Yousef Farhaoui · Alvaro Rocha
Zouhaier Brahmia · Bharat Bhushab
Editors

Artificial Intelligence and Smart Environment

ICAISE'2022

Springer

Editors
Yousef Farhaoui
Department of Computer Science
Moulay Ismail University, Faculty
of Sciences and Techniques
Errachidia, Morocco

Zouhaier Brahmia
University of Sfax
Sfax, Tunisia

Alvaro Rocha
ISEG
Universidade de Lisboa
Lisbon, Cávado, Portugal

Bharat Bhushab
School of Engineering and Technology (SET)
Sharda University
Greater Noida, Uttar Pradesh, India

ISSN 2367-3370 ISSN 2367-3389 (electronic)
Lecture Notes in Networks and Systems
ISBN 978-3-031-26253-1 ISBN 978-3-031-26254-8 (eBook)
https://doi.org/10.1007/978-3-031-26254-8

This Springer imprint is published by the registered company Springer Nature Switzerland AG
The registered company address is: Gewerbestrasse 11, 6330 Cham, Switzerland

Preface

Data is becoming an increasingly decisive resource in modern societies, economies, and governmental organizations. Data science, artificial intelligence and smart environments inspire novel techniques and theories drawn from mathematics, statistics, information theory, computer science, and social science. This book reviews the state of the art of big data analysis, artificial intelligence, and smart environments. It includes issues which pertain to signal processing, probability models, machine learning, data mining, database, data engineering, pattern recognition, visualization, predictive analytics, data warehousing, data compression, computer programming, smart city, etc. Papers in this book were the outcome of research conducted in this field of study. The latter makes use of applications and techniques related to data analysis in general and big data and smart city in particular.

The book appeals to advanced undergraduate and graduate students, post-doctoral researchers, lecturers, and industrial researchers as well as anyone interested in big data analysis and artificial intelligence.

Chairman of ICAISE’2022

Sebastian Garcia	Czech Technical University in Prague, Czech Republic
Younes Asimi	University of Ibn Zohr, Agadir, Morocco
Samaher Al_Janabi	University of Babylon, Iraq
Safa Brahmia	University of Sfax, Tunisia
Hind Hamrouni	University of Sfax, Tunisia
Anass El Haddadi	University of Abdelmalek Essaadi, Morocco
Abdelkhalek Bahri	University of Abdelmalek Essaadi, Morocco
El Wardani Dadi	University of Abdelmalek Essaadi, Morocco
Ahmed Boujraf	University of Abdelmalek Essaadi, Morocco
Ahmed Lahjouji El Idrissi	University of Abdelmalek Essaadi, Morocco
Aziz Khamjane	University of Abdelmalek Essaadi, Morocco
Nabil Kannouf	University of Abdelmalek Essaadi, Morocco
Youness Abou El Hanoune	University of Abdelmalek Essaadi, Morocco
Mohamed Addam	University of Abdelmalek Essaadi, Morocco
Fouzia Moradi	University of Abdelmalek Essaadi, Morocco
Hayat Routaib	University of Abdelmalek Essaadi, Morocco
Imad Badi	University of Abdelmalek Essaadi, Morocco
Mohammed Merzougui	FSJES, Oujda, Morocco
Asmine Samraj	Quaid-E-Millath Government College, India
Esther Asare	Anhui University of Science and Technology, China
Abdellah Abarda	FEM-Hassan 1 University, Morocco
Yassine El Borji	University of Abdelmalek Essaadi, Morocco
Tarik Boudaa	University of Abdelmalek Essaadi, Morocco
Ragragui Anouar	University of Abdelmalek Essaadi, Morocco
Abdellah Benami	FSTE-UMI, Errachidia, Morocco
Abdelhamid Zouhair	University of Abdelmalek Essaadi, Morocco
Mohamed El Ghmary	FSDM-USMBA, Morocco
Youssef Mejdoub	ESTC-UH2, Casablanca, Morocco
Amine Tilioua	FST-UMI, Errachidia, Morocco
Azidine Geuzzaz	EST Essaouira, University of Cadi Ayyad Marrakech, Morocco
Said Benkirane	EST Essaouira, University of Cadi Ayyad Marrakech, Morocco
Ali Ouacha	FS-Mohammed V University, Morocco
Fayçal Messaoudi	ENCG, USMBA, Morocco
Tarik Chanyour	Ain Chock Faculty of Sciences-UH2, Morocco
Naercio Magaia	School of Engineering and Informatics, University of Sussex, UK
Imane Halkhams	UPE, Fez, Morocco
Elsadig Musa Ahmed	Faculty of Business Multimedia University, Malaysia

Contents

A Collaborative Fog-Based Healthcare Intrusion Detection Security Using Blockchain and Machine Learning

Maryam Douiba[1](✉), Said Benkirane[1], Azidine Guezzaz[1], and Mourade Azrour[2]

[1] Technology Higher School Essaouira, Cadi Ayyad University, Marrakesh, Morocco
maryam.douiba@ced.uca.ma

[2] Faculty of Sciences and Technics, IDMS Team, Moulay Ismail University of Meknès, Meknes, Morocco

Abstract. Over recent years, the outbreak of Covid-19 has infected more than a billion people. Due to this crisis, the healthcare industry is revolutionizing using the Internet of Health Things (IoHT). As a result, the increasing number of distributed connected objects, their heterogeneity, and mobility have led to a dramatic expansion in the volume of medical data, consequently, a considerable increase in cybercrime. However, the security of the healthcare system must be considered a top priority. According to the policy principles of cybersecurity intrusion detection systems (IDS) are effective and indispensable security tools. We propose in this paper a collaborative distributed fog-based intrusion detection system reinforced by using blockchain to ensure trust and reliability between Fog nodes, and machine learning (ML) approaches with the effective open-source Catboost benefiting from the GPU library to get a record detection and computation time.

Keywords: IoHT · Security · IDS · Blockchain · Fog-nod · Catboost

1 Introduction

The main objective of the healthcare system is to provide a reliable solution to improve the health of human society [1, 2]. Recently, a universal crisis caused by the COVID-19 pandemic influenced the healthcare sector [1, 4]. All countries suffer from this dramatic increase in medical patients, and access to primary care has become difficult for patients [1, 4]. Healthcare systems have greatly benefited by introducing the IoHT technology or Medcin 4.0. The connection between the patient and the doctors is achieved through wireless media to communicate with IoT-enabled biomedical intelligent devices equipped with sensors, actuators, and processors like pulse oximeters, thermometers, glucometer, blood pressure, accelerometer, insulin pump, galvanic skin response, airflow, electromyography, electrocardiography, and other different kinds of monitoring of body that gathering, analyzing, and disseminating the health-related information of patients in real-time, also enabling to diagnostic, treat and monitor patients remotely [1, 2, 4, 11, 14]. The objective of the IoHT is to improve the quality of life through personalized e-health services without constraints of location and limitations on time [2].

Y. Farhaoui et al. (Eds.): ICAISE 2022, LNNS 635, pp. 1–6, 2023.
https://doi.org/10.1007/978-3-031-26254-8_1

Various sensors, actuators, and controllers provide integration, calculation, and interoperability in IoHT [14]. These devices have limited capabilities due to their minimal memory, processor, and storage capacity. As a result of these factors, implementing security becomes a difficult task. Hence IoHT security must be considered a top priority. Some recent research proposes using blockchain technology to boost intrusion detection systems and increase threat detection to defend against network attacks and secure data [5, 7–10, 12, 13, 15]. However, they have not taken into consideration the new Cloud and fog based IoT architecture that represents the new era of distributed IoT systems. The contribution of this work can be summarized in three points. Firstly, develop an understanding of the use of Blockchain to improve the collaboration between the IDS of each node in decentralized and distributed systems. Secondly, proposed a distributed collaborative intrusion detection system model to secure the IoHT using Blockchain technology to ensure trust and reliability between fog nodes. The Blockchain will allow fog nodes to securely exchange and store the attacks detected by each fog node. Thirdly, the collaboration on Fog-based using our improved anomaly-based intrusion detection model [3, 5] with the efficient combination of GB and DT algorithms using Catboost and taking advantage of GPUs to achieve record detection and computation time [6].

The remainder of this paper is structured as follows. In Sect. 2, we present a review of the IoHT cyberattack and some related works in intrusion detection approaches, including ML and DL. Section 3 described the essential steps of the proposed IDS design and suggested solutions to validate our collaborative intrusion detection approach. Finally, the paper is achieved with a conclusion and future works.

2 Review and Related Work

Nowadays, various approaches are designed and implemented to secure and protect the IoHT environment from unauthorized access and attacks, but still insufficient because of such challenges as response time, latency, huge data volume. So, it is necessary to boost security mechanisms like IDS, SDN, and blockchain with emerging artificial intelligence, such as ML and DL algorithm to secure fog and cloud in IoHT environments [3–6, 11]. IDS solution is among the more successful solutions for monitoring network behavior and detecting exploited nodes. Recently, researchers discovered blockchain has a huge potential to respond to the challenges of maintaining trust between nodes while distributing attacks. The objective of this detection-based techniques is to identify and isolate attackers. IDS detects packets in a network and determines whether they are an affirmative or negative attack. In addition, IDS can be enhanced using ML and DL algorithms. Also, it contributes to the development of prevention systems by detecting the nature of the attack [4–6]. Recently, some research proposed blockchain technology to reinforce IDS and improve attack detection. This combination of IDS, Blockchain, and ML attempts to confront network attacks and weaknesses and protect critical healthcare data from attack. The history of distributed intrusion detection systems is enriching; researchers have strived to improve the security of decentralized networks and distributed networks by collaborating with multiple IDSs but not especially adequate for the current development of IoHT. On the same research axis in 2018, Alexopoulos et al. [7] discussed the intersection of blockchains with CIDS. They proposed architecture for incorporating blockchains into CIDS using this technology as a mechanism to improve trust between monitors. In 2019 Hu et al. [9] presented a new CIDS approach using blockchain for distributed intrusion detection with a multi-pattern proposal generation method in MMG systems. The consensus process and incentive mechanisms in blockchain are used to accomplish CIDS without a central manager or trusted authority. In 2019 Li et al. [8] CBSigIDS is a generic architecture of collaborative blockchain signature-based IDSs that construct and update a trusted signature in a collaborative IoT network. CBSigIDS offers a precise detection method in distributed systems with robustness and effectiveness without requiring a trusted intermediate. In 2019 Kolokotronis et al. [12] suggested a distributed CIDS in which any IDS securely communicates trust information about external nodes and hosts with other CIDS members. This information is securely aggregated based on source credibility, computed based on challenge responses, and preserved on a blockchain. In 2020 Alkadi et al. [10] presented a CIDS designed to provide security and privacy through distributed intrusion detection and a blockchain with smart contracts in IoT networks. Intrusion detection is provided with a bi-directional long-term memory DL algorithm (BiLSTM). In 2020 Liang et al. [15] examined the characteristic behavior of clustering data in a Blockchain network and demonstrated that relying on a single clustering characteristic for anomaly identification is inefficient. So, they proposed a collaborative intrusion detection technique for a Blockchain-based on clustering features. That can discover patterns of attack in Blockchain data transactions quickly. A Blockchain dataset's abnormal features are discovered after multiple rounds of mutual competition among clustering nodes, a weighted combination is performed, and the weighted coefficients are achieved. In 2020 Winanto et al. [14] designed PoAS, which is Signature-based CIDS capable to improved trust between the nodes. CIDS proposed also enhances the robustness and successfully produces trusted attack signatures. Recently in February 2022 Li et al. [13] designed a new framework of blockchain-based collaborative intrusion detection in SDN (BlockCSDN) that benefits robustness and security by combining SDN, CIDS, and blockchain. Findings proved the validity and effectiveness of the technique used in repelling insider threats, improving alarm aggregation resilience, and preserving network bandwidth. As a result, Blockchain technology can improve intrusion detection systems and increase threat detection to protect the network. Nevertheless, this cited research did not consider the new cloud and fog based IoHT architecture, which symbolizes the new wave of dispersed IoT systems and, more specifically, the architecture of healthcare systems.

3 Proposal Model

The aim of our proposed model is to secure the IoHT system by protecting the fog-node domains upon receiving the event of a malicious attack by integrating IDS [3, 5] in fog computing. IDS is an effective security solution that can be enhanced with ML and DL algorithms, but because of the distributed nature of IoHT, one IDS is not sufficient to effectively identify different attacks. In fact, we propose a collaborative distributed fog-based intrusion detection system using machine learning approaches with the effective open-source Catboost [6] to create an anomaly-based intrusion detection model to detect the attacks in each node, and Blockchain to exchange, and store attacks detected by fog nodes in complete security, also to ensure trust and reliability between Fog nodes. Our model IDS [3, 5] should be used at any fog node of the fog cloud architecture to solve the security problem as shown in Fig. 1. Furthermore, using GPU in benefiting from Catboost open sources helps to get a record detection and computation time [6]. Given the distributed nature of IoHT, the IDS should be linked to each fog computing node, and the alert should be propagated through blockchain integration. It is employed in networks to secure routing and forwarding information.

3.1 IDS Using Catboost

Using a classification that combines DTs and GB method, our suggested IDS will be able to classify data activity as normal or attack [3, 5]. To improve security, IDS should be employed at each fog node in the fog cloud architecture to monitor network behavior and detect exploited nodes. The use of a GPU and the digital conversion proposed by Catboost solve the problem of categorical characteristics and accelerate the detection well. Our proposed approach is able to reduce the bias of the gradient estimation and improves the generalization capability.

The Building of IDS: Data is prepared and understood. We identified and removed all inconsistent values, such as real and NaN values. Then the feature vector $(X_m^1, X_m^2, ..., X_m^n)$ and target label y_m $(y_1, y_2 ..., y_m)$ are defined and prepared. All features are used. Then the categorical values are encoded with (CatBoost Encoder) by using the TS in a greedy way on the whole dataset to reduce overfitting and avoid target leakage and normalization problem [6]. The ordering approach creates a strong predictor in each category based. The fundamental concept is to compute the TS for x_k on a subset $D_k \subset D \backslash \{x_k\}$ [6]:

$$\hat{x}_j^i = \frac{\sum_{x_j \in D_k} 1_{\{x_k^i = x_j^i\}} \cdot y_j + \alpha p}{\sum_{x_j \in D_k} 1_{\{x_k^i = x_j^i\}} \cdot y_j + \alpha} \tag{1}$$

where 1 is the indicator function and p is set to the average of the target value over the sample with the parameter $\alpha > 0$. Subsequently, the features are transformed and combined. The test set is used to fine-tune the different variables and choose the most efficient to obtain the best performance. The training and testing process is implemented using ML ensemble classifier Catboost and GPU processing. The building model is able to predict an attack as positive.

3.2 Collaboration with Blockchain

The IDS should be utilized at each fog node in the fog cloud architecture to monitor network behavior and detect exploited nodes to improve security. To improve detection capabilities, collaborative intrusion detection promotes IDS nodes to communicate essential information with each other. On the other hand, the blockchain is considered a transparent, distributed, decentralized, and ordered ledger. Because of its nature, recently,

Blockchain technologies have demonstrated their utility in ensuring trust and reliability in distributed and decentralized systems. It is an excellent way to solve the fundamental problems of CIDS because it replaces centralized processing with secure collaborative processing due to its decentralized nature.

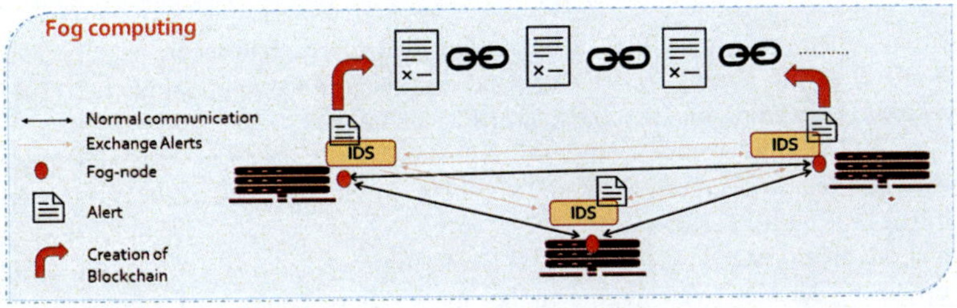

Fig. 1. The proposed collaborative intrusion detection system

The Building of Blockchain: Since the blockchain is created at the fog level, the analysis nodes will have the power they need to compute and process the data, but they can also be monitoring units that monitor and share the alert. All fog nodes in the blockchain network contribute to the detection and management of alerts, so all nodes participate fully as monitoring and analysis units. In our proposition, every alert detected by IDS is registered in a blockchain public. Then each alert generated must be broadcast and shared with all other nodes using the concept of Blockchain with P2P architecture to ensure that all attacks are exchanged. The blockchain's alerts are divided into blocks. Alerts and previous and current hash are all included in each block. The hash of the data in the block is like a key signature. The hash is cryptographically connecting each block. The Genesis Block is the first block. Each block is dependent on the one before it. The blockchain is shared between different nodes in the interest of collaborating. Furthermore, this exchange of alerts allows for evaluating the trust of other nodes. As a result, it can provide a solution for improving the trust calculation process. Hence, the enabling technologies of blockchains can make IoHT systems flexible for network design adoption resulting in prospects for a diverse range of applications; these technologies can be adapted to IoHT because they fit the notion of distributed fog and cloud architecture. Such blockchain can contribute to enhancing and securing IoHT.

4 Conclusion and Future Work

This work aims to develop an understanding of the use of Blockchain to enhance IDS and deal with decentralized and distributed IoHT systems. The proposed is a distributed collaborative intrusion detection system model to secure IoHT using an anomaly-based intrusion detection model with the effective combination of GB and DT algorithms from Catboost [6] open sources using GPU to get a record detection and computation time in each fog node and benefitting from Blockchain technology to ensure trust and reliability between nodes. It allows fog nodes to exchange, and store attacks detected by each fog node in complete security. Additional research must examine and test our proposal model at the fog nodes to evaluate how the collaborative IDS would work to secure IoHT network attacks.

References

1. Zhou, Z., Gaurav, A., Gupta, B.B., et al.: A statistical approach to secure health care services from DDoS attacks during COVID-19 pandemic. Neural Comput. Appl. 1–7 (2021). SN: 1433–3058. https://doi.org/10.1007/s00521-021-06389-6
2. Ahmadi, Z., Haghi Kashani, M., Nikravan, M., Mahdipour, E.: Fog-based healthcare systems: a systematic review. Multimedia Tools Appl. **80**(30), 36361–36400 (2021). https://doi.org/10.1007/s11042-021-11227-x
3. Douiba, M., Benkirane, S., Guezzaz, A., et al.: An improved anomaly detection model for IoT security using decision tree and gradient boosting. J. Supercomput. **79**, 3392–3411 (2022)
4. Bhardwaj, V., Joshi, R., Gaur, A.M.: IoT-based smart health monitoring system for COVID-19. SN Comput. Sci. **3**, 1–11 (2022). Epub 20 January 2022, PMID: 35079705; PMCID: PMC8772261
5. Douiba, M., Benkirane, S., Guezzaz, A., et al.: Anomaly detection model based on gradient boosting and decision tree for IoT environments security. J. Reliable Intell. Environ. (2022)
6. Prokhorenkova, L., Gusev, G., Vorobev, A., et al.: CatBoost: unbiased boosting with categorical features. In: Proceedings of the 32nd International Conference on Neural Information Processing Systems (2018)
7. Havarneanu, G., Setola, R., Nassopoulos, H., Wolthusen, S. (eds.): CRITIS 2016. LNCS, vol. 10242. Springer, Cham (2017). https://doi.org/10.1007/978-3-319-71368-7
8. Li, W., Tug, S., Meng, W., Wang, Y.: Designing collaborative blockchained signature-based intrusion detection in IoT environments. Future Gener. Comput. Syst. **96**, 481–489 (2019). ISSN 0167-739X
9. Hu, B., Zhou, C., Tian, Y., Qin, Y., Junping, X.: A collaborative intrusion detection approach using blockchain for multimicrogrid systems. IEEE Trans. Syst. Man Cybern. Syst. **49**(8), 1720–1730 (2019)
10. Alkadi, O., Moustafa, N., Turnbull, B., Choo, K.-K.R.: A deep blockchain framework-enabled collaborative intrusion detection for protecting IoT and cloud networks. IEEE Internet Things J. **8**(12), 9463–9472 (2021)
11. Hareesh, S.: Securing data in health care with block chain. J. Eng. Sci. **12**(08), 75–79 (2021). ISSN NO: 0377-9254
12. Kolokotronis, N., Brotsis, S., Germanos, G., Vassilakis, C., Shiaeles, S.: On blockchain architectures for trust-based collaborative intrusion detection. In: 2019 IEEE World Congress on Services (SERVICES), pp. 21–28 (2019)
13. Li, W., Yu, W., Meng, W., Li, J., Su, C.: BlockCSDN: towards blockchain-based collaborative intrusion detection in software defined networking. IEICE Trans. Inf. Syst. **E105.D**(2), 272–279 (2022). Released on J-STAGE 1 February 2022, Online ISSN 1745-1361, Print ISSN 0916-8532. https://doi.org/10.1587/transinf
14. Winanto, E.A., bin Idris, M.Y., Stiawan, D., Fatih, M.S.N., Sharipuddin: PoAS: enhanced consensus algorithm for collaborative blockchain intrusion detection system. In: 2020 3rd International Conference on Information and Communications Technology (ICOIACT), pp. 513–518 (2020)
15. Liang, W., Xiao, L., Zhang, K., Tang, M., He, D., Li, K.-C.: Data fusion approach for collaborative anomaly intrusion detection in blockchain-based Systems. IEEE Internet Things J. https://doi.org/10.1109/JIOT.2021.3053842

A Combined Approach of Computer Vision and NLP for Documents Data Extraction

Ali Benaissa[1(✉)], Abdelkhalak Bahri[1], and Ahmad El Allaoui[2]

[1] L-LST, T-DSCI, ENSA Al Hoceima, Abdelmalek Essaadi University, Tetouan, Morocco
ali.benaissa@etu.uae.ac.ma, abahri@uae.ac.ma

[2] L-STI, T-IDMS, FST Errachidia, Moulay Ismail University of Meknes, Meknes, Morocco
a.elallaoui@umi.ac.ma

Abstract. In recent years, computer vision and NLP became a major topic in computer studies drawing attention from researchers and computer scientists from the entire world. Deep learning technologies present the most outstanding performance in computer vision and NLP issues. In this article, we present an approach based on computer vision and NLP, for documents data extraction, we start from collecting data to predicting the documents objects, while using the NLP, meanwhile, we train the model based on NER technologies, to make the system intelligent.

Keywords: Computer vision · NLP · NER · Documents data extraction · Deep learning

1 Introduction

Among the difficulties facing technological progress is the analysis, is understanding and the extracting the valuable data from documents, considering the fact that unstructured data form more than 85% [1]. In this regard; Computer Vision (CV) [2] and Neural Languages Processing (NLP) [3] technologies, which belong to the deep learning field, provide us large libraries software to manipulate the data from the documents, such as OpenCv [4], Tesseract [1] and spaCy [5]... However, the difference in the structure of these documents constitutes an obstacle to these techniques. Therefore, in the order to bypass this issue we present this approach.

Computer Vision (CV) enables us to extract and manipulate objects contained in any file as documents or images, to take advantage of this technology; we used one of the most important libraries that play this role, called OpenCV, this library provides many programming languages, basically, this library is used for images processing tasks and Deep Learning tasks. Neural Languages Processing (NLP) to make us understand more about what CV gives us as data, we used spaCy library to do this task, this library can understand the data and give it a meaning or a definition, this sub domain of NLP called Named-entity Recognition (NER) [5].

In this article, we will present an approach based on deep learning, we can divide our proposed approach as: (1) Collect Data, (2) Extract text from documents, (3) Labeling text, (4) Text cleaning, (5) Predict using NER model.

Y. Farhaoui et al. (Eds.): ICAISE 2022, LNNS 635, pp. 7–13, 2023.
https://doi.org/10.1007/978-3-031-26254-8_2

2 Problematic Statement

In the era of digital transformation that the world is witnessing, which is rather reflected in the abandonment of what is paper for what is digital, and in light of the huge number of paper documents that exist with companies and the public sector as well. This is why it is a nightmare to lose papers, since most of the paper in our offices are not organized.

This research comes in this context, where we convert what is paper to digital. We structure and extract the specific items of value from the document, and give it a meaning or a definition.

3 Related Work

Many studies are done in the same context of Documents Data Extraction, which was knows by OCR, in the previous studies, the searchers usually used Computer Vision techniques only, as we will discuss in this section. In a survey article, Shantanu Kumar [6], discusses the importance of Relation Extraction, and the role of building the huge datasets, to achieve this goal, a supervised and semi-supervised learning methods are performed on CNNs models. Francesco Adamo et al. [7] present an automatic document processing system for medical data extraction, for extraction, they based on an open OCR engine, and the results discussion showed the presented system's advantages. Nancy Xin Ru Wang et al. [8] presented a system called "TableLab", which can extract tables from documents. The system based on fine-tuning technique to train the model, and this work directed to ordinary users and for AI engineers as well.

Our research comes due to the lack of studies that combine both CV and NLP techniques.

4 Dataset Construction

4.1 Dataset

Our objective is to extract information value from documents, to achieve this result, we had to build a new dataset based on documents type; that we want to extract data from. In this order, and to make a real use case, we decide to apply our approach on Business Cards; therefore, we collect Business Cards from the internet and from some small businesses near us, as images in JPEG file format, to build the dataset, Fig. 1 (in the end of this section) shows samples of the dataset.

4.2 Labeling

Most computer-assisted text analysis methods, including NER, labels the data (segmenting the text into tokens), this step comes after splitting the sum of the raw text extracted from dataset, into small pieces, called tokens, where we give each token what it represents in the documents, as Fig. 2 shows.

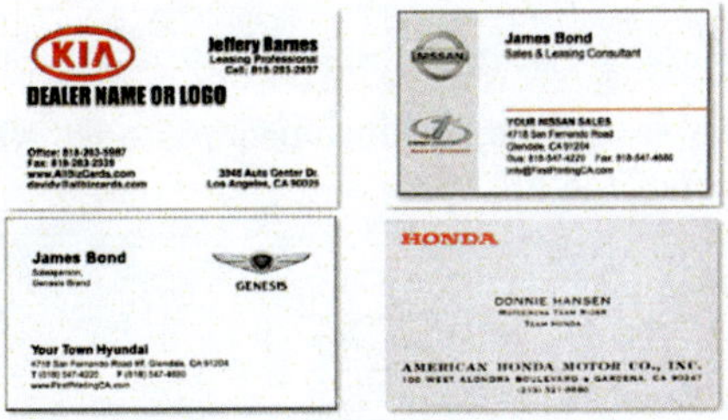

Fig. 1. Dataset samples.

	A	B	C
1	id	text	tag
2	000.jpeg	stu	O
3	000.jpeg	qcv	O
4	000.jpeg	BENAISSA	B-NAME
5	000.jpeg	Ali	I-NAME
6	000.jpeg	Researche	B-JOB
7	000.jpeg	636787	B-PHONE
8	000.jpeg	Izemouren	B-LOC
9	000.jpeg	ENSA Al H(	B-DES
10	000.jpeg	[]iudhgf	O

Fig. 2. Samples of labeling.

5 Proposed Approach

In this section, we will present our approach, and how can we make a combination between CV and NER.

Global View: Using CV we will scan our documents, identify location of each objects in our documents and finally we will extract the text to be able to manipulate it, with NLP we will extracts objects from given text and make clear for the objects and finally give a meaning for each object. As it is showing in Fig. 3.

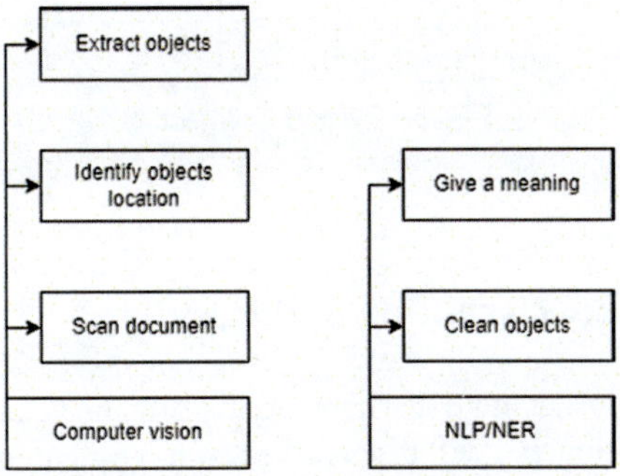

Fig. 3. Global view of the approach.

Steps: To achieve our goal by give a meaning for each object in documents, we will follow the following steps: (1) Data preparation, (2) Extract objects location as coordinates, (3) Label the extraction data, (4) Data pre-processing, (5) Train the model, (6) Prediction, as Fig. 4 explain.

5.1 Architecture

In this section, we will present the approach architecture; the Fig. 3 explains how the procedure of getting objects from documents goes on. First, we scan our document then

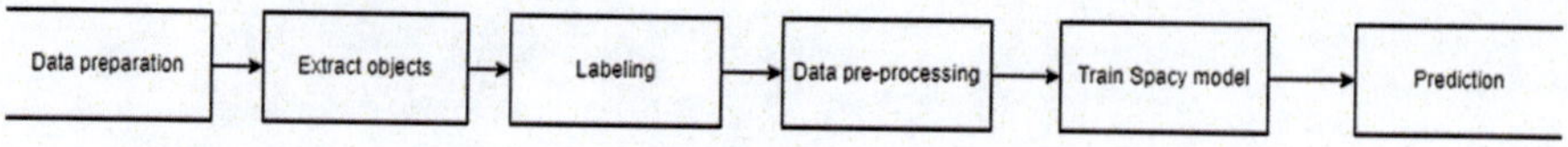

Fig. 4. Approach steps.

we extract the data, after the cleaning of the objects done, the document passed to NLP model to give meaning for each object as the table in Fig. 5 show.

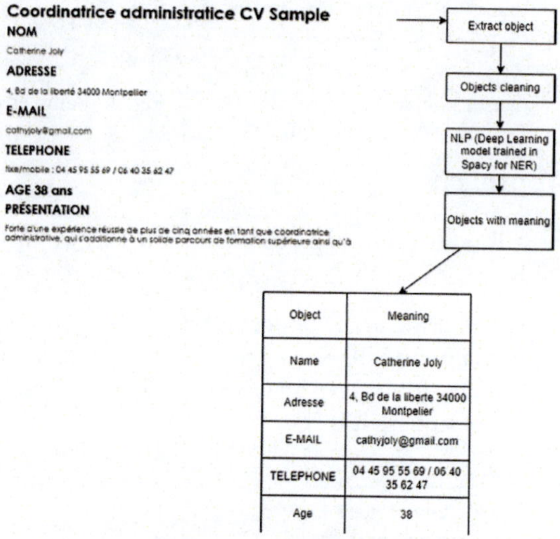

Fig. 5. System architect.

6 Experiments Model Training and Results

In this section, we will present our experimentation using the dataset we built, and based on our approach. In addition, we will discuss the results compared to the other proposed methods/approach. The presented experiments conducted on a machine with CPU of 2.6 GHz Intel i5 1145G7 with 16384 MB of RAM.

To train our spaCy pipelines, based on CPU, we passed the following parameters as presented in Table 1:

After this, we shuffled our dataset, then we divided it into 80% of train and 10% of test, then we trained our model based on the mention parameters in Table 1, if we looked at results of efficiency from the outside, we got medium results, and this is mainly due to the size of the dataset. Overall, the result presented in Fig. 6 are subject to change for the better, compared to the dataset's size, and the efficacy of labeling. Inside model, it is quite different, when we evaluate our model; we got high results, which was 100% for the tokens, and for entity we got 96% precision, 94% recall and 95% F-score, the table present the evaluation values (Table 2).

Table 1. Parameters value.

Parameter	Value
Language	English
Pipeline	Tok2vec and NER
Batch size	1000
Hidden width	64
Maxout pieces of NER	2
Rows	[5000,2500]
Width = 96	96
Depth = 4	4
Window size = 1	1
Dropout	0.1
Optimizer	AdamV2
Learn rate	0.001

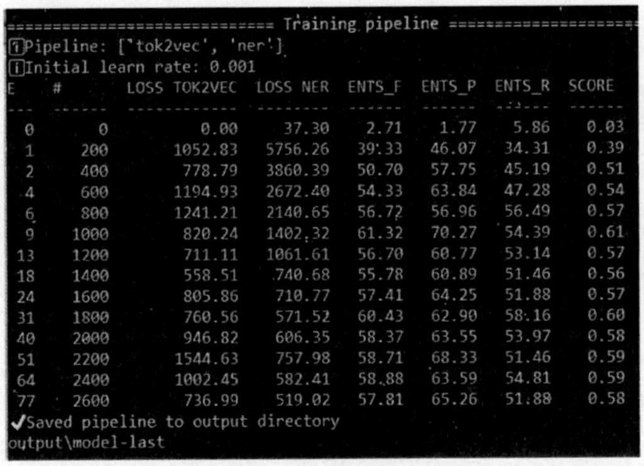
============================ Training pipeline ====================
ⓘ Pipeline: ['tok2vec', 'ner']
ⓘ Initial learn rate: 0.001

E	#	LOSS TOK2VEC	LOSS NER	ENTS_F	ENTS_P	ENTS_R	SCORE
0	0	0.00	37.30	2.71	1.77	5.86	0.03
1	200	1052.83	5756.26	39.33	46.07	34.31	0.39
2	400	778.79	3860.39	50.70	57.75	45.19	0.51
4	600	1194.93	2672.40	54.33	63.84	47.28	0.54
6	800	1241.21	2140.65	56.72	56.96	56.49	0.57
9	1000	820.24	1402.32	61.32	70.27	54.39	0.61
13	1200	711.11	1061.61	56.70	60.77	53.14	0.57
18	1400	558.51	740.68	55.78	60.89	51.46	0.56
24	1600	805.86	710.77	57.41	64.25	51.88	0.57
31	1800	760.56	571.52	60.43	62.90	58.16	0.60
40	2000	946.82	606.35	58.37	63.55	53.97	0.58
51	2200	1544.63	757.98	58.71	68.33	51.46	0.59
64	2400	1002.45	582.41	58.88	63.59	54.81	0.59
77	2600	736.99	519.02	57.81	65.26	51.88	0.58

✓ Saved pipeline to output directory
output\model-last

Fig. 6. Training pipeline.

Table 2. Evaluation values for tokens and entities.

	Token accuracy	Token precision	Token recall	Token F-score	Entity precision	Entity recall	Entity F-score
Value	1.0	1.0	1.0	1.0	0.96	0.94	0.95

If we go deeper into the performance of the model, and we evaluated each entity, compared to the dataset, which is small, the results was perfect, the Table 3 present the precision, recall and f-score for each entity.

Table 3. Each entity evaluation.

	Precision	Recall	F-score		Precision	Recall	F-score
B-PHONE	0.98	0.93	0.96	I-MAIL	0.96	0.90	0.93
I-PHONE	0.95	0.93	0.94	B-NAME	0.95	0.92	0.94
B-ORG	0.97	0.92	0.94	I-NAME	0.99	0.94	0.96
I-ORG	0.94	0.98	0.96	I-DES	0.96	0.90	0.93
B-WEB	0.96	0.96	0.96	B-MAIL	0.98	0.97	0.98
I-WEB	0.84	0.84	0.84	I-MAIL	0.96	0.90	0.93
B-MAIL	0.98	0.97	0.98				

6.1 Results Discussion

Although the studies done on the Documents Data Extraction using CV and NLP are not available, we compare our results to the existing OCR engine/systems and to the proposed approach based on CV and CNN, to compare between the existing method and our method, we will discuss the different classifiers.

- Convolution neural network (CNN) achieved a high accuracy in image recognition problems, but always needed huge amount of training data, as an example, we have Devanagari Character Recognition [9] which achieved a 98.19% of accuracy, and Arabic Character Recognition [10] achieved a 97.4% of accuracy.
- Neural networks (NN) could train with any variety of data points, inputs and layers, but as the number of neurons increases, the network becomes more complex, for example, a Gujarati Character Recognition [11] achieved an accuracy of 74.16%.
- SVM is easily integrated with different distance functions to quickly categorize new functions and a very small object training sample size, as an example, Devanagari Character Recognition [12] achieve a 98.62% of accuracy.
- CV/NLP is the approach that we present in this work, as we get in our experiments results, firstly we do not need a huge dataset to get a high precision. We achieved 100% for all entities we had tested in our case. Therefore, this method has had achieved promising results.

7 Conclusion and Future Work

One of the important tasks in Deep Learning is Document Data Extraction; in this article, we present a combined approach of CV and NLP/NER to improve the Document Data Extraction, whatever the form and structure of the document is.

In future work, we will focus on applying our approach, to the official Moroccan documents, especially the Arabic ones.

References

1. Kay, A.: Tesseract: an open-source optical character recognition engine. Linux J. **159**, 2 (2007)
2. Bird, S., et al.: Natural Language Processing with Python. O'Reilly Media, Sebastopol (2009)
3. Bradski, G.: The OpenCV Library. Dr Dobb's J. Softw. Tools **25**(11), 120–123 (2000)
4. Honnibal, M., Montani, I.: spaCy 2: natural language understanding with Bloom embeddings, convolutional neural networks and incremental parsing (2017)
5. Nasar, Z., Jaffry, S.W., Malik, M.K.: Named entity recognition and relation extraction: state-of-the-art. ACM Comput. Surv. **54**(1), Article 20, 39 (2022). https://doi.org/10.1145/3445965
6. Kumar, S.: A survey of deep learning methods for relation extraction. arXiv preprint arXiv:1705.03645 (2017)
7. Adamo, F., Attivissimo, F., Di Nisio, A., et al.: An automatic document processing system for medical data extraction. Measurement **61**, 88–99 (2015)
8. Wang, N.X.R., Burdick, D., Li, Y.: TableLab: an interactive table extraction system with adaptive deep learning. In: 26th International Conference on Intelligent User Interfaces-Companion, pp. 87–89 (2021)
9. Mehrotra, K., et al.: Unconstrained handwritten Devanagari character recognition using convolutional neural networks. In: Proceedings of the 4th International Workshop on Multilingual OCR. ACM (2013)
10. Ashiquzzaman, A., Tushar, A.K.: Handwritten Arabic Numeral Recognition using Deep Learning Neural Networks (2011)
11. Nasriwala, J.V., Patel, B.C.: Offline handwritten Gujarati character recognition based on water reservoir and radial histogram. Int. J. Electr. Electron. Comput. Sci. Eng. **2**, 353–379 (2016)
12. Jangid, M., et al.: SVM classifier for recognition of handwritten Devanagari numeral. In: 2011 International Conference on Image Information Processing (ICIIP). IEEE (2011)

Check for updates

A Comprehensive Analysis of Novel Intrusion Detection Systems for Internet of Things Networks

Zouhair Chiba(✉), Noreddine Abghour, Khalid Moussaid, Seddiq El Kasmi Alaoui, Tarik Chanyour, and Oumaima Lifandali

LIS Labs, Faculty of Sciences Aïn Chock, Hassan II University of Casablanca, Casablanca, Morocco

{ZOUHAIR.CHIBA,NOREDDINE.ABGHOUR,KHALID.MOUSSAID, SEDDIQ.ELKASMIALAOUI,TARIK.CHANYOUR}@univh2c.ma

Abstract. In recent years, the Internet of Things (IoT) paradigm has shown massive embracing by various industries notably the medical sector, vehicle manufacturers, home appliances manufacturers, etc. The acceptance of IoT technology has tremendously changed the manner we live. Hence, IoT is commonly referred to as the Internet of People (IoP) given that it is employed roughly by everyone in their daily lives, from individuals to institutions. There is a perpetual increase in connected devices around the globe. Nevertheless, IoT devices are a soft target and prone to attacks due substantially to their resource limitations, and the nature of their networks. Thus, IoT device protection against intrusions and attacks is a serious issue and an enormous source of disquietude for institutions and countries. To alleviate cyber threats, lots of IoT networks begin implementing an intrusion detection system (IDS) with a view to detecting malevolent events. Thus, an IDS plays a crucial role in preserving the security and privacy of IoT Networks. In this paper, we provide a comprehensive study and deep analysis of several modern and relevant IoT-specific IDS, where we underline primarily the suggested mechanisms, strengths, and limitations or shortcomings. This assists to mark the concerns that still require to be handled, well outlines the mainstream of research direction, and clears the way for new avenues of research for forthcoming researchers. Finally, we deliver a guide or support for scientists interested with IDS for IoT networks.

Keywords: Intrusion Detection System (IDS) · Internet of Things (IoT) · Networks

1 Introduction

With the emerging of the "Internet plus" era, the Internet of Things (IoT) is smoothly penetrating into various domains such education, surveillance, healthcare, agriculture, military, commerce, etc., and the scale of its equipment is also showing an explosive growing trend. The age of the "Internet of Everything" is coming [1]. IoT considered as the next generation interconnection mode, is a ubiquitous network built on the Internet.

Y. Farhaoui et al. (Eds.): ICAISE 2022, LNNS 635, pp. 14–26, 2023.
https://doi.org/10.1007/978-3-031-26254-8_3

In the age of digital transformation, it has become a popular technology, which connects a huge number of micro devices (smart phones, electricity meters, etc.) with the Internet according to the agreed protocol, fulfilling the smart communication between people, people and things, and things and things. It is the high-tech behind smart city, intelligent residence, and self-driving. The Internet of Things is known as the third industrial transition. It is defined as "interconnection, through the Internet, from computer equipment embedded in everyday objects, allowing it to send and receive data". There exist several ways IoT nodes connect to the Internet and this includes communication protocols such as the Transmission Control Protocol and the Internet Protocol (TCP/IP) using Message Queue Telemetry Transport (MQTT), Modbus TCP, Cellular, Long-Range Radio Wide Area Network (LoRaWAN), etc. As shown in Fig. 1, according to the flow of information collection, transmission and processing, the hierarchical structure of IoT can be separated into 3 layers namely, the perceptual layer, the network layer, and the application layer. Additionally, IoT gives a connected and machine-to machine communication environment for varied types of devices, which can be fruitful for supporting decision-making process in numerous organizations and companies, e.g., energy industry. It is presently developing at a quick speed by transforming a large number of physical elements through digital intelligence, making business faster, cleverer, and more effective. With the gradual scale of the IoT, its industrial value will be 30 times larger than the Internet, and it will become the next trillion level information industry business [1]. The market for IoT is increasing at a spectacular rate, commencing with 2 billion artifacts in 2006, a forecast rise of 200 billion by 2022 out of 200% [2].

A great number of conventional appliances in digital transformation, roughly have no synchronous configuration of defense capabilities, affecting the security, integrity, and reliability of the IoT. Moreover, owing to the integration and diversity variety of IoT terminals and applications, it carries more security uncertainty to IoT business. The ever-widening diversity of IoT interconnected devices offers attackers a broad network intrusion portal, leading to the IoT facing substantial security issues. Furthermore, Internet of Things devices have small size, hence, they commonly have limited resources, low processing capacity, and limited memory. As IoT devices resource limitations, this turns it arduous to implement security mechanisms. Security breach cases are prevalent in IoT systems, as security often does not receive due attention in the conception of these systems. In 2016, lots of IoT devices were hacked by means of Mirai malware to accomplish denial of service attacks [2]. In fact, Each layer of IoT architecture displayed in Fig. 1, is prone to numerous kinds of attacks and intrusions that could jeopardize the systems within the IoT. Some common attacks and intrusions on the IoT ecosystem encompass Data corruption breaches, Access control attacks, Spoofing attacks, Sybil attacks, Denial of Service (DoS) attacks, Distributed DoS, betrayal attacks, Operating System (OS) attacks, Jamming attacks, etc.

To tackle these malicious attacks and to guarantee that the active nature of IoT nodes and the security of IoT networks are preserved, loads of organizations and corporates are implementing Intrusion Detection Systems (IDSs). Furthermore, these IDSs can be configured at any layer in Fig. 1. An IDS plays a vital role in the IoT by insuring that the integrity, security, and privacy of data transmitted across its network are kept up. An IDS can prevent, detect, react and turn up any attacks or malevolent activities that have the potency to cripple an IoT network [3, 4].

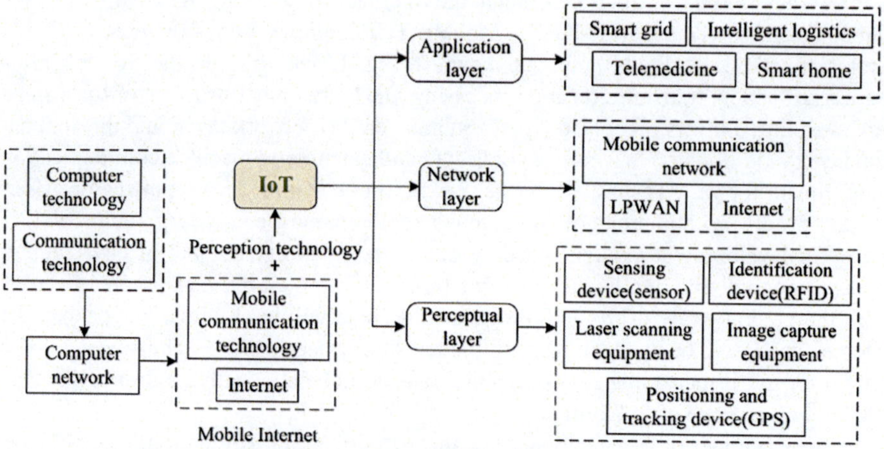

Fig. 1. Typical IoT architecture

The rest of this paper is organized as follows: Sect. 2 provides a deep and critical analysis of recent IDS for IoT. While Sect. 3 is dedicated to the discussion of the analyzed works, offering new guidelines and avenues of research. Finally, Sect. 4 ends with conclusions.

2 Analysis of Intrusion Detection Systems for IoT

This section gives a critical analysis of ten novel and relevant IDSs conceived for IoT networks. As displayed by Table 1, for each one, we provide the class of the methodology adopted, the mechanisms employed to detect attacks and malevolent activities, and the datasets employed to build and assess the IDS proposed. Besides, we underline the strong points and the shortcomings of these IDSs.

Table 1. Analysis of intrusion detection systems for IoT networks

Ref.	Method	Detection mechanism	Dataset	Characteristics/Strengths	Limitations/Challenges
[3]	Machine Learning	Random Forest (RF), Decision Tree (DT), Extra-Trees (ET), and Extreme Gradient Boosting (XGB) algorithms	UNSW-NB15 dataset	- For the binary modeling process, the GA-RF outperformed the baseline model LR and others Tree-based algorithms with a Test Accuracy (TAC) of 87.61% - For the multiclass modeling process, the GA-ET outperformed the base-line model NB and others Tree-based algorithms with TAC of 77.64%	- The proposed IDS underperformed for some minority classes such as Worms, Backdoor, and Analysis. Hence, TAC of 77.64% attained by GA-ET needs improvement - Both GA-RF and GA-ET yielded bad prediction time of 2.2 s and 3.8 s respectively. They were surpassed by DT and BN with 18.3 ms and 7.96 ms respectively
[5]	Deep Learning+Biological Heuristic Algorithm	APSO-CNN: Convolu-tional Neural Network optimized by Adaptive Swarm Optimization	Real dataset: the public IoT data set created by Meidan et al. [6]	- APSO-CNN offers best performance in terms of intrusion detection than three popular algorithms, namely SVM, FNN, and R-CNN - APSO-CNN achieves 96% of Accuracy and 97% of Average Precision	- According to [7], the Convolutional architectures of neural networks are not suitable for intrusion detection in IoT devices, due to the use of a large number of parameters in these models concerning the limited processing resources in IoT devices. It is need a lightweight architecture - Assessment of the impact of resources consumed by the APSO-CNN based IDS on the operation of IoT devices has not been performed

(*continued*)

Table 1. (*continued*)

Ref.	Method	Detection mechanism	Dataset	Characteristics/Strengths	Limitations/Challenges
[1]	Machine Learning	One-class SVM (OCSVM)	UNSW-NB15 dataset	- Compared with CMLW, TSDL and Threshold-optimized CNNs, this IDS achieves higher DR of Backdoor, Shellcode and Worms, the three low frequency attacks, reached 51.28%, 64.47% and 77.78%, Respectively - Low cost time; the IDS can detect 4472 records per second	- The FAR of 10.62% yielded is high. This performance indicator needs to be reduced - Low DR for some low frequency attacks, such as DoS, reconnaissance, and analysis
[2]	Machine Learning+Deep Learning	Extra Tree (ET), Random Forest (RF), and Deep Neural Network (DNN)	BoT-IoT, IoTID20, NSL-KDD, and CICIDS2018 datasets	Experimental outcomes obtained based on Bot-IoT, IoTID20, NSL-KDD, and CICIDS2018 datasets show that the IDS developed achieves high rates (between 99% and 100%) of DR, Recall, Precision, and Balanced Accuracy	- The prediction time is high compared to KNN, RF, and NB techniques. It should be reduced - U2R and infiltration attacks were hard to detect by this model; DR attained regarding these classes are 68.75% and 13.50%, respectively. While those attacks are perilous - Conception and implementation of countermeasure module are missing - Robustness of that IDS against network routing attacks such sinkhole, wormhole and selective forwarding was not assessed

(*continued*)

Table 1. (*continued*)

Ref.	Method	Detection mechanism	Dataset	Characteristics/Strengths	Limitations/Challenges
[7]	Deep learning	Parallel Deep Auto-Encoder (PDAE)	KDDCup99, CICIDS2017, and UNSW-NB15 datasets	- Lightweight and efficient architecture of NN - Few numbers of parameters - High Accuracy of 99.37%	- For evaluating the Performance of the proposed IDS, false alarm rate was not utilized. A better NIDS should attain a high DR and low FPR - To assess the low computational complexity of the proposed model, the detection time metric was not used. The number of parameters is not a meaningful metric for measuring time complexity
[8]	Deep Learning	Deep Neural Networks	ToN_IoT dataset [9]	The methodology proposed attains an accuracy of 94.17% for malware attacks detection, while it reaches an accuracy of 97.08% for the classification	The ability to recognize password attacks is very weak, it achieves 0% of precision, and 0% of recall

(*continued*)

Table 1. (*continued*)

Ref.	Method	Detection mechanism	Dataset	Characteristics/Strengths	Limitations/Challenges
[10]	Machine Learning	Random Forest (RF) and Extreme Gradient Boosting XGBoost)	BoT-IoT (IoT-based dataset) [2]	- Distributed Anomaly IDS using fog computing to detect DDoS attacks against blockchain-enabled IoT Network - Effective in de-tecting IoT based attacks with 99.99% Accuracy, 99.8% DR and 99.99% F1-score - Detection system takes less processing time for training and testing	- There is no cooperation between the IDSs deployed in fog nodes - The proposed IDS is anomaly IDS, so detection of a previous attack occurred again requires the same computational cost in the IDS. If the signature based detection is applied prior to anomaly detection, that will reduce computational cost. Anomaly IDS has to detect only unknown attacks, because known attacks are already detected by signature IDS and denied - The manual management of abnormal transactions can lead to the loss of certain transactions which can go unnoticed by the admin especially in the event of the enormous flow of these transactions

(*continued*)

Table 1. (*continued*)

Ref.	Method	Detection mechanism	Dataset	Characteristics/Strengths	Limitations/Challenges
[11]	Deep Learning	Convolutional Neural Network (CNN)	The NID (Network Intrusion Detection) Dataset and BoT-IoT dataset	Accuracy of 99.51% and 95.55% based on NID and BoT-IoT datasets, respectively	- Implementing sophisticated security mechanisms such CNN based IDS and software on IoT devices is often difficult due to its limited energy - Evaluation of impact of deploying the IDS on IoT device is missing - Accuracy obtained based on BoT-IoT needs improvement
[12]	Machine Learning	XGBoost, CatBoost, KNN, SVM, QDA, and NB algo-rithms	UNSW-NB15 dataset	The proposed machine leaning model achieves high performance compared to state-of-the-art works with an accuracy of 99.9% and MCC of 99.97%	- Computation cost, energy consumption and required memory for the ML model in the IoT context were not studied - Assessment of the effectiveness of that ML model by means of a specialized dataset containing contemporary attacks targeting IoT devices, namely BoT-IoT is needed

(*continued*)

Table 1. (*continued*)

Ref.	Method	Detection mechanism	Dataset	Characteristics/Strengths	Limitations/Challenges
[1]	Signature matching+Statistical approach+Blockchain+IPFS	Blacklist packet filter composed of a blacklist (all blacklisted IP addresses) and a look-up table (IDS signatures based on the blacklisted IP addresses)	Real datasets	Regarding betrayal attack (insider threat), the proposed filter achieves a smaller false rate than others. Whereas, concerning DDoS threat (external attack), it keeps the network bandwidth better than other filters	- The IDS performance was examined under DDoS attack and betrayal attack, whereas some other attacks could be considered notably advanced insider attacks such as fingerprinting attack or a hybrid attack - The blockchain can bring numerous benefits in securing the data integrity, but it may also induce more delay and overhead during the consensus process. Authors should deploy the filter in an even larger blockchain environment, considering a large number of nodes, e.g., over 100 nodes to validate the scalability concern

3 Discussion

In In the present work, we analyze thoroughly ten state-of-the-art research released between 2021 and 2022 (two published in 2021, and eight published in 2022), offering innovative and relevant IDS for IoT networks. We provide their pivotal specifics such as the class of the methodology, the mechanisms leveraged to develop an IDS and hence enabling the detection of attacks aiming at IoT networks, and the datasets employed for building and appraisal of the predictive model proposed. Additionally, the strengths and shortcomings of these works are emphasized and exposed clearly. This assists to pinpoint the issues that still need to be tackled, well defining the mainstream research direction, and preparing the ground for new avenues of research for upcoming scientists.

Based on our current study, for example, some researchers can focalize on optimization of a few works analyzed towards resolving the flaws or surmounting the challenges aforementioned. Furthermore, they can carry out a mixture of two approaches or works with the purpose of take the benefits of each approach and alleviate their drawbacks. As a result, the two proposed lines of research will surely improve the performing of IDS/IPS drawn up for IoT networks. Below, we offer pertinent recommendations and the primary conclusions harvested from the current review:

1. The UNSW-NB15 and BoT-IoT are advanced datasets [1–3, 10, 12] for IDS research which are broadly utilized in the literature. Likewise, it would be good to deem other fresh IoT-specific datasets that is to say IoTID20 [2] or ToN_IoT [9].
2. Assessment of an IDS using only KDD CU99 or NSL-KDD datasets is not persuasive. These datasets do not comprise recent types of network attacks, the traffic patterns contained are like to be outdated and unable to enlighten on the effectiveness of new techniques. Indeed, these datasets are quite old, they suffer from various issues emphasized by McHugh [13] and may not be an ideal representative of contemporary real networks. Thus, they can only be employed for a baseline appraisal. 2–3 more novel datasets are needed to validate the efficiency of a provided model.
3. Nothing is wholly secure in IoT appliances. This enables the users to retrieve their data without impediment in the IoT environment. Nonetheless, it produces an unsecured and vulnerable climate for the intruders to compromise any network segment. Users should be aware of all IoT appliances' security flaws to protect themselves against cyber threats. Numerous approaches and structures to mitigate the network threats have been built up. For example, machine learning can aid to detect and categorize attacks on a wide network.
4. Proficient intrusion detection in IoT networks asks the hugging of novel smart approaches such as Machine Learning (ML) algorithms. ML is a branch of Artificial Intelligence (AI) that endows numerous systems with the potential and the capacity to learn from experience and boost their decision-making process without any explicit programming. One of the principal ML techniques that have been successfully exploited in resolving complex practical issues is DNN. DNNs have the capability to overcome countless problems faced by the other existing methods used in intrusion detection [5].

5. Owing to resource constraints and complexity, conventional intrusion detection techniques are not secure and safe enough for the Internet of Things (IoT). Accordingly, there is a necessity to conceive and set up smart intrusion detection systems for IoT devices, to defend them against intrusions and assaults launched from compromised IoT devices.
6. IoT objects are constrained in terms of energy consumption, memory capacity, and processing capability. Due to these limitations, classical IDS approaches are not convenient to implement directly in IoT devices. It is required lightweight solutions. In effect, Basati & Faghih [7] suggested a lightweight architecture for IDS for IoT networks.
7. As aforementioned, in IoT networks, there are inherent limitations in memory, energy, and processing resources. Sadly, our analysis together with the survey done by Al-Taleb et al. [14] show the drawback of several works of don't take into consideration both performance and security concerns during the conception of IoT-specific IDS. Owing to these constraints, we advise surmounting these issues during the designing of IoT-specific IDS.
8. The hybrid detection method for building IDS is the best method enabling to heighten detection accuracy. In fact, combining signature-based detection (SD) and anomaly detection (AD) methods enables to leverage the strengths of each technique and relieve their shortcomings. We recommend construction IDS for IoT based on the hybrid approach [15].
9. Many approaches suggest employing machine learning techniques or deep learning techniques but most of these techniques necessitate elevated computation cost such as in [11] due to the iterations, which IoT devices can't afford because it will impact the power consumption and IoT appliances are limited resource devices.
10. Memory capacity is another exigency for some of the proposed approaches such as in [5], where the predictive model uses a large number of parameters but some of the IoT devices don't support sophisticated memory capacity and only provide a few kilobytes. Also, real-time detection is a significant issue since real-time detection is substantial in IoT applications due to the sensitive data and it will aid in lessening the damages.

4 Conclusions

Internet Currently, it is recognized that the countless IoT entities being linked overall will in general grow on the day by day premise. Accordingly, the IoT paradigm invades numerous sectors such as education, surveillance, healthcare, agriculture, military, and commerce. However, each layer of IoT architecture is prone to various types of attacks and intrusions that could jeopardize the systems within the IoT. To alleviate cyber threats, many IoT networks start implementing an intrusion detection system (IDS) to detect malicious events.

In this In this paper, firstly, we provide a comprehensive analysis of twelve state-of-the-art and relevant IDS developed between 2021 and 2022 for IoT networks, issuing their key specifics, strengths, limitations, and challenges with a view to spot the concerns that still require to be addressed, well point out the mainstream of research direction, and clear the way for new avenues of research for coming scientists. Secondly, we give a guide or support for scientists interested with IDS for Internet of Things networks. To sum up, three main conclusions are drawn; first of all, a significant rise in the use of advanced datasets UNSW-NB15 and BoT-IoT for building and evaluation of IDSs designed. We advise using these datasets as they hold a hybrid of real modern normal activities and synthetic contemporary attack behaviors. In the second place, due to resource constraints and complexity, classical intrusion detection approaches are unsuitable and not secure enough for the Internet of Things (IoT), hence, an efficient intrusion detection in IoT Networks asks embracing of novel and smart techniques such as Machine Learning and Deep Learnings algorithms. Lastly, as in IoT networks, there are inherent limitations in energy, memory, and processing resources, researchers should think about both performance and security matters during the designing and assessment of IoT-specific IDS.

References

1. Liu, J., Yang, D., Lian, M., Li, M.: Research on intrusion detection based on particle swarm optimization in IoT. IEEE Access **9**, 38254–38268 (2021)
2. de Souza, C.A., Westphall, C.B., Machado, R.B.: Two-step ensemble approach for intrusion detection and identification in IoT and fog computing environments. Comput. Electr. Eng. **98**, 107694 (2022)
3. Kasongo, S.M.: An advanced intrusion detection system for IIoT based on GA and tree based algorithms. IEEE Access **9**, 113199–113212 (2021)
4. Oktay, U., Sahingoz, O.K.: Proxy network intrusion detection system for cloud computing. In: 2013 the International Conference on Technological Advances in Electrical, Electronics and Computer Engineering (TAEECE), pp. 98–104. IEEE, Konya (2013)
5. Kan, X., et al.: A novel IoT network intrusion detection approach based on adaptive particle swarm optimization convolutional neural network. Inf. Sci. **568**, 147–162 (2021)
6. Meidan, Y., et al.: N-BaIoT—network-based detection of IoT botnet attacks using deep autoencoders. IEEE Pervasive Comput. **17**(3), 12–22 (2018)
7. Basati, A., Faghih, M.M.: PDAE: efficient network intrusion detection in IoT using parallel deep auto-encoders. Inf. Sci. **598**, 57–74 (2022)
8. Jamal, A., Hayat, M.F., Nasir, M.: Malware detection and classification in IoT network using ANN. Mehran Univ. Res. J. Eng. Technol. **41**(1), 80–91 (2022)
9. ToN_IoT Datasets. https://www.unsw.adfa.edu.au/unswcanberra-cyber/cybersecurity/ADFA-ton-iot-Datasets/. Accessed 3 July 2022
10. Kumar, R., Kumar, P., Tripathi, R., Gupta, G.P., Garg, S., Hassan, M.M.: A distributed intrusion detection system to detect DDoS attacks in blockchain-enabled IoT network. J. Parallel. Distrib. Comput. **164**, 55–68 (2022)
11. Saba, T., Sadad, T., Rehman, A., Mehmood, Z., Javaid, Q.: Intrusion detection system through advance machine learning for the internet of things networks. IT Prof. **23**(2), 58–64 (2021)
12. Saheed, Y.K., Abiodun, A.I., Misra, S., Holone, M.K., Colomo-Palacios, R.: A machine learning-based intrusion detection for detecting internet of things network attacks. Alex. Eng. J. **61**(12), 9395–9409 (2022)
13. McHugh, J.: Testing intrusion detection systems: a critique of the 1998 and 1999 Darpa intrusion detection system evaluations as performed by Lincoln laboratory. ACM Trans. Inf. Syst. Secur. **3**(4), 262–294 (2000)

14. Al-Taleb, N., Saqib, N.A.: Attacks detection and prevention systems for IoT networks: a survey. In: 2020 International Conference on Computing and Information Technology (ICCIT-1441), pp. 1–5. IEEE, Tabuk (2020)
15. Chiba, Z., Abghour, N., Moussaid, K., El Omri, A., Rida, M.: A survey of intrusion detection systems for cloud computing environment. In: 2016 International Conference on Engineering & MIS (ICEMIS), pp. 1–13. IEEE, Agadir (2016)

A Coupled Graph Theoric and Deep Learning Approaches for Nonrigid Image Registration

Omaima El Bahi[1](✉), Youssef Qaraai[1], and Ahmad El Allaoui[2]

[1] Mathematical and Computer Modelling of Systems Team, Applied Mathematics, Computing and Systems Laboratory, Faculty of Sciences and Techniques Errachidia, Moulay Ismail University of Meknes, Meknes, Morocco
omaimaelbahi1l@gmail.com

[2] Decisional Computing and Systems Modelling Team, Engineering Sciences and Techniques Laboratory, Faculty of Sciences and Techniques Errachidia, Moulay Ismail University of Meknes, Meknes, Morocco
a.elallaoui@umi.ac.ma

Abstract. Image registration is one of the essential elements in computer vision and several practical domains. Its objective is to recover a spatial transformation that aligns images. This is frequently formulated as an optimization problem based on an objective function that measures the quality of transformation with respect to the picture data and some prior information. This article revisits the concept of graph cuts as an effective optimization technique for image registration. We present a combination of graph cuts with superpixels segmentation via convolutional neural network (CNN), which produces a meaningful graph representation that can overcame the high computation cost.

Keywords: Image registration · Graph cuts · Nonrigid registration · Discrete optimization · Superpixels · Convolutional neural networks

1 Introduction

Image registration, also referred to as image fusion, is the process of determining correspondence between points or regions of two or more images of the same scene, taken at different times, from different viewpoints or by different sensors. Registration is widely used in remote sensing, medical imaging, cartography, computer vision, etc. A comprehensive review of image registration can be found in previous surveys [1–3] which are among the earliest articles published in this field.

Due to the diversity of the images to be registered, and due to the different types of deformations in the image (rigid and nonrigid deformations), it is impossible to develop a universal registration method. In this research, and based on the idea of applying graph theory in image processing due to its simplicity to provide a discrete and mathematical representation of different types of images, we aim to present a nonrigid image registration method that combines graph cuts and CNN-based superpixels image representation.

Y. Farhaoui et al. (Eds.): ICAISE 2022, LNNS 635, pp. 27–32, 2023.
https://doi.org/10.1007/978-3-031-26254-8_4

This paper is organized as follows: In Sect. 2 we present the proposed method and the basics and fundamental components. Section 3 describes the results of the proposed method. Section 4 concludes this study and paves the way for future re-search through recommending new areas of inquiry.

2 Theory and Methods

2.1 Nonrigid Image Registration

Nonrigid image registration aims to find the best deformable mapping between two images. In this type of registration, one of the images is considered fixed (I), and another is considered floating (J), which is transformed to the fixed image. The goal is to find the optimal correspondence map φ from I to J that maximizes the similarity between I and $Jo\varphi$ [5]. In the literature, the objective function is generally given as a combination of two components, one is the image similarity E_{sim} between the fixed and moving image impacted by the transformation map E, while the other term is a smoothing term E_{smooth} that regularizes the transformation, this latter is provided by:

$$\varphi_{opt} = argmin_{\varphi}(E_{sim}(I, Jo\varphi) + E_{smooth}(\varphi)) \tag{1}$$

where $Jo\varphi$ expresses the transformation of J according to φ.

According to [4], nonrigid image registration can be decomposed into three phases: The transformation model, the similarity measure and the optimization method. In our case, we concentrate on the optimization phase, especially the discrete optimization which is becoming widely used in image analysis.

2.2 Graph Cuts for Image Registration

Graph-based methods have been effectively applied in many areas with different forms of signals, such as computer vision, sensor networks and traffic. This is due to the fact that a graph is a simple and flexible model to structurally represent different types of data. The application of graph theory in image processing was proposed in the early 1970s [6]. Graph cuts is one of the graph-based methods that has been successfully applied to image segmentation and stereo matching. In this work we focus on the using of this method in image registration [8, 9].

In nonrigid image registration, the transformation functions used are nonlinear. One of the oldest algorithms based on graph cuts [7, 8] proposes a method where the registration of nonrigid images is considered as a discrete labeling problem. The optimization can be formulated on an undirected graph with a set of nodes S, and used for solving a labeling problem, where every node $s \in S$ must be attributed a label f_s from a finite set of labels L. Here, the labels represent a displacement vector, which allows the use of MRFs in image registration algorithms [12]. The MRF labeling problem can be posed as an optimization process, composed of data and piecewise smoothing terms:

$$E(f) = Edata(f) + \kappa Esmooth(f) \tag{2}$$

where $Edata(f)$ corresponds to the disagreement between the labeling f and the observed data, $Esmooth(f)$ is a piecewise smoothing term and κ is a weight coefficient that determines the impact of the smoothing term.

2.3 Proposed Approach

2.3.1 Superpixel Based Image Representation via Convolutional Neural Networks

Superpixels (see Fig. 1) have been frequently employed in various computer vision tasks [10, 11, 13]. Superpixels segmentation is widely used as a preprocessing for image processing tasks and it is used to reduce the dimensionality of the discrete registration [12].

A convolutional neural network (CNN) is a deep learning algorithm. This network has been widely used in computer vision for object classification and detection because of the features inspired by the visual cortex. CNNs are characterized by convolution and pooling layers. These layers introduce partial links to reduce the number of parameters and enhance the sharing of common features. In [14] CNN are used for superpixels segmentation. This method generates superpixels via the CNN through a single label-free image by minimizing a suggested objective function for superpixel segmentation in inference time. The method has three advantages over different available methods. First, it exploits a CNN image prior for superpixel segmentation, second it adaptively changes the number of superpixels based on the provided images, and thirdly it controls the superpixel property by adding an auxiliary cost to the objective function. By using this method, we can extract the superpixels from the images. An example of CNN based segmentation on one of the road images is illustrated in Fig. 1.

2.3.2 Graph Construction

The optimization problem is posed on an undirected graph, which is defined by the adjacency of the CNN based extracted superpixels. Each pair of directly adjacent superpixels is connected by an edge. The edges values are calculated on the basis of the absolute difference between the means intensities of the superpixels using the minimum spanning tree (MST) [12]. This approach can eliminate only the edges that connect superpixels of incoherent regions, while conserving all the connections that are in coherent structures.

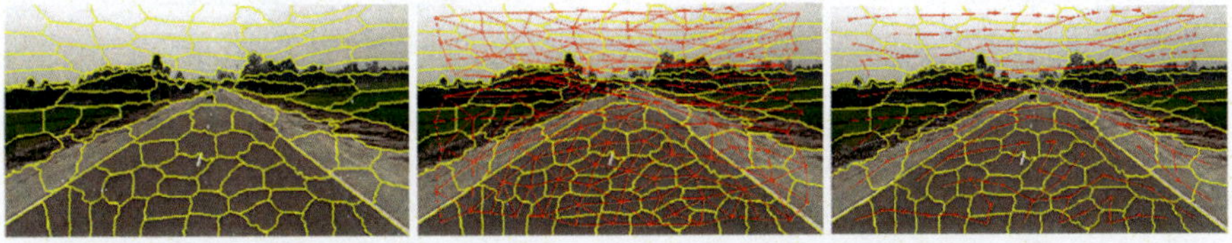

Fig. 1. Example of the graph construction of a 2D road image: the first image (left) shows the corresponding superpixels representation via CNN, with the borders of the superpixels drawn in yellow. The second shows the graph extracted with the edges connecting adjacent superpixels in red and the MST-based simplification of the graph is shown in the last image (right).

2.3.3 Similarity Measure

Several measures of similarity can be considered for intensity-based nonrigid image registration. The measure used in practice depends highly on the application and particularly on the modalities of the images which means how the images were acquired. The most commonly used similarity measures, based only on the differences between the intensities of the corresponding pixels, such as the sum of absolute differences (SAD) or the sum of squared differences (SSD). For us we consider SAD as a measure of similarity.

$$S_{SAD}(I, J) = \frac{1}{|\omega|} \sum\nolimits_{p \in \omega} |I(p) - J(p)| \tag{3}$$

2.3.4 Graph Cuts Optimization

The data cost term $Edata(f)$ is formulated as an average error computed for all pixels x in I and J grouped into a certain superpixels represented by a node p, for the applied displacement f_p:

$$Edata(f) = \Sigma p||I(xp) - J(xp + fp)|| \tag{4}$$

We suggest using the piecewise smoothing term in [8] for $Esmooth(f)$. For this kind of problem, we use the α-β-swap variant of the graph cut method [15] to optimize the energy function. We start by extracting the superpixels, then we create a resulting graph. After that, we simplify this graph by the MST method. We apply the graph cuts algorithm in order to optimize the computed energy.

3 Results and Discussion

Our proposed method is applied on 2D images in traffic flow area, the used images are converted to grey scale. We start by extracting superpixels, the number of super-pixels depends on the image dimension (we use 200 × 200 images) then we create the graph from them. The graph will be simplified according to MST method. We use the graph cuts in order to optimise the computed energy (see Fig. 2).

To evaluate our method, we compute the difference between the fixed image and the moving image, this variance is compared with the difference between the same fixed image and the registered image Fig. 3. We compute the absolute intensity difference (mean ± SD) before and after registration and we find (31.17 ± 34.98) before registration and (3.7 ± 10.34).

From the figure (Fig. 2 and Fig. 3) we can observe how the graph cuts method minimizes the dissimilarity between images after registration. The use of superpixels reduces the dimensionality of the problem and the use of CNN superpixels extraction [14] gives good results if the number of superpixels is lower than 100. This problem impacts the quality of our method which makes the computation cost longer. As a solution of this problem we propose the use of SLIC method [11, 12], which is one the most popular methods in this field due to the fast execution and direct control of the number of extracted superpixels.

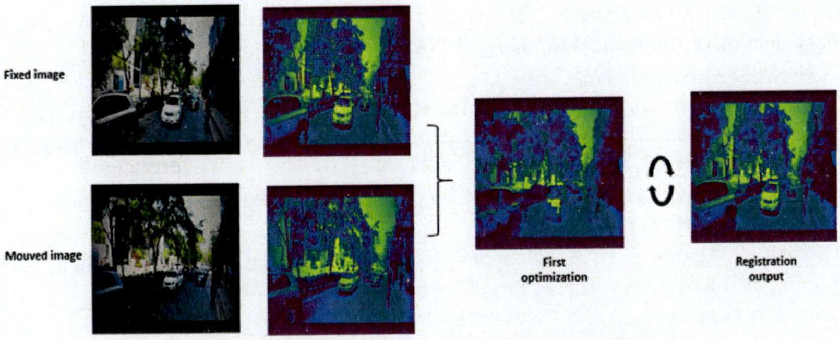

Fig. 2. Registration results of 2D road images.

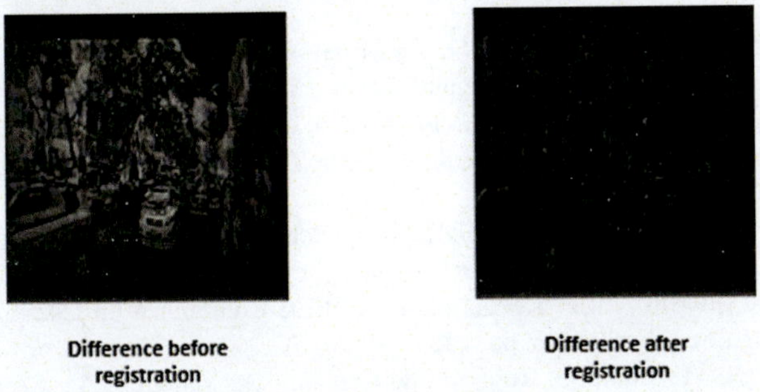

Fig. 3. Difference between fixed and moving images before and after registration.

4 Conclusion and Next Directions

In the literature graph cuts is a robust method for nonrigid registration with a better registration accuracy. For that we have proposed a nonrigid image registration, which combine CNN based superpixels representation and graph cuts as a method for discrete optimization. The reason to use CNN to extract superpixels is the changing of the number of superpixels according to the given images, but the cost of calculation is long, which has a direct effect on the registration. Actually, we are looking to reformulate the proposed work by using another technique for superpixels image representation, such as SLIC method, to accelerate the registration. Also, we are planning in the future to use deep graph cuts as a method to minimize the energy in discrete registration.

References

1. Brown, L.G.: A survey of image registration techniques. ACM Comput. Surv. (CSUR) **24**(4), 325–376 (1992)
2. Maintz, J.A., Viergever, M.A.: A survey of medical image registration. Med. Image Anal. **2**(1), 1–36 (1998)
3. Zitova, B., Flusser, J.: Image registration methods: a survey. Image Vis. Comput. **21**(11), 977–1000 (2003)

4. Sotiras, A., Ou, Y., Paragios, N., Davatzikos, C.: Graph-based deformable image registration. In: Handbook of Biomedical Imaging, pp. 331–359. Springer, Boston (2015). https://doi.org/10.1007/978-0-387-09749-7_18
5. Pham, N., Helbert, D., Bourdon, P., Carré, P.: Spectral graph wavelet based nonrigid image registration. In: 2018 25th IEEE International Conference on Image Processing (ICIP), pp. 3348–3352. IEEE, Athens (2018)
6. Zahn, C.T.: Graph-theoretical methods for detecting and describing gestalt clusters. IEEE Trans. Comput. **100**(1), 68–86 (1971)
7. Akshaya, R., Menon, H.P.: A review on registration of medical images using graph theoretic approaches. Indonesian J. Electr. Eng. Comput. Sci. **12**(3), 974–983 (2018)
8. Tang, T.W.H., Chung, A.C.S.: Non-rigid image registration using graph-cuts. In: Ayache, N., Ourselin, S., Maeder, A. (eds.) MICCAI 2007. LNCS, vol. 4791, pp. 916–924. Springer, Heidelberg (2007). https://doi.org/10.1007/978-3-540-75757-3_111
9. Lézoray, O., Grady, L.: Image Processing and Analysis with Graphs. CRC Press, Boca Raton (2012)
10. Liu, Z., Zhang, X., Luo, S., Le Meur, O.: Superpixel-based spatiotemporal saliency detection. IEEE Trans. Circuits Syst. Video Technol. **24**(9), 1522–1540 (2014)
11. Achanta, R., Shaji, A., Smith, K., Lucchi, A., Fua, P., Süsstrunk, S.: SLIC superpixels compared to state-of-the-art superpixel methods. IEEE Trans. Pattern Anal. Mach. Intell. **34**(11), 2274–2282 (2012)
12. Szmul, A., Papiez, B.W., Bates, R., Hallack, A., Schnabel, J.A., Grau, V.: Graph cuts-based registration revisited: a novel approach for lung image registration using supervoxels and image-guided filtering. In: Proceedings of the IEEE Conference on Computer Vision and Pattern Recognition Workshops, pp. 152–159 (2016)
13. Van den Bergh, M., Boix, X., Roig, G., de Capitani, B., Van Gool, L.: Seeds: superpixels extracted via energy-driven sampling. In: Fitzgibbon, A., Lazebnik, S., Perona, P., Sato, Y., Schmid, C. (eds.) ECCV 2012. LNCS, vol. 7578, pp. 13–26. Springer, Heidelberg (2012). https://doi.org/10.1007/978-3-642-33786-4_2
14. Suzuki, T.: Superpixel segmentation via convolutional neural networks with regularized information maximization. In: ICASSP 2020–2020 IEEE International Conference on Acoustics, Speech and Signal Processing (ICASSP), pp. 2573–2577. IEEE (2020)
15. Boykov, Y., Veksler, O., Zabih, R.: Fast approximate energy minimization via graph cuts. IEEE Trans. Pattern Anal. Mach. Intell. **23**(11), 1222–1239 (2001)

A Deep Convolutional Neural Networks for the Detection of Breast Cancer Using Mammography Images

Nourane Laaffat[1](✉), Ahmad Outfarouin[2], Walid Bouarifi[1], and Abdelilah Jraifi[1]

[1] Mathematics Computer Science and Communication Systems Research Laboratory, National School of Applied Sciences, Cadi Ayyad University, 46000 Safi, Morocco
nourane.laafat@gmail.com

[2] Management and Decision Support Laboratory, National School of Business and Management, Ibn Zohr University, 73000 Dakhla, Morocco

Abstract. Breast cancer (BC) is one of the malignant diseases that can affect women in the first degree. Its early diagnosis is necessary and effective in its early detection, which will contributes to its complete elimination, or slow down and delay its progression, and increases the success of treatment. Medical imaging plays an important role in the early diagnosis of diseases. In the case of BC, mammographic imaging allows clear visualization of the breast and discovery of BC in its early stages. Its classification is a technique that effectively helps radiologists in the analysis of medical images, it allows to detect pathologies and categorize them according to its appropriate stage. Deep convolutional neural network (DCNN) models have proven to be widely used in the medical field in recent years and achieve great feats in image classification, especially in terms of high performance and robustness. In this paper, we propose two approaches based on the Transfer Learning (TF) technique to classify BC using mammography images from the Mini-MIAS dataset. The first one consists in performing the classification process by fine-tuning the adopted pre-trained model with hyperparameters adjustment, while the second approach consists in using the pre-trained model as feature Extractor, then the classification phase is performed by Random Forest machine learning classifier. The DCNN model adopted in this work is VGG-19. Data augmentation technique is a necessary preprocessing step for the improvement of our small dataset. The experimental results show that the first approach classification give the highest accuracy of 97%.

Keywords: Breast cancer · Deep convolutional neural networks · Transfer learning · Mammography imaging · Data augmentation · Classification

1 Introduction

The initial attention and prominence given to the issue of breast cancer is due to the fact that it is the most common and widespread disease among all types of cancer in the world and that a very high cure rate and declining mortality rate can be achieved if

Y. Farhaoui et al. (Eds.): ICAISE 2022, LNNS 635, pp. 33–41, 2023.
https://doi.org/10.1007/978-3-031-26254-8_5

detected at a very early stage [1, 2]. Therefore, we try to focus on this subject in order to contribute in protecting the lives of many patients with this disease. According to the World Health Organization [3], in 2020, BC affected 2.3 million women and caused 685,000 deaths worldwide. In addition, in the last five years, 7.8 million women have been diagnosed with BC. Globally, women lose more years of life to breast cancer than to any other type of cancer. Breast cancer occurs in every country in the world and affects women of all ages from puberty. Mammography [1, 4, 5] is a radiographic examination that uses X-rays. It is based on the principle of x-raying each breast from the front and from the side in order to visualize the entire mammary gland. Mammography is designed specifically for the women's breast and aims to detect abnormalities if they exist before any symptoms appear [6], diagnose and examine symptoms of breast pathology, such as breast swelling, breast discharge, change in breast size and shape or skin wrinkles, nipple inversion, itching, crusty sores or rash around the breast, etc., and detect BC. Each year, a large volume of mammographic images must be analyzed, which requires intense work, enormous time and several interventions by different radiologists in order to help each other in the decision-making process, especially since this is a sufficiently delicate and essential task for the survival or not of a woman. To this end, several research studies have been directed toward automating mammography reading and decision making. Early work on automatic mammography image processing systems was aimed at providing radiologists with a second interpretation to help them detect/diagnose abnormal lesions at an early stage, regardless of their nature [7]. Specially, Computer-Aided Diagnosis (CADx) systems refer to a complete mammographic image processing system from preprocessing to classification and decision making. It aims to characterize and classify the detected anomaly as benign or malignant or as abnormal or normal.

Artificial intelligence is achieving great success in the medical field. For the classification of breast lesions, several Machine Learning (ML) and Deep Learning (DL) algorithms exist. DCNN is used in recent years in an intense way due to its success in the efficient classification of anomalies.

This work gives a presentation on the method of application of (DCNN) on the Women's breast dataset using the transfer learning technique in two different ways. The first one consists in performing the classification process by fine-tuning the adopted pre-trained model with hyperparameters adjustment, while the second approach consists in using the pre-trained model as Feature Extractors, then the classification phase is performed by Random Forest machine learning classifier. The DCNN model adopted in this work isVGG-19. Data augmentation technique is a necessary preprocessing step for the improvement of our small dataset.

The remainder of this document goes through the following structure. Section 2 is a presentation of related work. Section 3 is the methodology. Section 4 deals with the experimental results. Section 5 presents a discussion of the experimental results obtained. Finally, Sect. 6 presents a conclusion of the paper.

2 Related Work

Image classification in DL involves extracting image features directly from raw images by adjusting convolution and pooling layer parameters. DL models have made great progress in medical image classification and preventing disease such as BC, Cervical Cancer, Pneumonia, Skin disease…, especially with the emergence of TL techniques and the data augmentation that have evolved the performance rates of classification models. Transfer learning techniques have been able to remedy several problems, among which we cite the lack of data (medical images) available, since it is difficult to have large datasets, therefore, DCNN find themselves unable to better learn small datasets and end up finding themselves in the phenomenon of overfitting. The transfer learning exploits the pre-trained DCNN in different ways. Either to extract the features, and keep the important knowledge and pass it to a classification model to use it. Either to make specific modifications to it in order to obtain good results.

Alantari and Kim [8] contribute with a breast lesion detection and classification system. The detection phase is performed through the YOLO detector and obtains F1 scores of 99.2% and 98.02%, successively, for the DDSM and INbreast datasets. The classification phase is performed using regular feedforward CNN, ResNet-50, and InceptionResNet-V2 classifiers. For DDSM and INbreast datasets, successively, CNN classification models achieve an accuracy of 94.5% and 88.7%, ResNet-50 achieves 95.8% and 92.5%, and InceptionResNetV2 achieves 97.5% and 95.3%.

N. S. Ismail and C. Sovuthy [9] apply VGG16 and ResNet50 on the IRMA dataset and compare their performance to find the best breast cancer detector and classifier. VGG16 proves to be the top classifier, its accuracy is 94% while that of ResNet50 is 91.7%. In reference [10], the authors contribute with a CNN-based breast cancer mass classifier. They compare the performance criterion of the different models AlexNet and GoogleNet, which vary according to the design and hyperparameters, to find the best classifier using images from four mammogram databases named CBIS-DDSM, INbreast, MIAS and images from Egyptian National Cancer Institute (NCI). The classification models of AlexNet and GoogleNet, achieve accuracies of 100%, and 98.46%,respectively, for the CBISDDSM database, and 100%, and 92.5%, respectively, for the INbreast database, and 97.89% with AUC of 98.32%, and 91.58% with AUC of 96.5%, respectively, for NCI images, and 98.53% with AUC of 98.95%, and 88.24% with AUC of 94.65%, respectively, for MIAS database. Thus, AlexNet model proves to be the most accurate classifier. In reference [11], the authors present a new system idea to segment and classify breast cancer images. They performed the classification into malignant and benign using the InceptionV3, DenseNet121, ResNet50, VGG16 and MobileNetV2 models on MIAS, DDSM and CBIS-DDSM datasets. The proposed model with the classification phase performed by the Inception v3 model, especially with the DDSM dataset, achieves the best result with an accuracy of 98.87%, an AUC of 98.88%, a sensitivity of 98.98%, a precision of 98.79%, 97.99% of F1-score and 1.2134 s for computation time.

In reference [12], the authors contribute with a Covid-19 classifier through pre-trained DCNN models Xception, InceptionV3, MobileNetV2, VGG16, VGG19 and NasNetLarge. The study is done on a dataset consisting of chest x-ray images and formed by three classes: Normal, Covid and Pneumonia. During the training, the hyper-parameters used are: learning rate (1e-3), epochs (35) and batch size (8).The fine-tuned pre-trained VGG-16 model succeed to achieve the highest accuracy of 95.88%.

3 Methodology

Our proposed system consists of 4 steps, we illustrate its architecture in Fig. 1 (Fig. 1). The first step is to preprocess the mammographic images of women's breast from the Mini-Mias dataset which is freely available for scientific research. The goal of this step is to improve the size of the dataset and the quality of its images. Then, the second step is to divide our dataset into three (training, testing, and validation). Then, step 3 which concerns feature extraction and performed by the pre-trained DCNN model VGG-19 through the transfer learning technique. Finally, the fourth step is performed in two different ways. The first way, which presents the first approach, is to refine the pre-trained models, while implementing the softmax function at the last layer of the classification network (FC) to specify the predicted output probabilities for mammography class determination, in this case there are two outputs/classes: normal or abnormal. The second way is to classify the extracted features by an ML classifier, in this case RF, which results in approach 2.

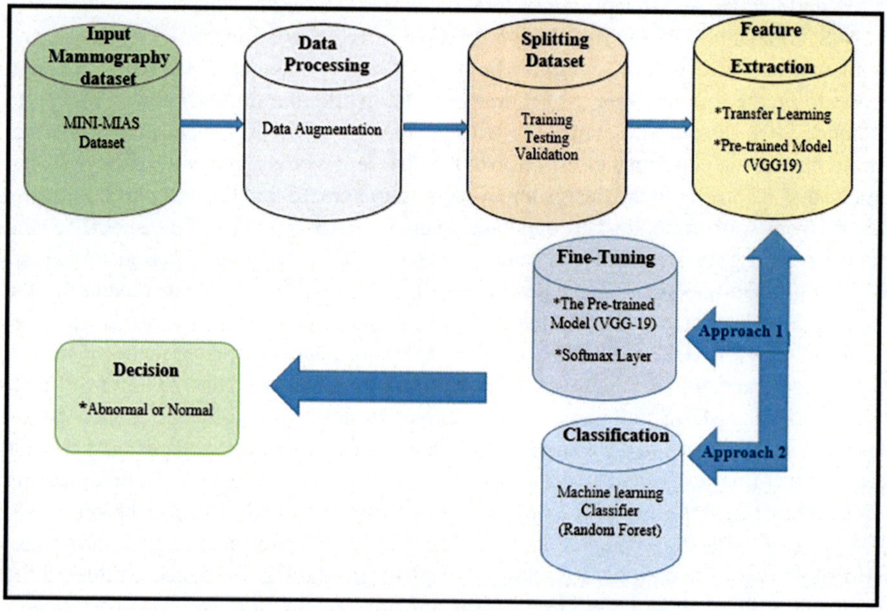

Fig. 1. Overview of our proposed system.

For the input stream of our system, we use the Mini-MIAS dataset [13] which has been made available free of charge for scientific research, through the Pilot European Image Processing Archive (PEIPA) of the University of Essex. It was collected and generated by a group of researchers from the UK at the Mammographic Image Analysis Society (MIAS), who are interested in understanding mammographic images and detecting breast lesions. The dataset consists of 322 mammographic images of the mediolateral oblique views of 1024*1024 pixels, including 113 images with abnormalities that may be benign or malignant and 209 without abnormalities (Normal) and divided into 2 classes (35, 1% abnormal and 64, 91% normal). Given the small size of our dataset, we necessarily go through the preprocessing technique to overcome the overfitting and improve the learning capability of the proposed model. For this, we use data augmentation to upsurge the samples number, especially those of the training set. It includes random image rotation, zooming and shearing, horizontal and vertical random flipping.

4 Experimental Results

It is always a question of finding powerful and robust models that will allow an accurate diagnosis as much as possible. We thus evaluate the performance of the models adopted, through the parameters measuring the effectiveness and efficiency. For BC classification, performance evaluation is performed by some commonly used measures such as accuracy, precision, recall and sensitivity [14]. We mention that for the first model based on fine-tuning, we fix, during the training process, the different hyper parameters used, as follows:

- Optimizer = Adam,
- learning rate = 0.00005,
- Batch size = 32,
- Epoch = 10.

4.1 Effectiveness

This section evaluates the efficiency of all adopted classifiers through the model speed prediction, correctly classified instances, misclassified instances and accuracy. Our Testing set consists of 32 images. We present the results in Table 1.

Simulation errors are also taken into account to better evaluate the performance of the classifiers. In this study, the classifiers effectiveness is also evaluated through:

- Kappa statistic (KS),
- Mean Absolute Error (MAE),
- Root Mean Squared Error (RMSE),
- Relative Absolute Error (RAE),
- Root Relative Squared Error (RRSE).

KS, MAE and RMSE are in numeric values. RAE and RRSE are in percentage. The results are shown in Table 2.

Table 1. Performance of the classifiers.

Evaluation criteria	Classifiers	
	Proposed model 1	Proposed model 2
Speed of prediction (s)	0.53	0.02
Correctly classified instances	31	29
Incorrectly classified instances	1	3
Accuracy (%)	97	91

4.2 Efficiency

This section evaluates the adopted classifiers efficiency through precision, recall, TPR (true positive rate), TNR (true negative rate), FPR (false positive rate), and FNR (false negative rate) for all proposed classifiers, directly after building the predictive model. We summarize the results of this comparison in Table 3 and Table 4. We present the confusion matrices that also serve as evaluators of the classifiers. Each row in Table 5 represents the rates in an actual class while each column refers to the predicted class.

Table 2. Simulation error.

Evaluation criteria	Classifiers	
	Proposed model 1	Proposed model 2
Kappa Statistic (KS)	0,93	0,78
Mean Absolute Error (MAE)	0,03	0,09
Root Mean Squared Error (RMSE)	0,18	0,31
Relative Absolute Error (RAE) %	6,92	20,77
Root Relative Squared Error (RRQE) %	37,22	64,46

Table 3. Comparison of TP, TN, FP, RN rates for the two proposed models.

Classifiers	Evaluation criteria			
	TPR	TNR	FPR	FNR
Proposed model 1	1	0,95	0,04	0
Proposed model 2	0.9	0,91	0,09	0,1

5 Discussion

From Table 1, we can see that the first proposed model takes about 0.53 s to make its predictions on the test dataset, unlike the second proposed model which takes 0.02 s.

Table 4. Comparison of accuracy measures for the two proposed models.

Classifiers	Evaluation criteria				Class
	Precision	Recall	F1-score	AUC	
Proposed model 1	1.00	0.91	0.95	0.95	Abnormal
	0.95	1.00	0.98		Normal
Proposed model 2	0.90	0.82	0.86	0.88	Abnormal
	0.91	0.95	0.93		Normal

Table 5. Confusion matrices

Classifiers	Actual class		Predicted class
	Abnormal	Normal	
Proposed model 1	10	1	Abnormal
	0	21	Normal
Proposed model 2	9	2	Abnormal
	1	20	Normal

On the other hand, the accuracy obtained by the first model (97%) is better than the accuracy obtained by the second model (91%). It can also be easily seen that the first model correctly classifies the largest number of instances with the smallest number of misclassified instances compared to the second model.

From Table 2, we can see that the first model has more chance of having the best classification according to the value of kappa statistic (0.93%) with the lowest error rate (0.03). The second model highest value of error rate explains its incorrectly classified instances (3) comparing to the first model (1).

It is possible to determine whether a model works well or not by its ability to detect the anomaly. For this, the values of TPR and TNR must be high, and those of FPR and FNR must be as low as possible. The capacity of correct identification of the cases with the anomaly is produced by TPR which represents the sensitivity, as well as that of the cases without the anomaly is produced by TNR which represents the specificity. Table 3 shows that the first model obtained the highest value of TPR and TNR (100% and 95% respectively) and the lowest value of FPR and FNR (0.04 and 0 respectively). The second model obtained the lowest value of TPR and TNR (90% and 91% respectively) and the highest value of FPR and FNR (0.09 and 0.1 respectively).

The area under the curve (AUC) measures how well a classifier is able to discern between existing classes. The chance of having a model that performs well and has a high ability to distinguish between normal and abnormal or benign and malignant or negative and positive anomaly (different classes) depends on a high AUC value. From Table 4, we can see that the first model has the highest AUC value (0.95) unlike the second model (0.88).

Table 5 shows the confusion matrix, and from which we compare the actual class and the predicted results obtained. The first model correctly predict 31 instances out of 32 instances (10 Abnormal instances that are indeed Abnormal and 21 Normal cases that are in fact Normal), and 1 poorly predicted case (1 case of Normal class predicted as Abnormal and 0 cases of Abnormal class predicted as Normal). This justifies why the accuracy of the first model is better than the second classifier, with lower error rate value.

In summary, the first proposed model was able to distinguish itself through its effectiveness and efficiency compared to the second model. Returning to the speed of prediction shown in Table 1, we can say that the first model takes longer given its architecture and its high power in making the prediction with more accuracy.

6 Conclusion

To analyze biomedical images in order to detect anomalies and classify them, there are several techniques and models in Deep learning and Machine learning. The challenge still remains to achieve a robust model: Fast, stable and accurate. In this work, we proposed an architecture of a CADx for diagnosis of Breast Cancer using the Mini-MIAS dataset which consists of mammographic images. It is based on two different proposed approaches and uses the transfer learning technique of VGG-19 being a pre-trained model. The first model proposed which relies on the fine-tuning of the pre-trained VGG-19 model achieved 97% of accuracy, while the second model where the pre-trained model is used as a feature extractor and the classification is done by using Random Forest being a machine learning classifier, achieved 90% of accuracy. The first model surpasses the second in terms of performance. In general, the Fine-tuning technique is well suited for small datasets.

References

1. NCI (National Cancer Institute). Annual Report to the Nation: Cancer Death Rates Continue to Decline; Increase in Liver Cancer Deaths Cause for Concern, 9 March 2016. https://www.cancer.gov/news-events/pressreleases/2016/annual-report-nation-1975-2012
2. Kavitha, T., et al.: Deep learning based capsule neural network model for breast cancer diagnosis using mammogram images. Interdisc. Sci. Comput. Life Sci. **14**, 113–129 (2021). https://doi.org/10.1007/s12539-021-00467-y
3. WHO Homepage. https://www.who.int/news-room/fact-sheets/detail/breast-cancer
4. Rampun, A., et al.: Breast pectoral muscle segmentation in mammograms using a modified holistically nested edge detection network. Med. Image Anal. **57**, 1–17 (2019). https://doi.org/10.1016/j.media.2019.06.007
5. Punitha, S., Amuthan, A., Joseph, K.S.: Benign and malignant breast cancer segmentation using optimized region growing technique. Future Comput. Inform. J. **3**(2), 348–358 (2018). https://doi.org/10.1016/j.fcij.2018.10.005
6. Cheikhrouhou, I.: Description et classification des masses mammaires pour le diagnostic du cancer du sein. (Description and classification of breast masses for the diagnosis of breast cancer). Doctoral thesis. University of Évry Val d'Essonne, France, (2012). https://dblp.org/rec/phd/hal/Cheikhrouhou12

7. Astley, S., Gilbert, F.: Computer-aided detection in mammography. Clin. Radiol. **59**(5), 390–399 (2014)
8. Al-antari, M. A., Kim, T.-S.: Evaluation of deep learning detection and classification towards computer-aided diagnosis of breast lesions in digital X-ray mammograms. Comput. Methods Programs Biomed. **196**, Article 105584 (2020)
9. Ismail, N.S., Sovuthy, C.: Breast cancer detection based on deep learning technique. In: 2019 International UNIMAS STEM 12th Engineering Conference (EnCon), pp. 89–92 (2019). https://doi.org/10.1109/EnCon.2019.8861256
10. Hassan, S.A., Sayed, M.S., Abdalla, M.I., Rashwan, M.A.: Breast cancer masses classification using deep convolutional neural networks and transfer learning. Multimedia Tools Appl. **79**(41–42), 30735–30768 (2020). https://doi.org/10.1007/s11042-020-09518-w
11. Salama, W.M., Aly, M.H.: Deep learning in mammography images segmentation and classification: automated CNN approach. Alexandria Eng. J. **60**(5), 4701–4709 (2021). ISSN 1110-0168, https://doi.org/10.1016/j.aej.2021.03.048
12. Makris, A., Kontopoulos, I., Tserpes, K.: COVID-19 detection from chest x-ray images using deep learning and convolutional neural networks. In: Proceedings of the 11th Hellenic Conference on Artificial Intelligence (2020). https://doi.org/10.1101/2020.05.22.20110817
13. http://peipa.essex.ac.uk/info/mias.html
14. Murtaza, G., et al.: Deep learning-based breast cancer classification through medical imaging modalities: state of the art and research challenges. Artif. Intell. Rev. **53**(3), 1655–1720 (2019). https://doi.org/10.1007/s10462-019-09716-5

A Dynamic Panel Data Analysis on the Impact of Quality of Institutions on the Economic Growth: Evidence from Wu-Hausman Specification Test

El Houssaine Fathi[1](✉), Ahlam Qafas[2], and Youness Jouilil[3]

[1] Department of Economics, Faculty of Economics and Management, Ibn Tofail University of Kenitra, Kenitra, Morocco
e.fathi91@gmail.com

[2] National School of Commerce and Management, Ibn Tofail University of Kenitra, Kenitra, Morocco

[3] Department of Economics, Faculty of Economics and Social Sciences Hassan II, University of Casablanca, Casablanc, Morocco

Abstract. In this paper research, we have tried to modulate and analyze the real impact of the quality of institutions on economic growth through a dynamic panel data analysis. Indeed, Stata Wu-Hausman's test revealed that the fixed effect model is the most appropriate model for our case study. Hence, econometric findings showed that, for all other things being equal, there is a positive and significant nexus between the quality of institutions and economic growth ($p < 5\%$).

The results obtained confirm that the quality of the institutional environment can have a stimulating effect on economic growth. In this context, policymakers should implement policies that strengthen these institutions in order to expect sustainable economic growth.

Keywords: Mathematical modeling · Panel analysis · Fixed effect · Random effect · Wu-Hausman test

1 Introduction

Economic growth with different rates has always been a subject of debate for economic researchers. Indeed, economists have always tended to explain the difference in growth through factors mainly related to the economic and geographical assets of countries, especially the availability of natural resources, human capital, the opening of the economy, and geographical position. However, this observation is being questioned with the emergence of economies that do not have these assets.

To this end, since the 1970s, researchers have sought to introduce other factors that explain economic growth, particularly the quality of institutions, most notably the regulatory system. Thus, the role of institutions has been for several economists a key

Y. Farhaoui et al. (Eds.): ICAISE 2022, LNNS 635, pp. 42–47, 2023.
https://doi.org/10.1007/978-3-031-26254-8_6

factor for growth. North and Tomas (1973) were probably the first to attempt to examine the relationship between institutional quality and economic growth.

In this succession of studies (Olaoye et al., 2019; Kim et al., 2018; Gurvich, 2016; Hall and Ahmad, 2012; Slesman et al., 2015; Breeny and Gillanders, 2010; Acemoglu et al., 2001; Acemoglu and Johnson, 2005; Keefer and Knack, 1997; North, 1991; North and Thomas, 1973) there is evidence that institutional quality is a better explanation of economic growth than neoclassical (see Solow, 1956; Cass, 1965) and endogenous (see Lucas, 1988) theory.

In this paper, we will contribute to this strand of studies by trying to analyze the real impact of the quality of institutions approached by the legal system on economic growth through a panel data analysis.

The manuscript will be structured as follows. In the first part, we will expose the mathematical model related to our study. In the second one, we will estimate the degree of causality between our exogenous factors namely the quality of institutions, and the endogenous ones, especially the Economic growth using STATA software. Lastly, we will discuss the results and conclude.

2 Mathematical Modeling

The objective of this section is to present the econometric model that associate the variable of interest to the set of explanatory variables (including the variable that approaches the quality of institutions). Indeed, it is a longitudinal model with two indices; time and individual (country).

Hence, let Y_{it} be a quantitative variable at time t ($t \in Z$) for the individual i ($i \in N^*$). Let's consider the following regression model:

$$\begin{aligned}
&Y_{it} = \alpha_{it} + \beta_0 QL_{it} + X'_{it}\beta + \mu_{it} \\
&\mu_{it} = \alpha_i + \zeta_{it} \\
&i = 1, \ldots, n \\
&t = 1, \ldots, T
\end{aligned} \quad (1)$$

The previous model can also be expressed by the following equation:

$$\begin{aligned}
&Y_{i,t} = \alpha_{it} + + \sum_{j=1}^{K} \beta_j X_{i,t,j} + \mu_{i,t} \\
&\zeta_{i,t} \sim i.i.d\left(0, \sigma^2\right) \\
&\mu_{i,t} = \alpha_i + \zeta_{i,t} \\
&i = 1, \ldots, n \\
&\zeta_{i,t} \sim i.i.d\left(0, \sigma^2\right)\sigma > 0
\end{aligned} \quad (2)$$

Regarding our case study, our dynamic panel econometric model can be expressed as follows:

$$\begin{aligned} &\mathrm{GDP}_{\mathrm{i,t}} = \alpha_{it} + \beta_0 QL_{\mathbf{i,t}} + \sum_{j=1}^{K} \beta_j X_{\mathrm{i,t,j}} + \mu_{\mathrm{i,t}} \\ &\mu_{\mathrm{i,t}} = \alpha_i + \zeta_{i,t} \\ &\mathrm{i} = 1, \ldots, \mathrm{n} \\ &\mathrm{t} = 1, \ldots, \mathrm{T} \\ &\zeta_{i,t} \sim i.i.d\left(0, \sigma^2\right)\sigma > 0 \end{aligned} \quad (3)$$

where:

- $\mathbf{GDP_{i,t}}$ is the dependent variable for the country i at time t. It represents the gross domestic product growth.
- $\boldsymbol{QI}$ refers to the quality of institutions which has been approached by the quality of regulation (REG_qual).
- $\mu_{\mathbf{it}}$ is the error term that is assumed to be independent and identically distributed ($\mu_{\mathrm{it}} \sim i.i.d\left(0, \sigma^2\right)$).
- α_{it} represents the individual effect.
- $\boldsymbol{\beta} = (\beta_1, \beta_2, \ldots \beta_K)R^K$ the set of K parameters.
- $\boldsymbol{X}_{\mathbf{i,t}} = \left(X_{i,t}^1, X_{i,t}^2 \ldots X_{i,t}^K\right)$ the set of K control variables

It should be mentioned that X includes the Inflation rate (INF), Government Expenditure (GOV_expend), Population Growth (P_growth), Opening of the Economy index (OE_index), Human Capital (CAPIH), and the initial gross domestic product per capita (GDP_init) that was taken by normal logarithm.

3 Estimation and Discussion

3.1 Estimation

To estimate our econometric model, we have used STATA 16 to obtain the parameters shown in Table 1. The results revealed that the variable of interest and the other explanatory variables generally produce effects in the expected directions.

Note:

Standard errors in parentheses

*** $p<0.01$, ** $p<0.05$, * $p<0.1$

FE indicates the fixed effect model

RE indicates the random effect model

Table 1. Fixed effect model versus random effect model

Variables	FE	RE
REG_qual	1.249134 ***	1.230605 *
	(0.001)	(0.058)
Log_GDP_init	−.405652***	−.0000345
	(0.017)	(0.305)
CAPIH	2.637706***	2.884457
	(0.036)	(0.141)
P_growth	−.0327754	−.1229468
	(0.402)	(0.459)
GOV_expend	−.1462637***	−.1352333
	(0.003)	(0.178)
INF	−.0235889***	−.1785004 ***
	(0.039)	(0.000)
OE_index	−.6063311	−2.002039**
	(0.116)	(0.028)
Constant	5.833783***	8.224214***
	(0.000)	(0.000)

Source: Authors' calculation using STATA software

3.2 Fixed or Random Effect Model

In order to decide what model to use - fixed effect or a random effect - it is necessary to apply the Wu-Hausman test (Griliches, Z. and Hausman, J.A., 1986). This test is based on the examination of these two hypotheses.

The first one implies that there is no correlation between the error term and the independent variables in our econometric model (The appropriate model is Random effects) while the other hypothesis implies that the correlation between the error term and the independent variables is statistically significant (the appropriate model is Fixed effects).

Statistically, we can write:

$$H_0 : \mathrm{Cov}(\alpha_i, X_{it}) = 0 \mathrm{versus} H_1 : \mathrm{Cov}(\alpha_i, X_{it}) = 0 \tag{4}$$

Wu-Hausman test is calculated from / based on the following expression:

$$\mathrm{H} = (\widehat{\beta^{RE}} - \widehat{\beta^{FE}}), (\mathrm{Var}(\widehat{\beta^{RE}}) - \mathrm{Var}(\widehat{\beta^{FE}})), (\widehat{\beta^{RE}} - \widehat{\beta^{FE}}) \tag{5}$$

The table below (Table 2) shows the STATA result of the Wu-Hausman test. It appears that the fixed effect is the most appropriate in our case since the associate P-value is superior to five percent (0.0357).

Table 2. Wh-Hausman test

Chi2 (6)	Prob > Chi2
12.48	0.0357

Source: Authors' calculation using STATA software

3.3 Discussion of the Findings

Regarding the control variables, we found that the coefficient associated with the population growth is negative but not significant (p-value = 0.402), which supports the hypothesis that the population explosion is occurring at the expense of economic growth. Population growth in this region has stabilized significantly since the late 1990s and early 2000s. The quality of human capital indicator has a significate positive association with economic growth in the MENA region. It is worth reminding that the level of education has been used as a proxy for the quality of human capital, which confirms the favorable impact of education on the economic growth of countries.

The coefficient associated with the inflation parameter is significantly negative, which indeed proves its detrimental impact on economic growth in the MENA region. It affects the macroeconomic distribution of income in MENA countries, and economic actors are unable to align their income growth with that of inflation. Therefore, the coefficient on initial GDP per capita is negative, which confirms the theory of conditional convergence.

The coefficient on government expenditure is negatively significant. This shows the unfavorable impact of increased public spending in the MENA region on its economic growth. In general, the rate of public expenditure in this region directly impacts its economic growth rate through the weight of public investment, especially investment in infrastructure or investment created by public enterprises. In terms of trade openness, measured by the opening of the economy index (OE_index), it was expected to have a positive effect on the economic growth rate, but the empirical study did not confirm our expectations.

As for our variable of interest, which represents the quality of the institutions, we found that Regulatory System (REG_qual) has a significant positive impact on the economic growth rate. This result is in accordance with the research of Olson et al. (1998), who proved that economic growth is directly correlated with governance effectiveness, thus (Kauffman et al., 2005) confirmed empirically that governance effectiveness positively stimulates the economic growth of a given economy.

4 Conclusion and Policy Implications

The empirical study of the relationship between institutional quality and economic growth was the subject of this paper. We tested our basic hypothesis using a sample of 10 countries over a 20-year time series, with an indicator of institutional quality and numerous control variables. The results obtained confirm that the quality of the institutional environment can have a stimulating effect on economic growth. In this context, policymakers should implement policies that strengthen these institutions in order to expect sustainable economic growth.

In spite of our theoretical reflection and empirical investigations, the results of our research work cannot be ideal. Any attempt at reality research admits in absolute terms, criticisms relative to the conditions of its execution some of them are due to the luck of the dataset and the others are due to the heterogeneity of our target population.

In addition, this result should be examined very closely since the intercept was highly significant which means that other factors explaining the variations in our target were not considered in our econometric model.

References

Ahn, S.C., Low, S.: A reformulation of the hausman test for regression models with pooled cross sectional time series data. J. Econom. **71**, 309–319 (1996)

Andersen, T., Tarb, F.: Financial liberalization, financial development and economic growth. J. Int. Dev. **15**, 189–209 (2003)

Arellano, M.: On the testing of correlated effects with panel data. J. Econom. **59**, 87–97 (1993)

Arestis, P., Demetriades, K.: Financial development and economic growth : the role of stock markets. J. Money Credit Bank **33**(1), 16–41 (2001)

Arestis, P., Demetriades, P.: Financial development and economic growth: assessing the evidence. Econ. J. **107**, 783–799 (1997)

Barro, R.: Economic growth in a cross section of countries. Q. J. Econ. **106**(2), 407–443 (1991)

Barro, R., Sala-i-martin, X.: Technological diffusion, convergence, and growth. J. Econ. Growth 2(1) (1995)

Belfqih, et al.: Investigating the nexus between FDI and institutional quality: Evidence from Morocco. Global J. Emerging Market Econ. 1–29 (2021)

Blundell, R., Bond, S.: Initial conditions and moment restrictions in dynamic panel data models. J. Econom. **87**, 115–143 (1998)

Demetriades, P., et al.: Finance, institutions and economic growth. University of Leicester, Discussion Papers in Economics 04/5 (2004)

Durbin, J.: Errors in Variables. Rev. Int. Statis. Institute **22**, 23–32 (1954)

Golub, G.H., Van Loan, C.F.: Matrix Computations, third edition, North Oxford Academic Press (1983)

Griliches, Z., Hausman, J.A.: Error in variables in panel data. J. Econom. **31**, 93–118 (1986)

Lechheb, H., Ouakil, H., Jouilil, Y.: Economic growth poverty and income inequality: implications for lower- and middle-income countries in the era of globalization. J. Private Equity (2019)

Youness, J.: The individual determinants of financial inclusion in Morocco: an exploratory study in the case of a credit application In: 2022 2nd International Conference on Innovative Research in Applied Science, Engineering and Technology (IRASET), pp. 1–6 (2022)

Youness, J., Driss, M.: An ARIMA model for modeling and forecasting the dynamic of univariate time series: the case of moroccan inflation rate. In: 2022 International Conference on Intelligent Systems and Computer Vision (ISCV), pp. 1–5 (2022)

Kuran, T.: Why the Midle east is economically underdeveloped: historical mechanism of institutional stagnation. J. Econ. Perspectives **18**(3), 71–90 (2004)

Kaufmann, et al.: Governance Matters IV: Governance Indicators for 1996–2004, World Bank (May 2005)

Platteau, J.-P.: Institutions, Social Norms, and Economic Development. Harwood, Amsterdam (2000)

Jouilil, Y., Lechheb, H.: Effect of Moroccan health insurance on individuals' healthcare utilization and expenditures. J. Private Equity **22**(1), 87–90 (2018)

A Literature Review of Solar Cell Overheating Control

Imad Laabab[1(✉)], Said Ziani[2], and Abdellah Benami[1]

[1] Laboratory of Mechanics, Energy Efficiency and Renewable Energies (M3ER), Department of Physics, Faculty of Sciences and Techniques of Errachidia (FSTE), Moulay Ismail University, Meknes, BP-509 Boutalamine, 52 000 Errachidia, Morocco
im.laabab@edu.umi.ac.ma, a.benami@fste.umi.ac.ma

[2] Laboratory of Networks, Computer Science, Telecommunication, Multimedia (RITM), Department of Electrical Engineering, High School of Technology (ESTC), Hassan II University, Casablanca, Morocco
said.ziani@univh2c.ma

Abstract. Energy efficiency is increased by optimizing buildings. Two potential energy sources are solar panels and thermal sensors. Methods for managing solar panel performance have been developed and advanced for decades, and they are now becoming ingrained in our daily lives and starting to have a substantial impact on sustainability. This research offers a cutting-edge evaluation of several methods for solar panel overheating protection that improves their functionality and lengthens their lifespan. Following an introduction to protection strategies, it covers recent patent applications that deal with this issue with an emphasis on how each strategy should be applied. The conclusion of this essay is therefore advised to include a consideration of the benefits and drawbacks of many defense methods. It's hoped that the results described in this review paper will serve as a helpful tool for researchers.

Keywords: Energy efficiency · Solar panel · Sustainability · Overheating · Protection

1 Introduction

The term overheating is used a lot by specialists to describe the evaporator, but it doesn't mean what it means. In fact, you can measure how hot something is. It is the difference in temperature between two places. The difference between the actual vapor temperature of the refrigerant and the saturation temperature of the refrigerant at the same point is used to measure superheat. There are two kinds of superheat at the bottom of the system: superheat from the evaporator and total superheat. Evaporator superheat starts when the vapor going into and out of the evaporator is 100% saturated. The change from liquid to gas depends on this concentration. This is a pressure and temperature chart showing the temperature at this point in the process. At the evaporator outlet is where the remote bulb for the thermostatic expansion valve is. Technicians usually put a thermocouple or thermistor at the evaporator's outlet to ensure the temperature stays

Y. Farhaoui et al. (Eds.): ICAISE 2022, LNNS 635, pp. 48–53, 2023.
https://doi.org/10.1007/978-3-031-26254-8_7

the same. This outlet has a pressure gauge that can measure how hot the saturated steam is. Some evaporator manufacturers put a Schrader fitting near the bulb of the evaporator so that the pressure can be measured. Currently, one of the most promising sources for supplying humanity's energy demands is solar energy. The benefit of this energy is that it is clean and renewable, and Morocco is regarded as one of the solar energy industry's most promising nations. Summer is the best time for solar panels. This is where they generate the most energy. But it's also the time of year when people go on vacation and leave their homes unoccupied. Solar collectors suffer from this major problem which is the phenomenon of overheating. Therefore, it's prudent to consider the risks and difficulties related to overheating in photovoltaic and thermal panels. In this work, we present a review of the methods [1], which are protecting solar panels from the overheating phenomenon, to evaluate each type and what is positive or negative. Also, we will figure out the new possible area of research as our main objective in this overview.

2 Autonomous Technologies for Preventing Overheating in Solar Collectors

Hajji A. figured out that overheating happens when a solar water heater absorbs more solar energy than its principal heat transfer fluid circuit can handle. This leads to excessive temperatures at the absorber. In the same way, for a photovoltaic panel, it is observed when the conditions are unfavorable (important solar radiation, over-dimensioning…) [2].

During periods of intense sunlight, it may occur on a solar installation for at least two main reasons:

- When a pump fails, an exchanger becomes clogged, or simply when there is a power failure, the recovered solar energy is no longer delivered to the solar storage volume.
- When the collected solar energy is not used, for instance when the building is uninhabited, or the solar installation is oversized.

Solutions were presented in the 2020s by Hajji A., Lahlou Y., and Abbou A. It's part of the theory and practice of variable structure systems whose principal aims are to protect solar panels from overheating [3].

This automatic system serves two purposes: it guards solar collectors against overheating and dust. This technique employs a blind that adjusts to the weather by rising and falling. The solar collector's efficiency is improved by this system.

The solar water heater was modeled using MATLAB/SIMULINK to illustrate the impact of overheating on a PV collector in the case of stagnation.

Based on our literature review [1,2,4], we can conclude that this autonomous protection system has excellent performance in terms of adaptability to different weather conditions. However, this solution has some shortcomings such as the complexity due to the over-parameterization and the inaccurate estimation of the maximum heat limit of the system components.Slaman M. and Griessen R. have shown that the prismatic form is a simple yet effective protection method against overheating. A prismatic structure can easily make using extrusion. A straightforward yet efficient self-limiter for the stagnation temperature is also provided by the switching fluid's evaporation. It is not necessary to significantly change the design of solar collectors to adopt a switchable prismatic form [5].

3 Innovative Smart Selective Coating

High-efficiency solar thermal systems often experience stagnant conditions where solar panel temperatures can reach as high as 190–200 °C. One of the main issues with solar thermal systems is known to be their stagnant environments. Stagnant solar thermal systems can be reduced to 150 °C because of a new intelligent method, which is distinguished by a significant boost in the infrared emissivity thermochromic effect [6]. Hot water production is ensured by the new thermochromic thermal solar system's high performance because of its low emissivity (6%) at low temperatures, high solar absorption factor (>94%), and the onset of the thermochromic effect at about 70 °C. This example shows the value of smart selective coating. The properties of a recent generation of selective coatings established on Alumina and Vanadium mixtures were discussed as reported in [6].

Images captured by an infrared camera and Fourier Transform Infrared Spectroscopy evidently demonstrated a sharp rise in emissivity at temperatures above 70 °C. With regard to the newest versions using thermochromic solar panels, aging behavior (elevated temperature, humidity, and temperature variations) was also shown to assure a minimum lifespan of 25 years.

4 The Usage of Supercritical Fluids

Due to supercritical fluid's capacity to work at temperatures higher than thermal oil, researchers are investigating its use in parabolic solar collectors. The utilization of several supercritical fluids in parabolic trough solar collectors was investigated in [7].

Zaharil H. A. examined and studied the effectiveness of various supercritical fluids.

The commercial software EES was used to construct, validate, and solve a 1-D mathematical model. At common inlet temperatures, the results demonstrated that water outperforms the other fluids in terms of energy performance. At an inlet temperature of 477 °C, all liquids displayed their highest energetic performance, with 46.46%, 46.24%, and 45.76% for Water, Sulphur hexafluoride, and Nitrogen, respectively [7].

5 Nanofluids in Solar Thermal Collectors

The majority of solar collectors could benefit from employing nanofluids to increase overall efficiency [8,9]. To highlight the effect of using nanofluids on solar thermal collectors, modeling of this product in a numerical simulation was done. This also appears to be applicable to synchronous motors [10,11].The use of thermophysical parameters from the heat transfer fluid in the collector flow models is one of the issues with the numerical simulation of nanofluids. The estimation of the manifold flow and heat transfer would be inaccurate because of any mistakes in the model of the thermo-physical characteristics of the nanofluid.

6 Hybrid Solar Collectors and Overheating

Thermal collector and photovoltaic technologies have gotten a lot of interest because they resolve the problem of unwanted overheating of solar panels [12,13].

A photovoltaic module for converting solar energy into electrical energy and a high-efficiency thermal conversion module that use a thermal fluid are combined to create a photovoltaic-thermal hybrid solar collector. The main objective of this advancement in solar technology is to cool PV panels to maximize electricity production while simultaneously producing valuable thermal energy. The study [14] examines the evolution and global outlook of photovoltaic-thermal technology.

These systems are designed to transport heat away from photovoltaic cells, allowing them to cool and increase efficiency by lowering resistance [15]. Within, a photovoltaic and thermal collector has been developed and confirmed, and the results indicate strong concordance of the databases. The results show that just cooling the solar cells can easily boost the power source of the photovoltaic-thermal mechanism. The same PV analysis techniques were applied to investigate hotspots that could exist in solar cells. In this scenario, overheating has been noticed in one of the PV module's cells when compared to the rest of the PVT cells. Solar cells may develop faults and damage the entire PV module if such problems are not immediately resolved. PVT systems are projected to have enormous potential shortly, despite their lack of mainstream use.

7 Discussion

Finally, we noticed in our study that high consideration was given to the overheating phenomenon. First, the autonomous protection system of the solar collectors against overheating gives us a new solution: safe and sustainable. Then, on one side, overheating protection was provided by prismatic features in the thermal solar collector. On the other side, circumstances of stagnation might have disastrous impacts on solar thermal. When evaluating the upsides and downsides of different solutions, it's important to consider the level of security provided. For example, when there is no heat transfer to the solar collector loop or system due to stagnation in the solar collector, a high degree of system stagnation protection can be achieved in comparison to procedures that subject system elements to high temperatures or pressures. Various methods have been invented for reducing the impacts of stagnation. Some of those methods offered aren't appropriate for all system designs and situations. Furthermore, the expense of anti-stagnation measures should be reduced, and passively functioning systems, which don't rely on controller or user input or needs extra power to operate, appear to have considerable durability advantages.In general, techniques for stagnation control differ by climate and were influenced by other system design factors like freeze security solutions, system design temperatures, collector type, and loads.

8 Conclusion

Solar collectors suffer from a major problem which is the phenomenon of overheating. Thus, this study is essentially a review of overheating protection. It showed the availability of many works concerning solar panel protection against overheating. Since many articles are published, it is satisfying to search for suitable literature in scientific journals. In contrast, patents are rare.In this overview, we conclude that despite the remarkable advantages of each method, on one hand, we notice the deficiency of solutions that consider not only the phenomenon of overheating but also the protection against freezing. On the other hand, the scarcity of water leads us to think about other responsible solutions to produce electricity. This study's findings also emphasize the need to use wavelets [16,17,18,19], integrated with data analysis techniques often employed in biomedical signal processing, such as ICA-NMF-SVD-PCA [20,21,22] to further improve the aforementioned techniques' efficacy. One of our primary concerns in this literature research is how to extend our effort to address these issues. Moreover, reducing costs and extending the life of solar power systems through the use of new components and materials is a major challenge for academics.

References

1. Laabab, I., Ziani, S., Benami, A.: Solar panels overheating protection - a review. Indonesian J. Electrical Eng. Comput. Sci. (In press) (Accepted 03 September 2022)
2. Hajji, A.: Techniques innovantes pour la gestion et l'optimisation de l'énergie consommée dans les bâtiments. PhD thesis in engineering sciences and techniques, Mohammadia School of Engineering EMI/ University Mohammed V Rabat (2022)
3. Hajji, A., Lahlou, Y., Abbou A.: Système autonome de protection des collecteurs solaires contre la surchauffe et la poussière. Application for invention n°50822- Filed on September 04, 2020- Notification date February 26 (2021)
4. Zaharil, H.A.: An investigation on the usage of different supercritical fluids in parabolic trough solar collector, Higher Institution Centre of Excellence (HICoE), UM Power Energy Dedicated Advanced Centre (UMPEDAC), Level 4, Wisma R&D, University of Malaya, Jalan Pantai Baharu, 59990, Kuala Lumpur, Malaysia. Renew. Energy **168**, 676–691 (2021)
5. Sajid, M.U., Ali, H.M.: Recent advances in application of nanofluids in heat transfer devices: A critical review. Renew. Sustain. Energy Rev. **103**, 556–592 (2019)
6. Achour, H.B., Said, Z., Chaou, Y., El Hassouani, Y., Daoudia, A.: Permanent magnet synchronous motor PMSM control by combining vector and PI controller. WSEAS Trans. Syst. Control **17**, 244–249 (2022)
7. Chaou, Y., Said, Z., Achour, H.B., Daoudia, A.: Nonlinear control of the permanent magnet synchronous motor PMSM using backstepping method. WSEAS Trans. Syst. Control **17**, 56–61 (2022)
8. Ghasemiasl, R., Hoseinzadeh, S., Javadi, M.A.: Numerical analysis of energy storage systems using two phase-change materials with nanoparticles. J. Thermophys. Heat Transf. **32**, 440–448 (2018)
9. Michael, J.J., Iniyan, S., Goic, R.: Flat plate solar photovoltaic– thermal (PV/T) systems: A reference guide. Renew. Sustain. Energy Rev. **51**, 62–88 (2015)
10. Sahota, L., Tiwari, G.N.: Review on series-connected photovoltaic thermal (PVT) systems: Analytical and experimental studies. Sol. Energy **150**, 96–127 (2017)

11. Jun, Y.-J., Park, K.-S., Song, Y.-H.: A study on the structure of Solar/ Photovoltaic Hybrid system for the purpose of preventing overheat and improving the system performance. Sol. Energy **230**, 470–484 (2021)
12. Lämmle, M., Oliva, A., Hermann, M., et al.: PVT collector technologies in solar thermal systems: A systematic assessment of electrical and thermal yields with the novel characteristic temperature approach. Sol. Energy **155**, 867–879 (2017)
13. Ramos C.A.F. et al.: Photovoltaic-thermal (PVT) technology: review and case study. In: International Conference on New Energy and Future Energy System (2019). IOP Conf. Ser.: Earth Environ. Sci. **354.** 012048
14. Ouhadou, M., El Amrani, A., Ziani, S., Messaoudi, C.: Experimental modeling of the thermal resistance of the heat sink dedicated to SMD LEDs passive cooling. In: Proceedings of the 3rd International Conference on Smart City Applications (2018)
15. Ouhadou, M., El Amrani, A., Messaoudi, C., Ziani, S.: Experimental investigation on thermal performances of SMD LEDs light bar: Junction-to-case thermal resistance and junction temperature estimation. Optik **182** (2019)
16. Ziani, S., El Hassouani, Y.: A new approach for extracting and characterizing fetal electrocardiogram. Traitement du Signal **37**(3), 379–386 (2020). https://doi.org/10.18280/ts.370304G
17. Ziani, S., El Hassouani, Y., Farhaoui, Y.: An NMF based method for detecting RR interval. In: International Conference on Big Data and Smart Digital Environment. Springer (2019)
18. Ziani, S., El Hassouani, Y.: Fetal-maternal electrocardiograms mixtures characterization based on time analysis. In: 2019 5th International Conference on Optimization and Applications (ICOA), pp. 1–5 (2019). https://doi.org/10.1109/ICOA.2019.8727619
19. Ziani, S., Jbari, A., Bellarbi, L.: QRS complex characterization based on non-negative matrix factorization NMF. In: 2018 4th International Conference on Optimization and Applications (ICOA), pp. 1–5 (2018). https://doi.org/10.1109/ICOA.2018.8370548
20. Ziani, S., Jbari, A., Bellarbi, L., Farhaoui, Y.: Blind maternal-fetal ECG separation based on the time-scale image TSI and SVD-ICA methods, Proc. Comput. Sci. **134** (2018)
21. Ziani, S., El, Y.: Fetal electrocardiogram analysis based on LMS adaptive filtering and complex continuous wavelet 1-D. In: Farhaoui, Y. (ed.) BDNT 2019. LNNS, vol. 81, pp. 360–366. Springer, Cham (2020). https://doi.org/10.1007/978-3-030-23672-4_26
22. Ziani, S., Jbari, A., Belarbi, L.: Fetal electrocardiogram characterization by using only the continuous wavelet transform CWT. In: International Conference on Electrical and Information Technologies (ICEIT), vol. 2017, pp. 1–6 (2017). https://doi.org/10.1109/EITech.2017.8255310

A Machine Learning-Based Model for Energy Efficiency Classification of an Unmanned Aerial Vehicle

Joseph Bamidele Awotunde[1](✉), Micheal Olaolu Arowolo[2], Agbotiname Lucky Imoize[3], Yousef Farhaoui[4], and Abidemi Emmanuel Adeniyi[5]

[1] Department of Computer Science, Faculty of Information and Communication Sciences, University of Ilorin, Ilorin 240003, Kwara, Nigeria
awotunde.jb@unilorin.edu.ng

[2] Department of Computer Sciences, Landmark University, Omu-Aran, Nigeria
arowolo.olaolu@lmu.edu.ng

[3] Department of Electrical and Electronics Engineering, Faculty of Engineering, University of Lagos, Akoka, Lagos 100213, Nigeria
aimoize@unilag.edu.ng

[4] STI Laboratory, T-IDMS Faculty of Sciences and Techniques, Moulay Ismail University of Meknès, Errachidia, Morocco
y.farhaoui@fste.umi.ac.ma

[5] Department of Computer Sciences, Precious Cornerstone University, Ibadan, Nigeria
adeniyi.emmanuel@lmu.edu.ng

Abstract. One of the most promising technologies for next-generation wireless networks is unmanned aerial vehicles (UAVs). They make excellent prospects for a range of applications due to their accessibility and capacity for establishing line-of-sight interactions with users. Because of the enormous amount of data available, artificial intelligence is rapidly developing and succeeding tremendously. As a result, a sizeable segment of the academic community has started integrating intelligence into the core of UAV networks by using machine learning (ML) approaches to address a range of drone-related problems. ML-based framework is designed to provide solutions to the different issues that have been time invariant when UAVs are utilized for communication. UAV-based delivery systems, real-time multimedia streaming, and intelligent transportation systems are all examples of this. The problems with wifi and security are revealed by this study. ML-based solution techniques are created to address these problems, enabling real-time secure operation of UAVs while leveraging wireless system resources in a flexible manner.

Keywords: Machine learning · Dimensionality reduction · Unmanned aerial vehicles · Classification

Y. Farhaoui et al. (Eds.): ICAISE 2022, LNNS 635, pp. 54–63, 2023.
https://doi.org/10.1007/978-3-031-26254-8_8

1 Introduction

Unmanned aerial vehicles (UAVs) have recently garnered considerable interest from academics and researchers. The usage of UAVs was discovered to be advantageous in a variety of application fields, including intelligent transportation systems, disaster management, surveillance, and wildlife monitoring [1]. A cluster head (CH) is chosen by the other nodes in a cluster, it is a reputable energy-saving strategy. At the same time, a significant problem in the UAV networks is scene identification from the remote sensing images of spatial precision that UAVs have taken. UAV often enables line-of-sight (LoS) connectivity and has the best link capabilities. UAVs are useful in a variety of fields, including better farming, disaster management, routine environmental surveillance, and disaster warning, thanks to their effective maneuverability, elastic innovation, and affordability [2]. UAVs are also used as mobile relays to increase capacity and network availability. As a result, UAV provides an aerial setting that has been extensively used in wireless communication systems since it allows users in locations without terrestrial buildings to communicate via aerial means [3].

As a result of the UAV's motion, the on-board cameras usually create blurry, noisy photos with low quality. Due to the need for realistic performances, the identification task is quite difficult in many UAV scenarios [4]. Numerous studies on UAVs have been conducted in an effort to track and locate certain items, people, including moving pedestrians, landmarks, landing places, and vehicles [5]. However, because various target image segmentation is crucial for many UAV applications, there are certain works that assume multiple detection and recognition [6]. Two significant limitations account for the discrepancy between the technological capacities and the application's specifications: (i) It is difficult to create and store different target classification algorithm; and (ii) even for a single object, successful object detection requires a lot of processing resources.

Nevertheless, there are a number of difficult problems and crises with the UAV communication paradigm. One of the main problems when using UAV for battery maintenance or recharging is the restricted battery, which reduces the UAV's durability. Finally, one of the measures of efficiency for UAV wireless communication is energy efficiency (EE) in bits/J. It is noted that UAVs demand maximal acceleration power, which is necessary for communication, in contrast to conventional terrestrial systems. The trajectory approach for UAV transmission is one of the most important when using the EE.

Machine learning models have been applied in various fields for classification, prediction, and detection of weather [7], images [8], diseases [9, 10], and energy efficient scene. This research introduces novel cluster-based UAV networks that conserve energy by employing ML-based model scene classification to address the EE and scene classification issues. The outcomes validated the superiority of the given model over the compared approaches. The study key contributions are summarized as follow:

1. the study proposes a ML-based model.
2. the study depends on the degree of the UAV, the RE, and the distance to neighboring UAVs.
3. it uses an ML-based approach to classify scenes.

The paper is oeganized as follows: Sect. 2 presents the methodology employ in the course of this study. Section 3 discusses the experiemental results with detailed discussion. Finally, Sect. 4 concludes the study with future direction towards the study.

2 Materials and Methods

The process that was employed to put the suggested model into practice is as follows.

2.1 Gathering Information and Defining Variables

This study aims to utilize ML-based methods to enhance different design and functional elements of UAV-based connectivity. By gathering traffic flows from eight distinct sorts of consumers, you can model channels, manage resources, control placement, and maintain security, use a laptop from DELL with an Intel Corporation Wireles NIC that is freely avaliable online and released in 2020 by the Carnegie Mellon University. This study's objective was to scientifically assess how much energy compusption by the UAVs, and efficiently improve the energy proficiency. Autonomously operating a DJI Matrice 100 (M100) drone, it carried a range of payload weights in a triangle pattern before landing. Using a variety of discrete variations, the specified parameters were changed between flights. These possibilities included payloads of 0 g, 250 g, and 500 g; cruise altitudes of 25 m, 50 m, and 75 m; and 100 m; and cruise speeds of 4, 6, 8, 10, and 12 m/s.

We simultaneously gather information from numerous on-board devices. The following onboard devices were utilized to get the data: * Wind sensor: pre-calibrated, UAV-mountable FT Technologies FT205 ultrasonic wind sensor with 0.1 m/s accuracy and a 10 Hz refresh rate; * Position sensor: GNSS/INS sensor kit 3DM-GX5–45 The GPS and IMU data are combined using a Kalman filtering method that is already present in these sensors. The sensor features a 10 Hz maximum output rate, a 2 m$ RMS horizontal accuracy, and a 5 m$ RMS vertical accuracy. Current and voltage sensor, Mauch Electronics PL-200. This sensor can capture voltages and currents up to 33 V and 200 A, respectively. The analog inputs digital expressions were created from the sensor. Using an 8 channels Analog to digital converter with 17 bits.

Data synchronization and recording were accomplished using the low-power Raspberry Pi Zero W and the Robot Operating System (ROS). The Raspberry Pi's microSD card was used to store the data. The data from each sensor was timed to a resonant speed of about 5 Hz using the Robot Operating System's Approximate Time message filter policy (ROS).

There were 196 flights with different operational parameters (payload, altitude, and speed). There were also 13 recordings made to evaluate the drone's supplementary power and hover circumstances [11].

2.2 Model Design

Before procedures like classification and clustering can be performed, feature selection should be achieved. Reduce symmetry in a dataset's (dimensionality) feature space as part of feature selection to prepare data (preprocess). By deleting noisy symmetrical features while maintaining the most useful ones, feature selection decreases the complexity of a dataset. Features with a high connection to other symmetry features are among the noisy features (redundant) and characteristics with a tenuous relationship to the target class (label of the instance) irrelevant.

Relief-F is utilized in this work to generate an initial filter for the feature database's high-dimensional features. The feature subset is created after the sorted features have been eliminated. We used our recommended Relief-F algorithm technique to the categorization learning algorithms are: "SVM and KNN" to examine the performance of UAV classifiers.

The proposed approach was employed in this work for data collecting, model construction, feature selection, data preparation, evaluation, and assessment. Figure 1 represents the methodology's proposed steps in their completeness.

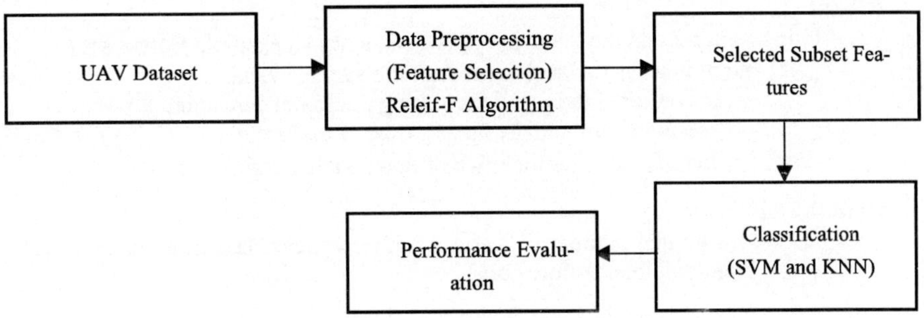

Fig. 1. The flowchart for the proosed methodology

2.3 Improved Releif-F Algorithm

An variation of the Relief technique called Relief-F is effective at evaluating feature strength, but it's only good for two-class situations. While the complexity remains the same, this approach can systematically compute probabilities and determine distances between sample distributions. Manage partial and multiclass data sets, as well. When dealing with a variety of challenges, the Optimization technique does not always choose the k-NN sample from all available sample sets. For continuous data, such as regression issues, it chooses the NN selection from each set of observations; The capacity to distinguish the closest distance between any two classes is used to determine the significance of a characteristic.

Another method for estimating a feature's value is Relief-F, and in numerous feature quality test scenarios, it has proven to be superior. In order to create the weight vector w that represents the quality of each feature, the Relief-F method employs training examples with feature values and class values that are randomly selected from the data. Depending on the feature's likelihood of class discrimination, the Relief-F technique's weight feature assessment criterion was determined greater feature significance for the classes being indicated by a larger predicted weight. All weights w[A] are initially set to zero, then the nearest hit Hand and nearest miss M are sought using an arbitrarily selected instance Ri. When utilizing the attribute A to distinguish between two situations of the same class is not desired, the reliability estimate was decreased by w[A]. When the characteristic A was made capable of classifying two instances into different values, the quality estimation w[A] increased [12].

Algorithm 1: Releif-F [15]

Input

S represents a first sample feature set, and $g_f^{n(f)}, g_C^{n(C)}, g_\gamma^{n(\gamma)}$ the initial population, the feature set is encoded by f, C and γ and those parameters have been encoded.

Output

characteristics that were extracted using the Relief-F technique

Repeat

1. Using Relief-F and the feature weights, sequence an example feature set t (ωit) and m times is updated to produce the average value.
2. Activation of the population using the Relief-F method
3. Determine individual's fitness. Cost of the feature is $1/n\sum_{i=1}^{n} f_i X C_i$ where C_i signifies the price of functionality and $f_i = 1, 0$

Until the termination test is met

4. a limited number of features with excellent classification accuracy and minimal feature cost

---

2.4 K-Nearest Neighbor (KNN)

KNN unsupervised ML model that excludes the need for a predictor variables to predict the results of a specific data set because of how well the classifier performs. The model is given ample training data, and we then leave it to determine which neighborhood a certain data point belongs to. The KNN model determines the separation of a fresh data point and its nearest neighbors, and K determines how many of its neighbors' votes are in favor; The class with the closest distance is given the new data point if K is 1. The following mathematical equations can be used to determine how far two locations are from one another [13].

KNN categorizes new places based on the majority of the neighboring k's sounds in relation to them. The closest neighbors K, who share the allotted position in the class, are strongly mutually exclusive is determined by the significance of distance.

KNN Pseudocode [14]

Sort (X, Y, and x);
Training data X, class labels Y, unidentified sample X, and class labels of Y
For i = 1 to *m*
do
Measure the distance d(Xi, x)
end for
Set i of indices for the k lowest distances should be calculated (Xi, x)
return
majority label for {Yi where i∈I}
end

The fundamental concepts of SVM are found in the research of Cortes and Vapnik. Due to how well it handles high-dimensional and exponentially non-separable datasets, In order to maximize the distance between the closest positive values, SVM provides the optimum hyperplane. And unfavorable samples. Two distinct groupings are represented by the white and black points. The two samples that are closest to the categorization line, H, are H1 and H2, respectively, and the classification interval is the space between them. With the separation margin optimized, the ideal classification hyperplane guarantees accurate classification [15].

In the SVM classifier, a penalty parameter allows for some misclassification. By dividing the hyperplane with maximal margin, more than two classes are classified using the non-linear SVM classifier. The separation hyperplane can be calculated without having for the patterning to be carried into feature space by using kernel functions. The mathematical versions of the linear, polynomial, radial basis kernels, and sigmoid are represented by Eqs. (1), (2), (3), and (4), respectively [16];

1. Linear Kernel: If the original input space contains linearly separable data, No need to move it into a high-dimensional space is necessary. In this scenario, the linear kernel can be used, it represents the two vectors' dot product in the initial space.

$$k\ (x, y) = (x, y) \tag{1}$$

2. Polynomial Kernel: The definition of polynomial kernels of degree p is

$$k(x, y) = ((x, y) + 1)^p \tag{2}$$

   where p is the degree of the polynomial.
3. Radial Basis Kernel

$$k(x, y) = \exp\frac{-||x - y||^2}{2\sigma^2} \tag{3}$$

   where is the radius control parameter and > 0.
4. Sigmoid Kernel

$$k(x, y) = tanh(m(x, y) + c),\ with\ some\ (not\ all) m > 0\ and\ c > 0 \tag{4}$$

2.5 Evaluation Measures

The study's evaluation measures included accuracy, precision, recall, and f-measure. The true positives (TPs) represented news that was projected to be false and actually was false news, whereas the true negatives (TNs) indicated news that was anticipated to be true but actually wasn't false news. False negatives (FNs) indicated news that was expected to be phony but wasn't, while false positives (FPs) indicated news that was expected to be fake but wasn't [17]:

3 Results and Analysis

The UAV dataset used in this work was fed into the Relief-F method, and the subset results were then provided onto the KNN and SVM classifiers. The results of all the experiments are compiled, the confusion matrix's output is displayed, and a summary of the performance evaluation is shown in Table 1 as well.

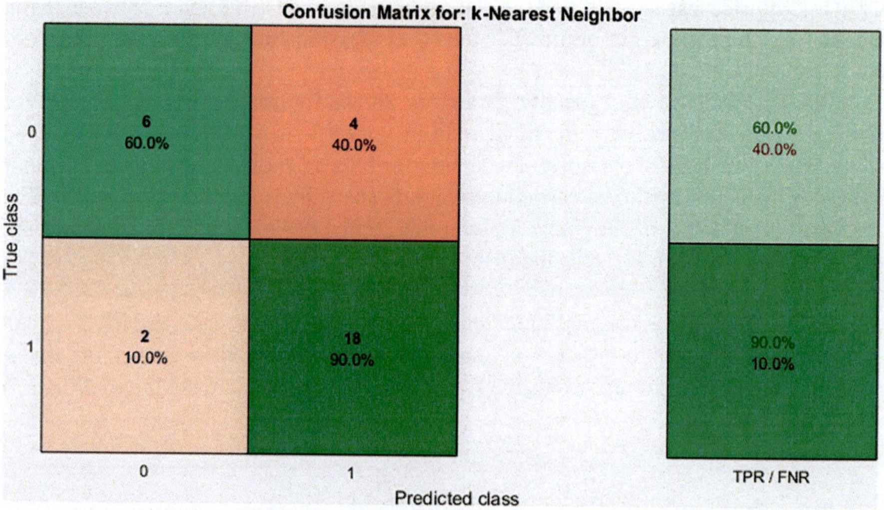

Fig. 2. Confusion matrix showing the dataset with KNN TP = 6, TN = 18, FP = 2, FN = 4

The evaluation method for each technique on the taken-into-account datasets is shown in Table 1. It is evident that the Relief-F + SVM method was able to attain a maximum accuracy of 90% on the UAV Dataset. Relief-F + KNN classifier had an accuracy rate of 80%. SVM performs better than KNN in terms of sensitivity and f-measure, among other performance metrics. The overfitting issue can be reduced and misclassified points can be reliably identified.

In this study, a feature selection and classification approach was carried out on a UAV generated dataset, using Relief-F a feature selection algorithms and KNN and SVM were utilized as classification techniques to extract pertinent information from the provided data. The result shows that Relief-F + SVM outperforms, in terms of accuracy, then the

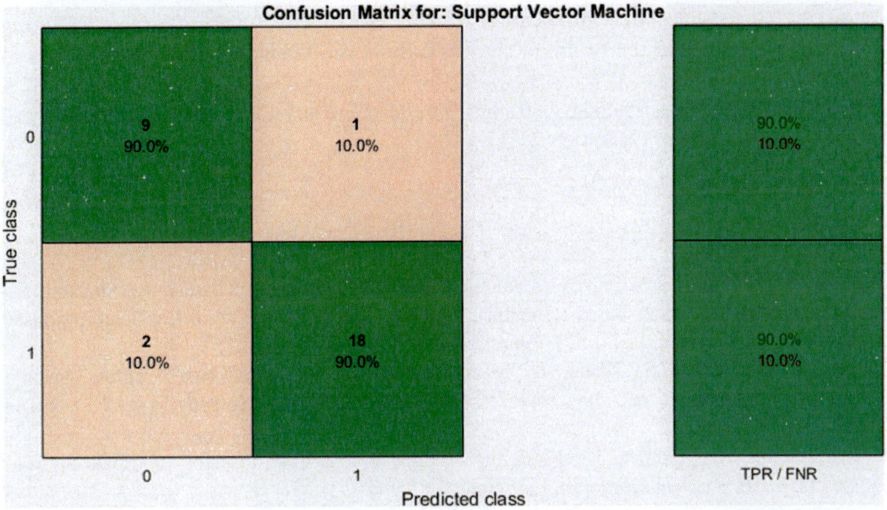

Fig. 3. Confusion matrix showing the dataset with SVM TP = 9, TN = 18, FP = 2, FN = 1

Table 1. Summary of prediction performance evaluation

Performance metrics	Relief-F + KNN	Relief-F + SVM
Accuracy (%)	80	90
Sensitivity (%)	60	90
Specificity (%)	90	90
Precision (%)	75	81.82
F-measure (%)	66.67	85.71

classifier is shown in the results are tabulated in Table 1. However, this study has shown a relevant output results.

4 Conclusion

This study proposes a Relief-F classifier for feature selection and KNN and SVM models for the classification of the dataset used for the performance evaluation. The UAV can initially record images using the suggested approach. The videos are then split up into a collection of frames. The video frames are then preprocessed to provide room for additional processing. After that, the training phase is conducted utilizing the classifiers for the classification and Relief-F for feature selection. The proposed model is run, and the input image is effectively classified into a scene. In comparison to all the previous approaches, the suggested method has the fastest computation time (70 s) and the highest accuracy (90.0%). The simulation results demonstrate the effectiveness of the suggested model. This work can be further enhanced by hyper-parameter-tuning modeling techniques in the future to significantly minimize calculation time.

Data Avaliability. https://kilthub.cmu.edu/articles/dataset/Data_Collected_with_Package_Delivery_Quadcopter_Drone/12683453.

References

1. Pustokhina, I.V., Pustokhin, D.A., Kumar Pareek, P., Gupta, D., Khanna, A., Shankar, K.: Energy-efficient cluster-based unmanned aerial vehicle networks with deep learning-based scene classification model. Int. J. Commun Syst **34**(8), e4786 (2021)
2. Xiang, T.Z., Xia, G.S., Zhang, L.: Mini-unmanned aerial vehicle-based remote sensing: Techniques, applications, and prospects. IEEE Geosci. Remote Sensing Mag. **7**(3), 29–63 (2019)
3. Al-Hourani, A., Gomez, K.: Modeling cellular-to-UAV path-loss for suburban environments. IEEE Wireless Commun. Lett. **7**(1), 82-85 (2017)
4. Jordan, S., et al.: State-of-the-art technologies for UAV inspections. IET Radar Sonar Navig. **12**(2), 151–164 (2018)
5. De Smedt, F., Hulens, D., Goedemé, T.: On-board real-time tracking of pedestrians on a UAV. In: Proceedings of the IEEE Conference on Computer Vision and Pattern Recognition Workshops, pp. 1–8 (2015)
6. Kapania, S., Saini, D., Goyal, S., Thakur, N., Jain, R., Nagrath, P.: Multi object tracking with UAVs using deep SORT and YOLOv3 RetinaNet detection framework. In: Proceedings of the 1st ACM Workshop on Autonomous and Intelligent Mobile Systems, pp. 1–6 (January 2020)
7. Abdulraheem, M., Awotunde, J.B., Adeniyi, A.E., Oladipo, I.D., Adekola, S.O.: Weather prediction performance evaluation on selected machine learning algorithms. IAES Int. J. Artifi. Intell. **11**(4), 1535–1544 (2022)
8. Kwak, G.H., Park, N.W.: Impact of texture information on crop classification with machine learning and UAV images. Appl. Sci. **9**(4), 643 (2019)
9. Awotunde, J. B., Oluwabukonla, S., Chakraborty, C., Bhoi, A.K., Ajamu, G.J.: Application of artificial intelligence and big data for fighting COVID-19 pandemic. In: Decision Sciences for COVID-19, pp. 3–26 (2022)
10. Abiodun, M.K., Misra, S., Awotunde, J.B., Adewole, S., Joshua, A., Oluranti, J.: Comparing the performance of various supervised machine learning techniques for early detection of breast cancer. In: Abraham, A., et al. (eds.) HIS 2021. LNNS, vol. 420, pp. 473–482. Springer, Cham (2022). https://doi.org/10.1007/978-3-030-96305-7_44
11. Rodrigues, T.A., Patrikar, J., Choudhry, A., Feldgoise, J., Arcot, V., Gahlaut, A.: Data Collected with Package Delivery Quadcopter Drone. Carnegie Mellon University. Dataset, Carnegie Mellon University (2020). https://doi.org/10.1184/R1/12683453
12. Ma, L., et al.: Evaluation of feature selection methods for object-based land cover mapping of unmanned aerial vehicle imagery using random forest and support vector machine classifiers. ISPRS Int. J. Geo Inf. **6**(2), 51 (2017). https://doi.org/10.3390/ijgi6020051
13. Flórez, J., Ortega, J., Betancourt, A., García, A., Bedoya, M., Botero, J.S.: A review of algorithms, methods, and techniques for detecting UAVs and UAS using audio, radiofrequency, and video applications. TecnoLógicas **23**(48), 269–285 (2020). https://doi.org/10.22430/22565337.1408
14. Khanam, Z., Alwasel, B.N., Sirafi, H., Rashid, M.: Fake news detection using machine learning approaches. IOP Conf. Ser. Mater. Sci. Eng. **1099**(1),(2021)

15. Zhou, Y., Zhang, R., Wang, S., Wang, F.: Feature selection method based on high-resolution remote sensing images and the effect of sensitive features on classification accuracy. Sensors **18**(7), 2013 (2018). https://doi.org/10.3390/s18072013
16. Vasantha, V.K., Keesara, V.R.: Comparative study on crop type classification using support vector machine on UAV imagery. In: Jain, K., Khoshelham, K., Zhu, X., Tiwari, A. (eds.) UASG 2019. LNCE, vol. 51, pp. 67–77. Springer, Cham (2020). https://doi.org/10.1007/978-3-030-37393-1_8
17. Arowolo, M.O., Adebiyi, M.O., Adebiyi, A.A., Okesola, O.J.: Predicting RNA-seq data using genetic algorithm and ensemble classification algorithms. Indonesian J. Electr. Eng. Comput. Sci. 21(2), 1073 (2021). https://doi.org/10.11591/ijeecs.v21.i2.pp1073-1081

A New CBIR Search Engine with a Vision Transformer Architecture

Smail Zitan[1](✉), Imad Zeroual[2], and Said Agoujil[1]

[1] MMIS, MAIS, FST Errachidia, Moulay Ismail University of Meknes, Meknes, Morocco
zitansmail@gmail.com

[2] L-STI, T-IDMS, FST Errachidia, Moulay Ismail University of Meknes, Meknes, Morocco

Abstract. The use of images is starting to grow in the majority of applications, including image search systems. However, the traditional text-based methods used to perform the search process have some limitations. Therefore, CBIR has emerged as an alternative and robust solution. The CBIR systems use the image's features to perform the search; however, most CBIR systems usually extract low-level features like texture, shape, and color. Such systems have not managed to give an ideal solution that can be effective and fast. Therefore, we propose a newly designed CBIR system that can extract more accurate and meaningful features while responding in a minimum time. Our CBIR system, presented in this paper, uses a transformer-based model to extract the features from images, called ViT (Vision Transformer). Then, we used the principal component analysis (PCA) method to minimize the dimensionality of the feature vector space. Finally, we implemented the Annoy library for similarity searches. The experiments indicate that our CBIR system is more accurate and faster than relevant CBIR systems.

Keywords: CBIR search engine · ViT transformer · Annoy · Principal component analysis

1 Introduction

At present, many applications still use different search engine systems as a solution to their needs using basic methods based on text queries or suggestions. However, due to the increase in the amount of data, numerous techniques are not very effective, especially for visual searches that look for images, rather than textual content. This makes the search process unsatisfactory in terms of speed, accuracy, and flexibility of the user experience.

With the advent of Content-Based Image Retrieval (CBIR), which is based on the image as a search query, the characteristics of an ideal search engine become potentially achievable but still a challenge. Current CBIR-based search engines mainly extract low-level features like texture, shape, and color. Consequently, the biggest challenge faced by these systems is the semantic gap issue [1]. Therefore, much effort is being made to extract high-level image features without a high computational burden to handle this issue.

Most CBIR approaches use convolutional neural networks as deep learning architecture to extract the image's features. Such CBIR models lead to satisfactory results

Y. Farhaoui et al. (Eds.): ICAISE 2022, LNNS 635, pp. 64–69, 2023.
https://doi.org/10.1007/978-3-031-26254-8_9

despite their high computational cost (e.g., [2]). However, in this paper, we propose an efficient and effective CBIR approach that can be applied to curry out the feature extraction task using a visual model based on a transformer architecture. This model is called the Vision Transformer (ViT), which is originally designed for a text-based content; yet, it was successfully used for image processing and outperformed familiar convolutional architectures [3]. Furthermore, we implemented ANNOY (Approximate Nearest Neighbors Oh Yeah) as a search indexing algorithm. The advantage of using the Annoy algorithm is that it is faster and leads to better results compared to alternatives methods such as faiss and genism [2]. For the same reason, we applied the Principal Component Analysis (PCA) as a data dimensionality reduction method [4]. Moreover, we used the Corel-1k, a publicly available corpus that comprises 1,000 images. These later are categorized into ten classes. At last, we evaluated the performance of the proposed CBIR system using standard measures, i.e., precision and recall. Finally, we selected two relevant CBIR systems from the literature to compare our findings. These two CBIR systems are proposed by Radha et al. [5] and Joseph et al. [6]. Noted that these CBIR systems are trained and tested using the Corel-1K dataset, also called WANG dataset [7].

The paper is organized in four sections besides this introduction. The second section describes the dataset used in this study and illustrates the architecture of our proposed CBIR. The Third section shows the results scored from the evaluations alongside a discussion. The paper is concluded in Section four, providing our perspectives for future works.

2 Materials and Methods

2.1 Dataset

To evaluate the CBIR system proposed, we used a subset of the Corel dataset, which consisted of 1,000 distinct images. Each image in the dataset is 384×256 or 256×384. Besides, the data is regrouped over ten semantic classes, and each class comprises 100 images. These categories are buses, mountains, beaches, elephants, food, flowers, Africa, horses, dinosaurs, and buildings. Several studies have exclusively used these categories to investigate the efficiency of their CBIR approaches, thereby enabling comparison with other methods. Figure 1 depicts samples of images of each class of the Corel dataset.

Fig. 1. Samples from the Corel-1K dataset.

2.2 Pre-processing

The dataset is composed of a thousand images but of two different sizes. Therefore, we performed a pre-processing process to normalize the dimension of the pictures. Since the ViT model was pre-trained using a resolution of 224 × 224, we resized the Corel-1K images into 224 × 224 pixels.

2.3 Methods

After the pre-processing task, our approach is based on the ViT architecture for extracting features from the Corel images. Then, we used the PCA as a features selector to minimize the dimensionality. Finally, we implemented the Annoy algorithm for similarity searches. Noted that those tasks are applied for both training and querying images. Figure 2 depicts our overall methodology.

During the feature extraction phase, the image is segmented into N square patches in raster order (left to right, up to down); then, the patches are flattened, resulting in N line vectors. These flattened patches are multiplied by a trainable embedding tensor which learns to linearly project each flat patch. A learnable CLS token is prepended to the sequence of patch embeddings. A trainable positional embedding tensor is added to the concatenated sequence of projections. This tensor learns 1D positional information for each of the patches to add a spatial representation of each patch within the sequence. Figure 3 illustrates those steps.

The result goes as input to the stacked transformer encoder (see Fig. 4). This later represents the second component of the ViT architecture. The transformer is made of L layers, each with two main sublayers. The first one is the Multi-Headed Self Attention, which assigns a self-attention operation to different projections of the input tokens. The second sublayer is a Feed-Forward Network. These two sublayers are preceded by a normalization layer and also ended with a skip connection layer [3].

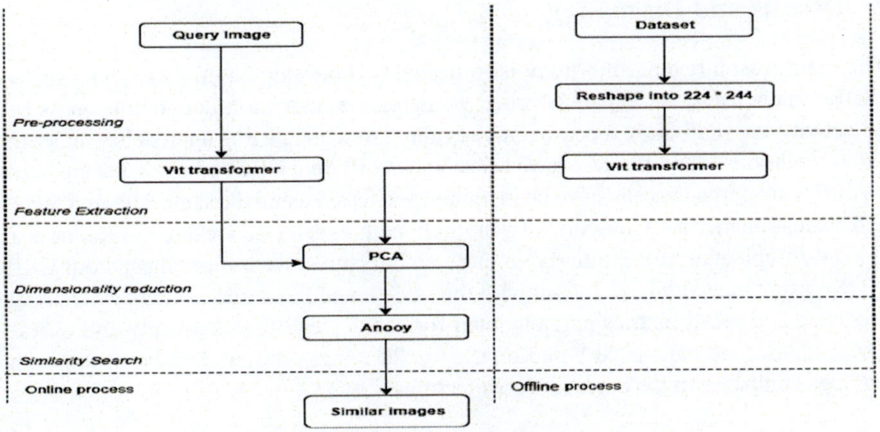

Fig. 2. Our overall CBIR methodology.

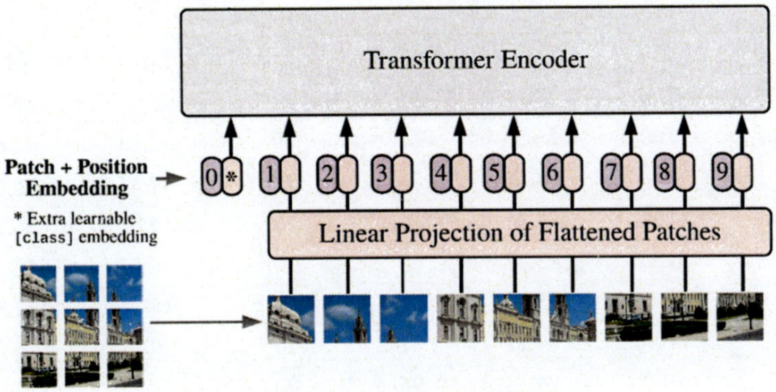

Fig. 3. Vision transformer architecture.

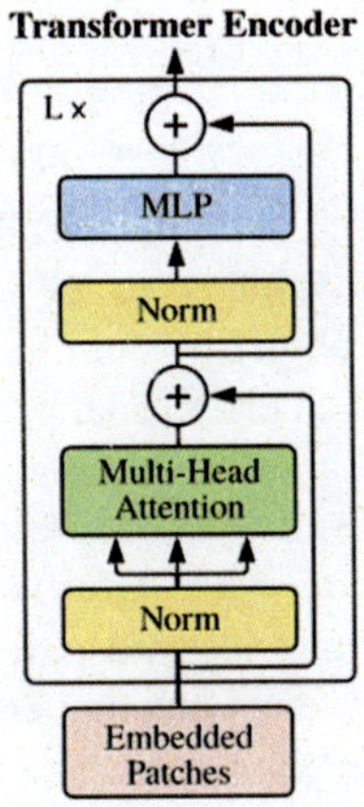

Fig. 4. Modelling the transformer encoder architecture.

3 Results and Discussion

The main reason behind this study is to design and develop an efficient and effective CBIR system using the ViT transformer as its features extractor. For completeness and objectivity, we shall make a comparative evaluation with existing approaches under the same conditions. Therefore, we have selected two different CBIR systems that are based on different approaches and have been trained and tested using the same dataset, classes, and evaluation metrics. Consequently, the architectures selected are those of Radha et al. [5] and Joseph et al. [6]. Similarly, to these architectures, we have evaluated our CBIR system using the Cored-1K dataset with the same ten aforementioned classes; then, the precision and recall metrics are computed for each category, individually and overall. These metrics are calculated based on the top 20 images retrieved. Table 1 shows the findings compared to the other two architectures.

Table 1. Our average precision and recall values compared to the existing results.

Category	Precision (%)			Recall (%)		
	Our	[5]	[6]	Our	[5]	[6]
Buses	100	100	81	20	20	16.2
Mountains	100	86.58	68	20	17.64	11.6
Beach	90	84.65	78	18	17.44	15.6
Elephants	100	91.45	81	20	17	16.2
Food	100	93.77	60	20	16	12
Flowers	100	95.87	89	20	18	17.8
Africa	100	84.11	81	20	19	16.2
Horses	100	94.13	95	20	18.44	19
Dinosaurs	100	100	100	20	20	20
Buildings	100	84.55	80	20	16	11
Average	**99**	91.32	81.3	**19.8**	17.94	16.2

According to Table 1, we achieved the best overall performance result for precision (i.e., 99%) and recall (i.e., 19.8%) compared to the other CBIR systems, Radha et al. (91.31% and 17.94%) and Joseph et al. (81.3% and 16.2%). Similarly, our CBIR system scored equal or highest results for each class separately.

These results can be explained based on two main factors. The first one is the ViT transformer extractor which provides a better representation of an image through its architecture that is based on attention. This latter is the essential element for vision networks to reach more increased robustness. Also, it is a computational primitive that permit quantifying pairwise entity interactions that enable a network to learn the hierarchies and alignments given in the input data. The second factor will be the search algorithm implemented (Annoy algorithm) to find similar images. The Annoy proves its effectiveness by retrieving accurate results with highest precision for most classes.

4 Conclusions and Perspectives

The ViT transformer is a recently introduced model in the research area of image retrieval, especially Content-Based Image Retrieval. So, it is still not evaluated extensively. Therefore, this article proposed a novel CBIR system that is based on the ViT transformer architecture as a features extractor; then, we applied the PCA method to reduce the vector space dimensionality; while using the Annoy algorithm as a search index similarity. This CBIR approach was next trained and tested using the Corel-1K dataset to compare the findings with the existing results in the literature. Based on the comparisons presented here, our approach outperforms the other models in terms of precision and recall for most classes individually as well as for overall performance.

Regarding our perspectives for future works, we plan to evaluate the CBIR system proposed using additional datasets over different fields, especially that the effectiveness and the rapidity of the ViT transformer as a features extractor is still not well investigated using big datasets.

References

1. Salih, S.F., Abdulla, A.A.: An effective bi-layer content-based image retrieval technique. J Supercomput. (2022). https://doi.org/10.1007/s11227-022-04748-1
2. Zitan, S., Zeroual, I., Agoujil, S.: Performance investigation of a proposed CBIR search engine using deep convolutional neural networks. In: Fakir, M., Baslam, M., and El Ayachi, R. (eds.) Business Intelligence. pp. 41–49. Springer International Publishing, Cham (2022). https://doi.org/10.1007/978-3-031-06458-6_3
3. Dosovitskiy, A., et al.: An Image is Worth 16x16 Words: Transformers for Image Recognition at Scale, http://arxiv.org/abs/2010.11929, (2021). https://doi.org/10.48550/arXiv.2010.11929
4. Karamizadeh, S., Abdullah, S.M., Manaf, A.A., Zamani, M., Hooman, A.: An overview of principal component analysis. JSIP. **04**, 173–175 (2013). https://doi.org/10.4236/jsip.2013.43B031
5. Radha, M.K., Sudha, R.V., Meena, M., Jayavadivel, R., Kanimozhi, S., Prabaharan, P.: Modified cuckoo search algorithm: feature subset selection & shape, color and texture features descriptors for content-based image retrieval. J. Univ. Shanghai Sci. Technol. **23**, 525–541 (2021). https://doi.org/10.51201/jusst/21/121046
6. Joseph, A., Rex, E.S., Christopher, S., Jose, J.: Content-based image retrieval using hybrid k-means moth flame optimization algorithm. Arab. J. Geosci. **14**(8), 1–14 (2021). https://doi.org/10.1007/s12517-021-06990-y
7. Wang, J.Z., Li, J., Wiederhold, G.: SIMPLIcity: semantics-sensitive integrated matching for picture libraries. IEEE Trans. Pattern Anal. Mach. Intell. **23**, 947–963 (2001). https://doi.org/10.1109/34.955109

A New Coronary Artery Stenosis Detection Method with a Hybrid LSTM-CNN Model

Xavier Lessage[1,2(✉)], Michal Nedoszytko[3], Saïd Mahmoudi[1], Lilian Marey[4], Olivier Debauche[1,5], and Sidi Ahmed Mahmoudi[1]

[1] Faculty of Engineering, Infortech-Numediart, ILIA, UMONS, Mons, Belgium
xavier.lessage@gmail.com
[2] Applied Research Centre, CETIC, Charleroi, Belgium
[3] Department of Cardiology, CHR-Mons-Hainaut, Mons, Belgium
[4] Department of Computer Science, ENSAE, Paris, France
[5] Liège - GxABT, TERRA/DEAL, Gembloux, Belgium

Abstract. Screening for coronary artery disease is a major health issue, knowing that the most common cause of death in industrialized countries is cardiovascular pathology (coronary artery disease, stroke, other cardiovascular diseases). Computer Aided Diagnosis systems (CADx) can assist cardiologists to and play a key role in detecting abnormalities and treating coronary arteries. In this paper we propose a deep learning classification method based on a new Hybrid CNN-LSTM Architecture. The aim of our method is to detect the presence of stenosis in the coronary arteries and to classify the type of arteries. Our experiments have been conducted using an anonymized database from a Belgian hospital (CHR Mons-Hainaut) thanks to a retrospective study.

Keywords: Coronary arteries · Stenosis detection · Classification · Convolutional neural networks · Recurrent neural network · Hybrid CNN-LSTM Architecture · Explainable artificial intelligence

1 Introduction

The diagnosis of coronary diseases is a health issue given that the most frequent cause of death in industrial countries is cardiovascular disease (coronary disease, stroke, other cardiovascular diseases), and that atherosclerosis is involved in 75% of cardiovascular deaths (Baumgartner[11]). The research problem addressed in this work consists in evaluating the performance of an automatic coronary angiography analysis system to detect the presence of disease (stenosis) using algorithms based on the latest advances in artificial intelligence. Our model is built on Convolutional Neural Networks (CNN) and Long Short-Term Memory (LSTM) networks, an artificial Recurrent Neural Network (RNN) architecture used in the field of Deep Learning (DL).

Y. Farhaoui et al. (Eds.): ICAISE 2022, LNNS 635, pp. 70–77, 2023.
https://doi.org/10.1007/978-3-031-26254-8_10

2 Related Works

Du et al. (Du[1])developed in 2018, an artificial intelligence method that uses deep neural networks for lesion detection based on coronary X-ray angiography images to assist the interventional cardiologist in percutaneous cardiovascular interventions. They trained it on a dataset consisting of 5,701 coronary angiography images. For the lesion detection task including stenosis diameter, calcification, thrombus, and dissection, the sensitivity or recall were 88.2%, 82.6%, 85.5%, 85.8% respectively. On the other hand, for the diseased segment prediction task, the precision (percentage of correct pixels) and sensitivity or recall (percentage of relevant pixels recovered) are 86.3% and 78.0% respectively. In (Lee[2])Lee Paul et al. used convolutional neural networks that they embedded in a smartphone to interpret coronary angiograms in real-time with the goal of reducing interobserver. To quickly maximize training data, coronary angiography images were obtained via web scans of Percutaneous Coronary Intervention (PCI) cases, reviews, and textbooks and anonymised. The final dataset consists of 4980 images, split in a ratio (6.8:2.9:3) for training, validation and testing. They used VGG16, VGG19, ResNet, DenseNet and Inception-ResNet V2 architectures which were pre-trained on ImageNet. They obtained with the VGG16-based CNN network which performed better than other models, 83% accuracy on the test dataset for artery anatomy and 74% accuracy on the test dataset for stenosis. Poorly predicted images mainly occurred in examples with overlapping arteries, making the topology difficult. Emmanuel Ovalle-Magallanes et al. proposed in 2020 (Ovalle-Magallanes[3]) a network-cut and fine-tuning hybrid method for stenosis detection in Xray Coronary Angiography (XCA) using patched-based stenosis detection. Their implementation was based on 20 distinct setups for the pre-trained (on the ImageNet dataset) VGG16, ResNet50, and Inception-v3 architectures. The pretrained Inception-v3 with only the three top blocks as feature extractor reaches an accuracy of 0.95%, precision of 0.93, sensitivity of 0.98, F1-score of 0.95, and specificity of 0.92 for stenosis detection.

3 Proposed Approach

We present and compare two types of architectures: CNN VGG16, Hybrid CNNLSTM models to detect the presence of stenoses in coronary arteries. In order to be able to explain the decision taken by the Deep Neural Network (DNN), we use a technique to visualize the pixels responsible for the classification. We have chosen the occlusion method. To obtain the best possible results, we use two techniques of data augmentation: (1) Over-Sampling which consists of rebalancing the dataset by increasing the number of instances of the minority class (with multiple copies of some instances of the minority class); (2) Multiple transformations, where we apply classical transformations such as rotation, rescale, shear, and shift. In general, binary classification is implemented with a CNN architecture and not with an LSTM architecture. But, this study it is quite special because the dataset is in fact, a sequence of images from a video (angiography). And the information from the previous sequences should be taken into account when classifying an artery as normal or not.

3.1 VGG16 Model

The VGG16 model consists of multiple layers, containing 13 convolution layers and 3 fully connected layers. The architecture of VGG16 is illustrated in Fig. 1.

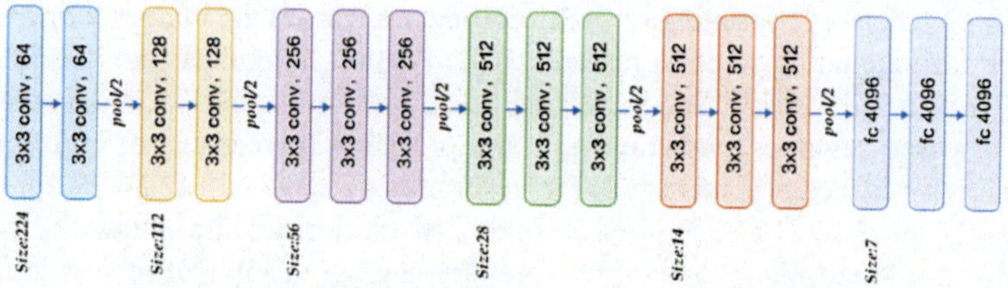

Fig. 1. Illustration of the VGG16 architecture (VGG[10]).

The proposed method exploits a pre-trained architecture on ImageNet with several modifications (removal of the last layer, addition of the 5 dense layers and setting the number of classes of the last layer to 2 (Normal, Disease). To avoid oscillations of accuracy and loss values between epochs during the training process, we suggest using the rate decay method that allows to optimize the learning rate value and thus accelerate the convergence and ensure accurate results.

3.2 Hybrid CNN-LSTM Model

A LSTM is a deep learning model derived from Recurrent Neural Networks. RNN are widely used in AI as soon as a temporal notion intervenes in the data, as in our case, with the image sequences. The idea behind the LSTM is to keep in memory previous inputs that would improve the prediction, while ensuring the stability of the network optimization phase. A LSTM perceptron is a cell composed of an input gate, an output gate and a forget gate, each managed by specific functions. These three gates present areas of calculation that regulate the flow of information (by performing specific actions). There are also two types of outputs (called states). A hybrid CNN-LSTM network is in fact a CNN whose final output is sent directly to an LSTM which will then make its prediction. For our proposed model hybrid CNN-LSTM model, in the first branch, we have a convolution layer and a pooling layer, in the second branch, we have an LSTM layer. These two branches end up with a dense layer to result in two outputs of the same size. These outputs are then concatenated through a dense one-layer network. The last layer provides the prediction: Disease (suspicion of stenosis) or Normal (no abnormality). This proposed mixed architecture is presented in Fig. 2.

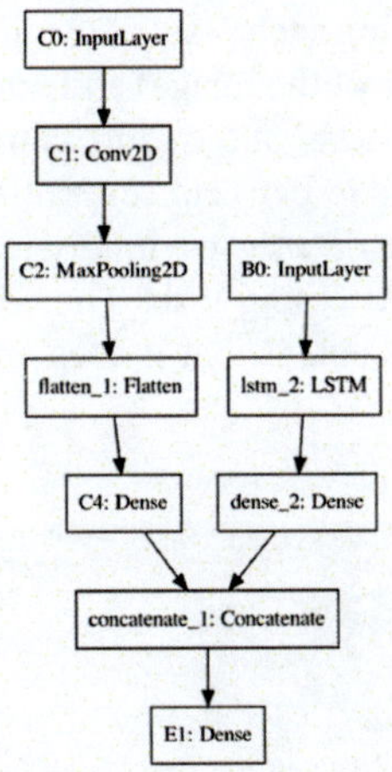

Fig. 2. The proposed CNN-LSTM hybrid schema

3.3 Explainable Deep Learning

To interpret the results provided by our models used for the classification of coronary artery anatomy and for the diagnosis of coronary artery disease we propose to apply the occlusion visualization technique (Fong[5]), (Mahmoudi[7]). Occlusion is mainly used in image classification and developed by Zeiler and Fergus (Zeiler[6]), who proposed to replace a square of input pixels by a gray square. The occluded pixels are important if the class probability drops significantly.

4 Evaluation Dataset

The angiographies acquired were pseudo-anonymised with a cryptographic hash algorithm (SHA3–256 with salting) at the hospital. Thanks to a specifically developed application, the images from the videos (angiography X-rays) were tagged by an experienced cardiologist based on typical 70% stenosis cutoff value.

4.1 Dataset

The database is composed of 1,468,067 images, which have been annotated beforehand by Dr. Nedoszytko, interventional cardiologist. From this database, we constructed a dataset comprising 2 classes (Disease, Normal) and consisting of 275,390 images. Unfortunately, our dataset have the disadvantage of being unbalanced with 194,620 and 80,970 images respectively for the Disease and Normal classes.

4.2 Pre-processing Phase

We pre-processed the raw data from the hospital to keep only the data with contrast. In fact, only the arteries containing contrast medium will be visible. The method that gave the best results consisted in scanning each pixel of the image (based on several diagonals starting from the lower left corner of the image) and retaining only those images whose pixel area corresponds to the grey level of the contrast medium. The grey level threshold is configurable, which allowed us to perform several tests and to retain only the most relevant images. Out of the entire dataset, we finally retained only 33% of the images. The Fig. 3 illustrates this pre-processing task. The advantage of measuring the grey levels present only on these diagonals is that it does not take into account the catheter present on the image.

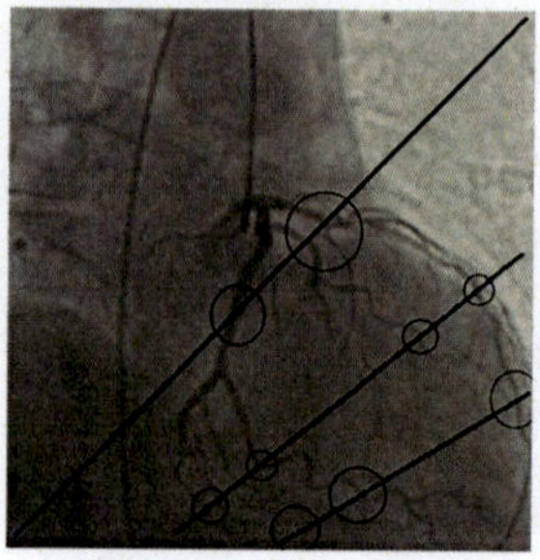

Fig. 3. Illustration of the concept to keep only the images with contrast product

5 Experimental Results

We used only the images with contrast product (about 91,800 images) thanks to the pre-processing phase. In order to overcome the class imbalance, we opted for image replication (Oversampling) to increase the images of the the minority class. Oversampling is performed only on the training base, while keeping the test and validation sets balanced.

5.1 Sensitivity (Recall) and Specificity (Precision)

We can notice from Table 1 that the hybrid model gives better results than VGG16 model in terms of F1 scores.

Table 1. F1 Score of Hybrid and VGG16 architecture (binary classification)

Hybrid Architecture			VGG16 Architecture		
Specificity	Sensitivity	F1 Score	Specificity	Sensitivity	F1 Score
89.9%	55.3%	68.5%	65.3%	0.1%	17.1%

5.2 Confusion Matrix

From Tables 2 and 3, we can also observe that the hybrid model is more efficient than VGG16 model.

Table 2. Confusion matrix for Hybrid architecture

TN	*FN*	*FP*	*TP*
0.100	0.380	0.053	0.470

Table 3. Confusion matrix for VGG16 architecture

TN	*FN*	*FP*	*TP*
0.490	0.430	0.025	0.470

5.3 Training and Testing Curves

We noticed that the curves obtained with VGG16 showed an absence of learning because the loss rate only increased. In contrast to the hybrid model, where the loss decreased as soon as the learning process started. The Fig. 4 shows the result obtained with the hybrid architecture.

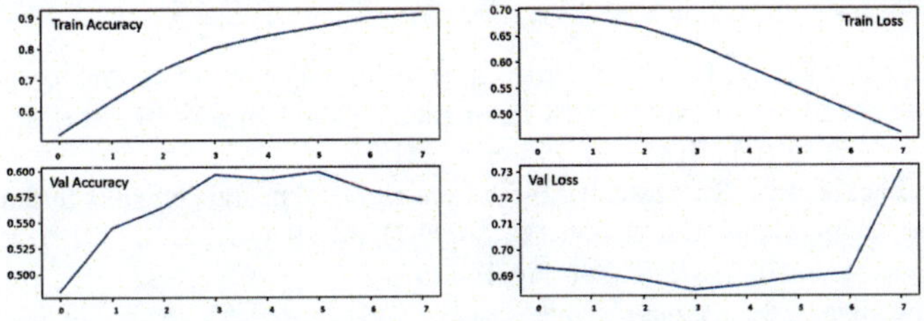

Fig. 4. Training and validation curves with the hybrid architecture

5.4 Explainability

The Fig. 5 shows the result obtained with sensitivity occlusion and we can notice that the green areas clearly target certain elements of the artery. On the other side, the most fluorescent point targets probable stenosis. Also, as a side effect, the catheter is sometimes highlighted which is logical because it is always present.

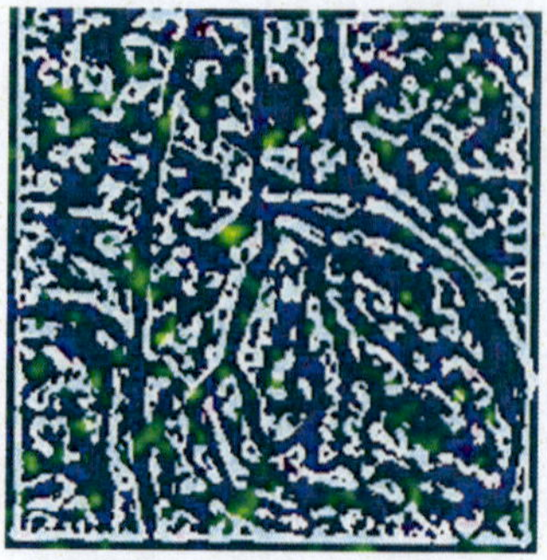

Fig. 5. Example of occlusion

6 Conclusion and Future Work

In this paper, we proposed two deep learning architectures: VGG16 and Hybrid CNN-LSTM to detect stenoses in coronary arteries. After processing the data, we managed to obtain relevant and efficient models: relevant because the curves obtained show real statistical learning, and efficient because their performance is superior to that of the references (VGG16 and the classical network). However, the performance achieved is, in absolute terms, are not those expected: barely 61% accuracy. Actually, in a real clinical context, this would not really help a cardiologist in his diagnosis of coronary artery disease. This can be explained by the fact that the data set on which we conducted our experiments should be annotated more finely by the cardiologist. This work is currently underway. by the latter and will enable us to obtain better results.

References

1. Du, T., Liu, X., Zhang, H., Xu, B.: Real-time lesion detection of cardiac coronary artery using deep neural networks. In: 2018 International Conference on Network Infrastructure and Digital Content (IC-NIDC), pp. 150–154 (2018)
2. Lee, P.C., Lee, N., Pyo, R.: Abstract 12950: Convolutional neural networks for interpretation of coronary angiography. Circulation **140**(Suppl 1), A12 950–A12950 (2019). https://www.ahajournals.org/doi/10.1161/circ.140.suppl1.12950
3. Ovalle-Magallanes, E., Avina-Cervantes, J.G., Cruz-Aceves, I., Ruiz-Pinales, J.: Transfer learning for stenosis detection in x-ray coronary angiography. Mathematics **8**(9) (2020). https://www.mdpi.com/2227-7390/8/9/1510
4. Resta, M., Monreale, A., Bacciu, D.: Occlusion-based explanations in deeprecurrent models for biomedical signals. Entropy **23**(8) (2021). https://www.mdpi.com/1099-4300/23/8/1064
5. Fong, R.C., Vedaldi, A.: "Interpretable explanations of black boxes by meaningful perturbation. In: IEEE International Conference on Computer Vision (ICCV), vol. 2017, pp. 3449–3457 (2017)
6. Zeiler, M.D., Fergus, R.: Visualizing and understanding convolutional networks. In: Fleet, D., Pajdla, T., Schiele, B., Tuytelaars, T. (eds.) Computer Vision – ECCV 2014: 13th European Conference, Zurich, Switzerland, September 6-12, 2014, Proceedings, Part I, pp. 818–833. Springer International Publishing, Cham (2014). https://doi.org/10.1007/978-3-319-10590-1_53

7. Mahmoudi, S.A., Stassin, S., El Habib, M., Daho, X.L., Mahmoudi, S.: Explainable deep learning for Covid-19 detection using chest X-ray and CT-scan images. In: Garg, L., Chakraborty, C., Mahmoudi, S., Sohmen, V.S. (eds.) Healthcare Informatics for Fighting COVID-19 and Future Epidemics, pp. 311–336. Springer International Publishing, Cham (2022). https://doi.org/10.1007/978-3-030-72752-9_16
8. Liang, G., Hong, H., Weifang, X., Zheng, L.: Combining convolutional neural network with recursive neural network for blood cell image classification. IEEE Access **6**, 36188–36197 (2018)
9. Antczak, K., Liberadzki, L.: Stenosis detection with deep convolutional neural networks. In: MATEC Web of Conferences, vol. 210, p. 04001 (2018). EDP Sciences
10. VGG et Transfer Learning - datacorner par Benoit Cayla. https://datacorner.fr/vgg-transfer-learning/. (Accessed 15 June 2022)
11. Baumgartner, C., Roffi, M., Perrier, A., Carballo, S.: La maladie coronarienne stable ou asymptomatique: un nouveau paradigme: Médecine Interne Générale. Revue médicale suisse, vol. 6(267) (2010)

A New Four-Ports MIMO Antenna for 5G IoT Applications

Ouafae Elalaouy[1](✉), Mohammed El Ghzaoui[2], and Jaouad Foshi[1]

[1] Faculty of Sciences and Techniques, Moulay Ismail University, Errachidia, Morocco
ou.elalaouy@edu.umi.ac.ma
[2] Sidi Mohamed Ben Abdellah University, Fes, Morocco

Abstract. Unlike other mobile network technologies, 5G is a disruptive technology. Because 5G can address not only the general public but also major economic sectors for B2B (Business to Business) uses. So that, the implementation of 5G system is harsh task. Antenna is one of the important element to implement 5G technology. In this regard, this paper deals with the design of defected ground structure Multiple Input Multiple Output (MIMO) antenna functioning in the millimeter-wave 5G band for the future IoT applications. The proposed MIMO antenna consists of four MIMO elements with eight identical patches and it was derived after multiple iterations. The bandwidth of the proposed antenna is about 2.5 GHz ranging from 26.8 GHz to 29.3GHz. This high bandwidth is obtained by the use of slots in the first iteration and defected ground in the final iteration. The MIMO antenna was simulated via HFSS. The study's major findings are as follows: First, the proposed antenna covers the band allocated to 5G in USA 27.5–28.35 GHz and Japan (27.5–28.28 GHz). The second finding is lower isolation less than −35 dB which mean the coupling between antenna can be ignored. Finally, the peak gain at the operating band is about 18.25dBi. The above-mentioned results made the MIMO antenna that covers 27.9 GHz (26.8–29.3 GHz) useful in 5G band handsets.

Keywords: MIMO · 5G · Gain · Bandwidth · Handsets

1 Introduction

MIMO technology has witnessed rapid development within the field of mobile communication [1–3]. The 5G communication system is expected to handle the growth in forecasted data traffic by embracing the MIMO antenna as an emerging key technique. Due to multipath fading, researchers are highly focusing on the use of many antennas both transmitting and receiving ends, to enhance the data rate and to provide decent link reliability through spatial diversity and spectral efficiency via multiplexing techniques. Although MIMO is utilized to overcome various drawbacks, it still suffers from mutual coupling, especially among closest spaces. Therefore, a number of approaches have been adopted to solve this hitch namely defected ground structure and antenna separation.

To realize MIMO antenna that can be used for smartphones, multi-antennas have been proposed in literature [4–10]. A 4-ports UWB MIMO antenna system has been

Y. Farhaoui et al. (Eds.): ICAISE 2022, LNNS 635, pp. 78–85, 2023.
https://doi.org/10.1007/978-3-031-26254-8_11

proposed in [4]. The antenna has high bandwidth along with high gain. In [5], a MIMO antenna with high isolation is designed for smartwatch applications. In [6], MIMO antenna operating at 28/38 GHz has been proposed for 5G smartphones. The suggested antenna offers a peak gain of 9dBi. Aghoutane et al. [7] have proposed a MIMO antenna for 5G application. The authors used a new geometry of ground plan with array elements to form the MIMO antenna. The proposed antenna has dual band properties, it covers the two bands (27.50–28.35 GHz) and 37 GHz (37–37.6GHz). In [8] a MIMO antenna array has been introduced for 5G application. The antenna covers the band (25.5–29.6 GHz); however, antenna gain is lower of about 8.3dBi which isn't efficient for 5G mobile applications. In [9], a slot patch antenna has been reported for 28/38 GHz bands. The antenna is dual band and offers a maximum peak gain of about 9dBi at 28 GHz. In [10], a new form of antenna is proposed. The proposed MIMO antenna with an end-launch connector model for 5G Millimeter Wave Mobile Applications was analyzed. The suggested antenna in [11] has an impedance bandwidth of 2.6 GHz (27.4–30.0 GHz) and 3.3 GHz (36.7–40.0 GHz). Besides, the antenna offers peak gains of 18 and 14.5 dB at 28.5 and 38.8 GHz respectively. The works carried out in the above cited references helped us to have a clear picture about the status of antenna for 5G application as well as the design requirements for the implementation of printed antennas for mobile applications. Many others research work have been reported in literature [12–15].

In the present work, a four ports MIMO antenna was proposed using defected ground structure to cope with the mutual coupling. The proposed MIMO antenna operates at 27.9 GHz and has a bandwidth of 2.5 GHz (26.8–29.3 GHz) with a peak gain of 18.25 dBi. Furthermore, it provides an isolation less than −35 dB. As a result, the proposed MIMO array antenna can be deemed a capable mm-wave antenna for upcoming 5G networks.

2 Design Flow

The presented MIMO antenna has been achieved after numerous spteps. In the first step, a single resonator is designed to work in 26.8–29.3 GHz. In the second, a two-element array antenna is built. Finally, the latest design is extended to four-element MIMO. Therefore; the subsequent sections give and explore the design evolution from a single element to the suggested MIMO configuration.

2.1 Single Element Antenna

The design flow of the proposed single input single output (SISO) has been initiated from a rectangular patch antenna [17–20]. The initial rectangular patch shows a narrowband. Hence, some modifications have been implemented to lead at the desired band such as etching the rectangular patch and adding a slot to get the 2nd of Fig. 1. The final version of the single antenna as shown in Fig. 1 was used to develop the MIMO antenna.

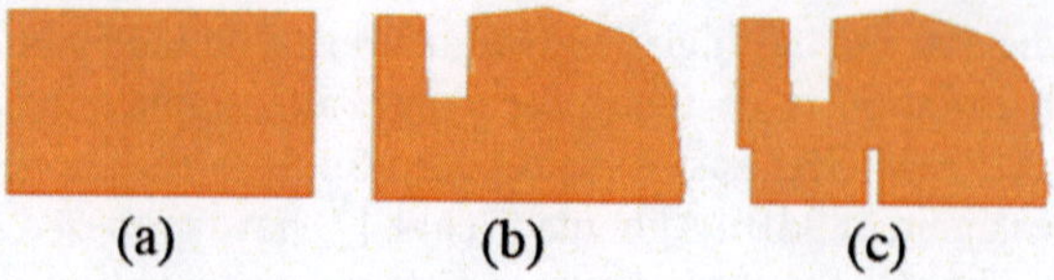

Fig. 1. Design flow of single radiator, (a) rectangular patch (patch 1), (b) patch2, (b) patch 3.

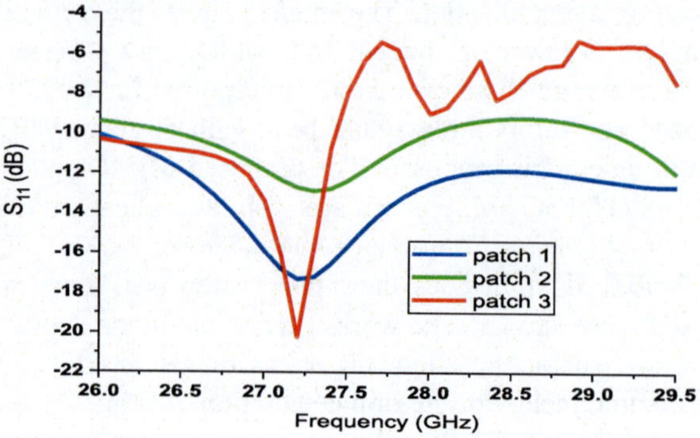

Fig. 2. Return loss versus frequency plot of different patches.

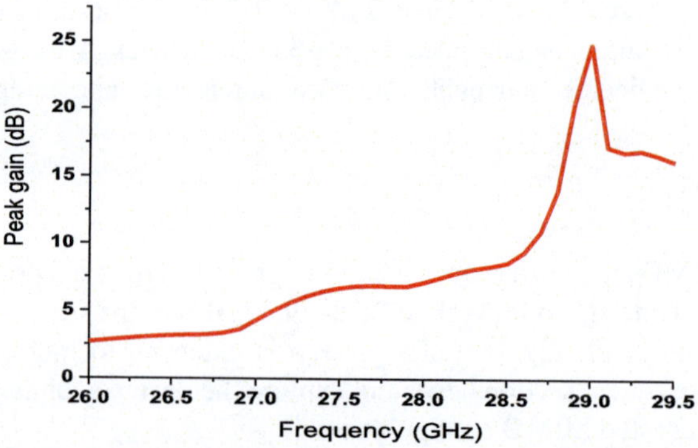

Fig. 3. Gain versus frequency plots for the proposal antenna.

The comparison of simulated return loss S11 versus frequency plots of the single element antenna is depicted in Fig. 2. It is contemplated from Fig. 2 that 1st antenna works in 26–27.9 GHz, whereas the 2nd antenna resonates at 27.3 GHz with a bandwidth from 26.4 GHz to 27.9 GHz and shows a bad reflection coefficient of −10.3 dB. Hence, the 3rd antenna is designed to improve the bandwidth and s11 respectively to 26 GHz–29.5 GHz and −17.21 dB. In Fig. 3 it is shown that the antenna has a peak gain of 7.19 dBi in both American and Japanese band that are respectively 27.5–28.35 GHz and 27.5–28.28 GHz.

2.2 Two-Element Array

During the second stage of the design, identical single-element patches were combined to create a two-element array using a T-junction power divider. The two-element array's structural arrangement, as well as its feeding network, are shown in Fig. 4.

Fig. 4. Two-element array antenna design.

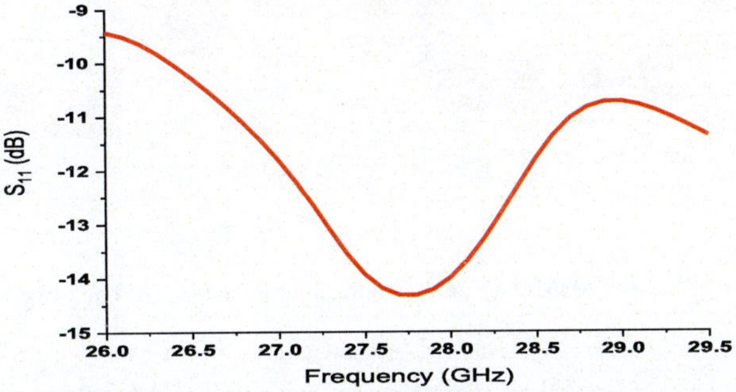

Fig. 5. Return loss versus frequency plot of two-element array antenna.

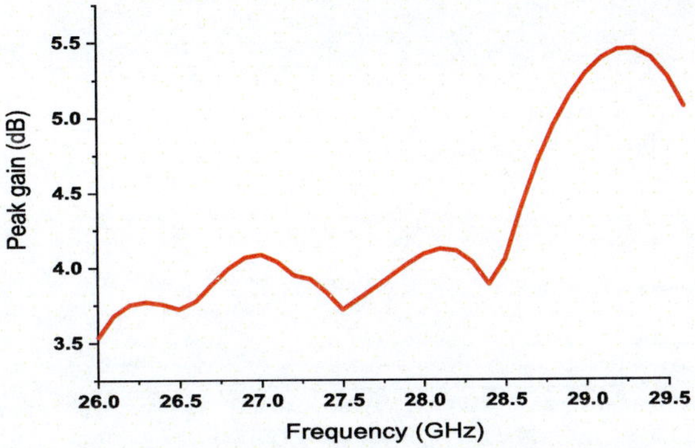

Fig. 6. Gain versus frequency plot of two-element array antenna.

In Fig. 5 we depicted the S_{11} of the basic array antennas. From Fig. 5, one can notice that the antenna has a good matching with a minimum value = –14.33 dB along with a −10 dB impedance bandwidth ranging from 26.4 GHz to 29.5 GHz with a central frequency of 27.8 GHz. As highlighted in Fig. 6, the two-element array peak gain equals 4.04 dBi in both American and Japanese band that are respectively 27.5–28.35 GHz and 27.5–28.28 GHz.

2.3 MIMO Antenna

To achieve the final version of MIMO antenna, the aforementioned developed array antenna was utilized as displayed in Fig. 7. The top plane of the MIMO antenna consists of eight identical patch antennas. The back plane consists of a defected ground structure to mitigate the mutual coupling. The whole design was built on a low-cost FR4 epoxy substrate of thickness h = 1.6 mm with dielectric constant $\varepsilon_r = 4.4$ and the resonant frequency $f_r = 27.9$ GHz.

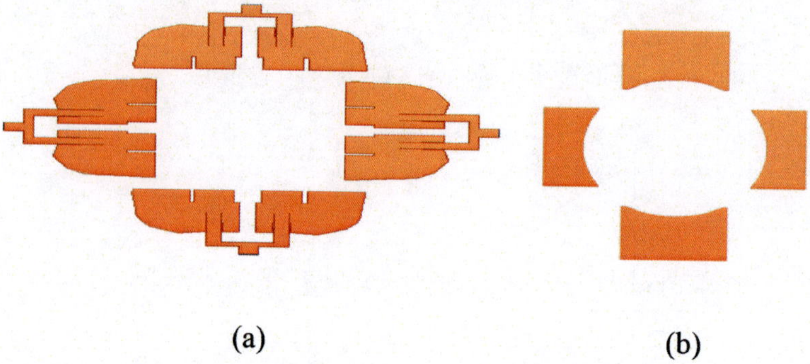

Fig. 7. Design of the MIMO antenna, (a) top plane design, (b) back plane design.

As shown in Fig. 8, the MIMO operates in 26.8–29.3 GHz and the return loss is – 14.2 dB. Further. The MIMO antenna achieves 18.25 dBi, in both American and Japanese band that are respectively 27.5–28.35 GHz and 27.5–28.28 GHz, as shown in Fig. 9.

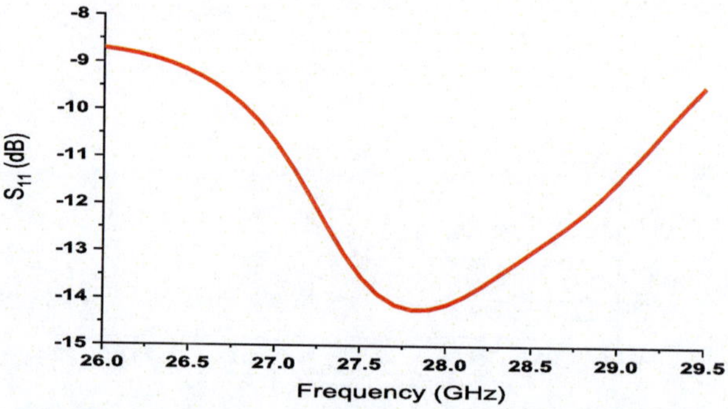

Fig. 8. Return loss versus frequency plot of MIMO antenna.

3 Comparison of Results

The study of increasing elements among MIMO antennas is much valued as it offers a detailed understanding into the enhancement of the characteristics parameters. The

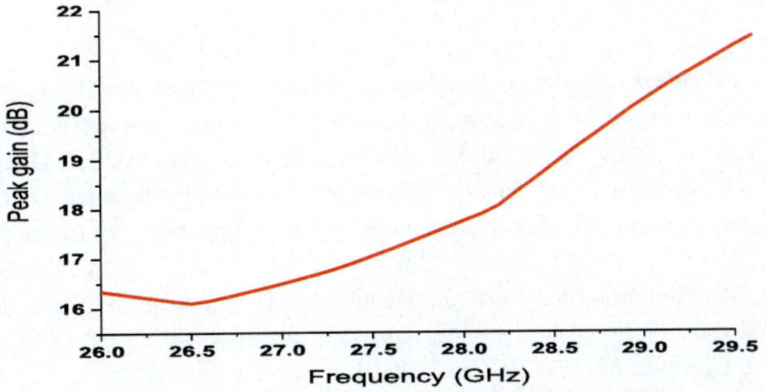

Fig. 9. Gain versus frequency plot of MIMO array antenna.

Table 1. Performance of Single, 2 Elements array and MIMO antenna.

Structures	Bandwidth (GHz)	S11 (dB)	Peak gain (dB)
Single element	[26–29.5] = 3.5	−17.21	7.19
2 elements array	[26.4–29.5] = 3.1	−14.33	4.04
MIMO with DGS	[26.8–29.3] = 2.5	−14.2	18.25

single-element antenna does indeed function at 27.3 GHz with a bandwidth of 3.5 GHz, but it does not provide satisfactory results in terms of gain. Furthermore, array configuration is built to increase gain for the proposed 5G applications. The two-element array antenna has a bandwidth of around 3.1 GHz (26.4–29.5) at 27.8 with a low return loss of roughly 14.33 dB. Compared to a single-element array, the gain provided by the two-element array is lower. Hence, the array antenna's lower gain is insufficient to meet 5G requirements. Yet it provides satisfactory results in terms of bandwidth. For the sake of improving antenna radiation, a four ports MIMO antenna array is projected 5G IoT applications. Table 1 shows that the considered MIMO array antenna outperforms other antenna proposed in literature. The peak gain obtained in this work is enhanced to 18.25 dBi.

4 Conclusion

In this paper, A four-port MIMO array antenna system operating in (26.8–29.3) with the modified ground structure was designed to meet 5G mobile phone requirements. The simulated MIMO antenna exhibits a bandwidth of 2.5 along with a high peak gain of 18.25 dBi. These results are obtained thanks to the suggested geometrical modification in the structure. So, the proposed MIMO antenna has proven to be useful in 5G applications.

References

1. Sangwan, M., Panda, G., Yadav, P.: A literature survey on different MIMO patch antenna. In: 2020 International Conference on Inventive Computation Technologies (ICICT), Coimbatore, India, pp. 912–918 (2020). https://doi.org/10.1109/ICICT48043.2020.9112566
2. Mestoui, J., El Ghzaoui, M., Fattah, M. et al. : Performance analysis of CE-OFDM-CPM Modulation using MIMO system over wireless channels. J. Ambient Intell. Hum. Comput. 11, 3937–3945 (2020). https://doi.org/10.1007/s12652-019-01628-0
3. El Ghzaoui, M., Hmamou, A., Foshi, J., Mestoui, J.: Compensation of non-linear distortion effects in MIMO-OFDM systems using constant envelope OFDM for 5G applications. J. Circuits Syst. Comput. **29**(16), 2050257 (2020)
4. Parchin, N.O., Basherlou, H.J., Al-Yasir, Y.I.A., Abdulkhaleq, A.M., Abd-Alhameed, R.A.: Ultra-wideband diversity MIMO antenna system for future mobile handsets. Sensors **20**, 2371 (2020). https://doi.org/10.3390/s20082371
5. Wen, D., Hao, Y., Wang, H., Zhou, H.: Design of a MIMO antenna with high isolation for smartwatch applications using the theory of characteristic modes. IEEE Trans. Antennas Propag. **67**(3), 1437–1447 (2019)
6. Farahat, A.E., Hussein, K.F.A.: 28/38 GHz dual-band yagi-uda antenna with corrugated radiator and enhanced reflectors for 5G MIMO antenna systems. Progress Electromagnetics Res. C **101**, 159–172 (2020). https://doi.org/10.2528/PIERC20022603
7. Aghoutane, B., Das, S., EL Ghzaoui, M., Madhav, B.T.P., El Faylali, H.: A novel dual band high gain 4-port millimeter wave MIMO antenna array for 28/37 GHz 5G applications. AEU – Int. J. Electron. Commun. **145**, 154071 (2022)
8. Khalid, M., et al.: 4-Port MIMO Antenna with Defected Ground Structure for 5G Millimeter Wave Applications. Electronics **9**(1), 71 (2020). https://doi.org/10.3390/electronics9010071
9. Liu, P., Zhu, X., Zhang, Y., Wang, X., Yang, C., Jiang, Z.H.: Patch antenna loaded with paired shorting pins and H-Shaped Slot for 28/38 GHz dual-band MIMO applications. IEEE Access **8**, 23705–23712 (2020). https://doi.org/10.1109/ACCESS.2020.2964721
10. Mahmoud, K.R., Montaser, A.M.: Synthesis of multi-polarised upside conical frustum array antenna for 5G mm-Wave base station at 28/38 GHz. IET Microwaves Antennas Propag. **12**(9), 1559–1569 (2018). https://doi.org/10.1049/iet-map.2017.1138
11. Aghoutane, B., El Ghzaoui, M., Das, S., Ali, W., El Faylali, H.: A dual wideband high gain 2 × 2 multiple-input-multiple-output monopole antenna with an end-launch connector model for 5G millimeter-wave mobile applications. Int. J. RF Microwave Comput.-Aided Eng. **32**(5) (2022)
12. Njogu, P., Sanz-Izquierdo, B., Elibiary, A., Jun, S.Y., Chen, Z., Bird, D.: 3D printed fingernail antennas for 5G applications. IEEE Access **8**, 228711–228719 (2020). https://doi.org/10.1109/ACCESS.2020.3043045
13. Sun, Y.-X., Wu, D., Fang, X.S., Yang, N.: Compact quarter-mode substrate-integrated waveguide dual-frequency millimeter-wave antenna array for 5G applications. IEEE Antennas Wirel. Propag. Lett. **19**(8), 1405–1409 (2020). https://doi.org/10.1109/LAWP.2020.3003305
14. Hussain, N., Awan, W.A., Ali, W., Naqvi, S.I., Zaidi, A., Le, T.T.: Compact wideband patch antenna and its MIMO configuration for 28 GHz applications. AEU – Int. J. Electron. Commun. **132** (2021). https://doi.org/10.1016/j.aeue.2021.153612
15. Chandra, R., et al.: IEEE 3rd 5G World Forum (5GWF). Bangalore, India **2020**, 514–518 (2020). https://doi.org/10.1109/5GWF49715.2020.9221471

16. Hakim, M.L., Uddin, M.J., Hoque, M.J.: 28/38 GHz dual-band microstrip patch antenna with DGS and stub-slot configurations and its 2×2 MIMO antenna design for 5G wireless communication. In: 2020 IEEE Region 10 Symposium (TENSYMP), Dhaka, Bangladesh, 2020, pp. 56–59 (2020). https://doi.org/10.1109/TENSYMP50017.2020.9230601
17. Aghoutane, B., Das, S., El Faylali, H., Madhav, B.T.P., El Ghzaoui, M., El Alami, A.: Analysis, design and fabrication of a square slot loaded (SSL) millimeter-wave patch antenna array for 5G applications. J. Circuits Syst. Comput. **30**(5), 2150086 (2021)

A New Proof of the Sobolev Logarithmic Inequality on the Circle R/2πZ

M'hammed Ouyahia(✉), Ali Hafidi, and Moulay Rchid Sidi Ammi

Department of Mathematics, AMNEA Group, Faculty of Sciences and Techniques, Moulay Ismail University, B. P. 509, Errachidia, Morocco
mhammed.ouyahia@usmba.ac.ma, hafidiali@gmail.com, rachidsidiammi@yahoo.fr

Abstract. We consider the Brownian motion semi-group defined on the circle R/2πZ, equipped with the uniform law $d\mu = \frac{dx}{2\pi}$. Via a method involving probabilistic techniques and by using the semigroup thechniques, we present a simple and clear demonstration of the Sobolev logarithmic inequality.

Keywords: Brownian motion semigroup · Sobolev logarithmic inequality

1 Introduction

For all smooth functions f on R^n, Gross's logarithmic Sobolev inequality [1] states that:

$$\int_{\mathbb{R}^n} f^2 \ln\left(f^2\right) d\gamma_n - \left(\int_{\mathbb{R}^n} f^2 d\gamma_n\right) \ln\left(\int_{\mathbb{R}^n} f^2 d\gamma_n\right) \leq 2 \int_{\mathbb{R}^n} |\nabla f|^2 d\gamma_n, \tag{1}$$

with dγn denotes the Gaussian normalized measure on$\surd$R^n:

$$d\gamma_n(dx) = (\sqrt{2\pi})^{-n} \exp\left(-|x|^2/2\right) dx,$$

and the usual gradient's length is $|\nabla f|$.

In actuality, this inequality is a strengthened version of the original Poincaré inequality.

$$\int_{\mathbb{R}^n} f^2 d\gamma_n - \left(\int_{\mathbb{R}^n} f d\gamma_n\right)^2 \leq \int_{\mathbb{R}^n} |\nabla f|^2 d\gamma_n. \tag{2}$$

In 1980, This inequality was provided on the circle by F. Weissler [3] using a direct method but with a complicated technique, and by O. Rothaus [2] in 1980 by a variationnel method.

In 1987 M. Emry and J. E. Yakich [4] provide a simple and direct proof of the Sobolev logarithmic inequality on the unite circle.

Our aim in this paper is to show a simple proof of the Sobolev logarithmic inequality on the circle R/2πZ.

These types of integral inequalities have a strong connection to the aspects of the long-time behavior of the parabolic PDEs [5].

Y. Farhaoui et al. (Eds.): ICAISE 2022, LNNS 635, pp. 86–89, 2023.
https://doi.org/10.1007/978-3-031-26254-8_12

2 Preliminaries

To keep the paper's organization in order and reasonable self-contained, in this section, we summarize the fundamental notion that will be applied throughout this work.

We consider on the circle S = R/2πZ the Brownian motion semi-group defined by:

$$P_t f(x) = \int_S f(x+y) \sum_{n \in Z} \frac{1}{\sqrt{4\pi t}} \exp^{-(y+2\pi n)^2/4t} dy, \ f \in C^\infty(S).$$

Acting on the space of Hilbert $L^2(S,d\mu)$.

The heat semi-group $(P_t)_{t>0}$ has the properties:

- $\frac{dP_t f}{dt} = (P_t f)'' = P_t(f'')$;
- $P_t f \to \int_S f \, d\mu$ on $L^2(S, d\mu)$ as $t \to \infty$ (ergodicity);
- $P_t f \to f$ as $t \to 0$.

3 Main Results

In this section, we will establish our proof of the Sobolev logarithmic inequality on the circle S = R/2πZ.

Theorem

For all smooth function f on the circle S = R/2πZ. S equipped with the uniform law dμ = $\frac{dx}{2\pi}$ the Sobolev Logarithmic Inequality on S states that:

$$\int_S f^2 \ln\left(f^2\right) d\mu - \left(\int_S f^2 d\mu\right) \ln \left(\int_S f^2 d\mu\right) \leq 2 \int_S f'^2 d\mu. \tag{3}$$

It is true for all f such that $f' \in L2(S, d\mu)$ and needs to verify $\int f' d\mu = 0$.

Proof of Theorem

In the fact, we will show that for all function f strictly positive in $C^\infty(S)$:

$$\int_S f \ln(f) d\mu - \left(\int_S f d\mu\right) \ln \left(\int_S f d\mu\right) \leq \frac{1}{2} \int_S \frac{f'^2}{f} d\mu. \tag{4}$$

Remplacing f by $f^2 + \varepsilon$ and letting ε to zero, we get the inequality (2).

To prove (4), we define for $t \in R^+$ the function $\phi(t)$ by

$$\varphi(t) = e^{2t} \int_S \frac{P_t\left(f'\right)^2}{P_t f} d\mu,$$

For all $f > 0$, we make the successive substituations $g = P_t f$ and $h^2 = 2g$, we have

$$\frac{d\varphi(t)}{dt} = e^{2t} \int_S \left(\frac{2g'^2}{g} + \frac{2g^{(3)}g'}{g} - \frac{g'^2 g''}{g^2} \right) d\mu$$
$$= 4e^{2t} \int_S \left(h'^2 + h' h^{(3)} + \frac{2h'^2 h''}{h} - \frac{h'^4}{h^2} \right) d\mu.$$

By integration by parts, we have

$$\int_S h' h^{(3)} d\mu - \int_S h''^2 d\mu \quad and \quad 3 \int_S \frac{h'^2 h''}{h} d\mu = \int_S \frac{h'^4}{h^2} d\mu.$$

We develop h into a Fourier series, we get

$$\int_S h''^2 d\mu \geq \int_S h^2 d\mu.$$

We deduce

$$\frac{d\varphi(t)}{dt} = -4e^{2t} \int_S \left(h''^2 - h'^2 + \frac{h'^4}{3h^2} \right) d\mu \leq 0.$$

.

For all $t \in \mathrm{R}^+$, this shows that ϕ is non-increasing function so

$$\int_S \frac{P_t (f')^2}{P_t f} d\mu \leq e^{-2t} \int_S \frac{P_0 (f')^2}{P_0 f} d\mu = e^{-2t} \int_S \frac{f'^2}{f} d\mu.$$

Let $\psi_p(x) = \frac{-x^p + x}{p-2}$ for $p \in [0, +\infty[$, $p \# 2$ be a smooth function, by Fubini's theorem, we get

$$\int_S \psi_p(f) d\mu - \psi_p \left(\int_S f d\mu \right) = \int_S \psi_p(P_0 f) d\mu - \int_S \psi_p(P_\infty f) d\mu$$
$$= - \int_0^\infty \frac{d}{dt} \int_S [\psi_p(P_t f) d\mu] dt$$
$$= - \int_0^\infty \int \left[(P_t f)'' \psi_p'(P_t f) d\mu \right] dt$$
$$= \int_0^\infty \int_S \left[(P_t f)' \psi_p'(P_t f) d\mu \right] dt$$
$$= \int_0^\infty \int_S \left[(P_t f')^2 \psi_p''(P_t f) d\mu \right] dt$$
$$= \frac{2}{p^2} \int_0^\infty \int_S \left[(P_t f')^2 (P_t f)^{\frac{2(1-p)}{p}} d\mu \right] dt$$

For $p \to 2$, we have $\psi_2 = \frac{1}{2}xln(x)$. Consequently

$$\int_s \psi_2(f)d\mu - \psi_2(\int_s fd\mu) = \frac{1}{2}\int_0^\infty \left[\frac{(P_tf')^2}{P_tf}d\mu\right]dt$$

$$\leq \frac{1}{2}\int_0^\infty e^{-2t}\int_S\left[\frac{(P_0f')^2}{P_0f}d\mu\right]dt$$

$$= \frac{1}{4}\int_S \frac{f^2}{f}d\mu.$$

This is the conclusion, which completes the proof.

References

1. Ané, C., et al.: Sur les inégalités de Sobolev logarithmiques, vol. 10. Société Mathématique de France, Paris (2000)
2. Rothaus, O., Logarithmic, S.: Sobolev inequalities and the spectrum of SturmLiouville operators. J. Func. Anal. **39**, 42–56 (1980)
3. Weissler, F.: Logarithmic Sobolev inequalities and hypercontractive estimates on the circle. J. Func. Anal. **37**, 218–234 (1980)
4. Emery, M., Yukich, J.E.: A simple proof of the logarithmic Sobolev inequality on the circle. In: Azéma, J., Yor, M., Meyer, P.A. (eds.) Séminaire de Probabilités XXI. LNM, vol. 1247, pp. 173–175. Springer, Heidelberg (1987). https://doi.org/10.1007/BFb0077632
5. Pazy, A.: Semigroups of Linear Operators and Applications to Partial Differential Equations. Springer, New York (1983). https://doi.org/10.1007/978-1-4612-5561-1

A New Robust Adaptive Control for Variable Speed Wind Turbine

Sanae El Bouassi(✉), Zakaria Chalh, and El Mehdi Mellouli

Laboratory of Artificial Intelligence, Data Sciences and Emerging Systems, School of Applied Sciences, Sidi Mohamed Ben Abdellah University, Fez, Morocco
{sanae.elbouassi,Zakaria.chalh}@usmba.ac.ma, mellouli_elmehdi@hotmail.com

Abstract. In This paper, we introduce an approximation of a sliding mode control for variable speed wind turbine. In the first step, we present the model of the system of the wind turbine. The second step consist to approximate the expression of the additive expression in the law control. This approach is related to the control law of this wind turbine, it leads to the approximate expression of the gain of the auxiliary expression in the additive term of the law control, in order to attenuate the external disturbances and compensate the error modeling and to overcome the charting problem. Finally, the stability and the efficiency of this control law are proved and in the end, we present the simulation, this one shows that the proposed approach is efficient and more precise and accurate in terms of settling time, tracking accuracy and the energy consumption.

Keywords: Variable speed wind Turbine · Sliding mode control · Lyapunov stability

1 Introduction

Nowadays, the human activities have diversified, specifically in production and in the improvement of the quality of life. However, that the increase in electricity consumption has affected the environment and the word's reserves. These causes pushed human to introduce a new source of energy: renewable sources.

Based on the mechanical study of the wind turbine already made in [12, 15] and [3]; which aims to maximize the energy absorbed from the wind using sliding mode control (SMC). The SMC design theory was introduced by S.V. Emelyanov and V.I. Utkin [1], the development of this theory is shown in [2].

Furthermore, the adaptive law using Lyapunov synthesis based on stability and convergence analysis for satisfying a specified tracking performance can tune the parameters.

This paper presents an approximate study, which makes it possible to find an approximate expression of the additive part in the control law more precisely the unknown positive gain existing in this expression in order to minimize the charting problem.

Y. Farhaoui et al. (Eds.): ICAISE 2022, LNNS 635, pp. 90–96, 2023.
https://doi.org/10.1007/978-3-031-26254-8_13

Nomenclature

Symbols

Ta	Aerodynamique torque	*Kls*	Shaft damping coefficient
Kr	Rotor external damping	*wg*	Generator speed
P	Air density	*Bls*	Stiffness coefficient
Jr	Rotor inertia	*kg*	Generator friction
wt	Angular speed of the rotor	*Tem*	Generator electromagnetic torque
Kr	Rotor external damping	λ	Tip-speed ratio
β	Blade pitch angle	*Cpopt*	Optimal wind energy
R	Rotor radius	*v (t)*	Wind speed
Jg	Generator inertia	*Cp*	Power coefficient

2 The System of Wind Turbine

The Aerodynamique torque's expression presented in Fig. 1 is given by [12]:

$$\text{Ta} = \frac{1}{2}\rho\pi R^3 \frac{Cp(\lambda, \beta)}{\lambda(t)} v(t)^2 \tag{1}$$

R is the radius of the rotor and v(t) is the speed of wind, ρ is the density of the air and Cp is the coefficient of power, which is given by:

$$\lambda(t) = R.w_t / v(t) \tag{2}$$

The following equation characterizes the dynamics of the rotor:

$$J_r \dot{w}_t = T_a - T_{ls} - k_r w_t \tag{3}$$

The generator speed is given by:

$$J_g \dot{w}_g = T_{hs} - k_g w_g - T_{em} \tag{4}$$

where T_{hs} is the high-speed shaft torque, Jg is the inertia of the generator and kg is the generator friction coefficient.

The transmission n_g is:

$$n_g = \frac{T_{ls}}{T_{hs}} \tag{5}$$

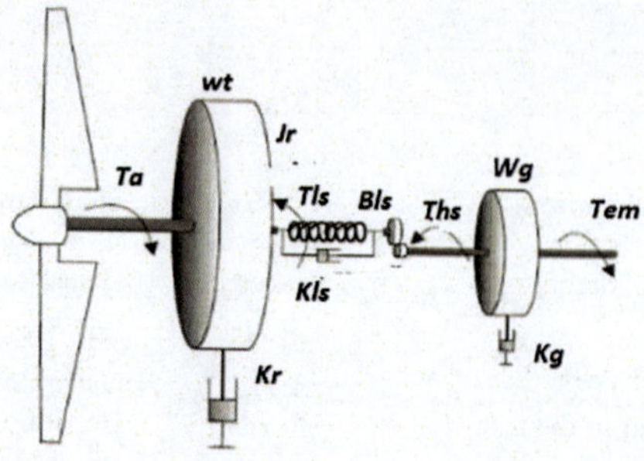

Fig. 1. Two-mass model of a wind turbine [21]

3 Problem Statement

The problem is to obtain the rotor speed w_t for tracking optimal reference y_r given by:

$$y_r = w_{topt} = \frac{v(t).\lambda_{opt}}{R} \tag{6}$$

$$\text{And } \lambda_{opt} = \frac{R.w_{topt}}{v(t)} \tag{7}$$

The system in Fig. 1 is represented by:

$$\begin{cases} \dot{x_1} = x_2 \\ \dot{x}_2 = f(x, t) + g.u(x, t) \\ y_1 = x_1 \end{cases} \tag{8}$$

where u = T_{em}, y =w_t, , and g = $(-\frac{1}{Jr} * \frac{k_{ls}}{n_g J_g})$ and f(x) is the function of the nonlinear system:

$$\text{f(x, t)} = \left(\frac{-k_g}{J_g} - \frac{J_g}{n_g * J_g}\right) * x_2 - \frac{k_r}{n_g * J_g} * x_1 + \frac{1}{J_g * n_g} T_a \tag{9}$$

The Tracking errors e and $\dot{e}$ are defined as:

$$e = w_{topt} - w_t \tag{10}$$

$$\dot{e} = \dot{w_{topt}} - \dot{w_t}$$

Define a sliding surface [1, 3] in the space of the error state as with k is a positive constant:

$$\sigma = \dot{e} + k * e \tag{11}$$

The surface derivative (10) is given by:

$$\dot{\sigma} = \ddot{e} + k * \dot{e} \tag{12}$$

Using (11), (12), the following theorem is proposed:

$$\dot{\sigma} * \sigma \prec 0 \tag{13}$$

The equivalent command is given by:

$$\dot{\sigma(t)} = 0$$

$$u_{eq}(t) = \frac{1}{g} * (\ddot{w_{topt}} - f(x) + k * \dot{e}) \tag{14}$$

In order to absorb disturbances and satisfy better robustness, the switching control law is given as:

$$\mu_{sw} = -k_d * sign(\sigma) \tag{15}$$

In this part, we can use the function arctg to eliminate the problem of charting.

Where k_d is a positive gain. This positive gain k_d can be approximate using adaptive law as [6]:

$$K_d = w_o + w_1|e| + w_2|\dot{e}| \tag{16}$$

where $w_n(n = 0, 1, 2)$ positive coefficients, e and $\dot{e}$ are the errors.

4 Stability Analysis

Choosing the Lyapunov function candidate to be [1, 3, 6]:

$$v(t) = \frac{1}{2} * \sigma^2 + \frac{1}{2}\sum_{i=0}^{2} \frac{\tilde{\omega}_i^2}{\varepsilon_i} \tag{17}$$

Therefore using (12), we find:

$$\dot{v}(t) = \dot{\sigma}\sigma + \sum_{i=0}^{2} \frac{\hat{\dot{w}}_l\tilde{\omega}_i}{\varepsilon_i} \tag{18}$$

$$= \ddot{\sigma}[e + k * \dot{e}] + \sum_{i=0}^{2} \frac{\hat{\dot{w}}_l\tilde{\omega}_i}{\varepsilon_i}$$

Using the time derivative of the tracking errors and using (4) then $\dot{v}(t)$ can be expressed as:

$$\dot{v}(t) \ddot{=} \sigma[w_{iopt} - \frac{1}{J_g} * (T_{hs} - k_g * w_g - u) + k * \dot{e}] + \sum_{i=0}^{2} \frac{\hat{\dot{w}}_l\tilde{\omega}_i}{\varepsilon_i} \tag{19}$$

The control law is given by:

$$u(t) = ueq(t) + \hat{u}sw(t) \tag{20}$$

Using (13), (14) and (19) into (18) we find the following expression:

$$\dot{v}(t) \leq (\frac{-T_{hs}}{J_g} - (w_0 + w_1|e| + w_2|\dot{e}|) \\ |\sigma|) + \sum_{i=0}^{2} \frac{\hat{w}_I\tilde{\omega}_i}{\varepsilon_i} \tag{21}$$

According to Lyapunov's theory, we find an approximate expression of the derivatives of the estimated values [6]:

$$\begin{cases} \widehat{\dot{w_0}} = \mathcal{E}_0|\sigma| \\ \widehat{\dot{w_1}} = \mathcal{E}_1|\sigma||e| \\ \widehat{\dot{w_2}} = \mathcal{E}_2|\sigma||\dot{e}| \end{cases} \tag{22}$$

where $\widehat{w}_0$, $\widehat{w}_1$ and $\widehat{w}_2$ are the estimates of w_o, w_1, and w_2.

5 Simulation Results

The parameters of the system two mass model are [12]:
R = 21.65 m, ρ = 1.29 Kg/m^3,$n_g = 43.156$, $J_r = 3.25 * 10^5$ kgm^2,
$B_{ls} = 2.691 * 10^5$ Nm/rad,
$k_g = 0.2$ Nm/rad/s, $\lambda_{opt} = 8.5$, $k_r = 27.36$ Nm/rad/s,

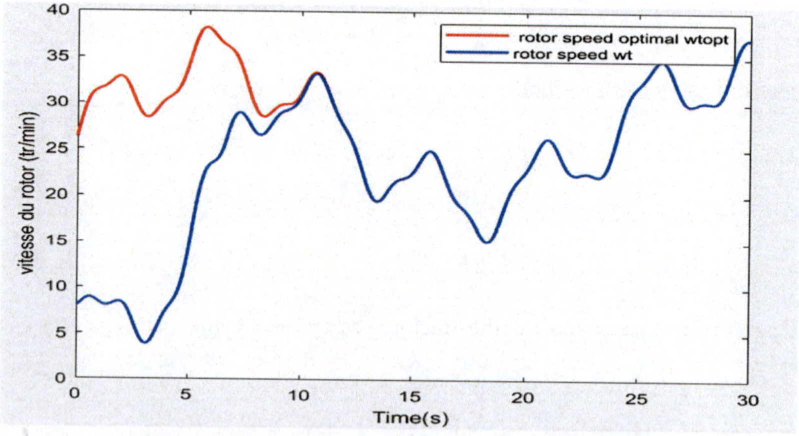

Fig. 2. Responses of the normal and optimal rotor speed

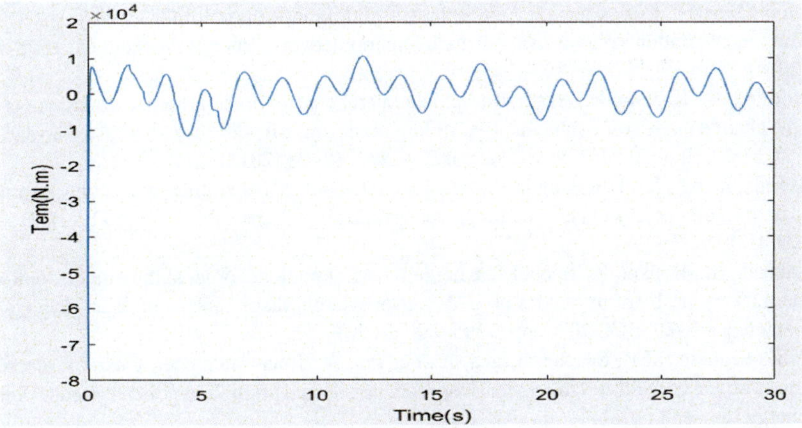

Fig. 3. The response of the control law

6 Conclusion

In this article, we were able to find an approximation of the positive gain existing in the second term in the law of control using adaptive laws, which allowed us to obtain more correct and effective results in term of certainty, this was clear in the results of simulation. In this optics, we can make and look for other approximation for other quantity in order to have more improvement in the functioning of our system.

References

1. Utkin, J., Guldner, J.: Shi, Sliding Mode Control in Electromechanical Systems, 2nd edn. CRC Press, Taylor and Francis, Boca Raton, London (2009)
2. Shtessel, Y., Edwards, C., Fridman, L., Levant, A.: Sliding Mode Control and Observation. Birkhäuser, New York (2014)
3. Mellouli, E.M., Chalh, Z., Alfidi, M.: A new robust adaptive fuzzy sliding mode controller for a variable speed wind Turbine. Int. Rev. Autom. Control **8**(5), 338 (2015)
4. Mellouli, E.M., Sefriti, S., Boumhidi, I.: Combined fuzzy logic and sliding mode approach's for modelling and control of the two link robot. In: 2012 IEEE International Conference on Complex Systems (ICCS), pp. 1–6
5. Naoual, R., Mellouli, E.M., Sefriti, S., Boumhidi, I.: Fuzzy sliding mode control for the two-link robot. Int. J. Syst. Control Commun. **6**(1), 84–96 (2014)
6. Mellouli, E.M., J Boumhidi, I Boumhidi ,Using fuzzy logic for eliminating the reaching phase on the fuzzy H∞ tracking control. Int. J. Modell. Identification Control **20**(4), 398–406 (2013)
7. Mellouli, E.M., Alfidi, M., Boumhidi, I.: Fuzzy sliding mode control for three-tank system based on linear matrix inequality. Int. J. Autom. Control **12**(2), 237–250 (2018)
8. El Hajjami, L., Mellouli, E.M., Berrada, M.: Robust adaptive non-singular fast terminal sliding-mode lateral control for an uncertain ego vehicle at the lane-change maneuver subjected to abrupt change. Int. J. Dynamics Control **9**(4), 1765–1782 (2021)
9. Sellami, A., Arzelier, D., M'hiri, R., Zrida, J.: A sliding mode control approach for systems subjected to a norm-bounded uncertainty. Int. J. Robust Nonlinear Control **17**(Issue 4), 327–346 (2007)

10. Chih-Ming, H., Cong-Hui, H., Fu-Sheng, C.: Sliding mode control for variable-speed wind Turbine generation systems using artificial neural network. Energy Procedia **61**, 1626–1629 (2014)
11. Norouzi, A., Kazemi, R., Azadi, S.: Vehicle lateral control in the presence of uncertainty for lane change maneuver using adaptive sliding mode control with fuzzy boundary layer. Proc. Inst. Mech. Eng. Part I: J. Syst. Control Eng. **232**, 12–28 (2018)
12. Mérida, J., Aguilar, L.T., Dávila, J.: Analysis and synthesis of sliding mode control for large scale variable speed wind turbine for power optimization. Renewable Energy **71**(11), 715–728 (2014)
13. Labbadi, M., Boukal, Y., Taleb, M., Cherkaoui, M.: Fractional order sliding mode control for the tracking problem of quadrotor UAV under external disturbances. In: European Control Conference 2020, ECC 2020, pp. 1595–600 (2020)
14. Pablo-Romero, M.P., Sánchez-Braza, A., Galyan, A.: Renewable energy use for electricity generation in transition economies: Evolution, targets and promotion policies. Renew Sustain Energy Rev. **138** (2021)
15. Aissaoui, A.G., Tahour, A., Abid, M., Essounbouli, N., Nollet, F.: Power Control of Wind Turbine based on Fuzzy Controllers, The Mediterranean Green Energy Forum 2013 (MGEF-13) (2013)
16. Chatri, C., Ouassaid, M., Labbadi, M., Errami, Y.: Integral-type terminal sliding mode control approach for wind energy conversion system with uncertainties
17. Yao, X., Guo, C., Xing, Z., Li, Y., Liu, S.: Variable speed wind turbine maximum power extraction based on fuzzy logic control. In: 2009 International Conference on Intelligent Human-Machine Systems and Cybernetics
18. Djilali, L., Badillo-Olvera, A., Rios, Y.Y., López-Beltrán, H., Saihi, L.: Neural high order sliding mode control for doubly fed induction generator based wind Turbines. IEEE Latin Am. Trans. **20**(2), February 2022
19. Li, Y., Xu, Z., Meng, K.: Optimal power sharing control of wind Turbines. IEEE Trans. Power Syst. **32**(1), January 2017
20. Hussain, J., Mishra, M.K.: An Efficient Wind Speed Computation Method Using Sliding Mode Observers in Wind Energy Conversion System Control Applications. IEEE Trans. Ind. Appl. **56**(1) (2020)
21. https://www.researchgate.net/figure/Two-mass-model-of-wind-turbine-system_fig5_322245010

A New Version of the Alkhalil Morpho Sys 2 Analyzer for Contemporary Documents

Samir Belayachi(✉) and Mazroui Azzeddine

Faculty of Sciences, Mohammed First University, Oujda, Morocco
samirbelayachi@gmail.com

Abstract. Morphological analysis is an important step in several applications of natural language processing. The richness of Arabic morphology makes the task of morphological analysis very complex. Arabic context-free morphological parsers widely used in applications suffer from several limitations such as poor coverage and the high average number of solutions per parsed word that negatively impacts the disambiguation phase. In this paper, we have developed a new version of the Alkhalil Morpho Sys analyzer dedicated to contemporary Arabic documents. This version is characterized by a reduced database built on a representative corpus of the standard Arabic language. The tests carried out show the high speed and the good coverage of this new version.

Keywords: Morphological analysis · Disambiguation · Standard Arabic language · Natural language processing

1 Introduction

The widespread use of the Internet has generated an exponential production of text data. The analysis of these texts generally goes through three levels: morphological, syntactic and semantic. Syntactic and semantic analyzes often use the results of the morphological analysis and their qualities are strongly impacted by that of the morphological analysis. The latter can be carried out without using the context of the words, and in this case, to an analyzed word can correspond to several analyses. It can also exploit the context to offer a single analysis per word.

The absence of diacritical marks in the majority of the Arabic texts greatly amplifies the ambiguity. Therefore, the word " فهم " /fhm[1] may correspond, depending on the context, to the verb " فَهَّمَ " /faham~a/ (he gets ready) or to the name " فَهْمٌ " /fahomN/ (understanding).

Several out-of-context morphological analyzers have been developed for the Arabic language. A comparative study between Sama [5] and Alkhalil Morpho

[1] Buckwalter transliteration: http://www.qamus.org/transliteration.htm/.

Y. Farhaoui et al. (Eds.): ICAISE 2022, LNNS 635, pp. 97–102, 2023.
https://doi.org/10.1007/978-3-031-26254-8_14

Sys 2 [3] analyzers has been carried out in [3]. This comparative study revealed that the Sama analyzer has low coverage and is relatively slow [3]. Similarly, the richness of the database of the Alkhalil Morpho Sys 2 analyzer, which represents a strong point in the analysis of classical texts, has the disadvantage of providing a very high average number of analyses for words in contemporary texts [3], and this negatively impacts the disambiguation phase. Thus, in this work we have developed a new version of the Alkhalil Morpho Sys 2 analyzer by filtering its database to keep only the vocabulary of the contemporary Arabic language.

The rest of the paper is organized as follows. In the following section, we present a state of the art on Arabic morphological analyzers. Section 3 is devoted to a description of the methodology adopted to build the new database. Then, in the next section, we used this database to develop the new version of the Alkhalil Morpho Sys. The second part of this section presents the results of a comparative study between the Alkhalil Mrpho Sys 2 analyzer and the new version. We end the paper with a conclusion and some thoughts on future work.

2 State of Art

We distinguish two types of morphological analysis. Out-of-context analysis in which words are analyzed without taking into account their contexts, and as a result several solutions are proposed for each analyzed word. While the second type of analyzers considers the context of words, and associates to each word a single analysis. We recall below some of the most cited morphological analyzers in the literature.

2.1 Out of Context Analyzers

- SAMA [5] is an improved version of a previous version developed for the Linguistic Data Consortium by Tim Buckwalter [4]. This analyzer provides all possible segmentations of the word. Its coverage is wider than the old version and the number of proposed solutions has increased considerably.
- Alkhalil Morpho Sys 2 [3] is an improved version of the Alkhalil Morpho Sys 1. The correction of errors in the database of the first version has increased the coverage and accuracy of this new version. Moreover, the reorganization of its database has considerably accelerated the analysis process, and the features have been enriched with the word lemma and its pattern.

2.2 Analyzers in Context

- MADAMIRA [6] is an open source morphological analyzer. It starts with an out-of-context morphological analysis using the SAMA [5] analyzer. Then, it uses in the disambiguation step an SVM classifier and language models. The information provided by Madamira is the segmentation of the word into proclitic + stem + enclitic, its diacritical form, its lemma and its POS tag.

- FARASA [1] is an open source morphosyntactic disambiguation system developed by the Arabic Language Technologies Department of Qatar Computing Research Institute. It is composed of a segmentation module based on a SVM classifier using linear kernels, a diacritization system, a POS tagger and a dependency parser based on the randomized glutton algorithm.
- Alkhalil Toolkit [2] has been developed by the Computer Science Research Laboratory of Mohammed First University of Oujda. It is an open source morphological disambiguation system that operates in two steps. In the first step, it performs an out-of-context analysis using the Alkhalil Morpho Sys 2 analyzer, and thus obtains several solutions for each analyzed word. Then, a disambiguation phase is performed based on hidden Markov models or splines and the Viterbi algorithm. Alkhalil Toolkit provides for each analyzed word its segmentation into proclitic + stem + enclitic, its diacritic form, its stem, its lemma, its root and its POS tag.

3 Building a Contemporary Vocabulary of Arabic

3.1 Collection and Pre-processing of the Contemporary Corpus

Corpora are indispensable resources for natural language processing applications, such as morphological analysis, machine translation and sentiment analysis. The exponential evolution of the data available on the Web, and the diversification of its topics make it a very important source for the construction of corpora. Indeed, 51% of corpora are built totally or partially from the Web [7].

In order to identify the vocabulary used in the Standard Arabic Language (MSA), we collected contemporary Arabic documents from the Web (see Table 1). The sites and digital libraries we have chosen are known for their rigorous quality of spelling of the texts they produce.

Table 1. Resources of the corpus

Sources	Type
Al Yaoum24; Al Jazeera; BBC Arabic; CNN Arabic Hespress; Al Bayan; Aresco; Nature Research Journal	Sites Web
Hindawi[2]; Ashamela[3]	Digital libraries

We have tried to cover a large number of themes, and to diversify the geographical origin of the resources. We have made sure that the books in the digital libraries we have chosen have been written by contemporary authors, and we proceeded to a pre-processing phase which consists in eliminating redundant or non-Arabic documents. Table 2 presents the statistics corresponding to the corpus divided by theme.

[2] https://www.hindawi.org.
[3] https://shamela.ws.

Table 2. Statistics on the corpus

Topic	Number of documents	Number of words	Vocabulary size
Culture	3564	2439435	252436
Politic	29786	10662091	258740
Sport	45167	11066958	216636
Sciences	25653	11215501	263702
Society	16109	4631100	176369
Sociology	111	5838694	274323
Religion	432	25984988	415084
Economy	27898	9321691	192013
Literature	337	13928747	595120
Health	21293	7154819	182215

3.2 Morphological Analysis of the Corpus

We recall that the first objective of this study is the identification of the vocabulary used in the MSA. In order to accomplish this task, we assumed that the collected corpus is representative of the MSA. Then, we used the Alkhalil Toolkit analyzer to perform a morphological analysis in the context of this corpus. The canonical morphological forms, provided by this analysis, that will allow us to identify the vocabulary of the MSA are the root, the lemma and the stem (see [3] for the definition of these forms).

Since the Alkhalil Morpho Sys 2 databases cover most of the vocabulary of the Arabic language (MSA and classical), we compared these databases with the canonical forms of the corpus in order to identify Arabic words that are no longer used in contemporary Arabic.

Table 3. Comparison between the lists of canonical forms of Alkhalil Morpho Sys 2 and those of the corpus

	Racine	Lemma	Stem
Corpus	6045	72311	282829
Alkhalil	7716	155238	3126354
Rates	78.34%	46.58%	9.05%

In analyzing these results (Table 3), we raise the following points:

- Among the 7716 roots in the Alkhalil Morpho Sys 2 database, only 6045 roots (78.34%) are present in the list of roots in the corpus. So, 21.66% of roots with their lemmas and stems are no longer used in contemporary texts.

- Only 46.58% of the lemmas in the AlKhalil database are present in the list of lemmas of the corpus.
- A good part of the inflected forms of the lemmas are absent. Indeed, to a lemma corresponds on average about 20 stems in the AlKhalil database, whereas a lemma in the corpus is associated on average with only 4 stems.

4 New Version of the Alkhalil Analyzer for the MSA

The objective of this work is to develop a new version of the Alkhalil Morpho Sys analyser that is more adapted to contemporary Arabic language documents.

4.1 Construction of the New Database of the Alkhalil Analyzer

In order to develop a new version of the Alkhalil analyser more suitable for contemporary Arabic documents, we proceeded to a filtering of the old database of the Alkhalil Morpho Sys 2 following these steps:

a) Eliminate from the list of roots in the Alkhalil Morpho Sys 2 database the roots that do not belong to the list of roots obtained from the morphological analysis of the corpus. Thus, the new list of roots contains only 6045 roots.
b) Eliminate from the Alkhalil Morpho Sys 2 database the lemmas and stems whose corresponding roots do not belong to the new list of roots.
c) For the roots belonging to the new list, keep in the list of corresponding lemmas only those that belong to the list of lemmas obtained from the morphological analysis of the corpus.
d) All the inflected forms of the lemmas in the new lemma list will constitute the new stem list. The new list is thus composed of 3290743 stems.

4.2 Comparison Between Alkhalil Morpho Sys 2 and the New Version

We first developed a new version of the Alkhalil analyzer by replacing the database of the Alkhalil Morpho Sys 2 analyzer with the one composed of the new lists of roots, lemmas and stems built in the previous section. Then, we collected from Al Jazeera a new corpus containing one million words and covering all the topics in the Table 2. We analyzed this corpus by both versions of Alkhalil, and we calculated the following indicators reported in Table 4:

- **Coverage-R**: rate of the words of the test corpus analyzed by the system.
- **Speed**: number of words that the system analyzes per second.
- **Nb-Lemma**: average number of lemmas proposed by the system.
- **Nb-Stem**: average number of stems proposed by the system.

We note that despite the limited size of the database of the new version of the Alkhalil analyzer, the coverage rates of the two analyzers are very close. On the other hand, the new version is faster than the Alkhalil Morpho Sys 2. Finally, and as expected, the average number of lemmas and stems per word proposed by this new version are lower than those relative to the Alkhalil Morpho Sys 2.

Table 4. Comparison between Alkhalil Morpho Sys 2 and the new version of Alkhalil

	Alkhalil2	Alkhalil3
Coverage-R	96.506	96.478
Speed	1218	1408
Nb-Lemma	3.38	3.06
Nb-Stem	3.73	3.43

5 Conclusion

In this paper, we have built a new version of the Alkhalil Morpho Sys 2 that is dedicated to contemporary Arabic documents. This version is built from a representative corpus of the MSA. The tests carried out show the high speed and the good coverage of this new version. In the future, we plan to enrich the database with new proper names and to integrate this new version into the Alkhalil Toolkit disambiguation system.

References

1. Abdelali, A., Darwish, K., Durrani, N., Mubarak, H.: Farasa: a fast and furious segmenter for Arabic (2016). https://doi.org/10.18653/v1/N16-3003
2. Boudchiche, M., Mazroui, A.: Spline functions for Arabic morphological disambiguation. Appl. Comput. Inform. (2020)
3. Boudchiche, M., Mazroui, A., Bebah, M.O.A.O., Lakhouaja, A., Boudlal, A.: Alkhalil morpho sys 2: a robust Arabic morpho-syntactic analyzer. J. King Saud Univ. Comput. Inf. Sci. **29**(2), 141–146 (2017)
4. Buckwalter, T.: Buckwalter Arabic morphological analyzer (BAMA) version 2.0. linguistic data consortium (LDC) catalogue number ldc2004l02. Technical report (2004). ISBN1-58563-324-0
5. Maamouri, M., Graff, D., Bouziri, B., Krouna, S., Bies, A., Kulick, S.: LDC standard arabic morphological analyzer (SAMA) version 3.1 ldc2010l01. Web Download Philadelphia: Linguist. Data Consortium (2010)
6. Pasha, A., et al.: MADAMIRA: a fast, comprehensive tool for morphological analysis and disambiguation of Arabic. In: Proceedings of the Ninth International Conference on Language Resources and Evaluation (LREC 2014), Iceland (2014)
7. Zeroual, I., Lakhouaja, A.: Data science in light of natural language processing: an overview. Procedia Comput. Sci. **127**, 82–91 (2018)

A Proposed Machine Learning Model for Intrusion Detection in VANET

Sara Amaouche[1](✉), Said Benkirane[1], Azidine Guezzaz[1], and Mourad Azrour[2]

[1] Technology Higher School Essaouira, Cadi Ayyad University, Marrakech, Morocco
saraa.amaouche232@gmail.com

[2] IDMS Team, Faculty of Sciences and Technics, Moulay Ismail University of Meknès, Meknes, Morocco

Abstract. In order to improve securities on VANET security still remains a delicate research subject. Actually, intrusion detection systems (IDSs) are an effective security tool which can be enhanced by ML algorithms. This paper presents an optimized intrusion detection approach using realistic dataset called ToN-IoT derived from a large-scale heterogeneous IoT network, to achieve our model we used the mutual information technique for feature selection and the synthetic minority oversampling technique (SMOTE) for class balancing Then to compare we tested various ML methods Logistic regression (LR), k-Nearest Neighbor (kNN), decision tree (DT), Random Forest (RF) and Support Vector Machine (SVM) for VANET Security.

Keywords: VANET security · Attacks · Machine learning · Intrusion detection

1 Introduction

VANET are a new type of application of Mobile Ad-hoc networks MANETs [1, 2, 31]. The role of the VANET network is to guarantee communication between vehicles, and allow access to information in real time [1]. The VANET network is exposed to security attacks by adversaries like DOS attack [3, 4]. At the beginning, VANETs used dedicated Short-Range Communication DSRC that uses a bandwidth of 75 MHz [5]. The existing VANET standards are not adequate for the services and applications expected [6, 7]. The added communication technologies increase the scope of VANETs and their connection with other enabling technologies such as the Internet of Things (IoT) [37], cloud computing [8, 9, 38]. Securing vehicles against intrusion is a challenging problem. Traditional practical security countermeasures such as encryption techniques and access control are no longer relevant for autonomous vehicles [10–12]. The rest of this paper is organized as follows: in Sect. 2, we introduce the background of VANET architecture, VANET security, intrusion detection approaches and we introduce other recent related works in intrusion detection approaches that include ML, DL and ensemble learning techniques. Section 3 describes the proposed solution for this new approach. The results obtained are discussed in Sect. 4. Finally, the paper is reached with a conclusion.

Y. Farhaoui et al. (Eds.): ICAISE 2022, LNNS 635, pp. 103–108, 2023.
https://doi.org/10.1007/978-3-031-26254-8_15

2 Related Works

In order to address the increasing of intelligent transport system applications, the IEEE 802.11p working group formed in 2004 to provide enhancements to the IEEE 802.11 standard for supporting wireless access in Vehicular Environments (WAVE) [13]. The IEEE 802.11p standard that was published in 2010 allows the 5.9 GHz ITS band to be used to enable V2V communications between vehicles and V2I communications between vehicles and RSUs. It should be note that the IEEE 802.11p standard only gives specifications for the physical base layer (PHY) and medium access control (MAC) [13]. In recent times, the IEEE 1609 working group [14] was formed to define additional layers (Above the IEEE 802.11p PHY/MAC layers), the resulting IEEE 1609 family of standards [15, 16].

VANETs are self-configuring networks composed of moving vehicles. so it causes the rapid topology change [17, 18]. Therefore, protocols developed for VANETs must be able to handle Link maintenance, frequent neighborhood change, and high mobility. In addition, network outages [18] are numerous, as the link between two vehicles can quickly disappear due to the highly dynamic network topology. In addition, another feature is the mobility pattern [18] because vehicles follow a particular mobility pattern that depends on the roads, traffic, speed, and driver [19]. In complement, the characteristic of the propagation model [18] As commonly known in wireless networks, the waves may face several obstacles such as buildings, trees, as well as interference from other vehicles, since the preparation model in VANET is not free space, then these disadvantages must be considered. Finally, the characteristic of onboard sensors [18], in VANETs, communicating vehicles are equipped with sensors to provide routing information. To address the security challenges of VANET, various studies have been conducted on machine learning algorithms to enhance the reliability of VANET by detecting intrusions and making predictions, and they have finally led to convincing results [20]. An Intrusion Detection System (IDS) is a mechanism to identify unusual activity on a network. IDS solutions are proposed for detecting internal attacks (which are attacks by compromised nodes) that cryptographic solutions cannot detect [21]. An IDS is typically used as a second line of protection after cryptographic systems [34,35,36]. Basically, an intrusion detection system is formed in three phases: a data collection phase that is followed by an analysis phase and a response phase to minimize the impact of the attack on the system. The IDS is installed on some specific nodes called monitors or monitor nodes [21]. The deployment of these nodes depends on the type of protocol and the architecture of the IDS. IDSs are grouped into three categories. First, signature-based on systems [21]. Second, the anomaly detection system [21]. Finally, System based on specifications [21]. Researchers have proposed IDS based on machine learning (ML) [22, 23, 24, 39] or deep learning (DL) methods [40, 41].

Many researchers have presented detection methods related to IDSs in the ad hoc network. Al-Jarrah et al. [26] combined RF with forward and reverse ranking feature selection methods. They used KDD-CUP99 dataset. In terms of results, RF-FSR achieved the best classification accuracy 99.90% and RF/BER 99.88%. Grover et al. [27] have proposed a methodology that uses classification algorithms supported by WEKA. The model had 99% of accuracy. Kumar et al. [28] proposed a new classifier to detect any form of malicious behavior on the network. The results obtained indicate that T-CLAIDS outperforms other existing schemes. The proposed model showed an accuracy of 96%. Wahab et al. [29] proposed a multidecision intelligent detection model named CEAP and they employed SVM. The results of the model in function of the network density scenario are 99,13% for Linear kernel, 99,04% for Multilayer percepton kernel, 99,13% for Quadratic kernel, 99,35% for Polynomial kernel and 99,67% Gaussian radial basis function kernel. Abdalla and Ahmed [25] proposed an IDS for VANET based on the ToN-IoT network dataset. The proposed model obtained these results: LR 86,5% of accuracy, NB 75,6%, DT 97,6%, RF 97,6%, Adaboost 90,8%, Knn 98,5%, SVM 86,5% and XGBoost 98,6%. H. Bangui, M. Ge, B. Buhnova [30] proposed an approach that employs RF for network intrusion detection. The suggested method achieved a maximum accuracy of 96.93%, while the lowest results are obtained with Bayesian coresets (82.4%), CNN (95.14%) and SVM (85.2%). A new routing approach ECRDP has been proposed in [32]. A new clustering-based routing protocol was proposed in [33].

3 Proposed Work

Since VANETs are directly attached to the Internet and due to the large number of messages shared between vehicles every second, VANETs are more vulnerable to many types of attacks. Detecting intrusion among vehicles is one of the major challenges, and that will help us to increase the security of VANET networks, that's why we present a model that will help us to solve this major problem, our intrusion detection model is based on the following ML techniques LR, kNN, DT, RF and SVM (Fig. 1).

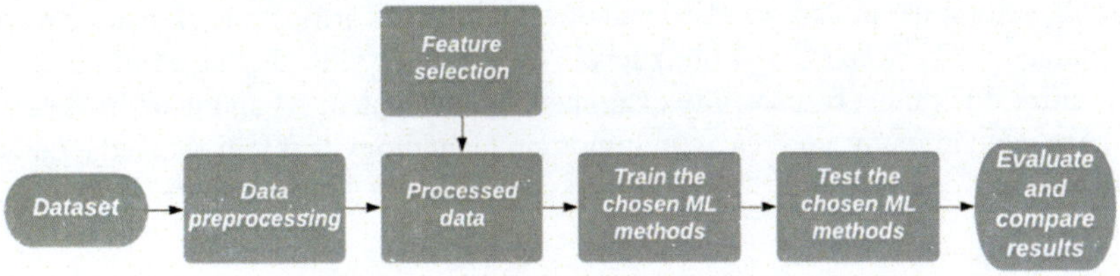

Fig. 1. Scheme of proposed approach

To have useful analysis these data must be processed correctly, the missing values in our model will be replaced by the most current value. For the conversion of categorical values into numeric values we use the one-hot encoding technique. To solve the problem of class imbalance, we use the SMOTE technique to balance the attack class, which is an oversampling technique that aims to enlarge the minority class.

The goal of feature selection in machine learning is to determine the best set of features to develop useful models of the phenomena under study to detect intrusion detection.

4 Evaluation Study

This section discusses the effectiveness of the ML methods used for VANTE by analyzing the results of the ToN-IoT dataset.

Models	Accuracy	Precision	Recall	F1-score
LR	0.960	0.963	0.960	0.959
KNN	0.996	0.996	0.996	0.996
DT	1.0	1.0	1.0	1.0
RF	0.999	0.998	0.998	0.998
SVM	0.989	0.988	0.988	0.988

As presented in the results table for the network ToN-IoT dataset. A 10-factor cross-validation was applied to all selected ML methods. Accuracy, precision, recall and also F1 score are presented to evaluate the selected ML methods. In general, DT shows significant results based on various feature engineering techniques that were applied to the dataset. First, we impute the missing values and convert the categorical features using one-shot encoding, then we apply the Min-Max normalization technique, thereafter we solve the imbalance classes problem by the smote technique, the results show that the accuracy is 1, the recall is 1, the precision is 1 and the F1 score is 1.

5 Conclusion and Future Works

VANETs are complex and time-critical systems due to the high mobility of vehicles in which physical safety of road users and transportation service efficiency are directly linked to the provision of cyber security. Although there is an upgrade of VANETs standards, setting the global standard and developing a security strategy is not a reality. The interactions between the various levels of VANETs with the world around them (smart cities, IoT) must be carefully examined and evaluated. AI and machine learning have been cited in many sources as an important technology that will shape the outlook for VANETs.

References

1. Hamdi, M.M., Audah, L., Rashid, S.A., Mohammed, A.H., Alani, S., Mustafa, A.S.: A review of applications, characteristics and challenges in vehicular Ad Hoc Networks (VANETs). In: 2020 International Congress on Human-Computer Interaction, Optimization and Robotic Applications (HORA). IEEE (2020)
2. Yousefi, S., Mousavi, M.S., Fathy, M.: Vehicular ad hoc networks (VANETs): challenges and perspectives. In: 2006 6th International Conference on ITS Telecommunications. IEEE (2006)
3. Biswas, S., Mišić, J., Mišić, V.: DDoS attack on WAVE-enabled VANET through synchronization. In: 2012 IEEE Global Communications Conference (GLOBECOM). IEEE (2012)
4. Zaidi, T., Faisal, S.: An overview: Various attacks in VANET. In : 2018 4th International Conference on Computing Communication and Automation (ICCCA). IEEE (2018)
5. Zeadally, S., Hunt, R., Chen, Y.S., Irwin, A., Hassan, A.: Vehicular ad hoc networks (VANETS): status, results, and challenges. Telecommunication Systems (2012)

6. Meireles, R., Boban, M., Steenkiste, P., Tonguz, O., Barros, J.: Experimental study on the impact of vehicular obstructions in VANETs. In: 2010 IEEE Vehicular Networking Conference. IEEE
7. Dar, K., Bakhouya, M., Gaber, J., Wack, M., Lorenz, P. : Wireless communication technologies for ITS applications [Topics in Automotive Networking]. IEEE Communications Magazine (2010)
8. Rasheed Hussain, F. A., Son, J., Eun, H., Oh, H.: Privacy-aware route tracing and revocation games in VANET-based clouds. In: IEEE 9th International Conference on Wireless and Mobile Computing, Networking and Communications (WiMob) (2013)
9. He, W., Yan, G., Da Xu, L.: Developing vehicular data cloud services in the IoT environment : IEEE Trans. Ind. Inf. (2014)
10. Loukas, G., Vuong, T., Heartfield, R., Sakellari, G., Yoon, Y., Gan, D.: Cloud-based cyber-physical intrusion detection for vehicles using deep learning. IEEE Access (2017)
11. Maduranga, M.W.P., Abeysekera, R.: Machine learning applications in IoT based agriculture and smart farming: a review. Int. J. Eng. Appl. Sci. Technol (2020)
12. Moustafa, N.: ToN-IoT Dataset (2020). https://cloudstor.aarnet.edu.au/plus/s/ds5zW91vdgjEj9i
13. Ben Hamida, E., Noura, H., Znaidi, W.: Security of cooperative intelligent transport systems: Standards, threats analysis and cryptographic countermeasures. Electronic (2015)
14. WG—Dedicated Short Range Communication Working Group. IEEE Standards Development Working Group. http://standards.ieee.org/develop/wg/1609_WG.html
15. Zhang, J., Ren, M., Labiod, H., Khoukhi, L. : Link duration prediction in VANETs via AdaBoost. In: GLOBECOM 2017–2017 IEEE Global Communications Conference. IEEE (2017)
16. ETSI—Intelligent Transport Systems. http://www.etsi.org/technologiesclusters/technologies/intelligent-transport
17. Taleb, A.A.: VANET routing protocols and architectures: an overview. J. Comput. Sci. (2018)
18. Joshi, P.K.: A survey of vanet routing protocols. J. Anal. Comput. (JAC) (2019)
19. Hu, J.G.J.Y.C., David, S.P.A.K.S., Johnson, B.: Design and evaluation of a metropolitan area multitier wireless ad hoc network architecture. In: Proceedings of the Fifth IEEE Workshop on Mobile Computing Systems & Applications (2003)
20. Talpur, A., Gurusamy, M.: Machine learning for security in vehicular networks: a comprehensive survey. IEEE Commun. Surv. Tutorials (2021)
21. Erritali, M., El Ouahidi, B.: A survey on VANET intrusion detection systems. In: Proceedings of the 2013 International Conference on Systems, Control, Signal Processing and Informatics (2013)
22. Almseidin, M., Alzubi, M., Kovacs, S., Alkasassbeh, M.: Evaluation of machine learning algorithms for intrusion detection system. In: 2017 IEEE 15th International Symposium on Intelligent Systems and Informatics (SISY). IEEE (2017)
23. Segal, M.R.: Machine learning benchmarks and random forest regression (2004)
24. Mahesh, B.: Machine learning algorithms-a review. In: International Journal of Science and Research (IJSR) (2020)
25. Gad, A.R., Nashat, A.A., Barkat, T.M.: Intrusion detection system using machine learning for vehicular ad hoc networks based on ToN-IoT dataset. IEEE Access (2021)
26. Al-Jarrah, O.Y., Siddiqui, A., Elsalamouny, M., Yoo, P.D., Muhaidat, S., Kim, K.: Machine-learning-based feature selection techniques for large-scale network intrusion detection. In: 2014 IEEE 34th International Conference on distributed Computing Systems Workshops. IEEE (2014)
27. Grover, J., Prajapati, N.K., Laxmi, V., Gaur, M.S.: (Machine learning approach for multiple misbehavior detection in VANET. In: International Conference on Advances in Computing and Communications. Springer, Heidelberg (2011)

28. Kumar, N., Chilamkurti, N.: Collaborative trust aware intelligent intrusion detection in VANETs. Computers & Electrical Engineering (2014)
29. Wahab, O.A., Mourad, A., Otrok, H., Bentahar, J.: CEAP: SVM-based intelligent detection model for clustered vehicular ad hoc networks. Expert Systems with Applications (2016)
30. Bangui, H., Ge, M., Buhnova, B.: A hybrid data-driven model for intrusion detection in VANET. Procedia Computer Science (2021)

A Quadratic Observer for Sensorless Drive System Controller

Ben Achour Hafid[1](✉), Ziani Said[2], and El Hassouani Youssef[1]

[1] Department of Physics, Faculty of Science and Technology Errachidia, Moulay Ismail University, BP-509 Boutalamine, Errachidia 52000, Morocco
h.benachour@edu.umi.ac.ma

[2] Laboratory of Networks, Compter Science, Telecommunication, Multimedia (RITM), Department of Electrical Engineering, High School of Technology ESTC, Hassan II University, Casablanca, Morocco

Abstract. This study introduces the challenge of observability and the synthesis of observers for the synchronous motor without a mechanical sensor. Then, we concentrate on using the Luenberger observer to estimate the location and velocity of the PMSM and implement it in vector control using a conventional PI controller. Tests have been done under the Matlab/Simulink system.

Keywords: PMSM · Nonlinear control · Luenberger observer · PI controller

1 Introduction

In the fields of automated control and electrical engineering, the subject of the power of synchronous machines with permanent magnets in the absence of a mechanical sensor has been extensively explored. The submitted technique tackles the specific issue of sensorless low-speed control of smooth pole machines. This paper aims to provide a method for controlling the position of an electromechanical actuator employed in an aeronautical setting (here for ailerons). A theoretical analysis of the two families of machines, salient and non-salient, revealed a variation in the observability of the models of the MSAP depending on the speed of the machine, with a loss of observability at low speed for machines with smooth poles in particular. To compensate for this loss, new models are built that account for the machine's vibrations, which are prompted by the injection of signals. Applying a novel analysis to these models makes it feasible to ensure the machine's observability across its whole speed range, including while it is stationary. We thus developed a method for observing location and velocity based on these novel models and a well-chosen injection. Unlike traditional approaches based on salience, this strategy applies to all machine kinds. It has been validated using an expanded Kalman observer on many test benches. Numerous tests have been conducted on an industrial applications-specific benchmark. The outcomes demonstrated the suggested observers' effectiveness, robustness, and limits. We were also interested in the machine's speed control without a mechanical sensor, for which we presented two ways

Y. Farhaoui et al. (Eds.): ICAISE 2022, LNNS 635, pp. 109–113, 2023.
https://doi.org/10.1007/978-3-031-26254-8_16

of observer synthesis. The first is based on the polytopic synthesis of a robust LPV observer. The second utilizes a slip mode observer of second order with adaptive gains. This demonstrates the experimental viability of these two observers. [1, 2] and [3, 4].

2 Luneberger Observer

An observer is a mathematical development allowing to reconstitute the internal states of a system from only the accessible data, which means the imposed inputs and the measured outputs.

The observation is performed in two phases; the first is an estimation stage while the second is a correction stage. The estimation is done by calculating the state quantities using models close to the system, and the correction is done by adding or subtracting the difference between the estimated states and those measured (estimation error) which is multiplied by a gain (L) [5]. This gain governs the dynamics and robustness of the observer. Therefore, its choice is important and must be adapted to the properties of the system whose states are to be observed. Let the system model be:

$$\begin{cases} \dot{x} = Ax + Bu \\ y = Cx \end{cases} \tag{1}$$

The Luenberger observer model is presented as follows:

$$\begin{cases} \dot{\hat{x}} = A\hat{x} + Bu + L(y - \hat{y}) \\ \hat{y} = C\hat{x} \end{cases} \tag{2}$$

Since $A\hat{x} + Bu$ represents the copy of the system dynamics and $L(y - \hat{y})$ represents the corrective term. The structure of a state observer is based on the system model, called a predictor or an estimator, which operates in open loop. To correct the error between the output of the system and the output of the model, we creat a structure a structure in which we introduce a locked-loop between the observer and the system. The aim of an observer is converging the estimated state to the true state value. This can be written as follows: Let the error described by $\tilde{x} = x - \hat{x}$. The goal of the observer being to guarantee the convergence of the error to zero which gives: $\lim_{t \to +\infty} \tilde{x} = 0 \to \lim_{t \to +\infty} x - \hat{x} = 0$.

The following figure illustrates the state model of the system in association with the observer. In order to converge the error to zero, the matrix must have negative eigenvalues, it is of great importance to note that the observer's dynamics must be faster than the dynamics of the system (Fig. 1).

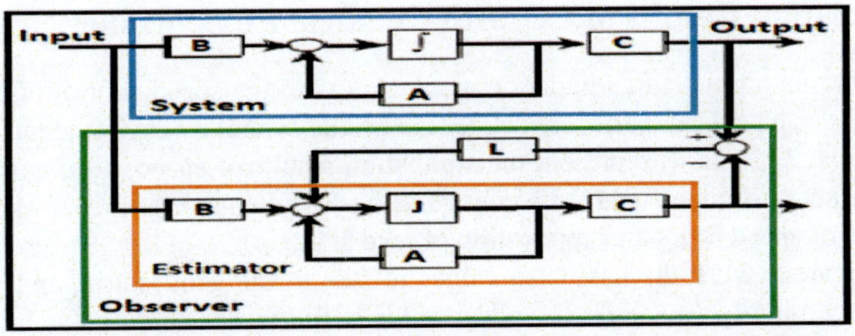

Fig. 1. State model of the system in association with the observer

3 Application of Luenberger Observer (LO) to PMSM

The state model of the PMSM in the rotate frame (d-q) is presented through the following system:

$$\frac{d}{dt}\begin{bmatrix} i_d \\ i_q \\ \Omega \\ \theta \end{bmatrix} = \begin{bmatrix} -\frac{R_s}{L_d} i_d + p\frac{L_q}{L_d}\Omega i_q \\ -\frac{R_s}{L_q} i_q - p\frac{L_d}{L_q}\Omega i_d - p\frac{\Omega\varphi_f}{L_q} \\ \frac{p}{J}(L_d - L_q) i_d i_q - \frac{f}{J}\Omega + \frac{P}{J}\varphi_f i_q - \frac{C_r}{J} \\ p\Omega \end{bmatrix} + \begin{bmatrix} \frac{1}{L_d} & 0 \\ 0 & \frac{1}{L_q} \\ 0 & 0 \end{bmatrix}\begin{bmatrix} v_d \\ v_q \end{bmatrix} \quad (3)$$

The equation-of-state model of the MSAP simplified to unmeasured values is written:

$$\frac{d}{dt}\begin{bmatrix} \theta \\ \omega \\ C_r \end{bmatrix} = \begin{bmatrix} 0 & 1 & 0 \\ 0 & -\frac{f}{J} & -\frac{1}{J} \\ 0 & 0 & 0 \end{bmatrix}\begin{bmatrix} \theta \\ \omega \\ C_r \end{bmatrix} + \begin{bmatrix} 0 \\ \frac{p\varphi_f}{J} \\ 0 \end{bmatrix} i_q \quad (4)$$

With

$$A = \begin{bmatrix} 0 & 1 & 0 \\ 0 & -\frac{f}{J} & -\frac{1}{J} \\ 0 & 0 & 0 \end{bmatrix}; B = \begin{bmatrix} 0 \\ \frac{p\varphi_f}{J} \\ 0 \end{bmatrix}; C = [100]; u = i_q$$

The gain matrix L can be chosen as follows:

$$\begin{cases} \frac{d\hat{\theta}}{dt} = \hat{\omega} \\ \frac{d\hat{\omega}}{dt} = -\frac{1}{J}C_r - \frac{f}{J}\hat{\omega} + l_2(\omega - \hat{\omega}) + l_1(\theta - \hat{\theta}) \\ \frac{d\widehat{C_r}}{dt} = l_3(\theta - \hat{\theta}) \end{cases} \quad (5)$$

$$\text{where} : \mathrm{L} = \begin{bmatrix} 0 & 0 & 0 \\ l_2 & l_1 & 0 \\ l_3 & 0 & 0 \end{bmatrix}$$

The characteristic equation of the observer is:

$$\det(sI - (A - LC)) = s^3 + s^2(\frac{f}{J} + l_1) + sl_2 - \frac{l_3}{J} = 0 \quad (6)$$

4 Simulation Results of LO with Classical PI Controller

To validate this observer in simulation associated with the speed control of a PMSM with a PI, tests have been performed under the Matlab/Simulink environment.

The Figs. 2, 3 and 4) represent the simulation results of speed, id-iq currents and electromagnetic torque of PMSM's control using Luenberger observer in the case of conversion of speed at 1 s and application of load at 2.5 s.

The previous work discussed the values related to the gain matrix, and provided results concerning l_1, l_2, and l_3. The added value of this work is offering an interval for each value that we can edit based on the desired need of damping. These values can be presented as follows: $l_1 = [20 - 60]$, $l_2 = [150 - 300]$, $l_3 = [0.1 - 0.3]$.

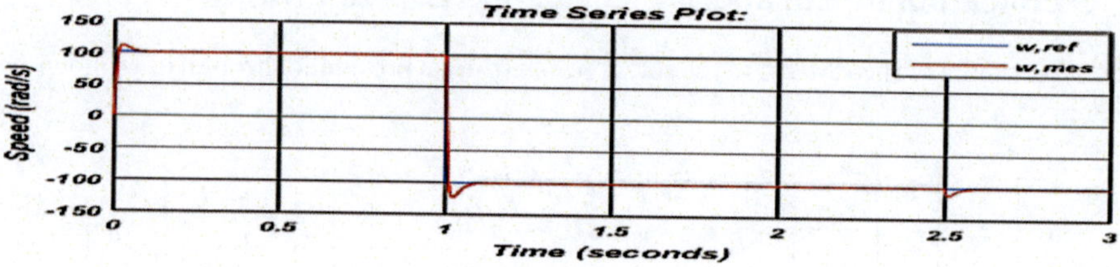

Fig. 2. Behavior of the observed speed of PMSM

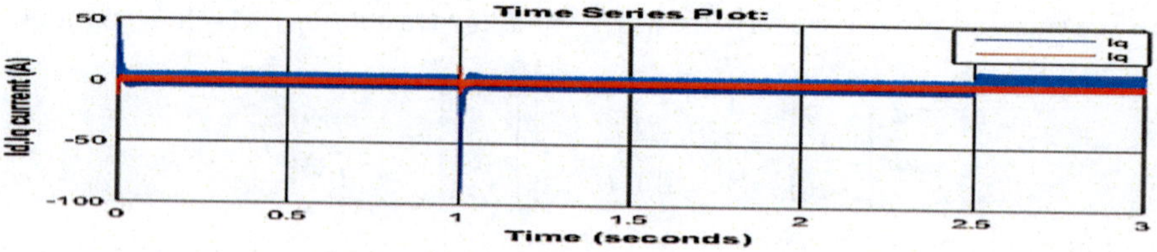

Fig. 3. i_d, i_q currents behavior of PMSM

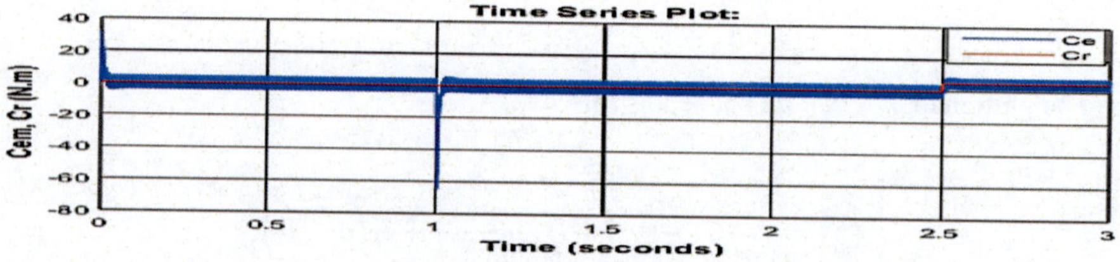

Fig. 4. Electromagnetic torque behavior of PMSM

These simulation results show good performance of the observer. The difference between the estimated speed and its measurement is very small. An error occurred when the load and speed have been changed [5]. Concerning the estimation of position and speed, a good tracking between the measurement and the observation is observed in the first part, but at the microscopic scale the phase shift between the real and the observed speed becomes notable during the zero crossing of the speed, a clear loss of observability of the latter with the estimation error which becomes important as this loss persists.

5 Conclusion

In this research, we discussed the use of the Luenberger observer coupled with the PI controller of the machine. Following its application, it was determined that this observer had a satisfactory dynamic response to the velocity and excellent disturbance rejection. Its observational resilience is limited during parametric change, and its estimation of the position in the unobservable zone is unstable.

This study's findings also emphasize the need to use wavelets [6–8], integrated with data analysis often employed in biomedical signal processing, such as PCA to further improve the aforementioned techniques' efficacy.

References

1. Ben Achour, H., Ziani, S., Chaou, Y., El Hassouani, Y., Daoudia, A.: Permanent Magnet Synchronous Motor PMSM Control by Combining Vector and PI Controller. WSEAS Trans. Syst. Control **17**, 244–249 (2022)
2. Zhu, G., Kaddouri, A., Dessaint, L.A., Akhrif, O.: A nonlinear state observer for the sensorless control of a permanent-magnet AC machine. IEEE Trans. Ind. Electron. **48**(6), 1098–1108 (2001)
3. Zhao, M., An, Q., Chen, C., Cao, F., Li, S.: Observer based improved position estimation in field-oriented controlled pmsm with misplaced hall-effect sensors. Energies **15**, 5985 (2022)
4. Ellis, G.: Observers in control systems, A pratical gide. Academic Press, An imprint Elsevier Science, Copyright 2002, Elsevier Science (USA)
5. Said, Z., El Hassouani, Y.: A new approach for extracting and characterizing fetal electrocardiogram. Traitement du Signal, Vol. 37, No. 3, pp. 379–386 (2020). https://doi.org/10.18280/ts.370304
6. Ziani, S., El Hassouani, Y.: Fetal-maternal electrocardiograms mixtures characterization based on time analysis. In: 2019 5th International Conference on Optimization and Applications (ICOA), 2019, pp. 1–5. https://doi.org/10.1109/ICOA.2019.8727619
7. Ouhadou, M., El Amrani, A., Messaoudi, C., Ziani, S.: Experimental investigation on thermal performances of SMD LEDs light bar: Junction-to-case thermal resistance and junction temperature estimation, Optik, vol. 182 (2019)
8. Ziani, S., Jbari, A., Bellarbi, L.: QRS complex characterization based on non-negative matrix factorization NMF. In: 2018 4th International Conference on Optimization and Applications (ICOA), pp. 1–5 (2018). https://doi.org/10.1109/ICOA.2018.8370548

A Review of the Application of Artificial Intelligence for Weather Prediction in Solar Energy: Using Artificial Neural Networks

Imad Laabab[1](✉), Said Ziani[2], and Abdellah Benami[1]

[1] Laboratory of Mechanics, Energy Efficiency and Renewable Energies (M3ER), Department of Physics, Faculty of Sciences and Techniques of Errachidia (FSTE), Moulay Ismail University, Meknes, BP-509, 52 000 Boutalamine, Errachidia, Morocco
im.laabab@edu.umi.ac.ma, a.benami@fste.umi.ac.ma

[2] Laboratory of Networks, Computer Science, Telecommunication, Multimedia (RITM), Department of Electrical Engineering, High School of Technology (ESTC), Hassan II University, Casablanca, Morocco
said.ziani@univh2c.ma

Abstract. All our activities are linked in one way or another to energy. Thanks to the control of energy conversion technologies, humans have been able to considerably increase their level of comfort and meet their increasingly varied needs. Our daily lives are now being impacted by artificial intelligence (AI), which is starting to have a big impact on sustainable development. AI can help humans predict weather and slope data to increase building occupants' safety and comfort. Algorithms and computer programs have been created for data mining and analysis. In several activities, Big Data (BD) is reviving AI methods and applications. This literature review presents the basic notions that are essential for any research on the use of AI in photovoltaic (PV) systems. The introduction suggests a review of the fundamental ideas of AI before considering the applications of artificial neural networks (ANN) to solar systems. This study confirms that BD and AI work well together to boost PV system's energy efficiency so that they use as little energy as possible while yet maintaining a comfortable environment.

Keywords: Artificial intelligence · Sustainable development · Artificial neural networks · Energy efficiency

1 Introduction

The modeling, analysis, and prediction of performance benefit greatly from the use of AI approaches. Modeling, control, and prediction techniques for energy system performance include challenging differential equations and take a lot of time to solve. Instead, AI algorithms don't need to worry about the mathematical relationships between the inputs and outputs of the systems they are studying in order to understand the key models of those systems. Then again, long-term meteorological data sets, such as those pertaining to solar radiation, temperature, or wind statistics, are necessary for the design,

Y. Farhaoui et al. (Eds.): ICAISE 2022, LNNS 635, pp. 114–119, 2023.
https://doi.org/10.1007/978-3-031-26254-8_17

management, and performance of solar panels. These long-term remedies are typically lacking in most tourist destinations or when they exist, they have a number of disadvantages. One of the best solutions to these problems seems to be using AI techniques. This paper briefly introduces some artificial intelligence approaches (artificial neural networks -ANN) often used in the field of photovoltaic energy. Two uses of AI include the forecasting and modeling of temperature and solar irradiance. First, in need of definitions, it is mandatory to identify several terms related to our subject:

- Artificial intelligence aims to make machines intelligent by creating a system that behaves like a human being [1]. It involves the use of smart objects that are aware of their surroundings and operate in ways that increase their chances of success [2].
- Machine Learning is a subset of AI. It tries to create algorithms that can gain knowledge from past information and enhance systems over time. They can change their internal programming to better carry out particular jobs by providing data to algorithms [3].
- Data science is a branch of science that focuses on all things data-related, including data collection, storage, analysis, cleansing, visualization, interpretation, and decision-making [4].

2 Methods

Neural networks are the AI method most frequently utilized for this subject, according to an assessment of the literature. The learning parameters are typically combinations of long-term meteorological measurements (temperature, sunshine, humidity, wind speed, precipitation, etc.) and geographic information about the study sites (altitude, latitude, longitude…). We provide a survey of the literature on this subject in the paragraphs that follow.According to Elizondo et al. [5], a back-propagation neural network employing the gradient descent approach for training and the square error as an accuracy criterion was developed so as to forecast the daily solar radiation. Daily precipitation value, daily minimum temperature, daily maximum temperature, daily sunshine under clear skies, day length, and day of the year, were utilized as the inputs for six parameters (calculated from the latitude and date of the day to be predicted). Estimates of day duration and sunshine were made depending on the solar constant, latitude, sun angle, and day of the year. In addition, Mohandes et al. [6] employed altitude, longitude, latitude, and sun duration as inputs for a multilayer perceptron neural network (MLP) and a neural network (RBF: Radial Basis Function) in order to forecast global solar radiation for 41 sites dispersed throughout the Kingdom of Saudi Arabia (KSA). The neural networks were accomplished using the data from 31 stations and tested using the data from the remaining 10 stations. The MLP network means absolute error as a percentage was 12.61%, while the RBF neural network was 10.09%.The ability of neural networks to forecast the monthly daily average sunshine of Al-Ain city in United Arab Emirates was proved by Maitha H. Al-Shamisi et al. [7]. Rajesh Kumar et al. [8] employed an artificial neural network to approximate the two Angstrom regression coefficients to determine the global daily solar radiation of a site. The accuracy of the anticipated radiation values is improved by using a neural network because these parameters are typically evaluated empirically. The altitude and latitude of the sites were used by the authors as two parameters to be input into the network (in northern India). Two months of measurements were utilized for validation, two months for testing, and eight months of measurements were used to train the network.Furthermore, K. Chiteka and C.C. Enweremadu [9] developed an artificial neural network to forecast horizontal irradiance for significant areas in Zimbabwe. Geographical information such as altitude, latitude, and longitude as well as meteorological information such as pressure, humidity, average temperature, and clear index were used to predict the global horizontal irradiation. The proportion of ground radiation to extraterrestrial radiation determines the brightness index. According to the research, this artificial neural network model may likewise be used to forecast global horizontal irradiance for the day before and the day after, with mean relative errors of 5.9% and 7.6%, respectively. For the same objectives, the authors later created a neural network using the brightness index as the sole input. It is possible to forecast insolation with reasonable accuracy using simply the brightness index, as demonstrated by this network's ability to predict worldwide horizontal irradiation with an average relative error of 5%.

2.1 Analysis

Research until now is based basically on statistical techniques. We then go through prediction methods for various temporal horizons.

Figure 1 presents the motivations for establishing prediction studies for optimizing energy.

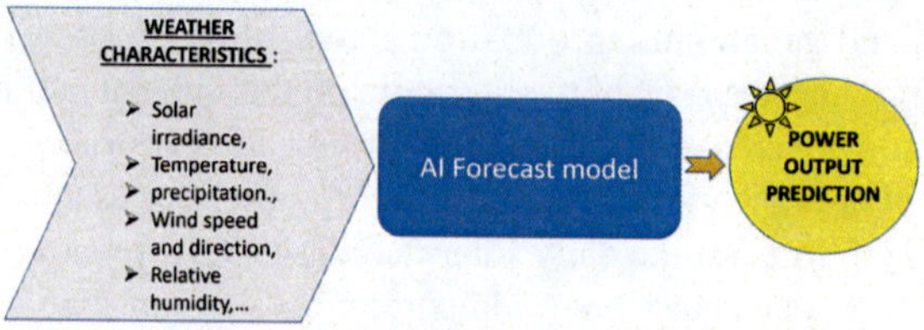

Fig. 1. AI forecast model principle

The two main categories of prediction methods are numerical weather prediction techniques, which depend on statistical methods and modeling of physical events in atmospheric dynamics.There are statistical forecasting techniques that don't use data from dynamic circulation models. Since no information on numerical weather prediction was available at the time, these simple methods are considered classics. These techniques are now employed for short-term and medium-term forecasting when numerical weather prediction information is insufficiently precise or frequent. Combining statistical techniques with numerical techniques for weather prediction is an important application of statistical techniques in weather prediction. In order to improve the dynamic prediction model outputs for specific quantities (probabilities) or places not covered by this one, statistical forecasting models are applied. The three categories of general prediction are short, medium, and long-term. The first prediction horizon is between one hour and one week in the future. The second prediction horizon is between one week and one year. It should be noted that nowcasting techniques are now possible due to improved resolution in satellites.

3 Results and Discussion

This overview gives a summary of the articles in this area that have been found in the literature. Even though this analysis is neither extensive nor full, it does a good job of laying out how much research has approached this issue and demonstrates the viability of using intelligent techniques to solve various weather forecasting issues. Conventional control approaches are not hindered by them; on the contrary, they can help AI techniques create more effective hybrid controls. The decision between intelligent modeling techniques and conventional methods is not always clear-cut because both have drawbacks, just like any modeling technique.

Artificial intelligence notably through its potential for prediction has a role to play in addressing these concerns. These challenges are made more interesting, though, by the fact that they are directly tied to time, which AI must assist us in learning to manage. It is projected that more AI technology will be used in the energy industry as systems become more complex and dispersed [10]. The energy transition may someday leverage AI applications, even though they are currently intermittent on networks.

The following Table 1 gives a summary comparison between two methods exposed in [11, 12], and [13].

Table 1. Comparison between three methods.

	Advantages	*Disadvantages*
Method 1: Bin Du et al. 2022, [11]	– This ANN and data clustering are combined – Clustering involved three different kinds of weather	– Not tested on other forms of solar energy systems for example photovoltaics
Method 2: S. Shanmuga Priya et al. 2015, [12]	– The global solar radiation is estimated and predicted using this ANN-based technique	– Only areas in India are acceptable for using this model to predict solar radiation

As illustrated in Table 1, we can mention that due to the inaccuracy of conventional or empirical theories, ANN approaches have been applied for prediction. The discussion of the results provides evidence that each method has its strengths and weaknesses.

4 Conclusion

We conclude that the energy sector can apply AI in a variety of ways. Its predictive capability mainly facilitates two important activities. The bibliographic analysis conducted reveals that many different fields use AI techniques, including control, prediction, and optimization. These techniques have an advantage over traditional methods because they typically don't call for an understanding of the internal parameters of the systems being examined or the intricate mathematical relationships regulating them. These techniques have an advantage over traditional methods because they typically don't call for an understanding of the internal parameters of the systems being examined or the intricate mathematical relationships regulating them. This optimization method can also be used to avoid the heating of LEDs in optics [9, 10]. It is sufficient to have experimental or projected measurements of a carefully designed mathematical model depicting the real performance and behavior of the researched system in order to create these intelligent models. This study's findings also emphasize the need to use wavelets [9, 11–13], integrated with data analysis techniques often employed in biomedical signal processing, such as ICA-NMF-SVD-PCA [14–16] to further improve the aforementioned techniques' efficacy, This also appears to be applicable to synchronous motors [17, 18].

References

1. Zahraee, S.M., Khalaji Assadi, M., Saidur, R.: Application of artificial intelligence methods for hybrid energy system optimization. Renew. Sustain. Energy Rev. **66**, 617–630 (2016). https://doi.org/10.1016/j.rser.2016.08.028
2. Martin Nascimento, G.F., Laranjeira, T., Thi-Tuyet-Hong Vu, Shadid, M.S., Wurtz, F.: Machine learning on buildings data for future energy community services. Benoît Delinchant. SGE 2021, Symposium de Génie Electrique 2021, Nantes 6–8 July 2021
3. Data Science and Energy: Some Lessons From Europe on Higher Education Course Design and Delivery, HDSR, January 2022
4. Elizondo, D., Hoogenboom, G., McClendon, R.W.: Development of a neural network model to predict daily solar radiation. Agric. For. Meteorol. **71**, 115–132 (1994)
5. Kumar, R., Aggarwal, R.K., Sharma, J.D.: Comparison of regression and artificial neural network models for estimation of global solar radiations. Renew. Sustain. Energy Rev. **52**, 1294–1299 (2015)
6. Chiteka, K., Enweremadu, C.C.: Prediction of global horizontal solar irradiance in Zimbabwe using artificial neural networks. J. Clean. Prod. (2016). https://doi.org/10.1016/j.jclepro.2016.06.128
7. Du, B., Lund, P.D., Wang, J.: Improving the accuracy of predicting the performance of solar collectors through clustering analysis with artificial neural network models. Energy Rep. **8**, 3970–3981 (2022). ISSN 2352–4847, https://doi.org/10.1016/j.egyr.2022.03.013
8. Shanmuga Priya, S., Iqbal, M.H.: Solar radiation prediction using artificial neural network. Int. J. Comput. Appl. (0975 – 8887), vol 116 – No. 16, April 2015
9. Ouhadou, M., El Amrani, A., Messaoudi, C., Ziani, S.: Experimental investigation on thermal performances of SMD LEDs light bar: Junction-to-case thermal resistance and junction temperature estimation, Optik, vol. 182 (2019)
10. Ouhadou, M., El Amrani, A., Ziani, S., Messaoudi, C.: Experimental modeling of the thermal resistance of the heat sink dedicated to SMD LEDs passive cooling. In: Proceedings of the 3rd International Conference on Smart City Applications (2018)
11. Said, Z., El Hassouani, Y.: A new approach for extracting and characterizing fetal electrocardiogram. Traitement du Signal **37**(3), 379–386 (2020). https://doi.org/10.18280/ts.370304
12. Said, Z., El Hassouani, Y.: Fetal Electrocardiogram Analysis Based on LMS Adaptive Filtering and Complex Continuous Wavelet 1-D. In: Farhaoui, Y. (eds) Big Data and Networks

Technologies. BDNT 2019. Lecture Notes in Networks and Systems, vol 81. Springer, Cham (2020). https://doi.org/10.1007/978-3-030-23672-4_26

13. Said, Z., El Hassouani, Y.: Fetal-maternal electrocardiograms mixtures characterization based on time analysis. In: 2019 5th International Conference on Optimization and Applications (ICOA), 2019, pp. 1–5 (2019). https://doi.org/10.1109/ICOA.2019.8727619
14. Ziani, S., El Hassouani, Y., Farhaoui, Y.: An NMF based method for detecting RR interval. In: Farhaoui, Y., Moussaid, L. (eds.) ICBDSDE 2018. SBD, vol. 53, pp. 342–346. Springer, Cham (2019). https://doi.org/10.1007/978-3-030-12048-1_35
15. Said, Z., Jbari, A., Bellarbi, L.: QRS complex characterization based on non-negative matrix factorization NMF. In: 2018 4th International Conference on Optimization and Applications (ICOA), 2018, pp. 1–5. https://doi.org/10.1109/ICOA.2018.8370548
16. Said, Z., Jbari, A., Belarbi, L.: Fetal electrocardiogram characterization by using only the continuous wavelet transform CWT. In: 2017 International Conference on Electrical and Information Technologies (ICEIT), pp. 1–6 (2017). https://doi.org/10.1109/EITech.2017.8255310
17. Chaou, Y., et al.: Nonlinear control of the permanent magnet synchronous motor PMSM using backstepping method. WSEAS Trans. Syst. Control **17**, 56–61 (2022)
18. Achour, H.B., Ziani, S., Chaou, Y., El Hassouani, Y., Daoudia, A.: Permanent magnet synchronous motor PMSM control by combining vector and PI controller. WSEAS Trans. Syst. Control **17**, 244–249 (2022)

A Sorting-AB Approach for an Enhanced IDS in Cloud Environment

Hanaa Attou[1(✉)], Azidine Guezzaz[1], Said Benkirane[1], and Mourade Azrour[2]

[1] Technology Higher School, Cadi Ayyad University, Essaouira, Morocco
hanaa.attou@ced.uca.ma

[2] Faculty of Sciences and Technics, Moulay Ismail University, Errachidia, Morocco

Abstract. Cloud computing (CC) is currently the most popular option for companies due to its flexibility and pay-per-use services. Nonetheless, privacy and security issues are important hurdles in the successful adoption of CC because of its distributed openness of CC, traditional intrusion detection technologies are often ineffective. An intrusion detection system (IDS) is an important and necessary component in ensuring network security and protecting network resources and infrastructure. In this research work, we effectively use an IDS based on sorting and AdaBoost (AB) classifier. We verify the effectiveness and the feasibility of the proposed IDS sorting-AB model by experiments on the NSL-KDD and CICIDS 2017 datasets. The operational results show our model's superiority over different techniques and its capacity to detect intrusions with high detection accuracy (ACC) using a minimum number of features.

Keywords: Cloud security · Intrusion detection system · AdaBoost · Sorting-approach

1 Introduction

CC has played a significant role in meeting rising storage and infrastructure demands [1]. The cloud model comprises 5 basic qualities, 3 cloud service models, and 3 deployment types [2, 7]. Despite the efficiency of services provided by these applications, there are still security concerns [3, 8]. Therefore, to secure cloud environment IDS is used [3]. Recently, several contributions show that ensemble learning methods have been applied to increase the detection rate of IDS [9, 10]. The main goal of this article is to confirm an improved IDS based on the sorting-AB approach, which combines the sorting and AB classifier. As a result, our suggested method increases ACC while reducing the number of features required. When we compared existing models, experimental findings on the NSL- KDD and CICIDS 2017 show that our proposed model performs well in terms of ACC and precision. The remainder of this paper is outlined below. In part 2, we review the background of CC architectures, IDS approaches, Machine Learning (ML) methodologies, and some recent domain-related work. In Sect. 3, we show and discuss our model, then, the outcomes of our model are presented in Sect. 4. The report ended with a conclusion.

Y. Farhaoui et al. (Eds.): ICAISE 2022, LNNS 635, pp. 120–125, 2023.
https://doi.org/10.1007/978-3-031-26254-8_18

2 Background and Related Works

This section provides an overview of CC architecture and intrusion detection methodologies and references to some recent IDS that use ML and DL algorithms to improve cloud security. Cloud delivery models are infrastructure as a Service (IaaS), platform as a service (PaaS), and software as a service (SaaS) [2, 5, 6, 14, 15] deployed in CC deployment models: Public, hybrid, and private cloud [6, 7, 11–13]. Despite the benefits of cloud services, security issue remains a challenge for cloud providers. Detecting all types of attacks using the traditional method is incredibly difficult [16, 26–28]. Recently, IDS is used to monitor network traffic, Anomaly base and misuse base are the two types of NIDS [4, 36, 39]. There are several limits of IDS [16]. Then to improve IDS, ML and ensemble learning methods were used [17, 37, 38]. In [19], the authors propose network-based intrusion detection architecture for protecting cloud services and networks from diverse threats. Authors in [20] have suggested that the monitoring of networks will be carried out using a novel detection approach. Authors [18] explain an ensemble-based machine learning approach that employs four classifiers. The system achieves 97.24%. In [21], the RF outperforms the various strategies employed Naive Bayes (NB), Support Vector Machine (SVM) and K-Nearest Neighbors (KNN). To identify distributed denial of service (DDoS), [22] employ SVM-based expert systems DDoS. Authors in [23] suggested a classification-based ML strategy to identify DDoS attacks in CC using three methods KNN, RF and NB. In [17], ML techniques are used to feed IDS and detect malicious network traffic, ISOT-CID is the dataset used to evaluate the system. In [32] Z. Liu et al. proposed a hybrid IDS using a GA-Based method and RF. In [33] the model based on ANN achieves 98.86% of ACC.

3 Proposed Sorting-AB IDS Framework

Our contribution aims to propose and implement an optimized model improving ACC and minimizing the number of used features. This section explains the proposed model.

Our model is separated into 3 crucial steps, as shown in Fig. 1: Firstly the dataset is preprocessed by founding and eliminating the NaN values, then we use one-hot encoding and normalization process. The preprocessed dataset is then divided into two groups: Training and testing. In the training step, 70% are employed and 30% are used for the test step. Secondly, we apply the sorting approach which trains the model on each variable that exists in the list of the best-selected variables from the earlier works. Then it returns the variables that have an ACC higher than 80%. Then we test them to find the minimum of variables that work well for our model with the AB classifier. The classifier set of complexity associated with keeping a set of weights over training samples and adapting these weights after each Boosting iteration: the weights of the training instances that are incorrectly identified by the existing element classifier will be enhanced while the weights of the training instances that are correctly classified will be reduced [35]. Adaboost's weight update has been suggested to be implemented in several different ways [35].

Thirdly, the attack can be predicted by the build model as positive. Based on metrics of confusion matrix like ACC, recall, precision and F1-score. ACC is the proportion of

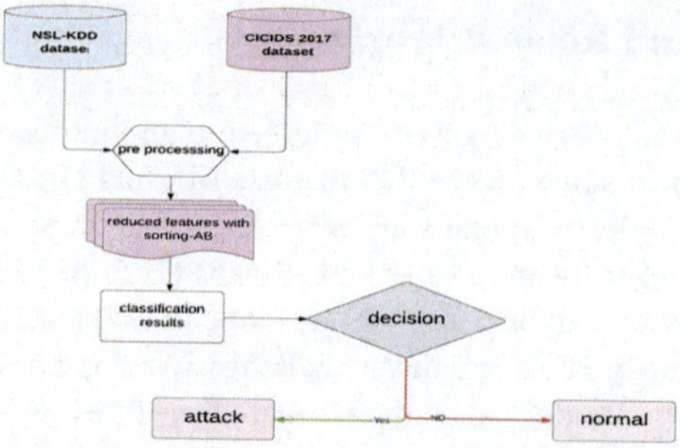

Fig. 1. Scheme of the sorting-AB model

correctly predicted outcomes compared to all cases. The recall is employed to figure out the proportion of correctly identified positive patterns. Precision measures the number of precisely correctly predicted features in a positive class out of all the normal behavior. F1-score the harmonic mean of the precision and recall.

4 Results and Discussions

The confusion matrix generates performance indicators that will be used to evaluate the suggested IDS. We trained our model using the NSL-KDD and CICIDS 2017 datasets. Table 1 presents the obtained results.

Table 1. Obtained metrics

Dataset	Recall(%)	F1-score(%)	Precision(%)	ACC(%)
NSL-KDD	95.13	96	98	95.13
CICIDS 2017	100	100	100	99.84

The evaluation shows that the proposed method accurately identifies cyber-attacks with 96.5% of ACC with 5 features, 96% of recall, 97% of f1-score and 98% of precision. 3 variables are selected for the final model from the subset of CICIDS 2017. Our proposed IDS achieves 99.84% of ACC. Figure 2 (a)(C) show that our proposed model performs well compared to the results found by GA-based J48 and GA-based NB classifiers [34] using NSL-KDD in terms of ACC.

5 variables are also used in the development of our model. Figure 2 (b) shows that our proposed model performs well compared to the results found by GA-RF [32] in terms of ACC. As well as for the construction of our model we used only 5 variables. This shows that the sorting-AB approach can detect intrusions with fewer variables. We use CICIDS 2017 to confirm the performance of the model, as shown in Fig. 2 sorting-AB with this dataset performs well the other models.

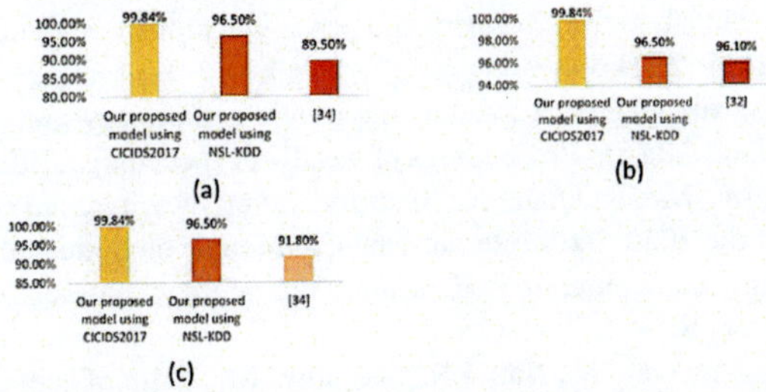

Fig. 2. Performance comparison between our proposed model and other models.

5 Conclusion

The effective approach for improving cloud security against attacks is intrusion detection. In this research, we provide an optimized intrusion detection model for cloud security. The results of this experiment highlight how well the system identifies attacks and distinguishes between normal and abnormal behavior improving IDS accuracy and reducing the number of employed features. The suggested strategy selects the lowest number of features possible, just 5 out of 41 were chosen from the NSL-KDD datasets. It also reduced the dimensionality of CICIDS 2017 to 3 by adopting the same model.

References

1. AlSalihi, M., Naji, A.M.: The impact of cloud computing on information technology in the Basra oil company as a model for the study. Turk. J. Comput. Math. Educ. **12**, 3644–3652 (2021)
2. Final version of NIST Cloud Computing definition published (2013). http://www.nist.gov/itl/csd/cloud-102511.cfm. Accessed 25 Aug 2013
3. Abusitta, A., Bellaiche, M., Dagenais, M., Halabi, T.: A deep learning approach for proactive multi-cloud cooperative intrusion detection system. Future Gener. Comput. Syst. **98**, 308–318 (2019)
4. Buczak, L., Guven, E.: A survey of data mining and machine learning methods for cyber security intrusion detection. IEEE Commun. Surv. Tutorials **18**, 1153–1176 (2016)
5. Singh, A., Chatterjee, K.: Cloud security issues and challenges: a survey. J. Netw. Comput. Appl. **79**, 88–115 (2016)
6. Ramachandra, G., Iftikhar, M., Khan, F.A.: A comprehensive survey on security in cloud computing. In: The 3rd International Workshop on Cyber Security and Digital Investigation (2017)
7. Patel, B., Kansara, N.: cloud computing deployment models: a comparative study. Int. J. Innovative Res. Comput. Sci. Tech. **9**, 45–50 (2021)
8. Gago, F., Pearson, C., Derrico, S., Alnemr, M., Pulls, R., Oliveira, T.: Cloud workshop: accountability in the cloud, In: Proceedings of the IFIP International Summer School on Privacy and Identity Management (2015)
9. Gu, J., Wang, L., Wang, H., Wang, S.: A novel approach to intrusion detection using SVM ensemble with feature augmentation. Comput. Secur. **86**, 53–62 (2019)
10. Tama, B.A., Rhee, K.H.: HFSTE: hybrid feature selections and tree-based classifiers ensemble for intrusion detection system. IEICE Trans. Inf. Syst. **100**, 1729–1737 (2017)

11. Abdulsalam, Y.S., Hedabou, M.: Security and privacy in cloud computing: technical review. Future Internet **14**, 11 (2021)
12. Wang, C., Ren, K., Yu, S., Urs, K.: Achieving usable and privacy-assured similarity search over outsourced cloud data. In: Proceedings of the 2012 Proceedings IEEE INFOCOM (2012)
13. Zhou, W., Zhang, M., Xie, R. Qian, W.: Security and privacy in cloud computing: a survey. In: Proceedings of the 2010 Sixth International Conference on Semantics (2010)
14. Zhang, Q., Cheng, L., Boutaba, R.F.: Cloud computing: state-of-the-art and research challenges. J. Internet Serv. **1**, 7–18 (2010)
15. Srinivasan, M.K., Sarukesi, K., Sai, P.R., Revathy, M.: State-of-the-art cloud computing security taxonomies: a classification of security challenges in the present cloud computing environment. In: ICACCI (2012)
16. Almeida, M., Alzubi, M., Kovacs, S. Alkasassbeh, M.: Evaluation of machine learning algorithms for intrusion detection system. In: IEEE 15th International Symposium on Intelligent Systems and Informatics (2017)
17. Alshammari, A., Aldribi, A.: Apply machine learning techniques to detect malicious network traffic in cloud computing. J. Big Data **8**(1), 1–24 (2021). https://doi.org/10.1186/s40537-021-00475-1
18. Singh, P., Ranga, V.: Attack and intrusion detection in cloud computing using an ensemble learning approach. Int. J. Inf. Technol. **13**, 565–571 (2021). https://doi.org/10.1007/s41870-020-00583-w
19. Mehibs, S.M., Hashim, S.: Proposed network intrusion detection system based on fuzzy c mean algorithm in cloud computing environment. J. Univ. Babylon Pure Appl. Sci. **26**, 27–35 (2018)
20. Guezzaz, A., Asimi, A., Asimi, Y., Tbatous, Z., Sadqi, Y.: A global intrusion detection system using PcapSocKS sniffer and multilayer perceptron classifier. Int. J. Netw. Secur. **21**, 438–450 (2019)
21. Ahmad, F.B., Ghulam, M., Nawaz, A., Kiani, A.A., Ali, T.: Securing cloud data: a machine learning based data categorization approach for cloud computing. Research square (2022)
22. Mubarakali, A., Srinivasan, K., Mukhalid, R., Jaganathan, S.C., Marina, N.: Security challenges in internet of things: distributed denial of service attack detection using support vector machine-based expert systems. Comput. Intell. **36**, 1580–1592 (2020)
23. Mishra, A., Gupta, B.B., Perakovic, D., Jose, F., Penalvo, G., Hsu, C.H.: classification based machine learning for detection of DDOS attack in cloud computing. In: IEEE International Conference on Consumer Electronics (2021)
24. Singh, A.: Security concerns and countermeasures in cloud computing: a qualitative analysis. Int. J. Inf. Technol. **11**, 683–690 (2018). https://doi.org/10.1007/s41870-018-0108-1
25. Kaur, P., Gosain, A.: Issues and challenges of class imbalance problem in classification. Int. J. Inf. Technol. **14**, 539–545 (2018). https://doi.org/10.1007/s41870-018-0251-8
26. Shafiq, M., Tian, Z., Sun, Y., Xiaojiang, D., Guizani, M.: Selection of effective machine learning algorithm and Bot-IoT attacks traffic identification for internet of things in smart city,. J. Pre-proof (2020)
27. Koroniotis, N., Moustafa, N., Sitnikov, E., Turnbull, B.: Towards the development of realistic botnet dataset in the internet of things for network forensic analytics: Bot-IoT dataset. Future Gen. Comput. Syst. **100**, 779–796 (2019)
28. Shafiq, M., Tian, Z., Bashir, A.K., Du, X., Guizani, M.: CorrAUC: a malicious Bot-IoT traffic detection method in IoT network using machine learning techniques. IEEE Internet Things J. **8**, 3242–3254 (2021)
29. Verma, P., Anwar, S., Khan, S.: Network intrusion detection using clustering and gradient boosting. In: Proceedings of the International Conference on Computing, Communication and Networking Technologies (2018)

30. Hamamoto, A., Carvalho, L., Sampaio, L., Abrao, T., Proenc, A.: Network anomaly detection system using genetic algorithm and fuzzy logic. Expert Syst. Appl. **92**, 390–402 (2018)
31. Besharati, E., Naderan, M., Namjoo, E.: LR-HIDS: logistic regression host-based intrusion detection system for cloud environments. J. Ambient. Intell. Humaniz. Comput. **10**(9), 3669–3692 (2018). https://doi.org/10.1007/s12652-018-1093-8
32. Zhiqiang, L., Yucheng, S.: A hybrid IDS using GA-based feature selection method and random forest. Int. J. Mach. Learn. Comput. **12**, 43–50 (2022)
33. Subba, B., Biswas, S., Karmakar, S.: A neural network based system for intrusion detection and attack classification. IEEE 2016
34. Desale, M., Ade, M.: Genetic algorithm based feature selection approach for effective intrusion detection system. In: International Conference on Computer Communication and Informatics (2015)
35. Xuchun, L., Wang, L., Sung, E.: AdaBoost with SVM-based component classifiers. Eng. Appl. Artif. Intell. **21**, 785–795 (2007)
36. Guezzaz, A., Benkirane, S., Azrour, M.: A novel anomaly network intrusion detection system for internet of things security. In: Azrour, M., Irshad, A., Chaganti, R. (eds.) IoT and Smart Devices for Sustainable Environment. EAI/Springer Innovations in Communication and Computing. Springer, Cham (2022). https://doi.org/10.1007/978-3-030-90083-0_10
37. Guezzaz, A., Asimi, A., Asimi, Y., Azrour, M., Benkirane, S.: A distributed intrusion detection approach based on machine leaning techniques for a cloud security. In: Gherabi, N., Kacprzyk, J. (eds.) Intelligent Systems in Big Data, Semantic Web and Machine Learning. AISC, vol. 1344, pp. 85–94. Springer, Cham (2021). https://doi.org/10.1007/978-3-030-72588-4_6
38. Azrour, M., Mabrouki, J., Farhaoui, Y., Guezzaz, A.: Security analysis of nikooghadam et al.'s authentication protocol for cloud-IoT. In: Gherabi, N., Kacprzyk, J. (eds.) Intelligent Systems in Big Data, Semantic Web and Machine Learning. AISC, vol. 1344, pp. 261–269. Springer, Cham (2021). https://doi.org/10.1007/978-3-030-72588-4_18
39. Guezzaz, A., Asimi, A., Mourade, A., Tbatou, Z., Asimi, Y.: A multilayer perceptron classifier for monitoring network traffic. In: Farhaoui, Y. (ed.) BDNT 2019. LNNS, vol. 81, pp. 262–270. Springer, Cham (2020). https://doi.org/10.1007/978-3-030-23672-4_19

A SWOT Analysis for Healthcare Using Machine Learning

Zineb Sabouri[1](✉), Noreddine Gherabi[1], Hakim El Massari[1], Sajida Mhamedi[1], and Mohamed Amnai[2]

[1] Lasti Laboratory Khouribga, Sultan Moulay Slimane University, ENSA Khouribga, Beni Mellal, Morocco
zineb.sabouri@usms.ac.ma
[2] Laboratory of Computer Sciences, Faculty of Sciences, IbnTofail University, Kenitra, Morocco

Abstract. Machine learning (ML) applications are used in variety of real-world situations, and they are proving a big impact on the healthcare sector. It has become a reality in clinical practice. ML algorithms are applied to examine disease predictions and medical records. The healthcare data can be utilized to find the best trial sample, acquire additional data points, evaluate ongoing data from study participants, and avoid data-based errors. Many recent efforts have aided the adoption of machine learning techniques in the medical sector to allow healthcare providers to focus on patient care rather than searching information. Most of the existing ML techniques have external and internal drawbacks that prevent their ultimate implementation in the clinical domain and various concepts related to ML need to be implemented in the medical studies so that health care practitioners can effectively interpret and guide research in this field. For this reasons, this paper analyzes the present state of ML, as applied to healthcare in a strengths, weaknesses, opportunities and threats (SWOT) analysis.

Keywords: SWOT analysis · Machine learning · Healthcare

1 Introduction

Clinical data is critical in the treatment of health problems, and ML systems can aid in disease diagnosis [1]. ML is a fast dominant force in the healthcare system. It transforms the manual health care system into an automated system and has the potential to dramatically improve patient care and reduce healthcare costs.

Increasingly, ML is seen as a way to enhance the quality, accuracy, efficiency, and safety of medical field by providing point-of-care decision support, facilitate new approaches and enabling routine quality measures to become reality. ML methods play an important role in the health-care process, including disease diagnosis, decision-making assistance, and early detection [2].

The techniques of ML falls under three types including supervised, unsupervised, and reinforcement learning [3]. Supervised learning is a sub category of ML in which machines use labeled data to predict the result [4]. Unsupervised learning uses methods

Y. Farhaoui et al. (Eds.): ICAISE 2022, LNNS 635, pp. 126–131, 2023.
https://doi.org/10.1007/978-3-031-26254-8_19

to find patterns in datasets that are not tagged or classified. Reinforcement learning is a type of ML where intelligent machine learns how to react in a given environment by doing actions and observing the outcome of actions.

The objective of this study was to lay the groundwork for a perspective on the potential impact of ML on the medical sector. Therefore, we adopted a SWOT analysis to identify strength, opportunities, threats and drawbacks currently confronting the implementation of ML algorithms in medical area.

To achieve the above goals the remainder of this article is structured into five sections: Sect. 2 gives an overview of machine learning in healthcare. Section 3 describes SWOT analysis and the methodology used. In Sect. 4 the results are discussed. Overall, conclusion and future work appear in the last section.

2 Literature Review

2.1 Applications of ML in Healthcare

There are a growing number of applications of ML in healthcare, with many benefits for both patients and doctors. Some of the popular applications of ML in the medical field include automating healthcare charging, medical decision support, and creating medical care rules. There are various examples of ML and healthcare approaches being used in medical [5], and wide range of applications such as diagnostic imaging, smart medical records, enhanced radiation therapy, individualized treatment, crowdsourcing data collection, machine learning-based behavioral change, clinical trials, and research [6].

ML algorithms were utilized to simulate the prediction and diagnosis of a thyroid patient. In the proposed model, an attempt is made to examine naive Bayes, k-nearest neighbor, and Support Vector Machine (SVM) for multiclass thyroid dataset classification [7, 8]. ML classification algorithms such as the random forest, decision tree, and K.N.N were implemented to predict heart disease [9, 10]. We can use ontology classification for data segmentation [7] after that we implement ML classification algorithms to predict chronic disease. According to the results, the random forest classifier algorithm outperforms the other classifier algorithms in terms of accuracy [11].

3 Research Methodology

3.1 SWOT Analysis

SWOT is an effective technique used to determine opportunities and threats of the context and identify the weaknesses and strengths of a product or service [12]. The first two aspects are linked to external factors of the environment, whereas strengths and weaknesses related to internal factors. These factors are the most crucial for the success of organizations in the future. SWOT method is frequently used as a technique for analyzing internal and external elements in order to develop a methodical strategy and provide help for dealing with a problem. It is a powerful technique for making decisions. By determining the strengths, weaknesses, opportunities, and threats, Companies can develop a strategy based on the strengths, eradicate weaknesses, exploit opportunities, and utilize an option to transferred threats into opportunities.

- Strength can be viewed as a resource that allows ML to attain its predetermined objectives.
- weakness can be considered as fault, limitation, or defect of ML to reach a special target.
- Opportunity can be regarded as a beneficial condition that the ML algorithm has the potential to profit on or employ to its benefits.
- Threat can be viewed as an undesirable situation that could have a negative effect by posing a barrier or restriction to attain goals.

Table 1. SWOT analysis.

Strengths (S)	Weaknesses (w)
• S_1: Early detection and better predictions of diseases • S_2: Improve clinical decision-making • S_3: Improve population health • S_4: Enhance the prediction of clinical outcomes and response to treatment • S_5: Boost data quality with less exposure, cheaper devices, and minimal intervention • S_6: Allow the integration of complex features • S_7: Optimize the standardization of processes • S_8: Allow objective and through comparison to populations • S_9: Fast execution • S_{10}: High accuracy • S_{11}: Save time and reduce costs	• W_1: Risk putting patients' lives in jeopardy from system errors • W_2: Lack of consideration for the user's perspective • W_3: Safety, efficacy, privacy, information and consent, the costs, and access time-, cost-, and resource-consuming • W_4: Lack of clarity for some ML algorithms • W_5: Privacy issues for data used for ML model training • W_6: Need a large volume of training data • W_7: Slow ML Adoption • W_8: Inconsistent results and lack of reproducibility • W_9: Lack formal ML competence • W_{10}: Black box models

(continued)

Table 1. (*continued*)

Strengths (S)	Weaknesses (w)
Opportunities (O)	Threats (T)
• O_1: Reducing human intervention • O_2: Creation of larger and better organized medical databases • O_3: Perform an automatic identification and recommendations of clinical terms • O_4: Extraction of semantic links among medical entities • O_5: Help solve critical healthcare problems • O_6: Improve clinical outcomes • O_7: Stimulate the man/machine collaboration • O_8: Reach diagnosis in a shorter time • O_9: Help in the organization of healthcare-diagnosis, risk assessment and urgency assessment • O_{10}: Discharge healthcare practitioners from repetitive tasks	• T_1: Harm patients if wrong decisions are taken • T_2: Obtain unfair results if training data is biased • T_3: Information privacy • T_4: Ethical and legal implication • T_5: Need validation by professionals • T_6: Absence of the integration of ML algorithms in clinical workflow • T_7: Need of advanced and efficient storage systems for dealing with large volumes of data • T_8: Lack of a fully established agreement between researchers, clinicians, and vendors for standard labeling of data • T_9: Depend on data quality

At the end of this session, the most crucial factors in each SWOT category was determined (see Table 1).

4 Discussion of Results

The SWOT analysis looks at the strengths of ML, suggesting that it will provide an early detection and better predictions of diseases. ML models can range from difficult-to-find chronic diseases in the early stages to other transmitted illnesses by maximizing data information. It optimizes the standardization of processes by analyzing images, detecting problems, and supplement doctor interpretation to improve diagnosis accuracy and patient care.

However, the SWOT analysis reflect a weaknesses that ML risk endangering the lives of patients by providing erroneous or inadequate data to ML algorithms. The effectiveness of an ML algorithm depends on the quality of the examples used for training; experts must classify ground truth examples and compile groups of individual patients, medical photos, or imaging reports in order to provide training data. This procedure takes a long time, costs a lot of money, and uses many resources. Furthermore, ML has several ethical issues including right to decide, privacy, efficacy, safety, costs and access that arise when ML techniques are used in healthcare sector.

As for opportunities, ML reduces human intervention and perform an automatic identification and recommendations of clinical terms particularly in time-consuming traditional tasks. This new vision is linked to advancements in computer science, covering both software and hardware elements, as well as the construction of larger and more organized medical databases.

A major threat is that ML can harm patients if wrong decisions or interpretations are taken. Consequently, the results of ML need to be interpreted intuitively by the clinician and validated much more comprehensively, which is one of the main reasons why ML is not being implemented in routine clinical practice. Moreover, there are inherent external limits resulting from the requirement for modern and efficient storage systems to deal with massive volumes of data that ML algorithms require as input.

Finally, Integration in a clinical setting, interpretability, and validation still require a significant amount of work, which will necessitate prospective randomized clinical studies. In this regard, there is still a long way to go before these tools are ready to be used in everyday patient care.

5 Conclusion

This study proposes a SWOT analysis method for evaluating the potential of ML technology in medical sector. Based on the strengths revealed, ML can be a powerful technology for clinicians, doctors and researchers. It is required to make decisions in treatment and diagnosis of patients before they happen and engaging patients more efficiently in their rehabilitation process, in order to reduce medication errors and saving people lives. However, existing limitations of ML including ethical implications, do not explicitly distrust the adoption and usefulness of ML methods in the healthcare industry, as long as better results and a high success percentage are achieved.

This SWOT technique can be enhanced by determining which SWOT components should be treated first, examining the eigenvalues, and find out the best match to the external and internal factors. This can be accomplished by combining SWOT and AHP [13] method called hybrid method A'WOT [12] Exploring that enable quantization of priorities to support decision of adoption of ML in medical domain [14].

References

1. Nayyar, A., Gadhavi, L., Zaman, N.: Chapter 2 - machine learning in healthcare: review, opportunities and challenges. In: Singh, K.K. Elhoseny, M. Singh, A. Elngar, A.A. (eds.) Machine Learning and the Internet of Medical Things in Healthcare, pp. 23–45. Academic Press (2021). https://doi.org/10.1016/B978-0-12-821229-5.00011-2
2. Sunarti, S., Rahman, F.F., Naufal, M., Risky, M., Febriyanto, K., Masnina, R.: Artificial intelligence in healthcare: opportunities and risk for future. Gac. Sanit. **35**, S67–S70 (2021). https://doi.org/10.1016/j.gaceta.2020.12.019
3. Verma, V.K., Verma, S.: Machine learning applications in healthcare sector: an overview. Mater. Today Proc. **57**, 2144–2147 (2022). https://doi.org/10.1016/j.matpr.2021.12.101
4. Sabouri, Z., Maleh, Y., Gherabi, N.: Benchmarking classification algorithms for measuring the performance on maintainable applications. In: Maleh, Y., Alazab, M., Gherabi, N., Tawalbeh, L., Abd El-Latif, A.A. (eds.) ICI2C 2021. LNNS, vol. 357, pp. 173–179. Springer, Cham (2022). https://doi.org/10.1007/978-3-030-91738-8_17
5. Musen, M.A., Middleton, B., Greenes, R.A.: Clinical decision-support systems. In: Shortliffe, E.H., Cimino, J.J. (eds.) Biomedical Informatics, pp. 795–840. Springer, Cham (2021). https://doi.org/10.1007/978-3-030-58721-5_24

6. Kalaiselvi, K., Deepika, M.: Machine learning for healthcare diagnostics. In: Jain, V., Chatterjee, J.M. (eds.) Machine Learning with Health Care Perspective. LAIS, vol. 13, pp. 91–105. Springer, Cham (2020). https://doi.org/10.1007/978-3-030-40850-3_5
7. Abbad Ur Rehman, H., Lin, C.-Y., Mushtaq, Z., Su, S.-F.: Performance analysis of machine learning algorithms for thyroid disease. Arab. J. Sci. Eng. **46**(10), 9437–9449 (2021). https://doi.org/10.1007/s13369-020-05206-x
8. Massari, H.E., Gherabi, N., Mhammedi, S., Sabouri, Z., Ghandi, H.: Ontology based decision tree model for prediction of cardiovascular disease. INDJCSE **13**(3), 851–859 (2022). https://doi.org/10.21817/indjcse/2022/v13i3/221303143
9. El Massari, H., Mhammedi, S., Sabouri, Z., Gherabi, N.: Ontology-based machine learning to predict diabetes patients. In: Maleh, Y., Alazab, M., Gherabi, N., Tawalbeh, L., Abd El-Latif, A.A. (eds.) ICI2C 2021. LNNS, vol. 357, pp. 437–445. Springer, Cham (2022). https://doi.org/10.1007/978-3-030-91738-8_40
10. Massari, H.E., Sabouri, Z., Mhammedi, S., Gherabi, N.: Diabetes prediction using machine learning algorithms and ontology. J. ICT Stand. **10**, 319–338 (2022). https://doi.org/10.13052/jicts2245-800X.10212
11. Ali, M.M., Paul, B.K., Ahmed, K., Bui, F.M., Quinn, J.M.W., Moni, M.A.: Heart disease prediction using supervised machine learning algorithms: performance analysis and comparison. Comput. Biol. Med. **136**, 104672 (2021). https://doi.org/10.1016/j.compbiomed.2021.104672
12. Bourhim, E.M.: Augmented reality for fire evacuation research: an A'WOT analysis. In: Abraham, A., Gandhi, N., Hanne, T., Hong, T.-P., Nogueira Rios, T., Ding, W. (eds.) ISDA 2021. LNNS, vol. 418, pp. 277–285. Springer, Cham (2022). https://doi.org/10.1007/978-3-030-96308-8_25
13. Bourhim, E.M., Cherkaoui, A.: Selection of optimal game engine by using AHP approach for virtual reality fire safety training. In: Abraham, A., Cherukuri, A.K., Melin, P., Gandhi, N. (eds.) ISDA 2018 2018. AISC, vol. 940, pp. 955–966. Springer, Cham (2020). https://doi.org/10.1007/978-3-030-16657-1_89
14. Nejjahi, R., Gherabi, N., Marzouk, A.: Towards classification of web ontologies using the horizontal and vertical segmentation. In: Noreddine, G., Kacprzyk, J. (eds.) ITCS 2017. AISC, vol. 640, pp. 70–81. Springer, Cham (2018). https://doi.org/10.1007/978-3-319-64719-7_7

A Systematic Review on LiDAR-Based 3D Object Detection

Adnan Anouzla(✉), Mohamed Bakali El Mohamadi, Nabila Zrira, and Khadija Ouazzani-Touhami

ADOS Team, LISTD Laboratory, Ecole Nationale Supérieure des Mines de Rabat, Rabat, Morocco
{adnan.anouzla,m.bakali-elmohamadi,zrira,ouazzani}@enim.ac.ma

Abstract. Light Detection And Ranging (LiDAR) is a popular sensor for providing 3-Dimensional (3D) information about a point cloud to localize and characterize its shape. Also, object detection in point clouds is critical in several robotic and computer vision tasks. Object detection is gaining popularity in both industry and research because of its vast range of applications in a variety of sectors. We can either make a fixed number of grid forecasts (i.e., single-stage) or use a proposal network to discover items, then a second network to fine-tune these proposals and produce a final prediction (i.e., two-stage). The purpose of this paper is to review single-stage and two-stage approaches in terms of used datasets, advantages, limitations, hyperparameters, and performance results.

Keywords: 3D object detection · Deep learning · Point clouds · LiDAR

1 Introduction

3D object detection has recently gotten a lot of attention and is now a hot area in Deep Learning (DL) research. Moreover, there are numerous real-world applications, such as self-driving cars [1], augmented/virtual reality [2], robots [3], and intelligent traffic systems that require 3D object detection. The point cloud, which is recorded by 3D sensors, is the most popular sort of data representation for 3D object detection to provide reliable information and display the 3D scene (e.g., LiDAR sensors). For autonomous driving, sensors must be able to detect and track moving objects in real-time, including vehicles, pedestrians, and cyclists. To do that, a LiDAR employs a laser scanner to determine the distance to the surroundings. However, due to the complexity of point clouds, detecting 3D objects in LiDAR remains a challenging problem. Objects like pedestrians, bikers, and traffic cones are frequently represented by a small number of points, making point cloud recognition difficult.

The most accurate first and second-stage 3D object identification techniques will be introduced and compared in this paper. Moreover, a comparison between these methods in terms of precision on three levels (Easy, Moderate, and Hard), hyperparameters, type of Graphics Processing Units (GPU), experimental results, and used datasets is provided. The objective of this benchmark aims to select the best LiDAR-based 3D object detection

Y. Farhaoui et al. (Eds.): ICAISE 2022, LNNS 635, pp. 132–137, 2023.
https://doi.org/10.1007/978-3-031-26254-8_20

approach that will be used further in navigation. This is to help blind people in their daily lives as they can move around in an independent way in an indoor environment since all the objects are detected and localized.

The rest of this paper is structured as follows. The most existing LiDARbased 3D object detection methods are introduced in Sect. 2. Section 3 presents the advantages and limitations of each method. Section 4 contains a discussion. Section 5 brings the paper to a close.

2 LiDAR-Based 3D Object Detection

In this section, the most recent LiDAR-based 3D detection methods which can be divided into single-stage and second-stage techniques are presented. Due to its straightforward architecture, the first category performs better than the second, which uses a second stage to refine the first category's predictions using features that are in line with region proposals. Table 1 summarizes each method category's hyperparameters, GPUs used for training networks, and public datasets.

Table 1. Comparison of the used datasets, GPU, and hyperparameters

Method	Dataset	GPU	Learning rate	Batch size	Epochs	Loss
PointPillars [4]	KITTY	GTX 1080 Ti	0.0002	2 for val 4 for test	160	- Focal loss - Sine-Error loss
CIA-SSD [5]	KITTY	TITAN Xp	Cosine annealing	4	60	- Focal loss - Smooth-L1 loss - Cross-entropy loss
Voxel R-CNN [6]	KITTY WAYMO	RTX 2080 T	0.01	30 for KITTI 32 for WAYMO	80 for KITTY 30 for WAYMO	- Focal Loss - Huber Loss
PV-RCNN++ [7]	KITTY WAYMO	Titan RTX	0.01	2	30	- The region proposal loss Lrpn - Keypoint segmentation loss Lseg - The proposal refinement loss Lrcn

2.1 Single-Stage Methods

Lang et al. [4] suggested a single-stage approach to detect 3D objects. PointPillars is a novel encoder that learns a point clouds representation structured in vertical columns using PointNets to predict the forms of 3D objects. For feature extraction, the authors divided a point cloud into pillars rather than voxels and then used a 2-Dimensional (2D) convolutional detection architecture for object localization. Additionally, pillars are very effective since all essential operations can be described as 2D convolutions that can be computed quickly on a GPU. Moreover, it is not essential to change the vertical binning manually by using pillars instead of voxels. In the PointPillars network, the input is point clouds, after which it estimates the 3D shape, and it comprises three steps: (i) A feature encoder for converting a point cloud into a pseudo-sparse image. (ii) The pseudo-image is processed into a high-level presentation using a 2D convolutional column. (iii) The 3D boxes are detected and regressed by a detection head. They evaluated the model using the Karlsruhe Institute of Technology and Toyota Technological Institute (KITTI) dataset. The PointPillars network runs at 62 Hz and is trained using just LiDAR point clouds.

Zheng et al. [5] provided a method called Confident Intersection over UnionAware Single-Stage Object Detector (CIA-SSD). It is based on the discovery that anchor-feature-based IoU predictions can distinguish between precise and imprecise bounding box regressions. The Spatial-Semantic Feature Aggregation (SSFA) module was created to integrate advanced abstract semantic data with low-level spatial features to improve the accuracy of bounding box and classification confidence predictions. It also includes two types of convolutional layers (spatial and semantic) as well as an attention-grabbing fusion module. There are three steps in the pipeline: (i) The sparse convolutional network (SPConvNet) for encoding point cloud inputs. (ii) The SSFA module for robust spatial-semantic feature extraction. (iii) A multi-tasking head with a confidence function for rectifying classification scores and Distance-variant IoU-weighted NMS (DI-NMS) for post-treatment. The model achieved performance according to the official ranking criteria (moderate average precision of 80.28%) and inference speed of more than 30.76 ms on the KITTI test set.

2.2 Two-Stage Methods

Deng et al. [6] suggested a voxel-based framework for 3D object detection that is simple but effective, named Voxel R-CNN. Three core modules make up Voxel R-CNN: (1) a 3D backbone network, (2) a 2D backbone network followed by the Region Proposal Network, and (3) a detection head with a new voxel RoI pooling operation. Voxel RoI pooling is used to extract region features from 3D voxel features after voxel R-CNN uses the input voxels to generate dense area proposals using Bird's-Eye-View (BEV) feature representations.

Shi et al. [7] proposed an advanced framework that is more than twice as fast as PV–RCNN. Moreover, PV-RCNN++ represents a significant advancement in developing more realistic 3D detection system with enhanced performance, lower resource utilization, and faster processing speed. The method provided two new components that increase the PV-RCNN framework's accuracy and efficiency. The first is the VectorPool aggregation module, which allows for more effective and efficient local feature aggregation from large-scale point clouds and the second is the sectorized proposal-centric technique, which allows for far quicker and better keypoint sampling. The set abstraction procedure in both the RoI-grid pooling module and the voxel set abstraction layer is replaced by the suggested VectorPool aggregation in the PV-RCNN++ detection system. The PV-RCNN++ framework consumes significantly less memory and computes resources than the PV-RCNN framework while achieving better 3D detection performance due to the VectorPool aggregation mechanism. The PV-RCNN++ framework is trained from the ground up using the ADAM optimizer with a learning rate equal to 0.01 and a cosine annealing learning rate decline method. The WAYMO open dataset was used for the evaluation. The PV-RCNN++ framework can provide state-of-the-art performance at 10 FPS for a detection range of 150 m $\times$ 150 m ($\times 3$ quicker than the PV-RCNN framework). On the KITTI dataset, it achieved state-of-the-art performance at 16 Frame Per Second (FPS) for a 70 m $\times$ 80 m detection area.

3 Advantages and Limitations

3.1 Advantages

PointPillars can use the full information represented by the point cloud by learning features rather than relying on fixed encoders. Furthermore, because pillars are used instead of voxels, no manual vertical direction binning is needed. Finally, pillars are particularly efficient since all critical operations may be represented as 2D convolutions that can be quickly computed on a GPU.

The spatial semantic feature aggregation (SSFA) module for obtaining strong spatial-semantic features for object predictions and the development of a confidence function to correct the classification score and address the misalignment between localization accuracy and classification confidence. Thus, the key contributions of CIA-SSD are the DI-NMS for smoother results and removing duplicate (zero-IoU) false positives. To avoid spatial information loss, SSFA keeps the spatial feature's dimensions the same as the input.

The voxel representation's spatial context in Voxel R-CNN is nearly sufficient for 3D object detection, and it is more efficient for feature extraction.

PV-RCNN++ progresses toward a more practical 3D detection system with performance improvements, lower resource consumption, and faster processing speed. The model makes use of two modules. The proposal-centric technique is the first module, this enables much faster and more accurate keypoint sampling. The VectorPool aggregation module, on the other hand, enables more efficient and effective local feature aggregation from large-scale point clouds.

3.2 Limitations

PointPillars can reach 105 Hz with minimal accuracy loss. While a LiDAR normally works at 20 Hz, it may be argued that this is an excessively long runtime. With further box augmentation, the detection performance for pedestrians declined dramatically. The idea is that by introducing ground truth sampling, per-box augmentation will be reduced.

Due to missing points around the boundary, candidate boxes in 3D detection by CIA-SSD may not have excellent object coverage, making it difficult to construct well-aligned boxes with the weighted average in DI-NMS.

The disadvantage of voxel-based is that voxelization frequently results in the loss of exact position information. Also, point-based approaches are less efficient in general because it is more expensive to find the nearest neighbor with a point representation for point set abstraction.

Performance suffers a bit in PV-RCNN++ when the discrete local kernel weights are replaced with shared kernel weights for relative position encoding.

4 Discussion

The Adam optimizer is used to completely train each model from scratch. During training, PointPillars used random mirroring flip, scaling and global rotation. Finally, they used x, y, and z to do a global translation. In CIA-SSD, four types of data augmentation are used to improve the model's generalization capacity. First, the entire point cloud is globally augmented, including scaling, flipping, and random rotation. The second type is a local augmentation that includes random rotation and translation on a part of the point cloud surrounding a groundtruth object. The third type is ground-truth augmentation. Last, omit objects of varied difficulty. Random mirroring flip, global rotation, scaling and translation, and other techniques are used in Voxel R-CNN for data augmentation. In PV-RCNN++, they used widely used data augmentation algorithms for 3D object detection, such as global rotation around the Z-axis,global scaling, random scene flipping, and ground-truth sampling augmentation. On the BEV object detection, the CIA-SSD model outperforms previous state-of-the-art methods in Moderate Average Precision (AP), i.e., 89.84%, and the Voxel-RCNN model ranks in the 1st place of the car category, as seen in Table 2, in terms of easy and hard AP, i.e., 94.85% and 86.13%, respectively. For the most essential 3D object detection benchmark in the car class, the PV-RCNN++ approach outperforms previous state-of-the-art methods in terms of moderate and difficult AP, i.e., 81.88% and 77.15%, respectively. On the easy level, Voxel R-CNN performs best, by 90.90% in AP.

Table 2. Comparison of performance on the KITTI test set in terms of Average Precision 3-Dimensional, and Bird's Eye View (AP3D, and APBEV) for car class

Method	AP3D (%)			APBEV (%)		
	Easy	Moderate	Hard	Easy	Moderate	Hard
PointPillars [4]	82.58	74.31	68.99	90.07	86.56	82.81
CIA-SSD [5]	89.59	80.28	72.87	93.74	**89.84**	82.39
Voxel R-CNN [6]	**90.90**	81.62	77.06	**94.85**	88.83	**86.13**
PV-RCNN++ [7]	90.14	**81.88**	**77.15**	92.66	88.74	85.97

5 Conclusion and Perspectives

This paper considered several methods for detecting objects in point clouds, including PointPillars, CIA-SSD, Voxel R-CNN, and PV-RCNN++. PV-RCNN++ outperformed previous 3D detection algorithms on the open datasets KITTI and WAYMO, achieving new state-of-the art performance. In the moderate and hard 3D detection challenges, the PVRCNN++ method performed best. The comparison of 3D object detection methods is one of our goals to select the most suitable approach. PV-RCNN++ will be used in our future work for 3D object detection in indoor navigation thanks to its excellent results. This will help blind people locate all the objects which are present in their surroundings.

References

1. Sautier, C., Puy, G., Gidaris, S., Boulch, A., Bursuc, A., Marlet, R.: Image-to-lidar self-supervised distillation for autonomous driving data. arXiv (2022). https://doi.org/10.48550/arXiv.2203.16258
2. Luleci, F., Li, L., Chi, J., Reiners, D., Cruz-Neira, C., Catbas, F.N.: Structural health monitoring of a foot bridge in virtual reality environment. Procedia Struct. Integr. **37**, 65–72 (2022). https://doi.org/10.1016/j.prostr.2022.01.060
3. Ismail, H., Roy, R., Sheu, L.-J., Chieng, W.-H., Tang, L.-C.: Exploration-based SLAM (e-SLAM) for the indoor mobile robot using lidar. Sensors **22**(4), 4 (2022). https://doi.org/10.3390/s22041689
4. Lang, A.H., Vora, S., Caesar, H., Zhou, L., Yang, J., Beijbom, O.: PointPillars: fast encoders for object detection from point clouds (2019), pp. 12697–12705. Consulté le: 1 octobre (2022). https://openaccess.thecvf.com/content_CVPR_2019/html/Lang_PointPillars_Fast_Encoders_for_Object_Detection_From_Point_Clouds_CVPR_2019_paper.html
5. Zheng, W., Tang, W., Chen, S., Jiang, L., Fu, C.-W.: CIA-SSD: confident IoU-aware single-stage object detector from point cloud. In: Proceedings of the AAAI Conference on Artificial Intelligence, vol. 35, no. 4, Art. no. 4 (2021). https://doi.org/10.1609/aaai.v35i4.16470
6. Deng, J., Shi, S., Li, P., Zhou, W., Zhang, Y., Li, H.: Voxel R-CNN: towards high performance voxel-based 3D object detection. In: Proceedings of the AAAI Conference on Artificial Intelligence, vol. 35, no. 2, Art. no. 2 (2021). https://doi.org/10.1609/aaai.v35i2.16207
7. Shi, S., et al.: PV-RCNN++: Point-voxel feature set abstraction with local vector representation for 3D object detection. arXiv (2022). https://doi.org/10.48550/arXiv.2102.00463

Check for updates

Adaptive Neural Fuzzy Inference System (ANFIS) in a Grid Connected-Fuel Cell-Electrolyser-Solar PV-Battery-Super Capacitor Energy Storage System Management

Ismail Elabbassi[1(✉)], Naima Elyanboiy[1], Mohamed Khala[1], Youssef El Hassouani[2], Omar Eloutassi[1], Sara Teidj[3], Monsif Ben Messaoud Layti[4], and Choukri Messaoudi[1]

[1] Optoelectronics and Applied Energy Techniques, Faculty of Science and Technology, Moulay Ismail University of Meknes, Errachidia, Morocco
ism.elabbassi@edu.umi.ac.ma

[2] New Energies and Materials Engineering, Faculty of Science and Technology, Moulay Ismail University of Meknes, Errachidia, Morocco

[3] Faculty of Science and Technology, Moulay Ismail University of Meknes, Errachidia, Morocco

[4] New Technology Trends (NTT), ENSATé, Abdelmalek Essaadi, Tetouan, Morocco

Abstract. The machine learning methods for the hybrid micro-grid management is claimed to be more efficient for a smart storage system management, also to minimize the greenhouse emission through the integration of the fuel cell as an alternative source in micro grids. The proposed method is the Adaptive Neural Fuzzy Inference System (ANFIS) that combine Fuzzy Logic (FL) and Artificial Neural Network (ANN) and contribute to improve the storage system management for micro-grid. Therefore, the available fuel cell power can be used to serve the critical loads during battery-SC-PV-Grid shortage condition also to charge the battery in the Normal State of Charge (SOC). The results show that the ANFIS artificial intelligence method is a good management process during production, consumption and storage of energy.

Keywords: Machine learning · Adaptive neural fuzzy inference system · Storage system · Energy storage system management · Fuel cell

1 Introduction

The demand for electricity is growing as the population increases around the world, besides the production of fossil resources is jeopardising the earth by producing poisonous gas like sulphur and nitrogen oxides that deteriorate the ozone layer [1]. The need to develop sophisticated and nature-friendly intelligent energy systems is essential; currently the priority in the management of electrical energy systems is not limited to meeting the needs of the population but also to minimize the emission of greenhouse gases. Greenhouse effect and hydrogen consumption reduction in hybrid systems [2]. Fuel cell can be indeed a solution for the intermittency of renewable energy hence its

Y. Farhaoui et al. (Eds.): ICAISE 2022, LNNS 635, pp. 138–143, 2023.
https://doi.org/10.1007/978-3-031-26254-8_21

great potential and cleanliness [3]. The main aim of this study is to propose an artificial intelligence method to manage the electrical energy in a micro-grid system at the level of consumption, production and storage to reduce energy consumption from fossil fuels, therefore strengthening the autonomy of the hybrid micro-grid system by extending the lifetime of the battery because the coupling batteries and super-capacitors in the form of hybrid energy storage systems seems to be the most appropriate solution to increase the lifetime of batteries [4].The remaining part of this work has been arranged as follow: Sect. 2 presents the proposed method based on Artificial neural network and fuzzy logic, Sect. 3 provides the aim results and discussion of the proposed method for micro-grid storage system management, Sect. 4 concludes the summary of the whole paper along some future prospects.

2 Mathematical Modelling and Description of System Components

The management system depends mainly on the behaviour of the load, the SOC of the battery, the SOC* of the super capacitor and the level of the hydrogen H2 tank, the power grid, the output of the PV generator and the fuel cell. The following section describes the mathematical model of each component.

2.1 PV Model

The photovoltaic module with a single diode is one of the elements which constitute the solar cell; this is defined as a source of light current Iph generated according to solar irradiation see Fig. 1. The solar cell output current expression used in is showed in (1):

$$I = I_{ph} - I_0\left[\left(e^{\left(\frac{V+R_s}{\alpha * V_T}\right)}\right) - 1\right] - \left(\frac{V + I * R_s}{R_{SH}}\right) \tag{1}$$

where: I is the generated current, Iph is the light current according to the solar irradiance, I0 denotes the reverse saturation current, V is the voltage, RS and RSH are the series and parallel resistance respectively, V0 denotes the output voltage, a is the ideality factor of the diode generally taken between {1, 2, 3} and VT is the thermal voltage [5].

2.2 Fuel Cell

Proton exchange membrane FC type feed by Hydrogen as a fuel is considered. There are several advantages of using FCs as a power source such as their high efficiency, compact size, silent or low noise operation and friendliness [6]. For a single FC, the overall voltage will be expressed as follows [7]:

$$V_{fc} = E_{Nernst} - V_{act} - V_{ohmic} - V_{con} \tag{2}$$

where E_{Nernst} is the raw output voltage generated from the chemical reaction, is expressed as follows [8]:

$$\mathrm{E_{Nernstt}} = 1.229 - 8.46 \times 10^{-4}(\mathrm{T} - 298.15) + 4.31 \times 10^{-5}\mathrm{T}\left(\ln(\mathrm{PH_2}\sqrt{\mathrm{PO_2}}\right)) \tag{3}$$

where T is temperature, P_{O2} and P_{H2} are the oxygen and hydrogen pressures respectively, V_{act} represents the activation over potential voltage, V_{ohmic} represents the ohmic voltage drop.

2.3 Pack Battery

The modelling is done for a modified and improved model of the Lithium-ion battery [9, 10].This improved model is used to maximize the accuracy of the model [11]. The lithium battery equivalent *SOC* is estimated by measuring the discharging current of a battery and integrating them over time [12]. The *SOC* is calculated by the following equation:

$$\mathrm{SOC(t) = SOC_0(t_0)}\frac{\eta}{\mathrm{C_n}}\int_{\mathrm{t_0}}^{\mathrm{t}}\mathrm{I(t)dt} \tag{4}$$

where $SOC_0(t_0)$ is the original state of charge, η denotes the Colombian efficiency, C_n is the nominal capacity, *I(t)* is the instantaneous discharge current of the battery.

2.4 Super Capacitor Model

The super-capacitor model is constructed using Debye polarization equivalent circuit, the electrical behaviour of the Debye polarization equivalent circuit model can be expressed as following equations [13]:

$$\frac{dV_{DA}}{dt} = -\frac{1}{R_L C_{DA}} V_{DA} + \frac{1}{R_L C_{DA}} V_N \tag{5}$$

$$\frac{dV_{DA}}{dt} = -\frac{1}{R_L C_N} V_{DA} + \frac{1}{R_L C_N} V_N - \frac{1}{C_N} I \tag{6}$$

$$V_T = V_N - IR_{ESR} \tag{7}$$

where RESR is the equivalent series resistor, RL is the leakage current resistance, CDA is the Debye adsorption capacitance, and CN is the nominal capacitance [14].

3 Results and Discussion

In the present investigation, a micro-grid consisting of PV, battery-super capacitor for the energy store management study under the variable climate conditions i.e. irradiance of 0 to 1000 W/m2 and the fixed temperature of the PV array at 26 °C [7]. The maximum power of elements constituting the micro-grid in this work for the simulation are: the PV generator with 14 KW, the battery with 4.5 KW, super-capacitor with 4.2 KW and the Electrolyser can consume up to 5.7 KW to produce during its production the hydrogen. When the PV generator output and the demand of the load are close to each other thus, P-demand (Is the difference between the PV generator output and the load demand) is very low, SOC is high, SOC* and H_T_level are all in Medium in which case the Battery supplies far more than the fuel cell to power the relevant load.

Once the charge reaches the maximum the battery can supply, the ANFIS Controller stops the charge of the storage system. Figure 2 illustrates the key findings of the study, which is supported by the Hybrid Artificial Intelligence Method as a combination of

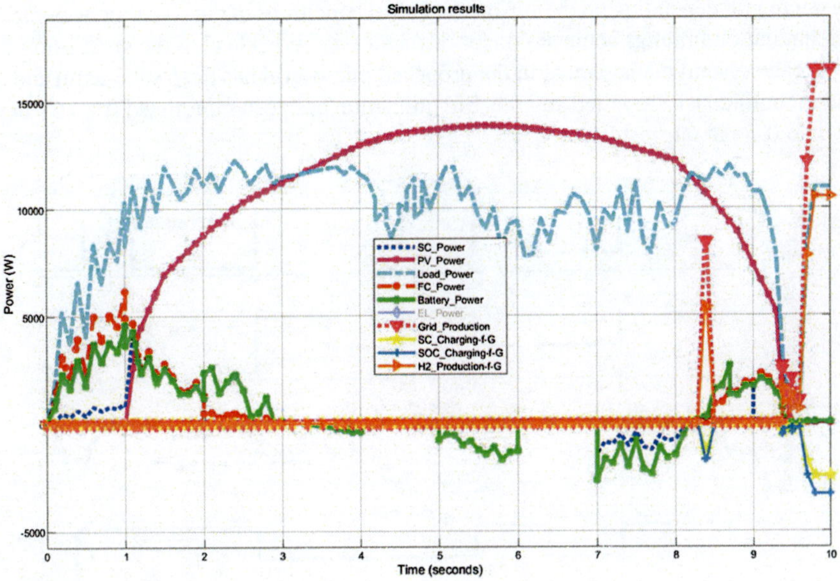

Fig. 1. System simulation results obtained from ANFIS energy management system

Artificial Neural Network and Fuzzy Logic [15], in view of ensuring a good management of electric power in a micro-grid system.

The simulation of the results is carried out over a time range of 10 s under conditions as shown in Fig. 1, it is for this reason that it does not matter the duration of simulation. As illustrated in Fig. 1, if SOC, SOC*, H2–tank-level are in the High state and the PV array has no power production, and there is a load demand from the load between t = 0 s and t = 1 s, in this case the power will be provided by the fuel cell, battery and the super-capacitor. As long as the PV array provides non-zero power, the contribution of the micro-grid elements to supply the load will be reduced from t = 1s to t = 2 s. Between t = 2 s and t = 3 s the SOC, SOC* and H2–tank-level are in medium state as is the load at maximum load demand in parallel with large PV production; in that case the ANFIS controls the Battery and the Fuel Cell in view of the difference between the demand of the Load and the lack of production of the PV generator. From t = 3 s to t = 4 s the SOC, SOC* maintains the same conditions as the preceding state but the H2–tank-level becomes High with 90%; in this case the PV production is enough to meet the load requirement, as quickly as an excessive load occurs in that case ANFIS controller controls the charging of the Battery to keep it in good condition, if the components constituting the micro-grid are in a low state as well as the PV array has no production, that is can be in night, and the charge has a significant need; in that case the power distribution network will intervene to ensure the load need and the charge of the storage system the proposed ANFIS control as illustrated in Fig. 2 in the range of t = 9.4 s and t = 10 s. Based on the above results shown in Fig. 2 under the ANFIS controller inputs adopted in this work as show in Fig. 2, it can be concluded that the neural adaptive fuzzy inference system is capable of providing good power management in a micro grid depending on the various technical conditions of the constituent elements. The simulation findings achieved under different conditions lead to the conclusion that the storage system management in the proposed micro-grid has been successful and the load is continuously supplied and the grid utilization has been minimized from 6 s in [6] down to 0.7 s in our results.

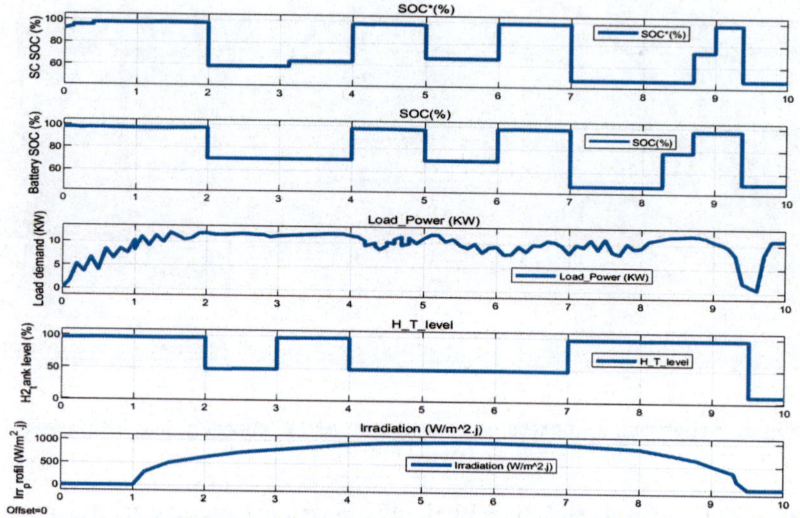

Fig. 2. ANFIS controller inputs adopt in this work

4 Conclusion and Future Work

The ANFIS method adopted in this work to manage the electrical energy between its production, consumption and storage in a micro-grid with a variable load of 12 KW power, under variable irradiation from 0 W/m2 to 1000 W/m2 is satisfactory in terms of ensuring the state of charge of the battery at the normal. The results show that the ANFIS artificial intelligence method is good management processes that contribute to improve the storage system management for micro-grid system.

The next works relate to:

- The integration of another energy sources in the micro-grid taking into account the fuel cell optimization using the Artificial Neural Networks method,
- Monitoring of energy storage system in real time using the Internet of Things.

References

1. Rezk, H., et al.: Comparison among various energy management strategies for reducing hydrogen consumption in a hybrid fuel cell/super-capacitor/battery system. Int. J. Hydrogen Energy **46**(8), 6110–6126 (2021)

2. Kouchachvili, L., Yaïci, W., Entchev, E.: Hybrid battery/super-capacitor energy storage system for the electric vehicles. J. Power Sour. **374**, 237–248 (2018)
3. Li, Y., Yang, J., Song, J.: Structure models and nano energy system design for proton exchange membrane fuel cells in electric energy vehicles. Renew. Sustain. Energ. Rev. **67**, 160–172 (2017)
4. Diana, L., et al.: Battery-super capacitor energy storage systems for electrical vehicles a review. Energies **15**, 5683 (2022)
5. Jha, V.: Comprehensive modeling and simulation of PV module and different PV array configurations under partial shading condition. Iran. J. Sci. Technol. Trans. Electr. Eng. **46**, 503–535 (2022). https://doi.org/10.1007/s40998-022-00494-5
6. Apiyo, E., Maina, C., Ngoo, L.: Modelling and energy management strategy in a grid connected solar PV-battery energy storage system. J. Agric. Sci. Technol. **20**(3), 4–15 (2021)
7. Chippada, D., Reddy, M.: Mathematical modelling and simulation of energy management in smart grid. Int. J. Smart Grid Clean Energy **9**(4), 746–755 (2020)
8. Kayla Kilgoa, M., et al.: Metal leaching from Lithium-ion and Nickel-metal hydride batteries and photovoltaic modules in simulated landfill leachates and municipal solid waste materials. Chem. Eng. J. **431**(2), 133825 (2022)
9. Zubi, G., Dufo- Lopez, R., Carvalho, M., Pasaoglu, G.: The lithium-ion battery: state of the art and future perspectives. Renew. Sustain. Energ. Rev. **89**, 292–308 (2018)
10. Bracco, S., Delfino, F., Trucco, A., Zin, S.: Electrical storage systems based on sodium/nickel chloride batteries: a mathematical model for the cell electrical parameter evaluation validated on a real smart microgrid application. J. Power Sources **399**, 372–382 (2018)
11. Chen, W., et al.: A review of lithium-ion battery for electric vehicle applications and beyond. Energy Procedia **158**, 4363–4368 (2019)
12. Dickson, N., et al.: State of charge estimation for lithium-ion batteries using model-based and data-driven methods: a review. IEEE Access **7**, 136116–136136 (2019)
13. Yüksek, G., Muratoğlu, Y., Alkaya, A.: Modelling of super-capacitor by using parameter estimation method for energy storage system. Adv. Eng. Days **1**, 29–31 (2022)
14. Benmouna, A., et al.: Efficient experimental energy management operating for FC/battery/SC vehicles via hybrid artificial neural networks-passivity based control. Renew. Energy **178**, 1291–1302 (2021)
15. Arcos-Aviles, D., Pascual, J., et al.: An energy management system design using fuzzy logic control: Smoothing the grid power profile of a residential electro-thermal microgrid. IEEE Access **9**, 25172–25188 (2021)

AI Applications in Smart Cities Between Advantages and Security Challenge

Mohamed Abdedaime[1](✉), Chaimaa Hazman[3], Ahlam Qafas[1], Mounir Jerry[2], and Azidine Guezzaz[3]

[1] National School of Business and Management, Ibn Tofail University, Kenitra, Morocco
abdedaimemohamed@gmail.com
[2] Faculty of Economics and Management, Ibn Tofail University, Kenitra, Morocco
[3] Higher School of Technology Essaouira, Cadi Ayyad University, Marrakech, Morocco

Abstract. In the last decades, smart cities are growing rapidly as a potential solution to several urban development difficulties such as traffic congestion, infrastructure development, pollution, economic growth, and a deficit in natural resources. This wave is being driven by fast-advancing technology and the increase in people's expectations. Nevertheless, in this work, we intend to examine the opportunities and challenges that cities may encounter while embracing Artificial Intelligence (AI). Consequently, to better understand the importance of this technology, we will discuss in this paper the roles and the applications area of AI in smart cities. The purpose of this paper is to provide a theoretical overview of AI solutions for smart cities that will assist these cities in developing better infrastructure and conditions to meet their everyday demands while also improving the quality of life in urban spaces. Hence, we will study some applications of AI solutions in the smart city to illustrate the benefits of this framework through an exemplary case study. In this article, we look particularly at the most recent research initiatives aimed at AI analytics in a smart-city environment. Indeed, the main aim of this work is to examine the opportunities and challenges that cities may face when adopting these technologies.

Keywords: AI · Smart cities · Privacy · IoT · Security · Machine learning

1 Introduction

In recent decades, there has been a major increase in urbanization, which has increased both the share of the world population living in cities and the share of global Gross Domestic Product (GDP) created in cities. Therefore, by 2050, more than 68% of the world population, or 9.7 billion people will live in urban areas [1]. This urban growth generates a variety of issues, namely air pollution, mental health issues, crime, loss of public space, land consumption, pollution, deteriorated infrastructure, and recession [2–4]. Simultaneously, these issues will present considerable barriers to smart cities and growth. In this context, AI, Big Data and IoT opportunities have refocused attention on the role of smart cities in enhancing smart life conditions such as transportation, smart

Y. Farhaoui et al. (Eds.): ICAISE 2022, LNNS 635, pp. 144–155, 2023.
https://doi.org/10.1007/978-3-031-26254-8_22

communities, access to services and technology, security, wastewater recycling, and better living conditions. However, smart cities are becoming more ubiquitous in politics. Furthermore, these solutions may be exploited in many aspects of city life, including urban issue resolution, natural resource management, real-time data management, and crime prevention, among others. Moreover, these new technologies may provide new social, ethical, and security issues. Furthermore, rather than extending the literature, this analysis aims to attract the attention of researchers to AI applications in smart cities. Also, this research tends to analyze whether smart cities could assume a higher quality of life as well as to investigate the smart AI solutions that could influence the smart city's implementation and efficiency. To have a comprehensive overview of the AI paradigm in a smart-city context, this article examines the latest recent research in AI smart-city applications. First, we presented AI- smart-cities solutions. Second, we discussed the advantages of AI solutions for future smart cities. In addition, we aim to highlight the challenges that may limit the effectiveness of AI applications in any smart-city scenario. Finally, open topics are addressed to provide research suggestions for the future.

2 Research Methodology

In this research, we adopted a systematic literature review approach to answer the research question. In fact, this method is considered as a tool for understanding state-of-the-art in the research that are related to technology. As result, this methodology is also considered as the most used for synthesizing knowledge since it is based on a rigorous and transparent process to identify studies. In addition, it allows us to identify critical knowledge gaps by highlighting the discrepancy between what is currently known, what needs to be known. Finally, we chose to adopt this research methodology to answer our research question: what are the opportunities and the challenges that cities may face when adopting AI solutions?

3 Background and Related Works

3.1 Smart City: An Emergent Paradigm

Transforming a city into a successful smart city has become an aspiration for many modern cities all over the world. Moreover, the population growth and urbanization trends are not only accelerating global resource and environmental degradation, as well as food, water, and energy insecurity, but they are also aggravating social inequities and turning our cities largely ungoverned [5].

In fact, a smart city was described as a city that employs digital data and technology to achieve efficiency to promote economic development, quality of life, and sustainability [6]. Subsequently, it was comprehended that digital data and technology are not the only key ingredients in the making of smart cities [7, 8]. Hence, this smart city definition is employed in this study because this new approach focuses on urban smartness not only in terms of technology but also other essential characteristics that constitute smart city transition capability [9]. Additionally, the factors affecting smart cities are still an understudied area of research. Besides theorical works, there is limited empirical evidence in the literature on how we could assume a successful transformation of cities into smart spaces by developing AI applications. As a result, the smart city is characterized as an infrastructure network constructed on very diversified technologies.

In this context, the smart city concept has emerged as a first attempt to use the enormous potential of ICTs to assist the development of public values via participatory democracy [10], smart cities, on the other hand, include the substantial and comprehensive application of ICT to various domains of city activity, necessitating the identification of specific aspects of cities for evaluation with a ranking system [11]. However, it is commonly accepted that due to [11]'s six primary aspects of smart dimensions (smart economy, smart people, smart governance, smart mobility, smart environment, and smart living) are the most common indicators of smart cities. In our case, the focus of this work will be on the application of technology in different city domains. As stated, AI is a critical technology for the emergence of smart cities. Consequently, it's interesting to study the roles of AI applications in the development of smart cities.

It's clear, that the concept of the smart city is gaining popularity among research labs, universities, governments, politicians, and ICT corporations [12–15]. It is a highly interdisciplinary field, which makes the concept of smart cities quite ambiguous and difficult [12, 16, 17]. In contrast, there is no common measure for assessing a smart city's level of smartness [18, 19]. Meanwhile, a review of the literature reveals that the smartness level of a smart city may be classified into six stages (see Fig. 1): ecological dimension, social dimension, governmental dimension, economic dimension, urban dimension, technological dimension. Consequently, we could confirm that is critical to exploit the enormous capabilities of ICT to support unique smart concepts and methods for tackling the inherent major issues and problems of modern cities, which has resulted in the development of smart cities.

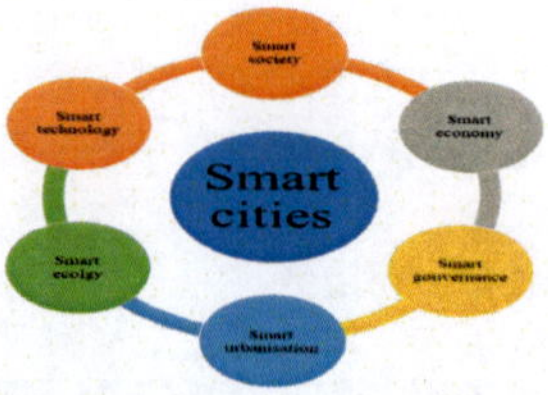

Fig. 1. Smart city's dimensions

[20] defines the smart city as a city that monitors and integrates conditions of all of its critical infrastructures, including roads, water and communications, can better optimize its resources, plan its preventive maintenance activities, and monitor security aspects while maximizing services to its citizens. Therefore, the United Nations International Telecommunication Union (ITU) used the definition: "A smart sustainable city employs information and communication technologies to improve the quality of life, the efficiency of urban activities and services, and the attractiveness of the city. It addresses the economic, social, and environmental demands of future generations". On the contrary,[21] has stated that there is no unique and shared definition of the smart city.

To summarise, smart cities are a new paradigm that incorporates varied communications infrastructures and devices producing a huge volume of data based on Big data, IoT and Artificial intelligent applications and solutions. Indeed, a smart city development not only entails the adoption of intelligent computing technologies in citizen life and city governance but also depends on human capital and support from institutions. In addition, smart city solutions are focused in the fields of healthcare, education, the environment, energy, and transportation. Their ultimate goal is to improve citizens' quality of life by delivering intelligent services.

3.2 AI as a New Support for Smart City

The use of artificial intelligence (AI), which has had a resurgence in recent years due to progress in technologies such as natural language processing, machine learning, computer vision, and robotics, is an emerging trend in the usage of ICT. Thus this is further boosted by the rise in available data, i.e., big data, as well as the significant gain in computing power [22, 23].

These AI applications provide automation and efficiency for smart city networks and services, such as autonomous cars for transportation and criminal identification and tracking for public safety. However, there is the possibility of negative impacts during the development and deployment of AI for smart cities, such as service discrimination, privacy, legal, and ethical difficulties [23, 24]. Furthermore, by building a framework of principles for AI governing in smart cities, we aim to better understand and address the difficulties of AI regulation in smart cities. As a consequence, this can assist academics and practitioners in comprehending the principles of regulation smart city AI applications and examining or developing public policies accordingly.

Research in AI topics is not a new field of study. Its origins may be traced back to the 1950s [25–27] when it began as an exploration into the nature of intelligence. In theory,this technology is becoming increasingly important in many aspects of smart cities today [28, 29]. Hence, we identified two purposes of definitions presented to characterize the idea of artificial intelligence from our review of the literature. Indeed, the first wave [30–32] defines artificial intelligence as the replication of human intelligence and aptitudes to develop intelligent machines. The second wave [33, 34] considers AI as a simulation of human intelligence vs computer capacities to handle controversial problems. From this perspective, [35] defined AI as the study of creating intelligent entities that can execute tasks and think like humans. Similarly, [32] claims that AI is a new field that "involves employing approaches based on human intelligence to address complicated issues" [31]. Due to the diversity of artificial intelligence definitions, there is no unique globally recognized definition of AI [36].

With regards to smart technology, AI refers to the machines or software that attempt to think and behave humanly and rationally [23]. So, the aim of AI not only includes the successful fidelity to human performance, but also the achievement of rational or ideal performance. Hence AI has largely been applied to reasoning, problem-solving, knowledge representation, natural language processing, perceptions (e.g., face recognition, speech recognition, computer vision), and motion manipulation (e.g., robotics) [37].

4 AI and Smart Cities: Applications and Benefits

After providing this theorical framework, we claim to propose an integrated approach for smart city AI governance that addresses AI solutions. Our approach contributes to the research on smart cities and artificial intelligence field. In fact, the study of dimensions of a smart city and the components of AI systems permit to highlight the important of AI on smart cities efficiency and a better understanding on interactions between AI solutions and smart cities implementation. In this perspective, in next sections, below we will discuss each dimension of smart city and its relation with AI solutions.

4.1 Smart Transport

This innovation is employed in the mobility field to develop new models of movement solutions, ensure intelligent vehicle tracking, provide smart parking, and improve security levels through the use of road sensors [38, 16], manage traffic, and ensure a smart display on road conditions to reduce pollution, optimize people mobility, and safety [39]. Additionally, mobility services must include not just traditional public transportation, but also, shared bicycles, cars, and Parking as well as electric charging stations. Therefore, AI is used in transportation systems to solve several problems and ensure intelligent transport [39, 40]. For example, AI is used to: develop autonomous cars, predict traffic risks and congestion zones [40, 41], ensure intelligent parking management, develop intelligent solutions to reduce accidents, etc.

4.2 Smart Healthcare

Such as technology allows for remote monitoring of patients (24/7) [42], linking doctors, patients, and nurses via smart device, and each entity can roam freely, collecting data about patients' states in real-time, locating patients and improving their safety, ensuring management of medical emergencies, management of transfusion information management, and real-time health monitoring [42, 43]. Moreover, the use of AI in healthcare provides several opportunities to solve real-world healthcare problems [44]. For example, it can ensure a correct diagnosis, a precise analysis, an exact prediction of diseases (such as cardiovascular disease, cardiac events, etc.), and an efficient treatment of hospital data [45, 46]. Finally, this technology can also help to determine an intelligent analysis of complex medical data to provide an effective prediction of diseases.

4.3 Urban Environment

Nowadays, cities consume more than 75% of global energy and reject more than 80% of carbon dioxide into the atmosphere. Also, the world's 600 largest metropolises are expected to contribute 65% of global growth. In this context, the AI technology can be used to control natural resource systems, temperature, air quality, environment, humidity, weather, rain, carbon dioxide level and harmful gases, and waste management to improve the urban environment. For instance, AI might contribute to domestic waste collection efficiency. Also, modern technologies of machine learning could enhance and expedite data collection. As result, when implemented in this method, AI technology will result in lower operating costs and a better response to any environmental issues.

4.4 Smart Home

The use of AI makes homes smarter and able to anticipate our actions and our preferences. Clearly, this technology allows, for example, to inform a resident when it is time to take their medication, alert the hospital if the resident has fallen, close the water, or turn off the oven, etc. Hence, AI will render homes intelligent and able to adapt to the needs of the habitants.

4.5 Smart Government

AI is used by governments to provide intelligent and personalized services to ensure a smart public administration [47–49]. Thus, the use of AI in governments will improve, for example, participation, transparency, efficiency, accountability, and inclusion to ensure an effective government.

4.6 Smart Learning

AI integrated into different education systems will help to improve the education level, make classrooms smart, support teachers, and students with the use of virtual assistants that can play the role of "tutor" for students, automate the different administrative tasks and provide personalized help [50]. Since the COVID-19 crisis, E-learning has emerged as an alternative to traditional classrooms; as a result, AI applications in education are being developed to not only design solutions for future crises but also to assist the aspirations of the new digital generation that has grown up with technology.

4.7 Smart Security

The use of AI can improve the security level in different areas of smart cities. This technology integrated into different objects can provide capabilities to monitor urban areas through the analysis of movements and the prediction of crimes, ensure monitoring of device networks to detect cyberattacks and to improve cybersecurity [51], or secure physical sites such as homes or public buildings.

5 Challenges of AI in Smart Cities

The increase in urban population generates more challenges for city managers. In fact, projections indicate that more than 70% of the world's population will be urbanized in the next decade. Consequently, this urbanization will occur new and increasing issues in the management of public service. Whereas, the importance of AI and the potential that it provides for communities, smart cities may confront a variety of AI-related issues. Indeed, from our review of the literature, we identified many societal, regulatory, and moral questions linked with AI applications.

5.1 Social Challenges

AI allows us to simulate human behavior and accomplish a variety of complex tasks. However, various researchers have affirmed the issue of unemployment, in which machines will replace workers in certain sectors [31]. Due to its advantages in reducing the cost of production within a high quality, in the next decade the work will tend to use machines more and more in different sectors including home activities and hospital services.

5.2 Economic Challenges

The development of AI applications requests a huge investment and a sophistical infrastructure. According to AI analysts, investments in AI industry are estimated to be valued at more than 1500 billion dollars by 2030, up from 62 billion dollars in 2022. Whatever the smart city market forecasts are, one thing is certain: they all expect massive gains. Meanwhile, poor countries and small companies are excluded to reach these attractive markets. In fact, only big companies are struggling to develop and investing in this industry. For instance: Amazon.com (AMZN), Microsoft (MSFT), Tesla (TSLA), Apple (AAPL), Meta Platforms (META), IBM (IBM), and Alphabet (parent company of Google) (GOOGL). Consequently, integrating AI solutions for smart cities impose enormous investments that are not within reach of most cities's manger.

5.3 Legal Issues

The great majority of AI solutions are concerned about individual freedoms and pose serious threats to information security. In addition, with so much public data still to be valued, intelligent cities face several problems. Hence, rather than its importance and presence in the daily life of citizens, tell now there is already no formal structure that organizes the roles and functions of AI [52]. For example, [53] had asked if "Will an autonomous automobile cause an accident"? Thus, [54] discovered that as artificial intelligence advances, robots will become future citizens of smart cities. Consequently, governments are called to anticipate those "digital citizens" and their impacts in the real life.

5.4 Ethical and Privacy Challenges

These challenges are among the most difficult for cities to address [55, 56]. Indeed, AI has the potential to: imitate the human brain, analyze human behavior [57], track people and perform facial recognition, or even create people's dependency on AI applications [31], all of which affect the confidentiality of human privacy and raise several ethical concerns. Undoubtedly, AI may be used to govern individuals and communicate their sensitive data. As a result, one of the crucial issues that AI must address is: "To what extent is privacy protected?" To result from numerous sensitive data sets might be acquired by sensing devices and shared without authorization with other sensors or third parties.

5.5 Security Issues

According to several authors [58–60], 70% of AI devices in a smart city are susceptible to attacks due to sufficient vulnerabilities such as free permissions or access, inadequate software protections, or weakly encrypted communication protocols. In addition, [60–63] stated that the implementation of smart cities networks depends on efficiency of data quality techniques in data preparation stage and its transformation to empowering and making a robust security approach for smart cities solutions. However, recognizing that these challenges represent a huge challenge, it is important to note that future development of AI can bring more complicated challenges. In this context, several authors [52, 55] indicated that with the next evolution of AI that will be a strong AI, we will witness several existential risks [57], where [32] illustrates that: "A powerful AI could overwhelm any human resistance" [32]. We support that AI applications for smart cities are being implemented or will be deployed in numerous parts of a smart city. However, large-scale AI technology deployment, such as data-driven prediction systems, may have unanticipated negative social implications.

In resume, in various scenarios, activities of individual users e.g., individual users' behaviors, such as their purchasing and preference patterns, health information, online and offline transactions, and their images, are utilized as training data. Clearly, this raises concerns about the use of privacy and the security of individual data obtained, stored, processed, shared, and produced from AI applications in smart cities. Additionally, to compliance with the core principles, the design of AI solutions also needs to follow the specific principles of data privacy, algorithm accountability, and transparency.

6 Conclusion and Future Works

The new hyper-connected environment made possible by AI within the smart cities improves efficiency and reduces costs, but it also creates a situation in which city services have no awareness of which devices to secure. In this paper, after examining smart city dimensions and definitions, we had highlighted AI smart city applications and their challenges. Hence, we released that personal data privacy has emerged as significant challenge. Although, AI solutions raise several ethical and technological considerations. As result, governments and corporations have to develop a consolidated approach to the design and deploy their AI solutions in order to deal with issues and potential negative consequences.

Future research may focus on identifying governance structures for smart city AI applications and integrating them into our conceptual generated framework to gain a better understanding of smart city AI governance. In future work, three key study pathways are proposed. First, we will expand our review of the literature to get a better understanding of how AI applications are used in smart cities. Second, we believe it is necessary to formulate a new paradigm for understanding how to get the most out of artificial intelligence solutions to ensure that these technologies are utilized responsibly and beneficially for city's citizens.

References:

1. United Nations, Department of Economic and Social Affairs, Population Division (2019). World Urbanization Prospects 2018: Highlights (ST/ESA/SER.A/421). https://population.un.org/wup/Publications/. Accessed 14 Aug 2022
2. Chourabi, H., Nam, T., Walker, S., et al.: Understanding smart cities: an integrative framework. In: Proceedings of the 45th Hawaii international conference on system sciences. Washington, DC. Computer Society, IEEE (2012).
3. De Paz, J.F., Bajo, J., Rodríguez, S., Villarrubia, G., Corchado, J.M.: Intelligent system for lighting control in smart cities. Inf. Sci. **372**, 241–255 (2016)
4. Hui, T.K., Sherratt, R.S., Sánchez, D.D.: Major requirements for building smart homes in smart cities based on Internet of Things technologies. Futur. Gener. Comput. Syst. **76**, 358–369 (2017)
5. Yigitcanlar, T., Butler, L., Windle, E., Desouza, K.C., Mehmood, R., Corchado, J.M.: Can building "artificially intelligent cities" safeguard humanity from natural disasters, pandemics, and other catastrophes? An urban scholar's perspective. Sensors **20**(10), 2988 (2020)
6. Mora, L., Bolici, R., Deakin, M.: The first two decades of smart-city research: a bibliometric analysis. J. Urban Technol. **24**(1), 3–27 (2017). https://doi.org/10.1080/10630732.2017.1285123
7. Araral, E : Why do cities adopt smart technologies? Contingency theory and evidence from the United States. Cities **106,** 102873 (2020). https://doi.org/10.1016/j.cities.2020.102873. ISSN: 0264–2751
8. Yigitcanlar, T., Foth, M., Kamruzzaman, M.: Towards post-anthropocentric cities: reconceptualizing smart cities to evade urban ecocide. J. Urban Technol. **26**(2), 147–152 (2019)
9. Yigitcanlar, T., Desouza, K.C., Butler, L., Roozkhosh, F.: Contributions and risks of artificial intelligence (AI) in building smarter cities: Insights from a systematic review of the literature. Energies **13**(6), 1473 (2020)
10. Moore, M.H.: Recognizing Public Value. Harvard University Press, Boston, MA (2013)
11. Giffinger, R., Fertner, C., Kramar, H., Kalasek, R., Pichler-Milanovic, N., Meijers, E.: Smart cities. Ranking of European medium-sized cities, Final Report. Vienna UT, Austria: Centre of Regional Science (2007)
12. Bibri, S.E., Krogstie, J.: Smart sustainable cities of the future: an extensive interdisciplinary literature review. Sustain. Cities Soc. **31**, 183–212 (2017)
13. Cocchia, A.: Smart and digital city: A systematic literature review. In: Smart City, pp. 13–43. Springer, Cham, Switzerland (2014). https://doi.org/10.1007/978-3-319-06160-3_2
14. Dohler, M., Vilajosana, I., Vilajosana, X., Llosa, J.: Smart cities: an action plan. In: Proceedings of Barcelona Smart Cities Congress, Barcelona, Spain (pp. 1–6), December 2011
15. Zhuhadar, L., Thrasher, E., Marklin, S., de Pablos, P.O.: The next wave of innovation review of smart cities intelligent operation systems. Comput. Hum. Behav. **66**, 273–281 (2017)
16. Hause, M., & Hummell, J. (2016, July). Making smart cities smarter–MBSE driven IoT. In INCOSE International Symposium (Vol. 26, No. 1, pp. 1675–1690)
17. Hollands, R.: Will the real smart city please stand up? Creative, progressive or just entrepreneurial? City: Anal. Urban Trends Culture, Theory, Policy, Action **12**, 303–320 (2008). https://doi.org/10.1080/13604810802479126
18. Albino, V., Berardi, U., Dangelico, R.M.: Smart cities: definitions, dimensions, performance, and initiatives. J. Urban Technol. **22**(1), 3–21 (2015)
19. Balakrishnan, A.: Cloud security concerns in 2018: Data breaches, security misconfigurations, AI and botnets. Retrieved October 25, 2019 from (2018). https://www.imperva.com/blog/2018-cloud-security-concerns-data-breaches-security-misconfigurations-ai-and-botnets/

20. Hall, R.E., Bowerman, B., Braverman, J., Taylor, J., Todosow, H., Von Wimmersperg, U.: The vision of a smart sustainable city. Scitech Connect: US. Dept. Energy Off. Sci. Tech. Inf. (OSTI) **28** (2009). http://www.osti.gov/scitech/servlets/purl/773961. Accessed 8 Feb 2014
21. Abdedaime, M., Qafas, A., Jerry, M., Guezzaz, A.: A KNN-based intrusion detection model for smart cities security. In: Gupta, D., Khanna, A., Hassanien, A.E., Anand, S., Jaiswal, A. (eds.) International Conference on Innovative Computing and Communications. Lecture Notes in Networks and Systems, vol. 492. Springer, Singapore (2023). https://doi.org/10.1007/978-981-19-3679-1_20
22. Russell, S.J., Norvig, P.: Artificial Intelligence: A Modern Approach. Pearson Education Limited, Malaysia (2016)
23. Stoica, I., et al.: A Berkeley view of systems challenges for AI. arXiv preprint arXiv:1712.05855 (2017)
24. Stoyanovich, J., Abiteboul, S., Miklau, G.: Data, responsibly: fairness, neutrality and transparency in data analysis. In: Proceedings of the 19th International Conference on Extending Database Technology, Bordeaux, France (2016)
25. Kurzweil, R.: The Age of Spiritual Machines: When Computers Exceed Human Intelligence. Penguin, New York (2000)
26. Simon, H.A.: The sciences of the Artificial. MIT Press, Cambridge (1996)
27. Turing, A.M.: Intelligent machinery, a heretical theory. Philos. Math. **4**(3), 256–260 (1996)
28. Augusto, J.C., Nakashima, H., Aghajan, H.: Ambient intelligence and smart environments: a state of the art. In: Handbook of Ambient Intelligence and Smart Environments, pp. 3–31. Springer, Boston (2010). https://doi.org/10.1007/978-0-387-93808-0_1
29. Nigon, J., Glize, E., Dupas, D., Crasnier, F., Boes, J.: Use cases of pervasive artificial intelligence for smart cities challenges. In: Ubiquitous Intelligence & Computing, Advanced and Trusted Computing, Scalable Computing and Communications, Cloud and Big Data Computing, Internet of People, and Smart World Congress (UIC/ATC/ScalCom/CBDCom/IoP/SmartWorld), 2016 Intl IEEE Conferences, pp. 1021–1027. IEEE, July 2016
30. Cath, C., Wachter, S., Mittelstadt, B., Taddeo, M., Floridi, L.: Artificial intelligence and the 'good society': the US, EU, and UK approach. Sci. Eng. Ethics **24**(2), 505–528 (2018)
31. Coppin, B.: Artificial Intelligence Illuminated. Jones and Bartlett Publishers, Sudbury (2004)
32. Yudkowsky, E.: Artificial intelligence as a positive and negative factor in global risk. Global Catastrophic Risks **1**(303), 184 (2008)
33. Aimé, X., Charlet, J., Maillet, D., Belin, C.: L'intelligence artificielle à la rencontre de la neuropsychologie: mémoire sémantique, vieillissement normal et pathologique. Gériatrie et Psychologie Neuropsychiatrie du Vieillissement **13**(1), 88–96 (2015)
34. Lauterbach, B.A., Bonim, A.: Artificial intelligence: a strategic business and governance imperative. NACD Directorship, 54–57 (2016)
35. David, B., Chuantao, Y., Zhou, Y., Xu, T., Zhang, B., Jin, H., Chalon, R.: SMART-CITY: problematics, techniques and case studies. In: Proceedings 8th International Conference on Computing Technology and Information Management (ICCM), vol. 1, pp. 168–174 (2012)
36. Cook, D.J., Augusto, J.C., Jakkula, V.R.: Ambient intelligence : technologies, applications, and opportunities. Pervasive Mob. Comput. **5**(4), 277–298 (2009)
37. Flasiński, M.: Introduction to Artificial Intelligence. Springer, New York (2016)
38. Behrendt, F.: Why cycling matters for smart cities. internet of bicycles for intelligent transport. J. Transp. Geogr. **56**, 157–164 (2016)
39. Tsaramirsis, G., Karamitsos, I., Apostolopoulos, C.: Smart parking: an IoT application for smart city. In: 2016 3rd International Conference on Computing for Sustainable Global Development INDIACom, pp. 1412–1416. IEEE, March 2016
40. Dias, G.M., Bellalta, B., Oechsner, S.: Predicting occupancy trends in Barcelona's bicycle service stations using open data. arXiv preprint arXiv:1505.03662 (2015)

41. Mathur, S., Modani, U.S.: Smart city-a gateway for artificial intelligence in India. In: 2016 IEEE Students' Conference on Electrical, Electronics and Computer Science SCEECS, pp. 1–3. IEEE, March 2016
42. Su, K., Li, J., Fu, H.: Smart city and the applications. In: 2011 International Conference on Electronics, Communications and Control ICECC, pp. 1028–1031. IEEE, September 2011
43. Boulos, M.N.K., Al-Shorbaji, N.M.: On the internet of things, smart cities and the WHO Healthy Cities. Int. J. Health Geogr. **13**, 10 (2014)
44. Zang, Y., Zhang, F., Di, C.A., Zhu, D.: Advances of flexible pressure sensors toward artificial intelligence and health care applications. Mater. Horiz. **2**(2), 140–156 (2015)
45. Patel, V.L., et al.: The coming of age of artificial intelligence in medicine. Artif. Intell. Med. **46**(1), 5–17 (2009)
46. Speight, P.M., Elliott, A.E., Jullien, J.A., Downer, M.C., Zakzrewska, J.M.: The use of artificial intelligence to identify people at risk of oral cancer and precancer. Br. Dent. J. **179**(10), 382 (1995)
47. Adadi, A., Berrada, M., Chenouni, D., Bounabat, B.: Ontology based composition of e-government services using AI planning. In: 2015 10th International Conference on Intelligent Systems: Theories and Applications (SITA, pp. 1–8. IEEE, October 2015
48. Pereira, G.V., Macadar, M.A., Luciano, E.M., Testa, M.G.: Delivering public value through open government data initiatives in a smart city context. Inf. Syst. Front. **19**(2), 213–229 (2016). https://doi.org/10.1007/s10796-016-9673-7
49. Schedler, K., Guenduez, A.A., Frischknecht, R.: How smart can government be? Discussing the barriers of smart government adoption. In: 2017 IPMN Conference, Shanghai, China (2017)
50. Holmes, W., Bialik, M., Fadel, C.: Artificial Intelligence in Education. Center for Curriculum Redesign, Boston (2019)
51. Sharbaf, M.: Artificial intelligence and cybersecurity. Bus. Strategy Artif. Intell. Econ. **5** (2018)
52. Bostrom, N.: Superintelligence. Dunod, Paris (2017)
53. Gurney, J.K.: Sue my car not me: products liability and accidents involving autonomous vehicles. U. Ill. JL Tech. & Pol'y, 247 (2013)
54. O'grady, M., & O'hare, G.: How smart is your city? Science **335**(6076), 1581–1582 (2012)
55. Frankish, K., Ramsey, W.M. (eds.): The Cambridge Handbook of Artificial Intelligence. Cambridge University Press, Cambridge (2014)
56. Hawking, S., Russell, S., Tegmark, M., Wilczek, F.: Stephen Hawking:\'Transcendence looks at the implications of artificial intelligence-but are we taking AI seriously enough?\'. The Independent, vol. 2014, no. 05-01, p. 9313474 (2014)
57. Brundage, M.: Taking superintelligence seriously: Superintelligence: Paths, dangers, strategies by Nick Bostrom (Oxford University Press, 2014). Futures **72**, 32–35 (2015)
58. Mehmood, Y., Ahmad, F., Yaqoob, I., Adnane, A., Imran, M., Guizani, S.: Internet-of-things-based smart cities: recent advances and challenges. IEEE Commun. Mag. **55**(9), 16–24 (2017)
59. Vattapparamban, E., Güvenç, İ., Yurekli, A. İ., Akkaya, K., Uluağaç, S.: Drones for smart cities: Issues in cybersecurity, privacy, and public safety. In: 2016 International Wireless Communications and Mobile computing Conference (IWCMC), pp. 216–221. IEEE, September, 2016
60. Azrour, M., Mabrouki, J., Guezzaz, A., Farhaoui, Y.: New enhanced authentication protocol for internet of things. Big Data Mining and Analytics. **4**(1), 1–9 (2021)
61. Guezzaz, A., Asimi, Y., Azrour, M., Asimi, A.: Mathematical validation of proposed machine learning classifier for heterogeneous traffic and anomaly detection. Big Data Min. Analytics **4**(1), 18–24 (2021)

62. Guezzaz, A., Benkirane, S., Azrour, M., Khurram, S.: A reliable network intrusion detection approach using decision tree with enhanced data quality. Secur. Commun. Netw. **2021**, 8, 1230593 (2021)
63. Azrour,M., Mabrouki, J., Guezzaz, A., Kanwal, A.: Internet of things security: challenges and key issues. Secur. Commun. Netw. **2021**, 11, 5533843 (2021)

An Analysis of NDN-IoHT Congestion Control Policies

Asmaa El-Bakkouchi[1](✉), Mohammed El Ghazi[1], Anas Bouayad[1], Mohammed Fattah[2], and Moulhime El Bekkali[1]

[1] Artificial Intelligence, Data Sciences and Emerging Systems Laboratory, Sidi Mohamed Ben Abdellah University, Fez, Morocco
asmaa.elbakkouchi@usmba.ac.ma
[2] IMAGE Laboratory, Moulay Ismail University, Meknes, Morocco

Abstract. The Internet of Health Things (IoHT) is gaining increasing attention from researchers because of its extensive use in healthcare. IoHT describes medical devices that aim to transfer health data securely and without loss between devices and health care personnel. Therefore, IoHT requires a sophisticated stack of protocols that allows communication between various types of devices with various service requirements. Named Data Net-working (NDN) is a potential internet architecture for IoHT because of its characteristics that make it the most appropriate solution for an IoHT environment. Since the data transmitted in IoHT is sensitive and non-sensitive data which is transferred concurrently and frequently, it increases the possibility of congestion and therefore packet loss. A congestion control policy that can effectively manage the traffic and adapt the transmission rate is needed. This paper presents an overview of different congestion control mechanisms proposed in NDN, discusses and evaluates their relevance for the IoHT domain. It also presents the main problems related to these mechanisms in IoHT.

Keywords: IoHT · NDN · Congestion control

1 Introduction

In recent years, the Internet of Things (IoT) has received great attention from researchers because of its objectives to connect several smart devices to the internet and share data between them [1]. IoT has been used in the health domain (IoHT) to make it smart by ensuring communication, exchanging and sharing of health data between different medical devices, applications and health workers [2]. As the transmitted data of healthcare circulate frequently in the network. This raises the risk of congestion and consequently packet loss, which is highly undesired in IoHT systems [3]. Nevertheless, IoHT uses the internet architecture IP for the transmission of health data which highlights the limitations of this architecture related to the heterogeneity of IoHT devices and the amount of data exchanged as well as to the requirements of each device which change from device to another between reliable unreliable transmission, high speed transmission or low speed

Y. Farhaoui et al. (Eds.): ICAISE 2022, LNNS 635, pp. 156–161, 2023.
https://doi.org/10.1007/978-3-031-26254-8_23

transmission. Therefore, IoHT needs an architecture that can handle the requirements of different devices and the amount of data exchanged. Recently, many architectures have been proposed for the future internet that change the communication paradigm by focusing on content rather than IP addresses. The characteristics of NDN architecture [4] have made it the most suitable architecture for the IoHT. NDN consists of three elements; Content Store (CS), Pending Interest Table (PIT) and Forwarding Interest Base (FIB) and exchanges two types of packets; interest and data. The consumer demands a content by sending an interest packet and the producer responds to this demand with a data packet. This content can be hidden in the cache of router to reply to future request for the same content without the need to go to the producer, which saves time especially in IoHT where time is very important. Network congestion is one of the main problems faced by researchers in every computer network evolution [5]. Since the IoHT exchanges and shares health data frequently this increases the risk of congestion. To deal with this problem, a congestion control policy that can effectively manage the traffic and adapt the transmission rate consequently is needed. This paper proposes an analysis of NDN congestion control mechanisms, their advantages and disadvantages, and their compatibility with the IoHT domain. It also presents the main problems related to these mechanisms in the IoHT domain. The remainder of the paper is organized as follows, Sect. 2 presents congestion in IoHT, Sect. 3 discusses the challenges related to IoHT in terms of congestion control and Sect. 4 concludes the paper.

2 Congestion in IoHT

To implement IoHT, a set of protocols is needed to handle communication over the internet. NDN has proven to be the most promising architecture for IoHT. As the sensors used by IoHT transfer health data frequently and simultaneously, this raises the possibility of congestion and therefore packet loss, which is highly undesired in an IoHT environment. Moreover, IoHT needs a congestion control mechanism that has the ability to be modified according to its requirements, to start with a slightly good rate to take advantage of the available bandwidth, to prevent when congestion occurs to adjust the sending rate of packets of interest and to stabilize the sending rate when the network resources are stable. In the literature, congestion control mechanisms in NDN are classified according to their congestion control mode [6].

2.1 Receiver-Based Control

This type of control uses consumer nodes to control network congestion and uses delay as an indicator of congestion. An increase in delay implies an increase in congestion and vice versa. To deal with congestion, the authors calculated many types of delay such as RTT, RTO and queue delay and most of these mechanisms use the AIMD mechanism to regulate the rate at which packets of interest are sent.

ICP [7] is considered as the first mechanism in this category that relies on RTT expiration as an indicator of congestion. **SIRC** [8] measures the arrival time of data packets to adjust the size of its cwnd, SIRC also proposes an estimation of RTO which takes into account the multihoming of content. **RAAQM** [9] calculates an RTT per path and maintains an RTO per path and the degree of congestion is estimated based on the calculated RTO value. **CCTCP** [10] relies on anticipated interest to predict chunk locations, detects congestion by Timeout and maintains a separate RTO for each considered source. The authors of [11] proved that using RTT value as an indication of congestion is not efficient, especially along the path of interest or data path, because it is not easy to determine when congestion is increasing or decreasing. For this purpose, **DCP** [11] relies on the value of the queue delay as an indication of congestion. This delay is calculated by measuring the delay time returned by the producer or intermediate nodes along the path of the transmitted data packets to decide whether the link is congested or not.

2.2 Hop-by-Hop Control

This type of control uses routers nodes to control network congestion and relies on detecting congestion, either by calculating the queue length or by calculating the interest packet rate. For the first category, **HobHIS** [12] monitors the queue length of data blocks as an indicator of congestion and depending on the occupancy of this queue and other available resources, the router adjusts the sending rate of interest packets. **HCCM** [13] calculates the queue length of interest packets of the output interface and uses two thresholds qmax and qmin of queue occupancy. When the queue length reaches one of the thresholds qmax or qmin, the router sends a notification to the downstream node to inform it of the congestion state and the regulation of the sending rate of interest packets. **IACCM** [14] calculates the length of the output queue of an interface n times during a period T and adjusts the size of its cwnd according to the state of the link; non-congested, mildly congested or severely congested. For the second category, **HIS** [15] calculates the shaping rate of incoming interest packets and adjusts its cwnd by calculating the difference between the optimal interest rate and the actual incoming interest rate. **Stateful Forwarding** [16] calculates the limit rate of interest packets to detect congestion, if this is detected, the router generates a NACK packet containing the link state so that the downstream router adjusts the cwnd size using AIMD.

2.3 Hybrid Control

This type of control uses consumers and routers nodes to control network congestion and relies on detecting congestion, either by measuring packet sojourn time or monitoring the queuing length. The first category is characterized by the use of CoDel-AQM to detect congestion by calculating the sojourn of packets in the router queue as in **PCON** [17], which uses this time to detect congestion and then marks the packets concerned to inform the consumer of the link situation to regulate the size of its cwnd using BIC mechanism. Similarly for **EC-Elastic** [18], which also calculates the sojourn time of packets to detect congestion and then marks the affected packets to inform the consumer of the link situation to regulate the size of its cwnd by a mechanism based on Elastic-TCP [19]. The same for **WinCM** [20] which uses the AIAD algorithm for its cwnd adjustment and controls the transfer rate of packets of interest by interface prefix in multiple paths. For the second category, it is characterized by calculating the queuing length of packets. Some works proposed to monitor the queue length of interest packets to detect congestion as in **ECP** [21] which depending on this value, it divides the link into three levels (Free, Busy and Congested) and returns the link state in a NACK packet to the consumer which is responsible for regulating the sending rate of interest packets using the MIAIMD algorithm. Other works proposed to monitor the queue length of data packets to detect congestion as in **BCON** [22] that calculates on each router the queue length of data packets. In case of congestion, the current interest limit will be halved and a NACK will be sent to the consumer to halve its cwnd. Otherwise, cwnd is increased by 1.

3 Discussion

In IoHT, each healthcare device has its specific requirements regarding transmission speed, transmission reliability, delay, and loss tolerance [5]. As the number of connected health devices increases, the amount of health data transmitted increases and consequently the risk of congestion also increases. Considering the explosive increase of intelligent devices in IoHT to make healthcare field intelligent, congestion control is an open challenge that need to be studied.

Based on the analysis of various mechanisms in NDN [7–22], we can conclude that most receiver-based mechanisms use RTT, RTO, queue delay and congestion source prediction for each flow to control congestion, which is not appropriate for NDN-IoHT because of the massive quantity of traffic generated by IoHT and the characteristics of NDN "multipath, multi-source and caching". For hop-by-hop and hybrid mechanisms, the authors use the delay between the interest and the corresponding data which is variable in NDN and the application of this type of control on each node makes its cost very high.

Some works have used caching in routers which is advantageous in IoHT since recovery time of data is very important in this area. However, the quantity of data transmitted by IoHT can quickly saturate the cache memory if it is limited, which requires cache management. The authors assumed schemes of a simple network (a few consumers and producers) and did not address the case where we have a large-scale network with several consumer and producer nodes involved in data transmission (the case of IoHT), which limits the use of these mechanisms. Moreover, all these works assumed that a consumer node always requests content by sending an interest packet and that a producer node responds with a corresponding data packet. However, they neglect the case of an architecture in which each node can both request content and respond to the request, i.e. a node that functions as both a consumer and a producer, which is present in IoHT devices.

In IoHT, health data is sensitive patient data that requires high protection from malevolent and unauthorized persons which is absent in all existing congestion control mechanisms in NDN that need to be considered to avoid such problems.

In IoHT, some scenarios use devices that rely on high-speed transfer of health data without reliability using a low-power protocol. However, other scenarios use devices that rely on fast and reliable transfer. Therefore, IoHT needs a mechanism that can handle the requirements of different devices. Managing the massive quantity of health data produced by IoHT devices is also a challenge, as each IoHT device produces an enormous volume of data that needs to be held in another IoHT storage device. With an increasing number of IoHT devices connected, the amount of data produced increases and the need for storage increases, posing a serious challenge for IoHT that NDN needs to take into consideration.

As a conclusion, despite the characteristics of NDN that make it the most suitable architecture for IoHT, and the congestion control mechanisms proposed by NDN that can be adapted in IoHT to ensure secure and lossless communications, several issues still need to be taken into consideration that are necessary. IoHT needs a robust mechanism that can be adapted to its communications and provide services according to the requirements of different devices.

4 Conclusion

The congestion problem is one of the principal problems encountered by researchers in every evolution of the Internet, which also continues in the deployment of the IoHT. The IoHT strategy aims to make the medical domain intelligent by ensuring the communication, exchange and transfer of health data over the internet. Therefore, if the amount of health data exchanged increases, the risk of congestion also increases. This paper provides an overview of different congestion control mechanisms proposed in NDN, discusses and evaluates their relevance for the IoHT domain and also presents the main problems related to these mechanisms in the IoHT domain.

As a conclusion, despite the characteristics of NDN that make it the most suitable architecture for IoHT, a robust congestion control mechanism that meets both the requirements of IoHT and the characteristics of NDN still an open issue.

References

1. Das, A.K., Zeadally, S., He, D.: Taxonomy and analysis of security protocols for Internet of Things. Future Gener. Comput. Syst. **89**, 110–125 (2018). https://doi.org/10.1016/j.future.2018.06.027
2. Aroosa, U.S.S., Hussain, S., et al.: Securing NDN-based Internet of Health Things through cost-effective signcryption scheme. Wirel. Commun. Mob. Comput. **2021**, 1–13 (2021). https://doi.org/10.1155/2021/5569365
3. El-bakkouchi, A., Ghazi, M.E.L., Bouayad, A., et al.: Average delay-based early congestion detection in named data of health things. Int. J. Adv. Comput. Sci. Appl. **13**, 335–342 (2022)
4. Zhang, L., Estrin, D., Burke, J., et al.: Named Data Networking (NDN) Project Named Data Networking (NDN) Project (2010)
5. Mishra, N., Verma, L.P., Srivastava, P.K., Gupta, A.: An analysis of IoT congestion control policies. Procedia Comput. Sci. **132**, 444–450 (2018). https://doi.org/10.1016/j.procs.2018.05.158
6. Ren, Y., Li, J., Shi, S., et al.: Congestion control in named data networking – a survey. Comput. Commun. **86**, 1–11 (2016). https://doi.org/10.1016/j.comcom.2016.04.017
7. Carofiglio, G., Gallo, M., Muscariello, L.: ICP: Design and evaluation of an Interest control protocol for content-centric networking. In: 2012 Proceedings IEEE INFOCOM Workshops. IEEE, pp 304–309 (2012)
8. Amadeo, M., Molinaro, A., Campolo, C., et al.: Transport layer design for named data wireless networking. In: 2014 IEEE Conference on Computer Communications Workshops (INFOCOM WKSHPS). IEEE, pp 464–469 (2014)

9. Carofiglio, G., Gallo, M., Muscariello, L., Papali, M.: Multipath congestion control in content-centric networks, pp. 363–368 (2014). https://doi.org/10.1109/infcomw.2013.6970718
10. Saino, L., Cocora, C., Pavlou, G.: CCTCP: a scalable receiver-driven congestion control protocol for content centric networking. In: Proceedings of IEEE ICC 2013 - Next-Generation Networking Symposium (ICC13 NGN), pp. 3775–3780 (2013)
11. Albalawi, A.A., Garcia-Luna-Aceves, J.J.: A delay-based congestion-control protocol for information-centric networks. Int. Conf. Comput. Netw. Commun. (ICNC) **2019**, 809–815 (2019). https://doi.org/10.1109/ICCNC.2019.8685491
12. Rozhnova, N., Fdida, S.: An effective hop-by-hop Interest shaping mechanism for CCN communications. In: 2012 Proceedings IEEE INFOCOM Workshops. IEEE, pp 322–327 (2012)
13. Mejri, S., Touati, H., Malouch, N., Kamoun, F.: Hop-by-hop congestion control for named data networks. In: Proceedings of IEEE/ACS International Conference on Computer System Application. AICCSA 2017-Oct, pp. 114–119 (2018). https://doi.org/10.1109/AICCSA.2017.36
14. Zhu, Y., Luo, Q., Tao, Y., Huang, R.: A congestion control mechanism based on identity authentication for named data networking. Eng. Lett. **28**, 231–237 (2020)
15. Wang, Y., Rozhnova, N., Narayanan, A., et al.: An improved hop-by-hop interest shaper for congestion control in named data networking. ACM SIGCOMM Work Inf. Centric Netw. **1**, 55 (2013)
16. Yi, C., Afanasyev, A., Moiseenko, I., et al.: A case for stateful forwarding plane. Comput. Commun. **36**, 779–791 (2013). https://doi.org/10.1016/j.comcom.2013.01.005
17. Schneider, K., Yi, C., Zhang, B., Zhang, L.A.: Practical Congestion Control Scheme for Named Data Networking. https://doi.org/10.1145/2984356.2984369
18. El-Bakkouchi, A., El Ghazi, M., Bouayad, A., et al.: EC-elastic an explicit congestion control mechanism for named data networking. Int. J. Adv. Comput. Sci. Appl. **12**, 594–603 (2021). https://doi.org/10.14569/ijacsa.2021.0121168
19. Alrshah, M.A., Al-Maqri, M.A., Othman, M.: Elastic-TCP: flexible congestion control algorithm to adapt for high-BDP networks. IEEE Syst. J. 1–11 (2019). https://doi.org/10.1109/jsyst.2019.2896195
20. Wang, M., Yue, M., Wu, Z.: WinCM: a window based congestion control mechanism for NDN, pp. 80–86 (2019). https://doi.org/10.1109/hoticn.2018.8606039
21. Ren, Y., Li, J., Shanshan, Shi., et al.: An explicit congestion control algorithm for named data networking. In: 2016 IEEE Conference on Computer Communications Workshops (INFOCOM WKSHPS). IEEE, pp 294–299 (2016)
22. Agarwal, A., Tahiliani, M.P.: BCON: Back pressure based congestion avoidance model for named data networks. In: 2016 IEEE International Conference Advanced Networks Telecommunications Systems, (ANTS 2016) (2017). https://doi.org/10.1109/ANTS.2016.7947823

An Enhanced Artificial Hummingbird Algorithm for Workflow Scheduling in Cloud

Adnane Talha(✉), Anas Bouayad, and Mohammed Ouçamah Cherkaoui Malki

Sidi Mohamed Ben Abdellah University, Fez, Morocco
{adnane.talha,anas.bouayad,ocamah.cherkaoui}@usmba.ac.ma

Abstract. The execution of complex, extensive scientific applications has been made much easier through workflows. Although it has many unresolved issues, the cloud serves as an appropriate platform for executing them. These issues must be resolved for an efficient workflow scheduling. Numerous algorithms have been put out for workflow scheduling; however, the majority of them fall short of taking into account the critical components of the cloud, such as heterogeneous resources, a pay-per-usage business model and elasticity. In this work, we present a hybrid approach based on the new Artificial Hummingbird optimization Algorithm (AHA) for modelling and optimizing a workflow-scheduling problem in cloud environment. The Predict Earliest Finish Time Algorithm (PEFT) is a heuristic model intervenes in the generation of the initial population. A good seed speeds up the process of discovering the best solution. The experimental outcomes for 100 tasks confirm that our method outperforms other cutting-edge algorithms.

Keywords: Cloud · PEFT · AHA · Metaheuristic algorithm · Task scheduling

1 Introduction

The importance of scientific computing is rising across a wide range of research fields. A typical application could have numerous interdependent tasks expressed as a workflow, needing effective task scheduling. Today, scientific applications made up of thousands of interconnected tasks can be run using cloud computing. The workflow paradigm is an effective technique to build such intricate applications in this environment. It is constituted of directed acyclic graph nodes (DAG) that represent application tasks, and connection lines between nodes that establish interdependence between these tasks. The purpose of the task scheduling challenge in the cloud is to allocate tasks to virtual machines while maintaining performance standards. There have been various attempts over the past ten years to employ bio-inspired scheduling algorithms to develop an optimal or nearly optimal schedule of the tasks on a given set of resources, in an effort to reduce the total schedule length and maximize resource utilization. The majority of research on workflow scheduling in the cloud has concentrated on reducing the overall cost of using the computing resources that cloud providers offer, rather than just the total execution time of workflows. However, the solution space for mapping tasks to associated resources grows exponentially as the tasks rises. Therefore, mechanisms to

Y. Farhaoui et al. (Eds.): ICAISE 2022, LNNS 635, pp. 162–172, 2023.
https://doi.org/10.1007/978-3-031-26254-8_24

improve search strategies of bio-inspired scheduling algorithms are needed in order to produce better scheduling solutions with fewer iterations and shorter execution times. In this study, we propose a multi-objective algorithm based workflow scheduling using the heuristic algorithms (PEFT) [1] and the bio-inspired Artificial Hummingbird Optimization [2] Algorithm (AHA) that uses the least amount of time and cost while maximizing the resource utilization. In order to obtain better solutions using AHA, we propose to first identify a schedule using heuristic algorithms and then apply these as initial solutions in the search space.

2 Related Work

This paper [3] develop a new systematic approach that takes into account both interactions and task security requirements when placing secure tasks in the cloud. Authors develop a model for task scheduling and suggest a heuristic approach based on the work's completion time and security criteria in order to respect security and performance. They also offer an additional attack response strategy to minimize some cloud security risks. In this article [4], authors presents a new resource provisioning method and a workflow scheduling strategie, titled Greedy Resource Provisioning and modified HEFT (GRP-HEFT), for shortening the computation time of a chosen workflow under the constraints of a budget for the hourly-based cost analysis of contemporary IaaS clouds. The goal of this research [5] is to investigate the cost optimization technique of deadline constrained workflows on cloud environement and to present two list scheduling strategie to address the issue look back workflow scheduling lbws and structure. The authors of this paper [6] present a hybrid metaheuristic (HBMMO) for multi objective optimization. The method enhance the multi objective scheduling of several scientific workflows in the cloud environment. The suggested algorithm incorporates Predict Earliest Finish Time algorithm (PEFT) to improve the convergence and variety in terms of reduced makespan, minimized cost, and efficient load balancing of the Virtual Machines, strengthening the potent global exploration ability of the nature-inspired metaheuristic.

3 The Problem Formulation

The scheduling of workflows is an NP-complete problem. The structure of workflow is frequently described as a DAG (Fig. 1) that has no circuits and whose arcs are directed. The illustration is W = (E, T), where T = $\{t_1, t_2 \ldots t_m\}$, denotes the collection of n tasks in a workflow. The interdependence of data and control between t_i and t_j is described by the sequence of directed edges $\{e_{i,j} | (t_i, t_j) \in E\}$. The task t_j can only continue with its execution if all of its predecessor tasks have completed their work and delivered any relevant data or parameters to t_j. The source task is the parent task of an edge, while the target task is the child task. Pred(t_j) and Succ(t_j) stand for the respective successor and predecessor of t_j. An output task is a task without a child, whereas an input task is a task without a parent. A typical workflow DAG scheme is depicted in Fig. 1.

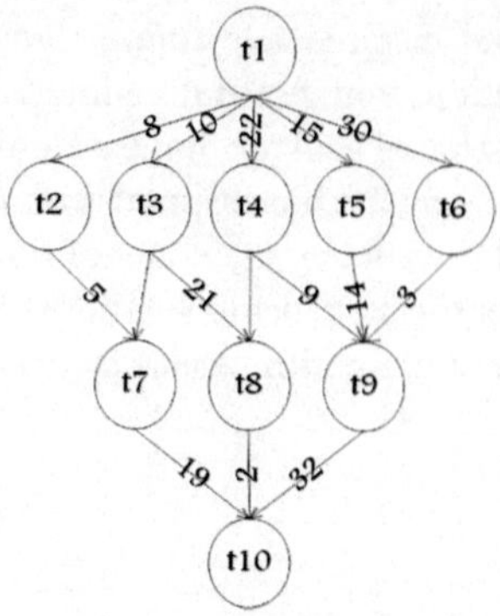

Fig. 1. Sample workflow DAG scheme

A Fitness Function

The aim of this study is to decrease task execution time and cost while maximizing resource utilization across all VMs in a cloud environment. In this case, three goals are taken into consideration. The first is lowering the time needed to execute a task (Mw). The second one is lowering the cost of execution (TCw), and the third is making the best use of available resources. We must precisely construct a fitness function that satisfies this in order to accomplish these objectives.

In light of this, the function is established using the following equations:

$$F = \min\left[\alpha_1(\text{Mw}) + \alpha_2(\text{TCw}) + \alpha_3(1 - \text{AVGuse})\right]$$

where $\boldsymbol{\alpha_i}$ represents the Weight value and $1 \leq i \leq 3$.

The Makespan

The makespan, which is defined as follows, expresses the total amount of time required to complete the workflow:

$$\mathbf{Mw} = \text{Max}\{\mathbf{Finish\_Time}(t_i, v_k)\text{i} = 1, .., \text{n k} = 1, .., \text{m} \tag{1}$$

The Total Cost (TCw)

Cost guidelines are specified for CSP services. The above term refer to the execution cost:

$$\mathbf{TCw} = \frac{((\boldsymbol{FT_k}) - (\boldsymbol{ST_k}))}{\boldsymbol{\gamma}} * \mathbf{LPT_k} \tag{2}$$

where $\mathbf{LPT_k}$ is the rental cost per unit time for $\boldsymbol{v_k}$, $\boldsymbol{FCT_k}$ ***and*** $\boldsymbol{SCT_k}$ are respectively the unit time to release the $\boldsymbol{v_k}$ (end-of-lease time) and the start-of-lease time during the execution of a workflow.

The Resource Utilization

By keeping resources as occupied as possible to generate the most profit, maximizing resource utilization is another critical necessity for cloud providers. The average resource utilization is calculated by the formula below:

$$\text{AVUse} = \frac{\sum_{k=1}^{p} Execution\ \ time\ \ of\ \ v_k(ET_k)}{Makespan*p} \tag{3}$$

where p is the number of VMs.

4 The Proposed Algorithm

The major aim of the suggested methodology is to allocate the task to a virtual machine (VM) employing (AHA), which reduces the overall execution time and cost although balancing the load. The load is a crucial factor in scheduling, which is the procedure of distributing the load among multiple nodes of a shared system in order to increase resource utilization and task response time and prevent a situation in which some of the VMs are heavily loaded while other nodes are idle or performing very few operations. The multi-objective algorithm that is developed in this study using two known algorithms PEFT and AHA is used to get around this issue.

1. **Predict Earliest Finish Time Algorithm (PEFT)**

The OCT is seen as a matrix, with the columns denoting the resources and the rows denoting the tasks. The OCT value is determined recursively using a backwards approach in accordance with (4). It will indicate how much it will cost to run each task that is a current task's child up until the exit node.

$$\text{OCT}(t_\text{i}, v_j) = \max_{t_\text{j} \in \text{succ}(t_\text{i})}[\min_{v_w \in P}\{\ \text{OCT}(t_\text{j}, v_w)\mathbf{ET}(\mathbf{t_j}, \mathbf{v_w}) + \mathbf{TT}_{\mathbf{i,j}}\}] \tag{4}$$

where $\mathbf{TT}_{i,k} = 0$, if$v_j = v_w$ for the exit task $\text{OCT}(t_{exit}, v_j) = \mathbf{ET}(\mathbf{t_{exit}}, v_j)$.

- **Prioritizing Phase**

The cumulative OCT defines the rank of every task as in (5). Tasks are ordered in the list on the basis of decreasing order of the rankOCT.

$$rank_{OCT}(t_\text{i}) = \frac{\sum_{j=1}^{P} OCT(t_\text{i}, v_j)}{P} \tag{5}$$

- **VM Selection Phase**

Optimistic (O_{EFT}) is computed utilizing (6) to choose the resource for the task. Here, resource allocation is not determined by the task's earliest completion time (EFT), because we anticipate that the tasks in the following steps will be completed in less time. The aim is to verify if the tasks are ahead to finished more rapidly.

$$O_{EFT}\left(t_\text{i}, v_j\right) = \text{EFT}(t_\text{i}, v_j) + \text{OCT}(t_\text{i}, v_j) \tag{6}$$

Algorithm 1 : The PEFT Algorithm
1. Formulate the OCT matrix using the formula given in Eq. 11.
2. Using the $rank_{OCT}$, prioritize the tasks and put them in a list according to the decreasing order of $rank_{OCT}$, as in Eq. 12
3. Repeat steps 4 to 6, until the list is empty.
4. For all the resources do
5. pick the top most task and compute its EFT using insertion based policy[1]
6. Calculate the OEFT as described in (13)
7. Allocate the task to the resource, which gives the minimum OEFT.
8. Return the optimal schedule.

The Peft algorithm

B Artificial Hummingbird Algorithm (AHA):

A hummingbird investigates various elements to select an appropriate source from a variety of food sources, such as nectar quantity and quality of particular flowers, as well as the nectar-refilling mechanism. This algorithm, which differs from previous algorithms in terms of search domain variety, was motivated by hummingbirds' exceptional flying abilities and accurate foraging techniques for accessing food sources.

The algorithm has a high likelihood of exploitation and strong exploration potential because to the various flight patterns. The visit table is also included to imitate the hummingbird's memory for identifying appropriate food sources. Hummingbirds use three different foraging technique and three different flight movements to get food from sources. Axial, diagonal, and omnidirectional are the three different flying patterns, whereas guided, territorial and migration foraging are the three different search tactics. The visit table of food sources is constructed using Eq. (7):

$$\mathrm{VT}_{ij} = \begin{cases} 0 \ \ if \ \ i \neq j \\ null \ \ i = j \end{cases}, \ \mathrm{i} = 1, \ldots, \mathrm{N}, \ \mathrm{j} = 1, \ldots, \mathrm{N} \tag{7}$$

where for i = j, VT_{ij} = null stands for the food taken by a hummingbird at its specific food source. In addition, i ≠ j, $\mathrm{VT}_{ij} = 0$ stands for the hummingbird i visiting j food source.

- **Guided Foraging**

Three flight behaviors are included in this foraging model (omnidirectional, diagonal, and axial flight). These three flight patterns are shown in 3D space in Figs. 2 and 3.

The axial flight can be defined as:

$$D_i = \begin{cases} 1 \ \ if \ \ i = randi([1, d]) \\ \quad 0 \ \ else \end{cases} \mathrm{i} = 1, \ldots, \mathrm{d} \tag{8}$$

The diagonal flight is formulated as:

$$D_i = \begin{cases} 1 \quad if \quad i = P(i), j[1, k], P = randperm(k), k \in [2, r_1(d-2)+1] \\ \quad 0 \quad else \end{cases} \text{i = 1... d} \tag{9}$$

The omnidirectional flight is represented as:

$$D_i = 1, i = 1, \ldots, d, \tag{10}$$

where randi([1, d]) stands for a random integer from 1 to d, randperm(k) represents a random permutation of integers from 1 to k, and r1 ∈ [0, 1] denotes a random number. The guided foraging behavior can be formulated as

$$V_i(t+1) = X_{i,t}(t) + a{*}D{*}\big(X_i(t) - X_{i,t}(t)\big), a \in N(0, 1) \tag{11}$$

where $X_{i,t}(t)$ represents the food source i at iteration t. $X_{i,t}(t)$ is the target food source that visited by ith hummingbird. The value of X_i can be update using the following equation:

$$X_i(t+1) = \begin{cases} X_i(t) if\, f(X_i(t)) \le f(V_i(t+1)) \\ \quad V_i(t+1)\, otherwise \end{cases} \tag{12}$$

where f stands for the fitness value.

- **Territorial Foraging**

The hummingbirds' territorial foraging strategy's local search is represented by the following equation:

$$V_i(t+1) = X_{i,t}(t) + b{*}D{*}(X_i(t)), b \in N(0, 1) \tag{13}$$

where b denotes the territorial factor.

- **Migration Foraging**

The following equation describes the migration of a hummingbird from the source with the lowest rate of nectar replenishing to a novel source chosen at random:

$$X_w(t+1) = L * r * (U - L) \tag{14}$$

where U and L are the upper and lower limit ranges, respectively, and X_w indicates the food source with the worst population rate of nectar replenishment.

C The Proposed Algorithm

Algorithm2: The hybrid PEFT-AHA algorithm

Input: Tasks T , Virtual Machines V
Output: assign tasks to the VMs
Begin
1. define Max_iter, n,
2. Initialize the set of population using schedule obtained from PEFT algorithm
3. Calculate the fitness value for each element.
4. While t< Max_iter
5. For each population update, calculate the direction switch vector D
6. If rand<1/3 Following diagonal flight using equation (16)
7. Else if rand <2/3
8. Follow omnidirectional flight using equation (17)
9. Else
10. follow axial flight using equation (15)
11. End if
12. End for
13. For each population update foraging behavior
14. If rand<1/2
15. Follow guided foraging using equation (19)
16. Else
17. Follow territorial foraging using equation (20)
18. If t =2n //
19. Follow migration foraging using equation (21)
20. End if
21. End for
22. Update position
23. Return the best fitness value
24. t = t+1
25. end while
26. End

Fig. 2. The PEFT-AHA algorithm

5 Simulation Results

We are doing the simulations with WorkflowSim in order to assess the performance of our suggested method. Montage, **Cybershake** and **Inspiral** are three scientific workflows that we take into consideration. These procedures demand varying amounts of data and computation, as well as varied structural characteristics. We conducted a set of experiments with existing well-known algorithms for 100 and 1000 tasks. The suggested algorithm is developed in Java and runs on the Eclipse platform. The tests have been carried on a workstation with a window OS and 2.5 GHz Intel Core i5 processor and 4 GB of RAM. The proposed PEFT-AHA approach is based on a population of 100 elements, which is sufficient to achieve reasonable convergence rate. Table 1, 2, 3, 4, 5 and 6 show the results of a series of studies using Montage workflow, Cybershake workflow and inspiral process to compute the makespan, cost and resource utilization accordingly.

Table 1. The execution time of applications for 100 tasks

	Montage	Cybershake	Inspiral
AHA	350,4	992,7	2145
PSO	503,9	825,2	1990,46
HEFT	410,56	927,5	2355,7
PEFT	318,96	1096,02	2490,23
PEFT-AHA	328,03	855	1782,31

Table 2. The execution time of applications for 1000 task

	Montage	Cybershake	Inspiral
AHA	2410,2	6325,95	17002,8
PSO	2256,03	6722,17	17878,29
HEFT	2688,45	6731,8	17400,65
PEFT	2257,69	6970,21	18754,36
PEFT-AHA	2196,57	6109,55	16832,33

Table 3. The total execution cost of applications for 100 tasks

	Montage	Cybershake	Inspiral
AHA	2006,5	42254,32	55325,47
PSO	2087,66	42915,25	55958,2
HEFT	1928,23	41320,98	59912,36
PEFT	1990,14	41450,36	59910
PEFT-AHA	1827,3	49310,75	55802,45

Table 4. The total execution cost of applications for 1000 tasks

	Montage	Cybershake	Inspiral
AHA	32800,12	124788,63	272114
PSO	33205,45	137836,87	267720,01
HEFT	34150,68	129930,25	276890,49
PEFT	33052,33	138990,56	269910,6
PEFT-AHA	32120,32	119829,2	261160,95

Table 5. Resource utilization of applications 100 tasks

	Montage	Cybershake	Inspiral
AHA	0,698	0,74	0,62
PSO	0,7353	0,68	0,78
HEFT	0,624	0,79	0,56
PEFT	0,705	0,76	0,59
PEFT-AHA	**0,732**	**0,85**	**0,85**

Table 6. Resource utilization of applications 1000 task

	Montage	Cybershake	Inspiral
AHA	0,61	0,8	0,74
PSO	0,75	0,723	0,842
HEFT	0,85	0,765	0,79
PEFT	0,79	0,769	0,866
PEFT-AHA	0,84	0,82	0,932

We compare our work with the other three algorithms namely PEFT [1], HEFT [7], AHA [2] and Particle swarm algorithm (PSO) [8]. The performance analysis based on the makespan, cost and percent of resource utilization for the suggested technique is shown in Figs. 4, 5, 6, 7 and 8 for 100 and 1000 tasks.

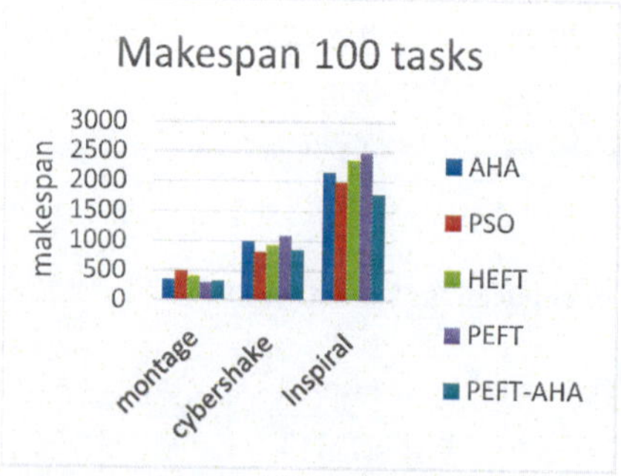

Fig. 3. Simulation results plot of the makespan for 100 tasks

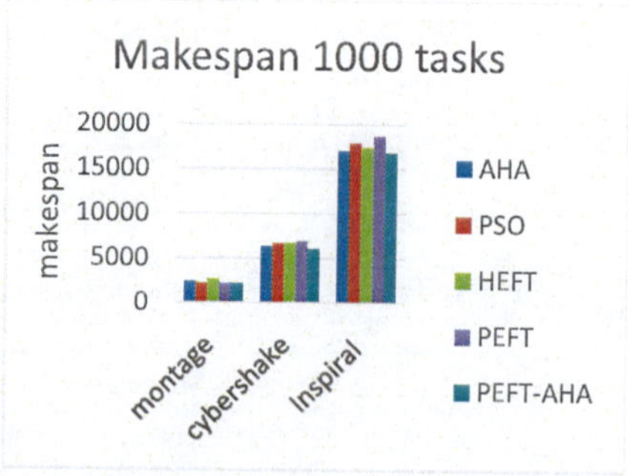

Fig. 4. Simulation results plot of the makespan for 1000 tasks

Figure 4 and 5 demonstrates that for the benchmark workflows with 100 task, our algorithm give a good result in comparison to the other techniques. For workflow with 1000 tasks, our method outperforms the three comparison algorithms. We can noticed that PEFT-AHA can complete tasks in less time. The PEFT-AHA improvement rate in terms

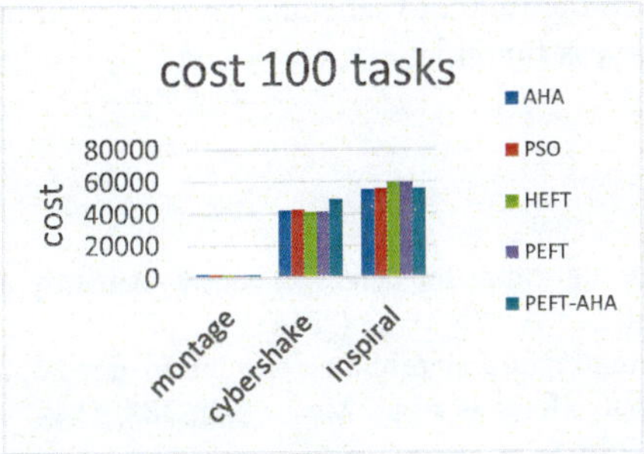

Fig. 5. Simulation results plot of the total cost for 1000 tasks

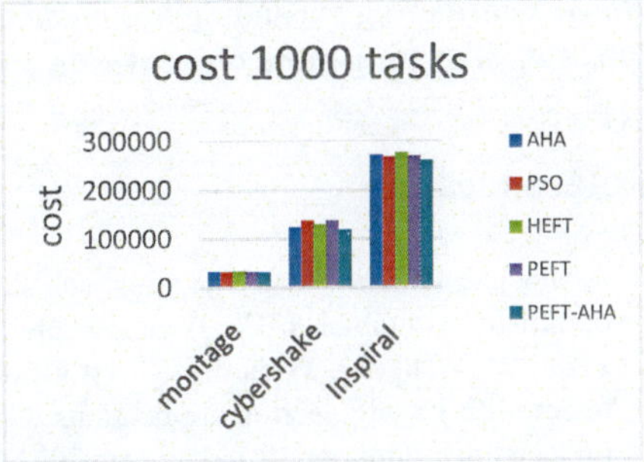

Fig. 6. Simulation results plot of the total cost for 100 tasks

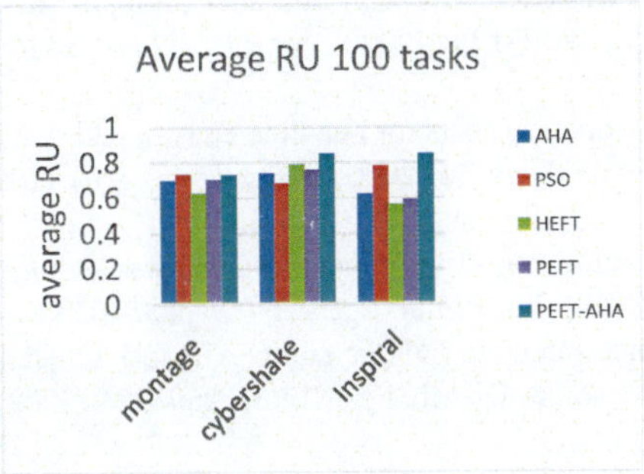

Fig. 7. Simulation results plot of the average resource utilization for 100 tasks

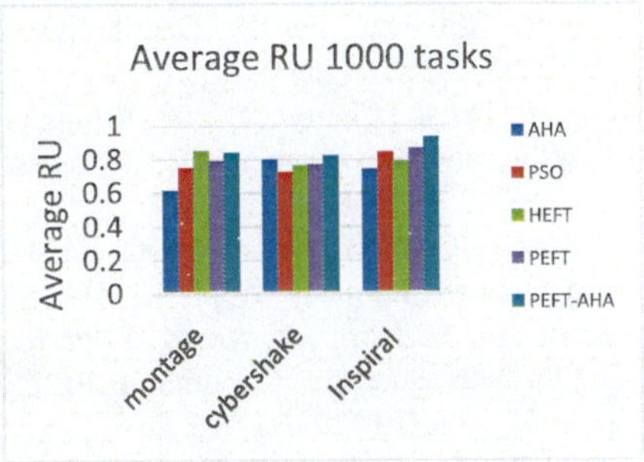

Fig. 8. Simulation results plot of average the resource utilization for 1000 tasks

of the cost of the workflow applications is depicted in Fig. 5 and 6. Results show that the proposed PEFT-AHA had the lowest cost, notably when a big workflow with 1000 tasks was being handled. This finding can be attributed to the hybrid algorithm's intelligent task allocation, which maximizes resource usage. Additionally, it selects appropriate VMs to carry out every task with the lowest costs implementation. In terms of resource usage (Fig. 7 and 8), the PEFT-AHA outperforms the AHA, PEFT, HEFT and PSO by 7.26%, 14.04%, and 4.54% respectively, while employing the Montage, CyberShake, and Inspiral processes.

6 Conclusion

In this paper, we propose a hybrid approach in which PEFT heuristic algorithms were used to obtain initial solutions in the search space. These initial solutions then serve as starting points for obtaining better solutions using the new artificial hummingbird optimization algorithm (AHA) to schedule tasks from scientific applications. This study try to compares the effectiveness of our hybrid technique with those of other approaches.

We are considering expanding this algorithm in the future to schedule several workflow apps and evaluating more objectives in our fitness function.

References

1. Arabnejad, H., Barbosa, 'J.G.: List scheduling algorithm for heterogeneous systems by an optimistic cost table', IEEE Trans. Parallel Distrib. Syst. **25**(3), 13 (2014)
2. Zhao, W., Wang, L., Mirjalili, 'S.: Artificial hummingbird algorithm: a new bio-inspired optimizer with its engineering applications. Comput. Methods Appl. Mech. Eng. **388**, 114–194 (2022). https://doi.org/10.1016/j.cma.2021.114194
3. Abazari, F., Analoui, M., Takabi, H., Fu, S.: MOWS: multi-objective workflow scheduling in cloud computing based on heuristic algorithm. Simul. Model. Pract. Theory 93, 119–132 (2019). https://doi.org/10.1016/j.simpat.2018.10.004
4. Faragardi, H.R., Saleh Sedghpour, M.R., Fazliahmadi, S., Fahringer, T., Rasouli, N.: GRP-HEFT: a budget-constrained resource provisioning scheme for workflow scheduling in IAAS clouds. IEEE Trans. Parallel Distrib. Syst. **31**(6), 1239–1254 (2020). https://doi.org/10.1109/ACCESS.2020.2971351
5. Han, P., Du, C., Chen, J., Du, X.: Minimizing monetary costs for deadline constrained workflows in cloud environments. IEEE Access **8**, 25060–25074 (2020). https://doi.org/10.1109/ACCESS.2020.2971351
6. Anwar, N., Deng, H.: A hybrid metaheuristic for multi-objective scientific workflow scheduling in a cloud environment'. Appl. Sci. **8**(4), 538 (2018). https://doi.org/10.3390/app8040538
7. Topcuoglu, H., Hariri, S., Wu, M.-Y.: Performance-effective and low-complexity task scheduling for heterogeneous computing. IEEE Trans. Parallel Distrib. Syst. **13**(3), 260–274 (2002). https://doi.org/10.1109/71.993206
8. Kennedy, J., Eberhart, R.: Particle swarm optimization. In: Proceedings of ICNN 1995 - International Conference on Neural Networks, p. 7 (1995)

An Enhanced Medical Diagnosis System for Malaria and Typhoid Fever Using Genetic Neuro-Fuzzy System

Joseph Bamidele Awotunde[1](✉), Agbotiname Lucky Imoize[2], Dotun Patrick Salako[1], and Yousef Farhaoui[3]

[1] Department of Computer Science, Faculty of Information and Communication Sciences, University of Ilorin, Ilorin 240003, Kwara State, Nigeria
awotunde.jb@unilorin.edu.ng

[2] Department of Electrical and Electronics Engineering, Faculty of Engineering, University of Lagos, Akoka, Lagos 100213, Nigeria
aimoize@unilag.edu.ng

[3] STI Laboratory, T-IDMS Faculty of Sciences and Techniques, Moulay Ismail University of Meknès, Errachidia, Morocco
y.farhaoui@fste.umi.ac.ma

Abstract. The major challenge of inadequate healthcare services in developing countries has been linked to the unavailability of qualified medical personnel and inefficient diagnostic techniques adopted. The advent of technology in medicine has aided the improvement of medical processes including diagnosis. Many computerized systems have been implemented over the years for medical diagnosis some of which included machine learning techniques. But the existing methods lack the power of selecting the most relevant signs and symptoms of malaria and typhoid fever. Therefore, this study develops a system capable of Malaria and Typhoid Fever diagnosis using a Genetic Algorithm, a Neuro-Fuzzy Inference System called GENFIS. In order to train the NFIS, the GA module determines the optimal set of network parameters, saves them, and then distributes them to the appropriate hidden layer nodes. The study made use of the MATLAB environment to test and evaluate the system. The accuracy performance of the proposed model revealed an accuracy of 97.2%, thus performing better than some of the existing systems. If completely adopted, the proposed approach has the potential to lessen the major issues with NFIS systems. Furthermore, it might be used to address difficult issues in different fields.

Keywords: Typhoid fever · Genetic algorithm · Neural networks · Fuzzy logic · Diagnosis · Healthcare systems

1 Introduction

One of the most common deadly diseases in the world, malaria is brought on by one or more Plasmodium species [1]. Over 50% of the world's population suffers from malaria,

Y. Farhaoui et al. (Eds.): ICAISE 2022, LNNS 635, pp. 173–183, 2023.
https://doi.org/10.1007/978-3-031-26254-8_25

with Sub-Saharan Africa causing the majority of cases and mortality [2]. According to the 2011 Global Malaria Report, in 2010, there were an estimated 655,000 fatalities from malaria, with an estimated 216 million cases [3]. Roughly 68% of the population of Ethiopia lives in malaria-risk areas and it is the most communicable disease in the country [4]. Serious illness and death are brought on by Plasmodium falciparum, a parasite found primarily in native communities in Ethiopia and other sub-Saharan nations [5].

Typhoid fever is an extensive long-lasting febrile infection caused by certain Salmonella serotypes. *Salmonella enterica serotype Typhi* is the most familiar type of salmonella that give rise to typhoid fever. In 2010, an estimated 13.5 million people were infected with typhoid disease around the world [3]. Malaria and typhoid share circumstances which are vital to their transmission.

The authors in [6] revealed that 17% of 200 fever patients had both malaria and typhoid fever, based on the bacterial confirmed diagnosis. Therefore, persons living in endemic or aboriginal areas are at risk of developing both diseases at the same time or an acute infection on top of a chronic infection. A high index of suspicion and reliable diagnostic method is necessary to diagnose a co-infection as most people are used to linking every sign and symptom to a single pathology [7].

Medical establishments like hospitals, rehabilitation centres, clinics etc. are the recognized standard options for diagnosis and treatment of most diseases and illnesses including malaria and typhoid. The absence of standard medical establishments in certain regions as well as the unavailability of well-trained and qualified medical personnel are what has led to the advent of computerized systems which in turn has made it possible for numerous patients to access proper medical diagnosis and receive correct treatment procedures online.

If meaningful quality improvements are to be made, information technology must play a key role in the transformation of the healthcare system [8]. The use of technology in the treatment and management of illnesses has been linked to reductions in hospitalizations, readmissions, lengths of stay, and costs, as well as improvements in some physiological measures, a high level of satisfaction, and improved medication compliance [9]. Technologies in medicine include; Expert systems for terminal and chronic illnesses or diseases, Online Medical Diagnosis systems (OMDS), Electronic Health Records (EHR), and Machine Learning Techniques/Systems, to mention a few [10].

Fuzzy Logic Systems (FLS) are programs which transfer inconsistently, a subquery output dataset from an input dataset. It is a type of logic with several values that are concerned with estimated rather than precise reasoning. Due to its adaptability, the FL concept, Fuzzy Logic Systems has gained popularity in contemporary Information technology, manufacturing techniques, medical diagnosis, pattern recognition, data mining, and decision-making to mention a few [11].

Neural Networks (NN) are a class of qualitative methods for learning based on organic neural networks (the brain and animals' central nervous systems) that are often used to calculate functions with many parameters. NN is a fantastic tool for creating systems that can implement tasks involving information processing in the same way that our brain does [12, 13].

The genetic Algorithm (GA) is motivated by the research into genetics which is founded on the concept of natural selection [12]. GA is principally laid based on evolution operation which focuses on solving population as opposed to other search strategies that work on a single answer [14, 15].

The Genetic Neuro-Fuzzy system is a hybrid model of a Fuzzy Logic (FL) component that deals with erroneous and missing medical data, a Genetic Algorithm that handles optimization of connection weights to train the Neural Network and the parameters that drive the Membership Functions (MF) are generated using Neural Networks (NN).

Therefore, this study proposes a hybrid Genetic-Neuro-Fuzzy based system (GENFIS) where; the Fuzzy Logic (FL) module deals with sloppy and partial medical data, while the Neural Networks (NN) module produces the parameters that control the Membership Functions automatically (MF), and the Genetic Algorithm is used to evolve the optimum connection weights for the training of the Neural Networks. The system is also aimed at providing consultancy options for patients who need direct access to a medical practitioner.

The purpose of this study is to design and develop a medical diagnosis and consultancy system for malaria and typhoid fever. The key contributions of this study are as follows: Design the proposed system; Implement the proposed system; Evaluate the medical diagnostic system in terms of correctness, functionality and usability.

2 Materials and Methods

The proposed system comprises of a base of knowledge (KB), a GENFIS, a DSE and an interface for the user that allows you to enter diagnosis variables and see the results of your diagnosis. The knowledge base (KB) contains both organized and unstructured information on the problem area, as well as operational data that has to be processed. The DSE contains filters, both Emotional and Cognitive that handle the medical practitioner's objective and intuitive feelings towards a patient (Fig. 1).

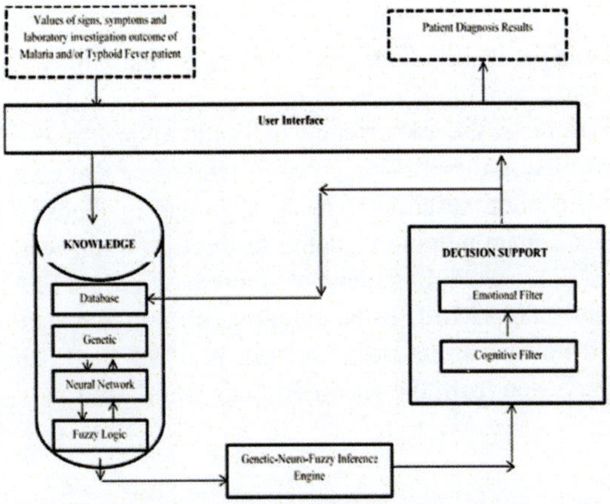

Fig. 1. The proposed genetic-neuro-fuzzy system framework

2.1 Data Collection

This dataset contains information on the prevalence, and attitudes that operate as risk factors in the diagnosis and the treatment options of malaria and malaria-typhoid co-infection in a medical facility of a higher institution in south-east Nigeria. It consists of the demographic, diagnostic and management information collected from June to November 2020. This study included 236 febrile volunteers, 104 males and 132 females from the medical center of Akanu Ibiam Federal Polytechnic Unwana, Afkpo Ebonyi state, Nigeria. Data from laboratory diagnoses as well as responses to a standardized questionnaire about the diagnostic and management techniques of malaria-typhoid co-infection are presented in Excel spreadsheets. Thick blood film microscopy was used to measure the density of malaria parasites whereas the typhoid diagnosis was based on the Widal test.

2.2 Fuzzy Logic Component

The FL component consists of a fuzzifier, a fuzzy inference engine, and a defuzzifier. The fuzzifier transforms clean input values into fuzzy ones. Assume that v is a jumbled collection of diagnostic parameters. In the V (Universe of Discourse) and that it denotes a component in v., then v is: $v = \{(x_i, \mu_v (x_i)) \mid x_i \in V, m\mu_v (x_i) \in [0,1]\}$ where μ_v (xi) is the function of membership (MF) of (xi) in v, and it shows the degree to which (xi) contributes to the Malaria/Typhoid Fever evaluation output. The fuzzy group that was used in this study to define the characteristics of Malaria and Typhoid Fever diagnosis consist of; mild, moderate and severe. A linguistic phrase from the defined fuzzy set is used to represent each diagnosis attribute, with a numeric value associated with each linguistic term. The linguistic phrase for the ith diagnosis variable, for example, is defined by the equation:

$$LT_i = \{\{\text{"Mild"}, \text{ if } x_i = 1\}, \{\text{"Moderate"}, \text{ if } x_i = 2\}, \{\text{"Severe"}, \text{ if } x_i = 3\}\}.$$

The language name for the ith diagnosis variable is denoted by LTi, and has = 1, 2, 3, 14; and the diagnostic variable's value of ith is denoted by xi (Ci).

The Rule Base reflects the experience of medical professionals who diagnose Malaria/Typhoid fever using the "If-Then" construct. NR = Kn, where Kn is a number. K, n, the number of different variables in each diagnosis attribute classification, and n, the number of linguistic parameters evaluated in the fuzzy set, were used to compute the number of rules (NR) in each diagnostic characteristic group. FEV and HAD are the inputting numbers, and "TF is Mild" is the diagnosis output, hence the rule base is made up of rules with the following structure. The corresponding linguistic term for both the inputs and outputs is chosen from the specified fuzzy set (Table 1).

Table 1. Attributes for diagnosing typhoid and malaria fever.

S/N	Category	Malaria/TF diagnosis attributes	Code
1	Patient Laboratory Investigation (PLI)	Elevated Liver Function Test	C1
2		Blood test	C2
3		Stool test	C3
4		Urine test	C4
5	Patient Medical History (PMH)	Fever	C5
6		Headache	C6
7		Abdominal pain	C7
8		Stomach pain	C8
9		Myalgia	C9
10		Lassitude	C10
11		Loss of appetite	C11
12		Vomiting	C12
13	Patient Physical Examination (PPE)	Body temperature	C13
14		Pulse rate	C14

2.3 Genetic Neuro-Fuzzy Inference Engine

The GENFIS is a deductive method that combines Genetic Algorithms, Neural Networks, and Fuzzy Logic components to supply a versatile system that adapts to its surroundings for processing ambiguous and erroneous data for the treatment of Malaria and Typhoid Fever. Active nodes make up the hidden and output layers, where computation occurs while the input layer consists of inactive/docile nodes. The feed-forward propagation approach is used by the inference system, which is made up of seven layers of neurons.

The reasoning process in the inference engine is driven by production rules based on Mamdani's inference engine. The first out of the seven layers is made up of nodes that are currently active which designate inputs to the system. The Numeric values are used as inputs denoting how mild/severe a patient feels. This layer produces language phrases that correlate to each input value as its outputs. Adaptive nodes make up the second layer that takes linguistic terms (the first layer's output) as input and output a membership grade set that looks like this:

$$L_2(xi) = \mu Ai(xi) \tag{1}$$

The triangle Membership Function is used to calculate the fuzzy value for each variable. (MF), which is set to:

$$\mu Ai(xi) = (xi - b)/(a - b) \tag{2}$$

where *a and b* are the variables that determine the shape of the triangular MF so that $b \leq xi \leq a$. The actions of the third layer's multipliers are fixed. These nodes compute the following firing strengths of the related rules:

$$L_3(xi) = \mu Ai(xi) * \mu Bi(xi) * \mu Ci(xi) \tag{3}$$

Nodes are fixed at the fourth layer. A kth rule's normalized strength is calculated as follows:

$$L_4(xi) = Wk/\sum 3j = 1Wj \tag{4}$$

The fifth layer involves the product of a rule's normalized firing strength and its corresponding output value to determine the variables' contribution to the diagnosis process:

$$L_5(xi) = L_4(xi) * L_3(xi) \tag{5}$$

The GENFIS ultimate output is represented by the sixth layer, which is made up of a single fixed node. It calculates the total number of incoming signals as follows:

$$Y = \sum ni = 1L_5(xi) \tag{6}$$

The system's output, which displays the patient's diagnostic, is as follows:

$$\text{Output} = \{\text{Mild; } 0.2 \le \text{Y} \le 0.4, \text{ Moderate; } 0.4 \le \text{Y} < 0.6, \text{ Severe; } 0.6 \le \text{Y} < 0.8\}$$

2.4 Decision Support Engine Component

The outcome of the GENFIS goes into the DSE system, which consists of filters, both cognitive and emotional. These filters handle both the physician's subjective and objective emotions towards the patient. The Cognitive filter helps to know certain medical information about the patient like allergies and current health state, while the Emotional filter provides certain information to determine what form of treatment is to be administered. It calculates the total sum of all incoming signals as follows: In conclusion, the DSE enhances the whole system's function and, as a result, aids in the efficient administration of therapy.

2.5 The Performance Metrics Used for the Proposed System

Precision: is a term used to show the amount of accuracy in the digit of a number. It refers to the amount of information that is conveyed by a number in terms of its digits. MATLAB uses 16 digits of precision by default and is further calculated using the symbolic math toolbox. Precision also shows the closeness between two or more measurements. It is defined by:

$$Precision = \frac{TP}{TP + FP} * 100\% \tag{7}$$

where TP is True Positive, FP is False Positive.

Sensitivity Analysis: The study of how the uncertainty in the result of a mathematical model or system might be separated and attributed to distinct sources of uncertainty in its input values is known as sensitivity analysis. It's also known as a method for discovering under a set of assumptions, how alternative values of an independent variable affect a given dependent variable. It is defined by:

$$Specificity = \frac{TN}{TN + FP} * 100\% \tag{8}$$

False Positive Rate (FPR): can be described as a measure of accuracy for a certain type of test such as a medical diagnostic test or a machine learning model. It is defined by:

$$FPR = \frac{FP}{TN + FP} * 100\% \tag{9}$$

Therefore overall accuracy of the system is defined by:

$$\text{OverallAccuracy} = \frac{TN + TP}{The\ total\ number\ of\ test\ items} \tag{10}$$

3 Results and Discussion

A high-performance language for technical computing is called MATLAB. For the solution of ordinary differential equations, direct algebra, statistics, Fourier analysis, filtering, optimization, numerical integration, and numerical integration, provide a huge library of functions. Additionally, it features graphics built-in for data visualization. The core of MATLAB is the MATLAB language, a matrix-based language that represents computer mathematics most intuitively.

3.1 Training the Model

The dataset for the model is loaded onto the interface (Figs. 2 and 3);

3.2 Testing the System

The chosen dataset is used to test and evaluate the system (Figs. 4 and 5).

3.3 The Proposed System Performance

Outlier, noisy, and inconsistent samples and cases were eliminated using the GA, and the mean was used to fill in for missing values. The algorithm was utilized as a feature selection tool for classification, with a 10-fold cross-validation approach, and NFIS received its output. Different features from the initial set of attributes were chosen for each run using GA, and their classification accuracy for each run was recorded. The experiment was repeated 50 times until the desired result could be obtained. The evaluation metrics gleaned from the suggested system are displayed in Table 2. The results from the experiment showed an accuracy of 96.7%, precision of 95.9%, and sensitivity of 96.3% respectively.

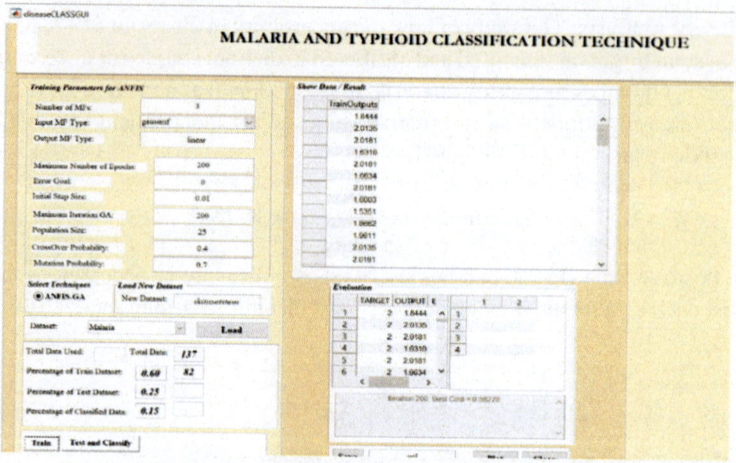

Fig. 2. Graphical user interface showing training phase

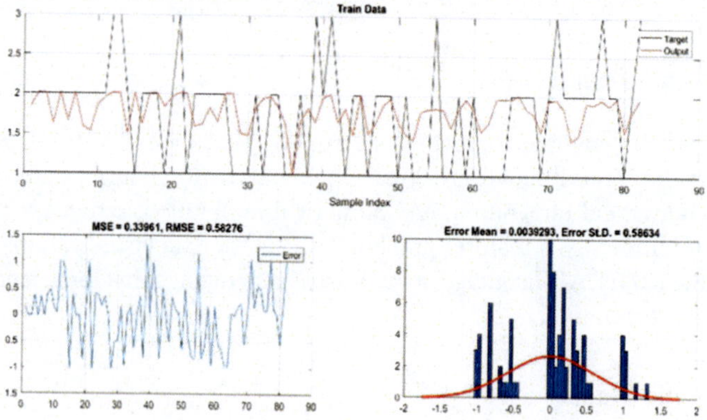

Fig. 3. Train validation showing mean square error, root mean square error, and error mean

3.4 The Comparison of the Proposed Model with ANN, and Fuzzy Logic

Table 3 shows the accuracy of the system's diagnosis of typhoid fever and malaria. The outcomes demonstrate that the hybridized system outperformed the single approach.

4 Conclusion

This study focused on the medical diagnosis of Malaria and Typhoid Fever using Genetic algorithms, Neural Networks and Fuzzy Logic in a MATLAB environment. It was designed to aid and eventually eliminate the need for manual medical diagnosis. The Genetic Neuro-Fuzzy system is a hybrid model of a Fuzzy Logic (FL) component which consists of a fuzzifier, a fuzzy inference engine and a defuzzifier used to analyse imprecise or medical information that is not complete, Neural Networks (NN) component that

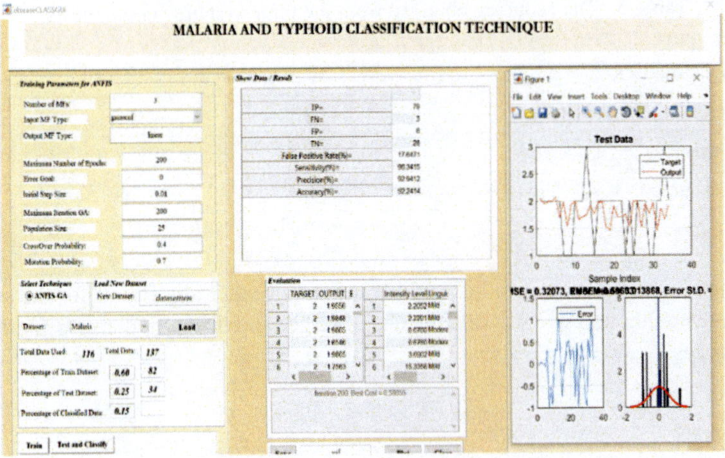

Fig. 4. Graphical user interface showing testing phase

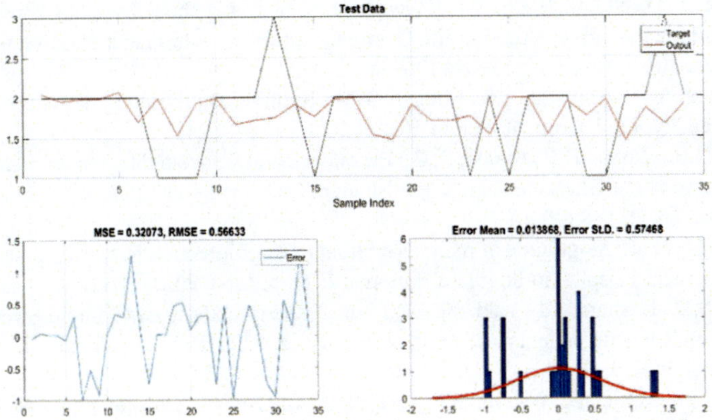

Fig. 5. Test validation showing mean square error, root mean square error, and error mean

Table 2. Confusion matrix

TP	FN	FP	TN	FPR(%)	Sensitivity(%)	Precision(%)	Accuracy(%)
79	3	6	28	17.65	97.3	96.9	98.4

spontaneously generates the values that drive the Membership Functions (MF) and a Genetic Algorithm that handles optimization of connection weights to train the Neural Network using attributes from the medical history, physical examination, and laboratory testing of the patient. This study also demonstrates the use of the genetic algorithm in a neuro-fuzzy system for effective selection of the weights of connections required to train the neural network which saves time and increases the accuracy of the system. The

Table 3. The accuracy obtained from the three systems for diagnosis

Performance metrics	In percentage
Fuzzy logic	90.7
ANN accuracy	94.5
GENFIS accuracy	98.4

recommendations for future studies include the use of other algorithms for the selection of optimum connection weights and the use of larger and different datasets for the types of all diseases.

References

1. Awotunde, J.B., Jimoh, R.G., Oladipo, I.D., Abdulraheem, M.: Prediction of malaria fever using long-short-term memory and big data. Commun. Comput. Inf. Sci. **1350**, 41–53 (2021)
2. Tessema, B., Ferede, G., Endris, M., Enawgaw, B.: Malaria, typhoid fever and their coinfection among febrile patients at a rural health Centre in Northwest Ethiopia: a cross-sectional study, Adv. Med. **2014**, Article ID 531074 (2014)
3. Maidabara, A.H., Ahmadu, A.S., Malgwi, Y.M., Ibrahim, D.: Expert system for diagnosis of malaria and typhoid. Comput. Sci. IT Res. J. **2**(1), 1–15 (2021)
4. Gontie, G.B., Wolde, H.F., Baraki, A.G.: Prevalence and associated factors of malaria among pregnant women in Sherkole district, Benishangul Gumuz regional state West Ethiopia. BMC Infect Dis. **20**(1), 1–8 (2020)
5. Alelign, D., et al.: Magnitude of malaria-typhoid fever coinfection in febrile patients at Arba Minch General Hospital in Southern Ethiopia. J. Trop. Med. **2022,** 8p (2022)
6. Vasava, S.D., Lakhani, S.J., Lakhani, J.D.: Prevalence of malaria and typhoid fever and coinfection in all febrile patients attending at tertiary care hospital in Vadodara. J. Appl. Biol. Biotechnol. **9**(5), 1–4 (2021)
7. Cheong Mun Keong, B., Wahinuddin, S.: Typhoid and Malaria Co-infection – An Interesting Finding in the Investigation of a Tropical Fever, Department of Medicine, Hospital Taiping, Jalan Taming Sari, 3400 Taiping, Perak, Malaysia (2003)
8. Institute of Medicine (US): Committee on Quality of Health Care in America. Crossing the Quality Chasm: A New Health System for the 21st Century. Washington (DC): National Academies Press (US) (2001). PMID: 25057539
9. Nesbitt T.S.: The Evolution of Telehealth: Where Have We Been and Where Are We Going? National Academies Press, Washington, DC (2012)
10. Ayo, F.E., Misra, S., Awotunde, J.B., Behera, R.K., Oluranti, J., Ahuja, R.: A mobile-based patient surgical appointment system using fuzzy logic. In: Proceedings of Third International Conference on Computing, Communications, and Cyber-Security, pp. 193–207. Springer, Singapore (2021). https://doi.org/10.1007/978-981-19-1142-2_15
11. Asogbon, M.G., Samuel, O.W., Omisore, M.O, Awonusi, O.: Enhanced neuro-fuzzy system based on genetic algorithm for medical diagnosis. J. Med. Diagn. Meth. **5,** 205 (2016)
12. Oladipo, I.D., Babatunde, A.O., Awotunde, J.B., Abdulraheem, M.: An improved hybridization in the diagnosis of diabetes mellitus using selected computational intelligence. In: International Conference on Information and Communication Technology and Applications, pp. 272–285 (2020). https://doi.org/10.1007/978-3-030-69143-1_22

13. Saramourtsis, A., Damousis, J., Bakirzis, A., Dokopoulos, P.: Genetic algorithm solution to the economic dispatch problem-application to the electrical power grid of crete island. Aristotle University of Thessaloniki, Department of Electrical and Computer Engineers Power System Laboratory, 540 06 Thessaloniki, Greece (1999)
14. Uzoka, F.M., et al.: Risk factors for some tropical diseases in an African country. BMC Public Health **21**(1), 1–10 (2021)
15. Arji, G., et al.: Fuzzy logic approach for infectious disease diagnosis: a methodical evaluation, literature and classification. Biocybernet. Biomed. Eng. **39**(4), 937–955 (2019)

An Image Encryption Algorithm Based on Substitution and Diffusion Chaotic Boxes

Younes Qobbi[1](✉), Abdellah Abid[1], Mariem Jarjar[1], Samir El Kaddouhi[3], Abdellatif Jarjar[2], and Abdelhamid Benazzi[1]

[1] MATSI Laboratory, Mohamed First University, Oujda, Morocco
qobbi.younes@ump.ac.ma
[2] High School Moulay Rachid, Taza, Morocco
[3] IEVIA Team, IMAGE Laboratory, Department of Sciences, Ecole Normale Supérieure, Moulay Ismail University of Meknès, Meknes, Morocco

Abstract. In this article, a new method for constructing a substitution table will be proposed for implementation in an image encryption algorithm controlled by chaotic maps. The substitution and diffusion operations are based on new functions. After loading the original image, this image is transformed into three vectors. The encryption of an image goes through two phases. In the first phase, a change of the value of a pixel is established using S-box. In the next phase, a CBC diffusion operating mode is used by integrating a new diffusion function and a second substitution table denoted D-box. This last operation protects our system from any differential attack.

Keywords: S-box · D-box · Substitution · Diffusion · Chaotic map

1 Introduction

Due to the massive transfer of image and multimedia data through unsecure channels, it is more and more important to make this kind of data private during transmission. So, the efficient way to achieve this requirement is encryption. Indeed, several classical encryption schemes such as IDEA, DES, 3DES and AES Have been suggested [1–4]. However, due to some intrinsic features of digital images, such as redundancy, high correlation of adjacent pixels and bulky capacity of data, the traditional algorithms are not suitable, especially exhibit some drawbacks and weakness [5].

Due to specific characteristics of chaotic systems, such as sensitivity to initial condition, and control parameters in one hand. The close relationship between the chaotic system and cryptography in other hand, the chaotic systems are considered good for image encryption, especially in confusion and diffusion processes using the substitution Boxes. Recently, many researchers have been focalised their works on different types of image encryption algorithms based on substitution and diffusion using chaotic systems [6–10].

In our method we propose a new technique based on two chaotic maps to generate two large S-boxes denoted S-box and D-box. Which are used in our image encryption for substitution and diffusion process respectively.

Y. Farhaoui et al. (Eds.): ICAISE 2022, LNNS 635, pp. 184–190, 2023.
https://doi.org/10.1007/978-3-031-26254-8_26

2 Chaotic Sequences

Our approach based on discrete chaotic maps for generation of pseudo-random numbers used in the construction of our chaotic boxes. These chaotic maps are described by the following equations:

PWLMCM map	Logistic map
$x_{n+1} = \begin{cases} \frac{x_n}{p} \; si\, 0 \le x_n \le p \\ \frac{x_n - p}{0.5-p} \; si\, p \le x_n \le 0.5 \quad (1) \\ f(1 - X_n, p) Otherwise \end{cases}$ $p \in]0, 0.5[$ and $x_0 \in]0, 1[$	$y_{n+1} = \mu_1 y_n (1 - y_n)$ (2) $\mu 1 \in [3.57, 4]$ *and* $y_0 \in [0.5, 1[$

3 Encryption Setting Design

3.1 S-Box and D-Box Construction

After loading the plain image of size (N, M). where N and M represent the height and width of plain image respectively. This image is transformed into three vectors (VR), (VG) and (VB) of size $(1, N \times M)$.The construction of S-box (SB) and D-box (DB) goes through the following steps:

Step 1: Set of initial conditions and control parameters of the logistic and PWLCM maps.

Step 2: Generate two sequences of pseudo-random numbers $\boldsymbol{x}$ and $\boldsymbol{y}$ of size $N \times M$.

Step 3: Generate two chaotic integer sequences (U) and (V) of size (1 × 256) by using the following algorithm.

```
For i = 0 to 255
U(i) = mod(x(i)×10^10 , 256)
V(i) = mod(y(i)×10^10 , 256)
Next i
```

Step 4: An ascending sort of elements of (U) and (V) provides two permutation vectors P1 and Q1 of size (1 × 256).

Step 5: The elements of (P1) and (Q1) are used to set two first rows of (SB). The other rows are obtained by using the following algorithm:

```
For i = 2 to 255
For j = 0 ... 255
SB(i,j) = SB(i-1, SB(i-2, j))
Next j, Next i
```

Step 6: A descending sort of elements of (U) and (V) provides two other permutation vectors (P2) and (Q2) of size (1 × 256).

Step 7: The elements of (P2) and (Q2) are used to set the two first rows of **(DB)**. The other rows are generated by the following algorithm:

```
For i = 2 to 255
For j = 0 … 255
DB(i,j) = DB(i-2, DB(i-1, j))
Next j, next j
```

3.2 Initialization Vector Computation

The initialization vector (Iv) of size (1, 3). This vector is computed from the original image by using the following algorithm:

```
Iv(o)=Iv(0) ⊕ VR(i)
Iv(1)=Iv(1) ⊕ VG(i)
Iv(2)=Iv(1) ⊕ VB(i)
```

3.3 Chaotic Vector Generation

Generate an integer sequences T of size (1, N × M) by using the equation as follows:

$$T(i) = \text{mod}(x(i) \times 10^{10}, 256) \tag{3}$$

This vector will be used in the new encryption functions.

4 Encryption and Decryption Process

4.1 Encryption Process

The improved S-box and D-box based image encryption algorithm in this work includes the following steps:

Step 1: the first original pixel value is changed by using the initialization vector (Iv) and bitwise operation.

Step 2: the result pixel input in the S-box for substitution phase using the new encryption function as follows:

```
VR(i)= SB(T(i), mod(a×VR(i)+b,256))
VG(i)= SB(T(i), mod(a×VG(i)+b,256))
VB(i)= SB(T(i), mod(a×VB(i)+b,256))
```

Where a is an odd number in Z/256 Z. This condition ensures that the number a is invertible in Z/256Z.

Step 3: The output pixel from the S-box is going through D-box and linked with the next original pixel by using the new function of diffusion as follows:

$$\mathtt{VR(i+1)} = \mathtt{DB(T(i),\ mod(c \times VR(i) + d,256))} \oplus \mathtt{VR(i)}$$
$$\mathtt{VG(i+1)} = \mathtt{DB(T(i),\ mod(c \times VG(i) + d,256))} \oplus \mathtt{VG(i)}$$
$$\mathtt{VB(i+1)} = \mathtt{DB(T(i),\ mod(c \times VB(i) + d,256))} \oplus \mathtt{VB(i)}$$

Where c is an odd number in Z/256Z. This condition ensures that the number c is invertible in Z/256Z. The diagram below (Fig. 1) represents the encryption process of our proposed method.

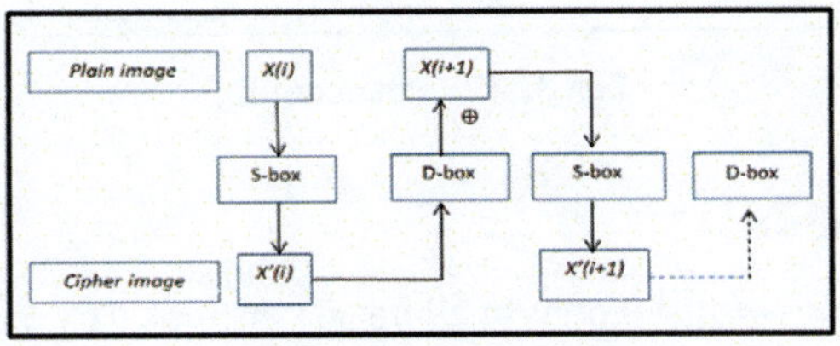

Fig. 1. Encryption process

4.2 Decryption Process

The proposed algorithm in this paper is based on Substitution and Diffusion, which require starting a decryption process from the last encrypted pixel to the first. It is goes through the following steps:

Step 1: Transform the cipher image of size (N, M) into three vectors (VR), (VG) and (VB) of size $(1, \mathrm{N} \times \mathrm{M})$.

Step 2: Reverse substitution phase using the inverses S-box denoted BS.

Step 3: reversion diffusion process to find the original pixel using the inverse function of diffusion.

5 Experimental Simulation and Security Analysis

To test the security of the proposed algorithm against known attacks, some simulations are carried out by our image encryption on several images, such as Lena (256 × 256) and Peppers (512 × 512. The plain images and their corresponding encrypted images are displayed in the Fig. 2.

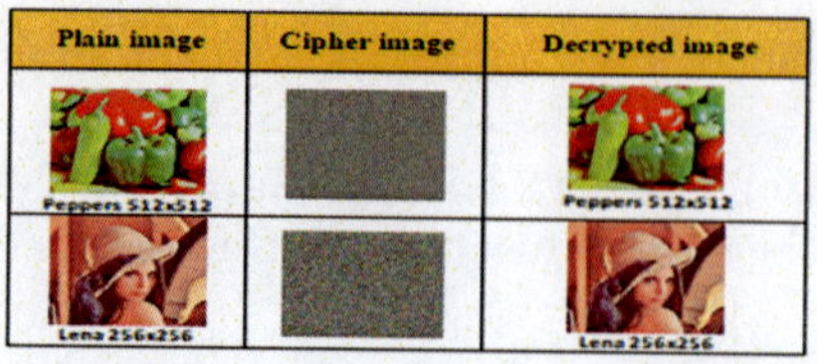

Fig. 2. Plain, cipher and decrypted images

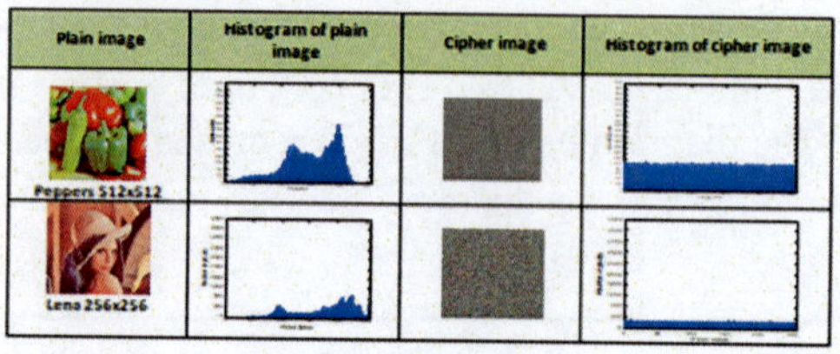

Fig. 3. Histogram result

5.1 Statistical Analysis

Histogram Analysis. The histograms of all images encrypted by our algorithm are flat as shown in the figure (Fig. 3). Which prove that our system is efficient against statistical attacks.

Information Entropy. The equation for calculating the information entropy H(m) of an image is as follows:

$$H(m) = -\sum_{i=1}^{255} \Pr(m_i) \times \log \Pr(m_i) \tag{4}$$

The information entropy values of an encrypted image by our algorithm are shown in the Table 1.

Correlation Coefficient Analysis. The adjacent pixels of a plain image are highly correlated in the three directions, and there should be no correlation between adjacent pixels in the cipher image.

The correlation between adjacent pixels of the plain images and the encrypted images of Lena (256 × 256) and Baboon (512 × 512) are shown in the Table 1.

5.2 Differential Analysis

The sensitivity is evaluated through a NPCR (Number of Pixels Change Rate) and UACI (Unified Average Changing Intensity). The following equations give expression of NPCR and UACI:

$$NPCR = \frac{\sum_1^w \sum_1^h D_{ij}}{w * h} \times 100\% \tag{5}$$

$$UACI = \frac{1}{w \times h} \frac{\sum_{ij} |IC1_{ij} - IC2_{ij}|}{255} \times 100\% \quad (6)$$

The result of NPCR and UACI corresponding to the slightly changed plaintext image in our system are shown in Table1 .

Table 1. Correlation Coefficients, Entropy, NPCR and UACI values

Image	HC	VC	DC	Entropy	NPCR	UACI
Cipher Lena (256 × 256)	– 0.0091	– 0.0022	– 0.0023	7.9992	100%	33.52%
Cipher Peppers (512 × 512)	0.0025	– 0.0010	– 0.0065	7.9998	100%	33.68%

6 Comparison with Other Works

See Table 2

Table 2. Comparison with other methods

Method	Entropy		NPCR %		UACI %	
	Lena	Peppers	Lena	Peppers	Lena	Peppers
Our	7.9992	7.9998	100	100	33.5204	33.684
Ref [9]	–	7.9843	–	99.59	–	33.6839
Ref [10]	7.9993	–	99.6149	–	33.4554	

7 Conclusion

This work exposes a new improvement of the classical method of Vigenère. The proposed method ensured by two new substitution tables S-box and D-box, the construction of two boxes is based on chaotic permutations and translations, attached to new substitution functions in the used ring. The use of a new diffusion technique protects our image encryption algorithm from any differential attack. The simulation results and security analysis presented by a correlation coefficient close to zero and a uniform histogram of the ciphered images ensure entropy close to theoretical value, which prove that our system is efficient against any known attack.

References

1. Habutsu, T., Nishio, Y., Sasase, I., Mori, S.: A secret key cryptosystem by iterating a chaotic map. In: Davies, D. W. (ed.) EUROCRYPT 1991. LNCS, vol. 547, pp. 127–140. Springer, Heidelberg (1991). https://doi.org/10.1007/3-540-46416-6_11
2. Schneier, B.: Applied Cryptography: Protocols, Algorithms, and Source Code in C. Wiley, Hoboken (2007)
3. Coppersmith, D.: The data encryption standard (DES) and its strength against attacks. IBM J. Res. Dev. **38**(3), 243–250 (1994)
4. Patidar, V., Pareek, N.K., Purohit, G., Sud, K.K.: Modified substitution–diffusion image cipher using chaotic standard and logistic maps. Commun. Nonlinear Sci. Numer. Simul. **15**(10), 2755–2765 (2010)
5. Li, S., Chen, G., Zheng, X.: Chaos-based encryption for digital image and video. In: Multimedia Encryption and Authentication Techniques and Applications, pp. 129 (2006)
6. Farah, M.A.B., Farah, A., Farah, T.: An image encryption scheme based on a new hybrid chaotic map and optimized substitution box. Nonlinear Dyn. **99**(4), 3041–3064 (2019). https://doi.org/10.1007/s11071-019-05413-8
7. Ullah, A., Jamal, S.S., Shah, T.: A novel scheme for image encryption using substitution box and chaotic system. Nonlinear Dyn. **91**(1), 359–370 (2017). https://doi.org/10.1007/s11071-017-3874-6
8. Qobbi, Y., Jarjar, A., Essaid, M., Benazzi, A.: Development of large chaotic s-boxes for image encryption. In: Motahhir, S., Bossoufi, B. (eds.) ICDTA 2021. LNNS, vol. 211, pp. 847–858. Springer, Cham (2021). https://doi.org/10.1007/978-3-030-73882-2_77
9. Ullah, A., Jamal, S.S., Shah, T.: A novel construction of substitution box using a combination of chaotic maps with improved chaotic range. Nonlinear Dyn. **88**(4), 2757–2769 (2017). https://doi.org/10.1007/s11071-017-3409-1
10. Hasanzadeh, E., Yaghoobi, M.: A novel color image encryption algorithm based on substitution box and hyper-chaotic system with fractal keys. Multimedia Tools Appl. 79, 1–19 (2019). https://doi.org/10.1007/s11042-019-08342-1

An Image Encryption Scheme Based on DNA Sequence Operations and Chaotic System

Mariem Jarjar[1], Abdellah Abid[1], Younes Qobbi[1], Samir El Kaddouhi[3(✉)], Abdellhamid Benazzi[1], and Abdellatif Jarjar[2]

[1] MATSI Laboratory Mohamed First University Oujda, Meknes, Morocco
{qobbi.younes,a.benazzi}@ump.ac.ma
[2] Hight School Moulay Rachid Taza, Taza, Morocco
[3] IEVIA Team, IMAGE Laboratory, Department of Sciences, Ecole Normale Supérieure, Moulay Ismail University of Meknès, Meknes, Morocco
s.elkaddouhi@umi.ac.ma

Abstract. This work presents a new method for encrypting medical and color images. This technique is based on the use of genetic operations acting on DNA sequences controlled by chaotic maps. To overcome the problem of uniform images, the original image is confused with a chaotic image. Then to prepare the image for genetic operations, a passage to the nucleotide sequence is established. The second phase of image encryption consists of applying the mutation and inversion operations. Several simulations performed on a large number of images of different sizes and formats ensure that our method is not subject to any known attacks.

Keywords: Vigenere grid · Chaotic map · Broadcast function · Genetic operator

1 Introduction

With the rapid development of information and communication technology, multimedia information needs to be extended and transmitted through the network. Due to the opening and sharing of the network, the security of multimedia information, in particular the security of the digital image, has become an important issue. One of the methods of protecting images is to use encryption algorithms, which convert the original image into an encrypted image that is not easy to understand [1, 2].

On the one hand, different kinds of image encryption methods based on different technologies have presented, among them, chaos-based techniques are more interested in recent years [3–5]. Chaotic encryption was characterized by high sensitivity to initial conditions, pseudo-random behavior, non-periodicity, simplicity of hardware and software implementation [3]. Therefore, it offers researchers the possibility to improve some classical techniques, like Hill [6, 7], Vigenère [8, 9], Feistel [10, 11].

Y. Farhaoui et al. (Eds.): ICAISE 2022, LNNS 635, pp. 191–198, 2023.
https://doi.org/10.1007/978-3-031-26254-8_27

On the other hand, genetic algorithms are widely used in the field of image encryption. They use selection, crossover and mutation operators inspired by evolutionary biology. The majority of these algorithms involve a calculation on the DNA by assigning static values to the nucleotides (A = 0, C = 1, T = 2, G = 3) [12–15]. However, in our work, we proposed to associate chaotic values to nucleotides from a table generated by permutations.

2 The Proposed Method

2.1 Chaotic Sequences Development

In order to build a new algorithm using a single-encryption key, we will use the 2D logistics map [16]. This choice is due to the simplicity of its development and its high sensitivity to the initial parameters. From the map used, we generate five chaotic vectors **XL**, **XR**, **YL** **CL** and **ML** in Z/256Z and a control binary vector **BV**.

2.2 Plain Image Preparation

The original image of size (n, m) is decomposed into three one-dimensional vectors **VR**, **VG** and **VB** of size (1, nm). Then, VR, VG and VB are concatenated into a single vector **X** (x1, x2,….., x3 nm) of size (1, 3 nm).

An initialization value (**IV**) will be calculated from the original image for a boot block modification and start the encryption process.

2.3 Diffusion Process

Algorithm 1 : Diffusion process

```
X(0)=X(0) ⊕ IV
For i=1 to 3nm-1
   X(i)=X(i) ⊕ YL(i)
If (i<3nm-1) then
      X(i+1)=X(i) ⊕ X(i+1) ⊕ XL(i)
Next i
```

2.4 Passage to Binary Vector

The Matrix **TB** gives the binary writing of the numbers between 0 *and* 3.

For the pixel $X(i)$, the transformation into (XB) binary follow the algorithm below

```
Algorithm 2 : Transformation into binary vector
For i=0 to 3nm-1
    G=E(X(i)/16) ; D= X(i)-16*G;
    a=E(G/4) ; b=G-4*a ; c=E(D/4) ; d=D-4*c;
    XB(8*i)=TB(a,0)
    XB(8*i+1)=TB(a,1)
    XB(8*i+2)=TB(b,0)
    XB(8*i+3)=TB(b,1)
    XB(8*i+4)=TB(c,0)
    XB(8*i+5)=TB(c,1)
    XB(8*i+6)=TB(d,0)
    XB(8*i+7)=TB(d,1)
Next i
```

XB is subdivided into two blocks XB1 and XB2 of size (1.12 nm).

2.5 Transition to Nucleotid Vector

Let be the Table 1 **(PN)** of some exhibited chaotic values of the generated chaotic vectors.

Table 1. PN table

1	$XL(n)$	$XL(m)$	$XR(nm)$	$YL(n+m)$
2	$XR(n)$	$YL(m)$	$XL(nm)$	$YL(m)$

- (P_1) permutation obtained by a descending sort on the values of the first row of the table (PN), indicates the value of the nucleotide
- (P_2) permutation obtained by a descending sort on the values of the second row of the table (PN), indicates the value of the complementary of the nucleotide

The Table 2 **(TN)** represents the new chaotic values of the nucleotides.

Table 2. TN table

Rank		0	1	2	3
1	1	A	C	T	G
2	(P_1)	$P_1(A)$	$P_1(C)$	$P_1(T)$	$P_1(G)$
3	(P_2)	$P_2(A)$	$P_2(C)$	$P_2(T)$	$P_2(G)$

Table 3. Entropy values

Name	Image	Entropy of cipher image	Ref [14]
Image 1	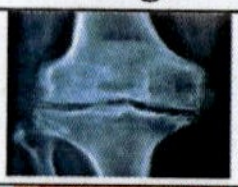	7.9995	------
Image 2		7.9992	7.9992

Table 4. Correlation values of adjacent pixels

Name	Horizontal correlation	Vertical correlation	Diagonal correlation
Image 1	0.0015	0.0025	0.0027
Image 2	0.0007	0.0045	0.0025

Table 5. NPCR and UACI values

	Our Method		Ref [14]	
Name	NPCR %	UACI %	NPCR %	UACI %
Image 1	99.61	33.45	----	----
Image 2	99.63	33.46	99.61	33.36

2.6 Generation of a Chaotic Nucleotide Vectors

We generate two nucleotide tables (**TC** and **TM**) by the following algorithm:

```
Algorithm 3 : Chaotic nucleotide vectors
For i=0 to 12nm-1
   TC(i)=TN(0,CL(i))
   TM(i)=TN(1,CL(i))
Next i
```

2.7 Transformation of the Binary Image to Nucleotide Image

A chaotic transition table (*TA*) of size (12 nm, 3) is generated by the following process.

- Q1: permutation obtained by an ascending sort on the first 12 nm values of the vector (*XL*), defining the first column of the table (*TA*)
- Q2: permutation obtained by an ascending sort on the first 12 nm values of the vector (*XR*), defining the second column of the table (*TA*)

- Q3: permutation obtained by an ascending sort on the first 12 nm values of the vector (*YL*), defining the third column of the table (*TA*)

The vector (*XN*) is given by the algorithm below.

```
Algorithm 4 : Passage to nucleotide level
For i=0 to 12nm-1
   S=XB1(TA(i,0))*2+XB2(TA(i,1))
   XN(TA(i,2))=TN(1,s)
Next i
```

2.8 Genetic Operations

Reversion Operation. The reversion operation consists in the change of a nucleotide into its complement (**XC**) according to the value of the control binary vector BV This transaction is described by the following algorithm

```
Algorithm 5 : Reversion operation
For i=0 to 12nm-1
If BV(i)=0 then
    XC(i)=XN(i) ⊕ TC(i)
Else  XC(i)= TN(1,XN(i)) ⊕ TM (i)
Next i
```

Adapted Genetic Mutation. We subdivide the vector XC into two vectors B and C of size (1, 6 nm). B and C are subdividing into **M** sub-blocks of 6 n nucleotides. This mutation is entirely controlled by the vector BV. If $BV(i) = 0$ then exchange between block B(i) and C(i)

2.9 Decryption Process

Our algorithm is a symmetrical encryption system; therefore, the reciprocal functions will be used in the recovery of the original image. The steps of decryption process are:

- Encrypted image loading.
- Transformation into single vector.
- Passage in DNA sequence
- Reverse mutation and reversion operations
- Passage into binary code
- Passage into Z/256Z ring.
- Reverse confusion.
- Construction of the original image.

3 Simulation and Security Analysis

3.1 Key-Space Analysis

The chaotic sequence used in our method ensures strong sensitivity to initial conditions, and can protect it from any brutal attacks. If we use single-precision real numbers $\mathbf{10^{-10}}$ to operate, the total size of the key will greatly exceed $\approx \mathbf{2^{180}} \gg \mathbf{2^{110}}$.

3.2 Secret Key's Sensitivity Analysis

Our encryption algorithm has a very high sensitivity to initial conditions, which means that a small degradation of a single parameter used will automatically make a huge difference to the encrypted image. This ensures that without the real encryption key, the original image cannot be restored.

3.3 Histogram Analysis

All images tested by our algorithm have a uniformly distributed histogram.

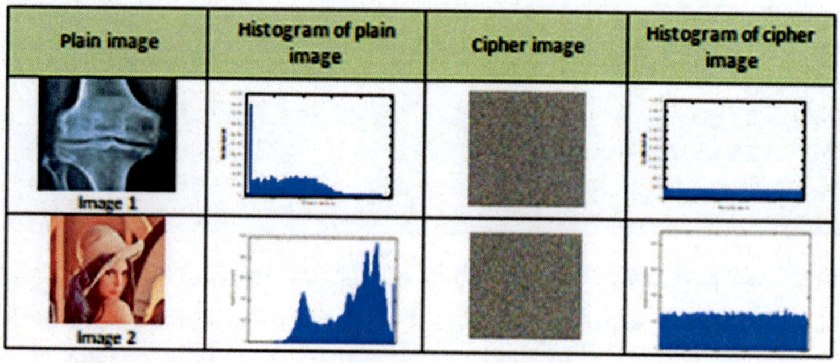

Fig. 1. Plain and Cipher image histograms

3.4 Entropy Analysis

The entropy expression is determined by the equation below:

$$H(MC) = \frac{1}{s}\sum_{i=1}^{s} - p(i)log_2(p(i)) \tag{1}$$

All the entropy values of the images tested by our algorithm are close to 8. This proves that the method is far from a statistical attack.

3.5 Correlation Analysis

Correlation is a technique that compares two images to estimate the displacement of pixels in one image relative to another reference image. The relevant expression is defined by the following equation

$$r = \frac{cov(x, y)}{\sqrt{V(x)}\sqrt{V(y)}} \tag{2}$$

3.6 Differential Analysis

Let be two encrypted images, whose corresponding free-to-air images differ by only one pixel, from $(\boldsymbol{C}_1)$ and $(\boldsymbol{C}_2)$, respectively. The *NPCR* and UACI mathematical analysis of an image are given by the Eq. 4 and 5.

$$NPCR = \left(\frac{1}{nm}\sum\nolimits_{i,j=1}^{nm} D(i, j)\right) * 100 \tag{3}$$

$$UACI = \left(\frac{1}{nm}\sum\nolimits_{i,j=1}^{nm} Abs(C_1(i, j) - C_2(i, j))\right) * 100 \tag{4}$$

$$WithD(i, j) = \begin{cases} 1 \; if \;\; C_1(i, j) \neq C_2(i, j) \\ 0 \; if \;\; C_1(i, j) = C_2(i, j) \end{cases}$$

4 Conclusion

In this article, we have proposed a method for encrypting color and medical images of various sizes. Before starting the encryption process, the original image is subjected to several transformations, namely a binary transformation and then DNA sequence transformation. The encryption process goes through a phase of confusion with a chaotic image then a phase of inversion and mutation operations controlled by chaotic maps. The results of several security tests prove the robustness of our approach against different types of attacks.

References

1. Kumari, M., Gupta, S., Sardana, P.: A survey of image encryption algorithms. 3D Res. **8**(4), 1–35 (2017). https://doi.org/10.1007/s13319-017-0148-5
2. Puteaux, P., Ong, S.Y., Wong, K.S., Puech, W.: A survey of reversible data hiding in encrypted images – The first 12 years. J. Vis. Commun. Image Represent. 77, (2021). https://doi.org/10.1016/j.jvcir.2021.103085
3. Zia, U., McCartney, M., Scotney, B., Martinez, J., AbuTair, M., Memon, J., Sajjad, A.: Survey on image encryption techniques using chaotic maps in spatial, transform and spatiotemporal domains. Int. J. Inf. Secur. **21**(4), 917–935 (2022). https://doi.org/10.1007/s10207-022-00588-5

4. Wang, X., Teng, L., Qin, X.: A novel colour image encryption algorithm based on chaos. Signal Process. **92**, 1101–1108 (2012). https://doi.org/10.1016/j.sigpro.2011.10.023
5. Brahim, A.H., Pacha, A.A., Said, N.H.A.D.J.: A new image encryption scheme based on a hyperchaotic system & multi specific S-boxes. Inf. Secur. J. 00, 1–17 (2021). https://doi.org/10.1080/19393555.2021.1943572
6. Essaid, M., Akharraz, I., Saaidi, A., Mouhib, A.: Image encryption scheme based on a new secure variant of Hill cipher and 1D chaotic maps. J. Inf. Secur. Appl. **47**, 173–187 (2019). https://doi.org/10.1016/j.jisa.2019.05.006
7. Hraoui, S., Gmira, F., Abbou, M.F., Oulidi, A.J., Jarjar, A.: A new cryptosystem of color image using a dynamic-Chaos Hill cipher algorithm. Procedia Comput. Sci. **148**, 399–408 (2019). https://doi.org/10.1016/j.procs.2019.01.048
8. Li, S., Zhao, Y., Qu, B., Wang, J.: Image scrambling based on chaotic sequences and Veginère cipher. Multimed. Tools Appl. **66**, 573–588 (2013). https://doi.org/10.1007/s11042-012-1281-z
9. Touil, H., Akkad, N.E.L., Satori, K.: Text Encryption: Hybrid cryptographic method using Vigenere and Hill Ciphers. In: 2020 International Conference Intelligent System Computation Vision, ISCV 2020 (2020). https://doi.org/10.1109/ISCV49265.2020.9204095
10. Yao, W., Zhang, X., Zheng, Z., Qiu, W.: A colour image encryption algorithm using 4-pixel Feistel structure and multiple chaotic systems. Nonlinear Dyn. **81**(1–2), 151–168 (2015). https://doi.org/10.1007/s11071-015-1979-3
11. Zhang, X., Zhou, Z., Niu, Y.: An image encryption method based on the Feistel network and dynamic DNA encoding. IEEE Photonics J. **10**, 1–14 (2018). https://doi.org/10.1109/JPHOT.2018.2859257
12. Wang, X., Zhao, M.: An image encryption algorithm based on hyperchaotic system and DNA coding. Opt. Laser Technol. **143**, 107316 (2021). https://doi.org/10.1016/j.optlastec.2021.107316
13. Zhang, S., Liu, L.: A novel image encryption algorithm based on SPWLCM and DNA coding. Math. Comput. Simul. **190**, 723–744 (2021). https://doi.org/10.1016/j.matcom.2021.06.012
14. Bhat, J., Moon, A.H.: Color image encryption and authentication using dynamic DNA encoding and hyper chaotic system. Expert Syst. Appl. **206**, 117861 (2022). https://doi.org/10.1016/j.eswa.2022.117861
15. Mahmud, M., Atta-ur-Rahman, Lee, M., Choi, J.Y.: Evolutionary-based image encryption using RNA codons truth table. Opt. Laser Technol. **121**, 105818 (2020). https://doi.org/10.1016/j.optlastec.2019.105818
16. Gao, X.: Image encryption algorithm based on 2D hyperchaotic map. Opt. Laser Technol. **142**, 107252 (2021). https://doi.org/10.1016/j.optlastec.2021.107252

An Improvement to the Cloud Service Research and Selection System's Usage of the Skyline Algorithm

Imane El Khammar[1](✉), Mohamed El Ghmary[2], and Abdellah Idrissi[1]

[1] Department of Computer Science, Faculty of Sciences, Mohammed V University, Rabat, Morocco
{imane_elkhammar,a.idrissi}@um5.ac.ma
[2] Department of Computer Science, FSDM, Sidi Mohamed Ben Abdellah University, Fez, Morocco
mohamed.elghmary@usmba.ac.ma

Abstract. The popularity of cloud computing has led to a growth in the number of cloud services, consumers now have a wide range of cloud services to choose from. Selecting the ideal cloud service from a plethora of commonly available options has become extremely complex, especially for non-computer science consumers. It is difficult for clients to choose cloud service based on which best meets their needs. This research focuses on a cloud service selection mechanism that allows consumers to establish their perception of quality criteria in this huge amount of service. Our approach is based on a skyline algorithm which is known for his power in selecting from large databases to define the most dominant cloud service, and this will increase the speed of our tool to calculate the Skyline and to have a more powerful, general and efficient tool. In this work we discuss our algorithm and some experimental findings that demonstrate the great potential of our strategy.

Keywords: Cloud computing · Cloud service selection · Skyline Algorithm · Branch-and- bound skyline algorithm

1 Introduction

Cloud computing as an idea is rapidly gaining traction as internet technologies evolve [1]. It allows users to store and manage data, install models and architectures, and access resources on the Internet by providing computing services. Accordingly, it strives to make big data secure, reliable, scalable and shareable using web technologies [2]. Cloud computing allows computing devices to collaborate and reduces the need for platform adequacy. Clients often employ a third-party cloud platform for their computing and storage needs, paying exclusively for the products they actually use. Cloud computing has countless advantages over traditional computing in terms of price, effectiveness, and dependability [3].

Customers or vendors can use cloud computing to access a variety of services, including: (SaaS), (PaaS), (IaaS), and (DaaS) [4].

Y. Farhaoui et al. (Eds.): ICAISE 2022, LNNS 635, pp. 199–205, 2023.
https://doi.org/10.1007/978-3-031-26254-8_28

Public, private, community, and hybrid deployment methods are the most often used ones for cloud services [5, 6].

One of the most difficult aspects of choosing a cloud service is the variety of cloud computing, which makes comparing one cloud service to another tough. Another barrier for cloud service selection is a lack of awareness of a service's nonfunctional qualities. Because numerous cloud providers give similar services and multiple services are supplied by cloud provider with varying quality characteristics, a technique to assist cloud customers in selecting the best services is required.

Processing skyline queries has received a lot of attention lately [7]. Several applications, like information suggestion, multi-criteria decision making, and intelligent data analysis used to resolve issues in research like web service choosing, depend heavily on skyline outcomes, query processing over uncertain data [8]. The Skyline operator may manage several competing items at once and return all of the ones that are not in control of the others [9].

In our BB-CSRSS tool, we have generalized work of [10] too deals with the large databases, we opted for an approach based on finding the nearest neighbor with BBS skyline algorithm.

The remainder of the paper is structured as follows. Associated research initiatives are covered in Sect. 2. Section 3 goes over our suggested approach. Results and comments are included in Sect. 4. And a conclusion is found in Sect. 5.

2 Related Work

In the database literature, many types and variations of skyline approaches have been documented. The majority of these skyline approaches aimed to increase the speed with which skyline queries were processed, however most of them do not handle cases where the database is huge. In the following, we present the related approaches for processing skyline queries and the approaches based on clustering.

in this work [10] they development a Web Agent utilizing the Skyline approach to identify which Cloud services best satisfy users' needs, this algorithm is based on the BNL Skyline algorithm. But using the BNL algorithm they have neglected the multidimensional databases and the large databases.

In this work [11], they directly tackled the Input/Output operations used to calculated skyline to address the issue of performance, both MinMax and Euclidean distance were used to initialize the input list for the skyline, and Euclidean distance was also used to sort the input list.

In this research [12], they present a two-phase technique for computing skylines, based on Quickhull, which divides the computation into many phases by finding the supported no dominated points and BNL algorithms, which compute the unsupported no dominated points.

As a solution for the skyline queries' time cost, which rises exponentially as the quantity and the number of skyline queries increase. In order to efficiently react, [13] they suggested a cost-based (CA) algorithm that does not require the implementation of skyline queries based on enormous base data. Instead, it merely pre-stores an ideal collection of skyline views. In two stages approximate calculation and adjustment.CA employs the map/reduce distributed computing approach; can swiftly build the ideal collection of skyline views.

they proposed [14], a combination of an outranking method (ELECTRE IS) system and selection services with the DENCLUE-IM clustering algorithm which showed promising results but still based on data which is not large.

To increase the effectiveness of the selecting process, [15] they employed a strategy based on the hierarchical clustering technique and skyline, Only the best candidates are provided to the selection module by skyline based on the clustering findings and mapping of end user requests to a cluster.

This work [16], uses the improved Skyline algorithm to locate the point in the dataset closest to the origin and confirms that point to be the SP point to filter out and eliminate dominance checks. The real SP services are then separated from the set of suspicious SP services using the bitmap approach.

Although these works addressed the challenge of looking for and choosing Cloud services, the majority of them dealt with the issue of scaling down the skyline set. The majority of research that has been done ignores the selection issue in a huge database. We outline our strategy for handling this aspect below.

3 Our Proposed Approach

Regardless of the quality of their service, cloud providers use a range of marketing strategies to entice customers to use their services, therefore cloud services are increasing day by day. To deliver superior cloud services, we applied our approach BB-CSRSS it's a generalization or an improvement of CSRSS [10], they used the BNL algorithm, it is a powerful algorithm but in small databases, we adapted the BBS algorithm, it is the most effective algorithm for handling large data sets [17], which is an important criteria to have a preferment Cloud Service selection Methods (CSSMs) [18].

BBS only travels through the R-tree once, in contrast to NN. Data is organized in R-tree. Each node in the R-tree can host three entries. Additionally, all leaf nodes have the ability to hold three input items too. Similar to NN, where data points are arranged in accordance with their minimum distances from an origin point (mindist) or minimum bounding rectangles (MBR), Calculating the mindist of a point involves adding up its coordinates, while calculating the mindist of an (MBR) entails measuring the distance between the (MBR)'s lower-left corner and the origin. The method selects the nearest tree points to the origin among all the unexplored places at each stage. Additionally, it maintains these discovered points in a set list in memory for the dominance validation phase (Fig. 1).

The BBS algorithm's pseudo code is then presented [19].

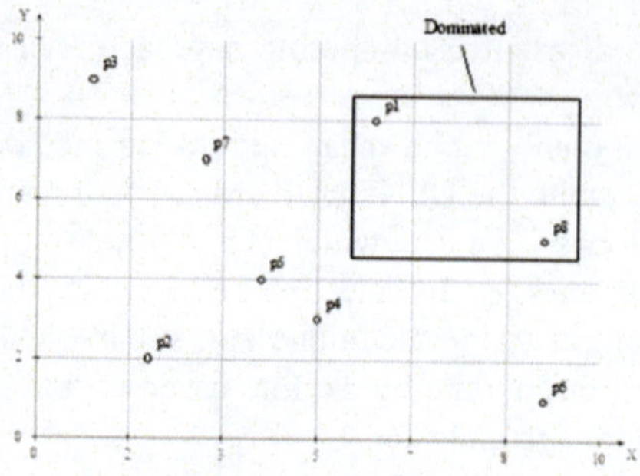

Fig. 1. BBS algorithm illustration.

Input: A data (R-tree).
Output: The skyline points of dataset B.

```
L=Ø // L is a list of dominant points
add all entries of the root D in the heap.
while B not empty yet do
    remove d// d is the top entry from D
    if d is dominated by some point in L do remove d
      else // d is not dominated
        if d is an intermediate entry then
          for each child di of d
              if di is not dominated by some point in L insert di into heap
            else // d is a data point
              insert di into L
            end
          end
        end
    end
end while
```

The BBS algorithm outperforms other skyline techniques by guaranteeing a low input/output cost, a low number of R-tree node accesses, and a short Process duration [20].

4 Experimentation and Results

The DELL workstation that we utilized for the studies has an Intel Core i7 3.60 GHz CPU, 8 GB of main memory, Windows 10 Professional as the operating system, and Oracle Database as the database management system. Python is used to implement our algorithm.

For each dimension, we generated over 150000 Cloud services using random values. We ran our software with a range of input sizes, from 1000 to 150000 cloud services, and a range of dimension numbers, from 1 to 6. Next, we took measurements of the Skyline's size and execution time. The outcomes are displayed in Figs. 2 and 3.

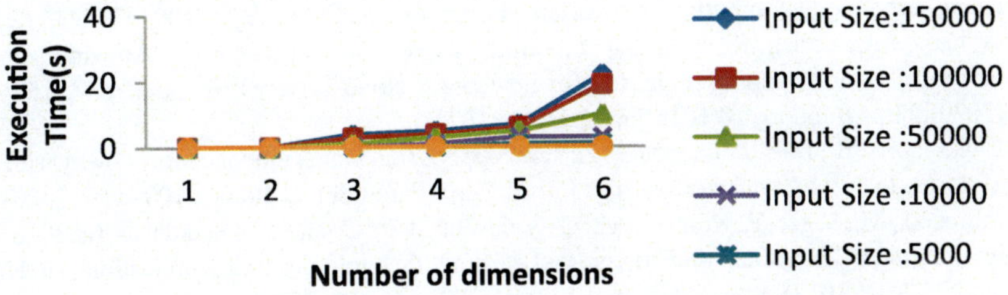

Fig. 2. Execution time/number of dimensions for various input sizes.

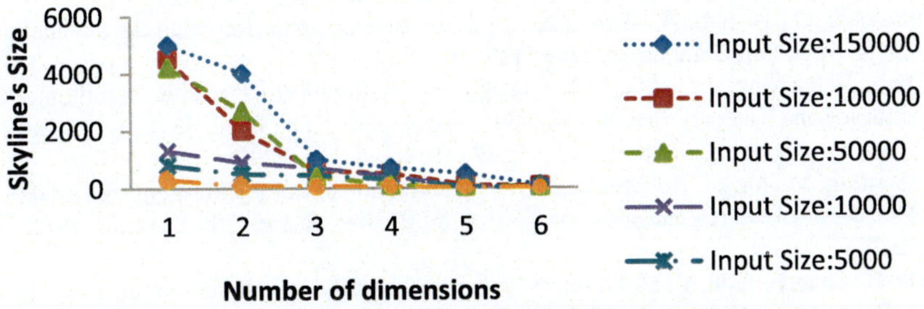

Fig. 3. Skyline's size/number of dimensions for various input sizes.

When there are fewer than three dimensions or a smaller-than-50,000 input, the execution time doesn't vary significantly. When computing a 6-dimensional skyline for 150000 cloud services, the maximum execution time is 23 s. When there are more than five dimensions, the Skyline size, which is relatively modest compared to the input size, starts to converge for all sizes.

Based on these results and comparing them with the results of the other authors, we find that our method excelled and gave much better results, especially in terms of the amount of Input Sizes.

5 Conclusion

Finding the best cloud service rapidly from a large dataset is the aim of the selection cloud service challenge. In this work, we developed an algorithm that enables searching and choosing Cloud services, based on the BBS Skyline algorithm. The experimental findings demonstrate that we can handle a sizable amount of data Skyline using our method in less than 20 s. The conclusion is clear that our strategy produces very positive outcomes. Be aware that our algorithm is general and can be applied to any problem of a similar nature. In the future, this strategy will be investigated.

References

1. Bypour, H., Farhadi, M., Mortazavi, R.: An Efficient Secret Sharing-based Storage System for Cloud-based Internet of Things. International Journal of Engineering 32(8), 1117–1119 (2019)
2. De la Prieta, F., Rodríguez-González, S., Chamoso, P., Corchado, J.M., Bajo, J.: Survey of agent-based cloud computing applications. Future Gener. Comp. Syst. **100**, 223–236 (2019)
3. Wahab, O.A., Cohen, R., Bentahar, J., Otrok, H., Mourad, A., Rjoub, G.: An endorsement-based trust bootstrapping approach for newcomer cloud services. Inf. Sci. **527**, 159–175 (2020). https://doi.org/10.1016/j.ins.2020.03.102
4. Chakraborty, B., Das, S.: Introducing a new supply chain management concept by hybridizing TOPSIS, IoT and Cloud Computing. J. Inst. Eng. (India) Ser. C **102**(1), 109–119 (2020)
5. Moravcik, M., Segec, P., Kontsek, M.: Overview of Cloud Computing Standards. In: 2018 16th International Conference on Emerging eLearning Technologies and Applications (ICETA). pp. 395–402 (2018). https://doi.org/10.1109/ICETA.2018.8572237
6. Aryotejo, G., Kristiyanto, D.Y., Mufadhol: Hybrid cloud: bridging of private and public cloud computing. J. Phys.: Conf. Ser. **1025**, 012091 (2018)
7. Cheng, Y., Morimoto, Y.: Bull. Netw. Comput. Syst. Softw. **8**, 81–86 (2019)
8. Belkasmi, D., Hadjali, A., Azzoune, H.: On fuzzy approaches for enlarging skyline query results. Appl. Soft Comput. **74**, 51–65 (2019)
9. Özyer, T., Zhang, M., Alhajj, R.: Integrating multi-objective genetic algorithm based clustering and data partitioning for skyline computation. Appl. Intell. **35**, 110–122 (2011)
10. Idrissi, A., Abourezq, M.: Skyline in Cloud Comput. **60**, 12 (2005)
11. Abourezq, M., Idrissi, A., Rehioui, H.: An amelioration of the skyline algorithm used in the cloud service research and selection system. Int. J. High Perform. Syst. Archit. **9**, 136–148 (2020)
12. Du, H., Shao, L., You, Y., Li, Z., Fu, D.: A two phase method for skyline computation. In: Jia, Y., Du, J., Zhang, W. (eds.) CISC 2019. LNEE, vol. 592, pp. 629–637. Springer, Singapore (2020). https://doi.org/10.1007/978-981-32-9682-4_66
13. Huang, Z., Xu, W., Cheng, J., Ni, J.: An efficient algorithm for skyline queries in cloud computing environments. China Commun. **15**, 182–193 (2018)
14. Rehioui, H., Idrissi, A., Abourezq, M.: The research and selection of ideal cloud services using clustering techniques: track: big data, data mining, cloud computing and remote sensing. In: Proceedings of the International Conference on Big Data and Advanced Wireless Technologies. pp. 1–6. Association for Computing Machinery, New York, NY, USA (2016). https://doi.org/10.1145/3010089.3010138
15. Purohit, L., Kumar, S.: Clustering based approach for web service selection using skyline computations. In: 2019 IEEE International Conference on Web Services (ICWS), pp. 260–264 (2019). https://doi.org/10.1109/ICWS.2019.00052
16. Liang, X., Lu, Q., Li, M.: Research on web service selection based on improved skyline algorithm. In: 2019 IEEE International Conference on Parallel & Distributed Processing with Applications, Big Data & Cloud Computing, Sustainable Computing & Communications, Social Computing & Networking (ISPA/BDCloud/SocialCom/SustainCom) (2019)
17. Wang, H., et al.: Integrating reinforcement learning and skyline computing for adaptive service composition. Inf. Sci. **519**, 141–160 (2020). https://doi.org/10.1016/j.ins.2020.01.039

18. Alabool, H., Kamil, A., Arshad, N., Alarabiat, D.: Cloud service evaluation method-based multi-criteria decision-making: a systematic literature review. J. Syst. Softw. **139**, 161–188 (2018). https://doi.org/10.1016/j.jss.2018.01.038
19. Fariss, M., Asaidi, H., Bellouki, M.: Comparative study of skyline algorithms for selecting Web Services based on QoS. Procedia Comp. Sci. **127**, 408–415 (2018)
20. Fariss, M., El Allali, N., Asaidi, H., Bellouki, M.: Prefiltering approach for web service selection based on QoS. In: 2019 International Conference on Systems of Collaboration Big Data, Internet of Things & Security (SysCoBIoTS), pp. 1–5 (2019)

An Innovative Approach for Supervised Link Prediction Using Feature Embedding Methods

Mohamed Badiy[1](✉), Fatima Amounas[2], and Saleh Bouarafa[3]

[1] RO.AL&I Group, Faculty of Sciences and Technics, Moulay Ismail University of Meknes, Errachidia, Morocco
m.badiy@edu.umi.ac.ma

[2] RO.AL&I Group, Computer Sciences Department, Faculty of Sciences and Technics, Moulay Ismail University of Meknes, Errachidia, Morocco
f.amounas@umi.ac.ma

[3] LRIT, Associated Unit to CNRST (URAC No 29), Faculty of Sciences, Mohammed V University in Rabat, Rabat, Morocco

Abstract. Link prediction is a fundamental task for graph-structured data with many applications, including social network analysis, recommendation systems, bioinformatics, computer science, and so on. Due to the numerous applications in various fields, the link prediction problem is a great challenge. To address this issue, several link prediction approaches have been suggested. Recently, supervised machine learning has become one of the most important approaches to solve the link prediction problem. Although several research works have been shown that this approach provide the best results, there is still need to improve the previous approaches. In this paper, we introduce a new approach to supervised link prediction for achieving better performances using feature learning methods. The main focus of this work is to do the supervised link prediction by using embedding-based methods. To do so, we consider a set of embedding methods as the feature vectors for training machine learning classifiers. The experimental results on the Facebook page dataset revealed satisfactory results, which encourage us to further analysis. Moreover, the use of feature embedding methods will achieve better performance.

Keywords: Link prediction · Social network · Machine learning · Embedding methods · Feature

1 Introduction

In the last decade, social networks have become growingly crucial in people's lives due to the wide variety of applications in several parts of a person's life, including social relationships, educational topics, and commercial encounters. The social network is represented as a graph collection of nodes with the relationships between them. The node represents a person or another entity, and the edge represents the relationships between them. At present, link prediction is the current trend to analyze the social network. It exploits the existing networks information to predict the likely links that will emerge in

Y. Farhaoui et al. (Eds.): ICAISE 2022, LNNS 635, pp. 206–211, 2023.
https://doi.org/10.1007/978-3-031-26254-8_29

the future [1]. The basic goal of link prediction is to predict future links between entities in the network [2, 3]. The three basic methods for link prediction are similarity techniques, embedding methods, and supervised learning approaches. The first category defines a similarity measure for each pair of unconnected vertices, assuming that the more similar the pair of vertices, the more likely there is a connection between them. The similarity-based methods assign a similarity score to a pair of nodes based on the structure of the network [4–7]. The second category know n as embedding-based techniques, tries to map network data to a lower dimensional space while preserving network neighborhood information as much as possible [8–10]. As result, in embedded space, similar nodes are close to each other. These methods can be classified into five categories: matrix factorization based methods, edge reconstruction based methods, graph kernel based methods, deep learning based methods and generative model based methods. The third approach is a learning-based method, which develops a model based on training data and observes predicted new links [11–13]. As compared to similarity-based algorithms, these techniques provide better performance. Many experiments with this approach have shown promising results, but the choice of a set of features for the training of a classifier remains a major challenge. In this context, we propose an approach of supervised link prediction using feature embedding methods. The main idea of this work is to improve performance by using embedding-based methods as a feature vectors for supervised learning algorithms. The rest of this paper is structured as follows. In Sect. 2, the proposed approach based on feature embedding methods is described in detail. The experimental results are discussed in Sect. 3. Conclusions are drawn in Sect. 4.

2 Proposed Approach

In this work, we suggest an improved supervised link prediction approach with feature embedding methods. Here, we adopt the following supervised machine learning methods, as described in Sect. 2.2. Our goal is predict the likelihood of a link's existence in the social graph by considering Facebook Page dataset. Whereas the primary achievement of this work was to identify an optimal subset of features in order to achieve the superior performance of the classifiers. Fig. 1 depicts the block diagram of the proposed methodology. The proposed model is composed of two main phases which include: Feature extraction, and Classification.

2.1 Feature Extraction

This section explains the features used in the proposed approach. In fact, feature selection is a crucial step in the supervised link prediction task. It very important to choose the right features for any classifier. In this paper, we utilize three different feature embedding methods. In what follows, we briefly explain these features.

– **Deepwalk:** represents an embedding method for learning latent representations of vertices in a network. This method uses local information obtained from truncated random walks to learn latent representations.

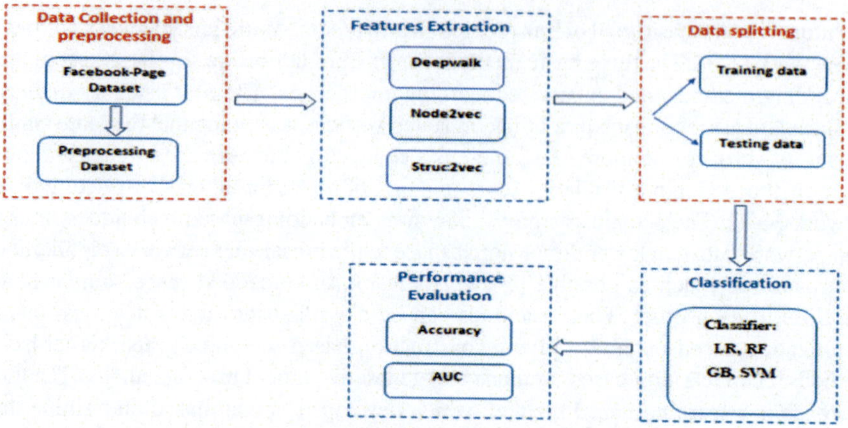

Fig. 1. Block diagram of the proposed approach

- **Node2vec:** is an embedding method that transforms networks into numerical representations [14]. This method uses graph embedding approach and generates features by simulating biased random walks in a vector representation, and storing each node in a d-dimensional space according to their features.
- **Struc2vec:** is one of the strongest node embedding methods which can effectively capture each node's structural features and embed it into a vector [15]. This method constructs a new graph according to each node's degrees of different layers and uses the skip-gram to make the nodes with similar structure obtain similar embeddings.

2.2 Supervised Machine Learning Algorithms Extraction

The features discussed above serve as input for a supervised machine learning model that predicts whether or not a pair of currently disconnected nodes will connect in the future. The machine learning algorithms adopted in our case are l isted below:

- **Logistic Regression (LR):** is a supervised machine learning which can be used in classification task. This algorithm allows to calculate or predict the probability of a binary (yes/no) event occurring.
- **Random Forest (RF):** is another supervised machine learning classifierused to solve regression and classification problems, which combines the output of multiple decision trees to reach a single result.
- **Gradient Boosting(GB):** is one of the most popular algorithms in the field of machine learning. It is powerful enough to produce a prediction model in the form of an ensemble of weak prediction models.
- **Support vector machine (SVM):** is one of the powerful algorithms that can be used for classification task. It is a discriminated classifier that basically separates the given data into classes by finding the best separable hyperplane that categorizes the given training data.

3 Results and Discussion

In this section, we explain how we performed our experiments on a i5 processor having 3.2 GHz clock rate by using Google Colab as tool. In order to evaluate the performance of the different algorithms, we used the Facebook Page data set in our experiments. Table 1 shows the statistics of the dataset considered. In this work, we considered the concatenated embedding vectors of node pairs obtained by applying three different embedding methods: Deepwalk, Node2vec and Struc2vec as input to the supervised machine learning algorithms. For validation purposes, we use standard evaluation metrics such as accuracy and AUC. Table 2 shows the performance of learning models built using learned embedding vectors. The graphical representation of the table described below is shown in Fig. 2.

Table 1. Detail of the dataset

Dataset	Nodes	Edges	Average degree
Facebook page	620	2102	6

Table 2. Comparison of different models in terms of accuracy and AUC values

		Performance	
Feature	**Model**	**Accuracy**	**AUC**
Deepwalk	LR	0.931	0.787
	RF	0.937	0.912
	GB	0.934	0.880
	SVM	0.942	0.938
Node2Vec	LR	0.930	0.811
	RF	0.935	0.919
	GB	0.933	0.881
	SVM	0.944	0.946
Struc2vec	LR	0.933	0.799
	RF	0.933	0.814
	GB	0.931	0.784
	SVM	0.937	0.856

From the results, we make the following observations. In all ensemble learning models, Logistic Regression (LR), Random Forest (RF), Gradient Boosting (GB) and Support vector machine (SVM), trained with the Facebook Page dataset, Node2vec

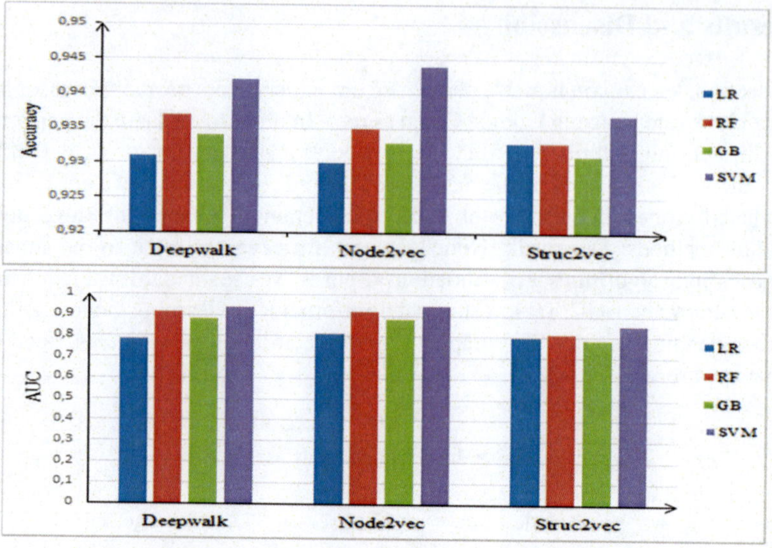

Fig. 2. Performance analysis of different link prediction models.

representation achieved slightly more successful results in terms of accuracy and AUC values compared to Deepwalk and Struc2vec representation. In experiments, the best performance was observed in SVM model trained with Node2vec representations, with 0.944 in terms of accuracy and 0.946 in terms of AUC values. The Node2vec method increased the accuracy and AUC values of the SVM model by approximately 1% and 9% respectively, compared to Struc2vec.

4 Conclusion

Machine learning has made remarkable progress in many fields, such as the social network analysis field. Currently, supervised link prediction is the main challenge in this area. In this paper, we present a new approach to supervised link prediction based on embedding methods, which predict the future link in a social network. We focused on the task of link prediction and carried out experiments on several typical models in the Facebook page network. Here, we considered the embedding methods as the feature vectors for training the supervised machine learning classifiers. Through the analysis of the experimental results, we found that the use of different embedding methods is a very effective feature for supervised link prediction. For the future work, our goal is to combine the mentioned methods in order to greatly improve the accuracy of the different models.

References

1. Martínez, V., Berzal, F., Cubero, J.C.: A survey of link prediction in complex networks. ACM Comput. Surv. **49** 69 (2016). https://doi.org/10.1145/3012704

2. Haghani, S., Keyvanpour, M.R.: A systemic analysis of link prediction in social network. Artif. Intell. Rev. **52**(3), 1961–1995 (2017). https://doi.org/10.1007/s10462-017-9590-2
3. Kumar, A., Singh, S.S., Singh, K., et al.: Link prediction techniques, applications, and performance: a survey. Physica A: Stat. Mech. Appl. **553**, 124289 (2020). https://doi.org/10.1016/j.physa.2020.124289
4. Zhou, K., et al.: Attacking similarity-based link prediction in social networks, arXiv preprint arXiv:1809.08368 (2018). https://doi.org/10.48550/arXiv.1809.08368
5. Mumin, D., Shi, L.-L., Liu, L.: An efficient algorithm for link prediction based on local information: considering the effect of node degree. Concurr. Comput.: Pract. Exp. **34** e6289 (2022). https://doi.org/10.1002/cpe.6289
6. Yu, C., Zhao, X., An, L., et al.: Similarity-based link prediction in social networks: a path and node combined approach. J. Inform. Sci. **43**(5), 683–695 (2017). https://doi.org/10.1177/0165551516664039
7. Mallek, S., Boukhris, I., Elouedi, Z., et al.: Evidential link prediction in social networks based on structural and social information. J. Comput. Sci. **30**, 98–107 (2019). https://doi.org/10.1016/j.jocs.2018.11.009
8. Goyal, P., Ferrara, E.: Graph embedding techniques, applications, and performance: a survey. Knowl. Based. Syst. **151**, 78–94 (2018). https://doi.org/10.1016/j.knosys.2018.03.022
9. Wu, C., et al.: Link prediction based on graph embedding method in unweighted networks. In: 2020 39th Chinese Control Conference (CCC). IEEE (2020). https://doi.org/10.23919/CCC50068.2020.9188785
10. Ozge, K.A.R.T.: Network embedding for link prediction in bipartite networks. Avrupa Bilim ve Teknoloji Dergisi **27**, 311–317 (2021). https://doi.org/10.31590/ejosat.937722
11. Pecli, A., Cavalcanti, M.C., Goldschmidt, R.: Automatic feature selection for supervised learning in link prediction applications: a comparative study. Knowl. Inf. Syst. **56**(1), 85–121 (2017). https://doi.org/10.1007/s10115-017-1121-6
12. Deepanshu, M.: and Rinkaj Goyal, "Supervised-learning link prediction in single layer and multiplex networks." Mach. Learn. Appl. **6**, 100086 (2021). https://doi.org/10.1016/j.mlwa.2021.100086
13. Kumari, A., Behera, R.K., Sahoo, K.S., et al.: Supervised link prediction using structured-based feature extraction in social network. Concurr. Comput. Pract. Exp. **34**, e5839 (2022). https://doi.org/10.1002/cpe.5839
14. Aditya, G., Jure, L.: node2vec: Scalable feature learning for networks. In: Proceedings of the 22nd ACM SIGKDD International Conference on Knowledge Discovery and Data Mining, pp. 855–864, (2016). https://doi.org/10.1145/2939672.2939754
15. Ribeiro, L., Saverese, P.H.P., Figueiredo, D.R.: Struc2vec: learning node representations from structural identity. ACM **17** 13–17 (2017). https://doi.org/10.1145/3097983.3098061

Check for updates

An Intelligent Monitoring Approach Based on WiFi Sensing for Smart Hospital

Hicham Boudlal(✉), Mohammed Serrhini, and Ahmed Tahiri

Faculty of Sciences, Mohammed First University, BV Mohamed VI, BP 717, 60000 Oujda, Morocco
hicham.boudlal5@gmail.com

Abstract. Traditional sensing techniques for smart hospitals are recognized to be labor- and technically-intensive for practical adoption. As a result, researchers are employing alternate sensing techniques. One of the major technologies enabling connection for smart hospital services is WiFi. With its sensitivity to environmental dynamics, WiFi signal is now widely used for a variety of sensing applications, such as gesture recognition and fall detection, in addition to its primary usage for communication. WiFi-based sensing offers non-contact sensing in a private mode, simultaneous sensing and data transmission without additional connection infrastructure, and remote sensing without handheld sensors. This paper aims to investigate existing systems and methods that use WiFi sensing, conduct experiments to analyze how different environments as well as system parameters affect data and the ability of monitoring human activity without the need for sensors on their bodies.

Keywords: WiFi sensing · Channel state information · Intelligent monitoring · Human activity recognition · Smart hospital · Multi-environment

1 Introduction

Recognition of human gestures and activities gained significant interest during the last few years through various applications for monitoring human movement and behavior in indoor areas. Its applications involve health monitoring as well as fall detection for the elderly, contextual awareness, and numerous other applications based on the Internet of Things [1]. In the existing systems, the person is required to wear a device containing motion sensors, including an accelerometer and gyroscope. Sensing data are locally treated at the wearing device or transferred to a server to extract features, followed by the use of supervised learning algorithms for classification. Such monitoring is called active monitoring. However, at all times wearing a device is cumbersome as well as might not be practical within numerous passive activity recognition applications, where the individual may not be wearing a sensor or wireless device [2]. Furthermore, when installed, it has non-negligible expenses and high overhead [3]. Although camera-based models could be employed to recognize passive activity, the required line-of-sight (LOS) is a main limit for such models. In addition, camera-based approaches had confidentiality

Y. Farhaoui et al. (Eds.): ICAISE 2022, LNNS 635, pp. 212–223, 2023.
https://doi.org/10.1007/978-3-031-26254-8_30

concerns and they could not be used for a wide range of environments. For this reason and to overcome this limitations, passive monitoring system based on wireless signal, that doesn't infringe privacy of individuals, was desired [4]. Recently, WiFi technology has been the focus of much research on activity recognition due to its ubiquitous availability in indoor environments. It involves a WiFi AP (access point) along with one or more WiFi enabled devices located at various points in the environment. When an individual initiates an activity, body movement impacts the wireless signals and modifies the system's multipath profile [5].

The paper is structured as follows: In Section 2, we first provide some background information before going through the details of the technology. Then in Section 3, the design details and overview system architecture are presented. The experiment environment, hardware/software setup and implementation are described in Section 4. Section 5 provides the conclusions and directions for future research.

2 Background Information

Using WiFi radios as sensors is a new concept known as WiFi sensing [6]. The WiFi signal is described in this section along with some background information [7]. We specifically describe two digital sensor readings: the RSSI (received signal strength indicator) and CSI (channel state information).

2.1 Received Signal Strength Indicator (RSSI)

RSSI measures the overall amplitude of the received signal, which is averaged over all subcarriers. Most WiFi devices can get the received signals, which show the path loss of wireless signals relative to a specific distance and may be calculated using the Log-normal Distance Path Loss (LDPL) model [8]:

$$P(d) = P(d_0) + 10\gamma log\frac{d}{d_0} + X_\delta \tag{1}$$

where $P(d)$ stands for the RSSI measurement, which shows the path loss at distance d in decibels (dB). Path loss at reference distance d_0 is denoted by $P(d_0)$. X_δ is a zero average normal noise brought on by flat fading, and γ is the exponent of the path loss.

The LDPL method describes the variation of received signal strength across distance caused by path loss and shadowing. Path loss is the result of transmission power dissipation in the propagation channel, whereas shadowing is a result of barriers attenuating the signal power by reflection, absorption, scattering and diffraction. Scattering, and diffraction [8]. Although RSSI is a basic and simple approach that can be easily produced in any fundamental WiFi device without the need for extra hardware, it is challenging to record signal fluctuations in real time. Additionally, it was demonstrated that RSSI is unreliable in many application contexts since its stability is not guaranteed even in a static indoor environment [9].

2.2 Channel State Information (CSI)

Wireless devices in the IEEE 802.11 n/ac standards utilize MIMO (Multiple Input Multiple Output) systems. Through the utilization of MIMO technology, the diversity gain, matrix gain, and multiplexing gain can be increased, as well as decreasing co-channel interference [10]. Orthogonal frequency division multiplexing (OFDM) was widely implemented by several WiFi standards for coding and sending information from a transmitter to a receiver among the numerous modulation techniques [11].

The CSI describes the channel from the transmitter to the receiver in a more fine-grained way. For OFDM subcarriers, CSI information represents the amplitude and phase information. The received signal could be represented as follows:

$$y = H \times x + n, \tag{2}$$

where y is the received signal, x is the transmitted signal, n is the channel noise, and H is the CSI, a matrix of complex values that represents the channel frequency response (CFR) for each specific subcarrier within each spatial stream. Consequently, CSI is a $m \times n \times w$ matrix, where m is the number of transmitter antennas, n is the number of reception antennas, and w is the number of subcarriers, for all subcarriers and special streams. Such a fine-grained matrix might accurately represent the channel's temporal and spectral characteristics as well as any changes brought on by a multipath effect at tiny scales. The CSI components H of every received packet may be represented as follows, supposing that there are m transmitters and n receivers in the MIMO systems:

$$H = \begin{bmatrix} H_{1,1} & \cdots & H_{1,n} \\ \vdots & \ddots & \vdots \\ H_{m,1} & \cdots & H_{m,n} \end{bmatrix} \tag{3}$$

In OFDM, the whole frequency spectrum is split into 56 or 114 frequency subcarriers, depending on whether the channel bandwidth is 20 MHz or 40 MHz. Each subcarrier's CSI is, where $|h|$ denotes the amplitude and θ the phase:

$$h = |h|e^{j\theta} \tag{4}$$

3 System Design

This section describes the challenges, details, and overview of our system design.

3.1 Challenges

Our objective is to use CSI measurement from a single pair of WiFi devices to identify human movement. A variety of issues must be overcome in order to construct such a system in real environments like a typical hospital scenario.

Reliability in a Real-World Setting. WiFi equipment placement in actual situations is subject to alter over time, and human gestures vary from person to person. Under these difficult circumstances, which include varying distances between the AP and WiFi devices and the presence of walls obstructing line-of-sight between WiFi devices, our system should be able to offer accurate gesture monitoring (NLOS scenarios).

Sensing with a Single AP and WiFi Device Pair. Assuming there is just one wireless link (between AP and device) through the human body, our system is intended to operate using existing WiFi infrastructure. The system should also consume WiFi bandwidth as little as feasible, for example, by solely using existing beaconing traffic.

3.2 System Architecture

With the use of variations in the correlation between CSI sequence frames that are influenced by human movement, our system aims to detect movement. When there are periods of greater fluctuation, it may be determined that the subject's movements have an effect on the RF environment. It is possible to detect movement by measuring this.

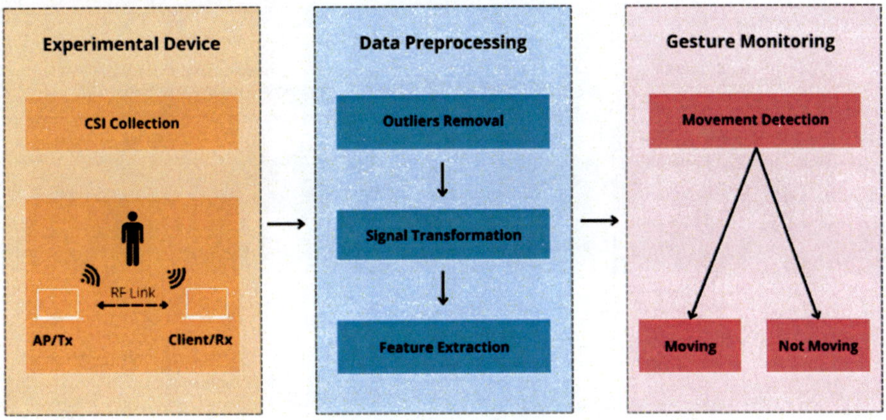

Fig. 1. Overall system diagram.

As illustrated in Fig. 1, this system may be split in six sections. Data are collected using CSI hardware and either delivered to the system as a batch or buffered on a real-time system. The data are then pre-processed to eliminate errors and noise.

3.3 Preprocessing Data

Although with the best equipment and configuration, CSI data includes a lot of noise, making it impossible to use raw data. First, the surroundings and reflected radio waves are to cause, followed by potential hardware and software issues. In order to create a reliable and accurate human activity detection model using WiFi CSI data, data preprocessing is a crucial component.

Outliers Removal Technique. CSI can produce a lot of undesirable noise since it is extremely sensitive to environmental changes. To detect small activities, it is therefore required to filter out the noise. In order to eliminate high frequency signal elements, we use a low pass filter [12]. After that, the amplitudes and phases of the CSI contain the noise produced by internal changes of state, such as transmission power, rate adaptations, and thermal noise in the devices, etc. It changes the signal and contains erroneous values which are not caused by human presence. We apply the Hampel filter to address this issue [13]. Furthermore, to further reduce high frequency noise, we also apply a running mean filter, often known as a moving average [12]. Figure 2 illustrate the results with comparison of the CSI amplitude before and after applying filtering.

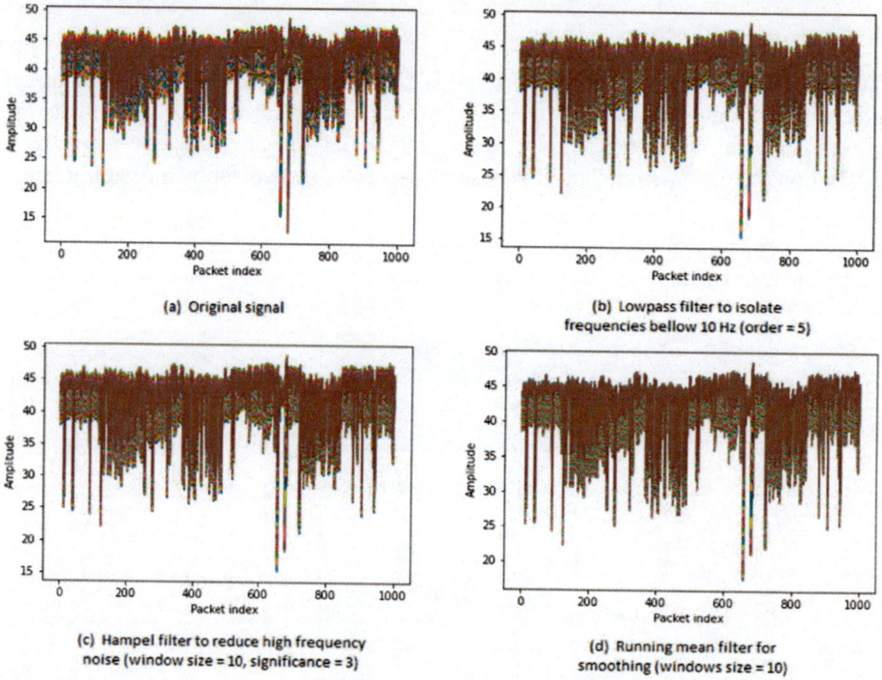

Fig. 2. Illustrate the results with comparison of the CSI amplitude before and after applying filtering.

Signal Transformation Based on Short Time Fourier Transform. A time series of CSI measurements is analyzed using signal transformation methods for time-frequency analysis. In order to compute the Fast Fourier Transform (FFT) coefficients individually on each segment, the Short-Time Fourier Transform (STFT) separates the input into shorter segments of equal length, as indicated in Eq. 5. By displaying the time series data in both the time and the frequency domains, STFT can detect changes in dominating frequencies over time [14]. Figure 3 shows the short time Fourier transform.

$$X(t,k) = \sum_{n=-\infty}^{\infty} x[n]w[n-t]e^{-jkn} \quad (5)$$

where w is the window function, k is the frequency index, and t is the time index.

Signal Extraction with Principal Component Analysis. Signal extraction is used to retrieve the signal targets from raw or preprocessed measurements of CSI. PCA is utilized generally for blind signal separation and feature extraction. A matrix is converted into a set of principal components using PCA, which employs an orthogonal transformation. The input is expected to be a set of variables that may be correlated, but the principal components are a collection of linearly uncorrelated variables. Figure 4. Illustrate the principal component analysis (PCA) for CSI dimension reduction.

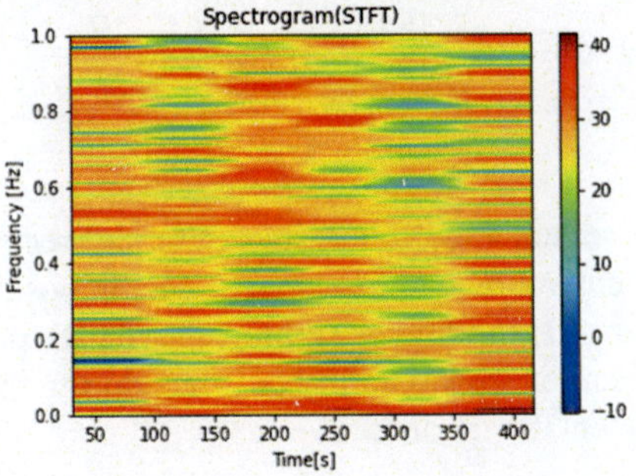

Fig. 3. Shows the short time Fourier transform.

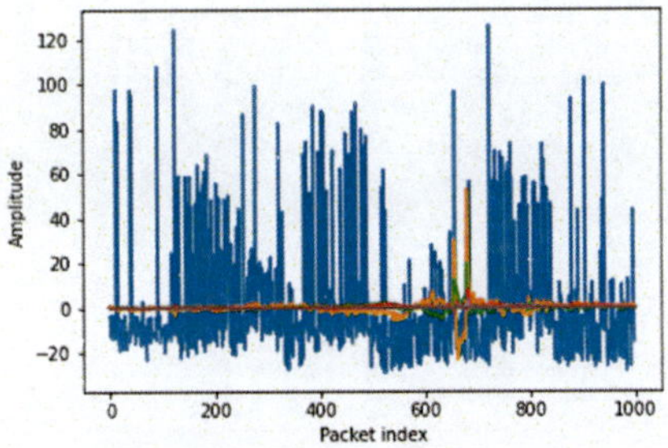

Fig. 4. Illustrate the principal component analysis (PCA) for CSI dimension reduction.

4 Experimental Study

This section describes both the environment where this experiment is conducted as well as the hardware and software configuration specifications.

4.1 Experiment Environment

The experiments are conducted in two different environments and their dimensions and key signal processing parameters, as shown in Table 1:

Table 1. Summary of experimental parameters.

Characteristics and Parameters	Environment	
	Laboratory	Office
Dimensions	6.9 m * 8.7 m	2.95 m * 3.9 m
Occupancy	1 Person	1 Person
Bandwidth	20 MHz	20 MHz
Channel	2437 MHz	2437 MHz
Frequency	2.417 GHz	2.417 GHz
Antennas	1Rx * 2Tx	1Rx * 2Tx
Subcarriers	56	56

Two computers that serve as the transmitter (Tx) and receiver (Rx) were used to instrument the two test environments. The Tx-Rx pair is positioned so that each test environment's center is inside the line of sight of the Tx-Rx pair. There is no change in the physical environment of the laboratory or office during the experiment. Figure 5 illustrate the two experimental environment.

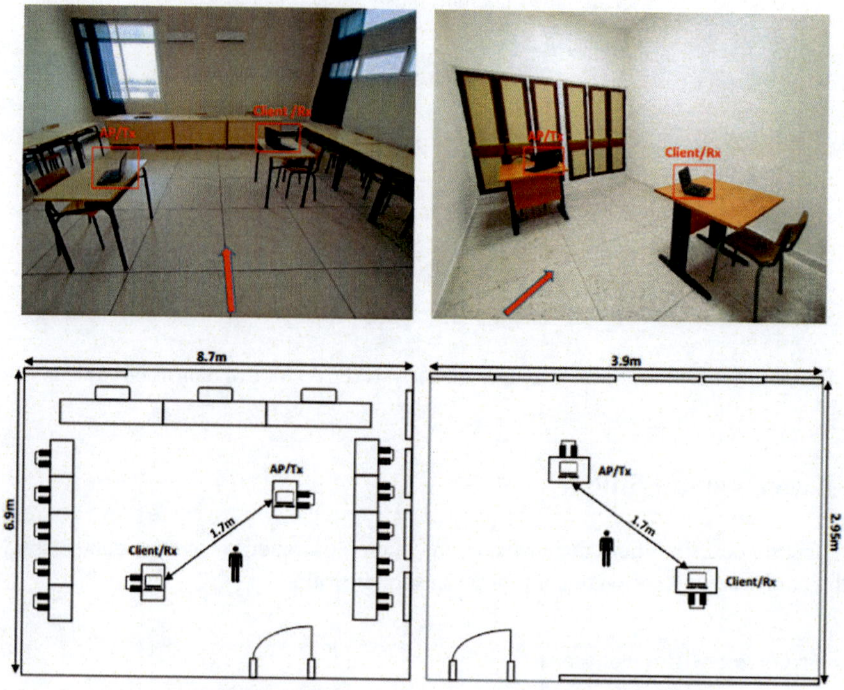

(a) Picture of the laboratory (b) Picture of the office

Fig. 5. Two experimental environment and scenario. The height of the WiFi device is 75 cm.

4.2 Hardware Configuration

Atheros CSI tool is an 802.11n experimental and measurement tool that is open source for CSI data collecting. It enables the extraction of complete physical layer wireless communication data from Atheros WiFi network cards, including CSI, received packet payload, data rate, timestamp, RSSI of each antenna, and more [15]. A large community has developed around it, testing various Atheros NICs and working on projects to make getting started easier. Moreover, there are a lot of resources accessible including MATLAB programs that analyze data quickly. We made the decision to employ it in our study in light of that.

The receiver and transmitter were two PCs equipped with Atheros Wi-Fi NICs. To facilitate the extraction of physical layer information, a customized CSI tool was installed there. Since they operate in AP-Client mode, the transmitter and receiver are set up to function in AP and client modes, respectively. We utilize Hostapd to produce AP. The client computer then has to connect to the newly formed AP in a similar manner to how a phone connects to a WiFi network. They then join the same network and are free to communicate.

The system was then built and put into operation, with the transmitter sending data to the receiver, which processed it, computed CSI, and logged it for further examination. It was built based on [16] and [17]. In order to manage, store and visualize the CSI in real time, the configuration consists of three computers. It operates as follows:

- The sending computer (A) sends data to the receiving computer (B).
- The receiving computer (B) receives the data and calculates the CSI.
- The receiving computer (B) sends the raw CSI to the user's computer (C).
- The user's computer (C) manages the data, stores it and visualizes it in real time.

The hardware computers utilized for the experiments are shown in Table 2.

Table 2. List of experiment hardware.

Device name	Role	Hardware	Specifications
Computer (A)	AP/Tx	Packard Bell EasyNote TE11HC	Atheros CSI tool is installed and implemented on Ubuntu 14.04 LTS (64-bit) with custom Linux kernel 4.1.10 +, based on Qualcomm Atheros AR5BWB222 network card. It is in 802.11n AP mode using a Hostapd
Computer (B)	Client/RX	Ordinateur HP Mini 110 4117sk	Atheros CSI tool is installed and implemented on Ubuntu 14.04 LTS (64-bit) with custom Linux kernel 4.1.10 +, based on Atheros network card. It is in 802.11n client mode
Computer (C)	CSI visualizer	HP ProBook 450 G3	Anaconda installed to compile python scripts. It is in 802.11n client mode

4.3 Experiments

For the experiment, a desktop Tx with ping generates WiFi frames, and in order to gather CSI, we set the Tx to send 1000 packets with 10^4 microseconds between each transmission. Rx monitors and processes the CSI for the channel between Tx and Rx while actively receiving the ping packets issued by Tx. All 56 OFDM subcarriers' CSI are extracted by Atheros and sent to MATLAB. The receiver then uses the UDP protocol to deliver raw data to the server. As a result, the system consists of a real-time monitoring system and computer-based CSI transmission.

The gesture for the lab and office were made in the line of sight of the computers Tx-Rx, which were placed 1.7 m apart. One volunteer was instructed to walk briskly between Rx and Tx in both layouts as part of an experiment to assess the robustness of recognition. The laboratory and office experiment examined the single user case. Figure 6 shows the configuration of device used for data collection.

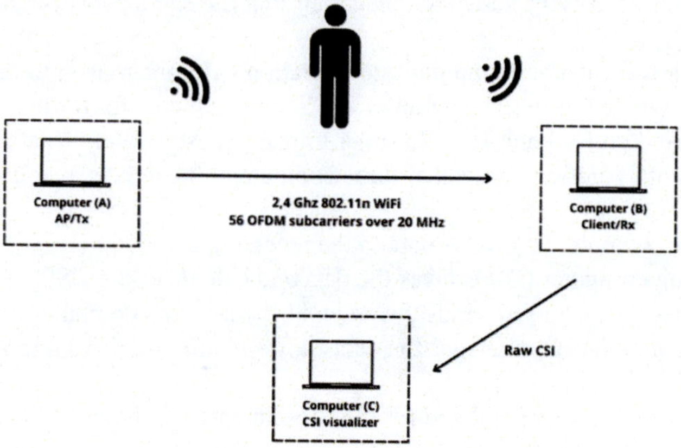

Fig. 6. CSI capture device configuration.

With the use of WiFi CSI data, this experiment aims to demonstrate how the system described in Section 3 can detect gestures. CSI data will be used to evaluate the feasibility of identifying a pattern of WiFi movements. How various settings and the people in them effect data is another consideration.

Gesture Monitoring. When observable physical movements are used to convey a certain message, it is known as gesture (For example, Okay symbol or clapping hands, tapping fingers, tilting head to one side). There are many commonly used gestures, so we focused on the brisk walking. The purpose is to investigate if there is a pattern of gestures in the data we collected for the experiment.

These experiments performed a brisk walking gesture: small pause and brisk walking movement, a small pause. Based on the results shown in Fig. 7, the data has a specific pattern when the person briskly walk, therefore we can differentiate them.

Figure 7 shows the 3D CSI amplitude of subcarriers over time, with time on the y-axis, amplitudes [dBm] on the z-axis, and 56 subcarrier frequencies on the x-axis. A separate frequency is represented by each subcarrier, which provides as an additional source of CSI data. The Oz axis contains the amplitude designates the quality of the signal in a given subcarrier. The maximum amplitude is reached when the signals do not encounter any interference. The Ox axis contains the index subcarriers, each subcarrier is received with an independent quality which helps us to understand what is happening to the signal during its transmission. The amplitude variations demonstrate changes that can be linked to environmental occurrences. The person's brisk walk may be the reason of this pattern. This discovery strongly suggests that by utilizing the CSI from off-the-shelf WiFi equipment, we may be able to accomplish device-free fine-grained body gesture monitoring.

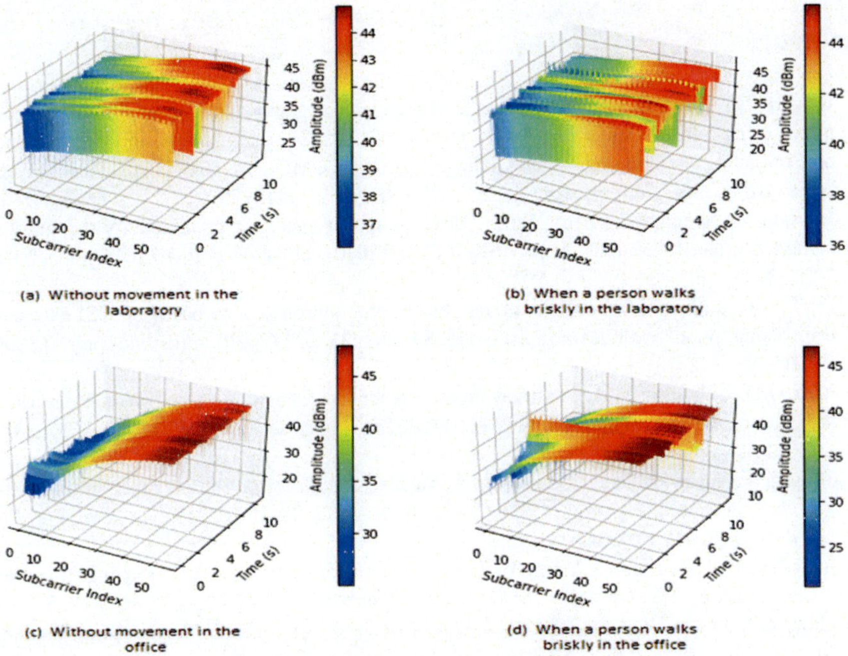

Fig. 7. Illustrate CSI amplitude of subcarriers over time in 3D.

Impact of Environment and System Parameters. The influence of the test environment and system parameter on data performance is assessed. The WiFi transmitter and receiver are placed side by side on a table at one end, as shown in Fig. 5. We encounter that the CSI sensing's coverage increases with the distance between Tx and Rx. There is usually just one person working at the office. The detection coverage is then extended by positioning the Tx and Rx far apart. Since TX and RX are placed close to one another in the lab, no outside influences may significantly affect CSI, and only user activity affects the signals.

5 Conclusion and Future Work

The paper presents a new approach to the use of WiFi signals in smarts hospitals and generally in all indoor environments. CSI can be used to monitor patients without the need to wear special equipment, for example, to monitor the movements of sick people in a care facility. By locating patients, we can protect people with Alzheimer's disease when they want to leave the facility. To build such a system based on WiFi, it is enough to use the pre-installed routers (WiFi devices) without the need for additional hardware. In contrast to conventional wearable sensors and systems that need close proximity, WiFi sensing with CSI can provide consumers a high accuracy contactless monitoring service. Finally, other ideas as well as initial results are motivating more research in this area.

References

1. Banjarey, K., Prakash Sahu, S., Kumar Dewangan, D.: A Survey on Human Activity Recognition using Sensors and Deep Learning Methods. In: 2021 5th International Conference on Computing Methodologies and Communication (ICCMC), pp. 1610–1617. IEEE, Erode, India (2021). https://doi.org/10.1109/ICCMC51019.2021.9418255
2. Hussain, Z., Sheng, M., Zhang, W.E.: Different approaches for human activity recognition: a survey. J. Netw. Comput. Appl. **167**, 102738 (2020). https://doi.org/10.1016/j.jnca.2020.102738
3. Li, T., Shi, C., Li, P., Chen, P.: A novel gesture recognition system based on CSI extracted from a smartphone with nexmon firmware. Sensors. **21**, 222 (2020). https://doi.org/10.3390/s21010222
4. Lara, O.D., Labrador, M.A.: A survey on human activity recognition using wearable sensors. IEEE Commun. Surv. Tutor. **15**, 1192–1209 (2013). https://doi.org/10.1109/SURV.2012.110112.00192
5. Yousefi, S., Narui, H., Dayal, S., Ermon, S., Valaee, S.: a survey on behavior recognition using wifi channel state information. IEEE Commun. Mag. **55**, 98–104 (2017). https://doi.org/10.1109/MCOM.2017.1700082
6. Jiang, H., Cai, C., Ma, X., Yang, Y., Liu, J.: Smart home based on wifi sensing: a survey. IEEE Access. **6**, 13317–13325 (2018). https://doi.org/10.1109/ACCESS.2018.2812887
7. Zhou, Z., Chenshu, W., Yang, Z., Liu, Y.: Sensorless sensing with WiFi. Tsinghua Sci. Technol. **20**, 1–6 (2015). https://doi.org/10.1109/TST.2015.7040509
8. Seidel, S.Y., Rappaport, T.S.: 914 MHz path loss prediction models for indoor wireless communications in multifloored buildings. IEEE Trans. Antennas Propag. **40**, 207–217 (1992). https://doi.org/10.1109/8.127405
9. Liu, J., Liu, H., Chen, Y., Wang, Y., Wang, C.: Wireless sensing for human activity: a survey. IEEE Commun. Surv. Tutor. **22**, 1629–1645 (2020). https://doi.org/10.1109/COMST.2019.2934489
10. Xiao, Y.: IEEE 802.11n: enhancements for higher throughput in wireless LANs. IEEE Wirel. Commun. **12** 82–91 (2005). https://doi.org/10.1109/MWC.2005.1561948
11. Nee, R.V., Prasad, R.: OFDM For Wireless Multimedia Communications. Artech House, London, UK (2000)
12. Forbes, G.: CSIKit: Python CSI processing and visualisation tools for commercial off-the-shelf hardware (2022). https://github.com/Gi-z/CSIKit
13. Chowdhury, T.Z.: Using Wi-Fi Channel State Information (CSI) for Human Activity Recognition and Fall Detection. 139

14. Ma, Y., Zhou, G., Wang, S.: WiFi sensing with channel state information: a survey. ACM Comput. Surv. **52**, 1–36 (2020). https://doi.org/10.1145/3310194
15. Xie, Y., Li, Z., Li, M.: Precise power delay profiling with commodity Wi-Fi. IEEE Trans. Mob. Comput. **18**, 1342–1355 (2019). https://doi.org/10.1109/TMC.2018.2860991
16. YaxiongXie: xieyaxiongfly/Atheros-CSI-Tool-UserSpace-APP, https://github.com/xieyaxiongfly/Atheros-CSI-Tool-UserSpace-APP Accessed 23 Apr 2022
17. NovelSense/Atheros-CSI-Tool-UserSpace-APP, https://github.com/NovelSense/Atheros-CSI-Tool-UserSpace-APP. Accessed 23 Apr 2022

Analysis of Packet Health Fields on a WSN MicaZ-Crossbow Platform

Karim Lahma(✉) and Mohamed Hamroui

RITM Laboratory, CED Engineering Sciences, ESTC, Hassan II University of Casablanca, Casablanca, Morocco
karim.lahma@gmail.com

Abstract. A wireless sensor network is made up of a large number of autonomous sensors capable of collecting and working together to transmit data to the base station. In this work, we analyze the fields of Packet Health for better decision making. Indeed, we set up a comparison between a normal sensor node and a disturbed one. To do this, we use the wireless sensor network MicaZ-Crossbow experimental platform composed of two sensor nodes.

Keywords: WSN · MicaZ · Packet Health

1 Introduction

Wireless sensor networks (WSNs) are used today in many applications. They are a category of wireless networks with a large number of nodes and they are characterized by a very dense and large-scale deployment.

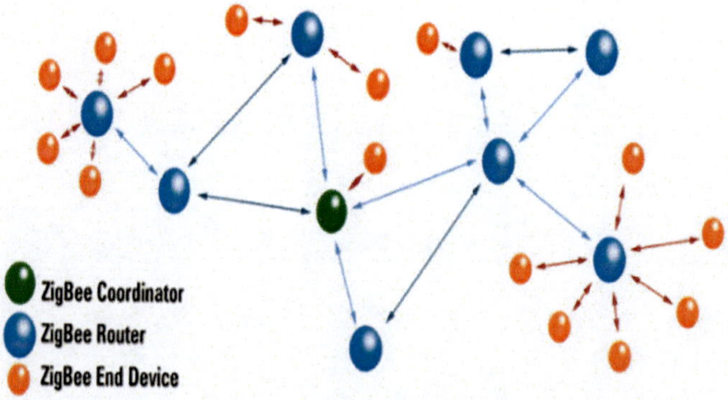

Fig. 1. Architecture of the wireless sensor network

In recent years, WSNs have attracted much attention from industry and the research community [1], particularly in the areas of monitoring, tracking, disaster monitoring, home automation, industrial control and battlefield surveillance. In general, WSNs are

Y. Farhaoui et al. (Eds.): ICAISE 2022, LNNS 635, pp. 224–229, 2023.
https://doi.org/10.1007/978-3-031-26254-8_31

composed of many self-organizing disposable electronic devices equipped with a short-range wireless transceiver. Also, they are equipped with a limited power supply (usually batteries), a detection unit and processing and memory resources [2]. When deployed over an area of interest, the sensors can collect scalar information such as humidity, pressure, temperature, brightness, and seismic variations, depending on application needs and nature of the detection unit provided to each source node [3].

When transmitting data packets over wireless links, the packets may become corrupted and require retransmission. Indeed, the sensor nodes generally have a single radio antenna and share the same transmission channel. Moreover, the simultaneous transmission of data from several sensors can produce collisions and thus a loss of the information transmitted. Retransmitting lost packets can result in a significant loss of power.

The use of RCSF-based systems has drawbacks, mainly the verification of the information received since it goes through several phases, including collection, transmission and storage. To remedy this, many works exist around the notion of reliability, among others the authors [4] propose a model for modeling the imprecision of epistemic and random uncertainty. We highlight another work [5], which focuses on data fusion using evidence theory (also called belief theory) in the field of automatic language identification.

Several parameters serve as guidelines to improve the proposed routing algorithms and protocols, we cite:

- Fault tolerance;
- Scalability;
- Operating environment;
- Network topology;
- Material constraints;
- Transmission medium;
- Energy consumption.

In the following part, we will give an overview of the WSN MicaZ-Crossbow platform and describe the structure of the 'Packet Health' framework. Then we will compare the results obtained using a WSN composed of two sensor nodes. Finally, we will interpret the results in order to contribute to better decision-making.

2 WSN MicaZ-Crossbow Platform and Packet Health

2.1 WSN MicaZ-Crossbow Platform

The MicaZ sensor nodes we used (MDA100) are very small in size and resource limited (Fig. 2). They are autonomous and capable of processing and transmitting information via radio waves (ZigBee). They are programmable, in particular conditions as often in embedded computing. Programming cannot be done directly on the module. Consequently, they are interfaced to the PC via USB thanks to a support on which they do not care: the MIB 520 (Fig. 3).

Fig. 2. MicaZ MDA100 sensor node

Fig. 3. MicaZ MIB520 base station

2.2 Packet Health

Packets Health allows the user to monitor the health of the XMesh network. They are enabled or disabled from the user's application. If activated, they are transmitted upstream to the base station. These packets contain useful information about the performance of the Mote in the mesh with respect to radio traffic.

The precompiled XMesh applications provided with the monitoring interface of the 'MoteView' platform generate Health Packets regularly. These Health Packets encapsulate the state of the wireless mesh over time with display of the latest Health Packet readings received for each node in the network.

There are two types of Health Packet in XMesh:

- Statistics Health Packets: Packets containing statistics on the state of the sensor node.
- Neighbor Health Packets: Packets containing information about neighboring sensor nodes.

In this paper, we will exploit the information given by the statistical Packet-Health (Fig. 4) in order to describe the quality of the information exchanged and that of the sensor node. Based on this information, we will have:

- Visibility on sensor nodes that have a large number of lost packets, this is possible from the 'Dropped' field;
- Information on the quality of the link level between the different sensor nodes as well as their links with the base station;
- Visibility on the number of transmissions needed to send a packet from a node to the base station, this is possible from the 'Path-Cost' field;
- Visibility into the 'Signal Strength Indicator Level' received from a sensor node's radio, this information is stored in the 'Link quality' field.

All this data and others can be used for better visibility of the state of the sensor nodes as well as the quality of the information exchanged.

3 Interpretation of Results

Packets in WSNs can be discarded for many reasons, such as congestion, node failure, and optimization mechanisms [6]. Additionally, multiple accesses to the communication

Data Fields	Description	
uint16_t seq_num	This is the sequence number of the health packet.	
uint16_t num_node_pkts	Number of application packets that have been generated locally.	
uint16_t num_fwd_pkts	Number of radio packets that the mote has forwarded in the mesh.	
uint16_t num_drop_pkts	Number of radio packets that the mote has dropped due to failed link level retransmissions.	
uint16_t num_rexmits	This is the number of times that the message was re-transmitted.	
uint8_t battery_voltage	Battery reading.	
uint16_t power_sum	Power statistics in low power radio stack (not used)	
uint8_t rsvd_app_type	Reserved MEMSIC field to identify sending application.	
hn_quality nodeinfo	A record containing information about the link to the present parent	
	uint16_t node_id	Parent's ID
	uint8_t link_quality	Link quality to parent
	uint8_t path_cost	The routing path cost to the base station
	uint8_t radio_link_indicator	rssi or lqi value of the parent

Fig. 4. Structure of the 'Packet Health'

medium can also lead to packet collision, requiring a MAC protocol to manage shared access to the channel and to retransmit colliding packets. In this case, many works have proposed solutions to recover lost data or to minimize the impact of packet loss on the final quality of data received at the receiver [7] [8].

In this work, we use an experimental WSN Micaz-Crossbow platform composed of two sensor nodes in order to visualize the transmitted packets and interpret the results in order to contribute to a better decision-making. The software used for the visualization of the Packet Health fields is the 'XSniffer'.

Figure 5 illustrates the variation of the 'link quality' field between two sensor nodes, the first is intact (normal) and the second is disturbed in order to see the effect of this on the fields of this packet. We noticed that the variation of the link quality of the second sensor node is unstable compared to the variation of the first sensor node. This is clear from minute 42:26.

Figure 6 illustrates the variation of the 'Dropped' field. We notice that the second sensor node has lost several packets, this is due to the disturbance we performed.

Contrary to some works [9][10][11][12] which do not take into account the particular characteristics that can negatively affect the estimation of the state of a sensor node, our work takes into consideration the information from the different fields given by Packet Health. Based on this information, we will have, among other things, visibility on the creation of shared data [13][14], on the number of lost packets used and on the quality of the Emitted/Transmitted signal. All this information facilitates the interpretation of the results as well as contributing to decision-making.

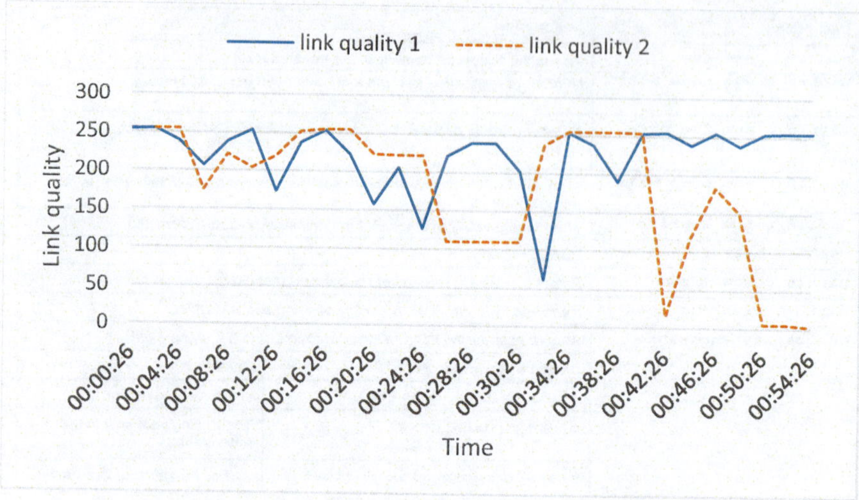

Fig. 5. Representation of the link quality variation for two sensor nodes

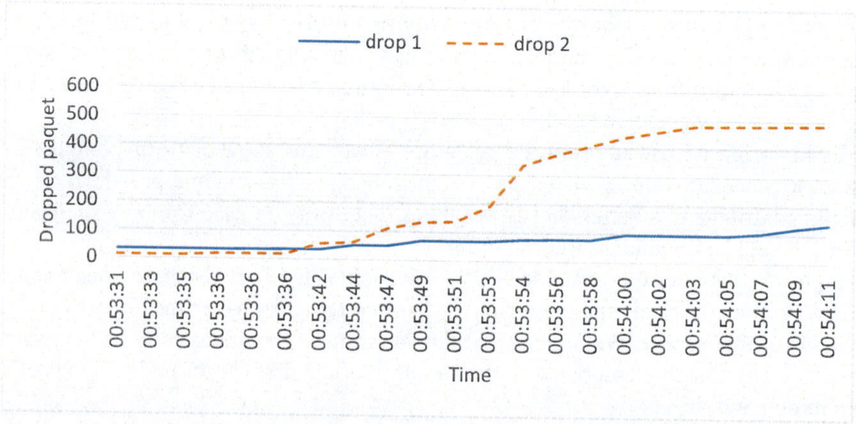

Fig. 6. Representation of the 'Dropped' variation

4 Conclusion

The sensor nodes of the WSN Crossbow MicaZ (ZigBee) experimental platform that we used emit packets called 'Packet Health'. These packets contain several information about the state of the sensor nodes. In this work, we exploited this information in order to contribute to the decision-making on the credibility of the information exchanged by the WSN. Indeed, we used tools to generate, visualize and then process the different information shared by this WSN platform.

References

1. Swetha, R., Santhosh, V., Anita, S.: Wireless Sensor Network : A Survey. Int. J. Adv. Res. Comput. Commun. Eng. **2018**, 71124 (2018)
2. Jino Ramson, S.R., Moni, D.J.: Applications of wireless sensor networks — A survey. In: 2017 International Conference on Innovations in Electrical, Electronics, Instrumentation and Media Technology (ICEEIMT), pp. 325–329 (2017)
3. Martinho, J., Prates, L., Costa, J.: Design and Implementation of a wireless multi parameter patient monitoring system. Procedia Technology, pp. 542–549 (2014)
4. Simon, C., Weber, B.: Fiabilité imprécise par les réseaux de capteurs de croyance. Nancy Université, CNRS, Congrès de Maîtrise des Risques et de Sûreté de Fonctionnement (2008)
5. Arturo, J., Celaya, G.: Fusion d'informations en identification automatique des langues. Thèse de l'Université Paul Sabatier Toulouse III, Formation Doctorale Informatique de l'Image et du Langage, Juillet (2005)
6. Costa, D., Guedes, L.: A: Survey on multimedia-based cross-layer optimization in visual sensor networks. Sensors **11**, 5439–5468 (2011)
7. Almalkawi, I., Zapata, M., Al-Karaki, J., Morillo-Pozo, J.: Wireless multimedia sensor networks: Current trends and future directions. Sensors **10**, 6662–6717 (2010)
8. Sasikumar, G. Vignesh Ramamoorthy, H., Natheem Mohamed, S.: An analysis on animal tracking system using wireless sensors. Int. J. Adv. Res. Computer Sci. Softw. Eng. **4**, 8 (2014)
9. Chen, J., Kher, S., Somani, A.: Distributed fault detection of wireless sensor networks. In: Workshop on Dependability issues in wireless ad hoc networks and sensor networks, 2006, New York, USA, pp. 65–72 (2006)
10. Zhang, L., Zhang, F., Yuan, J., lan Weng, J., hua Zhang, W.: Sensor fault diagnosis and location for small and medium-scale wireless sensor networks. In: Natural Computation (ICNC), 2010 Sixth International Conference on, vol. 7, pp. 3628 –3632 (2010)
11. Mao Huang, R., song Qiu, X., li Ye, L.: Probability-based fault detection in wireless sensor networks. In: Network and Service Management (CNSM), 2010 International Conference on, pp. 218 –221 (2010)
12. Krishnamachari, B., Iyengar, S.: Distributed bayesian algorithms for fault-tolerant event region detection in wireless sensor networks. Comput. IEEE Trans. **53**, 241–250 (2004)
13. Lahma, K., Hamraoui, M.: Information credibility modeling in a wsn via leach routing protocol. IJACT Int. J. Adv. Comput. Technol. **7**(10), 2838–2842 (2018)
14. Lahma, K., Hamraoui, M.: Evaluation de la crédibilité de l'information d'une chaine de nœuds de capteurs : RCSF. Mediterr. Telecommun. J. **6**(2), 5 (2018)

Artificial Intelligence Applications in Date Palm Cultivation and Production: A Scoping Review

Abdelaaziz Hessane[1(✉)], Ahmed EL Youssefi[1], Yousef Farhaoui[1], Badraddine Aghoutane[2], and Youssef Qaraai[1]

[1] STI Laboratory, T-IDMS Faculty of Sciences and Techniques of Errachidia, Moulay Ismail University of Meknès, Errachidia, Morocco
{a.hessane,ah.elyoussefi}@edu.umi.ac.ma, {y.farhaoui, y.qaraai}@fste.umi.ac.ma

[2] IA Laboratory, Department of Computer Science, Faculty of Sciences, Moulay Ismail University of Meknès, Meknès, Morocco
b.aghoutane@umi.ac.ma

Abstract. Date palm cultivation is considered one of the most important levers in the economy of many countries, especially in North Africa and the Middle East. This tree and the fruits it produces are of great importance in terms of their nutritional and medicinal value, as well as their uses in some biochemical applications. In the last decade, artificial intelligence and its applications in the field of precision agriculture constituted a fertile field for research. Consequently, the date palm sector is emerging with new and modern technologies to satisfy global sustainability standards. This article provides an overview of studies on the application of artificial intelligence (AI) in the date palm agriculture industry throughout the past decade (2012–2021). After applying exclusion criteria, the scoping review was constructed to answer four (4) predetermined research questions by analyzing 43 publications. Based on the defined research questions, the analysis of the examined literature included the yearly and geographical distribution of papers, the most widely adopted algorithms, and the research trends in the field of AI applications for date palm cultivation and production. Investigations have indicated that AI is underutilized in certain applications such as yield estimation and diseases and pest control and management. Nonetheless, intelligent systems using machine vision and artificial intelligence are evolving to improve the date palm agricultural industry. This article explains the importance of artificial intelligence in the date palm agricultural industry and presents insight for future study in this field.

Keywords: Date palm · Phoenix dactylifera · Date fruit · Artificial intelligence · Machine learning · Deep learning

1 Introduction

The transition from traditional agriculture to a smart one based on new technologies is becoming a necessity to overcome various challenges principally related to food

Y. Farhaoui et al. (Eds.): ICAISE 2022, LNNS 635, pp. 230–239, 2023.
https://doi.org/10.1007/978-3-031-26254-8_32

and water security. In this context, Artificial Intelligence (AI) is used and had shown promising results regarding the improvement of agricultural decision support systems [1]–[3]. The variety of AI-based approaches and the methods that are implemented in the context of Precision Agriculture (PA) made it possible to have multiple applications such as real-time monitoring of crops, crop health assessment, automatic irrigation, and plant disease detection.

Date palm is considered as one of the most important crops in the world, not only because of its nutritional and medicinal value [4, 5] but also because of its traditional and socioeconomic importance to many countries, especially in North Africa and the Middle East regions [6]. In 2020, the global production of dates reached 9.45 million tons with an increase of about 2.3 points from 2010 [7]. To meet the rising demand for date fruit in global markets, the producers face many challenges, including the necessity to have enough high-quality yields to attain self-sufficiency and to improve their export capacity. Moreover, the date palm industry must be emerged to meet the agriculture 4.0 standards which is characterized with the automatization of processes, adaptation to climate changes and the reasonable allocation of resources [8].

In the literature, few authors have reviewed the application of AI in date palm production and cultivation field. For instance, Eldin et al. conducted a review on methods used to identify Red Weevil within palm trees [9]. The AI-based Robotic Systems mainly designed for the quality assessment of date fruits are reviewed in [10]. The literature review demonstrates the absence of a review of current research trends and the scope of AI use for date palm. To the best of the authors' knowledge, no comprehensive review has previously investigated diverse aspects of the date palm domain within the context of AI. This review aimed to analyze significant advancements, current trends, research gaps, and innovative ideas to stimulate research on date palms using AI. Consequently, high-quality recommendations are envisaged to be produced and presented to researchers based on a meticulous examination of the available literature.

The remainder of the article is structured as follows. Section 2 presents the review methodology. It details the procedure that was followed during the design and execution phases. Furthermore, the results are presented in Sect. 3, while the conclusion and perspectives are provided in Sect. 4.

2 Review Methodology

From a problem-solving standpoint, this paper examines previous research that employ AI to assist the date palm farming industry. To ensure a comprehensive investigation based on current research trends, we started with the topic selection phase, then we arranged the empirical findings and answers to the "research questions" based on the review protocol depicted in Fig. 1. Finally, the information extraction and reporting step was done accordingly.

2.1 Topic Selection and Search String Definition

After a thorough review of the current literature on latent advancement in date palms with the use of AI, a less intact research topic was carefully chosen to undertake this

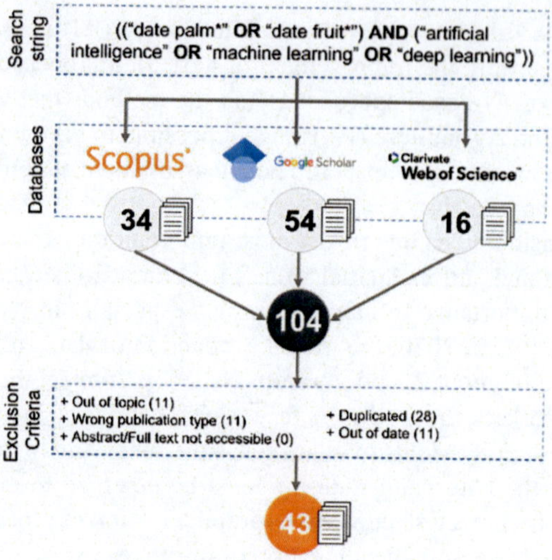

Fig. 1. Proposed review protocol

investigation. This review topic is ideal to analyze numerous dimensions in the quest to explore the studies that have been published in the realm of date palm and AI.

The fundamental search was conducted automatically using the Rayyan tool [11]. The initial search input was "date palm" AND "artificial intelligence." Articles from different databases (Web of science, Google scholar and Scopus) were retrieved, and abstracts were reviewed to determine the keywords' synonyms. A more complex searching string was built based on the findings of the earlier step with the aim to not miss relevant studies. The final search string was as follows: ("date palm*" OR "date fruit*") AND ("artificial intelligence" OR "machine learning" OR "deep learning").

2.2 Review Boundary and Exclusion Criteria

The existing literature on the subject was gathered during a ten-year timeframe (2012–2021). Only technical publications proposing the use of AI to investigate date palm were examined. Given that AI applications exists in many fields that are irrelevant to our domain-specific investigation, all articles that did not specifically apply AI to agricultural date palm were excluded. In addition, several exclusion criteria were established to limit the scope of the review. In the systematic article selection process, the database was filtered using the five exclusion criteria (EC) outlined below.

- EC 1: Publications that are not related to the agricultural sector of date palm with the application of artificial intelligence (Out of topic)
- EC 2: Publications that are survey/review articles reviewing old publications (Wrong publication type)
- EC 3: Full text/abstract of the publication is unavailable.
- EC 4: Publications that are already retrieved (Duplicated)
- EC 5: Work that is published before 2012 and after 2021 (Out of date frame)

2.3 Research Questions

The purpose of the review was determined by the answers to four research questions (RQ). The first question (RQ 1) relates to the annual rate of publications that apply AI to date palm. Second question (RQ 2) addresses the research trends, while the third (RQ 3) investigates the approaches, algorithms and methods that have been highly used during the last decade to serve date palm agriculture industry. Final question (RQ 4) focuses on the geographical distribution of retrieved articles.

3 Results and Discussion

This section summarizes the findings of the review methodology proposed in the previous section. Considering the development of the number of published research papers on the topic over the past decade, it is evident that it has been in a state of continuous growth, except for the years between 2014 and 2016. However, the number of research remains relatively low, reaching only 15 papers in 2021. Figure 2 illustrates the yearly distribution of research publications. This result addresses the first research question (RQ 1).

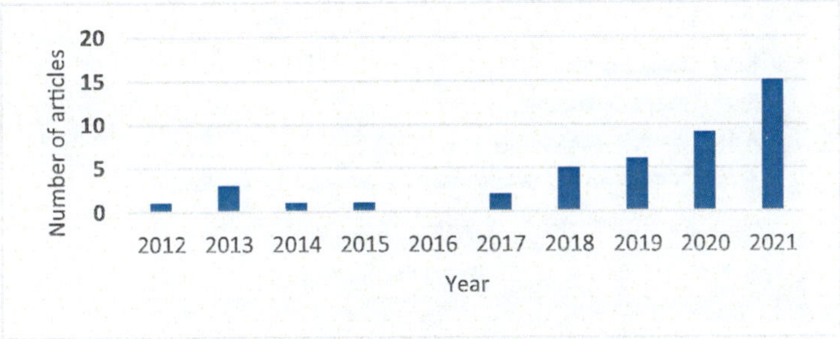

Fig. 2. Annual distribution of reviewed articles

The findings related to the second research question (RQ 2) provide additional understanding of the current research trends in the addressed domain. The results indicate that around 95% of the published work concentrated on classification approaches, while the remaining 5% focused on regression analysis for prediction. Important research topics in the classification category included date fruit detection and classification [12]–[29], palm tree identification [30]–[38], date fruit quality assessment [39]–[45], and date fruit pests and diseases management and control [46]–[51]. In contrast, regression analysis focuses exclusively on yield estimation [52, 53]. Figure 3 depicts the relative proportions of the research trends.

We further evaluated the reviewed database to determine the most prevalent AI approaches and methods (referring to RQ 3). Approximately 96% of the analyzed publications utilized the supervised learning approach. In contrast, Unsupervised learning is utilized infrequently, and the reinforcement learning strategy is never employed. Moreover, 58% of publications employed deep learning-based methods, such as Convolutional Neural Network (CNN: 42.3%) and Artificial Neural Network (ANN: 15,83%).

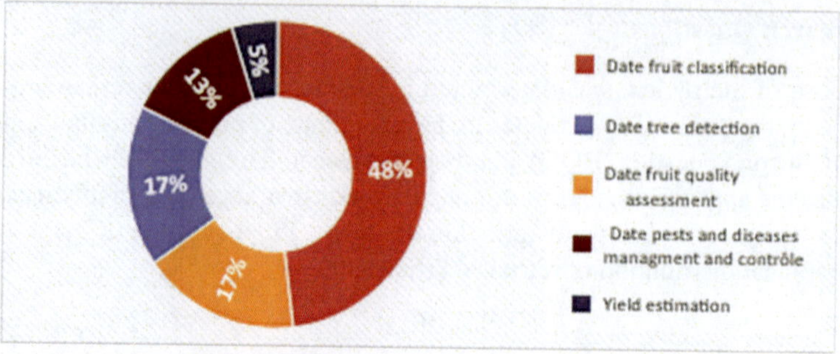

Fig. 3. Article's objective-based distribution.

The percentage of utilization of Support Vector Machine (SVM) and K-Nearest Neighbor (KNN) algorithms is 13.46% and 9.62%, respectively. Other methods are used less frequently than those described above. We only examined methods with a frequency of use greater than or equivalent to the average, which is 4.7 publications. Figure 4 shows the highly adopted methods and their usage frequency in the context of this discussion.

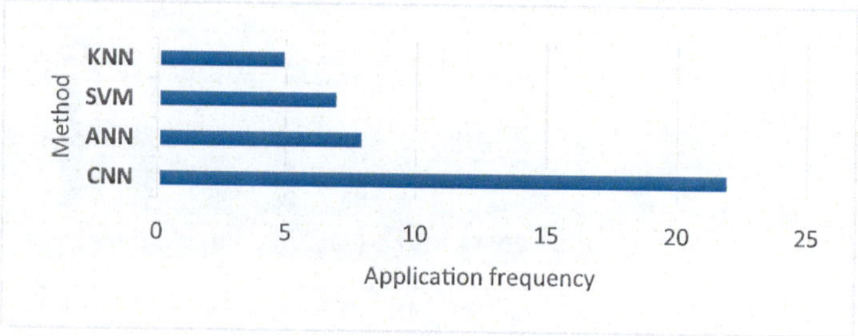

Fig. 4. Highly adopted algorithms

To address the fourth research question (RQ 4), we retrieved the first author's affiliation to determine the geographical distribution of the reviewed articles. The results indicate that the countries most interested in adopting AI in date palm agriculture are the world's leading producers, namely Saudi Arabia, Oman, Iran, and Algeria. However, just 20% of countries have published more than or equal to the average number of publications per country, which is 2.15 papers, indicating that the field remains fertile and promising. Figure 5 presents the contribution of researchers from all countries regarding the number of publications.

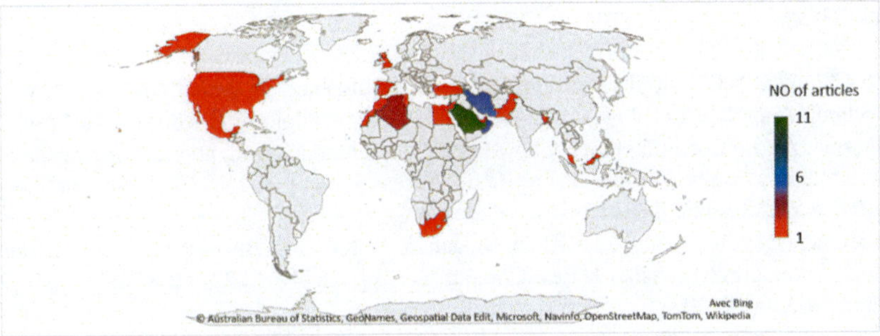

Fig. 5. Geographical distribution of reviewed publications

4 Conclusion and Perspectives

This study examined research conducted within the last decade (2012–2021) on the application of artificial intelligence to date palm. The reviewed studies were separated into two categories for clarity: classification and regression analysis. The study listed the primary objectives in the classification category as date fruit classification, palm tree detection, date quality assessment, and date diseases management and control. In contrast, just the estimation of date yield was included in the regression analysis. Surprisingly, fruit yield forecasting, and the disease detection models have not been widely implemented. The results demonstrate that the most widely used algorithms were supervised deep learning techniques such as CNN and ANN. In comparison, methods based on machine learning are less prevalent. Due to the limited number of researchers from a small number of countries, important research dimensions were lacking, such as soil classification to identify appropriate land for a specific variety of date palm, automated pest and disease recognition, identification the symptoms of climate, nutrient, and water stress in date palm crop, optimization of fertilizers and so on. Importantly, research trends revealed that existing and recent techniques such as Computer Vision (CV), Unmanned Aerial Vehicles (UAV), and Remote Sensing (RS) were coupled with machine learning and deep learning approaches to provide effective decision-making systems within the context of the date palm agriculture industry. Current research is insufficient to build practically useful instruments capable of raising yields, enhancing quality, and ensuring the long-term viability of plantations in an environmentally responsible manner.

This study gave a lucid understanding of the advancement of the date palm in relation to artificial intelligence and can encourage scholars to identify pertinent issues in this field. This work expands the present research on "date palm and artificial intelligence" from multiple vantage points aiming to pave the road for the automation and intelligence-based development of the date palm. Future study should concentrate on a technical analysis of the performance of AI-based models applied to the date palm agriculture area. In addition, additional research concerns must be addressed, such as the availability of datasets and sources to examine date palms using ML/DL, critical input features, and how the models are applied and evaluated.

References

1. Eli-Chukwu, N.C.: Applications of artificial intelligence in agriculture: a review. Eng. Technol. Appl. Sci. Res. **9**(4), 4377–4383 (2019). https://doi.org/10.48084/etasr.2756
2. Sharma, A., Jain, A., Gupta, P., Chowdary, V.: Machine learning applications for precision agriculture: a comprehensive review. IEEE Access **9**, 4843–4873 (2021). https://doi.org/10.1109/ACCESS.2020.3048415
3. Bharman, P., Ahmad Saad, S., Khan, S., Jahan, I., Ray, M., Biswas, M.: Deep learning in agriculture: a review. Asian J. Res. Comput. Sci., pp. 28–47 (2022) https://doi.org/10.9734/ajrcos/2022/v13i230311
4. Temitope Idowu, A., Osarumwense Igiehon, O., Ezekiel Adekoya, A., Idowu, S.: Dates palm fruits: a review of their nutritional components, bioactivities and functional food applications. AIMS Agric. Food, **5**(4), 734–755 (2020) https://doi.org/10.3934/agrfood.2020.4.734
5. Al-Mssallem, M.Q.: The role of date palm fruit in improving human health. J. Clin. DIAGNOSTIC Res. (2020). https://doi.org/10.7860/JCDR/2020/43026.13442
6. El Hadrami, A., Al-Khayri, J.M.: Socioeconomic and traditional importance of date palm. Emirates J. Food Agric. **24**(5), 371–385 (2012)
7. "FAOSTAT.[Production/Yield quantities of Dates in World + (Total)]," 2020. https://www.fao.org/faostat/Accessed 16 Jun 2022
8. Yahya, N.: Agricultural 4.0: Its Implementation Toward Future Sustainability, pp. 125–145 (2018)
9. Eldin, H.A., Waleed, K., Samir, M., Tarek, M., Sobeah, H., Salam, M.A.: A Survey on Detection of Red Palm Weevil Inside Palm Trees: Challenges and Applications. In: ACM International Conference Proceeding Series, pp. 119–125 (2020) https://doi.org/10.1145/3436829.3436861
10. Scaria, B., Aziz, N.A., Siddiqi, M.A.M.: AI Based Robotic Systems for the quality control of Date Palm Fruits - A Review. In: Proceeding of 2019 International Conference on Digitization: Landscaping Artificial Intelligence, ICD 2019, pp. 227–231 (2019) https://doi.org/10.1109/ICD47981.2019.9105743
11. Ouzzani, M., Hammady, H., Fedorowicz, Z., Elmagarmid, A.: Rayyan—a web and mobile app for systematic reviews. Syst. Rev. **5**(1), 210 (2016). https://doi.org/10.1186/s13643-016-0384-4
12. Alhamdan, W.S.N., Howe, J.M.: Classification of Date Fruits in a Controlled Environment Using Convolutional Neural Networks. In: Hassanien, A.-E., Chang, K.-C., Mincong, T. (eds.) AMLTA 2021. AISC, vol. 1339, pp. 154–163. Springer, Cham (2021). https://doi.org/10.1007/978-3-030-69717-4_16
13. Koklu, M., Kursun, R., Taspinar, Y.S., Cinar, I.: Classification of date fruits into genetic varieties using image analysis. Math. Probl. Eng., **2021** (2021)https://doi.org/10.1155/2021/4793293
14. Hossain, M.S., Muhammad, G., Amin, S.U.: Improving consumer satisfaction in smart cities using edge computing and caching: a case study of date fruits classification. Futur. Gener. Comput. Syst. **88**, 333–341 (2018). https://doi.org/10.1016/j.future.2018.05.050
15. Aiadi, O., Kherfi, M.L.: A new method for automatic date fruit classification. Int. J. Comput. Vis. Robot. **7**(6), 692 (2017). https://doi.org/10.1504/IJCVR.2017.087751
16. Manickavasagan, A., Al-Shekaili, N.H., Al-Mezeini, N.K., Rahman, M.S., Guizani, N : Computer vision technique to classify dates based on hardness. J. Agric. Mar. Sci. [JAMS], **22**(1), 36 (2018). https://doi.org/10.24200/jams.vol22iss1pp36-41
17. Muhammad, G.: Date fruits classification using texture descriptors and shape-size features. Eng. Appl. Artif. Intell. **37**, 361–367 (2015). https://doi.org/10.1016/j.engappai.2014.10.001

18. Manickavasagan, A., Al-Shekaili, H.N., Thomas, G., Rahman, M.S., Guizani, N., Jayas, D.S.: Edge detection features to evaluate hardness of dates using monochrome images. Food Bioprocess Technol. **7**(8), 2251–2258 (2013). https://doi.org/10.1007/s11947-013-1219-0
19. Nozari, V., Mazlomzadeh, M.: Date fruits grading based on some physical properties. **9**(7), 1703–1713 (2013)
20. Benam, M.P.M.R., Mazloumzadeh, S.M.: Using adaptive neuro-fuzzy inference system for classify date fruits. J. Agric. Technol. 9(5), 1309–1318 (2013).
21. Haidar, A., Dong, H., Mavridis, N.: Image-based date fruit classification. In: 2012 IV International Congress on Ultra Modern Telecommunications and Control Systems, Oct., pp. 357–363 (2012) https://doi.org/10.1109/ICUMT.2012.6459693
22. Khayer, M.A., Hasan, M.S., Sattar, A.: Arabian date classification using CNN algorithm with various pre-trained models. In: Proceedings of the 3rd International Conference on Intelligent Communication Technologies and Virtual Mobile Networks, ICICV 2021, pp. 1431–1436 (2021) https://doi.org/10.1109/ICICV50876.2021.9388413
23. Pérez-Pérez, B.D., García Vázquez, J.P., Salomón-Torres, R.: Evaluation of Convolutional Neural Networks. Hyperparameters with Transfer Learning to Determine Sorting of Ripe Medjool Dates. Agriculture **11**(2), 115 (2021) https://doi.org/10.3390/agriculture110201215
24. Khriji, L., Ammari, A.C., Awadalla, M.: Artificial intelligent techniques for palm date varieties classification. Int. J. Adv. Comput. Sci. Appl. **11**(9), 489–495 (2020). https://doi.org/10.14569/IJACSA.2020.0110958
25. Abi Sen, A.A., Bahbouh, N.M., Alkhodre, A.B., Aldhawi, A.M., Aldham, F.A., Aljabri, M.I.: A classification algorithm for date fruits. In Proceedings of the 7th International Conference on Computing for Sustainable Global Development, INDIA Com 2020, pp. 235–239 (2020) https://doi.org/10.23919/INDIACom49435.2020.9083706
26. Khriji, L., Ammari, A.C., Awadalla, M.: Hardware/software co-design of a vision system for automatic classification of date fruits. Int. J. Embed. Real-Time Commun. Syst. **11**(4), 21–40 (2020). https://doi.org/10.4018/IJERTCS.2020100102
27. Faisal, M., Alsulaiman, M., Arafah, M., Mekhtiche, M.A.: IHDS: Intelligent harvesting decision system for date fruit based on maturity stage using deep learning and computer vision. IEEE Access **8**, 167985–167997 (2020). https://doi.org/10.1109/ACCESS.2020.3023894
28. Aiadi, O., Kherfi, M.L., Khaldi, B.: "Automatic date fruit recognition using outlier detection techniques and gaussian mixture models", elcvia electron. Lett. Comput. Vis. Image Anal. **18**(1), 52 (2019). https://doi.org/10.5565/rev/elcvia.1041
29. Altaheri, H., Alsulaiman, M., Muhammad, G.: Date fruit classification for robotic harvesting in a natural environment using deep learning. IEEE Access **7**, 117115–117133 (2019). https://doi.org/10.1109/ACCESS.2019.2936536
30. Alburshaid, E.A., Mangoud, M.A.: "developing date palm tree inventory from satellite remote sensed imagery using deep learning. In: 2021 3rd IEEE Middle East and North Africa COMMunications Conference. MENACOMM **2021**, 54–59 (2021). https://doi.org/10.1109/MENACOMM50742.2021.9678262
31. Gibril, M.B.A., Shafri, H.Z.M., Shanableh, A., Al-Ruzouq, R., Wayayok, A., Hashim, S.J.: Deep convolutional neural network for large-scale date palm .mapping from uav-based images. Remote Sens. **13**(14) (2021). https://doi.org/10.3390/rs13142787
32. El Hoummaidi, L., Larabi, A., Alam, K.: Using unmanned aerial systems and deep learning for agriculture mapping in Dubai. Heliyon **7**(10), e08154 (2021). https://doi.org/10.1016/j.heliyon.2021.e08154
33. Rhinane, H., Bannari, A., Maanan, M., Aderdour, N.: Palm Trees Crown Detection and Delineation From Very High Spatial Resolution Images Using Deep Neural Network (U-Net). In: 2021 IEEE International Geoscience and Remote Sensing Symposium IGARSS, Jul. 2021, pp. 6516–6519. https://doi.org/10.1109/IGARSS47720.2021.9554470

34. Culman, M., Delalieux, S., Tricht, K.V.: Palm Tree Inventory from Aerial Images Using Retinanet. In: 2020 Mediterranean and Middle-East Geoscience and Remote Sensing Symposium, M2GARSS 2020 - Proceedings, 2020, pp. 314–317 (2020) https://doi.org/10.1109/M2GARSS47143.2020.9105246
35. Cousin, R., Ferry, M.: Automatic localization of phoenix by satellite image analysis. Arab J. Plant Prot. **37**(2), 83–88 (2019). https://doi.org/10.22268/AJPP-037.2.083088
36. Al-Ruzouq, R., Shanableh, A., Gibril, M.A., Al-Mansoori, S.: Image segmentation parameter selection and ant colony optimization for date palm tree detection and mapping from very-high-spatial-resolution aerial imagery. Remote Sens. **10**(9) (2018) https://doi.org/10.3390/rs10091413
37. Djerriri, K., Ghabi, M., Karoui, M.S., Adjoudj, R.,: Palm trees counting in remote sensing imagery using regression convolutional neural network. In: International Geoscience and Remote Sensing Symposium (IGARSS), vol. 2018-July, pp. 2627–2630 (2018). https://doi.org/10.1109/IGARSS.2018.8519188
38. Almaazmi, A.: Palm trees detecting and counting from high-resolution WorldView-3 satellite images in United Arab Emirates. In: Proceedings of SPIE - The International Society for Optical Engineering, 2018, vol. 10783, (2018). https://doi.org/10.1117/12.2325733
39. Shoshan, T., Bechar, A., Cohen, Y., Sadowsky, A., Berman, S.: Segmentation and motion parameter estimation for robotic Medjoul-date thinning. Precision Agric. , 1–24 (2021). https://doi.org/10.1007/s11119-021-09847-2
40. Keramat-Jahromi, M., Mohtasebi, S.S., Mousazadeh, H., Ghasemi-Varnamkhasti, M., Rahimi-Movassagh, M.: Real-time moisture ratio study of drying date fruit chips based on on-line image attributes using kNN and random forest regression methods. Measurement **172**(1), 108899 (2021)
41. Faisal, M., Albogamy, F., Elgibreen, H., Algabri, M., Alqershi, F.A.: Deep learning and computer vision for estimating date fruits type, maturity level, and weight. IEEE Access **8**, 206770–206782 (2020). https://doi.org/10.1109/ACCESS.2020.3037948
42. Raissouli, H., Ali, A., Mohammed, S., Haron, F., Alharbi, G.: Date Grading using Machine Learning Techniques on a Novel Dataset. Int. J. Adv. Comput. Sci. Appl., **11**(8), (2020). https://doi.org/10.14569/IJACSA.2020.0110893
43. Hakami, A., Arif, M.: Automatic Inspection of the External Quality of the Date Fruit. Procedia Computer Science **163**, 70–77 (2019). https://doi.org/10.1016/j.procs.2019.12.088
44. Nasiri, A., Taheri-Garavand, A., Zhang, Y.-D.: Image-based deep learning automated sorting of date fruit. Postharvest Biol. Technol. **153**, 133–141 (2019). https://doi.org/10.1016/j.postharvbio.2019.04.003
45. Alavi, N.: Quality determination of Mozafati dates using Mamdani fuzzy inference system. J. Saudi Soc. Agric. Sci. **12**(2), 137–142 (2013). https://doi.org/10.1016/j.jssas.2012.10.001
46. Al-Shalout, M., Mansour, K.: No Title. In: 22nd International Arab Conference on Information Technology (ACIT), 2021, pp. 1–5, [Online]. Available: https://ieeexplore.ieee.org/abstract/document/9677103
47. Karar, M.E., Reyad, O., Abdel-Aty, A.-H., Owyed, S., Hassan, M.F.: Intelligent IoT-Aided early sound detection of red palmWeevils. Comput. Mater. Contin. **69**(3), 4095–4111 (2021). https://doi.org/10.32604/cmc.2021.019059
48. Wang, B., et al.: Towards Detecting Red Palm Weevil Using Machine Learning and Fiber Optic Distributed Acoustic Sensing. Sensors **21**(5), 1592 (2021). https://doi.org/10.3390/s21051592
49. Alaa, H., Waleed, K., Samir, M., Tarek, M., Sobeah, H., Salam, M.A.: An intelligent approach for detecting palm trees diseases using image processing and machine learning. Int. J. Adv. Comput. Sci. Appl. **11**(7), 434–441 (2020). https://doi.org/10.14569/IJACSA.2020.0110757

50. Magsi, A., Mahar, J.A., Razzaq, M.A., Gill, S.H.: Date Palm Disease Identification Using Features Extraction and Deep Learning Approach. In: Proceedings of the 2020 23rd IEEE Int. Multi-Topic Conference INMIC 2020, no. November (2020). https://doi.org/10.1109/INMIC50486.2020.9318158
51. Teena, M., Manickavasagan, A., Al-Sadi, A.M., Al-Yahyai, R., Deadman, M.L., Al-Ismaili, A.: Near infrared imaging to detect Aspergillus flavus infection in three varieties of dates. Eng. Agric. Environ. Food **11**(4), 169–177 (2018). https://doi.org/10.1016/j.eaef.2018.04.002
52. Heyns, K.: Estimation Methods for Date Palm Yield – a Feasibility Study," no. March (2021)
53. Husain, M., Khan, R.A.: Date Palm Crop Yield Estimation a Framework. SSRN Electron. J. (2019). https://doi.org/10.2139/ssrn.3509195

Check for updates

Artificial Intelligent-Based System for Thermal Comfort Control in Smart Building

Youssef Boutahri(✉) and Amine Tilioua

Research Team in Thermal and Applied Thermodynamics (2.T.A.), Mechanics, Energy Efficiency and Renewable Energies Laboratory (L.M.3.E.R.), Department of Physics, Faculty of Sciences and Techniques Errachidia, Moulay Ismaïl University of Meknès, B.P. 509, Boutalamine, Errachidia, Morocco

yo.boutahri@edu.umi.ac.ma, a.tilioua@umi.ac.ma

Abstract. Buildings consume a significant amount of the world's energy. An important portion of this energy is utilized to maintain building thermal comfort. Since Heating Ventilation Air Condition (HVAC) systems consume the bulk of a building's energy, it is vital to optimize this energy while preserving thermal comfort for occupants. The Predict Mean Vote (PMV) model is considered as one of the most important models describing the building's thermal comfort. In this context, the accurate and real-time prediction of the PMV index is crucial. This paper proposes an artificial intelligence-based model to be integrated into HVAC systems to predict Predicted Mean Vote (PMV) using two Machine Learning (ML) algorithms, namely Random Forest (RF) and Artificial Neural Network (ANN). The performance of the investigated methods was tested using several evaluation metrics such as Root Mean Square Error (RMSE) and Correlation Coefficient (R2). The obtained results have revealed that random forest is more accurate with an RMSE = 0.075 and R2 = 0.965, in comparison with Artificial Neural Network (ANN) which shows an RMSE = 0,171 and R2 = 0.801. The proposed work will be an essential stage in the deployment of HVAC systems, to maintain thermal comfort of occupants and optimize energy consumption in smart buildings.

Keywords: Energy · HVAC Systems · Thermal comfort · Machine learning · Artificial neural network · Smart building

1 Introduction

Buildings consume a significant amount of word's energy. Most of this energy is consumed by HVAC systems. Therefore, reducing the energy demand of these systems is essential for total energy conservation and greenhouse gas reduction. Commercial and industrial building lose up to 30% and 25% of their heat through ventilation, respectively [1]. Its potential to improve energy efficiency by increasing occupant comfort and decreasing energy consumption, the operation of HVAC systems has received a significant attention over the past decade [2]. To decide the layout of building systems based on the occupants' comfort whish reflect their opinions, collective models, namely the PMV model, have traditionally been used in the design and management

Y. Farhaoui et al. (Eds.): ICAISE 2022, LNNS 635, pp. 240–246, 2023.
https://doi.org/10.1007/978-3-031-26254-8_33

of HVAC systems. The PMV model, which combines two human-related parameters, clothing level and metabolic rate, to make indoor spaces thermally tolerable for most occupants, has been standardized by the American Society of Heat, Refrigeration, and Air-conditioning Engineers (ASHRAE) [5]. ASHRAE 55 [3] suggests a PMV of ± 0.5 for a suitable building environment. The PMV score may be quantitatively determined utilizing six comfort parameters (four interior environment elements, including Temperature (T), Humidity (H), Mean Radiant Temperature (MRT), and Air Velocity (Va), and two essential components, including Metabolic Rate (Met) and Clothing Insulation (Clo)). Recent researchers have also attempted the forecasting of the association between sub munching and the PMV score using ML. Optimizing energy consumption in buildings and modeling thermal comfort using ML methods are among the most popular topics addressed by researchers in recent years.

Liu et al. [4] investigated the use of an ANN and a gradient boosting decision tree (GBDT) to develop a predictive models of occupant air conditioning use behavior. Their results showed that the models used outperformed logistic regression. Tien et al. [1] proposed a method to recognize and detect manual windows used in buildings using a deep learning framework, with an accuracy of 97.29% for classification and 95.5% for detection. Abdulgader et al. [5] proposed an energy-efficient thermal comfort model for HVAC systems in smart buildings to improve the thermal comfort of building occupants while reducing energy consumption. Their results show that applying a large database to the Support Vector Machine (SVM) model can significantly improve the accuracy. Gao et al. [6] suggested using deep reinforcement learning (DeepComfort), to improve thermal comfort control in buildings. They used Deep Neural Networks (DNN) and Deep Deterministic Policy Gradients (DDPG) methods in this scenario. They were able to reduce energy consumption by 4.31% and improve thermal comfort by 13.6%. Zhang et al. [7] studied the impact of DNN topology on the control of a building's HVAC system. According to their results, the addition of hidden layers and neurons improves the modeling performance. Zhang et al. [8] used ML algorithms such as linear regression (LR) and decision tree (DT), to improve the thermal comfort system. Bhattacharya et al. [9] discussed an ML-based framework to effectively evaluate the impact of non-ideality sensor for building systems. Gan et al. [10] presented a Building Information Modeling (BIM)-based computational framework and data-driven ML models to estimate the best thermal comfort of indoor spaces with natural ventilation. Evaluation of their results shows that the average errors range from 0.03 to 0.13 for the PMV and from 0.35 to 5.52 for the predicted percentage of dissatisfied (PPD).

In this work, we propose an artificial intelligence-based framework to build a thermal comfort model that can be integrated into HVAC systems. Compared to the mentioned literature research, our model showed some performance that can be explained by the use of environmental and personal variables related to the occupants. The rest of this paper is described as follows. Section 2 presents a thermal comfort system architecture that introduces the environmental and personal factors of thermal comfort. Section 3 proposes the methodology, experiments, data processing and adopted ML algorithms. In Sect. 4, we show the results and performance of the evaluation of ML methods.

2 Proposed System Architecture

In this section we aim to present a ML approach using an available open access dataset [11]. Designing effective thermal control methods for intelligent buildings will undoubtedly optimize the energy consumption of HVAC systems while maintaining occupant comfort at acceptable levels. Our goal in this study is to design thermal comfort system that can provide occupant thermal comfort with optimal energy consumption, Fig. 1 describes the overall architecture of the proposed thermal comfort system.

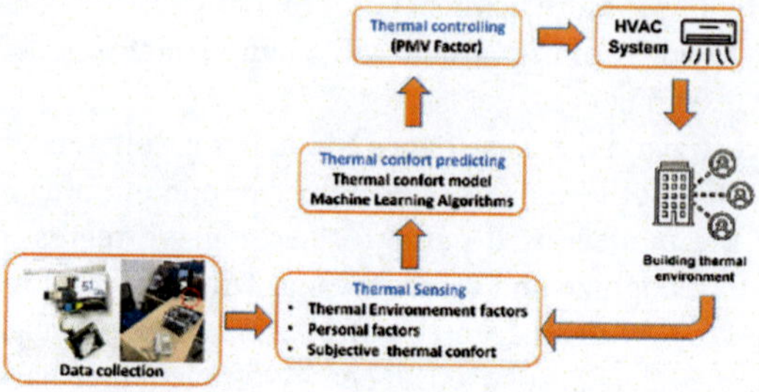

Fig. 1. Proposed architecture of the thermal comfort control system

3 Methodology and Experiment

3.1 Data Set

The dataset used in this work was collected from summer 2017 to winter 2019 by Elnaklah [11]. The data was generated using a raspberry pi based embedded system connected to several sensors in a total of 13 office buildings. In addition to thermal comfort indicators, the dataset contains absenteeism and presenteeism.

3.2 ML Models

RF model: Rf is a ML technique based on the theory of decision trees that uses fits for multiple sub-samples of the initial dataset at the training level to create computation and tree-arranging decision trees [12].

ANN model: The numerous linked neurons in the ann have varying weights for their connections. With extensive parallel processing, dispersed information storage, and strong self-organizing capabilities, it is a sort of abstraction, simplification, or simulation of the human brain. Theoretically, neural network algorithms may approximate any function, particularly nonlinear ones [13].

4 Results and Discussion

The prediction of PMV values is essential for the optimization of energy consumption in buildings. This prediction allows for better control of HVAC systems. Optimal control of HVAC systems offers many economic and ecological benefits.

According to Fig. 2 (a) 50% of the values taken by the PMV index are between 0 and 0.5 while 25% of the values are located between 0.5 and 1.25. The remaining 25% of the values are located between −0.7 and 0.0. The value of PMV is influenced by several variables such as (temperature, relative mean temperature, relative humidity, air velocity, activity rate, tissue insulation….). For this purpose, the correlation of PMV with the other variables in our data was studied. Figure 2 (b) shows the correlation matrix using Pearson's method. The correlation matrix shows that there is a strong dependence between the values of PMV and (PPD, Tr, Ta, Rh, Met, Clo, Av), Table 1 shows the values of Pearson coefficients.

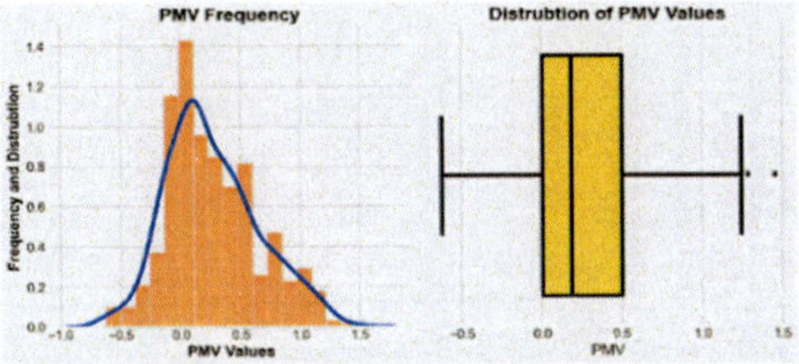

(a)

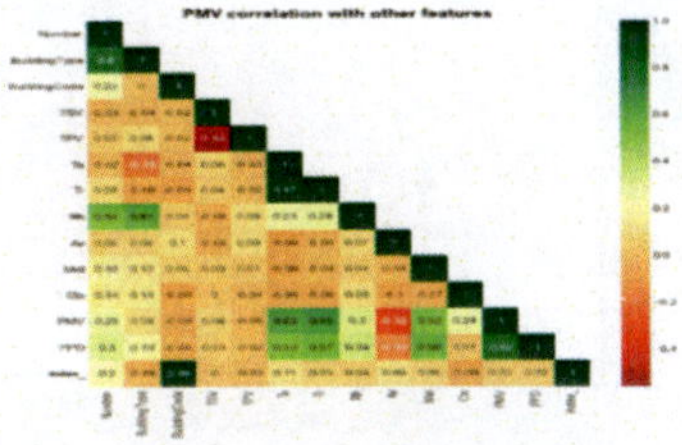

(b)

Fig. 2. (a) Frequency and distribution of the PMV index values; (b) Correlation Matrix

Table 1. PMV correlation coefficient by the Pearson method

	PPD	Tr	Ta	Met	Rh	Clo	Av
Pearson coefficient	0.82	0.65	0.62	0.52	0.30	0.24	−0.32

The model was implemented and tested using ML algorithms such as RF and ANN on the basis of collected data. Table 2 shows the evaluation result of predicted PMV model algorithms for the training phase and testing phase for the different algorithms. The predicted values of PMV were compared to the actual PMV of the database using four evaluation metrics, MSE, RMSE, MAE and R2 Score.

Both algorithms predict PMV more accurately in the training phase than in the testing phase (Table 2). This is because 75% of the dataset was utilized for training and 25% for testing. RF method has 96.5% accuracy and 0.075 in the RMSE, greater than ANN's 80.1% accuracy and 0.171 in RMSE. Figure 3 and Fig. 5 shows the anticipated and real PMV index using RF and ANN. Figure 4 illustrates the difference between the PMV index predicted by RF and its real value. We observe that this difference is in the range 0.05 for most samples. We also observe the presence of 8 peaks in the pattern, which means a higher error in these 8 points. However, these peaks represent less than 6.50% of

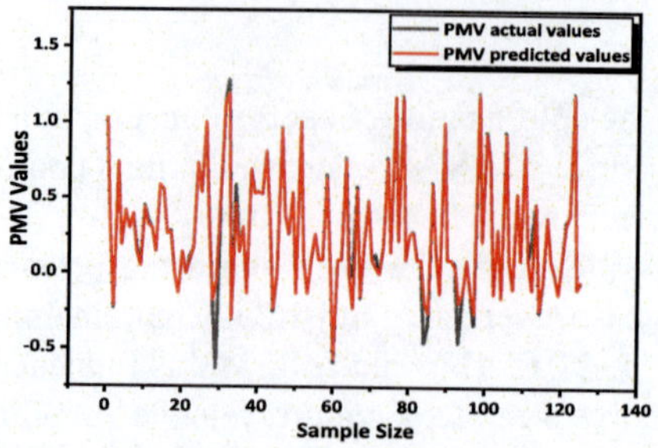

Fig. 3. Difference between the value of the PMV index predicted by RF and its real value

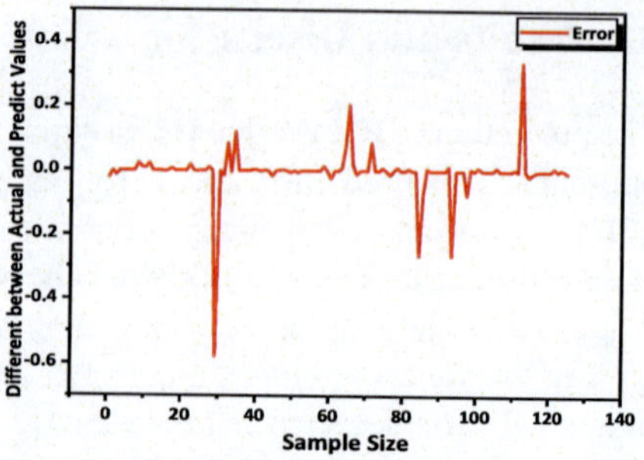

Fig. 4. Error variation using RF algorithm

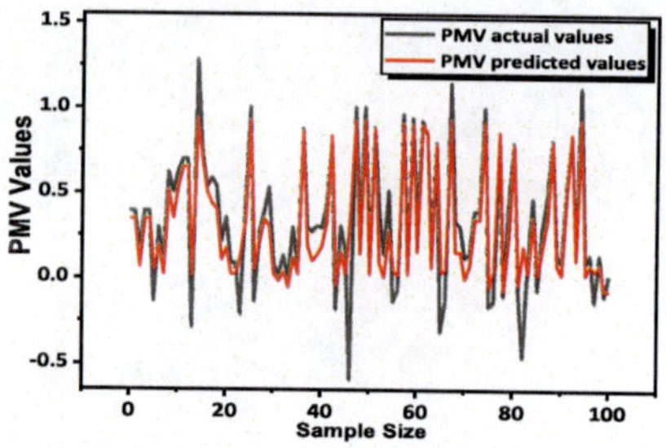

Fig. 5. Difference between the value of the PMV index predicted by ANN and its real value

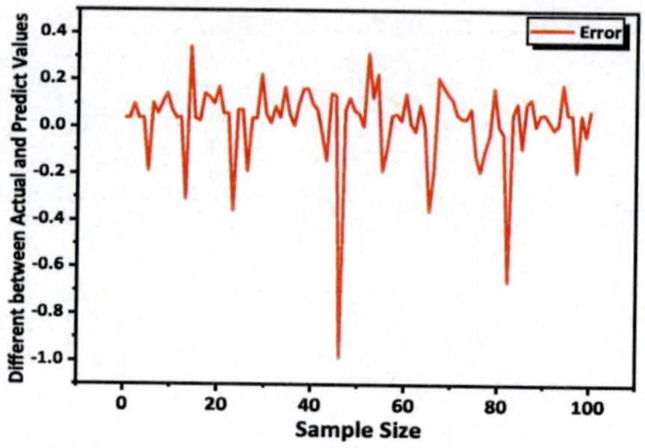

Fig. 6. Error variation using ANN algorithm

Table 2. Metrics results for both models during training and testing phases

	Phase	R2	MAE	MSE	RMSE
RF	Train	0.992	0.01	0.001	0.037
	Test	0.965	0.021	0.006	0.075
ANN	Train	0.76	0.114	0.037	0.191
	Test	0.801	0.102	0.029	0.171

the totality of samples, which explains the high value of R2 despite the presence of these 8 peaks. Figure 4 and Fig. 6 reveal an inaccuracy in the negative PMV predictions. Thus, our model isn't trained enough to forecast the PMV's negative peaks; upon watching the PMV's trajectory, we discover it's rich in positive peaks and misses negative peaks. This explains why the ANN model can only track positive peaks. By limiting the length of our dataset, the model error increases.

Figure 6 displays PMV real and anticipated values. The discrepancy between the predicted and real PMV index for each sample reveals that the ANN model is less accurate.

5 Conclusion

In this paper, we propose a framework based on artificial intelligence to build a model that guarantees thermal comfort by ensuring energy efficiency in buildings. The objective is to study the use of ML algorithms, such as RF and ANN, to predict the PMV index, with the aim of integrating it into HVAC systems. Firstly, we described the architecture of the thermal comfort system and its impact on the energy efficiency of buildings. Then, we trained the dataset using ML algorithms to predict the PMV of the thermal comfort. The obtained results were evaluated using evaluation metrics such as MAE, RMSE, and R2 coefficient. We notice that the RF algorithm shows better performance with 0.965 for R2 coefficient and 0.075 for RMSE, while ANN shows a value of 0.801 for R2 coefficient and 0.171 for RMSE. In our future research, we will extend our studies in this field by investigating the use of other deep learning methods such as DNN, CNN, DRN… to predict thermal comfort in the control phase.

References.

1. Tien, P.W., Wei, S., Liu, T., Calautit, J., Darkwa, J., Wood, C.: A deep learning approach towards the detection and recognition of opening of windows for effective management of building ventilation heat losses and reducing space heating demand. Renew. Energy **177**, 603–625 (2021)
2. Jung, W., Jazizadeh, F., Diller, T.E.: Heat flux sensing for machine-learning-based personal thermal comfort modeling. Sensors **19**(17), 3691 (2019)
3. ASHRAE: Thermal Environmental Conditions for Human Occupancy; ASHRAE: Atlanta, GA, USA (2017)
4. Liu, H., Sun, H., Mo, H., Liu, J.: Analysis and modeling of air conditioner usage behavior in residential buildings using monitoring data during hot and humid season. Energy Build. **250**, 111297 (2021)
5. Abdulgader, M., Lashhab, F.: Energy-efficient thermal comfort control in smart buildings. In: 2021 IEEE 11th Annual Computing and Communication Workshop and Conference (CCWC), pp. 0022–0026. IEEE (2021)
6. Gao, G., Li, J., Wen, Y.: Energy-efficient thermal comfort control in smart buildings via deep reinforcement learning. arXiv preprint arXiv:1901.04693 (2019)
7. Zhang, W., Hu, W., Wen, Y.: Thermal comfort modeling for smart buildings: a fine-grained deep learning approach. IEEE Internet Things J. **6**(2), 2540–2549 (2018)
8. Zhang, W., Wen, Y., Tseng, K.J., Jin, G.: Demystifying thermal comfort in smart buildings: an interpretable machine learning approach. IEEE Internet Things J. **8**(10), 8021–8031 (2020)
9. Bhattacharya, S., Sharma, H., Adetola, V.: Towards Learning-Based Architectures for Sensor Impact Evaluation in Building Controls. In: Proceedings of the Twelfth ACM International Conference on Future Energy Systems, pp. 493–498 (2021)
10. Gan, V.J., Luo, H., Tan, Y., Deng, M., Kwok, H.L.: BIM and data-driven predictive analysis of optimum thermal comfort for indoor environment. Sensors **21**(13), 4401 (2021)

11. Elnaklah, R.: Dataset for "Indoor environment quality and work performance in 'green' office buildings in the Middle East." University of Bath Research Data Archive, Bath (2020)
12. Boutahir, M.K., Farhaoui, Y., Azrour, M.: Machine learning and deep learning applications for solar radiation predictions review: morocco as a case of study. In: Digital Economy, Business Analytics, and Big Data Analytics Applications, Studies in Computational Intelligence, p.1010 (2022)
13. Russel, S., Norvig, P.: Artificial Intelligence: A Modern Approach, Pearson Education Limited, Malaysia (2016)

Assessing Hydrologic Impacts of Future Land Cover Change Scenarios in a Mediterranean Watershed

S. El Harche(✉), M. Chikhaoui, M. Naimi, M. Seif-Ennasr, and A. Chaaou

Department of Natural Resources and Environment, Hassan II Institute of Agronomy & Veterinary Medicine, Rabat, Morocco
s.elharche@iav.ac.ma

Abstract. This paper presents a study in which future land cover scenarios were examined for their impact on sediment production and surface runoff, using (KINEROS2) in combination with (AGWA) tool. The baseline land cover reference was modified to reflect the perspective of stakeholders in 2040 by three possible future scenarios, which include technical differences related to conservation, existing planning trends, and open development. The most significant increase was associated with plans with agricultural land cover and urban areas under the urbanization and development scenario, assumed to be the primary environmental stressors affecting the condition of the Tleta watershed. While the largest decrease was associated with plans with agroforestry use with less urban spread under the Moderate Conservation scenario.

Keywords: Land cover · Erosion · Scenarios · KINEROS2 · AGWA · Tleta Watershed

1 Introduction

The socio-economic activities and the non-irrigated agricultural sector depend highly on water resources and rainfall [1–3]. Soil losses affect agricultural production and resulting in an accelerated siltation rate in reservoirs mainly for drinking water supply and irrigation. The complex interactions between land cover and climate make it difficult to predict the impacts of global change on runoff and erosion [4]. The GIS system and remote sensing are two systems that highlight environmental degradation in the case of soil erosion by integrating physical variables and human activities [5, 6]. The use of hydrological models with climate scenarios remains the most common approach to study the hydrological impact of climate and land cover change, generally corrected according to the area of interest [7, 8]. In this study, the potential impacts of three scenarios relative to current conditions in the Tleta Basin are modeled in a geographic information system (GIS) by KINEROS2 or (K2). To apply this model on an operational basis, it is essential to have automated procedures that can be repeated, accurate and relatively simple. In this sense. AGWA, developed by EPA, the Agricultural Research Service of the United States Department of Agriculture (USDA) and the University of Arizona, to provide

Y. Farhaoui et al. (Eds.): ICAISE 2022, LNNS 635, pp. 247–253, 2023.
https://doi.org/10.1007/978-3-031-26254-8_34

qualitative estimates of runoff and erosion in relation to landscape change [10]. The coupled model K2/AGWA has been used as a tool for modeling flows and solids fluxes in Tleta basin [11], the use of different scenarios of land cover will help to identify areas at risk in the Tleta watershed. The main objectives of this study are: (1) assess how the changes in the landscape through forecasting scenarios affect processes of runoff and sediments yield. And (2) deduce the best scenario that affects the hydro sedimentary regime of the Tleta watershed.

2 Methodology

2.1 Study Area

The Tleta watershed (180 Km^2) is located in the Moroccan rifaine chain, between latitudes 35°33' and 35°39' N, longitudes 5°42' and 5°37' W (Fig. 1). The soil characteristics of the watershed have been determined from previous studies [12, 13]. The rainfall data was obtained from the ibn Batouta station in the outlet of the watershed [14].

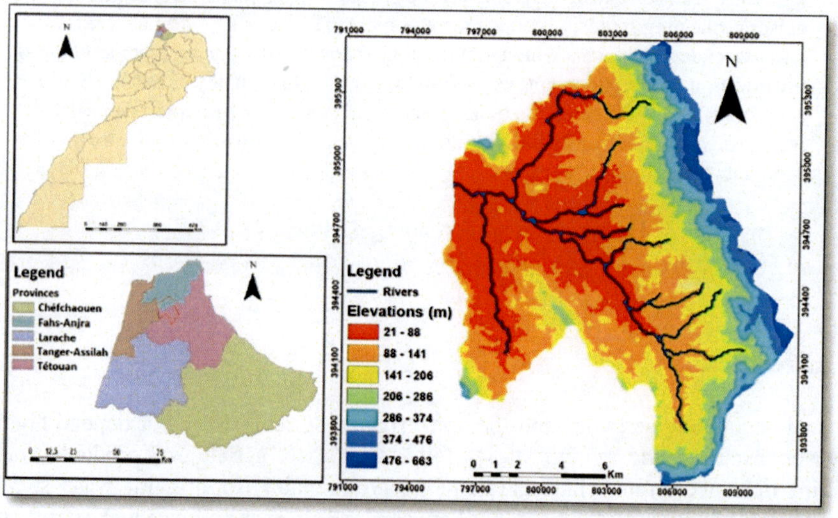

Fig.1. Location of the Tleta catchment area

2.2 Modeling Approach

KINEROS2 is an event physical model describing the processes of, surface runoff, interception, dynamic infiltration, and erosion of watersheds and erosion in agricultural, rangeland and urban watersheds [9, 15]. AGWA is a Geographic Information System (GIS) interface designed to facilitate the management and analysis of water quantity and quality in watersheds [10]. The first step required to conduct watershed analyses or assessments is the delineation of the watershed [15]. The second step is the spatial complexity of the watershed subdivision which controlled by Contributing Source Area (CSA), and corresponds to the heterogeneity of the watershed and the spatial resolution of the simulated processes [16]. K2 assumes that runoff is generated by a Hortonian mechanism [18], and the Smith and Parlange equation to calculate the infiltration rate [19]. The third step is the parameterization of the watershed element which must be characterized by its hydraulic geometry, the length of the flow, and the unique properties of the land cover and soil [18]. The last step is the rainfall data required to run the Kineros2 model, which is taken from daily climate data records from ibn batouta station for the period (1998–2018). In the K2 manual [9, 19], and documentation on the K2 website, the are tables to provide initial estimates for K2 parameters [15, 19].

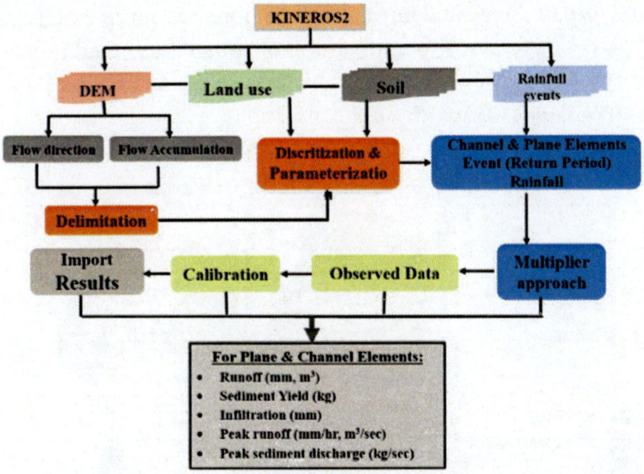

Fig.2. Methodology used in soil erosion assessment for KINEROS2-AGWA

2.3 Alternative Future Scenarios by 2040

The simulations were performed using parameters optimized from baseline conditions (2018)[20]. The assumptions based on modeling approaches in each prospective work led to uncertainties in climate projections and socioeconomic scenarios. This is almost certainly due to the high quality and availability of the data, as well as the simulation models used. Future developments will necessarily require the testing of a broader range of scenarios as well as the quantification of the uncertainties associated with the results provided on expected changes.

3 Results and Discussion

3.1 Analysis of Land Cover Change

Expansion of urbanization and the conversion of natural vegetation into crops are major markers of the influence of human activities. In terms of the entire size of the basin, there

Table 1. Alternative future scenarios in the Tleta watershed in the year 2040.

Scenarios 2040	Description
Urbanisation & Development (S1)	Increasing the rate of urbanization, hight industrial and commercial activity, degraded some crops land
Moderate Conservation (S2)	Protection of ecosystems, converting some crops land to agroforestry, moderate urban area expansion
Development balance (S3)	Low extension of the urban area, local development of community (develop the local products sector)

are relatively few urban areas, and agricultural regions are more extensive in the basin valley's center part. Conversions to agricultural or urban areas tend to have less impact at higher elevations. To establish input parameters, calibrated K2 was used for each of the three land cover scenarios for 2040.

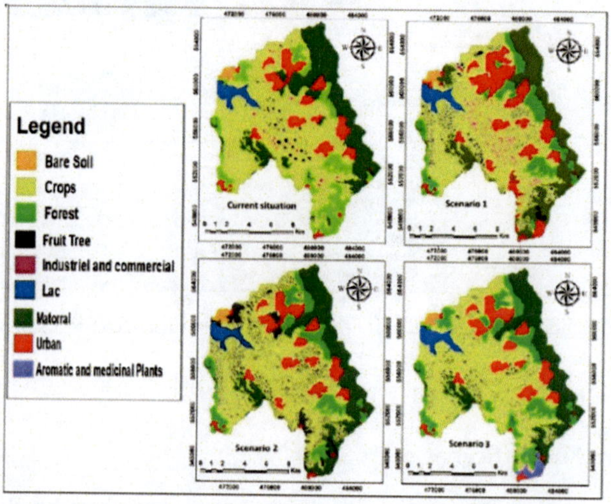

Fig.3. Land cover map for current situation vs future scenarios

3.2 Temporal and Spatial Distribution of Hydrological Processes

The Development scenario experienced the most change, because runoff is typically directed as quickly as possible to natural watercourses, floods are stronger and faster.

The runoff produced by the sealing acts as a catalyst, amplifying the erosion phenomenon. Simulated increases in surface runoff occur primarily within Plans located in the watershed's northern, the greatest impact appears to be also centered in the middle parts, where the majority of new development is planned. as well as the stream near the dam. The conversion of natural land use into artificial infrastructure reduces soil natural

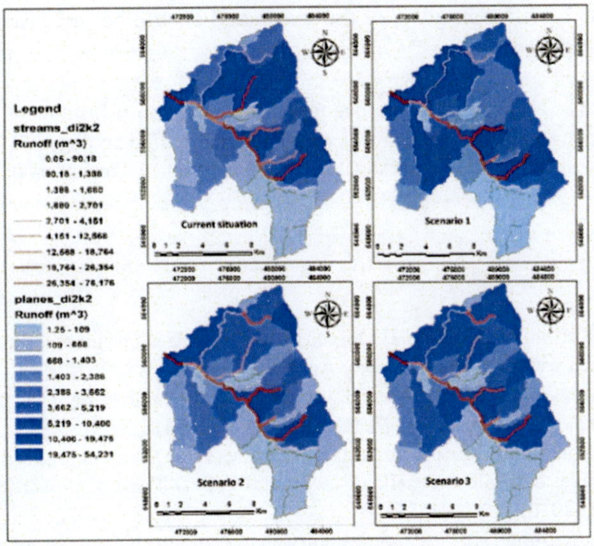

Fig.4. Change in surface runoff for current situation vs future scenarios

capacity to store water and rapidly converts precipitation to surface runoff; stormwater systems prevent any infiltration into the soil, altering river water regimes. There is a minor rise in runoff rate and erosion in scenario 2 due to moderate urbanization and an increase in the surface area of fruit trees, which has favored interception and infiltration. Average increases are also observed for Scenarios 3 as a result of a decrease in matorral land and fruit trees to the benefit of cultivated land, which is consistent with the results of the swat model in the same watershed [21].

Table. 2. Simulated outputs for current situation vs future scenarios (S1), (S2), (S3)

	Baseline 2018	Simulated change by 2040		
		S1	S2	S3
Runoff (m^3)	618,550	1,212,310	634,163	646,883
Sediment yield (Kg/ha)	95,146	241,651	99,267	105,978

4 Conclusions

The scenario 1 have the greatest detrimental influence on surface hydrology and resulting in a larger simulated area. Surface runoff and sediment yield are dramatically reduced in alternative scenarios 2 by reducing agricultural land in favor of greater fruit tree cover was successful in lowering the rate of soil erosion. The loss of fruit tree plantations in favor of cultivated land increased the rate of runoff and sediment in scenario 3, but less than in scenario 1. Different areas of the watershed should be organised according to specific local conditions.

Acknowledgments. The authors would like to thank the Hassan II Institute of Agronomy and Veterinary Medicine and the Hassan II Academy of Science and Technology (GISEC Project) for their manifold support. we also like to extend our warm thanks to the reviewers of the manuscript.

References

1. Schilling, J., Korbinian, P.F., Elke, H., Jürgen, S.: Climate change, vulnerability and adaptation in North Africa with focus on Morocco. Agric. Ecosyst **156**, 12–26 (2012)
2. Born, K., Fink, A.H., Paeth, H.: Dry and wet periods in the northwestern Maghreb for present day and future climate conditions. Meteorol. Z. **17**, 533–551 (2008)
3. Driouech, F.: Distribution des précipitations hivernales sur le Maroc dans le cadre d'un changement climatique: descente d'échelle et incertitudes (PhD Thesis). Université de Toulouse, Toulouse 163 (2010)
4. Tomer, M.D., Schilling, K.E.: A simple approach to distinguish land-use and climate-change effects on watershed hydrology. J. Hydrol. **376**(1), 24–33 (2009)
5. El Garouani, A., Hao, C., Lawrence, L., Tribak, A., Abharour, M.: Cartographie de l'utilisation du sol et de l'erosion nette a partir d'images satellitaires et du sig idrisi au nord-est du maroc. Télédétection **8**(3), 193–201 (2008)
6. Bonn, F.: La spacialisation des model d'erosion des sols a l'aide de la teledetection et le SIG: possiblité, erreure et limites. Science et changements planétaires/sécheresse **9**(3), 185–192 (1998)
7. Senatore, A., Mendicino, G., Smiatek, G., Kunstmann, H.: Regional climate change projections and hydrological impact analysis for a Mediterranean basin in Southern Italy. J. Hydrol. **399**, 70–92 (2011)
8. Ruelland, D., Ardoin-Bardin, S., Collet, L., Roucou, P.: Simulating future trends in hydrological regime of a large SudanoSahelian catchment under climate change. J. Hydrol. **216**, 424–207 (2005)
9. Smith, R.E., Goodrich, D.C., Woolhiser, D.A., Unkrich, C.L.: KINEROS - A kinematic runoff and erosion model. chapter 20. Computer Models of Watershed Hydrology, (Ed. by Singh, V. J.) Water Resour. Pub., highlands Ranch, Colo 697–732 (1995)
10. Miller, S., et al.: The automated geospatial watershed tool. Environ. Model. Softw. **22**, 365–377 (2007)
11. El Harche, S., Chikhaoui, M., Naimi, M.: Estimation Of Runoff And Erosion Rates Using Agwa - Kineros2 Model: Application To A Mediterranean Watershed European Scientific Journal. ESJ **16**(36), 1 (2020)
12. Ezzine, H., Naimi, M., Zhim, S.: Développement et couplage d'un sigweb-sol au modèle impelero pour le diagnostic et l'évaluation de l'érosion hydrique à l'echelle parcellaire dans le bassin versant de tleta. HTE N°149/150 (2011)
13. Inypsa, M.: Projet Intégré de développement Agricole de Tanger-Tétouan, Secteur de Tétouan : Etude des sols au 1/100.000 (Edition au 1/50.000). S.A. et Direction Provinciale de l'Agriculture de Tétouan (1987)
14. ABHL: Etude d'actualisation du plan directeur d'aménagement intégré des ressources en eau des bassins Loukkos, Tangérois et côtiers Méditerranéens. Mission 1, Sous Mission 1.1 : Analyse et synthèse des connaissances actuelles dans le domaine des ressources en eau, Ressources en eau superficielles, ABHL, Tétouan, Maroc, p. 186 (2013)

15. Goodrich, DC., et al.: Kineros2/agwa: Model use,calibration, and validation. The Soil & Water Division of ASABE (2012)
16. Lane, L.J., Woolhiser, D.A., Yevjevich, V.: Influence of simplifications in watershed geometry in simulation of surface runoff. Hydrol. Paper **81**, 80 (1975)
17. Kalin, L., Govindaraju, R.S., Hantush, M.M.: Effect of geomorphologic resolution on modeling of runoff hydrograph and sedimentograph over small watersheds. J. Hydrol. **276**(1–4), 89–111 (2003)
18. .Miller, S., et al.: Integrating landscape assessment and hydrologic modeling for land cover change analysis. American Water Resources Association **38**(4), 915-929 (2002)
19. Woolhiser, D.A., Smith, R.E., Goodrich, D.C.: KINEROS, a kinematic runoff and erosion model. Documentation and user manual. U.S. Department of Agriculture, Agricultural Research Service 77, p. 133 (1990)
20. Hérivaux, C., Vinatier, F., Sabir, M., Guillot, F., Rinaudo, J.: Combining scenario workshops, quantitative approaches and land use change modeling to design plausible future land use scenarios in the Tleta catchment (Morocco). In: International Conference "Vulnerability of the Mediterranean soils to water erosion: State of knowledge and adaptation strategies in the face of global change" (2018)
21. Choukri, F., et al.: Distinct and combined impacts of climate and land use scenarios on water availability and sediment loads for a water supply reservoir in northern Morocco. International Soil and Water Conservation Research **8**(2), 141–153 (2020)

Assessing the Improvements Brought by Artificial Intelligence on the Prediction of Aerodynamic Coefficients

Jad Zerouaoui[1(✉)], Altaf Alaoui[2], Badia Ettaki[1,2], and ElMahjoub Chakir[1]

[1] Laboratory of Material and Subatomic Physics, Faculty of Sciences, Ibn Tofail University, Campus Universitaire, BP 133, Kenitra, Morocco
jadzerouaoui@icloud.com

[2] LyRICA: Laboratory of Research in Computer Science, Data Sciences and Knowledge Engineering, School of Information Sciences Rabat, Rabat, Morocco

Abstract. The implementation of Artificial Intelligence in the aerospace field is fairly new and various domains of the aerospace discipline are to be explored. This paper provides a general review on the upbringing and improvements of Artificial Intelligence as a pilar of the next generation of scientific research. Applications and developments brought by Artificial Intelligence and Deep Learning to the aerospace industry generally, mainly the prediction of aerodynamic coefficients and design optimization using intricate neural networks replicating the process of the human brain. Scientific developments participated in improving major areas in the aerospace field. Due to the expensive and time-consuming process related to the complexity of the geometry and the precision of the results required, researchers integrated surrogate models based on Deep Learning in order to improve the calculation process of the aerodynamic coefficients. On one hand, this paper introduces and discusses the general architecture of several networks used to enhance the calculation ability of the machine, while presenting the process of each work along with recounting the different methods employed. On the other hand, a discussion regarding the challenges presented with utilizing Computational Fluid Dynamics while assessing and comparing the results brought by surrogate models based on efficiency and time consumption is also tackled.

Keywords: Artificial intelligence · Deep learning · Aerospace · Prediction · Aerodynamic coefficient · Computational fluid dynamic

1 Introduction and Background

Artificial Intelligence (AI) by definition is a scope of numerical and engineering approaches that mimic mental processes, for example training, thinking and critical thinking. As reported by John McCarthy, AI is the "*... science and engineering of making intelligent machines, especially intelligent computer programs ...*" [1]. Numerous ways of characterizing AI ranges from broad power to usefulness, basic procedures (statistical interference, symbolic, Machine Learning (ML), etc.).

Y. Farhaoui et al. (Eds.): ICAISE 2022, LNNS 635, pp. 254–263, 2023.
https://doi.org/10.1007/978-3-031-26254-8_35

ML is considered one of AI methods that is being used effectively in various cases. ML is often joined with progressed examination disciplines for example, activities exploration and information science in modern applications. Deep Learning (DL) is a specific ML roused by how the human mind proceeds information by recreating and simulating several layers of artificial neurons [2].

Neural networks are a multivariate, non-parametric training algorithm; Computational Neural Networks (CNN) comprise of a tangled gathering of artificial neurons that studies and analyzes data simultaneously utilizing an auto associative procedure to compute. A neural network is a way of demonstrating and modeling complicated connections among inputs and outputs or to find similarities in the data [3].

CNN have demonstrated a high ability in understanding and representing the geometry and have been applied in the analyses of visual imagery.

In the phase of aerodynamic design for aircrafts, the computation of the aerodynamic coefficients of airfoils has been a critical step in the process of aerodynamic optimization of the aircraft. High-fidelity Computational Fluid Dynamics (CFD) simulations have always been a computationally expensive process and time overwhelming. The recent shine of ML and AI in numerous domains have brought the attention of the scientific community on the abilities and the applications of these methods in multitude of research fields namely the prediction of aerodynamic coefficients.

Recently, several researchers applied ML and very often neural networks in diverse ways to provide help, bring improvements on computational fluid dynamics.

The following part of the review will discuss the general architecture of several networks and each step that were detrimental to understand and complete the purpose of each topic. Secondly, a deliberation about the general techniques used in each of the experiments that lead to somewhat similar results with an obvious increase in performance. The third part or section confers about the experiments that lead to every result discussed in the following review. In the fourth chapter several CFD experiment results will be discussed. And in the final and sixth chapter we will analyze and get through with a conclusion.

2 General Architecture of the Networks

2.1 Discuss the Similarities and Dissimilarities

LeNet-5, is a CNN firstly introduced by Lecun et al. [4], comprised of seven layers, excluding the inputs (which are trainable parameters). Its wide utilization in the field of image recognition made this network the basis of each work discussed in this article. Upon this, Yuan Zelong et al. [5], favored this network due to its utility in the field of recognition of computer vision. Especially since this convolutional network obtains effectively, even the most complex characteristics of any image. In the same context, Hai Chen et al. [6], used the same prediction model established in LeNet-5, and has changed the activation function from a sigmoid to a rectified linear unit due to its faster convergence rate. On the other hand, Hassan Moin et al. [7] have introduced a new method, which uses Keras API in python to design the neural networks, an adam optimizer have been used for the optimization scheme and Rectified Linear Unit (ReLU) was utilized as the activation function in all layers due to its effectiveness and its computational simplicity.

Yao Zhang et al. [8] profited of the latter architecture being one of the cores of the computational structures allowing an important part of modern DL techniques. Meanwhile, Ricardo Wallach et al. managed to discuss and present their chosen network which was a Multi-Layer Perceptron (MLP) since it is one of the most employed networks for nonlinear functions approximation as well as the Functional-Link Network (FLN) and the Radial Basis networks (RBN) which were harder to implement in Matlab compared to the MLP [9].

Improving and deepening the CNN has been the aim of Boping Yu et al. [10]. As for deepening, the whole network, it had been deepened from convolutional layers, pooling layers to fully connected layers. And in order to expedite and boost the training speed and also reduce internal covariate shift between the previous pooling layer and the next convolutional layer, a Batch Normalization (BN) is used following each pooling operation. The dropout technique is used to avoid over-fitting problems as the dropout rate deployed was 0.7.μ

The discussed articles share the use of a ReLU activators due to its advantages as shown by Krizhevsky et al. [11], whereas each article differs in which way the ConvNet is utilized. Both ReLU and tanh activators were being used [5]. Even though the tanh activator was criticized for its disappearing gradients in the training process. As for the architecture, the activation and loss function of the regression layer is entirely linear and Mean Square Error (MSE) function respectively. While keeping the same structure as LeNet-5 and changing the activation function from a sigmoid to a ReLU function, the output layer was altered to a regression using MSE [6]. Both articles share the use of only two convolutional and pooling layers, in order to avoid and thwart any overfitting and guaranteeing prediction accuracy.

The discussed architecture is modified version of LeNet-5 since the aerodynamic meta-modeling requires various alterations, and flow conditions need to be implemented [8]. And following the same output model as discussed previously, it needs to be a continuous regression instead of a discrete classification.

Using an Artificial Neural Network (ANN), the researchers designed the networks employing Keras API in Python [7]. A trial-and-error procedure was utilized for the test cases since the characteristics of the hidden layers must be defined as a compromise between the network size, the accuracy, precision of the generated output and the training time, as well as overtraining and oscillatory behavior avoidance. Deepening and improving their network induced a CNN of 13 hidden layer that have been constructed, including three convolutional layers, three pooling layers, three batch normalization layers and four fully connected layers [10].

2.2 Discussion of the Different Methods Employed

Results of different methods have shown enormous autonomy and improvement on the results obtained by using ML and AI surrogate models. In this section we will discuss various results obtained from the calculations and compare them to each other.

In the process of aerodynamic coefficients prediction, the discussed methods, networks and setups were efficient, reliable and time saving in comparison with the traditional CFD solver. Back propagation and the employment of a ReLU activator are some of the characteristics that are common and mutual between all the networks. The training process for the network employed consisted of the usage of the adaptive moment estimation (Adam optimization algorithm) [5]. The latter has shown great results when dealing with computing very noisy or sparse gradients on non-convex operations [12]. Hicks-Henne bump functions are employed to parameterize the geometric shape of airfoils, this type of shape parametrization method is efficient to get plenty of accurate airfoil shapes for the training dataset [13]. In order to miniconvolmize the time-consuming calculations, surrogate models are used, these models are considered as data-driven approximation models for expensive and complex numerical simulations. Kriging [14], as a Bayesian interpolation method which provide statistic prediction of a black box function by minimizing its MSE, this model is employed to compare its results with the network utilized. The training process of the next network includes forward calculation that extracts image features with the convolutional and pooling layers, and the differences between predicted results and actual results were predictions errors [6]. Error back propagation transfers the prediction errors backwards using the algorithms [15]. In the latter paper, the training algorithm selected Stochastic Gradient Descent with Momentum (SGDM). The network produced in the subsequent article [7] requires an accurate airfoil shape for the training results, therefore the UIUC Airfoil Data Site [16] is being used. In this work three network structures are being tested, first comes the MLP network that serves as a baseline architecture, next comes the architecture that provides and serves as an intermediate architecture between the MLP and the fully convoluting third and last network. The third network employs a full image-like 2D array combining the airfoil shape and the airflow conditions. All three architectures have been implemented in the Python environment using a modern, open-source library for DL.

Relying on a totally different network, ANN have been used, but the learning part of any ANN is mutual with the previously discussed articles, which is done through its back propagation algorithm. This algorithm minimizes the loss function using common optimization algorithms such as SGDM or the ADAM optimizer [8]. The data set employed was generated using Javafoil using macros. Six different datasets were created.

The method employed in the prediction model of [9] is rather for General Transport aircraft.

3 General Techniques Used

In order to assure an efficient working and productivity from each network several techniques have been used.

3.1 Geometry Representation

In order to translate the information provided into datasets, or more specifically decide on the inputs, one must denote the geometry information of airfoils. The latter can be represented in numerous ways, SDF was implemented, which is a method that was proven to work effectively [17]. The airfoil profile was considered to be a closed region boundary described

$$Z = \left\{ \boldsymbol{\Gamma}_{\boldsymbol{\Omega}}(x, y) = 0 | \boldsymbol{\Omega} \subset \boldsymbol{R}^2, (x, y) \in \boldsymbol{R}^2 \right\} \quad (1)$$

where $\boldsymbol{\Omega}$ is the domain enclosed by an airfoil profile, $\boldsymbol{\Gamma}_{\boldsymbol{\Omega}}$ is the closed boundary of $\boldsymbol{\Omega}$. $\Gamma(x, y) < 0$ if and only if *(x y,)* is within the boundary and $\Gamma(x, y) > 0$ is the opposite situation.

An SDF $D(i, j)$ associated to a profile boundary presents the nearest distance from a given (x_i, y_j) to the boundary

$$D(i,j) = min\left\| \boldsymbol{r}(x_j, y_j) - \boldsymbol{r}(x_\Gamma, y_\Gamma) \right\| \mathrm{sign}\left(\Gamma\left(x_i, y_j\right)\right), (i = 1, 2, \ldots, m, j = 1, 2, \ldots, n) \quad (2)$$

where $\boldsymbol{r}(x, y)$ is the location of a Cartesian XY coordinate (x, y), (x_Γ, y_Γ) is the nearest point from the airfoil boundary. The K-D Tree method [18] is employed to investigate the closest discrete point of the airfoil profile for a given coordinate.

On the other hand, the geometry representation of the CNN was an airfoil image which was a grayscale image, most of the airfoils are flat, and its thickness and length differed by an order of magnitude, the thickness of the airfoil was magnified ten times in this paper in order to rally the prediction model accuracy [6]. The flow conditions needed to be implemented, therefore they had to be converted into images to be recognized by the network. The conversion process of these images discussed in this paper consisted of convolving the airfoil images with the respected flow conditions, then, combined with the original airfoil image, a CAIs are generated, these serve as input for the CNN.

Each architecture in the next article is intended to process a unique input data structure, therefore, involving different process for its data preparation [7]. Firstly, the MLP makes use of the freestream Mach number, Reynolds number, angle of attack and 100 y coordinates. The second network employs the same parameters; however, the y-coordinates have a form of the 2D array. The last network employs a full image-like 2D array fusing the airfoil shape and the flow conditions.

A wide range of NACA 4- and 5-digits airfoils is generated in *Javafoil* using macros, these coordinates are then spaced to capture the leading and trailing edge shapes by having denser points around these areas [8]. Based on the specifications discussed, six different datasets were created (three for each series), hence a total of 176400 and 352800 samples in each 4- and 5-digit datasets respectively.

Specified airfoils templates were selected in the following study [9], since it is indicated it is for general transport airplanes.

Hicks-Henne fitting method [13] is used to generate the training and testing airfoil datasets. In this context, the user defined OpenFOAM solver was chosen as the simulation tool to obtain the aerodynamic coefficients. The simulating Mach number is 0.4, Reynolds number is 3.63×10^6 and the angle of attack is 6.13°.

3.2 Network Training

The training process of each network is somewhat similar in every article mentioned. One can deduce that the training process examined in the first article [5], back propagation method is employed to renew the weights and biases, and the optimization algorithm uses Adam algorithm.

The formulas employed to update the parameters

$$\begin{aligned} \boldsymbol{m}_t &= \beta_1 \boldsymbol{m}_{t-1} + (1-\beta_1)\boldsymbol{g}_t \\ \boldsymbol{s}_t &= \beta_2 \boldsymbol{s}_{t-1} + (1-\beta_2)\boldsymbol{g}_t^2 \\ \hat{\boldsymbol{m}}_t &= \frac{\boldsymbol{m}_t}{1-\beta_1^t}, \hat{\boldsymbol{s}}_t = \frac{\boldsymbol{s}_t}{1-\beta_2^t} \\ \boldsymbol{\theta}_{t+1} &= \boldsymbol{\theta}_t - \frac{\eta}{\sqrt{\hat{s}_t}+\varepsilon}\hat{\boldsymbol{m}}_t \end{aligned} \quad (3)$$

where $\boldsymbol{\theta}_t$ is the weights and the biases of the network. m_t and s_t are the estimated values of the mean and variance of $\boldsymbol{\theta}_t$'s gradients respectively. $\hat{\boldsymbol{m}}_t$ and $\hat{\boldsymbol{s}}_t$ are bias corrected values of $\hat{\boldsymbol{m}}_t$ and $\hat{\boldsymbol{s}}_t$, β_1 and β_1 are the decay rates, η is the learning rate and $\mathcal{E}$ is a small positive constant for numerical stabilizations.

One can also notice the use of the back propagation and also the forward calculation method in the cited article [6]. Where the forward calculation isolates image features of the airfoil, where the author managed to obtain prediction values through the output layer. Error back propagation transmits prediction errors backward by the algorithms such as gradient descent and updates the network weights and biases. In the latter article the training algorithm selected is SGDM.

The following work discussed demonstrates that the usage of several networks in chain including the MLP at the start induce faster training and reduction in the amount o simulation data required for the prediction accuracy [6]. In the MLP architecture, the learning capability can be increased by additional hidden layers and/or additional hidden units in each hidden layer. A total of 133 sets of 2D airfoil geometry are used as training and validation datasets. All three networks' biases and weights provided in the discussed article were trained using AdaDelta Algorithm [19].

4 Computational Fluid Dynamics Simulations

Establishing several simulations using conventional CFD simulations, is extremely costly in both calculations and time. Simulating flight conditions in order to determine the aerodynamic coefficients of a single airfoil can take up to hours, depending on the number of iterations entered by the user.

Simulations were used on the airfoil NACA 0010-35 (see Fig. 1), highlighting the velocity and the pressure contours around the airfoil. High precision is mandatory, since these results are used as training sets for the network, for more accuracy, efficiency and rapidity. Therefore, overcoming all the drawbacks introduced by traditional CFD calculations and simulations.

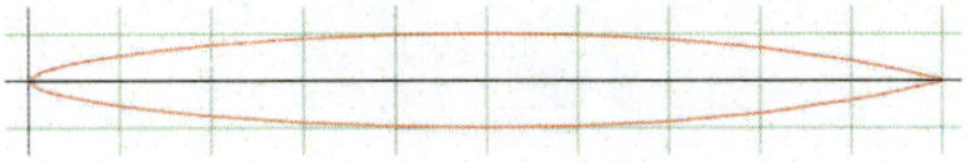

Fig. 1. NACA 0010-35 on a plane.

Ansys [21], is a pioneer software used to simulate or recreate physical phenomena while most importantly respecting the principal laws of physics and fluid dynamics. In Fig. 2, it is apparent that Ansys can deal with several analysis systems, the *Fluid Flow (Fluent)* category of analysis can deal efficiently with the aerodynamic simulations needed.

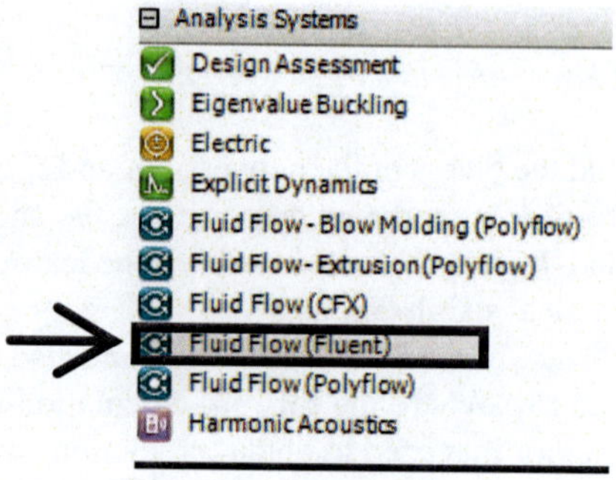

Fig. 2. Ansys homepage

The geometry part consists of generating the airfoil and the envelope that simulates the flying conditions (see Fig. 3). Therefore, correctly identifying the boundary conditions (see Fig. 4), is critical, in order to obtain accurate results.

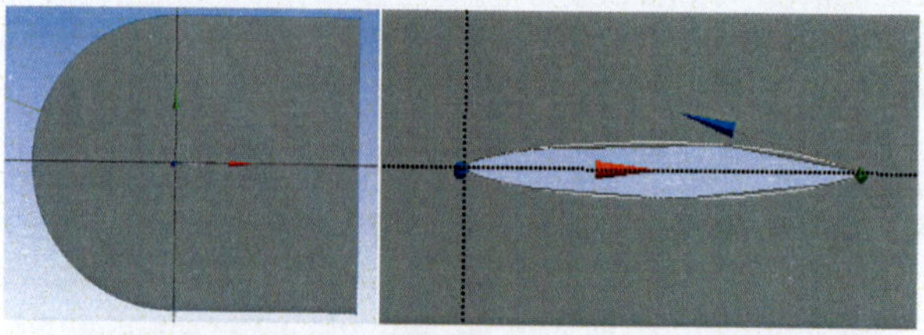

Fig. 3. Boolean operations to re-create flying conditions

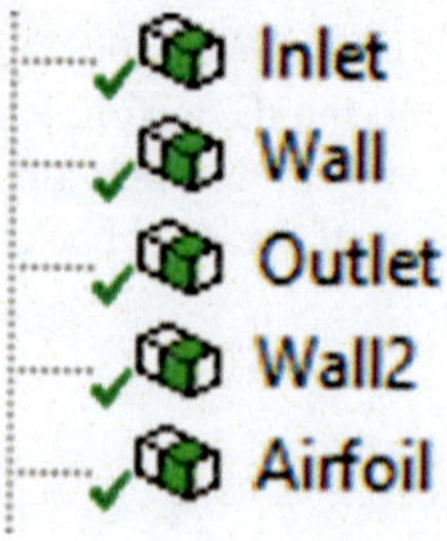

Fig. 4. Boundary conditions.

The meshing part of the process is very crucial. Since the elements where the software should do the calculations must be precise and finite for optimal results. Therefore, it is recommended to use the *All-triangles method* for better accuracy and also reducing the element size (see Fig. 5). Various methods are used on the mesh such as, *edge sizing* and *refinement* in order to obtain a smooth and accurate mesh (see Fig. 6).

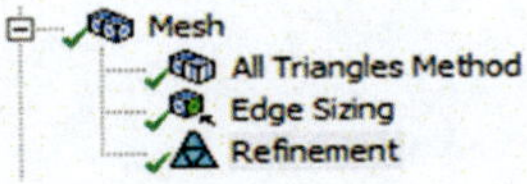

Fig. 5. Methods used to ensure a reliable mesh

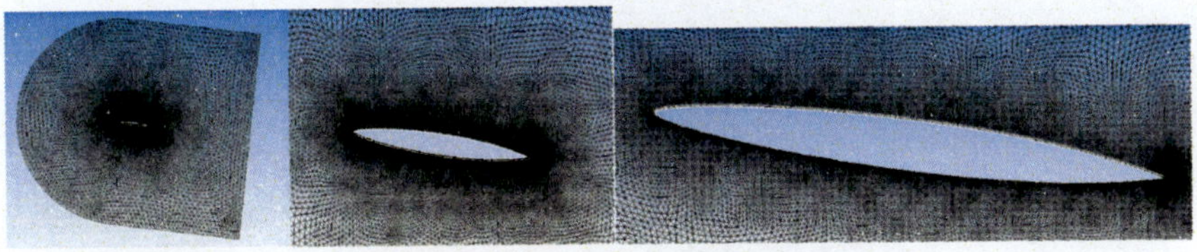

Fig. 6. The precision of the mesh

After ensuring a good and accurate mesh, one should set the boundary conditions of the simulation. The input was taken as a velocity input with 100 m/s, while the outlet was set to pressure outlet and the velocity on the walls is set to Zero due to the no-slip condition of the fluids (see Fig. 7).

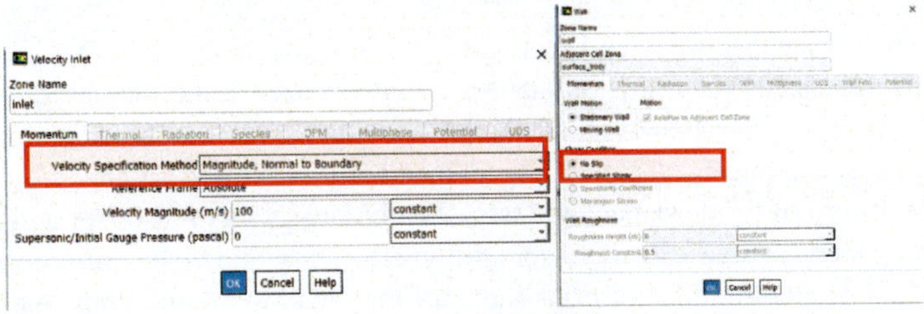

Fig. 7. Boundary conditions

A set of iterations was chosen to be two hundred iterations, the following chart shows the continuity of the calculations and also mapping the x and y velocities. After, completing the calculations, the software automatically opens the *results* part (see Fig. 8). One can easily manipulate the controls in order to create several contours on several planes. The velocity contour is shown in Fig. 9.

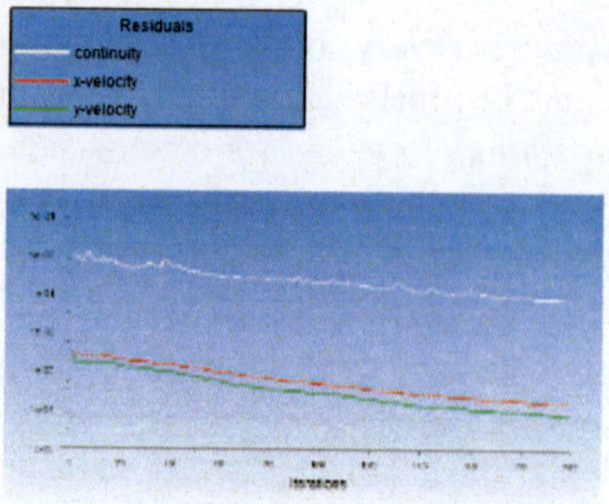

Fig.8. Residuals of the calculations

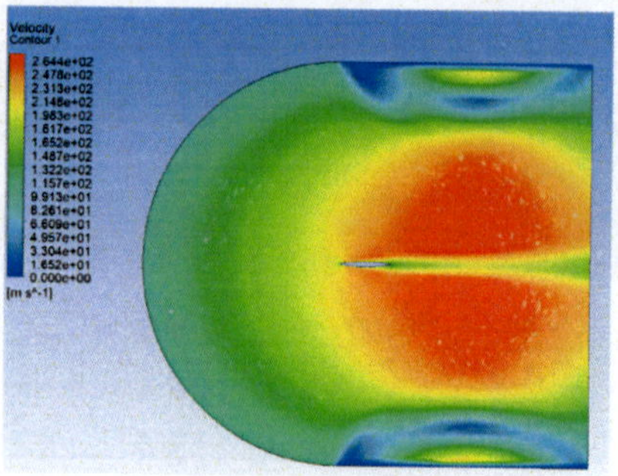

Fig. 9. Velocity contour

5 Conclusion and Perspectives

The following paper discusses and evaluates the improvements brough by AI in the domain of Aerospace. Efficiency and accuracy have proven to be increased as well as the time consumption has been reduced, the latter being the most frowned upon problem throughout the aerodynamic analysis. The debated and examined articles prove that AI is indeed the future of the aerospace industry. Which brings us to discover and examine closely several areas of studies which were difficult to grasp due to the time consumption problem. The introduction of AI in this industry may lead to autonomous maintenance with less to no human interference. As CNN are replicas of human neuron, smart training, innovative product designs are now easy to fetch and attain. Looking ahead, there are many opportunities for AI and ML in the aerospace industry.

References

1. Sharma, L., Garg, P.K. (Eds.): Artificial Intelligence: Technologies, Applications, and Challenges (1st ed.). Chapman and Hall/CRC, Boca Raton (2021)
2. Topór, T.: Machine learning in oil and gas exploration: Machine learning era (n.d.). https://www.researchgate.net/publication/356064327

3. Sharmila, L., Leoni Sharmila, S., Dharuman, C., Venkatesan, P., Professor, A.: A comparative study of neural network and fuzzy neural network for classification a comparative study of neural network and fuzzy neural network for classification view project a comparative study of neural network and fuzzy neural network for classification. Int. J. Sci. Eng. Res. **10**(1) (2019)
4. Lecun, Y., Bottou, E., Bengio, Y., Haffner, P.: Gradient-based learning applied to document recognition (1998)
5. Yuan, Z., Wang, Y., Qiu, Y., Bai, J., Chen, G.: Aerodynamic coefficient prediction of airfoils with convolutional neural network. In: Zhang, X. (eds.) The Proceedings of the 2018 Asia-Pacific International Symposium on Aerospace Technology (APISAT 2018). APISAT 2018. LNEE, vol. 459, pp. 34–46. Springer, Singapore (2019). https://doi.org/10.1007/978-981-13-3305-7_3
6. Chen, H., He, L., Qian, W., Wang, S.: Multiple aerodynamic coefficient prediction of airfoils using a convolutional neural network. Symmetry **12**(4) (2020)
7. Zhang, Y., Sung, W.J., Mavris, D.: Application of convolutional neural network to predict airfoil lift coefficient. In: AIAA/ASCE/AHS/ASC Structures, Structural Dynamics, and Materials Conference (210049) (2018)
8. Moin, H., Khan, H.Z.I., Mobeen, S., Riaz, J.: Airfoil's Aerodynamic Coefficients Prediction using Artificial Neural Network (2021). http://arxiv.org/abs/2109.12149
9. Wallach, R., Mattos, B., Girardi, R., Curvo, M.: Aerodynamic coefficient prediction of transport aircraft using neural network. In: 44th AIAA Aerospace Sciences Meeting and Exhibit (2006)
10. Yu, B., Xie, L., Wang, F.: An improved deep convolutional neural network to predict airfoil lift coefficient. In: Jing, Z. (eds.) Proceedings of the International Conference on Aerospace System Science and Engineering 2019. ICASSE 2019. LNEE, vol. 622. Springer, Singapore (2020). https://doi.org/10.1007/978-981-15-1773-0_21
11. Krizhevsky, A., Sutskever, I., Hinton, G.E.: ImageNet classification with deep convolutional neural networks. Commun. ACM **60**(6), 84–90 (2017)
12. Alom, M.: Adam Optimization Algorithm (n.d.). https://towardsdatascience.com/adam-latest-trends-in-deep-learningoptimization-
13. Poole, D.J., Allen, C.B., Rendall, T.C.: S Metric-based mathematical derivation of efficient airfoil design variables. AIAA J. **53**(5), 1349–1361 (2015)
14. Nielsen, H.B., Lophaven, S.N., Søndergaard, J.: DACE - A Matlab Kriging Toolbox. Computer programme, Informatics and Mathematical Modelling, Technical University of Denmark, DTU (2002)
15. Oh, S.H.: Error back-propagation algorithm for classification of imbalanced data. Neurocomputing **74**(6), 1058–1061 (2011)
16. https://m-selig.ae.illinois.edu/ads/coord_database.html
17. Luo, H., Wang, X., Lukens, B.: Variational analysis on the signed distance functions. J. Optim. Theory Appl. **180**(3), 751–774 (2018). https://doi.org/10.1007/s10957-018-1414-2
18. Devroye, L., Jabbour, J., Zamora-Cura, C.: Squarish k-d trees. SIAM J. Comput. **30**, 1678–1700 (2000). Ahmad AL Smadi et al. Deep convolutional neural network-based system for fish classification.
19. Huang, L., Yang, D., Lang, B., Deng, J.: Decorrelated batch normalization (2018)

Attitude Control of LEO Satellite via LQR Based on Reaction Wheels Versus Magnetorquer

Taha Ennaciri[1](✉), Ahmed El Abbassi[1], Nabil Mrani[2], and Jaouad Foshi[1]

[1] Team Renewable Energies and Information Processing and Transmission Laboratory, Faculty of Sciences and Technologies, Errachidia, Moulay, Ismail University, Meknes, Morocco
t.ennaciri@edu.umi.ac.ma

[2] Team Technologie d'Information et du Multimedia (TIM) High School of Technology, Meknès, Moulay Ismail University, Meknes, Morocco

Abstract. In this paper, attitude control of a LEO microsatellite via LQR is discussed using reaction wheels and magnetorquer. A comparative study is established between the two approaches showing the advantages of implementing a magnetorquer over reaction wheels in terms of time response and accuracy.

Keywords: Linear quadratique regulator · Reaction wheels · Magnetorquer · LEO satellite

1 Introduction

An attitude control system of a satellite generally includes an angular momentum storage device in order to absorb the effects of external disturbing torques acting on the satellite's body (such as atmospheric drag, solar radiation pressure, and more.).

Conventionally, such an angular momentum storage device includes inertial actuators such as: gyroscopes, reaction wheels, magnetorquers, and more. They are implemented to create an attitude control torques which its effects is to compensate for the effects of external disturbing torques [1].

A comparison of reaction wheels and magnetorquer actuators has been studied based on a dynamic model of a LEO satellite via LQR. In this study a perturbed system of a LEO satellite equipped by three reaction wheels on one hand, and equipped by magnetorquer on the other hand, the last one shows both a better response time, and accuracy by using less energy [2, 3].

2 Satellite Attitude Dynamic Model

2.1 With Reaction Wheels

The satellite attitude dynamic, Kinematics and reaction wheels Model related to our system are previously detailed in our Work comparative study between linear quadratic

Y. Farhaoui et al. (Eds.): ICAISE 2022, LNNS 635, pp. 264–269, 2023.
https://doi.org/10.1007/978-3-031-26254-8_36

regulator and fractional-order robust controller for the three-axis attitude control system of LEO satellite using reaction wheels [4].

Furthermore, the study will be based on the system written as bellow:

Where:

$$\begin{cases} I_x\ddot{\phi} = \omega_0\dot{\psi}(I_z - I_y + I_x) + \dot{L}_x + \dot{\psi}\dot{\theta}(I_y - I_z) + \phi\left[4(I_z - I_y)\omega_0^2 - \omega_0\dot{\theta}(I_z - I_y)\right] \\ I_y\ddot{\theta} = 3(I_x - I_z)\omega_0^2\theta + \dot{\psi}\phi(I_z - I_x) + \dot{L}_y + \dot{\psi}\psi\omega_0(I_x - I_z) + \phi\left[\psi\omega_0^2(I_x - I_z) + \omega_0\dot{\phi}(I_z - I_x)\right] \\ I_z\ddot{\psi} = \psi\left[\omega_0^2(I_x - I_y) + \dot{\theta}\omega_0(I_y - I_x)\right] + \dot{\phi}\left[(I_y - I_x - I_z)\omega_0 + \dot{\theta}(I_x - I_y)\right] + \dot{L}_z \end{cases} \quad (1)$$

Canonical form of the model:

$$\dot{x} = Ax(t) + Bu(t),\ y = Cx(t) + Du(t) \quad (2)$$

where:

- θ, ϕ and ψ are Roll, Pitch and Yaw angles respectively.
- I is the inertia matrix of the satellite.
- ω_0 is the orbit angular velocity.
- L is the total moment vector of reaction wheel $L = \left[L_x, L_y, L_z\right]^T$

$$A = \begin{pmatrix} 0 & 0 & 0 & 1 & 0 & 0 \\ 0 & 0 & 0 & 0 & 1 & 0 \\ 0 & 0 & 0 & 0 & 0 & 1 \\ 4\omega_0^2\frac{I_z - I_y}{I_x} & 0 & 0 & 0 & 0 & \omega_0\frac{I_x + I_z - I_y}{I_x} \\ 0 & 3\omega_0^2\frac{I_x - I_z}{I_y} & 0 & 0 & 0 & 0 \\ 0 & 0 & \omega_0^2\frac{I_x - I_y}{I_z} & \omega_0\frac{I_y - I_x - I_z}{I_z} & 0 & 0 \end{pmatrix}$$

$$B = \begin{pmatrix} 0 & 0 & 0 \\ 0 & 0 & 0 \\ 0 & 0 & 0 \\ \frac{1}{I_x} & 0 & 0 \\ 0 & \frac{1}{I_y} & 0 \\ 0 & 0 & \frac{1}{I_z} \end{pmatrix};\ C = \begin{pmatrix} 1\,0\,0\,0\,0\,0 \\ 0\,1\,0\,0\,0\,0 \\ 0\,0\,1\,0\,0\,0 \\ 0\,0\,0\,1\,0\,0 \\ 0\,0\,0\,0\,1\,0 \\ 0\,0\,0\,0\,0\,1 \end{pmatrix};\ D = \begin{bmatrix} 0\,0\,0 \\ 0\,0\,0 \\ 0\,0\,0 \\ 0\,0\,0 \\ 0\,0\,0 \\ 0\,0\,0 \end{bmatrix};\ x = \begin{pmatrix} \phi \\ \theta \\ \psi \\ \dot{\phi} \\ \dot{\theta} \\ \dot{\psi} \end{pmatrix};\ u = \begin{bmatrix} L_x \\ L_y \\ L_z \end{bmatrix}.$$

2.2 With Magnetorquer

The satellite is equipped with magnetometers, measuring the satellite axis compo-nents of the Earth's magnetic field. Angular position sensors and angular speed measurement gyroscopes are associated with on-board electronics, make it possible to develop the torque functions, and therefore the currents to be injected into three coils following the 3 axes to obtain the control magnetic moment M.

Using magnetorquer the satellite dynamic model becomes [5]:

$$\begin{cases} I_x\ddot{\phi} + (I_x + I_z + I_y)\omega_0\dot{\psi} + 4(I_y - I_z)\omega_0^2\phi = M_{cx} + M_{px} \\ I_y\ddot{\theta} + 3(I_x - I_z)\omega_0^2\theta = M_{cy} + M_{py} \\ I_L\ddot{\psi} - (I_x + I_z - I_y)\omega_0\dot{\phi} + (I_y - I_x)\omega_0^2\psi = M_{cz} + M_{pz} \end{cases} \quad (3)$$

Canonical form of the model:

$$\dot{x} = Ax(t) + Bu(t), y = Cx(t) + Du(t)$$

where:

- $M_c = \begin{bmatrix} M_{cx} \; M_{cy} \; M_{cz} \end{bmatrix}^T$ is the control torque associated with gravitational force.
- $M_p = \begin{bmatrix} M_{px} \; M_{py} \; M_{pz} \end{bmatrix}^T$ is the total moment of external non-gravitational disturbances.

$$A = \begin{bmatrix} 0 & 1 & 0 & 0 & 0 & 0 \\ 4\frac{I_z-I_y}{I_x}\omega_0^2 & 0 & 0 & 0 & 0 & \frac{I_x+I_y-I_z}{I_x}\omega_0^2 \\ 0 & 0 & 0 & 1 & 0 & 0 \\ 0 & 0 & 3\frac{I_z-I_x}{I_y}\omega_0^2 & 0 & 0 & 0 \\ 0 & 0 & 0 & 0 & 0 & 1 \\ 0 & \frac{I_x+I_z-I_y}{I_y}\omega_0 & 0 & 0 & \frac{I_x-I_y}{I_z}\omega_0^2 & 0 \end{bmatrix}$$

$$B = \begin{bmatrix} 0 & 0 & 0 \\ \frac{1}{I_x} & 0 & 0 \\ 0 & 0 & 0 \\ 0 & \frac{1}{I_y} & 0 \\ 0 & 0 & 0 \\ 0 & 0 & \frac{1}{I_z} \end{bmatrix}; C = \begin{bmatrix} 1\,0\,0\,0\,0\,0 \\ 0\,1\,0\,0\,0\,0 \\ 0\,0\,1\,1\,0\,0 \\ 0\,0\,0\,1\,0\,0 \\ 0\,0\,0\,0\,1\,1 \\ 0\,0\,0\,0\,0\,1 \end{bmatrix}; D = \begin{bmatrix} 0\,0\,0 \\ 0\,0\,0 \\ 0\,0\,0 \\ 0\,0\,0 \\ 0\,0\,0 \\ 0\,0\,0 \end{bmatrix} x = \begin{bmatrix} \phi \\ \dot{\phi} \\ \theta \\ \dot{\theta} \\ \psi \\ \dot{\psi} \end{bmatrix}; u = \begin{bmatrix} M_{cx} + M_{px} \\ M_{cy} + M_{py} \\ M_{cz} + M_{pz} \end{bmatrix}$$

3 Controllability and Observability

They are an essential step to successfully design a controller for the system.

The literature as well as our previous works has provided us with a detailed study regarding this important step, therefore a summarized version is to be presented [4].

The system must be:

- Completely state controllable
- Observable

We opted for Kalman's method to study the controllability and the observability of the system using controllability matrix $C_n = \begin{bmatrix} B \; AB \; A^2B \ldots A^{n-1}B \end{bmatrix}$ and the observability matrix $o = \begin{bmatrix} C \; CA \; CA^2 \ldots CA^{n-1} \end{bmatrix}^T$.

where:

n = 6.

Since rank (C_n) = rank (O) = 6 = futll rank, the system is controllable and observable.

4 Linear Quadratic Regulator Controller

Considering the system (2):

The goal of such controller is to find a solution $u(t)$ which minimizes the cost function J.

With J is written as [4–6]:

$$J = \frac{1}{2}\int_{t_0}^{T}\left(x^T(t)Qx(t) + u^T(t)Ru(t)\right)dt \tag{4}$$

R and Q are positive definite matrices.
The solution given for such a problem is:

$$u(t) = -Kx(t) \tag{5}$$

With K is the gain matrix found by resolving the algebraic Riccatti equation.

5 Simulations and Results

Set the parameters for Satellite Simulations:
Initial condition:

$$x(0) = (1^\circ, 1^\circ, 1^\circ, 0.1^\circ/s, 0.1^\circ/s, 0.1^\circ/s,)$$

Satellite weight = 120 kg;
Orbit angular velocity: $\omega_0 = 0.0010764\ \frac{\text{rad}}{\text{s}}$.
Satellite Inertia Matrix: $I_x = 9.8194\ \text{kgm}^2$; $I_y = 9.7030\ \text{kgm}^2$; $I_z = 9.7309\ \text{kgm}^2$..
Orbit = 686 km.

Figures 1, 2 and 3 represent a comparison between reaction wheels versus magnetorquer via LQR controller for the evolution of the Roll, Pitch and Yaw angles versus time.

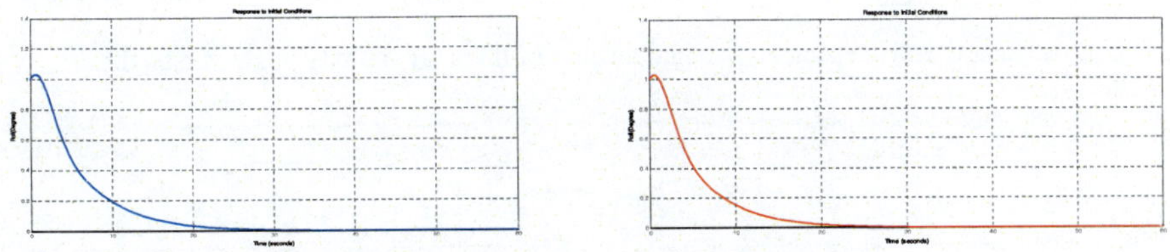

Fig. 1. Reaction wheels against magnetorquer for the Roll angle versus time.

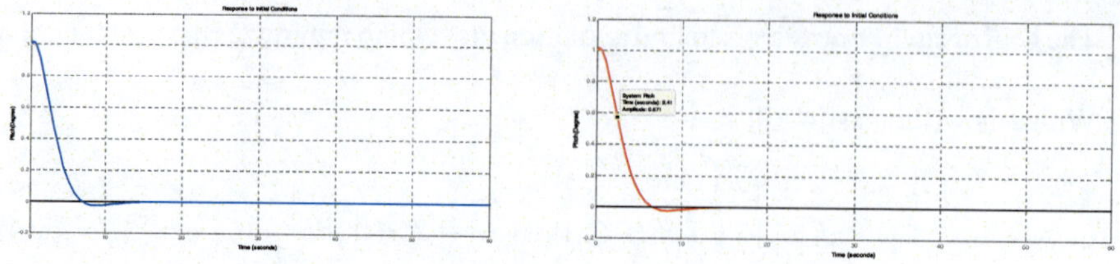

Fig. 2. Reaction wheels against magnetorquer for the Pitch angle versus time.

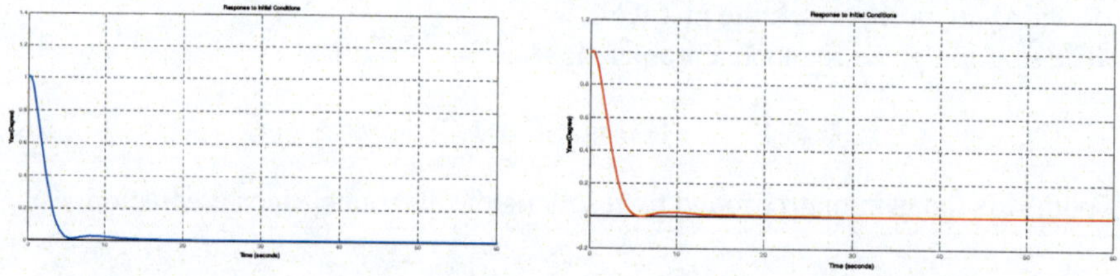

Fig. 3. Reaction wheels against magnetorquer for the Yaw angle versus time.

Discussion:

As shown based on Fig. 1 to Fig. 3 it is noted that the magnetorquer shows that it is quicker in term of response time by comparing the evolution of the angles versus time, taking for example the Roll angle, the reaction wheels needs roughly 32 s to settle down compared to 30 s for the magnetorquer.

Figures 4, 5 and 6 represent a comparison between reaction wheels versus magnetorquer via LQR controller for the velocity of the Roll, Pitch and Yaw angles ver-sus time.

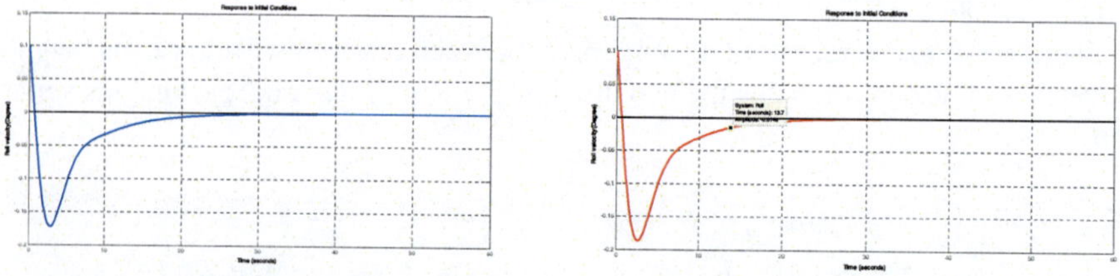

Fig. 4. Reaction wheels against magnetorquer of the Roll velocity angle versus time.

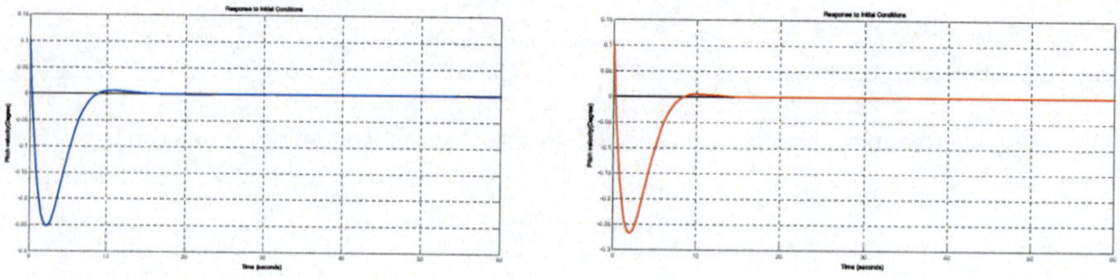

Fig. 5. Reaction wheels against magnetorquer of the Pitch velocity angle versus time.

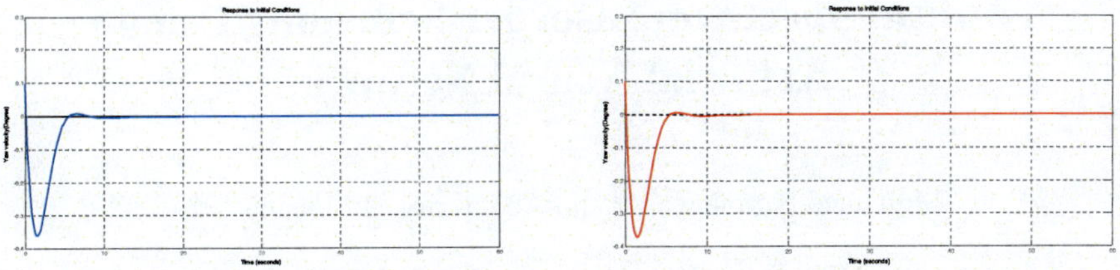

Fig. 6. Reaction wheels against magnetorquer of the Yaw velocity angle versus time.

Discussion:
Both reaction wheels and magnetorquer have an overshooting value for Pitch and Yaw angles witch tend to settle quickly hence no significant impact on overall performance. The magnetorquer shows again rapidity to settle down against reaction wheels for all the angles but not by a big margin taking for example the Roll angle: it takes for the magnetorquer 37 s to settle down versus 38 s for reaction wheels.

6 Conclusion

In this paper, a comparison between Reaction wheels and magnetorquer actuators for the three axis LEO satellite via linear quadratic regulator controller has been established. The simulation results shows clearly that the magnetorquer has a slightly better response time than his counterpart. Even though the deference is insignificant but it will make a radical change depending on the mission assigned to the LEO satellite.

References

1. Jovanovic, N., et al.: Design of magnetorquer-based attitude control subsystem for FORESAIL-1 satellite. IEEE J. Miniat. Air Space Syst. **2**(4), 220–235 (2021)
2. Kukreti, S., et al.: Genetic algorithm based LQR for attitude control of a magnetically actuated CubeSat. In: AIAA Infotech@ Aerospace, p. 0886 (2015)
3. Arefkhani, H., Sadati, S.H., Shahravi, M.: Satellite attitude control using a novel constrained magnetic linear quadratic regulator. Control Eng. Pract. **101**, 104466 (2020)
4. Ennaciri, T., Mrani, N., Foshi, J.: Comparative study between linear quadratic regulator and fractional-order robust controller for the ThreeAxis attitude control system of LEO satellite using reaction wheels. In: 2022 2nd International Conference on Innovative Research in Applied Science, Engineering and Technology (IRASET), pp. 1–5. IEEE, March 2022
5. Mrani, N.: Fractional order controller for satellite attitude stabilization based on magnetorquer. Int. J. Adv. Res. Sci. Technol. **4**(4), 406–411 (2015)
6. Charles, U., Christian, C., James, O.: Design of linear quadratic regulator for the three-axis attitude control system stabilization of microsatellites. Int. J. Sci. Eng. Res. **7**(6), 834–845 (2016)

Autism Spectrum Disorder Screening Using Artificial Neural Network

Mohamed Ikermane[1](✉) and Abdelkrim El Mouatasim[2]

[1] Laboratory LABSI, Faculty of Sciences, Ibn Zohr University, Agadir, Morocco
ikermane.mohamed@gmail.com

[2] Department of Mathematics and Management, Faculty of Polydisciplinary Ouarzazate (FPO), Ibn Zohr University, Ouarzazate, Morocco
a.elmouatasim@uiz.ac.ma

Abstract. Autism spectrum disorders, a neuro-developmental illness marked by deficits in social, communication, and behavioral development and associated with high healthcare expenses. Autism has no cure, and the purpose of treatment is to enhance your child's functional abilities by lowering autism spectrum disorder symptoms. As a result, early diagnosis can significantly reduce its symptoms and promote the child's growth and learning. The process of diagnosing autism can be lengthy and expensive, and the increasing number of Autistic cases around the world shows an urgent need for a rapid, easy, and reliable self-administered Autism screening tool that can be used by professionals, parents, and caregivers to ensure if the subject exhibits some of the typical symptoms of autism, and whether they should pursue formal clinical diagnosis or not. This paper presented an Autism spectrum disorders (ASD) screening tool using a dataset consisting of 20 features, 10 behavioral features from the brief self-administered ASD screening methods (AQ-10-Adult), and ten personal information that have been proven to be beneficial in detecting ASD patients. Our ASD classifier model had an accuracy rate of 98.3% when using artificial neural networks (ANN).

Keywords: Autism diagnosis · ASD · Artificial neural network · Self-administered autism test · AQ-10-Adult · Keras

1 Introduction

Autism spectrum disorders (ASD) is a lasting, multifaceted, neuro-developmental illness characterized by deficiencies in social and communication skills development as well as repetitive or limited behaviors and interests [1]. There is no doubt that autism is on the increase, with prevalence rates growing at an alarming pace [2]. An estimated 1.5% of the global population is on the autistic spectrum, and it is believed that many more individuals with ASD are going undiagnosed [3, 4]. ASD early detection may significantly minimize the related healthcare and educational expenses and parents may better understand their child's requirements [6, 7]. The existing diagnostic processes are time-consuming and expensive, which results in long wait periods for an ASD diagnosis, for example in the United Kingdom the average waiting time is 42 months [8, 9]. Therefore, the need

Y. Farhaoui et al. (Eds.): ICAISE 2022, LNNS 635, pp. 270–275, 2023.
https://doi.org/10.1007/978-3-031-26254-8_37

for self-administered diagnostic services has grown in response to expanding public awareness of ASD [5], thereby numerous self-administered diagnostic tests, such as the Autism-Spectrum Quotient Test (AQ) [15] and the Quantitative Checklist for Autism in Toddlers (Q-CHAT) [16], are intended to quantify the expression of Autism Spectrum features in a person subjectively. Machine learning is a branch of study that combines fields of study like mathematics, and Computer science in order to develop "Intelligent" prediction models that are both accurate and dependable. Artificial Neural Networks (ANNs) are a subset of machine learning algorithms that are made up of linked nodes inspired by the simplicity of neurons in the brain [14]. Works on machine learning and autism aimed to speed up diagnosis times and increase diagnostic accuracy in order to better serve people with ASD [10]. Many studies using machine learning approaches attempted to identify research on ASD diagnosis speed, accuracy, and ASD's key dataset features [11–13].

Tabath [17] previously suggested "ASDTests," a mobile-based application that may be used as a self-administered Autism screening tool based on the AQ and Q-CHAT screening Questionnaires and personal characteristics that have shown to be useful in ASD detection. In addition to ten personal questions such as age, sex, and ethnicity... Subject must also answer ten questions of the self-administrated Q-CHAT method corresponding to the subject age category (infant, child, adolescent, and adult) in order to utilize this mobile app. Fourth months later a dataset of 1452 cases was collected. Due to the imbalanced data (96% non-ASD) the infant dataset was eliminated. Later two supervised learning algorithms, Logistic Regression, and Naïve Bayes were applied to the three datasets Child, Adolescents, and Adults. The results clearly revealed that the dataset features acquired by the ASD app are strongly influential in detecting ASD and that a high degree of prediction accuracy was achieved using these features. In this present work we implemented the Artificial neural network techniques by making use of the Python open-source neural-network framework Keras [19] to build an accurate, and reliable self-administered ASD prediction tool.

2 Materials and Methods

2.1 ASD Screening Dataset

The dataset employed in this current work was obtained from the mobile app known as "ASDTests" [17] and was only restricted to the Adult category, The publicly available dataset [18] consists of 703 different occurrences, 189 of them identified as Autistic (86 males and 103 females), and the remaining 515 are non- autistic (281 Males, 234 Females). Participants came from 67 different countries, and their ages varied from 17 to 64 years old, with a standard deviation of 9.8 years. The used dataset contains 20 features, 10 of them were obtained based on user answers to the AQ-10-Adult Questions Table 1, a shortened adult version of the original AQ [20].

Table 1. AQ-10-Adult dataset screening Questions

AQ-10-Adult Questionnaire Item
I often notice small sounds when others do not
I usually concentrate more on the whole picture, rather than the small details I find it easy to do more than one thing at once
If there is an interruption, I can switch back to what I was doing very quickly I find it easy to "read between the lines" when someone is talking to me
I know how to tell if someone listening to me is getting bored
When I'm reading a story, I find it difficult to work out the characters' intentions I like to collect information about categories of things (e.g., types of cars..)
I find it easy to work out what someone is thinking or feeling just by looking at their face I find it difficult to work out people's intentions

Also, 10 features were collected based on the subject personal information, and which has been proven to be strongly influential in detecting ASD Table 2.

Table 2. Personal features used in the autism screening questionnaire

Attribute	Description
Age	Age in years (number)
Gender	Male or Female (M or F)
Ethnicity	List of common ethnicities (Latino, Asian,..)
Born with jaundice	Whether the case was born with jaundice (yes or no)
Family member with PDD	A family member has pervasive developmental disorders (yes or no)
Who is completing the test	Self, parent…
Country of residence	Choose from the list of countries
Used the screening app before Previously used this app (yes or no) AQ-10-Adult score	For each "Yes" answer we score 1 point
ASD class	Whether the subject is an Autistic (yes or no)

2.2 Artificial Neural Network Model

One of the most critical concerns in the use of artificial neural networks is determining the number of hidden layers and the number of neurons in each layer. Using Keras Tuner

[21], you may iterate over all the possible combinations of the number of hidden layers, and the width of each hidden layer (number of neurons) in a pre-defined range in order to determine the number and width of hidden layers that best fit your dataset. Because ANNs may cope with non-linear issues through the use of an activation function, one of the most commonly employed activation functions in multi-layer neural networks is the Rectified Linear Unit (ReLU). Additionally, due to the categorization nature of our model and that the probability of anything occurring only between 0 and 1, the Sigmoid is the best choice as an activation function for the output layer.

To prevent the ANN model from overfitting, we used the dropout technique [22], which is a method that removes neurons from the hidden layers temporarily during the training phase, the goal here is to avoid having the model learn too much information and noise from the training data, which would have a negative influence on the model's performance on unseen data.

3 Results and Discussion

In this model, we added a dropout layer after each of the two hidden layers, with P, the probability of a neuron, in the hidden layer, being excluded from the network equal to 0.4 and 0.2, respectively. Each time our model's gradient is adjusted, various units are excluded depending on a probability hyperparameter P. This approach enables dropout to decrease overfitting, which occurs when the model attempts to learn too many details in the training data while also accounting for noise in the training data. As a consequence, model performance on unseen or test datasets is quite low. As illustrated in the Fig. 1, and by analyzing the optimization Learning and performance learning curves. we can see that the model has a good fit.

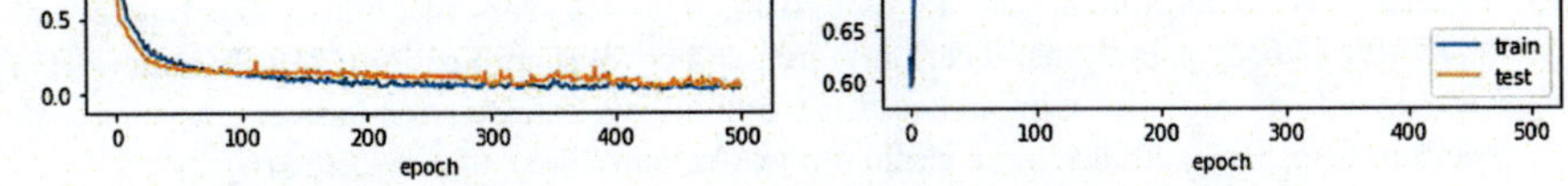

Fig. 1. Performance (right fig) and Optimization (left fig) Learning Curves

As shown in Table 3 below. Even with an imbalanced dataset and a reasonable number of instances, the results are impressive. The model has an accuracy rate of 98.7%. It's convenient to say that those dataset features are important to the ANN model's prediction of ASD presence and that we can rely on them using this self-administrated ASD questionnaire and the Artificial neural network techniques.

Table 3. Autism artificial neural network classification report

Class	Precision	Recall	f1-score	Support
Non Autistic	0.976	1.000	0.988	123
Autistic	1.000	0.943	0.971	53
Accuracy	0.983			
Precision	1.000			
Recall	0.943			
f1-score	0.971			

4 Conclusion

Autism diagnostic services are frequently overburdened, making it difficult for families in need of an evaluation to get an appointment. A self-administrated Autism screening method can help parents determine if their children exhibit some of the typical behaviours and thoughts of adults with autism. Using it, parents and caregivers may be encouraged to take their suspected autistic children to get a proper diagnosis. As part of this study, we developed a reliable and accurate ASD Prediction Model using the self-administered AQ-10-Adult and some personal characteristics that have been shown to be important in Autism detection. By implementing an artificial neural network model, the ASD classifiers have achieved a 98% accuracy rate.

References

1. Gambrill, E.: The diagnostic and statistical manual of mental disorders as a major form of dehumanization in the modern world. Res. Soc. Work. Pract. **24**, 13–36 (2014). https://doi.org/10.1177/1049731513499411
2. Baio, J., et al.: Prevalence of autism spectrum disorder among children aged 8 years—autism and developmental disabilities monitoring network, 11 sites, United States, 2014. MMWR Surveill. Summ. **67**(6), 1 (2018)
3. Brugha, T.S., et al.: Epidemiology of autism spectrum disorders in adults in the community in England. Arch. Gen. Psychiatry **68**(5), 459–465 (2011)
4. Fitzgerald, M.: The clinical gestalts of autism: over 40 years of clinical experience with autism. In: Autism-paradigms, recent research and clinical applications. IntechOpen (2017)
5. Russell, A.J., et al.: The mental health of individuals referred for assessment of autism spectrum disorder in adulthood: a clinic report. Autism **20**(5), 623–627 (2016)
6. Buescher, A.V., Cidav, Z., Knapp, M., Mandell, D.S.: Costs of autism spectrum disorders in the United Kingdom and the United States. JAMA Pediatr. **168**(8), 721–728 (2014)
7. Leigh, J.P., Du, J.: Brief report: forecasting the economic burden of autism in 2015 and 2025 in the United States. J. Autism Dev. Disord. **45**(12), 4135–4139 (2015)
8. Crane, L., Chester, J.W., Goddard, L., Henry, L.A., Hill, E.: Experiences of autism diagnosis: a survey of over 1000 parents in the United Kingdom. Autism **20**(2), 153–162 (2016)
9. Bishop, D.: Definition, diagnosis & assessment in a history of autism by A. Feinstein (2010)
10. Thabtah, F.: Machine learning in autistic spectrum disorder behavioral research: a review and ways forward. Inform. Health Soc. Care **44**(3), 278–297 (2019)

11. Li, M.S., Shanavas, A.M.: A study on Autism spectrum disorders using classification techniques. Int. J. Soft Comput. Eng. **4**(5), 88–91 (2014)
12. Bone, D., Goodwin, M.S., Black, M.P., Lee, C.C., Audhkhasi, K., Narayanan, S.: Applying machine learning to facilitate autism diagnostics: pitfalls and promises. J. Autism Dev. Disord. **45**(5), 1121–1136 (2015)
13. Wall, D.P., Kosmicki, J., Deluca, T.F., Harstad, E., Fusaro, V.A.: Use of machine learning to shorten observation-based screening and diagnosis of autism. Transl. Psychiatry **2**(4), e100–e100 (2012)
14. Yegnanarayana, B.: Artificial Neural Networks. PHI Learning Pvt. Ltd., New Delhi (2009)
15. Baron-Cohen, S., Wheelwright, S., Skinner, R., Martin, J., Clubley, E.: The autism-spectrum quotient (AQ): evidence from Asperger syndrome/high functioning autism, males and females, scientists and mathematicians. J. Autism Dev. Disord. **31**(1), 5–17 (2001)
16. Allison, C., et al.: The Q-CHAT (Quantitative Checklist for Autism in Toddlers): a normally distributed quantitative measure of autistic traits at 18–24 months of age: preliminary report. J. Autism Dev. Disord. **38**(8), 1414–1425 (2008)
17. Thabtah, F.: An accessible and efficient autism screening method for behavioural data and predictive analyses. Health Inform. J. **25**(4), 1739–1755 (2019)
18. Thabtah, F.: Autism screening adult. UCI Machine Learning Repository (2017). https://archive.ics.uci.edu/ml/datasets/Autism+Screening+Adult
19. Chollet, F.: Keras. GitHub (2015). https://github.com/fchollet/keras
20. Allison, C., Auyeung, B., Baron-Cohen, S.: Toward brief “red flags” for autism screening: the short autism spectrum quotient and the short quantitative checklist in 1,000 cases and 3,000 controls. J. Am. Acad. Child Adolesc. Psychiatry **51**(2), 202–212 (2012)
21. O’Malley, T., et al.: github.com/keras-team/kerastuner (2019). Accessed 2 Apr 2022
22. Srivastava, N., Hinton, G., Krizhevsky, A., Sutskever, I., Salakhutdinov, R.: Dropout: a simple way to prevent neural networks from overfitting. J. Mach. Learn. Res. **15**(1), 1929–1958 (2014)

Backstepping Control of the Permanent Magnet Synchronous Generator (PMSG) Used in a Wind Power System

Chaou Youssef[1](✉), Ziani Said[2], and Daoudia Abdelkarim[1]

[1] Department of Physics, Faculty of Science and Technology Errachidia, Moulay Ismail University, BP-509, Boutalamine, 52 000 Meknes, Errachidia, Morocco
yo.chaou@edu.umi.ac.ma

[2] Department of Electrical Engineering, High School of Technology, Hassan II University, Casablanca, Morocco

Abstract. We investigate the complex regulation of a windy turbine's energy motor synchronous generator. The primary contribution of this research is testing the dynamical control approach for Turbine-PMSG systems utilizing the back-stepping technique. Numerous simulation results of Turbine-PMSG systems are shown and studied to illustrate the performance of the nonlinear output feedback approach. The response characteristics to variations in rotation speed show the algorithm's efficiency.

Keywords: Wind turbine · Nonlinear control · Turbine-PMSG systems · Backstepping technique

1 Introduction

The world's consumption of electrical energy is constantly increasing due to industrialization and the daily needs of mankind. The main source of electricity production comes from fossil fuels, which are still the main cause of global warming and various environmental problems [1]. Morocco is 95% reliant on imports for its energy requirements. The Moroccan government's goal is to reduce this dependency to 85% by 2020. By this date, renewable energy sources will account for 40% of overall power output. The kingdom plans to use this renewable and limitless energy source, which has significant potential. Through the operations of the national office, the state is the primary driver of the development of renewable energies in Morocco (ONEE). ADEREE contributes to policy implementation. Morocco's estimated wind potential is almost 25,000 MW, including 6,000 MW in investigated locations. Wind-generated 6.7% of installed electrical power in 2013. The government of Morocco intends to treble this number by 2020. Currently, the most used generators in the production of wind energy are based on the asynchronous generator and the synchronous generator with permanent magnets [2]. Faced with the problems of wind or water power production, the permanent magnet synchronous generator has advantages which are among others examples no excitation

Y. Farhaoui et al. (Eds.): ICAISE 2022, LNNS 635, pp. 276–281, 2023.
https://doi.org/10.1007/978-3-031-26254-8_38

circuit, low maintenance. In this paper, the authors propose a modeling study and behavioral simulations of a wind energy production system based on a permanent magnet synchronous machine (PMSG). The studied system is presented (see Fig. 1). This system can be controlled by the classical PI controller and vector control [3, 4], but can't ensure adequate performances such as stability and protection from perturbations. To solve this issue, a variety of nonlinear control approaches have been designed and tested to command the Turbine-PMSG systems such as input-output linearization control.

1.1 Model of the Turbine and the Permanent Magnet Synchronous Machine

The theoretical power applied to the turbine is given by (1), where ρ is the density of air, S Is the surface swept by the turbine, β is the pitch angle of the blade, V is the wind speed in (m/s).

$$P_t = \frac{1}{2} C_p(\lambda, \beta).\rho.S.V^3 \tag{1}$$

The ratio between the turbine speed and the wind speed is expressed by (2), where Ω_t is the speed of rotation of the turbine and R_t is the radius of the blade.

$$\lambda = \frac{\Omega_t.R_t}{V} \tag{2}$$

The power coefficient (C_p) has a theoretical limit of 0.59 called the "Betz limit". This limit is never attained in practice [11]. This coefficient is estimated by using (3),

$$C_p(\lambda, \beta) = A_1\left[\left(\left(\frac{A_2}{\lambda_i}\right) - A_3\beta - A_4\right)e^{\frac{A_5}{\lambda_i}} + A_6\lambda\right] \tag{3}$$

The torque of the wind turbine C_g is given by (4) which can be obtained from the mechanical power.

$$C_g = \frac{P_t}{\Omega_t} \tag{4}$$

The system's mechanical equation is represented by (5), with J_t and J_g are the moments of inertia of turbine and alternator, respectively, f_t and f_g are coefficients of the frictional viscosity of turbine and alternator, respectively, Ω_{mec} is the rotation speed of the alternator, and G the speed multiplier ratio.

$$C_g - C_{em} = J\frac{d\Omega_{mec}}{dt} + f\Omega_{mec} \tag{5}$$

With: $J = J_g + \frac{J_t}{G^2}$ and $f = f_g + \frac{f_t}{G^2}$.

2 PMSG Control by Backstepping Technique

The electrical model in Park reference (d, q) of the PMSM in generator operation is reproduced from the model of the machine in motor operation, by reversing the direction of the currents i_d and i_q in the Park marks [5, 6, 7]. The model global of Turbine-PMSG systems thus can be written:

$$\begin{bmatrix} \frac{di_d}{dt} \\ \frac{di_q}{dt} \\ \frac{d\Omega}{dt} \end{bmatrix} = \begin{bmatrix} -\frac{R_s}{L_d} i_d + \frac{L_q}{L_d} \omega i_q \\ -\frac{R_s}{L_q} i_d - \frac{L_d}{L_q} \omega i_d - \frac{\varphi_f}{L_q} \omega \\ -P\frac{\varphi_f}{J} i_d - \frac{P(L_d - L_q)}{J} i_d i_d - \frac{f}{J}\Omega \end{bmatrix} + \begin{bmatrix} \frac{-1}{L_d} & 0 & 0 \\ 0 & \frac{-1}{L_q} & 0 \\ 0 & 0 & \frac{1}{J} \end{bmatrix} \begin{bmatrix} v_d \\ v_d \\ C_g \end{bmatrix} \tag{6}$$

Note that $\Omega = \Omega_{mec}$ for driving a PMSG by the wind turbine where R_s is the stator resistance, L_d and L_q are the inductances in the (d, q) frame, i_d and i_q are the stator currents, ω is the electrical velocity of the and φ_f is the PMSG remanent flux.

3 Designed of Backstepping Controller

The application of the Backstepping synthesis for the PMSG used in a wind turbine by keeping the same general structure of the system while ensuring a regulation and a limitation of the currents. We follow the same reasoning, same procedure and same calculation in order to determine the PMSM control variables of the PMSG.

$$\begin{cases} i_{dref} = 0 \\ v_{dref} = L_d\left[-K_1 e_1 - \dot{i}_{dref} - \frac{R_s}{L_d} i_d + \frac{\omega L_q}{L_d} i_q\right] \\ i_{qref} = \left(-k_2 e_2 - \dot{\omega}_{ref} - \frac{f}{J}\omega + \frac{1}{J} C_r\right)\left(\frac{J}{P\varphi_f}\right) \\ v_{qref} = L_q\left[K_3 e_3 + \dot{i}_{qref} + \frac{R_s}{L_q} i_q + \frac{P}{L_q}(L_d i_d + \varphi_f)\right] \end{cases} \tag{7}$$

The (Fig. 1) below represents the global scheme of the backstepping control of the. PMSG including in wind system.

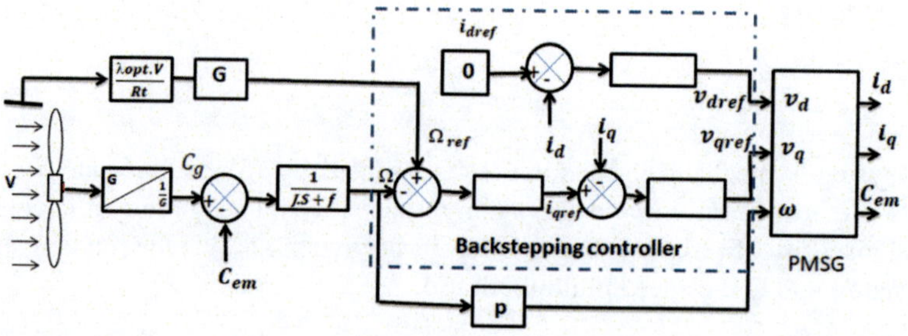

Fig. 1. Schematic representation of the backstepping control of the Turbine-PMSG systems

3.1 Simulations Results

The adopted control is based on the Backstepping method applied to a GMSM, whose model is nonlinear and multi-variable, the Table 1 below shows the parameter values used to test the system through numerical simulation.

Table 1. The Turbine-PMSG parameters.

Parameters	Values	Parameters	Values
$\boldsymbol{P}$	4	f	0.0014
R_s	$0.6[\Omega]$	C_{pmax}	0.39
L_d	0.0014 [$\boldsymbol{mH}$]	β	2°
L_q	0.0028 [$\boldsymbol{mH}$]	ρ	1.22 [kg°/m^3]
φ_f	0.2 [$\boldsymbol{Web}$]	λ_{opt}	10
R_t	3 [m]	G	6
j	$0.02\left[N.mS^2/rad\right]$		

3.2 Simulations and Discussion

The first test is intended to evaluate the robustness of the Backstepping control on the wind energy conversion system. Obviously, the wind profile for this test does not accurately represent actual circumstances. Of course, the wind profile for this test doesn't reflect actual conditions, however, it does attest to the performance of the selected control with respect to abrupt changes. For proper operation of the complete system, we must adopt the reference current $i_d = 0$, which provides the highest electromagnetic torque of the machine with a less stator current. The (Fig. 2) represent respectively the results generated by simulating the wind pattern and the angular velocity of the machine. The application of the Backstepping command gives a good performance in terms of speed and quality of the generated energy. They are clearly visible through the current-voltage, driving torque C_g and the electromagnetic torque C_{em} figures shown in (Figs. 3).

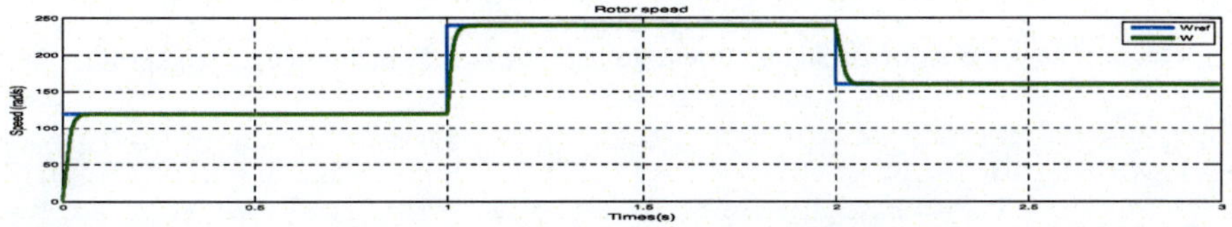

Fig. 2. Response of the reference speed (drive) tracking by the PMSG in case of ideal wind speed profile

The Second Test. Is in view of validating the Backstepping command, a second test with a random wind pattern more similar to the reality evolution. The real wind is implemented, to adapt it to the slow dynamics of the studied system. The applied control shows a good follow-up of the instructions and the quality of the generated current-voltage. Thus the degree of tracking of the point at maximum power and the effectiveness of the speed control provided by the backstepping technique. The simulation results obtained are shown in (Fig. 4, Fig. 5 and Fig. 6).

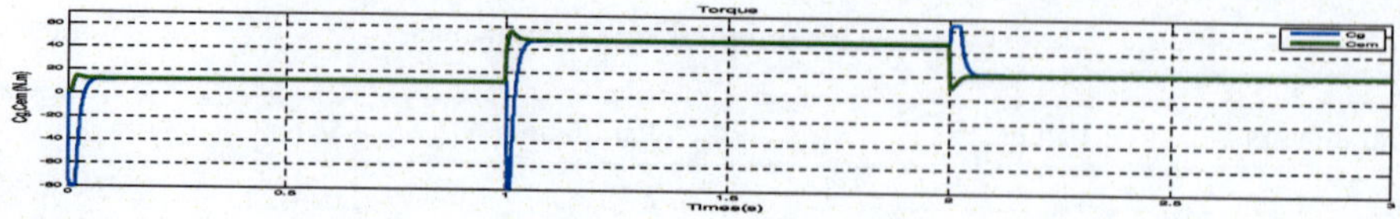

Fig. 3. The behavior of the driving torque C_g and the torque C_{em} in case of ideal wind speed profile

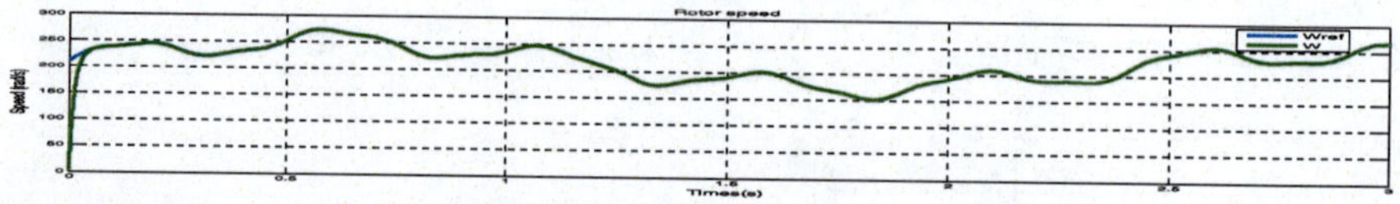

Fig. 4. Response of the reference speed tracking (drive) by the PMSG in case of random wind speed profile

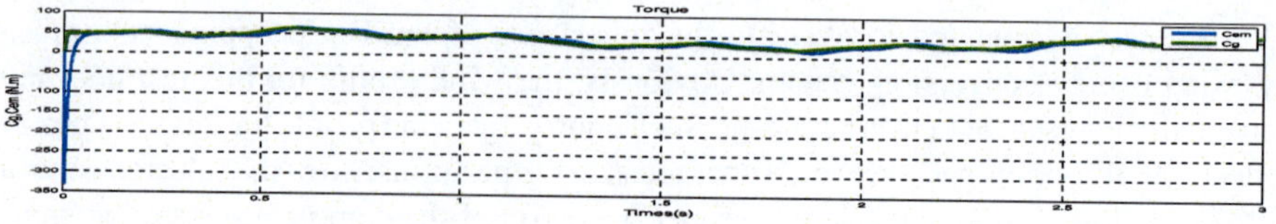

Fig. 5. The behavior of the driving torque C_g and torque C_{em} in case of random wind speed prof

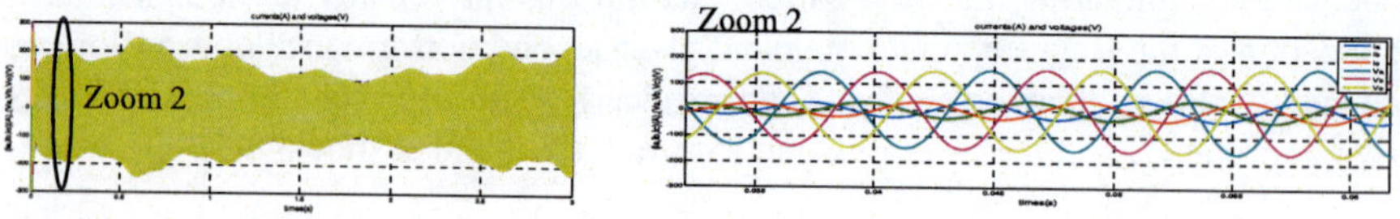

Fig. 6. The behavior of the stator current-voltage in case of random wind speed profile

4 Conclusion

This work is dedicated to the Backstepping control applied to a PMSG based variable speed wind energy conversion system. The modeling of the wind energy conversion system, specifically, the turbine and PMSG, as well as the Backstepping control theory are developed in this work. The simulation and the implementation of the control are presented. The results obtained allow to conclude that the Backstepping technique gives better control and performances for different wind profiles. This optimization method can also be used to avoid the heating of LEDs in optics [9, 10] and this study's findings also emphasize the need to use wavelets [11, 12, 13, 14], integrated with data analysis techniques often employed in biomedical signal processing, such as ICA-NMF-SVD-PCA [15, 16], to further improve the aforementioned techniques' efficacy.

References

1. Sorensen, B.: Renewable Energy—Its Physics, Engineering, Environmental Impacts. Economics & Planning (2004)
2. Ghedamsi, K., Aouzellag, D., Berkouk, E.: Control of wind generator associated to a flywheel energy storage system. Renew. Energy **33**(9), 2145–2156 (2008)
3. Singh, A., Roy, O.: Performance analysis of a PMSM drive using PID controllers. Electron. Inf. Plan. **37**(3), 80–87 (2010)
4. Rebouh, S., et al.: Nonlinear control by input-output linearization scheme for EV permanent magnet synchronous motor. In: 2007 IEEE Vehicle Power and Propulsion Conference. IEEE (2007)
5. Hamida, M.A., et al.: Robust adaptive high order sliding-mode optimum controller for sensorless interior permanent magnet synchronous motors. Math. Comput. Simul. **105**, 79–104 (2014)
6. Chaou, Y., Ziani, S., Achour, H.B., Daoudia, A.: Nonlinear control of the permanent magnet synchronous motor PMSM using backstepping method. WSEAS Trans. Syst. Control **17**, 56–61 (2022)
7. Achour, H., Ziani, S., Chaou, Y., El Hassouani, Y., Daoudia, A.: Permanent magnet synchronous motor PMSM control by combining vector and PI controller. WSEAS Trans. Syst. Control **17**, 244–249 (2022)
8. Ouhadou, M., El Amrani, A., Messaoudi, C., Ziani, S.: Experimental investigation on thermal performances of SMD LEDs light bar: junction-to-case thermal resistance and junction temperature estimation. Optik **182** (2019)
9. Ouhadou, M., El Amrani, A., Ziani, S., Messaoudi, C.: Experimental modeling of the thermal resistance of the heat sink dedicated to SMD LEDs passive cooling. In: Proceedings of the 3rd International Conference on Smart City Applications (2018)
10. Ziani, S., El Hassouani, Y.: A new approach for extracting and characterizing fetal electrocardiogram. Traitement du Signal **37**(3), 379–386 (2020). https://doi.org/10.18280/ts.370304G
11. Ziani, S., El Hassouani, Y., Farhaoui, Y.: An NMF based method for detecting RR interval. In: Farhaoui, Y., Moussaid, L. (eds.) Big Data and Smart Digital Environment. ICBDSDE 2018. Studies in Big Data, vol. 53. Springer, Cham (2019). https://doi.org/10.1007/978-3-030-12048-1_35
12. Ziani, S., El Hassouani, Y.: Fetal-maternal electrocardiograms mixtures characterization based on time analysis. In: 2019 5th International Conference on Optimization and Applications (ICOA), pp. 1–5 (2019). https://doi.org/10.1109/ICOA.2019.8727619
13. Ziani, S., Jbari, A., Bellarbi, L.: QRS complex characterization based on non-negative matrix factorization NMF. In: 2018 4th International Conference on Optimization and Applications (ICOA), pp. 1–5 (2018). https://doi.org/10.1109/ICOA.2018.8370548
14. Ziani, S., Jbari, A., Bellarbi, L., Farhaoui, Y.: Blind maternal-fetal ECG separation based on the time-scale image TSI and SVD-ICA methods. Procedia Comput. Sci. **134** (2018)
15. Ziani, S., El Hassouani, Y.: Fetal electrocardiogram analysis based on LMS adaptive filtering and complex continuous wavelet 1-D. In: Farhaoui, Y. (ed.) BDNT 2019. LNNS, vol. 81, pp. 360–366. Springer, Cham (2020). https://doi.org/10.1007/978-3-030-23672-4_26
16. Ziani, S., Jbari, A., Belarbi, L.: Fetal electrocardiogram characterization by using only the continuous wavelet transform CWT. In: 2017 International Conference on Electrical and Information Technologies (ICEIT), pp. 1–6 (2017). https://doi.org/10.1109/EITech.2017.8255310

BDIV: Healthcare Blockchain Data Integrity Schemes Verification on Storage Cloud

Soumia Benkou[1](✉) and Ahmed Asimi[2]

[1] Department of Mathematics and Computer Science, Ibn Zohr University, Agadir, Morocco
benkou.soumia2015@gmail.com, soumia.benkou@edu.uiz.ac.ma
[2] Department of Mathematics, Ibn Zohr University, Agadir, Morocco
asimiahmed@uiz.ac.ma

Abstract. The health data provided by an individual's various health care or other organizations is very sensitive and private. Firstly, this sensitive electronic health data to be stored and shared on a private or community site, in terms of privacy and security. Sometimes, existing security mechanisms in health systems find problems in preserving data on the Internet. Cloud Computing (CC) provides a solid infrastructure for health IT (HIT) on the Internet with his speed, flexibility, rapid deployment, resource sharing, infrastructure efficiency, scalability and the energy savings. The health sector needs to meet current and future demands while keeping costs to a minimum to enable collaboration between Information and communication technologies (ICT) in health care facilities.

Keywords: Healthcare · Integrity · Taxonomy

1 Introduction

Cloud data storage issues include data integrity issues, especially in healthcare. Traditional auditing uses a central, internal or external trustworthy entity to verify whether the stored data is intact or not. This delegation leads to data corruption or a problem in performing this task voluntarily or involuntarily. For this, Blockchain technology is implemented and used to guarantee the confidentiality and protection of the patient's personal medical data. Save data, which is only visible to the patient or authorized users (family, doctor or other data subjects) (Mettler 2016).

Blockchain is a distributed database consisting of a chain of blocks using a peer-to-peer network. Genesis (Park and Park 2017) is the parent node that does not reference a previous block. In the continuation of this investigation, we present some Blockchain context and background in Sect. 2. Then, in Sect. 3, we present a scheme of Blockchain Data integrity on Cloud Storage and finally the conclusion.

Y. Farhaoui et al. (Eds.): ICAISE 2022, LNNS 635, pp. 282–286, 2023.
https://doi.org/10.1007/978-3-031-26254-8_39

2 Blockchain Background and Related Work

2.1 Blockchain Background

The global crisis of 2008–2009 was used on cryptocurrency concepts such as Bitcoin using the Blockchain as a protocol. The characteristics of the Blockchain are summarized in: *Decentralization, autonomy, transparency, immutability and cryptographic link.*

There are three types of Blockchain Auditing: Public Blockchain, which is completely open and anyone can participate without any permission to anyone willing to participate. Private Blockchain used by private organizations by formulating the read-write access authorization and accounting qualification according to the rules of the private organization. Consortium Blockchain: Consortium Blockchain is a hybrid Blockchain allowing limited participation of authorized members (organization or group to which each node of the Blockchain refers). Proof of work (PoW) and IOTA are protocols of Blockchain (Fig. 1).

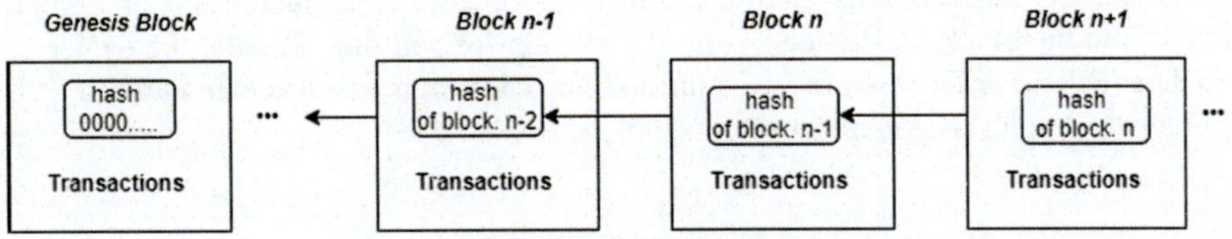

Fig. 1. Structure of blockchain (Wen 2022)

2.2 Related Works

(Wang et al. 2019) propose a personal health records sharing scheme that combines the cloud storage, searchable symmetric encryption, attribute-based encryption, Blockchain technology(Ethereum) and smart contract to assure privacy protection, keyword search and data integrity verification. The data integrity verification contract is implemented in a new scheme and the contract is deployed to the official Ethereum test network Rinkebey (Benil and Jasper 2020) introduces a new scheme which consists of EC-ACS verification and auditing using Blockchain technology enabled in MCS to provide auditing, data integrity and secure data tracking to show the change records of medical data.

(Sharma et al. 2021) introduces Blockchain as a secure and efficient architecture for medical big data management by encoding and processing healthcare data in IPFS (Inter Planetary File System). (Xie et al. 2021) introduced a Blockchain-based decentralized data integrity service ensuring service reliability. Through Blockchain technology using Proof-of-Work (PoW) consensus as an algorithm (Mardiansyah et al. 2022) introduces a new personal health records system that verifies data integrity and security through cryptographic hash processing of files and messages during the transaction, data identification and event password verification.

3 Proposed Method

3.1 System Model

The scheme of data integrity verification using Blockchain is showed on the figure above. The roles in this model are:

- Patient and doctor who store the data on the Cloud.
- A CSP is a company providing pay-as-you-go cloud services and especially storage and computing power.

Patient and doctor store its data on the CSP with a data integrity using the Blockchain as an audit. This verification corresponds to each patient, doctor or other users and the CSP a node. The blocks delegated by the Blockchain for auditing store auditing proofs that can be accessed by any node and. Each time an audit request is made by the patient and the physician to the CSP, the CSP then generates an audit proof and distributes it to the integrity verification system using the Blockchain, which includes it in a block to add it to the blocks delegated by the Blockchain for auditing. Finally, the owner of the data (Patient or Doctors) receives the audit proof from the Blockchain and can then decide whether his data is intact or not (Fig. 2).

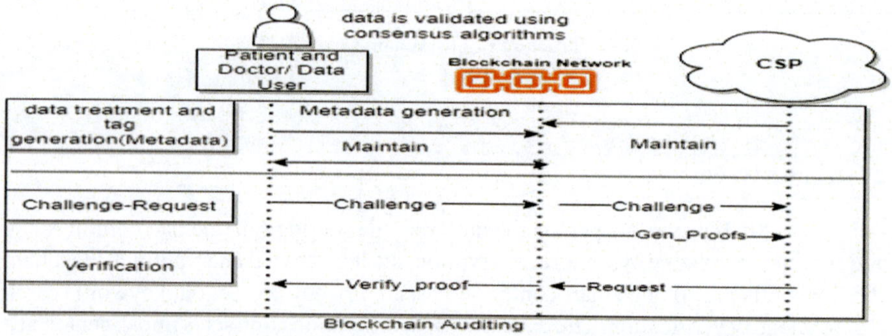

Fig. 2. Blockchain data integrity verification on storage cloud

3.2 Discussion

In our scheme, we have: **Node selection o**f master node and some slave node is required. **Public/private keys**: by Patient or Doctors due to the use of asymmetric encryption in Blockchain technology. **Metadata generation**: Patient and doctor generates metadata (such as HVTs) for their files. **Challenge- Response**: from the Patient and doctor se to the CSP to probabilistically to verify the data blocks in the Cloud via **Consensus** after communication between the nodes that manage the Blockchain together. **The Decision**: The Patient and Doctor make right decision about the corruption of data (Table 1).

Table 1. Comparison between proposed scheme and other schemes

Scheme	Dynamic Data	Public Auditing	Privacy Preserving	Authentication	Use of Blockchain
(Wang et al. 2019)	No	Yes	Yes	No	No
(Kim et al. 2020)	No	No	Yes	Yes	Yes
(Sharma et al. 2021)	Yes	No	Yes	Yes	Yes
(Margheri et al. 2020)	Yes	No	Yes	Yes	Yes
(Xie et al. 2021)	Yes	Yes	Yes	Yes	Yes
(Mardiansyah et al. 2022)	Yes	No	No	No	Yes
(Chen et al. 2022)	No	Yes	Yes	Yes	Yes
(El Balmany et al. 2022a, b)	Yes	Yes	Yes	Yes	No
Our scheme	Yes	Yes	Yes	Yes	Yes

3.3 Limitations on Blockchain Data Integrity Verification and Future Trends

Despite the advantages of adapting Blockchain to store, manage and audit health data in the cloud, many challenges and limitations in general are generated. We cite scalability, data privacy in healthcare and Consensus algorithms used in Blockchain. As a challenge of BDIV,we cite storage and communication efficiency and security attacks like.

4 Conclusion

In this article, we present the verification of previous health data integrity verification using Blockchain and proposed a scheme that can improve the reduce computational efficiency of the whole system. Through Blockchain traceability, data verification is possible by the patient, doctor or even data users at any time. This shows that we can find different type of Blockchain according to its usefulness in the field of health and especially the latest version of Blockchain Consortium which can provide a solution ensuring high immutability, batch auditing, secure and privacy-preserving. at a lower cost.

References

Balmany, C.E., Asimi, A., Bamarouf, M., Tbatou, Z.: Dynamic proof of retrievability based on public auditing for coded secure cloud storage. Multimed. Tools Appl. 1–21 (2022)

Benil, T., Jasper, J.J.: Cloud based security on outsourcing using blockchain in E-health systems. Comput. Netw. **178**, 107344 (2020)

Chen, C.-L., Yang, J., Tsaur, W.-J., Weng, W., Wu, C.-M., Wei, X.: Enterprise data sharing with privacy-preserved based on hyperledger fabric blockchain in IIOT's application. Sensors **22**, 1146 (2022)

El Balmany, C., Tbatou, Z., Asimi, A., Bamarouf, M.: Secure virtual machine image storage process into a trusted zone-based cloud storage. Comput. Secur. **120**, 102815 (2022)

Kim, M., Yu, S., Lee, J., Park, Y., Park, Y.: Design of secure protocol for cloud-assisted electronic health record system using blockchain. Sensors **20**, 2913 (2020)

Mardiansyah, V., Sari, R.F., et al.: Lightweight blockchain framework for medical record data integrity. J. Appl. Sci. Eng. **26**, 91–103 (2022)

Margheri, A., Masi, M., Miladi, A., Sassone, V., Rosenzweig, J.: Decentralised provenance for healthcare data. Int. J. Med. Inform. **141**, 104197 (2020)

Mettler, M.: Blockchain technology in healthcare: the revolution starts here. In: 2016 IEEE 18th International Conference on E-Health Networking, Applications and Services (Healthcom), pp. 1–3 (2016)

Park, J.H., Park, J.H.: Blockchain security in cloud computing: use cases, challenges, and solutions. Symmetry **9**, 164 (2017)

Sharma, P., Borah, M.D., Namasudra, S.: Improving security of medical big data by using Blockchain technology. Comput. Electr. Eng. **96**, 107529 (2021)

Wang, S., Zhang, D., Zhang, Y.: Blockchain-based personal health records sharing scheme with data integrity verifiable. IEEE Access **7**, 102887–102901 (2019)

Wen, H.A.: A blockchain-based privacy preservation scheme in mobile medical. Secur. Commun. Netw. (2022)

Xie, G., Liu, Y., Xin, G., Yang, Q.: Blockchain-based cloud data integrity verification scheme with high efficiency. Secur. Commun. Netw. **2021** (2021)

Big Data and the Effectiveness of Tourism Marketing: A Prospective Review of the Literature

Naoual Bouhtati[1], Majda Kamal[2], and Lhoussaine Alla[1](✉)

[1] LAREMEF Laboratory, Sidi Mohamed Ben Abdellah University of Fez, Fez, Morocco
lhoussaine.alla@usmba.ac.ma

[2] Higher International Institute of Tourism of Tangier, Tangier, Morocco

Abstract. The emergence of big data brings a new wave of marketing strategies by supporting the personalization and adaptation of sales, service and customer service. Particularly in the field of tourism, marketing needs big data to improve the customer experience, in particular the personalization and adaptation of its offer.

The objective of this research is to answer the following question: **to what extent does the anchoring of Big Data techniques promote the improvement of the marketing effectiveness of tourism companies?**

We have identified the main concepts and their interactions, to suggest a framework for the adoption of data analytics in tourism marketing. The main socle of this study was an exploratory literature review.

The research highlighted the gradual use of Big Data Transactional Data/Non-transactional/Social Data tools by tourism professionals in the management of data relating to both supply and demand. Tourism, so as to improve their marketing effectiveness in terms of customer relations (personalization customization, empowerment, interactivity, etc.), customer intelligence (needs, habits, attitudes, behaviors, sensitivities, etc.), campaigns and promotions (segmentation, targeting, direct marketing, etc. and ROI (turnover, market share, outstanding customer, etc.).

Keywords: Big data · Marketing data analytics · Tourism marketing · Marketing strategy · Customer relationship management · Marketing effectiveness

1 Introduction

Today, and more than ever, consumers, and in particular tourists, are constantly informed on merchant websites, comparators and social networks, they exchange and transmit data on their interests and their experiences [1] in complete spontaneity. Thanks to the Internet, the data shared in the last decade has exceeded that of the entire history of humanity.

In all sectors and particularly tourism, recent technological developments (Internet, websites, social media, mobile applications, chatbots, robotics, self-service kiosks, etc.)

Y. Farhaoui et al. (Eds.): ICAISE 2022, LNNS 635, pp. 287–292, 2023.
https://doi.org/10.1007/978-3-031-26254-8_40

and the arrival of artificial intelligence have created a technological layer important in the interaction between companies in the travel, tourism and hospitality sector and their customers.

This revolution could not be possible without Big data which is at the heart of artificial intelligence applied to the tourism industry. From now on, companies operating in this sector can easily obtain valuable information to forecast tourism demand, make better decision-making, manage knowledge flows and customer interaction, and provide the best service in a more efficient and effective way. This can translate into improved productivity, increased customer satisfaction, personalized and effective marketing campaigns, and above all adequate customer relationship management. However, research questions on the role of Big Data in improving the customer experience in the tourism sector remain open [1].

The paper presents a theoretical and conceptual framework of the subject, the methodology, the results and their discussion, as well as the conclusions and research perspectives.

2 Theoretical and Conceptual Framing

2.1 Big Data and its Application in Marketing

Nowadays, Big data is one of the most important areas of future technology, which is rapidly spreading across many industries and businesses. Through multiple tools and techniques, revolutionizing its 7Vs (Value, Volume, Variability, Visualization, Veracity, Variety, Velocity) [3], Big Data can bring added value to companies by finding models and relationships inter data, using a range of techniques (Transactional Data, Non-transactional Social Data) [2].

Progress in analytical techniques, information technology, technological cost reduction, and attempts to capture and retain customers, are driving the growing use of Big Data. Companies have used this technology to gain a competitive advantage. In relation to marketing, Big Data can enable companies to build more effective relationships with their customers, creating opportunities for automating relational processes and managing useful business information. These technologies can cause greater interactivity between actors, with more possibilities for collecting, processing, sharing and disseminating information (Netflix [3] and Starbucks cases for illustration [4]).

2.2 Analytics Big Data and Reinventing Tourism Marketing

Marketing, being a key function of the tourism business, is the ideal basis for the application of Big Data Analytics, with results that improve reputation, customer loyalty, financial performance, and business position. Compared to the competition. Over the past decade, traditional tools were based on a predetermined strategy schedule and supported the mass market mentality, maintaining the focus on the person and their needs. Recently, the rapid development of technologies and the consolidation of the Internet as a key gateway to markets have led to drastic changes in the balances and given more power and control to the consumer [5]. Big Data has enhanced the ability to connect directly with tourists and facilitates maximum personalization in any communication effort [2].

New trends in tourism marketing focus on "marketing intelligence" through the use of data to make marketing-related decisions. Recent data shows that the 5Ps (people, product, promotion, price and place) of the marketing mix can be used to manage big data, creating the conditions for innovations involving significant changes in product design, promotion and price adjustment.

Tourism is a phenomenon characterized by variability, spontaneity, risk, adventure and expectations are affected by the changing desires of tourists. The immediate and targeted adaptation of approach strategies acquires a personal and unique character thanks to Big Data. As Big Data technologies flood the tourism industry worldwide, large corporations are making the most of the dynamics of data [6].

Likewise, from the data provided by Tourists through the various transactions carried out, any tourist unit could benefit from the possibilities offered by Big Data in the management and understanding of its customers, as well as a better configuration of its advertising campaigns.

Finally, Big Data tools allow tourists to better manage their stay (information, location, booking, comparison, experience, appreciation, recommendations, etc.) and professionals to better develop their offer (location, services, prices, experiences, simulations, …) and personalize the tourist's travel itinerary [7].

3 Methodology

3.1 Research Protocol

To carry out our exploratory literature review, we have generally complied with the basic protocol of the systematic literature review [8], via a rigorous framework by the research problem, by the primary delimitation of the research and the definition of the collection tools and data extraction. Indeed, the interest of this documentary research is to provide some answers to our main question: to what extent does the anchoring of Big Data techniques promote the improvement of the marketing effectiveness of tourism companies?

To better operationalize this research, we have defined our preliminary scope of investigation by setting the inclusion and exclusion criteria for references, in terms of general information (context, year of publication, etc.), methodological information (theoretical framework, qualitative design, quantitative design, etc.) and documentary information (nature of the document, languages, etc.).

Then, for the collection of references, the search was conducted via Google Scholar search engines, as well as through scientific databases: Sciencedirect, Researchgate and Citeseerx. The operationalization of the research was done in French and English, using the following keywords: "Marketing Big Data", "Marketing Data Analytics", "Tourism Marketing Big Data" and "Tourism Marketing Data Analytics". Thus, 15 primary studies were sectioned, before finally retaining only 5, after purification. The data extraction is carried out through the in-depth examination of the content of the references in order to extract the main Big Data Analytics tools adopted and the marketing effectiveness metrics generated.

3.2 Characteristics of the Sample of References Examined

The references examined (Table 1) are very recent (post 2015), more exploratory (4/5), more qualitative (100%) and less contextualized. This seems to reflect an embryonic level of research and thus partial and provisional results, very far from elaborate analysis models and confirmatory studies.

4 Results and Discussion

4.1 Results

Our literature review is based on the content analysis of the main works identified in the literature dealing with the issue of the anchoring of Data Analytics tools and its impact on tourism marketing effectiveness (Table 1).

Table 1. Summary of the literature review

Authors	Mobilized big data tourism mechanisms	Tourism marketing effectiveness indicators selected
[9]	Inventory data/Loyalty programs Bookings/User ID data Web analytics data/Search data Travel reviews /	Audience segmentation Announcement frequency Intensive ad targeting Personalization customization
[10]	User ID data/Search data Inventory data/Operations	Reputation/Customer Loyalty Customer empowerment Competitive positioning
[11]	Social media data Bookings Searchdata	Accessibility and relevance of information Behavior prediction Purchase incentive/Customer experience
[12]	Search data/Web analytics data	Indexing by search engines Relevance of information
[13]	Social media data Web analytics data	Choice of tourist destinations Attractiveness of the destination Promoting the destination

The big data of Tourism operators consists of the huge data sets collected by internal and external digital platforms. This helps them monitoring the customers/prospects engagement behavior in terms of content (web and application analysis), the process of managing their staying (search, reservation, promotion, feedback, etc.), and interactivity with the main stakeholders in the tourism ecosystem, this could be also expanded to the international level.

4.2 Discussion and Perspectives

The results of our theoretical study are consistent with multiple research works on the issue [14]. On the one hand, the marketing managers of tourism companies, as well as institutional managers of the development of the tourist strategy, are aware of the importance of the use of techniques and tools of Big Data Analytics, in particular Social media. Data [15] to boost the operational efficiency of their commercial strategy and make it a lever for innovation and demarcation, prediction and effectiveness of marketing communication [16].

Finally, the intelligent and intelligible adoption of Analytics Data Marketing tools is necessary for tourism professionals in order to meet the challenge of improving the effectiveness of their marketing strategy and the satisfaction and loyalty of their customers., via value co-creation focused on reducing transaction costs, research efficiency, personalization and customer relationship management [17], via the mobilization and integration of the 7Ps (Product, Process, People, Promotion, Physical evidence, Place, Price) [11]. While these concerns are increasingly understood by experts and consultants, with more conceptual mastery and mapping of good practices [17], academic research in this area remains limited and still exploratory [14].

5 Conclusion

This paper seeks to understand the state of the art of reflection and practices concerning the adoption of technologies related to big data in the field of tourism marketing and the implications generated on the marketing effectiveness of these establishments.

The general finding of the study reveals a big trend towards the use of Big Data as a lever for the development of competitive advantage, Knowledge management and optimization of databases use, etc… In terms of marketing, the studies examined, had reflected a massive use of multiple big data tools, in order to improve customer relationship management, to further develop customer intelligence, the main objective is effectively predict their needs and desires, in order to eventually stand out significantly among competitors.

While in principle the remarkable contribution made by Big Data techniques and solutions to improving the marketing efficiency of the company, the realization of these gains comes up against a reluctant state of mind of certain researchers and tourism professionals, as well as the deficit in terms of dissemination of good practices in this area, particularly in the Moroccan context [18, 19]. As a research perspective, we envisage the study of the impact of Social media data on the commercial performance of the company, particularly in terms of customer intelligence, the effectiveness of the targeting of advertising campaigns, the development of company offer, etc. [20].

References

1. Maghraoui, S., Belghith, E.: The customer experience: what contributions of artificial intelligence technologies, vol. 8 (2019)

2. Fan, C., Yan, D., Xiao, F., Li, A., An, J., Kang, X.: Advanced data analytics for enhancing building performances: from data-driven to big data-driven approaches. Build. Simul. **14**(1), 3–24 (2020). https://doi.org/10.1007/s12273-020-0723-1
3. Kallinikos, J., Nucci, F. (nd): Netflix: a data and media hybrid, vol. 70 (2020)
4. Rijmenam, M.: Starbucks, the world largest coffee shop, grinds a lot of data datafloq (2016). Accessed 18 July
5. Middleton, V., Fyall, A.: Marketing for Travel and Tourism. Butterworth Heinemann, Oxford (2009)
6. Ren, S., Zhang, Y., Liu, Y., Sakao, T., Huisingh, D., Almeida, C.: A comprehensive review of big data analytics throughout product lifecycle to support sustainable smart manufacturing: a framework, challenges and future research directions. J. Cleaner Prod. **210**, 1343–1365 (2018)
7. Fararni, K.A., Nafis, F., Aghoutane, B., Yahyaouy, A., Riffi, J., Sabri, A.: Hybrid recommender system for tourism based on big data and AI: a conceptual framework. Big Data Min. Anal. **4**(1), 47–55 (2021)
8. Petticrew, M., Roberts, H.: Systematic Reviews in the Social Sciences: A Practical Guide. Blackwell Publishing, USA (2006)
9. Bank, AD.: Big data for better tourism policy, management, and sustainable recovery from COVID-19. Asian Development Bank (2021)
10. Mantas, C., Rossidis, L.., Malik, S., Belias, D.: The use of big data in tourism: current trends and directions for future research. RICHTMANN (2021)
11. Liu, C.H., Horng, J.S., Chou, S.F., Yu, T.Y., Huang, Y.C. Lin, J.Y.: Integrating big data and marketing concepts into tourism, hospitality operations and strategy development. Qual. Quant. 1–18 (2022).https://doi.org/10.1007/s11135-022-01426-5
12. Soualah-Alila, F., Coustaty, M., Faucher, C., Wannous, R.: TourinFlux project: contribution of semantic web technologies for tourism data management. In: 6th edition of the multidisciplinary conference AsTRES: Association Tourism Research and Higher Education, pp.12. University of Western Brittany, Quimper, France (2016)
13. Miah, S.J., Vu, H.Q., Gammack, J., McGrath, M.: A big data analytics method for tourist behavior analysis. Inf. Manag. **54**(6), 771–785 (2017)
14. Li, H., Wang, Q., Zhang, L., Cai, D.: Big data in china tourism research: a systematic review of publications from English journals. J. China Tourism Res. **18**, 1–19 (2022)
15. Zhao, Z., Zhu, M., Hao, X.: Share the gaze: representation of destination image on the Chinese social platform WeChat moments. J. Travel Tourism Mark. **35**(6), 726–739 (2018)
16. Mariani, M.: Big data and analytics in tourism and hospitality: a perspective article. Tourism Rev. **75**(1), 299–303 (2020)
17. Li, J., Xu, L., Tang, L., Wang, S., Li, L.: Big data in tourism research: a literature review. Tour. Manag. **68**, 301–323 (2018)
18. Alaei, A.R., Becken, S., Stantic, B.: Sentiment analysis in tourism: capitalizing on big data. J. Travel Res. **58**(2), 175–191 (2019)
19. Centobelli, P., Ndou, V.: Managing customer knowledge through the use of big data analytics in tourism research. Curr. Issue Tour. **22**(15), 1862–1882 (2019)
20. Boulaalam, O., Aghoutane, B., Ouadghiri, D.E., Moumen, A., Cheikh Malinine, M.L.: Proposal of a big data system based on the recommendation and profiling techniques for an intelligent management of moroccan tourism. Procedia Comput. Sci. **134**, 346–351 (2018). https://doi.org/10.1016/j.procs.2018.07.200

Big Data Application in Education: Overview

Hanae Aoulad Ali[1(✉)], Chrayah Mohamed[2], and Bouzidi Abdelhamid[1]

[1] TIMS LABORATORY, FS, Abdelmalek Essaadi University, Tetuan, Morocco
hanae.aouladali@etu.uae.ac.ma

[2] TIMS LABORATORY, ENSATE, Abdelmalek Essaadi University, Tetuan, Morocco

Abstract. As the Internet, computers and smartphones increasingly become part of the environment in which today's young people grow and learn, educational institutions and education systems are urged to take advantage information and communication technology education. The development of technologies in education has brought out many issues, including the improvement of learning experiences. In this context, Big Data technology offers valuable information on the effectiveness of teaching. Teachers can mobilize them to fight against school dropout.

Keywords: Big data in education · Educational data mining · Data mining tools · Big data applications

1 Introduction

The COVID-19 crisis has exacerbated existing inequalities and exposed the deficient capacities of our fundamental social and educational institutions. Changes in the educational system have been interpreted in the world at large [11–14]. Especially since we have the ambition to meet the challenges and build a sustainable future. As education has been shaped according to COVID-19, has it really met the need of the 21st century age?

Indeed, the crisis has prompted innovation in the education system to allow the continuation of teaching activities and ensure educational continuity. But these innovations have also brought to light promising new perspectives open to pedagogy and rapid changes to teaching methods that can only be sustainable if no one is left behind.

In fact, online education was an idea that many institutions had conceived, but it was not studied much. Thanks to COVID-19, the transition of education to online platforms could not be achieved step by step. Education has undergone a compulsory change without sufficient study time. For such a radical change, it must be done step by step and it is a process that takes time. There was no opportunity to see the shortcomings of the process of changing the education system [13–15]. The world has suddenly abandoned traditional education, and modernized education according to age. Moreover, this sudden change in the educational system has brought many problems. With online education, we believe that there will be some skill gaps in prospective students. This can affect productivity in business life in the future [1].

Y. Farhaoui et al. (Eds.): ICAISE 2022, LNNS 635, pp. 293–300, 2023.
https://doi.org/10.1007/978-3-031-26254-8_41

Currently, with this crisis that has shaken the world, universities are adapting the capacities of their university servers and are working intensively on new didactic concepts and new information platforms. The current exceptional situation also serves as a test case for many innovative digital teaching and learning formats that have long been waiting to be implemented in our universities [2]. Actors from academia, politics, business and society have long since networked within the framework of the Higher Education Forum on Digitization and generated many impulses. The support of this initiative for the development of higher education is currently proving its worth. For this, the university opts for a new form of teaching which is online learning and consequently the exploitation of learning strategies via the Net and ICT for the purposes of higher education.

Moreover, it is clear that the future of higher education needs to be rethought in many ways. Ensuring educational continuity during the closure of establishments has become the priority of States all over the planet. Indeed, an increasing number of people have turned to digital, which has forced teachers to deliver their lessons through the Internet, including our country which has favored various methods of distance education, depending on the levels of classes. In areas where Internet access is limited, the state has made more use of traditional means of distance education, often combining television and radio broadcasting of educational programs and distribution of printed materials [3].

Learner results, grades, quality of online participation… Education generates massive volumes of digital data, the analysis of which (often carried out by artificial intelligence) makes it possible to improve pedagogical approaches. This is where Big Data comes in, which is about extracting useful information from large amounts of data. Combined with "increased storage capacities and real-time analysis tools", these data now offer "unparalleled possibilities for exploiting information", according to the CNIL (National Commission for Computing and Liberties) [4]. Raising questions of digital sovereignty and the protection of private data, "regulated" Big Data can also help support student success.

In education, Big Data corresponds to both the compilation of structured (predefined and formatted) and unstructured data (generated by online activities: messaging, multimedia files, etc.). Since they provide valuable information on the activities and progress of learners, their processing enriches their "phygital" pathways. "Learning analytics, which can be found in LMSs or LXPs, make it possible to better follow the paths of students in a learning context and to make a relevant adaptation of educational content", explains Anthony Hié, director of transformation at Excelia, as part of a webinar organized by Blackboard [5].

Big Data also makes promises in the fight against school dropout. By bringing out indicators such as a higher failure rate or a longer time to pass an exercise, this approach makes it possible "to identify the weak points of the students, to offer them suitable revision modules and thus to promote their success" says Anthony Hié. It thus contributes to the hyper-personalization of learning methods, greatly facilitated by the hybridization of lessons. This strategy is part of the more global approach of adaptive learning, which uses computers as a teaching medium to refine learning experiences according to the unique needs of each learner [6].

Faced with increasingly large volumes of data, schools must better manage relations with prospects, students, alumni, prescribers, partners, etc. For them, the use of this data is therefore a means of creation of value to obtain competitive advantages.

The purpose of this study is to overviews of BIG DATA application in education environment. Second section present related work. Furthermore Big Educational Data as well as application of big data application in education Finally,we present a general conclusion.

2 Related Work

This study uses the [10] approach to locate relevant literature, which consists of three phases: planning, doing the review, and recording the results of the review. First, in order to define the articles, including the keyword "Big Data for Education," this research relies on a search in the Saudi Digital Library SDL with the years of publication customized from 2015 to 2020. There were about 521,716 objects discovered. According to [10], the second phase is to perform the review, which attempts to find relevant studies, select major studies, assess their quality, and synthesize data. The final stage is to document the outcome of the evaluation. The order of the outcomes in this study is determined by the date. [9] provide applications for educational big data mining to demonstrate how big data can be used to solve educational problems. [11] They discuss the challenges of dealing with large amounts of data, as well as the most popular techniques for educational data mining, such as regression, nearest neighbor, clustering, and classification, and the top tools used, such as Hadoop and map-reduce, and how those technology can be used in a variety of ways in learning analytics. The investigation also showed a small number of studies concentrating on measuring learning outcomes using natural language text analysis. This study only analyzed a small number of accessible papers; and the need for a thorough literature review on the main issues from 2010 to 2019.

This study analyzed just a small number of accessible papers, and there is a need for a comprehensive analysis of literature on big data in education and learning from 2010 to 2019. [12] Examines the current situation of stakeholders in learning analytics, including as students, teachers, and researchers. They discussed the advantages and disadvantages that these stakeholders confront. [13] They investigate how cloud-based big data mining might be used to Indian education and research; they then overcome the hurdles and reap the rewards. They use the phrase "educational intelligence" to refer to concepts such as learning analytics, academic analytics, and big data analytics [14].

They clarify Big Data's Positioning in Education, followed by theoretical research.

They use the phrase "educational intelligence" to refer to concepts such as learning analytics, academic analytics, and big data analytics. [14] They clarify the Big Data Positioning in Education, then highlight that theoretical research and practical implementation of big data for education are still in the early stages. He believes that big data will have a significant impact on schooling. He highlights the benefits of integrating big data in education in two areas: learning analytics and educational policy. He also emphasizes that big data for education is still in its early stages.

Emphasizes map-based management and visual analysis methods, which they feel would assist consumers and scholars that use big data in education. He says that there have been relatively few big data studies on education policy. He talks about how to approach big data and how to use big data in education policy research.

They highlight to data mining methodologies and applications used in education, and they describe how learning analytics are used in colleges and universities. They conduct a literature review to define the advantages of educational data mining (EDM) and learning analytics (LA), as well as the value of big data in education. Then they bring out the difficulties. They talk about big data in Nurse Education. They discuss the advantages of big data in nurse education, as well as the hurdles and possible hazards. [13] They offer PABED (Project Evaluating Big Education Data), an educational intelligence tool for analyzing large amounts of educational data. N. Xu and colleagues. They studied big data utilizing education and applied their findings to the MOOC, which is a training platform. Suggested a Map Reduce-based deep learning analysis technique for dealing with college data investigate the elements that influence online learning students' behavior in a big data environment. The aspects that influence a student's knowledge number of internet education and offer control measures based on those characteristics.

3 Big Educational Data

3.1 History of Big Data

Although the concept of Big Data is relatively new, big data sets date back to the 60s and 70s, when the world of data was just getting started with the first data centers and the development of the relational database.

In 2005, there was an awareness of the amount of data that users generated on Facebook, YouTube and other online services. Hadoop (an open source infrastructure created specifically to store and analyze big data games) was developed that same year. NoSQL also began to be used more and more at this time.

The development of open source frameworks such as Hadoop (and, more recently, Spark) has been critical to the growth of big data, as they make it easier to use big data and reduce storage costs. Since then, the volume of Big Data has exploded. Users still generate huge amounts of data, but it's not just humans who use it.

With the advent of the Internet of Things (IoT), more and more objects and devices are connected to the Internet, collecting data on customer usage patterns and product performance. The emergence of machine learning has produced even more data.

While Big Data has come a long way, its usefulness is only just beginning to be felt. Cloud computing has further increased its possibilities. The cloud offers tremendous scalability, developers can simply quickly run dedicated clusters to test a subset of data. Additionally, graph databases are becoming increasingly important, with their ability to display huge amounts of data in ways that make analyzes quick and comprehensive.

4 The 5 V's of Big Data

As for the last solution, it speeds up the request processing time.

The volume corresponds to the mass of information produced each second. According to studies, to get an idea of the exponential increase in the mass of data, it is considered that 90% of the data was generated during the years when the use of the Internet and social networks experienced strong growth [7]. The set of all data produced from the beginning of time until the end of the year 2008, would now fit the mass of those that are generated every minute. In the business world, the volume of data collected every day is of vital importance.

Velocity
Velocity is the speed with which new data is developed and deployed. For example, if you post messages on social media, they can go "viral" and spread in no time. This involves analyzing data along its lineage (sometimes called in-memory analysis) without it being essential that this information be stored in a database [7].

Variety
Only 20% of the data is structured and then stored in relational database tables similar to those used in accounting management. The remaining 80% are unstructured. This can be images, videos, texts, voices, and many more... Big Data technology allows analysis, comparison, recognition, and classification of data of different types such as conversations or messages on social networks, photos on different sites etc. These are the different elements that make up the variety offered by Big Data [7].

Veracity
Veracity concerns the reliability and credibility of the information collected. As Big Data allows to collection of an indefinite number and several forms of data, it is difficult to justify the authenticity of the contents, if we consider the Twitter posts with the abbreviations, colloquial language, the hashtag, typos etc. However, computer geniuses are developing new techniques that should facilitate the management of this type of data [7].

Value
The notion of value corresponds to the profit that can be derived from the use of Big Data. These are usually the companies that are starting to get incredible benefits from their Big Data. According to managers and economists, companies that are not serious about Big Data risk being penalized and dismissed. Since the tool exists, not using it would lead to losing a competitive privilege [7].

5 Applications of Big Educational Data

Here is a summary of the applications from linked publications that may be obtained by examining data collected in education:

Predicting student performance: It is the earliest and most widely used use of Data Mining in education. Using prediction to estimate value defines the student's status, such as dropout or their duties. Data visualization: It may aid in a variety of ways, including assessing student behaviour and statistically visualizing it. Using big data mining techniques, we may provide specifics about the educational environment and extract knowledge like statistical indicators on student involvement in forums, the sequence in which students study topics, and the amount of materials students utilize, among other things Also, display the data for usage in visual representations to assist people comprehend the examined data [8]

Providing smart feedback: For example, comments on how to improve student learning efficiency, arrange course resources and then assist students in taking necessary action. Giving comments also aids in the extraction of fresh, hidden, and intriguing ideas from data [8]. Different data mining tools, such as clustering, classification, and association rule mining, are utilized to offer feedback, allowing teachers to automatically obtain input from the learning process.

Big data provides insights and recommendations: It can aid in the provision of insights to aid in the educational process [38]. Furthermore, it can propose to the student based on their behaviours, websites visited, the next assignment to be completed, and so on. Also, the ability to design material, interfaces, or customised exercises for each individual learner.

Big data modelling: Include modelling of their abilities and knowledge in the development of student's cognitive models. In order to automate the student models, it is also necessary to automatically incorporate student attributes such as motivation, contentment, learning behaviour, and so on. For example, association-rule algorithms have been used in online education to model the personality traits of students.

Detecting undesired student behaviours: Identify students who exhibit some form of issue or atypical behaviour, such as erroneous activities, low motivation, and so on. To detect these sorts of behaviours, several DM approaches, like as classification and grouping, are applied. That is, to give proper assistance [8].

Grouping students to highly personalized activities.: This entails grouping students based on, for example, features or personal qualities that assist teachers in grouping students and building a personalized learning system, in order to construct successful group learning. Classification and clustering are classification algorithms used for this [8].

Constructing course content: The goal is to assist instructors and developers in developing courses and learning content automatically. It also encourages the interchange of existing learning resources between systems—the clustering DM approach used to produce customised courseware by constructing a personalized web tutor tree [8]. Social Relationship mining: Rather than researching individual features or characteristics, this is used to investigate interactions between persons. A social network is a collection of individuals who are linked together by connections such as friendship. Collaborative or social filtering is the most popular DM strategy used to mine social networking sites in education. It is a means of creating predictions about a student's interests by obtaining information about what they favour [8]. Text data from forums, chat rooms, and social networks can be used.

Scheduling and planning: The goal is to improve conventional education by preparing future courses, such as course scheduling, resource allocation, admission and counselling assistance, curriculum development, and so on [8], for example, in academic planning, to investigate the impacts of changes in admissions, and so on.

Furthermore, it was utilized to produce strategies, such as creating a successful an educational strategy that designs Intelligent Systems, which respond to student behaviour in the near term and improves the learning process in the long term. Association rules are a data mining approach used in higher education for task planning through classification, estimate, and visualization [8].

Skill estimation: Estimate the student's ability in the educational setting.
Foreign language learning.

6 Conclusions

Digitization represents, in our time, a necessity for the development and competitiveness of the majority of sectors. It is a process for transforming objects, professions, methods and tools into more efficient and more efficient digital code. The higher education sector is no exception to this digital transformation. Thus, over the past two decades, digital pedagogy has become one of the most appropriate forms of learning for the current generation of higher education students who prefer a training model with smart universities.

Big data represents one of the technological advancements that is currently being developed and required. Big data will also be capable of storing large amounts of data and integrate data from various sources. We can use correct and timely data from everywhere safely and conveniently with big data.

In this study, We cover what big data means to the field of education.

References

1. Amirhosein, M., et al.: Ensemble boosting and bagging based machine learning models for groundwater potential prediction. Water Resour. Manag **35**(1), 23–37 (2021)
2. Ali, H.A., et al.: A course recommendation system for Moocs based on online learning. In: 2021 XI International Conference on Virtual Campus (JICV). IEEE (2021)
3. Rincy, T.N., Gupta, R.: Ensemble learning techniques and its efficiency in machine learning: a survey. In: 2nd International Conference on Data, Engineering and Applications (IDEA). IEEE (2020)
4. Polikar, R.:Ensemble learning. In: Zhang, C., Ma, Y. (eds.) Ensemble machine Learning. Springer, Boston, MA, pp. 1–34 (2012). https://doi.org/10.1007/978-1-4419-9326-7_1
5. Dalipi, F., Imran, A.S., Kastrati, Z.:MOOC dropout prediction using machine learning techniques: review and research challenges. In: 2018 IEEE Global Engineering Education Conference (EDUCON). IEEE (2018)
6. Edalati, M., Imran, A. S., Kastrati, Z., Daudpota, S. M.: The potential of machine learning algorithms for sentiment classification of students' feedback on MOOC. In: Arai, K. (ed.) IntelliSys 2021. LNNS, vol. 296, pp. 11–22. Springer, Cham (2022). https://doi.org/10.1007/978-3-030-82199-9_2
7. Taha, E.A., Kamal, E.K., Eddine, C.M.: Toward a new framework of recommender memory based system for MOOCs. Int. J. Electr. Comput. Eng. (IJECE) **7**(4), 2152–2160 (2017)
8. Bai, X., et al.: Educational big data: predictions, applications and challenges. Big Data Res. **26**, 100270 (2021)
9. Costa, E.B., et al.: Evaluating the effectiveness of educational data mining techniques for early prediction of students' academic failure in introductory programming courses. Comput. Hum. Behav. **73**, 247–256 (2017)
10. Hew, K.F., et al.: What predicts student satisfaction with MOOCs: a gradient boosting trees supervised machine learning and sentiment analysis approach. Comput. Educ. **145**, 103724 (2020)

11. Ali, H.A., et al.: COVID-19 and distance learning. In: 2022 XII International Conference on Virtual Campus (JICV). IEEE (2022)
12. Ali, H.A., et al.: Prediction MOOC's for student by using machine learning methods. In: 2021 XI International Conference on Virtual Campus (JICV). IEEE (2021)
13. El Habti, F.E., Chrayah, M., Bouzidi, A., Ali, H.A.: Blended learning platform model. In: 2022 XII International Conference on Virtual Campus (JICV), pp. 1–4. IEEE (2022)
14. Ali, H.A., Mohamed, C., Abdelhamid, B.: Challenges of E-learning in Covid 19 pandemic. In: 2022 XII International Conference on Virtual Campus (JICV), pp. 1–3. IEEE (2022)
15. Ali, H.A., et al.: A comparative evaluation use bagging and boosting ensemble classifiers. In: 2022 International Conference on Intelligent Systems and Computer Vision (ISCV). IEEE (2022)

BPMN to UML Transformation for MDA Approach to Represent an EDM Acquisition Functionality

Soufiane Hakkou[1](✉), Redouane Esbai[1], Mohamed Achraf Habri[1], and Lamlili El Mazoui Nadori Yasser[2]

[1] Department of Computer Science, Pluridisciplinary Faculty of Nador, Mohammed the 1st University, Nador, Morocco
soufianehakkou@yahoo.fr

[2] MATSI Laboratory, Mohammed the 1st University, Oujda, Morocco

Abstract. EDM as a system stands for a different digital process that helps manage, track, store documents, and then reduce the use of paper. One of the main functionalities of EDM is the digitization part which is a critical step in the EDM system as it reflects the real file on paper in an electronic file; The user needs to be sure that the real content of his file is identical to the digitized one before saving it or using it in another process. In this paper we propose a novel transformation of the BPMN (Business Process Model and Notation) representation to the UML (Unified modeling language) representation, using QVT (query views transformation) language, in which the user will benefit from both the simplicity and readability of BPMN and the stability and precision of UML. Moreover, this solution applied on an MDA (model-driven architecture) framework will not only save time and cost but also improve the software quality.

Keywords: BPMN · EDM · MDA · QVT · UML

1 Introduction

With the expansive growth of the technology and its use in many industries [1] and companies, electronic document management becomes an obligation, in order to keep up the digitization and reduce the use of paper as it turns into an outdated method to manage or, stock information since it is more difficult to track data in it comparing to digital files that could be indexed, and then searched easily. In this context we thought of using one of the main functionalities of an EDM [2] system which is the digitization of files, to help the user access information easily.

Since it becomes a necessity to find a simple representation that is easy to understand by the different staff of an organization to ensure the best collaboration and develop the productivity inside the enterprise, we had the idea of using a modeling language that is easy to implement, use and understand by the different actors in a company which is the BPMN [3], it is a standard modeling language adopted by the OMG [4] (Object

Y. Farhaoui et al. (Eds.): ICAISE 2022, LNNS 635, pp. 301–306, 2023.
https://doi.org/10.1007/978-3-031-26254-8_42

Management Group) considering its efficiency, simplicity to use since it is based on clear graphical representations.

The OMG created the MDA [5] standard to facilitate the migration of the Information system from an old one with old or outdated technologies to a new one with new technologies without losing data, time, or energy in that migration.

The MDA as architecture is composed of different levels such as the CIM (Computing independent model) and the PIM (Platform independent model) levels, it is recommended to use the UML modeling in the PIM level, it could also be used in the CIM one, but in our case, we are proposing to use the BPMN model since it will simplify the understanding of the first phase to users from different specialties in a company.

2 Related Work

In [7] Farah Deeba et al. proposed a data transformation of UML class diagram by using MDA, the authors used ANDROMDA source code generator following the MDA patterns, on a multi-level marketing, the proposed approach by the authors will save time and effort, but the authors focus on generating the code without taking in consideration the EDM functionalities or including the data part.

In [8] the authors propose a new modeling approach using BPMN and ODM as CIM models in the first step then transform the representation into an Interaction Flow Modeling Language diagram, the authors applied the proposed model-based development process on a case study of e-health. However, the authors focus on the abstract part for modeling taking in consideration the CIM metamodels and the PIM metamodels.

In [9] authors Wiem Khlif et al. proposed a transformation model from BPMN model to aligned UML over an MDA approach, the business process model at the CIM level was modeled with the BPMN and the PIM level was modeled through a UML use case diagram, the transformation proposed on this approach will offer an enterprise efficiency and simplicity of developing a new information system or migrating to old one in order to develop it.

In [10] The authors propose a transformation from BPMN into UML, in order to convert from the business process (BP) to the activity diagram (Ad), this transformation will preserve the semantics meaning of the context applied on, it will also bring many benefits to the business process and the modeling.

3 Background Knowledge

3.1 EDM Overview

EDM standsforElectronicDocumentManagement [12], it refers to the document management process in a company. This process includes different steps (Fig. 1):

1. Document acquisition
2. Document indexing and classifying
3. Documents storage
4. Document access and distribution of documents
5. Document archiving

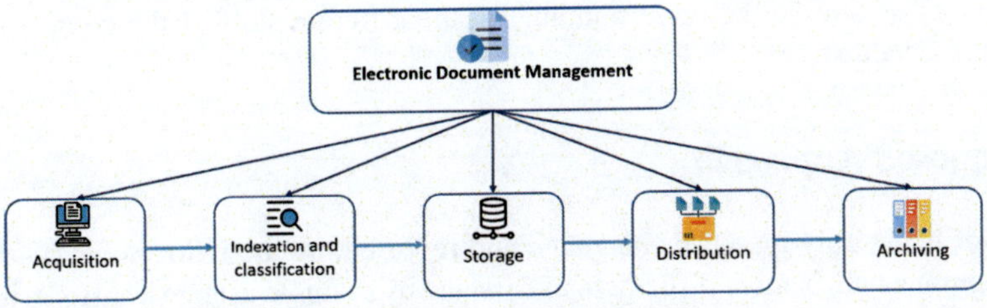

Fig. 1. Main steps in EDM system

3.2 Mda

Model Driven Architecture (MDA) [15] is an approach for software development, adopted by OMG,it provides instructions and guidelines to organize software specifications that are presented as models. The MDA is composed of 3 different models:

CIM: or independent model computation the first level in the MDA framework it helps describe different actors and actions in the system.
PIM: the platform-independent model, it defines the business-oriented model.
PSM: or the platform-specific model, it describes the technical part of the system, to understand and facilitate the implementation of the platform (Fig. 2).

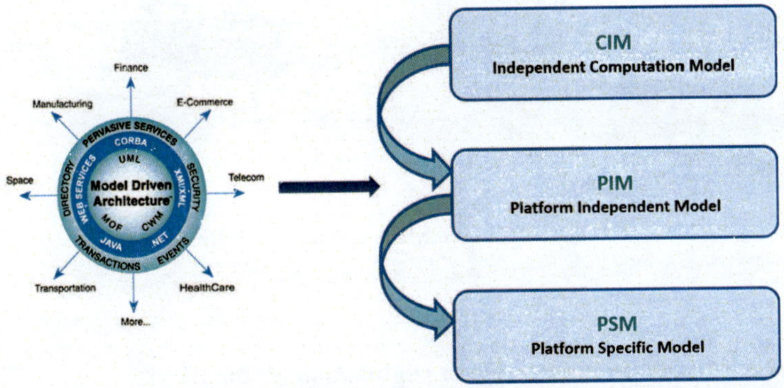

Fig. 2. MDA models

In this paper, we are focusing on the first and the second models (CIM and PIM), to describe the transformation between the two of them, by implementing the BPMN on the CIM level and the UML on the PIM.

3.3 BPMN Standard

BPMN [17] stands for a standard Business Process Model and Notation, it enables the organization to simply understand their internal plan, since it provides graphical notations, these simple the understanding on the different staff of the company from different specialties.

4 Proposed Approach

Instead of using only the UML language and representing the CIM model by the use case diagram which is more difficult to understand by a simple user, to transform it into a PIM defined by the class diagram, we propose a representation by combining 2 different modeling languages, adopted by the OMG, which are the BPMN and the UML [18].

We choose to study the case of a main functionality in the EDM: document acquisition.

First, we represented in a clear and understandable process diagram for the different collaborators using the BPMN language as the source model for the input, after the transformation, we come to our PIM level which is the UML class diagram for the output model.

This transformation will be achieved with the help of the QVT language [19] which had the advantage of being a standard of the OMG (Fig. 3).

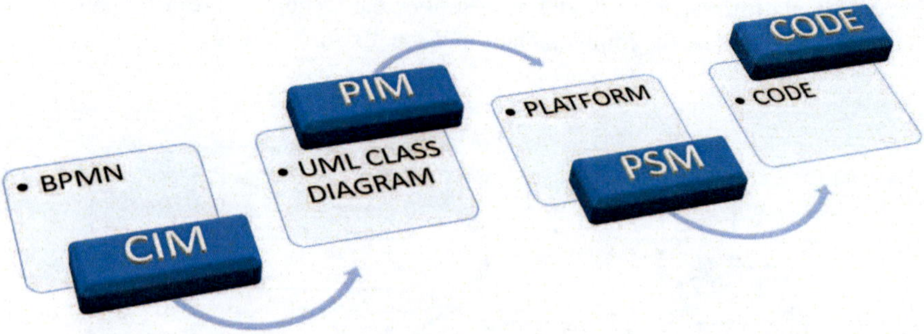

Fig. 3. Proposed MDA approach models

4.1 Use Case Study on the Process of Digitization by the BPMN

Our BPMN diagram shows the collaboration between two main collaborators that affect the digitization mechanism: the user and the executed machine.

In the input, the user that is a member of a group have the right to digitize a paper document, he will authenticate and ask for the digitization of a file, by sending a request to the machine; this last will execute the request and send the result to the user to validate the quality and the integrity of the document.

In the output, if the digitized document doesn't meet the requirements of the user, the action ends and we need to restart the operation from the beginning, if not, the user will send a request asking the machine to save and classify the file in a specific placement as an electronic document (Fig. 4).

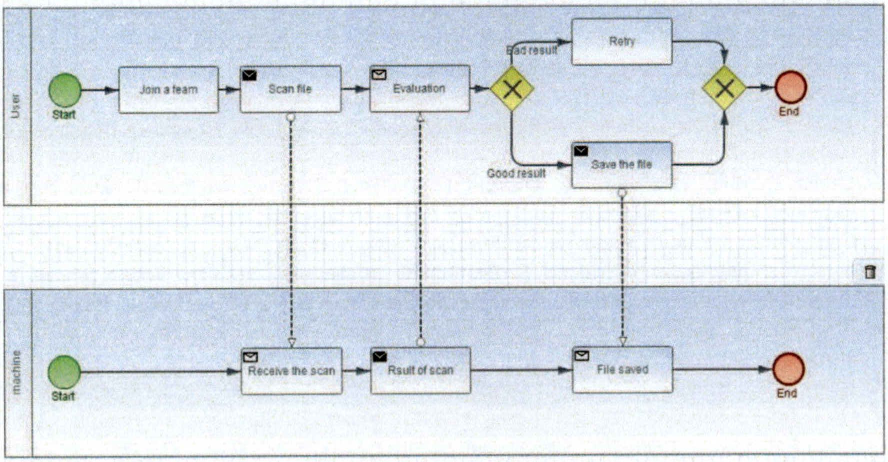

Fig. 4. BPMN diagram of our use case study

5 Conclusion

In this article, we have proposed a new approach for adopting BPMN for the MDA framework to ensure the transformation between the CIM level on which we used the BPMN, and the PIM level on which we adopted the UML class diagram recommended by the MDA framework, studying the first functionality of the EDM system which is document acquisition this proposal will help facilitate the understanding of the software for simple users since the BPMN will enable that by its graphical simple representation and the UML part will be the part for a more experienced user in the field of IT, the work was restricted to the CIM and PIM level and we did not generate code, for our next study we are going to use the QVT to generate the code to automatically allow the transformation between the CIM and the PIM, this will enable us to work on the next level of the MDA framework which is the PSM, ultimately we are going to work on more complicated Functionalities of the EDM other then the document acquisition part.

References

1. Guo, F., Jahren, C.T., Turkan, Y.: Electronic document management systems for the transportation construction industry. Int. J. Constr. Educ. Res. **17**, 52–67 (2021)
2. Kurganova, M.V.: Enterprise knowledge base based on EDM system. In: Ashmarina, S.I., Horák, J., Vrbka, J., Šuleř, P. (eds.) IES 2020. LNNS, vol. 160, pp. 179–188. Springer, Cham (2021). https://doi.org/10.1007/978-3-030-60929-0_24
3. Chinosi, M., Trombetta, A.: BPMN: an introduction to the standard. Comput. Stan. Interfaces **34**, 124–134 (2011)
4. Colomb, R., et al.: The object management group ontology definition metamodel. In: Calero C., Ruiz F., Piattini, M. (eds.) Ontologies for Software Engineering and Software Technology, pp. 217–247. Springer, Heidelberg (2006). https://doi.org/10.1007/3-540-34518-3_8

5. Bucchiarone, A., Cabot, J., Paige, R.F., Pierantonio, A.: Grand challenges in model-driven engineering: an analysis of the state of the research. Softw. Syst. Model. **19**(1), 5–13 (2020). https://doi.org/10.1007/s10270-019-00773-6
6. Petrasch, R., Kongkachandra, R.: Towards enterprise data management: data model mapping. In: IEEE International WIE Conference on Electrical and Computer Engineering (WIECONECE), pp. 72–74 (2018)
7. Deeba, F., Kun, S., Shaikh, M., Dharejo, F.A., Hayat, S., Suwansrikham, P.: Data transformation of UML diagram by using model driven architecture. In: IEEE 3rd International Conference on Cloud Computing and Big Data Analysis (ICCCBDA), pp. 300–303 (2018)
8. Laaz, N., Kharmoum, N., Mbarki, S.: Combining domain ontologies and BPMN models at the CIM Level to generate IFML models. Procedia Comput. Sci. **170**, 851–856 (2020)
9. Khlif, W., Elleuch Ben Ayed, N., Ben-Abdallah, H.: From a BPMN model to an aligned UML analysis model, pp. 623–631 (2018)
10. Habri, M.A., Esbai, R., Lamlili El Mazoui Nadori, Y.: BPMN to UML class diagram using QVT. In: Ben Ahmed, M., Teodorescu, H.-N.L., Mazri, T., Subashini, P., Boudhir, A.A. (eds.) Networking, Intelligent Systems and Security. SIST, vol. 237, pp. 593–602. Springer, Singapore (2022). https://doi.org/10.1007/978-981-16-3637-0_42
11. Rhazali, Y., Hadi, Y., Mouloudi, A.: Third World Conference on Complex Systems (WCCS) (2015)
12. Keating, P., Murray, J., Schenkel, K., Merson, L., Seale, A.: Electronic data collection, management and analysis tools used for outbreak response in low- and middle-income countries: a systematic review and stakeholder survey. London School of Hygiene and Tropical Medicine, London, United Kingdom (2021)
13. Boujrad, M., Lamlili El Mazoui Nadori, Y.: A new artificial intelligence-based strategy for digital marketing reinforcement. In: Ben Ahmed, M., Rakıp Karaş, İ., Santos, D., Sergeyeva, O., Boudhir, A.A. (eds) SCA 2020. LNNS, vol. 183, pp. 689-699. Springer, Cham (2021). https://doi.org/10.1007/978-3-030-66840-2_52
14. Anh, P.V., Khanh, N.D.T., Dat, T.M., Van Dan, P.: Improved OCR quality for smart scanned document management system. J. Sci. Tech.-Le Quy Don Tech. Univ. **210**, 9 (2020)
15. https://www.mda.org/
16. Montalvo, A., Parra, P., Rodriguez Polo, O., et al.: Model-driven system-level validation and verification on the space software domain. Softw. Syst. Model, 1–28 (2021)
17. https://www.bpmn.org/
18. Sbai, R., Arrassen, I., Meziane, A., Erramdani, M.: QVT transformation by modeling. (IJACSA) Int. J. Adv. Comput. Sci. Appl. **2(5)** (2011)
19. Esbai, R., Elotmani, F., Belkadi, F.: Model-driven transformations: toward automatic generation of column-oriented nosql databases in big data context. Int. J. Online Biomed. Eng. (iJOE), **15**(09) (2019)

Business Intelligence Models for E-Government in Mauritania: A Survey

Mohamed El Moustapha El Arby Chrif[1], Moustapha Mohamed Saleck[2](✉), Aichetou Cheikh Mohamedou N'Diaye[1], and El Benany Mohamed Mahmoud[1]

[1] Department of Mathematics and Computer Science, University of Nouakchott, Nouakchott, Mauritania

[2] LAROSERI Laboratory, Computer Sciences Department, Faculty of Sciences, University of Chouaib Doukkali, El Jadida, Morocco

saleck.moustapha@gmail.com

Abstract. Today, In Mauritania social media such as Facebook and Tweeter are widely used not only for family or friendly communication but also for sharing, advertising, especially used in the field of social policy, therefore a large amount of information produced and reflects the intentions of these users. This data must be used in an optimal way in order to help decision-makers when making decisions. The aim of this survey is to compare the existing methods of analyzing data flows in real time applying to social networks. We therefore seek to compare the strategic BI models of real-time social media data analysis applied to government information in Mauritania. This model will allow to give on the one hand a clear and timely vision, by grouping the topical subjects by their importance according to the audiences of the public, also indicating their feelings towards the publications by classifying them according to their polarity. On the other hand, only relevant data can be saved in the data warehouse which will then be searchable by government sectors with the possibility of accessing only the data that concerns it, which will lead to the speed of BI analysis process.

Keywords: Business intelligence · Big data · Sentiment analysis · Social media · e-Gov

1 Introduction

Due to the technological evolution today, especially in the field of information technology and in a context where information sources are fragmented, voluminous and complex, there is a real need to consolidate and analyze them in order to have a global vision and to exploit the company's information assets. However, "too much information kills information"… The objective of BI is to help managers in their decision making and in the analysis of the performance of their company.

The BI process essentially revolves around the ability to collect a large amount of similar data in a single "unified data reference" and then in turn load the restructured data streams that will impact the quality of decision [1] to data stores and then display them to end users.

Y. Farhaoui et al. (Eds.): ICAISE 2022, LNNS 635, pp. 307–312, 2023.
https://doi.org/10.1007/978-3-031-26254-8_43

The objective of this work is to study the feasibility of real time in BI and to address this issue from a performance, cost and maintainability point of view. The goal is to create a business intelligence dashboard to observe the performance of government news topics or channels posted on social media accounts such as Facebook and Twitter [2]. Our BI model of real-time supervision will be useful in informing current policy discussions in many areas of government service delivery in Mauritania. This brief introduction is followed by a section that presents an overview of the basic business intelligence literature, a second section that discusses the state of the art of the topic at hand and a conclusion at the end.

2 Overview of the Literature

The emergence of the data warehouse as a repository, advances in data cleansing, and increased hardware and software capabilities are combining to create a richer business intelligence environment than ever before [3].

2.1 Business Intelligence

According to [4] Business intelligence is a set of operational decision support systems based on an integrated database to facilitate access to information. Business Intelligence as a sub-type of sentiment analysis has evolved with the advancement of technology on the Web, from Web 1.0, 2.0 to 3.0 [5].

2.2 Data Warehouse

Data Warehouse as its name indicates a data warehouse allows to store historical information of the company in a structured non-volatile way and classified by subject.

According to [7] A data warehouse is the main source of information that feeds the analytical processing within an organization. It is the target location for the integration of data from multiple sources, both internal and external to an organization.

2.3 Datamart

According to [7] DataMart is a subject-oriented data reference, similar in structure to the data warehouse, but it contains the data needed for decision support for a specific department or group within an organization.

Bill Inmon finds that a Data Mart is a stream of data from the Data Warehouse that brings together specialized data, gathered for a particular business. And according to Ralph Kimball, the DataMart that make up the Data Warehouse, Kimball defines the DataMart as a subset of the Data Warehouse made up of detailed, interrelated tables [8].

2.4 ETL

ETL is the acronym for "Extract Transform Load" ETLs are software programs that allow for the transformation of raw data extracted from multiple sources into a format adopted by a data warehouse [9, 10]. It consists on [4] extracting, transforming and loading data from heterogeneous sources. It also allows to:

- discover, analyze and extract data from heterogeneous resources
- clean and standardize data
- load data into a Data warehouse or Database

2.5 Dashboard

Dashboard is a dashboard that gathers several visualizations, in other words, it is an interface designed for managers to show them the indicative state of each element in the system to which they apply through a simple and understandable illustration. Dashboard Presents output of a BI system.

2.6 e-Government

Definition. According to [11] E-Government is the use of technology to enhance the access to and delivery of government service to benefit citizens, business partner and employees. E-Government is a broad and vague concept. Therefore, there is no universally accepted definition, but it can simply be defined as the use of the World Wide Web and other information and telecommunications technologies to provide government information and services to citizens [4].

3 Related Works

Social media are used as quality input data in business intelligence research [12]; Therefore, it appears useful to take it into account during the process of data analysis that can be defined as a subfield of static data description concerns on which is based the decision-making. In the following, with the methods of this analysis process we try to give the links that can exist between the different data and to draw the main information contained in these data.

3.1 Social Media Data Analysis

Today, social media are widely used. Indeed, they produce voluminous data that constitute a large part of Big Data. However, social media analytics is about collecting and analyzing this data with analytical tools to make decisions.

Existing Studies

Approach [13], this article contributes to the information systems literature by presenting a summary of the main challenges and difficulties faced by researchers in the stages of the social media analysis research process that precede data analysis: discovery, collection and preparation. As a second contribution to the literature, they also point researchers to possible solutions to these challenges. To be able to classify these, they rely instead on theory from the big data literature. In particular, they use the four Vs: volume, velocity, variety, and veracity.

Approach [14], in this paper, they proposed a deep learning approach to identify emotions expressed through Hindi-English mixed language in various social media platforms such as Twitter and Facebook. To implement the detection model, they collected and cleaned 12,000 Hindi-English mixed sentences from different sources that contain emotions such as happy, sad, and angry.

3.2 Sentiment Analysis

Sentiment analysis refers to the exploration of opinions. It is sometimes characterized as a subcategory of text mining [5] whose goal is to determine the inclination or attitude of a communicator through the contextual polarity of their writing or speaking.

Existing Studies

Approach [2] sought to analyze social media content using several data mining methods to shed light on the information for decision makers. Indeed, have selected the best among the three algorithms namely Naïve Bays, decision Tree and SVM to classify the social media texts which will be very useful information for the implementation of a data warehouse.

Approach [15] proposed a methodology that combines two analysis methods, machine learning and lexicon-based approach, to provide a better understanding of the application of sentiment analysis in social media and improve the result compared to using a single approach. The machine learning approach based on algorithms to extract and detect the sentiment of a data, while the lexicon-based approach works by counting the positive and negative words related to the data.

Approach [10] Proposed a contribution presented by an improvement of the traditional sentiment lexicon-based approach; the said approach takes into consideration the intensity of sentiments and punctuation, it can detect emojies and argost. This study contributes to the evolution of research on digital interactions between government and citizens.

3.3 Synthesis

Two approaches have been discussed in the section one to describe the difficulties encountered when analyzing social media data, which today represents an important source of information.

First, the approach [13] describes the major challenges and problems encountered in the research phase that precedes the data analysis phase. It proposes as a solution the combination of computer science techniques and social science theories.

Secondly, the approach [14] proposes a detection model based on deep learning to identify the emotions expressed in mixed Hindi-English language. According to the author, the most successful and tested classification model with a classification accuracy of 83.21% is CNN-BiLSTM.

On section two, extracting useful information from social media and loading it into the data warehouse is the main objective of the latter approaches. The first approach orders Naïve Bays, Decision Tree and SVM algorithms according to their performance in text classification and then selected the best among the three for social media content analysis, the results given by this study are shown next (Table 1):

Table 1. Algorithms according to their performance in text classification [2].

Algorithms	Accuracy
SVM	78.99%
Naïve Bayes	74.67%
Decision Tree	55.66%

In the second approach, the researchers argue that the two methods of analysis machine learning and lexicon-based have very similar performance in terms of accuracy, thus recommended to combine these two methods as they complement each other and the results improve compared to the use of one alone.

The third approach is all simple an improvement of the lexicon-based approach, which takes into account the intentions of feelings and punctuation.

4 Conclusion

According to our research we have found that most of the studies that have been done in the field of business intelligence are generally based on the analysis of historical data; In other words the BI batch, however a great amount of useless flows mixed with relevant information this set is stored in the data warehouse, on the one hand naturally every information whatever is surrounded by others that in the majority of cases help to null of the all, on the other hand the storage of these data before being analysed has also many negative aspects as for example the increase of the stored quantities and this on its side can delay even damage the process of the analysis.

From our point of view and according to the situation indicated above and the place occupied by BI today in the business market worldwide and in Mauritania precisely. We seek to raise the problem caused by the classical BI batch processing method by studying the feasibility of real-time data analysis in order to optimize resources and save time. Our objective is to study an approach to real-time sentiment analysis applied to e-Gov.

References

1. Wieder, B., Ossimitz, M.L.: The impact of business intelligence on the quality of decision making - a mediation model. Procedia Comput. Sci. **64**, 1163–1171 (2015). https://doi.org/10.1016/j.procs.2015.08.599

2. Kurnia, P.F., Suharjito.: Business intelligence model to analyze social media information. Procedia Comput. Sci. **135**, 5–14 (2018). https://doi.org/10.1016/j.procs.2018.08.144
3. Negash, S.: Business intelligence. Commun. Assoc. Inf. Syst. **13**(Feburary 2004) (2004).https://doi.org/10.17705/1CAIS.01315
4. Oumkaltoum, B., Mohamed Mahmoud, E.B., Omar, E.B.: Toward a business intelligence model for challenges of interoperability in egov system: transparency, scalability and genericity. In: 2019 International Conference on Wireless Technologies, Embedded and Intelligent Systems (WITS), pp. 1–6 (2019). https://doi.org/10.1109/WITS.2019.8723756
5. Al-Kharusi, M.I., Usman, A.I., Awwalu, J.: Application of sentiment analysis in business intelligence. Int. J. Knowl. Innov. Enterepreneurship **3**(3), 51–60 (2015)
6. Rokade, P.P., Aruna, K.D.: Business intelligence analytics using sentiment analysis-a survey. Int. J. Electr. Comput. Eng. **9**(1), 613–620 (2019). https://doi.org/10.11591/ijece.v9i1.pp613-620
7. Loshin, D.: Data warehouses, online analytical processing, and metadata. Bus. Intell. **1**, 75–90 (2003). https://doi.org/10.1016/b978-155860916-7/50007-5
8. Ben Rejab, F.: Chapitre 2 data warehouse conventional (2021)
9. N'Diaye, A.C.M., Chrif, M.E.M.E.A., El Mahmoud, B.M., El Beqqali, O.: Apply sentiment analysis technology in social Media as a tool to enhance the effectiveness of e-government : application on Arabic and Mauritanian dialect 'HASSANIYA.' In: 2021 Fifth International Conference On Intelligent Computing in Data Sciences (ICDS), pp. 1–5 (2021). https://doi.org/10.1109/ICDS53782.2021.9626766
10. Georgiadou, E., Angelopoulos, S., Drake, H.: Big data analytics and international negotiations: sentiment analysis of Brexit negotiating outcomes. Int. J. Inf. Manag. 51(December 2019), 102048 (2020). https://doi.org/10.1016/j.ijinfomgt.2019.102048
11. Novakouski, M., Lewis, G.A.: Interoperability in the e-Government Context. no. January (2012)
12. Choi, J., Yoon, J., Chung, J., Coh, B.-Y., Lee, J.-M.: Social media analytics and business intelligence research: a systematic review. Inf. Process. Manag. **57**(6), 102279 (2020). https://doi.org/10.1016/j.ipm.2020.102279
13. Stieglitz, S., Mirbabaie, M., Ross, B., Neuberger, C.: Social media analytics – challenges in topic discovery, data collection, and data preparation. Int. J. Inf. Manage. **39**, 156–168 (2018). https://doi.org/10.1016/j.ijinfomgt.2017.12.002
14. Sasidhar, T.T., Premjith, B., Soman, K.P.: Emotion detection in Hinglish (Hindi+English) code-mixed social media text. Procedia Comput. Sci. **171**(2019), 1346–1352 (2020). https://doi.org/10.1016/j.procs.2020.04.144
15. Drus, Z., Khalid, H.: Sentiment analysis in social media and its application: systematic literature review. Procedia Comput. Sci. **161**, 707–714 (2019). https://doi.org/10.1016/j.procs.2019.11.174

CF Recommender System Based on Ontology and Nonnegative Matrix Factorization (NMF)

Sajida Mhammedi[1](✉), Hakim El Massari[1], Noreddine Gherabi[1], and Mohamed Amnai[2]

[1] Lasti Laboratory, National School of Applied Sciences, Sultan Moulay Slimane University, Khouribga, Morocco
sajida.mhammedi@usms.ac.ma
[2] LARIT Laboratory, Faculty of Sciences, IbnTofail University, Kenitra, Morocco

Abstract. Recommender systems are a kind of data filtering that guides the user to interesting and valuable resources within an extensive dataset. by providing suggestions of products that are expected to match their preferences. However, due to data overloading, recommender systems struggle to handle large volumes of data reliably and accurately before offering suggestions. The main purpose of this work is to address the recommender system's data sparsity and accuracy problems by using the matrix factorization algorithm of collaborative filtering based on the dimensional reduction method and, more precisely, the Nonnegative Matrix Factorization (NMF) combined with ontology. We tested the method and compared the results to other classic methods. The findings showed that the implemented approach efficiently reduces the sparsity of CF suggestions, improves their accuracy, and gives more relevant items as recommendations.

Keywords: CF · Matrix factoring · NMF · Ontology · Sparsity · Accuracy

1 Introduction

A recommender system is a fully automated system that analyzes user preferences and predicts user behavior. The research interest in this field is still very high, mainly because of the practical significance of the problem. Among the approaches of recommender systems, we distinguish the collaborative filtering approach that relies on the users' ratings to find the most similar ones. CF is believed to be the most effective strategy for dealing with the problem of overloaded data in the context of e-commerce. However, collaborative filtering-based recommender systems typically encounter issues such as data sparsity, scalability, prediction inaccuracy, and recommendation accuracy [1, 2], for instance, compares recommendation methodologies, whereas [3, 4] classifies recommender systems that use AI algorithms. Furthermore, [4, 5] categorizes the techniques based on suggestion factors. To increase the accuracy of CF recommendations, the researchers construct a semantic-level information model based on an ontological notion to address the issues mentioned above [6, 7]. Furthermore, ontology recommender systems are classified according to the tool, ontology type, and ontology representative language in

Y. Farhaoui et al. (Eds.): ICAISE 2022, LNNS 635, pp. 313–318, 2023.
https://doi.org/10.1007/978-3-031-26254-8_44

[3, 4, 8]. Ontologies have been widely used in various fields and contexts because they are incredibly valuable constructs [9, 10]. The work in [11] created a hybrid RS with better semantics by combining the reasoning of ontology-based semantic similarity with conventional elements-based CF. To solve the difficulties of cold start and data sparsity, authors in [12] developed a method that includes the item's semantic domain knowledge as well as the user's social trust network. To enhance the accuracy of a library book recommender, Liao [13] used the Chinese library categorization system as a reference. Ranjbar [14] constructed ANFIS, a multi-standard recommender system, by merging semantic information of items based on ontology with user population statistics. This indicated that incorporating semantic information may increase the prediction accuracy of a multi-standard RS. There are a variety of extensive analyses due to the increased research focus on ontology for recommendation systems. For our research proposal, we are interested in hybrid recommender systems, which are slightly more advanced forms of traditional RS but based on CF. Two problems influenced by RS arise the sparsity of the user-element scoring matrix and the lack of recommendation accuracy when the data is significant.

The objective of our work is to remedy these two problems. For this reason, we use the matrix factorization algorithm using one of its methods: the Nonnegative Matrix Factorization (NMF), which is one of the methods used to accelerate the search for content recommendations for users. We also use the conceptualization of items based on ontology. This paper begins by discussing research that is pertinent to the current study. Then moves on to present the background related to our work. The proposed system design is then detailed in-depth. Experiment findings and their evaluation are then presented. Finally, a conclusion was conducted.

2 New Hybrid CF Recommender Approach

2.1 System Architecture

The suggested system of the CF recommendation algorithm based on semantic modeling of items (ontology) and dimensionality reduction approach using NMF is depicted in Fig. 1. In the beginning, the tree of the hierarchical structure of items was built using ontology conception. The semantic similarity between two concepts was computed utilizing the overlapping connections of the concepts' words. The user-item evaluation matrix was then filled in, and certain missing values in the sparse matrix were estimated using semantic similarity. Furthermore, the semantic similarity threshold was adjusted to improve the content of the matrix data while preserving the source matrix's fundamental attribute characteristics.

2.2 Similarity Calculation Process

Conception of Ontology and Concept Semantic Similarity

We should mention that ontologies are employed in various domains and settings since they are extremely useful constructs. As a result, their role differs depending on the

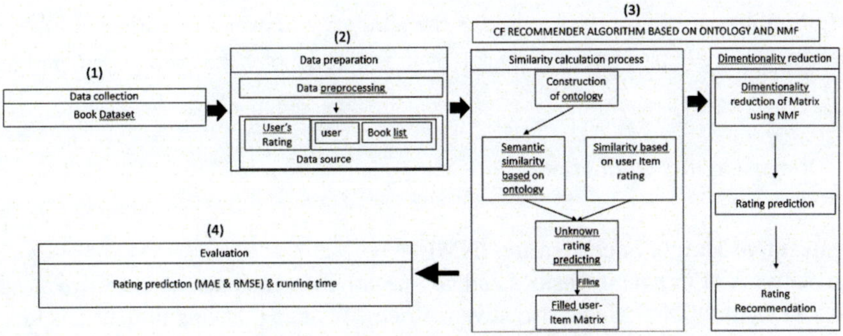

Fig. 1. Hybrid System Architecture.

framework. This section is inspired by the work done in [15], which suggests an automatic construction of large ontology from a JSON file, according to its domain of use. To use the stated system, we'll use retool utilities to convert our CSV file to a JSON file (https://retool.com/utilities/convert-csv-to-json). We believe that a vocabulary's concepts may be structured into a hierarchical tree, with the most specific terms linked to the more general ones via a parent-child relationship.

The notion of distance between items is based on the similarity of their meaning or semantic content, and semantic similarity is a measure established on a set of documents or words. It's a mathematical method for estimating the strength of a semantic link between language units, concepts, or instances using a numerical description generated by comparing evidence to support or describe their nature. The similarity between two concepts, C1 and C2, is calculated using the ontology's hierarchical structure. We assume the following two assumptions:

- The more similar ancestors any two family members share, the closer they are.
- A family element's distance from a common ancestor affects its distance from other family elements.

The number of common ancestors to the genealogy (i.e., the set of concepts from the root to the concept) of two concepts C1 and C2, is the ratio of their similarities [16]. Its definition is as follows:

$$\text{SemSimilarity}\left(\text{Cl}_i, \text{Cl}_j\right) = \frac{2\ \text{deep}\left(\text{nearP}\left(\text{Cl}_i, \text{Cl}_j\right)\right)}{\left[\text{L}\left(\text{Cl}_i,\ \text{nearP}\left(\text{Cl}_i, \text{Cl}_j\right)\right) + \text{L}\left(\text{Cl}_j,\ \text{nearP}\left(\text{Cl}_i, \text{Cl}_j\right)\right) + 2\text{deep}\left(\text{nearP}\left(\text{Cl}_i, \text{Cl}_j\right)\right)\right]}$$

Predicting Unknown Rating

To predict rating, we apply the following equation:

$$\left(\frac{P(U_u}{R_{U_u}}, b\right) = \frac{\sum_{v\in N(U_u)} \text{SimSimilarity}\ (U_u, U_v)\times\left(R_{U_v}(b) - R_{U_v}\right)}{\sum_{v\in N(U_u)} \text{SimSimilarity}\ (U_u, U_v)}$$

where SimSimilarity (U_u, U_v) indicates the similarity between two users. $\left(\frac{P(U_u}{R_{U_u}}, b\right)$ represents the prediction rating of book b by user u, R denotes the average rating of user u and user v.

2.3 Dimensionality Reduction

Non-negative Matrix Factorization (NMF)
For a matrix A of m x n dimensions and non-negative coefficients, i.e., positive or null, NMF can factor it into two non-negative matrices, W and H, having dimensions m x k, k x n respectively. Here, the matrix A is defined as:

$$A_{mx\,n} = W_{mx\,k}H_{kx\,n}$$

where, A is the original Input Matrix (Linear combination of W & H), W is the Matrix of Features, H refers to coefficient Matrix (Weights associated with W), and k indicates low rank approximation of A (k ≤ min(m, n)) The goal of NMF is dimensionality reduction and feature extraction. Thus, when we define the lower dimension as k, the goal of NMF is to find two matrices W R^{mk} and H R^{nk} having only nonnegative elements. Therefore, using NMF makes it possible to obtain factorized matrices with dimensions significantly smaller than the product matrix. Intuitively, NMF assumes that the original input consists of a set of hidden features represented by each column of the matrix W. Each column of the matrix H represents the "coordinates of a data point" in the matrix W. In simple terms, it contains the weights associated with the matrix W. In this context, each data point represented as a column in A can be approximated by an additive combination of non-negative vectors, defined as columns in W. This technique has since been widely used in many fields with the aim to study the structure of huge sparse matrices.

3 Experimental Evaluations Results

3.1 Dataset

The BookCrossing (taken from the BookCrossing.com book RS) is an online free book club. This dataset is based on notes rather than tags. Indeed, users are encouraged to evaluate the books they share by assigning a number between 1 and 10, with the higher the score, the better the book's worth. The purchase of books is solely perceived as a sign of contentment, referred to as implicit voting. The dataset we employed for our analysis is openly available and comprises three tables of data: 278858 users who produced 1149780 ratings on 271379 books. We performed a preprocessing phase before utilizing the Bookcrossing dataset, which included the following steps:

- Removing implicit (zero) ratings is the first step. This experiment exclusively looks at explicit evaluations because implicit ratings tend to add noise to the acquired data.
- Eliminate users who have never rated books. Unrated books will never be rated since the suggested system only offers ratings to books that neighbor have rated. And we remove the picture URL field, in the book's data.

3.2 Implementation and Evaluation

To measure the accuracy of these predictions, the estimated interactions are compared with the actual interactions, i.e., those created by the user. The suggested Hybrid recommender system is evaluated using two precision criteria: MAE and RMSE. Both of these metrics allow for easy interpretation because they are on the same scale as the original scores. We employed a 5-fold cross-validation to calculate the different criteria for each testing subgroup, splitting rating data into five parts and using one as the testing set and the other as the training set at each iteration. Finally, we compared our innovative hybrid strategy's with CF, CB, and CF + NMF approaches. Figure 2. Ontology-based conceptualizing and matrix factorization have significantly influenced the development of CF recommendation systems. Furthermore, matrix factorization decreased the size of the matrix and led to faster response time, but it also increased the suggestion performance.

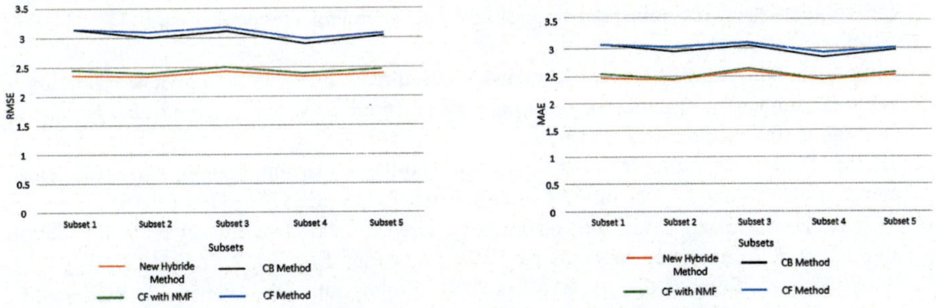

Fig. 2. The performances in terms of RMSE and MAE.

4 Discussion

Sparsity and scalability are two significant problems when creating RS. As a result, in this work, efforts were made to solve these challenges to enhance the performance of a RS using dimensionality reduction techniques and ontology in the CF. The method is put to the test using real-world data. According to MAE and RMSE measurements, the innovative hybrid CF recommender method effectively improved the performance of CF recommender system. Our findings indicated that hybrid strategies might be used to address the sparsity and scalability issues that face RS. We also found that the technique surpasses previous methods that rely purely on CF, CB, or CF + NMF. Because the recommended technique employs semantic similarity relations for the item in the item-based CF, the suggested approach improves the accuracy of the recommendations. As an out-come, using semantic similarity improves the accuracy of the hybrid approach's suggestions.

5 Conclusion and Future Work

In our work, we have implemented a book RS based on ontology integration and matrix factorization, using (NMF) algorithm. The proposed approach improves the CF system's data sparsity and predictive accuracy. Results proved that ontology-based conceptualization and matrix factorization played an essential role in the RS based on CF. Moreover, the matrix factorization reduced the matrix size, which improved recommendation performance. In the future, we plan to integrate more ML approaches in an attempt to get better results and to improve the performance of recommendations even further.

References

1. He, C., Parra, D., Verbert, K.: Interactive recommender systems: a survey of the state of the art and future research challenges and opportunities. Expert Syst. Appl. **56**, 9–27 (2016)
2. Bhareti, K., Perera, S., Jamal, S., Pallege, MH., Akash, V., Wiieweera, S.: A literature review of recommendation systems. In: 2020 IEEE International Conference for Innovation in Technology (INOCON), pp 1–7 (2020)
3. Javed, U., Shaukat, K., Hameed, I., Iqbal, F., Alam, T.M., Luo, S.: A review of content-based and context-based recommendation systems. Int. J. Emerg. Technol. Learn. **16**, 274–906 (2021)
4. Tarus, J.K., Niu, Z., Mustafa, G.: Knowledge-based recommendation: a review of ontology-based recommender systems for e-learning. Artif. Intell. Rev. **50**(1), 21–48 (2017). https://doi.org/10.1007/s10462-017-9539-5
5. Truong, H.M.: Integrating learning styles and adaptive e-learning system: current developments, problems and opportunities. Comput. Hum. Behav. **55**, 1185–1193 (2016)
6. Mohana, H., Suriakala, D.: A study on ontology based collaborative filtering recommendation algorithms in E-commerce applications. IOSR J. Comput. Eng. **19**, 2278–2661 (2017)
7. Bassiliades, N., Symeonidis, M., Meditskos, G., Kontopoulos, E., Gouvas, P., Vlahavas, I.: A semantic recommendation algorithm for the PaaSport platform-as-a-service marketplace. Expert Syst. Appl. **67**, 203–227 (2017)
8. George, G., Lal, A.M.: Review of ontology-based recommender systems in e-learning. Comput Educ **142**, 103642 (2019)
9. El Massari, H., Gherabi, N., Mhammedi, S., Ghandi, H., Qanouni, F., Bahaj, M.: Integration of ontology with machine learning to predict the presence of covid-19 based on symptoms. Bull. Electr. Eng. Inform. **11**, 2805–2816 (2022)
10. El Massari, H., Gherabi, N., Mhammedi, S., Sabouri, Z., Ghandi, H.: Ontology-based decision tree model for prediction of cardiovascular disease. Indian J. Comput. Sci. Eng. **13**, 851–859 (2022)
11. Al-Hassan, M., Lu, H., Lu, J.: A semantic enhanced hybrid recommendation approach: a case study of e-Government tourism service recommendation system. Decis. Support Syst. **72**, 97–109 (2015)
12. Shambour, Q., Lu, J.: A trust-semantic fusion-based recommendation approach for e-business applications. Decis. Support Syst. **54**, 768–780 (2012)
13. Liao, S., Kao, K., Liao, I., Chen, H., Huang, S.: PORE: a personal ontology recommender system for digital libraries. Electron. Libr. **27**, 496–508 (2009)
14. Ranjbar Kermany, N., Alizadeh, S.H.: A hybrid multi-criteria recommender system using ontology and neuro-fuzzy techniques. Electron. Commer. Res. Appl. **21**, 50–64 (2017)
15. Mhammedi, S., El Massari, H., Gherabi, N.: Composition of Large Modular Ontologies Based on Structure. In: Maleh, Y., Alazab, M., Gherabi, N., Tawalbeh, L., Abd El-Latif, A.A. (eds.) ICI2C 2021. LNNS, vol. 357, pp. 144–154. Springer, Cham (2022). https://doi.org/10.1007/978-3-030-91738-8_14

Circuit Analysis of Series and Shunt Rectifier Topologies for RF Energy Harvesting Applications at 5.80 GHz

Salah Ihlou[1(✉)], Ahmed El Abbassi[1], Abdelmajid El Bakkali[1], and Hafid Tizyi[2]

[1] Department of Physics the Faculty of Science and Technology, Moulay Ismail University Errachidia, Errachidia, Morocco
ihlou.salah@gmail.com

[2] STRS Laboratory, Rabat, Morocco

Abstract. A study of RF-DC converter circuit topologies built for wireless energy harvesting at a frequency of 5.80 GHz is presented in this work, which includes experimental results. A study of two rectification circuits is being carried out at the conversion-power level, including research, development, simulation, and comparison. In most cases, the rectifier circuit consists of an association of one or more Schottky diodes mounted in series or in parallel. When comparing the series and shunt topologies to the parallel topologies at 5.80 GHz, the power efficiency of the series and shunt topologies is 70% and 74%, respectively. Detailed investigations have been conducted into each of these topologies, as well as evaluations of their overall performance. The results of the rectifier circuit simulation must be received using the Advanced Design System-ADS Software. The incident power has been set for the tests at 10 dBm. The shunt topology has demonstrated good results in terms of RF-DC con-version efficiency.

Keywords: Microstrip low pass filter · Rectifier circuit · Rectenna · Schottky diode

1 Introduction

Low-power current wireless communication systems, battery life and miniaturization are the two most important design considerations. There is a substantial investigation into the feasibility of recycling ambient electromagnetic energy[1], particularly in heavily populated urban areas.

Therefore, low-power, low-voltage (Low-V) circuits, small energy-consumption footprints, low-cost manufacturing, and high integration power have all received considerable attention [2, 3]. Reduced power dissipation results in a lower battery size, which allows it to endure for a longer period.

A rectenna is an important component of a wireless energy transmission system.

Three major components of a rectenna design are an antenna, a matching network, and a rectifier circuit. A typical block design is seen in Fig. 1. In this study, the topologies of series and shunt rectifier circuits are compared and contrasted for comparison. The

Y. Farhaoui et al. (Eds.): ICAISE 2022, LNNS 635, pp. 319–324, 2023.
https://doi.org/10.1007/978-3-031-26254-8_45

efficiency of the rectifier's RF-to-DC conversion is improved by the use of a harmonic balancing (HB) simulation.

1.1 Rectenna Design

This is a Microstrip patch antenna that was printed on FR4 substrate with permittivity εr = of 4.3 and thickness (h) of 1.6 mm. There are two tapered microstrip lines, one Schottky diode, a matching microstrip line, and an output low pass filter in this rectifier.

A singu-lar frequency voltage source was used as the rectifier's primary input source.

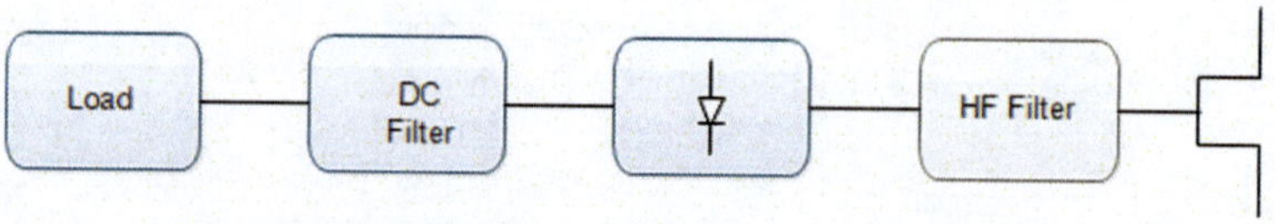

Fig. 1. General block diagram of rectenna.

The accuracy of the rectenna estimate is governed by the quantity of output voltage produced by the antenna and the efficiency with which it performs RF-DC conversion [2, 4]. The efficiency of RF-to-DC conversion is calculated as follows:

$$\eta = \mathrm{PDC\,out}/\mathrm{PRF\,int} \tag{1}$$

1.2 Rectifying Circuit Background

1.2.1 Series mounted diode

The first circuit built that uses an HSMS 2850 Schottky diode in a series topology configuration (Fig. 2). This allows the recovery of the DC element in the circuit.

The low pass filter is made up of an inductor in series and a capacitor in parallel.

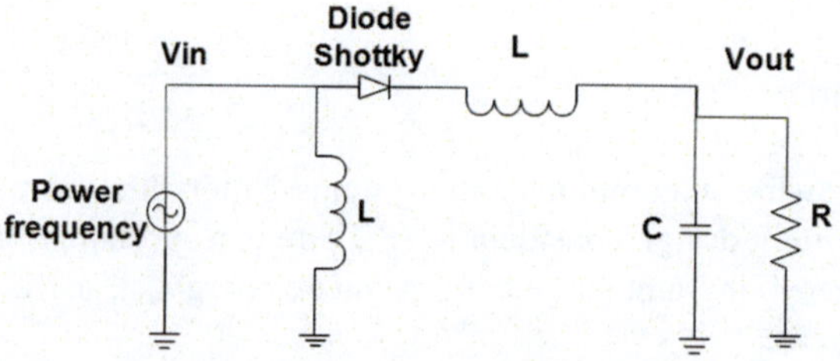

Fig. 2. Single series mounted diode.

Figure 3 shows the power levels before and after the rectifier circuits. The power is transferred to DC at the output, as can be observed.

Figure 4 depicts the variation in efficiency as a function of the incident RF power level.

At 5.80 GHz and with an incidence power of 8 dBm, we can see that the rectification circuit has the highest efficiency possible, with 72% at the highest frequency.

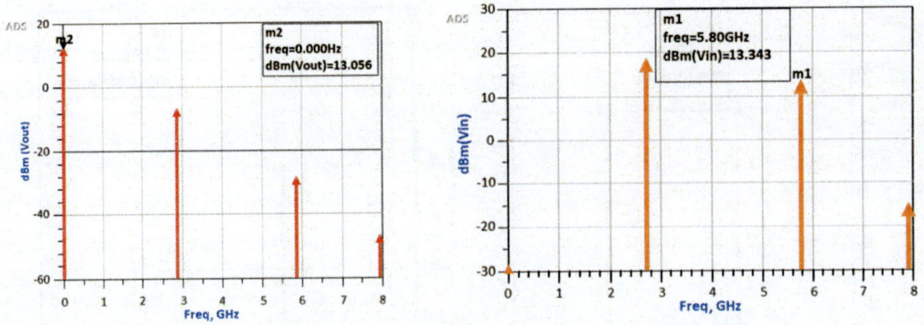

(a) (b)

Fig. 3. (a) Input Spectrum power and (b) Output Spectrum power.

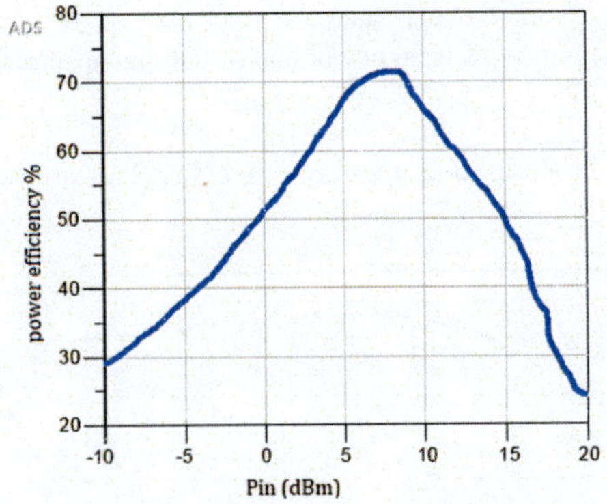

Fig. 4. Power efficiency versus input power.

1.2.2 Shunt Mounted Diode

Figure 5 displays a second rectifier arrangement for consideration. The diode is shunt inserted and is similar to the diodes that are used in series topologies in terms of performance.

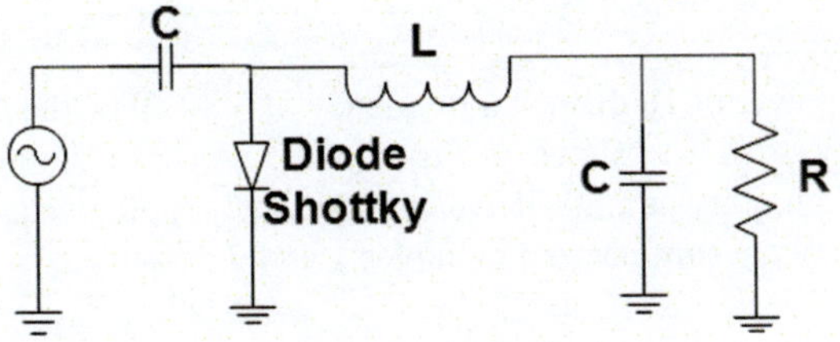

Fig. 5. Single shunt mounted diode

Figure 6 depicts the power levels prior to and after the rectifier circuit. The power is converted to DC at the output, as can be seen.

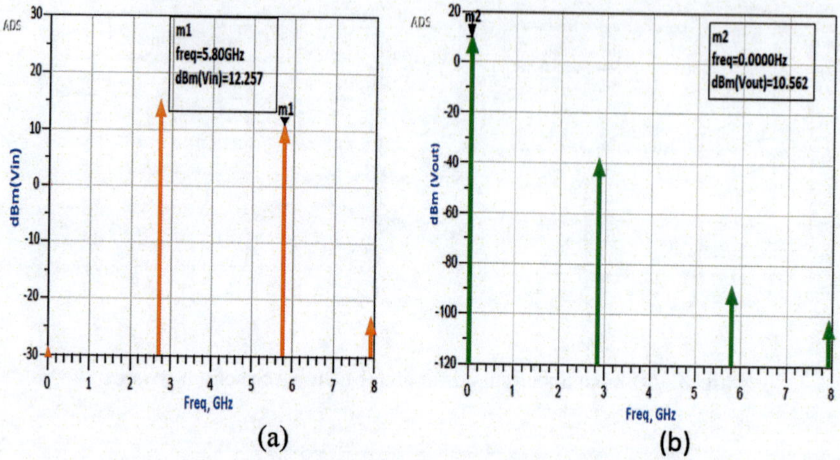

Fig. 6. (a) Input spectrum power and (b) Output spectrum power.

The best efficiency is attained at a voltage of 1.8 KΩ, as shown in Fig. 7.

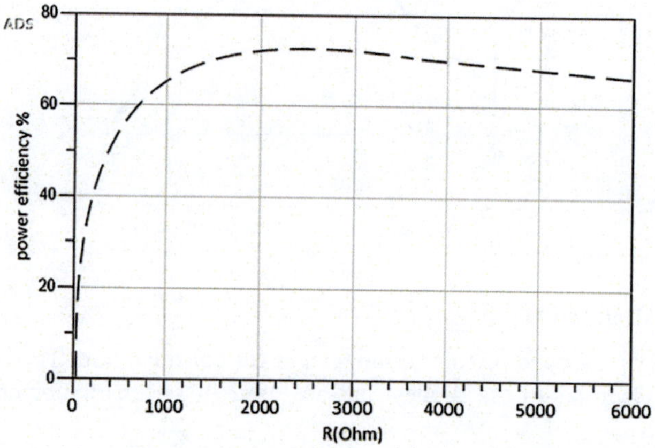

Fig. 7. Power efficiency versus load impedance.

Using an input power of 10 dBm at a frequency of 5.80 GHz, the structure achieves maximum efficiency of 74% as shown in Fig. 8.

The results of an inquiry at 5.80 GHz using the same topologies as a previous study at the same frequency are summarized in Tables 1 and 2.

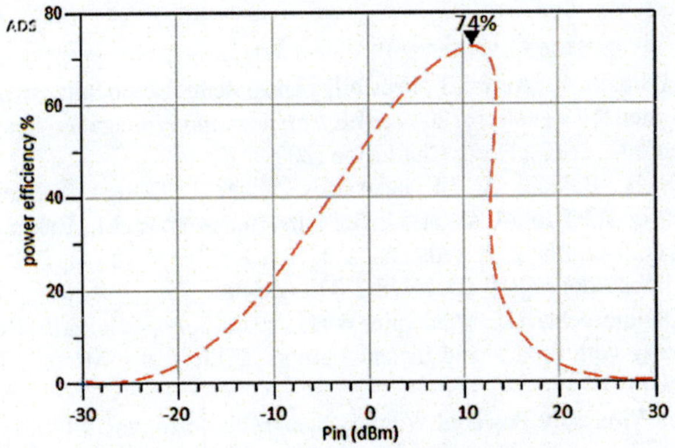

Fig. 8. Power efficiency versus input power.

Table 1. Comparison with anterior rectifier using diode serie.

Ref	Topology	Diode	Input power (dBm)	Efficiency (%)	Freq (GHz)
This work	Diode in series	HSMS 2860	8	71	5.80
[5]	Diode in series	HSMS 282C	20	80	2.45

Table 2. Comparison with anterior rectifier using diode shunt.

Ref	Topology	Diode	Input power (dBm)	Efficiency (%)	Freq (GHz)
This work	Diode in shunt	HSMS 2860	10	74	5.80
[6]	Diode in shunt	HSMS 2860	10	60	2.45

As seen in the Table 1 and 2, this research is somewhat efficient when compared to other studies, although it has the advantage of having a low input power. The optimum value for this task is found at 10 dBm as the efficiency rises with input power.

2 Conclusion

In this study, we discussed two different rectifier topologies for use in an RF energy harvesting system. The efficiency of the serial rectifier's RF-to-DC conversion is around 71%. When the HSMS2860 diode is used in the shunted rectifier, the RF-DC conversion efficiency is 74%, and the frequency of operation is 5.80 GHz, which is quite high.

For the sake of these experiments, the incident power is set at 10 dBm. In terms of RF-DC conversion efficiency, the shunt topology has shown promising results.

References

1. Kawahara, Y., Tsukada, K., Asami, T.: Feasibility and potential application of power scavenging from environmental RF signals. In: 2009 IEEE Antennas and Propagation Society International Symposium, pp. 1–4. IEEE, North Charleston (2009)
2. Ihlou, S., Tizyi, H., Bakkali, A., El Abbassi, A., Foshi, J.: Design of a new CP microstrip patch antennas for WPT to a UAV at 5.8 GHz. In: Ben Ahmed, M., Rakıp Karaş, İ, Santos, D., Sergeyeva, O., Boudhir, A.A. (eds.) SCA 2020. LNNS, vol. 183, pp. 992–1002. Springer, Cham (2021). https://doi.org/10.1007/978-3-030-66840-2_75
3. Hagerty, J.A., Helmbrecht, F.B., McCalpin, W.H., Zane, R., Popovic, Z.B.: Recycling ambient microwave energy with broad-band rectenna arrays. IEEE Trans. Microw. Theory Tech. **52**, 1014–1024 (2004)
4. Paing, T., et al.: Wirelessly-Powered Wireless Sensor Platform, vol. 4 (2007)
5. Liu, C., Tan, F., Zhang, H., He, Q.: A novel single-diode microwave rectifier with a series band-stop structure. IEEE Trans. Microw. Theory Tech. **65**(2), 600–606 (2017)
6. ken-system: A Study of Energy Harvesting from Ambient RF Sources, https://www.ieice.org/ken/paper/20110909k08P/eng/. Accessed 12 Apr 2022

CNN-Based Deep Features with Ensemble Learning for COVID-19 Classification

Youssra El Idrissi El-Bouzaidi(✉) and Otman Abdoun

Laboratory of Information Security, Intelligence Systems and Applications-ISISA, Faculty of Science, Abdelmalek Essaadi University, Tetouan, Morocco
youssra.elidrissi.elbouzaidi@gmail.com, Abdoun.otman@gmail.com

Abstract. In this paper, we propose a hybrid system that can automatically detect coronavirus disease and speed up medical image analysis processes by using artificial intelligence technique. Our system consists of two parts: First, to perform feature extraction, we used a deep convolutional network that is based on the transfer learning technique, in this step, we include eight well-known convolutional neural networks for comparison purposes. In the second part, a voting classifier is considered, combining three classifiers, including random forest (RF), support vector machine (SVM), and K-nearest neighbor (KNN), to classify radiological images into three classes: COVID-19, normal, and pneumonia, collected from two public medical repositories. The results show that deep learning and radiological images are able to retrieve relevant COVID-19 features with an accuracy of 96.87%.

Keywords: COVID-19 · Chest X-ray · Artificial intelligence · Medical imaging

1 Introduction

The fast propagation of the coronavirus worldwide is leading to containment which penalizes human life and disables industries. The highly transmissible nature of this disease, insofar as detection occupies an important place in the fight against this disease and in taking the preventive plan seriously. According to the WHO, on July 14, 2022, more than 556,897,312 confirmed cases have been reported worldwide, at the result in the death of 6,356,812 patients [1]. Within only 2 months, the contamination spread from Wuhan to the rest of China to other countries such as Morocco. Since the first case detected on 3nd January 2020, on 14th July 2022, Morocco has recorded 1,244,892 confirmed cases, and 16,165 deaths. To date, only 12,080,887 tests have been performed in Morocco [2], while the Moroccan population is estimated at 36,683,720 until July 15, 2022 [3].

The sensitivity that characterizes COVID-19 tests poses a major challenge. Radiologists are usually able to identify COVID-19 cases through the review of radiological images; on the other hand, these tasks are performed manually. To address a need for more rapid interpretations a number of artificial intelligence (AI) systems using deep learning were proposed with promising results for the detection of patients infected with COVID-19, focusing primarily on medical images [4, 5].

Y. Farhaoui et al. (Eds.): ICAISE 2022, LNNS 635, pp. 325–330, 2023.
https://doi.org/10.1007/978-3-031-26254-8_46

We have recently published a work [6, 7] where we proposed a model based on deep CNN transfer learning using pre-trained DensNet121, to classify x-ray images as COVID-19, normal and pneumonia, the proposed model achieved an accuracy of 96.52 for the three classes. Another work published by us [8] In this paper, to extract features from radiological images, a deep transfer learning approach was suggested, and subsequently for classification of the obtained features, we used SVM. An accuracy of 96.09% and f1 score of 97%, were achieved by the proposed classification model, for COVID-19 detection based on three classes.

Our main objective in this paper, is to increase diagnostic performance given the limited amount of data available in open access. To achieve our goals, we proposed CovDec a deep transfer learning approach using convolutional neural networks, based on the pre-trained CNN DenseNet121 model to classify COVID-19 and other classes, the transfer learning method has been applied to achieve superior performance. The features extracted from the convolution layers were classified using a combination of the following three algorithms: Support Vector Machine (SVM), Random Forest (RF), and K-nearest neighbor (KNN).

This paper is presented along the following lines: Sect. 2: introduces the proposed methodology, while Sect. 3 lists in detail the experimental work, the dataset, and devoted to a discussion of the results and Sect. 4 conclusion.

2 The Proposed Methodology

2.1 First Phase: X-ray Images Pre-processing

This preliminary step prepares the data with the respective deep transfer neural network employed for every image. Depending on the architecture, each of these different CNNs needed to be sized differently. As part of the preprocessing process. For example: a 229 × 229 image size is required for Xception, whereas the size of Dense-Net121 is 224 × 224. We employed a simple and commonly used bilinear interpolation function. Similarly, normalization was performed on all images based on their respective architectures. The Min-Max-Scalar function was used for normalizing an i-th image in the interval [0, 1]. The data were split by 80% and 20% for the training and validation, respectively. To compare the performance of the different networks, the same training and validation data sets were prepared and selected for all networks.

Data augmentation involves artificially creating new training data using modifications to our existing images. Essentially, by making a number of modifications and using different changes in training set, including random vertical and horizontal shift that can move all pixels in the X-ray image in the horizontal or vertical direction and keep the same dimensions, as well as vertical and horizontal flip that inverts all pixel lines and columns for vertical or horizontal flip, and finally random zooming and rotation, we create new augmented images.

2.2 Second Phase - Feature Extraction

In this step, to extract features, we used an existing set of pre-trained convolutional neural network architectures, providing excellent results obtained using various classification tasks as alternate to building an architecture from scratch. Therefore, the DenseNet121 CNN architecture [9] was considered as an effective way to perform the extraction of features using its weights which are made in the ImageNet dataset [10]. The output results in the form of a matrix with columns representing the number of classes (COVID-19, Normal, Pneumonia) and lines representing an image count.

As part of our work, to compare performance, we reviewed eight pretrained CNNs, that are well known: 1- Xception, 2- ResNet101V2, 3- ResNet50V2, 4-DenseNet201, 5- VGG19, 6- VGG16, 7- InceptionResNetV2, and 8-MobileNetV2.

2.3 Third Phase: Classification with Machine Learning

Finally, the last and third step is the classification. We propose a voting classifier from the aggregation of various classifiers such as Support Vector Machine (SVM), Random Forest (RF), and K-nearest neighbor (KNN) to improve the overall accuracy. After extracting the relevant features by a convolutional neural network previously trained in the second phase, these classifiers are trained to obtain optimal weights.

3 Experimental Results

3.1 Dataset

During the development stage, the investigation concentrated to obtain X-ray images for three classes as COVID-19 (+), pneumonia (+), and normal X-ray images. The dataset was obtained from GitHub, the public source repository that Dr. Joseph et al. has shared [11], and from the Kaggle "Chest XRay Images (Pneumonia)" [12].

3.2 Results

In this section, a detailed evaluation of the performance of the CovDec model using the test data are given in Fig. 1.

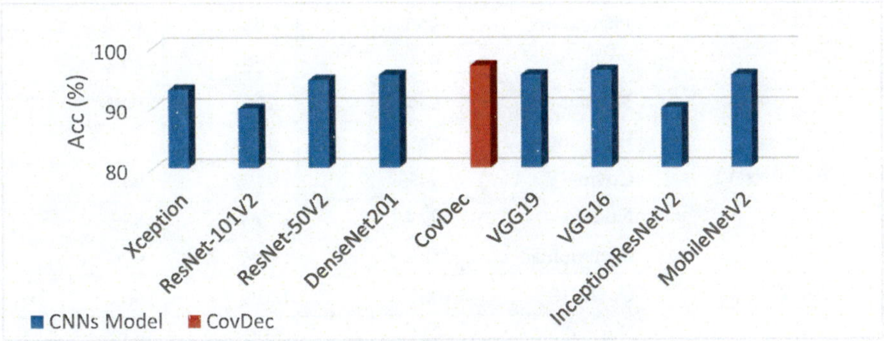

Fig. 1. Details of the performance results obtained from CovDec and eight CNNs in diagnosis of Covid-19 disease in training and validation datasets.

We have obtained the best performance as the accuracy of 96.87%, recall of 100%, and specificity value of 100%, and F1-score of 100% for the CovDec model. The lowest performance values have been yielded an accuracy of 89.84%, for InceptionResNet v2 and ResNet101 v2. As a result, The CovDec with DenseNet121 pre-trained ImageNet CNN model for feature extraction is therefore superior to other models, both for training and for testing as shown in Table 1. In addition, MobileNet v2, DenseNet121, DenseNet201, and ResNet101 v2, demonstrated a high performance for Covid-19 Class.

Table 1. Confusion matrix of the used CNNs.

CNN models	Class	Precision	Recall	F1-score
Xception	Covid-19	100%	88%	93%
	Normal	97%	86%	91%
	Pneumonia	90%	99%	94%
ResNet-101V2	Covid-19	100%	100%	100%
	Normal	91%	76%	83%
	Pneumonia	87%	96%	91%
ResNet-50V2	Covid-19	100%	94%	97%
	Normal	97%	88%	93%
	Pneumonia	92%	99%	95%
DenseNet201	Covid-19	100%	100%	100%
	Normal	97%	88%	93%
	Pneumonia	93%	99%	96%
CovDec	Covid-19	100%	100%	100%
	Normal	97%	93%	95%
	Pneumonia	96%	99%	97%
VGG19	Covid-19	89%	100%	94%
	Normal	100%	88%	94%
	Pneumonia	95%	99%	97%
VGG16	Covid-19	94%	100%	97%
	Normal	97%	90%	94%
	Pneumonia	96%	99%	97%
InceptionResNetV2	Covid-19	94%	94%	94%
	Normal	92%	79%	85%
	Pneumonia	88%	96%	92%
MobileNetV2	Covid-19	100%	100%	100%
	Normal	95%	90%	93%
	Pneumonia	94%	97%	96%

This proposed model provided an automatic detection process for COVID-19 from radiography images. The experimental studies were implemented using Python with Google Colaboratory [13], commonly known as "Google Colab", a scientific initiative aimed at prototyping machine learning (ML) models over high-performance device technologies of GPUs. For deep feature extraction we utilized a deep learning architecture DensNet121, these deep transfer learning approaches allowed for training network weights on large datasets, as well as on small datasets, fine-tuning pre-trained network weights. The resulting deep features from deep model are classified using machine learning (ML) algorithms into three classes COVID-19, pneumonia and normal. The objective of the second phase of the methodology was to test which algorithm best classifies the features extracted with the deep transfer learning method (first phase) that represent the COVID-19 disease. For the classification experiment performed, 10-fold cross-validation was used. To ensure that the results of the experiment are not unstable, the experiment was repeated 10 times. In order to compare the results, nine well-known CNNs were used to provide a comprehensive view of the role of artificial intelligence in this task. The results showed that deep learning could distinguish COVID-19 from pneumonia and healthy patients with high performance. The DensNet121 architecture was effective in this task. Based on these results, features are extracted from the convolution layers of the DenseNet121 architecture to be ready for the next step which includes the final classification.

This last scenario allows the performance of the proposed classification algorithm to be evaluated. According to our analyses, the proposed voting classifier system would achieve an accuracy of 96.87% (see Table 1). Experimental comparison of the results obtained by the voting classifier proves that the proposed model for identifying COVID-19 patients through x-ray images has improved efficiency. Thus, it is possible to enhance diagnostic capabilities while avoiding the significant burden on radiologists associated with the early detection of COVID-19.

4 Conclusion

In this paper, we utilized artificial intelligence (AI) techniques, namely, deep transfer learning (CNN) and machine learning (ML) algorithms in building detection model that can discriminate between COVID-19, Pneumonia and Normal from chest X-ray images. In the experiments, an accuracy of 96.87% was achieved using a voting classification between SVM, KNN and RF algorithms, the results can be improved by obtaining a huge dataset. In the future, we can use various types of classifiers besides different types of features such as texture features to describe the texture of the chest. In addition, there are some pre-processing and data augmentation methods that can be applied for improving image viewability in chest radiography.

References

1. Home. https://www.who.int
2. http://www.covidmaroc.ma/Pages/AccueilAR.aspx

3. Démographie & population | Site institutionnel du Haut-Commissariat au Plan du Royaume du Maroc. https://www.hcp.ma/Demographie-population_r142.html
4. Benbrahim, H., Hachimi, H., Amine, A.: Deep transfer learning with apache spark to detect COVID-19 in chest X-ray images, vol. 13 (2020)
5. Abbas, A., Abdelsamea, M.M., Gaber, M.M.: Classification of COVID-19 in chest X-ray images using DeTraC deep convolutional neural network. Appl. Intell. **51**(2), 854–864 (2020). https://doi.org/10.1007/s10489-020-01829-7
6. El Idrissi El-Bouzaidi, Y., Abdoun, O.: DenTcov: deep transfer learning-based automatic detection of coronavirus disease (COVID-19) Using chest X-ray images. In: Motahhir, S., Bossoufi, B. (eds.) ICDTA 2021. LNNS, vol. 211, pp. 967–977. Springer, Cham (2021). https://doi.org/10.1007/978-3-030-73882-2_88
7. El Idrissi, E.B.Y., Otman, A.: Transfer learning for automatic detection of COVID-19 disease in medical chest X-ray Images. http://www.iaeng.org/IJCS/issues_v49/issue_2/IJCS_49_2_09.pdf
8. Youssra, E.I.E.-B., Otman, A.: Application of artificial intelligence to X-ray image-based coronavirus diseases (COVID-19) for automatic detection. In: Lazaar, M., Duvallet, C., Touhafi, A., Achhab, M.A. (eds.) BDIoT 2021. LNNS, vol. 489, pp. 208–220. Springer International Publishing, Cham (2022). https://doi.org/10.1007/978-3-031-07969-6_16
9. Huang, G., Liu, Z., Van Der Maaten, L., Weinberger, K.Q.: Densely connected convolutional networks. In: 2017 IEEE Conference on Computer Vision and Pattern Recognition (CVPR), pp. 2261–2269 (2017). https://doi.org/10.1109/CVPR.2017.243
10. Deng, J., et al.: ImageNet: a large-scale hierarchical image database. In: 2009 IEEE Conference on Computer Vision and Pattern Recognition, pp. 248–255 (2009). https://doi.org/10.1109/CVPR.2009.5206848
11. Cohen, J. P. IEEE8023/Covid-chestxray-dataset (2020)
12. Chest X-Ray Images (Pneumonia) | Kaggle. https://www.kaggle.com/paultimothymooney/chest-xray-pneumonia
13. Bisong, E.: Building Machine Learning and Deep Learning Models on Google Cloud Platform: A Comprehensive Guide for Beginners. Apress (2019). https://doi.org/10.1007/978-1-4842-4470-8

Combination of Renewable Energy Source and Battery for Quality of Service of Connected Objects

Wafaa Ennabirha[1](✉), Ahmed Moutabir[2], and Abderraoouf Aboudou[1]

[1] High School of Technology, Laboratory in Computer Networks, Telecommunications and Multimedia, Hassan 2nd University, Casablanca, Morocco
wafaa.ennabirha.doc2020@ensem.ac.ma

[2] Aïn Chok Faculty of Science (FSAC), Laboratory of Electrical and Industrial Engineering, Information Processing, IT and Logistics (GEITIIL), Hassan 2nd University, Casablanca, Morocco

Abstract. Internet of Things (IoT) technology has attracted a lot of attention in recent years in all fields. The operational safety of this system and data security is a necessity. The power supply of its components, especially the wireless sensors, is a priority. This article aims to present a hybrid system composed of a photovoltaic solar generator and a battery.

Our objective is to propose an autonomous sensor solution based on a solar energy recovery system that can be implemented in different applications. In this article, we focus on combining sources such as solar and battery to provide a backup solution for all energy needs. First, a mathematical of single diode PV panel model is presented and simulated. Then, the Perturb and Observe algorithm to ensure the MPPT objective is presented. Simulated models have been developed and tested in MATLAB/ SIMULINK.

Keywords: Internet of Things · PV generator · Battery · Single diode · MPPT

1 Introduction

The Internet of Things (IoT) is defined as a network of interconnected objects using different communication protocol. The IoT enables objects to be detected or operated directly via existing network technology, enabling a closer connection between the real world and computer-based networks and enhancing performance [1, 2]. IoT is applicable in a large field of science and technology [3]. This concept started to develop with RFID (Radio Frequency Identification), a wireless communication technology that marked the beginning of communicating objects. Then, many communication protocols appeared such as Bluetooth, Zigbee, WiFi, Sigfox, and LoRa [3, 4]. Since then, there has been a significant increase in the number of connected objects with fully energy-autonomous.

IoT security is therefore essential given the emergence of this technology in our daily lives. This security has three aspects: technological, human and systematic.

Y. Farhaoui et al. (Eds.): ICAISE 2022, LNNS 635, pp. 331–338, 2023.
https://doi.org/10.1007/978-3-031-26254-8_47

Technology protection concerns the security of data, communications and network infrastructures and their functionalities.

The energy autonomy of connected objects will certainly contribute to their development. The use of renewable energies such as solar can be an interesting solution. To overcome the problems of intermittency of these energies, they can be combined with other sources of storage.

The use of hybrid energy systems can guarantee the permanent power supply of the wireless sensors used.

Extracting the maximum power from the PV generator is the most desired performance then to guarantee the tracking of the maximum power point requires the use of algorithms. Several MPPT algorithms [5] have been developed such as the perturb and observe [6] and Incremental conductance [7] methods.

In this paper, the hybrid system is presented, the PID control is developed and the perturb and observe algorithm is used to track the maximum power of the PV array. The model of the photovoltaic generator is given and in the end, we illustrate the performance of the system by numerical simulations Using Simulink Matlab.

2 Hybrid System Scheme

The circuit in Fig. 1 represents the hybrid system composed of the PV generator and a battery. Two DC-DC converters are used which are the Boost converter associated with the PV generator and the Buck-Boost converter associated with the battery. DC-DC converters conform to the known principle of pulse-width modulation (PWM).

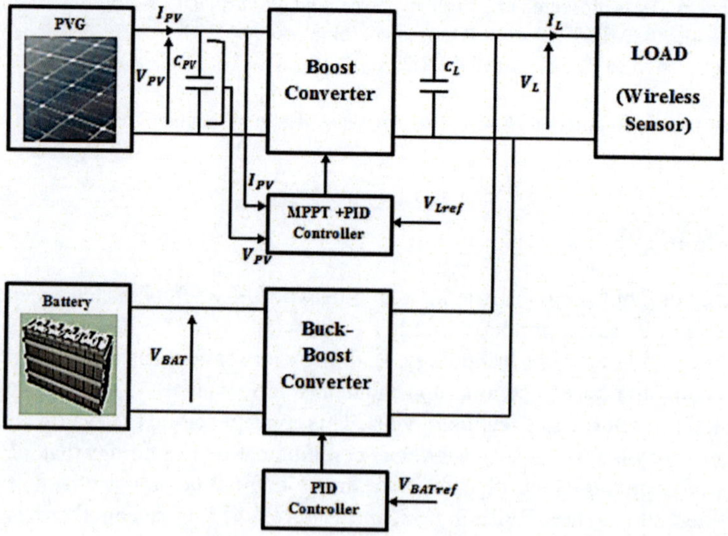

Fig. 1. Hybrid system scheme

Where:

- I_{PV} and I_L are respectively the PV and the load currents.
- V_{PV}, V_L and V_{BAT} denotes the PV, the load and the battery voltages.

3 PV Panel Model

The PV cell is made of semiconductor materials and directly transforms light energy into electrical energy.

The PV cell can be modeled in several ways. Depending on the physical phenomena involved and the structure of the cell, several models have been defined. Single-diode and double-diode models are the most widely used. These models are improved if the different losses are taken into account [8].

The single diode PV cell model adopted in this work is shown in Fig. 2.

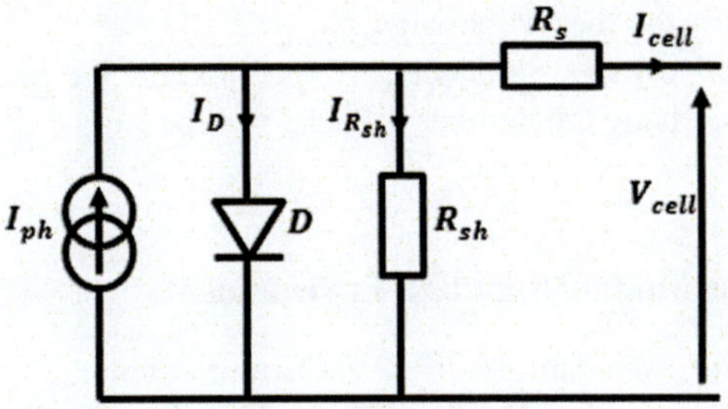

Fig. 2. Single diode PV cell model

The PV panel is defined by the following equations:

$$I_{cell} = I_{ph} - I_D - I_{R_{sh}} = I_{ph} - I_S\left(e^{\left(\frac{V_{cell}+R_sI_{cell}}{V_T}\right)} - 1\right) - \frac{V_{cell} + R_sI_{cell}}{R_{sh}} \quad (1)$$

$$I_{ph} = \frac{G}{1000}[I_{scr} + K_T(T - 298{,}5)] \quad (2)$$

$$I_s = I_{rr}\left[\frac{T}{298{,}5}\right]^3\left[exp\frac{qE_G}{\gamma K_B}\left(\frac{1}{298{,}5} - \frac{1}{T}\right)\right] \quad (3)$$

The PV array voltage and current are defined by:

$$\begin{cases} V_{PV} = N_s V_{cell} \\ I_{PV} = N_p I_{cell} \end{cases} \quad (4)$$

The PV power is given by:

$$P = I_{PV} V_{PV} \quad (5)$$

The different parameters used in the equations are defined as follows:

- $I_{cell}, I_{ph}, I_D, I_{R_{sh}}, I_S, I_{rr}, I_{scr}$ and I_{PV} are respectively the PV cell current, the photonic current, the diode current, the shunt resistor current, the saturation current, the reverse saturation current, the schort circuit current and the PV panel current.
- V_{cell} and V_{PV} are respectively de PV cell and the PV array voltages.
- q is the charge of electron $(1, 6.10^{-19}C)$, K_B is the Boltzmann's constant $(1, 38.10^{-23})$, γ is the diode ideality factor for diffusion current, E_G is a bandgap energy of a semiconductor, V_T is the thermal voltage.
- G and T are respectively the irradiance (kW/m^2) and the PV cell temperature (K)
- N_s and N_p are respectively the number off cells in series and the number off modules in parallel.

3.1 Modelling in the Matlab/ Simulink Environment

Using Matlab software, we obtained the PV Current – voltage and Power – Voltage curves at 25 °C and at many level of irradiance. These characteristics are respectively shown in Fig. 3 and Fig. 4.

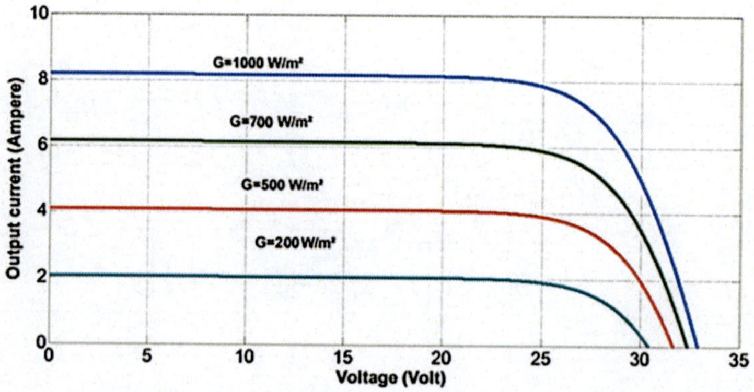

Fig. 3. Current-voltage curves (curves obtained at 25 °C and at many levels of irradiance)

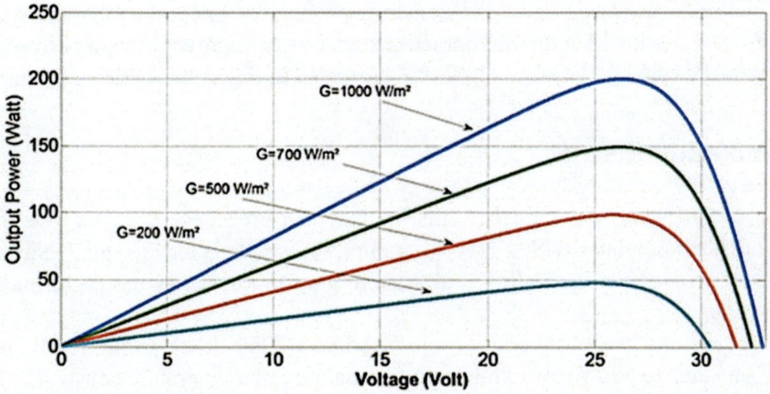

Fig. 4. Power – voltage curves (curves obtained at 25 °C and at many levels of irradiance)

4 Controller Design

There are three operational control objectives:

- Ensure a MPPT control.
- Regulating the output voltage v_L to a reference value v_{Lref}.
- Ensure the global stability of the system.

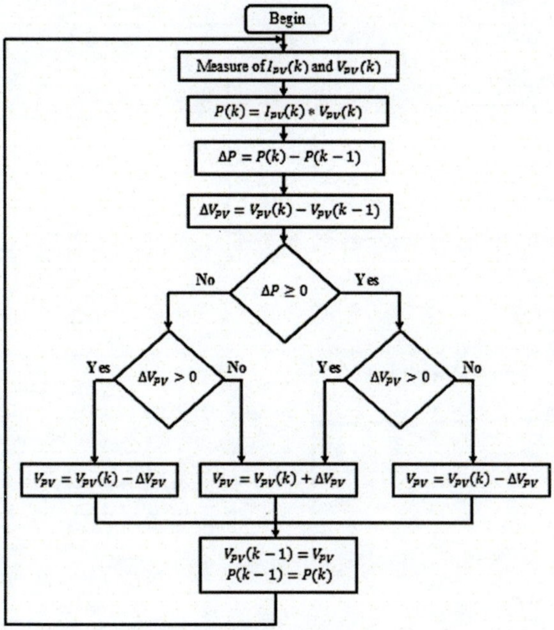

Fig. 5. The Perturb and Observe algorithm Current

- The controller used is a PID which ensures tracking by integral action and stability by derivative action. A judicious choice of parameters improves system performance.
- Using the Perturb and Observe algorithm presented in Fig. 5, the MPPT is guaranteed.

5 Simulation Results

In order to verify the theoretical results predicted in precedent Sections, the Hybrid system has been simulated with the set of parameter values depicted in Table 1. Numerical simulations were made in the Simulink/Matlab platform to verify the performance of the controller.

The variations were applied to the irradiance (Fig. 6-a) from $400\,\mathrm{W/m^2}$ to $1000\,\mathrm{W/m^2}$ then to $600\,\mathrm{W/m^2}$ while maintaining the cell temperature constant at 25°C. The PV power is shown at Fig. 6-b.

The resulting control performances are illustrated by Fig. 6-c to Fig. 6-e.The reference DC bus voltage value is held at $v_{Lref} = 100\,\mathrm{V}$.

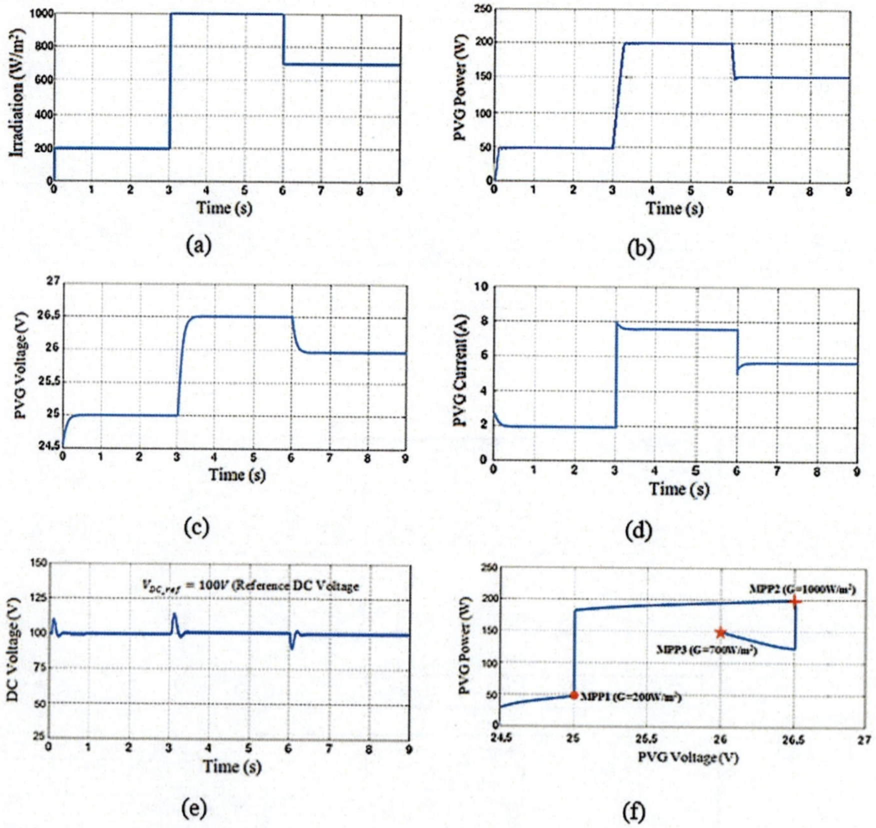

Fig. 6. (a) Variation of irradiations. (b) PV power. (c) PV array voltage. (d) PV array current.(e) DC bus voltage. (f) PV array power – voltage characteristic.

From these Figures, it is clear that the maximum power point is always reached very quickly with excellent accuracy and good performances according to atmospheric condition changes.

The DC bus voltage is regulated to its desired value. In Fig. 6-f, it is clear that the tracking of the maximum power point is guaranteed.

Table 1. Parameter values for the Hybrid system.

Parameters	Values	Parameters	Values
R_s	$0, 1124\,\Omega$	I_{scr}	$3, 45$ A
R_{sh}	$6500\,\Omega$	I_{rr}	$4, 842\,\mu$A
γ	$1, 7404$	N_s	36
N_p	1	*Reference*: v_{BAT}	24
Reference: v_L	100 V	Frequency	10 kHz
Capacitors	$C_{PV} = 4700\,\mu$F $C_{dc} = 470\,\mu$F	Swiches	IGBT

6 Conclusion

This paper presents a study of a Hybrid system. The perturb and observe algorithm is designed to extract maximum power from the PV array. To track the designed trajectory and regulating the load voltage, a PID controller is developed to modulate the duty cycle of the boost and buck-boost converters. The proposed controller has the advantages of robustness, accurate tracking, fast response and good performance. Simulation results, obtained under Matlab/Simulink environment, show the control performance and dynamic behavior of hybrid system provides good results and show that the control system is robust and efficiency.

This work opens the horizons to several perspectives at the level of recovery, optimization and storage of energy for communicating sensors. At the level of energy recovery and optimization, we aim at characterizing the solar energy to feed the sensor and overcome the temporary unavailability of energy sources. At the energy storage level, we propose to model the energy storage in multi-source. Then, we model the consumption of the collector to converge towards its neutral operating point and thus extend its life.

References

1. Tiwary, A., Mahato, M., Chidar, A., Chandrol, M.K., Shrivastava, M., Tripathi, M.: Internet of Things (IoT): research, architectures and applications. Int. J. Future Revolution Comput. Sci. Commun. Eng. **4**, 23–27 (2018)

2. Puar, V.H., Bhatt, C.M., Hoang, D.M., Le, D.-N.: Communication in internet of things. In: Bhateja, V., Nguyen, B., Nguyen, N., Satapathy, S., Le, D.N. (eds.) Information Systems Design and Intelligent Applications. Advances in Intelligent Systems and Computing, vol. 672, pp. 272–281. Springer, Singapore (2018). https://doi.org/10.1007/978-981-10-7512-4_28
3. Zeebaree, S.R., Shukur, H.M., Hussan, B.K.: Human resource management systems for enterprise organizations: a review. Period. Eng. Nat. Sci. (PEN) **7**, 660–669 (2019)
4. Evans, D.: The internet of things: how the next evolution of the internet is changing everything. Cisco white paper (2011)
5. Saravanan, S., et al.: Maximum power tracking algorithms for photovoltaic system. Renew. Sustain. Energy Rev. **57**, 192–204 (2016). https://doi.org/10.1016/j.rser.2015.12.105
6. Kamran, M., et al.: Implementation of improved Perturb & Observe MPPT technique with confined search space for standalone photovoltaic system. J. King Saud Univ. Eng. Sci. **32**, 432–441 (2020). https://doi.org/10.1016/j.jksues.2018.04.006
7. Shahid, H., et al.: Implementation of the novel temperature controller and incremental conductance MPPT algorithm for indoor photovoltaic system Sol. Energy **163**, 235–242 (2018). https://doi.org/10.1016/J.SOLENER.2018.02.018
8. Nguyen, X.H., Nguyen, M.P.: Mathematical modeling of photovoltaic cell/module/arrays with tags in Matlab/Simulink. Environ. Syst. Res. **4**(1), 1–13 (2015). https://doi.org/10.1186/s40068-015-0047-9

Covid-19 Dataset Analysis: A Systematic Review

Anoual El Kah[1(✉)] and Imad Zeroual[2]

[1] Faculty of Sciences, Mohamed First University, Oujda, Morocco
elkah.anoual.mri@gmail.com
[2] L-STI, T-IDMS, FST Errachidia, Moulay Ismail University of Meknes, Meknes, Morocco

Abstract. Due to their need to be connected to the rest of the world, people started to use social networks extensively to share their feelings and be informed, especially during the Covid-19 pandemic and its lockdown. The tremendous growth of content in social media increased the frequency of researchers' work on natural language understanding, text classification, and information retrieval. Unfortunately, not all languages have benefited equally from this interest. Arabic is an example of such languages. The main reason behind this gap is the limited number of datasets that addressed Covid-19-related topics. To this aim, we performed the first-of-its-kind systematic review that covered, to the best of our knowledge, the most Arabic Covid-19 datasets freely available or access granted upon request. This paper presents these 15 datasets alongside their features and the type of analysis conducted. The general concern of the authors is to direct researchers to reliable and freely available datasets that advance the progress of Arabic Covid-19-related studies.

Keywords: Covid-19 · Dataset analysis · Text classification · Arabic language · Systematic review

1 Introduction

During the Covid-19 pandemic, the use of virtual social networks significantly increased since those social networks were among the very few spaces left for people to share their thoughts and opinions, and be connected with each other. This tremendous growth of content in social media increased the frequency of researchers' work on natural language understanding, text classification, and information retrieval, especially those related to the Covid-19 pandemic. Unfortunately, not all languages have benefited equally from this interest. Arabic is an example of such languages.

Due to the scarcity of freely available datasets, the progress in Arabic Covid-19-related studies, especially text classification, lucks behind compared to other languages like English, Spanish, and Chinese. Some relevant efforts have been done and the main purpose of this paper is to shed light on it. Following the guidelines of the updated 2020 version of the Preferred Reporting Items for Systematic reviews and Meta-Analyses (PRISMA) [1], we performed a systematic review to cover the most currently freely available or access granted upon request Arabic Covid-19-related datasets. Nowadays, more than 10.5 million Arabic tweets are published per day, making Arabic among the top

Y. Farhaoui et al. (Eds.): ICAISE 2022, LNNS 635, pp. 339–344, 2023.
https://doi.org/10.1007/978-3-031-26254-8_48

five dominant languages on Twitter [2]. Consequently, several studies have performed different kinds of text classification using Arabic tweets, such as sentiment analysis [3], misinformation detection [4], and authorship attribution [5]. Similarly, most Arabic Covid-19 datasets are built using Arabic tweets. The paper is organized in three sections besides this introduction. The second section describes the methodology used to perform this systematic review. The third section summarizes the findings related to the datasets included in our PRISMA review. The fourth section concludes the paper.

2 Methodology

We performed our systematic review following the updated 2020 PRISMA guidelines [1]. This systematic review is performed using relevant online databases (e.g., Preprints, Arxiv, IEEE Xplore, and Google Scholar), catalogues (i.e., Masader[1] and ELRA[2]), repositories (e.g., Github) and Google dataset search engine. Then, we evaluated the quality and reliability of the works covered, including the subject investigated and the dataset availability. Figure 1 displays the illustrative flow diagram.

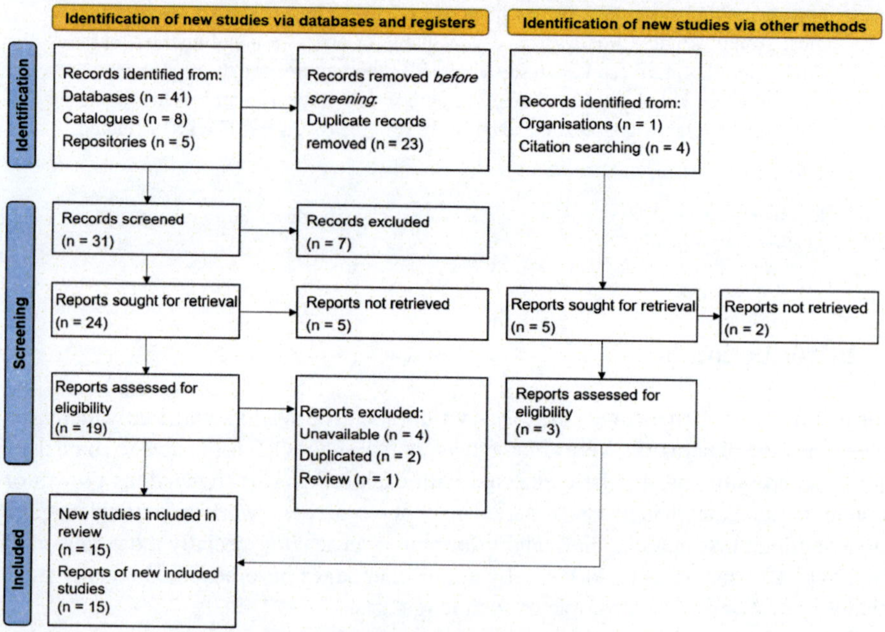

Fig. 1. PRISMA 2020 flow diagram.

At first, we identified 54 records via databases, catalogues, and repositories, while only five records are identified via citation searching and our private university database. Before screening the records, 23 duplicated records are removed. Then, 14 records are

[1] https://arbml.github.io/masader/.
[2] http://catalogue.elra.info/en-us/.

removed after the screening task, in which seven records were unrelated and seven more records were inaccessible. We were left with 22 records assessed for eligibility. After analyzing these records, one review was removed, two records that used similar datasets were also removed, and four records were removed since their datasets are not available. Thus, the final records included in our review are 15 records.

3 Results and Discussion

As aforementioned, all the 15 datasets included in this review are freely available or access granted upon request. Except for one dataset (i.e., AlRiyadh[3]) that comprises Arabic newspaper articles addressing Covid-19-related events, the other 14 datasets contain Arabic tweets that were compiled from Twitter during the Covid-19 pandemic. Each dataset was initially collected for one or more specific purposes. We have gathered and listed those purposes bellow:

- P1: Statistics about Covid-19-related issues.
- P2: Misinformation of fake news detection.
- P3: Sentiment analysis.
- P4: Topic identification.
- P5: Rumors detection.
- P6: Source prediction.
- P7: Emotions detection.
- P8: Symptoms detection.
- P9: Hate-Speech detection.
- P10: Sarcasm detection.

The global time frame used to collect the data is from the first of January 2020 till March 2021. However, some datasets are periodically updated. In addition, most of these builders have implemented their datasets in text classification tasks to prove their reliability. Therefore, we have summarized the different classification tasks that involved the datasets covered in this review. This summary also included the relative conditions of each study that led to its highest performance, namely the dataset size, the number of classes, and the classifier implemented. Table 1 presents the features of these datasets alongside the experiments conducted.

Table 1. Features of Arabic Covid-19 datasets.

Dataset	Purpose	Size (Nb. of docs.)	Nb. of classes	Classifier	Best performance
Alqurashi et al. [6]	P1	6,086,085	None	None	None
COVID-19-FAKES [4]	P2	216,209	2	None	None

(continued)

[3] https://github.com/alioh/AlRiyadh-Newspaper-Covid-Dataset.

Table 1. (*continued*)

Dataset	Purpose	Size (Nb. of docs.)	Nb. of classes	Classifier	Best performance
Alhumoud et al. [3]	P3	416,292	3	Enssemble model	90.21% acc.
Alsudias et al. [7]	P4	1,048,575	6	K-means	None
	P5	2,000	5	LR & SVM	84.03% acc.
	P6		3	LR	77% acc.
Addawood et al. [8]	P1, P3, P6	3.8M	3	SVM	98% acc.
AraEmoCorpus [9]	P7	5.5M	6	LSTM	82.8% F-score
	P8		2		75% F-score
ArCOV-19 [10]	P1, P4, P6	3,140,158	Multiclass	None	None
ArCOV19-Rumors [11]	P5	9,414	3	MARBERT	75.7% acc.
AlRiyadh	P2	20,571	33	None	None
ArCorona [12]	P2	8K	2	SVM	85.4% acc.
			13		62.8% acc.
ARACOVID19-MFH [13]	P2, P3, P6, P9	10,828	10	AraBERT	97.31% F-score
ARACOVID19-SSD [14]	P3	5,162	2	AraBERT	95.97% F-score
	P10		3	SVM	92.26% F-score
NLP4IF [15]	P2	4,056	2	Enssemble model	76.3% acc.
ArCovidVac [16]	P4	10K	2	AraBERT	86.4% acc.
			10		75.4% acc.
	P3		3	SVM	82.5% acc.
Alam et al. [17]	P2	4,966	2	XLM-RoBERTa	84,27% acc.
			Multiclass		65,58% acc.

As observed, the smallest dataset is the NLP4IF [15]. This latter comprises around 4k tweets. On the other hand, the largest dataset, i.e., Alqurashi et al. [6], included more than 6M tweets. Hence, the average size of Arabic Covid-19 datasets is 1.26M documents per dataset. Most datasets have 2 or 3 categories since they are primarily used for binary (e.g., True or False) and ternary classification (e.g., positive, negative, and neutral). Those datasets that have multiclass are usually used for topics classification task. The only dataset that is not classified is Alqurashi et al. [6], however, its authors claimed that it can be adjusted to be used for automatic text classification.

According to their original published papers, three datasets have not been yet implemented and evaluated for any kinds of classification problems. Those datasets are Alqurashi et al. [6], COVID-19-FAKES [4], and ArCOV-19 [10]. However, they might have been used for classification tasks, and the experiments are published separately from their original papers. The other 12 datasets have been used to train and test several machine learning-based classifiers. The classifiers that achieved the highest performance are Support Vector Machine (SVM), Logistic Regression (LR), Long Short-Term Memory neural networks (LSTM), and pretrained transformer-based models for the Arabic language like AraBERT, MARBERT, and XLM-RoBERTa.

4 Conclusion

This review was performed according to the PRISMA guidelines, which covers 15 freely available Arabic Covid-19 datasets. Most of these datasets are built by crawling Arabic tweets during the Codiv-19 pandemic. Through this papers, relevant dataset-related statistics were presented. Moreover, the primarily reasons behind building those datasets are illustrated given the classification problems addressed for each dataset and the best performance achieved. Most publishers of those datasets claimed that their datasets might be used for other kinds of classification problems and they will periodically be updated. Other researchers have claimed that their datasets will be available as well. Our general concern is to direct researchers to reliable and freely available datasets that advance the progress of Arabic Covid-19-related studies. However, the need for more datasets is urgent, especially from other sources like Facebook and Newspaper websites.

References

1. Page, M.J., et al.: PRISMA 2020 explanation and elaboration: updated guidance and exemplars for reporting systematic reviews. bmj **372**, 1–36 (2021)
2. Alshaabi, T., et al.: The growing amplification of social media: measuring temporal and social contagion dynamics for over 150 languages on Twitter for 2009–2020. EPJ Data Sci. **10**(1), 1–28 (2021). https://doi.org/10.1140/epjds/s13688-021-00271-0
3. Alhumoud, S.: Arabic sentiment analysis using deep learning for covid-19 twitter data. Int. J. Comput. Sci. Netw. Secur. **20**(9), 132–138 (2020)
4. Elhadad, M.K., Li, K.F., Gebali, F.: COVID-19-FAKES: a Twitter (Arabic/English) dataset for detecting misleading information on COVID-19. In: Barolli, L., Li, K.F., Miwa, H. (eds.) INCoS 2020. AISC, vol. 1263, pp. 256–268. Springer, Cham (2021). https://doi.org/10.1007/978-3-030-57796-4_25
5. El Kah, A.E., Zeroual, I.: A new classification model with a multi-layer dimensionality reduction approach. In: 2022 11th International Symposium on Signal, Image, Video and Communications (ISIVC), pp. 1–6 (2022). https://doi.org/10.1109/ISIVC54825.2022.9800211
6. Alqurashi, S., Alhindi, A., Alanazi, E.: Large Arabic twitter dataset on COVID-19. ArXiv Prepr. arXiv:2004.04315 (2020)
7. Alsudias, L., Rayson, P.: COVID-19 and Arabic Twitter: how can Arab world governments and public health organizations learn from social media? In: Proceedings of the 1st Workshop on NLP for COVID-19 at ACL 2020 (2020)
8. Addawood, A., et al.: Tracking and understanding public reaction during COVID-19: Saudi Arabia as a use case. In: Proceedings of the 1st Workshop on NLP for COVID-19 (Part 2) at EMNLP 2020 (2020)
9. Al-Laith, A., Alenezi, M.: Monitoring People's emotions and symptoms from Arabic tweets during the COVID-19 pandemic. Information **12**, 86 (2021). https://doi.org/10.3390/info12020086

10. Haouari, F., Hasanain, M., Suwaileh, R., Elsayed, T.: ArCOV-19: the first Arabic COVID-19 Twitter dataset with propagation networks. In: Proceedings of the Sixth Arabic Natural Language Processing Workshop. pp. 82–91. Association for Computational Linguistics, Kyiv (2021)
11. Haouari, F., Hasanain, M., Suwaileh, R., Elsayed, T.: ArCOV19-rumors: Arabic COVID-19 twitter dataset for misinformation detection. ArXiv Prepr. arXiv:2010.08768 (2020)
12. Mubarak, H., Hassan, S.: ArCorona: analyzing Arabic tweets in the early days of coronavirus (COVID-19) pandemic. In: Proceedings of the 12th International Workshop on Health Text Mining and Information Analysis, pp. 1–6. Association for Computational Linguistics (2021)
13. Ameur, M.S.H., Aliane, H.: AraCOVID19-MFH: Arabic COVID-19 multi-label fake news & hate speech detection dataset. Procedia Comput. Sci. **189**, 232–241 (2021)
14. Ameur, M.S.H., Aliane, H.: AraCOVID19-SSD: Arabic COVID-19 sentiment and sarcasm detection dataset (2021). https://doi.org/10.48550/arXiv.2110.01948
15. Qarqaz, A., Abujaber, D., Abdullah, M.: R00 at NLP4IF-2021 fighting COVID-19 infodemic with transformers and more transformers. In: Proceedings of the Fourth Workshop on NLP for Internet Freedom: Censorship, Disinformation, and Propaganda, pp. 104–109. ACL (2021). https://doi.org/10.18653/v1/2021.nlp4if-1.15
16. Mubarak, H., Hassan, S., Chowdhury, S.A., Alam, F.: ArCovidVac: analyzing arabic tweets about COVID-19 vaccination. ArXiv Prepr. arXiv:2201.06496 (2022)
17. Alam, F., et al.: Fighting the COVID-19 infodemic: modeling the perspective of journalists, fact-checkers, social media platforms, policy makers, and the society. In: Findings of the Association for Computational Linguistics: EMNLP 2021, pp. 611–649. Association for Computational Linguistics, Punta Cana (2021). https://doi.org/10.18653/v1/2021.findings-emnlp.56

Deep Learning Algorithms for Skin Cancer Classification

Mariame Oumoulylte[1], Ahmad El Allaoui[2](✉), Yousef Farhaoui[2], Fatima Amounas[3], and Youssef Qaraai[4]

[1] L-LSA, T-SDIC, ENSA Al Hoceima, ABdelmalek Essadi of Tetouan, Tetouan, Morocco
[2] L-STI, T-IDMS, FST Errachidia, Moulay Ismail University of Meknes, Meknes, Morocco
a.elallaoui@umi.ac.ma
[3] RO.AL&I Group, FST Errachidia, Moulay Ismail University of Meknes, Meknes, Morocco
f.amounas@umi.ac.ma
[4] L-MAIS, T-MIS, FST Errachidia, Moulay Ismail University of Meknes, Meknes, Morocco

Abstract. Deep learning algorithms, in particular convolutional networks, have rapidly become a methodology of choice for analyzing medical images. The aim of this work is to demonstrate two critical binary classification of skin lesions using Deep Learning algorithms. The proposed method based on a custom convolutional Neural Network trained end-to-end from images directly, using only pixels and disease labels as inputs, we train a CNN using a dataset of 3600 images. A deep learning based method convolutional neural network classifier is used for the stratification of the extracted features. After evaluation of model we find that the accuracy scores up to 72.73%, it achieved 76.15% for sensitivity, 83.06% for specificity, 77.40% for precision and 76.76% for F1 score.

Keywords: Skin cancer · Deep learning · Convolutional neural network · Medical imaging

1 Introduction

One of the deadliest diseases that currently afflict humankind is cancer. Skin cancer is one of the most common and the deadliest among all types of cancer. Among these, melanoma is the most aggressive and deadly form [1] of cancer. Skin cancer are of two major sorts, namely nonmelanoma and melanoma (Merkel cell carcinomas, squamous cell, basal cell, and so on) [2] as shown in Fig. 1. Melanoma is the dangerous skin cancer, however, if it is detected in the early stages, it can be curable, but progressive melanoma is deadly. Therefore, it is well known that the early treatment and detection of skin cancer can minimize the morbidity. Computer vision with Deep Neural Networks have achieved superior performance in areas such as segmentation, image classification, object detection, pose estimation and activity recognition. With this rise in computer vision and the wider scope of deep learning, there has been increased attention in the application of these technologies within the healthcare segment. Automated classification of skin lesions using images is a challenging task owing to the fine-grained variability in the appearance of skin lesions [3].

Y. Farhaoui et al. (Eds.): ICAISE 2022, LNNS 635, pp. 345–351, 2023.
https://doi.org/10.1007/978-3-031-26254-8_49

This paper is structured as follows: in the second part, we give background and cite related works, in the third we treated the methodology used, in the fourth section, we will discuss the methods used, and the construction of the dataset. In the fifth section we will present our experiments, and the results, in final section we finish with a conclusion, and we open about upcoming works.

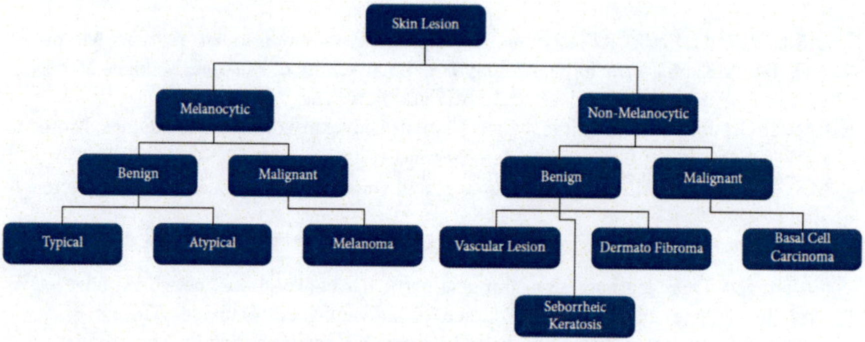

Fig. 1. Diagnosis trees structure for skin lesion.

2 Background and Related Work

The aim of this work is to demonstrate two critical binary classification of skin lesions using Deep Learning algorithms. The proposed method based on a custom convolutional Neural Network trained end-to-end from images directly, using only pixels and disease labels as inputs. We train a CNN using a dataset of 3600 images consisting of two different classes of skin cancer: malignant moles versus benign moles.

Dermatologists does the diagnosis of the skin cancer. They can access the images of cancer patients and analyze the result whether the patient has cancerous cells or not. Because of having cancerous cells, dermatologist suggest it as malignant melanoma and benign on vice versa. The issue with this framework is, it sets aside a lot of time to process a ton of patients and furthermore it takes a great deal of labor to expand the rate of recognition which makes the cost go up. The developing computerized system can automate this skin cancer detection process that will assist the dermatologists, and makes their works easier and faster. The previous research [4] was developed using an edge detection method with K-NN and C-NN algorithms provide an accuracy of 75% and 75.6% respectively to classify skin disorders that potentially benign cancer and skin disorders that have the potential to be malignant cancer by using the International Skin Imaging dataset Collaboration (ISIC). In research [5], using the ISIC dataset for the conditions of skin cancer and skin benign tumors, an automatic skin disease classification system was developed based on deep learning with PNASNet-5-Large architecture which gives the best performance accuracy of 76%. Furthermore, other studies using CNN for the detection of skin diseases [6–8] provide performance accuracy of 80.52%, 86.21%, 87.25% respectively. To increase the amount of data and to improve the performance of the skin cancer detection system, the ISIC data augmentation process was performed [9]. The best accuracy was obtained by 95.91% using Alexnet.

3 Materials and Methods

3.1 Dataset

Our dataset comes from the ISIC (International Skin Image Collaboration) archive [10]. It consists of 1800 pictures of benign moles and 1800 pictures of malignant classified moles. The pictures have all been resized to low resolution (224 × 224 × 3) RGB and divided into two classes, "benign" and "malignant". We load in the pictures and turn them into numpy arrays using their RGB values. As the pictures do not have any labels, these need to be created as shown in Fig. 2. Finally, the pictures are added together to a big training set and shuffeled.

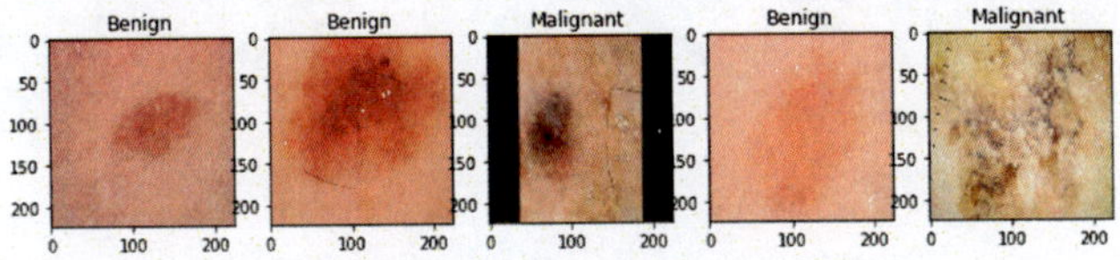

Fig. 2. Labels of images.

3.2 Proposed Deep Learning Model

The labeled images "benign" and "malignant" were used in this system. Images were put into dataset relying upon their analysis mark which has been extracted from the metadata of the pictures. The dataset has been organized into two classes one containing all the cancerous images and other containing non cancerous images. In our proposed system, there exist three layers. First layer is the input layer where the data sets are trained on. Input layer collects data that are delivering and adding some weight with it that goes to hidden layers. Theneurons of hidden layer separate the features from the data to find out a pattern. The pattern is then used as basis to output layers that selects to appropriate classes. Finally, binary classification are used which appropriately select class 1 and class 0. For our case, class 0 means that the images are benign and class 1 means that the images aremalignant. How our system is implemented, using convolutional neural network is depicted in Figs. 3. The characteristics of the layers for our custom CNN are summarized in Table 1.

3.3 Steps of the System

We followed the steps bellow to classify the images into benign and malignant images.

Step1: Initializing all the images and all the parameters that are needed for system. The huge size of the images is a big problem in computer vision because the very large input data requires considerably more computing capacity to save multiple images, which is time consuming and wastes memory. In our work, we resize all the pictures in order to process them with less memory and graphical computational power and we normalize the pictures to prevent overfitting.

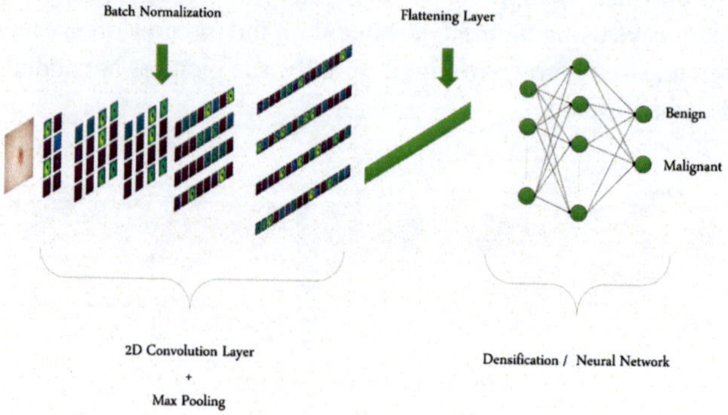

Fig. 3. Example of a skin lesion image is processed by the custom CNN model.

Table 1. Summary of our model layers

Layer (type)	Output shape	Param
Convo2D	(done, 224, 224, 64)	1792
Maxpol2D	(done, 112, 112, 64)	0
dropout	(done, 112, 112, 64)	0
Convol2D	(done, 112, 112, 64)	36928
Maxpol	(done, 56, 56, 64)	0
Droput	(done, 56, 56, 64)	0
Flatten	(done, 200704)	0
Dense	(done, 128)	25690240
Dense	(done, 2)	258
Total params: 25, 729,218; Trainable params: 25,729,218 Non-trainable params: 0		

Step 2: The system loada training and test images, make dictionary of images, takes them as input and saves them into the system. Each of the preprocessed images are saved

in the record along with their classes. From the dataset, benign and malignant images are taken for further processing. The recorded images are used to feed to a convolutional Neural Network.

Step 3: The system build a convolutional neural network model and finds out the prediction. Feeding the preprocessed data to convolutional neural network (CNN). Three types of layers are present in a convolutional Neural Network. That are given bellow:

- Convolution layer: it is the first layer that will process the image as an input system model. The image will be convoluted with a filter to extract feature from the input image that is called the feature map. In our model, we used the 2D convolution with a kernel of 3 * 3 and 64 filters.
- Pooling layer: pooling layers are typically used to reduce the image size, control over-fitting and increase the computation speed. They will be usually inserted in the CNN method after several convolution layer. Pooling layer is used to reduce data using max-pooling or mean pooling, in our model we used the MaxPoling2D with kernel 2 * 2 which Calculate the maximum value for each patch of the feature map.
- Fully connected layer: it is the layer at the end of the architecture used, This layer will connectall the neuron of the previous activation layer. In this stage, all neuron in the input layer need to be transformed into one-dimensional data(flatten). After that softmax activation can be used to classify the pictures.

Step 4: Training with the convolutional neural network that are generated in step 3. We trained our model for 50 epochs with a batch size of 64. Every times the loss of the system decreases to a certain level and the accuracy increase with the epochs. While training epochs is approximately 40, then we don't remark any change in loss so we have to stop our iterations at 50.

Step 5: Testing model on test data. We used k-fold cross validation technique (in our example k = 3) for testing our model, we evaluate the result with the standard evaluation metrics as accuracy, precision, recall, and f1 score.

Step 6: Save the model to file JSON and serialize weight to HDF5 into the system for further testing purposes.

3.4 System Performance

To evaluate and to determine the performance of the model, accuracy, recall, precision, specificity and f1 score metrics are utilized.

$$\text{Accuracy} = \frac{TP + FN}{P + N + FP + FN} \tag{1}$$

$$Recall/Sensitivity = \frac{TP}{TP + FN} \tag{2}$$

$$Specificity = \frac{TN}{TN + FP} \tag{3}$$

$$Precision = \frac{TP}{TP + FP} \tag{4}$$

$$F - Score = \frac{2 * \text{Precision} * \text{Recall}}{\text{Precision} + \text{Recall}} \tag{5}$$

4 Results and Discussion

4.1 Model Accuracy and Loss

We plot in Fig. 4a the variation of the accuracy for the CNN during training. The figure shows an increase in accuracy with the number of epochs for both the training and the test datasets. The final value reported for the accuracy is 0.75 for training dataset, while the accuracy is 0.69 for testing dataset. We show in Fig. 4b the loss during training, the value of loss decrease with the number of epochs for both the training and the test datasets.

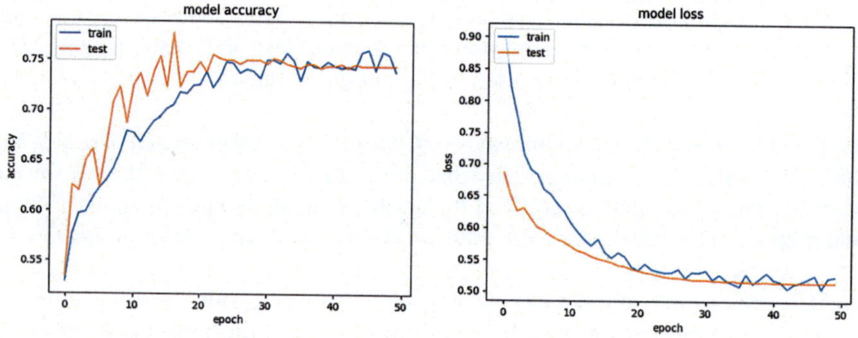

(a) (b)

Fig. 4. **(a)** Variation of the accuracy for the CNN during training, **(b)** The loss during training

4.2 Cross-Validation Model

The cross-validation is a procedure used to avoid overfitting and estimate the skill of the model on new data. It is a very useful technique for assessing the effectiveness of the model, in particularly in cases where we need to mitigate over-fitting. The procedure has a single parameter called k that refers to the number of groups that a given data sample is to be split into. In regards to the accuracy of our cross-validation model for the images classification of skin cancer, we find an accuracy of 73.15%, 74.86%, 70.19% and the accuracy of the model is the average of the accuracy of each fold which is equal to 72.73%. Table 2 showing the result of recall, precision, specificity and F1 Score.

Table 2. Accuracy of 3-Fold cross-validation

Parameter	Result
Recall	76.15
Precision	77.40
Specificity	83.06
F1 score	76.76

5 Conclusion

In this work, we implemented a Convolutional Neural Networks approach for skin lesions classification. We have developed a system that can help patients and doctors to be able to detect or identify skin cancer classes whether it is benign or malignant. The experimental and evaluation part showed that our model achieves good performance in terms of different metrics.

References

1. Jinnai, S., Yamazaki, N., Hirano, Y., Sugawara, Y., Ohe, Y., Hamamoto, R.: The development of a skin cancer classification system for pigmented skin lesions using deep learning. Biomolecules **10**(8), 1123 (2020)
2. Hekler, A., Jochen, S.U., Alexander, H.E., et al.: Superior skin cancer classification by the combination of human and artificial intelligence. Eur. J. Cancer **120**, 114–121 (2019)
3. Sun, R., Deutsch, E., Fournier, L.: INSERM U1030, Département de Radiothérapie, Gustave Roussy, université Paris-Saclay, F-94805, Villejuif, France (2021)
4. Savera, T.R., Suryawan, W.H., Agung, W.S.: Deteksi Dini KankerKulit Menggunakan K-NN dan convolutional neural network. J. JTIIK. **7**(2), 373–378 (2020)
5. Milton, M.A.A.: Automated skin lesion classification using ensemble of deep neural networks in ISIC: skin lesion analysis towards melanoma detection challenge (2018)
6. Jianu, S.R.S., Ichim, L., et al.: Automatic diagnosis of skin cancer using neural networks. In: The XIth International Symposium on Advanced Topics in Electrical Engineering, 28–30 March 2019, Bucharest, Romania (2019)
7. Hameed, N., et al.: Multi-class skin diseases classification using deep convolutional neural network and support vector machine (2018)
8. Zhang, X., et al.: Towards improving diagnosis of skin diseases by combining deep neural network and human knowledge. BMC Med. Inform. Decis. Making **18**(Suppl 2), 59 (2018)
9. Hosny, K.M., et al.: Classification of skin lesions using transfer learning and augmentation with Alex-net. PLOS One **14**, e0217293 (2019)
10. Bechelli, S., Machine, D.J.: Learning and deep learning algorithms for skin cancer classification from dermoscopic images. Bioengineering **9**, 97 (2022). https://doi.org/10.3390/bioengineering9030097

Design of an Adaptive Neuro-Fuzzy Inference System for Photovoltaic System

Salma Benchikh(✉), Tarik Jarou, Elmehdi Nasri, and Lamrani Roa

Advanced Systems Engineering Laboratory, National School of Applied Sciences, Ibn Tofail University, Kenitra, Morocco
{Salm.benchikh,Tarik.jarou,Roa.lamrani}@uit.ac.ma, elmehdi.nasri@gmail.com

Abstract. A photovoltaic module directly converts solar radiation into electrical energy. It has a crucial characteristic known as the maximum power point, which is a single operational point where the maximum of power is given (MPP). The temperature and irradiance of the solar cells in the module determine the voltage and current at this point. An extraction of the maximum power possible of the photovoltaic modules is necessary and this by a tracking mechanism called Maximum Power Point Tracking. The most commonly adopted MPPT techniques are Perturbation et Observation (P&O), Incremental Conductance (IC) et Hill Climbing (HC). They are used for their speed of convergence and simplicity of implementation. However, the maximum power point (MPP) in these methods is affected by environmental conditions. To deal with the above problem, one of the solutions is the utilization of one of the soft computing methods Adaptive Neuro-Fuzzy Inference System is required.

Keywords: MPPT · ANFIS · Fuzzy logic · ANN · Photovoltaic

1 Introduction

A photovoltaic generator can only provide a maximum power for specific current and voltage values. MPPT techniques are used in photovoltaic systems to get the maximum power point. However, this is not easy to achieve. In this paper, an intelligent and innovant approach is developed, that combines two intelligent techniques, namely Fuzzy logic and Artificial Neural Network. The objective of the method is to identify the point of maximum power whatever the variation of the climatic conditions (irradiation, temperature) even in the most unfavorable case, as well as to improve the results obtained with classical algorithms already developed.

An ANFIS (Adaptive Network Based Fuzzy Inference System) is one of the MPPT techniques, it is a hybrid method that integrates the advantages of the two techniques mentioned above. The advantages of fuzzy logic controllers are their capacity to operate with uncertain inputs, with no need for a specific mathematical model, and their capacity to process non-linear data [1]. On the other hand, ANN learning algorithms are capable to operate under imprecise and time-varying conditions with minimal human interaction.

Y. Farhaoui et al. (Eds.): ICAISE 2022, LNNS 635, pp. 352–358, 2023.
https://doi.org/10.1007/978-3-031-26254-8_50

2 Conception of Photovoltaic System

MPPT controller for photovoltaic array is connected to load via DC-DC converter, as shown below in Fig. 1.

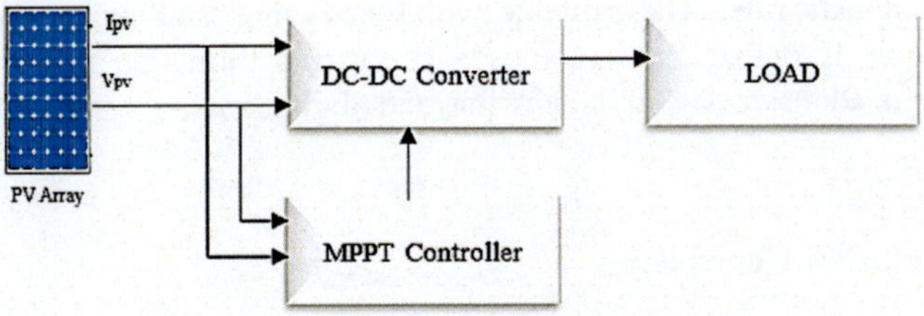

Fig. 1. The block diagram of photovoltaic system

2.1 Characteristics I-V and P-V of Photovoltaic Panel

PV panel is characterized by its I-V and P-V characteristics that change by solar irradiation and temperature. Solar irradiation mainly affects the output current, which increases with the increase of solar irradiation. At the same time, the temperature affects the output voltage, the latter increases with the lower temperature.

3 Improvement of ANFIS for MPPT Controller for Photovoltaic System

(See Figs. 2 and 3).

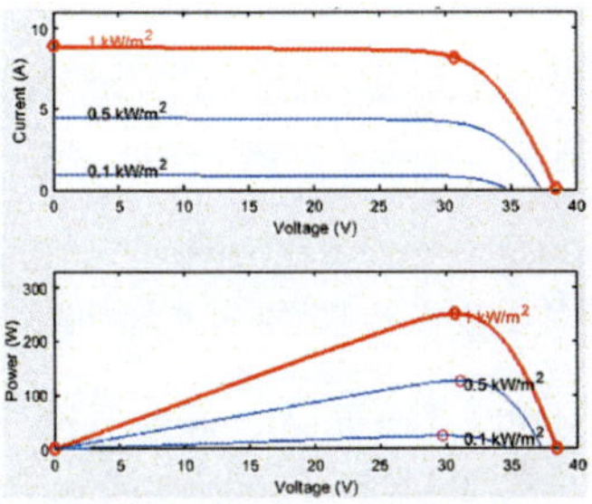

Fig. 2. The PV panel at 25 °C and the specified irradiation.

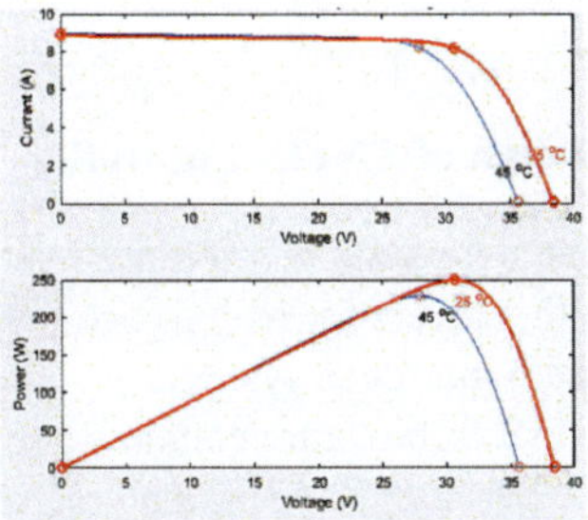

Fig. 3. The PV panel at 1000W/m^2 and the specified temperature.

3.1 Design of Fuzzy Logic Controller

The first part in fuzzy logic is fuzzification, the objective of fuzzification is to define the membership functions for the different variables which allows to fuzzify the input variables. The goal of the fuzzification is to transform the input variables into linguistic variables or fuzzy variables. The second part is Inferences, it links input and output variables by linguistic rules. These rules are combined using "and" and "or" connections.

The last one is defuzzification, it is a fuzzy transformation into a determined information. In addition, signal processing and digital/analog conversion are often required.

3.2 Design of ANN Controller

One of the artificial intelligent MPPT systems that can resolve nonlinear issues is the ANN controller. The input, hidden, and output layers made up the three layers of the ANN controller. The input layer is made up of two neurons. The hidden layer has ten neurons, the output layer contains one neuron, the two input neurons are connected to the hidden layer neurons, and each hidden layer neuron was linked to the output neuron via weighted linking. [2] (Fig. 4).

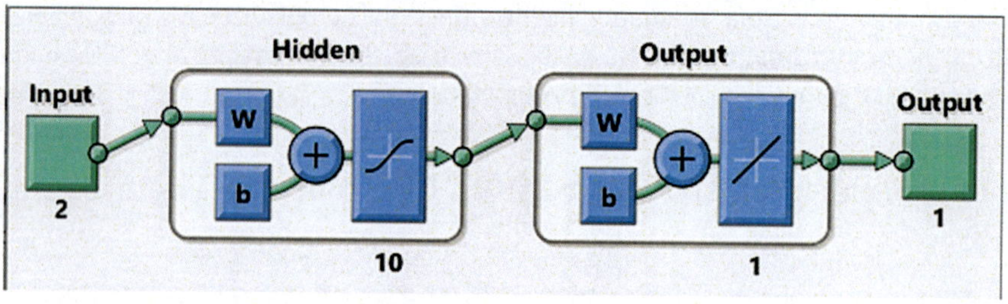

Fig. 4. The ANN design

3.3 Design of ANFIS Controller

Adaptive neuro-fuzzy inference system consists in using a neural network of the MLP type with 5 layers for which each layer is designed to perform a step of a Takagi Sugeno type fuzzy inference system.

ANFIS architecture contains five layers Fig. 5 shows. The square nodes represent an adaptive part they have parameters, and the circle nodes represent non-adaptive sections they do not have parameters. However, each node applies a function to its input signals. The parameters of the adaptive nodes will be modified throughout the training process of ANFIS [4] (Fig. 6).

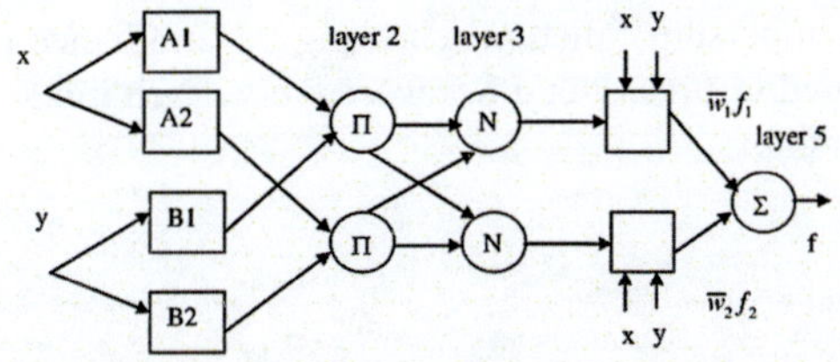

Fig. 5. ANFIS architecture for two rules.

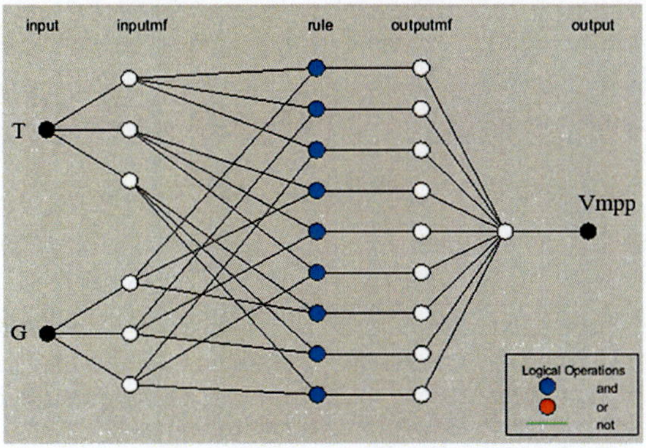

Fig. 6. Architecture of ANFIS system.

The ANFIS model's first layer is intended to fuzzification. The second one is for fuzzy rules; each neuron in this layer corresponds to a fuzzy rule, which is calculated as a function of the activation of the neurons in the first layer's outputs. Each neuron calculates the normalized degree of truth of a particular fuzzy rule in the third layer; the resulting number is the fuzzy rule's contribution to the outcome, hence the name normalization. Defuzzification is the fourth layer, and summation is the last one. The last one is summation; this layer contains of a single neuron that includes the output of ANFIS by summing the outputs of all defuzzification neurons [3].

4 Simulation and Results

The proposed ANFIS controller disposes of two inputs; the solar irradiation (G) and ambient temperature (T) and one output maximum voltage (Vmpp). The ranges of temperature and irradiation are 15 °C to 35 °C and 0W/m^2 to 1000W/m^2 respectively. For each set of T and G value given as input to the PV system, the corresponding Vmpp is measured. Through the simulation, A total of 1,000 data sets were collected for randomly selected irradiance and temperature levels. Each data set consists of T, G and corresponds to a value of Vmpp [6].

The two input membership functions showing the irradiance and the temperature variations are illustrated in Figs. 7 and 8, respectively. Each input and output of fuzzy system has 3 memberships.

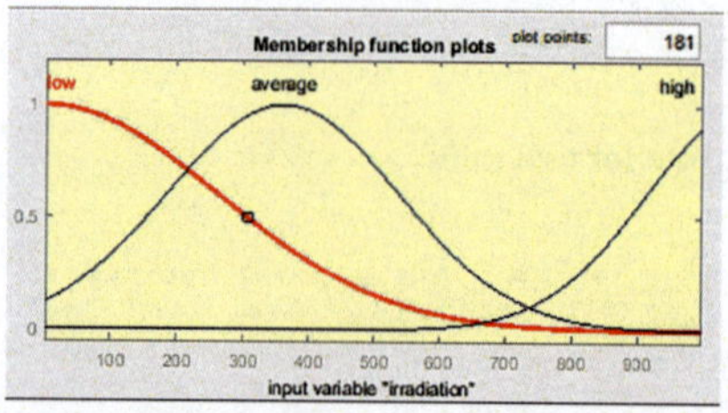

Fig. 7. Membership function plots of irradiation input.

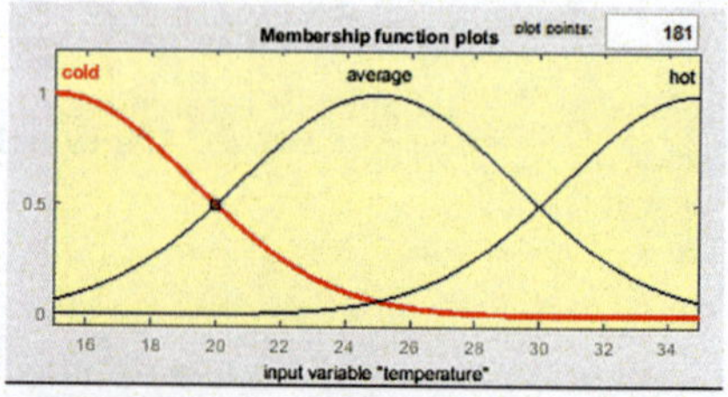

Fig. 8. Membership function plots of temperature input.

The irradiation has as membership functions low, average and high, for the temperature, it has cold, average and hot as membership functions.

Figure 9 demonstrate that by moving the red line of each irradiation or temperature rule, the Vmpp output changes correspondingly.

For example, the output results of the controller, for an irradiation of 499W/m^2 and a temperature of 25 °C, the MPP Voltage equal to 30.7 V.

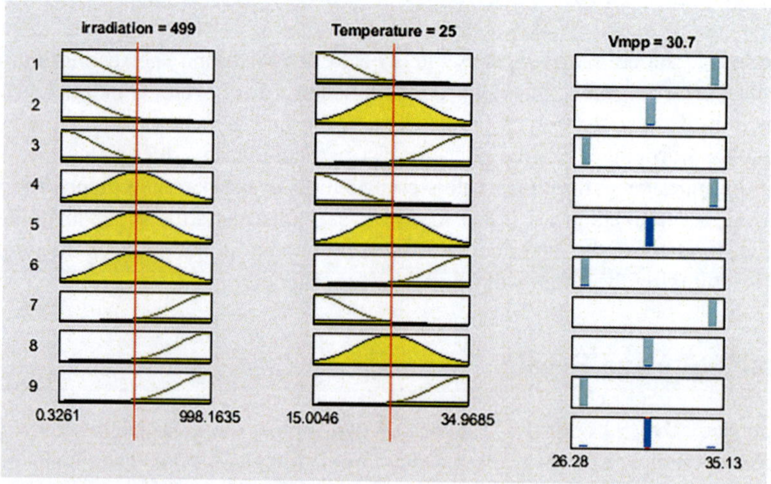

Fig. 9. Input and output relationships of the proposed ANFIS controller.

Figure 9 demonstrate that by moving the red line of each irradiation or temperature rule, the Vmpp output changes correspondingly.

For example, the output results of the controller, for an irradiation of 499 W/m^2 and a temperature of 25 °C, the MPP Voltage equal to 30.7 V (Figs. 10 and 11).

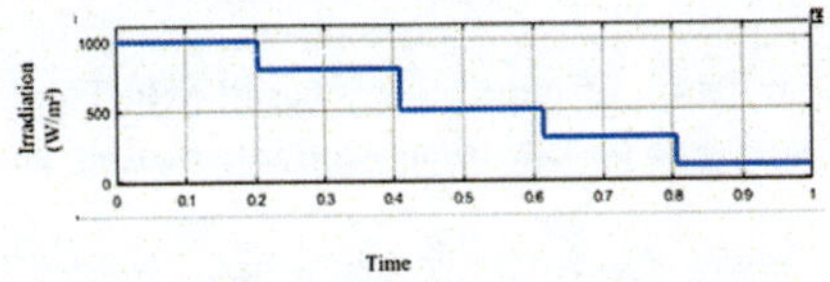

Fig. 10. Variation in irradiation level.

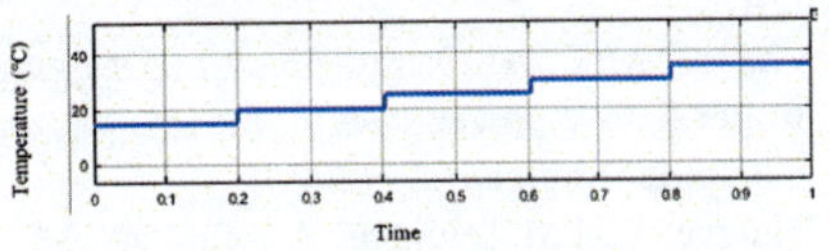

Fig. 11. Variation in temperature level.

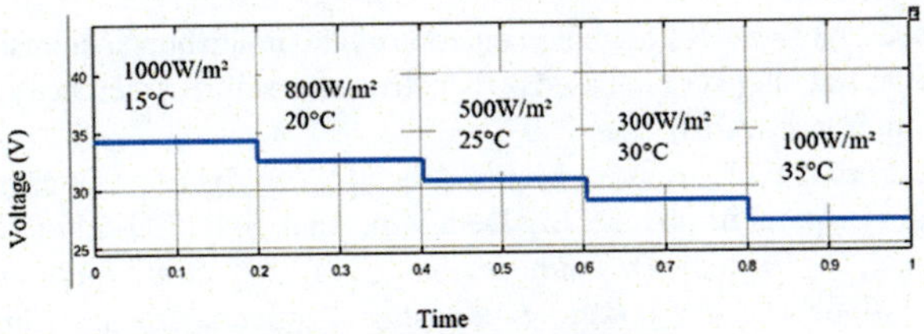

Fig. 12. Vmpp under changing temperature and irradiation levels

Figure 12 presents that the MPP voltage is affected by the temperature, its value changes with the variation of the temperature, however, the ANFIS controller had the rapidity of response and reacts rapidly to changing climate. Comparing the ANFIS controller to other artificial intelligence controllers ANN or FL, the ANFIS controller presents a robustness, a rapidity of response and less oscillations during the metrological change, it has better tracking MPP than the other controllers.

5 Conclusion

To ensure the operation of a photovoltaic generator at its maximum point, MPPT controllers are frequently used, which are specifically developed to extract the maximum power in different climatic changes. In this paper a design of ANFIS controller is presented, the inputs of the controller are the solar irradiation and the ambient temperature, and the output is the MPP voltage, the results of simulation presented in this paper prove the performance of ANFIS controller under all meteorological conditions, as well as the ability to adjust system parameters.

ANFIS controller collect advantages of FL and ANN, it is simple to understand and to implement, it has a higher efficiency under climatic conditions changing, and it does not necessitate too many mathematical calculations or equations.

The results obtained in this work encourage further research in this direction and to improve it. In the future work a simulation of this model proposed will applicated for remote and isolated installation.

Acknowledgment. Benchikh Salma would thank the CNRST for financing research activities.

References

1. Khosrojerdi, F., Taheri S., Cretu, A.M.: An adaptive neuro-fuzzy inference system-based MPPT controller for photovoltaic arrays. In: 2016 IEEE Electrical Power and Energy Conference (2016)
2. Heelan, M.Y., Al-Qrimli, F.A.M.: Design and simulation of neuro-fuzzy based MPPT controller for PV power system. In: International Conference on Electrical, Communication, and Computer Engineering (2020)
3. Ndiaye, E.H.M., Ndiaye, A., Tankari, M.A., Lefebvre, G.: Adaptive neuro-fuzzy inference system application for the identification of a photovoltaic system and the forecasting of its maximum power point. In: The 7th International Conference on Renewable Energy Research and Applications (2018)
4. Abadi, I., Imron, C., Noriyati, R.D.: Implementation of maximum power point tracking (MPPT) Technique on solar tracking system based on adaptive neuro-fuzzy inference system (ANFIS). In: E3S Web Conference (2018)
5. Boutahir, M.K., Farhaoui, Y., Azrour, M., El Allaoui, A.: Effect of feature selection on the prediction of direct normal irradiance. Big Data Min. Anal. **5**(4), 309–317 (2022)
6. Sheik Mohammed, S., Devaraj, D., Imthias Ahamed, T.P.: Maximum power point tracking system for stand-alone solar PV power system using adaptive neuro-fuzzy inference system. In: Biennial International Conference on Power and Energy Systems: Towards Sustainable Energy (2016)

Disturbance Observer-Based Adaptive Sliding Mode Control for Autonomous Vehicles

Rachid Alika[1](✉), El Mehdi Mellouli[2], and El Houssaine Tissir[1]

[1] LISAC Laboratory, Department of Physics, Faculty of Sciences Dhar El Mehraz, University Sidi Mohammed Ben Abdellah, Fez, Morocco
{rachid.alika,elhoussaine.tissir}@usmba.ac.ma

[2] LISA Laboratory, National School of Applied Sciences, University Sidi Mohammed Ben Abdellah, Fez, Morocco

Abstract. This study focuses on the lateral control of an autonomous vehicle, the super-twisting sliding mode controller was developed on this system, the disturbances were considered and collected by an observer. The optimization of the control parameters of the super-twisting mode using the Particle Swarm Optimization (PSO) algorithm has been performed. We minimize the lateral position error of the autonomous vehicle in order to follow its reference trajectory. In this autonomous vehicle, the steering angle is the control input and the lateral displacement is the output. The simulations are carried out by MATLAB and show that the results of the control by super-twisting sliding mode with optimization of the PSO parameters and observation of the disturbances are better.

Keywords: Autonomous vehicles · Lateral dynamics · STSMC · PSO · Disturbance observer

1 Introduction

Several articles discuss the control of the autonomous vehicle to ameliorate the lateral dynamics of the autonomous vehicle. Different control methods, observers, artificial intelligence methods have been introduced in order to make the system more efficient. In ref [1], we have developed a strategy for controlling the lateral dynamics of an autonomous vehicle using the particle swarm optimization and neural network (RBFNN). In [2], we approximate the equivalent control of STSMC by fuzzy logic to control the autonomous vehicle. Optimization of STSMC parameters by PSO to control vehicle lateral dynamics was investigated in [3]. An in-depth study has been achieved on the lateral control of the autonomous vehicle in [4]. Ref [5] developed a new output feedback control methodology for nonlinear active suspension systems. The ref [6] is concerned in the control H_∞ of the active suspensions of the vehicle.

The sliding mode control used for the autonomous system in [2, 3–7]. The sliding mode control was applied to the three-tank system in [8, 9] and was also used for a two-link robot system in [10].

Y. Farhaoui et al. (Eds.): ICAISE 2022, LNNS 635, pp. 359–366, 2023.
https://doi.org/10.1007/978-3-031-26254-8_51

The Observer is one of the tools that help the control to achieve its operation, the Disturbance Observer is widely employed to attenuate and estimate disturbances. Ref. [11] realizes a new observer-based controller design for input disturbance systems.

The lateral dynamics of an autonomous vehicle controlled by a STSMC in this paper, furthermore the disturbances were considered and collected by an Atsuo disturbance observer. Optimization of the STSMC parameters by PSO algorithm was performed. The objective of this work is minimized the lateral error of the system in order to follow its reference path. A comparison between the use of the standard parameters of the SMC [7] and the parameters optimized by PSO only and also with the parameters optimized by PSO combined with the observation of disturbances. The illustrations show the robustness and the precision of the proposed method.

This paper has been organized in the manner described below. We presented the autonomous vehicle's dynamic model and the reference trajectory model in Sect. 2. We developed a strategy of control, optimized the controller parameters and observed the disturbances in Sect. 3. We present in Sect. 4, the proposed method and the results obtained. The conclusions and future work presented in Sect. 5.

2 Vehicle Dynamics Model and Reference Trajectory

2.1 Vehicle Bicycle Model

The dynamic bicycle model, shown in Fig. 1, allows representing the lateral behavior of the autonomous vehicle [7]. In this model, the vehicle is considered as symmetrical and the sideslip angles are equal (same axle). The force of the tire used represented by a linear model, the bicycle model composed of lateral and yaw dynamics, the angles are supposed small, V_x represents velocity longitudinal.

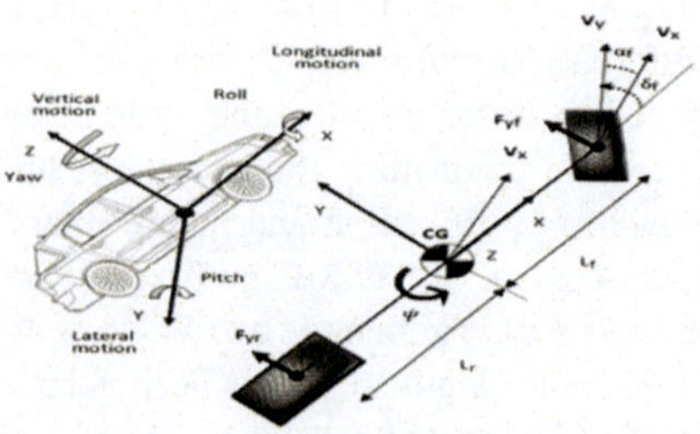

Fig. 1. Bicycle model vehicle.

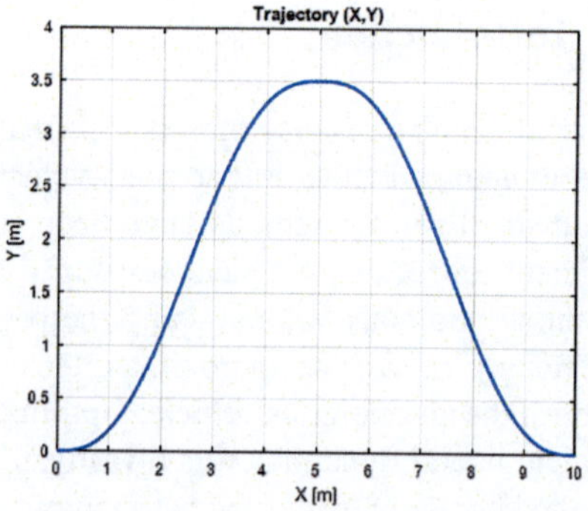

Fig. 2. Double lane change maneuver.

The bicycle model shown as follows,

$$\begin{cases} y = f_1(x,t) + g_1(x,t)u(t) + d_1(x,t) \\ \psi = f_2(x,t) + g_2(x,t)u(t) + d_2(x,t) \end{cases} \quad (1)$$

where y is the lateral position and ψ is the yaw angle.

$$f_1(x,t) = -\frac{(C_f + C_r)}{mV_x}\dot{y} - \left(\frac{L_fC_f - L_rC_r}{mV_x} + V_x\right)\dot{\psi}$$

$$f_2(x,t) = -\frac{L_fC_f - L_rC_r}{I_zV_x}\dot{y} - \frac{L_f^2C_f + L_r^2C_r}{I_zV_x}\dot{\psi}$$

$$g_1(x,t) = \frac{C_f}{m},\ g_2(x,t) = \frac{L_fC_f}{I_z},\ u(t) = \delta$$

$$d_1(x,t) = 0.1\ f_1(x,t),\ d_2(x,t) = 0.1\ f_2(x,t)$$

where the state vector is $x = [\dot{y},\ \dot{\psi}]^T$, the control input is $u(t) = \delta$, $f(x,t) = [f_1(x,t), f_2(x,t)]^T$ and $g(x,t) = [g_1(x,t),\ g_2(x,t)]^T$ are functions of dynamics system. The disturbances considered to constitute 10% of the autonomous vehicle dynamics, the disturbances are $d(x,t) =,\ [d_1(x,t),\ d_2(x,t)].^T$.

2.2 Reference Path Model

In this study, the maneuver trajectory corresponds to double lane changes, the one illustrated in Fig. 2. The lane change trajectory [2], represented by a five-degrees polynomial function (Quintic), where the time conditions are satisfied in (2) as follows,

$$y(t) = at^5 + bt^4 + ct^3 + dt^2 + et + f \tag{2}$$

3 Control Strategy

3.1 Super-Twisting Control

The super-twisting control has two components [3], u_1 is a function of the sliding variable and function u_2 is determined by its derivative. The control law given by,

$$u_{ST}(t) = u_1u_2 \begin{cases} u_1 = -\alpha|s|^{\tau}\mathrm{sign}(s),\ \tau \in]0, 0.5] \\ \dot{u}_2 = -\beta\mathrm{sign}(s) \end{cases}$$

3.2 Cose of Disturbance Observer

We chose the disturbances observer suggested by Atsuo et al. in [12] as,

$$\hat{d}_1 = k_1\left(\dot{y} - \hat{\theta}\right) \tag{3}$$

$$\hat{\theta} = \hat{d}_1 + g_1(x,t)u(t) + k_2\left(\dot{y} - \hat{\theta}\right) + f_1(x,t) \tag{4}$$

where $\hat{d}_1$ and $\hat{\theta}$ are estimation value of d_1 and $\dot{y}$ respectively, $k_1 > 0, k_2 > 0$.

We chose lyapunov function as,

$$V_1 = \frac{1}{2k_1}\tilde{d}_1^2 + \frac{1}{2}\tilde{\theta}^2 \tag{5}$$

where $\tilde{d}_1 = d_1 - \hat{d}_1$ and $\tilde{\theta} = \dot{y} - \hat{\theta}$.

From the Lyapunov function (5), we can obtain its time derivative by,

$$\dot{V}_1 = \frac{1}{k_1}\tilde{d}_1\dot{\tilde{d}}_1 + \tilde{\theta}\dot{\tilde{\theta}} = \frac{1}{k_1}\tilde{d}_1\left(\dot{d}_1 - \dot{\hat{d}}_1\right) + \tilde{\theta} = \left(\ddot{\lambda} - \dot{\hat{\theta}}\right).$$

Replace (3) and (4), we have,

$$\dot{V}_1 = \frac{1}{k_1}\tilde{d}_1\dot{d}_1 - \frac{1}{k_1}\tilde{d}_1\left(-k_1\left(\hat{\theta} - \dot{y}\right)\right) + \tilde{\theta}(f_1(x,t) + g_1(x,t)\,u(t) + d_1(x,t))$$
$$-\left(\hat{d}_1 + g_1(x,t)\,u(t) - k_2\left(\hat{\theta} - \dot{y}\right) + f_1(x,t)\right)$$
$$\dot{V}_1 = \frac{1}{k_1}\tilde{d}_1\dot{d}_1 - \tilde{d}_1\left(\tilde{\theta}\right) + \tilde{\theta}\left(d_1 - \left(\hat{d}_1 - k_2\left(\hat{\theta} - \dot{y}\right)\right)\right)$$

We chose k_1 relatively large we can get $\frac{1}{k_1}\dot{d}_1 \approx 0$, also we chose k_2 relatively large we can get,

$$\dot{V}_1 = \frac{1}{k_1}\tilde{d}_1\dot{d}_1 - k_2\tilde{\theta}^2 \le 0 \tag{6}$$

3.3 Design of SMC Control Based Observer

The disturbance observer-based controller allows minimizing lateral error while maintaining the desired path for the autonomous vehicle. The steering angle is the control input and the lateral position is the output of the autonomous vehicle. The design of the SMC is established, the sliding surface is given by,

$$s = \dot{e}_y + \lambda e_y \tag{7}$$

where e_y is the lateral error, y_{ref} is the desired lateral position, y is the actual position and λ denotes a positive design parameter. The lateral position error is,$e_y = y - y_{ref}$

We have the time derivative of (7) is,

$$\dot{s} \underset{y}{=} e + \lambda\dot{e}_y = a_y - a_{y_{ref}} + \lambda\dot{e}_y = y + V_x\dot{\psi} - a_{y_{ref}} + \lambda\dot{e}_y \tag{8}$$

where a_y is the current lateral acceleration and $a_{y_{ref}}$ is desired lateral acceleration. Since $a_y = y + V_x\dot{\psi}$ [4].

Substituting Eq. (1) into Eq. (8), the following is obtained,

$$s. = f_1(x,t) + g_1(x,t)u(t) + d_1 + V_x\dot{\psi} - a_{y_{ref}} + \lambda\dot{e}_y \tag{9}$$

The lateral control signal consists of equivalent and switching control laws as follows,

$$u(t) = \frac{1}{g_1(x,t)}\left[u_{eq}(t) + u_{sT}(t)\right] \tag{10}$$

The equivalent control law is defined as,

$$u_{eq}(t) = a_{y_{ref}} - f_1(x,\ t) - V_x\dot{\psi} - \hat{d}_1 - \lambda\dot{e}_y \tag{11}$$

and the switching control law is,

$$u_{sT}(t) = -\alpha|s|^{1/2}\text{sign}(s) - \beta\int \text{sign}(s)dt \tag{12}$$

where α and β are positive design parameters.

The Eqs. (11) and (12) are replaced in the Eq. (10), we get the following,

$$u(t) = \frac{1}{g_1(x,t)}\left[a_{y_{ref}} - f_1(x,t) - V_x\dot{\psi} - \hat{d}_1 - \lambda\dot{e}_y - \alpha|s|^{1/2}\text{sign}(s) - \beta\int sign(s)dt\right] \tag{13}$$

where $\tilde{d}_1 = d_1 - \hat{d}_1$.

We chose lyapunov function as, $V_2 = \frac{1}{2}s^2$. The time derivative of V_2 can be obtained

$$\dot{V}_2 = s\dot{s} = s((f_1(x,t) + g_1(x,\ t)\ u(t) + d_1 + V_x\dot{\psi} - a_{y_{ref}} + \lambda\dot{e}_y) \tag{14}$$

Substituting Eqs. (13) into Eq. (14), we obtained,

$$\dot{V}_2 = s(-\hat{d}_1 + d_1 - \alpha|s|^{1/2}\text{sign}(s) - \beta\int \text{sign}(s)dt) = s\tilde{d}_1 - \alpha|s|^{1/2}|s| - |s|\int\beta\ dt \leq 0 \tag{15}$$

the stability of this system is assured if α and β must satisfy, $\alpha \geq \left|\tilde{d}_1\right|$ and $\beta \geq 0$ or $\alpha \geq 0$ and $\beta \geq \left|\tilde{d}_1\right|/t$, where t_0 the initial time is greater than 0.

The closed system lyapunov function can be represented as follows,

$$V = V_1 + V_2 = \frac{1}{2k_1}\tilde{d}_1^2 + \frac{1}{2}\tilde{\theta}^2 + \frac{1}{2}\ s^2.$$

we can demonstrate a clear from Eq. (6) and (15) that the system is stable $\dot{V} \leq 0$.

The work is done at medium speed, the yaw angle effect is low for this the yaw dynamics is not taken into consideration.

3.4 Particle Swarm Optimization

This technique is inspired by a social model which represents the swarm theorem [3]. The update of each particle is based on Eq. (16),

$$\begin{cases} v_i(j+1) = wv_i(j) + c_1r_1(p_i(j) - x_i(j)) + c_2r_2\big(p_g(j) - x_i(j)\big) \\ x_i(j+1) = x_i(j) + v_i(j+1) \end{cases} \tag{16}$$

where j is the number of iterations and the inertia weight isw. $c_1$and c_2 are the random numbers between [0,1]. The velocity of the particle is represented by v_i and the position of the particle is x_i. $p_i is$ personal best and p_g is global best.

The objective function is given by,

$$F_{obj} = e_y + \lambda \dot{e}_y \tag{17}$$

according to (17) and (8), $F_{obj} = \dot{s}$, with $J_{min}(\lambda, \alpha, \beta) = \dot{s}$, find λ, α and β when the minimum global value of $\dot{s}$ is found, where λ, α and β are the variables of this function.

4 Simulation Results

In this work we realized, the optimization of the STSMC controller's parameters (λ, α, β) by the PSO algorithm, as well as the observation of the disturbances by the Atsuo observer. We have added disturbances that can be affected the autonomous vehicle, the disturbances considered constitute 10% of the dynamics of the system. The longitudinal speed used is $V_x = 13.5$ (m/s).

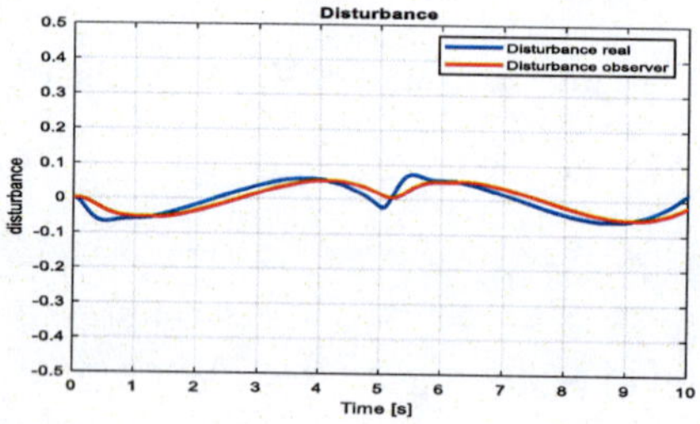

Fig. 3. Disturbances real and observed.

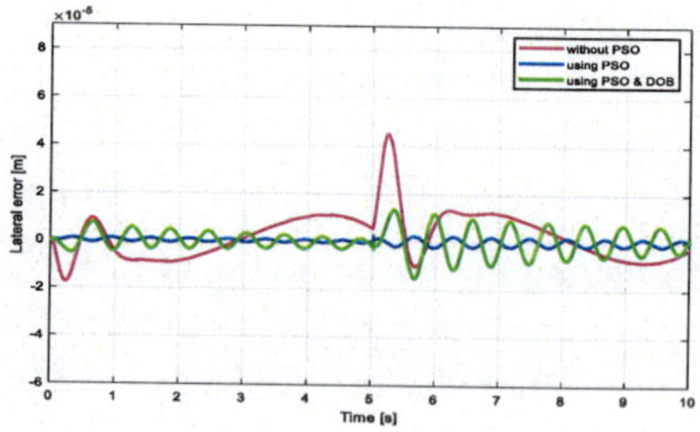

Fig. 4. Lateral error.

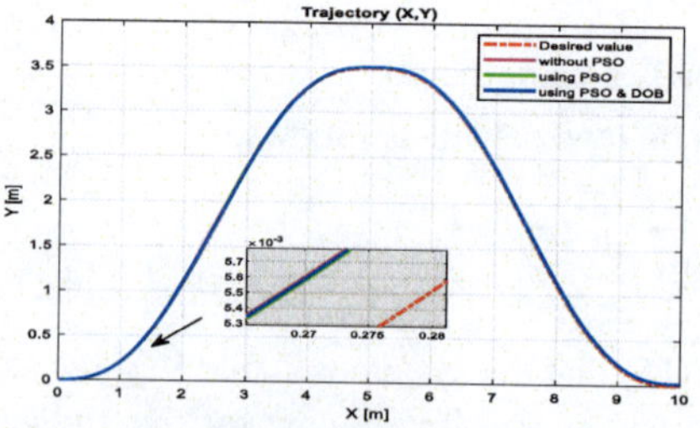

Fig. 5. Trajectory.

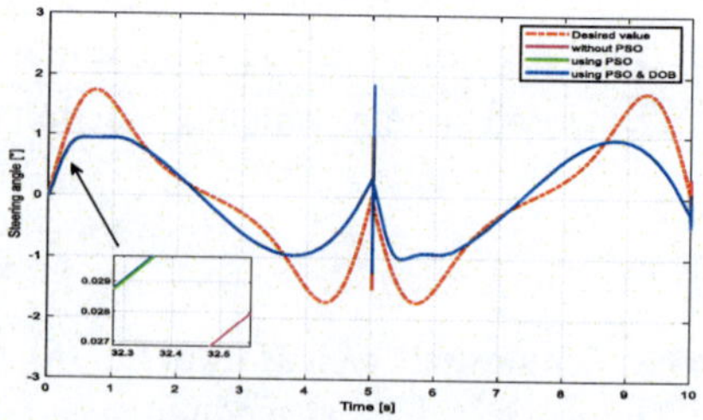

Fig. 6. Steering angle.

The STSMC controller robustness is compared with the standard parameters of the SMC controller used in [7] and with the optimization parameters λ, α and β shown in Figs. 3, 4, 5 and 6. We have clearly noticed the results obtained by optimizing STSMC parameters with PSO are more acceptable than those find by the standard SMC parameters. And the results obtained by STSMC combined with PSO and disturbance observer are also acceptable than the results obtained by the SMC controller alone.

Figure 3 shows the variations of the disturbances affecting the system and the disturbances observed. The observer is efficient, and the disturbances observed are almost the same disturbances considered in the system.

Table 1 displays the results of the optimal parameters found by the PSO algorithm.

Table 1. STSMC parameter values found by particle swarm optimization (PSO).

Parameters	Value	Fitness function
λ	400	−0.1591
α	0.2	
β	2.9	

Table 2. Maximum error value obtained by the different methods.

Methods	Max error value
STSMC Without PSO	$4.54\ 10^{-5}$
STSMC With PSO	$0.23\ 10^{-5}$
STSMC with PSO & DOB	$1.34\ 10^{-5}$

The lateral error represented in Fig. 4 shows that the STSMC is suitable for following the reference path with low error in all cases. The large value of the lateral displacement error is $4.54\ 10^{-5}$ (with normal SMC parameters) and $0.23\ 10^{-5}$ in the second case (with SMC parameter optimization by PSO) in transient mode and does not exceed $1.34\ 10^{-5}$ in the case of the use of the STSMC combined with the PSO and the disturbance observer, the values that we compared are shown in Table 2.

Figure 5 illustrates the different trajectories, it can be seen that all the control methods allow the autonomous vehicle to follow the desired path. And the STSMC controller modifies better than others.

The progression of the steering angle appears in Fig. 6. The dynamic variables obtained by STSMC are better in all cases. The steering angle variation find by STSMC-PSO is lower than the steering angle variation find by SMC alone. And the steering angle variation find by STSMC with PSO and the disturbance observer is greater than the steering angle variation find by SMC alone.

5 Conclusions

In this document, the dynamic lateral of an autonomous vehicle is controlled by a disturbance observer-based STSMC. Disturbances affecting the system are observed. The parameters of this controller are optimized. The different tests have clearly shown the strength of the proposed control (robustness, precision), that appear from the weak amplitude of lateral error during the transient periods. The comparison of the robustness of the STSMC controller with PSO without and with a disturbance observer, and of the SMC controlled only was made. The robustness of STSMC controller with PSO is very excellent compared to SMC controller alone and better than STSMC controlled with PSO based on Atsuo Disturbance Observer.

An in-depth study of artificial intelligence combined with SMC will be developed.

References

1. Alika, R., Mellouli, El M., Tissir, E.H.: Adaptive modified super-twisting sliding mode control based on PSO with neural network for lateral dynamics of autonomous vehicle. Int. J. Model. Identif. Control **42**(4) (2023)
2. Alika, R., El Mellouli, M., Tissir, E.H.: Adaptive higher-order sliding mode control based fuzzy logic TS for lateral dynamics of autonomous vehicles. In: 12th International Conference on ICICS, Valencia, Spain, pp. 358–363 (2021). https://doi.org/10.1109/ICICS52457.2021.9464623
3. Alika, R., El Mellouli, M., Tissir, E.H.: Optimization of higher-order sliding mode control parameter using particle swarm optimization for lateral dynamics of autonomous vehicles. In: 1st International Conference on IRASET, Morocco, 16–19 April (2020). https://doi.org/10.1109/IRASET48871.2020.9092119
4. Rajamani, R.: Vehicle Dynamics and Control. Springer, New York (2012)
5. Mrazgua, J., Chaibi, R., Tissir, E.H., Ouahi, M.: Static output feedback stabilization of T-S fuzzy active suspension systems. J. Terramech. **97,** 19–27 (2021). https://doi.org/10.1016/j.jterra.2021.05.001, (2021)
6. Mrabah, L., Mrazgua, J., Ouahi, M., Tissir, E.L.H.: $H_{\infty }$ control for half vehicle active suspension systems in finite frequency domain. In: Saka, A., et al. (eds.) CPI 2019. LNME, pp. 1–12. Springer, Cham (2021). https://doi.org/10.1007/978-3-030-62199-5_1
7. Tagne, G., Talj, R., Charara, A.: Higher-order sliding mode control for lateral dynamics of autonomous vehicles, with experimental validation. In: IEEE Intelligent Vehicles Symposium (IV), Gold Coast: Australia (2013). https://doi.org/10.1109/IVS.2013.6629545
8. El Mellouli, M., Boumhidi, I.: Direct adaptive fuzzy sliding mode controller without reaching phase for an uncertain three-tank-system. Int. J. Model. Identif. Control **25**(4), 335–342 (2016)
9. El Mellouli, M., Alfidi, M., Boumhidi, I.: Fuzzy sliding mode control for three-tank system based on linear matrix inequality. Int. J. Autom. Control **12**(2), 237–250 (2018)
10. Naoual, R., El Mellouli, M., Sefriti, S., Boumhidi, I.: Fuzzy sliding mode control for the two-link robot. Int. J. Syst. Control Commun. **6**(1), 84–96 (2014). https://doi.org/10.1504/IJSCC.2014.062808
11. El Fezazi, N., Lamrabet, O., El Haoussi, F., Tissir, E.H.: New observer-based controller design for delayed systems subject to input saturation and disturbances. Iranian J. Sci. Technol. Trans. Electr. Eng. **44**(3), 1081–1092 (2019). https://doi.org/10.1007/s40998-019-00298-0
12. Atsuo, K., Hiroshi, I., Kiyoshi, S.: Chattering reduction of disturbance observer based sliding mode control. IEEE Trans. Ind. Appl. **20**(2), 456–461 (1994)

Electric Vehicle Backstepping Controller Using Synchronous Machine

Chaou Youssef[1](✉), Ziani Said[2], and Daoudia Abdelkarim[1]

[1] Department of Physic, Faculty of Science and Technology Errachidia, Moulay Ismail University, BP-509 Boutalamine, 52 000 Errachidia, Morocco
yo.chaou@edu.umi.ac.ma

[2] Department of Electrical Engineering, High School of Technology, Hassan II University, Casablanca, Morocco

Abstract. This paper's contribution is mainly centered on the nonlinear control method, evaluated for the power train system of single-engine electric cars, primarily employing the feedback control approach based on the nonlinear model. The computations of the investigated system are provided, evaluated, and highlighted to demonstrate and highlight the performance and resilience of the control approach utilized in the face of disturbances and parametric fluctuations.

Keywords: Electric vehicle · Nonlinear control · Traction chain · Backstepping technique

1 Introduction

The concept of utilizing electric energy to propel cars is not new; it was conceived as soon as the first electric motors began to operate, with the discovery of batteries preceding this development. Prototype railway cars were propelled by electromagnetic motors driven by batteries as early as the 1840s. Several electric vehicles powered by electrochemical accumulators were created around the end of the twentieth century [1, 2]. To increase the performance of PMSM and, by extension, the performance of electric vehicle drives, robust and nonlinear control approaches with the capacity to reject disturbances have been implemented. These nonlinear control approaches include input-output linearization control [3], sliding mode control [4], backstepping control [5, 6], and DTC. In this paper, the authors provide a backstepping nonlinear control approach for the drive train of single-motor electric cars, precisely, the permanent magnet synchronous machine (PMSM).

2 Modeling of the Elements of the Traction Chain

2.1 Vehicle Dynamics

The vehicle is modeled as a moving mass subjected to the driving force F_t [9]. The different forces to which the moving vehicle is subjected:

Y. Farhaoui et al. (Eds.): ICAISE 2022, LNNS 635, pp. 367–373, 2023.
https://doi.org/10.1007/978-3-031-26254-8_52

According to Newton's second law, the acceleration of the vehicle can be written as:

$$\frac{dV}{dt} = \frac{F_t - F_R}{M} \tag{1}$$

Then

$$F_t = MgC_{rr} + \frac{1}{2}\rho S_f C_x (V - V_{vent})^2 + Mg\sin(\alpha) + M.\frac{dV}{dt} \tag{2}$$

With
F_t: The total traction force of the vehicle
F_R: The total resistance force of the vehicle
$F_r = MgC_{rr}$: Rolling resistance force of the wheels
$F_a = \frac{1}{2}\rho S_f C_x (V - V_{vent})^2$: Aerodynamic force
$F_g = Mg\sin(\alpha)$: Tilting force
$F_{acc} = M.\frac{dV}{dt}$: Acceleration force
Note that the M is the total mass of the vehicle, g is the acceleration of gravity, ρ is air density (kg/m^3), S_f is frontal section of the vehicle, V is the vehicle speed, V_{vent} is wind speed and C_{rr} is rolling coefficient.

2.2 Model of the Electric Motorization of the Electric Vehicle

The mathematical model of the PMSM is given as that in [6]. The dynamic part of the vehicle to be studied is coupled to the electric motorization through a gearbox, whose n is the speed reduction ratio. The relationship between the wheel speed and the motor speed is given by:

$$\Omega = n.\Omega_{wleel} \tag{3}$$

The dynamic equation for the rotation of each drive wheel (two-wheel vehicle) is given by:

$$J_r\dot{\Omega}_{wheel} = C_t - R\frac{F_R}{2} \tag{4}$$

$$J_r\dot{\Omega}_{wheel} = C_t - \frac{R}{2}(M\frac{R}{n}\dot{\Omega}_{wheel} + F_R) \tag{5}$$

From this Eq. (5), we find the traction torque of the wheel:

$$C_t = \frac{J_r}{n}\dot{\Omega} + \frac{R}{2}(M\frac{R}{n}\dot{\Omega} + F_R) \tag{6}$$

The motor rotation equation is expressed as follows:

$$J_m\dot{\Omega} + f\Omega = C_{em} - \frac{C_t}{n} \tag{7}$$

From this Eq. (6) and (7), we find finally:

$$J\dot{\Omega} + f\Omega = C_{em} - C_r \tag{8}$$

With: $J = J_m + \frac{J_r}{n^2} + M\frac{R^2}{2n^2}$ and $C_r = \frac{R}{2n}F_R$

Considering the electrical equation of the PMSM motor, the global motorization model of the electric vehicle is written:

$$\begin{bmatrix} \frac{di_d}{dt} \\ \frac{di_q}{dt} \\ \frac{d\Omega}{dt} \end{bmatrix} = \begin{bmatrix} -\frac{R_s}{L_d}i_d + \frac{L_q}{L_d}\omega i_q \\ -\frac{R_s}{L_q}i_d - \frac{L_d}{L_q}\omega i_d - \frac{\varphi_f}{L_q}\omega \\ P\frac{\varphi_f}{J}i_d - \frac{P(L_q - L_d)}{J}i_d i_d - \frac{f}{J}\Omega \end{bmatrix} + \begin{bmatrix} \frac{1}{L_d} & 0 & 0 \\ 0 & \frac{1}{L_q} & 0 \\ 0 & 0 & \frac{-1}{J} \end{bmatrix} \begin{bmatrix} v_d \\ v_d \\ C_r \end{bmatrix} \quad (9)$$

Note that $\Omega = \Omega_{mec}$ is the PMSM rotor speed, R_s is the stator resistance, L_d and L_q are the inductances in the (d,q) frame, i_d and i_q are the stator currents, ω is the electrical speed and φ_f is the PMSM remanent flux. The Eq. (9) represents the dynamic model of a nonlinear system whose general form is the following:

$$\dot{X} = F(X) + G(X)U \quad (10)$$

3 Backstepping Control Design

This technique uses one state as a virtual controller to another state since the system is in triangular feedback form. The design of backstepping control, nonlinear systems or subsystems of the form (9).

$$\begin{aligned} \dot{x}_1 &= f_1(x_1) + g_0(x_1)x_2 \\ \dot{x}_2 &= f_2(x_1, x_2) + g_0(x_1, x_2)x_3 \\ &\vdots \\ \dot{x}_n &= f_n(x_1, \ldots, x_n) + g_0(x_1, \ldots, x_n)u \end{aligned} \quad (11)$$

where: $x = [x_1, x_2, \ldots, x_n]^T \in \Re^n, u \in \Re$

We wish to make the output $y = x$ follow the reference signal y_{ref} supposed to be known. The system being of order n, the design is done in n steps.

4 Designed of Backstepping Controller for EV Speed Control

We follow the same reasoning, same procedure and same calculation as in [6] in order to determine the PMSM control variables references (10) of the electric vehicle.

$$\begin{cases} i_{dref} = 0 \\ v_{dref} = L_d[K_1 e_1 + \dot{i}_{dref} + \frac{R_s}{L_d}i_d - \frac{\omega L_q}{L_d}i_q] \\ i_{qref} = (k_2 e_2 + \dot{\omega}_{ref} + \frac{f}{J}\omega + \frac{1}{J}C_r)(\frac{J}{P\varphi_f}) \\ v_{qref} = L_q[K_3 e_3 + \dot{i}_{qref} + \frac{R_s}{L_q}i_q + \frac{P\Omega}{L_q}(L_d i_d + \varphi_f)] \end{cases} \quad (12)$$

5 Simulation Results and Discussion

5.1 Simulations Results

The adopted parameter values: $K_1 = K_2 = 10000$ and $K_3 = 5000$. The electric vehicle parameters used for the simulations are given in Table 1.

Table 1. The electric vehicle parameters.

Parameters	Values	Parameters	Values
P	4	M	1000 [kg]
R_s	0. 6 [Ω]	S_f	1.9 [m^2]
L_d	0.0014 [mH]	C_x	0.25
L_q	0.0028 [mH]	g	9.8
φ_f	0.2 [Web]	ρ	1.23 [kg /m^3]
J_m	$0.2 \left[N.mS^2/rad\right]$	C_{rr}	0.01
J_r	$0.14 \left[N.mS^2/rad\right]$	n	4
f	0.3	R	0.3 [m]

5.2 Simulations and Discussion

The First Test. This test treats the problem of trajectory tracking. We consider that the vehicle traces a straight path at a reference speed fixed at 80km/h speed step (see Fig. 2) with the slope of 30% at (t = 2.5 s) and the wind speed is fixed at 10 m/s will be applied on the road profile. We notice from the (Fig. 2), that the speed of the vehicle tracking well the reference speed, with a response time 1,8 s. After 1.8 s, it is the permanent regime. At the moment of passage by the slope, the vehicle kept the same speed. The electromagnetic torque was also represented in the (Fig. 3, Fig. 4).

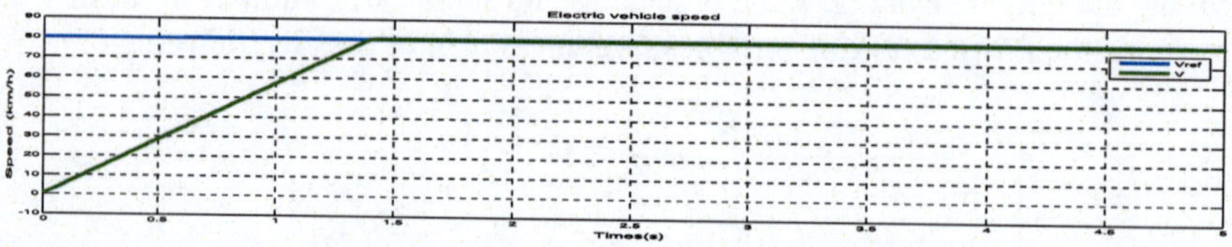

Fig. 2. Speed tracking response for reference for constant speed with the slope of 30% at (t = 2.5 s)

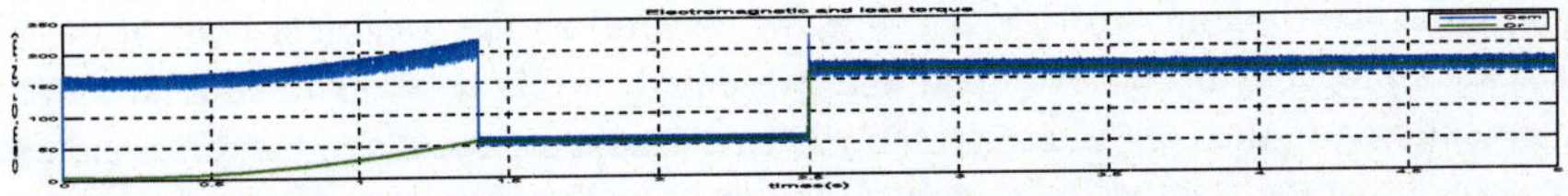

Fig. 3. The behavior of the driving torque C_r and the electromagnetic torque C_{em} in case with the slope of 30% at (t = 2.5 s)

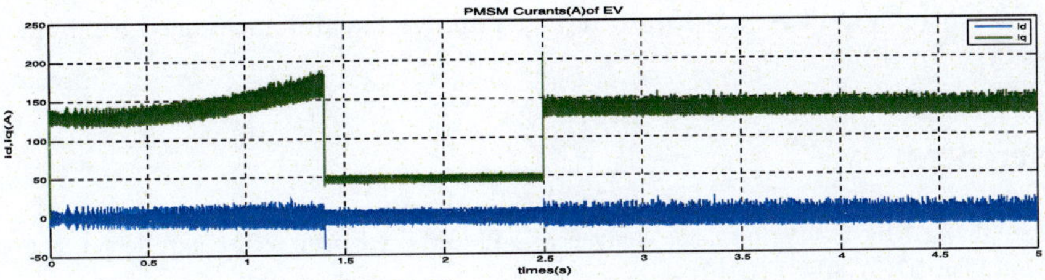

Fig. 4. The behavior of the (id,iq) current in case of constant speed and 30% at (t = 2.5 s)

The Second Test. This test treats the problem of trajectory tracking. It is considered that the vehicle traces a straight path with a different reference speed step, acceleration and deceleration of VE (see Fig. 5). The variation of the electromagnetic torque was also represented in the (Fig. 6).

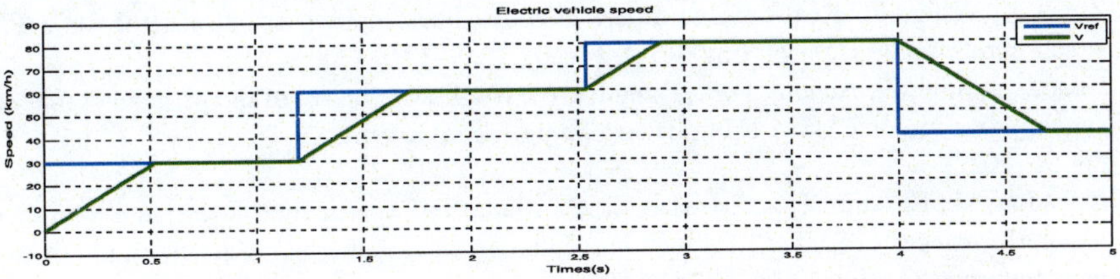

Fig. 5. Speed tracking response for a different reference variable speed

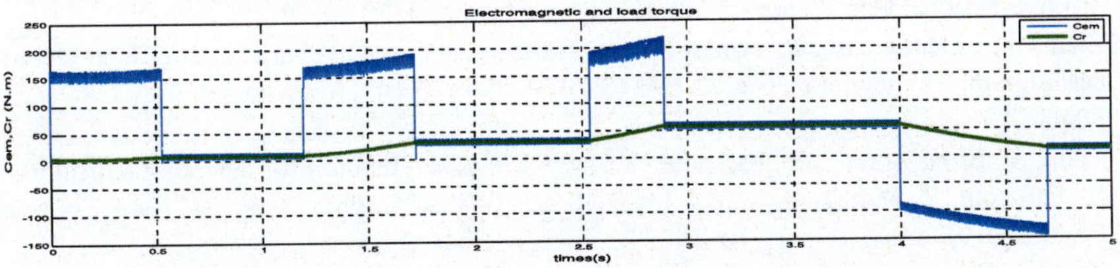

Fig. 6. The behavior of the driving torque C_r and the electromagnetic torque C_{em}

6 Conclusion

This work is focused on the Backstepping control applied to an electric vehicle powertrain or motorization system using the PMSM. The modeling of the EV powertrain system, in particular the dynamics of EV and PMSM, as well as the nonlinear control theory of Backstepping has been developed and presented in this work. The simulation and implementation of the control are presented. The simulation results of the different tests have shown that the Backstepping control is robust and provides a very good trajectory trackand, this optimization method can also be used to avoid the heating of LEDs in optics [7, 8]. This study's findings also emphasize the need to use wavelets [9–12], integrated with data analysis techniques often employed in biomedical signal processing, such as ICA-NMF-SVD-PCA [13–15], to further improve the aforementioned techniques' efficacy.

References

1. Singh, A., Roy, O.: Performance analysis of a PMSM drive using PID controllers. Electron. Inf. Plan. **37**(3), 80–87 (2010)
2. Achour, H.B., Ziani., S., Chaou, Y., El Hassouani, Y., Daoudia, A.: Permanent magnet synchronous motor PMSM control by combining vector and PI controller. WSEAS Trans. Syst. Control **17,** 244–249 (2022)
3. Rebouh, S., et al.: Nonlinear control by input-output linearization scheme for EV permanent magnet synchronous motor. In 2007 IEEE Vehicle Power and Propulsion Conference. IEEE (2007)
4. Hamida, M.A., et al.: Robust adaptive high order sliding-mode optimum controller for sensorless interior permanent magnet synchronous motors. Math. Comput. Simul. **105**, 79–104 (2014)
5. Zhou, J., Wang, Y.: Real-time nonlinear adaptive backstepping speed control for a PM synchronous motor. Control. Eng. Pract. **13**(10), 1259–1269 (2005)
6. Chaou, Y., Said, Z., Achour, H.B., Daoudia, A.: Nonlinear control of the permanent magnet synchronous motor PMSM using backstepping method. WSEAS Trans. Syst. Control **17,** 56–61 (2022)
7. Ouhadou, M., El Amrani, A., Messaoudi, C., Ziani, S.: Experimental investigation on thermal performances of SMD LEDs light bar: Junction-to-case thermal resistance and junction temperature estimation. Optik **18**2 (2019)
8. Ouhadou, M., El Amrani, A., Ziani, S., Messaoudi, C.: Experimental modeling of the thermal resistance of the heat sink dedicated to SMD LEDs passive cooling, In: Proceedings of the 3rd International Conference on Smart City Applications (2018)
9. Ziani, S., El Hassouani, Y.: A new approach for extracting and characterizing fetal electrocardiogram. Traitement du Signal. **373** (3), 379–386 (2020). https://doi.org/10.18280/ts.370304G
10. Ziani, S., El Hassouani, Y., Farhaoui, Y.: An NMF based method for detecting RR interval. In: Farhaoui, Y., Moussaid, L. (eds.) ICBDSDE 2018. SBD, vol. 53, pp. 342–346. Springer, Cham (2019). https://doi.org/10.1007/978-3-030-12048-1_35
11. Ziani, S., El Hassouani, Y.: Fetal-maternal electrocardiograms mixtures characterization based on time analysis. In: 2019 5th International Conference on Optimization and Applications (ICOA), pp. 1–5 (2019). https://doi.org/10.1109/ICOA.2019.8727619
12. Ziani, S., Jbari, A., Bellarbi, L.: QRS complex characterization based on non-negative matrix factorization NMF. In: 2018 4th International Conference on Optimization and Applications (ICOA), pp. 1–5 (2018), https://doi.org/10.1109/ICOA.2018.8370548
13. Ziani, S., Jbari, A., Bellarbi, L., Farhaoui, Y.: Blind maternal-fetal ECG separation based on the time-scale image TSI and SVD-ICA methods. Procedia Comput. Sci.**134** (2018(

14. Ziani, S., El Hassouani, Y.: Fetal electrocardiogram analysis based on LMS adaptive filtering and complex continuous wavelet 1-D. In: Farhaoui, Y. (ed.) BDNT 2019. LNNS, vol. 81, pp. 360–366. Springer, Cham (2020). https://doi.org/10.1007/978-3-030-23672-4_26
15. Ziani, S., Jbari, A., Belarbi, L.: Fetal electrocardiogram characterization by using only the continuous wavelet transform CWT. In: 2017 International Conference on Electrical and Information Technologies (ICEIT), pp. 1–6 (2017). https://doi.org/10.1109/EITech.2017.8255310

Emotion Detection in Real-Time Video Using Deep Learning

Moulay Lhabib El Hadi[1] (✉) and M'barek Nasri[2]

[1] MATSI EST, Oujda, Morocco
mlhabibelhadi@gmail.com
[2] University Mohammed I, BP 717, 60000 Oujda, Morocco

Abstract. Challenges such as emotion recognition can not typically be solved through classical machine learning techniques and these results are limited in the context of processing very large amounts of data. A new line of research is being developed with the emergence of the concept of deep learning and large databases. The aim of this work is to describe the survey of Face Expression Recognition (FER) techniques, which includes the three major stages such as preprocessing, feature extraction, and classification, and develop the basis that leads to Convolutional Neural Networks (CNN) to create a classifier that recognizes the facial expression of a human in real-time. The classification is performed on the publicly available FER-2013 dataset of over 35,000 face images with an in-the-wild setting of 7 distinct emotions. The proposed method is competitive with existing methods in terms of training time, testing time, and recognition accuracy.

Keywords: Emotion recognition · Face expression recognition · Machine learning · Deep learning · Preprocessing · Feature extraction · Classification

1 Introduction

We,humans, have a few diverse facial feelings by which we can comprehend someone else's emotions, speak with one another, and understand his aim in accomplishing something. Over the past decade, the computer vision and scientific research communities envisioned developing systems capable of recognizing facial expressions in videos or static images. Most of these facial expression analysis systems attempt to classify expressions into a few broad emotional categories, such as joy, sadness, anger, surprise, fear, and disgust.

The main objective of this work is to analyze the emotions of a human being and to give estimates for each basic emotion. The analysis of emotions is done according to facial expressions.

Thanks to the emotion detector, it is possible to avoid accidents caused by mood changes (such as anger).

Y. Farhaoui et al. (Eds.): ICAISE 2022, LNNS 635, pp. 374–381, 2023.
https://doi.org/10.1007/978-3-031-26254-8_53

This main objective can be divided into 3 sub-objectives which are:

- Detect the face in a video stream.
- Place the characteristic points (or "landmarks") [1].
- Find the emotion by analyzing the movement of the characteristic points.

In this work, we carried out experiments for 7 emotional states with the Fer2013 dataset. We obtained a test set accuracy of 67%.

The outline of this paper is as follows: Sect. 2 presented the methodology of the FER system. Section 3, for the application and discussion of the dataset, gave the architecture of the CNN we have utilized in our work, then gave the methodology to deploy our trained model on image and webcam stream, and finally, we discussed our research challenges. Finally, Sect. 4 presents the conclusion.

2 Methodology of the FER System

The overview of the FER system is illustrated in Fig. 1. The FER system includes the major stages such as face image preprocessing, feature extraction and classification.

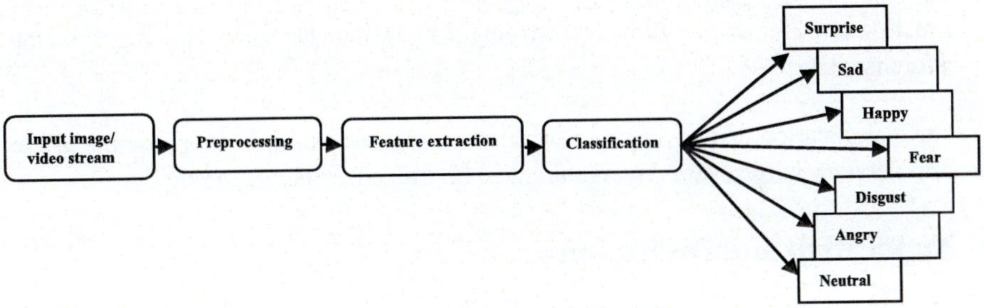

Fig. 1. Overview of the FER system

2.1 Preprocessing

Face detection is the preprocessing step for recognizing facial expressions. It's a process that can be used to improve the performance of the FER system, and it can be carried out before the feature extraction process [2]. Image preprocessing includes different types of processes such as image clarity and scaling, contrast adjustment, and additional enhancement processes to improve the expression frames.

2.2 Feature Extraction

The feature extraction process is the next stage of the FER system. Feature extraction converts pixel data into a higher-level representation of the shape, motion, color, texture, and spatial configuration of the face or its components. Feature extraction generally reduces the dimensionality of the input space. The reduction procedure should retain essential information as it is an important task in pattern recognition systems. Feature extraction can be done using various techniques (Fig. 2).

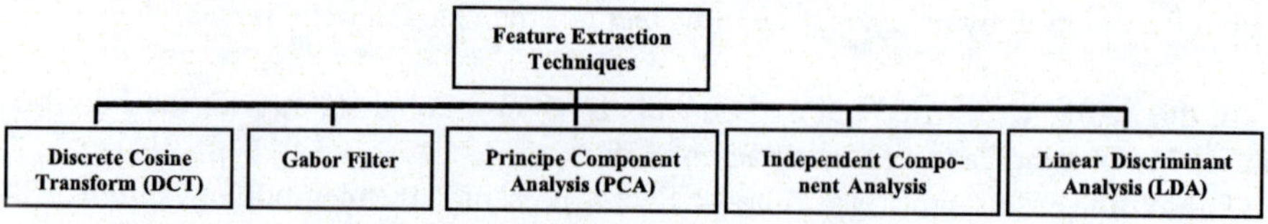

Fig. 2. Some techniques of feature extraction

2.3 Classification

Classification is the final stage of the FER system in which the classifier categorizes the expressions such as smiles, sad, surprise, anger, fear, disgust, and neutral. It is divided into two sub-steps:

1. Standardized data and creation of reference models by extracting the characteristics and launching a learning process.
2. Creation of the classifier allows classifying the new data by comparing it against the reference data.

The classification can be done in several ways [3]. For example, it is possible to use a neural network with several layers that must be trained at the beginning.

3 Application and Discussion

3.1 Facial Emotion Recognition Dataset (FER-2013)

For such an attractive feature, many datasets have been developed for facial emotion recognition tasks. Created in 2013, FER-2013, the data consists of 48x48 pixel grayscale images of faces. The task is to categorize each face based on the emotion shown in the facial expression into one of seven categories. The training set consists of 28,709 examples and the public test set consists of 3,589 examples.

3.2 Architecture of Our CNN

We started by training a classical convolutional neural network. The network has 440000 trainable parameters. The architecture of our first model is presented in Fig. 3.

When plotting the accuracy and the loss in terms of the epochs (Fig. 4), on both the training and the validation set, we observe a rather clear overfitting that occurs after approximately 20 epochs.

For this reason, several approaches and techniques have been developed and implemented. In the next sections, we will cover the main techniques, and the outcomes of these different techniques, as well as the final model that has been implemented.

Layer (type)	Output Shape	Param #
conv2d_2 (Conv2D)	(None, 48, 48, 32)	320
max_pooling2d_2 (MaxPooling2	(None, 24, 24, 32)	0
batch_normalization_1 (Batch	(None, 24, 24, 32)	128
conv2d_3 (Conv2D)	(None, 22, 22, 32)	9248
max_pooling2d_3 (MaxPooling2	(None, 11, 11, 32)	0
batch_normalization_2 (Batch	(None, 11, 11, 32)	128
conv2d_4 (Conv2D)	(None, 11, 11, 32)	9248
max_pooling2d_4 (MaxPooling2	(None, 5, 5, 32)	0
conv2d_5 (Conv2D)	(None, 5, 5, 32)	9248
flatten_1 (Flatten)	(None, 800)	0
dense_1 (Dense)	(None, 512)	410112
dense_2 (Dense)	(None, 7)	3591

Total params: 442,023
Trainable params: 441,895
Non-trainable params: 128

Fig. 3. Keras model summary

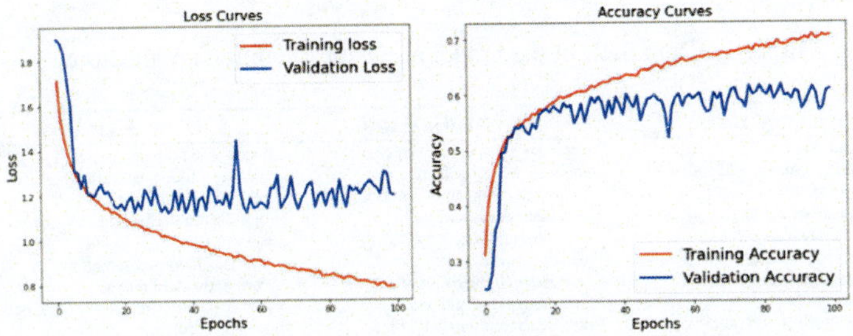

Fig. 4. Accuracy and loss on training and validation set

A) Dimension Reduction through Auto-Encoding

The first step that was implemented was to auto-encode the images in order to reduce the dimension [4]. The implemented auto-encoder proposes a dimension reduction of more than 95%, by reducing the input image to a dimension of 12 * 12 pixels (Fig. 5).

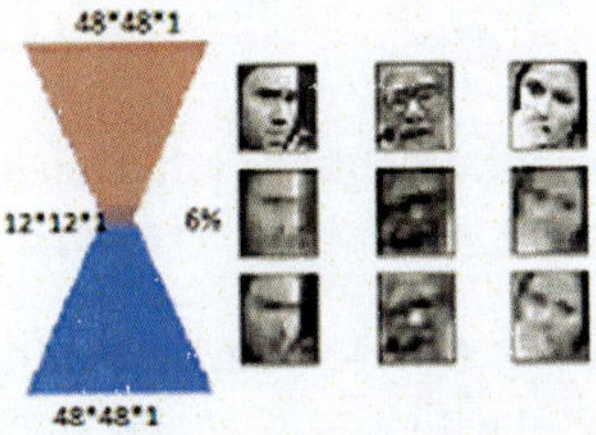

Fig. 5. Auto-Encoding the input images

B) XCeption and Depthwise Separable convolutions

Xception is a deep convolutional neural network architecture that involves depth-separable convolutions [5]. It was developed by Google researchers. The Xception architecture has overperformed VGG-16, ResNet, and Inception V3 in most classical classification challenges. It is an efficient architecture that relies on two main points:

- Depthwise Separable Convolution
- Shortcuts between convolution blocks as in ResNet

Depthwise Separable Convolutions are alternatives to classical convolutions that are supposed to be much more efficient in terms of computation time [6].

In order to limite overfitting and increase the accuracy of our model, we modified our first model using the exception architecture described in Table 1.

Table 1. Definition of the model besed on XCeption architecture

Entry flow	Midle Flow	Exit flow
def entry_flow(inputs) : x = Conv2D(32, 3, strides = 2, pad ding='same')(inputs) x = BatchNormalization()(x) x = Activation('relu')(x) x = Conv2D(64,3,padding='same')(x) x = BatchNormalization()(x) x = Activation('relu')(x) previous_block_activation = x for size in [128, 256, 728] : x = Activation('relu')(x) x = SeparableConv2D(size, 3, pad-ding='same')(x) x = BatchNormalization()(x) x = Activation('relu')(x) x = SeparableConv2D(size, 3, pad-ding='same')(x) x = BatchNormalization()(x) x = MaxPooling2D(3, strides=2, padding='same')(x) residual = Conv2D(size, 1, strides=2, padding='same')(previous_block_activation) x = keras.layers.Add()([x, residual]) previous_block_activation = x return x	Def middle_flow(x, num_blocks=8) : previous_block_activation = x for _ in range(num_blocks) : x = Activation('relu')(x) x = SeparableConv2D(728, 3, padding='same')(x) x = BatchNormalization()(x) x = Activation('relu')(x) x = SeparableConv2D(728, 3, padding='same')(x) x = BatchNormalization()(x) x = Activation('relu')(x) x = SeparableConv2D(728, 3, padding='same')(x) x = BatchNormalization()(x) x = keras.layers.Add()([x, revious_block_activation]) previous_block_activation = x return x	def exit_flow(x, num_classes=7) : previous_block_activation = x x = Activation('relu')(x) x = SeparableConv2D(728, 3, pad-ding='same')(x) x = BatchNormalization()(x) x = Activation('relu')(x) x = SeparableConv2D(1024, 3, pad-ding='same')(x) x = BatchNormalization()(x) x = MaxPooling2D(3, strides=2, pad-ding='same')(x) residual = Conv2D(1024, 1, strides=2, padding='same')(previous_block_activation) x = keras.layers.Add()([x, residual]) x = Activation('relu')(x) x = SeparableConv2D(728, 3, pad-ding='same')(x) x = BatchNormalization()(x) x = Activation('relu')(x) x = SeparableConv2D(1024, 3, pad-ding='same')(x) x = BatchNormalization()(x) x = GlobalAveragePooling2D()(x) x = Dense(num_classes, activa-tion='softmax')(x) return x

We arrive at the phase of quantification and evaluation of the results of the developed approach, basing ourselves on two parameters: the recognition rate (Accuracy) and the rate of loss or error (Loss).

We compiled our CNN using ADAM optimizer with loss function as categorical_crossentropy and evaluation metrics as accuracy, Batch_size as 128, and training the CNN on the complete dataset.

From Fig. 6, We obtained a test set accuracy of 67% and a training set accuracy of around 93%.

To show the robustness of our model we will compare it with some research on the same FER2013 database as shown in the table.

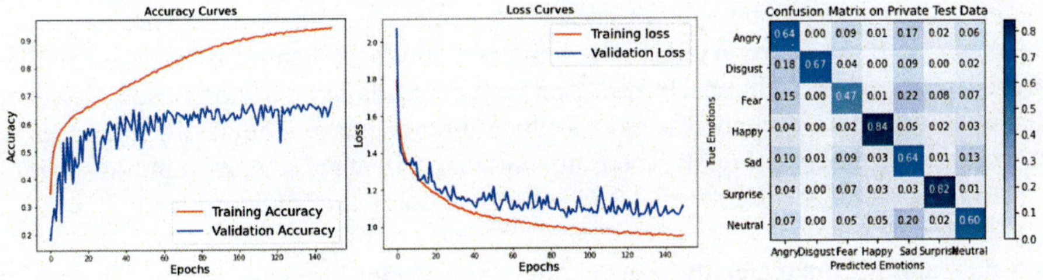

(A) (B)

Fig. 6. **(A)** Accuracy on training and validation set. **(B)** Normalized confusion matrix of our model

Table 2. Standalone-based neural network (SBNN) results comparison

Related works	Proposed method	Test accuracy
[7]	CNN + Batch Normalization	60.12%
[8]	CNN + Batch Normalization + Varying number of filters	65%
[9]	Custom CNN	66.67%
	One-vs-All (OVA) SVM	45.95%
[10]	VGG-16 + GAP	69.40%
[11]	New architecture based on attentional CNN	70.02%
Our model	**CNN based on Xception architecture**	**67%**

From the Table 2, our proposed model also achieves similar, albeit lower, test accuracy as compared to the other approaches.

3.3 Emotion Classification in Real-Time Video

In the context of human sentiment analysis, a key component is to be able to understand emotions from video input, not only from pictures; the methodology to deploy our trained model on a webcam stream is the following (Fig. 7):

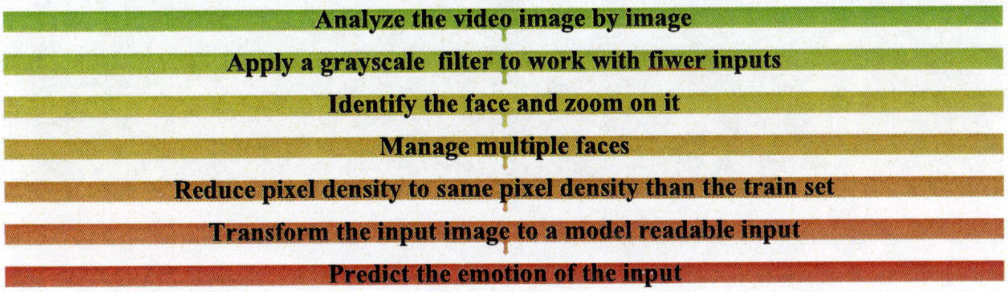

Fig. 7. Methodology to deploy our trained model on a webcam stream

Face detection is done with a Cascade Classifier. Cascade classifiers are based on the concatenation of several classifiers, using all the information from the output of the previous classifier as additional information for the next classifier in the cascade. Cascade classifiers are typically applied to regions of interest in an image. The classification seems to work well in practice (Fig. 8). There are, however, still many sources of improvement.

To enhance the visualization, we added a lot of sources of improvements:

- Add features from manually selected filters (e.g. Gabor filters)
- Take into account the frequency of eye blinks
- Take into account the symmetry of the key points on a face
- Display all the key points of the face
- Align the face by scaling the facial features.

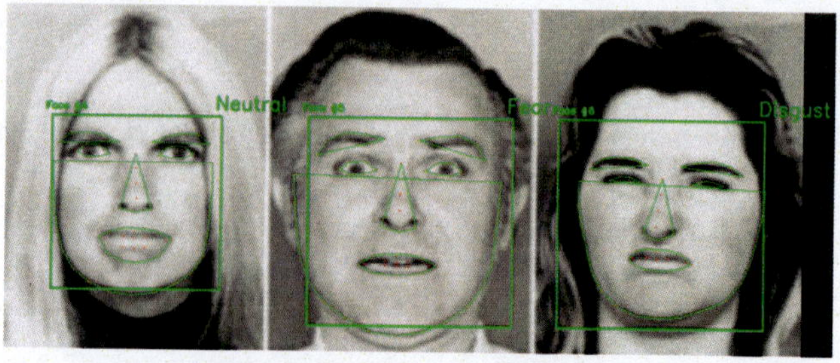

Fig. 8. Image emotion recognition

Our model works well, and the accuracy reaches 67%. We can illustrate the accuracy of the algorithm with a concrete example.

Additional work has also been done to allow for multi-face real-time emotion recognition (Fig. 9).

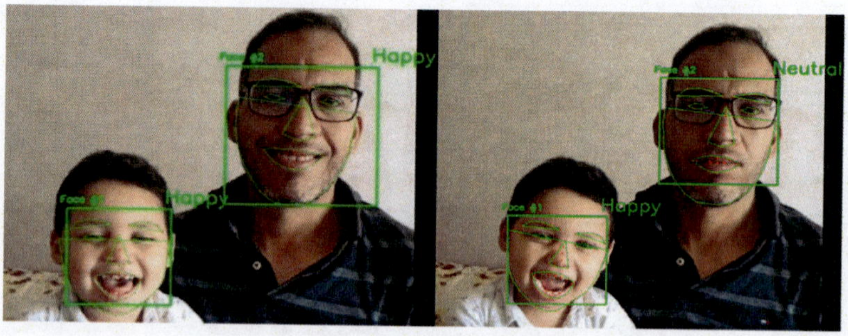

Fig. 9. Multi-Faces emotion recognition

We came to the conclusion that the XCeption model allows a shorter training time on GPUs, more image processing per second in real-time prediction, and prevents overfitting. Thanks to the XCeption architecture, we limit the number of parameters. Our model is among the best and can be used to obtain a good results.

4 Conclusion

Emotions are important pieces of information to retrieve and process. The face of the person is one of the carriers of this information. In this work, we have described the survey of FER techniques, and we have proposed and tested a CNN model based on the Xception architecture on a very large database, which is that of FER2013. Our proposed model has been built systematically to prevent overfitting, reduce the number of parameters, and more. Our classifier allows for shorter GPU training time and more images processed per second in real-time prediction. In the carried out experiments, for 7 emotional states, we achieved a good classification accuracy of emotions—67% with the Fer2013 dataset. Which allows us to say that our proposed model also achieves similar, albeit lower, test accuracy as compared to the other approaches.

References

1. Kazemi, V., Sullivan, J.: One millisecond face alignment with an ensemble of regression trees. In: 2014 IEEE Conference on Computer Vision and Pattern Recognition (2014). https://doi.org/10.13140/2.1.1212.2243
2. Poursaberi, A., Noubari, H.A., Gavrilova, M., Yanushkevich, S.N.: Gauss – Laguerre wavelet textural feature fusion with geometrical information for facial expression identification. EURASIP J. Image Video Process, pp. 1–13 (2012)
3. Tan, P.-N., Steinbach, M., Kumar, V.: Introduction to Data Mining, p. 17, 54. Pearson Education, London (2006)
4. Wang, Y., Yao, H., Zhao, S.: Auto-encoder based dimensionality reduction. Neurocomputing **184**, 232–242, ISSN 0925-2312, (2016).https://doi.org/10.1016/j.neucom.2015.08.104
5. Chollet, F.: Deep learning with depthwise separable convolutions. In: Proceedings of the IEEE Conference on Computer Vision and Pattern Recognition (CVPR), pp. 1251–1258 (2017)
6. Hua, B.-S., Tran, M.-K., Yeung, S.-K.: Pointwise convolutional neural networks. In: Proceedings of the IEEE Conference on Computer Vision and Pattern Recognition (CVPR), pp. 984–993 (2018)
7. Talegaonkar, I., Joshi, K., Valunj, S., Kohok, R., Kulkarni, A.: Real time facial expression recognition using deep learning. SSRN Electron. J. (2019). https://doi.org/10.2139/ssrn.3421486
8. Agrawal, A., Mittal, N.: Using CNN for facial expression recognition: a study of the effects of kernel size and number of filters on accuracy. Vis. Comput. **36**(2), 405–412 (2019). https://doi.org/10.1007/s00371-019-01630-9
9. Quinn, M., Sivesind, G., Reis, G.: Real-time Emotion Recognition From Facial Expressions (2017)
10. Kusuma, G.P., Jonathan, A.P.L.: Emotion recognition on FER-2013 face images using fine-tuned VGG-16. Adv. Sci. Technol. Eng. Syst. J. **5**(6), 315–322 (2020)
11. Minaee, S., Abdolrashidi, A.: Deep-Emotion: Facial Expression Recognition Using Attentional Convolutional Network (2019)

Energy Efficiency Optimization Techniques for the 5G Cellular Networks

Macoumba Fall[1](✉), Younes Balboul[1], Mohammed Fattah[2], Said Mazer[1], Moulhime El Bekkali[1], and Ahmed Ahmed Dooguy Kora[3]

[1] IASSE Laboratory, Sidi Mohamed Ben Abdellah University, Fez, Morocco
macfallm@gmail.com

[2] IMAGE Laboratory, Moulay Ismail University, Meknes, Morocco

[3] EDMI, Cheikh Anta Diop University, Dakar, Senegal

Abstract. The new generation of cellular networks called 5G has been defined in order to satisfy several use cases summarized in 3 main implementations: eMBB, mMTC and uRLLC. The technologies used in these different use cases will induce high energy consumption in this next-generation network. The optimization of energy efficiency has thus become a major challenge for this new generation and several strategies to improve energy efficiency have been proposed in the literature.

This paper presents a review and a comparative analysis of these different strategies, divided into four groups, those relating to the base station, those based on the organization of the network, those based on Software Defined Networks and those based on machine learning.

Keywords: 5G · Energy efficiency · Ecology · Energy · Optimization · Mobile networks

1 Introduction

Conceived as a disruptive technology, the 5th generation of mobile networks was created to meet several use cases divided into three main scenarios: eMBB (Enhanced Mobile Broadband), mMTC (massive Machine Type Communication) and uRLLC (Ultra-Reliable Low Latency Communication) representing respectively a high-speed use (greater than 10 Gbps), a high number of connected objects (1million objects per km^2) or a low-latency use (less than 1 ms) [1, 2].

The 5th generation network is created with the aim of solving 4G network saturation, mainly in urban areas, and enabling the digital transition of industry and services. The different specifications of 5G are based on a new radio architecture and other major technological advances which can however lead to greater energy consumption of mobile networks [3].

An increasing energy curve has been observed generation after generation, but for this new generation, the energy consumption may reach a critical value due to the forecasts of the increase in traffic and the number of connected objects. Thus, this ecological issue makes this generation one of several controversial, despite the technological leap and the transformation of uses that it promises.

Y. Farhaoui et al. (Eds.): ICAISE 2022, LNNS 635, pp. 382–387, 2023.
https://doi.org/10.1007/978-3-031-26254-8_54

In order to solve this major problem of 5G, a lot of research has been directed towards optimizing energy efficiency in order to slow down this energy curve and make it less energy consuming.

This article begins by introducing the ecological challenges facing this new technology, then it summarizes the optimization techniques and strategies that have been proposed in the literature to reduce energy consumption, and concludes with a comparative analysis of these strategies in terms of implementation complexity, potential energy gain and deployment cost.

2 Ecological Challenge of 5G

In order to meet the different requirements of 5G, especially regarding the number of users and expected data volumes, which will grow exponentially, telephone operators are required to increasingly densify their core and access networks; with the use of heterogeneous networks, base stations are more and more numerous.

The authors of [4], as well as [5] reported that 3% of the world's electrical energy is consumed by the information and communication technology (ICT) sector and that it is responsible for 2% of global CO2 emissions. And, in terms of ict consumption, most of it is made by wireless technologies, with mobile networks as the main contributor.

One of the factors of the environmental impact of mobile networks and the carbon footprint and energy consumption related to the production of equipment and mainly smartphones.

In [6], it is reported that the manufacture of a smartphone releases 30 kg of CO_2. Worldwide, more than 1.4 billion smartphones were produced during the year 2021 reveals us [7]. This emits greenhouse gases and pollutes water, air and soil.

3 Energy Efficiency Optimization Strategies

In recent years, several researches have been conducted to propose strategies to optimize energy efficiency given the challenge that this represents in the implementation of mobile networks or 5th generation systems.

This research was mainly oriented towards 7 topics that present the following diagram (Fig. 1):

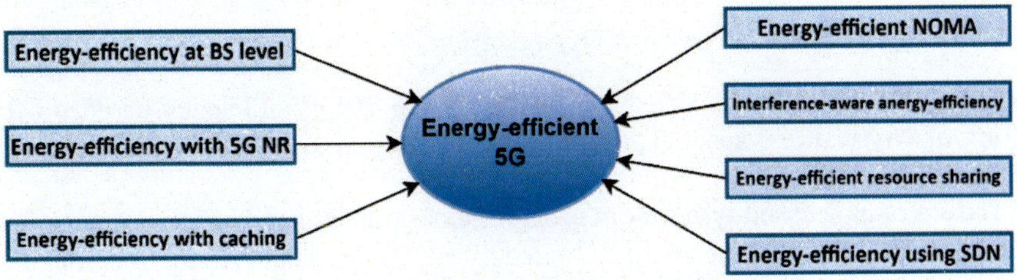

Fig. 1. Key research topics on optimizing energy efficiency in 5G networks [8]

These different topics mentioned in the literature make it possible to organize the strategies into three groups: base station, organization of the network and Machine Learning.

3.1 Base Station Energy Efficiency

Analysis of energy consumption within mobile networks has shown that stations are the entity that consumes most of the energy. They consume about 57% of the network's energy [9].

A large part of the research has thus focused on the energy optimization of the base station. Among the strategies explored we find:

- C-RAN: which describes both centralized RAN and Cloud RAN. The first describes the ability to centralize the BBUs of the base stations of a given area on a single site while the Cloud RAN goes further by virtualizing certain network functions at the server level. In [10], the authors discussed the usefulness of removing most processing functionality from the baseband to a CPU.
- Cell standby: technical or strategic measures have been proposed to put the small cells into standby based on cell load, cell selection or traffic modelling. This is called an On-Off switching strategy. In [11] and [12], the authors reviewed the proposed On/Off switching strategies for 5G. Better interference management: heterogeneous networks lead to an increase in interference, deteriorating energy efficiency.

  [13] proposes the use of variable transmission power levels. [14] provides a robust distributed algorithm to reduce the adverse effects of computational complexity and noise on resource allocation. The authors of [15] proposed a resource allocation technique to minimize interference on the EU side.
- Control of the DRR connection: this strategy mainly impacts the EU, whose permanent signalling exchanges with the base station promote faster battery depletion. [16] discusses this exhaustion and proposes for 5G the removal of constant monitoring of the PDCCH channel for incoming transmissions.
- Proactive data caching, depending on the popularity of the data, has also been proposed, for small cells. Several researches have focused on organizing content based on users' locations and constantly adjusting clusters based on the distribution of data popularity, instead of having the same caching on the network. [17] Relies on small caching that maximizes energy efficiency.

3.2 Strategies Based on Network Organization

Several studies have been carried out on improving energy efficiency based on the sharing of physical and spectral resources between operators through the sharing of infrastructure.

The use of unlicensed systems or tapes for indoor use has been mentioned in several researches. [18] proposes the use of EU-rated distributed antenna system for indoor use, with also the possibility of using a band without a license.

The virtualization of certain network functions, which previously were on dedicated hardware, is an avenue increasingly explored in research. The virtual network architecture described in [19] presents the interconnection between multiple virtual units as well as the interconnected physical units to form a larger system.

3.3 Machine Learning

Machine learning techniques are increasingly present in ICTs. The authors highlight a transmission with an intelligent antenna that can, for example, adjust its power or frequency band according to learning from frequency signals.

The authors of [20] proposed a policy of enabling and disabling small cells in order to optimize power consumption via distributed Q-learning applied to incoming traffic requests.

The authors of [21] also use machine learning techniques for energy-efficient resource allocation in the heterogeneous 5G cloud radio access network. Cloud radio access networks are considered a key component of 5G in order to provide higher data rates and lower intercellular interference. The authors developed a resource allocation scheme to maximize the energy efficiency of the UEs served by the radio heads while minimizing interference.

[22] highlighted spectrum detection techniques based on mathematical analysis to reduce interference problems exacerbated by random deployments of femto BS. The authors formulated objective functions to define an optimal power allocation for users, in order to maximize energy efficiency.

4 Comparative Study of Strategies

The different strategies proposed, have various advantages and disadvantages, related to the complexity of the implementation, the expected gain or the costs that it would require. We have thus defined the following table ("+" represents a low level, "++" a relative or medium level and "+++" a high level), which summarizes these different aspects:

Consuming nearly 60% of the energy of the network, the stations are naturally of major interest in the quest to improve the energy efficiency of the network. They thus concentrate the greatest diversity of strategies. However, some areas are increasingly explored, such as C-RAN, SDN or virtualization, which occupy an important place in the definition of future technologies (Table 1).

Table 1. Comparative table of energy efficiency optimization strategies

Group	Strategy	Complexity	Gain	Cost
Base station	C-RAN	++	+++	++
	On/Off switching	++	+++	+

(*continued*)

Table 1. (*continued*)

Group	Strategy	Complexity	Gain	Cost
	Reduced interference	++	+	+
	RRC connection control	+	+	+
	Data caching	++	+	++
Organization of the network	Sharing resources	+++	+++	+++
	NOMA	+++	+++	++
	Use of tapes without a license	++	++	+
	Virtualization	++	+++	++
Machine learning	Putting cells to sleep	++	++	+
	Resource allocation	++	+	+
	Reduced interference	++	+	+

5 Conclusion

Several strategies have been proposed in the literature to improve the energy efficiency of 5G and thus reduce the energy consumption of networks. The analysis of these different strategies, in terms of complexity, potential gains and implementation costs, has allowed us to identify some, which are more or less interesting than others in terms of results.

However, these different groups of strategies can be considered more as complementary tracks than competing tracks. In order to more effectively achieve the ecological objectives of 5G, all these aspects must be taken into account in order to sustainably define this so-called disruptive technology.

References

1. Reddy, P.K., et al.: 5G new radio key performance indicators evaluation for IMT-2020 radio interface technology. IEEE Access **9**, 112290–112311 (2021)
2. Erunkulu, O.O., Zungeru, A.M., Lebekwe, C.K., Mosalaosi, M., Chuma, J.M.: 5G mobile communication applications: a survey and comparison of use cases. IEEE Access **9**, 97251–97295 (2021)
3. Agiwal, M., Roy, A., Saxena, N.: Next generation 5G wireless networks: a comprehensive survey. IEEE Commun. Surv. Tutor. **18**(3), 1617–1655 (2016)
4. Xu, X., Yuan, C., Chen, W., Tao, X., Sun, Y.: Adaptive cell zooming and sleeping for green heterogeneous ultradense networks. IEEE Trans. Veh. Technol. **67**(2), 1612–1621 (2018)
5. Boumaiz, M., et al.: Energy harvesting based WBANs: EH optimization methods. Procedia Comput. Sci. **151**, 1040–1045 (2019)
6. Moutaib, M., Fattah, M., Farhaoui, Y.: Internet of things: energy consumption and data storage. Procedia Comput. Sci. **175**, 609–614 (2020)
7. Usama, M., Erol-Kantarci, M.: A survey on recent trends and open issues in energy efficiency of 5G. Sensors **19**(14), 3126 (2019)
8. Tan, R., Shi, Y., Fan, Y., Zhu, W., Wu, T.: Energy saving technologies and best practices for 5G radio access network. IEEE Access **10**, 51747–51756 (2022)

9. Younes, B., Fattah, M., Said, M., Bekkali, M.E.: 5G uplink interference simulations, analysis and solutions: the case of pico cells dense deployment. Int. J. Electr. Comput. Eng. **11**(3), 2245 (2021)
10. Feng, M., Mao, S., Jiang, T.: Base station ON-OFF switching in 5G wireless networks: approaches and challenges. IEEE Wirel. Commun. **24**(4), 46–54 (2017)
11. Beitelmal, T., et al.: Sector and site switch-off regular patterns for energy saving in cellular networks. EEE Trans. Wireless Commun. **17**(5), 2932–2945 (2018)
12. Mamane, A., et al.: The impact of scheduling algorithms for real-time traffic in the 5G femto-cells network. In: 9th International Symposium on Signal, Image, Video and Communications, ISIVC 2018, pp. 147–1512 (2018)
13. Mamane, A., Fattah, M., El Ghazi, M., Balboul, Y., El Bekkali, M., Mazer, S.: Proportional fair buffer scheduling algorithm for 5G enhanced mobile broadband. Int. J. Electr. Comput. Eng. **11**, 4165–4173 (2021)
14. Mamane, A., Fattah, M., Ghazi, M.E., Bekkali, M.E., Balboul, Y., Mazer, S.: Scheduling algorithms for 5G networks and beyond: classification and survey. IEEE Access **10** (2022)
15. Yen, C., Chien, F., Chang, M.: Cooperative online caching in small cell networks with limited cache size and unknown content popularity. In: 3rd International Conference on Computer and Communication Systems (ICCCS), Nagoya, Japan (2018)
16. Georgakopoulos, A., Margaris, A., Tsagkaris, K., Demestichas, P.: Resource sharing in 5G contexts: achieving sustainability with energy and resource efficiency. IEEE Veh. Technol. Mag. **11**(1), 40–49 (2016)
17. Xu, B., Chen, Y., Carrión, J., Zhang, T.: Resource allocation in energy-cooperation enabled two-tier NOMA HetNets toward green 5G. IEEE J. Sel. Areas Commun. **35**(12), 2758–2770 (2017)
18. Ng, D.W.K., Breiling, M., Rohde, C., Burkhardt, F., Schober, R.: Energy-efficient 5G outdoor-to-indoor communication: SUDAS over licensed and unlicensed spectrum. IEEE Trans. Wirel. Commun. **15**(5), 3170–3186 (2016)
19. Chafi, S.-E., Balboul, Y., Mazer, S., Fattah, M., Bekkali, M.E.: Resource placement strategy optimization for smart grid application using 5G wireless networks. Int. J. Electr. Comput. Eng. **12**, 3932–3942 (2022)
20. Nguyen, X.-N., Saucez, D., Barakat, C., Turletti, T.: OFFICER: a general optimization framework for OpenFlow rule allocation and endpoint policy enforcement. In: IEEE Conference on Computer Communications (INFOCOM) (2015)
21. Miozzo, M., Giupponi, L., Rossi, M., Dini, P.: Switch-on/off policies for energy harvesting small cells through distributed Q-learning. In: IEEE Wireless Communication and Networking Conference Workshops (WCNCW), San Francisco, CA, USA (2017)
22. AlQerm, I., Shihada, B.: Enhanced machine learning scheme for energy efficient resource allocation in 5G heterogeneous cloud radio access networks. In: IEEE 28th Annual International Symposium on Personal, Indoor, and Mobile Radio Communications (PIMRC), Montreal, QC, Canada (2017)

Energy Management Strategy Based on Neural Network for Hybrid Renewable System

Benzaouia Mohammed[1](✉), Hajji Bekkay[1], Rabhi Abdelhamid[2], and Mellit Adel[3]

[1] Renewable Energy, Embedded System and Data Processing Laboratory, National School of Applied Sciences, Mohamed First University, Oujda, Morocco
m.benzaouia@ump.ac.ma

[2] Laboratory of Modelization, Information and Systems, University of Picardie Jules Verne, 3 Rue Saint Leu, 80039 Amiens Cedex, France

[3] Renewable Energy Laboratory, Faculty of Sciences and Technology, Jijel University, Jijel, Algeria

Abstract. The exploitation of several renewable energy sources is the optimal and promising solution to provide water for agricultural applications mainly irrigation. Generally, these agricultural areas are isolated where access to the grid is difficult and requires a significant investment. This paper presents a Neural Network-based energy management system (ANN-EMS) for a hybrid system consisting of a wind turbine, photovoltaic panels as the primary sources and batteries as secondary source, supercapacitors are used to satisfy the starting current peaks of the motor-pump unit as the batteries are unable to satisfy them. The system has been designed and simulated using Matlab/Simulink environment. The obtained results proved the efficiency of the proposed ANN-EMS algorithm under random weather conditions and load variations.

Keywords: Energy management strategy (EMS) · ANN · PV · Wind turbine · Batteries · Supercapacitors

1 Introduction

Traditionally, irrigation has been based on the use of conventional resources such as diesel or propane generators which consume huge amounts of expensive fuels that cause environmental problems and increase air pollution due to the emission of $CO2$ in the air [1]. Alternative solutions are available to reduce the dependence on fossil resources, namely the exploitation of wind and solar energies, which are clean and friendly to environment [2, 3] and [4]. Several researches have been carried out on these systems, and due to their intermittent nature, a great attention has been placed on hybrid systems, which are a combination of these two sources (photovoltaic and wind) with hybrid storage systems using batteries and supercapacitors. However, in order to ensure a reliable operation of the system it is necessary to integrate an intelligent and efficient energy management system (EMS) that takes into account the availability of power production of each source, the load demand and finally the states of charge of the batteries and

Y. Farhaoui et al. (Eds.): ICAISE 2022, LNNS 635, pp. 388–394, 2023.
https://doi.org/10.1007/978-3-031-26254-8_55

supercapacitors [5] and [6]. In [7], the authors proposed an energy management strategy based on fuzzy logic to guarantee the distribution of the energy production of a solar-powered hybrid space ship of 5000 cars. The manager takes into account the production of solar energy, the power supplied by the battery and the demand of the ship's electrical load, to decide the direction of the flow between the systems. Other researchers have proposed a decentralized energy management strategy based on predictive controllers for a medium voltage DC photovoltaic electric vehicle charging station [8]. The obtained results showed the efficiency of the manager, while keeping a fixed Bus voltage with system expansion. In our last study [9], a hybrid autonomous system was proposed, which consists of a photovoltaic (PV) source and batteries as a secondary energy source, plus a management system based on a TOR (on/off) algorithm to manage the flow between the two sources, while keeping the batteries' state of charge between 20% and 80%.

In this research paper, a system composed by a wind turbine based on a permanent magnet generator (PMSG), PV panels, Li-ion batteries with supercapacitors and an energy management system based on the use of artificial intelligence precisely neural networks is presented and evaluated under different scenarios.

This paper is organized as follows: Sect. 2 presents the different stages of the studied system; Sect. 3 is devoted to the description of the proposed neural network energy management system. The discussion of the obtained simulation results is provided in Sect. 4. Finally, the last section concludes this document and presents the improvement prospects.

2 Presentation and Modeling of Studied System

The structure of the studied system is represented in Fig. 1; it contains a variable speed wind turbine, a photovoltaic generator, batteries and supercapacitors, which are all connected to the DC bus by unidirectional or bidirectional DC-DC converters, and a PMDC motor connected to a centrifugal pump.

2.1 Wind Turbine Modeling

The mathematical model of the adopted wind turbine is given by equations [1–5], the output power produced of the wind turbine is described as follow:

$$P_w = \frac{1}{2}\rho\pi R^2 C_p(\lambda, \beta)v^3 \tag{1}$$

where: R is the turbine radius, v is the wind speed and $C_p(\lambda, \beta)$ is the power coefficient which depends on the tip speed ratio λ and the blade pitch angle β, it is expressed by the following equation:

$$\frac{1}{\gamma} = \frac{1}{\lambda + 0.08\beta} - \frac{0.035}{\beta^3 + 1} \tag{2}$$

$$\lambda = \frac{\omega_t R}{v} \tag{3}$$

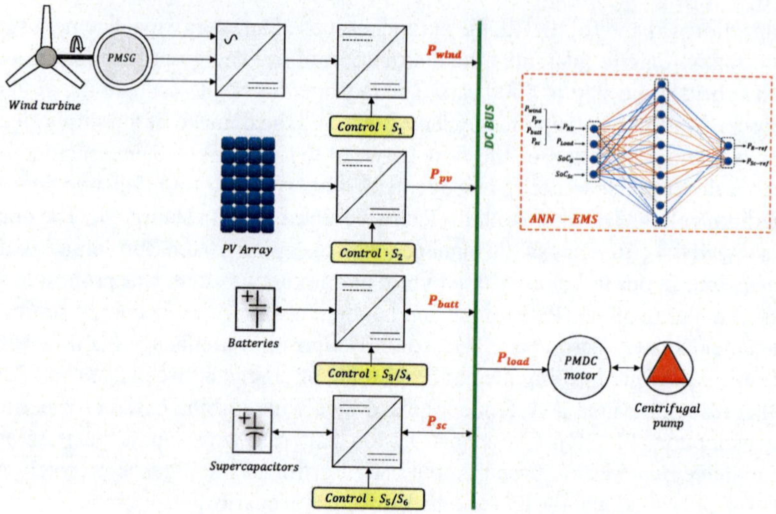

Fig. 1. Stand-alone water pumping system description.

With, $C_1 = 0.5176$, $C_2 = 116$, $C_3 = 0.4$, $C_4 = 5$, $C_5 = 21$ and $C_6 = 0.0068$. The aerodynamic torque T_a is formulated as follows:

$$T_a = \frac{P_w}{\omega_t} = \frac{1}{2}\rho SRC_t(\lambda, \beta)v^2 \tag{4}$$

$$C_t(\lambda, \beta) = \frac{C_p(\lambda, \beta)}{\lambda} \tag{5}$$

where, $C_t(\lambda, \beta)$ is the torque coefficient.

2.2 Photovoltaic Panels Modeling

The $I - V$ and $P - V$ characteristics that reflect the non-linear operation of any photovoltaic generator is given by the following equations:

$$I_{PV} = I_{ph} - I_{\mathrm{d}} - I_{sh} \tag{6}$$

$$I_{PV} = I_{ph} - I_{\mathrm{s}} e^{\left(\frac{q(V_{PV} + I.R_s)}{nkT}\right) - 1} - \frac{(V_{PV} + I.R_s)}{R_{sh}} \tag{7}$$

where: I_s is the saturation current, q is the electron charge, k is Boltzmann constant, n is the diode ideality factor, T is the PV cell temperature (K).

2.3 Batteries Modeling

The electrical model of a battery is given by the following Equation:

$$V_{Batt} = E_B \pm I_{Batt} R_i \tag{8}$$

where: E_B is the voltage source, R_i the internal resistance and I_{Batt} is the current of the battery. The battery capacity C_{Batt} is determined by:

$$C_{Batt} = \frac{E_d.N_d}{V_{Batt}.DOD.\eta_{Batt}} \tag{9}$$

where: E_d is the daily electrical energy required by the load (PMDC motor and the centrifugal pump), N_d number of autonomy days, V_{Batt} is the battery voltage, DOD is the depth of discharge and η_{Batt} is the battery performance. The remaining accessible capacity in the battery is given by the following state of charge formula:

$$SoC(t) = SoC(t-1) - \int_{t-1}^{t} \frac{I_{Batt}}{C_{Batt}} dt \tag{10}$$

2.4 Supercapacitors Modeling

The supercapacitor (SC) model is based on a voltage source and a resistor. The charge/discharge process and the voltage of the supercapacitor are expressed by the Equations:

$$V_{SC} = \frac{N_s Q_t d}{N_p N_e \varepsilon \varepsilon_0 A_i} + \frac{2N_s N_s RT}{\mathrm{F}} \sinh^{-1}\left(\frac{Q_t}{N_p N_e^2 A_i \sqrt{8RT\varepsilon\varepsilon_0 c}}\right) - R_{SC} i_{SC} \tag{11}$$

With:

$$Q_t = \int i_{SC} dt \tag{12}$$

The self-discharge behavior is expressed as follows ($i_{SC} = 0$):

$$Q_t = \int i_{Self\_dis} dt \tag{13}$$

where, V_{SC} is the SC voltage. i_{SC} is the SC current. N_p is the number of parallel SC's. N_s is the number of series SC's. N_e is the number of layers of electrodes. Q_t is the electric charge. R is ideal gas constant. T is the work temperature. F is Faraday constant. d is the molecular radius. A_i is the boundary region between electrolyte and electrodes. ε, ε_0, are the permittivity of the material, the permittivity of free space and the molar concentration, respectively.

3 Proposed ANN-EMS

To guarantee an optimal power control of the subsystems and to ensure a balance between production and load demand, an energy management system (EMS) based on the ANN has been integrated, the inputs being the power described by the equation, the load demand, the states of charge of the batteries and supercapacitors. The outputs being the reference powers of the batteries and supercapacitors. The ANN manager allows the flexible switching between the operating modes described in Fig. 2, with a high precision compared to the classical EMS systems. The structure of the ANN used is given by the Fig. 3. In order to obtain the best results, a database of 800 values has been used and divided as follows: 70% of the data are used for training, 15% for validation, and finally, 15% for testing.

	Modes	Description
$P_{RE} > P_{load}$	SoC_b<80% and SoC_{sc}<80%	Charging batteries and charging or discharging SCs
	SoC_b<80% and SoC_{sc}≥80%	Charging batteries and discharging SCs
	SoC_b<80% and SoC_{sc}<50%	Charging batteries and charging SCs
	SoC_b≥80% and 50% ≤ SoC_{sc}≤80%	isolate batteries and charging or discharging SCs
	SoC_b≥80% and SoC_{sc}≥80%	isolate batteries and discharging SCs
	SoC_b≥80% and SoC_{sc}≤50%	isolate batteries and charging SCs
$P_{RE} < P_{load}$	SoC_b>20% and SoC_{sc}<80%	Charging batteries and charging or discharging SCs
	SoC_b>20% and SoC_{sc}≥80%	Charging batteries and discharging SCs
	SoC_b>20% and SoC_{sc}<50%	Charging batteries and charging SCs
	SoC_b≤20% and 50% ≤ SoC_{sc}≤80%	isolate batteries and charging or discharging SCs
	SoC_b<20% and SoC_{sc}≥80%	isolate batteries and discharging SCs
	SoC_b<20% and SoC_{sc}≤50%	isolate batteries and charging SCs

Fig. 2. Operating modes of ANN-EMS.

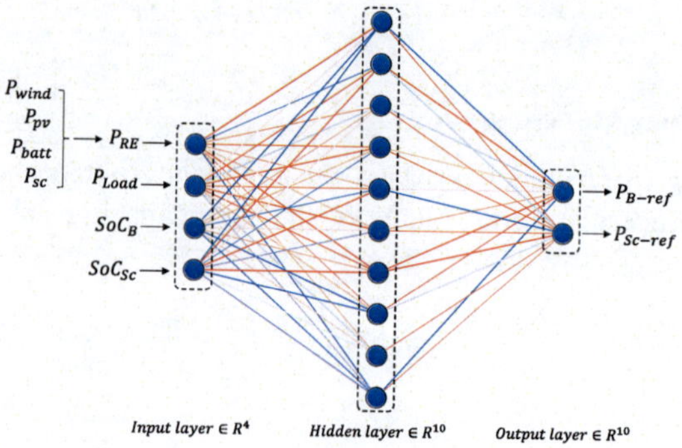

Fig. 3. Neural network architecture for the proposed EMS.

4 Simulation and Results

The studied hybrid system and the proposed ANN-EMS has been simulated in the Matlab/Simulink environment. In order to verify the efficiency and robustness of the model, a solar irradiance profile varying from [1000, 700, 900, 1000, 650 w/m2], a wind speed varying from [8, 9, 10, 7, 6 m/s], with a fixed temperature of 25 °C were used.

The obtained power profiles under these conditions are shown in Fig. 4 (a). Note that these powers are maximized by MPPT algorithms. Figure 4 (b) represents the load profile required by the motor-pump unit. During [0:1s], [2s:4s] and [3s:4s], the energy production is higher than the demand so the excess is stored in the batteries, for the rest of the time ([2s:3s] and [4s:5s]) the batteries provide the required power difference, while keeping the SoC_B and SoC_{sc} between the minimum and maximum margins 20% and 80% respectively (Fig. 4 (e) and Fig. 4 (f)). It is also noticeable that the supercapacitors absorb the current peaks (transient regime), and the batteries maintain the permanent regime of the demand. The ANN-EMS allowed also to maintain a fixed DC bus at 50V (Fig. 4 (d)).

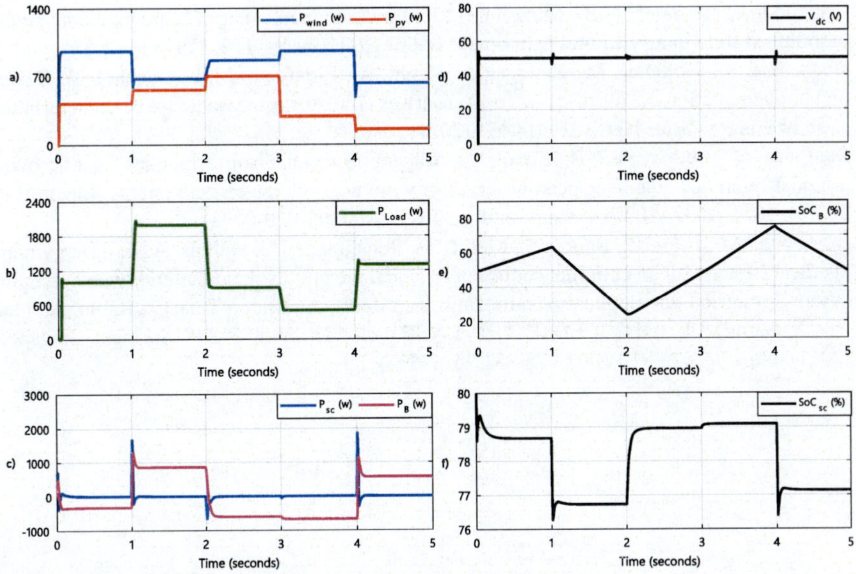

Fig. 4. **(a)** PV and wind turbine power, **(b)** Load demanded, **(c)** Supercapacitors and batteries power, **(d)** DC bus voltage, **(e)** SOC of batteries, **(f)** SOC of supercapacitors.

5 Conclusion

In this paper, an energy management system for a hybrid system composed by a wind turbine, PV panels, batteries and supercapacitors has been presented and simulated in the MATLAB/*Simulink* environment under different scenarios. The use of artificial intelligence tools for energy management ensures a flexible permutation between the operating modes, as well as a smooth transmission of the energy flows between the sub-systems.

References

1. Zaouche, F., Rekioua, D., Gaubert, J.P., Mokrani, Z.: Supervision and control strategy for photovoltaic generators with battery storage. Int. J. Hydrogen Energy **42**(30), 19536–19555 (2017)
2. Benzaouia, M., et al.: Design and performance analysis of a photovoltaic water pumping system based on DC-DC boost converter and BLDC motor. In: 2019 7th International Renewable and Sustainable Energy Conference (IRSEC), pp. 1–6. IEEE (2019)
3. Soufyane, B., Abdelhamid, R., Smail, Z.: Signed-distance fuzzy logic controller adaptation mechanism based MRAS observer for direct-drive PMSG wind turbines sensorless control. In: 2020 American Control Conference (ACC), pp. 4083–4089. IEEE (2020)
4. Shakeri, M., et al.: Implementation of a novel home energy management system (HEMS) architecture with solar photovoltaic system as supplementary source. Renew. Energy **125**, 108–120 (2018)
5. Roumila, Z., Rekioua, D., Rekioua, T.: Energy management based fuzzy logic controller of hybrid system wind/photovoltaic/diesel with storage battery. Int. J. Hydrogen Energy **42**(30), 19525–19535 (2017)

6. Khiareddine, A., Salah, C.B., Mimouni, M.F.: Power management of a photovoltaic/battery pumping system in agricultural experiment station. Sol. Energy **112**, 319–338 (2015)
7. Benzaouia, S., Mokhtari, M., Zouggar, S., Rabhi, A., Elhafyani, M.L., Ouchbel, T.: Design and implementation details of a low cost sensorless emulator for variable speed wind turbines. Sustain. Energy Grids Netw. **26**, 100431 (2021)
8. Soufyane, B., Abdelhamid, R., Smail, Z.: Adaptation mechanism techniques for improving a model reference adaptive speed observer in wind energy conversion systems. Electr. Eng. **102**(3), 1621–1637 (2020). https://doi.org/10.1007/s00202-020-00984-x
9. Benzaouia, M., Hajji, B., Rabhi, A., Mellit, A., Benslimane, A., Dubois, A.M.: Energy management strategy for an optimum control of a standalone photovoltaic-batteries water pumping system for agriculture applications. In: Hajji, B., Mellit, A., Marco Tina, G., Rabhi, A., Launay, J., Naimi, S.E. (eds.) ICEERE 2020. LNEE, vol. 681, pp. 855–868. Springer, Singapore (2021). https://doi.org/10.1007/978-981-15-6259-4_89

Estimation of Global Irradiation on Horizontal Plane Using Artificial Neural Network

Mohammed Benchrifa[1(✉)], Jamal Mabrouki[2], and Rachid Tadili[1]

[1] Laboratory of Solar Energy and Environment, Mohammed V University in Rabat, Faculty of Sciences, B.P. 1014, Rabat, Morocco
simobenchrifa@gmail.com

[2] Laboratory of Spectroscopy, Molecular Modeling, Materials, Nanomaterial, Water and Environment, CERNE2D, Mohammed V University in Rabat, Faculty of Science, Avenue Ibn Battouta, BP 1014, Agdal, Rabat, Morocco

Abstract. All studies concerning solar thermal or photovoltaic systems require the knowledge of solar radiation data but measurements of these radiation are in most cases not available. Consequently, the objective of this work is to establish a prediction model of the global solar irradiation on a daily scale using the artificial neural network (ANN). For this purpose, using the meteorological data collected in the Solar Energy and Environment Laboratory (LESE), five models were tested in order to find the combinations of input variables that lead to an efficient prediction. As a result, it was found that the fifth model (M5) is the most valid one because the obtained errors are 1.32 for RMSE, 1.2 for MAE and 0.94 for R^2.

Keywords: Global solar radiation · Artificial Neural Networks (ANN) · Prediction · Multi Layer Perceptron (MLP)

1 Introduction

The growth in global energy demand, the inevitable depletion of fossil fuels in the longer term, and the environmental degradation caused by these types of energy, has prompted the development of renewable sources of energy, thus ensuring sustainability and environmental protection [1]. The use of solar photovoltaic energy seems to be a necessity for the future. Indeed, solar radiation is the most abundant energy resource on earth. The amount of energy released by the sun (captured by the planet earth) during one hour could be enough to cover the world's energy needs for one year [2].

In order to exploit this energy and to optimize its collection by the photovoltaic and thermal collectors, it is necessary to know the distribution of solar irradiation at the photovoltaic installation location, under different orientations and inclinations.

However, solar irradiance is one of the most difficult meteorological parameters to estimate because it is a function of several geographical and astronomical parameters and is dependent on meteorological and atmospheric conditions [3]. Nevertheless, several estimation models on different time scales (hour, day, and month) from the available meteorological data are developed [4–10]. Among these models, we can cite physical

Y. Farhaoui et al. (Eds.): ICAISE 2022, LNNS 635, pp. 395–400, 2023.
https://doi.org/10.1007/978-3-031-26254-8_56

models, empirical models, artificial neural network models, hybrid models using a combination of several methods, and the models based on the satellite images data. Thus, the physical models are based on physical parameters, which estimates the global radiation from insolation duration, meteorological variables, and spatial variables in different sky conditions [20–24, 26–30].

The objective of our work is to use a mathematical modelling tool to predict solar irradiation. To this end, we are interested in the application of artificial neural networks to find the most efficient model to predict the daily global solar irradiation received on a horizontal plane for the region of Rabat, based on meteorological data that we were able to retrieve from the solar energy and environment laboratory of the Faculty of Sciences of Rabat, and astronomical data calculated from existing mathematical relationships.

2 Materials and Methods

2.1 Artificial Neural Networks (ANN)

The capabilities of a single neuron are limited. The implementation of complex functions requires the integration of several neurons, operating in parallel, in the form of a network with a specific topology. The neuron inputs are either the inputs of the global network or the outputs of other neurons. The connections between the neurons that make up the network describe the topology of the model. It can be arbitrary, but most often it is possible to distinguish a certain regularity. The most important parameters of this model are the synaptic coefficients. They build the resolution model according to the information given to the network. We must therefore find a mechanism that allows us to calculate them from quantities that we can acquire from the problem. This is the fundamental principle of learning. Once the adjustment of the synaptic weights has been made, the neural network constitutes a non-linear statistical model [14]. The advantage of neural networks over conventional regression methods is that they generally require a smaller number of adjustable parameters to obtain a non-linear model of a given accuracy [15]. Two types of neural networks can be distinguished: non-looped and looped networks. Looped networks are often used for signal processing, control or regulation. Non-looped networks are used for recognition, classification or prediction.

2.2 Multilayer Perceptron (MLP)

The architecture of a neural network is the organisation of the neurons in the same network, i.e. the way they are ordered and connected. Most neural networks use the same type of neurons. Some rarer architectures are based on didactic neurons. The architecture of a neural network depends on the task to be learned (problem to be solved). The Multi-Layer Perceptron (MLP) (Fig. 1) is probably the simplest and best known neural network, most commonly used for approximation, classification and prediction problems [17]. It consists of several layers of fully connected neurons. The activation function used is mainly the sigmoid function. This type of network is in the general family of forward propagation networks.

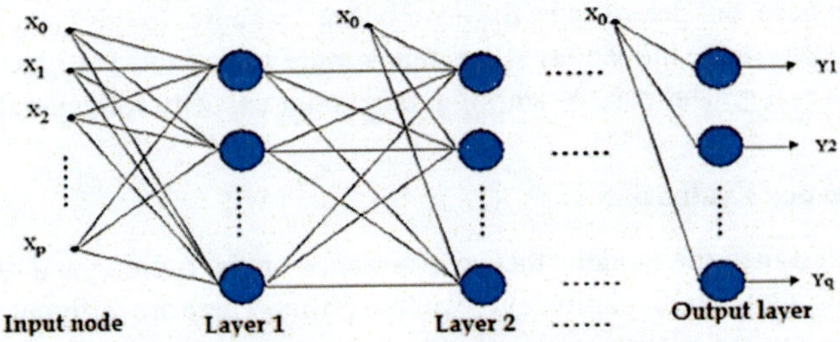

Fig. 1. Structure of a multilayer perceptron

3 Results and Discussion

3.1 Model Development

The objective of our study is the use of the neural network (MLP) as a tool for the prediction of solar irradiation by supervised learning. For this, the input data are: the insolation S (h) the theoretical day length S0 (h), the extraterrestrial solar irradiation G0 (wh.m-2.day-1), the maximum temperature Tmax (°C) and the minimum temperature Tmin (°C).

Several models were tested in order to find the combinations of input variables that lead to an efficient prediction. The inputs used for these models were chosen based on the calculation of the correlation coefficient between solar irradiation and each of the mentioned variables We report in the following table the 5 models that we considered to be the most interesting (Table 1).

Table 1. Various models used for training the neural network

Models	Input values					
	S	S0	G0	T_{max}	T_{min}	ΔT
M_1	X	X				
M_2	X		X			
M_3	X	X	X			
M_4				X	X	
M_5	X	X	X	X	X	X

For each model, the number of neurons in the hidden layer was varied from 1 to 10, and 500 training sessions were carried out, with a different initialization for each of these architectures; then the value of the synaptic weights that gives the minimum RMSE on the training base was saved.

We noticed that from 3 neurons the learning phenomenon starts to appear. The indication of the presence of overlearning is shown by the decrease in performance in the test phase; the successive addition of neurons in the hidden layer improves the error on the training base, but deteriorates the error on the test base. In this case, the network, with more neurons in the hidden layer, has many more parameters (weights and biases), which introduces a greater ease of learning, but a poor capacity for generalization.

3.2 Performance Evaluation

After having designed the models, their performance on the training and test data must be evaluated through the calculation of the output errors. There are different error criteria in the literature, but the criteria used in this work are: Correlation coefficient (R^2), Mean absolute percentage error (MAPE) and Root mean square error (RMSE) calculated in Table 2.

Table 2. Models performance evaluation

Modèles	R2	MAE	RMSE	R^2	MAE	RMSE
	Training			Testing		
M_1	0.924	0.56	0.95	0.86	1.26	1.44
M_2	0.92	0.69	1.06	0.85	1.33	1.61
M_3	0.95	0.55	0.88	0.9	1.2	1.4
M_4	0.921	0.84	1.16	0.88	1.27	1.69
M_5	0.989	0.33	0.63	0.94	1.1	1.32

As can be seen in Table 2 for the different models, the best performance is obtained by the fifth model according to the three error criteria. The error obtained is 1.32 for RMSE, 1.2 for MAE and 0.94 for R^2. As results, the characteristics of this model (weights and bias) can be kept for a possible real implementation to estimate the global solar irradiation on horizontal plane from Rabat's meteorological data.

4 Conclusion

The presented study was motivated by the need of global solar irradiation measurement. This data is very important for the study and design of solar systems, especially photovoltaic systems. This study confirms the ability of neural networks to predict solar irradiance accurately, and the predicted data can therefore be used in the absence of measurements. The results indicate that neural network modelling shows great potential for estimating the solar resource in remote areas or areas without measuring stations.

Thus, in the present work, 5 models based on different variables were compared, and the comparison between them showed that the fifth model is the most suitable for the chosen site and the quality of the data generated by this model is very satisfactory ($R = 0.97$; $RMSE = 1.32$), and exceeds that estimated from empirical methods [6].

Finally, our model can provide synthetic data series for the global component of solar radiation which can be used in the dimensioning and optimal design of solar thermal and photovoltaic energy systems [20–22].

References

1. Scheffran, J., Felkers, M., Froese, R.: Economic growth and the global energy demand. In: Green Energy to Sustainability: Strategies for Global Industries, pp. 1–44 (2020)
2. Mabrouki, J., Bencheikh, I., Azoulay, K., Es-Soufy, M., El Hajjaji, S.: Smart monitoring system for the long-term control of aerobic leachate treatment: dumping case Mohammedia (Morocco). In: International Conference on Big Data and Networks Technologies, pp. 220–230 (2019)
3. Benchrifa, M., Tadili, R., Idrissi, A., Essalhi, H., Mechaqrane, A.: Development of new models for the estimation of hourly components of solar radiation: tests, comparisons, and application for the generation of a solar database in Morocco. Int. J. Photoenergy (2021)
4. Benchrifa, M., Essalhi, H., Tadili, R., Bargach, M.N., Mechaqrane, A.: Development of a daily databank of solar radiation components for Moroccan territory. Int. J. Photoenergy (2019)
5. Benchrifa, M., Essalhi, H., Tadili, R., et al.: Estimation of daily direct solar radiation for rabat. In: Sayigh, A. (ed.) Sustainable Energy Development and Innovation, pp. 629–634. Springer, Cham (2022). https://doi.org/10.1007/978-3-030-76221-6_69
6. Yorukoglu, M., Celik, A.N.: A critical review on the estimation of daily global solar radiation from sunshine duration. Energy Convers. Manage. **47**(15–16), 2441–2450 (2006)
7. Teke, A., Yıldırım, H.B., Çelik, Ö.: Evaluation and performance comparison of different models for the estimation of solar radiation. Renew. Sustain. Energy Rev. **50**, 1097–1107 (2015)
8. Ahmad, M.J., Tiwari, G.N.: Solar radiation models—a review. Int. J. Energy Res. **35**(4), 271–290 (2011)
9. Bencheikh, I., et al.: The adsorptive removal of MB using chemically treated artichoke leaves: parametric, kinetic, isotherm and thermodynamic study. Sci. Afr. **9**, e00509 (2020)
10. Karabacak, K., Cetin, N.: Artificial neural networks for controlling wind–PV power systems: a review. Renew. Sustain. Energy Rev. **29**, 804–827 (2014)
11. Koley, E., Shukla, S.K., Ghosh, S., Mohanta, D.K.: Protection scheme for power transmission lines based on SVM and ANN considering the presence of non-linear loads. IET Gener. Transm. Distrib. **11**(9), 2333–2341 (2017)
12. Mocanu, E., Nguyen, P.H., Gibescu, M., Kling, W.L.: Deep learning for estimating building energy consumption. Sustain. Energy Grids Netw. **6**, 91–99 (2016)
13. Dorvlo, A.S., Jervase, J.A., Al-Lawati, A.: Solar radiation estimation using artificial neural networks. Appl. Energy **71**(4), 307–319 (2002)
14. El Badaoui, H., Abdallaoui, A., Chabaa, S.: Using MLP neural networks for predicting global solar radiation. Int. J. Eng. Sci. **2**(12), 48–56 (2013)
15. Mabrouki, J., Azrour, M., Farhaoui, Y., El Hajjaji, S.: Intelligent system for monitoring and detecting water quality. In: Farhaoui, Y. (ed.) BDNT 2019. LNNS, vol. 81, pp. 172–182. Springer, Cham (2020). https://doi.org/10.1007/978-3-030-23672-4_13
16. Benchrifa, M., Mabrouki, J.: Simulation, sizing, economic evaluation and environmental impact assessment of a photovoltaic power plant for the electrification of an establishment. Adv. Build. Energy Res. **16**, 1–18 (2022)
17. Benchrifa, M., Mabrouki, J., Elouardi, M., et al.: Detailed study of dimensioning and simulating a grid-connected PV power station and analysis of its environmental and economic effect, case study. Modeling Earth Syst. Environ. 1–9 (2022)
18. Mabrouki, J., Fattah, G., Al-Jadabi, N., et al.: Study, simulation and modulation of solar thermal domestic hot water production systems. Modeling Earth Syst. Environ. 1–10 (2021)
19. Azrour, M., Mabrouki, J., Fattah, G., et al.: Machine learning algorithms for efficient water quality prediction. Modeling Earth Syst. Environ. **8**(2), 2793–2801 (2022)

20. Mabrouki, J., et al.: Smart system for monitoring and controlling of agricultural production by the IoT. In: Azrour, M., Irshad, A., Chaganti, R. (eds.) IoT and Smart Devices for Sustainable Environment, pp. 103–115. Springer, Cham (2022). https://doi.org/10.1007/978-3-030-90083-0_8
21. Azrour, M., Farhaoui, Y., Ouanan, M., Guezzaz, A.: SPIT detection in telephony over IP using K-means algorithm. Procedia Comput. Sci. **148**, 542–551 (2019)
22. Azrour, M., Mabrouki, J., Guezzaz, A., et al.: Internet of things security: challenges and key issues. Secur. Commun. Netw. **2021** (2021)
23. Mabrouki, J., Azrour, M., Hajjaji, S.E.: Use of internet of things for monitoring and evaluating water's quality: a comparative study. Int. J. Cloud Comput. **10**(5–6), 633–644 (2021)
24. Mabrouki, J., Benbouzid, M., Dhiba, D., El Hajjaji, S.: Simulation of wastewater treatment processes with Bioreactor Membrane Reactor (MBR) treatment versus conventional the adsorbent layer-based filtration system (LAFS). Int. J. Environ. Anal. Chem. 1–11 (2020)

Evaluating the Impact of Dataset Size on Univariate Prediction Techniques for Moroccan Agriculture

Rachid Ed-daoudi[1(✉)], Altaf Alaoui[1], Jad Zerouaoui[1], Badia Ettaki[1,2], and Jamal Zerouaoui[1]

[1] Laboratory of Engineering Sciences and Modeling, Faculty of Sciences, Ibn Tofail University, Campus Universitaire, BP 133, Kenitra, Morocco
rachid.ed-daoudi@uit.ac.ma

[2] LyRICA: Laboratory of Research in Computer Science, Data Sciences and Knowledge Engineering, School of Information Sciences Rabat, Rabat, Morocco

Abstract. Learning models used for prediction are mostly developed without taking into account the size of datasets that can produce models of high accuracy and better performance. Although, the general believe is that, large dataset is needed to construct a predictive learning model. To describe a data set as large in size depends on the circumstances and context of prediction. This means that what makes a dataset to be considered as being big or small is controversial. In this paper, the ability of the predictive model to adapt to a particular size of data in training is examined. The study experiments on three different sizes of Moroccan agricultural data using a variety of statistical and Machine Learning techniques, to create predictive models with a view to establishing if the size of data has any effect on the accuracy of a model. The output of each model is measured using the Mean Absolute Error (MAE) and r-squared, and comparisons are made. The results of training the models through the three partitioned dataset show that, the models trained with the smallest and largest size of training data appear to be less accurate, while the models trained with a medium sized dataset delivers a much better results.

Keywords: Prediction · Statistical techniques · Machine learning · Agriculture data · Data size

1 Introduction

Crop yield prediction has the great importance in food production in Morocco. Policy makers in rely on accurate predictions to make timely import and export decisions to strengthen national food security [1, 2]. The productivity of crops can be predicted by using various techniques [3].

There are several techniques used in modelling dataset in the purpose of making predictions in many human areas of activity [4]. Times series, machine learning and neural networks are among the successful techniques for their processing capabilities.

Y. Farhaoui et al. (Eds.): ICAISE 2022, LNNS 635, pp. 401–407, 2023.
https://doi.org/10.1007/978-3-031-26254-8_57

These techniques have been reported to have performed well in the construction of predictive model in several studies [5–8]. These widely used techniques are based on fitting a curve through the data, which mainly involve finding a relationship from the predictors to the predicted.

The present study modelled datasets in order to determine the effect of using dataset of various sizes for model construction in order to predict crops yield in Morocco. Some of the questions the paper would answer include: What differences does it make to use small or large dataset to develop a predictive model? How can the accuracy of a model be determined and what can be done to improve the accuracy of a predictive model?

In order to address these questions, the study uses a variety of prediction techniques to experiment on three different sizes of dataset in order to create a comparison between models capable of predicting the crop production based on the relationships established from the input data. Findings from this study unveils the direct effect of model construction using different sizes of the dataset.

This paper is organized as follows: in the next section, used data is presented. Some predictive learning models reported in the literature follow this, and the section that follows presents the proposed approaches and the performance of the training process is illustrated. The experimental results are presented and discussed in a separate section, while the study is concluded in the final section.

2 Data

Food and Agriculture Organization of the United Nations (FAOSTAT) [9] provides data related to food and agriculture for two hundred countries. In this study, we use two datasets related to Morocco: monthly eggs production and yearly soft wheat production, starting from 1962 to 2019. The collected data is structured in two columns (unit of time and production values).

3 Prediction Methodology

Prediction methodologies are used with the aim of forecasting the behavior that a given indicator will have. It is almost always necessary to make predictions with more than one methodology, in order to compare and observe which is the most accurate and robust prediction [10].

3.1 Statistical Models

The Box-jenkins methodology considers that the time series to be predicted is generated by a stochastic process, the nature of which can be characterized by means of a model [11]. An autoregressive (AR) model is a representation of a random process, in which the variable of interest depends on its past observations. Specifically, the variable of interest, or output, depends linearly on its previous values [12].

The technical indicator "Moving Average" (MA) shows the average value of an indicator over a certain period of time. When Moving Average is calculated, the value of the indicator is mathematically averaged over a given period of time [13].

An ARIMA model is a class of linear models that utilizes historical values to forecast future values. ARIMA stands for Autoregressive Integrated Moving Average, each of which technique contributes to the final forecast. Let's understand it one by one [14].

3.2 Exponential Smoothing

The Exponential Smoothing prediction methodology uses the Holt-Winters exponential smoothing method to decompose the time series into seasonal and trend components to effectively predict future time periods. Its maximum effectiveness is obtained when the values of the time series follow a gradual trend and present a seasonal behavior in which the values follow a repeated cyclical pattern over a given number of time periods [15].

3.3 Regression

Regression analysis is a statistical method that shows the relationship between two or more variables. Usually expressed in a graph, the method evaluates the relationship between a dependent variable with the independent variables [16].

4 Application and Results

In this section, the development of predictive models based on different sizes of the dataset and the evaluation of these models is presented. In this work, proportions of datasets are created under three sizes: small (59 observations) related to the annual production of soft wheat in Morocco between 1962 and 2019, Medium (700 observations) relating to the monthly production of eggs in Morocco between 1991 and 2004 and large (2005 observations) relative to the daily production of eggs in Morocco between 2010 and 2016. Then we will apply the prediction techniques discussed in the previous part and note the application process and the result.

4.1 Experiment I

Small dataset (59 observations) of annual production of soft wheat in Morocco between 1962 and 2019.

In this experiment, the dataset is partitioned into a training set representing 60% of data, while the leftover data sets are partitioned to the validation set and testing set in equal percentage. While the training set learn the relationship between the input attributes and the target, the validation set track the error that occurs during the training.

The Mean Square Error (MSE) is the error computed during this process using the formula in Eq. 1:

$$MSE = \frac{1}{n}\sum\nolimits_{i=1}^{n}(\widehat{Y}i - Yi)^2 \tag{1}$$

where n is the number of samples, ŷi is the network output and yi is the target value.

MAE measures the average magnitude of the errors in a set of predictions. It's the average over the test sample of the absolute differences between prediction and actual observation where all individual differences have equal weight as gives the Eq. 1:

$$MAE = \frac{1}{n}\sum_{j=1}^{n} |Yj - \widehat{Y}i| \tag{2}$$

where n is the number of samples, ŷi is the network output and yi is the target value.

Table 1 presents the evaluation measures for experiment 1.

Table 1. Evaluation measures for experiment I

Technique	MAE	MSE	R^2
Holtwinters	1566	45.59	0.006
MA	887	59.45	0.08
AR	878	80.39	0.08
AutoARIMA (0, 0, 3)	530	83.64	0.26
Randomforest	334	43.08	0.23
SVR	121	56.03	0.11
Tensorflow LSTM	80.93	50.45	0.38

4.2 Experiment II:

Medium Dataset (700 lines) of monthly production of eggs in Morocco between 1991 and 2004.

In the second experiment, the dataset used for the model construction is increased to 700 lines. Similarly, using the same technique as in Experiment I, this dataset is partitioned into training set, validation set and testing set. The validation sets perform the role as described in experiment I.

Table 2. Evaluation measures for experiment II

Technique	MAE	MSE	R^2
Holtwinters	1566	45.59	0.20
MA	887	66.2	0.34
AR	878	65.32	0.36
AutoARIMA (0, 0, 3)	530	64.71	0.5
Randomforest	334	59.45	0.16
SVR	121	80.39	0.20
Tensorflow LSTM	80.93	64.71	0.66

4.3 Experiment III

Larger dataset (2005 observations) of daily eggs production in Morocco between 2010 and 2016.

In the third experiment, the dataset used for model construction is increased to 2005 lines. The partitioning of dataset to three portions also conforms to 60% for training, 20% for validation and 20% for testing. These sets of data, perform roles as described in Experiment I. Table3 presents the evaluation measures for experiment 3.

Table 3. Evaluation measures for experiment III

Technique	MAE	MSE	R^2
Holtwinters	1566	126	0.2
MA	887	113	0.18
AR	878	91	0.14
AutoARIMA (0, 0, 3)	530	105	0.28
Randomforest	334	102	0.346
SVR	121	10	0.489
Tensorflow LSTM	80.93	121.98	0.57

5 Discussion

The study experimented three different sizes of Moroccan agricultural data, with a view to establish if the size of data has any effect on the accuracy of a predictive model.

From the experiments, it can be seen that the final MSE value for Experiment 2 appears small, while it appears bigger for Experiment 1 and 3. In general, all the models trained well as no over-fitting is shown up. The response of each model to an untrained input dataset, however, gives much better clarifications on their level of accuracies. The model in experiment 3 has the least MAE, while model 1 (constructed using 59 data entries) shows the MAE of highest value. These results have shown that, simulating a predictive model using over 50% of an untrained input data of a small proportion dataset can affect the prediction accuracy.

However, the values of r^2 shown in Table 3 (constructed using 2500 data entries) are the less significant. This shows that larger training sets can lead to an over training of the model, and cause a loss of information through prediction [17].

Table 2 of Experiment 2 show that a compromise can be found between size of the data and the proportion of training and validation set. An acceptable approach could be to stop training when performance on a validation dataset starts to degrade. The values of r^2 through the respective experiments tend to be the highest for machine learning techniques. We observe a very suitable training for the LSTM and SVR techniques, especially in experiment 2.

6 Conclusion

This paper evaluates and presents the resulting outputs of modelling different sizes of the dataset for prediction purposes. In the course of experimentations, three different sizes of the dataset are modelled using various prediction techniques. The accuracy of the simulated outputs is measured using mean absolute error and the comparison shows the degree of error associated with each trained model, while the value of r^2 determines the best techniques performances for the case study.

The predictive model constructed with the smallest dataset records highest error when simulated with untrained inputs, while other models constructed using more dataset records better accuracy MSE. Thus, it can be inferred from the results of this study that, using sufficient data set for predictive model construction can lead to better accuracy and the model's ability to generalize. Although, due to vagueness that surrounds the size of the dataset, it is difficult to say precisely when a dataset can be considered to be big enough for the model to be accurate without it to be over fitted.

References

1. Bandyopadhyay, G., Chattopadhyay, S.: Single hidden layer artificial neural network models versus multiple linear regression model in forecasting the time series of total ozone. Int. J. Environ. Sci. Technol. **4**, 141–149 (2007). https://doi.org/10.1007/BF03325972
2. Basavanhally, A., Doyle, S., Madabhushi, A.: Predicting classifier performance with a small training set: applications to computer-aided diagnosis and prognosis. Paper presented at the 2010 IEEE International Symposium on Biomedical Imaging: From Nano to Macro (2010)
3. Dobbin, K.K., Simon, R.M.: Sample size planning for developing classifiers using high-dimensional DNA microarray data. Biostatistics **8**(1), 101–117 (2007)
4. Ed-daoudi, R., Alaoui, A., Ettaki, B., Zerouaoui, J.: A review of prediction techniques in some domains of human activity. J. Comput. Sci. Submitted in 2022
5. Haykin, S.: Neural Networks and Learning Machines. Pearson Education, Upper Saddle River (2009)
6. McArthur, D.P., Encheva, S., Thorsen, I.: Predicting with a small amount of data: an application of fuzzy reasoning to regional disparities. J. Econ. Stud. **41**, 12–28 (2013)
7. Mukherjee, S., Tamayo, P., Rogers, S., Rifkin, R., Engle, A.: Estimating dataset size requirements for classifying DNA microarray data. J. Comput. Biol. **10**, 119–142 (2003)
8. Oladokun, V., Adebanjo, A., Charles-Owaba, O.: Predicting students' academic performance using artificial neural network: a case study of an engineering course. Pac. J. Sci. Technol. **9**(1), 72–79 (2008)
9. Osmanbegović: Data mining approach for predicting student performance (2012)
10. Özel, T., Karpat, Y.: Predictive modeling of surface roughness in hard turning using regression and neural networks. Int. J. Mach. Tools Manuf. **45**, 467–479 (2005)
11. Skillicorn, D.: Understanding datasets: data mining with matrix decompositions (2007)
12. Suh, S.C.: Practical Applications of Data Mining. Jones & Bartlett Learning, Burlington (2012)
13. van der Ploeg, T., Austin, P.C., Steyerberg, E.W.: A simulation study for predicting dichotomous endpoints. BMC Med. Res. Methodol. **14**(1), 137 (2014)
14. Witten, I.H., Frank, E., Hall, M.A.: Data Mining Practical Machine Learning Tools and Techniques, 3rd edn. Morgan Kaufmann, San Francisco (2011)

15. Cuesta, E., Kirane, M., Malik, S.A.: Image structure preserving denoising using generalized fractional time integrals. Signal Process. **92**, 553–563 (2012)
16. Shen, S., Liu, F., Anh, V., Turner, I.: Detailed analysis of a conservative difference approximation for the time fractional diffusion equation. J. Appl. Math. Comput. **22**, 1–19 (2006)
17. Alikhanov, A.A.: A new difference scheme for the time fractional diffusion equation. Comput. Phys. **280**, 424–438 (2015)

Experimental Assessment of MPPT Based on a Neural Network Controller

Benzaouia Mohammed[1] (✉), Hajji Bekkay[1], Rabhi Abdelhamid[2], and Benzaouia Soufyane[3]

[1] Renewable Energy, Embedded System and Data Processing Laboratory, National School of Applied Sciences, Mohamed First University, Oujda, Morocco
m.benzaouia@ump.ac.ma

[2] Laboratory of Modelization, Information and Systems, University of Picardie Jules Verne 3, rue Saint Leu, 80039 Amiens Cedex 1, France

[3] Laboratory of Systems and Information Sciences of Aix Marseille University, Marseille Cedex 20, France

Abstract. Improving the efficiency of photovoltaic systems depends mainly on extracting the maximum power output from the panels with high accuracy in tracking the maximum power point (MPP) despite variations in load and weather conditions (temperature and solar irradiance). In this paper, a neural network (ANN) based maximum power point tracking (MPPT) controller has been developed and experimentally validated under real conditions. The objective is to overcome the drawbacks of the classical MPPT algorithms (P&O and INC), namely the response time and the oscillation around the MPP. The obtained results show a high accuracy of the MPP tracking.

Keywords: PV · DC-DC Boost converter · ANN · MPPT

1 Introduction

Given the continuous growth of the energy demand in the different sectors of industry, production and agriculture, the recourse to the exploitation of renewable energy sources such as solar, wind and hydraulic energy represents the best solution to get rid of the dependence of conventional energy resources and their undesirable impact on the environment [1]. PV energy is classified as a green and friendly energy, which presents several advantages such as the easiness of installing in different areas and the low maintenance cost due to the absence of moving parts [2]. However, the main drawback is the low conversion efficiency, which is around 9–17% under normal conditions. To maximize the efficiency and improve the overall performance of the PV system, several techniques have been implemented to track the MPP. These techniques consist in adjusting the duty cycle of the DC-DC converter according to the variations of temperature and solar irradiance. In the literature three different types of control are distinguished, direct, indirect and AI-based control, differentiated in terms of simplicity of implementation, response time, reliability and accuracy of tracking [3]. The open circuit voltage (OCV) and short

Y. Farhaoui et al. (Eds.): ICAISE 2022, LNNS 635, pp. 408–414, 2023.
https://doi.org/10.1007/978-3-031-26254-8_58

circuit current (*SCC*) method are some of the indirect control techniques [4]. The idea is to determine the MPP voltage and current values (V_{MPP}, I_{MPP}) through a linear equation by multiplying the values of V_{OC} and I_{SC} with a variable gain K [5]. Yet, these methods lack accuracy for tracking the MPP under variable and rapid weather conditions. Among the direct control methods, *P&O* and *INC* which are widely used because of their simplicity of implementation, the few sensors used and their high reliability and efficiency [6]. The integration of AI has allowed to surmount several inconveniences of direct and indirect methods such as response time and oscillations around the MPP. In [7], the authors proposed a technique based on fuzzy logic with two inputs, error and error variation, twenty-five fuzzy rules were used to adjust the variation of the duty cycle, the system was tested under temperature conditions fixed at 25 °C and a profile of solar irradiance varying from 600 w/m^2 to 1000 w/m^2. Other research has focused on the development of hybrid MPPT control such as FL-P&O and FL-INC to reduce the number of fuzzy rules on one hand and to integrate the advantages of FL and P&O - INC algorithms and finally to facilitate the implementation part of the MPPT controller [8].

In this research paper, a PV system connected to a DC-DC Boost converter and a variable load is presented. A neural network (ANN) based MPPT technique has been developed and implemented in real time to maximize the power under variable conditions (weather and load). This paper is organized as follows: the modeling of the different stages of the system is presented in the second section. In the third section, a presentation of the classical MPPT techniques as well as the proposed ANN MPPT approach. Experimental results are presented in the fourth section and finally a conclusion.

2 Presentation and Modeling of Studied System

The structure of the studied system is depicted in Fig. 1; it contains a photovoltaic panel connected to a DC-DC Boost converter, which is controlled by a neural network (ANN) controller to extract the maximum power. The PWM signals are generated from a dSPACE MicroLabBox 1202. At the output a variable load has been connected to simulate real demand profiles.

2.1 Photovoltaic Panel Model

The non-linear behavior of PV panel is given by Eqs. (1 and 2). The effect of solar irradiance and temperature on photovoltaic panel operation is shown in Fig. 2. The different parameters of the chosen PV panel are summarized in Table 1.

$$I_{PV} = I_{ph} - I_{\mathrm{d}} - I_{sh} \tag{1}$$

$$I_{PV} = I_{ph} - I_{s}e^{(\frac{q(V_{PV}+I.R_s)}{nkT})-1} - \frac{(V_{PV} + I.R_s)}{R_{sh}} \tag{2}$$

where: I_s is the saturation current, q is the electron charge, k is the Boltzmann constant, n is the diode ideality factor, and T is the PV cell temperature (K).

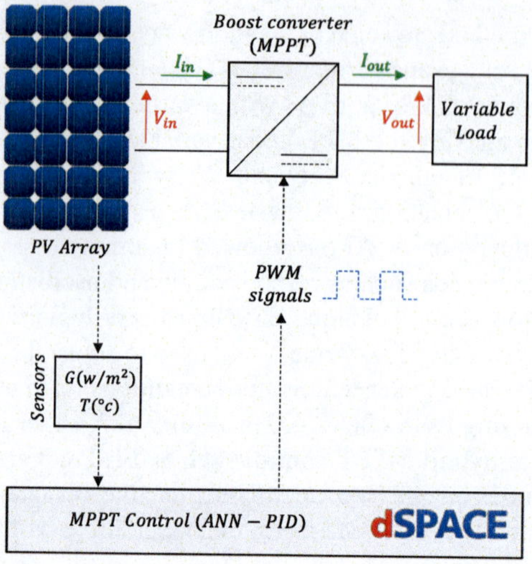

Fig. 1. Stand-alone water pumping system description.

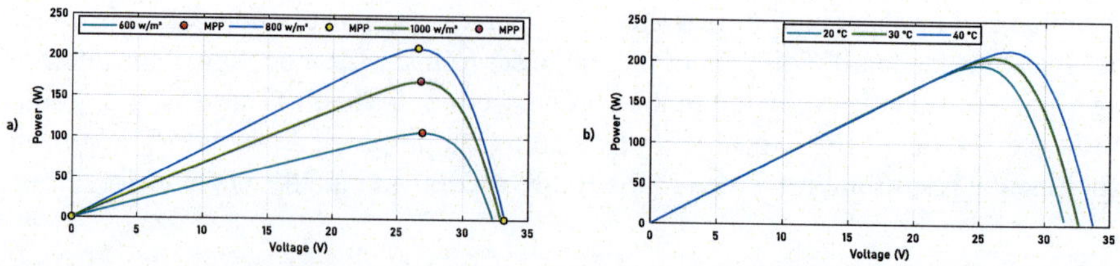

Fig. 2. a) $P - V$ characteristic for different values of solar irradiance, b) $P - V$ characteristics for different values of temperture.

Table 1. The PV module parameters.

PV module Kyocera Solar KD210GX-LFBS	
Parameters	Values
Numbers of cells in a module, N_c	54
Open circuit voltage, V_{OC}	33.2 V
Short circuit current, I_{SC}	8.58 A
Maximum voltage at MPP, V_{MPP}	26.6 V
Maximum current at MPP, I_{MPP}	7.9 A
Maximum Power at MPP, P_{MPP}	210.14 W

2.2 DC-DC Boost Converter Model

For MPP operation, the DC-DC converter is used as an adaptation stage between the PV panels and the load. It is composed by an inductor (L), a controlled switch (IGBT), a diode (D) and a capacitor (C). The design of the converter is based on Eqs. (3 and 4):

$$L = \frac{V_i(V_O - V_i)}{\Delta I_L.f_s.V_O} \tag{3}$$

$$C = \frac{I_O.D}{\Delta V_O.f_s} \tag{4}$$

where, V_i is the input voltage, V_o is the output voltage, I_O is the output current, f_s is the switching frequency, ΔI_L is the estimated inductor ripple current, ΔV_O is the estimated output ripple voltage, and D is the duty cycle.

3 Maximum Power Point Tracking (MPPT)

3.1 Perturb & Observe (P&O)

The simplicity of implementation and the reliability of MPP tracking made this technique very useful in many systems. The P&O algorithm consists in perturbing the system and observing its effect on the output power then adjusting the duty cycle with a well-defined step according to Eq. (5):

$$D_{k+1} = D_k \pm \Delta D \tag{5}$$

where, D_{k+1} is the next variation of duty cycle, D_k is the precedent variation of duty cycle, and ΔD is the step size.

3.2 Incremental Conductance (INC)

The INC algorithm uses the instantaneous conductance I/V and the incremental conductance dI/dV to determine the MPP of the $P - V$ curve characteristic of PV panels. Depending on the relationship between the two values, the location of the operating point is determined.

3.3 Neural Network (ANN)

Figure 3 shows the block diagram of the proposed three-layer feedforward ANN applied to our system. The ANN controller is used to estimate the output voltage V_{MPP} for each value of solar irradiance and temperature, and it is composed of three essential layers:

- Input layer: The ANN controller contains two input neurons which correspond to the measured values of the climatic variables influencing the operation of the panel (temperature and solar irradiance).
- Hidden layer: In this layer, a function applies weights to the input values and links them through an activation function as an output.
- The output layer represented by a single neuron, which is the output voltage value V_{MPP}.

The used database has been extracted by simulating the mathematical model of the panel (i.e. finding the V_{MPP} and I_{MPP} for a temperature range from 0 °C to 45 °C), and divided as follows: 70% of the data are used for training, 15% for validation, and finally, 15% for testing.

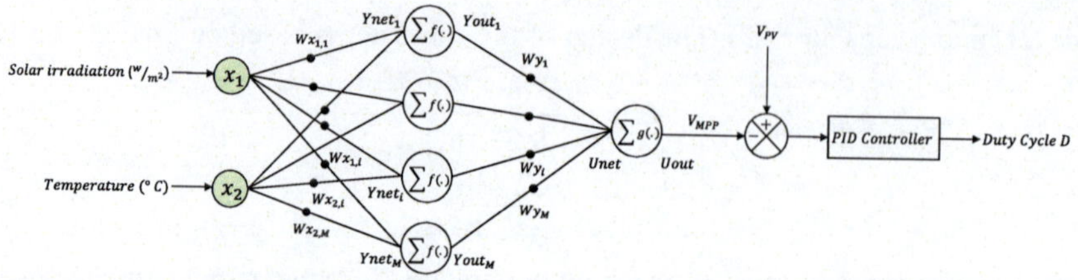

Fig. 3. Flowchart of the ANN based MPPT algorithm.

4 Experimental Results and Discussion

The laboratory experimental results are realized on the basis of the existing MIS laboratory test bench, as shown in Fig. 4. The output power produced by the PV panel under a solar irradiance value of G = 323 w/m^2 and a temperature of 37.3 °C (Fig. 5. (a) and Fig. 5. (b)), being 76 w (Fig. 5. (c)). Comparing this result to the simulated characteristics of the PV panel (Fig. 5. (e)), it is clear that the controller has allowed to reach the maximum operating point with great accuracy and precision. The noticed peaks are related to the change of the load, and as already detailed, it is necessary to track the MPP even under the load changes, this is achieved in our case by the ANN controller, while keeping an accurate tracking of the V_{MPP} (obtained by the characteristic) and the $V_{Measured}$ as shown in Fig. 5. (d) with less chattering percentage (Fig. 5. (f)).

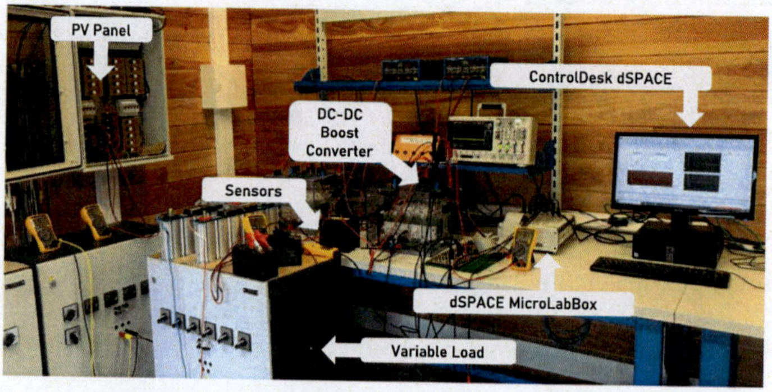

Fig. 4. Experimental setup (MIS laboratory).

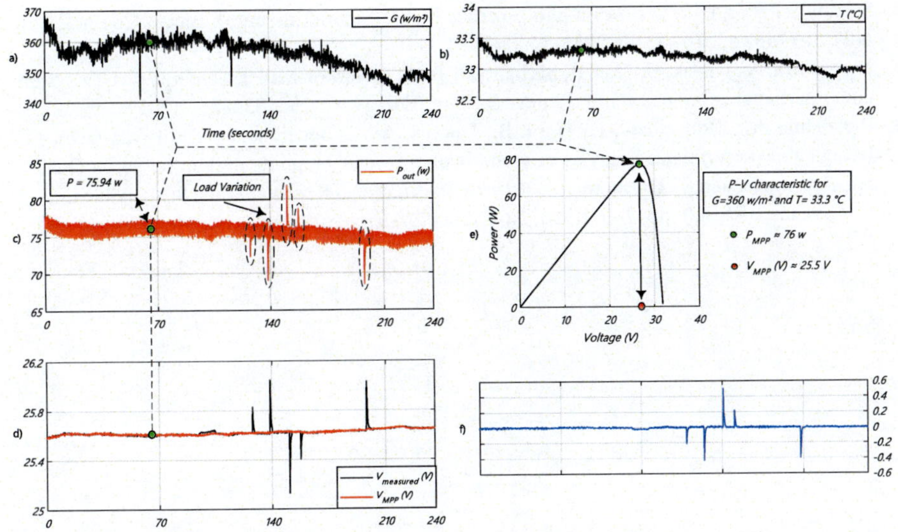

Fig. 5. **(a)** Solar irradiance (w/m^2), **(b)** Temperature (°C), **(c)** PV power (w), **(d)** Measured voltage and V_{MPP} (V), **(e)** $P - V$ characteristic, **(f)** Tracking error.

5 Conclusion

This paper discusses the implementation and experimental validation of a neural network controller to extract the maximum power from PV panels in order to improve the response of a photovoltaic system under real variable atmospheric conditions. Among the advantages of the controller, the accuracy of tracking the maximum PV power point with minimal response time compared to classical MPPTs. In addition, the low oscillation around the MPP ensures stable operation.

References

1. Soufyane, B., Abdelhamid, R., Smail, Z.: Adaptation mechanism techniques for improving a model reference adaptive speed observer in wind energy conversion systems. Electr. Eng. **102**(3), 1621–1637 (2020). https://doi.org/10.1007/s00202-020-00984-x
2. Benzaouia, M., Hajji, B., Rabhi, A., Mellit, A., Benslimane, A., Dubois, A.M.: Energy management strategy for an optimum control of a standalone photovoltaic-batteries water pumping system for agriculture applications. In: Hajji, B., Mellit, A., Marco Tina, G., Rabhi, A., Launay, J., Naimi, S.E. (eds.) ICEERE 2020. LNEE, vol. 681, pp. 855–868. Springer, Singapore (2021). https://doi.org/10.1007/978-981-15-6259-4_89
3. Karami, N., Moubayed, N., Outbib, R.: General review and classification of different MPPT Techniques. Renew. Sustain. Energy Rev. **68**, 1–18 (2017)
4. Benzaouia, S., Mokhtari, M., Zouggar, S., Rabhi, A., Elhafyani, M.L., Ouchbel, T.: Design and implementation details of a low cost sensorless emulator for variable speed wind turbines. Sustain. Energy Grids Netw. **26**, 100431 (2021)
5. Mellit, A., Kalogirou, S.A., Hontoria, L., Shaari, S.: Artificial intelligence techniques for sizing photovoltaic systems: a review. Renew. Sustain. Energy Rev. **13**(2), 406–419 (2009)

6. Selvan, S., Nair, P.: A review on photo voltaic MPPT algorithms. Int. J. Electr. Comput. Eng. **6**(2), 567–582 (2016). ISSN 2088-8708
7. Mohapatra, A., Nayak, B., Das, P., Mohanty, K.B.: A review on MPPT techniques of PV system under partial shading condition. Renew. Sustain. Energy Rev. **80**, 854–867 (2017)
8. Benzaouia, M., Bouselham, L., Hajji, B., Dubois, A.M., Benslimane, A., El Ouariachi, M.: Design and performance analysis of a photovoltaic water pumping system based on DC-DC boost converter and BLDC motor. In: 2019 7th International Renewable and Sustainable Energy Conference (IRSEC), pp. 1–6. IEEE, November 2019

Explainable Prediction of Intelligent DTN Routing

El Arbi Abdellaoui Alaoui[1(✉)], Khalid Nassiri[2], Stéphane Cédric Koumétio Tékouabou[3], and Said Agoujil[4]

[1] IEVIA Team, Department of Sciences, Ecole Normale Supérieure, Moulay Ismail University of Meknès, Meknés, Morocco
e.abdellaouialaoui@umi.ac.ma

[2] Perception, Robotics, and Intelligent Machines Research Group (PRIME), University of Moncton, Moncton, Canada

[3] Center of Urban Systems (CUS), Mohammed VI Polytechnic University (UM6P), Hay Moulay Rachid, 43150 Ben Guerir, Morocco

[4] École Nationale de Commerce et de Gestion, Moulay Ismail University of Meknes, El Hajeb, Morocco

http://www.abdellaoui.net

Abstract. The objective of this work is to create an intelligent routing system based on machine learning methods, the usage of which opens up yet another opportunity to classify bundles as having successfully reached their destination or not. These networks periodically do an assessment, enabling the selection of the sort of routing appropriate for a certain circumstance. After that, it conducts data categorization and reduces the entries' extraneous information. Despite the issues mentioned, our task is to create an intelligent routing system that can distinguish between bundles that have reached their destination and those that have not. Machine learning is the primary technology used in constructing the intelligent routing system. Several machine learning approaches, including boosting and bagging, have been employed to classify whether bundles have successfully reached their destination or not. Machine learning improves accuracy now that we can learn directly from data rather than human knowledge. The results demonstrate that the Random Forest (RF) and ExtraTrees Classifier have an overall accuracy of 80% (ET).

Keywords: Delay tolerant networks · Data collection · Machine learning · Shap value

1 Introduction

A set of mobile nodes known as a Delay Tolerant Network (DTN) may dynamically exchange information with one another without the requirement for an established infrastructure [4,5,13]. In DTN, a node is a physical device that is intermittently connected to another. If there is contact between two nodes, then a message exchange will be created between these two nodes. This exchange

Y. Farhaoui et al. (Eds.): ICAISE 2022, LNNS 635, pp. 415–420, 2023.
https://doi.org/10.1007/978-3-031-26254-8_59

requires a DTN routing protocol. Many routing classes for DTN can be used for this purpose [9]. We have many difficulties since the nodes are mobile and may enter or exit the network at any moment, especially regarding security, QoS, and routing systems [14]. There are now a number of suggestions and studies trying to find solutions to these problems. In addition, we have noticed recently that DTN protocol design is increasingly adapting to new technologies, such as the Internet of Things (IoT) [1,12]. We need a "facade" in the IoT environment so that items may connect and interact. The application protocols are employed here and create what is referred to as an application interface [8]. In fact, things are interconnected through the Internet and may seek services from one another as well as provide services to other items. As a consequence, information is shared across the many objects [10]. As a result, an object that does not already possess the information the user has requested might ask the objects it is related to get the needed data. Even if the service in issue does not have the required information in its database [3], the user may still get it thanks to this mechanism. Additionally, whether the communication system can predict when the bundles will arrive at the destination. This is how we can make our intelligent and participative routing system a reality. We will use machine learning as a potential approach to do this [15].

In this paper, we offer an intelligent DTN routing system based on standard routing protocols and the intelligent Machine Learning (ML) technique for predicting if a bundle has arrived or not. By dynamically altering existing routing protocols, our approach aims to solve problems affecting modern technologies like the concept of a smart city [7] and the internet of things (IoT) [11] while continuing to improve network performance. In fact, the architecture of the Internet of Things (IoT) requires communication between objects, especially when sharing information between linked objects.

2 Methodology and Approach

In this part, we outline our strategy and lay out the chosen course of action. We begin with a summary of the procedure, describing the many steps using explanatory illustrations. The database we created is then highlighted, with each of its attributes described. In the last part of this section, we will develop the system of bundle classification using machine learning methods and end with the explicability of bundle classification models.

To illustrate the prediction structure in this context, we give Fig. 1, in which we display the crucial components of the learning system, such as preprocessing, model development, etc.

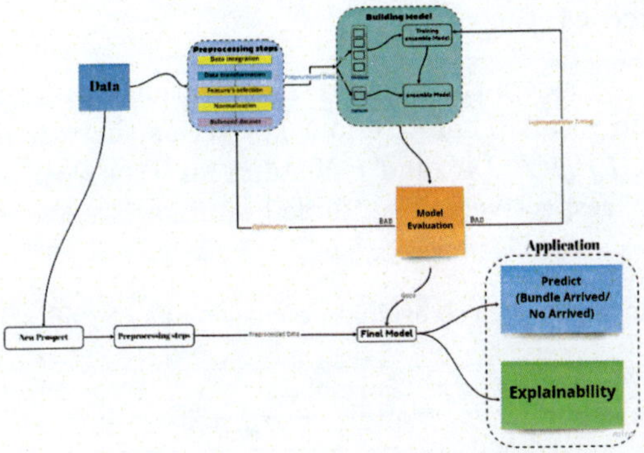

Fig. 1. Architecture of proposed model for classification of the bundles

2.1 Model Interpretation for Machine Learning

How can we comprehend the choices that these models recommend so that we may have faith in them? Transparency is one of the most often discussed ethical issues in the contemporary world. Indeed, it is vital to be able to explain the AI's forecast or decision to a client or basic user. In addition, explaining a method for complicated learning enables optimization of the model by identifying the critical variables used to classify the bundles beforehand.

Lundberg and Lee's Shap (SHapley Additive exPlanations) technique [6]. It enables all sorts of models to provide a posterior explanations for each variable's contribution. To calculate the Shapley values of features for a given instance, we simulate various combinations of feature values, keeping in mind that some of these combinations may be completely devoid of a feature. We compute the difference between the predicted value and the mean of the predictions based on the actual data for each combination. The usual form of the Shapley value φ_i expression is:

$$\varphi_i = \sum_{S \subseteq N \setminus \{i\}} \frac{|S|!(M - |S| - 1)!}{M!} (f_x(S \cup i) - f_x(S)), \tag{1}$$

such as M Number of features, S features set, f_x the prediction function at time x, $f_x(S) = E[f(x)|x_S]$, i is the ith feature.

2.2 Experimental Study and Results Analysis

In this part, utilizing two separate datasets for both DTN protocols, we conduct comprehensive experiments to assess the performance of the suggested system. The results are then examined and discussed.

2.3 Performance Metrics

Evaluation metrics to assess the bundles classification quality are accuracy, F1-score, precision, and recall. There are four key terms that are necessary to compute these metrics: TP, FP, TN, and FN stand for True Positive, False Positive, and True Negative, respectively. The Table 1 summarises these metrics:

Table 1. Metrics to evaluate our system's classification

N	Metric	Formula
1	Accuracy	$\frac{TP+TN}{TP+TN+FP+FN}$
2	Precision	$\frac{TP}{TP+FP}$
3	Recall	$\frac{TP}{TP+FN}$
4	f-score	$\frac{2*(Recall*Precision)}{Recall+Precision}$

2.3.1 Epidemic Dataset Predictive Performance. The Table 2 displays the Epidemic protocol-based data's predictive classification performance for the arrived and rejected bundles. These results show that the ET algorithm performs best across all parameters, with scores of 83.63%, 83,92%, 83,58% and 83,63% correspondingly for accuracy, precision, f1-score, and recall. The performances of RF, BC, and XGB will follow immediately after these ones. The worst performances have always come from DT and KNC.

Table 2. Epidemic protocol data classification performance in terms of Accuracy, Precision, F1-score, and Recall

Classifier	Epidemic protocol			
	Accuracy	Precision	F1-score	Recall
RF	0.8345	0.8373	0.8355	0.8357
BC	0.8274	0.8291	0.8270	0.8273
ET	**0.8363**	**0.8392**	**0.8358**	**0.8363**
XGB	0.8175	0.0.8191	0.8173	0.8176
DT	0.8012	0.8010	0.8036	0.8064
KNC	0.6937	0.6735	0.7106	0.7525

2.4 Model Explainability

Figure 2 shows how the Shap values determine the variables' global importance. It is a way to figure out which parameter or set of parameters affected the prediction (Classification). Each point represents a Shap value. For example, red points represent high values of the variable, and blue points represent low values of the variable. With the bar plot (??), which shows how important each feature is, the size of the bundle and the distance between the source node and the destination node are the ones that help predict the most.

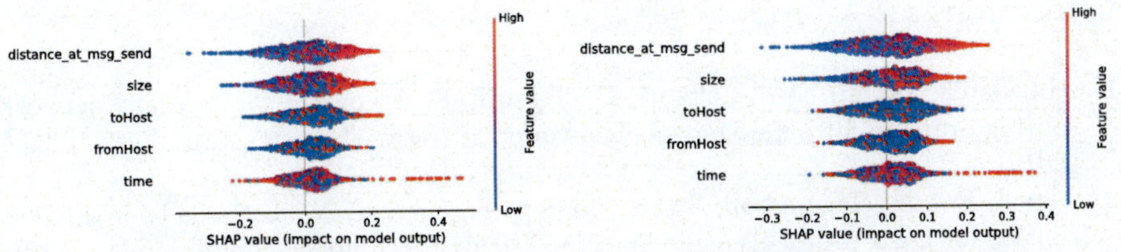

(a) Shap-Epidemic (b) Shap-Spray and Wait

Fig. 2. SHAP summary graphic shows DTN protocol classification feature relevance. (Color figure online)

2.5 Comparative Analysis

In this section, we compared the model developed in this study with other ML models [2]. To further validate the efficiency of our model, which demonstrated the best performance against the base classifiers as demonstrated in Table 3.

Table 3. Comprehensive comparison of the Epidemic protocol for classification of bundles.

Classifier	Epidemic protocol							
	Model in [2]				Our model			
	Accuracy	Precision	F1-score	Recall	Accuracy	Precision	F1-score	Recall
RF	0.8087	0.8045	0.7940	0.7879	0.8345	0.8373	0.8355	0.8357
BC	0.8021	0.7970	0.7870	0.7813	0.8274	0.8291	0.8270	0.8273
ET	**0.8119**	**0.8097**	**0.7964**	**0.7893**	**0.8363**	**0.8392**	**0.8358**	**0.8363**
XGB	0.7604	0.7531	0.7392	0.7331	0.8175	0.8191	0.8173	0.8176
DT	0.8012	0.7910	0.7914	0.7917	0.8012	0.8010	0.8036	0.8064
KNC	0.6338	0.6248	0.6247	0.6300	0.6937	0.6735	0.7106	0.7525

3 Conclusion

This paper has developed a system of intelligent DTN routing using machine learning techniques. The crucial task of the system is to classify the bundle arrived and not arrived for a destination node. In addition, we used the SHAP values to analyze the decision that was produced by our model. Because of this, decision-makers are more able to trust the model and understand how to weigh its suggestions with other decision-making considerations.

References

1. Abdellaoui Alaoui, E.A., Zekkori, H., Agoujil, S.: Hybrid delay tolerant network routing protocol for heterogeneous networks. J. Netw. Comput. Appl. **148**, 102456 (2019)
2. Alaoui, E.A.A., Tekouabou, S.C.K., Maleh, Y., Nayyar, A.: Towards to intelligent routing for DTN protocols using machine learning techniques. Simul. Model. Pract. Theory **117**, 102475 (2022)
3. Chowdhury, A., Raut, S.A.: A survey study on internet of things resource management. J. Netw. Comput. Appl. **120**, 42–60 (2018)
4. Le, T.: Multi-hop routing under short contact in delay tolerant networks. Comput. Commun. **165**, 1–8 (2021)
5. Li, W., Galluccio, L., Bassi, F., Kieffer, M.: Distributed faulty node detection in delay tolerant networks: design and analysis. IEEE Trans. Mob. Comput. **17**(4), 831–844 (2017)
6. Lundberg, S.M., Lee, S.I.: A unified approach to interpreting model predictions. In: Advances in Neural Information Processing Systems, pp. 4766–4775 (2017)
7. Manogaran, G., Nguyen, T.N.: Displacement-aware service endowment scheme for improving intelligent transportation systems data exchange. IEEE Trans. Intell. Transp. Syst. 1–11 (2021)
8. Naeem, M.A., Nguyen, T.N., Ali, R., Cengiz, K., Meng, Y., Khurshaid, T.: Hybrid cache management in IoT-based named data networking. IEEE Internet Things J. **9**(10), 7140–7150 (2021)
9. Panda, M., Ali, A., Chahed, T., Altman, E.: Tracking message spread in mobile delay tolerant networks. IEEE Trans. Mob. Comput. **14**(8), 1737–1750 (2015)
10. Park, T., Saad, W.: Distributed learning for low latency machine type communication in a massive internet of things. IEEE Internet Things J. **6**(3), 5562–5576 (2019)
11. Serhani, A., Naja, N., Jamali, A.: AQ-routing: mobility-, stability-aware adaptive routing protocol for data routing in MANET-IoT systems. Clust. Comput. **23**(1), 13–27 (2020)
12. Seyhan, K., Nguyen, T.N., Akleylek, S., Cengiz, K., Islam, S.K.H.: Bi-GISIS KE: modified key exchange protocol with reusable keys for IoT security. J. Inf. Secur. Appl. **58**, 102788 (2021)
13. Sun, X., et al.: Performance of DTN protocols in space communications. Wirel. Netw. **19**(8), 2029–2047 (2013)
14. Wang, E., Yang, W.-S., Yang, Y.-J., Jie, W., Liu, W.-B.: An efficient message dissemination scheme for minimizing delivery delay in delay tolerant networks. J. Comput. Sci. Technol. **33**(6), 1101–1124 (2018)
15. Yang, H., Xie, X., Kadoch, M.: Machine learning techniques and a case study for intelligent wireless networks. IEEE Network **34**(3), 208–215 (2020)

Facial Emotion Recognition Using a GoogLeNet Architecture

Imane Bouslihim(✉) and Walid Cherif

Laboratory of Computer & Data and Information Sciences (Dc-ISDI),
School of Information Sciences (ESI), Rabat, Morocco
{imane.bouslihim,wcherif}@esi.ac.ma

Abstract. Convolutional Neural Networks are widely used for Facial Emotion Recognition (FER) by virtue of their inherent feature extraction mechanism from images. In this paper, we built a Facial Emotion Recognition system which performs emotion classification on the FER- 2013 dataset. We carried out an analysis of the architectures presented in the literature before proposing a Transfer Learning model based on a pre-trained GoogLeNet architecture. The obtained results showed that the suggested model accomplishes the FER task effectively using GoogLeNet, and achieved an accuracy of about 63.39%.

Keywords: Emotion recognition · Face detection · Convolutional neural network · Transfer learning · GoogLeNet

1 Introduction

Nowadays, due to the vast number of facial images available on the internet, Facial Expression Recognition (FER) is now one of the most popular Artificial Intelligence technologies to learn about a person's emotional state. Also known as Facial Emotion Recognition, this technology aims to classify human emotions such as neutral, angry, disgusted, fearful, happy, sad, and surprised from facial image datasets. Recently, Facial Emotion Recognition has gained a lot of interest in both academic and industrial areas. Many fields have opened up to the application of FER such as education, marketing, security, and medicine [6].

In order to reliably classify emotions displayed by humans using computers, typical FER systems can be created using Transfer Learning with Deep Learning approaches such as Convolutional Neural Network (CNN), that attempt to interpret facial movements and changes [1]. These systems usually consist of three main stages: face detection, feature extraction and emotion classification.

Deep Learning-based algorithms are the most popular methods used to classify human emotions. The primary goal of this work is to present a pre-trained Deep Learning model based on Transfer Learning using a GoogLeNet-Convolutional neural network in order to accurately predict emotion. In this study, we'll train our facial Emotion Recognition model using labeled facial expression images from the FER-2013 dataset.

Y. Farhaoui et al. (Eds.): ICAISE 2022, LNNS 635, pp. 421–426, 2023.
https://doi.org/10.1007/978-3-031-26254-8_60

The rest of this paper is organized as follows: Sect. 2 cites some related works. Section 3 describes specifics about the proposed model. Section 4 reports and discusses the experimental results. Section 5 concludes the paper.

2 Related Works

This section provides a brief overview of some related works on Facial Emotion Recognition. In Ref. [1] the authors proposed a very deep DCNN model using Transfer Learning technique in which a pre-trained DCNN model was adopted by replacing its upper dense layer(s) compatible with FER. The proposed system was verified on KDEF and JAFFE datasets and different pre-trained DCNN models such as DenseNet-161. On the KDEF and JAFFE test sets, DenseNet- 161 achieved the highest FER accuracies of 96.51% and 99.52%, respectively.

Tian et al. [7] proposed a deep learning model for Teacher expression recognition using a GoogLeNet-InceptionV3 convolutional neural network (CNN). The classification accuracy rate achieved after training was 81.4%. This model was also verified using teachers' lecture videos and achieved an accuracy rate of 90%. Basbrain et al. [2] presented a FER system for eyeglass detection and feature extraction using GoogleNet-CNN. The proposed system included a glasses detector, CNNs and a multiple classifier system (MCS). The proposed system was verified using the USTC-NVIE (NVIE) database and the accuracy achieved was 99.7%. The authors also applied the neural network approach in the MCS for score fusion, which have increased the classification rates of about 10%.

3 Proposed Model

3.1 Transfer Learning with CNN

Transfer Learning (TL) technique is a popular way to quickly develop models, its fundamental principle is to take a model that has been trained on a large dataset and apply its knowledge to a smaller dataset. In this regard, in order to create an accurate FER system, the model proposed in this paper suggests a Deep Convolutional Neural Network (DCNN) modeling through Transfer Learning using a pre-trained CNN model. The idea is that low-level basic features are common for most images and only high-level features need to be tuned [1]. Therefore, a pre-trained CNN model will be accurate by replacing its upper layers with the dense layers to make it compatible with FER data.

A Convolutional Neural Network (CNN), for the most part, consists of three types of layers: convolutional layers, max-pooling layers, and fully connected layers. The first two types of layers are responsible for feature extraction, introducing non-linearity into the network, and reducing the dimension of the features. While the fully connected layer is responsible for classifying the input image based on the features previously extracted by the other layers [4].

3.2 GoogLeNet

GoogLeNet is a 22 layers deep convolutional neural network, counting only layers with parameters, while the total number of layers used is almost 100 (see Fig. 1). GoogLeNet has been trained on over a million images and can classify images into 1000 object categories, it takes an image as input and outputs a label for the object in the image together with the probabilities for each of the object categories. Its architecture contains parts of the network which are executed in parallel, these parts are called "Inception modules". The last layer applies a global average pooling and a soft-max activation function to produce a prediction [3]. The architecture of GoogLeNet shown in Fig. 1 takes 224 × 224 size images with RGB color channels and Rectified Linear Units (ReLU) serve as the activation functions for each convolution layer [3].

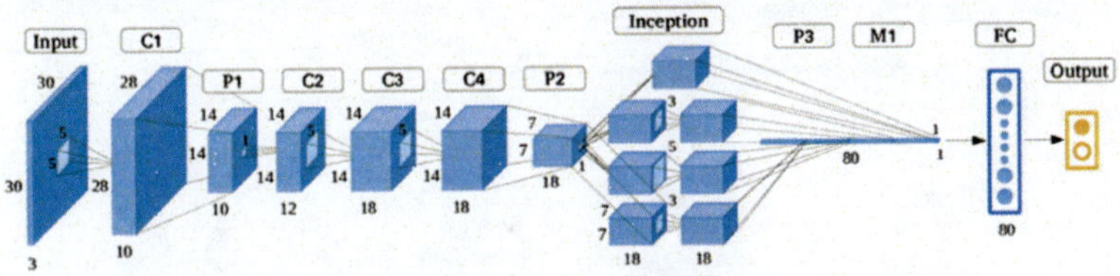

Fig. 1. Basic architecture of GoogLeNet

3.3 Dataset

For training our proposed FER model, we used FER-2013 dataset provided by Kaggle, introduced at the International Conference on Machine Learning (ICML) in 2013 by Pierre-Luc Carrier and Aaron Courvill [8], It is a widely used dataset for benchmark and assessing the performance of systems and approaches. The FER-2013 dataset contains nearly 36,000 face RGB photos of various emotions with a size limit of 48×48, and its primary labels are categorized into 7 types: 0 indicates angry, 1 disgust, 2 fear, 3 happy, 4 sad, 5 surprised, and 6 neutral. The disgust expression has the fewest images (less than 600) compared to the other labels, which each have almost 5,000 examples (see Fig. 2).

Fig. 2. Examples of facial images from the FER-2013 Dataset

3.4 FER Using Transfer Learning with Pre-trained GoogLeNet-CNN

In Fig. 3, we describe the training and the testing stages of the proposed Facial Emotion Recognition model based on Transfer Learning (TL) using pre-trained GoogLeNet-CNN. As shown in Fig. 3, in the training stage, the architecture is defined, by replacing the last dense layers of the pre-trained model with the new dense layer(s) to recognize the emotion expressed in a given facial image of the training dataset. Fine-tuning of the model architecture is then performed using the convolutional base of the pre-trained model and the added density layer(s). Face image pre-processing such as resizing and cropping is performed to prepare a clean dataset which is used for fine-tuning. Finally, in the testing phase, a cropped image is used as input to the system, and the highest probability output emotion is taken as the predicted class.

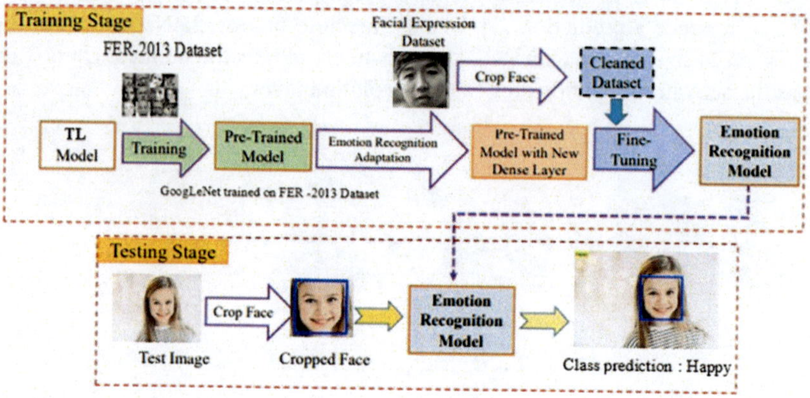

Fig. 3. Training and testing stages of proposed FER model with GoogLeNet-CNN

4 Test and Results

The experimental section of this work is evaluated using MATLAB, a widely used platform for Facial Emotion Recognition. It provides a large library of Transfer Learning Models as well as multiple Deep Learning toolboxes.

As mentioned before FER-2013 dataset contains 48×48 facial images and GoogLeNet only supports images with size 224 × 224 and RGB color channels.

Therefore, we had to convert the size of the FER-2013 dataset images using a simple MATLAB script. Then, we applied our FER model on the FER-2013 dataset and obtained an accuracy of 63.39%. The training process was designed to go through 40180 iterations and with the initial learning rate of 0.003, promising results were received from the first iterations. Figure 4 shows examples of the FER-2013 dataset recognized using proposed model.

The proposed model has demonstrated successful emotion classifications as it has also encountered difficulties. The FER task was quiet challenging especially when facial expressions were confusing or emotions were weakly expressed in images. In Fig. 5, the correct expression for the first misclassified image is disgust but predicted as angry. In this image, the hand is put on the nose and the eyebrows and eyes are frowning just like an angry face. Also, the misclassification of the second image as fear is logical as the mouth is open and eyes are closed, so the expression appears to be fear more than

Fig. 4. Images from FER-2013 dataset classified using GoogLeNet algorithm

sad. For the third image the eyes look shrunken and the mouth looks frowning, which is similar to sad while the correct expression is angry. Finally, in the fourth image the mouth is too normal to be classed as sad, this expression is almost indistinguishable by the algorithm.

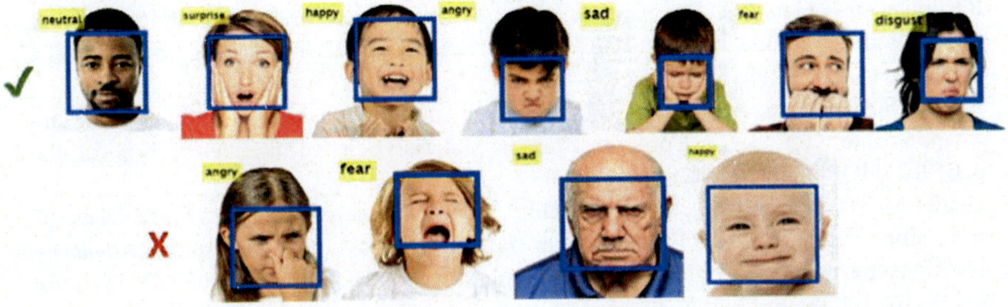

Fig. 5. Examples of images classified by proposed GoogLeNet-CNN model

4.1 Results Analysis

We further evaluate the proposed model by comparing it to some related works which used existing algorithms mainly AlexNet and ResNet-18 for Facial Emotion Recognition using FER-2013 dataset. We note from Table 1 that the work incorporating GoogLeNet and AlexNet [4] is shown to have achieved the best FER accuracy of 83%. On the other hand, our proposed method achieved an accuracy of 63.39%. In addition, we can notice that our model achieved better results comparing to the AlexNet [9] and the ResNet-18 [5] methods which achieved respectively an accuracy of 61.1% and 52.1%

Table 1. Accuracies of proposed model and existing works on FER-2013 dataset

Author	Method used	Accuracy
Panagiotis Giannopoulos [4]	GoogLeNet and AlexNet	83%
Proposed model	GoogLeNet	63.39%
Shuwen Zhao [9]	AlexNet	61.1%
Satnam Singh [5]	ResNet-18	52.1%

5 Conclusion

In this paper, we presented a Facial Emotion Recognition model based on Transfer Learning using a pre-trained GoogLeNet-Convolutional Neural Network architecture. According to results, testing the proposed model on the FER-2013 dataset that contains seven types of emotion expressions (happy, angry, disgust, fear, sad, surprised, and neutral), the accuracy obtained for Facial Emotion Recognition is 63.39%. Further, using the 48×48 original image dimensions for training the GoogLeNet algorithm, might enhance the classification accuracy.

References

1. Akhand, M., Roy, S., Siddique, N., Kamal, M.A.S., Shimamura, T.: Facial emotion recognition using transfer learning in the deep CNN. Electronics **10**(9), 1036 (2021)
2. Basbrain, A.M., Gan, J.Q., Sugimoto, A., Clark, A.: A neural network approach to score fusion for emotion recognition. In: 2018 10th Computer Science and Electronic Engineering (CEEC), pp. 180–185. IEEE (2018)
3. Fatima, S.A., Kumar, A., Raoof, S.S.: Real time emotion detection of humans using mini-Xception algorithm. In: IOP Conference Series: Materials Science and Engineering. vol. 1042, p. 012027. IOP Publishing (2021)
4. Giannopoulos, P., Perikos, I., Hatzilygeroudis, I.: Deep learning approaches for facial emotion recognition: a case study on FER-2013. In: Hatzilygeroudis, I., Palade, V. (eds.) Advances in Hybridization of Intelligent Methods. SIST, vol. 85, pp. 1–16. Springer, Cham (2018). https://doi.org/10.1007/978-3-319-66790-4_1
5. Singh, S., Schicker, D.: Seven basic expression recognition using ResNet-18. arXiv preprint arXiv:2107.04569 (2021)
6. Smirnov, D.: Emotion recognition using facial feature extraction. Rowan University (2019)
7. Tian, Y., Han, T., Wu, L.: Teacher facial expression recognition based on GoogLeNet-InceptionV3 CNN model. In: Wang, W., Wang, G., Ding, X., Zhang, B. (eds.) Artificial Intelligence in Education and Teaching Assessment, pp. 69–78. Springer, Singapore (2021). https://doi.org/10.1007/978-981-16-6502-8_8
8. Zahara, L., Musa, P., Wibowo, E.P., Karim, I., Musa, S.B.: The facial emotion recognition (FER-2013) dataset for prediction system of micro-expressions face using the convolutional neural network (CNN) algorithm based Raspberry Pi. In: 2020 Fifth International Conference on Informatics and Computing (ICIC), pp. 1–9. IEEE (2020)
9. Zhao, S., Cai, H., Liu, H., Zhang, J., Chen, S.: Feature selection mechanism in CNNs for facial expression recognition. In: BMVC, p. 317 (2018)

Fake Profile Identification Using Machine Learning

Fatna El Mendili[1(✉)], Fatima Zahra Louhab[1], Nisrine Berros[2], Younes Filaly[2], Hamza Badri[2], Younès El Bouzekri El Idrissi[2], and Mohammed Fattah[1]

[1] Image Laboratory, School of Technology, Moulay Ismail University of Meknes, Meknes, Morocco
f.elmendili@gmail.com

[2] Engineering Sciences Laboratory, National School of Applied Sciences, Ibn Tofail University, Kenitra, Morocco

Abstract. Nowadays, remarkable growth in the use of social networks, as well as the manner of using them has fundamentally changed the way of using the Internet. This increases the quantity and quality of shared data in these social networks. In addition, the security and confidentiality of data become an important concern to the significance of detecting malicious profiles in OSNs (Online Social Networks), the social network security has offered multiple techniques for detecting malicious profiles and content spam. In this paper, the authors test a new system to detect malicious profiles. First, they propose the identification of false profiles by the security approach based on social honeypots. Second, they suggest detecting those that are similar to the profiles detected by the approach of honeypot using a similarity function based on collaborative filtering. This work uses a hybrid system based on honeypots and a recommender system algorithm for detecting malicious users and those are similar. The approach has greatly improved the quality of abstraction in terms of performance and design. The algorithm is also fast and simple to implement. Experimental results show the stability and accuracy (over 99,99%) of our approach.

Keywords: Malicious users · Social honeypots · Collaborative filtering · Recommender system

1 Introduction

Twitter is a great platform for communication and sharing, it attracts profiles while providing services to disseminate messages of 140 characters [1].

Every month, over 42 million new accounts are created on Twitter; the openness of Twitter also leads to the popularity of spamming activities on Twitter [1].

Unfortunately, the attackers have made their attention on the OSN and exploit them in carrying out various types of attacks like phishing [2–4], the injection of malicious codes, and the dissemination of malicious software [2, 3, 5]. These malicious behaviors can cause serious privacy and economic problems. User's private data are popular on the black market and access to them may lead to economic crimes.

Y. Farhaoui et al. (Eds.): ICAISE 2022, LNNS 635, pp. 427–432, 2023.
https://doi.org/10.1007/978-3-031-26254-8_61

Detecting reconnaissance activities is very difficult since usually it is performed outside of the organization's premises and without direct interaction with the organizational resources. At some point, the reconnaissance phase enables the attacker to find an entry point into the organization leading to the next phases [6].

Social media is already ripe with threats: between 8%–10% of all social media, profiles are malicious in nature [7]. Twitter is one of the most popular social media, which has 313 million active monthly users who post 500 million tweets a day [8]. This popularity attracts the attention of spammers who use Twitter for malicious purposes such as phishing of legitimate users or the distribution of malicious software and advertises through URLs shared in tweets, aggressively track/untrack legitimate users and divert topics with a tendency to attract their attention, spreading pornography. In August 2014, Twitter revealed that 8.5% of its active monthly users, which is equivalent to about 23 million users automatically, have contacted their servers for regular updates. Thus, the detection and filtering of spammers from legitimate users are mandatory in order to provide a spam-free environment on Twitter [9].

Due to the different use agendas on the platform, it can often be difficult to identify users that can be mined for meaningful information.

However, the main purpose of this article is to propose a new system for detecting malicious profiles (fake users). The system is based on social honeypots and collaborative filtering.

The contribution of this article is:

- Propose a layer based on the social honeypots to detect the malicious profiles. The social honeypots are used in our approach to attract the profiles of the system and extract the characteristics of malicious profiles.
- Propose a layer based on the collaborative filtering to calculate the similarity between the malicious profiles detected by the Layer based on honeypots and the layer of collaborative filtering with the aim to present the profiles that are similar to the malicious profile.
- Draw the new system for the detection of false profiles.

The rest of this paper is structured as follows. In Sect. 2 we provide the previous related work. In Sect. 3, we present the proposed hybrid system. And finally, our conclusions, research limitations, and future works are summarized in Sect. 4.

2 Related Work

Many efforts have been made to develop spam detection techniques on Twitter in the last decade. In this section, we present the state-of-the-art techniques for detecting malicious users in social network [1]. Jasek et al. [10] suggested the general concept of using honeypots to detect activities associated with APTsfically targets social networks and takes into consideration their logistical concerns. Several previous studies [11, 12] have focused on the identification of spammers that use social honeypots and the creation of classifiers in order to distinguish social malicious users from legitimate profiles. Spammers generally write tweets that contain a hashtag and URL according to the following research studies that analyzed commonly used hashtags and URL: COMPA [13] detected compromised accounts that wrote spam tweets based on the tweeting language of the user's account, the tweeting time window, the URL, and the mention "receiver". H.L. Gururaj et al. [14] are proposing an natural language processing (NLP) technique to find suspicious users based on the daily conversations between the users. They demonstrated the behaviour of each user through their anomaly activities. Another machine learning technique called support vector machine (SVM) classifiers to detect the toxic comments in the comments blog.

Zijie Yang et al. [15] present the first comprehensive study of growing-up behaviors of malicious accounts in WeChat, one of the major PC-MSNs with billions of daily active users across the globe. The analysis reveals that the behavior patterns of growing-up accounts are very similar to that of benign users, and yet quite different from typical malicious accounts. Based on this observation, they design Muses, a detection system that can automatically identify subtle yet effective behaviors (features) to distinguish growing-up accounts before they engage in obvious malicious campaigns. Edward Kwadwo Boahen, et al. [16] propose an improved machine learning approach, Word Embedding and KNN (WE-KNN), which addresses the limitations encountered by the techniques used. They detailed their proposed WE-KNN for feature extraction, OSN user behavior selection, and classification. They evaluated their proposed model using the standard benchmark datasets, namely KDD Cup '99 and NSL-KDD, and performed the implementation in WEKA. Xingfa Shen et al. [17] propose a trust-based detection framework to detect malicious users from different types of data from a real dating site. In particular, they developed a user trust model to distinguish malicious users from legitimate users. In addition, they proposed a new data balancing method to improve the recall rate of malicious user detection. Extensive experiments have been conducted on real-world datasets. The results show that the proposed approach yields a precision of up to 59.16% and a recall rate of up to 73%.

3 Proposed Hybrid System

This Section presents details about the proposed hybrid system architecture. Can be divided into three parts:

- Security layer based on the social honeypots
- Security layer based on collaborative filtering
- Classification layer.

3.1 Security layer based on the social honeypots

This layer of security based on the social honeypots allows detecting malicious profiles based on the characteristics of the honeypots and characteristics of profiles. The process of this layer is the following:

- The deployment of social honeypots for harvesting information of malicious profiles.

– Analysis of the characteristics of these malicious profiles and those of deployed honeypots for creating classifiers that allow to filter the existing profiles and monitoring the new profiles.

This layer for detecting malicious profiles based on Social Honeypot deploys and maintains social honeypots for trapping evidence of malicious profile behavior. In our system, it inserts a honeypot that plays a role of a malicious profile and legitimate, if the honeypot attracts a profile, this last is malicious, then the honeypot detects the malicious activities of this profile, (e.g., by crawling the profile of the user sending the unsolicited friend request plus hyperlinks from the profile to pages on the Web-at-large). As honeypots collect the characteristics of malicious profiles, (for example spam Number of friends, the text on the profile, age, etc.), it is easy to detect in a community the characteristics of legitimate profiles with the aim of classifying the malicious profile with spam that propagate in the social networks. This is called a type of strategy by "Feature-based strategy.". A new method used in our approach to improve our classification and increase the ability to detect an attacker on social networks is the "honeypot feature-based strategy", this strategy uses the whole characteristics of honeypots that interact with users to refine our ranking. The whole data collected is becoming an integral part of the training of a classifier of malicious profiles. By an iterative refinement of selected characteristics using a set of algorithms for automatic classification, which are implemented on "Weka Machine Learning. Toolkit" the authors can explore the wider space of malicious profiles.

Figure 1 presents the deployment and collection data.

First, the researchers have developed a program for the automatic creation of profiles in the form of social honeypots, they created 200 profiles for social honeypots and 100000 Twitter profiles, which contains the malicious and legitimate profiles.

After collecting the characteristics of profiles, the classification is a necessary step to classify the legitimate profiles and those who are malicious. The researchers chose four types of classification algorithm to make a comparison between them and use the one that gives good results.

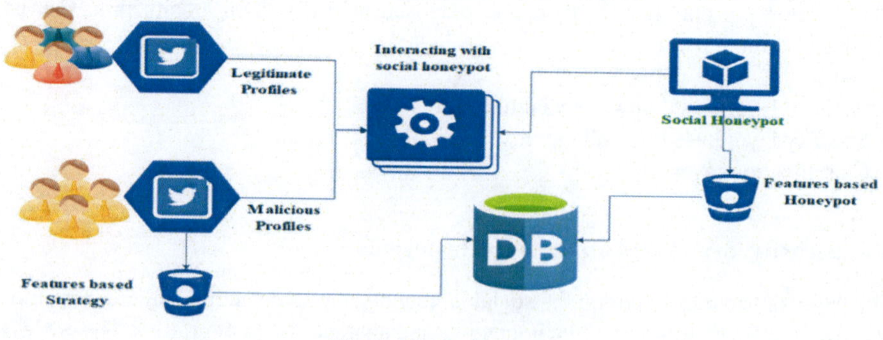

Fig. 1. Collection data

This step allows you to evaluate the performance of our classification and compare our proposal to the other current approaches. They use the recall, precision-measure,

TP rate, FP rate confusion matrix. Recall (sensitivity) is defined as the ratio of correctly classified spam in total real spam. Presents the third step of the first layer of the system proposed (Fig. 2).

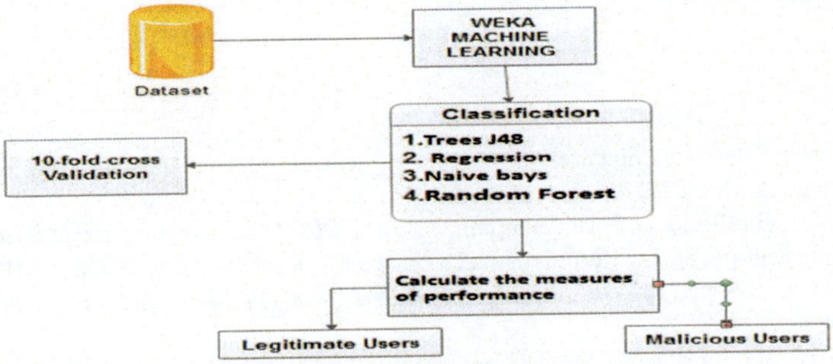

Fig. 2. Classification

3.2 Security Layer Based on Collaborative Filtering

This layer is operated as follows:

The development of the systems layer of recommendation is based on the Algorithm: collaborative filtering to detect similar profiles that is legitimate or malicious well, the idea is to find identical profiles to the layers based on the social honeypot's and collect the maximum of users who share the same behaviors, tastes, and operations on the platforms of communication including Twitter.

All first the collaborative filtering seeks similar profiles to an active user, this last one is a malicious profile produced by the first layer, then the system will suggest items, pages, or content that is already appreciated by users similar to the malicious Profile "User Active".

The result will give similar profiles to the malicious profile and their suggestions which gives a global vision on the interest of malicious profiles and which can be of content, which contains spam, url, and Malware.

We take the set of users who have tweeted the tweet j as the training set for K-NN and each user who has not tweet the tweet as a test point. For each user who has not tweet the tweet (test point), we compute the similarity to users who have tweet the tweet, and assign an estimated rating based on the known ratings of the neighbors.

4 Conclusion

In this paper, we presented, a system to detect malicious profiles on the social network Twitter. This system is based on social honeypots and collaborative systems. The hybrid system proposed allows detecting malicious profiles by the characteristics of the social honeypots and malicious accounts inserted in the system. Also detecting those who are similar by collaborative filtering. We used four types of algorithms to test the proposed system in a machine learning Weka. The classification algorithm via the regression, gives a better result (precision, a positive rate, negative rate, recall, and F-Measure equal 99,99%). Future work may include several aspects; such as conducting more theoretical studies on the outperformance of our methods to better understand the social honeypots based on malicious users' detection framework.

References

1. Wu, T., et al.: Detecting Spamming Activities in Twitter Based on Deep-Learning Technique. Wiley, New York (2017)
2. Eshete, B., Villafiorita, A., Weldemariam, K.: BINSPECT: holistic analysis and detection of malicious web pages. In: Keromytis, A.D., Di Pietro, R. (eds.) SecureComm 2012. LNICSSITE, vol. 106, pp. 149–166. Springer, Heidelberg (2013). https://doi.org/10.1007/978-3-642-36883-7_10
3. Eshete, B., Villafiorita, A., Weldemariam, K.: EINSPECT: evolution-guided analysis and detection of malicious web pages. Technical report, Fondazione Bruno Kessler (2012)
4. Aggarwal, A., Rajadesingan, A., Kumaraguru, P.: PhishAri: automatic realtime phishing detection on Twitter. In: eCrime Researchers Summit (eCrime), pp. 1–12. IEEE (2012)
5. Rahman, M.S., Huang, T.-K., Madhyastha, H.V., Faloutsos, M.: Efficient and scalable socware detection in online social networks. In: USENIX Security (2012)
6. Kemp, S.: Digital in 2017: Global Overview. https://wearesocial.com/special-reports/digital-in-2017-global-overview. Accessed 24 Jan 2017
7. Ahmad: How Many Internet and #SocialMedia Users are Fake? http://www.digitalinformationworld.com/2015/04/infographic-how-many-internetsusers-are-fake.html. Accessed 2 Apr 2015
8. Verma, M., Divya, Sofat, S.: Techniques to detect spammers in Twitter - a survey. Int. J. Comput. Appl. **85**(10), 27–32 (2014). ISSN 0975-8887
9. Webb, S., Caverlee, J., Pu, C.: Social honeypots: making friends with a spammer near you. Presented at the CEAS, California, CA, USA (2008)
10. Benevenuto, F., Magno, G., Rodrigues, T., Almeida, V.: Detecting spammers on Twitter. In: Collaboration, Electronic Messaging, Anti-abuse and Spam Conference (CEAS), vol. 6, p. 12 (2010)
11. Martinez-Romo, J., Araujo, L.: Detecting malicious tweets in trending topics using a statistical analysis of language. Expert Syst. Appl. **40**(8), 2992–3000 (2013)
12. Chakraborty, A., Sundi, J., Satapathy, S.: SPAM: a framework for social profile abuse monitoring. Technical report, Department of Computer Science, Stony Brook University, Stony Brook, NY 11794-4400, USA (2012)
13. Gururaj, H.L., Tanuja, U., Janhavi, V., Ramesh, B.: Detecting malicious users in the social networks using machine learning approach. Int. J. Soc. Comput. Cyber-Phys. Syst. **2**(3), 229–243 (2021)
14. Yang, Z., et al.: On detecting growing-up behaviors of malicious accounts in privacy-centric mobile social networks. In: Annual Computer Security Applications Conference (ACSAC 2021), pp. 297–310. Association for Computing Machinery, New York (2021)
15. Boahen, E.K., Bouya-Moko, W.C., Elvire, B.: Detection of compromised online social network account with an enhanced KNN. Appl. Artif. Intell. **34**, 777–779 (2020)
16. Shen, X., Lv, W., Qiu, J., Kaur, A., Xiao, F., Xia, F.: Trust-aware detection of malicious users in dating social networks. IEEE Trans. Comput. Soc. Syst. (2022). https://doi.org/10.1109/TCSS.2022.3174011

Faster RFID Authentication Scheme Based on ECC for Improving the Security in IoT Environment

Hind Timouhin[1](✉), Fatima Amounas[2], and Saleh Bouarafa[3]

[1] RO.AL&I Group, Faculty of Sciences and Technics, Moulay Ismail University of Meknes, Errachidia, Morocco
h.timouhin@edu.umi.ac.ma

[2] RO.AL&I Group, Computer Sciences Department, Faculty of Sciences and Technics, Moulay Ismail University of Meknes, Errachidia, Morocco
f.amounas@umi.ac.ma

[3] LRIT, Associated Unit to CNRST (URAC No 29), Faculty of Sciences, Mohammed V University in Rabat, Rabat, Morocco

Abstract. Today, the security of the Internet of Things (IoT) represents an emergent research field which has attracted much attention from research communities. Radio Frequency Identification (RFID) technology is an important part of the IoT environment. Therefore, a key concern for the security of the IoT is how to handle the security issue in RFID systems. The RFID authentication protocol is a crucial cryptographic mechanism for maintaining communication security between the tag and the server. Recently, RFID authentication protocols based on ECC (Eiliptic curve cryptography) have been widely studied because they could offer greater security features than traditional RFID authentication. This paper introduces a novel scheme with improved security level based on ECC for RFID authentication mechanism. The proposed method maintains security between different components of RFID system. The experimental results show that the proposed scheme is more secure and has less computational cost than other existing methods. Furthermore, the proposed scheme will not only enhance the security of information but also save computation time and reduce power requirements, which will find its suitability for most applications of IoT.

Keywords: Security · Internet of Things · Authentication · RFID · Elliptic curve cryptography · Multiplication scalar

1 Introduction

In the last decade, with the greater advancement in IoT enabling technologies starting with Radio Frequency Identification system (RFID), Cloud, and Big data analytics have been deployed in many applications such smart homes, smart cities, and healthcare [1]. The Internet of Things refers to the precise communication between all the connected objects around the world. These linked objects can be any type of device, such as personal computers, vehicles or surveillance system. The connectivity between the objects

Y. Farhaoui et al. (Eds.): ICAISE 2022, LNNS 635, pp. 433–439, 2023.
https://doi.org/10.1007/978-3-031-26254-8_62

is realized through the Internet to make them accessible globally. This paradigm integrates heterogeneous systems or networks such as sensor networks, the radio frequency identifier and smart cities. RFID is a promising technology of the future that is expected to connect billions of devices. It is widely utilized in industries as an active method to recognize objects. Applications of RFID technology include healthcare, e-passports, credit card, pharmaceutical, supply chain management, and so on. Due to the numerous applications in different areas, this technology has attracted much attention of the researcher. RFID technology involves two key elements called readers and tags. In the RFID network, the tag encrypts the identity details and transfers it to the reader. The reader can then authenticate the obtained data and the tag ID by relying on information available at the back-end server. RFID systems can be vulnerable to various attacks and we should find out the solution in the short term. So, the most interesting challenge of this technology will be protection issue. It can be achieved by authentication protocols. The authentication strategy is a crucial step in the RFID system. In the previous works [2, 3], the authentication protocols adopted the hash functions and the symmetric key cryptography in RFID system. Recently, the elliptic curve cryptography (ECC) is an effective approach for RFID authentication protocol, due to its good performance and low key size specifications. ECC is a public key cryptographic solution, which offers smaller key sizes and lower computing costs compared to other alternatives like RSA. It is widely implemented in many devices such as mobile phones, smart cards, biometric passports and some other important businesses [4, 5]. In this paper, we suggest a novel scheme with improved security level based on ECC for RFID authentication systems. This proposed protocol maintains security and has less computational cost compared to other existing methods. The remainder of this paper is organized as follows: In Sect. 2, some basic theories connected with elliptic curve cryptography are discussed. Section 3 is devoted to the proposed scheme, followed by its security strength analysis. Section 4 is dedicated to the performance analysis of the proposed work compared with the existing works. The last section presents a conclusion and future work.

2 An Overview of Elliptic Curve Cryptography

Elliptic curves are used in several types of cryptographic systems, including key exchange protocols, digital signature algorithms, and authentication protocols. Elliptic curve cryptography uses elliptic curves where the parameters are all constrained to elements of a finite field. Elliptic curves over two finite fields are mostly used, prime field $\mathbb{F}_p$, where p is a prime and binary field F_{2^m}, where m is a positive integer [6]. In this work, we introduce a novel authentication scheme with improved security level based on ECC for RFID systems. The mathematical operations of ECC are defined over the elliptic curve given by the following equation:

$$y^2 = x^3 + ax + b[p] \tag{1}$$

(where $p \neq 2, 3$ is a prime).

An elliptic curve E over a finite field $\mathbb{F}_p$ can be made into an abelian group by defining an additive operation on its points. ECC adopts scalar multiplication, which includes point doubling and adding operation. A scalar multiplication is achieved by a series of addition operation. The scalar multiplication kP can be computed by repeating the addition operation k times. The strength of an ECC cryptosystem depends on the difficulty of finding the number of times. This operation is known as the Elliptic Curve Discrete Logarithm Problem (ECDLP).

3 Proposed Authentication Protocol

Authentication service has a significant role in the fast development that recent technology has known recently. So, the main aim of this paper is to develop an efficient authentication scheme based on ECC to secure the transmission in the RFID system. The authentication procedure includes two stages, which are the initialization phase and the authentication phase. The flowchart of the proposed authentication scheme is shown in Fig. 1. Some of the notations related to the proposed protocol are illustrated in Table 1.

Table 1. Some of the notations related to the proposed protocol.

Notation	Description
$\mathbb{F}_p$	A finite field with p elements and p is a large prime a large prime
E	An elliptic curve over the finite field $\mathbb{F}_p$
a, b	Are the elements of $\mathbb{F}_p$ which are used to define the elliptic curve E
P	A base point on the elliptic curve E
K	A random point on the elliptic curve E
s, P_s	The private and public keys of the server
t, P_t	The private and public keys of the tag

3.1 Initialization

During the initial phase, we generate some public parameters: an elliptic curve over the finite field $\mathbb{F}_p$ and the base point, which having the largest order n such that $nP = \Omega$. In this phase, the private key and public key of the server and the tag are generated using elliptic curve parameters. The server chooses randomly an integer s and computes it's public key $P_s = sP$. The tag chooses his owner secret t and computes $P_t = tP$ as a public key. Assuming that each tag has identity information, after that, the server keeps its private and public keys in its database and the tag keeps its private key and its identity information in its memory.

3.2 Authentication Phase

The authentication procedure consists of the following steps:

- **Step 1**: The server randomly selects an point $K(k_1, k_2)$ on elliptic curve and computes: $R_1 = k_1P$. Then, the result point is transmitted to the tag.

- **Step 2:** After receiving R_1, the tag imbeds its own ID denoted r_1 as a point PT in E($\mathbb{F}_p$): $PT = r_1 R_1$. Then computes the parameter: $R_2 = r_1 P$. The tag sends the message (PT, R_2) to the server.

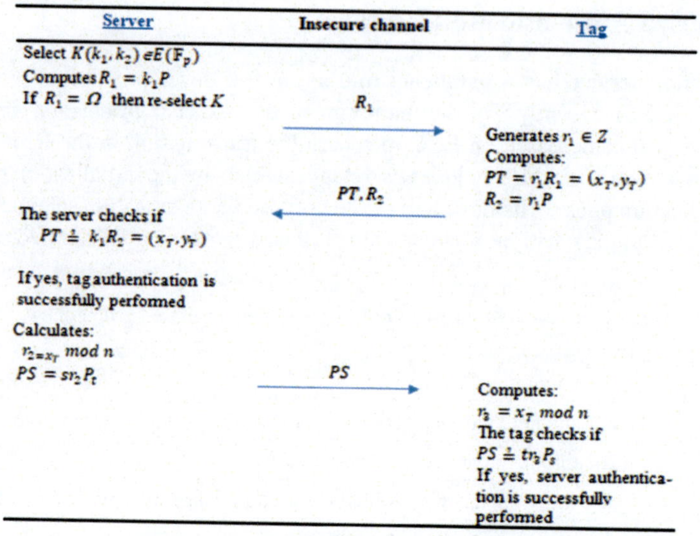

Fig. 1. Proposed authentication scheme

- **Step 3:** Upon receiving this message, with its private key, the server checks if: $PT = k_1 R_2$. If yes, the tag is successfully authenticated, else stop. In the second part of the current process, the server computes $r_2 = x_T \ mod \ n$, and uses its private key to calculate $PS = s r_2 P_t$. The message PS is sent to the tag.
- **Step 4:** After getting the message, the tag produces the corresponding secret $r_3 = x_T \ mod \ n$. The tag checks if the quantity $t r_3 P_s$ is equal to PS. If yes, the server is successfully authenticated, else the stop.

3.3 Required Security Services

The proposed scheme ensures the most RFID security services like mutual authentication, confidentiality and integrity.

- **Mutual authentication**

This service is achieved by two processes. The first one enables the server to authenticate the tag using its identity r_1. The server generates a random point in elliptic curve, and then sends an authentication request to the tag. When the tag received the message, he proceeds to generate its own ID and computes PT and R_2. The result points are sent to the server. After obtaining the message (PT, R_2) from the tag in the step 2, the server checks the validity of $PT = k_1R_2$. If yes, the tag is successfully authenticated. The second process allows to the tag to authenticate the server. The server produces $PS = sr_2P_t$ and sends the value PS to the tag. The value sr_2 cannot be obtained by an attacker due to the hardness of the ECDLP. Upon receiving this message, the tag verifies that the quantity tr_3P_s is equal to PS. If yes, the server is successfully authenticated, else the authentication is failed.

– **Confidentiality**

In our case, the identity information r_1 of the tag is embedded as a point into elliptic curve called PT and securely transmitted to the server. Then the tag sends the pair of point (PT, R_2) to the server. Here, if the attacker can get the point PT, he cannot extract the ID information of the tag due to the difficulty of solving the Elliptic Curve Discrete Logarithm Problem (ECDLP).

– **Integrity**

Our method ensures the integrity of the data exchanged between the tag and the server. Here, the secret quantities k_1 and r_1 are known only by the server and the tag respectively. These two values are used to compute the messages R_2, PT and PS, which are transmitted during the communication. If an attacker tries to modify the data transmitted between these two entities, the authentication process fails. In our case, the secret values cannot be transmitted directly over unsecure channel. According to ECDLP, the attacker cannot retrieve the secret key from the message.

4 Comparative Analysis

To evaluate the performance, we present the comparative analysis of our method with the protocols presented recently in [7–9] based on the computation cost in order to prove that the proposed protocol is efficient and lightweight. In the authentication scheme, the cost of a computation depends on how long its execution takes to perform the different operations involved. In the ECC approach, the RFID authentication protocol calculation time is proportional to the number of elliptic curve scalar multiplication operations (ECSM). In this work, we denote T_{SM} as the time required for the execution of the doubling point operation. According to [10], the execution time of one scalar multiplication is 64 ms. Table 2 presents the calculation cost comparisons with some associated works.

It can be clearly observed from Table 2 that the tag's runtime and the server's runtime are both 192 ms. As a result, the total time needed to execute our protocol is 384 ms. As a consequence; we can observe that, in comparison to other protocols, our suggested method requires less computing time to complete the necessary amount of scalar multiplication operations.

Table 2. Comparison of different schemes in term of Computation cost.

Scheme	Tag	Server	Total
Alamr et al. [7]	$4T_{SM}$	$5T_{SM}$	576
Naeem et al. [8]	$4T_{SM}$	$3T_{SM}$	448
Gasbi et al. [9]	$4T_{SM}$	$5T_{SM}$	512
Proposed scheme	$3T_{SM}$	$3T_{SM}$	384

5 Conclusion

Recently, elliptic curve cryptography is an effective solution that can be used in IoT applications to reduce the computational and to achieve reliable security for the transmitted data. To secure communications between the various RFID system components in an IoT context, several systems employ ECC. In this paper, a new authentication scheme using ECC for RFID systems in IoT is proposed. The obtained results show that the suggested approach is capable of completing the authentication between tag and server. The comparison between the proposed method and recent research works shows the effectiveness of our approach in terms of computing performance. In future work, the proposed protocol may be investigated and analyzed to be deployed in Healthcare systems and other similar fields.

References

1. Kislay, A., Singh, P., Shankar, A., Nayak, S.R., Bhoi, A.K.: A review on internet of things in healthcare applications. In: Mallick, P.K., Bhoi, A.K., Barsocchi, P., de Albuquerque, V.H.C. (eds.) Cognitive Informatics and Soft Computing. LNNS, vol. 375, pp. 387–394. Springer, Singapore (2022). https://doi.org/10.1007/978-981-16-8763-1_3
2. Radan, A., Samimi, H., Moeni, A.: A new lightweight authentication protocol in IoT environment for RFID tags. Int. J. Eng. Technol. **7**(4.7), 344–351 (2018). https://doi.org/10.14419/ijet.v7i4.7.23028
3. Mansoor, K., Ghani, A., Chaudhry, S.A., Shamshirband, S., Ghayyur, S.A.K., Mosavi, A.: Securing IoT-based RFID systems: a robust authentication protocol using symmetric cryptography. Sensors **19**, 4752 (2019). https://doi.org/10.3390/s19214752
4. Singh, A.K., Patro, B.D.K.: Elliptic curve signcryption based security protocol for RFID. KSII Trans. Internet Inf. Syst. **14**(1), 344–365 (2020). https://doi.org/10.3837/tiis.2020.01.019
5. Lamrani Alaoui, H., El Ghazi, A., Zbakh, M., Touhafi, A., Braeken, A.: A highly efficient ECC-based authentication protocol for RFID. J. Sens. **2021** (2021). https://doi.org/10.1155/2021/8876766. Article ID 8876766, 16 pages
6. Verri Lucca, A., et al.: A review of techniques for implementing elliptic curve point multiplication on hardware. J. Sens. Actuator Netw. **10**, 3 (2020). https://doi.org/10.3390/jsan10010003
7. Alamr, A.A., Kausar, F., Kim, J., Seo, C.: A secure ECC-based RFID mutual authentication protocol for internet of things. J. Supercomput. **74**(9), 4281–4294 (2016). https://doi.org/10.1007/s11227-016-1861-1

8. Naeem, M., Chaudhry, S.A., Mahmood, K., Karuppiah, M., Kumari, S.: A scalable and secure RFID mutual authentication protocol using ECC for Internet of Things. Int. J. Commun. Syst. **33**(13), e3906 (2019). https://doi.org/10.1002/dac.3906
9. Gabsi, S., Kortli, Y., Beroulle, V., Kieffer, Y., Alasiry, A., Hamdi, B.: Novel ECC-based RFID mutual authentication protocol for emerging IoT applications. IEEE Access (IF3.367) **9** (2021). https://doi.org/10.1109/ACCESS.2021.3112554
10. Dinarvand, N., Barati, H.: An efficient and secure RFID authentication protocol using elliptic curve cryptography. Wirel. Netw. **25**(1), 415–428 (2017). https://doi.org/10.1007/s11276-017-1565-3

Feature Selection Impact on Time Series Problems for Solar Radiation Forecasting

Hasna Hissou[1](✉), Said Benkirane[2], Azidine Guezzaz[2], and Abderrahim Beni-Hssane[1]

[1] Faculty of Science, Science and Technology Research Structure, Chouaïb Doukkali University, Avenue of Faculties, 24000 El Jadida, Morocco
hhissou@gmail.com

[2] Higher School of Technology, Computer Sciences Department, Cadi Ayyad University, Km 9, Road to Agadir, Essaouira Aljadida PB. 383, 44000 Ghazoua, Morocco

Abstract. Solar Radiation (Rs) received on our planet is measured in billions of watts. It is introduced into power grid networks and used for various applications. Even though it is still challenging to maintain a perfect balance between the production and consumption of electricity because of the radiation's intermittent and unpredictable nature. Accurate prediction of sun rays is thus critical in the solar industry. Many Machine Learning (ML) approaches and preprocessing models are used for Rs forecasting and complexity reduction to reduce these concern. In this paper, we present a model that shows the impact of feature selection approaches on Rs forecasting based on time series lag values. We use Recursive Feature Elimination (RFE) with random forest (RF), Decision Tree (DT), Logistic Regression (LR), Classification and Regression Tree (CART), Person (Pr) and Gradient Boosting models (GBM). The results show that our proposed approach gives reliable results in terms of some relevant parameters.

Keywords: Machine learning · Solar radiation · Feature selection · Forecasting

1 Introduction

Solar energy offers a lot of potential to serve the world's energy needs. It is one of the world's primary energy sources and it presents many advantages [1]. Solar radiation (Rs) is the amount of energy attaining the Earth's surface. It is influenced by the atmospheric layers. It affects the Earth's temperatures, global mean sea level, and climate extremes. It is also used for various applications and models [2]. Nowadays, many countries are considering introducing solar energy into their power grids. At all times, the electrical operators must ensure that the production and consumption of electricity are in perfect balance, which is very difficult to maintain as the Rs is sporadic, unexpected, and not controllable [3]. Therefore, accurate measurements of Rs's fluctuations are necessary. Various devices are used, but they are expensive [4]. To address these issues, several methods for Rs forecasting have been presented. The earliest techniques were: empirical models (simple with limited accuracy), statistical models (more accurate than empirical

Y. Farhaoui et al. (Eds.): ICAISE 2022, LNNS 635, pp. 440–446, 2023.
https://doi.org/10.1007/978-3-031-26254-8_63

models but unable to clearly express the nonlinear relationship between Rs and the other parameters), time series-based models, and so on [5]. Recently, ML models (neuron-based, Multilayer Perceptron (MLP) networks [6], kernel-based [7], tree-based [8], fuzzy logic, and ensemble learning models [9], etc.) are thought to be a viable approach due to their excellent forecasting accuracy. The advantage of using ML models is that they can address problems that are impossible to represent using explicit algorithms [3]. One of the challenging aspects of estimating a variable is determining the type and amount of variables that should be utilized as model inputs [10]. As the quality of the training data determines the accuracy, it is essential to decide on the best input for prediction models. This process is known as feature selection (FS)[11]. FS is utilized in the pre-processing section to simplify the models. It minimizes computing costs, increases training speed, and enhances prediction precision by removing unnecessary or redundant information and retaining only the most significant ones [12]. In this research article, we study the impact of FS models on improving the quality of Rs forecasting. We used RFE with different models: RF, DT, CART, LR, Pr, and GBM. We compare the feature importance score and the accuracy of each to the others.

In terms of this paper's structure, Sect. 2 discusses the background. Section 3 goes through the methodology followed in this study and presents the proposed model. Section 4 describes the environment of our work. It gives the results and discusses them. The most important findings and future works are highlighted in Sect. 5.

2 Related Works

This section summarizes Rs prediction methods related to various artificial intelligence methodologies [13, 14]. In terms of ML, it has been used by many academics to predict Rs. a lot of them present methods like Artificial Neural Networks (ANN), SVR, RF, regression tree (RT) or GB, etc. Others suggested hybrid or ensemble learning algorithms. Voyant et al., Meenal and Selvakumar, and Pang et al. evaluated a variety of ML algorithms. They showed that the ANN does not achieve high predictive results but suggests an algorithm improvement approach [3, 15]. Xue demonstrated, using a backpropagation neural network (BPNN), that the projected accuracy is affected by the selection of input data [16]. In general, the ANN and ARIMA approach present almost the same accuracy. SVM, random forests, and RT, all produce nearly equal error statistics and perform similarly [3]. Despite some research on Rs estimation, the literature on comparing alternative strategies for selecting inputs for calculating Rs is restricted. Rabehi et al. assessed Rs in the south of Algeria using ANN, Boosted Decision Tree (BDT), and a new combination of these models with Linear Regression (LR). A novel selection approach, Procrustes Analysis (PA), is proposed and demonstrated to lead to a better-compared selection of input variables for estimating Rs. PA has been utilized in a few research studies to optimize the amount and kind of model inputs [17]. Biazar et al. investigate a new input selection method, gamma test (GT), and PA with the help of non-linear models of ANN and SVM. He identifies the best input variables for estimating Rs using GT and PA. PA outperformed GT using fewer input variables and minimizing model uncertainty [10]. Table 1 covers some recent feature selection methods and related research.

Table 1. Recent feature selection methods and corresponding researches

References	Location	FS method
Boutahir et al., 2022 [18]	Errachidia, Morocco	RF, XGBoost, CatBoost, LightGBM
Zeng et al., 2020 [19]	130 sites, China	RF
Al maraashi, 2018 [20]	8 sites, Saudi Arabia	ReliefF, MonteCarlo uninformative variable elimination, random-frog, Laplacian score

3 Our Proposed Approach

This section shows the proposed model (Fig. 1) and presents the methodology.

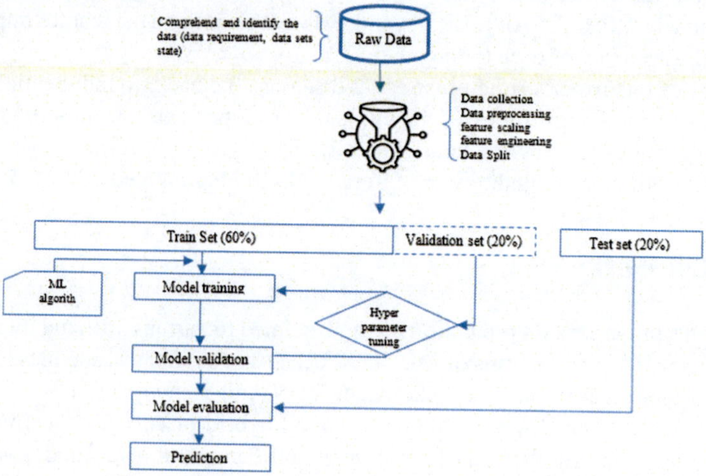

Fig. 1. Proposed design of our model

The first stage of data preprocessing of a time series model is to make it stationary. Seasonality was eliminated via differencing: We took the seasonal difference, resulting in a seasonally adjusted time series with 12-month period. We converted the dataset into supervised learning problem. We used the lag observations (i-1) as inputs and the actual observation (t) as output, moved the first 12 rows of data with NaN values, then generated a new time series with lag values of 12 months in order to forecast the current observation. The feature engineering phase minimizes the most relevant dimensions using RFE (a wrapped method that internally employs filter-based FS). It operates by building prediction models, weighing features, reducing those with the lowest weights, and repeating the procedure until the required features remain. Then we assess and train the model's attributes: The data is divided into training (60%), validating (20%), and test (20%) sets. We used RFE with RF, CART, DT, LT, Person, and GBM. We fit all the models successively on the same supervised learning view of the dataset. Finally, we evaluate the model's performance, refine its hyper-parameters, and validate it.

4 Experimental Study

4.1 Environment

The data set used is from The National Centers for Environmental Prediction (N.C.E.P). The data were collected from 1979 to 2014 (36 years). It has 12988 daily precipitations, minimum and maximum temperatures, wind, humidity, longitude, elevation, latitude, and sunshine records. Our research is processed on a laptop with a Core-i5 3437U CPU operating at 2.4 GHz and 16 GB of DDR3, running Windows 10 Professional 64 bits. Python version 3.9.7 is used to train the model. Repeated KFold to evaluate it with three repeats and ten folds across all repetitions and folds. The Mean Absolute Error (MAE) shown in Eq. (1), the mean and standard deviation of the model's accuracy are provided. The relative value and score of all input features for a specific model are quantified using feature importance. MAE is a measurement of the distinction in error between two examples of the same event occurring. Expected vs. observed comparisons are examples of Y against X, where x_i the actual value and y_i is the predicted.

$$MAE = \frac{\sum_{i=1}^{n} |y_i - x_i|}{n} = \frac{\sum_{i=1}^{n} |e_i|}{n} \tag{1}$$

The feature importance scores each feature based on its relevance. The Min-max normalization is conducted to prevent the negative influence problem caused by large weights. It maintains the correlations between the initial data values. It considers that $\min_X$ and $\max_X$ are the minimum and maximum values of an attribute X. It computes and then converts a value v_i of X to v'_i in the range [new_min_X, new_max_X]. Equation (2) calculates the range transformation formula being used for normalizing [21].

$$v' = \frac{v - min_X}{max_X - min_X}(new\_max_X - new\_min_X) + new\_min_X \tag{2}$$

4.2 Discussion of Results

After generating the new time series with lag values of 12 months, the Autocorrelation for each lag variable depicts the significant lag values with the most significant at 1, 2, and 12 months. As we used RFE with the different models, we fit all the models successively on the same supervised learning view of the dataset. The relative feature relevance ratings for each of the 12 lag observations are summarized. The importance scores are shown in Fig. 2, and the FS rank is shown in Fig. 3 for each input feature.

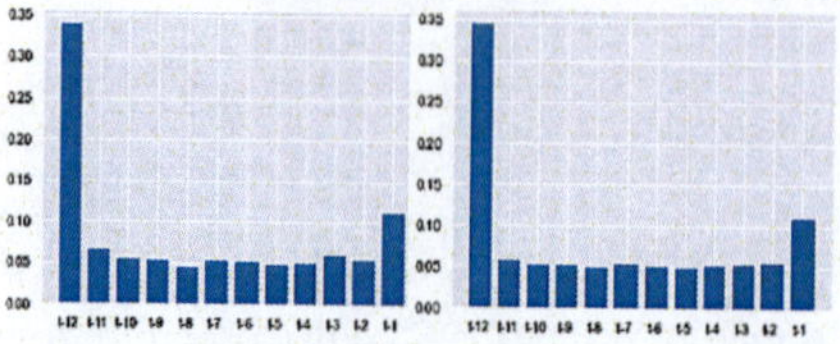

Fig. 2. Feature importance scores RF /DT

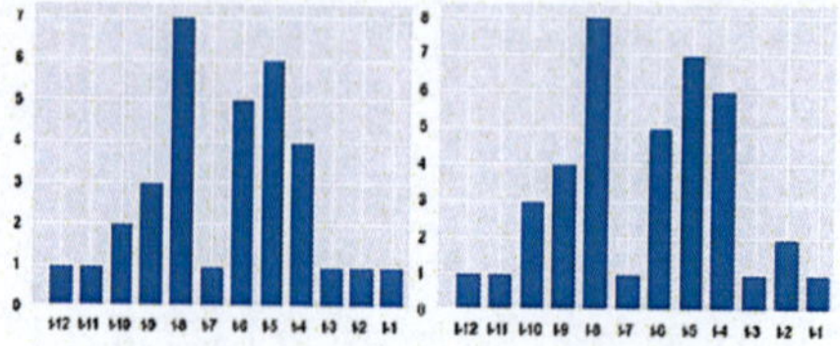

Fig. 3. FS rank using RF /DT

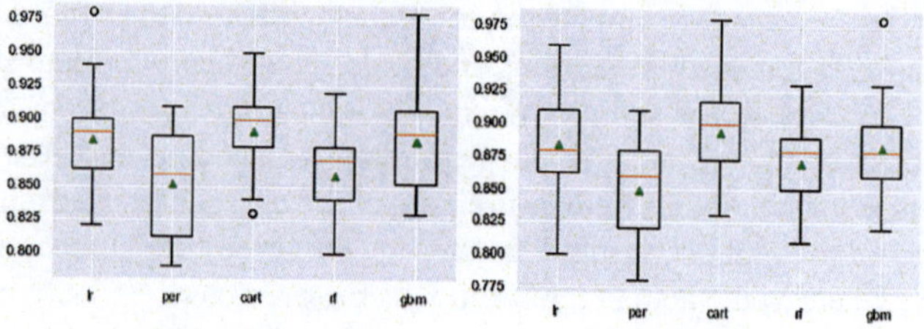

Fig. 4. Accuracy scores /std vs. nMAE /std

For the accuracy score distribution (left graph) and nMAE (right graph), a box and whisker graphic is generated for each model (Fig. 4). For RF, the features t–12, t–2, t–7, t–3, and t–1 successively present the highest scores. For the DT, we have a different selection of features (excluding the first and fourth), which are t–12, t–1, t–11, t–3, and t–7 successively. The ranking of the other features changes according to the model used.

5 Conclusion and Perspectives

We may state that LR, CART, and GBM outperform, followed by the remaining models. The model employed inside RFE impact significantly which features are picked and, as a result, the prediction performance. For the same data, we had different rankings of features depending on the model used. Nevertheless, the findings of this study will serve as the foundation for future research that will be enhanced. Thus, the future perspectives are first to investigate various FS methods to use the one best suited to the time series and then to assess the model's performance regarding the FS.

References

1. Ağbulut, Ü., Gürel, A.E., Biçen, Y.: Prediction of daily global solar radiation using different machine learning algorithms: evaluation and comparison. Renew. Sustain. Energy Rev. **135**, 110114 (2021). https://doi.org/10.1016/j.rser.2020.110114
2. Ozgoren, M., Bilgili, M., Sahin, B.: Estimation of global solar radiation using ANN over Turkey. Expert Syst. Appl. **39**(5), 5043–5051 (2012). https://doi.org/10.1016/j.eswa.2011.11.036
3. Voyant, C., et al.: Machine learning methods for solar radiation forecasting: a review. Renew. Energy **105**, 569–582 (2017). https://doi.org/10.1016/j.renene.2016.12.095

4. Huang, L., Kang, J., Wan, M., Fang, L., Zhang, C., Zeng, Z.: Solar radiation prediction using different machine learning algorithms and implications for extreme climate events. Front. Earth Sci. **9**, 596860 (2021). https://doi.org/10.3389/feart.2021.596860
5. Fan, J., et al.: Comparison of support vector machine and extreme gradient boosting for predicting daily global solar radiation using temperature and precipitation in humid subtropical climates: a case study in China. Energy Convers. Manag. **164**, 102–111 (2018). https://doi.org/10.1016/j.enconman.2018.02.087
6. Guezzaz, A., Asimi, A., Mourade, A., Tbatou, Z., Asimi, Y.: A multilayer perceptron classifier for monitoring network traffic. In: Farhaoui, Y. (ed.) BDNT 2019. LNNS, vol. 81, pp. 262–270. Springer, Cham (2020). https://doi.org/10.1007/978-3-030-23672-4_19
7. Chen, J.-L., Li, G.-S., Wu, S.-J.: Assessing the potential of support vector machine for estimating daily solar radiation using sunshine duration. Energy Convers. Manag. **75**, 311–318 (2013). https://doi.org/10.1016/j.enconman.2013.06.034
8. Tibshirani, S., Friedman, H.: The Elements of Statistical Learning: Data Mining, Inference and Prediction, p. 764. Springer, New York (2017). https://doi.org/10.1007/978-0-387-84858-7
9. Bounoua, Z., Chahidi, L.O., Mechaqrane, A.: Estimation of daily global solar radiation using empirical and machine-learning methods: a case study of five Moroccan locations. Sustain. Mater. Technol. **28**, e00261 (2021). https://doi.org/10.1016/j.susmat.2021.e00261
10. Biazar, S.M., Rahmani, V., Isazadeh, M., Kisi, O., Dinpashoh, Y.: New input selection procedure for machine learning methods in estimating daily global solar radiation. Arab. J. Geosci. **13**(12), 1–17 (2020). https://doi.org/10.1007/s12517-020-05437-0
11. Guezzaz, A., Benkirane, S., Azrour, M., Khurram, S.: A reliable network intrusion detection approach using decision tree with enhanced data quality. Secur. Commun. Netw. **2021**, 1–8 (2021). https://doi.org/10.1155/2021/1230593
12. Meenal, R., Michael, Prawin Angel, Pamela, D., Rajasekaran, E.: Weather prediction using random forest machine learning model. Indon. J. Electr. Eng. Comput. Sci. **22**(2), 1208 (2021). https://doi.org/10.11591/ijeecs.v22.i2.pp1208-1215
13. de Freitas Viscondi, G., Alves-Souza, S.N.: Solar irradiance prediction with machine learning algorithms: a Brazilian case study on photovoltaic electricity generation. Energies **14**(18), 5657 (2021). https://doi.org/10.3390/en14185657
14. Yu, Y., Cao, J., Zhu, J.: An LSTM short-term solar irradiance forecasting under complicated weather conditions. IEEE Access **7**, 145651–145666 (2019). https://doi.org/10.1109/ACCESS.2019.2946057
15. Meenal, R., Selvakumar, A.I.: Assessment of SVM, empirical and ANN based solar radiation prediction models with most influencing input parameters. Renew. Energy **121**, 324–343 (2018). https://doi.org/10.1016/j.renene.2017.12.005
16. Xue, X.: Prediction of daily diffuse solar radiation using artificial neural networks. Int. J. Hydrogen Energy **42**(47), 28214–28221 (2017). https://doi.org/10.1016/j.ijhydene.2017.09.150
17. Rabehi, A., Guermoui, M., Lalmi, D.: Hybrid models for global solar radiation prediction: a case study. Int. J. Ambient Energy **41**(1), 31–40 (2020). https://doi.org/10.1080/01430750.2018.1443498
18. Boutahir, M.K., Farhaoui, Y., Azrour, M., Zeroual, I., El Allaoui, A.: Effect of feature selection on the prediction of direct normal irradiance. Big Data Min. Anal. **5**(4), 309–317 (2022). https://doi.org/10.26599/BDMA.2022.9020003
19. Zeng, Z., et al.: Daily global solar radiation in china estimated from high-density meteorological observations: a random forest model framework. Earth Space Sci. **7**(2) (2020). https://doi.org/10.1029/2019EA001058

20. Almaraashi, M.: Investigating the impact of feature selection on the prediction of solar radiation in different locations in Saudi Arabia. Appl. Soft Comput. **66**, 250–263 (2018). https://doi.org/10.1016/j.asoc.2018.02.029
21. Al Shalabi, L., Shaaban, Z.: Normalization as a preprocessing engine for data mining and the approach of preference matrix. In: 2006 International Conference on Dependability of Computer Systems, Szklarska Poreba, pp. 207-214 (2006). https://doi.org/10.1109/DEPCOS-RELCOMEX.2006.38

Fetal Electrocardiogram Identification Using Statistical Analysis

Said Ziani(✉)

Laboratory in Computer Networks, Telecommunications and Multimedia, High School of Technology, Hassan II University, Casablanca, Morocco
SAID.ZIANI@univh2c.ma

Abstract. This research is an extension of previous work that focused solely on statistical models such as PCA to address the challenges of source separation. The recovery of fetal electrocardiograms is handled by deploying a combination of wavelet packets and statistical analysis. This investigation will demonstrate that this excessive number of channels is superfluous, assuming that the efficiency of the channel in terms of the energy and information it conveys is what is crucial. We advocate adopting the Continuously wavelet to select channels prior to performing the statistical algorithm. As a result of implementing this method, we can use only two channels as contrasted to four.

Keywords: Electrocardiogram ECG · Wavelet · Identification · JADE

1 Introduction

The input isolation is accomplished by using a mixing operator, who is also unknown, to pull out unknown source signals from only knowing their image, which is also called mixing or observation. This is why the separation is called blind. This is an awful way to set up a problem. So, the model has to be based on certain assumptions. Once these assumptions have been made clear, a stochastic optimization method can be used to solve this model. But when it comes to separating sources, there are some things that make it hard to tell them apart. At first, adding a separation criterion to a normalization criterion was the only way to do it. A method called "stochastic gradient" was used to find the minimum value. At the end of this study, after the gradient of this criterion was calculated, a new blind source separation algorithm was made and compared to an existing algorithm and a reference. Heart conditions are the primary cause of mortality in Morocco. cardiac illnesses account for more than 40% of mortality in the Kingdom, or four out of ten deaths. Figures that sound the warning bells... The risk of heart attack, cardiovascular accident (CVA), and mortality significantly increase due to the occasionally severe summer circumstances, including high heat. Those with cardiovascular disease are urged to be incredibly attentive during this season. High blood pressure (HTA) is Morocco's leading cardiovascular risk factor, followed by dyslipidemia (the presence of excessively high levels of lipids in the blood), obesity, diabetes, and smoking. It should be mentioned

Y. Farhaoui et al. (Eds.): ICAISE 2022, LNNS 635, pp. 447–453, 2023.
https://doi.org/10.1007/978-3-031-26254-8_64

that cardiac diseases do not discriminate against either the old or the young. The prevalence of congenital cardiac abnormalities is estimated to be 1 in 125 births annually. Fetal heart monitoring may assist pediatric cardiologists to diagnose and treat cardiac problems as soon as possible. Insight into the fetus's cardiac condition may be gleaned through an fECG [1]. Sensors placed on the mother's belly may measure this. Yet, the signal's brightness is modest and mingled with a lot of background noises. However, the maternal electrocardiogram (mECG) intensity is significantly larger than that of the fetus, rendering it the major source of infection. Thus, the fundamental issue is how to separate the fECG from the rest of the data [2, 3]. In this paper, we contribute by combining statistical techniques with the wavelet packet transform o minimize the original eight channels to only two. Wavelets arose due to the need for frequency and temporal analysis in particular academic disciplines1. Fourier analysis was the only technique that permitted the decomposition and reconstruction of a signal without information loss; however, it only provides frequency analysis and does not permit the temporal localization of abrupt changes, such as the appearance of a second musical note after the first has been played. In 1909, the first wavelet was composed of a short negative pulse followed by a short positive pulse. In 1946, a mathematician created a Fourier-like function transformation2 applied to a temporal frame expressing a Gaussian function. Eventually, the word wavelet entered the mathematical language.

2 Theoretical Background

2.1 Model

Figure 1 summarizes the issue of isolation [4]. A contrast function C{ν} is a function of the probability density of random[] sequence $\nu \in \mathbb{R}^{n \times T}$

$$\begin{cases} M \in \mathbb{R}^{n \times n}, C\{Ms\} \geq C\{s\}, \\ C\{Ms\} = C\{s\} ifandonlyifM \sim I_n \end{cases} \tag{1}$$

If B is a matrix that minimizes $C\{Bx\} = C\{BAs\}$
So $BA \sim I_n$
And consequently $B=A^{-1}$
So: Y=BX

2.2 Jade Algorithm

The original statement of the challenge of extracting distinct basic entities from composite signals was performed in 1984 by two researchers in neuroscience and signal processing to replicate the coding and decoding of human movement using a self-adaptive neuromagnetic network. This early research prepared the basis for the problem of blind source separation. In the second part of the 1980s, a community arose around this subject. The signal processing community chose a statistical formalism, but scholars sought to expand principal component analysis using a connectionist approach. The jade algorithm designed by Cardoso [5] aims at minimizing the following contrast function:

$$C^{\circ}_{JADE}\{y\} \stackrel{\text{def}}{=} \sum_{ijk} {c_{y_i y_j y_k y_l}}^2 \tag{2}$$

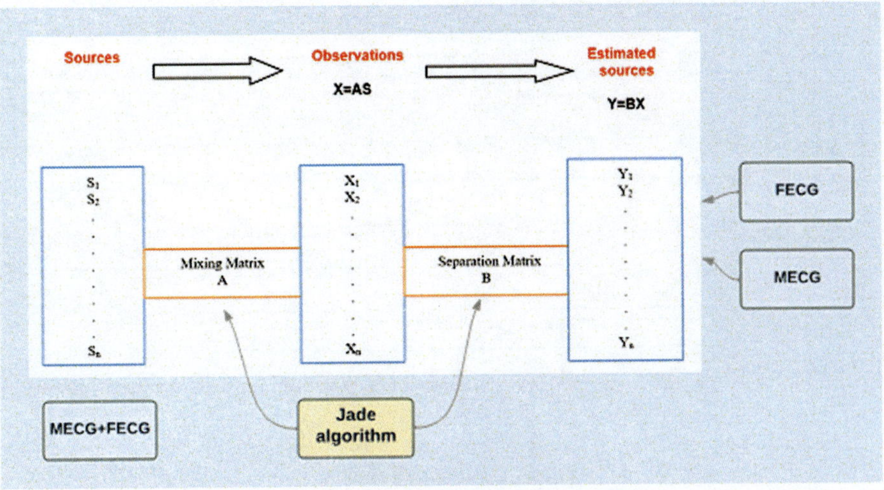

Fig. 1. Combining and separating sources

And this minimization is done via the simultaneous diagonalization of a matrix set of cumulants.

3 Methods

3.1 ECG Recording

We aim to get the FECG signal out of one lead using time-scale imaging segmentation. The data is acquired from an international database Daisy. Figure 2 illustrates the temporal representation of the recordings signals denoted f(t).

3.2 Algorithm

To address the issue of isolation, this study employs a distinctive methodology. Figure 3 depicts these sequential steps.:

4 Simulation Results

We begin by with CWT to find the most useful signals, we start by computing the continuous wavelets transform (CWT) of signals f1 through f8 using the Cgau wavelet [3]. After this, we'll demonstrate how Ti (a, b) changes over time which provided the following results (Fig. 4 a and b).

By adopting the scalogram representation, which consists of drawing the isocontour lines of the modulus of the wavelet field determined on a dyadic network, it is possible to display the "wavelet" energy of a given signal. Adopting the concept of wavelet frame, Letscan has created the original concept of super-resolution wavelet analysis, which

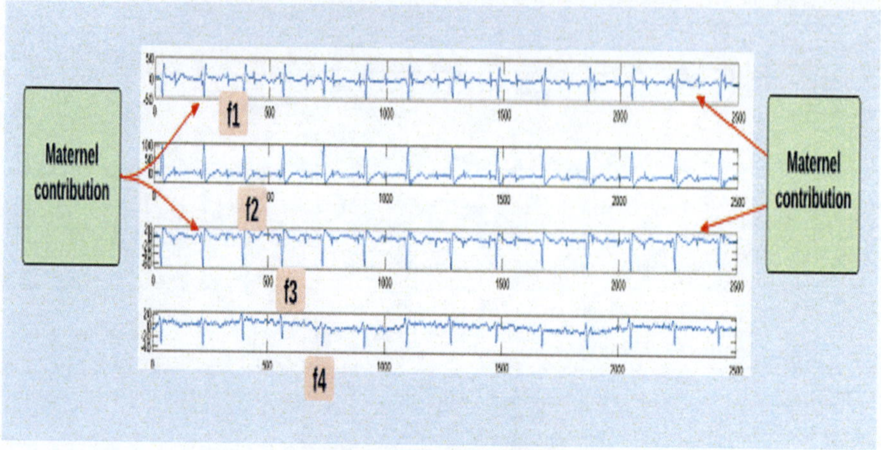

Fig. 2. Abdominals signals

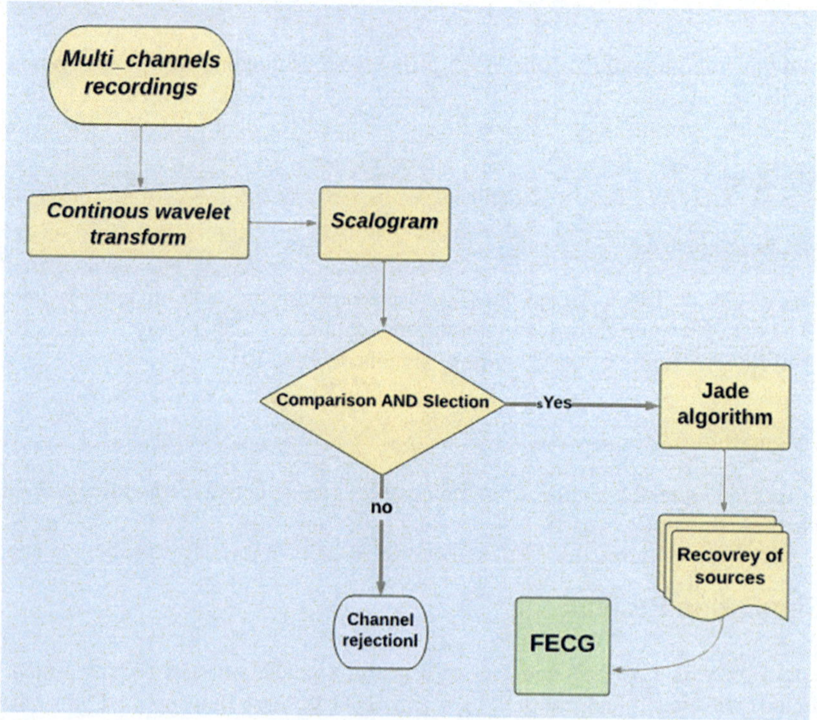

Fig. 3. Global algorithm

enables the study of the instantaneous frequencies of a signal decomposed on dense dyadic grids of the time-frequency plane with adaptive zoom capabilities. Based on the result shown in Fig. 4, we can see that the fetus's electrical activity is missing from the

(a)

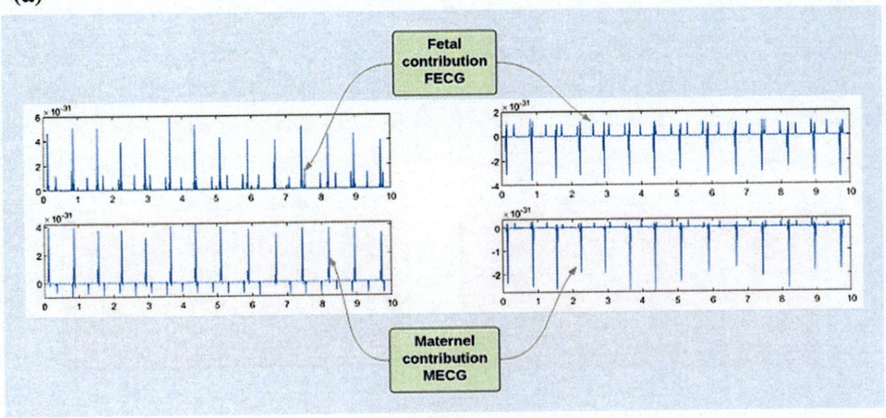

(b)

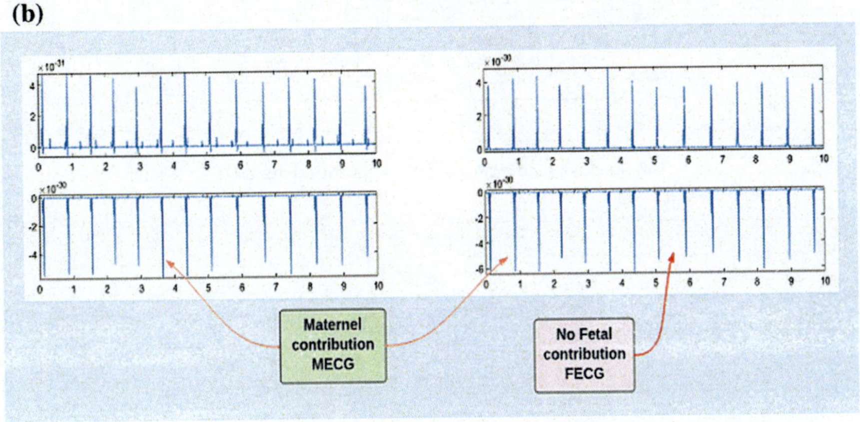

Fig. 4. Representation of the temporal evolution of CWT transform

signals f6, f7, and f8. So, instead of eight, there will only be five. Also, channel 1 has more details about how the mother's heart is beating. Because of this, you can't count on this channel. Our method comprises computing the continuous wavelets transform (CWT) of all channels to make an efficient selection of useful ones. Then, we'll display the time-scale space (Fig. 4). By inputting the four channels f1, f2, f3, and f4 to the JADE function in Matlab, the following separation result was obtained Figs. 5 and 6:.

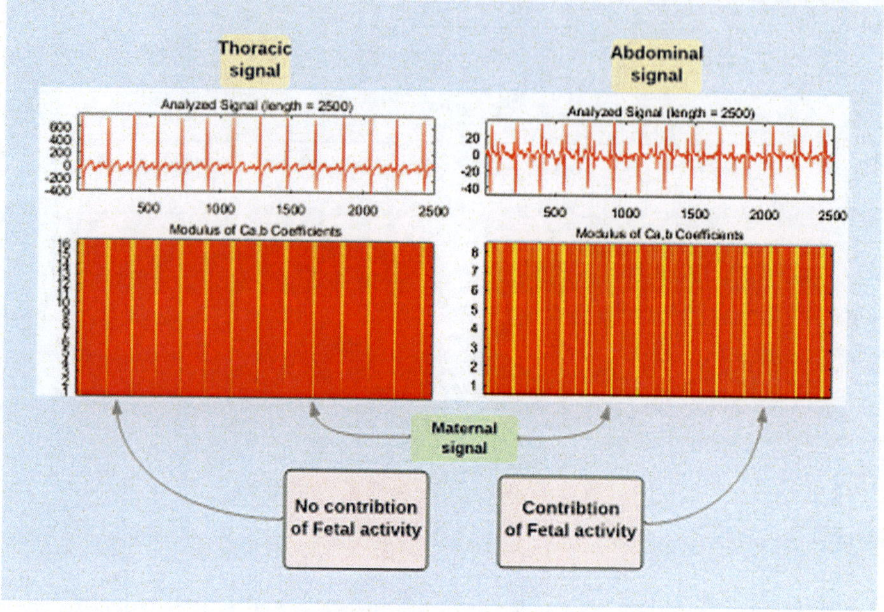

Fig. 5. CWT analysis of the Abdominal signal

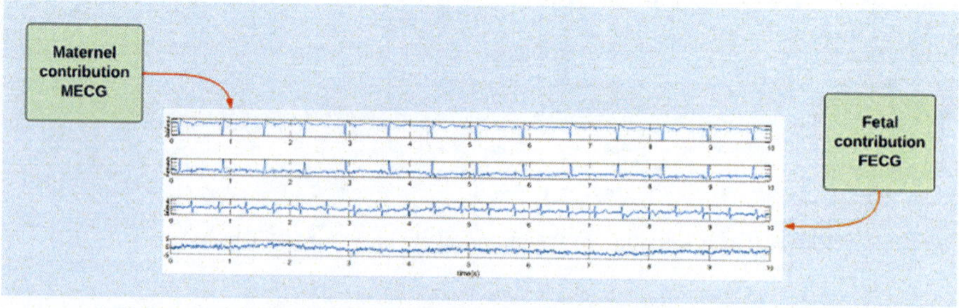

Fig. 6. Fetal and Maternal Contribution

5 Discussion and Conclusion

According to the results of the simulations presented in Fig. 5, blind source separation is effective. Moreover, the fECG signal is recovered with clearly distinguishable complex QRS. Ultimately, we will compare our findings to those in the literature[4, 5]: We observe that although the QRS complexes are evident in both circumstances, they are not identical. In conclusion, the usage of JADE as a blind source separation technique is not novel, but what is unknown in this article is the combination of JADE and the continuous wavelets transform CWT to minimize the number of channels from eight to four. Moreover, this combination leads to two significant and novel outcomes: First,- The energy criteria are more effective than the statistical independence requirement. Second,- A careful modification of JADE in constructing cumulant matrices can yield significant results, even when relying on a single channel. The contribution offered in this research can significantly enhance the work on a single-channel separation, particularly for those utilizing ICA in conjunction with EMD, NMF [6–8], SVD, and ISVD. This also appears to be applicable to synchronous motors [9].

References

1. Said, Z., El Hassouani, Y.: A new approach for extracting and characterizing fetal electrocardiogram. Traitement Sig. **37**(3), 379–386 (2020). https://doi.org/10.18280/ts.370304
2. Ziani, S., El Hassouani, Y.: Fetal electrocardiogram analysis based on LMS adaptive filtering and complex continuous wavelet 1-D. In: Farhaoui, Y. (ed.) BDNT 2019. LNNS, vol. 81, pp. 360–366. Springer, Cham (2020). https://doi.org/10.1007/978-3-030-23672-4_26
3. Said, Z., El Hassouani, Y.: Fetal-maternal electrocardiograms mixtures characterization based on time analysis. In: 2019 5th International Conference on Optimization and Applications (ICOA), pp. 1–5 (2019). https://doi.org/10.1109/ICOA.2019.8727619
4. Ouhadou, M., El Amrani, A., Messaoudi, C., Ziani, S.: Experimental investigation on thermal performances of SMD LEDs light bar: Junction-to-case thermal resistance and junction temperature estimation. Optik **182**, 580–586 (2019)
5. Ziani, S., El Hassouani, Y., Farhaoui, Y.: An NMF based method for detecting RR interval. In: Farhaoui, Y., Moussaid, L. (eds.) ICBDSDE 2018. SBD, vol. 53, pp. 342–346. Springer, Cham (2019). https://doi.org/10.1007/978-3-030-12048-1_35
6. Said, Z., Jbari, A., Bellarbi, L.: QRS complex characterization based on non-negative matrix factorization NMF. In: 2018 4th International Conference on Optimization and Applications (ICOA), pp. 1–5, (2018). https://doi.org/10.1109/ICOA.2018.8370548
7. Ouhadou, M., El Amrani, A., Ziani, S., Messaoudi, C.: Experimental modeling of the thermal resistance of the heat sink dedicated to SMD LEDs passive cooling. In: Proceedings of the 3rd International Conference on Smart City Applications (2018)
8. Said, Z., Jbari, A., Belarbi, L.: Fetal electrocardiogram characterization by using only the continuous wavelet transform CWT. In: 2017 International Conference on Electrical and Information Technologies (ICEIT), pp. 1–6, (2017). https://doi.org/10.1109/EITech.2017.8255310
9. Chaou, Y., et al.: Nonlinear control of the permanent magnet synchronous motor pmsm using backstepping method. WSEAS Trans. Syst. Control **17**, 56–61 (2022)

Fuzzy Logic Based Adaptive Second-Order Nonsingular Terminal Sliding Mode Lateral Control for Uncertain Autonomous Vehicle

Moussa Abdillah(✉), Ayoub Belkheir, Najlae Jennan, and El Mehdi Mellouli

Laboratory of Engineering, Systems and Applications LISA, National School of Applied Sciences, Fez, Morocco
{abdillah.moussa,ayoub.belkheir,najlae.jennan}@usmba.ac.ma

Abstract. We present in this article a T-S fuzzy logic theory combined with Adaptive Second-Order Nonsingular Terminal Sliding Mode Control ASONTSMC for autonomous vehicle modeling and lateral control. Firstly we present the dynamics of the vehicle by adopting the bicycle model which is difficult to study, secondly we show in details how to approximate a positive constant or a gain of the control law by using the T-S fuzzy logic algorithm associated with ASONTSMC, before concluding we confirm our approach by simulations performed on MATLAB.

Keywords: T-S fuzzy logic · ASONTSMC · Autonomous vehicle · Lateral control · Bicycle model · Gain

1 Introduction

Mobility and transport safety are being revolutionized by the mass production of self-driving cars. These vehicles have the ability to drive themselves without human intervention. This technology is not new, but it has progressed over time. For this reason, in the market there are many different models that differ from each other depending on the technology implemented. These technologies are called ADAS or simply systems that assist, help and replace human driving in certain situation.

In fact, the society of Automotive Engineers (SAE) has classified autonomous vehicles into six levels: No automation level, Driver assistance level, Partial automation level, Conditional automation level, High automation level and Full automation level [7].

To be able to drive alone, the vehicle must have a total or partial control with the ability to detect any kind of obstacles using sensors that replace the eyes, it must also be able to analyze and interpret information's received via sensors to plan a good trajectory but also control and command certain parameters to maintain on the chosen trajectory not to collide with other road users. In this context, several techniques and algorithms are used for the control and the command of which the lateral control is the subject of our study in this paper. In [1], an integral Sliding Mode Control SMC is used to get the wished torques that drives the car and converges its path to the wanted one [4–10]. A novel autonomous drift controller for a distributed drive electric vehicle whose configuration provides more possibilities for drift has exploited in [8].

Y. Farhaoui et al. (Eds.): ICAISE 2022, LNNS 635, pp. 454–461, 2023.
https://doi.org/10.1007/978-3-031-26254-8_65

So, the main objective of our work is the modeling and controlling of an autonomous vehicle model bicycle by combing fuzzy logic and adaptive second-order nonsingular terminal sliding mode control ASONTSMC [3] for the estimation of a positive constant appeared on the control low.

We have subdivided our work as followers: First, we present the dynamics of the vehicle, then, we give details of the control strategy, simulation results and discussions are given at the last to confirm our approach and we ended with a conclusion of the whole work. All simulations are done on MATLAB.

2 Mathematical Modeling of the Vehicle

There are two types of mathematical models of the vehicle, in literature: the bicycle model and the four- wheel model.

We employ in this work a dynamic bicycle model given by [2] to illustrate the lateral behavior of our study system. The Fig. 1 represents this bicycle model. We adopt this model as a means of representing the lateral behavior of the car, we also assume that it is symmetrical and that the drift angles are equals. We neglect the roll and pitch dynamics. Vx represents the velocity longitudinal.

The dynamic bicycle model shown as followers:

$$\begin{cases} \ddot{y} = -\left(\frac{C_f+C_r}{mV_x}\right)\dot{y} - \left(\frac{L_fC_f-L_rC_r}{mV_x} + V_x\right)\dot{\psi} + \frac{C_f}{m}\delta \\ \ddot{\psi} = -\left(\frac{L_fC_f-L_rC_r}{I_zV_x}\right)\dot{y} - \left(\frac{L_f^2C_f+L_r^2C_r}{I_zV_x}\right)\dot{\psi} + \frac{L_fC_f}{I_z}\delta \end{cases} \tag{1}$$

where y represents the lateral position, ψ is the yaw angle and δ the steering angle.

By posing:

$$X = [X_1X_2X_3X_4]^t = [y\,\dot{y}\,\psi\,\dot{\psi}]^t \tag{2}$$

We obtain the state representation of the following system [13]:

$$\begin{cases} \dot{X}_1 = X_2 \\ \dot{X}_2 = f_1(X,t) + g_1(X,t)U(t) \\ \dot{X}_3 = X_4 \\ \dot{X}_4 = f_2(X,t) + g_2(X,t)U(t) \end{cases} \tag{3}$$

$$\dot{X}_4 = f_2(X,t) + g_2(X,t)U(t)$$

where:

$f_1(X,t) = -\left(\frac{C_f+C_r}{mV_x}\right)X_2 - \left(\frac{L_fC_f-L_rC_r}{mV_x} + V_x\right)X_4$, $f_2(X,t) = -\left(\frac{L_fC_f-L_rC_r}{I_zV_x}\right)X_2 - \left(\frac{L_f^2C_f+L_r^2C_r}{I_zV_x}\right)X_4$

$$g_1(X,t) = \frac{C_f}{m},\ g_2(X,t) = \frac{L_fC_f}{I_z},\ U(t) = \delta$$

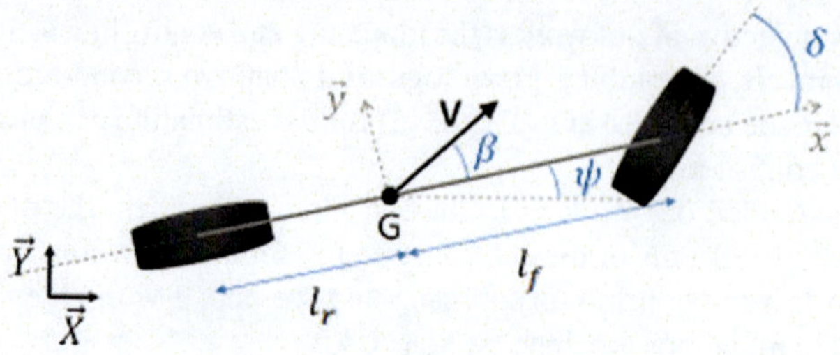

Fig. 1. Bicycle model vehicle

3 Controller Strategy Design

This section presents the controller strategy based on T-S fuzzy logic to approximate a gain that appeared on the control low. Firstly, we construct the ASONTSMC for determining the control low and secondly we fused a T-S Fuzzy logic method with the ASONTSMC constructed on the first steps to estimate the gain. The Stability of our system is assumed by using the Lyapunov theory.

3.1 Classical Adaptive Second-Order Nonsingular Terminal Sliding Mode Control [3] Design

The adaptive second-order nonsingular terminal sliding mode control [3] ASONTSMC is constructed by two terms: a proportional derivative combined other part with absolute value.

It's defined by the relation:

$$S = S_v + K\left|\dot{S}_v\right|^{\frac{p}{q}} sign\left(\dot{S}_v\right) \tag{4}$$

where K is a positive constant, p et q are positive odd integers satisfying 1 < p/q < 2.

S_v is the proportional derivative (5) and $\dot{S}_v$ his first derivative.

$$S_v = K_1\dot{e}_v + K_2 e_v \tag{5}$$

with e_v, $\dot{e}_v$ and $\ddot{e}_v$ represent respectively the lateral displacement tracking error, his first and second derivatives of which their formulas are given in [4]:

$$e_v = X_1 - X_{1d}$$

$$\dot{e}_v = X_1 + V_x e_\psi \tag{6}$$

$$\ddot{e}_v = f_1(X, t) + g_1(X, t)U(t) + V_x\dot{e}_\psi$$

X_{1d} represents the desired trajectory, $e_\psi = X_3 - X_{3d}$ is the yaw angle error and X_{3d} the desired yaw angle.

By deriving the sliding surface S, we obtain:

$$\dot{S} = \dot{S}_v + K\frac{p}{q}|\dot{S}_v|^{\frac{p}{q}-1}\ddot{S}_v = K\frac{p}{q}|\dot{S}_v|^{\frac{p}{q}-1}\left(\ddot{S}_v + \frac{q}{pK}|\dot{S}_v|^{2-\frac{p}{q}}sign(\dot{S}_v)\right) \tag{7}$$

To guaranty the stability of our system, let's take the Lyapunov function [1–13]:

$$V = \frac{1}{2}S^2 \tag{8}$$

In the case where the derivative of this function is negative, this confirms the stability of our system, which mains that:

$$\dot{V} = S\dot{S} < 0 \tag{9}$$

That's why we define the sliding surface derivative:

$$\dot{S} = -K_d sign(S) - \alpha S \tag{10}$$

where K_d is a positive constant and α is constant positive or a gain that will be approximated using fuzzy logic in the next section.

Based on (7) and the conditions (8) and (9), we obtain the control low as:

$$\dot{U}(t) = -\frac{1}{K_1 g_1(X,t)}\left[G + \frac{q}{pK}\left(|\dot{S}_v|^{2-\frac{p}{q}}sign(\dot{S}_v) + |\dot{S}_v|^{1-\frac{p}{q}}(K_d sign(S) + \alpha S)\right)\right] \tag{11}$$

where G is a function given as follower as:

$$G = \left[K_1\left(g_2(X,t)\left(2V_x + \frac{C_f l_f - C_r l_r}{mV_x}\right) - g_1(X,t)\frac{C_f + C_r}{mV_x}\right) + K_2 g_1(X,t)\right]U(t)$$
$$+\left(K_2 - K_1\frac{C_f + C_r}{mV_x}\right)f_1(X, t + K_1\left(2V_x + \frac{C_f l_f - C_r l_r}{mV_x}\right)f_2(X,t) + K_2 V_x X_4$$
$$-V_x\left(K_2\dot{X}_{3d} + K_1 \underset{3d}{X}\right) \tag{12}$$

3.2 T-S Fuzzy Logic Methodology

In this section, we use Fuzzy Logic algorithm [14–16] to approximate a positive constant α using $\hat{\alpha}$ (X, θ) approximate function.

The fuzzy technique is outlined by fuzzy IF-THEN rules [5].

This fuzzy technique is employed here to perform a mapping from an input vector $X = [X_1, X_2, X_3, \ldots .X_n]^t$ to an output $\hat{\alpha}$ (X, θ).

The fuzzy rule base consists of a collection of fuzzy rules IF-THEN in the following form [6–9]:

$$R^j \text{ if } X_1 \text{ is } A_1^j \text{ and} \ldots \text{ and } X_n \text{ is } A_n^j, \text{ then } \hat{\alpha}\ (\text{X}, \theta) \text{ is } \theta_j \tag{13}$$

where A_1^j, $i = 1, 2, .., n$ are fuzzy variables characterized by membership functions $\mu_{A_1^j}$ (Xi) and θ_j define value of the output singleton.

$\hat{\alpha}$ can be defined by [5–12]:

$$\hat{\alpha}(X,\theta) = \frac{\sum_{j=1}^{M} \theta_j(\prod_1^n \mu_{A_1^j}(Xi))}{\sum_{j=1}^{M}(\prod_1^n \mu_{A_1^j}(Xi))} = \theta^T \xi(X) \quad (14)$$

So,

$$\alpha(X,\theta) = \theta^* \xi(X) \quad (15)$$

where M is the total number of fuzzy rules, $\theta = [\theta_1, \theta_2, \ldots\ldots\ldots, \theta_n]^T$ represents an adjustable parameter vector and $\xi(X) = [\xi_1(X), \xi_2(X), \ldots\ldots\ldots\ldots \xi_n(X)]^T$ is the fuzzy basis functions defined as followers as:

$$\xi(X) = \frac{(\prod_1^n \mu_{A_1^j}(Xi))}{\sum_{j=1}^{M}(\prod_1^n \mu_{A_1^j}(Xi))} \quad (16)$$

The new control low is obtained so:

$$\dot{U}(t) = -\frac{1}{K_1 g_1(X,t)}\left[G + \frac{q}{pK}\left(|\dot{S}_v|^{2-\frac{p}{q}} sign(\dot{S}_v) + |\dot{S}_v|^{1-\frac{p}{q}}(K_d sign(S) + \hat{\alpha}S)\right)\right] \quad (17)$$

By substituting (17) in (7), where $\ddot{S}_v = K_1 g_1(X,t)\dot{U}(t) + G$, we obtain:

$$\dot{S} = -K_d sign(S) + \hat{\alpha}S \quad (18)$$

Adding and subtracting (18) by αS, we obtain:

$$\dot{S} = (\alpha - \hat{\alpha})S - (K_d sign(S) + \alpha S) \quad (19)$$

Substituting (14) and (15) in (19) and taking $\hat{\theta} = \theta^* - \theta$ we obtain:

$$\dot{S} = \hat{\theta}\xi(X)S - (K_d sign(S) + \alpha S) \quad (20)$$

To guaranty the stability of our system, let's take a Lyapunov function defined as follower as [6]:

$$V = \frac{1}{2}S^2 + \frac{1}{2\upsilon}\hat{\theta}^T\hat{\theta} \quad (21)$$

The first derivative is giving as:

$$\dot{V} = S\dot{S} + \frac{1}{\upsilon}\hat{\theta}^T\dot{\hat{\theta}} \quad (22)$$

By substituting (20) in (22) and by posing $\dot{\hat{\theta}} = -\dot{\theta}$ and $\dot{\theta} = \upsilon\xi(X)S^2$ we find:

$$\dot{V} = -S(K_d sign(S) + \alpha S) = -(K_d|S| + \alpha S^2) \le 0 \quad (23)$$

So, the system is stable. This confirms the stability theory of Lyapunov.

4 Results Obtained

The main objective of this section, is to show and confirm the efficiency of our approach strategy by simulations based on MATLAB.

Firstly, we define the membership functions as followers [6]:

$$\begin{cases} \mu_{\alpha i^1}(X_i) = \exp\left(-\left(\frac{X_i - b_n}{2\sigma}\right)^2\right) \\ \mu_{\alpha i^2}(X_i) = \exp\left(-\left(\frac{X_i}{2\sigma}\right)^2\right) \\ \mu_{\alpha i^3}(X_i) = \exp\left(-\left(\frac{X_i - b_p}{2\sigma}\right)^2\right) \end{cases} \tag{24}$$

where: $b_n = -\frac{\pi}{6}, b_p = \frac{\pi}{6}$ and $\sigma = 5 \quad i = 1, 2, 3, 4$

Secondly, we consider one lane change maneuver as the desired trajectory for the lateral displacement, a quintic function defines as follower [4]:

$$X_{1d} = \mathrm{C}_5\mathrm{t}^5 + \mathrm{C}_4\mathrm{t}^4 + \mathrm{C}_3\mathrm{t}^3 + \mathrm{C}_2\mathrm{t}^2 + \mathrm{C}_1\mathrm{t} + \mathrm{C}_0 \tag{25}$$

To determine the coefficients, we consider these initial conditions:

$$\text{for the initial time } t_i = 0, \text{ we have } : X_{1d} = 0, \dot{X}_{1d} = 0 \text{ and } \ddot{X}_{1d} = 0$$

$$\text{for the final time tf} = 10\,\text{s}, \text{ we have } : X_{1d} = X_{1dmax} = -5, \dot{X}_{1d} = 0 \text{ and } \underset{1d}{X} = 0$$

After calculations, we obtain:

$$X_{1d} = -0.0003\mathrm{t}^5 + 0.0075\mathrm{t}^4 + 0.0500\mathrm{t}^3 \tag{26}$$

We get the following results (see figures), after simulations.

From the Fig. 2, we can observe that the trajectory of vehicle for the ASONTSMC-FL is driven towards the reference more quickly and precisely. The two trajectories are almost similar with an error that tends to zero. This is confirmed by Fig. 3 which shows the error between the output and the desired output.

Figure 4 and Fig. 5 prove the efficiency and the robustness of the algorithm developed for the control of our system. Figure 4 shows behavior of a certain known positive constant $\alpha = 1500$ and his approximation value $\hat{\alpha}$ using the fuzzy logic algorithm and we can see that $\hat{\alpha}$ tends to α. Figure 5 shows the control lows obtained by ASONTSMC and ASONTSMC using fuzzy logic method and we notice that they have merged which proves the efficiency of our approach.

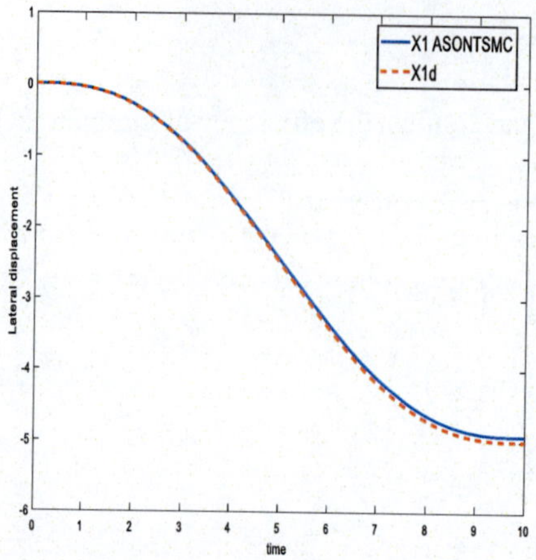

Fig. 2. The vehicle trajectory using ASONTSMC

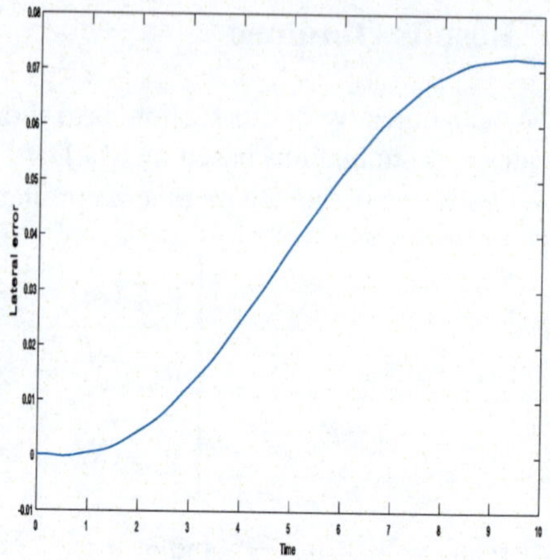

Fig. 3. The tracking error

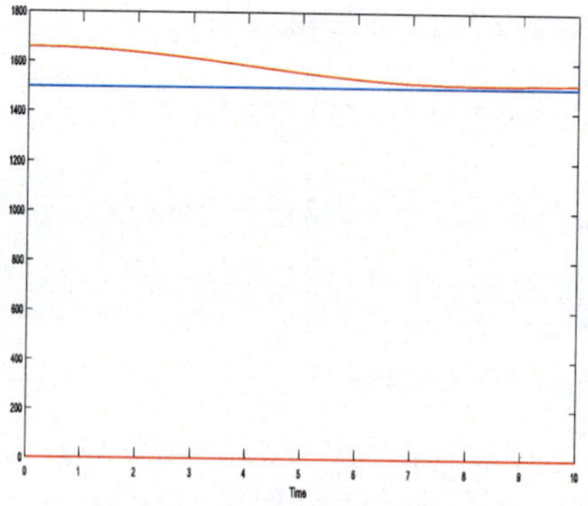

Fig. 4. The behavior of α and $\bar{\alpha}$

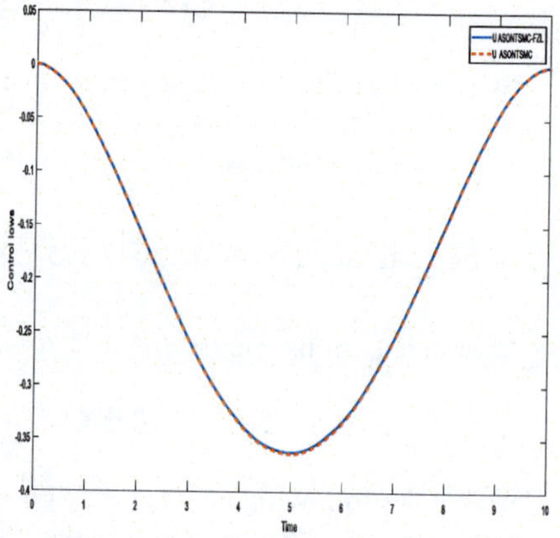

Fig. 5. The control lows generated by the 2 Controllers

5 Conclusion

For this work, we combined the TS- fuzzy logic algorithm with ASONTSMC for controlling an autonomous vehicle bicycle model and to estimate a gain. Results obtained have demonstrated the effectiveness of our proposed algorithm for the tracking trajectory but also the control low.

The efficiency and the robustness of our control strategy were proved by the simulation results. In the first time, we observed that the trajectory of vehicle for the ASONTSMC and ASONTSMC-FL was driven towards the reference more quickly and precisely. In addition, the approximation of the gain allowed us to degenerate a control law capable to stabilize the lateral dynamics of the vehicle and thus to pursue the desired trajectory.

We plan to design a control structure for our future work, based on Adaptive Fast Terminal Sliding Mode Control AFTSMC with a T-S Fuzzy Logic algorithm for modeling and controlling an autonomous vehicle.

References

1. Sabiha, A.D., Kamel, M.A., Said, E., Hussein, W.M.: ROS-based trajectory tracking control for autonomous tracked vehicle using optimized back steepingand sliding mode control. Rob. Autonom. Syst. J. (2022)
2. Fokam, G.T.: Commande et planification de trajectoires pour lanavigation de véhicules autonomes. Ph.D. thesis, Université de Haute Alsace Université de Technologie de Compiègne, , p. 11 (2014)
3. Yuan, J., Hui, J.: Adaptive second-ordernonsingular terminal sliding mode power levelcontrolfor nuclear power plants. Nuclear Eng. Technol. J. **54**, 1644–1651 (2021)
4. Hajjami, L.E., Mellouli, E.M., Baradar, M.: Robust adaptive non-singularfast terminal sliding-mode lateralcontrol for an uncertain ego vehicleat the lane-change maneuversubjected to abrupt change. Int. J. Dyn. Control **9**, 1–18 (2021)
5. Mellouli, E.M., Boumhidi, I.: Fuzzy tracking control design using sliding mode approach for nonlinear dynamic systems (2012)
6. Mellouli, E.M., Sefriti, S., Boumhidi, I.: Combined fuzzy logic and sliding mode approach's for modelling and control of the two link robot. In: IEEE International Conference on Complex Systems (ICCS), pp. 1–6 (2012)
7. Nagy, T.D., Ukhrenkov, N., Drexler, D.A., Takacs, A., Haidegger, T.: Enabling quantitative analysis of situation awareness: system architecture for autonomous vehicle handover studies. In: IEEE International Conference on Systems, Man and Cybernetics (SMC) Bari, Italy. 6–9 October 2019
8. Xiaohui, H., Junzhi, Z., Yuan, J., Weilong, L., Chengkun, H.: Autonomous drift controller for distributed drive electric with input coupling and uncertain disturbance. ISA Trans. **120**, 1–17 (2022)
9. Naoual, R., Mellouli, E.M., Boumhidi, I.: Adaptive fuzzy sliding mode control for the two-link robot. In: 9th International conference on Intelligent Systems: Theories et Applications (SITA 2014) (2014)
10. Naoual, R., Mellouli, E.M., Sefriti, S., Boumhidi, I.: Fuzzy sliding mode control for the two-link robot. Int. J. Syst. Control Commun. **6**, 84–96 (2014)
11. Mellouli, E.M., Chalh, Z., Alfidi, M.: A new robust adaptive fuzzy sliding mode controller for a variable speed wind turbine. Int. Rev. Autom. Control **8**(5), 338 (2015)
12. Mellouli, E.M., Naoual, R., Bouhmidi, I.: A new modified sliding mode controller based fuzzy logicc for a variable speed wind turbine. Int. J. Ecol. Dev. **32**(1), 44–53 (2017)
13. Hajjami, L.E., Mellouli, E.M., Berrada, M.: Neural network based sliding mode lateral control for autonomous vehicle. In: 1st International Conference on Innovative Research in Applied Science, Engineering and Technology (IRASET) (2020)
14. Rastelli, J.P., Penas, M.S.: Fuzzy logic steering control of autonomous vehicles inside roundabouts. Appl. Soft Comput. **35**, 662–669 (2015)
15. Hajyan, M., Nikoofard, A.: Fuzzy control of autonomous vehicle at non-signalized intersection in mixed traffic flow. In: IEEE Xplore, 9th Iranian Joint Congress on Fuzzy and Intelligent Systems (CFIS), April 2022
16. Taghavifar, H., Hu, C., Qin, Y., Wei, C.: EKF-neural network observer based type-2 fuzzy control of autonomous vehicles. IEEE Trans. Intell. Transp. Syst. **22**(8), 4788–4800 (2021)

Fuzzy Semantic Query Mapping and Processing

Salem Chakhar[1,2] and Zouhaier Brahmia[3](✉)

[1] Portsmouth Business School, Faculty of Business Law, University of Portsmouth, Portsmouth, UK
salem.chakhar@port.ac.uk
[2] Centre for Operational Research and Logistics, University of Portsmouth, Portsmouth, UK
[3] MIRACL Laboratory, Faculty of Economics and Management, University of Sfax, Sfax, Tunisia
zouhaier.brahmia@fsegs.rnu.tn

Abstract. Semantic data models are powerful conceptual modelling tools but they lack effective implementation mechanisms. Thus, fuzzy semantic data models are often mapped and implemented through relational or object-oriented database models. In a previous work, we have proposed a collection of mapping rules to map and implement the Fuzzy Semantic Model (FSM) as a Fuzzy Object-Relational database Model (FuzzORM). However, in such a work, only FSM schema mapping has been addressed. In this paper, we propose a middleware approach to map any FSM query into an equivalent query written in the object-relational query language SQL3.

Keywords: Fuzzy database · Fuzzy semantic model · Fuzzy object-relational database model · Query mapping · Query processing

1 Introduction

The Fuzzy Semantic Model (FSM) [1] is a semantic data model [2, 3]. It allows an entity (i.e., an instance) to be partially member of its class according to a given degree of membership that reflects the level to which the entity verifies the properties of this class. FSM (i) uses basic concepts of classification, association, generalization/specialization, composition, aggregation and grouping, which are commonly used in semantic modelling, and (ii) supports fuzziness, uncertainty and imprecision of real-world at attribute, entity, class and intra-/inter-classes relationships levels. Hence, it provides tools to formalize and conceptualize real-world within a manner adapted to its perception and representation by humans.

The semantic data models [2, 3] are powerful conceptual modelling tools but they lack effective implementation mechanisms. Thus, most of fuzzy semantic data models have been mapped and implemented through relational or object-oriented database models. However, the relational data models alone are not able to fully support all the concepts of semantic modelling. In addition, the object-oriented data models support several concepts of semantic modelling but their implementation is more complex than the relational ones. Thus, a collection of mapping rules has been proposed in [4] to map and

Y. Farhaoui et al. (Eds.): ICAISE 2022, LNNS 635, pp. 462–468, 2023.
https://doi.org/10.1007/978-3-031-26254-8_66

implement FSM as a Fuzzy Object-Relational database Model (FuzzORM). The Object-Relational DataBase Management System (ORDBMS) PostgreSQL has been used for the implementation of FuzzORM.

A first version of retrieve queries syntax for the FSM has been proposed in [1]. Such a version looks too complex and includes only a few types of retrieve queries. In addition, the work proposed in [4] addressed only the schema mapping with no support of query mapping or processing. Thus, based on these previous works, this research paper presents the conceptual architecture of FSM query mapping and processing tool. The latter relies on a middleware approach serving as dialogue system between the user and the host DBMS. The query mapping and processing tool allows the user to specify different data management and retrieve queries using FSM query language, into an equivalent query in the object-relational query language SQL3.

The rest of the paper is organized as follows. Section 2 overviews FSM and its retrieve query language. Section 3 presents the architecture of the proposed approach, along an illustrative query example. Section 4 concludes the paper.

2 Fuzzy Semantic Data Model

2.1 Principles of Fuzzy Semantic Data Model

Let E be the universe of discourse. A fuzzy entity e in E is a natural or artificial entity that one or several of its properties are fuzzy. At the extensional level, a fuzzy class K in E is a collection of fuzzy entities having some similar properties: $K = \{(e, \mu(e))$: $e \in E$ and $\mu(e) > 0\}$, where μ: $E \rightarrow [0, 1]$ is the membership function that maps the elements of E to the range [0, 1], and $\mu(e)$ represents the degree of membership (d.o.m) of the fuzzy entity e in class K. At the intensional level, a fuzzy class K is defined as a collection of attributes: $K = (Att_1, \ldots, Att_p, dom)$.

Each attribute FSM is basically characterized by its name, data type and domain. A data type may be exact (e.g. integer, char) or fuzzy. A set of fuzzy data types supported by FSM are detailed in [5]. The domain of an attribute is the set of values the attribute may take. Attributes may be single-valued, i.e., the attribute cannot have more than one value at a given time, or multi-valued, i.e., the attribute can have several values at a given time. In general, the values of a multi-valued attribute may be related with different logical connectors (e.g. AND, OR, XOR).

Each class is associated with one or several decision rules used to compute the DOM. Decision rules may be based on attributes or on common semantics. An attribute-based decision rule is a condition of the form <Attr> <op> <v>, where Attr is an attribute, op is a binary or a set operator; and v in D(Attr). Operator op may be crisp or fuzzy. A semantic decision rule is a semantic phrase used to specify the members of a fuzzy class. Semantic decision rules are mainly useful to define exact classes. The degree to which each decision rule determines fuzzy classes is not the same. To ensure this, each decision rule is associated with a non-negative weight reflecting its importance in deciding whether or not an entity e is a member of a given fuzzy class.

2.2 Fuzzy Semantic Data Model Query Language

The query language devoted to FSM-based databases has been proposed in [1]. This query language uses the concepts of perspective class and qualification and introduces thresholds in the FROM and WHERE clauses. The thresholds in the FROM clause correspond to the global degree of membership and may be mapped to the support of fuzziness at entity/class level. The thresholds in the WHERE clause correspond to the partial degree of membership and may be mapped to the support of fuzziness at the attribute level.

The notion of perspective class is simply defined as the class in which the user is primarily interested when formulating his/her query. It simplifies query formation and allows users with different interests to approach the database from points of view appropriate to their needs. The perspective class can be associated with an appropriate syntactic process, called qualification, allowing immediate attributes of other classes to be treated as if they were attributes of the perspective class. This process may be extended through the entity-valued attributes concept to the attributes related by more than one level of qualification. These attributes are called extended attributes. Apart from that, the notion of perspective class can be combined with generalization hierarchies to simplify query formation.

The generic syntax of a retrieve query is FSM is given in [1]. The following are two illustrative examples of data retrieval operations:

- Query 1: Retrieve the name and the type of supernova that have global d.o.m equal to or greater than 0.7 and have luminosity greater than 15 with partial d.o.m equal to or greater than 0.9:

```
FROM supernova WITH DOM ≥ 0.7
RETRIEVE SNovaName, TypeSNova
WHERE Luminosity > 15 WITH DOM  ≥ 0.9
```

- Query 2: Retrieve the name, the date of discovery and the discoverer of all supernovae which are not located in the milky-way galaxy with d.o.m not less than 0.5:

```
FROM supernova, discovery
RETRIEVE SNovaName, DiscoveryDate,
         PersonName OF Discoverer OF supernova
WHERE supernova.Location NOT IN (FROM galaxy
                                 RETRIEVE Location
                                 WHERE GalaxyName='milky-way')
WITH DOM  ≥ 0.5
```

3 Middleware Approach to Query Mapping and Processing

This section proposes a FuzzySeQL Sever devoted to translate user queries into a form compatible with the host DBMS, send the translated queries to the host DBMS, retrieve the results and present them to the user. FuzzySeQL is designed as a middleware between the user and the host DBMS.

The architecture of FuzzySeQL Server is given in Fig. 1. The FuzzySeQL Parser translates FSM queries into an algebra expression corresponding to the initial query. It assigns the different processing operations to the host DBMS and add transformation FuzzyToSQL operation indicating the final results should be transformed to FuzzySeQL Server. Then, for each algebraic operation identifies the best algorithm to computing the operation. The FSM2SQL component translates the operations that need to be executed in the DBMS into SQL. Then, it passes an execution-ready plan to the Execution Engine.

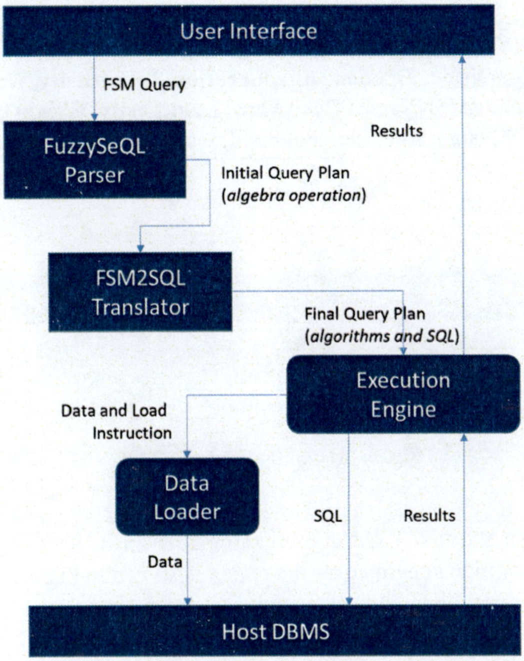

Fig. 1. Architecture of FuzzySeQL server

Before illustrating the working principles of FuzzySeQL Sever through a database example, we should mention that, for fuzzy query processing, we need to extend the binary and the set operators that may be used in the definition of the decision rules or the user query. The authors in [6] proposed an extension of a large set of operators to apply to fuzzy data. For illustration, we reproduce here the specification of FuzzyEQ (Fuzzy Equal) operator. The specifications of the other operators are available in [6].

The Fuzzy Equal operator models the equality concept for imprecise data values. Four different membership functions can be distinguished:

$\mu_{\simeq}(\tilde{x}, \tilde{y})$	y	$\tilde{y}$
x	1 if $x = y$; 0 otherwise	$= \sup_{(x,y)\in X\times Y} \min\{1, \pi_{\tilde{y}}(y)\}$
$\tilde{x}$	$\pi_{\tilde{x}}(y)$	$= \sup_{(x,y)\in X\times Y} \min\{p(x,y), \pi_{\tilde{x}}(x), \pi_{\tilde{y}}(y)\}$

where x and y are two crisp values, $\tilde{x}$ and $\tilde{y}$ are two fuzzy values, $p(x, y)$ is the proximity relation, and $\pi_{\tilde{x}}$ (y) and $\pi_{\tilde{y}}$ (x) are the possibility distributions on domains X and Y, respectively.

For illustration purposes, we consider the FSM database schema in Fig. 2, which is reproduced from [7]. The mapping of fuzzy class STAR according to the mapping rules proposed in [4] is given in Fig. 3.

GALAXY(<u>GalaxyName</u>, **Age, Location**);
STAR(<u>StarName</u>, *TypeStar*, **Age, Location, Luminosity, Weight**);
SUPERNOVA(<u>SNName</u>, *TypeSNova*, **Luminosity, Weight**);
PERSON(<u>Name</u>, **Age**, ResearchField);

Fig. 2. Database example

TypeStar	...	Age Value	| DataType	|ParametersList	...	DOM
NOVA	...	Young	Linguistic Label	(0.7,1.6,2.1,3)	...	0.75
SUPERNOVA	...	10.2	Real	()	...	1.0
SUPERNOVA	...	Very young	Interval	(0.01,0.85)	...	0.8

Fig. 3. Mapping of fuzzy STAR

Let first consider the first FSM query example given in Sect. 2.2 (Query 1). The principle of the query plan generated by the parser is shown in Fig. 4. This plan consists solely of algebraic operations and assigns all the processing to the DBMS. Algebraic operators in the initial plan include both regular and fuzzy operators. The fuzzy operators have their own algorithms for the middleware and need to be translated into regular SQL.

As advocated by the mapping rules designed in [4], the DOM of entities in FSM are mapped as columns in FuzzORM database model. This means the condition in FROM clause in Query 1 will be simply moved into the WHERE clause. The latter will be then composed of two conditions:

DOM $\geq$ 0.7 and

Luminosity > 15 WITH DOM $\geq$ 0.9

The first condition will be treated as in the classical SQL (since DOM is a crisp attribute). In the contrary, the second condition is fuzzy (since attribute Luminosity is fuzzy). Thus, a specific operation, FuzzyG (Fuzzy Greater), is used to compute the degree of satisfaction of Luminosity > 15. The result will be a value in the range [0, 1]. The next step consists in adding a temporary column called Luminosity_tmp that will contain the result of FuzzyG for all instance in the database. For this, the following two instructions should be executed:

```
ALTER TABLE STAR ADD Luminosity_tmp float;
UPDTAE STAR SET Luminosity_tmp = FuzzyEQ(Luminosity,15)
```

We note that the second instruction will be applied over all the instances in the relation STAR (since there is no restriction in this instruction).

The final SQL3 query that will be sent to the Execution Engine is as follows:

```
SELECT SNovaName, TypeSNova
FROM SUPERNOVA
WHERE DOM >= 0.7 and Luminosity_tmp > 0.9;
```

```
SELECT SNovaName, TypeSNova
FROM SUPERNOVA
WHERE DOM >= 0.7 and Luminosity_tmp > 0.9
↑
UPDATE SUPERNOVA SET Luminosity_tmp = FuzzyEQ(Luminosity,15)
                     ↑
ALTER TABLE SUPERNOVA ADD Luminosity_tmp float
↑
FROM SUPERNOVA WITH DOM ≥ 0.7
RETRIEVE SNovaName, TypeSNova
WHERE Luminosity > 15 WITH DOM  ≥ 0.9
```

Fig. 4. Principles of query plan

4 Conclusion

In this paper, we have proposed a middleware approach that allows to automatically map any fuzzy semantic query, specified under the FuzzySeQL query language, into an equivalent query written in the object-relational SQL3 language. In the near future, we will develop a tool that supports our approach. Such a tool will be used to experimentally evaluate the performances of our proposal. Besides, we are also planning to deal with FuzzySeQL query optimization.

References

1. Bouaziz, R., Chakhar, S., Mousseau, V., Ram, S., Temouldi, A.: Database design and querying within the fuzzy semantic model. Inf. Sci. **177**(21), 4598–4620 (2007)
2. Embley, D.W.: Semantic data model. In: Liu, L., Özsu, M.T. (eds.) Encyclopedia of Database Systems, 2nd edn. pp. 3391–3393. Springer, New York (2018). https://doi.org/10.1007/978-1-4614-8265-9_105
3. Ram, S.: Intelligent database design using unifying semantic model. Inf. Manag. **29**(4), 191–206 (1995)

4. Jandoubi, S., Bahri, A., Yacoubi-Ayadi, N., Chakhar, S., Labib, A.W.: Enhanced fuzzy object-relational database model for efficient implementation of the fuzzy semantic model. In: Proceedings of the 2015 IEEE International Conference on Fuzzy Systems (FUZZ-IEEE 2015), Istanbul, Turkey, 2–5 August 2015, pp. 1–8. IEEE (2015)
5. Jandoubi, S., Bahri, A., Yacoubi-Ayadi, N., Chakhar, S., Labib, A.W.: Imperfect information representation and implementation in fuzzy databases. In: Proceedings of the 2nd International Conference on Knowledge Management, Information and Knowledge Systems (KMIKS 2015), Hammamet, Tunisia, 16–18 April 2015, pp. 1–19 (2015)
6. Chakhar, S., Telmoudi, A.: Conceptual design and implementation of the fuzzy semantic model. In: Proceedings of 11th International Conference on Information Processing and Management of Uncertainty (IPMU 2006), Paris, France, 2–7 July 2006, pp. 2438–2445 (2006)
7. Jandoubi, S., Bahri, A., Yacoubi Ayadi, N., Chakhar, S., Labib, A.: Modelling, representation and implementation of imperfect information for an enhanced exploration of large databases. J. Decis. Syst. **26**(2), 152–169 (2017)

Geographic Information System for the Study of Water Resources in Chaâba El Hamra, Mohammedia (Morocco)

Jamal Mabrouki[1](✉), Mohammed Benchrifa[2], Mariem Ennouhi[1], Karima Azoulay[1], Imane Bencheikh[1], Toufik Rachiq[1], Khadija El-Moustaqim[1], Naif Al-Jadabi[1], Mourade Azrour[3], Abou-elaaz Fatima-zahra[4], and Souad El Hajjaji[1]

[1] Laboratory of Spectroscopy, Molecular Modeling, Materials, Nanomaterial, Water and Environment, CERNE2D, Mohammed V University in Rabat, Faculty of Science, Molecular Modeling Avenue Ibn Battouta, BP1014 Agdal, Rabat, Morocco
j.mabrouki@um5r.ac.ma

[2] Laboratory of Solar Energy and Environment, Mohammed V University in Rabat, Faculty of Sciences, B.P. 1014, Rabat, Morocco

[3] IDMS Team, Department of Computer, Faculty of Sciences and Technologies, Moulay Ismail University of Meknès, 52000 Errachidia, Morocco

[4] Scientific Institute, Geophysics, Natural Patrimony and Green Chemistry Research Centre (GEOPAC), Mohammed-V University in Rabat, Rabat, Morocco

Abstract. Adaptations in wetlands are exceptionally powerful and have been evaluated using a variety of information using a few strategies (examination of authentic maps, correlation of topographic maps, translation of photographs, etc.). The Chaâba El Hamra plain is deeply compromised by the water contamination identified with synthetic substances, waste water and high wastage discharges and the concentrated use of compost. It is useful to verify the nature of the water resources. This verification requires taking into account the guide for defence against groundwater contamination while relying on advanced information from the territory model and exogenous information, for example, terrain, and climate, soil and source information. Using the state of the hydrological equilibrium and Geographic Information Systems (GIS), we were able to plot the volumes of runoff and invasion in the watersheds of the Chaâba El Hamra plain and assess the effect of urban sprawl on the waters of the investigation area. The objective of this work is, on the one hand, to assess the impact of the release on the physico-chemical nature of the water in the wells of the investigation site and, again, to decide on the state of the water table in the area of the release. The results obtained showed that the groupings of the components considered (metals and contamination markers) fluctuate according to the surface area of the wells in the groundwater and that the groundwater was expelled to the discharge.

Keywords: Groundwater · Discharge · Mohammedia · GIS · Hydrogeology

Y. Farhaoui et al. (Eds.): ICAISE 2022, LNNS 635, pp. 469–474, 2023.
https://doi.org/10.1007/978-3-031-26254-8_67

1 Introduction

Following a general evolution on landfills in Morocco, in addition to the odors they emit, the juice of the landfills alter the quality of surface and underground waters due to their high pollution load (organic matter, minerals, and bacteriological pollution, indicated by the development of bacteria, viruses and algae, heavy metals, organochlorines, and inorganic salts) [1]. Groundwater assumes a key job in the solidness of provincial populaces in the BeniYakhlef area, Mohammedia, Morocco. They are misused by wells, springs and boreholes; depleted by different customary and present day procedures used to remove groundwater for drinking water supply and water system. The piezometric investigation of the shallow spring made it conceivable to determine the heading of groundwater stream that is by and large from east to west. It additionally shows an inventory of this sheet by the limestone massif, just as by the immediate penetration of water [2]. The delimitation of the watershed and the extraction of the hydrographic system just as the acknowledgment of the incline map by utilizing the DEM (advanced height models) pictures [3], had the option to decide the waste of surface water and their way in the bowl. The procedure utilized for this work has the disposition of recognizing the vertical weakness of a water table to a contamination originating from the outside of the ground, in the area of Mohammedia. In spite of this substance decent variety, water is commonly reasonable for human utilization, except for wells situated close to ranches. Additionally for water system [4–11], water presents a general decent quality.

2 Materials and Methods

The municipality of Chaâba El Hamra Ru-rale of BeniYakhlef, Mohammedia, Morocco located on the road P3313.ThE Chaâba El Hamra depends on the balanced stockpiling of strong waste so as to maintain a strategic distance from any danger of damage to human wellbeing and the earth. It is situated on a site of rather impermeable lithological nature, and intended to deal with five huge containers, which are burrowed as and when required, controlled, comprised of seepage system and assortment, permitting recuperating and treating the leachate shaped in the landfill re-specting the fixed qualities identifying with deliberation and water utilization [12–17].

In hydrogeology, and in the piezometric, the working tools are varied and more precise than the others. These tools allow an accurate and complete study based on measurements of latitudes, longitudes, altitudes and piezometric levels [18]. Piezometric level measurement is the main operation in the groundwater resource inventory. It corresponds to the altitude of the water level in natural balance (sounding or well). It is calculated by difference between the soil dimension (mark on the structure), the altitude Z and the depth of the water [19]. The static levels of each structure are measured using a piezometric probe. This consists of a probe attached to a conductive metric tape. When in contact with water, the probe emits an audible alarm, which allows us, thanks to the metric tape, to measure the height of the surface of the water table with respect to the surface of the ground, or a mark, whose height compared to the ground, is known. The equipment used to accomplish this field work is the Global Positioning System (GPS) for positioning the water points. The data processing is done by the software ARCGIS

10.3. Software Surfer allows to create Grids that will interpolate the irregular data of our points x, y, z so the ordinates.

3 Results and Discussion

The profundity of the water table in the investigation territory shifts from under 60 m in the nourishing territories close to the flaws, and in excess of 40 m in the inside, or significantly more than 65 m close to the shortcoming of Chaâba El Hamra. Underground stream is constrained by transverse and longitudinal shortcomings.

The piezometric bend shows numerous peculiarities, in particular despondency (elliptic bend with uniting flows) towards Oued N'fifikh that speaks to a zone of misfortune that implies that the last is encouraged by the water table. Notwithstanding the discouragements, there is the nearness of a large portion of a tenth of bulges (elliptic bend with dissimilar ebb and flow liners: vault piezometry) encompassing the release site comparing when all is said in done to huge feed locales from the surface (feed through precipitation water spills surface water by shortcoming). The guide introduced isopieze bend circular cone-molded discouraged with a variable dispersing that enables us to describe the Convergent Radial or Convex-type sheet whose concavity is situated downstream. For this situation, the stream segment diminishes toward stream. This sort of spring portrays the general seepage zones of the groundwater either superficially towards the conduit or at profundity.

As a component of this investigation, associates in groundwater estimations and inspecting were directed in December 2017. This is to get ready piezometric maps illustrative of the ebb and flow condition of the groundwater levels of the spring. It ought to be noticed that there is heterogeneity in the spatial dispersion of estimation focuses basically because of the topographical area of the various wells.

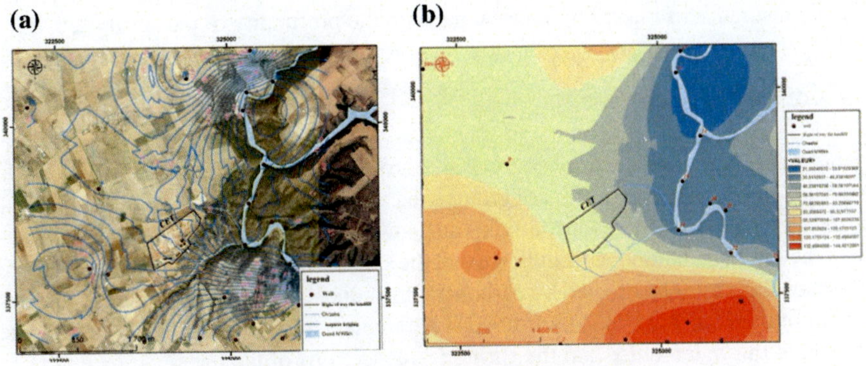

Fig. 1. **(a)** Piezometric map (b) piezometric surface of the aquifer wells

Using these different measurements, the piezometric water map was produced (Fig. 1 (a)). These maps allowed us to determine the overall direction of flow in the aquifer system. Figure 1 (b) shows the variety of the degree of water in the spring at wells

(P1, P2… P16) (circumstance of the wells). This leads us to reason that the variety of the piezometric surface of the water table in the area is reliant on the varia-tion of the precipitation and the parallel feedings of the groundwater. In this way, we see that the water level is high in progress, close to the bolstering regions and on everything beside flaws. The water level ascents with the arrival of precipitation and the heading of ground-water stream is toward the North.The stream of the web can be impacted by the geometry in tilted squares of its substratum, which causes the developing of the mio-pliocene downpour coat and the thickening of the spring.

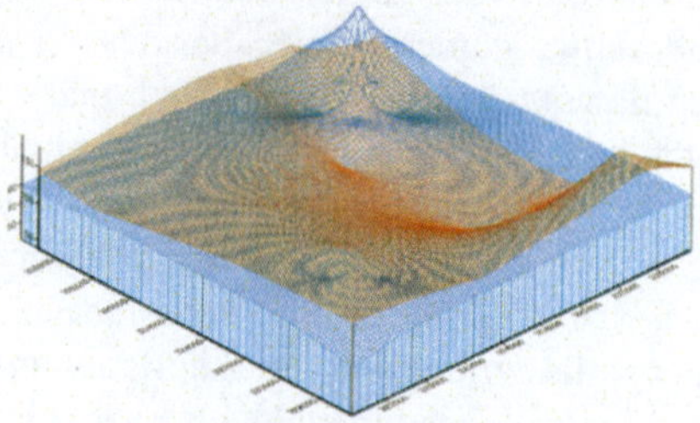

Fig. 2. 3D superposition of piezometric dimensions

Its understanding interprets a progression of the water table toward the north and toward the west. There are two watersheds in this seaside zone: the first in the southwest and the second in the east. The hydrographic system alludes to the topographical and physical qualities of the plain and waterways of a given area. The hydrographic system in the plain of Chaâba el Hamra is created, with a stream depleting this watershed, of which the greater parts are discontinuous streams. The N'fifikh aqueduct waterway is long. The profundity of the water table speaks to the vertical separation crossed by a contaminant at the outside of the ground to arrive at the water table. By and large, the potential insurance of the spring increments with the profundity of the spring [20–29].

4 Conclusion

Groundwater in fluvio-lacustrine and volcanic storages moves from SW to NE towards the ocean. This broad development of groundwater is guided by the degradation phenomena and structures that influence the site. The geographical setting and spatial appropriation of the synthetic components indicate that the synthesis of the groundwater substance of the source is strongly affected by carbonate decomposition, evaporitic striastic developments of the mud and by the hydrological parameters of the region; to be specific, the direction of the watercourse and the spring residence time of the level.

References

1. Mabrouki, J., Azrour, M., Farhaoui, Y., El Hajjaji, S.: Intelligent system for monitoring and detecting water quality. In: Farhaoui, Y. (ed.) Big Data and Networks Technologies, pp. 172–182. Springer, Cham (2020). https://doi.org/10.1007/978-3-030-23672-4_13
2. Azoulay, K., et al.: Adsorption mechanisms of azo dyes binary mixture onto different raw palm wastes. Int. J. Environ. Anal. Chem., 1–20 (2021)
3. Azrour, M., Mabrouki, J., Guezzaz, A., Kanwal, A.: Internet of things security: challenges and key issues. Secur. Commun. Netw. **2021**, 1–11 (2021)
4. Mabrouki, J., El Yadini, A., Bencheikh, I., Azoulay, K., Moufti, A., El Hajjaji, S.: Hydrogeological and hydrochemical study of underground waters of the tablecloth in the vicinity of the controlled city Dump Mohammedia (Morocco). In: Ezziyyani, M. (ed.) Advanced Intelligent Systems for Sustainable Development (AI2SD'2018), pp. 22–33. Springer, Cham (2019). https://doi.org/10.1007/978-3-030-11881-5_3
5. Farki, K., Zahour, G.: Contribution to the Understanding of the Sedimentary and Tectono-Volcanological Evolution of Oued Mellah (Coast Meseta, Morocco). Colloque International Conference of SIG Users, Taza GIS-Days, pp. 568–572 (2012)
6. Agence du Bassin Hydraulique du Bouregreg et de la Chaouia, Monographie des ressources en eau commune urbiane Bniyakhlaf (2011)
7. Hamid, W.: Contribution à l'étude tectono-volcanique et sédimentologique du bassin permien et triasique de l'Oued N'Fifikh (Meseta nord occidentale, Maroc). Ph. D. thesis, Université Hassan II Mohammedia. Casablanca (FSBM), p. 77 (2003)
8. Mabrouki, J., Bencheikh, I., Azoulay, K., Es-soufy, M., El Hajjaji, S.: Smart Monitoring System for the Long-Term Control of Aerobic Leachate Treatment: Dumping Case Mohammedia (Morocco). In: Farhaoui, Y. (ed.) Big Data and Networks Technologies, pp. 220–230. Springer, Cham (2020). https://doi.org/10.1007/978-3-030-23672-4_17
9. Azrour, M., Mabrouki, J., Fattah, G., Guezzaz, A., Aziz, F.: Machine learning algorithms for efficient water quality prediction. Model. Earth Syst. Environ. **8**(2), 2793–2801 (2022)
10. Zahour, G., Farki, K., Belkhattab, H.: Carrières de Mohammedia: impact environnemental et perspectives de réhabilitation. In: 4ème édition du colloque international: argiles et environnement, Oujda, pp. 7–8, 26–28 Novembre 2010
11. Mabrouki, J., Moufti, A., Bencheikh, I., Azoulay, K., El Hamdouni, Y., El Hajjaji, S.: Optimization of the coagulant flocculation process for treatment of leachate of the controlled discharge of the city Mohammedia (Morocco). In: Ezziyyani, M. (ed.) AI2SD 2019. LNEE, vol. 624, pp. 200–212. Springer, Cham (2020). https://doi.org/10.1007/978-3-030-36475-5_19
12. Bencheikh, I., et al.: The use and the performance of chemically treated artichoke leaves for textile industrial effluents treatment. Chem. Data Collections **31**, 100597 (2021)
13. Azrour, M., et al.: New efficient and secured authentication protocol for remote healthcare systems in cloud-IoT. Secur. Commun. Netw. **2021** (2021)
14. Bencheikh, I., et al.: The adsorptive removal of MB using chemically treated artichoke leaves: parametric, kinetic, isotherm and thermodynamic study. Sci. Afr. **9**, e00509 (2020)
15. Smida, H., Abdellaoui, C., Zairi, M., Ben Dhia, H.: Cartographie Des Zones Vulnérables A La Pollution Agricole Par La Méthode DRASTIC Couplée A Un Système D'information Géographique (SIG): Cas De La Nappe Phréatique De Chaffar (sud de Sfax, Tunisie). Science et changements planétaires / Sécheresse **21**(2), 131–146 (2010)
16. Mabrouki, J., et al.: Simulation of wastewater treatment processes with Bioreactor Membrane Reactor (MBR) treatment versus conventional the adsorbent layer-based filtration system (LAFS). Int. J. Environ. Anal. Chem., 1–11 (2020)

17. Mabrouki, J., Azrour, M., Hajjaji, S.E.: Use of internet of things for monitoring and evaluating water's quality: a comparative study. Int. J. Cloud Comput. **10**(5–6), 633–644 (2021)
18. Mabrouki, J., et al.: Study, simulation and modulation of solar thermal domestic hot water production systems. Model. Earth Syst. Environ. **8**(2), 2853–2862 (2022). https://doi.org/10.1007/s40808-021-01200-w
19. Abrouki, Y., et al.: Optimization and modeling of a fixed-bed biosorption of textile dye using agricultural biomass from the Moroccan Sahara. Desalin Water Treat **240**, 144–151 (2021)
20. Fattah, G., Ghrissi, F., Mabrouki, J., Kabriti, M.: Control of physicochemical parameters of spring waters near quarries exploiting limestone rock. In: E3S Web of Conferences, Vol. 234, p. 00018. EDP Sciences (2021)
21. Mabrouki, J., et al.: Smart system for monitoring and controlling of agricultural production by the IoT. In: Azrour, M., Irshad, A., Chaganti, R. (eds.) IoT and Smart Devices for Sustainable Environment, pp. 103–115, Springer, Cham (2022). https://doi.org/10.1007/978-3-030-90083-0_8
22. Rahmani, M., et al.: Adsorption of (methylene blue) onto natural oil shale: kinetics of adsorption, isotherm and thermodynamic studies. Int. J. Environ. Anal. Chem., 1–15 (2021)
23. Bencheikh, I., Mabrouki, J., Azoulay, K., Moufti, A., El Hajjaji, S.: Predictive analytics and optimization of wastewater treatment efficiency using statistic approach. In: Farhaoui, Y. (ed.) Big Data and Networks Technologies, pp. 310–319. Springer, Cham (2020). https://doi.org/10.1007/978-3-030-23672-4_22
24. Ghizlane, F., Mabrouki, J., Ghrissi, F., Azrour, M.: Proposal for a high-resolution particulate matter (PM10 and PM2. 5) capture system, comparable with hybrid system-based internet of things: case of quarries in the Western Rif, Morocco. Pollution **8**(1), 169–180 (2022)
25. Azrour, M., Mabrouki, J., Farhaoui, Y., Guezzaz, A.: Experimental evaluation of proposed algorithm for identifying abnormal messages in SIP network. In: Gherabi, N., Kacprzyk, J. (eds.) Intelligent Systems in Big Data, Semantic Web and Machine Learning, pp. 1–10. Springer, Cham (2021). https://doi.org/10.1007/978-3-030-72588-4_1
26. Fattah, G., Ghrissi, F., Mabrouki, J., Cherifi, H.: Impact of limestone deposit operation on the yield of surrounding agricultural land: case of the Western Rif, Morocco. In: E3S Web of Conferences, vol. 337, p. 02005. EDP Sciences, June 2022
27. Loukili, H., et al.: Combining multiple regression and principal component analysis to evaluate the effects of ambient air pollution on children's respiratory diseases. Int. J. Inf. Technol. **14**(3), 1305–1310 (2022)
28. Azrour, M., Mabrouki, J., Guezzaz, A., Farhaoui, Y.: Survey of high school students' usage of smartphone in Moroccan rural areas. Int. J. Cloud Comput. **10**(3), 265–274 (2021)
29. Azrour, M., Mabrouki, J., Farhaoui, Y., Guezzaz, A.: Security analysis of Nikooghadam et al.'s authentication protocol for cloud-IoT. In: Gherabi, N., Kacprzyk, J. (eds.) Intelligent Systems in Big Data, Semantic Web and Machine Learning, pp. 261–269. Springer, Cham (2021). https://doi.org/10.1007/978-3-030-72588-4_18

Gilbert Cell Down-Conversion Mixer for THz Wireless Communication

Abdeladim El Krouk[1](✉), Abdelhafid Es-Saqy[2], Mohammed Fattah[1], Said Mazer[2], Moulhime El Bekkali[2], and Mahmoud Mehdi[3]

[1] EST Moulay Ismail University, Meknes, Morocco
abdeladim.elkrouk@usmba.ac.ma
[2] AIDSES Laboratory, Sidi Mohamed Ben Abdellah University, Fez, Morocco
[3] Université Libanaise de Beyrouth, Beirut, Lebanon

Abstract. In this work we presented the design of a double balanced down-conversion frequency mixer in a receiving system using the 0.15 μm PH15NHF transistor of MMIC (Monolithic Micro-wave Integrated Circuit) technology in the Terahertz frequency band. This mixer based essentially on the Gilbert cell that achieves a conversion gain of 8.8 dB at the RF frequency of 140 GHz, with a low local oscillator power of 4 dBm, an average noise figure of about 8.3 dB and a 1 dB compression points of –12 dBm. We used two Baluns at the RF and LO gates to transform the two signals to differential signals.

Keywords: Conversion gain · Down-conversion mixer · Gilbert cell · 140 GHz

1 Introduction

Wireless communication systems in general and millimeter wave systems above 100 GHz and near 140 GHz in particular play a very important role in transmitting data at the highest possible rates [1]. This frequency band is of great importance, as it is located in one of the earth's atmospheric windows and situated between two peaks [2], one of oxygen absorption and the other of water absorption, which results in a low attenuation rate in the earth's atmosphere [4]. In addition, fundamental mode imaging and radar systems operating in the D band [110–170 GHz] are now readily available [5].

Therefore, in this work we proposed the design of a doubly balanced mixer based on a Gilbert cell with an RF frequency around 140 GHz in the architecture of the receiving chain as described in Fig. 1.

2 Circuit Designs

Figure 2 below shows the basic schematic of the Gilbert cell mixer.

This mixer allows to down-conversion from a 140 GHz RF frequency signal to a 1 GHz IF intermediate frequency, using the PH15NHF transistor of the UMS production line.

Y. Farhaoui et al. (Eds.): ICAISE 2022, LNNS 635, pp. 475–480, 2023.
https://doi.org/10.1007/978-3-031-26254-8_68

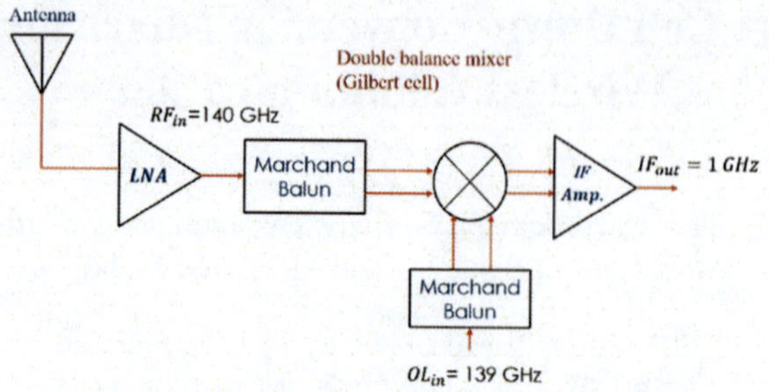

Fig. 1. Schematic diagram of the reception chain

All the gates (LO, RF and IF) of the "Gilbert cell" mixer are characterized by a differential topology. This characteristic prevents the transmission of LO and gives a good isolation between the gates [6].

In addition, the Gilbert mixer can have good conversion gain, good isolation between gates, and acceptable linearity [9].

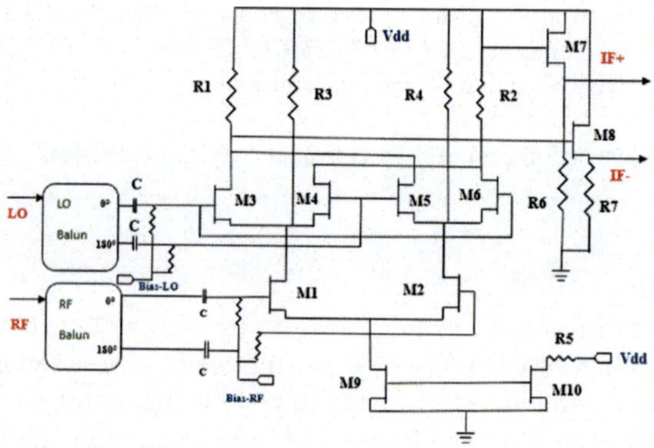

Fig. 2. Electrical diagram of the double balanced mixer (Gilbert cell).

The operation of this cell can be divided into three distinct stages RF, LO and IF.

The RF stage is an amplification stage constituting a differential pair formed by transistors (M1 and M2) whose role is to convert the differential voltage into a differential current.

The LO stage is a switching stage constituting two differential pairs formed by the transistors (M3, M4, M5 and M6), which makes it possible to carry out a multiplication between RF voltage and LO voltage.

The IF stage is an output stage composed of two transistors (M7, M8).

We used in this circuit the technique of charge injection (by adding two resistors R3 and R4). The role of this technique in this circuit is to inject current into the transistors of the RF stage and increase the conversion gain.

In the following, two Marchand baluns were used in the LO and RF ports to convert the input signals into differential signals, of equal amplitude and 180° phase difference.

A current mirror (transistors M9 and M10) is used to supply the current needed by our circuit.

Finally, to adapt the output of the mixer (IF+ and IF–) we add two transistors (M7 and M8) source trackers.

3 Simulations and Results

3.1 The Power of the Local Oscillator

The power of the local oscillator (LO) signal applied to the mixer must be carefully chosen to obtain the optimum value for the gain, compression point and noise figure.

Figure 3 shows the evolution in dB of the conversion gain as a function of the injected LO power in dBm, we choose the value of the power that allows to reach the highest gain (8.8 dB).

Figure 4 represents the evolution in dB of the noise factor as a function of the injected LO power. The value 4 dBm gives a minimum value of noise factor of about 8.3 dB.

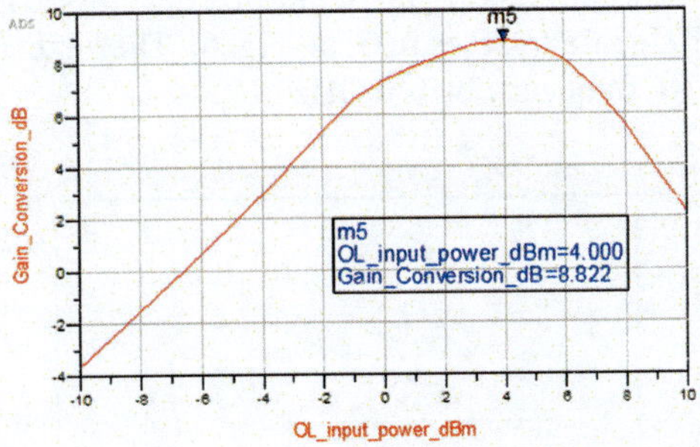

Fig. 3. Simulation of the conversion gain expressed in dB as a function of the OL power expressed in dBm.

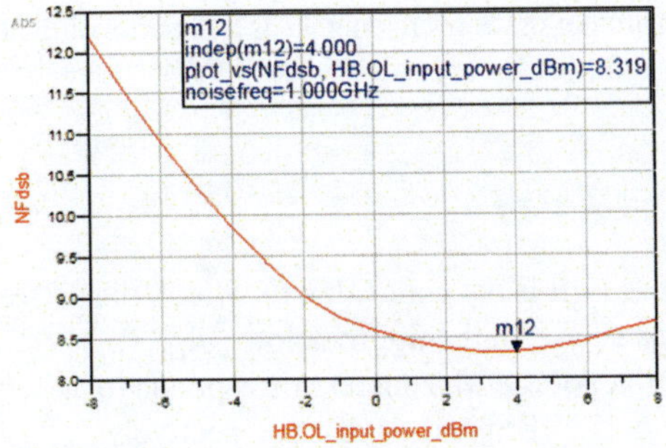

Fig. 4. Simulation of the DSB noise factor in dB as a function of the LO power in dBm.

Therefore, for all simulations, the optimal value of OL power and RF power are respectively set at 4 dBm and –30 dBm.

3.2 Conversion Gain

Figure 5 below shows the variation of the gain as a function of the RF power of the mixer. The gain reaches 8.8 dB for an RF power of –30 dB, at the RF frequency of 140 GHz.

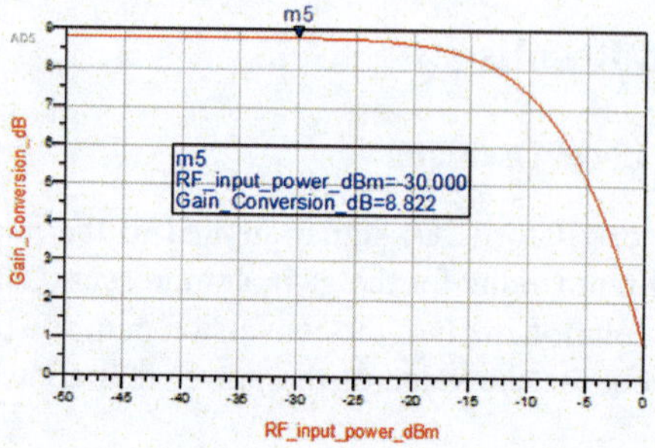

Fig. 5. The variation of the conversion gain as a function of the RF input power.

Then, we simulated the conversion gain of the mixer expressed in dB as a function of the frequency of the RF signal expressed in GHz (Fig. 6). The mixer reaches a conversion gain of 8.8 dB for the RF frequency of 140 GHz.

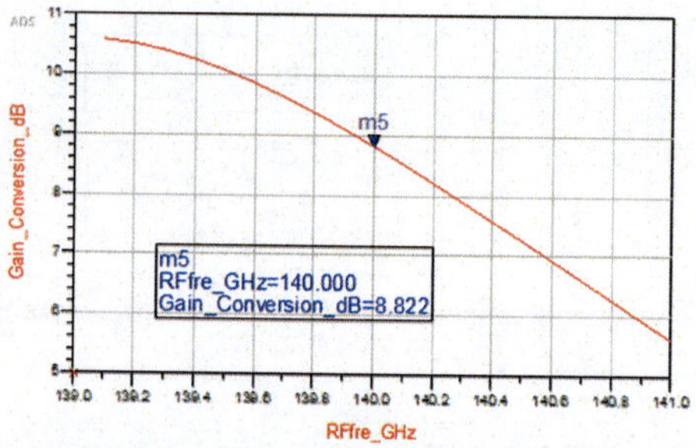

Fig. 6. The variation of the conversion gain as a function of the RF frequency.

3.3 Compression Point at 1 dB

The compression point at 1 dB is the gain value for which the conversion gain no longer follows its linear line, as shown in the figure below (Fig. 7).

The 1 dB compression point of this mixer corresponds to an RF input power and an IF output power equal to –12 dBm and –4 dBm respectively.

The summarized performance of the double-balanced mixer is shown in Table 1, with various published millimeter wave mixers. The proposed Gilbert mixer can achieve moderate to high conversion gain, and good isolation between the three ports, with competitive linearity.

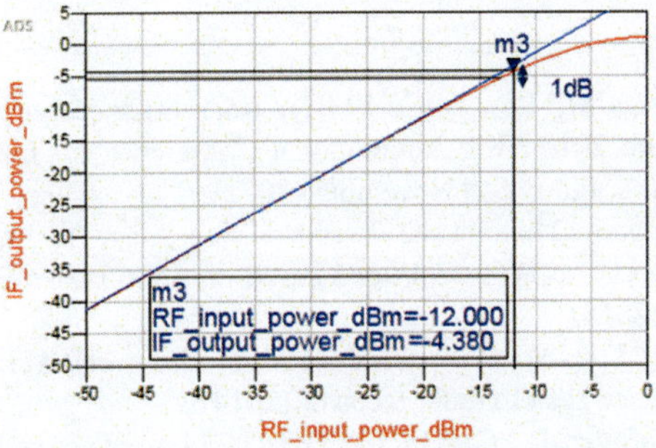

Fig. 7. The variation of the IF output power as a function of the RF input power in dBm.

Table 1. Performance comparison with recent mixers.

References	The work	[6]	[7]	[8]	[9]
process Technology	PH15 UMS	90-nm CMOS	90 nm CMOS	130 nm SiGe	130 nm SiGe
Circuit Topology	Double-Balanced Gilbert-Cell	Gilbert-cell	LC-tank oscillator	bottom-LO trans	Double Balanced Gilbert-Cell
RF fre. (GHz)	140	75 – 85	75–100	98–140	140 – 170
IF fre. (GHz)	1	0.1	NA	1–13	NA
CG (dB)	8.8	1.5	7.9	2.6	11
LO-RF iso. (dB)	68	49.2	44.7	45	NA
NF (dB)	8.3	23.3	15.4	NA	15
LO power (dBm)	4	5	0	5	NA
1-dB Compression point (dBm)	– 12	NA	NA	– 7.2	– 13

4 Conclusion

In this work, we presented a double balanced Gilbert mixer using the PH15NHF transistor from UMS foundry. This down-conversion mixer converts the RF signal from a 140 GHz frequency to a 1 GHz IF frequency using baluns in the RF and LO gates. It achieves an excellent CG of 8.8 dB with a low noise figure (NF) of 8.3 dB and has good isolation between the gates. In the future work, we will make some improvement of this mixer to reduce the noise factor and improve the conversion gain.

References

1. Eranna, U., Ravikumar, M.: Low power FMCW transceiver design with 140-ghz in CMOS 45 nm technology. In: 2020 IEEE International Conference for Innovation in Technology (INOCON), Bengaluru, India, 6–8 November (2020)
2. Daekeun, Y., Kiryong, S., Mehmet, K., Bernd, T., Jae-Sung, R.: An oscillator and a mixer for 140-GHz heterodyne receiver front-end based on SiGe HBT technology. J. Semicond. Technol. Sci. **15**(1) (2015)
3. Asad, J., Bashar, Z.: Low-Noise 24 GHz 0.15 μm GaAs pHEMT Gilbert cell mixer for intelligent transportation system radar receiver (2014)
4. Dong-Hyun, K., Jae-Sung, R.: CMOS 138 GHz low-power active mixer with branch-line coupler. Electron. Lett. **48**(10) (2012)
5. Ahmed, F., Furqan, M., Heinemann, B., Stelzer, A.: A SiGe-based broadband 140–170-GHz downconverter for high resolution FMCW radar applications. In: Proceedings IEEE MTT-S International Conference Microwaves for Intelligent Mobility, pp. 1–4, May 2016
6. Almohaimeed, H., Abdullah M.: Down-converter gilbert-cell mixer for WiMax applications using 0.15μm GaAs HEMT technology (2014)
7. Yo-Sheng, L., Chien-Chin, W., Jay-Ming, L.: 94 GHz down-conversion mixer with gain enhanced Gilbert cell in 90 nm CMOS, received. 2017LLC (2017)
8. Neda, S., Abdolreza, N., Sona, C., Zhongxia Simon He, M.B., Herbert, Z.: A 100–140 GHz SiGe-BiCMOS Sub-Harmonic Down-Converter Mixer. 978-2-87487-048-4 © (2017)
9. Yo-Sheng, L., Senior, M.: A W-band down-conversion mixer in 90 nm CMOS with excellent matching and port-to-port isolation for automotive radars. IEEE (2014)
10. Es-saqy, A., et al.: A 5G mm-wave compact voltage-controlled oscillator in 0.25 μm pHEMT technology. Int. J. Electric. Comput. Eng. (IJECE) **11**(2), 1036–1042 (2021)
11. Es-saqy, A., et al.: Very low phase noise voltage controlled oscillator for 5G mm-wave communication systems. In: 1st International Conference on Innovative Research in Applied Science, Engineering and Technology, IRASET (2020)
12. Younes, B., et al.: 5G uplink interference simulations, analysis and solutions: the case of Pico cells dense deployment. Int. J. Electric. Comput. Eng. (IJECE) **11**(3), 2245–2255 (2021)
13. Moutaib, M., et al.: Internet of things: energy consumption and data storage. Procedia Comput. Sci. **175**, 609–614 (2020)
14. Salah-Eddine, D., Imane, H., Mohammed, F., Younes, B., Said, M., Moulhime, EL.: Design of a microstrip antenna patch with a rectangular slot for 5G applications operating at 28 GHz. Journal of TELKOMNIKA Telecommunication Computing Electronics and Control. Vol. 20, No. 3, p. 527~536(2022)
15. Salah-Eddine, D., Imane, H., Mohammed, F., Younes, B., Said, M., Moulhime, EL.: Study and design of printed rectangular microstrip antenna arrays at an operating frequency of 27.5 GHz for 5G applications. J. Nano- Electron. Phys. **13**(6), 06035(5pp) (2021)

Heikin Ashi Candlesticks for Cryptocurrency Returns Clustering

Ahmed El Youssefi(✉), Abdelaaziz Hessane, Ahmad El Allaoui, Imad Zeroual, and Yousef Farhaoui

STI Laboratory, T-IDMS Faculty of Sciences and Techniques, University Moulay Ismail of Meknes, Errachidia, Morocco
{ah.elyoussefi,a.hessane,i.zeroual}@edu.umi.ac.ma, y.farhaoui@fste.umi.ac.ma

Abstract. Trading systems suggest several charting techniques allowing traders to better visualize the price movement of their assets. Heikin Ashi candlestick is one of the charting techniques that allows traders to visualize the price and trend movement of a cryptocurrency, which helps them in deciding the next move with their assets. Within this work, we used the KMeans clustering algorithm, to cluster Heikin Ashi candlesticks and logarithmic returns, to investigate the best number of classes to be used for logarithmic returns of a selected set of four cryptocurrencies retrieved from the Binance historical data center. We found that for the considered cryptocurrencies, the best k-values range from three to five. These results suggest the necessity of the clustering task before any classification task when dealing with cryptocurrency logarithmic returns forecasting, instead of an empirically predefined range of classes representing uptrend, downtrend, and no change in logarithmic returns.

Keywords: Heikin Ashi · Cryptocurrency · Trading · Clustering · KMeans

1 Introduction

1.1 Trading Cryptocurrencies

Traders use technical and fundamental analysis approaches to get feedback about the market dynamics and help them make decisions about their assets: holding (Hodling in the cryptocurrency terms [1]), buying or selling. While technical analysis uses historical trends of prices and volumes [2] and other regularly released information. Fundamental analysis relay on the information released about a blockchain project or cryptocurrency on the main media channels and on the press.

Trading systems offer the well-known Japanese Candlesticks visualization technique to help traders in visualizing the price movement of assets. Each Japanese candlestick represents the OHLC (open, high, low, close) data of a given period. Heikin Ashi candlesticks OHLC (OpenHA, HighHA, LowHA, CloseHA) data for a given period i, are

Y. Farhaoui et al. (Eds.): ICAISE 2022, LNNS 635, pp. 481–485, 2023.
https://doi.org/10.1007/978-3-031-26254-8_69

derived from the Japanese candlestick (For the same period) using the formulas: 1, 2, 3 and 4 below

$$OpenHA_i = \frac{1}{2}(Open_{i-1} + Close_{i-1}) \tag{1}$$

$$HighHA_i = \max(Open_i, Close_i, High_i) \tag{2}$$

$$LowHA_i = \min(Open_i, Close_i, Low_i) \tag{3}$$

$$CloseHA_i = \frac{1}{4}(Open_i + High_i + Low_i + Close_i) \tag{4}$$

Heikin ashi candlesticks give insights about the trend [3] of the market since it contains averaged data from previous periods.

1.2 Kmeans Clustering

The fast convergence and algorithm simplicity [4] made of KMeans clustering algorithm the goto algorithm for partitioning clustering, even though it requires specifying the number of clusters (k) in advance. To define these values, the elbow method could be used. Elbow method help in deciding the optimal number of clusters k [5] after which adding more clusters won't add sufficient information.

1.3 Logarithmic Returns

Symmetry and time additivity of logarithmic returns [6, 7, 8] encourages researchers to use them in cryptocurrency and stock returns instead of the simple returns. The formula for a logarithmic return of an asset i on a period p is:

$$\boldsymbol{Return_{i,p}} = \ln(\boldsymbol{Close_{i,p}}) - \ln(\boldsymbol{Close_{i,p-1}}) \tag{5}$$

Within this work we applied the KMeans clustering algorithm to a set of selected cryptocurrencies to define the clusters to be used for cryptocurrency logarithmic returns classification, using the Heikin Ashi technical indicator.

2 Data and Methods

2.1 Data Retrieval

We have selected empirically a list of four cryptocurrencies tradable on the USDT market within the Binance exchange. Table 1 gives more information about the set of selected pairs. The start date is different from one pair to another since the listing dates of the pairs within Binance exchange are different, and hence data availability dates are different.

Table 1. Information about selected trading pairs.

Cryptocurrency	Start date	End date
Bitcoin	August 2017	June 2022
Ethereum	August 2017	June 2022
Binance coin	November 2017	June 2022
Cardano	November 2019	June 2022

2.2 Data Preprocessing

After data retrieval, logarithmic returns and Heikin Ashi values have been calculated using the OHLC data. Standard normalization has been also applied to all the data columns.

2.3 Kmeans and Elbow Method

We used a range from 2 to 10 for the elbow technique to choose the best value of k. to plot the different inertia results for each value of k, we used yellowbrick [9] which is a machine learning visualization library offering different tools for data visual analysis.

3 Results

The results read that the optimal number of k varies from 3 to 5 for the four selected cryptocurrencies. The elbow plot for each selected cryptocurrency is shown in the Figs. 1, 2, 3 and 4 below.

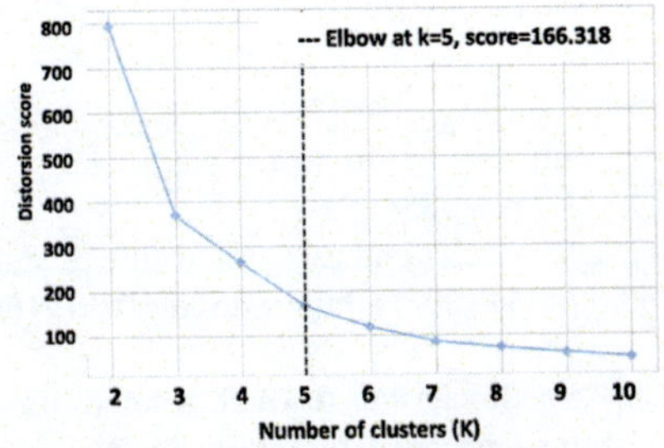

Fig. 1. Elbow method results for KMeans algorithm applied to the BTC/USDT pair

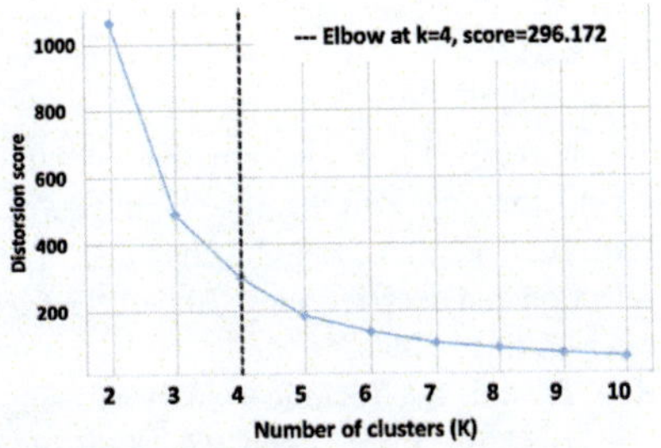

Fig. 2. Elbow method results for KMeans algorithm applied to the ETH/USDT pair

The obtained clustering results suggest that, instead of using empirically partitioned returns intervals, prior to any classification task. We need to consider clustering Heikin Ashi candlesticks to label the returns and to get more consistent classes.

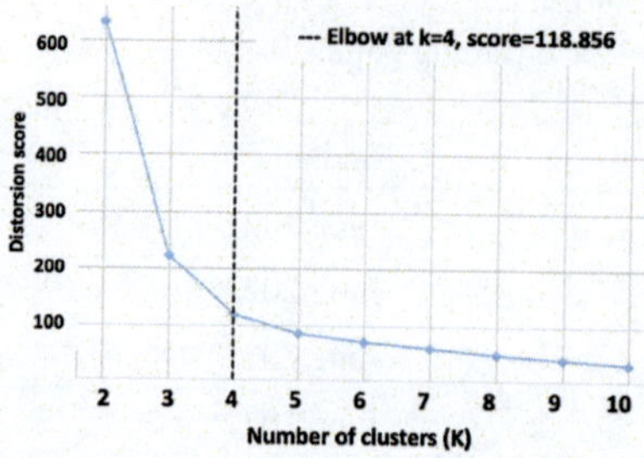

Fig. 3. Elbow method results for KMeans algorithm applied to the BNB/USDT pair

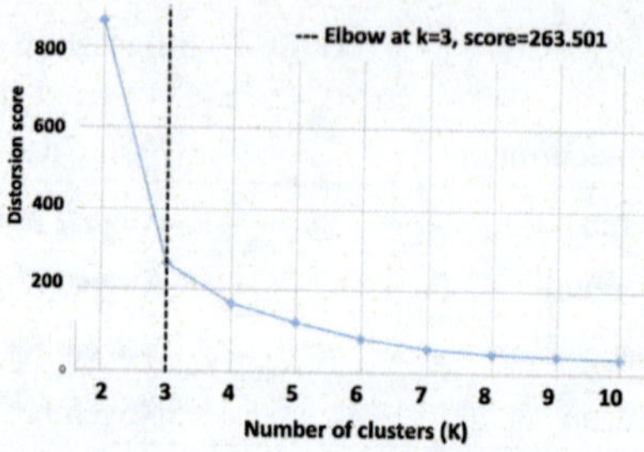

Fig. 4. Elbow method results for KMeans algorithm applied to the ADA/USDT pair

4 Conclusion and Perspectives

We conducted a clustering task of the heikin ashi candlestick of four cryptocurrencies prices data retrieved from Binance's historical data center. Our results suggest that we need to use three to five classes when we target a one-step logarithmic return of the considered cryptocurrencies.

The obtained results are limited to one Heikin Ashi candlestick, we need to consider conducting more research on a group of candlesticks to catch the sequence nature of the candlesticks.

We need also to compare the Heikin Ashi candlestick technique with other charting techniques which might give more insights into which one is the best when it comes to forecasting the price movement of an asset.

While we have opted for the KMeans algorithm and the elbow method for this research we need to test more clustering algorithms and more optimal k-value selection algorithms.

References

1. Trimborn, S., Li, Y.: Informative effects of expert sentiment on the return predictability of cryptocurrency. SSRN J. (2021). https://doi.org/10.2139/ssrn.3834279
2. Haq, A.U., Zeb, A., Lei, Z., Zhang, D.: Forecasting daily stock trend using multi-filter feature selection and deep learning. Expert Syst. Appl. **168**, 114444 (2021). https://doi.org/10.1016/j.eswa.2020.114444
3. Roy Trivedi, S.: Technical analysis using Heiken Ashi Stochastic: to catch a trend, use a HASTOC. Int. J. Fin. Econ. **27**, 1836–1847 (2022). https://doi.org/10.1002/ijfe.2245
4. Yuan, C., Yang, H.: Research on K-value selection method of K-means clustering algorithm. J. **2**, 226–235 (2019). https://doi.org/10.3390/j2020016
5. Madhulatha, T.S.: An overview on clustering methods (2012). https://doi.org/10.48550/ARXIV.1205.1117
6. Sung, S.-H., Kim, J.-M., Park, B.-K., Kim, S.: A Study on cryptocurrency log-return price prediction using multivariate time-series model. Axioms **11**, 448 (2022). https://doi.org/10.3390/axioms11090448

7. Yazdani, S., Hadizadeh, M., Fakoor, V.: Computational analysis of the behavior of stochastic volatility models with financial applications. J. Comput. Appl. Math. **411**, 114258 (2022). https://doi.org/10.1016/j.cam.2022.114258
8. Kim, J.-M., Cho, C., Jun, C.: Forecasting the price of the cryptocurrency using linear and nonlinear error correction model. JRFM **15**, 74 (2022). https://doi.org/10.3390/jrfm15020074
9. Bengfort, B., Bilbro, R.: Yellowbrick: visualizing the Scikit-learn model selection process. JOSS **4**, 1075 (2019). https://doi.org/10.21105/joss.01075

Health Surveillance and Management System Using WBSNs

Mohammed Moutaib[1](✉), Tarik Ahajjam[2], Mohammed Fattah[1], Yousef Farhaoui[2], Badraddine Aghoutane[3], and Moulhime El Bekkali[4]

[1] IMAGE Laboratory, Moulay Ismail University, Meknes, Morocco
m.moutaib@edu.umi.ac.ma
[2] L-STI, T-IDMS, University of Moulay Ismail, Faculty of Science and Technics, Errachidia, Morocco
[3] IA Laboratory, Moulay Ismail University, Meknes, Morocco
[4] IASSE Laboratory, Sidi Mohamed Ben Abdellah University, Fez, Morocco

Abstract. Wireless Body Sensor networks or WBSN are one of the technologies that have a great impact on human development. This technology is part of the IoT family, which today represents a major part of our daily lives. WBSNs have improved the medical field in such a way that each patient is monitored via a group of lightweight sensors. On the one hand, they allow the collection of data from the human body. On the other hand, they transmit them via the network with a specific transmission method, thus analyzing them in order to make decisions.

In this paper, a patient health monitoring and management system has been proposed with multiple master nodes based on cluster-tree topology, to avoid data retransmission and collision domains. First, a complete study with the architectural design is started in the first part. Second, an architecture implementation was performed in the human body to demonstrate the effectiveness of the solution. Finally, the result and the visualization of the data will be started in this step, which summarizes our work.

Keywords: WBSN · Energy consumption · IoT · Cloud computing · ECG · Data storage

1 Introduction

Today, the IoT or Internet of Things is one of the most efficient and powerful communication technologies of the 21st century [1]. In the IoT architecture, all objects are part of the Internet due to their communication and computing power. The Internet of Things is an important part of our daily life. Many smart objects in our daily life are connected and communicate with each other [2, 3]. This technology aims to achieve a dimension between the real world and the virtual world.

The Internet of Things encompasses all current areas such as smart cities, agriculture, connected vehicles, the health sector, which is our focus of study.

The effectiveness of the Internet of Things in the medical field relies on the fact that its objects analyze, communicate and process data autonomously [4]. This technology

Y. Farhaoui et al. (Eds.): ICAISE 2022, LNNS 635, pp. 486–491, 2023.
https://doi.org/10.1007/978-3-031-26254-8_70

is represented by a set of devices with their own identity and increasingly sophisticated computing and communication capabilities. IoT objects embed a very large number of actuators and sensors that will measure the environment and that can act on the physical world and the virtual world. However, the IoT poses several challenges in the medical field, with its great advantage, its dynamic nature, its data and these heterogeneous systems, which constitute powerful devices and various other advantages like the mobility of the sensors in the body, the power. [5]. These characteristics require methods to achieve a solution, in order to extract data from several data sources and store them in databases for analyzes.

Alternatively, the connectivity between these connected objects requires a lot of energy [6, 7].

These problems related to energy consumption considerably slow down the evolution and rapid deployment of this high technology [8]. The main causes of energy waste in body sensor networks are collisions and the retransmission process. However, the main cause of the collision occurred when two sensors or two nodes in network try to transmit data at exactly the same time and frequency, and the main cause of the retransmission process occurred when the collision occurs or data is not received correctly due to channel fading. These characteristics require appropriate tools and methods for the realization of applications capable of (i) avoiding collision domains and (ii) interacting with data retransmission as well.

The main objective of our work is to design a general architecture to collect data from a network of wireless sensors with a minimum of energy consumption. More specifically, our work aims to:

- Analyze the problem of the classic architecture of wireless sensor networks, as well as other architectures already proposed for IoTs in the medical field specifically, to extract the constraints to which a new architecture will have to submit to collect data.
- Propose a new architecture aimed at overcoming the drawbacks noted, and meeting the constraints that we have set.
- Search for solutions to make an implementation possible on sensors whose performance is reduced.
- Carry out an implementation of the data collection architecture in order not only to validate the feasibility of the proposed architecture, but also to test briefly the conditions of use of the architecture.

In this context, we propose a data flow management architecture. Our solution is composed of several parts ranging from the introduction to the result.

2 Literature Review

So far, several researches have been carried out on the energy consumption of connected objects. Among these researches [9] studied the cost function for body sensor networks. They described a hybrid resource block allocation control set which is subsequently derived and used to analyze the optimal power allocation on resource blocks. A solution has been proposed by [10] through a cognitive cooperative communication with two master nodes, his proposal is based on two main axes collision and retransmission of data which are the main causes of energy waste in body networks. They proposed a new WBSN network architecture, a dual master-slave network architecture. One of the master nodes is fixed on the human body. The other master node functions as a monitoring node and can receive data sent by the sensor. [11] proposed a distributed data flow management system for the Internet of Things, he brought a solution for access to data flows in the form of services and for the automatic deployment of continuous processing according to the device characteristics.

An application approach proposed by [12] with an IoT flow management application. This solution aims to demonstrate the usefulness of a centralized topology, by implementing this solution in a real case. A system of portable sensors is proposed by [13] to measure various physiological parameters such as temperature. The data collected from the sensors will be transmitted to a gateway server via a Bluetooth connection. The data received by the server will be transformed into an observation file and stored on a remote server in order to retrieve them later.A heuristic algorithm for programming several independent segments aimed at improving the degree of exchange and parallel transmission has been proposed by [14]. Cross-type which makes energy efficient and minimizes delays in WSN networks has been proposed [15].

3 System Architecture

Our solution aims to implant several sensors in the human body. These will provide data according to each organ to monitor the patient's condition. Each node is equipped with a processor and a sensor to collect data.

Figure 1 illustrates the system architecture for our health surveillance system, the main components of which we then describe:

Several wearable sensors that measure physiological biomarkers, such as ECG, temperature, respiratory rate, EMG muscle activity, perform data acquisition (Step1). The sensors connect to the network through an intermediate data aggregator or concentrator (Step2), which is usually a smartphone located near the patient.

The data transmission components of the system are responsible for transmitting patient data from the patient's home to servers that centralize the collected data (Step3) with assured security and confidentiality, ideally in real time. Typically, this collected data will be analyzed to give the patient's condition to the doctor and triggered an alert in case of an emergency patient condition.

3.1 Architecture Design

We started by designing our network of body sensors (Step1) based on several studies. Our solution was created with the tree topology (cluster-tree) which partitioned the nodes into groups called "clusters". A cluster consists of a particular node called the cluster head and other nodes. The latter only communicate with their cluster head. The latter is then in charge of forwarding the messages received to the base which will collect the information from all the nodes hierarchically. Finally send them to the network sink (access point) to be viewed via the network (Fig. 2).

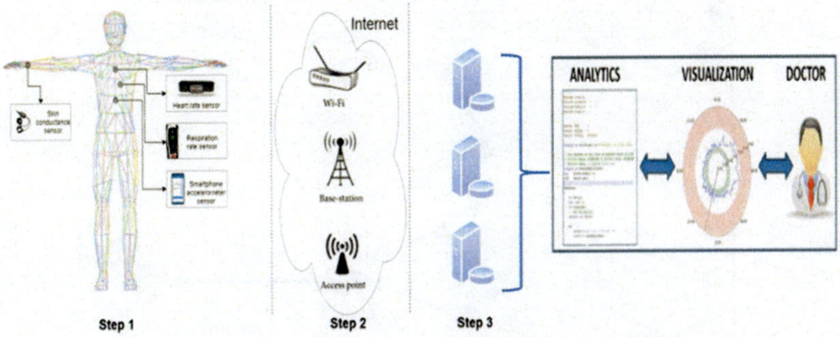

Fig. 1. Remote patient monitoring system architecture.

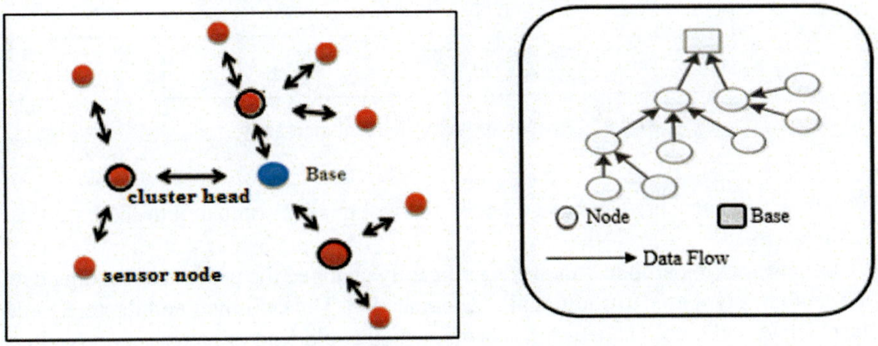

Fig. 2. Cluster tree network topology.

Our architecture is then prioritized according to the role of the network elements (the base, cluster heads, nodes). There may be a change in hierarchy between nodes and cluster heads depending on the energy available in each node. This balances the difference in energy levels of all nodes to avoid the disappearance or isolation of a node and better extend the life of the network. By adopting the cluster-tree topology, each node in the network is often provided with a device to estimate the battery level of the node.

3.2 Implementation

We implanted our solution in the human body (Fig. 3) and explained as follows:

In the architecture of Fig. 3, multiple sensors are evenly distributed over the human body for data collection.

Each of the sensors collects a set of information such as ECG, EEG signals and transfers them to the base. The Nodes are fixed on the human body.

The base is placed outside the human body and has the role of collecting data. Thus, it has a network-monitoring role. The distances between the nodes can also cause a major problem of data retransmission, because one can find nodes located at great distance or at short distance caused by the mobility of the human body.

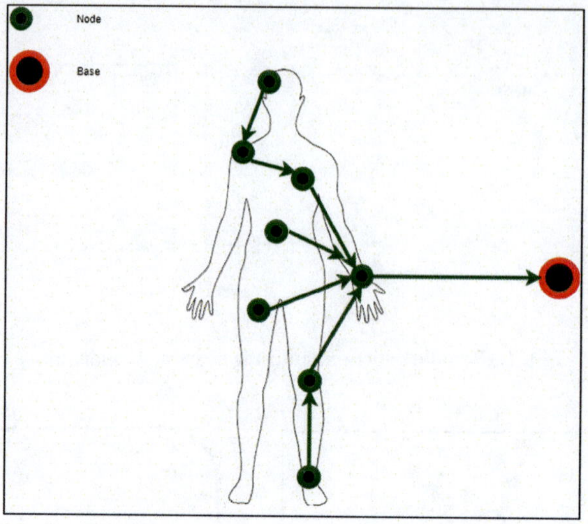

Fig. 3. Location of nodes in the human body.

Thus, several sensors can have a better channel quality compared to the base than the others can.

Our new network architecture has significantly changed the process of sending data to the database compared to traditional communication. Thus, this new architecture could reduce delays and conflicts between sensors, which could lead to better energy savings and efficient data collection.

4 Conclusion

In this paper, a novel architecture solution is proposed for patient health monitoring in the medical field. The proposed algorithm is based on cluster-tree topology for data collection with distributed method. The objective of the proposed solution can be summarized as follows: first, to reduce the process of sending data between the database and the nodes by using a distributed architecture (collection stage). Second, transmit the collected data through the network with a specific transmission method, and third, analyze the transmitted data and make decisions. The proposed solution has made it possible to obtain a better quality of transmission of critical data. In addition, it could significantly reduce the probability of losing the lives of patients caused by the variation of ECG signals and controlled patient condition. In this work we started by describing the first part of data collection, in the next works we will start the next two parts.

References

1. Weiser, M.: The computer for the 21st century. Sci. Am. **256**(3) (1991)
2. Gope, P., Hwang, T.: Untraceable sensor movement in distributed IoT infrastruc-ture. IEEE Sens. J. **15**(9), 53405348 (2015)

3. Boumaiz, M.: and all "Energy harvesting based WBANs: EH optimization methods." Procedia Comput. Sci. **151**, 1040–1045 (2019). https://doi.org/10.1016/j.procs.2019.04.147
4. Moutaib, M., Fattah, M., Farhaoui, Y.: Internet of things: energy consumption and data storage. Procedia Comput. Sci. **175**, 609–614 (2020)
5. Tarik, A., Farhaoui, Y.: Recommender system for orientation student BDNT 2019: big data and networks technologies, pp 367–370 (2019)
6. Moutaib, M., Ahajjam, T., Fattah, M., Farhaoui, Y., Aghoutane, B., el Bekkali, M.: Optimization of the energy consumption of connected objects. Int. J. Interact. Mob. Technol. (iJIM) **15**(24), 176–190 (2021). https://doi.org/10.3991/ijim.v15i24.26985
7. Moutaib, M., Ahajjam, T., Fattah, M., Farhaoui, Y., Aghoutane, B., el Bekkali, M.: Reduce the energy consumption of IOTs in the medical field. Digit. Technol. Appl., 259-268 (2022).https://doi.org/10.1007/978-3-031-02447-4_27
8. Moutaib, M., Ahajjam, T., Fattah, M., Farhaoui, Y., Aghoutane, B.: Reduce the energy consumption of connected objects. In: Proceedings of the 2nd International Conference on Big Data, Modelling and Machine Learning (2021). https://doi.org/10.5220/0010728900003101
9. Nazir, M., Sabah, A.: Cooperative cognitive WBSN: from game theory to population dynamism. In: 2011 3rd International Congress on Ultra Modern Telecommunications and Control Systems and Workshops (ICUMT), pp. 1–6, Singapore, Singapore (2011)
10. Alkhayyat, A., Thabit, A.A., Al-Mayali, F.A., Abbasi, Q.H.: WBSN in IoT health-based application: toward delay and energy consumption minimization. J. Sens. **2019**, 1–14 (2019). https://doi.org/10.1155/2019/2508452
11. Billet, B.: Système de gestion de flux pour l'Internet des objets intelligents. Calcul parallèle, distribué ET partagé [cs.DC]. Université de Versailles-Saint Quentin en Yvelines (2015)
12. Mottola, L., Picco, G.P.: Programming wireless sensor networks: fundamental concepts and state of the art. ACM Comput. Surv. **43**(3) (2011)
13. Babu, S., Chandini, M., Lavanya, P., Ganapathy, K., Vaidehi, V.: Cloud-enabled remote health monitoring system. In: International Conference on Recent Trends in Information Technology (ICRTIT), July 2013, pp. 702–707 (2013)
14. Gandham, S., Zhang, Y., Huang, Q.: Distributed time-optimal scheduling for convergecast in wireless sensor networks. Comput. Netw. **52**(3), 610–629 (2008)
15. Shi, L., Fapojuwo, A.O.: TDMA scheduling with optimized energy efficiency and minimum delay in clustered wireless sensor networks. IEEE Trans. Mob. Comput. **9**(7), 927–940 (2010)

How to Define a Functional Charges Copybook for an Agricultural Robot?

Najia Ait Hammou[1](✉), Hajar Mousannif[1], and Brahim Lakssir[2]

[1] Computer Science Department, Faculty of Sciences Semlalia, Cadi Ayyad University, Marrakech, Morocco
najia.aithammou@ced.uca.ma, mousannif@uca.ac.ma
[2] Digitalization and Microelectronics Smart Devices Department, MAScIR, Rabat, Morocco
b.lakssir@mascir.ma

Abstract. Weeds are undesirable plants that grow in the field. Their development is variable, hence the need to use means of struggle since they are uncontrollable. The difficulty of the weeding task has pushed farmers to think about its automation and robotization. It is a question of setting up autonomous robots capable of crossing the fields and removing weeds in a reduced time. Hence the need to carry out a functional analysis for the development of specifications dedicated to these agricultural robots. The functional analysis allows not only to determine the different behaviors of the robot in the field but also the essential equipment used. So, this study enable to define the objective behind designing an agriculture robot. This starts with the analysis of the need and ends with the tools needed to perform the weeding task. The purpose of this article is to help each designer define the basics of functional analysis first, so that he can draw his different diagrams, namely the Horned Beast, the Octopus, and the FAST (function analysis system technique).

Keywords: Agricultural robot · Weeds · Weeding · Functional analysis · Horned beast · Octopus · FAST

1 Introduction

In the management and production of agriculture, robotics is becoming increasingly important. In order to run farms effectively, agriculture needs time-saving and autonomous technologies. The current trend in agriculture is toward automation to increase production through the use of equipment and technology [1].

This automation calls for hard work in the design of the product in question, namely the study of its functional analysis to identify its utility a priori. In fact, functional analysis is a key tool in the design process that helps to define the architecture of new ideas and explore their potential. It is done to improve the functional specifications of the new product, to match its functions to the physical components, to make sure all required components are mentioned, and then establish links between the components of the new product [2]. The system life cycle's concept definition phase initiates in a

Y. Farhaoui et al. (Eds.): ICAISE 2022, LNNS 635, pp. 492–497, 2023.
https://doi.org/10.1007/978-3-031-26254-8_71

serious effort to define the physical and functional specifications of a new system (or a significant upgrade to an existing system) in order to satisfy operational needs and predetermined the essentials requirements. It is a commitment to adequately describing the system in order to make quantitative predictions about a number of factors, including its operational effectiveness, development time, and life cycle cost. Following this stage, system development entails translating the system concept into hardware and software, as well as developing it for operational and production use [3].

This paper begins with the definition of the role of functional analysis before moving on to a bibliographical study of all its diagrams, namely the horned beast diagram, octopus diagram, and FAST diagram. Then we try to apply these diagrams to the design of an agricultural robot. Finally, we use the results to summarize and point to other perspectives.

2 Related Works

2.1 What is the Role of Function Analysis?

Function analysis is a method for determining and analyzing the requirements of a project, product, or service. Function analysis encourages the solution of innovative problems by focusing attention on the necessary performance or need rather than the anticipated answer. It is a method for describing a complex system's operation as a function. The intention is to define the original function using more understandable functions. Intuitive research (brainstorming), environmental analysis, sequential analysis (process mapping), and movement and effort analysis are just a few of the methods utilized to pinpoint functions [4].

So, designers must follow the following steps during the functional analysis process as presented in chronological order:

- The needs analysis identifies the needs expressed in the horned beast diagram.
- The functional analysis of the need helps to identify the relationships between the product and its context of use in order to release service functions that are able to satisfy the need. It is presented in an octopus diagram.
- The technical functional analysis determines the technical functions necessary for the service functions and guides designers in the search for technical solutions. It is presented in the Function Analysis System Technique (FAST) diagram.

2.2 Horned Beast Diagram

A horned-beast diagram is an instrument for performing a functional analysis of a need. This graphic demonstrates whether the product is practical for the user and whether it satisfies his needs. It is the APTE (Application to Business Techniques) method's initial step. The reason for the diagram's nickname, Horned Beast, is that the line connecting the two upper bubbles resembles a bull's horn (see Fig. 1).

This diagram demonstrates the value of the new product and how it satisfies user needs. Three questions must be resolved by this diagram. Who does it serve? What does it accomplish? For what reason? [5].

2.3 Octopus Diagram

The connection between a good or service and its surroundings is depicted by an octopus diagram. It is a technique for analysis that is employed within the APTE methodology (Application to Business Techniques). The octopus diagram enables us to visualize a product's service functions, identify its primary and secondary functions, and show how each of these functions interacts with its environment. This diagram is a useful technique for displaying functional analysis graphically. It enables us to visually enhance and simplify a portion of the specs. The elements of the external environment are positioned around the product or service, which is at the center of the figure. According to their relationship, lines connect the various pieces (see Fig. 2).

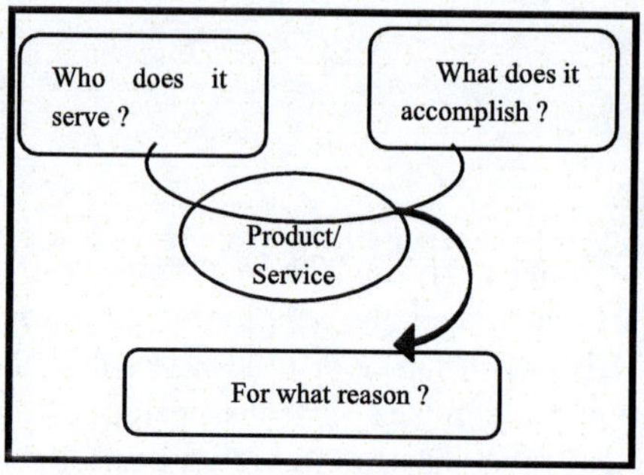

Fig. 1. Structure of a horned beast diagram.

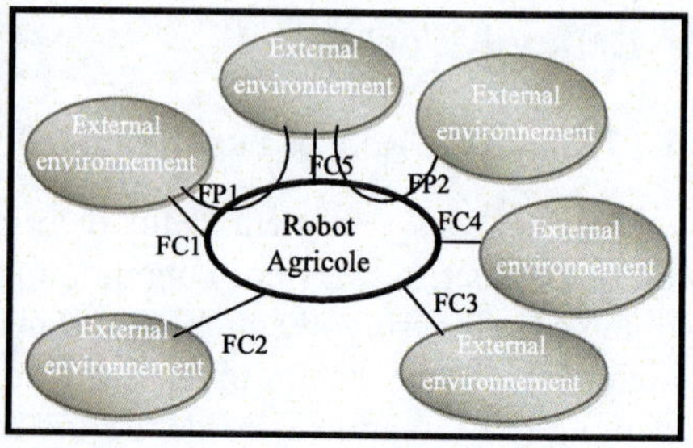

Fig. 2. Structure of an octopus diagram.

It is the second tool in the APTE method, which shows the functional specifications. It establishes the product's or service's value to its customers. In order to accomplish this, we must identify the many functions of the product and ascertain their limits as well as how they respond to their surroundings. In connection to the product and the surrounding environment, this figure illustrates the many functions. These two functions—the primary functions (FP) and the constrained functions (FC)—are required to create this diagram [6].

2.4 FAST Diagram

The Function Analysis System Technique (FAST) diagram is a graphic illustration of how functions are connected and why they work together in a system (product or process) to provide the desired goods or services [7].

A FAST Diagram answers three important questions:

How should this function be carried?
Why should this function be performed?
When does this function need to be accomplished? This question allows to identify the functions that occur at the same time (see Fig. 3) [8].

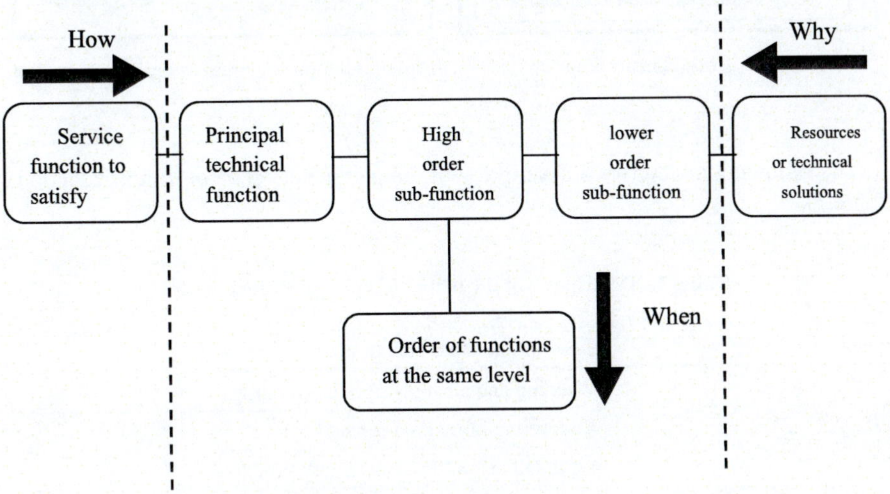

Fig. 3. Structure of a FAST diagram.

3 Results and Discussion

3.1 Horned Beast Diagram

We start the functional analysis with the analysis of the need. This last one allows us to obtain, as a result, the specifications of the need. The goal is to design an agricultural robot. The following diagram illustrates the needs of the product in the Horned Beast diagram (see Fig. 4).

As shown in the [9] for the walking device. We find in the first step the purpose of the design, which is an agricultural robot. In the second one, it defines the farmer for whom it will be beneficial. The third step is to determine weeding as the action that will be accomplished. The last step informs us about the goal of design, which is to reduce the drudgery of farmers' work and limit the use of herbicides and pesticides.

3.2 Octopus Diagram

We then continue with the functional analysis of the need. This step allows us to obtain, as a result, the functional specifications. (see Fig. 5).

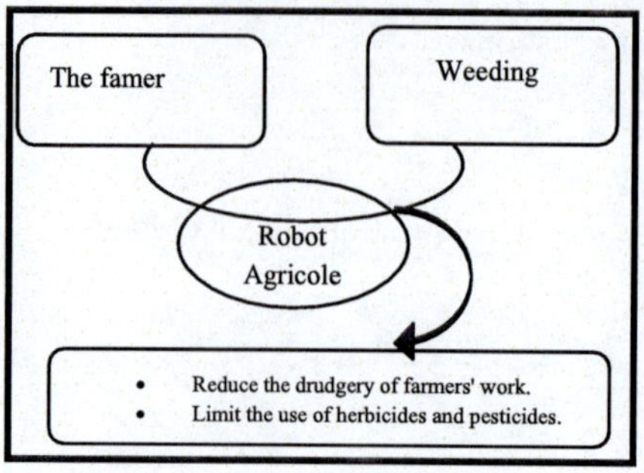

Fig. 4. Horned beast diagram

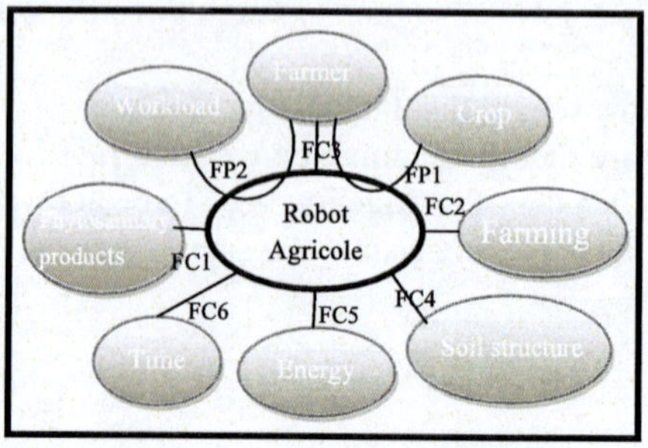

Fig. 5. Octopus diagram of agricultural robot.

The following table gives a description of the parameters and constraints functions (Table 1).

Table 1. Description of the main and constraints functions

Function	Description of the function
FP1	Weeding and cleaning agricultural crops
FP2	Reduce the farmer's workload
FC1	Limit exposure to phytosanitary products
FC2	Optimize the productivity of the farm
FC3	To be guided according to the need
FC4	Preserve soil structure
FC5	Save energy
FC6	Save time

As illustrated on the octopus diagram, the agricultural robot is connected to the external elements of its environment. This diagram define the functional parameters and the functional contraints related to the agriculture robot. The functional parameters are associated with two bodies. The functional constraints giving rise to the relationship with the agricultural robot are the constraints that it should have while performing the main purpose of the design [10].

As shown on the report [9] we can elaborate function parameters and function constraints:

Functions parameters: 1. Farmers are able to remove the weeds and clean the crops by using agricultural robot. 2. The agricultural robot is using to reduce the tediousness of weeding tasks.

Functions constraints: 1. The robot limit the use of herbicides and pesticides by using a mechanical instrument for weeding. 2. The robot can help farmers to focus on the other tasks. 3. The robot can cross the field as needed by the farmer. 4. The robot can protect the structure of the soil. 5. Farmers don't need more effort if the robot is here. 6. Farmers can manage their time when they use the robot.

4 Conclusion

A key technique in the design process for defining novel concepts' structures is functional analysis. A functional analysis is conducted to clarify the new product's functional requirements. Understanding the connections between the new product's components is important, as is making sure that all required components are specified and that none are extraneous [2].

The analysis of the need focuses on the need of the designer for the product, while the functional analysis of the need is used to identify the relationships between the context of use and this product to satisfy the need. The latter focuses on the functions of the product to be designed and not on the technical functions or technological solutions.

The definition of the service functions: main and constraints allows the mind to open up on the required technical functions, and it is what is normally dedicated to the technical functional analysis illustrated by the FAST diagram.

Acknowledgement. This work was supported by the Ministry of Higher Education, Scientific Research and Innovation, the Digital Development Agency (DDA) ant the CNRST for the realization of the 1st Moroccan agricultural robot.

References

1. (PDF) Status and Scope of Robotics in Agriculture (2022). https://www.researchgate.net/publication/312589560_Status_and_Scope_of_Robotics_in_Agriculture (consulté le 19 juillet 2022)
2. Viola, N., Corpino, S., Fioriti, M., Stesi, F.: Functional analysis in systems engineering: methodology and applications. In: Cogan, B. (ed.) Systems Engineering - Practice and Theory, InTech (2012). https://doi.org/10.5772/34556
3. Kossiakoff, A., Seymour, S.J., Flanigan, D.A., Biemer, S.M.: Systems Engineering Principles and Practice, 1st édn. Wiley (2020). https://doi.org/10.1002/9781119516699
4. Function Analysis Phase - Canadian Society of Value Analysis (2022). https://www.valueanalysis.ca/functionanalysis.php (consulté le 29 juin 2022)
5. Bête à cornes : définition, utilité et méthode | Lucidchart (2022). https://www.lucidchart.com/pages/fr/bete-a-cornes (consulté le 19 juillet 2022)
6. Qu'est-ce qu'un diagramme pieuvre ?, Lucidchart (2022). https://www.lucidchart.com/pages/fr/diagramme-pieuvre (consulté le 18 juillet 2022)
7. Borza, J.S.: FAST Diagrams: The Foundation for Creating Effective Function Models, p. 10 (2011)
8. Diagramme FAST : définition, utilité et exemples (2022). https://www.manager-go.com/gestion-de-projet/articles/analyser-un-probleme-dusage-avec-le-diagramme-fast (consulté le 18 juillet 2022)
9. Caro, S.: A Report on the design of a walking device to help elderly people walk and take off their own weight, p. 29 (2003)
10. Djami, A.N.N., Nzié, W., Yamigno, S.D.: Disassembly evaluation in design of a system using a multi-parameters index. Mod. Mech. Eng. **10**(1), Art. n° 1 (2020). https://doi.org/10.4236/mme.2020.101001

Check for updates

Hybrid Congestion Control Mechanism as a Secured Communication Technology for the Internet of Health Things

Asmaa El-Bakkouchi[1](✉), Mohammed El Ghazi[1], Anas Bouayad[1], Mohammed Fattah[2], and Moulhime El Bekkali[1]

[1] Artificial Intelligence, Data Sciences and Emerging Systems Laboratory, Sidi Mohamed Ben Abdellah University, Fez, Morocco
asmaa.elbakkouchi@usmba.ac.ma

[2] IMAGE Laboratory, Moulay Ismail University, Meknes, Morocco

Abstract. The Internet of Health Things (IoHT) has been proposed as an extended version of the Internet of Things (IoT) focusing on making the medical field intelligent by providing smart health services and sharing health data remotely in a secured way without data loss. However, IoHT uses the current internet architecture to transmit healthcare data, which highlights the limitations of this architecture due to the heterogeneity of IoHT devices and the volume and size of data exchanged. Named Data Networking (NDN) has emerged as a potential internet architecture for IoHT because of its characteristics that make it the optimal solution for IoHT. In addition, health data is transferred frequently in IoHT, which increases the possibility of congestion and therefore packet loss. To address this issue, we propose to use EC-Elastic, an NDN congestion control mechanism to control congestion in IoHT. EC-Elastic aims to maximize the use of available bandwidth, ensure data transmission with reasonable delay and minimize packet loss, which is highly required in IoHT systems. EC-Elastic has been implemented in ndnSIM and compared against PCON. The simulation results show that EC-Elastic can effectively improve the throughput compared to PCON with a reasonable delay while maintaining a low packet loss rate.

Keywords: IoHT · Congestion control · NDN

1 Introduction

The Internet of Things (IoT) is a new concept that consists of several intelligent devices capable of communicating with each other [1]. IoT has been deployed in several areas such as healthcare, smart grids, agriculture, smart cities and industry, etc. Healthcare is one of the main application areas covered by IoT. In this area, IoHT represents medical devices and applications capable of communicating and exchanging health data with each other and with healthcare personnel [2] with the objective of making healthcare services intelligent. Among the applications of IoT in the medical field, remote monitoring of patients using multiple sensors and the transmission of the data collected remotely to

Y. Farhaoui et al. (Eds.): ICAISE 2022, LNNS 635, pp. 498–503, 2023.
https://doi.org/10.1007/978-3-031-26254-8_72

the hospital servers or directly to the healthcare providers etc. The data transmitted can be sensitive data such as vital signs (heart-rate monitoring, glucose monitoring, etc.) and non-sensitive data such as movement signs (arm sensors, leg sensors, etc.). As the sensors used by IoHT transfer health data frequently and simultaneously, this raises the possibility of congestion and therefore packet loss, which is highly undesired in an IoHT environment especially for sensitive data that requires low delay and low packet loss. However, IoHT uses the current internet architecture to transmit health data, which highlights the limitations of this architecture due to the heterogeneity of IoHT devices, the volume and size of data exchanged. Several promising architectures have been proposed for the future internet, such as ICN (Information Centric Network) architectures [3] that are based on the use of content names over IP addresses. NDN [4] has been proposed as the most promising ICN architectures to solve the communication problems of the IoT environment. NDN supports two types of packets, interest and data, searches for desired content by name independently of its location linked to an IP address and uses a pull-based communication mode, meaning that to pull content, the consumer sends an interest packet and then the producer of the requested content responds with a data packet containing the requested content. Another important feature of NDN architecture is caching, i.e. caching the content in the content store of the intermediate router so that if there is a future request for the same content, the router will respond to the consumer's request rather than having to go to the producer, which is advantageous "time-saving" especially in an IoHT environment where time is important. NDN has three data transfer structures; Content Store (CS), Forwarding Information Base (FIB) and Pending Interest Table (PIT). CS is a memory for temporarily storing a copy of each transferred data packet. FIB allows packets of interest to be routed based on the longest prefix correspondence. PIT is a table that adds an entry when an interest packet is not satisfied and removes it when that interest packet gets its data packet. To solve the congestion problem in the healthcare domain, we propose to use EC-Elastic (an Explicit Congestion Elastic) that checks for congestion at the router and then reports this status to the consumer, which in turn adjusts the size of it congestion window (cwnd). EC-Elastic's objective is to reduce packet loss, improve utilization of available bandwidth with reasonable transmission delay. The following is the structure of this paper. The proposed algorithm is described in Sect. 2 and evaluated in Sect. 3, while the conclusion is presented in Sect. 4.

2 EC-Elastic Algorithm

EC-Elastic [5] has been proposed to control congestion in NDN using three steps; firstly, congestion detection which is done at the intermediate router by calculating the time that each packet takes in the queue. The queue used is Controlled Delay-Active Queue Management (CoDel-AQM) [6] which calculates the sojourn time of each packet in the queue and then compares it to a threshold called "TARGET" that is equal to 5 ms. If the calculated time is below this threshold, the link is considered normal, in this case no change in the congestion window size increase process. Otherwise, if the calculated time exceeds the threshold, the link is considered congested. In this case, EC-Elastic moves to the second step which consists of "Explicit signaling of detected congestion" where, after detecting the congestion by the router node, CoDel explicitly marks this packet and sends it to the consumer node that is responsible for adjusting the cwnd (third step). At the consumer node, cwnd is adjusted according to the type of packet received. If the received packet is normal, it increases its cwnd. Otherwise, if the received packet is marked, it decreases its cwnd. EC-Elastic uses two phases for increasing the size of its congestion window. It starts with the Slow Start phase which increases exponentially the cwnd by 1 according to the following formula:

$$\text{cwnd} = \text{cwnd} + 1 \tag{1}$$

This phase corresponds to cwnd below a predefined threshold (ssthresh) "cwnd < ssthresh". Once cwnd exceeds the threshold, EC-Elastic enters the congestion avoidance phase where the window increase becomes less aggressive. In this phase, the cwnd is increased by WWF/cwnd. It is based on the variation of the Window-correlated Weighting Function WWF [7], which has the objective of increasing the use of the available bandwidth. WWF is defined as follows:

$$WWF = \sqrt{\frac{RTT\ max}{RTT\ current} * cwnd} \tag{2}$$

where the current RTT is presented by RTTcurrent, the maximum RTT is presented by RTTmax and the current congestion window is presented by cwnd. For the reduction of the congestion window size which is done when the consumer receives a marked packet. EC-Elastic reduced it according to the phase where congestion has been detected.

$$\text{cwnd} = \text{cwnd} * \beta_1 \qquad \text{Slow Start} \tag{3}$$

$$\text{cwnd} = \text{cwnd} * \beta_2 \qquad \text{Congestion Avoidance} \tag{4}$$

where, β_1 and β_2 are the parameters for adjusting the congestion window size with $\beta_1 = 0.85$ and $\beta_2 = 0.9$. In an IoHT environment, the time to receive health data is important and if we initialize cwnd it will take a considerable time to meet the threshold again. For this, in the case of Timeout, we use a multiplicative decrease.

3 Performance Evaluation

EC-Elastic has been implemented in ndnSIM simulator [8] and evaluated against PCON (Practical Congestion Control scheme)[9]. The topology and parameters of the simulation are presented in Fig. 1 where it contains four consumers, two routers and four producers. The simulation time is fixed at 30 s.

3.1 Simulation Metrics

The performance evaluation of EC-Elastic and PCON is done in terms of their throughput, transmission delay and packet loss rate. Throughput is considered as an important metric for assessing network performance. It is the size of data successfully received compared to the interest demanded [10]. Delay refers to the delay between the sending of an interest packet and the reception of its data packet. Packet loss rate represents the number of packets dropped per second.

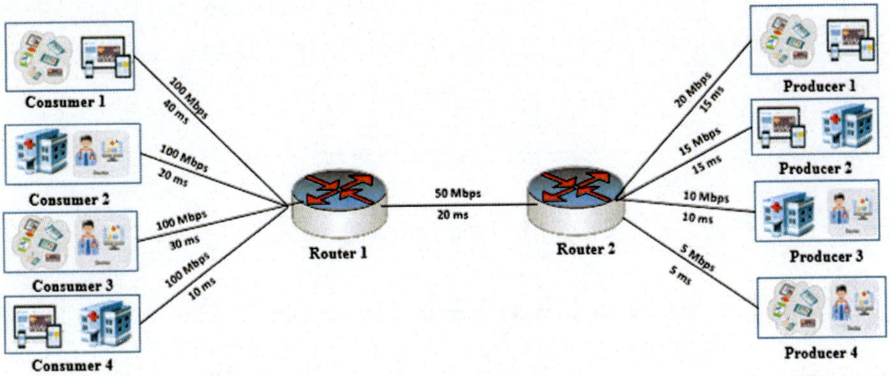

Fig. 1. The basic topology

3.2 Simulation Results

Figure 2 shows the simulation result of the throughput for both mechanisms. We observe that the best throughput was recorded by EC-Elastic and this is due to the method used to increase its congestion window size, which relies on using WWF in the congestion avoidance phase to use the available bandwidth fully, and the method used to reduce the congestion window which in case of congestion instead of dividing the size of the congestion window by two as in the case of PCON, it uses two parameters β_1 and β_2 according to which phase the congestion has been detected to avoid waiting for a long time before the congestion window reaches its maximum again because when the congestion window is increased the throughput increases and vice versa.

Figure 3 shows the delay measurement of both mechanisms. We observe that the delay results of EC-Elastic are quite close to those of PCON. The cases where the delay of EC-Elastic is higher than that of PCON are explained by the fact that when the congestion window reaches its maximum, several packets circulate at the same time in the link and consequently more time is needed to receive the data packets. However, this delay remains reasonable since EC-Elastic increases the use of the available bandwidth.

For the packet loss rate, the simulation results illustrated in Table 1 show a negligible loss rate for both mechanisms which validates the effectiveness of the method used by both mechanisms for congestion detection (CoDel-AQM) which allows to react quickly to the congestion event to avoid losing the data that are important in the health sector.

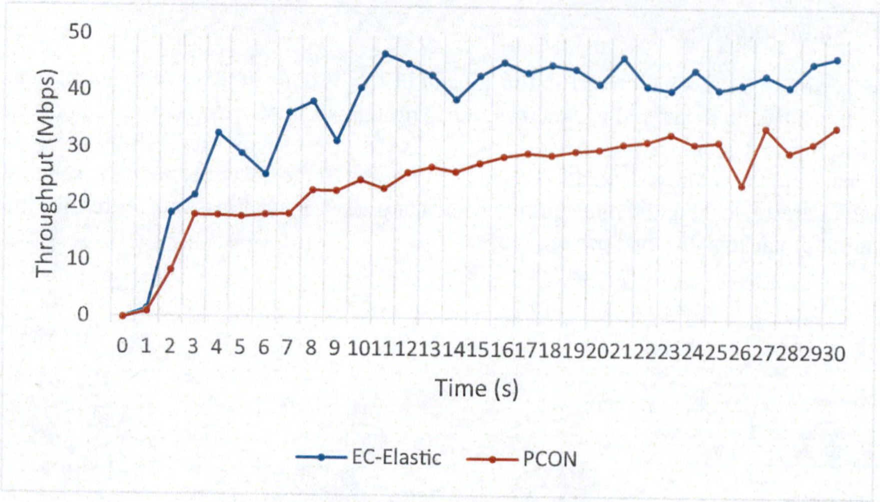

Fig. 2. Throughput measurement.

Table 1. Packet loss rate measurement.

Algorithms	EC-Elastic	PCON
Packet loss rate	0,0042	0,0078

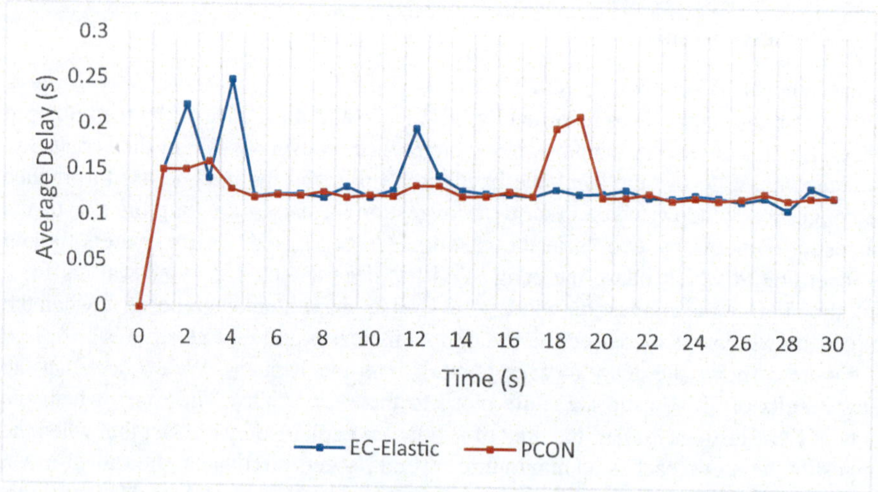

Fig. 3. Average delay measurement.

4 Conclusion

This paper presents EC-Elastic that has been proposed for NDN to control network congestion. In EC-Elastic, routers detect congestion using CoDel-AQM and then report this state to consumers which are responsible for adjusting the size of the congestion window. The performance evaluation showed that EC-Elastic can effectively improve throughput compared to PCON with reasonable delay and negligible packet loss rate demonstrating that our approach EC-Elastic has the potential to be a feasible solution to avoid and control congested network in IoHT systems where congestion is highly undesirable, especially for sensitive data that requires low delay and low packet loss.

References

1. Das, A.K., Zeadally, S., He, D.: Taxonomy and analysis of security protocols for Internet of Things. Futur. Gener. Comput. Syst. **89**, 110–125 (2018). https://doi.org/10.1016/j.future.2018.06.027
2. Aroosa, U.S.S., Hussain, S., et al.: Securing NDN-based internet of health things through cost-effective signcryption scheme. Wirel. Commun. Mob. Comput. **2021** (2021). https://doi.org/10.1155/2021/5569365
3. Ahlgren, B., Dannewitz, C., Imbrenda, C., et al.: A survey of information-centric networking. IEEE Commun. Mag. **50**, 26–36 (2012). https://doi.org/10.1109/MCOM.2012.6231276
4. Zhang, L., Estrin, D., Burke, J., et al.: Named Data Networking (NDN) Project Named Data Networking (NDN) Project (2010)
5. El-Bakkouchi, A., Ghazi, M. El., Bouayad, A., et al.: EC-elastic an explicit congestion control mechanism for named data networking. Int. J. Adv. Comput. Sci. Appl. **12**, 594–603 (2021). https://doi.org/10.14569/ijacsa.2021.0121168
6. Nichols, K., Jacobson, V.: Controlling queue delay. Commun. ACM **55**, 42 (2012). https://doi.org/10.1145/2209249.2209264
7. Alrshah, M.A., Al-Maqri, M.A., Othman, M.: Elastic-TCP: flexible congestion control algorithm to adapt for high-BDP networks. IEEE Syst. J. **1–11** (2019). https://doi.org/10.1109/jsyst.2019.2896195
8. Mastorakis, S., Afanasyev, A., Moiseenko, I., Zhang, L.: ndnSIM 2.0: A new version of the NDN simulator for NS-3. NDN Proj 1–8 (2015)
9. Schneider, K., Yi, C., Zhang, B., Zhang, L.: A practical congestion control scheme for named data networking. In: Proceedings of the 2016 Conference on 3rd ACM Conference on Information-Centric Networking - ACM-ICN 2016, pp 21–30. ACM Press, New York, New York, USA (2016)
10. El-Bakkouchi, A., Ghazi, M. El., Bouayad, A., et al.: Packet loss and delay measurement analysis of TCP variants in NDN congestion control. In: 2020 1st International Conference on Innovative Research in Applied Science, Engineering and Technology IRASET 2020 2–6 (2020). https://doi.org/10.1109/IRASET48871.2020.9092197

Impact of Feature Vectorization Methods on Arabic Text Readability Assessment

Safae Berrichi[1,2](✉), Naoual Nassiri[1,2], Azzeddine Mazroui[1], and Abdelhak Lakhouaja[1]

[1] Department of Computer Science, Faculty of Sciences, Mohammed First University, Oujda, Morocco
berrichi.safae@gmail.com

[2] Research Center of the School of Advanced Engineering Studies Oujda (CREHEIO), Oujda, Morocco

Abstract. Assessing the readability of texts is a crucial challenge in an information-intensive society. One of the many problems raised in this area is determining whether the models trained for this task are effective for low-resource languages. Since representing texts by vectors composed only of linguistic features is the most explored approach in the field of readability prediction of Arabic texts, word embedding can represent an attractive alternative in this field. It is another popular and efficient method for text classification that generates a vector representation of words based on semantic and syntactic similarities between linguistic units. In this study, we evaluated the impact of different word vectorizations on the readability of texts dedicated to readers of Arabic as a first foreign language. We also examined the effect of combining these representations with classical linguistic features. The test results show that the F-score of the combination-based approach outperforms the classical one by 8%.

Keywords: Readability · Arabic language · Words embedding · Random forest algorithm · TF-IDF · L2

1 Introduction

With the emergence of new techniques in Natural Language Processing (NLP) and Machine Learning (ML), attention has once again been focused on readability as a classification task [14]. Readability assesses the ease or the difficulty level of documents such as web articles, history, and education books. Automatic readability measurement has typically been performed using conventional mathematical readability formulas [7], based purely on lexical and syntactic features such as word length, sentence length and word difficulty. These approaches have been criticized for their reductive character, which overlooks word order and context information and separates the feature extraction process from the classification model design process. In recent years, another approach based on the

Y. Farhaoui et al. (Eds.): ICAISE 2022, LNNS 635, pp. 504–510, 2023.
https://doi.org/10.1007/978-3-031-26254-8_73

Term Frequency-Inverse Document Frequency (TF-IDF) algorithm has emerged. It consists in representing each word of a text by a numerical vector associated with a value indicating its "weight". Although this method is well adapted to many text mining tasks, the extremely high dimensionality of text represents a significant challenge for most classification algorithms.

Recently, deep neural networks have shown impressive performances in many NLP-related tasks. Indeed, some neural approaches have been proposed to predict text readability for languages with large linguistic resources, such as English [8]. Moreover, unlike traditional ML methods, text features in a neural model are automatically extracted following the learning process. Indeed, a first neural model (word embedding (WE) methods) represents the words of the text by continuous and dense numerical vectors. These methods include non-contextual WE techniques that consist in uniquely representing each word independently of the different contexts during the learning phase, as in the Skip-gram and the continuous bag of words (CBOW) techniques [10], and contextual WE techniques that learn semantics at the sequence level by considering the sequence of all the words in the documents such as the BERT algorithm [6].

However, learning these WEs requires the use of large corpora of text and is very time consuming. As an alternative, various pre-trained WE models are publicly available for NLP tasks. The proposed approach in this paper consists in exploiting the knowledge of the pre-trained model araVec[1] in the readability measurement domain. We also used the standard TF-IDF feature vectorization method. Thus, we first developed a readability measurement model based entirely on one of these vectorization methods. Then, we used the traditional ML Random Forest algorithm to develop a readability measurement model based on the combination of these vector representations with manually crafted linguistic features. The test results support that these combined methods are not only effective for Arabic, which is a low-resource language, but also provide information about the readability domain.

The rest of this paper is organized as follows. In Sect. 2, we present some previous researches on text readability measurement approaches. In Sect. 3, we provide details on the data used and the methodology adopted. We present our experiments on text readability measurement in Sect. 4, and we discuss the obtained results. We conclude this paper with some thoughts on future work.

2 Related Work

Since the beginning of the 21st century, ML-based text readability measurement methods have continuously incorporated various rich features to improve the performance of the readability model. For example, the authors of [4] were interested in predicting text comprehension, processing (reading time) and vocabulary familiarity for adult readers. They examined the use of features that address various levels of language (lexical, syntactic, and semantic). Highlighting that statistical ML models are highly language dependent and do not perform well for

[1] https://github.com/bakrianoo/aravec.

short texts, the authors of [11] introduced deep reinforcement learning models to assess text readability. Their contribution includes automating feature extraction, which reduces the dependence of the readability assessment task on the language under study, and requires using only a minimal portion of the text to assess its readability instead of the entire text.

More recent readability assessment models for well-resourced languages have used deep learning techniques. The model performance is often improved by acquiring more data, but readability annotations are time-consuming. In order to overcome the problem of limited size of readability data sets for deep learning models, some approaches have evaluated the impact of incorporating linguistic features. Indeed, reference [5] exploited the output of the transformer deep learning model and appended it with linguistic features that were then fed into a ML algorithm such as Support Vector Machine (SVM). Their results showed that while the training data is sufficient, linguistic features do not have a substantial advantage over deep learning methods.

Alternatively, reference [9] proposed a readability assessment approach in which they incorporated the knowledge of large pre-trained language models such as BERT. They combined its information-rich sentence embeddings as a distinct feature set for traditional ML algorithms with hand-collected linguistic features. Results on English and Filipino language data sets showed that the proposed method outperforms the supervised classification approaches commonly used in the readability assessment literature.

Regarding previous works on Arabic text readability measurement, they focused on hand crafted linguistic features in combination with traditional ML algorithms. The authors of [12] presented a study in which they collected a GLOSS[2] corpus. They introduced a total of 170 features to represent a text. They reported an F-score of 0.9 when testing on the training data. In an improvement of this first work, they exploited the previously constructed GLOSS corpus using 133 features instead of the 170 [13]. They achieved an accuracy of 100% when testing on the training data. Based on the review of the numerous studies related to the prediction of readability of Arabic texts [2], we found that these studies focused only on the evaluation of different ML classifiers using a set of predetermined features. The use of the WE vector representation of text by pre-trained language models has been addressed, to our knowledge, only recently. For this reason, we considered in this study to take into account vector representations and compare their impact to those used in the Arabic language readability evaluation.

3 Methodology

In this section we describe the different corpora we used as well as the methods we adopted to generate the different vector representations of the texts in order to use them as input for the ML algorithms.

[2] https://gloss.dliflc.edu/.

3.1 Data

In order to train and evaluate the Arabic readability prediction models, we used in this study educational corpora intended for learning Arabic as a first foreign language (L2). The texts of the latter are annotated by their difficulty levels according to the ILR scale [3], and we grouped them into three levels: Easy (level 1 and 1+), Intermediate (level 2), and Difficult (level 2+ and 3).

We collected data from three different corpora, namely Aljazeera-Learning[3] (AL), GLOSS Listening (GL), and GLOSS Reading (GR)[4]. Regarding the distribution of this corpora, we have 356 texts at easy level, 214 texts at intermediate level, and 249 texts at difficult level.

3.2 Feature Extraction

Different linguistic features are used in automatic readability measurement of the Arabic text. These features are grouped into two families according to the type of information they provide to the readability model (lexical and syntactic). In this work, we have used 70 features covering these two families and taken from those used in [14].

Representing a text by vectors based on a statistical measure (TF-IDF) or by WE is another prominent mechanism used to build ML models. In this work, we used two feature vectorization methods. The first one is TF-IDF which determines the importance of a word by weighting its frequency of occurrence in the document and by computing its frequency of occurrence in other documents. The second method used is Word2vec. It is a technique for generating WEs in the form of dense numerical vectors. The general idea is based on using the context of adjacent words to identify similar words. We adopted the pre-trained araVec model. This is built using the Gensim[5] tool, created to handle many common NLP tasks, and includes an implementation of the Word2Vec model. AraVec provides six different word embedding models related to one of three text domains (Tweets, WEB, and Wikipedia). For each of the three domains, two word embedding models are built, one of which uses the CBOW technique and the other uses the Skip-Gram technique. The models for these two techniques are generated with one of two vector dimensions 300 and 100. In this study, we used the two models trained on Wikipedia data with dimension equal to 300. In addition, since the readability prediction model is built on a set of texts annotated by difficulty levels, for each text we tried to compute the average of the WEs corresponding to it in order to provide a single vector per text. In addition to evaluate these models (TF-IDF/word2vec), we are interested in combining them with hand-crafted linguistic features through concatenation in order to retrieve the effective model that can achieve good performance. The proposed architecture is described in Fig. 1.

[3] https://learning.aljazeera.net/.
[4] https://gloss.dliflc.edu/.
[5] https://radimrehurek.com/gensim/about.html.

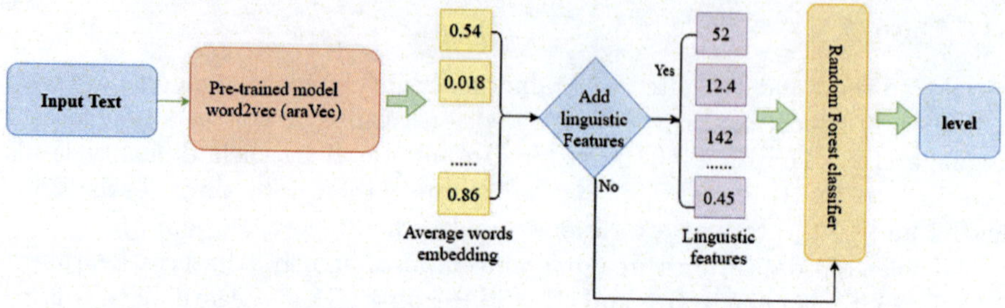

Fig. 1. AraVec model process with/without linguistic feature vectors

4 Experiments and Results

We performed a set of experiments to examine the impact of vector-based text representation methods on the Arabic readability assessment. We compared the performance of the six following models:

- Linguistic features (LF): original classification using only the 70 hand crafted linguistic features.
- TF-IDF: classification based on feature vectorization by the TF-IDF algorithm.
- AraVec-SG: classification using sentence embeddings from the araVec model via the Skipgram technique with a vector size of 300.
- AraVec-CB : classification using sentence embeddings from the araVec model via the CBOW technique with a vector size of 300.
- TF-IDF + LF : classification based on the concatenation of vectors obtained by TF-IDF and LF.
- AraVec-SG + LF : classification based on the concatenation of vectors obtained by AraVec-SG and LF.
- AraVec-CB +LF : classification based on the concatenation of vectors obtained by AraVec-CB and LF.

The generation of the text vector representations by TF-IDF and WE was performed after a data pre-processing phase. This phase consists in eliminating stop words using a list provided by the nltk library. Furthermore, for the model based on TF-IDF, we represented the words of the texts by their stems in order to keep contextual information on the word without increasing the dimensionality of the TF-IDF vectors. To assign a unique stem to each word in the text, taking into account its context, we used Alkhalil's stemmer [1].

The experiments were performed with the Random Forest classifier using 10-fold cross-validation. Table 1 shows the results, in terms of accuracy and F-score, of the different models.

As described in Table 1, the models trained using the non-contextual WEs (AraVec-SG and AraVec-CB) outperform the model based on the TF-IDF algorithm in terms of F-score, but their accuracies are worse. In addition, AraVec-SG

Table 1. Performance of the different classification models

Model	Accuracy	F-score
LF	74.76%	70.16%
TF-IDF	73.65%	63.92%
AraVec-SG	71.93%	67.76%
AraVec-CB	70.35%	65.95%
TF-IDF + LF	80.14%	77.43%
AraVec-CB + LF	80.14%	78.22%
AraVec-SG + LF	**80.63%**	**78.82%**

generated by the Skip-gram model performs better than the AraVec-CB model generated by CBOW. Moreover, these three models (TF-IDF, AraVec-SG, and AraVec-CB) perform less well than the LF model based only on hand crafted linguistic features. Indeed, the F-score of the LF model is respectively 2% and 4% higher than those of the AraVec-SG and AraVec-CB models. This can be explained by the limitations of the morphosyntactic information picked up by the WE-based models.

To further evaluate the impact of pre-trained WE and TF-IDF models in the readability domain, we trained the classification model by combining hand crafted linguistic feature with both non-contextual WE and TF-IDF models. The test results show that the combined training (TF-IDF, AraVec-SG+LF, AraVec-CB+LF) allows a substantial improvement in the quality of the readability measure compared to models using exclusively the vectorization methods (TF-IDF, AraVec-SG and AraVec-CB) or the linguistic features (LF). Indeed, we show an average increase of 8% in terms of F-score for the AraVec-CB+LF model. This combination provides the benefit of their complementarity, since WE captures semantic regularities that are not presented in NLP systems and therefore in the LF model. These first results encourage us to evaluate other pre-trained models of WE that take into account sequence-level semantics.

5 Conclusion and Future Work

In this paper, we proposed a new method to improve the accuracy of readability measurement of Arabic texts. We used the pre-trained araVec model based on CBOW and Skip-gram techniques for a dense vector representation of sequences. Then, we combined this representation with manually collected linguistic features to develop our models to measure the readability of Arabic texts. These models were evaluated on educational corpora dedicated to L2 learners and annotated by difficulty levels. The test results indicate a significant improvement in the readability accuracy.

As a future work, we plan to extend this work to contextual vector representations of words, and to conduct an in-depth analysis of their impact on the quality of readability measurement. Similarly, the evaluation of deep ML architectures, and the collection of additional data represent other directions of investigation for text readability measurement.

References

1. Boudchiche, M., Mazroui, A.: Spline functions for arabic morphological disambiguation. Applied Computing and Informatics (2020)
2. Cavalli-Sforza, V., Saddiki, H., Nassiri, N.: Arabic readability research: current state and future directions. Procedia Comput. Sci. **142**, 38–49 (2018)
3. Clark, J.L.D., Clifford, R.T.: The FSI/ILR/ACTFL proficiency scales and testing techniques: development, current status, and needed research. Stud. Sec. Lang. Acquisit. **10**(2), 129–147 (1988). https://doi.org/10.1017/S0272263100007270
4. Crossley, S.A., Skalicky, S., Dascalu, M., McNamara, D.S., Kyle, K.: Predicting text comprehension, processing, and familiarity in adult readers: new approaches to readability formulas. Discourse Process. **54**(5–6), 340–359 (2017)
5. Deutsch, T., Jasbi, M., Shieber, S.: Linguistic features for readability assessment. arXiv preprint arXiv:2006.00377 (2020)
6. Devlin, J., Chang, M.W., Lee, K., Toutanova, K.: Bert: pre-training of deep bidirectional transformers for language understanding. arXiv preprint arXiv:1810.04805 (2018)
7. Duffy, T.M.: Readability formulas: what's the use? In: Designing Usable Texts, pp. 113–143. Elsevier (1985)
8. Filighera, A., Steuer, T., Rensing, C.: Automatic text difficulty estimation using embeddings and neural networks. In: Scheffel, M., Broisin, J., Pammer-Schindler, V., Ioannou, A., Schneider, J. (eds.)European Conference on Technology Enhanced Learning. LNISA, vol. 11722, pp. 335–348. Springer, Cham (2019). https://doi.org/10.1007/978-3-030-29736-7_25
9. Imperial, J.M.: Bert embeddings for automatic readability assessment. arXiv preprint arXiv:2106.07935 (2021)
10. Mikolov, T., Chen, K., Corrado, G., Dean, J.: Efficient estimation of word representations in vector space. arXiv preprint arXiv:1301.3781 (2013)
11. Mohammadi, H., Khasteh, S.H.: Text as environment: a deep reinforcement learning text readability assessment model. arXiv preprint arXiv:1912.05957 (2019)
12. Nassiri, N., Lakhouaja, A., Cavalli-Sforza, V.: Arabic readability assessment for foreign language learners. In: Silberztein, M., Atigui, F., Kornyshova, E., Mtais, E., Meziane, F. (eds.) Natural Language Processing and Information Systems. LNISA, vol. 10859, pp. 480–488. Springer, Cham (2018). https://doi.org/10.1007/978-3-319-91947-8_49
13. Nassiri, N., Lakhouaja, A., Cavalli-Sforza, V.: Modern standard arabic readability prediction. In: Lachkar, A., Bouzoubaa, K., Mazroui, A., Hamdani, A., Lekhouaja, A. (eds.) Arabic Language Processing: From Theory to Practice, CCIS, vol. 782, pp. 120–133. Springer, Cham (2018). https://doi.org/10.1007/978-3-319-73500-9_9
14. Nassiri, N., Lakhouaja, A., Cavalli-Sforza, V.: Arabic l2 readability assessment: dimensionality reduction study. Journal of King Saud University - Computer and Information Sciences (2021). https://www.sciencedirect.com/science/article/pii/S1319157820306297

Impact of Landscape Change on Hydrological and Sediment Response of the Tleta Watershed in Northern of Morocco

S. El Harche(✉), M. Chikhaoui, and M. Naimi

Hassan II Institute of Agronomy andVeterinary Medicine, Rabat, Morocco
s.elharche@iav.ac.ma

Abstract. The landscape analysis can be used to characterize, test and assist in planning and considering solutions based on future impacts. In this study, two landscape are compared from past to current conditions (1998–2018) in the Tleta watershed for identifying the areas at risk of being most affected. In this context, AGWA tool will be coupled with the KINEROS2 model, to investigate the manner in which landscape changes over a 20-year period have affected runoff and erosion rate processes. According to the results, the modification of landscape are the main factors explaining increase of runoff observed (+6.08%) and sediments yield (+3.85%).

Keywords: Landscape · Runoff · Erosion rate · KINEROS2 · AGWA · Tleta watershed

1 Introduction

Landscape change as a fundamental land surface variable, directly affects soil erosion, resulting soil losses decreases the soil fertility consequently affecting agricultural production. Additionally, it also results in accelerated siltation rates reducing reservoir storage capacities which negatively impacts on drinking and irrigation water supply [1]. The dual influence of the Atlantic Ocean and the Mediterranean Sea means that the Rif Mountain range is threatened by degradation, especially by the western climate influenced by moisture from the Atlantic Ocean. Hence the need to assess the capacity of climate models to predict the climate in such complex areas [2]. The use of several model parameters may increase the forecast uncertainty but requires a complicated calibration procedure to reduce the uncertainty. In this study, the potential impacts of tow scenarios relative to past and current conditions in the Tleta catchement are modeled in a geographic information system (ARCGIS) by KINEROS2 or (K2) (The Kinematic Runoff and Erosion Model), that is a physical model describing the processes of interception, dynamic infiltration, surface runoff, and erosion in watersheds[3]. To apply this model on an operational basis, it is essential to have automated procedures that can be repeated, accurate and relatively simple. In this sense. AGWA, (Automated Geospatial Watershed Assessment tool) was developed by EPA, the Agricultural Research Service

Y. Farhaoui et al. (Eds.): ICAISE 2022, LNNS 635, pp. 511–517, 2023.
https://doi.org/10.1007/978-3-031-26254-8_74

of the United States Department of Agriculture (USDA) and the University of Arizona, to provide qualitative estimates of runoff and erosion in relation to landscape change. It facilitates the parameterization of the model [4]. The K2 model coupled with AGWA tool has been used as a tool for modeling flows and solids fluxes in Tleta basin [4], its use in different scenarios of landscape will help to identify areas at risk in the watershed. What interests this study is to estimate the effects of 20 years of landscape on runoff and sediments yields using the AGWA modeling tools coupling with the Kineros2 model.

2 Methodology

2.1 Study Area

The Tleta watershed is located in the north-western part of the Moroccan rifaine chain, occupying a surface area of about 180 km^2. It's characterized by a Mediterranean climate [5]. Downstream of the basin is the Ibn Batouta dam with an initial storage capacity corresponding to 43.6 million m^3 in 1977, decreased to reach a value of 29.1 Mm^3 in 2013. The reservoir has been subject to an annual silting up of about 0.4 Mm^3/year, which represents a loss of 33% of its initial capacity [6] (Fig. 1).

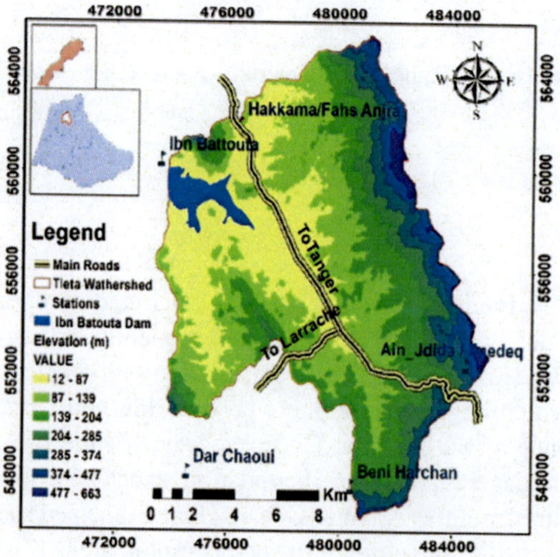

Fig. 1. Location of the Tleta watershed

2.2 AGWA-K2 Modeling Process

KINEROS2, is a largely updated version of KINEROS which is has been described in detail and published in 1990 [13], it's an event physical model describing the processes of, surface runoff, interception, dynamic infiltration, and erosion of watersheds and erosion in agricultural, rangeland and urban watersheds [7]. The model assumes that runoff is generated by a Hortonian mechanism and deals dynamically with infiltration, taking into account the relief, the river system, landscape and geology [10]. The model uses the Smith and Parlange equation to calculate the infiltration rate [11] and the kinematic wave equation for the Hortonian runoff it assumes [12]. The computation of Hortonian runoff on plans and channels is based on physical and mathematical equations as follows [3]:

$$\mathrm{q(x, t)} = \frac{\partial h}{\partial t} + \frac{\partial Q}{\partial x} \tag{1}$$

$$\mathrm{q(x, t)} = \frac{\partial A}{\partial t} + \frac{\partial Q}{\partial x} \tag{2}$$

where, h is the water height per width unit, Q the flow unit, A the wet section, t is the time, x the distance and q(x, t) the net inflow. The dynamic mass balance equation for erosion is giving by the following formula:

$$\mathbf{q_s(x,t)} = \frac{\partial (ACs)}{\partial t} + \frac{\partial (QCs)}{\partial x} - e(x,t) \tag{3}$$

where A is the cross-sectional area of flow, Cs is the local sediment concentration, Q is the water discharge, e is the surface erosion or deposition rate and qc is the rate of lateral sediment in flow for channels. The step required to conduct watershed analyses or assessments is the delineation [6], to involves generating an output for the watershed in question. The Ibn Batouta dam station was chosen as the outlet [4]. The spatial complexity of the watershed subdivision is controlled by Contributing Source Area (CSA) which corresponds to the heterogeneity of the watershed and the spatial resolution of the simulated processes [7, 8, 9]. The AGWA-KINEROS2 version 3.10.0 and reference guide are available from http://www.tucson.ars.ag.gov/agwa/. The parameterization of the watershed element must be characterized by its hydraulic geometry, the length of the flow, and the unique properties of the landscape and soil. The flow is controlled by the parameters of gradient, slope length, and soil hydraulic resistance [12]. The elements delimited by the AGWA are represented in K2 by a set of parameter values that are assumed to be uniform across the board, using weighting to estimate a mean value for each parameter [10]. In order to introduce the parameters of soil roughness and interception, the vegetation cover layer must be taken into account to help modify the hydraulic conductivity at saturation previously defined by the soil layer [13]. AGWA is designed to provide default K2 Parameters based on these tables and the properties of the topography, soils, and landscape available on GIS data layers [6, 12].

2.3 Input for K2 Simulation

The data required to run the K2 model, were taken from daily climate data records from the ibn batouta station for the period (1998–2018), including classified Landsat satellite image, confront with the reality of the terrain in the same year for 2018, soil map determined from previous studies [14], compatible with ArcGIS, and DEM digital elevation model extracted from the SPOT satellite data (Table 1).

Table 1. Input data, its resolution and source used in K2 model simulations

Type	Source	Description
DEM	SPOT (20 m)	Digital Elevation model
Soil	Inypsa (1987)	Soil survey (1/50 000)
Land cover	Landsat (30 m)	Image processing
Rainfall	ABHL	Daily rainfall (1998–2018)

3 Results and Discussions

3.1 Landscape Maps Evolution (1998–2018)

A comparison of the two maps derived for 1998 and 2018 in Fig. 2 shows that landscape has changed significantly as a result of strong anthropogenic actions.

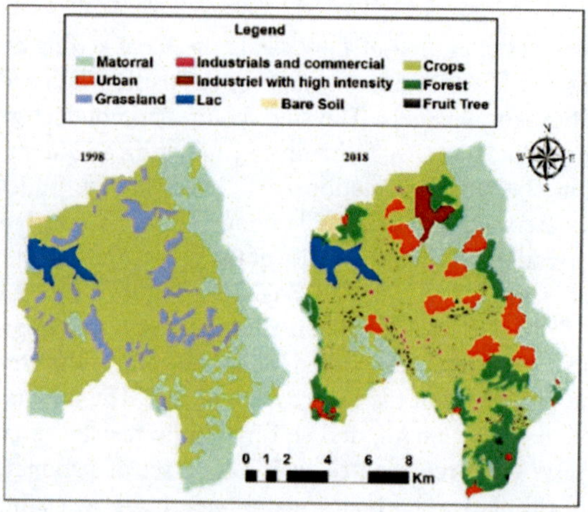

Fig. 2. Study area landscape change maps (1998–2018)

(Fruit tree), and (Forest) planted to increase plant biomass and promotes infiltration and interception. (Bare Soil), in the north-western part of the basin, (Grassland), replaced mostly by urban area. (Crops), located on the hills as well as the depressions, (Matorral), used as a rangeland, in the south-eastern part of the basin and more or less dense matorral on hills and slopes. (Urban), dispersed in the watershed. (Industrial with high intensity) with the new city of Cherrafat. (Industrial and Commercial) dispersed in the watershed in several small units. Finally, (Lac "Ibn Batouta dam").

3.2 Simulation Results

The increasing of output is due to increase in urban and cultivated area favoring the tilling of the soil that reduces its water retention capacity and reduces infiltration. Indeed, the cartography shows that the amount of land used for crops is relatively large. Addition to this, the topography shows some steep slopes. In previous study the K2 model has been tested in region of interest and has proven its robustness and reliability in predicting hydrological and erosion processes with R^2 and NSE > 0.8 for different precipitations events [4] (Table 2).

Table. 2. Simulated results (1998–2018)

Outputs	1998	2018 (% Change)
Runoff (m^3)	580919	618550 (+6.08)
SY (kg/ha)	91484	95146 (+3.85)

The simulation results showed that the runoff and sediment yield has increased (+6.08%, +3.85%). The response of the watershed to rainfall is longer in the presence of dense vegetation, the runoff is reduced compared to tilled soil or an urban area producing large volumes of overland runoff and sediments which is considered to be the major environmental stressor affecting the hydrological process [15] (Figs. 3 and 4).

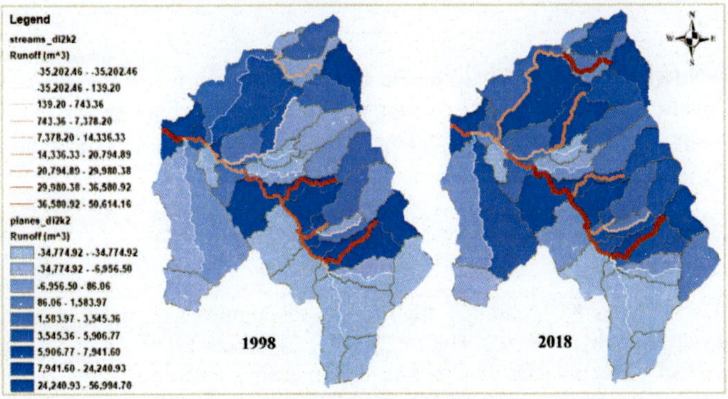

Fig. 3. The amount of Runoff during a precipitation event (1 h)

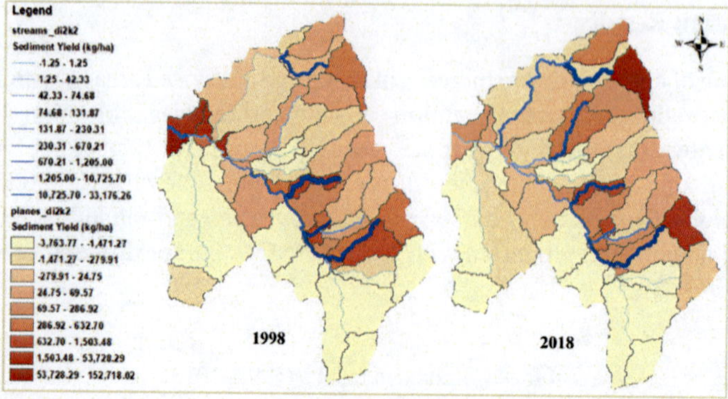

Fig. 4. The amount of sediments yield during a precipitation event (1 h)

4 Conclusion

The landscape changes associated with development is significantly altering the hydrological and erosion rate. The changes noted are mainly associated with increasing urbanization in favor of vegetated surfaces replaced by impervious surfaces. The most remarkable results can be summarised as follows: (i) Surface runoff and sediments yields has increased as urbanization, and Industrial increases in northern of the watershed; which leads to (ii) a decrease in the dam capacity from 35.6 Mm^3 in 1998 to 27.6 Mm^3 in 2018 with a total siltation around 8 Mm^3, (iii) The coupling AGWA/K2 model is a promising means of assessing, quantifying the effects of land cover change at the Tleta catchment level.

Acknowledgments. The authors would like to thank the Hassan II Institute of Agronomy and Veterinary Medicine and the Hassan II Academy of Science and Technology (GISEC Project) for their manifold support. We also like to extend our warm thanks to the reviewers of the manuscript.

References

1. Tomer, M., Schilling, K.: A simple approach to distinguish land-use and climate-change effects on watershed hydrology. J. Hydrol. **376**(1), 24–33 (2009)
2. Prein, A., et al.: Precipitation in the euro-cordex and simulations: high resolution, high benefits. Clim. Dyn. **46**, 383–412 (2015)
3. Smith, R.E., Goodrich, D.C., Woolhiser, D.A., Unkrich, C.L.: KINEROS - A kinematic runoff and erosion model. In: Singh, V.J. (ed.) Chapter 20. Computer Models of Watershed Hydrology, pp. 697–732. Water Resour. Pub., Highlands Ranch, Colo (1995)
4. El Harche, S., Chikhaoui, M., Naimi, M.: Estimation of runoff and erosion rates using agwa - kineros2 model: application to a mediterranean watershed european scientific journal. ESJ **16**(36), 1 (2020)
5. ABHL.: Etude d'actualisation du plan directeur d'aménagement intégré des ressources en eau des bassins Loukkos, Tétouan, Maroc, p. 186 (2013)
6. Goodrich, D.C., et al.: Kineros2/agwa: Model use, calibration, and validation. The Soil & Water Division of ASABE (2012)

7. Lane, L.J., Woolhiser, D.A., Yevjevich, V.: Influence of simplifications in watershed geometry in simulation of surface runoff. Hydrol. paper **81**, 80 (1975)
8. Thieken, A.H., Lücke, A., Diekkrüger, B., Richter, O.: Scaling input data by GIS for hydrological modelling. Hydrol. Process. **13**(4), 611–630 (1999)
9. Helmlinger, K.R., Kumar, P.E., Foufoula, G.: On the use of digital elevation model data for Hortonian and fractal analyses of channel networks. Water Resour. Res. **29**(8), 2599–2614 (1993)
10. Miller, S., et al.: Integrating landscape assessment and hydrologic modeling for landscape change analysis. Am. Water Resources Assoc. **38**(4), 915–929 (2007)
11. Smith, R.E., Parlange, J.Y.: A parameter-efficient hydrologic infiltration model. Water Resour. Res. **14**(3), 533–538 (1978)
12. Woolhiser, D.A., Smith, R.E., Goodrich, D.C.: KINEROS, a kinematic runoff and erosion model. documentation and user manual. U.S. Department of Agriculture. Agric. Res. Serv. **77**, 133 (1990)
13. Levick, L.R., et al.: Adding global soils data to the automated geospatial watershed assessment tool (AGWA). (Sustainability of Semi-Arid and Riparian Areas). International Symposium on Transboundary Water Manage, pp. 16–19 (2004)
14. Inypsa, M.: Projet Intégré de développement Agricole de Tanger-Tétouan, Secteur de Tétouan: Etude des sols au 1/100.000 (Edition au 1/50.000). Direction Provinciale de l'Agriculture de Tétouan (1987)
15. Rwanga, S., Ndambuki, J.: Accuracy assessment of land use/land cover classification using remote sensing and GIS. Int. J. Geosci. **08**(611), (2017)

Implementation of Artificial Intelligence Methods for Solar Energy Prediction

Abdellatif Ait Mansour(✉), Amine Tilioua(✉), and Mohammed Touzani

Department of Physics, Faculty of Sciences and Techniques Errachidia, Moulay Ismaïl University of Meknès, B.P. 509, Boutalamine, Errachidia, Morocco
ab.aitmansour@edu.emi.ac.ma, a.tilioua@umi.ac.ma

Abstract. In order to satisfy the growing world energy demand and decreasing the emission of greenhouse gases the requirement for more green energy has led to an increased focus on research related to forecasting solar energy recently. In this study we aim to develop forecast models, based on Artificial Neural Network and Random Forrest algorithms to predict daily solar energy based on daily historical meteorological data measured between 2019 and 2021. The accuracy and the performance of each model are compared using mean squared error, mean absolute percentage error, mean absolute error, max error and R-squared for evaluation. The prediction of daily solar energy from the daily maximum, minimum and average values of the metrological variables using Artificial Neural Network and Random Forest was carried out. The results obtained indicate that both models can predict daily solar energy with good accuracy (MAPE = 13%). On the one hand, the RF model showed excellent accuracy during the training phase (MAPE = 8%, R2 = 0.97), but it failed to show same results during the testing phase (MAPE = 13%, R2 = 0.79). On the other hand, the ANN was able to maintain the same results during training and testing (MAPE = 13%, R2 = 0.81).

Keywords: Solar energy prediction · Machine learning · Artificial intelligence · Random forest · Artificial Neural Network

1 Introduction

Nowadays, a large part of the world's energy production is based on exhaustible and polluting sources [1]. Despite serious efforts to reduce dependence on these sources by promoting renewables as an alternative, fossil fuels continue to contribute about 63.1% of global electricity generation in 2019 according to the International Energy Agency [2]. The current dependence on this polluting source has caused carbon dioxide (CO2) emissions, greenhouse gas (GHG) problems and environmental pollution [3, 4]. The human activities lead to an increase in GHG emissions and contribute to the increase of some substances that create disturbances on the planet. To satisfy the growing global energy demand [4] and decreasing GHG emissions, renewable energy sources (RES) can have a major role in the future. According to the European Renewable Energy Council in 2006, about half of the world's energy production will be from RES in 2040, and the entire world's energy production will be from RES by 2090.

Y. Farhaoui et al. (Eds.): ICAISE 2022, LNNS 635, pp. 518–523, 2023.
https://doi.org/10.1007/978-3-031-26254-8_75

Solar energy is one of the ecological energy alternatives that are inexhaustible at the human scale. It has grown rapidly in recent years [5]. Morocco as a developing country is in search of inexhaustible and ecological alternative energy sources. These sources can cover domestic energy needs and provide energy services without GHG emissions [3]. Among these sources, solar energy is favored in Morocco because of its exceptional potential. Morocco has one of the highest insolation rates in the world (over 2,200 kWh/m^2 in the southern regions) [6]. The strong dependence of solar energy on environmental parameters makes this energy flow difficult to predict. To this end, several research [7–10] have been conducted to predict the solar energy in the long term, short term, and real time to improve the energy control of this source and to make the energy flow control system predictive instead of been reactive, hence the accurate prediction of the solar energy is major key to increase the rentability of photovoltaic and solar concentrators systems.

The main objective of this paper is to study the possibility of predicting the daily solar energy from average, maximum and minimum metrological parameters values. With this aim in mind, we had firstly gathered the numerical data. Then, the data has been processed to handle missing and duplicate values and to determine the factors that are correlated with the solar energy. Finally, an Artificial Neural Network (ANN) and Random Forest (RF) methods have been used to predict and estimate daily solar energy.

The estimation in advance of the energy flow, allows to manage the electric production by benefiting to the maximum of this energy source. Many research has been carried out in this field. I. Jebli et al. [7] investigated the prediction accuracy of four different Machine Learning (ML) models: linear regression (LR), RF, support vector regression (SVR), and ANN, to predict solar energy, using dataset collected in Errachidia region between 2016 and 2018. The results obtained from their study showed that RF (MAE = 67.33; R2 = 0.58) and ANN (MAE = 30.3; R2 = 0.93) predict better compared to LR(MAE = 79.2; R2 = 0.31) and SVR(MAE = 77.4; R2 = 0,44), hence the results provided by ANN outperform all other models for daily solar energy predictions. M. Cococcioni et al. [10] have developed a prediction model of day ahead energy production in solar PV, using time series analysis and neural networks (NN). They implemented a Nonlinear Autoregressive with Exogenous input (NARX) time series analysis model using a feedforward NN with tapped delay lines. The authors showed that the proposed NN can accurately match the daily generated energy curve and predict the daily accumulated energy with less than 5% error. The focus on solar radiation prediction is continuously growing, several prediction methods have been used, these approaches are classified into two categories: Statistical approaches and numerical weather prediction approaches (NWP) I. Sansa et al. [11] proposed a hybrid model composed of autoregressive-moving-average (ARMA) and (NARX) to predict weekly averages of solar radiation. Their study shows that the performance of the ARMA model is not satisfactory when a large fluctuation in solar radiation is presented, therefore the NARX model was used when solar radiation exceeds 10% using temperature and historical solar radiation as inputs. They have evaluated the performance of the proposed hybrid model using MSE (=3%), MAE (=12%) and RMSE (=18%). S. S. Priya et al. [12] studied the feasibility of an ANN model, developed using the ANN consisting of two-layer feed-forward NN trained with the Lavenberg-Marquardt algorithm, to estimate and predict global solar radiation, using data collected in 4 cities: Bangalore, Thiruvananthapuram, Chennai and Hyderabad. In the training phase of the AI model, they used mean temperature, maximum temperature, minimum temperature, and altitude as input parameters. Their study indicates that the ANN model has less than 8.09% MAPE.

2 Data Collection and Methodology

The data used in this study has been collected between January 2019 and December 2021 in the region of Errachidia in Morocco. It contains the minimum, average and maximum values of weather parameters (Temperature, Humidity, Wind speed, THW index, Chill index, pressure,...) and the solar energy.

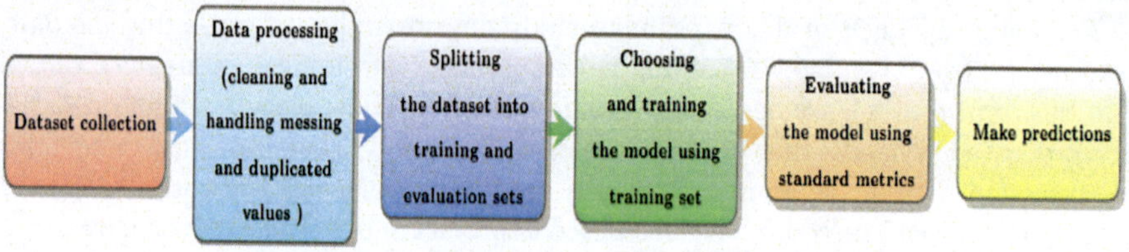

Fig. 1. The methodology used during the study

In this study we aim to predict daily solar energy using numerical weather data, the collected data were analysed to eliminate duplicated and missing values. To determine the factors that are correlated with the solar energy Pearson correlation method has been used. Then, we had splited the dataset in a certain ratio so that our model can learn from a fraction of data, and we could analyse its performance from the rest of that data. In this study two-thirds of the dataset were used to train the two ML models, while the remaining thirds of the data were used to test and evaluate the performance of the two models (see Fig. 1).

3 Results and Discussion

Figures 2 and 3 show the solar energy values measured daily during the three years, and solar energy predicted values using RF and ANN respectively. The RF and ANN predict the daily solar energy with a good accuracy (MAPE = 13%) according to [13]. This accuracy can be justified by the fact that these two ML models support non-linearity. Hence, they are very powerful tools when there is a complex and non-linear relationship between the data.

Table 1 shows the values of the different evaluation parameters. On the one hand, despite the extreme accuracy shown by the RF model during the training phase (R2 = 0.97 and MAPE = 8%), this model failed to show this accuracy during the evaluation phase (R2 = 0.79 and MAPE = 13%), these results are confirmed in Figs. 4 and Fig. 6. On the other hand, the ANN model shows identical results during the training and testing phases (see Fig. 5 and Fig. 7), these results show the robustness of the ANN

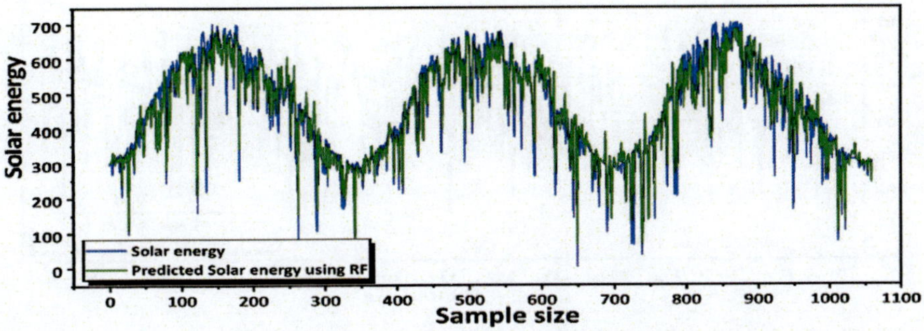

Fig. 2. Three years actual and predicted solar energy using RF

model. However, these results can be explained by the high self-adaptation, robustness and inference ability of this model according to [7], which surpasses other conventional mathematical models in terms of accuracy and adaptability.

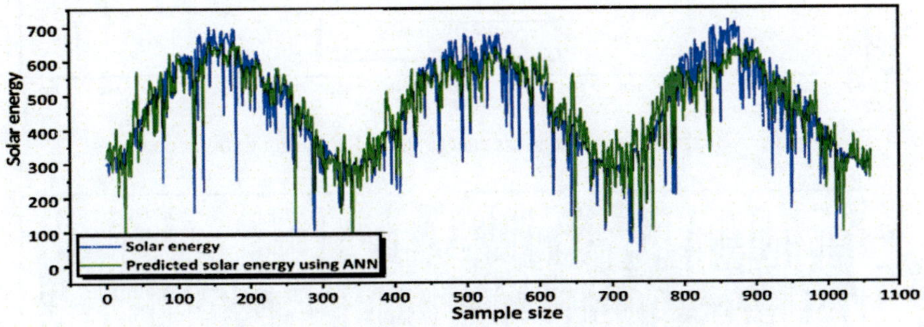

Fig. 3. Three years actual and predicted solar energy using ANN.

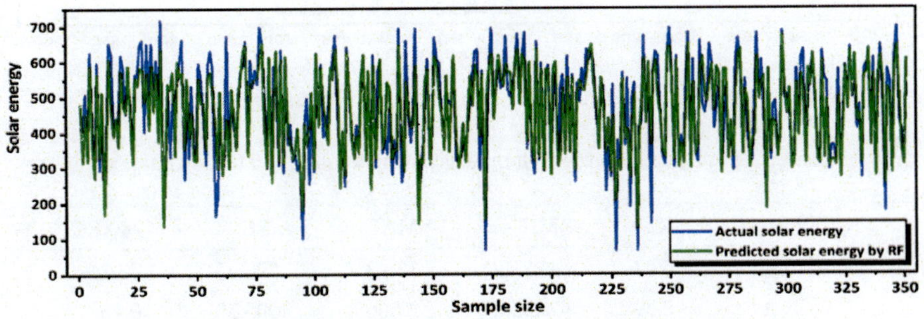

Fig. 4. Daily actual and predicted solar energy using RF during testing phase.

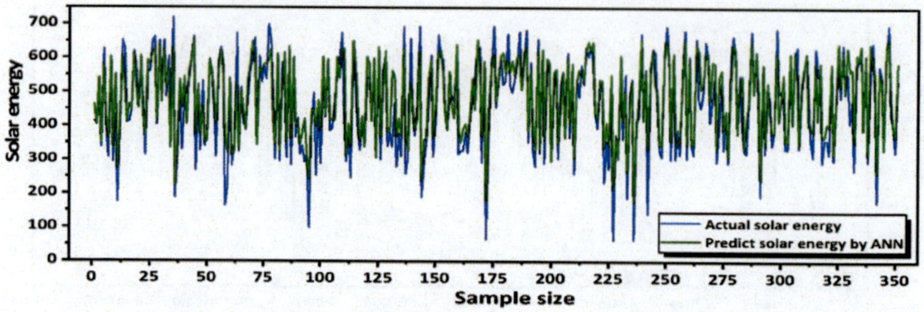

Fig. 5. Daily actual and predicted solar Energy using ANN during testing phase

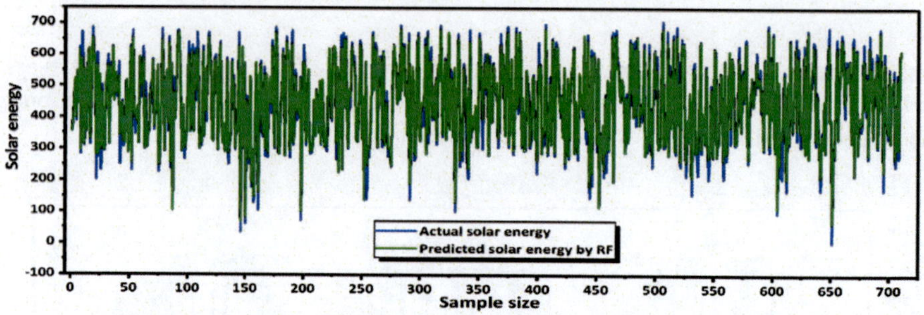

Fig. 6. Actual and predicted solar energy using RF during training phase

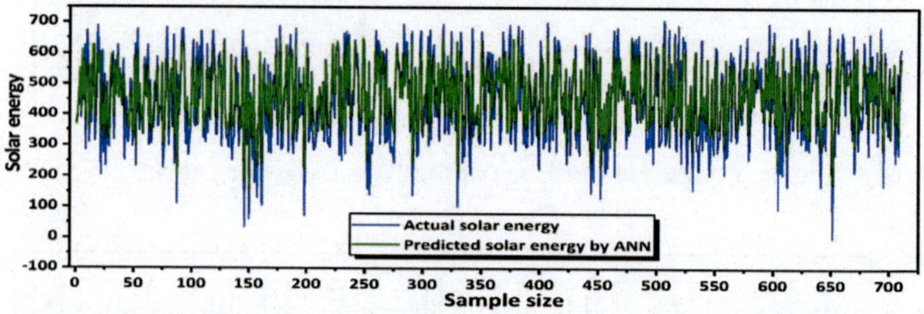

Fig. 7. Actual and predicted solar energy using ANN during training phase

Table 1. Metrics results for both models during training and testing phases

	Phase	R2	ME	MAE	MSE	MAPE
RF	Train	0.97	125.58	13.43	590.45	0.08
	Test	0.79	265.23	37.06	4005.33	0.13
ANN	Train	0.81	373.56	35.71	3700.86	0.13
	Test	0.81	247.23	37.83	3824.66	0.13

4 Conclusion

In this paper, we have examined the use of ANN and RF models to predict daily solar energy. The obtained result of this study indicates that the proposed ML models have a low MAPE (=13%), which means a good prediction accuracy. The models studied use meteorological parameters (daily average, maximum and minimum values) as training parameters to predict daily solar energy. A comparison of the values predicted by these two ML models with the measured data shows that the ML models can accurately predict the daily solar energy. In our future research we will further extend our studies in this field by investigating the use of other deep learning methods (supervised and unsupervised), and Time Series Analysis, as well as the combination of these forecasting methods (hybrid models), in order to improve the forecasting accuracy.

References

1. Trop, P., Goricanec, D.: Comparisons between energy carriers' productions for exploiting renewable energy sources. Energy **108**, 155–161 (2016)
2. Electricity - Fuels & Technologies – IEA. https://www.iea.org/fuels-and-technologies/electricity. Accessed 16 June 2022
3. Qazi, A., et al.: Towards sustainable energy: a systematic review of renewable energy sources, technologies, and public opinions. IEEE Access. **7**, 63837–63851 (2019)
4. Kralova, I., Sjöblom, J.: Biofuels–renewable energy sources: a review. J. Dispersion Sci. Technol. **31**(3), 409–425 (2010)
5. Wang, J., Zhong, H., Lai, X., Xia, Q., Wang, Y., Kang, C.: Exploring key weather factors from analytical modeling toward improved solar power forecasting. IEEE Trans. Smart Grid **10**, 1417–1427 (2019)
6. Solar resource maps of Morocco. https://solargis.com/maps-and-gis-data/download/morocco. Accessed 16 June 2022
7. Jebli, I., Belouadha, F.-Z., Kabbaj, M.I., Tilioua, A.: Prediction of solar energy guided by pearson correlation using machine learning. Energy **224**, 120109 (2021)
8. Aggarwal, S.K., Saini, L.M.: Solar energy prediction using linear and non-linear regularization models: a study on AMS (American Meteorological Society) 2013–14 Solar Energy Prediction Contest. Energy **78**, 247–256 (2014)
9. Detyniecki, M., Marsala, C., Krishnan, A., Siegel, M.: Weather-based solar energy prediction. In: 2012 IEEE International Conference on Fuzzy Systems, pp. 1–7 (2012)
10. Cococcioni, M., D'Andrea, E., Lazzerini, B.: 24-hour-ahead forecasting of energy production in solar PV systems. In: 2011 11th International Conference on Intelligent Systems Design and Applications, pp. 1276–1281 (2011)
11. Sansa, I., Boussaada, Z., Bellaaj, N.M.: Solar radiation prediction using a novel hybrid model of ARMA and NARX. Energies **14**, 6920 (2021)
12. Priya, S.S., Iqbal, M.H., Scholar, P.: Solar radiation prediction using artificial neural network, Int. J. Comput. Appl. **116**(16) (2015)
13. Lewis, C.: International and Business Forecasting Methods. London, Butterworths (1982)

Improved Parallel Genetic Algorithm for Fixed Charge Transportation Problem

Ahmed Lahjouji El Idrissi[1], Ismail Ezzerrifi Amrani[1], and Ahmad El Allaoui[2](✉)

[1] Abdelmalek Essaâdi University, Tetouan, Morocco
a.lahjoujielidrissi@uae.ac.ma
[2] L-STI, T-IDMS, FST Errachidia, Moulay Ismail University of Meknes, Meknes, Morocco
a.elallaoui@umi.ac.ma

Abstract. Genetic algorithm (GA) are a powerful method to different combinatorial and NP-Complete problems. This helps us find an optimal solution if the exact methods are not effective. Additionally, GA delivers the best solution in a reasonable time. But standard genetic algorithms require a rather high execution cost, especially for larger instances. However, the parallel genetic algorithm (PGA) is a mechanism that uses several genetic algorithms at the same time to solve the same task. In this work, we present a new approach to solve the FCTP problem using parallel genetic algorithms (PGA). Indeed, we propose a parallel algorithm-oriented multiprocessor architecture with distributed memory.

Keywords: Genetic algorithm · Parallel genetic algorithm · Fixed charge transportation problem

1 Introduction

The fixed charge transportation problem FCTP is a combinatorial academic and NP-hard problem which is classified among the resource allocation problem [1]. So, there are several optimization methods that already give significant results. So, we choose the genetic algorithms which they are the most used in multiple NP-Hard optimization problems like the FCTP problem [2]. Indeed, genetic algorithms are able to find solutions that may not be precise solutions, but are fast and efficient. We will also try to use parallel genetic algorithms in order to study the differences and extent of the effect of parallelism in accelerating the search for solutions [3]. This paper will be articulated in four parts; The first part generally presents the FCTP problem. In the second part we continued with the general presentation of standard genetic algorithms. In the third we examined at parallel genetic algorithms and some of their models. In particular, three models; Master-Slave Approach, Coarse-Grained Approach and the third is the combination between the two. Finally, in the last part we present the results found to solve the FCTP problem with the proposed approaches in comparison by standard genetic algorithms.

Y. Farhaoui et al. (Eds.): ICAISE 2022, LNNS 635, pp. 524–530, 2023.
https://doi.org/10.1007/978-3-031-26254-8_76

2 Problem Description

The Fixed Charge Transport Problem (FCTP) is an academic optimization problem. It was formulated by Hirsch and Dantzig [4]. Balinski modified the FCTP to make a linear integer problem [5]. More, Adlakha proposed a heuristic algorithm to solve the FCTP [6]. A destination group j = 1,…, n served by a group of production centers i = 1,…, m. while each producer has a given production capacity S_i and each destination has a demand to satisfy D_j. A variable transportation cost is charged for each product unit sent by the producers to the warehouses plus a fixed cost regardless of the quantity transported. The problem is to determine the amount of product to be sent from each production location for each warehouse to minimize the total fixed and variable costs to serve all destinations. The mathematical formulation of the FCTP is as follows:

$$Min\ Z = \sum_{i=1}^{m}\sum_{j=1}^{n}(c_{ij}x_{ij} + f_{ij}y_{ij})$$

$$y_{ij} = \begin{cases} 1, & x_{ij} \geq 0 \\ 0, & x_{ij} = 0 \end{cases}$$

s.t

$$\sum_{j=1}^{n} x_{ij} \leq S_i \qquad i = 1,2,\ldots,m$$

$$\sum_{i=1}^{m} x_{ij} \geq D_j \qquad j = 1,2,\ldots,n$$

$$x_{ij} \geq 0$$

$$i = 1,2,\ldots,m\ ;\ j = 1,2,\ldots,n$$

c_{ij} : variable cost from source i to destination j;
x_{ij} : quantity transported on the route (i,j);
f_{ij} : fixed cost associated with route (i,j);
y_{ij} : a binary variable y_{ij}=1 if x_{ij}>0 and 0 if x_{ij}= 0;
S_i : amount of supply at source i;
D_j : amount of demand at destination j;

3 Standard Genetic Algorithm

Genetic algorithms are very effective methods for solving the optimization problems [7]. It begins with the generation of an initial population. These individuals undergo several operators inspired by biological evolution, called genetic operators (selection, crossing, mutation and insertion) to produce a new population containing in principle a better individual than the previous ones. This population evolves with each iteration of the genetic algorithm until a stopping criterion is satisfied after a certain limited number of iterations or the optimal solution is found. Normally, after a certain number of iterations, the standard genetic algorithm finds the best solution.

3.1 Master-Slave Approach MS-GA

For the Master-Slave approach, the population is divided into subsets which are the subject of genetic operations (crossing and mutation) in parallel on several threads. Then we combine the subsets of the population so that it moves on to the next evolution. For our case, the population is shared between the 4 threads and each one starts the processes individually. Algorithm 2 represents the evolution algorithm in which a transmitted population is divided into subsets (Fig. 1).

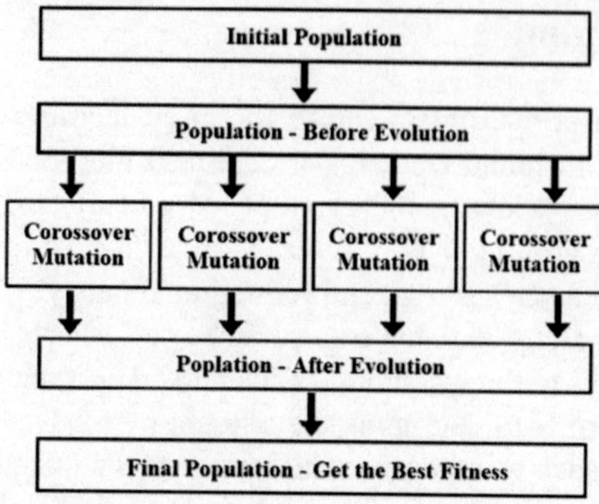

Fig. 1. A representative figure of Master-slave model.

Algorithm 1 : For each thread (Master-Slave Approach)

Begin
Get the size for old population
Create a new population with the same size to store result
For *counter j from 0 to size* ***do***
    Perform evolution on the population
    Perform crossover on the two parents
    Add child Tour to new population
End for
For *each Tour in the new population* ***do***
    Perform mutation on the Tour to add more genetic material
End for
End

Algorithm 2 : General Parallel genetic algorithm (Master-Slave Approach)

Begin
Create two lists with size 4 to store Population
Create 4 threads
Get the size for each sub-population = pop/thread-size(4)
Split the population into sub-populations
For *number of threads* **do**
  Create sub population
  Create threads and start
End for
Wait for all threads to finish using join()
Store evolved populations in an array
Create a new population of the original size
For *number of threads* **do** *(Combine sub-sets into one)*
  Add the new tour list to the new population
End *for*
Return the new population
End

3.2 Coarse-Grained Approach CG-GA

The Coarse-Grained approach consists of partitioning the population into subsets before performing the genetic evolution processes, such that each subpopulation performs the genetic evolution processes on its own and returns the best fitness of its population. After all subpopulations have completed operations, Coarse-Grained must manually choose the most optimal chromosome within the subpopulations. So, we divide the population into four sets of subsets of the population. So in this case Each initialized subpopulation consists of 25 chromosomes as long as the initial population and the maximum of iterations is 100 times. This is shown in Fig. 2.

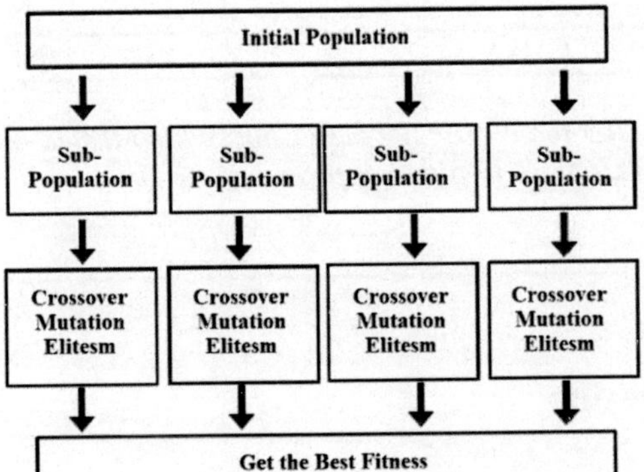

Fig. 2. A representative figure of Coarse-Grained model.

The Coarse-Grained approach is based on separating the population on the different threads before performing the genetic evolution operators, so that the threads only need to send their new generation to the algorithm common general genetics. Indeed, after an iteration of the parallel general algorithm, we can know the best chromosome according to their fitness function. In our case, the operation is done 100 times. Algorithm 3 represents the pseudo-code of the Coarse-Grained Approach.

Algorithm 3 : For each thread (Master-Slave Approach)

Begin
For *counter j from 0 to 100* ***do***
  Perform evolution on the population passed in
End for
End

In Algorithm 3, we present the Coarse-Grained model approach. Then, in this model the population is divided by the number of threads to ensure that the workloads of each thread are equivalent in order to optimize the capacity of the processors.

3.3 Hybrid Approach MSCG-GA

The performance improvement of the master-slave approach decreases when the size of the instances increases on the other hand and the performance of the Coarse-Grained approach is not efficient for small instances. Thus, we propose a hybridization approach MSCG-GA of two previous approaches in order to find new solutions for our problem. In addition, to gain more resolution time. In Algorithm 4, we propose a hybrid approach suitable for all FCTP problem instances. So the approach is changeable depending on the size of the instances. If the instance has a small size (smaller than 10 * 20), then the algorithm tries to apply the Master-Slave parallel approach. Otherwise, it applies the parallel Coarse-Grained approach. The proposed approach appears in Algorithm 4.

Algorithm 4 : MSCG-GA

Begin
If *the size is smaller than (10*20)* ***then*** *Choose the Master -Slave Approach*
Else *Choose the Coarse -Grained Approach*
End if
End

4 Numerical Results

After the results obtained for other similar problems we applied this approach on several instances of the FCTP problem already cited in previous articles using the priority-based representation [8]. In order to see the effectiveness of the proposed approaches, we compared them with the standard genetic algorithm using four test problems (Table 1).

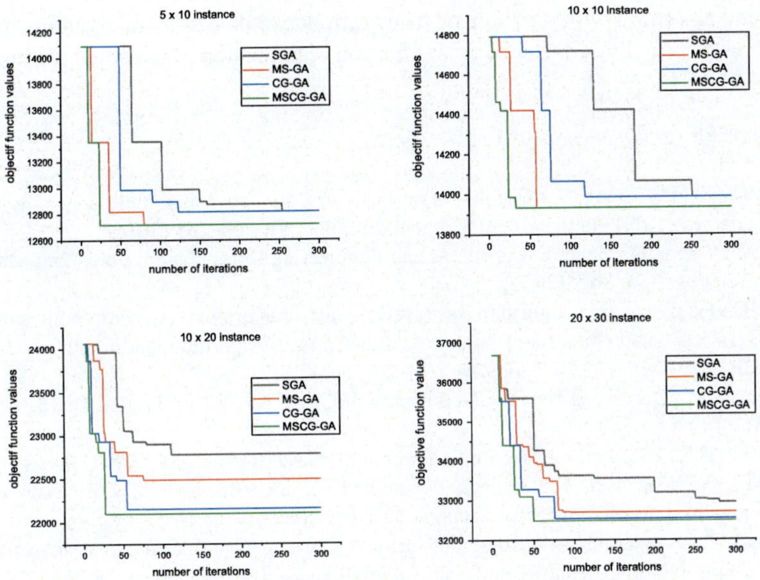

Fig. 3. Optimal solution based on the number of iterations for the FCTP problem.

Table 1. Best and average results by proposed approach and standard GA for FCTP problem.

Problem size	SGA	MS-GA	CG-GA	MSCG-GA
4 × 5	9291	9291	9291	**9291**
5 × 10	12734	**12718**	12734	**12718**
10 × 10	13987	**13934**	13987	**13934**
10 × 20	22258	22150	**22095**	**22095**
20 × 30	32936	32683	**32526**	**32471**
30 × 50	55450	55269	**55007**	**55007**

The obtained results show that our approaches which based on parallelism brings an improvement of the performances of GA. In effect; the Master-Slave MS-GA approach is advantageous for small instances. On the other hand, the Coarse-Grained approach is more effective for larges but the most important is the hybrid approach which brings together the characteristics of two approaches (see Fig. 3).

5 Conclusion

In this work, we introduced the principle of parallelism in the genetic algorithm to improve the performance. Three architectures are proposed, namely Master-Slave, Coarse-Grained and Hybrid Approach. The three approaches show the efficiency of the genetic algorithm with parallelism. They introduced diversity and dynamism in the population, which allowed to save more time and optimization of the objective function.

References

1. Nicholson, C.D., Zhang, W.: Optimal network flow: a predictive analytics perspective on the fixed-charge network flow problem. Comput. Ind. Eng. **99**, 260–268 (2016)
2. Gen, M., Cheng, R.: Genetic Algorithms and Engineering Optimization, Book, 2nd edn. John Wiley & Sons, New York (2000)
3. Peng, C.: Parallel genetic algorithm for travelling salesman problem. In: Proceedings of SPIE 12253, International Conference on Automation Control, Algorithm, and Intelligent Bionics (ACAIB 2022), vol. 122530Z, pp. 259–267 (2022)
4. Hirsch, W.M., Danzig, G.B.: The fixed charge problem. Naval Res. Logist. Q. **15**, 413–424 (1968)
5. Balinski, M.L.: Fixed-cost transportation problems. Naval Res. Logist. Q. **8**(1), 41–54 (1961)
6. Adlakha, V., Kowalski, K.: A heuristic method for more-for-less in distribution related problems. Int. J. Math. Educ. Sci. Technol. **32**, 61–71 (2001)
7. Holland, J.: Adaptation in natural and artificial systems: an introductory analysis with applications to biology, control, and artificial intelligence. (U. of M. Press, Ed.) (1975)
8. Lotfia, M., Tavakkoli, R.: A genetic algorithm using priority-based encoding with new operators for fixed charge transportation problems. Appl. Soft Comput. **13**, 2711–2726 (2013)

Increasing the Efficiency of Industry 4.0 Through the Use of the Digital Twin Concept

Kawtar Agouzzal(✉), Ahmed Abbou, and Abdelghani Hajji

Mohammed V University of Rabat, Mohammadia School of Engineers, Rabat, Morocco
agouzzalkawtar96@gmail.com, abbou@emi.ac.ma

Abstract. The ensuing specifications for the layout, setup, and operation of our facilities become essential for success. In the past, we have frequently made structures and control systems more sophisticated, which has led to rigid, monolithic production systems. However, the future must become "lean" in both organization and planning.

technology, too! We must create the technological tools necessary to reduce planning work, hasten setup and planning, and quickly adjust to the product modifications while operations are in progress. We should incorporate smart technologies into our daily lives in order to solve these difficulties.

The development of wireless communication will enable us to do away with cables. Numerous old control panels will be replaced by powerful computers or cellphones, and fragments will be replaced by abstract services byte controllable. These advances will lead not only to the mobility of people but also to the digitization of industrial systems which offers opportunities for innovation in terms of energy transition and cost reduction. To do this, autonomous systems will need to interact with their real environment, also known as a "digital twin".

The objective of this article is to evaluate the large increase in complexity to ensure the appropriate behavior of the autonomous system during production, in order to achieve the desired production goal. This goal can only be reasonably achieved through extensive use of model-based simulation, not only during design and planning but also during other phases of the life cycle, for purposes such as diagnosis and optimization of operations.

This article presents the functioning of the four aspects that determine the future of manufacturing: Modularity - Connectivity - Autonomy - Digital pairing.

Keywords: Digital twin · Synergies · Digitization · Energy saving · MBSE · CPPS

1 Introduction

Over the past few years, improvements in communications and microelectronics technology has become more and more prevalent in our daily lives [9]. Today's reality still keeps us a long way from this ideal. We place too much emphasis on technology and try to get a competitive edge by packing more features into each individual device. Despite all efforts to change this, production plants are also getting more and more sophisticated.

Y. Farhaoui et al. (Eds.): ICAISE 2022, LNNS 635, pp. 531–536, 2023.
https://doi.org/10.1007/978-3-031-26254-8_77

Longer planning phases and an intolerably long time to market follow as a result. On the basis of cutting-edge computer technology, it was hoped that completely automated plants would find solutions to challenges related to cost and quality [8]. At every level of product design, manufacture, and production, digitalization boosts efficiency across all industries, particularly those that use significant amounts of energy. Energy is needed for the operation of digital technologies and the new services they enable. And to the degree that digitalization increases energy use efficiency, it may also induce industry to utilize more energy, particularly through procedures that result in substantial cost reductions at negligible or no net cost. Manufacturing companies now have the chance to reach a completely new level of productivity because of the growing digitization of every stage of the production process. Modularity in product design and production modules is the first step, which boosts the effectiveness of production system engineering. With autonomy, the production system can react to unforeseen circumstances intelligently and effectively without the need for supervisory intervention. The digital twin, a notion in which the data created at each stage of the product life cycle is made transparently available to the next stages, is the means to accomplish all of these aims [10]. This article's goal is to assess the significant complexity increase and ensure that the autonomous system behaves appropriately during production to meet the targeted production goal. The model-based simulation must be heavily utilized to attain this goal, not only during the design and planning phases but also during other life cycle phases for tasks like diagnostic and operational optimization. The four elements that will affect manufacturing's future are highlighted in this article: Connectivity, Digital, Autonomy, and Modularity [1].

2 Methodology

The digital twin is a very dynamic idea that gets more complex over the course of its life. The digital twin, which is delivered concurrently with the product or even earlier, is built around the MBSE. It serves as the framework for the support systems while they are in use. These software solutions assist operators by making simulation-based forecasts and computing control and service decisions when used in conjunction with intelligent data techniques. Over the course of the system's or product's life, the models automatically change [5]. Autonomous systems must have as much knowledge as feasible about the system's general condition, the goods to be produced, the geometry and capabilities of the parts and tools to be employed, and their own configuration and capabilities. All of this historical information will be gathered via the idea of the digital twin and made available to the autonomous systems that are currently carrying out certain industrial procedures. Consequently, the digital twin depicts the entire status of the process and environment at any given time. This data and the models the digital twin provide will be used to predict the effects of the autonomous system's actions in a specific situation and enable the autonomous system to change its course of action in response to changes in the products, production volume, and automatic management of exceptions and errors without the need for manual supervision or reconfiguration.

3 Modeling Autonomous Systems

3.1 Cyber-Physical Production System (CPPS)

Consider a system comprising four CPPSs, called a "cyber-physical production system" (CPPS): *Robotic loading/unloading station (including a buffer carousel), *CNC drilling machine, *CNC milling machine and *Transport system (Fig. 2). Each production unit keeps a digital doppelganger of itself that includes data about the unit ID, its skills and capabilities, its configuration, its present states, and the pallets it now has. Each pallet maintains a local copy of the digital twin of the component it transports (Fig. 1) [10]. All the information required for production in this CPPS is contained in the digital twin, including the number of components in the pallet, the number of parts in the pallet, and more: *Room ID and type of room, *Production order number and priority, *Production flow with information such as the NC program number, *List of skills and tools for each step of the operation. *Current state and location, *NC Program Files, *Production history (e.g. which operation was performed on which machine).

While bulk data, like the NC program file, are kept in the contact memory, frequently used information, like part identification, is stored on the RFID. While bulk data, like the NC program file, are kept in the contact memory, frequently used information, like part identification, is stored on the RFID [11].

3.2 Model of Digital Twin Synchronization

Consider the movement of pieces through this CPPS:

At the loading/unloading station, a part is taken out of a tray and fastened to a pallet. The tray memory's digital copy of the part is read, and it is then written to the pallet memory.

- The raw part-carrying pallet is carried to the buffer first.
- The pallet is sent to the milling machine's input branch when it is ready.
- The part's digital twin is updated after milling. The next procedure is then awaited in the output segment.
- Important: The pallet can be moved to the carousel's buffer to clear space so as not to obstruct the machine.
- The pallet is brought in for processing as soon as the drill is available.
- Following drilling, the digital twin of the part is updated, and the pallet is then brought to the buffer.
- The part is transported from the buffer to the load/unload station, where it is taken off the pallet and put in a tray. After that, the part's digital twin is moved from the pallet memory to the tray memory.

3.3 Model of Autonomy in the Flow of Part

By giving the production units direct access to the digital twins of the parts, which are typically managed by a central manufacturing execution system, the production units are

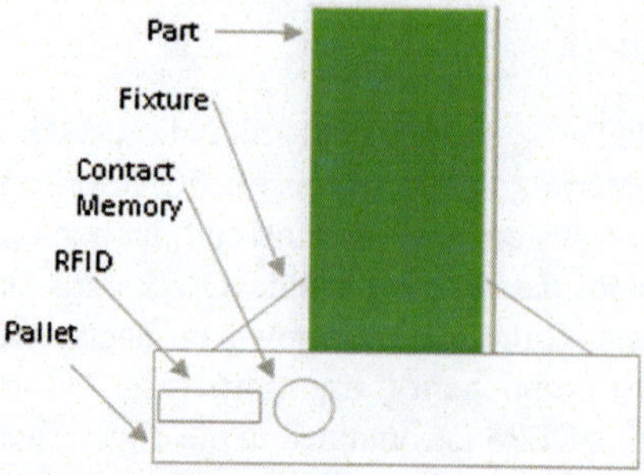

Fig. 1. Example of local memory on a pallet

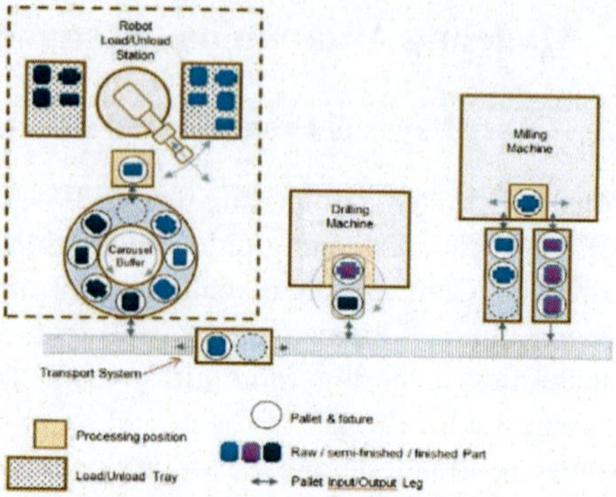

Fig. 2. Model of a cyber-physical production system

now able to arrange the component flow on their own, for instance through a negotiation process (Siemens, 2014) [12]. Let's use the recently finished milling process on the portion at location (4):

- The milling machine decides that drilling is the next operation after examining the digital twin of the component.
- He requests that all equipment volunteer to perform the drilling (i.e. to the drilling machine as well as to himself since he also has the drilling skill).
- It compares the costs and times of all returning offers according to predetermined criteria:
- **Case A:** The milling machine sends a transport order to have the part delivered to it, and the drilling machine receives the order (this should be the typical case by design).
- **Case B:** The milling machine receives the order (maybe because the drilling machine has a backlog of work), and it starts drilling the workpiece.
- **In case c,** the milling machine sends a transport order for the component to be buffered since no machine is able to bid (for example, the drilling machine is broken and the milling machine is lacking the necessary tools).

3.4 Model of Autonomy in Handling Breakdowns

Assuming that the milling machine in the CPPS example has the capacity to stamp the pallets in addition to milling and drilling. Think about the failure of a carousel buffer as an example. Since the finished part in the drill has nowhere to go, this would often result in a standstill in the part flow. The transport system temporarily substitutes the milling machine, which has the ability to buffer, for the carousel buffer. To make place for the semi-finished parts to be drilled, the finished drill bits can now be sent to the milling machine's infeed leg for buffering. By doing this, the buffer carousel would not need to be fixed until all the pieces had been machined by the CPPS. The delivery of raw parts is resumed, the finished parts in the milling machine are removed, and production resumes as soon as the buffer carousel is functioning once again.

4 Discussion of Results and Comparison

We've given a succinct summary of the digital twin knowledge base in this article. Without a doubt, the framework's effectiveness must be assessed in terms of the tasks of thinking and information it can learn. We have noted that this system excels at a variety of reasoning tasks, which distinguishes it from other automatic knowledge processing systems. These tasks require the acquisition of episodic mementos as well as reasoning based on simulation, visual reasoning using an interior global model, and reasoning visually. Our comparison of the digital twin concept with other systems shows that it appears to be the most successful. These mechanisms work in concert to greatly enhance the ability of automatic systems to carry out manipulation tasks. With the assistance of an online knowledge service for OPENEASE, many of these features can be tested.

5 Conclusion

The degree of technological use and digitization varies greatly between member states, regions, and corporate size. In this case, a thorough knowledge management of the production system's capabilities and state-of-the-art is necessary. The production system's collection, storage, and processing of all accessible (sensor) data as well as operational factors like order quantity primarily handle the first topic. Although some desirable quantities can be measured directly, it might be difficult to anticipate future behavior based solely on operating data, especially when a flexible production system is taken into account. On the other hand, combining real-world data with design simulation models can produce accurate predictions based on real-world data. In order to support operators and planners during routine operations as well as for maintenance and service using simulation-based forecasts, this enables the usage of simulation support systems. This process is made possible by the digital twin concept since all models and data are accessible in a unified and well-aligned environment. The sector is currently undergoing a dual digital and climate change, and there are crucial the potential for energy savings and decarbonization, among other synergies, between the two. However, the energy business could also be disrupted by digitalization. New regulations, however, based on a thorough comprehension of the energy and digital sectors, would minimize these negative effects and emphasize the positive ones.

References

1. Acatech, F.: Recommendations for implementing the strategic initiative INDUSTRIE 4.0, Final report of the INDUSTRUE 4.0 Working Group (2013)
2. Beckey, George, A.: Autonomous Robots. MIT Press (2005)
3. Shafto, M., et al.: DRAFT modeling simulation information technology & processing roadmap. Technol. Area **11** (2010)
4. Siemens, A worldwide production language, Siemens Industry Journal. siemens.com/industry (2014)
5. Tenorth, M., Beetz, M.: A knowledge processing infrastructure for cognition-enabled robots. Int. J. Robot. Res. (IJRR) **32**(5), 566–590 (2013)

6. Wahlster, W.: Digital product memory: embedded system keep a diary. In: Harting tec News, vol. 15, pp. 7–9 (2007)
7. Zühlke, D.: Smart factory - from vision to reality in factory technologies. In: Proceedings of the 17th IFAC World Congress (plenary paper), vol. 41, no. 2, pp. 14101–14108. Seoul (2008)
8. Yin, K.: Case Study Research and Applications: Design and Methods. SAGE, London (2017)
9. Zainal, Z.: Case study as a research method. J. Kemanusiaan, 1–6 (2007)
10. Kljajić, M., Škraba, A.: Simulation approach to decision assessment in enterprises. Syst. Theory Simul. **75**(4), 199–210 (2002)
11. Goossens, P.: Industry 4.0 and the Power of the Digital Twin, from Design News Direct, pp. 5–3 (2017)
12. Nankervis, A., Connell, J., Montague, A., Burgess, J. (eds.): The Fourth Industrial Revolution. Springer, Singapore (2021). https://doi.org/10.1007/978-981-16-1614-3
13. Thiers, G., Christian, M.: Automated production system simulations using commercial off-the-shelf simulation tools. In: Proceedings of the 2016 Winter Simulation Conference, pp. 1036–1047 (2016)

Explainable Machine Learning for Identifying Malicious Profiles in Online Social Networks

Amine Sallah[1(✉)], El Arbi Abdellaoui Alaoui[2], and Said Agoujil[3]

[1] Department of Computer Science, Faculty of Sciences and Techniques, Moulay Ismail University of Meknes, Errachidia, Morocco
aminefste@gmail.com

[2] Department of Sciences, Ecole Normale Supérieure, Moulay Ismail University of Meknes, Meknes, Morocco

[3] École Nationale de Commerce et de Gestion, Moulay Ismail University of Meknes, El Hajeb, Morocco

Abstract. According to Our World In Data, more than two-thirds of all Internet users regularly utilize numerous social networking networks. A fake virtual identity might result in issues including the transmission of spam, malware, and misinformation, so verifying a genuine profile is a subject of significant concern. In this study, we focus specifically on an Instagram dataset using machine learning (ML). How do we understand the decisions suggested by a fake account detection system in order that we can trust them? Here, we have included approaches for understanding complicated Machine Learning Models to understand the reasoning behind a model choice, such as SHAP values and Anchors. At the very least, explainability can facilitate understanding of various opaque models. Consequently, there has been a significant resurgence in recent years of scholarly interest in the topic of Explainable Artificial Intelligence (XAI), which is focused on creating novel techniques for interpreting and explaining machine learning models. In particular, the interpretation of the decisions made by the fake account detection system used in the Instagram dataset is the subject of this study's analysis of machine learning interpretability techniques.

Keywords: Fake account · Machine learning · Instagram · XAI · Classification · Interpretability · Explainability

1 Introduction

Artificial intelligence (AI) has much potential to improve private and public life. The automatic discovery of patterns and structures in vast troves of data is a fundamental component of data science, and it is now driving applications in various fields. It is better to have a precise model, but it is also better to get an explainable model, especially for making efficient and transparent decisions [15].

Y. Farhaoui et al. (Eds.): ICAISE 2022, LNNS 635, pp. 537–545, 2023.
https://doi.org/10.1007/978-3-031-26254-8_78

The interpretability of models is one of the most important flaws in current machine learning and deep learning algorithms, and it is addressed by the topic of explainable AI. Understanding how and why a prediction is formed is crucial as algorithms get more potent and provide forecasts with higher accuracy. Users would find it challenging to believe the forecasts made by real-world AI systems without interpretability and explainability. Explainability and interpretable machine learning are crucial for the following reasons:

- Interpretability is necessary in order to describe how the model functions from the standpoints of both diagnostic and debugging.
- To explain the model's judgments and the reasoning behind them to the end-user, we need to provide explanations.

In order to deploy a model, it is essential to investigate the biases present in most datasets or models. One method of identifying these biases in the model is interpretability. Interpretability of models is a necessity since many businesses, like banking and healthcare, have regulatory obligations regarding transparency, trust, explainability, and fidelity of models. For example, according to various interpretations of the GDPR (General Data Protection Regulation) rules in the European Union, AI systems must provide justifications for their findings. (According to some interpretations, persons need to be made aware when automated methods are used in choices that might impact them.) The fact that ML models are being used more and more in a variety of industries helps the reader understand the article's goal in particular. However, as methodologies become more common and complicated, business stakeholders at the very least are becoming more concerned about the limitations of models, data-specific biases, and other issues. Similar to this, data science practitioners sometimes rely on industry standards like SHAP [9] because they are unaware of new academic methodologies or find it difficult to understand the distinctions between various methods. In order to save space, we will concentrate on data-driven techniques, particularly machine learning models, with the primary objective of classifying Instagram accounts as authentic or fraudulent. We shall refer to explainable machine learning in the article as XAI for the sake of convenience, although we should stress that the AI community has a much wider perspective than only (statistical) classification problems [4]. The writers of this article [2] have provided a summary of explaibiliy approaches for developing precise and significant explanations, but effectively expressing them to a wide audience is also crucial. Last but not least, when it comes to terminology, we will primarily refer to a model as the underlying machine learning technique, such as convolutional neural networks, logistic regression, or random forests, and we will refer to an approach or method as an algorithmic pipeline that is used to explicitly simplify, interpret, or otherwise derive explanations from a model. The effectiveness of a machine learning model and its capacity to provide explicable and understandable predictions clearly trade off. On the one hand, there are what are referred to as "black-box models," which include ensembles [5,10] and deep learning [8]. The so-called "white-box" or "glass-box" models, in contrast, readily provide results that can be understood. Common examples are linear [16] and decision-tree based [13] models. Although easier to understand and use, the later models are less effective than the older and fall short of cutting-edge performance. Their economical design is to blame for both their subpar performance and their ease of interpretation and explanation. 381 distinct scientific publications published between 2004 and 2018 were gathered and examined as part of a thorough literature assessment by Adadi and Berrada [1]. They organized all of the research in the area of explainable AI along four primary axes and emphasized the need of introducing more formalism into the field of XAI as well as increasing interaction between people and machines. They advocated adopting explainability in other areas of machine learning after emphasizing the tendency of the community to examine explainability solely in terms of modeling. Finally, they made a study suggestion that may focus on the composition of already-used explainability approaches.

2 Approaches for Explainability

How do we understand the decisions suggested by these models in order that we can trust them?. One of the ethical issues that is very often raised in our modern world is that of transparency. Indeed, it is necessary to be able to explain the prediction or decision of an artificial intelligence to a customer or simple user of the AI [3]. In Moreover the explanation of an algorithm of complex learning allows also to optimize the model by predetermining the important variables to identify a fake account. For opaque models in particular, we might consider the following types of post-hoc explanations:

2.1 Rule-Based Learning

The core of rule-based learning is the intuitive act of creating rules that specify how a model produces its outputs. The resultant rules vary in complexity from simple "if-else" statements to fuzzy rules, or propositional rules that encode intricate connections between variables. These systems are often simple to comprehend, making them transparent models as people also use rules in daily life. Having said that, some design elements, such as the coverage (quantity) and specificity (length) of the created rules, determine the precise degree of transparency. The designers of LIME [12] noted that LIME was unable to accurately describe the model in various instances. As a result, they suggested Anchors [11] as a novel technique for doing model interpretation. In contrast to LIME, Anchors learns how to describe the model using the "local area." The phrase "local region" refers to an improved generation of data set for explanation. An anchor explanation is a rule that "anchors" the prediction enough locally so that changes to the instance's remaining feature values are unimportant. In other words, the forecast is (nearly) always the same for situations when the anchor holds. Any two- or more-class black box classifier may be explained using the anchor approach. The classifier just has to build a function that accepts raw text or a numpy array and delivers a prediction for us to consider it to be functional (integer). A well-liked technique for deciphering the logic underlying the conclusions of sophisticated machine learning (ML) classifiers is rule-based explanations. Anchors should put their attention on regional explanations based on if-then rules that are relevant in the area around a target instance. Although this has shown to be successful in providing trustworthy explanations, anchor-based explanations are not without its limitations. These include lengthy, excessively detailed regulations and imprecise explanations. Anchors are simple to understand, have very clear coverage, and are only applicable when all of the requirements of the rule are satisfied. If they do apply, they do so with great accuracy (by design).

2.2 Feature Relevance

The SHAP (SHapley Additive exPlanations) contribution is one of the most well-liked ones both here and in XAI as a whole [9]. Building a linear model around the instance that has to be explained in this situation, then interpreting the coefficients to represent the significance of the characteristic, is the goal. This concept is comparable to LIME; in fact, LIME and SHAP are connected; nonetheless, SHAP has a number of attractive theoretical aspects. Its mathematical foundations are based on Shapley values, notably coalitional Game Theory [14]. The average predicted marginal contribution to the model's choice, after all potential combinations have been taken into account, is roughly the Shapley value of a feature. The Shapley value φ_i has the following generic expression:

$$\varphi_i = \sum_{S \subseteq N \backslash \{i\}} \frac{|S|!(M - |S| - 1)!}{M!} (f_x(S \cup i) - f_x(S)), \tag{1}$$

such as M Number of features, S features set, f_x the prediction function at time x, $f_x(S) = E[f(x)|x_S]$, i is the ith feature.

SHAP approach is additive, so a prediction can be written as the sum of the different effects of the variables (shap value φ_i) added to the base value φ_0. The base value being the average of all the predictions of the dataset:

$$f(x) = y_{pred} = \varphi_0 + \sum_{i=1}^{M} \varphi_i z_i'. \tag{2}$$

with, y_{pred} the predicted value of the model for this example, $z' \in \{0,1\}^M$ when the variable is observed $z_i' = 1$ or unknown $z_i' = 0$.

2.3 Visual Explanations

Create visuals with the intention of making models easier to grasp. The established methodologies may aid in understanding the decision boundary or the way characteristics interact with one another, despite certain inherent difficulties (such as our incapacity to comprehend more than three dimensions). Because of this, visualizations are often used in conjunction with other tactics, particularly when speaking to non-expert audiences. PDP plots, or partial dependence plots, display the marginal influence of several factors on the result. As a model-agnostic approach, it does not rely on the kind of model and defines the relationship between model inputs and model outputs. communicating with a non-technical audience is simpler. The majority of the strategies are simple to use and logical.

3 Methodology

In this section, we describe our proposed approach. Specifically, we explain how each technique is used in the process of identifying fake profiles on Instagram. The process consists of three main mining steps. The methodology to detect a fake profile on Instagram is described in Fig. 1. First step, a dataset containing the relevant profile functionalities is available. In the second step, we have fitted Random Forest Classifier. In the last step, we show that XAI methods enable users to predict how a model would behave on unseen instances with less effort and higher precision.

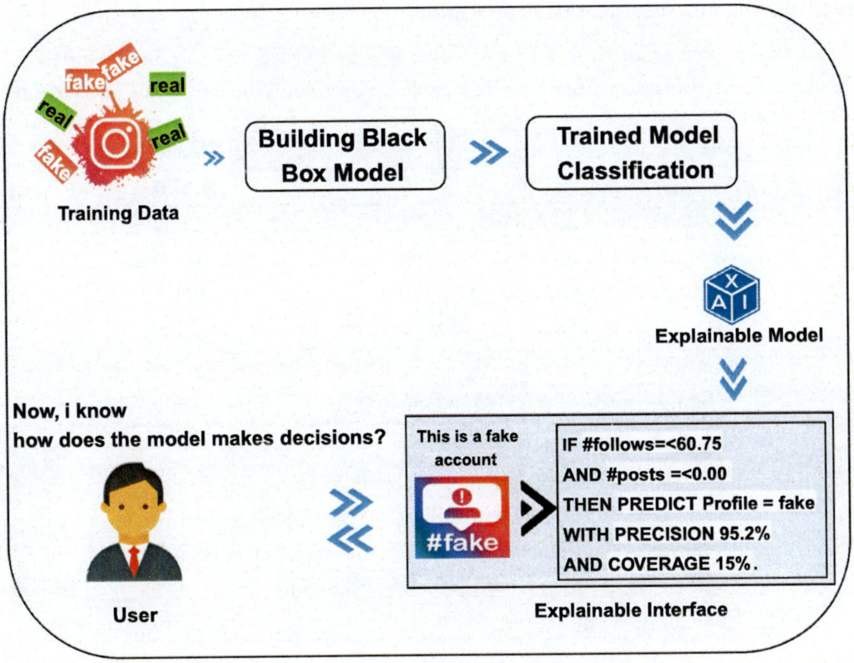

Fig. 1. Design approach to detect fake account

We applied our experience to Instagram account dataset, this Dataset was taken from Kaggle. The target variable named fake, which allows to know if it is a fake account or not, it takes two values 0 (real and 1 (fake).

Initially, Random Forests (RF) was suggested as a technique to enhance the accuracy of single decision trees, which often have poor generalization due to overfitting. In an effort to lower the variance of the resultant model and improve generalization, random forests seek to solve this problem by merging many trees together [6,7]. Each individual tree is trained on a separate portion of the training dataset in order to do this, collecting various aspects of the data distribution in order to provide an aggregate forecast. This process produces models that are very expressive and accurate, but it sacrifices interpretability since a full forest is much harder to grasp than a single tree, necessitating the employment of post-hoc explainability approaches to help the user comprehend the decision-making process.

4 Results and Analysis

In this section, we interpret the results of the classification models. After creating our Random Forest classifier using the training data sets, we apply the model to new data. We want to know how the black-box model makes decisions by exploiting various interpretation methods.

Table 1. Contributions Table-How has each feature contributed to the prediction?

Reason	Effect
Average of population	49.88%
profilepic = −1.5771810295587505	+12.52%
#followers = −0.10801175170356069	+10.23%
#posts = −0.28106920778535793	+9.31%
description length = −0.5942076966276106	+5.34%
nums/length username = −0.7813955851993025	−5.03%
fullname words = 0.5118653038645521	−4.99%
private = 1.317305628032864	−2.62%
external URL = −0.34385530815430465	+1.36%
nums/length fullname = −0.2947966769422395	−0.77%
#follows = -0.42559686710105077	+0.75%
name==username = −0.18257418583505533	−0.25%
Final prediction	75.73%

Table 1 shows the contribution that each individual feature has had on the prediction for a specific observation. The contributions (starting from the population average) add up to the final prediction. This allows to explain exactly

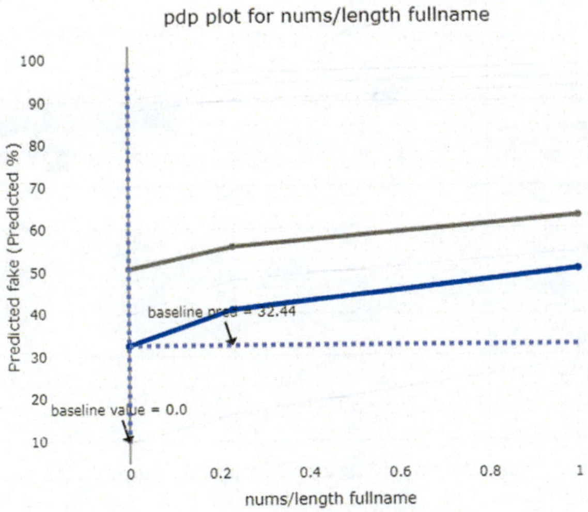

Fig. 2. PDP for fake account probability based on nums/length fullname

how each individual prediction has been built up from all the individual ingredients in the model. This plot shows us what are the main features affecting the prediction of a single observation, and the magnitude of the SHAP value for each feature. For example the feature *#profilepic*, has a positive impact, which pushes the predicted value higher.

We compute and visualize the partial dependence of the fake account probability on the feature *nums/length fullname* for the random forest. The PDP in the Fig. 2 show how the model prediction would change if you change one particular feature. The plot shows us a sample of observations and how these observations would change with this feature (gridlines). The average effect is shown in grey. The effect of changing the feature for a single Index is shown in blue. For *nums/length fullname*, it is clear that the number of numbers in the fullname increases the probability of being a fake account.

We now want to discover the reason behind the model's prediction that a certain person has an actual profile. The anchors algorithm offers a result justification similar to that in Fig. 3. The coverage of a rule is the proportion of examples covered by a rule, and The precision of a rule is the proportion of examples correctly classified by a rule. As you can see the corresponding anchors explanation is: $IF\#followers > 688.00$ AND $\#posts > 77.00$ THEN PREDICT Profile = real WITH PRECISION 97% AND COVERAGE 16%. The example shows how anchors can provide essential insights into a model's prediction and its underlying reasoning. The result shows which attributes were taken into account by the model, which in this case, is the $\#followers > 688.00$ and $\#posts > 77.00$. This rule may be used to confirm the model's behavior since humans are the ultimate arbiters of accuracy. In addition, the anchor informs us that it holds for 16% of cases in perturbation space. In certain circumstances, the explanation

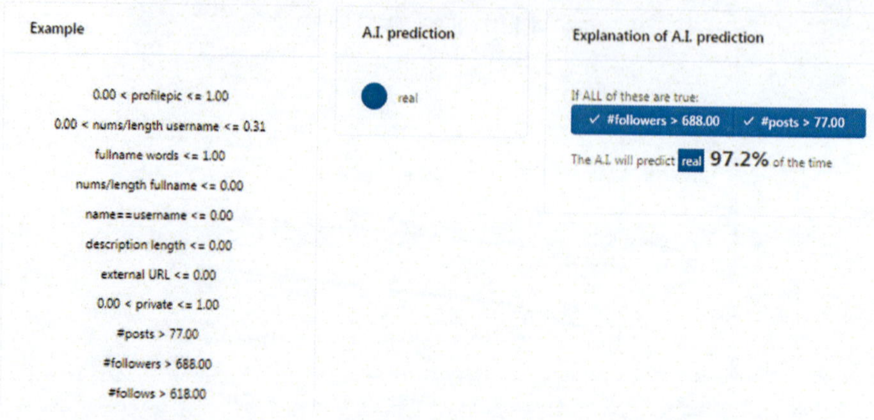

Fig. 3. Generated anchors for Tabular Instagram dataset

is 97% accurate, indicating that the expected result is nearly entirely due to the predicates shown.

Specific rules with a greater number of feature predicates are used for predictions that are close to decision boundaries, and as a result, their anchors may be complicated and provide insufficient coverage. We ensure (with a high chance) that precision will be over 0.95 since we set the threshold at 0.95, which means that predictions for situations in which the anchor holds will be accurate at least 95% of the time.

5 Conclusion

The objectives of the developed techniques are still somewhat unclear since XAI is a vast and young discipline of machine learning. The methods used in this research are representative of the variety of different explainability perspectives that are taken into account in the industry. For instance, model simplification approaches build comparatively simple models as proxies for the opaque ones, while visual explanations examine a model's inner understanding of a problem using graphical tools. Feature relevance approaches provide insights by measuring and quantitatively ranking the importance of a feature. It is important to note at this point that the kind of insights the user would want to get or possibly the types of explanations they are more at ease comprehending will directly affect the approach they choose for the application at hand.

Data Availability. The data used to support the findings of this study are available from the corresponding author upon request.

References

1. Adadi, A., Berrada, M.: Peeking inside the black-box: a survey on explainable artificial intelligence (XAI). IEEE Access **6**, 52138–52160 (2018)

2. Belle, V., Papantonis, I.: Principles and practice of explainable machine learning (2021)
3. Carvalho, D.V., Pereira, E.M., Cardoso, J.S.: Machine learning interpretability: a survey on methods and metrics (2019)
4. Chakraborti, T., Sreedharan, S., Grover, S., Kambhampati, S.: Plan explanations as model reconciliation–an empirical study. In: 2019 14th ACM/IEEE International Conference on Human-Robot Interaction (HRI), pp. 258–266. IEEE (2019)
5. Chen, C., Guestrin, T.: XGBoost: a scalable tree boosting system. In: Proceedings of the 22nd ACM SIGKDD International Conference on Knowledge Discovery and Data Mining, San Francisco, California (2016)
6. Kramer, O., et al.: On ensemble classifiers for nonintrusive appliance load monitoring. In: Corchado, E., Snášel, V., Abraham, A., Woźniak, M., Graña, M., Cho, S.-B. (eds.) HAIS 2012. LNCS (LNAI), vol. 7208, pp. 322–331. Springer, Heidelberg (2012). https://doi.org/10.1007/978-3-642-28942-2_29
7. Latha, C.B.C., Jeeva, S.C.: Improving the accuracy of prediction of heart disease risk based on ensemble classification techniques. Inform. Med. Unlocked **16**, 100203 (2019)
8. LeCun, Y., Bengio, Y., Hinton, G.: Deep learning. Nature **521**(7553), 436–444 (2015)
9. Lundberg, S.M., Lee, S.I.: A unified approach to interpreting model predictions. In: Advances in Neural Information Processing Systems, 2017-December(Section 2), pp. 4766–4775 (2017)
10. Polikar, R.: Ensemble learning. In: Zhang, C., Ma, Y. (eds.) Ensemble Machine Learning, pp. 1–34. Springer, Boston (2012). https://doi.org/10.1007/978-1-4419-9326-7_1
11. Ribeiro, M.T., Guestrin, C.: Anchors: high-precision model-agnostic explanations
12. Ribeiro, M.T., Singh, S., Guestrin, C.: "Why should i trust you?" Explaining the predictions of any classifier. In: Proceedings of the ACM SIGKDD International Conference on Knowledge Discovery and Data Mining, pp. 1135–1144, 13–17 August 2016
13. Safavian, S.R., Landgrebe, D.: A survey of decision tree classifier methodology. IEEE Trans. Syst. Man Cybern. **21**(3), 660–674 (1991)
14. Shapley, L.S.: A value for n-person games (no. rand-p-295). Rand Corp Santa Monica CA (1952)
15. Shohei Shirataki and Saneyasu Yamaguchi. A study on interpretability of decision of machine learning. In: Proceedings - 2017 IEEE International Conference on Big Data, Big Data 2017, January 2018, pp. 4830–4831 (2017)
16. Sanford Weisberg. Applied Linear Regression, vol. 528. Wiley (2005)

Intrusion Detection Framework for IoT-Based Smart Environments Security

Chaimae Hazman[1](✉), Said Benkirane[1], Azidine Guezzaz[1], Mourade Azrour[2], and Mohamed Abdedaime[3]

[1] Technology Higher School Essaouira, Cadi Ayyad University, Marrakech, Morocco
chaimaa.hazman@gmail.com

[2] IDMS Team, Faculty of Sciences and Technics, Moulay Ismail University of Meknès, Meknes, Morocco

[3] National School of Commerce and Management, Ibn Tofail University, Kenitra, Morocco

Abstract. Since of huge progress of Internet of things (IoT) and networking technologies, and the expanding number of devices linked to the Internet, security and privacy problems must be addressed in order to secure hardware and information networks. Real-time monitoring of network resources and information is required to guarantee security. Intrusion detection systems have been utilized to continuously monitor, detect, and notify on an intrusion incident. Indeed, intrusion detection systems (IDSs) are a powerful cybersecurity tool that can be improved with machine learning (ML) and deep learning (DP) algorithms. These are intended to improve standard quality criteria like as accuracy (ACC), recall, and precision, but they hardly take time processor performance into effect. A novel IDS for IoT-based smart environments that utilizes DL and Supervised ML, is presented in this paper. The framework typically offered an optimum anomaly detection model that combines deep extraction based on the stacked autoencoder and combining feature selection using Information gain (IG) and Genetic algorithms (GA) (DEIGASe) which employs Multi-layer Perceptron (MLP) and support vector machine (SVM), K-nearest neighbors (K-NN) for classification. The BoT-IoT dataset was used to validate the proposed model metrics when compared to previous IDS, the results show that the proposed approach produces excellent Accuracy (ACC), Recall, Roc-auc and F1-score performance metrics.

Keywords: Intrusion detection · Smart environments · IoT · ML · DL · BoT-IoT

1 Introduction

The IoT architecture has recently been used in the building of smart environments, such as smart cities and smart houses, with a wide range of application sectors and related services. The goal of creating such smart environments is to improve human life by addressing challenges such as the living environment, energy consumption, and industrial needs [2, 4, 13]. Many of the foregoing difficulties consider the current standards and approaches for IoT security into question. As a result, the IoT security issue is measured by a number of parameters, including verification, authorization, privacy, access

Y. Farhaoui et al. (Eds.): ICAISE 2022, LNNS 635, pp. 546–552, 2023.
https://doi.org/10.1007/978-3-031-26254-8_79

control, information storage, system architecture, and maintenance. [6] Many technologies, such as firewalls, antiviruses, and IDSs, were introduced to protect systems from intrusions and assaults [7, 9, 10]. Based on rules, signatures, or models, IDS are aimed to identify between normal and abnormal system activity [7, 19, 20]. However, they are still inadequate [3]. Hence, it is critical to strengthen security mechanisms with growing better artificial intelligence, such as ML and DL algorithms. Generally, IDSs are classified into three types: SIDS, AIDS, and hybrid IDS (HIDS). SIDS analyzes signatures or patterns to collected events [6, 7, 12] in order to detect known threats with a low FAR and high DR. Because no prior signature was kept for any new attacks, SIDS became less efficient as the number of zero-day attacks increased. As a consequence, AIDS is being investigated as a viable cure. It can identify zero-day attacks [3, 5, 7, 11] by deducing typical behavior profiles and treating abnormal deviations as assaults. SIDS and AIDS are combined in HIDS. They benefit from the increased DR of known attacks and lower the FAR for future attacks [3, 15].

This research work proposes new IDS for IoT-based smart environments that utilizes DL and Supervised ML to enhance detection rate and make reliable decisions. Typically, the proposed framework integrates an optimum ensemble- based intrusion detection approach that employs SVM, MLP, K-NN, and combines deep extraction with information gain (IG) and genetic algorithms (GA) feature selection (DEIGASe). Experiment findings on Bot-IoT datasets show that our approach performs well in regard to ACC, DR, and FAR. The rest of this paper is organized as follows. In the Sect. 2, we present a related works of IDS that integrate ML and DL to secure IoT environments. The section is achieved by a comparison study of various models. The novel framework is designed and described in detail is the Sect. 3. The Sect. 4 shows the obtained results and comparison with other previous works. Finally, the study ends with a conclusion and future work.

2 Related Works

This section provides an overview of IoT and intrusion detection approaches, while also references to some recent relevant research on IDS that use ML and DL algorithms to improve IoT security. In 2020 Sarker et al. [16] proposed an intrusion detection tree (IntruDTree) ML- based security model, and thus a feature selection based on security feature ranking. The intruDTree model is tree-like. Their proposed model was validated with an intrusion dataset that included both normal and intrusion data. It demonstrates that using feature selection methods assisted the model in scoring 98% of ACC. Moreover, Jabbar et al. [17] presented RFAODE, an ensemble classifier developed with RF and the average one-dependence estimator (AODE). They used AODE to overcome the descriptor dependency problem in the Nave Bayes classifier, and RF to enhance the accuracy and lower error rates. On the Kyoto dataset, they evaluated the effectiveness of their suggested ensemble classifier (RFAODE). They significantly outperform the NB, AODE, and RF algorithms with an accuracy of 90.51% and a FAR of 0.14. Furthermore, Chaabouni et al. [18] recommended an edge ML-based OneM2M IDS for IoT security. The experimental results show that the detection rate is 93.80%, the accuracy is 92.32%, the precision is 92.95%, the FPR is 1.53%, and the CPU training time is

9280 ms. LightGBM clearly achieves the best results. According to, Guezzaz et al. [5] develop an intrusion detection approach that uses PcapSocks and an MLP classifier to collect network data. The PcapSocks sniffer and MLP classifier are used to recognize and categorize events as either normal or invasive.

In 2021 Ullah et al. [19] proposed a deep learning model IDS based on a convolutional neural network (CNN) for binary and multicast classifications, with a minimum detection rate of roughly 99.7%.

In 2022 Saba et al. [1] proposed a deep learning IDS model based on a CNN technique for anomaly-based intrusion detection systems (IDS) that takes advantage of IoT power, offering attributes to efficiently scan all IoT traffic The suggested model is capable of detecting any potential intrusion and aberrant traffic behavior. The model was trained and evaluated using the NID Dataset and the BoT-IoT datasets, and it obtained 99.51% and 92.85% accuracy, respectively. Guezzaz et al. [14] uses ML approaches to provide a hybrid IDS for Edge-Based IIoT Security. This novel hybrid framework is based on the identification of abuse and anomalies through the use of K-Nearest Neighbor (K-NN) and Principal component analysis (PCA) approaches.The K-NN classifier, in particular, has been added to boost detection accuracy and make effective decisions, while the PCA is employed for better feature engineering and training. The collected findings demonstrated that our suggested framework has several benefits over other current models. On the NSL-KDD dataset, it achieves 99.10 percent ACC, 98.4% DR, and 2.7% false alarm rate (FAR), while on the Bot-IoT dataset, it achieves 98.2% ACC, 97.6% DR, and 2.9 percent FAR. Furthermore, Guezzaz et al. [8] proposed a network intrusion detection model based on DT and improved by data quality on the NSL-KDD and CICIDS2017 datasets. They tried to compare their model to others based on the same data sets. Their model achieved 99.42% accuracy on the NSL-KDD dataset and 98.80% on the CICIDS2017 dataset.

3 Proposed Intrusion Detection Framework

In this section, we present several techniques used to demonstrate our intrusion detection technique for IoT environment security. Using GPU, we will develop and implement an appropriate model that will increase detection rate, accuracy, and processing time. The goal of this approach is to evaluate a better IDS based on feature engineering methods and multiple ML classifiers (MLP, SVM, KNN). Figure 1 illustrates our optimized model process, which is separated into three critical components:

Data quality component: was separated into 2 phases, the first of which was data preparation. Data has been prepared and understood. As a result, we recognized and eliminated any inconsistencies, such as NaN values. The next phases will include feature engineering. We integrated a feature extraction using deep extraction with information gain, and then we used genetic algorithms on actual network traffic data to build a better training set while decreasing training time and processing cost. As a fitness function, the GA selects the greatest set of features with the greatest AUR-ROC score, highest ACC, and lowest FPR.

Classifier component: we integrate different model for classification, the objective of this component is to create a classifier model using transformed data from the data quality component as input. The classifier process will be separated into two important phases: model training and model validation. Half of the data is utilized to train an MLP, SVM, and KNN classifier, which are employed in our suggested technique, in the first stage, and the remaining data is used to evaluate our model in training set.

Predictive component: the design framework can forecast a positive attack. It is analyzed and verified using metrics from the confusion matrix such as ACC, recall, precision, FPR, FNR, AUC-ROC and f1-score. A classification mechanism that evaluates when incoming traffic is normal or intrusive. Thus, the aim of this component is to determine a binary output to validate the data with a yes or no answer. As a consequence, we classed both groups as numerical variables, with 0 indicating regular activities and 1 denoting intrusion. It is critical to remember that the degree of qualities must be chosen ahead of time. We employed a variety of approaches to partition the data into a training and test set for the validation stage of our model we proposed the kfold method.

3.1 DEIGASe Features Engineering Technique

In this technique we proposed the deep Extraction, IG and GA Selection (DEIGASe) in the feature engineering step, which extracts features using a deep-structured stacking autoencoder previous to feature selection based on the amount of mutual information provided between every feature and the target feature. The GA is then used to choose the optimum set of features that has the greatest AUR-ROC score, the highest ACC, F1-score, and the lowest FPR as a fitness function.

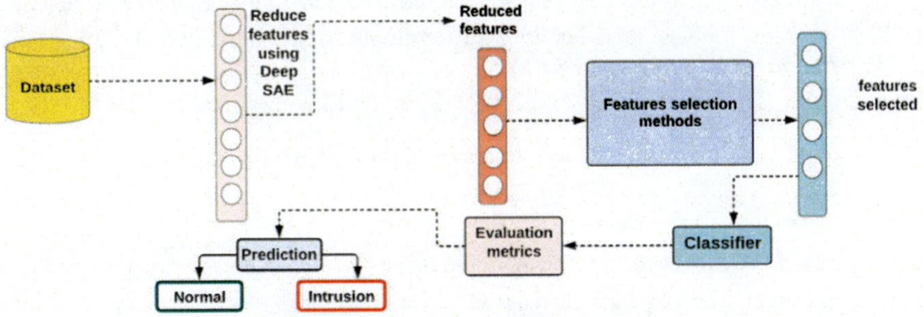

Fig. 1. Scheme of our framework

Experimental Evaluation and Results

3.2 Environment Description and Dataset

In general, IDS assessment is a key issue, the optimal performance parameters of any classifier are determined by the dataset used for model training and testing. This study makes use the datasets: BoT-IoT. it's been created and labeled for potential multiclass usage. An assault flow, a category of attacks, and a subclass were all designated by the label features. In truth, BoT-IoT has 99.99% more attacks than benign ones (0.01%), and it has 46 attributes, including the target variable. Our technique is tested experimentally on a kaggle machine with 13 GB of RAM, 15 GB of GPU memory, and a 64-bit operating system. Jupyter Lab and Python 3.9.7, as well as the pandas, numpy, and sklearn modules, were used to create the model. IDS assessment is a key topic. The dataset used for model training and testing affects the best performance parameters for each classifier.

3.3 Discussion of Results

Various metrics are generated to evaluate IDS performance based on the followed parameters listed in Table 1.

True positives (TP) are data points that have been correctly identified attack. True negative (TN) data items have been correctly categorized as normal.

False Positive (FP): the model predicts an assault but it does not happen. False Negative (FN): the model predicts normal yet it is not the reality.

We employ Bot-IOT dataset to check our performance of the model, our technique for feature selection still provides the greatest performance in each classifier. As stated previously, the findings obtained reflect the approach's performance in accuracy, precision, recall, and f1-score 99%. These results are described in Table 1.

The findings are similar, with 99% accuracy for SVM and K-means, 99% precision, 99% recall and 99% roc-auc for each except for MLP, which received 100% recall also has 99.99% for all metrics. The difference is seen in the training time, where K-means outperforms the other approaches by 0.07192 s seconds vs 50.950 s seconds for SVM and 45.29 s seconds for MLP. Furthermore, the MLP technique surpassed the other methods with 0% FPR, followed by K-means with 0.18% and SVM with 0.25%. When we compare the curves produced by plotting true positive rate (TPR) against FPR, we find that our models have the same capacity to differentiate between the two classifications, with 99.99% for MLP, K-NN, and SVM.

Table 1. The outcomes of performance indicators on the BoT-IoT dataset.

Model	Accuracy	Precision	Recall	F1-score	Auc-roc	Time training
MLP	0.999	0.999	1.00	0.999	0.999	45.29 s
SVM	0.98288	0.98078	0.985	0.988	0.992	50.950 s
K-NN	0.999	0.999	0.999	0.999	0.994	0.07192 s

4 Conclusion

Intrusion detection is suitable for boosting IoT security against threats, particularly when IDS is integrated with smart environments security. This research provides an enhanced intrusion detection model for IoT security based on an anomaly detection approach that improves IDS accuracy while reducing processing time. In this research, we provided an effective intrusion detection approach based on MLP, SVM and K-NN with feature engineering based on deep extraction and information gain, genetic algorithms for features selection approaches, GPU utilization benefits and influences the model. According to the findings of this study, the proposed model would aid in the development of an effective IoT NIDS with a high detection rate. The acquired findings show that our methodology may produce excellent results while significantly reducing training time, demonstrating the value of feature selection approaches in improving the effectiveness of intrusion detection systems. Furthermore, when compared to existing intrusion detection algorithms, our methodology exhibits encouraging results. As a result, while this article only explored binary classification, our future work will incorporate multi-class classification.

References

1. Saba, T., Rehman, A., Sadad, T., Kolivand, H., Bahaj, S.A.: Anomaly-based intrusion detection system for IoT networks through deep learning model. Comput. Electric. Eng. 107810 (2022)
2. Von Solms, R., Van Niekerk, J.: From information security to cyber security. Comput. Secur. **38**, 97–102 (2013)
3. Azrour, M., Mabrouki, J., Farhaoui, Y., Guezzaz, A.: Security analysis of Nikooghadam et al.'s authentication protocol for cloud-IoT. In: Gherabi, N., Kacprzyk, J. (eds.) Intelligent Systems in Big Data, Semantic Web and Machine Learning. AISC, vol. 1344, pp. 261–269. Springer, Cham (2021). https://doi.org/10.1007/978-3-030-72588-4_18
4. Kafle, V.P., Fukushima, Y., Harai, H.: Internet of things standardization in ITU and prospective networking technologies. IEEE Commun. Mag. **54**(9), 43–49 (2016)
5. Guezzaz, A., Asimi, A., Mourade, A., Tbatou, Z., Asimi, Y.: A multilayer perceptron classifier for monitoring network traffic. In: Farhaoui, Y. (ed.) BDNT 2019. LNNS, vol. 81, pp. 262–270. Springer, Cham (2020). https://doi.org/10.1007/978-3-030-23672-4_19
6. Guezzaz, A., Asimi, A., Asimi, Y., Azrour, M., Benkirane, S.: A distributed intrusion detection approach based on machine leaning techniques for a cloud security. In: Gherabi, N., Kacprzyk, J. (eds.) Intelligent Systems in Big Data, Semantic Web and Machine Learning. AISC, vol. 1344, pp. 85–94. Springer, Cham (2021). https://doi.org/10.1007/978-3-030-72588-4_6
7. Azrour, M., Mabrouki, J., Guezzaz, A., Farhaoui, Y.: A survey of high school students' usage of smartphone in moroccan rural areas. Int. J. Cloud Comput. **10**(3), 265-274 (2021)
8. Guezzaz, A., Benkirane, S., Azrour, M., Khurram, S.: A reliable network intrusion detection approach using decision tree with enhanced data quality. Secur. Commun. Netw. (2021)
9. Buczak, A.L., Guven, E.: A survey of data mining and machine learning methods for cyber security intrusion detection. IEEE Commun. Surv. Tutorials **18**(2), 1153–1176 (2016)
10. Guezzaz, A., Asimi, A., Asimi, Y., Tbatou, Z., Sadqi, Y.: A lightweight neural classifier for intrusion detection. Gen. Lett. Math. **2**, 57–66 (2017)
11. Azrour, M., Mabrouki, J., Guezzaz, A., Kanwal, A.: Internet of Things security: challenges and key issues. Secur. Commun. Netw. 5533843, 11 (2021)
12. Guezzaz, A., Asimi, A., Batou, Z., Asimi, Y., Sadqi, Y.: A global intrusion detection system using PcapSockS sniffer and multilayer perceptron classifier. Int. J. Netw. Secur. **21**(3), 438–450 (2019)
13. Koroniotis, N., Moustafa, N., Sitnikova, E.: A new network forensic framework based on deep learning for Internet of Things networks: A particle deep framework. Future Gen. Comput. Syst. **110**, 91–106 (2020)
14. Guezzaz, M. Azrour, S., Benkirane, M., Mohyeddine, H., Attou, M., Douiba. A.: Lightweight hybrid intrusion detection framework using machine learning for edge-based IIoT security. Int. Arab J. Inf. Technol. **19**(5) (2022)

15. Diro, A., Chilamkurti, N.: Distributed attack detection scheme using deep learning approach for Internet of Things. Future Gen. Comput. Syst. **82**, 761–768 (2017)
16. Sarker, I.H., Abushark, Y.B., Alsolami, F., Khan, A.I.: Intrudtree: a machine learning based cyber security intrusion detection model. Symmetry **12**(5), 754 (2020)
17. Jabbar, M.A., Aluvalu, R., Seelam, S.S.R.: RFAODE: a novel ensemble intrusion detection system. Procedia Comput. Sci. **115**, 226–234 (2017)
18. Chaabouni, N., Mosbah, M., Zemmari, A., Sauvignac, C.: A OneM2M intrusion detection and prevention system based on edge machine learning. In: IEEE/IFIP Network Operations and Management Symposium, pp. 1–7 (2020)
19. Ullah, I., Mahmoud, Q.H.: Design and development of a deep learning-based model for anomaly detection in IoT Networks. IEEE Access, **9**, 103906–103926 (2021)
20. Shafiq, M., Tian, Z., Sun, Y., Du, X., Guizani, M.: Selection of effective machine learning algorithm and Bot-IoT attacks traffic identification for internet of things in smart city. Futur. Gener. Comput. Syst. **107**, 433–442 (2020)

IoT Network Attack Types by Application Domains

Ouijdane Fadli[1(✉)], Younes Balboul[1], Mohammed Fattah[2], Said Mazer[1], and Moulhime Elbekkali[1]

[1] Artificial Intelligence and Data Science and Emerging Systems Laboratory, Sidi Mohamed Ben Abdellah University of Fez, Fez, Morocco
{ouijdane.fadli,moulhime.elbekkali}@usmba.ac.ma

[2] Image Laboratory University of Moulay Ismail of Meknes, Meknes, Morocco

Abstract. In a world dominated by digital technology, the Internet of Things (IoT) has occupied our daily lives beyond our knowledge. It has established an ecosystem that connects multiple systems to provide intelligent performance in each task of our lifestyles and it has made everything and everyone connected to the internet. From wearables to smart cities and smart industries, our hobbies, our movements are facilitated by this technology that greatly improves our quality of life and increases our comfort. Based on low-cost sensors, the IoT provides the possibility to render the several devices and objects around us addressable, identifiable, and locatable. Though the IoT brought infinite benefits, it creates several challenges, especially in security and privacy. These security challenges are complicated due to the variety of IoT protocols that are heterogeneous, serve various requirements, and are utilized in diverse application domains. However, because IoT is commonly deployed in hostile or remote environments and operate without monitoring, preventing attacks and protecting the privacy of sensitive collected data is critical for many IoT applications. This paper represents a detailed study of the general concepts on the security of IoT networks by application domains by describing their vulnerabilities, their security needs. The classification of the types of attacks on IoT networks by application domains is also discussed.

Keywords: Internet of Things · Attacks · Privacy · Confidentiality · Security challenges · Vulnerabilities

1 Introduction

Nowadays, the Internet of Things (IoT) is one of the most popular subjects amongst scientists and researchers [1]. It is viewed as a universal feature which enables everything around us to be connected to the internet with the ability to communicate with another objects, apps, and infrastructures with no need for human intervention [2, 3].

The range of connected objects increases considerably day by day. These objects are based on a common addressing system which makes it possible for them to interoperate and collaborate with each other in order to generate innovative applications and services like smart building, smart industry, smart grid, smart banking, etc. [4, 5].

Y. Farhaoui et al. (Eds.): ICAISE 2022, LNNS 635, pp. 553–558, 2023.
https://doi.org/10.1007/978-3-031-26254-8_80

IoT represents precious innovation, however, it may also be a major risk to the cybersecurity of diverse systems. That pose an important current cybersecurity threat, because there are multiple access points that can be breached and contain some security vulnerabilities. In general, one weakness on the system's security stack may constitute an overall security threat to the entire system and create opportunities for attackers [6].

Once IoT apps are utilized across sensitive and critical environments, severe cybersecurity threats intrinsically emerge [7]. In the case where cyberattacks occur on critical environments, the results will be catastrophic. Cutting off power to a hospital, changing the configuration of money transaction systems and exploiting the functionality of smart cars while they are moving are some of the devastating examples [8, 9].

Therefore, protecting the privacy of sensitive collected data is critical for many domains. Managing these problems and guaranteeing the safety and confidentiality of IoT services should be a fundamental priority. This paper represents a detailed study of the general concepts on the security of IoT networks by application domains by describing their vulnerabilities, their security needs. The classification of the types of attacks on IoT networks by application domains is also discussed.

2 IoT Application Domains

IoT applications offer a large variety of different solutions that significantly impact on individuals, companies, and organizations in their daily lives [10, 11]. The potential advantages expected facilitate its adoption by this variety of users.

IoT impacts several different domains, among those where this technology is proving to be very useful and can provide greater contributions, the following application areas can be cited: Healthcare; Building; Energy; Banking.

2.1 Security Vulnerabilities in the IoT Networks by Application Domains

High-speed connection of monitoring and control processes across various IoT application domains exposes Internet-enabled systems to a large variety of attacks and raises the probability and the risk of vulnerabilities and threats [12]. This section presents the various security vulnerabilities in IoT networks per application domains.

Medical and Healthcare Domain: Is identified as one of the critical sectors, that provides crucial and necessary services along with vital digital assets, accompanying with several difficulties, challenges, and intricacies [13]. The weaknesses within the medical domain that attackers are exploiting, variety and mostly involve the usage of Internet of medical things (IoMTs) to ensure proper supervision of patients both in clinic and at home. These devices generally use wireless communication which brings several security vulnerabilities at two distinct levels: attacking the route construction and maintenance, and attacking the useful data by injecting, modifying or removing packets.

Building Domain: Avoiding data loss and holding attackers off is critical, however, once that involves affecting the security and safety of our personal life at home through smartphone and IoT devices, it may be more vital to prioritize security first and foremost. The following are the most commonly exploited vulnerabilities in building domain.

Firstly, personal IoT equipment might be insecure. There are certain domestic IoT devices which are hurriedly commercialized, whose level of security can be not properly considered. Secondly, the domestic network can be unsecured and all available data can be easily accessed by hackers.

Energy Domain: IoT-based smart power energy systems are one of the most critical infrastructures, which are based on complicated architectures and involve sensitive equipment. They utilize critical communication systems which may cause large-scale damage to national safety, if the confidentiality, integrity, or availability of the communication is compromised [14]. The vulnerabilities within the power system domain that attackers are exploiting, variety and mostly involve [15]: Authentication system failure leading to corruption of existing appropriate procedures, Utilization of non-audited APIs, The use of wireless communication networks for data transmission which can falsify legitimate information if on node is breached, Using the same security keys for transmitting information, Usage of a common login/password on all equipment.

Banking Industry Domain: These days, using IoT frequently takes place through mobile devices and developing both mobile and online solutions was an important way for the banking community moved beyond the traditional fatiguing and time-consuming process that caused problems for users [16]. Safety threats in mobile and online banking, constitute an important issue for both bankers and users due to the technological improvements and the security vulnerabilities in each innovation. Among these vulnerabilities, we can mention: Usage of insecure Wi-Fi connectivity, Utilization of third-party applications (i.e., Applications developed by hackers with the same name as legitimate applications), Unencrypted data (i.e., Users store information on their PCs within their financial organization and online unencrypted), Weakness of passwords used in online banking applications.

2.2 Security Requirements for IoT Systems

Security is the worst nightmare of all IoT apps and services across different domains. Even though there are various novel advanced safety and security solutions, reaching complete security is almost unattainable [17]. However, every security solution must be able to fulfill the following security requirements.

Data Confidentiality: Refers to one of types of security feature designed to ensure that sensitive data and information are not available to non-authorized users. It ensures that private data are protected from abuse by hackers and can be accessed only by authorized persons. Various algorithms and protocols are being suggested to provide privacy to critical information in the IoT environment [18]. Data encoding, authentication and authorization processes are among the best ways to guarantee confidentiality [19].

Data Integrity: Guarantees that the data exchanged between devices is authentic and not manufactured or altered by hackers during communication. Encryption-based integrity controls represent one type of data integrity guarantee. Some of the widely used algorithms are AES128/256, MD5, SHA and S-box.

Data Availability: The primary objective of security systems is to guarantee the accessibility of information and services to the user at all times [20, 21]. Data availability indicates the ability of IoT-based systems to maintain its continuity of service even during denial-of-service attacks. To achieve continuous data and service accessibility, the following security objectives for availability should be taken into account [22]:

- Failure Resilience: The system should have the capacity to utilize the self-protection and the self-healing process upon a failure or an attack.
- Scalability: IoT devices can be structured hierarchically to handle the evolution.

Freshness: Replication attacks may be avoided through guaranteeing the freshness of receiving information. This involves checking that the incoming data from the IoT devices is timely and not duplicated. This technique is usually realized by utilizing sequence numbers and timestamps within the data packets transferred by the sensor [23].

2.3 Types of Attacks on IoT Networks by Application Domains

This part of our work represents the commonly existing attacks in the different IoT application domains [24] (Table 1).

Table 1. Types of attacks on IoT networks by application domains

Attacks	Description	Application domain
Denial of service or distributed attack (DDoS)	This attack focuses on flooding IoT systems with wrong data, in order to render its network unable to offer normal service either by attacking the network's bandwidth or its connectivity	Healthcare; Building; Banking; Energy
SQL injection attack	Hackers insert SQL statements inside the database and retrieve the necessary data, or remove and damage the entire database	Healthcare; Energy; Banking
Eavesdropping	The attacker regularly surveys the IoT systems, in order to capture critical information	Healthcare; Building
Malwares	Hackers use existing hardware and software weaknesses to attack the IoT system	Healthcare; Building; Banking; Energy

(*continued*)

Table 1. *(continued)*

Attacks	Description	Application domain
Man-in-the-Middle	The adversary installs a controlled system between the victim's system and an Internet resource that it uses to intercept, forge or manipulate any communication between them	Building; Agriculture
Device hijacking	An adversary dominates and gains control of a IoT device to falsifying information	Energy; Banking; Healthcare; Building
Phishing	Phishing attacks consist the transmission of false information appearing to be coming from a trusted source	Energy; Banking;
Spoofing	Includes all types of cyberattacks consisting of stealing digital identities like email addresses, domain names or IP addresses, in order to obtain confidential information	Energy; Banking;
False data injection	This attack refers to the situation in which an attacker compromises sensor record	Power;

3 Conclusion

This paper discusses different types of attacks on various smart IoT applications. This work started with a description of different vulnerabilities in some IoT application domains. In the second part, several of the important security requirements were described in detail. Then, we moved to the part presenting the types of IoT attacks on different application areas. In our future work, we plan to develop a secure approach to ensure the security and privacy of sensitive data in different IoT domains.

References

1. Damghani, H., et al.: Classification of attacks on IoT. In: 4th International Conference on Combinatorics, Cryptography, Computer Science and Computation (2019)
2. Resul, D.A.S., Gündüz, M.Z.: Analysis of cyber-attacks in IoT-based critical infrastructures. Int. J. Inf. Secur. Sci. **8**(4), 122–133 (2020)
3. Younes, B., et al.: 5G uplink interference simulations, analysis and solutions: the case of pico cells dense deployment. Int. J. Electr. Comput. Eng. **11**(3), 2245 (2021)
4. Chafi, S.E., Balboul, Y., Mazer, S., Fattah, M., El Bekkali, M.: Resource placement strategy optimization for smart grid application using 5G wireless networks. Int. J. Electr. Comput. Eng. **12**(4), 2088–8708 (2022)

5. Daghouj, D., et al.: UWB waveform for automotive short range radar. Int. J. Eng. Appl. (IREA) **8**(4), 158 (2020)
6. Liang, X., Kim, Y.: A survey on security attacks and solutions in the IoT network. In: 2021 IEEE 11th Annual Computing and Communication Workshop and Conference (CCWC), pp. 0853–0859. IEEE (2021)
7. Liang, X., Kim, Y.: A survey on security attacks and solutions in the IoT network. In: 2021 IEEE 11th Annual Computing and Communication Workshop and Conference (CCWC), pp. 853–859. IEEE (2021)
8. Jurcut, A. D., Ranaweera, P., Xu, L.: Introduction to IoT security. IoT Secur. Adv. Authentication, 27–64 (2020)
9. Atlam, H.F., Wills, G.B.: IoT security, privacy, safety and ethics. In: Farsi, M., Daneshkhah, A., Hosseinian-Far, A., Jahankhani, H. (eds.) Digital twin technologies and smart cities. IT, pp. 123–149. Springer, Cham (2020). https://doi.org/10.1007/978-3-030-18732-3_8
10. Moutaib, M., Fattah, M., Farhaoui, Y.: Internet of things: energy consumption and data storage. Procedia Comput. Sci. **175**, 609–614 (2020)
11. Mamane, A., et al.: Scheduling algorithms for 5G networks and beyond: classification and survey. IEEE Access (2022)
12. MacDermott, Á., et al.: Securing things in the healthcare internet of things. In: 2019 Global IoT Summit (GIoTS), pp. 1–6. IEEE (2019)
13. Kioskli, K., Fotis, T., Mouratidis, H.: The landscape of cybersecurity vulnerabilities and challenges in healthcare: security standards and paradigm shift recommendations. In: The 16th International Conference on Availability, Reliability and Security, pp. 1–9 (2021)
14. Gunduz, M.Z., Das, R.: Cyber-security on smart grid: threats and potential solutions. Comput. Netw. **169**, 107094 (2020)
15. Islam, S.N., Baig, Z., Zeadally, S.: Physical layer security for the smart grid: vulnerabilities, threats, and countermeasures. IEEE Trans. Industr. Inf. **15**(12), 6522–6530 (2019)
16. Yildirim, N., Varol, A.: A research on security vulnerabilities in online and mobile banking systems. In: 2019 7th International Symposium on Digital Forensics and Security (ISDFS), pp. 1–5. IEEE (2019)
17. Atlam, H.F., Alenezi, A., Alassafi, M.O., Alshdadi, A.A., Wills, G.B.: Security, cybercrime and digital forensics for IoT. In: Peng, S.-L., Pal, S., Huang, L. (eds.) Principles of internet of things (IoT) ecosystem: Insight paradigm. ISRL, vol. 174, pp. 551–577. Springer, Cham (2020). https://doi.org/10.1007/978-3-030-33596-0_22
18. Saad-Eddine, C., Younes, B.: Performance & energy consumption metrics of a data center according to the energy consumption models cubic, linear, square and square root. In: 2019 7th Mediterranean Congress of Telecommunications (CMT), pp. 1–5. IEEE (2019)
19. Deep, S., et al.: A survey of security and privacy issues in the Internet of Things from the layered context. Trans. Emerg. Telecommun. Technol. **33**(6), e3935 (2022)
20. Mamane, A., et al.: The impact of scheduling algorithms for real-time traffic in the 5G femto-cells network. In: 2018 9th International Symposium on Signal, Image, Video and Communications (ISIVC), pp. 147–151. IEEE (2018)
21. Chafi, S.-E., et al.: Cloud computing services, models and simulation tools. Int. J. Cloud Comput. **10**(5–6), 533–547 (2021)
22. Khanam, S., et al.: A survey of security challenges, attacks taxonomy and advanced countermeasures in the internet of things. IEEE Access **8**, 219709–219743 (2020)
23. Karunarathne, S.M., Saxena, N., Khan, M.K.: Security and privacy in IoT smart healthcare. IEEE Internet Comput. **25**(4), 37–48 (2021)
24. Anwar, R.W., Abdullah, T., Pastore, F.: Firewall best practices for securing smart healthcare environment: a review. Appl. Sci. **11**(19), 9183 (2021)

IoT-Based Intelligent System of Real-Time Data Acquisition and Transmission for Solar Photovoltaic Features

Naima Elyanboiy[1(✉)], Mohamed Khala[1], Ismail Elabbassi[1], Nourddine Elhajrat[1], Sara Teidj[2], Omar Eloutassi[1], and Choukri Messaoudi[1]

[1] Optoelectronics and Applied Energy Techniques, Faculty of Science and Technology, Moulay Ismail University of Meknes, Errachidia, Morocco
naima.elyanboiy@gmail.com, ism.elabbassi@edu.umi.ac.ma
[2] Faculty of Science and Technology, Moulay Ismail University of Meknes, Errachidia, Morocco

Abstract. The optimal operation of photovoltaic solar panels requires efficient energy monitoring, in order to ensure perfect monitoring of energy production and its affecting factors we develop a real-time data acquisition, transmission, and logging system. In this work, a database has been generated to record the collected data related to a photovoltaic solar panel. This system is based on a powerful ESP32 microcontroller which ensures the transmission of all the collected parameters to the No SQL database managed by MongoDB software. The hardware prototype of this system requires a set of sensors to measure data related to temperature, humidity, irradiation, voltage, and current of the adopted load, and related power, the ESP32 module is in charge of reading the sensors, adapting, and sending the data to the database. Therefore, this generated database allows energy managers and data scientists to make reliable studies on the effect of recorded parameters on energy production by applying artificial intelligence methods.

Keywords: Internet of Things · Acquisition system · Database · Solar photovoltaic · ESP32

1 Introduction

Actually, the global need for electrical energy is continuously increasing due to population growth, in parallel with the rarefication of natural fossil and fissile energy resources. Renewable energies are attributed to minimizing this challenge [1].

The solar photovoltaic system like any other industrial process can be submitted to various failures and anomalies during its operation; hence it has to be monitored to enhance system performances [2]. Those undesirable effects essentially due to environmental fluctuations and the intrinsic structure of the photovoltaic system reduce the system performance and its consistency. Various researches have been made to identify these inconveniences and their causes. Furthermore, accurate structure anomalies prediction involves additional effective maintenance and direct intervention. The main objective of this work is to comprehend an intelligent system for monitoring the functionality of

Y. Farhaoui et al. (Eds.): ICAISE 2022, LNNS 635, pp. 559–565, 2023.
https://doi.org/10.1007/978-3-031-26254-8_81

photovoltaic generators using a system based on the Internet of Things]. The Internet of Things (IoT) refers to the process of connecting physical objects to the Internet in order to make them intelligent and aim to communicate between various intelligent devices, it's also an excellent platform to develop a smart and effective monitoring system [3–5]. As discussed in [6], a smart IoT proposed system allows to providing real-time surface solar radiation levels, interconnection with wireless networks, and a specific web-based system for data monitoring.

Controlled and monitored using ADAFRUIT-programmed IoT connected devices Output from the DC microgrid, by mobile phone or Laptop or any remote device. It has been suggested in [7] that it can control and monitor using IoT connected devices programmed by ADAFRUIT Output from the DC microgrid, by cell phone or Laptop or any other remote device. However, the biggest limitation of IoT is ensuring the security of the application in large databases. In addition, an unintelligent IoT system will have limited capabilities and will be unable to scale with big data [8].

This project aims to develop a system to measure and monitor solar panel parameters using ESP32 technology and real-time transmission to MongoDb software. For this purpose, six parameters are considered: temperature, humidity, light intensity, voltage and current of the adopted load, and its power will be measured. This work will be organized into the following sections: In Sect. 2 we present the proposed work in which we describe the Hardware design tools. Section 3 focuses on the results discussion, and lastly the conclusion.

2 Proposed Work

The main objective of this work is to realize an effective and reliable monitoring system that allows the measurement of the parameters of the solar panel array using the data provided by the sensors. In this system, various parameters of solar panels are monitored. In this part, we describe the design of the IoT-based system of real-time data acquisition and transmission for solar photovoltaic features and its wiring links in Fig. 1.

2.1 System Overview

The proposed system allows measuring the different parameters of an electrical system powered by a solar panel using IoT, a set of sensors to give data for each of these parameters is used as shown in Fig. 2. A solar panel is used to produce electrical energy, which depends on the light intensity, the time of day, and the weather fluctuations. Hence, this system is designed using ESP32 to perform the necessary calculations, while the analogue output of the sensors is fed into the ADC channels of ESP32, which is a module that integrates a microcontroller with Wi-Fi designed by ESPRESSIF with high performance, low power consumption and a wide range of I/Os (inputs and outputs). An integrated product with extremely low power consumption with a 2.4 GHz Wi-Fi solution and several GPIOs (General Purpose Inputs and Outputs) that makes it ideal for IoT applications [9]. ESPRESSIF with high performance, low power consumption and a wide range of I/Os (inputs and outputs), that makes it ideal for IoT applications [9].

After the calculation, all these data are transmitted to the storage system (MongoDb and PhpMyAdmin) for statistical purposes and future studies.

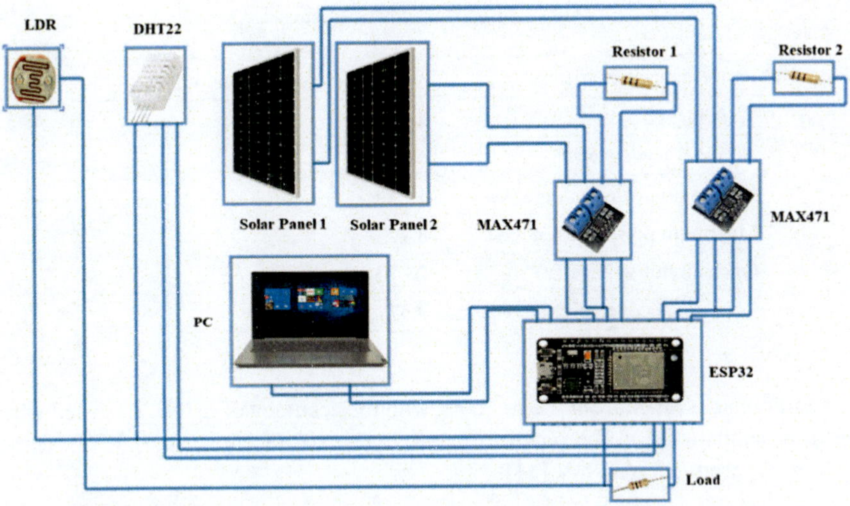

Fig. 1. Block diagram of the data acquisition and transmission system via IoT.

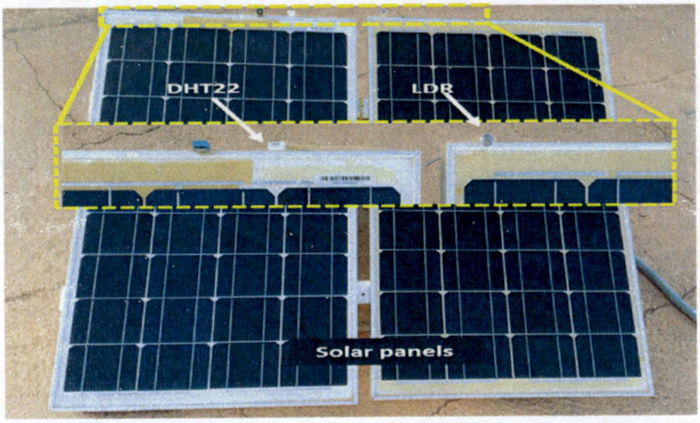

Fig. 2. DHT22 and LDR sensors fixed on the solar panels.

2.2 Hardware Design Tools

Characteristics of Photovoltaic Solar Panels

In this work two photovoltaic solar panels of the same type are used, Table 1. Illustrates their electrical characteristics, one is considered as a reference, and the other is on which we monitor.

Digital Humidity and Temperature Sensor DHT 22

DHT22 is more accurate and has a wider measuring range; the response time should not exceed 1 or 2 s [10]. From the DHT22 datasheet, we found that this sensor can measure

Table 1. PV panel characteristics.

Performance parameters	Values in standard conditions
Weight (kg)	4
Nominal power (watt)	60
Maximum power voltage (v)	19.3
Maximum power current (A)	3.12
Open circuit voltage (v)	23.10
Courant de court-circuit (A)	3.37

the humidity in the range from 0% to 100% with the accuracy of ± 2%. It is used when measuring temperature in the range from −40 to + 80° (accuracy ± 0.5 °C), with a power supply from 3.5 V to 6 V [11].

Light Dependent Resistor (LDR)
LDRs provide a voltage value that is proportional to the intensity of the light. The ESP 32 has a built-in Analog to Digital Converter (ADC), so when a voltage value is given to an Analog pin, an equivalent digital output is produced automatically [12].

MAX471 Voltage-Current Sensor
The MAX741 is a sensor that can measure voltage and current. All sensors mentioned above need a programming interface to process their operation in machine language, hence the interest in using the Arduino IDE software. The program code written for the Arduino board is also called a sketch [13].

3 Results and Discussions

The experimental device implemented in this work is a real-time data acquisition system. For this purpose. The figure Fig. 3 summarizes the algorithm of the acquisition system.

The system consists mainly of three essential steps:

Step 1: The implementation and mounting of two identical photovoltaic solar panels, then the sensors needed to measure the physical parameters. However, it is necessary to ensure protected wiring to avoid the effect of high temperatures in the study area.
Step 2: After connecting the measurement sensors, a proposal algorithm is written in C++ language on the Arduino platform, in which the program is able to read the measured values and transmit them to a database through a microcontroller. In order to send the data to the local web server, the chosen microcontroller as a solution in this experiment is the ESP32 module. This module supports Wi-Fi wireless communication and has a high-performance dual-core processor.
Step 3: Firstly, using phpMyadmin and when the code is compiled and uploaded, the web server interface sends the data to the requested client via HTTP (Hyper Text Transfer Protocol) and using the XAMPP environment where the local web server is managed

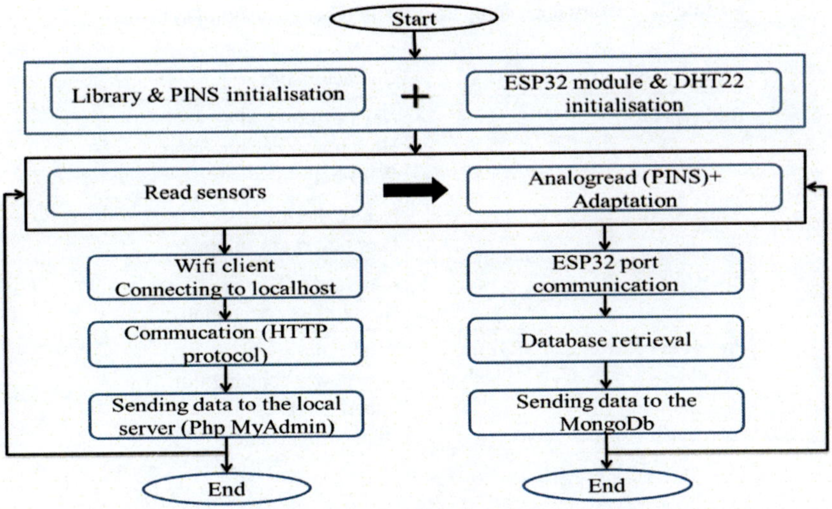

Fig. 3. Flowchart of the synchronized data recording process.

by an appropriate PHP code. XAMPP is a PHP development environment, containing a distribution of free software to set up a local web server. It is characterized by its flexibility of use and easy installation. On the phpMyAdmin software which is designed to manage MySQL administration on the web, while remain executing SQL statement directly [14], we create and manage the destination database where the transferred and visualized data will be stored. Secondly, using NoSQL database management system become more practical with the propagation of cloud-based applications and the Internet of Things (IoT) and for their ease and efficiency in processing high volumes of dynamic, heterogeneous and often unstructured data [15]. MongoDB allows data to be stored as documents and offers high performance and scalability options [16]. Figure 4 (a) shows the different parameters to be collected in the form of documents of data created and stored in MongoDb. However, Fig. 4 (b) shows the different parameters to be collected in the form of a table in phpMyAdmin database, and with a time delay of 120 s. Many IoT solutions for monitoring PV systems based on low-cost processing boards, such as Raspberry or Arduino boards have been frequently used. In this work, the ESP32 controller is selected to acquire, process, and transmit the data collected in real-time. Due to its high computational performance compared to the other modules as shown in Table 2. In [17], the microcomputer Raspberry Pi provides very high computing performance, but it needs external ADC devices to acquire analog measurements, which more expensive and consume a significant amount of power to operate than ESP32 board. As illustrated in [18, 19], the Arduino Mega2560 controller and the PIC16f877 require an Ethernet or GSM shield and the computing performance is lower while the ESP32 is characterized by its built-in wifi system and its high computing capacities.

Table 2. Performance comparison with other acquisition systems.

Data acquisition controllers	Characteristics
RasberryPi	Need external ADC High power consumption WIFI built in
Arduino Mega2560	Need Ethernet Low power consumption
PIC 16F877	Integrate ADC No wireless connection
Proposed controller (ESP32)	WIFI built in Low power consumption Integrate ADCs High computation

According to the manipulation of the two database management systems developed in this work. The NoSQL is more scalable, easier, and recommended for the management of Big-Data in comparison with the SQL database.

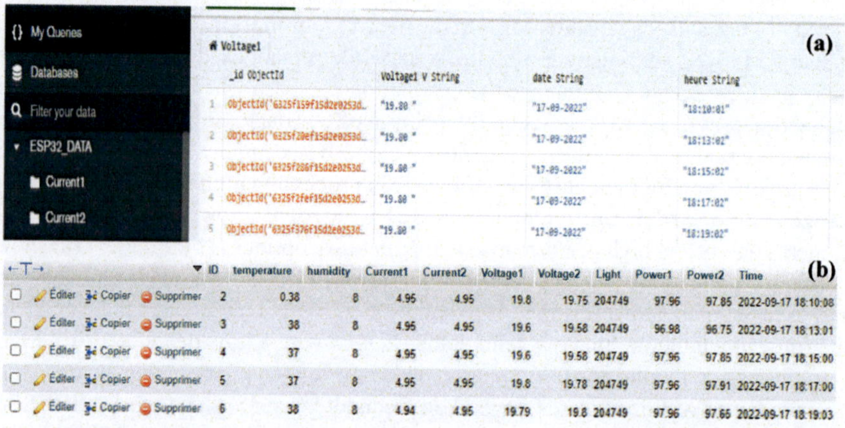

Fig. 4. Synchronized database table in MongoDb (a), and in phpMyAdmin (b).

4 Conclusion

The data acquisition system is a set of devices that are interconnected in order to automatically obtain a series of panel solar measurements. The analysis and the choice of the characteristics of an acquisition system are mainly linked to the aimed photovoltaic solar energy embedded equipment. The study performed in this work is focused on the achievement of photovoltaic solar energy embedded equipment that eases the data acquisition system for monitoring and control of the solar panel features. A real-time data acquisition system based on an ESP32 board is designed, connected by the following elements: temperature and humidity sensor dht22, current and voltage sensor MAX741, and light intensity sensor LDR The measured parameters are transmitted via the wireless Wi-Fi integrated in the ESP32 module to the No SQL database managed by the MongoDB software which is more efficient than using the phpMyAdmin software.

References

1. Triki-Lahiani, A., et al.: Fault detection and monitoring systems for photovoltaic installations: a review. Renew. Sustain. Energy Rev. **82**, 2680–2692 (2017)
2. Zbib, B., et al.: Fault detection and diagnosis of photovoltaic systems through I-V curve analysis. In: Proceedings of the 2nd International Conference on Electrical, Communication and Computer Engineering (ICECCE), pp.1–6. IEEE Explore (2020)
3. Laghari, A.A., et al.: A review and state of art of Internet of Things (IoT). Arch. Comput. Meth. Eng. 1–19 (2021).https://doi.org/10.1007/s11831-021-09622-6
4. Mellit, A., et al.: Artificial intelligence and internet of things to improve efficacy of diagnosis and remote sensing of solar photovoltaic systems, challenges, recommendations and future directions. Renew. Sustain. Energy Rev. **143**, 110889 (2021)
5. Ghosh, A., et al.: Artificial intelligence in Internet of things. CAAI Trans. Intell. Technol. **3**(4), 208–218 (2018). ISSN 2468–2322
6. Rocha, Á.B., et al.: Development of a real-time surface solar radiation measurement system based on the Internet of Things (IoT). Sensors **21**(11), 3836 (2021)
7. Tiwari, V., et al.: Monitoring and control of PV microgrid using IoT. In: 2021 7th International Conference on Electrical Energy Systems (ICEES), 292–298. IEEE Explore (2021)
8. Tellawar, M.P., et al.: An IoT based smart solar photovoltaic remote monitoring system. Int. J. Eng. Res. Technol. **8**(9), 235–240 (2019)
9. Bet, M.J.A., et al.: Design and implementation of an open-Source IoT and blockchain-based peer-to-peer energy trading platform using ESP32-S2 Node-Red and MQTT protocol. Energy Rep. **7**, 5733–5746 (2021)
10. Talshyn, K., et al.: Experience of connecting sensors to the controller based on the arduino board for use on multicopters. J. Theor. Appl. Inf. Technol. **100**(6) (2022). ISSN: 1992–8645
11. Mehmet, T., et al.: Real-time monitoring of indoor air quality with internet of things-based e-nose. Appl. Sci. **9**, 3435 (2019)
12. Soumyajit, M., et al.: Smart light for home with automatic direction and intensity adjustment using arduino. Int. J. Rec. Technol. Eng. **8**(4), 200–215 (2019)
13. Siti, M., et al.: Recovery of electrical energy from mechanical energy on a rotating playground bowl. e-Proceeding Appl. Sci. **6**(2), 3332 (2020)
14. Nur, k., et al: Design and Build an E-Catalog Application for Typical Jombang Products. Multidiscipline - International Conference (2021). E-ISSN: 2809–6142
15. Sabrina, S., et al.: Security & privacy issues and challenges in NoSQL databases. Comput. Netw. 108828 (2022)
16. Ameya, N., et al.: Type of NOSQL databases and its comparison with relational databases. Int. J. Appl. Inf. Syst. **5**(4), 16–19 (2013)
17. Pereira, R., et al.: IoT embedded linux system based on Raspberry Pi applied to real-time cloud monitoring of a decentralized photovoltaic plant. Measurement **114**, 286–297 (2018)
18. Rahman, M., et al.: Global modern monitoring systems for PV based power generation: a review. Renew. Sustain. Energy Rev. **82**, 4142–4158 (2018)
19. Benghanem, M.: Measurement of meteorological data based on wireless data acquisition system monitoring. Appl. Energy **82**(12), 2651–2660 (2009)

IoT-Enabled Smart Agriculture: Security Issues and Applications

Mouaad Mohy-eddine[1](✉), Azidine Guezzaz[1], Said Benkirane[1], and Mourade Azrour[2]

[1] Higher School of Technology, Cadi Ayyad University, 44000 Essaouira, Morocco
mouaadmohyeddine@gmail.com

[2] IDMS Team, Faculty of Sciences and Techniques, Moulay Ismail University of Meknès, Errachidia, Morocco

Abstract. The Internet of Things (IoT) harmonizes existing technologies, such as wireless sensors and cloud computing, to form a homogeneous and powerful environment. Hence, IoT has emerged in different areas of life, such as healthcare, industrial sectors, and agriculture. Agriculture is one of the major components of developing countries' economies, and it plays a vital role in sustaining human life. Nevertheless, the human ability to reproduce far exceeds the ability of the earth to secure the food necessary. Therefore, the emergence of IoT has seen significant growth to help agricultural production quantity and quality. This paper provides a mini-review on IoT security scenarios applied in smart farming. In addition, we investigate the applications and use cases, issues, and challenges of IoT in smart agriculture. By presenting this paper, we aim to draw guidelines to develop an intrusion detection system suitable for the smart agriculture networks.

Keywords: Smart farming · IoT · IDS · Authentication · Blockchain

1 Introduction

A study has warned the world that the population grows way quicker than its means of subsistence, which would inevitably result in famine and other disasters [1]. However, the agriculture technologies evolution is helping to fill the gap between population growth and food production. Smart farming or smart agriculture, is the adoption of IoT, blockchain, and Artificial intelligence technologies in the traditional agriculture [2].

IoT interconnects billions of embedded objects to serve a meaningful purpose [3, 4]. Smart agriculture applies IoT in many domains, such as farm management, harvesting system, soil cultivation, irrigation [2], animal health, activity monitoring [5], temperature monitoring, infrastructure management, controlling soil moisture, and monitoring humidity [7]. Various papers have been published presenting research, challenges, and smart farming applications [2, 5, 6]. The [2, 8] mentioned the IoT benefits in agriculture, such as quantity and quality of production that affect the population's health and lifespan. These devices do not have high computational and memory capabilities [9]. Smart farming is boosting the agriculture field's quality and productivity by decreasing human intervention and automating various processes. Smart agriculture ecosystems are

Y. Farhaoui et al. (Eds.): ICAISE 2022, LNNS 635, pp. 566–571, 2023.
https://doi.org/10.1007/978-3-031-26254-8_82

exposed to severe threats due to adopting IoT technologies [10, 11]. Exploiting smart farming vulnerabilities can lead to severe problems [12]. Sensors collect and send data to data lakes in the cloud, which makes maintaining data privacy, confidentiality, authenticity, and access critical. Considerable research on enhancing IoT security already exists, industrial IoT [13], smart city [14], and smart grid [15] can be adapted. Nevertheless, different IoT environments need well-adopted security mechanisms that can fit each environment's security issues.

We analyzed and condensed pertinent papers published between 2019 and 2022 to provide a baseline for future works. Our paper presents a comprehensive survey of smart agriculture applications and security issues. In Sect. 2, we provide a smart farming architecture and applications of IoT in agriculture. Section 3 presents smart farming security issues, and some works in security solutions are presented. The paper is concluded with a conclusion and future work. Architecture and applications.

1.1 Smart Farming Architecture

Smart farming architecture, as illustrated in Fig. 1, conserves the three basic three-layer architecture layers, which are the perception layer, the network layer, and the edge and cloud layer. Perception layer [2, 6, 12] contains the physical devices and sensors, including drones, autonomous tractors. These devices sense, gather information and help other devices perform smart farming missions. Network layer allows devices to guide and share data. It is the core of smart farming environments. It provides connectivity between the physical layer and edge and cloud layer [6, 12]. It facilitates the communication of heterogeneous devices [12]. Edge and cloud layer [12] contains two layers, the edge layer and the cloud layer. They receive the data gathered by the physical layer. The edge is between the end device and the end user and helps in the local computation and actuation. The cloud offers advanced computation, decisions and storage.

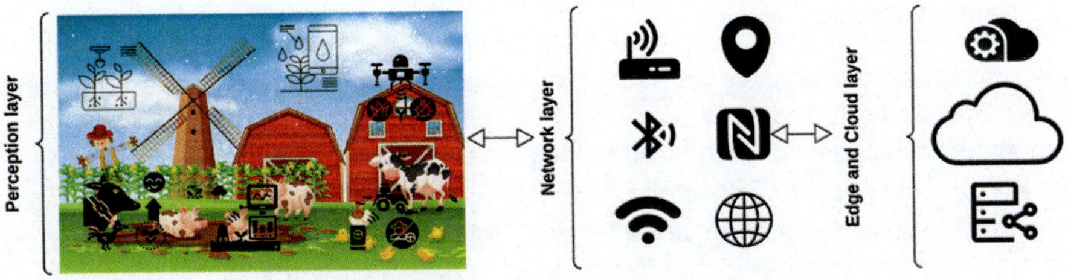

Fig. 1. Smart farming architecture.

1.2 Smart Farming Applications

IoT in the agriculture brings numerous benefits. It helps increase the quality and quantity of the produced food, reduce the cultivation cost, monitor and lower the wastage, optimize time, and make the agriculture industry more environmentally friendly. Precision agriculture: it ensures that crops and soil have just their need to remain healthy and result in maximum productivity. To reach that, smart agriculture uses IoT alongside other technologies. The [1] reviews smart monitoring and irrigation control strategies for irrigation scheduling.

Livestock farming: it is the automatic monitoring of the health and well-being of domestic animals to improve productivity and prevent health issues [3]. In [4], a review was presented on applied technologies for livestock management. Greenhouses: A closed area or a building designed to protect plants from bad weather. Smart greenhouses became more intelligent in providing the rightful climate for the plants without any manual intervention [5]. [6] presented a review on the role of IoT in smart farming and optimized greenhouse environments. IoT applied in other scenarios, like soil cultivation, irrigation, harvesting, etc. [3].

2 Smart Farming Security and Solutions

2.1 Security Issues

Understanding security issues is the first step to developing adequate solutions.

Maintaining Data security is considered one of the most motivations in research. In case of leakage, the big data generated could cause severe losses to farmers. Furthermore, most smart farms rely on third parties to analyze their data [12], which raises the risk of phishing attacks. The persistent need for real-time analysis and automatic actuation made data integrity as important as data privacy.

Network issues cannot be denied as it is the main channel between different architecture layers. In machine communication, ensuring that the senders and the receivers are the authorized and trusted entities is critical. In smart livestock, a buyer needs access to monitoring devices of a specific animal to analyze its data before buying [12], which raises trust and authorization and would lead to authentication of devices.

2.2 Smart Farming Attacks

Data attacks: Data leakage is the most devastating attack on a farmer because it harms farm productivity and exposes its strategy to competitors [16]. Intentionally or unintentionally, insider attacks could lead to huge losses in data, productivity, and finances. Usually, data centers across the globe store data as a cloud advantage, but it is vulnerable to distributed attacks. Attackers make the data stores unavailable for any user, falsifying the data, injecting misinformation into the network, which will affect real-time decisions, applying traffic analysis attacks, or man-in-the-middle attacks [17]. Device and network attack: Smart farming communication relies on sensors, RFID, meters, readers, and GPS devices [17]. Radio frequency communication is exposed to radio jamming attacks to disrupt and apply noise to interfere with the signals [12, 17]. DoS, malware injection, and spoofing attacks [10, 16, 18] are some attacks on conventional networks that are applied in smart agriculture environments [16, 18].

2.3 Countermeasures

After an extensive investigation, we have concluded that there is a lack of countermeasures in the literature for smart farming. It is due to the great focus given to developing smart systems and frameworks to enhance productivity in the agriculture field. [19] presented an intrusion detection system (IDS) to classify and identify intrusions in IoT-enabled smart irrigation in smart farming. A smart contract-based blockchain-envisioned authenticated key agreement mechanism in a smart farming environment was found in [20]. Autoencoder-based anomaly detection in the smart farming ecosystem was developed in [21] using an unsupervised machine learning model and achieved 98.98% accuracy. [22] suggested a blockchain architecture to protect drones and IoT sensors in agriculture by generating smart contracts. [23] proposes a platform based on blockchain for agriculture to ensure data integrity in fish farms. In [24] they created an algorithm to preserve privacy with a white Gaussian noise perturbation to ensure smart farm data's integrity. [25] enhanced authentication and exchanging protocol to detect various attacks on IoT environments. [26] proposed an anomaly detector to categorize heterogeneous traffic within an IoT network. To improve an IDS detection rate, [27] presented a model based on ML to mitigate security vulnerabilities in IoT. [28] designed architecture of mutual authentication protocol for distributed systems (Table 1).

Table 1. Literature review summary.

Reference	Year	Main theme	Threats	Countermeasures
Vangala et al. [2]	2022	Proposed a smart farming architecture, benefits, applications and security	-	√
Ullo et al. [5]	2021	Employment of IoT and smart sensors in agriculture applications	-	-
Quy et al. [6]	2022	Challenges of applying IoT in agriculture, architecture and application	√	-
Gómez et al. [8]	2019	Used software, devices, applications and benefits of IoT in agriculture	-	-
Demestichas et al. [11]	2020	Existing threats, vulnerabilities and mitigation measures in smart farming environments	√	√
Gupta et al. [12]	2020	Smart farming architecture, attacks, and solutions, security and privacy	√	√

3 Conclusion and Future Work

By providing this study, we look forward to building a baseline for future works of smart farming security. We began our paper with the smart farming architecture and some applications of IoT in the field. Then we provided smart farming security issues, attacks, and some countermeasures. Smart farming environment security is in critical need of contributions to mitigate security vulnerabilities. It is right that most IoT security works can be applied to smart farming security, but it needs some focused work for this field. For that, our future work will concentrate on developing a security mechanism alongside learning models for smart agriculture to mitigate security issues.

References

1. Thomas, R.M.: An Essay on the Principle of Population, as It Affects the Future Improvement of Society. J. Johnson, London (1798)
2. Vangala, A., Das, A.K., Chamola, V., Korotaev, V., Rodrigues, J.J.: Security in IoT-enabled smart agriculture: architecture, security solutions and challenges. Cluster Comput., 1–24, (2022). https://doi.org/10.1007/s10586-022-03566-7
3. Chanal, P.M., Kakkasageri, M.S.: Security and privacy in IoT: a survey. Wireless Pers. Commun. **115**(2), 1667–1693 (2020). https://doi.org/10.1007/s11277-020-07649-9
4. Azrour, M., Mabrouki, J., Guezzaz, A., Kanwal, A.: Internet of Things security: challenges and key issues. Secur. Commun. Netw. **2021**, 11 (2021)
5. Ullo, S.L., Sinha, G.R.: Advances in IoT and smart sensors for remote sensing and agriculture applications. Remote Sens. **13**(13), 2585 (2021)
6. Quy, V.K., et al.: IoT-enabled smart agriculture: architecture, applications, and challenges. Appl. Sci. **12**(7), 3396 (2022)
7. Haque, M.A., Haque, S., Sonal, D., Kumar, K., Shakeb, E.: Security enhancement for IoT enabled agriculture. Materials Today: Proceedings (2021)
8. Gómez-Chabla, R., Real-Avilés, K., Morán, C., Grijalva, P., Recalde, T.: IoT applications in agriculture: a systematic literature review. In: Valencia-García, R., Alcaraz-Mármol, G., Cioppo-Morstadt, J.D., Vera-Lucio, N., Bucaram-Leverone, M. (eds.) CITAMA2019 2019. AISC, vol. 901, pp. 68–76. Springer, Cham (2019). https://doi.org/10.1007/978-3-030-10728-4_8
9. Al Asif, M.R., Hasan, K.F., Islam, M.Z., Khondoker, R.: STRIDE-based cyber security threat modeling for IoT-enabled precision agriculture systems. In: Proceedings of the 2021 3rd International Conference on Sustainable Technologies for Industry 4.0, pp. 1–6. IEEE (2021)
10. Sontowski, S., et al.: Cyber-attacks on smart farming infrastructure. In: Proceedings of the 2020 IEEE 6th International Conference on Collaboration and Internet Computing (CIC) , pp. 135–143. IEEE (2020)
11. Demestichas, K., Peppes, N., Alexakis, T.: Survey on security threats in agricultural IoT and smart farming. Sensors **20**(22), 6458 (2020)
12. Gupta, M., Abdelsalam, M., Khorsandroo, S., Mittal, S.: Security and privacy in smart farming: challenges and opportunities. IEEE Access **8**, 34564–34584 (2020)
13. Huang, H., Ye, P., Hu, M., Wu, J.: A multi-point collaborative DDoS defense mechanism for IIoT environment. Digit. Commun. Netw. (2022)
14. Sangaiah, A.K., Javadpour, A., Jáfari, F., Pinto, P., Ahmadi, H., Zhang, W.: CL-MLSP: the design of a detection mechanism for sinkhole attacks in smart cities. Microprocess. Microsyst. **90**, 104504 (2022)

15. Dehalwar, V., Kolhe, M.L., Deoli, S., Jhariya, M.K.: Blockchain-based trust management and authentication of devices in smart grid. Cleaner Eng. Technol. **8**, 100481 (2022)
16. Sinha, B.B., Dhanalakshmi, R.: Recent advancements and challenges of Internet of Things in smart agriculture: a survey. Futur. Gener. Comput. Syst. **126**, 169–184 (2022)
17. Rouzbahani, H.M., et al.: Communication layer security in smart farming: a survey on wireless technologies. arXiv e-prints arXiv:2203.06013 (2022)
18. Kansal, N., Bhushan, B., Sharma, S.: Architecture, security vulnerabilities, and the proposed countermeasures in Agriculture-Internet-of-Things (AIoT) systems. In: Pattnaik, P.K., Kumar, R., Pal, S. (eds.) Internet of Things and Analytics for Agriculture, Volume 3. SBD, vol. 99, pp. 329–353. Springer, Singapore (2022). https://doi.org/10.1007/978-981-16-6210-2_16
19. Raghuvanshi, A., et al.: Intrusion detection using machine learning for risk mitigation in IoT-enabled smart irrigation in smart farming. J. Food Qual. (2022)
20. Vangala, A., Sutrala, A.K., Das, A.K., Jo, M.: Smart contract-based blockchain-envisioned authentication scheme for smart farming. IEEE Internet Things J. **8**(13), 10792–10806 (2021)
21. Adkisson, M., Kimmell, J. C., Gupta, M., Abdelsalam, M.: Autoencoder-based anomaly detection in smart farming ecosystem. In: Proceedings of the 2021 IEEE International Conference on Big Data (Big Data), pp. 3390–3399. IEEE (2021)
22. Chinnaiyan, R., Balachandar, S.: Reliable administration framework of drones and IoT sensors in agriculture farmstead using blockchain and smart contracts. In: Proceedings of the 2020 2nd International Conference on Big Data Engineering and Technology, pp. 106–111 (2020)
23. Hang, L., Ullah, I., Kim, D.-H.: A secure fish farm platform based on blockchain for agriculture data integrity. Comput. Electron. Agric. **170**, 105251 (2020)
24. Chukkapalli, S.S.L., Ranade, P., Mittal, S., Joshi, A.: A privacy preserving anomaly detection framework for cooperative smart farming ecosystem. In: Proceedings of the 2021 Third IEEE International Conference on Trust, Privacy and Security in Intelligent Systems and Applications (TPS-ISA), pp. 340–347. IEEE (2021)
25. Azrour, M., Mabrouki, J., Guezzaz, A., Farhaoui, Y.: New enhanced authentication protocol for Internet of Things. Big Data Min. Analytics **4**(1), 1–9 (2021)
26. Guezzaz, A., Asimi, Y., Azrour, M., Asimi, A.: Mathematical validation of proposed machine learning classifier for heterogeneous traffic and anomaly detection. Big Data Min. Analytics **4**(1), 18–24 (2021)
27. Guezzaz, A., Benkirane, S., Azrour, M.: A novel anomaly network intrusion detection system for Internet of Things security. In: IoT and Smart Devices for Sustainable Environment, pp. 129–138, (2022). https://doi.org/10.1007/978-3-030-90083-0_10
28. Tbatou, Z., Asimi, A., El Balmany, C., Asimi, Y., Guezzaz, A.: A novel architecture of a strong and mutual authentication protocol for distributed systems. Eng. Lett. **28**(2) (2020)

Learning Analytics in the Teaching of French as a Foreign Language (FFL) and Big Data: What Resources? For What Skills?

Sara Ouald Chaib(✉), Imane Joti, and Samira Khoulji

Information Systems Engineering Research Team (ERISI), Abdelmalek Essaadi University – ENSA of Tetouan, Tetouan, Morocco

sara.oualdchaib@etu.uae.ac.ma, {ijoti,skhoulji}@uae.ac.ma

Abstract. Today, Big Data technologies and the generalization of the use of digital tools by students allow us to build large bodies of data on their behavior.

After the arrival of the pandemic (Covid 19), learning analytics has become an emerging trend in Moroccan universities. It offers advantages and challenges for online learning. To answer questions related to the exploitation of this data which may provide useful information to language teachers in the context of specific lessons or learning environments. A quantitative analysis was conducted for the collection and evaluation of the student interaction process for distance language courses. We have found that the main potential of digital learning of French as a foreign language (FFL) lies in the provision of authentic linguistic resources accessible to learners.

Keywords: Big Data · Learning analytics · Online learning · FFL

1 Introduction

Higher education courses, especially at universities, tend to be less interactive due to larger enrolments. The teacher, having less feedback from his students, does not know the effectiveness of his methods until the results of the exams. With the emergence of advanced technologies, the digital learning sector is constantly evolving. New educational formats are emerging at the same time as new digital tools, thus promising the sector many educational innovations.

Education generates massive volumes of digital data (learner results, grades, quality of online participation, etc.), the analysis of which, often carried out by artificial intelligence, makes it possible to improve pedagogical approaches [1]. It's here where Big Data comes into play, which consists of extracting useful information from large amounts of data [2]. So, Big Data is all the data transmitted directly or indirectly by the learner during his training and this is through the various devices put in place [3]. In the case of language learning in a university context, the problem of the exploitation of Big Data is essential knowledge that the analysis of learning data aims to improve the overall effectiveness of training systems.

Y. Farhaoui et al. (Eds.): ICAISE 2022, LNNS 635, pp. 572–580, 2023.
https://doi.org/10.1007/978-3-031-26254-8_83

Other findings in terms of student habits can be seen when they test and visit multiple learning platforms and sites given their learning needs. In most cases, they find massive educational content created by multiple methods and scenarios. This led us to ask the following questions: What are the types of technological and educational platforms most appreciated by learners according to their expectations and what are their proposals for possible improvements?

In this context, our research work will be based on experimental studies and several types of interviews (questionnaires and computerized tests) which were carried out to determine the characteristics and training needs of the students (in terms of resources, functionalities, evaluations, and software).

2 Learning Analytics at the Service of FFL

Learning Analytics is the measurement, collection, analysis, and processing of data about learners and their environments with the aim of understanding and optimizing learning and the environments in which it occurs [4]. In language education research, the availability of large datasets related to social media posts, and traffic data to the ways and frequency with which learners interact with online resources, presents opportunities very interesting for teachers who have access to learner data beyond what is observable in the classroom, from grade point averages in school subjects to comparisons of performance between classes, courses, and teachers [5]. On the other hand, it requires teachers to develop new skills, new Big Data analysis techniques will have to be adopted to deal with the five characteristics of big data: volume, velocity, variety, veracity, and value [6]. It should be noted that the most important step is the analysis phase, which makes it possible to question how the training is designed and to identify areas for improvement to optimize the learner's training course [7]. In addition, this makes it possible to establish correlations between different indicators such as the duration of the training and the success rate.

Nowadays, researchers have started to realize that it is very important to analyze the data loads used and generated in foreign language teaching. Wang and Zhang [8] explored how Big Data methodology can help reinvent and reshape EFL (English as a Foreign Language) teaching. Using Big Data methodology, they statistically analyzed the frequency of change in the composition of learners and obtained scores, changes in vocabulary and sentence, and types of writing errors respectively. Also, with the trend of mobile learning, further research has been directed toward collecting and combining learning analytics data offered by modern smartphones [9]. This has contributed to the development of adapted and personalized learning systems for language learning [10].

Many universities use LMS-like online education platforms to manage a large mass of students. Data provided by the latter such as Moodle, mainly focus on administrative types of reporting data, such as the frequency of logins or the number of clicks on particular resources [11]. The visualization tools used tend to simplify the learning process and may not give a full picture of its complexity [12]. So, the effectiveness of Learning Analytics depends on a variety of factors and how the data was collected and processed.

In our research, we have listed a few performance indicators such as participation rate, and student interaction, which were collected and analyzed during an activity session.

These indicators make it possible to understand what pleases or, conversely, what repels the student [13]. For the trainer, this makes it possible to rethink, for example, the formats of his training by integrating new activities or by integrating more playful activities [14]. Therefore, meeting the needs of students by considering the constraints encountered by them.

3 Research Method

Since the arrival of the covid 19 pandemic in Morocco, the process of switching to distance education has been accelerated. This resulted in a real challenge for teachers and students who were not used to this new mode of Teaching/Learning. So, in a desire to support the commitment of students in a process of regulation, with the new training situation. We need to collect information to create tailor-made learning solutions and adapt the content of each training according to the needs of the learners [15].

In this context, we have developed an activity in the form of a questionnaire intended for scientific students in the faculties of science at Abdelmalek Essaadi University, which is disseminated on the university's Moodle platform and social networks. This questionnaire aims to collect all the data related to the expectations and needs of students using FFL digital learning systems. All this information will be collected, measured, used, and analyzed to improve and optimize the online learning process [16]. The observation needs identified by this questionnaire are grouped around these two axes:

- Identify the actions of university students according to the context of the use of online resources.
- Identify the actions of university students according to the nature of the resources used.

On the other hand, to develop distance resources according to student practices, several types of interviews were carried out with students and teachers of the Language and Terminology department to determine the characteristics and needs of students in FOP type training (French for Specific Purposes).

4 Results

The data below are based on the observation and responses of a sample of 450 SMAI (Mathematical Sciences and Applications/Mathematical and Computer Sciences) (Semester 1) students from the Faculty of Sciences (FS) of Tetouan, Abdelmalek Essaadi University who follow the Language and Terminology module training at a distance during the year 2021–2022.

The students concerned by this study, (97.1%) are new registrants.

The average age was 18 [range 17–27]. The breakdown by gender was (41%) women and (59%) men (Graph 1).

61.7% of students said they had a personal laptop, 91.2% had Internet access at home and all have electronic addresses (e-mail). 76.3% have smartphones.

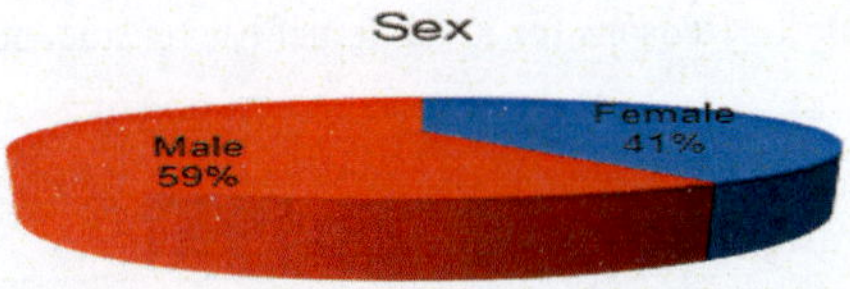

Graph 1. Gender.

Concerning the use of the Internet, 90.3% of the students used this tool within the framework of their studies; 75% with an average connection time of one hour or more per day (Graphs 2 and 3).

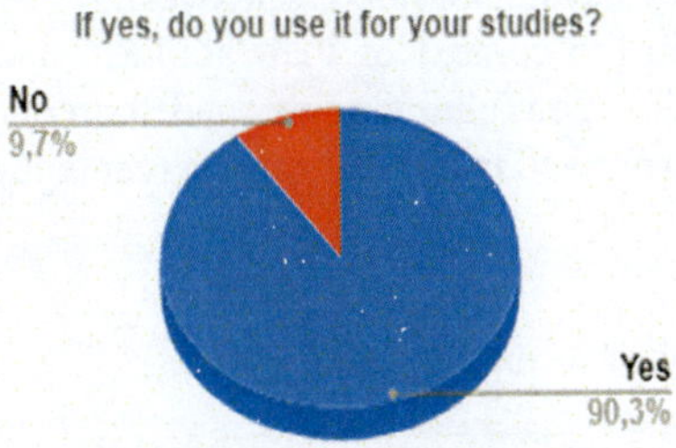

Graph 2. Use for studies.

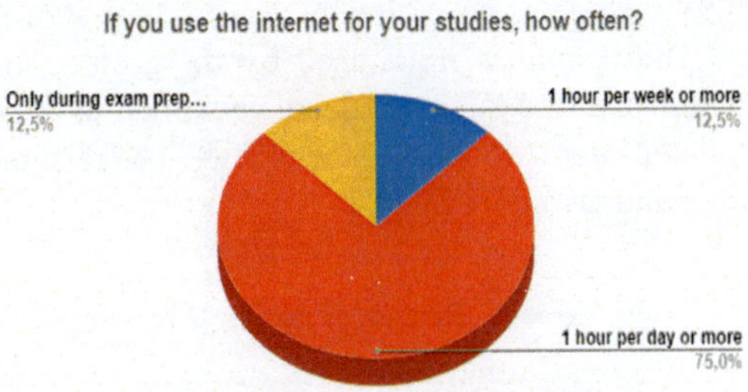

Graph 3. Frequency of use.

The reasons for this high rate of internet use during this year among students are linked to the current situation relating to the covid-19 pandemic. Higher education in Morocco has moved from all face-to-face to all remotely, which has forced students to follow their courses via the Internet.

However, (Graph 4) the reasons given by the students for not using the internet or not using it very much during their studies were: (Table 1).

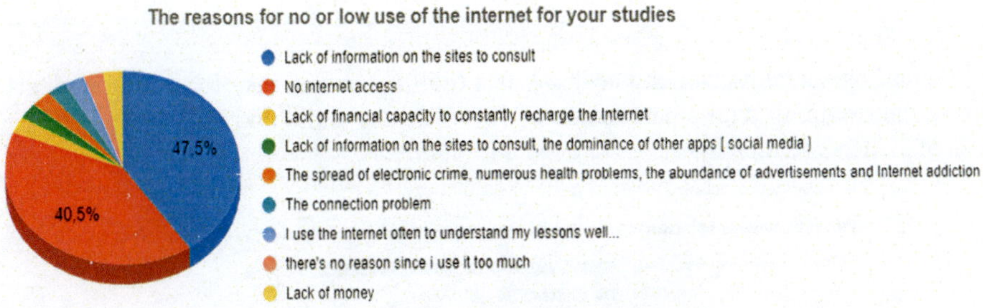

Graph 4. Reasons for low Internet use.

The platforms most used by these students during their studies were Microsoft Teams (50%), Moodle (30%), Google Classroom (10%), and Google Meet (10%). Most of the students have already used the Microsoft Teams platform during their studies in the previous year of the baccalaureate to be able to follow their courses remotely through

Table 1. Reasons for low Internet use by students.

Pattern	%
Lack of information on the sites to consult	47
No internet access	40.5
Lack of financial capacity to constantly recharge the Internet	8.7

virtual classes. It has been one of the solutions adopted by the Ministry of Education since the appearance of the covid-19 pandemic in public and private schools. A new remote work environment is adopted at the university, it is the Moodle platform, and 30% of students have already worked with it or they have discovered it this year.

The resources most used by these students in the context of their studies were chat/instant messaging, social networks, Wikipedia, YouTube, Google, and the educational sites (Moodle) of the faculty. The results expressed in numbers and percentages are summarized in Table 2.

Table 2. Resources and tools most used by FS students for their learning.

Attendance resources	%
The chat/instant messaging	70.61%
Social networks (e.g., Facebook)	64.44
Faculty pedagogical web (Moodle)	44.44
YouTube and Google	22.22
Forums and discussions	6.67
Wiki (ex: Wikipedia)	2.22
Software specific to your discipline	0

Regarding social networks, Facebook is a tool that seems easy to use for students during their training with a percentage of 50% And the audio and video resources by 50% of students (Graph 5).

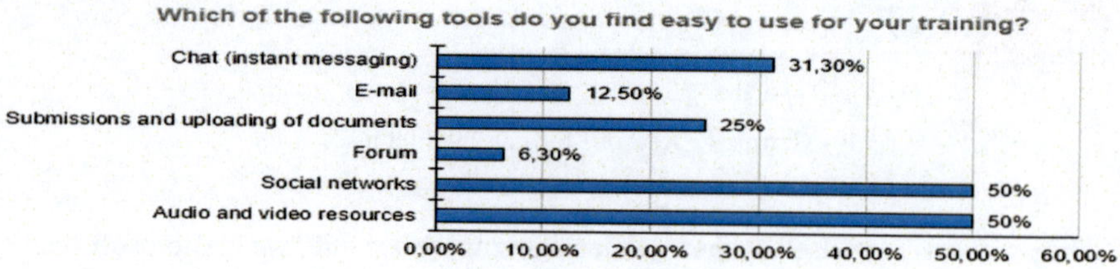

Graph 5. Easy-to-use tools.

In response to questions about MOOCs and e-Learning platforms:

- 82.4% of students are unaware of MOOCS,
- 63.3% say they do not know about e-Learning platforms, nor use them.

5 Discussion

5.1 Objective

This observational study allowed us to have objective data on the use that students make of online resources in their learning activities (especially after the covid-19 pandemic).

The idea was born from the observation and recognition of the "gap" between the desire of universities and teachers to innovate and promote the production of digital resources: courses, practical work, videos… And the actual use that the students do. On the other hand, to be able to design a product that meets the needs of the students of the FS of Tetouan.

5.2 Discussion of the Results

In this work, we noted that university students in the 1st year of the FS had little recourse to courses and educational web resources. However, during the current situation caused by the pandemic, things changed, students had good use of the educational resources made available to them on the university platform. Most students do not use software specific to their discipline. Although 91.9% of students are equipped with computers and have an Internet connection and the frequency of use for learning activities is generally good with an average connection of 1 h per day. Students prefer to use social networks such as Facebook groups (50%) for sharing course-related resources and as a means of accessing discussion forums. These interactions are interesting, but their quality remains abstract to the absence of supervision by teachers. To meet the needs of students, the head of the "Language and Terminology" module [17] has created a Facebook group for each year of study, for communication (scheduling of courses, exams), and for sharing of courses.

Most students lean towards audio and video resources (50%), rather than specific resources made available to them (such as uploading and downloading documents, email, etc.). As such, audiovisual resources are considered among the most valuable practices for learning, because it is faster and closer to face-to-face lessons to watch and hear than to enter a site or download lessons for reading.

A very large majority of students are also unaware of the MOOCs and other e-learning resources to which they could have access. In response to this state of affairs, the students mentioned as the main reason, the ignorance of the existence of these sites which are dedicated to them. Knowing that 50% of the students did not answer this question.

From this observation, we can only formulate other hypotheses about this low use of available resources. So, is this a real communication problem? teachers do not transmit information to students about these resources even if they are involved in the development of these resources which requires effort? we can even ask questions about the quality of these resources, whether on the site or the platform, in terms of ergonomics, user-friendliness, usability, satisfaction (attractiveness, speed of access), and efficiency. On the other hand, we can even mention the subject of the quality of content, design, relevance, and clarity of these educational resources put online, which mainly depends on the teaching skills of the teacher, his mastery of the IT tool, and its motivation.

On the other hand, what are the motives related to the student? we can cite the factor of non-mastery of computer tools also among students, but, if we talk about the technical side, most students, in recent years, are comfortable in the handling of digital tools. In addition, we have noticed that the student, instead of going to look for information on the sites that are made available to him, looks on other sites. This may be related to communication problems and ignorance of the existence of these sites or because they do not bring added value.

In summary, we can say that this work confirms that, in general, the use of new online teaching/learning technologies remains a powerful educational tool offering varied learning opportunities (Online courses, Videos, exercises, etc..) but it must be thought out, planned, and integrated into a structured set of educational activities to contribute to the achievement of learning objectives. In addition, other tasks must be handled carefully such as technical aspects, organizational, training, and even information issues. Considering the results, it should only be recalled that the digital resources used remotely remain a complement to face-to-face teaching/learning activities. And that we must think of tools that will create added value for students for their learning [18].

All these practices and data collected are pedagogically relevant elements that could be instrumented by Learning Analytics tools [19], hence the importance of analyzing these practices in detail to take them into account in the new solutions designed by teachers.

6 Conclusion

For the student, a better knowledge of his own learning strategies and an evaluation of their relevance is a major issue of higher education. What was presented in this work through a study that allowed us to identify the actions of university students according to the context of the use of online resources and their nature, is to develop resources that meet their needs.

Through this work, we tried to show the potential of big data and Learning Analytics in understanding the behaviors and practices of university students via e-learning technologies. Since they allowed us in this phase of collecting, quantifying, and analyzing this data to be able to determine what kind of resources should be considered by language teachers in their future distance practices, to increase student engagement and motivation.

From this research, we found that the gradual abandonment of communication tools that are part of the LMS such as discussion forums has increased. Students today are showing increasing use of instant messaging apps available for smartphones and are trying to create study or support groups for each subject. In this context, we plan, in an evolutionary perspective in the future, to develop and evaluate self-training resources based on conversational agent systems, taking into consideration all the needs and practices of students already studied to allow them to work at their own pace and preferences. And to create, on the other hand, an interface allowing remote feedback to be delivered in various ways to allow teachers to properly adapt the teaching offer.

References

1. Wen, Y., Song, Y.: Learning analytics for collaborative language learning in classrooms: From the holistic perspective of learning analytics, learning design and teacher inquiry. Educ. Technol. Soc. **24**(1), 1–15 (2021)
2. El Alfy, S., Marx Gómez, J., Dani, A.: Exploring the benefits and challenges of learning analytics in higher education institutions: a systematic literature review. Inf. Discov. Deliv. **47**(1), 25–34 (2019)
3. Gašević, D., Tsai, Y.-S., Drachsler, H.: Learning analytics in higher education – Stakeholders, strategy and scale. Internet High. Educ. **52**(2022), 100833 (2022)
4. Siemens, G., Long, P.: Penetrating the fog: analytics in learning and education. EDUCAUSE Rev. **46**(5), 30–40 (2011)
5. Lan, Y.-J., Chen, N.S., Sung, Y.-T.: Guest editorial: Learning analytics in technology enhanced language learning. Educ. Technol. Soc. **20**(2), 158–160 (2017)
6. Gandomi, A., Haider, M.: Beyond the hype: Big data concepts, methods, and analytics. Int. J. Inf. Manage. **35**(2), 137–144 (2015)
7. Pardos, Z.A., Fan, Z., Jiang, W.: Connectionist recommendation in the wild: on the utility and suitability of neural networks for personalized course guidance. User Model. User-Adap. Inter. **29**(2), 487–525 (2019)
8. Wang, Z., Zhang, Y.: Big data methodology and EFL teaching innovation — a case study of one million-odd EFL writings on the same topic. Comput. Assist. Foreign Lang. Educ. China **165**, 3–8 (2015)
9. Karakaya, K., Bozkurt, A.: Mobile-assisted language learning (MALL) research trends and patterns through bibliometric analysis: empowering language learners through ubiquitous educational technologies. System, 102925 (2022)
10. Wong, B.T.M., Li, K.C., Cheung, S.K.: An analysis of learning analytics in personalised learning. J. Comput. High. Educ., 1–20 (2022)
11. Atapattu, T., Falkner, K.: Impact of lecturer's discourse for student video interactions: video learning analytics case study of MOOCs. J. Learn. Analytics **5**(3), 182–197 (2018)
12. Gelan, A., et al.: Affordances and limitations of learning analytics for computer–assisted language learning: a case study of the VITAL project. Comput. Assist. Lang. Learn. **31**(3), 294–319 (2018)
13. DePaolo, C.A., Wilkinson, K.: Get your head into the clouds: using word clouds for analyzing qualitative assessment data. Tech. Trends **58**(3), 38–44 (2014)
14. Reinders, H., Lan, Y.J.: Big data in language education and research. Lang. Learn. Technol. **25**(1), 1–3 (2021)
15. Sointu, E., et al.: Learning analytics and flipped learning in online teaching for supporting preservice teachers' learning of quantitative research methods. Seminar.net **18**(1) (2022)
16. Kew, S.N., Tasir, Z.: Developing a learning analytics intervention in E-learning to enhance students' learning performance: a case study. Educ. Inf. Technol. **27**, 7099–7134 (2022)
17. Joti, I., OualdChaib, S.: Devices for self-training in FOS. In: Proceedings of the 1st International Colloquium on Hard and Soft Skills, Morocco, pp. 77–97 (2020)

18. Viberg, O., Wasson, B., Kukulska-Hulme, A.: Mobile-assisted language learning through learning analytics for self-regulated learning (MALLAS): a conceptual framework. Australas. J. Educ. Technol. **36**(6), 34–52 (2020)
19. da Silva, L.M., et al.: Learning analytics and collaborative groups of learners in distance education: a systematic mapping study. Inform. Educ. **21**(1), 113–146 (2022)

Lexical Simplification of Arabic Educational Texts Through a Classification Approach

Naoual Nassiri[1,2](✉), Safae Berrichi[1,2], Abdelhak Lakhouaja[1], Violetta Cavalli-Sforza[3], and Azzeddine Mazroui[1]

[1] Department of Computer Science, Mohamed First University, Oujda, Morocco
naoual.nassiri@gmail.com

[2] Research Center of the Higher Engineering Studies School, Oujda (CREHEIO), Oujda, Morocco

[3] School of Science and Engineering, AI Akhawayn University, Ifrane, Morocco
v.cavallisforza@aui.ma

Abstract. Readers wishing to learn a new language need texts adapted to their reading level. Indeed, readers with language or learning difficulties encounter challenges in understanding texts containing complex words. Therefore, identifying these words and generating easier synonyms and definitions can enhance text comprehension. In order to confront these problems, we introduce in this study a lexical simplification approach to educational Arabic texts based on readability measurements. This approach involves the task of identifying complex words and proposing substitution candidates followed by a process of selecting the best substitution and replacing it in the original text. The performance of the approach is evaluated on two different corpora dedicated to readers of Arabic as a first foreign language. The obtained results are very encouraging and constitute a first attempt to automate the simplification of Arabic texts in order to support new learners.

Keywords: Readability · Simplification · Arabic language · L2 · Education · Classification

1 Introduction

Since education is one of the main pillars of modern societies, learning a first foreign language (L2) is recommended in all educational systems. One of the factors that determine the success of teaching is the appropriateness of the educational content to the learners' levels. Thus, providing simplified text-based content will be beneficial for beginner learners as well as those with learning difficulties.

Text simplification is an active sub-field of natural language processing, due to the practical and theoretical problems it addresses, as well as its many practical applications. It consists in reducing the complexity of vocabulary and sentence structure of a text while preserving its original meaning, in order to improve

Y. Farhaoui et al. (Eds.): ICAISE 2022, LNNS 635, pp. 581–587, 2023.
https://doi.org/10.1007/978-3-031-26254-8_84

its readability (the ease with which a learner can decode a text). Text simplification is relevant for people with cognitive disabilities, as well as for illiterates and foreign language learners.

Among the various approaches used in this field, we distinguish the rule-based approaches, the statistical learning-based approaches, and the hybrid approaches which combine the two previous ones. These approaches can be classified according to four simplification levels, namely:

1. **Lexical level:** it consists of identifying and replacing complex words in a given text with simpler synonyms.
2. **Syntactic level:** it consists in identifying the grammatical complexities of a text and rewriting them in simpler structures.
3. **Explanation generation level:** this level consists of taking a difficult concept in a text and enriching it with additional information, which puts it in context and improves the user's understanding.
4. **Generating simplified sentences via automatic translation level:** the simplification process can be considered as a translation of complex texts into simple texts. The translation task is then handled as a conversion from the complex source language X to the same target language X, but simpler.

Given the unavailability of Arabic texts with their simplified versions, and the lack of a standard reference for complex words, simplifying a text and evaluating the degree of simplification continues to be a challenging task. Thus, in this study, we will exploit a text readability prediction model for lexical simplification. This system takes into account the fact that a text predicted as difficult can contain words of different difficulty levels (easy, medium, and difficult).

The rest of this paper is organized as follows. In Sect. 2, we present some previous research on text simplification. In Sect. 3, we provide details on the data and tools used. We present our approach to Arabic text simplification in Sect. 4, and we discuss the obtained results. Finally, we draw some conclusions and suggest some thoughts for future work.

2 Related Work

Although research in the field of text simplification has made significant progress in several languages, the simplification of Arabic texts has only recently attracted the attention of researchers and little work has been done in this domain.

The area of text simplification has been investigated since the 1990s s and is still developing and becoming more automated today. Text simplification has been the subject of much work, especially for second language learners. One of the earliest efforts at automatic simplification was a grammar and style checker developed for simplified English writers [6]. Subsequently, another work applied statistical phrase-based machine translation techniques to produce simplified English texts [10].

Furthermore, the authors in [8] proposed the use of unsupervised extraction techniques to automatically create learning corpora for multilingual simplification from raw website data. They evaluated their approach on the basis of simplification requirements in English, French, and Spanish.

In view of the growing need to produce personalized health information that is readable and understandable by patients, the authors in [7] presented a text simplification tool that helps medical content producers to produce accessible text. The tool uses features that provide concrete suggestions. This tool provides good coverage with suggestions on more than one-third of words and sentences when evaluated on a medical corpus. These suggestions are accurate to more than 40%, although the accuracy varies depending on the source of the text. The authors adopted both word-level and sentence-level approaches.

Regarding the Arabic language, Al Khalil et al. [2] described in their work the motivation and outline of a project called SAMER (Simplification of Arabic Masterpieces for Extensive Reading), which focuses on simplifying a collection of novels. The project involves rewriting the collection of novels in a simple and readable style for school-age learners. Several researchers in Arabic literature and pedagogy contributed to the project. The simplification task was carried out by experts, who followed a set of general guidelines to make changes in vocabulary, structure, and style.

3 Data and Tools

In this section, we provide a brief description of the data and tools we used in this study.

3.1 Data

We collected two corpora dedicated to reading Arabic as an L2 to evaluate our approach.

1. GLOSS-Reading[1]: is a collection of reading texts for maintaining and improving learners' language skills. The texts in this corpus are annotated by their difficulty levels according to the ILR scale [5]. We have used 90 difficult texts from this corpus.
2. HindaouiCorpus: is a collection of texts that we have collected from the Hindaoui Foundation[2]. The texts are derived mainly from children's stories, novels, and plays, and are not annotated by their difficulty levels. We have used 100 difficult texts from this corpus in this study.

The first corpus is used to test the performance of the simplification approach, based on the readability prediction model, on data similar to the one on which the model was trained. However, the second corpus (HindaouiCorpus) is used to analyze the behavior of the approach with texts different from the one the readability prediction model learned during the training phase.

One of the most important steps in the lexical simplification task is the selection of the most appropriate synonym. The selection of this synonym was

[1] https://gloss.dliflc.edu/.
[2] https://www.hindawi.org/.

made on the basis of a frequency dictionary. It is composed of the 5 000 most frequent words of the Arabic language. This list of words was compiled from a corpus containing millions of words. Such a list is useful to decide if a word is frequent and therefore simple. In this work we used the frequency dictionary developed in [9].

3.2 Tools

Since the frequency dictionary is accessed by the couple (Lemma, PoS), we need to annotate the texts of our corpora with the same morphological information in order to match the synonyms of the complex words with those of the frequency dictionary. To do this, we used Alkhalil-Lemmatizer [4] and Alkhalil-PoS-Tagger [1].

4 Methodology

In this section, we present our lexical simplification process. As illustrated in Fig. 1, for a given text, the process consists of:

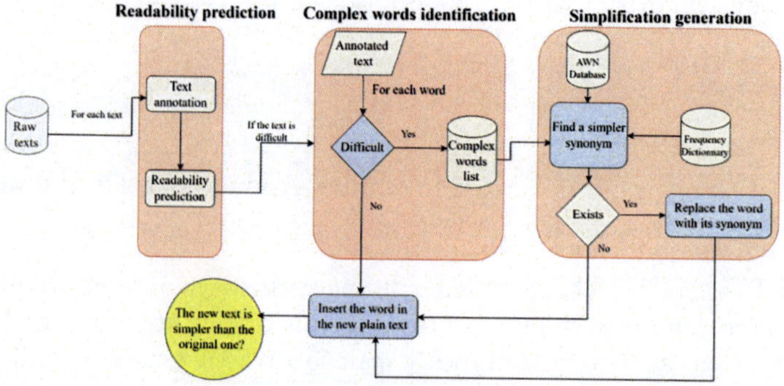

Fig. 1. Lexical simplification process

1. Predicting the difficulty level using a readability prediction model;
2. Simplifying lexically complex words;
3. Checking if the new text is less complex than the original one.

4.1 Difficulty Level Prediction

In order to measure the difficulty level of our texts, we need a readability prediction model. In this study, we have exploited the model dedicated to predict readability of Arabic L2 texts developed by [9]. As illustrated on Fig. 1, the first step of our process is a morphological annotation of the text words, followed by the application of the prediction model to retrieve the initial readability level of the text. Then, we keep only the texts with a difficult level to simplify them. Texts predicted by the medium or easy level will thus be eliminated in the following steps of the simplification process, as we assume that they do not require simplification.

4.2 Complex Words Identification

In this step, the system must be able to analyze the text and identify the list of complex words. The process followed for this step is as follows:

1. The inputs for this step are texts with a difficult level, and whose words have been morphologically annotated during the readability prediction phase;
2. The PoS annotation is then used to determine which words do not require simplification. Indeed, it is clear that it is useless to simplify some categories of words such as proper nouns and prepositions. Therefore, we started by filtering these words.
3. A second filtering step is then applied to the remaining words. It consists in eliminating the words belonging to the lexicon generated by the texts of lower levels (medium and easy), and whose words are supposed to be simple.

4.3 Substitution Generation

This step will be performed in two phases.

1. **First phase:** for each complex word, we produce a list of substitution candidates. To do so, we queried the linguistic database "Arabic WordNet" **(AWN)** [3] built manually by professionals, and in which a target word is associated with a set of synonyms.
2. **Second phase:** it consists in selecting the simplest synonym and substituting it for the complex word. A substitute is considered to be the simplest if it belongs to the lexicon of the lower levels, or otherwise if it belongs to the frequency dictionary with a higher frequency than the original word. Words that have no synonyms (the list of synonyms returned by AWN is empty) and words for which no synonym is simple will be kept without simplification.

4.4 Experiments and Results

In this section, we will apply the approach presented in Sect. 4.3 on a set of 190 difficult texts, 90 of which are drawn from the GLOSS-Reading corpus, and 100 drawn randomly from HindaouiCorpus. The *AWN* database, from which the synonyms were drawn, contains 11,269 words with their synonym lists. These words are classified according to four grammatical categories (verb, noun, adjective, and adverb). Table 1 shows some examples of words that have been replaced by simpler substitutions.

Table 1. Examples of simplified words

Complex word	Simple substitution	Translation
رمز /rmz/	علامة /ElAmp/	Sign
فجوة /fjwp/	فتحة /ftHp/	Gap
وكر /wkr/	عش /E$/	Nest
منوال /mnwAl/	طريقة /Tryqp/	Method

After the step of replacing the complex words with their simple synonyms, we recalculated the difficulty level of the new texts and we compared the results with the difficulty levels of the original texts. Table 2 shows the percentage of simplified texts for GLOSS-Reading and HindaouiCorpus, respectively.

Table 2. Results of the simplification

After simplification			
	Easy	Medium	Difficult
GLOSS-Reading	3	19	68
HindaouiCorpus	2	72	26

The obtained results show that 25% of the GLOSS-Reading corpus texts have been simplified whereas HindaouiCorpus reports a simplification percentage of 74%. By analyzing the simplified texts and some texts that have not been simplified by our approach, we have noticed the following:

1. The number of complex words replaced in a simplified text varies from 4 to 15 words.
2. The texts that have not been simplified are texts for which no synonyms for the difficult words are found in the AWN database. This lack of synonyms is due to the fact that AWN contains only 7,960 nouns and 2,538 verbs.

From these results, we can conclude that lexical simplification plays an important role in changing the difficulty levels of texts. These first performances encourage us to investigate other approaches of lexical simplification.

5 Conclusion and Future Work

This paper proposes an approach for lexical simplification of Arabic texts. We have tested this approach on two different corpora (GLOSS-Reading and HindaouiCorpus), and the obtained results are very encouraging.

Further research is needed in the area of text simplification. For example, in future work, we can improve this work by using a more representative thesaurus.

Exploiting pre-trained models like Bert for a vector representation of words can also improve the performance of the simplification model. Similarly, the use of the syntactic simplification level, which consists in breaking down the long and complex sentences and changing their syntax, will allow further simplification of the texts. Finally, the use of machine translation could be a possible direction for the simplification of texts.

References

1. Ababou, N., Mazroui, A.: A hybrid Arabic POS tagging for simple and compound morphosyntactic tags. Int. J. Speech Technol. **19**(2), 289–302 (2015). https://doi.org/10.1007/s10772-015-9302-8
2. Al Khalil, M., Habash, N., Saddiki, H.: Simplification of arabic masterpieces for extensive reading: a project overview. Procedia Comput. Sci. **117**, 192–198 (2017). Arabic Computational Linguistics
3. Black, W., et al.: Introducing the Arabic wordnet project. In: Proceedings of the Third International WordNet Conference, pp. 295–300. Citeseer (2006)
4. Boudchiche, M., Mazroui, A.: A hybrid approach for Arabic lemmatization. Int. J. Speech Technol. **22**(3), 563–573 (2018). https://doi.org/10.1007/s10772-018-9528-3
5. Clark, J.L.D., Clifford, R.T.: The fsi/ilr/actfl proficiency scales and testing techniques: development, current status, and needed research. Stud. Second. Lang. Acquis. **10**(2), 129–147 (1988). https://doi.org/10.1017/S0272263100007270
6. Hoard, J.E., Wojcik, R., Holzhauser, K.: An automated grammar and style checker for writers of simplified English, pp. 278–296. Springer, Netherlands, Dordrecht (1992). https://doi.org/10.1007/978-94-011-2854-4_19
7. Kauchak, D., Leroy, G.: A web-based medical text simplification tool (2020)
8. Martin, L., Fan, A., De la Clergerie, E., Bordes, A., Sagot, B.: Multilingual unsupervised sentence simplification (2020)
9. Nassiri, N., Lakhouaja, A., Cavalli-Sforza, V.: Arabic l2 readability assessment: dimensionality reduction study. J. King Saud Univ. Comput. Inf. Sci. (2021). https://doi.org/10.1016/j.jksuci.2020.12.021
10. Wubben, S., van den Bosch, A., Krahmer, E.: Sentence simplification by monolingual machine translation. In: Proceedings of the 50th Annual Meeting of the Association for Computational Linguistics (Volume 1: Long Papers), pp. 1015–1024. Association for Computational Linguistics, Jeju Island, Korea (2012)

Check for updates

License Plate Character Recognition System Using YOLOv5

Mohamed El Ghmary[1](✉), Younes Ouassine[2], and Ali Ouacha[2]

[1] Department of Computer Science, FSDM, Sidi Mohamed Ben Abdellah University, Fez, Morocco
mohamed.elghmary@usmba.ac.ma

[2] Faculty of Sciences, Department of Computer Science, Mohammed V University in Rabat, Rabat, Morocco
{younes.ouassine,a.ouacha}@um5r.ac.ma

Abstract. Recently, the field of artificial intelligence has seen many advances in deep learning and image processing. It is now possible to recognize images or even find objects in an image. One of the most interesting applications of image processing is automatic license plate recognition (ALPR). The Automatic License Plate Recognition system is not only focused on parking lots but can be used in all those facilities necessary to control, monitor, and have a record of all vehicles that pass through certain access. Example: private garages of companies, shopping centers, tolls, hospitals, etc. In this paper, we explain how we applied object detection methods to build a system capable of recognizing and extracting Moroccan license plates from images. We used transfer learning on YOLOv5 and trained it on 724 images of vehicles registered in Morocco to obtain a system capable of supporting a parking system.

Keywords: Plate detection · Character recognition · YOLOv5

1 Introduction

Reliable license plate recognition (LPR), also known as automatic license plate recognition (ANPR), has traditionally been expensive and only for a limited number of applications. The rapid development of IP-based cameras and deep learning is now enabling greater use of automated LPR solutions. Depending on your needs, they can help increase day-to-day efficiency, service levels and security. This task is a complex problem due to many factors, such as poor lighting conditions, blurred images, Arabic characters of license plate numbers in the Moroccan case, and variability of plates.

ALPR systems typically involve two steps: License plate (LP) detection and character detection in our approach, we decided to build two models separately instead of directly using libraries such as easyOCR due to their weaknesses in handling the variance of Moroccan license plate shapes and Arabic letters.

In our system; the first model has been trained to detect the license plate, which will then be cropped into the original image, before being passed to the second model,

Y. Farhaoui et al. (Eds.): ICAISE 2022, LNNS 635, pp. 588–594, 2023.
https://doi.org/10.1007/978-3-031-26254-8_85

which has been trained to detect the characters. The main objective of this work is to apply the YOLOv5 algorithm for each step of our ALPR system on Moroccan plates. The work done in our research is divided into several stages. We start with the stage of discussing the articles that deal with topics related to our work, and then we move to the next part, which is a paragraph that contains the data sources used in training the YOLOv5 architectures. For this algorithm, we detailed their architecture and operation in the next part. Concerning the last part made to discuss the obtained results.

2 Related Works

A license plate detection system is a system that recognizes each plate on a vehicle and then reads the characters on the plate. Automating plate reading will be facilitating many traditional jobs that are plagued by human error. Several papers have addressed this topic with a tendency to use two different neural networks for efficiency and accuracy problems. Many researchers have approached the plate detection stage using object detection methods based on convolutional neural networks.

In the paper [1], the authors noted that the Fast-YOLO model achieved a low recall rate when detecting plates without prior vehicle detection. Therefore, they used the Fast-YOLO model to detect first the frontal view of the cars and then their plates in the detected areas, which resulted in high precision and recall rates on a dataset containing Brazilian plates. For [2, 11, 12] the authors have partitioned the input image into subregions, forming an overlapping grid. A score for each region was produced using a convolutional neural network and the patches were detected by analyzing the outputs of neighboring subregions. The results for detecting Brazilian plates in images with multiple vehicles, reaching a recall rate of 83% on a public dataset by them.

The researchers of papers [3, 4, 13] used for their license plate character segmentation process projection profile method. They used a horizontal projection in [5, 14–16] this projection is a process that consists in finding the maximum peak or also known as the space between the characters by iteratively projecting the graph vertically on the other hand in [6–10] we find that the authors used a vertical projection for this method, the vertical axis is scanned in search of a black point in the images to detect the character.

3 Dataset

We used a dataset containing images of vehicles registered in Morocco collected by the Laboratory of Modelling, Simulation, and Data Analysis (MSDA) of the Polytechnic University Mohammed 6. The total data we used to train the two models was 724, 580 images were used as training data, and 144 images as validation data. The dataset is divided into two parts: a first one called segmentation contains the images of the vehicles and their annotation of the plates in XML and TXT format and a second one called OCR contains the images of the plates and their annotation of the characters in XML and TXT format (Fig. 1).

Fig. 1. Example of dataset

4 Architecure (YOLOv5)

This part will focus on the description of the YOLOv5 architecture used in our work (Fig. 2).

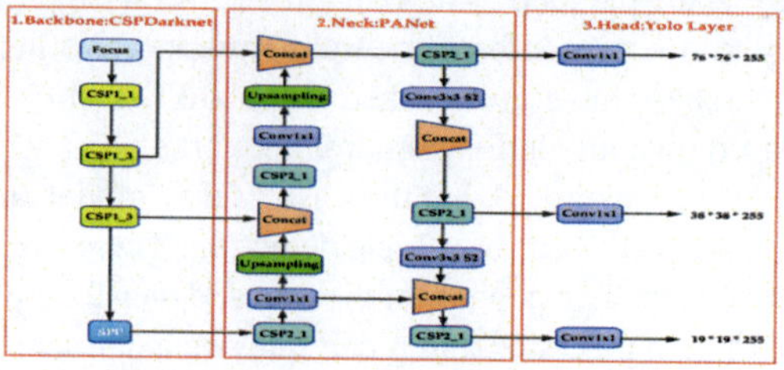

Fig. 2. YOLOv5 architecture

YOLOv5 is a new convolutional neural network (CNN) that detects objects in real-time with high accuracy. This approach uses a single neural network to process the entire image, then separates it into multiple parts and predicts bounding boxes and probabilities for each component. These bounding boxes are estimated based on the expected probability. The method looks at the image only once, in that it makes predictions after a single forward propagation of the neural network. It then provides the detected features after non-maximal deletion (which ensures that the object detection algorithm identifies each object only once).

These three sections can be assimilated into a set of layers, composing a functional unit. Each of these parts corresponds to a role, established by this set of layers, which were "pieces" of networks already existing in the literature and which were simply lightly modified.

- **Backbone:** The Backbone model is mainly used to extract the key features from an input image. CSPs (Cross Stage Partial Networks) are used as the backbone in YOLO v5 to extract rich and useful features from an input image.
- **Neck:** The Model Neck is used to create feature pyramids. Feature pyramids help models generalize successfully when scaling objects. They make it easier to identify the same object in different sizes and scales. Feature pyramids are very useful in helping models run efficiently on unseen data. Other models, such as FPN, BiFPN, and PANet, use different kinds of feature pyramid approaches. PANet is used as a move in YOLOv5 to obtain feature pyramids.

- **Head:** The Head model is primarily responsible for the final detection step. It uses anchor boxes to construct the final output vectors with class probabilities, objectivity scores, and bounding boxes.

YOLOv5 is available in four models, namely s, m, l and x, each offering different accuracy and detection performance, as shown below (Fig. 3).

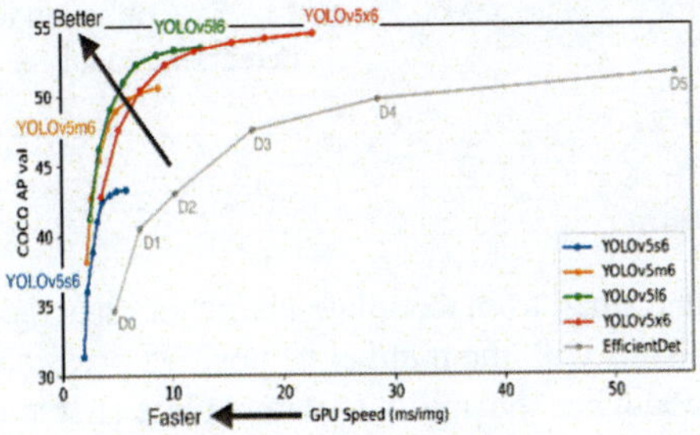

Fig. 3. YOLOv5 variants

5 Results and Analysis

For the last part of our work, we will discuss the detailed results for each part of the system.

5.1 License Plate Detection

- Precision: our models trained to detect vehicle plates has the ability to make True Positive (TP) comparisons with the number of positive predicted data of 89.50%.
- Inference time: our model has the ability to detect a vehicle plate on 8 ms (Table 1).
- Model size: The checkpoints generate after training takes 14.8 MiB from storage memory (Figs. 4 and 5).

Table 1. Experiment result of License Plate Detection

Detector	Precision	Recall	Inference time	Model size
YOLOv5	89.50%	90.6%	8 ms	14.8 MiB

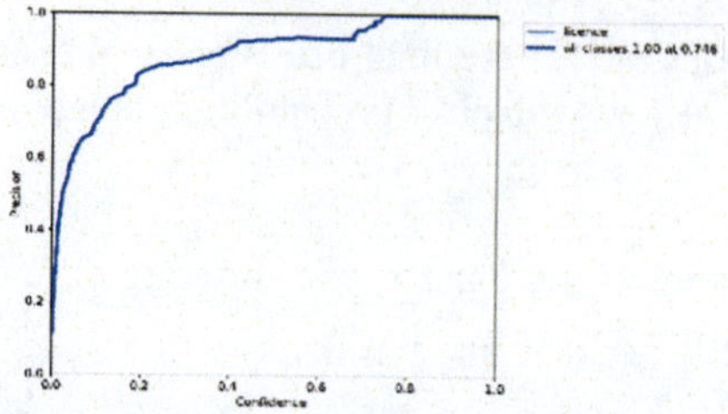

Fig. 4. Precision curve for the plate detection model

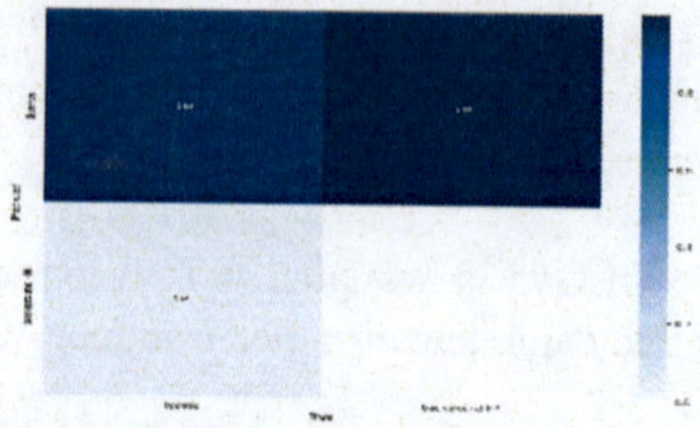

Fig. 5. Confusion matrix for the plate detection model

5.2 Character Detection

- Precision: our models trained to detect plate characters has the ability to make True Positive (TP) comparisons with the number of positive predicted data of 97.50%.
- Inference time: our model has the ability to detect plate characters on 9 ms (Table 2).
- Model size: The checkpoints generate after training takes 14.9 MiB from storage memory (Figs. 6 and 7).

Table 2. Experiment result of Plate Characters Detection

Detector	Precision	Recall	Inference time	Model size
YOLOv5	97.50%	95.1%	9 ms	14.9 MiB

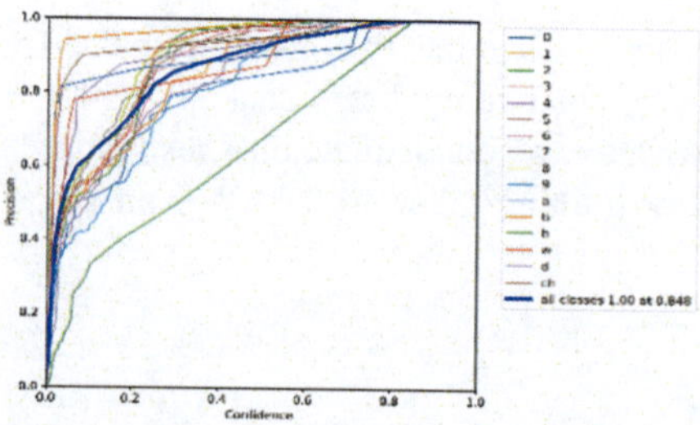

Fig. 6. Precision curves for the characters detection model

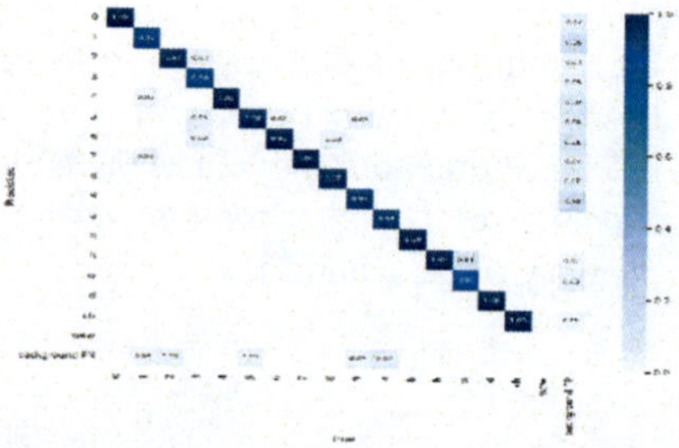

Fig. 7. Confusion matrix for the characters detection model

6 Conclusion

In this paper, we create a system to automate the process in parking systems, private garages of companies, shopping malls, tolls, and hospitals… Our system is focused on how to automatically recognize the characteristics of vehicle plates. This license plate recognition system consists of two parts, each part is equipped with an object detection model.

Our system composes of two object detection models for Object License Plate Detection, we used transfer learning on the YOLOv5 algorithm to build the Object License Plate Detector. Our model achieved a precision of 89.50% with an inference time of 8ms, In the Characters Detection model we also used transfer learning on the YOLOv5 algorithm and we get a precision of 97.50% with an inference time of 9ms.

Since YOLOv5 is initially implemented in PyTorch, it benefits from the established PyTorch ecosystem: support is simpler and deployment easier especially in surveillance cameras. Deployment on mobile devices is also simpler, as the model can be easily compiled into ONNX and CoreML.

References

1. Birmohan, S., Dalwinder, S., Manpreet, K.A., Gurwinder, S.: Automatic number plate recognition system by character position method. Int. J. Comput. Vis. Robot., January 2016
2. Vanshika, R.: Automatic number plate recognition. In: Proceedings of the International Conference on Innovative Computing and Communication (ICICC) 2021, July 2021
3. Praveen, R., Arihant, P.: Automatic license plate recognition for Indian roads using faster-RCNN. In: Proceedings of the International Conference on Advanced Computing (ICoAC) 2019, Chennai, India, December 2019
4. Palanivel Ap, N., Vigneshwaran, T., Madhanraj, R.: Automatic number plate detection in vehicles using faster R-CNN. In: Proceedings of the International Conference on System, Computation, Automation and Networking (ICSCAN), July 2020
5. Raj, S., Gupta, Y., Malhotra, R.: License plate recognition system using YOLOv5 and CNN. In: Proceedings of the 8th International Conference on Advanced Computing and Communication Systems (ICACCS), March 2022
6. Chen, J., Yang, S., Jiang, S: Face detection with the faster R-CNN. In: Proceedings of the International Conference on Electronic Measurement and Instruments, ICEMI, November 2019
7. Amatya, M.L., Sarvanan, K.N.: The state of the art – vehicle number plate identification a complete survey. Int. Res. J. Eng. Technol. (IRJET) **4**(2), 785–792 (2017)
8. Quanfu, F., Lisa, B., John, S.: A closer look at faster R-CNN for vehicle detection. In: IEEE Symposium on Intelligent Vehicle, June 2016
9. Liang, Z., Yuan, F., Jianqiao, S., Yicao, Z., Jin, S.: Research and implementation of database operation recognition based on YOLOv5 algorithm. In: Proceedings of the International Conference on Computer Information Science and Artificial Intelligence (CISAI), September 2021
10. Sung, J.Y., Yu, S.B.: Real-time automatic license plate recognition system using YOLOv4. In: Proceedings of the 2020 IEEE International Conference on Consumer Electronics-Asia (ICCE-Asia), November 2020
11. Andreas, C., Ruben, S.: An end-to-end optical character recognition pipeline for Indonesian identity card. In: Proceedings of the International Conference of Information and Communication Technology (ICoICT), August 2021
12. Hoang T, Ha Duyen, T., Nguyen, T.: YOLOv5 based deep convolutional neural networks for vehicle recognition in Smart University campus. In: Proceedings of the International Conference on Hybrid Intelligent Systems, March 2022

13. Nhu-Van, N., Christophe, R., Jean-Christophe, B.: Comic characters detection using deep learning. In: Proceedings of the 14th IAPR International Conference on Document Analysis and Recognition (ICDAR), November 2017
14. Tian-Hao, W., Tong-Wen, W., Ya-Qi, L.: Real-time vehicle and distance detection based on improved YOLOv5 network. In: Proceedings of the 3rd World Symposium on Artificial Intelligence (WSAI), June 2021
15. Upesh, N., Hossein, E.: Comparing YOLOv3, YOLOv4 and YOLOv5 for autonomous landing spot detection in faulty UAVs. Sensors **22**, 464 (2022)
16. Peiyuan, J., Daji, E., Fangyao, L., Ying, C., Bo, M.: A review of YOLO algorithm developments. In: Proceedings of the 8th International Conference on Information Technology and Quantitative Management (ITQM 2020 & 2021): Developing Global Digital Economy after COVID-19 (2022)

Lie-Trotter and Strang-Marchuk Methods for Modeling the 1D-Transport with Reaction Equation

Inasse El Arabi(✉), Anas Chafi, and Salaheddine Kammouri Alami

Industrial Engineering Department, Sidi Mohamed Ben Abdellah University, Fes, Morocco
Inasse.elarabi@usmba.ac.ma

Abstract. In this work, we will study the 1D transport of contaminants in a saturated porous medium which can be presented by different phenomena, such as advection, diffusion, and reaction. The system of the three equations linked together is considered a difficult system to solve since each equation has its stability and convergence condition. Therefore, our objective is to develop a new strategy that will allow us to solve this kind of problem and obtain more effective results, to compare the two procedures of this approach method and specify the best procedure for modeling this type of system. The method utilized here is the operator splitting method, which is a good method to solve these kinds of complicated models. The main idea behind this strategy is to split down a complex problem into smaller subsystems, known as division sub-problems, and solve each one individually using the appropriate numerical method. The effects of operator splitting methods on the solution of advection-diffusion-reaction are examined, within the context of this works two operator splitting methods, Lie-Trotter and Strang-Marchuk splitting methods were used and comparisons were made through various decomposition rate. Obtained results were compared with analytical solutions to the problems and available methods in the literature. It is seen that the Lie-Trotter splitting method has lower error norm values than the Strang-Marchuk splitting method. But, the Lie-Trotter splitting method produces accurate results for very small values of the numerical result for an application concerning the transport of a contaminant will be presented to enhance the value of our results, and prove the efficiency of the LTM (Lie-Trotter Method).

Keywords: Transport · Reaction · Saturated porous media · Finite different scheme · Splitting technique · Lie-Trotter · Strang-Marchuk

1 Introduction

Different phenomena are responsible for the transport of matter [1–5]. This matter can be transported by the movement of the fluid (advection), by diffusion from concentrated areas to areas of lower concentration, by reaction with other constituents present in the medium represented, or by other phenomena.

The advection-diffusion-reaction system (ADRS) is a combination of the advection, diffusion, and reaction equations [1, 4–6].

Y. Farhaoui et al. (Eds.): ICAISE 2022, LNNS 635, pp. 595–600, 2023.
https://doi.org/10.1007/978-3-031-26254-8_86

The ADRs modeling has attracted interest from a variety of fields, including medicine, biology, bioremediation, chemistry, water treatment, and engineering [1–4].

This modeling is treated as a coupling of three different terms: Advection, diffusion, and reaction.

The system of the three equations linked together is considered a difficult system to solve since each equation has its stability and convergence condition. As a result, much work has gone into developing an efficient and stable numerical technique, a new code to solve this kind of system, and comparing between procedures in terms of efficiency. Hence, to solve this coupled system, an operator splitting approach is given, also known as a splitting method, which allows the individual elements of the mathematical model to be treated separately [4–6].

Our objective is to develop a new code with two different operator splitting approaches and compared against them at the level of stability and efficiency.

In this work, an application in the environment was treated to compare the two procedures and prove our results.

Two operator splitting approaches, Lie-Trotter (LT) and Strang-Marchuk (SM) were utilized in the context of this work, and comparisons were done using various decomposition rates. The two operator splitting methods separates the transport (advection and diffusion) and reaction phases. Therefore, a finite difference approach is used to discretize the three terms of the ADR equation.

2 Methodology

We're dealing with the problem of one-dimensional transport with the reaction of matter in a saturated porous medium.

The equations which describe the advection-diffusion-reaction model are as follows:

$$\frac{\partial C}{\partial t} + V * \frac{\partial C}{\partial x} - D_x \frac{\partial^2 C}{\partial x^2} - \lambda * C = 0 \tag{1}$$

where t is time, x is spatial coordinate, C is the concentration of matter, V is the velocity of the flow, and Dx is diffusion coefficient, λ is the decomposition rate. Our system of Eq. (1) will be treated by two operator splitting approaches, Lie-Trotter (LT) and Strang-Marchuk (SM), which will be solved using the finite difference method.

2.1 Lie-Trotter Operator Splitting Method

The Lie-Trotter operator splitting method is a first-order operator splitting method and it is named a sequential splitting method. Applying this method to our ADR equation, from which the problem is divided into three sub-problems that will be treated by the finite difference method [4].

$$\frac{\partial C_1}{\partial t} + V \frac{\partial C_1}{\partial x} = 0 \quad C_1(t_n, x) = C(t_n, x) \quad t \in [t_n, t_{n+1}] \tag{2}$$

$$\frac{\partial C_2}{\partial t} = D_x \frac{\partial^2 C_2}{\partial x^2} \quad C_2(t_n, x) = C_1(t_{n+1}, x) \quad t \in [t_n, t_{n+1}] \tag{3}$$

$$\frac{\partial C_3}{\partial t} = -\lambda C_3 \quad C_3(t_n, x) = C_2(t_{n+1}, x) \quad t \in \left[t_n, t_{n+1}\right] \tag{4}$$

In this case, the problem is divided into three sub-problems, advection, diffusion, and reaction where C1 is the concentration in the advection process and C2 is the concentration in the diffusion of the process and C3 is the concentration in the reaction of the process. Equation (2), Eq. (3), and Eq. (4) represent the advection equation, the diffusion equation, and the reaction equation, respectively. In the solution process, Eq. (2) will be solved for a time interval of Δt using the initial condition of Eq. (1). The result obtained from there will be the initial condition of Eq. (3). Then, Eq. (3) will be solved for a time interval of Δt. The result obtained from there will be the initial condition of Eq. (4). Then, Eq. (4) will be solved for a time interval of Δt and the solution of Eq. (1) will be obtained for a time interval of Δt. Thus, the problems will be solved consecutively by combining them with the initial conditions [4].

Using the difference finite method, our algorithm can be rewritten as:

$$C_{1i} = C_i - \frac{V * \Delta t}{\Delta x} * (C_i - C_{i-1}) \tag{5}$$

$$C_{2i} = C_{1i} + \frac{D * \Delta t}{\Delta x} * \left(C_{1i+1} - 2C_{1i} + C_{1i-1}\right) \tag{6}$$

$$C_{3i} = \lambda * C_{2i} \tag{7}$$

2.2 Strang-Marchuk Operator Splitting Method

The Strang-Marchuk operator splitting method is a second-order operator splitting method. This method is used to divide our ADR equation into two sub-problems, namely advection and diffusion-reaction, similar to the Lie-Trotter operator splitting method. However, this time, the sub-problems will be solved in three steps in total [13, 15]. In a Strang-Marchuk operator splitting approach, we treat the advection equation twice at half time ($\frac{\Delta t}{2}$) and the diffusion-reaction equation only once at each time step (Δt). Which, the Strang Marchuk Splitting algorithm is as follows [4–6]:

$$\frac{\partial C_1}{\partial t} + V\frac{\partial C_1}{\partial x} = 0 \quad C_1(t_n, x) = C(t_n, x) \quad t \in \left[t_n, t_{n+\frac{1}{2}}\right] \tag{8}$$

$$\frac{\partial C_2}{\partial t} = D_x\frac{\partial^2 C_2}{\partial x^2} + \lambda C_2 \quad C_2(t_n, x) = C_1\left(t_{n+\frac{1}{2}}, x\right) \quad t\epsilon\left[t_n, t_{n+1}\right] \tag{9}$$

$$\frac{\partial C_1}{\partial t} + V\frac{\partial C_1}{\partial x} = 0 \quad C_1\left(t_{n+\frac{1}{2}}\right) = C_2(t_{n+1}, x) \quad t \in [t_{n+\frac{1}{2}}, t_{n+1}] \tag{10}$$

where Eq. (8) will be solved for a time interval of ($\frac{\Delta t}{2}$) using the initial condition of Eq. (1). The solution of Eq. (8) will be used as the initial condition of Eq. (9), this equation will be solved for a time interval of Δt. The result obtained from Eq. (9) will be the initial condition of Eq. (10). Finally, Eq. (10) will be solved for a time interval of $(\frac{\Delta t}{2})$. Thus, the solution of the Eq. (1) Will be obtained for a time interval of Δt [4]. Using the difference finite method, our algorithm can be rewritten as [4]:

$$C_{1i} = C_i - \frac{V * \Delta t}{2 * \Delta x} * (C_i - C_{i-1}) \tag{11}$$

$$C_{2i} = C_{1i} + \frac{D * \Delta t}{2 * \Delta x}(C_{1i+1} - 2C_{1i} + C_{1i-1}) + \lambda C_{1i} \tag{12}$$

$$C_{3i} = C_{2i} - \frac{V * \Delta t}{2 * \Delta x}(C_{2i} - C_{2i-1}) \tag{13}$$

3 Numerical Tests

In our work, we will deal with the case of the transport of a contaminant within a soil-groundwater system, by changing the value of λ.

The parameters used in the simulation are as follows [6]:

$$C^{exact} = exp^{\frac{-K_3 * X}{2}} (erfc\left(\frac{X - K_2 * t}{K_1}\right) + exp^{K_3 * X} erfc(\frac{X + K_2 * t}{K_1})$$

With: $K_1 = \sqrt{4 * D * t}$ $K_2 = \sqrt{V^2 + 4 * \lambda * D}$ $K_3 = \sqrt{\left(\frac{V}{2 * D}\right)^2 + \frac{\lambda}{D}}$.

With an initial, condition equal to: $C(X, 0) = 0$.

In addition, boundary condition equal to: $C(0, t) = 1$ and $C(L, t) = 0$.

In our case, the velocity of the flow and diffusion coefficient is taken as:

V = 0.1 and D = 0.05 With Cr = 0.01, Pe = 0.1, $\Delta x = 0.1$ and $\Delta t = 0.01$, as well as the length of the medium is chosen and is given as L = 10 [6].

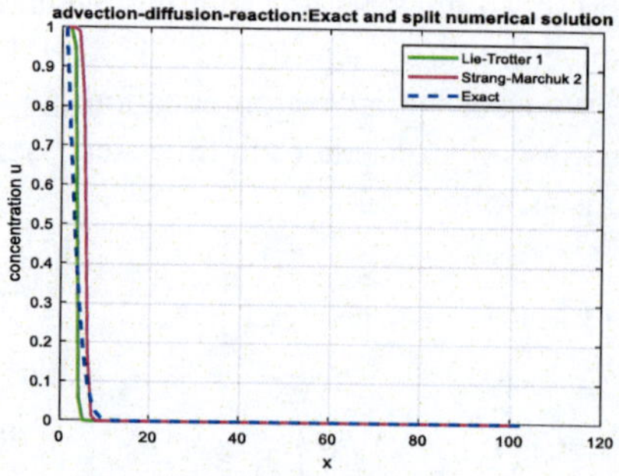

Fig. 1. ADR equation with Strang-Marchuk and Lie Trotter $\lambda = 10^{-6}$

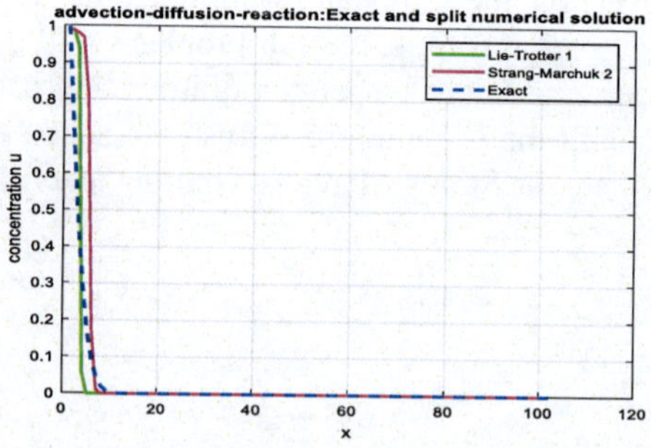

Fig. 2. ADR equation with Strang-Marchuk and Lie Trotter $\lambda = 10^{-3}$

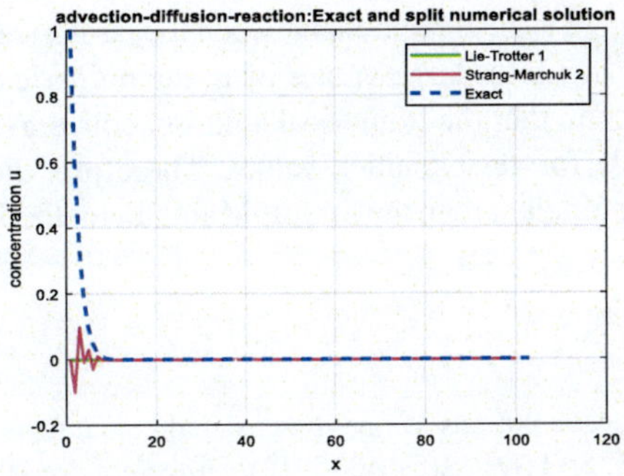

Fig. 3. ADR equation with Strang-Marchuk and Lie Trotter $\lambda = 1$

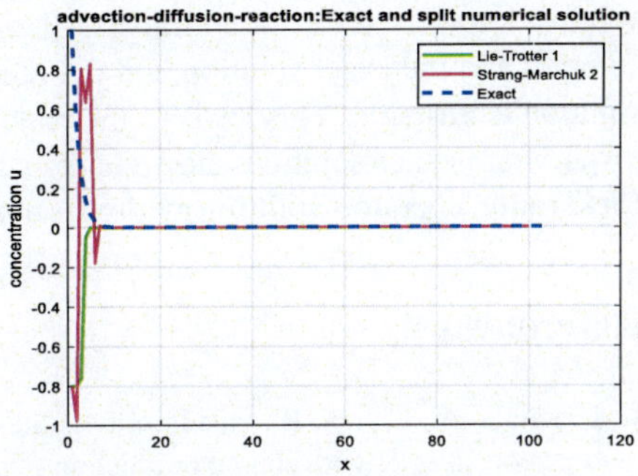

Fig. 4. ADR equation with Strang-Marchuk and Lie Trotter $\lambda = 1.8$

Table 1. Strang-Marchuk and Lie-Trotter ERROR

	Strang-Marchuk Error	Lie-Trotter Error
$\lambda = 10^{-6}$	0.1331	0.1131
$\lambda = 10^{-3}$	0.1332	0.1176
$\lambda = 1$	0.1784	0.1837
$\lambda = 1.8$	0.2401	0.2177

According to Table 1 and Figs. 1, 2, 3 and 4, we can see that when the λ value increases, the error increases as well. We can also see an instability of the patterns for large values of λ. As we can notice that the Strang-Marchuk approach is less efficient than the Lie-Trotter one.

In conclusion, Lie-Trotter is the more efficient procedure method to solve this kind of system.

4 Conclusion and Discussion

In this work, we have presented a numerical approach able to solve a coupled system of transport (Advection and diffusion) and reaction equation in the one-dimensional case. This approach is based on the splitting technique with two different procedures, named Lie-Trotter and Strand-Marchuk. Those procedures have the advantage of treating each subsystem that composes our mathematical model with the numerical method that adopted it. Thus, the advection, diffusion, and reaction problem were treated by the finite difference method, which proved to be very efficient in the different tests presented. Our result has been tested on a one-dimensional transport of a contaminant and biodegradation problem with first-order degradation kinetics. The error values of the two techniques were compared, and it was discovered that the approach of the Lie-Trotter has a lower error than the Strang-Marchuk method. It was also discovered that when the value of λ lowers, the error diminishes, then there is an agreement between the numerical solution and the exact solution. On the other hand, for large values of λ, the numerical solution is unstable. This leads to the conclusion that the numerical solution converges towards the exact solution and remains stable for very small λ values. Therefore, the Lie-Trotter, operator-Splitting method is more efficient than the Strang-Marchuk method.

References

1. Moranda, A., Cianci, R., Paladino, O.: Analytical solutions of one-dimensional contaminant transport in soils with source production-decay. Soil Syst. **2**(3), 40 (2018). https://doi.org/10.3390/soilsystems2030040
2. Verrall, D.P., Read, W.W.: A quasi-analytical approach to the advection-diffusion–reaction problem, using operator splitting. Appl. Math. Model. **40**(2), 1588–1598 (2016). https://doi.org/10.1016/j.apm.2015.07.023
3. Onate, E., Miquel, J., Nadukandi, P.: An accurate FIC-FEM formulation for the 1D advection–diffusion–reaction equation. Comput. Methods Appl. Mech. Eng. **298**, 34 (2016)
4. Bahar, E., Gürarslan, G.: Numerical solution of advection–diffusion equation using operator splitting method. Int. J. Eng. Appl. Sci. **9**(4), 76–88 (2017). https://doi.org/10.24107/ijeas.357237
5. El Arabi, I., Chafi, A., Alami, S.K.: Numerical simulation of the advection-diffusion-reaction equation using finite difference and operator splitting methods: application on the 1D transport problem of a contaminant in saturated porous media. In: E3S Web of Conferences, vol. 351, p. 01003. EDP Sciences (2022)
6. El Arabi, I., Chafi, A., Alami, S.K.: 1-D numerical simulation of transport and biodegradation in saturated porous media. Int. J. Differ. Equ. (IJDE) **17**(2), 177–187 (2022)

Machine Learning Application in Precision Marketing: A Systematic Literature Review and Comparative Study

Nouhaila El Koufi(✉), Abdessamad Belangour, and Mounir Sadiq

Laboratory of Information Technology and Modeling, Faculty of Sciences Ben M'sik, Hassan II University, Casablanca, Morocco
elkoufi.nouhaila1@gmail.com

Abstract. With the arrival of the big data era and the rapid advancement of artificial intelligence (AI), the traditional marketing strategy has become helpless and unsuitably for a company's long-term growth. The development of a precise and accurate marketing model became the interest of researchers and enterprises. Precision marketing based on AI techniques can provide personalized communication between customers and companies, attract prospective clients, and build targeted marketing recommendations for high-value clients. Therefore, this paper presents a systematic literature review on the current trends of machine learning applications in precision marketing. In this SLR, we have studied, analyzed, and discussed the new proposed solutions published between 2019 and 2022, as well as a comparative study of machine learning methods employed in this domain was performed. As results, we find seven ML methods provide better results in the literature: RF, SVM, CNN, KNN, XGBoost, SVM, and LG. This study provides a reference for searchers in this domain.

Keywords: Precision marketing · Machine learning · Big data · Systematic literature review

1 Introduction

With the fast technological revolution, the advancement of information technology, and the popularization of online services, consumption habits and lifestyles in the new generation have also changed and developed. Accordingly, companies search to change the traditional marketing strategy toward a powerful precision marketing method based on new and advanced to realize long-term growth. Marketing has long been a critical element in developing companies, engaging clients, and increasing revenue. It considers a communication channel between firms and customers, with the primary goal of attracting and influencing customers' demands and consumption to enhance sales. Moreover, businesses make decisions based on marketing mechanisms [1]. In order to stay competitive, the enterprises need to apply a developed marketing system that respect the new era needs. Developing an efficient marketing strategy requires the analysis of the behavior and purchase history of consumers. Accordingly, marketing techniques

Y. Farhaoui et al. (Eds.): ICAISE 2022, LNNS 635, pp. 601–607, 2023.
https://doi.org/10.1007/978-3-031-26254-8_87

require the ability to create correct predictions as well. Machine learning technology has advanced significantly since the dawn of the data era. Think to machine learning; it became possible to predict underlying patterns from data. Thus, the combination of machine learning with its ability to produce high-end accurate predictive and precision marketing strategies can develop and serve this field well. In this paper, we adopted the guidelines proposed by Keele et al. [2] to perform a systematic literature review (SLR).

The organization of the rest of this paper is as follows: Sect. 2 defines research objectives, SLR methodology, presents RQs and illustrates the search strategy. Section 3, defines exclusion and inclusion criteria, and presentes the quality assessment. Data strategy and synthesis is presented in Sect. 4. Section 5 presents the obtained results. We conclude by a general conclusion.

2 Research Questions and Search Strategy

The objective of this study is to identify and categorize the current state of the art in machine learning applications to precision marketing. We have analyzed, interpreted, and discussed the literature published between 2019 and 2022. In order to realize this purpose, a set of research questions has proposed as illustrated in Table 1.Our search technique involved all the well-known online scientific databases related to this field: Springer Link digital library, Scopus, IEEEXplore Digital Library, DBLP, and Science Direct. So, after a deep search of these electronic data sources, we have collected a set of primary studies. In order to boost the search results, we analyzed the bibliography of each selected paper. For choosing the search string, we have selected specific terms relying on common keywords used in the literature pertinent to this SLR for collecting the major available appropriate papers in the review. The string search used to build this SLR review mainly includes the union of the keyword ''precision marketing" and the specific terms related to this review: ("Precision Marketing") and ("Machine Learning" OR "Machine Learning techniques" OR "ML methods" OR "Machine Learning approaches"). ("PM") based on ("ML" OR "AI" OR "Machine learning methods" OR "Artificial Intelligence Methods").

3 Exclusion and Inclusion Criteria and Quality Assessment

Firstly, we have used predefined criteria such as the analysis of the (keywords, abstracts, conclusions, title, etc.) as initial factors to analyze studies. Then, we applied our proposed inclusion and exclusion criteria to select the primary studies the most pertinent to our research objectives and questions. The proposed inclusion and exclusion criteria were presented in Table 2.

In order to perform a good and efficient SLR, we have defined a quality assessment checklist presented in Table 3. We have affected a high weight to QA1 and QA2, which address the machine learning technique used, architecture of the proposed solution and comparison of the findings with others. In other hand, we have affected a low coefficient to QA3, QA4 and QA5, which address the theoretical and experimental results, the limitation of the proposed solution. The final quality score for each given paper is computed using the Eq. (1) defined as follow:

$$S = \sum_{i=1}^{2} \overline{W} * 0.7 + \sum_{i=3}^{5} \overline{W} * 0.3 \tag{1}$$

Table 1. Research questions

	Questions
RQ1	What ML methods can be used in precision marketing?
RQ2	What is the type of data utilized in precision marketing?
RQ3	How evaluate the performance of proposed solutions?
RQ4	Which machine learning techniques give an accurate result in this domain?

Table 2. Exclusion and inclusion criteria.

Inclusion	Exclusion
Researches published between 2019 to 2022	Articles not associated with our subject
Papers published at conference or journal	Papers less than five pages
Articles written in English language	Research not relates to our proposed research questions
Papers contains keywords relates to this domain	Workshop, technical reports and master thesis
Articles applied AI in precision marketing	Theoretical study

Table 3. Quality assessment.

	Quality assessment	Weight
QA1	Did the study compare the proposed solution with the existing solutions?	2
QA2	Are the obtained results clearly and deeply explained and discuss?	2
QA3	Are the theoretical and experimental results presented?	1
QA4	Are the limitation of the proposed solution mentioned?	0.5
QA5	Are the aim of study clearly defined?	1

Table 4. Data extraction method.

	Data extraction
DRQ1	ML techniques utilized in precision marketing field
DRQ2	The type of dataset used (Real or synthetic)
DRQ3	The evaluation metrics used
DRQ4	The performance results of ML methods used

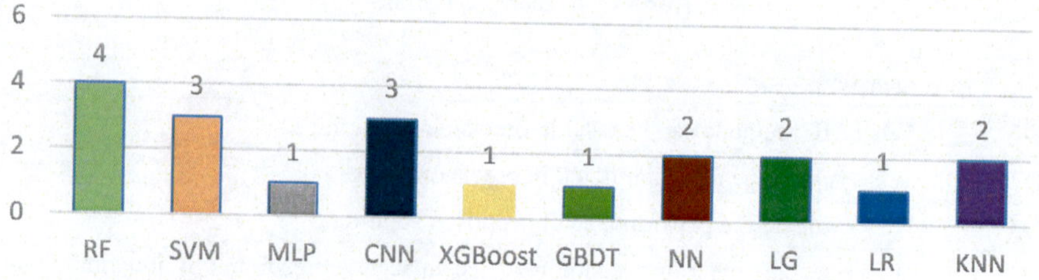

Fig. 1. Distribution of selected papers by ML techniques.

4 Data Extraction Strategy and Synthesis

Aiming to extract meaningful information and respond to the research questions and objectives (See Sect. 2), we have applied the data extraction list to the selected papers. The data extraction form involves general relevant points are the title of the research, date of publication, type of paper (journal article, conference paper), and authors names. Moreover, specific information related to research questions that we have defined in Table 4.The analytical stage combines two points the extraction of data and the summarization of results found from the selected studies to respond to research questions. Moreover, we analyzed the extracted data using statistical and descriptive narrative techniques. We utilized the data visualization method based on illustration tools like (box plots, histograms, etc.) for visualizing, clarifying, and making a clear presentation of the results.

5 Results of Protocol

We have found 100 papers published between 2019 and 2022 in this domain. The total search result was growth to 150 after the analysis of the bibliography of selected articles. Then, we have applied the inclusion and exclusion criteria in the chosen papers. As a result, the number of identified articles reduced from 150 to 29. Furthermore, we have measured the quality of opted studies using the quality assessment criteria that we already defined in Sect. 3 to ensure that we have collected the more relevant papers. Therefore, 11 papers selected as the final set that responds to our proposed criteria. On the other hand, since 2020, the rate of publishing in this field has increased dramatically.

5.1 Distribution of Selected Studies by Machine Learning Methods and Publication Venue

In Fig. 1, we have presented the most used ML approaches in the current literature (2019–2022) related to the research subject: RF, SVM, MLP, CNN, XGBoost, NN, LG, LR, and KNN. Among the previous list, the most ML techniques frequently adopted RF with 22%, SVM with 17%, CNN with 17%. As illustrated in Table 5, some papers combine several ML algorithms. It noted that the use of machine learning techniques in the selected research takes three ways: the use of the one ml method, the combination of various ML algorithms, or the combination of ML technique with other methods. We have illustrated the distribution of selected articles according to publication venues. That the opted papers publish in a variety of conferences and journals. After the search result analysis, we state that the highest number of papers related to our subject was published at the Conference with 57%. There are six conferences represented in the research articles. On the other hand, the articles came from four journals in total.

5.2 Comparison of Machine Learning Methods Using Performance Results

As illustrated in Table 5 and Fig. 1, many ML methods have been applied in precision marketing area. The performance of used ML techniques was measured using different performance metrics like precision, loss-log error, accuracy, recall, and F-measure. Table 5 represents a comparative study of ML approaches employed in PM. As illustrated in Table 5 and Fig. 1, many ML methods have been applied in precision marketing area. The performance of used ML techniques was measured using different performance metrics like precision, loss-log error, accuracy, recall, and F-measure. Table 5 represents a comparative study of ML approaches employed in PM. In [1], the authors compared the accuracy of MLP (84.20%), CNN (74.21%), SVM (76.34%), RF (72.15%), and LG (67.06%). In [3], RF realized accuracy of 73.9%. The results obtained in [5] show that the XGBoost, RF, and LG recorded an accuracy of 83.3%, 89.9%, and 89.9%, respectively. SVM registered an accuracy of 89.8% in [6]. The accuracy obtained by CNN is 93% in [7]. NN recorded the accuracy of 85%, 91% in [8, 9], respectively. The obtained results in [10] presented that LG and RF realized a good and same value of accuracy (99.3%). LR and SVM realized the accuracy of 72.10% and 85.1%, respectively. The performance of CNN was measured in [11] using the accuracy metric. As results, CNN recorded 83.21%. KNN. In [12], the authors measured the accuracy of KNN (94.2%). According to the comparison of results found in the literature, we conclude that SVM, CNN, KNN, XGBoost, and LG.

5.3 Data Extraction Results

After filtering the selected papers, we have obtained a set of 11 papers corresponding to proposed criteria, research questions, and quality assessment defined in previous sections. Table 5 represents a summary of the answers to our proposed research questions. It is include four columns, each column present a response to a specific research question. The first defines the title of chosen articles. The second column, the corresponding RQ1, represents the machine learning approaches used. The third column represents the data

Table 5. Summary of selected studies.

Study	Method	Type of data	Performance metrics	Evaluation results
[1]	MLP, CNN, SVM, RF and LG	RD	Accuracy	**Accuracy** SVM = 76.34%, MLP = 84.20%, CNN = 74.21%, RF = 72.15%, LG = 67.06%
[3]	RF	RD	Accuracy	**Accuracy** RF = 73.9%
[4]	KNN, SVM	RD	Accuracy	**Accuracy** KNN = 99.11%, SVM = 76.56%
[5]	XGBoost, RF and LG	RD	Accuracy, recall, AUC	**Accuracy** XGBoost = 83.3%, RF = 89.9%, LG = 89.9%
[6]	SVM	SD	Accuracy, Precision, Recall and F-measure	**Accuracy** SVM = 89.8%
[7]	CNN	RD	Accuracy	**Accuracy** CNNs = 93%
[8]	NN	RD	Accuracy	**Accuracy** NN = 85%
[9]	NN	RD	Accuracy, Loss values	**Accuracy** NN = 91%
[10]	LR and RF	RD	Accuracy, precision and recall	**Accuracy** LG = 99.3%, RF = 99.3%
[11]	CNN	RD	Accuracy	**Accuracy** CNN = 83.21%
[12]	KNN	RD	Accuracy	**Accuracy** KNN = 94.2%

type used in each selected paper (RQ2). Therefore, in the fourth column, the column responds to the question RQ3 and the final column answers to RQ4. It noted that the use of machine learning techniques in the selected research takes three ways: the use of one ml method, the combination of various ML algorithms, or the combination of ML approach with other methods. The results obtained from Table 5 illustrate that various selected studies in this area used real-time data with 95% of the total. This conduct that the opted papers present the application of their proposed solutions more realistic.

6 Conclusion

Through this systematic literature review, we have explored the trend of machine learning methods applications' in precision marketing. A set of 11 papers was selected after filtering and application of the criteria and research questions that we have defined in the planning stage. We have considered the papers published between 2019 and 2022 in this study in order to explore the current findings in this domain. In addition, it provides an overview of the practical implications and improvement of using machine learning in the precision marketing field. Moreover, this study gives a good reference for developing new solutions to improve precision marketing based on machine learning.

References

1. Zheng, Y., Li, L., Zhang, J., Xie, Q., Zhong, L.: Using sentiment representation learning to enhance gender classification for user profiling. In: Asia-Pacific Web (APWeb) and Web-Age Information Management (WAIM) Joint International Conference on Web and Big Data, pp. 244–258. Springer, Cham (2019). https://doi.org/10.1007/978-3-030-26075-0_1
2. Keele, S.: Guidelines for performing systematic literature reviews in software engineering. Tech. Rep. **5**(2), 1–57 (2007)
3. Liu, Y., Cheng, D., Pei, T., Shu, H., Ge, X., Ma, T.: Inferring gender and age of customers in shopping malls via indoor positioning data. Environ. Plann. B Urban Analytics City Sci. **47**(9), 1672–1689 (2020)
4. Zhang, S., Liao, P., Ye, H.Q., Zhou, Z.: Multiple resource allocation for precision marketing. J. Phys: Conf. Ser. **1592**(1), 1–18 (2020)
5. Qiongyu, S.: Prediction of O2O coupon usage based on XGBoost model. In: Proceedings of the 11th International Conference on E-business, Management and Economics, pp. 33–36. ACM, Beijing (2020).
6. Jiang, L.: Support vector machine (SVM) marketing strategy analysis method based on time series. IOP Conf. Ser. Mater. Sci. Eng. **750**(1), 1–6 (2020)
7. Chiu, M.C., Chuang, K.H.: Applying transfer learning to achieve precision marketing in an omni-channel system–a case study of a sharing kitchen platform. Int. J. Prod. Res. **59**(24), 7594–7609 (2021)
8. Liu, H.: Big data precision marketing and consumer behavior analysis based on fuzzy clustering and PCA model. J. Intell. Fuzzy Syst. **40**(4), 6529–6539 (2021)
9. Zhang, Y.: Prediction of customer propensity based on machine learning. In: Asia-Pacific Conference on Communications Technology and Computer Science (ACCTCS), vol. 2(5), pp. 5–9. IEEE (2021)
10. Chen, Q., Cai, S., Gu, X.: Research on the precise marketing method of goods based on big data technology. In: Proceedings of the 2nd International Conference on E-Commerce and Internet Technology (ECIT), vol. 2(5), pp. 364–367. IEEE (2021)
11. Liu, X.: E-commerce precision marketing model based on convolutional neural network. Sci. Program. **750**(1), 33–136 (2022)
12. Zou, J., Li, H.: Precise marketing of E-commerce products based on KNN algorithm. Comput. Intell. Neurosci. **750**(1), 33–136 (2022)

Mathematical Modeling of Financial Time Series Volatility: A GARCH Model

Jouilil Youness(✉) and Mentagui Driss

Laboratory of Partial Differential Equations, Spectral Algebra and Geometry, Department of Mathematics, Faculty of Sciences, Ibn Tofail University of Kenitra, Kenitra, Morocco
y.jouilil@gmail.com

Abstract. In the era of economic data modeling, machine learning algorithms, are increasingly suitable for big data, especially for univariate time series. The main aim of this contribution is to establish a step-by-step process of time series modeling using a special financial algorithm namely the generalized autoregressive conditional heteroscedasticity (GARCH) algorithm.

As a consequence, our analytic findings argued that the hybrid ARIMA-GARCH can reflect the specific of our time-series characteristics and have better predictive power than the simple ARIMA algorithm.

Keywords: Time series · GARCH model · Mathematical modeling

1 Introduction

In the era of economic data modeling, machine learning algorithms, are increasingly suitable for big data, especially for univariate time series.

The present article will provide the procedure to model and analyze univariate financial time series. The remainder of the paper will be divided as follows. The first part focuses on the ARCH-GARCH model. The second one exposes an analytical case study in which we will modulate the stock price volatility. In the last one, we will choose the best model.

2 GARCH Model

2.1 Model Identification: ARIMA Model

2.1.1 ARMA Model

An Auto Regressive Moving Average (ARMA) algorithm of p and q orders is a combination of the Auto Regressive (AR) algorithm and Moving Average (MA) one.

Mathematically, an ARMA (p,q) can be expressed as follows [1]:

$$y_t = \mu + \sum\nolimits_{j=1}^{P} \Phi_J y_{t-j} + \sum\nolimits_{j=1}^{q} \theta_J \mathrm{a}_{t-j} + \mathrm{a}_t \tag{1}$$

Y. Farhaoui et al. (Eds.): ICAISE 2022, LNNS 635, pp. 608–614, 2023.
https://doi.org/10.1007/978-3-031-26254-8_88

Using the backward shift operator (L), the reducing form of the algorithm can be expressed:

$$\Phi_p(L)y_t = \theta_J(L)a_t \tag{2}$$

where:

- L: indicates the backward shift operator
- $L^k y_t = y_{t-k}$; for each $k \in \{0,1, \ldots\}$
- $\theta(L) = 1 - \theta_1 L - \theta_2 L^2 - \ldots \theta_q L^q$
- $\Phi(L) = 1 - \Phi_1 L - \Phi_2 L^2 - \ldots \Phi_p L^p$

2.1.2 ARIMA Model

An ARIMA (p,d,q) algorithm takes the following form:

$$\Delta^d y_t = (1 - L)^d y_t \tag{3}$$

where Δ indicates to the difference operator and the process y_t follows an ARMA(p,q).

2.2 GARCH Model: A Brief Overview

In this part, we will expose the theoretical framework related to GARCH model. Hence, consider the following model:

The process a_t is a GARCH(p,q) process if it is stationary and if it satisfies, for all t and some strictly positive-valued process σ_t, the equations [2, 3].

$$a_t = \sigma_t z_t \tag{4}$$

$$\sigma_t^2 = \alpha_0 + \sum_{j=1}^{q} \alpha_J a_{t-j}^2 + \sum_{j=1}^{P} \beta_J \sigma_{t-j}^2 \tag{5}$$

where:

- $\alpha_0 > 0$
- $\alpha_i \geq 0, i = 1, \ldots, q,$
- $\beta_j \geq 0, j = 1, \ldots, p.$

3 Analytical Study: Application to Stock Price Volatility

In this part, we will present an application of GARCH model using a real-Word dataset. More specifically, our target series is stock price volatility. It covered the period from 1975 to 2020. The used data is available for download at the word bank platform.

3.1 Software

This part has been done using R software. In other words, our time series modeling was implemented in R language using the Keras deep learning library.

It should be mentioned that the dataset used in this GARCH-R article is stock price volatility.

3.2 Implementing of GARCH

The following code chunk shows our statistical manipulations using our time series (Tables 1, 2 and Fig. 1):

Table 1. Elementary manipulations using R language

```
setwd("C:/Users/LENOVO/Desktop/Finance")
myts <- read_excel("myts.xlsx")
head(myts)
myts$Index <- ts(myts$Index, start = c(1975,1), frequency = 1)
plot(myts$Index)
class(myts$Index)
start(myts$Index)
end(myts$Index)
summary(myts$Index)
```

Source: Author's manipulations using R language

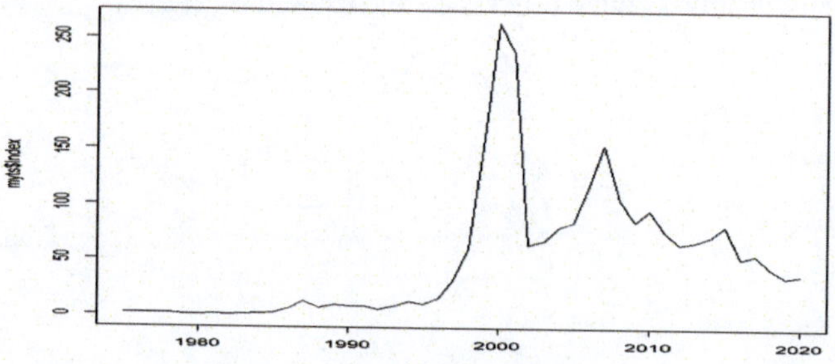

Fig. 1. Display the target series.

Table 2. Estimation ARIMA(0,1,3) model

	Coefficient
Ma1	0.3843
	(0.1729)
Ma2	-0.4095
	(0.1245)
Ma3	-0.4765
	(0.2192)
Sigma^2	986.3
Log likelihood	-217.95
AIC	443.91
AICc	444.91
BIC	451.13

Source: Author's manipulations using R language

Note:
Parentheses indicate standard errors.

3.3 Outputs

3.3.1 Estimation ARIMA Model

3.3.2 Estimation ARIMA-GARCH Model

Graphically, to check the homoscedasticity, we can use the Autocorrelation Function (ACF) and the Partial Autocorrelation Function (PACF) (Fig. 2).

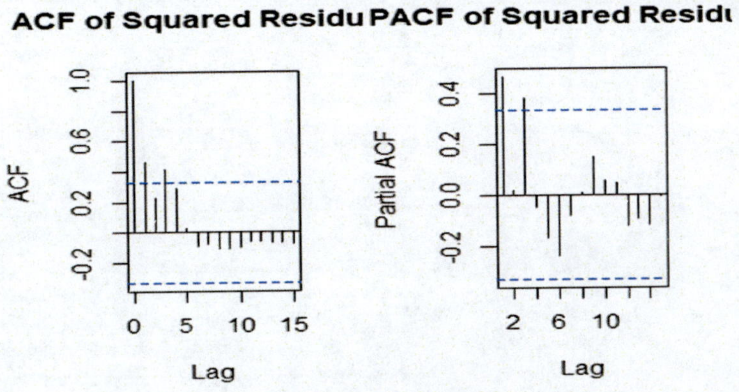

Fig. 2. ACF and PACF

Statistically, to check the homoscedasticity of our univariate time series, we have used Engle's ARCH Test. The results are presented in the figure bellow (Fig. 3):

```
> archTest(r)
Q(m) of squared series(LM test):
Test statistic:  22.91827  p-value:  0.01105232
Rank-based Test:
Test statistic:  99.97067  p-value:  0
```

Fig. 3. Engle's test

Since p values is less than 5 percent, we reject the null hypothesis. Therefore, we can conclude the presence of ARCH effects. Hence, based on previous results, here are the findings of the estimation of the standard GARCH model (Fig. 4).

```
*---------------------------------*
*          GARCH Model Fit        *
*---------------------------------*

Conditional Variance Dynamics
-----------------------------------
GARCH Model       : sGARCH(1,1)
Mean Model        : ARFIMA(1,0,1)
Distribution      : norm

Optimal Parameters
------------------------------------
        Estimate  Std. Error  t value Pr(>|t|)
mu       1.42677    0.961163   1.4844  0.13770
ar1      0.96540    0.034732  27.7958  0.00000
ma1      0.48083    0.060441   7.9555  0.00000
omega    1.39402    0.708316   1.9681  0.04906
alpha1   0.99900    0.160173   6.2370  0.00000
beta1    0.00000    0.001389   0.0000  1.00000

Robust Standard Errors:
        Estimate  Std. Error  t value Pr(>|t|)
mu       1.42677    0.209836  6.79948  0.00000
ar1      0.96540    0.035086 27.51532  0.00000
ma1      0.48083    0.044091 10.90536  0.00000
omega    1.39402    1.413041  0.98654  0.32387
alpha1   0.99900    0.226185  4.41674  0.00001
beta1    0.00000    0.001551  0.00000  1.00000

LogLikelihood : -131.1845
```

Fig. 4. Estimation ARIMA-GARCG model

3.3.3 Validation of the Model

The results of obtained model can be interpreted as follows.

The Criteria and Evaluation of Statistic Fit using many criteria such as the Akaike Information Criterion (AIC), The Mean Absolute Percentage Error (MAPE) and The Root Mean Squared Error (RMSE). Hence, the figure shows that the lowest AIC, MAPE, RMSE values are considered to be the best model for modeling properties.

Using our target series, we found that our series is ARIMA-GARCH. See above for more details.

4 Conclusion

Exploring GARCH model allows us to make robust modeling since it is the most powerful model, especially when we employ a financial dataset.

In fact, the ARIMA algorithm is commonly used together with the GARCH model, especially for non-stationary series. As a consequence, our analytic findings argued that the hybrid ARIMA-GARCH can reflect the specific of our time-series characteristics and have better predictive power than the simple ARIMA algorithm.

What would be the results in terms of modeling and forecasting if we compare ARIMA-GARCH to deep learning algorithms?

References

1. Black, F.: Studies of stock market volatility changes. In: Proceedings of the 1976 Meeting of the Business and Economic Statistics Section, American Statistical Association, Washington, DC, USA (1976)
2. Bollerslev, T., Engle, R.F., Nelson, D.B.: Arch models. Handb. Econ. **4**, 2959–3038 (1994)
3. Bollerslev, T.: Generalized autoregressive conditional heteroskedasticity. Journal of Econometrics **31**(3), 307–327 (1986)
4. Stock, J., Watson, M.: Introduction to Econometrics, Third Update, Global Edition. Pearson Education Limited (2015)
5. Yaziz, S.R., Azizan, N.A., Zakaria, R., Ahmad, M.: The performance of hybrid ARIMA-GARCH modeling in forecasting gold price. In: Proceedings of the 20th International Congress on Modelling and Simulation, Adelaide, Australia, 1–6 December 2013; pp. 1201–1207
6. Sun, K.: Equity return modeling and prediction using hybrid ARIMA-GARCH model. Int. J. Financ. Res. **8**, 154–161 (2017)
7. Ismail, M.T., Mustapa, F.: Modelling and forecasting S&P500 stock prices using hybrid ARIMA-GARCH model. J. Phys. Conf. Ser. **1366**, 012130 (2019)
8. Lechheb, H., Ouakil, H., Jouilil, Y.: Economic growth poverty and income inequality: implications for lower- and middle-income countries in the era of globalization. The Journal of Private Equity (2019)
9. Youness, J.: The individual determinants of financial inclusion in morocco: an exploratory study in the case of a credit application. In: 2022 2nd International Conference on Innovative Research in Applied Science, Engineering and Technology (IRASET), pp. 1- 6 (2022)
10. Youness, J., Driss, M.: An ARIMA model for modeling and forecasting the dynamic of univariate time series: the case of moroccan inflation rate. In: 2022 International Conference on Intelligent Systems and Computer Vision (ISCV), pp. 1–5 (2022)

11. Jouilil, Y., Lechheb, H.: Effect of Moroccan health insurance on individuals' healthcare utilization and expenditures. J. Private Equity **22**(1), 87–90 (2018)
12. Shumway, R.H., Stoffer, D.S.: Time Series Analysis and Its Applications With R Examples, 2nd ed.; Springer: Berlin/Heidelberg, Germany (2000).https://doi.org/10.1007/978-3-319-52452-8
13. Batrancea, L.: An econometric approach regarding the impact of fiscal pressure on equilibrium: evidence from electricity, gas and oil companies listed on the New York stock exchange. Mathematics **9**, 630 (2021)
14. Jouilil, Y., Mentagui, D.: LSTM deep learning vs ARIMA algorithms for univariate time series forecasting: a case study. In: 2022. 8th International Conference on Optimization and Applications (ICOA) (2022)
15. Sohaib, K., Driss, E., Jouilil, Y.: An econometric analysis of the determinants of small's cities' attractiveness: evidence from Moroccan case. In: 2022. 8th International Conference on Optimization and Applications (ICOA) (2022)

Mathematical Modeling of Monetary Poverty: Evidence from Moroccan Case

Yassine El Aachab(✉) and Mohammed Kaicer

Laboratory of Analysis Geometry and Applications, Department of Mathematics Faculty of Sciences, Ibn Tofail University, Kenitra, Morocco
y.elaachab@gmail.com, mohammed.kaicer@uit.ac.ma

Abstract. The objective of this study is to identify the variables that impact mostly house- holds poverty. Using the multiple logistic regression model, especially the logit model on the national survey on household consumption and expenditure dataset.

The R-probit results indicate that the household head's, household size, average annual expenditure, and household head's employment status are statistically significant determinants of household well-being ($p < 5\%$).

Keywords: Logistic regression model · Poverty · Morocco

1 Introduction

One of the multivariate analytic models that are most frequently employed in a variety of fields is logistic regression. It assesses the relationship between the likelihood that an event will occur (qualitative explained variable) in our case, poor or no poor, and, the elements that are most likely to affect it (explanatory variables).

In this sense, the target of this project paper is to study poverty using a model of regression, especially the multiple logistic regression model.

The rest of this paper will be as follows. In the first part, we will expose the theoretical framework related to poverty phenomena. The second one presents the multiple logistic regression model. The third shows some statistical descriptive results related to our target dataset. In the fourth one, we lead to see the significance and impact of our national household consumption and expenditure survey metadata variables on income poverty, using the multiple logistic regression model in R.

2 Theoretical Framework

Poverty is a difficult problem to discuss[1]. In connection with urban development and the widespread desire for a particular way of life, poverty presents itself in various ways and consequently takes on numerous shapes, both in the North and the South [2]. Poverty is a dynamic phenomenon that has been seen and manifested in various ways across time [3].

Y. Farhaoui et al. (Eds.): ICAISE 2022, LNNS 635, pp. 615–620, 2023.
https://doi.org/10.1007/978-3-031-26254-8_89

As a result, there is no agreement on the best way to investigate the phenomenon, and fresh conceptual and methodological insights are routinely made. To understand the new types of poverty that will unavoidably emerge in a generalized crisis where ecological limits are growing more and more significant, this conceptual work is still required [4].

3 Empirical Framework

For logistic regression, the logit model of the output variable y_i is a Bernoulli random variable (it can only take two values, 1 or 0) and [5]

$$P(y_i = 1|x_i) = S(x_i\beta) \tag{1}$$

where

$$S(t) = \frac{1}{1 + \exp(-t)} \tag{2}$$

is the logistic function, x_i is 1*K -dimensional input vector, and β is a K*1-dimensional coefficient vector.

Furthermore,

$$P(y_i = 0|x_i) = 1 - S(x_i\beta) \tag{3}$$

The parameter to be estimated using maximum likelihood is represented by the vector of coefficients β. We suppose that an IID sample with **N** data points is used to perform the estimation.

$$(y_i, x_i) \text{ for i} = 1,2,3 \ldots\ldots \text{N} \tag{4}$$

The formula for the likelihood of an observation (y_i, x_i) can be written as

$$L(\beta; y_i, x_i) = [S(x_i\beta)]^{y_i}[1 - S(x_i\beta)]^{1-y_i} \tag{5}$$

A represents the $N * 1$ vector of all outputs by y and represents the $N * K$ matrix of all inputs by X. Because the observations are IID, the likelihood of the entire sample is equal to the product of the likelihoods of the individual observations [5]:

$$L(\beta; y, X) = \prod_{i=1}^{N} [S(x_i\beta)]^{y_i}[1 - S(x_i\beta)]^{1-y_i} \tag{6}$$

The log-likelihood of the logistic model is

$$l(\beta; y, X) = \sum_{i=1}^{N} [-\ln 1 + exp(x_i\beta) + y_i x_i\beta] \tag{7}$$

The resolution of the maximization of the likelihood function can be written in the form:

$$\widehat{\beta_i} = \widehat{\beta_{i+1}} + (X^T W_{t-1} X)^{-1} X^T (y - \hat{y}_{t-1} \tag{8}$$

4 Data and Results

4.1 Data

The data used in this paper comes from the National Household Living Standards Survey Morocco, which is a survey conducted by the household survey division in the High Commission for Planning. In order to, carry out our experiment and analysis, we first carried out a pre-processing process, such as cleaning the data, transforming it, filtering it, etc. We selected all valid data in 12 regions of Morocco, for the 2014 survey year. In total, there are 15970 valid data.

4.2 Statistical Descriptive

````
```{r}
smp <- smp%>%mutate(Vulnérable  = as.factor(Vulnérable ))
smp <- smp%>%mutate(Milieu  = as.factor(Milieu ))
smp <- smp%>%mutate(Sexe_CM  = as.factor(Sexe_CM))
smp <- smp%>%mutate(Etat_matrimonial_CM    = as.factor(Etat_matrimonial_CM  ))
smp <- smp%>%mutate(Niveau_scolaire_agreg_CM   = as.factor(Niveau_scolaire_agreg_CM  ))
smp <- smp%>%mutate(Profession_agreg_CM  = as.factor(Profession_agreg_CM  ))
smp <- smp%>%mutate(Secteur_activité_agreg_CM  = as.factor(Secteur_activité_agreg_CM  ))
smp <- smp%>%mutate(Situation_profession_agreg_CM  = as.factor(Situation_profession_agreg_CM ))
smp <- smp%>%mutate(Type_activité_dominante_CM  = as.factor(Type_activité_dominante_CM   ))
smp <- smp%>%mutate(Diplôme_agregé_CM  = as.factor(Diplôme_agregé_CM  ))
smp <- smp%>%mutate(Lieunaissance_CM  = as.factor(Diplôme_agregé_CM  ))
````

Fig. 1. Transformation of some variables as qualitative variables Source: Author's manipulations using R.

If we summarize the data frame, we see that dichotomous data are treated as qualitative variables (Figs. 1, 2).

```
   N_ménage           DAM           Taille_ménage   Taille_agregée      Age_CM        Age_quin_CM      Région_12       Milieu
 Min.   :    1   Min.   :   4286   Min.   : 1.00   Min.   :1.000   Min.   :16.00   Min.   : 4.00   Min.   : 1.000   1:10380
 1st Qu.: 3993   1st Qu.:  39898   1st Qu.: 3.00   1st Qu.:3.000   1st Qu.:42.00   1st Qu.: 9.00   1st Qu.: 3.000   2: 5590
 Median : 7986   Median :  58098   Median : 5.00   Median :5.000   Median :52.00   Median :11.00   Median : 5.000
 Mean   : 7986   Mean   :  75068   Mean   : 4.74   Mean   :4.337   Mean   :52.63   Mean   :11.07   Mean   : 5.117
 3rd Qu.:11978   3rd Qu.:  88005   3rd Qu.: 6.00   3rd Qu.:6.000   3rd Qu.:62.00   3rd Qu.:13.00   3rd Qu.: 7.000
 Max.   :15970   Max.   :1224977   Max.   :26.00   Max.   :6.000   Max.   :98.00   Max.   :16.00   Max.   :12.000

 Pauvre    Vulnérable Sexe_CM   Lieunaissance_CM Etat_matrimonial_CM Niveau_scolaire_agreg_CM Diplôme_agregé_CM
 0:15471   0:14471    1:13068   1:11057          1:  570             0:7356                   1:11057
 1:  499   1: 1499    2: 2902   2: 3936          2:12886             1:  17                   2: 3936
                                3:  977          3:  515             2:3588                   3:  977
                                                 4: 1999             3:1582
                                                                     4:1476
                                                                     5: 992
                                                                     6: 959

 Type_activité_dominante_CM Profession_agreg_CM Secteur_activité_agreg_CM Situation_profession_agreg_CM
 1       :11587             4       :3180       5       :3635             0:  74
 8       : 1423             5       :3158       1       :3225             1:4001
 4       : 1071             6       :2046       4       :2022             2:5997
 7       :  899             2       :1565       3       :1546             3:5296
 10      :  437             3       :1421       2       :1425             4: 278
 2       :  306             (Other): 599        (Other): 116              5: 172
 (Other):  247              NA's    :4001       NA's    :4001             9: 152
```

Fig. 2. Statistical descriptive Source: Author's manipulations using R.
```

4.3 Econometric Results

```
```{r}
# Logit model
logit<- glm(Pauvre ~ Taille_ménage+ Age_CM+ Sexe_CM+ Etat_matrimonial_CM+ DAM+
Diplôme_agregé_CM+ Situation_profession_agreg_CM, family=binomial (link = "logit") ,
data = smp)
summary(logit)

```
```

Fig. 3. Conceptuel R programme Source: Author's manipulations using R.

Since we named our model (logit), R won't output any results from our regression because of this. We use the summary command to get the outcomes (Figs. 3, 4).

```
Deviance Residuals:
   Min       1Q  Median       3Q      Max
-1.795   0.000   0.000    0.000    3.351

Coefficients:
                                  Estimate Std. Error z value Pr(>|z|)
(Intercept)                     -7.011e-01  1.897e+01  -0.037   0.9705
Taille_ménage                    1.153e+01  1.232e+00   9.362  < 2e-16 ***
Age_CM                          -2.242e-03  1.833e-02  -0.122   0.9027
Sexe_CM2                         3.862e-01  7.538e-01   0.512   0.6084
Etat_matrimonial_CM2             3.976e+00  1.895e+01   0.210   0.8338
Etat_matrimonial_CM3             5.372e+00  1.900e+01   0.283   0.7773
Etat_matrimonial_CM4             4.388e+00  1.896e+01   0.231   0.8169
DAM                             -2.655e-03  2.833e-04  -9.372  < 2e-16 ***
Diplôme_agregé_CM2               3.192e+00  7.270e-01   4.391 1.13e-05 ***
Diplôme_agregé_CM3              -2.379e-01  1.230e+03   0.000   0.9998
Situation_profession_agreg_CM1 -2.563e+00  1.537e+00  -1.667   0.0954 .
Situation_profession_agreg_CM2 -2.496e+00  1.468e+00  -1.700   0.0891 .
Situation_profession_agreg_CM3 -2.496e+00  1.498e+00  -1.666   0.0958 .
Situation_profession_agreg_CM4 -3.845e+00  2.061e+00  -1.865   0.0622 .
Situation_profession_agreg_CM5 -3.026e+00  2.054e+00  -1.473   0.1407
Situation_profession_agreg_CM9 -4.774e+00  2.493e+00  -1.915   0.0555 .
---
Signif. codes:  0 '***' 0.001 '**' 0.01 '*' 0.05 '.' 0.1 ' ' 1

(Dispersion parameter for binomial family taken to be 1)

    Null deviance: 4441.2  on 15969  degrees of freedom
Residual deviance:  183.9  on 15954  degrees of freedom
AIC: 215.9

Number of Fisher Scoring iterations: 23
```

Fig. 4. Logit model estimators Source: Author's manipulations using R.

4.4 Discussion the Results

According to the previous results, we can say that the coefficients of the modality Age_CM,Etat_matrimonial_CM2, Etat_matrimonial_CM3,Etat_matrimonial_CM4, and Diplôme_agregé_CM2 impact positively the target variable (status of poverty), while the modalities of the variable Situation_profession_agreg_CM, Age_CM and negatively the Diplôme_agregé_CM3 modality's.

We compare the value of p for each term to your service threshold to evaluate the null hypothesis in order to ascertain whether the relationship between the answer and each term of the model is statistically significant. There being no correlation between the word and the response is the null hypothesis. A significance threshold (also known as alpha or) of 0.05 is often effective. A 0.05 service threshold indicates a 5% chance of drawing the incorrect conclusion that there is a link [6].

In these results, the predictors are the following modalities: Taille_ménage, DAM, Diplôme_agregé_CM2,Situation_profession_agreg_CM1,Situation_profession_agreg_CM2,Situation_profession_agreg_CM3,Situation_profession_agreg_CM4,Situation_profession_agreg_CM5 et Situation_profession_agreg_CM9.

Using our micro dataset retrieved from the national survey on household consumption, we have adopted the algorithms of logit regression in order to identify the modalities that impact positively or negatively the poverty. The R outputs showed that our predictions are perfect in solving problems of discrimination and enabling us to separate the data into two classes.

5 Conclusion

As we have seen, the present paper showed the poverty is determined by many factors such as social and economic dimensions. Indeed, living in a rural area enhance the probability of being poor comparatively to citizen one. An extension of this paper is highly recommended and could give more value of this issue by using advanced statistical techniques, especially machine learning algorithms.

References

1. Mowafi, M., Khawaja, M.: Poverty. Journal of Epidemiology and Community Health **59**(4), 260–4 (2005)
2. Lasida, E., Lompo, K.M., Dubois, J.-L. : La pauvreté: une approche socio-économique. Entretien avec Jean-Luc Dubois Institut Catholique de Paris | « Transversalités » 2009/3 N° 111, pages 35 à 47. https://sci-hub.hkvisa.net/https://doi.org/10.3917/trans.111.0035 or https://www.cairn.info/revue-transver-salites-2009%E2%80%933-page-35.htm
3. Train, K.: Discrete Choice Methods with Simulation, Cambridge University Press, 2e éd., 408 (2009)
4. Gelman, A., Hill, J.: Data analysis using regression and multilevel/hierarchical models, Cambridge University Press, coll. « Analytical Methods for Social Re- search », 1re éd., 648 (2006)

5. Taboga, M. : Logistic regression - maximum likelihood estimation. Lectures on probability theory and mathematical statistics. Kindle Direct Publishing. Online appendix (2021). https://www.statlect.com/fundamentals-of-statistics/logistic-model-maximum-likelihood
6. Fromont, M.: Tests Statistiques Rejeter, ne pas rejeter... Se risquer ?. Année univer sitaire 2015–2016. https://perso.univ-rennes2.fr/system/files/users/fromont_m/PolyTests.pdf

Miniaturized Patch Metamaterial Antenna for 5G 3.5 GHz Band

Abdel-Ali Laabadli(✉), Youssef Mejdoub, Abdelkebir El Amri, and Mohamed Tarbouch

RITM Laboratory, CED Engineering Sciences, Ecole Supérieure de Technologie, Hassan II, University of Casablanca, Casablanca, Morocco
aali.laabadli@gmail.com

Abstract. In this paper we present a compact patch antenna miniaturized with the metamaterial technique, this antenna resonate at 3.5 GHz and his band cover the one of 5 G, the band 3.5 GHz. Starting from a conventional patch antenna that resonate at 4.7 GHz we have shifted the resonance frequency to 3.5 GHz by etching two symmetric metamaterial (MTM) circular unit cells based on complementary split ring resonator (CSRR) on the ground plane of the conventional antenna. The conception was done with FR-4 dielectric as a substrate, its permittivity is 4.4 and its loss tangent is 0.0025, and cooper annulled as metal for radiating element and the ground. The simulation was done by CST solver. The conceived metamaterial antenna has a gain of 2.1dB and a bandwidth of 872.9 MHz from 3.1421 GHz to 4.015 GHz.

Keywords: Miniaturization · Complementary Split ring resonator (CSRR) · Metamaterial

1 Introduction

Metamaterials or "left-handed materials" are artificial structures whose electromagnetic properties have never been observed or realized in natural materials, for example, negative permittivity and/or permeability. Because of those particular properties, they are widely used in antenna miniaturization. In the following, some works reported in literature show that metamaterials present an effective solution for reducing the size of an antenna and improving its performances [1, 5]. In [1]: The authors introduced in their paper two metamaterial unit cells with negative permittivity and permeability above the patch antenna, they achieved a reduction of 43%. In [2]: The authors designed two metamaterial unit cells with negative permittivity and permeability in the vicinity of a half loop antenna; they shifted the resonance frequency by 20%. In [3]: The authors proposed a miniaturized patch antenna on substrate loaded with ring-shape metamaterial with permeability near to zero and permittivity negative, they miniaturized the antenna by 68%. In [4]: The authors proposed a miniaturized rectangular patch antenna loaded by six metamaterial unit cells with negative permeability, the conceived antenna resonates at two frequencies inferior to the frequency of the antenna without unit cells; that corresponds to a size reduction of 40% and 30%. In [5]: The authors present a miniaturized

Y. Farhaoui et al. (Eds.): ICAISE 2022, LNNS 635, pp. 621–626, 2023.
https://doi.org/10.1007/978-3-031-26254-8_90

square patch antenna by using concentric complementary split ring resonator (CSRR) structures in between the patch and the ground plane, the achieved size is 25% smaller than to the size of a conventional square patch antenna. In the next of this paper we present the steps followed to design a reduced size metamaterial antenna, resonating at 3.5 GHz.

2 Antenna Design Methodology

2.1 Design of the 4.7 GHz Conventional Patch Antenna

1) *Calculating the length and width of the patch*

Basing on the TLM (Transmission Line Model) explained in [6], the frequency resonance (fr) of the patch antenna, Fig. 3 (b), is approximated by the Eq. (1)

$$f_r = \frac{c}{2W\sqrt{\frac{\varepsilon r+1}{2}}} \tag{1}$$

c = 3.10e8 ms-1, speed of the light. ε_r = The relative permittivity of the dielectric. W: The width of the patch.

From Eq. (1), if fr = 4.7 GHz and ε_r = 4.4, the width of the patch is W = 19.41mm. The length of the patch, L = 14.68 mm was finded out by substituting the values of W, ε_r, fr and h (the thickness of the substrate) in the below formulas.

$$\varepsilon reff = \frac{\varepsilon r+1}{2} + \frac{\varepsilon r-1}{2} \times \left[1 + 12\frac{h}{W}\right]^{-\frac{1}{2}} \tag{2}$$

$$Leff = \frac{\lambda}{2\sqrt{\varepsilon reff}} \tag{3}$$

$$\lambda = \frac{c}{fr} \tag{4}$$

$$\frac{\Delta L}{h} = 0.412 \times \frac{(\varepsilon reff + 0.3) \times (\frac{W}{h} + 0.264)}{(\varepsilon reff - 0.258) \times (\frac{W}{h} + 0.8)} \tag{5}$$

$$Leff = L + 2 \times \Delta L \tag{6}$$

2) *Chosing the Feeding Modes*

The rectangular patch microstrip antennas have several feeding techniques, such as the microstrip line (with and without inset), coaxial probe, proximity coupling, and CPW.

Referring to the comparison between those feeding techniques done in [7], microstrip line with inset configuration is the most suitable for this work even its low bandwidth, it is simple to design and offers high gain.

Basing on the on the Eq. (7) [8], the width of the microstrip line that has an impedance of 50 Ω is 3.083

$$Z_0 = \frac{120\pi}{\sqrt{\varepsilon_r}\left(\frac{W}{h} + 1.393 + 0.667\ln\left(\frac{W}{h} + 1.44\right)\right)} \tag{7}$$

With: Z0: the characteristic impedance of the microstrip line.

W: the width of the microstrip line. / h: the high of the substrate.

To verify the previous theoretical study, the conventional microstip patch antenna with inset microstrip line feeding was simulated with CST. The antenna is printed on the substrate EPOXY FR4 with relative permittivity εr = 4.4, thang loss equal to 0.025 and a thickness of 1.6 mm. The other parameters are: antx = 19.41 mm, anty = 14.68 mm, sbx = 2*antx, sby = 2*anty, trx = 3.083 mm, insx = 1 mm and insy = 4.7 mm.

Figure 1 presents the S11 of the conventional patch antenna, the simulated resonance frequency is 4.674 GHz (green curve), it is lower than the resonance frequency given by TLM 4.7 GHz, CST and other commercial software based on a full wave methods like: Finite Element Method (FEM), Finite Integral Technique (FIT), and the Method of the Moment (MoM) are more accurate compared to the TLM method.

Hence, the design is optimized to have 4.7 GHz as the resonance frequency with the full wave method used in CST, The new dimensions of the patch are: antx = 19.29mm and anty = 14.59 mm. Figure 2 (Red curve) shows the S11 of the optimized antenna.

To demonstrate the effect of the inset length (insy) on the Impedance matching, a parametric study was done; Fig. 2 shows the variation of S11 versus insy. For 4.7 GHz the better impedance matching is obtained when insy = 4.7 mm.

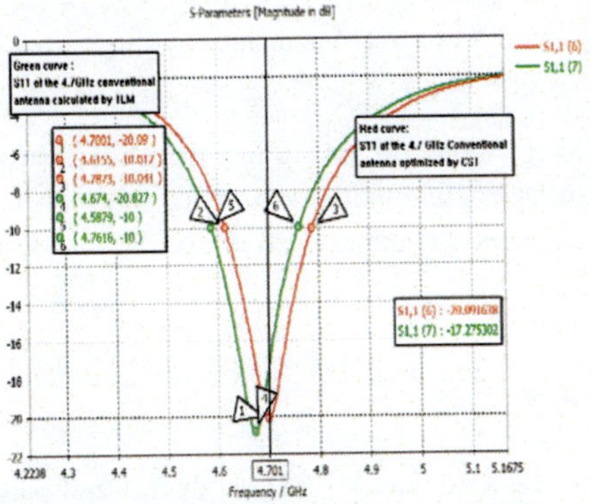

Fig. 1. S11 at 4.7 GHz of both conventional antennas:(Green) calculated with TLM, (Red) optimized by CST (full wave method

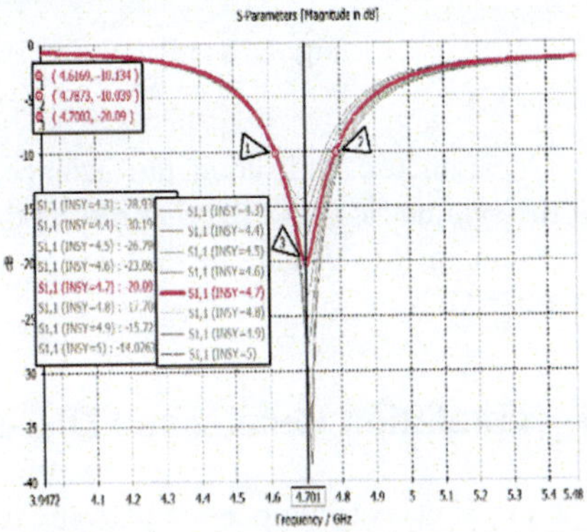

Fig. 2. Parametric study of S11versus insy

2.2 Design of the Metamaterial Antenna

As shown in Fig. 3(a), on the ground plane of the previous conventional patch antenna we etch two symmetric metamaterial unit cells, each unit cell is a circular complementary split ring resonator (CSRR), that their centres have a distance X1 from the center of the patch antenna, the parameters values of the new MTM antenna are in (mm): sbx = 38.58, sby = 29.18, antx = 19.29, anty = 14.59, trx = 3.083, insx = 1, insy = 4.7, G = W = S = 1, R0 = 7.8.

(a) (b)
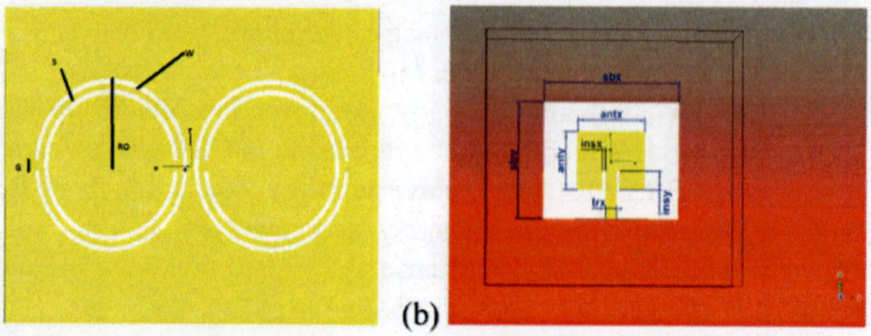

Fig. 3. Back view of the MTM patch antenna, (b): Top view of both antennas MTM antenna and 4.7 GHz conventional patch antenna.

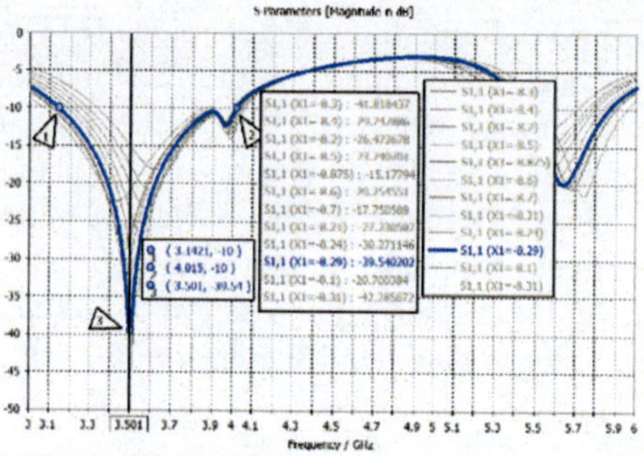

Fig. 4. S11 Parametric study about X1 of MTM antenna.

A parametric study about the distance X1 Fig. 4, was done to define the right position of the symmetric unit cells that results the suitable conclusion, we notice that at X1 = 8.29 mm, the MTM antenna resonate at 3.5 GHz and the bandwidth is from 3.1421 GHz to 4.015 GHz.

3 Simulation Results and Discussion

The Fig. 5 gives a comparison between the initial antenna (4.7 GHz conventional patch antenna (Sect. 2.1)) that was conceived without metamaterial unit cells on the ground and the new designed antenna, antenna with metamaterial unit cells that was etched on the ground of the 4.7 GHz conventional patch antenna (Sect. 2.2). As shown in the figure, the resonance frequency of the conventional antenna was shifted from 4.7 GHz to 3.5 GHz which means that the antenna size was reduced. We noticed also, that there is another generated resonance frequency of 5.64 GHz.

Figure 6 illustrates the 2D radiation patterns of the Metamaterial antenna at its resonance frequency fr = 3.5 GHz. We observe that the radiation pattern is bidirectional in both planes E-Plane (bleu graph) H-Plane (red graph).

To demonstrate that we have made a reduced size 3.5 GHz antenna, we will compare the dimensions of our conceived antenna to a 3.5 GHz conventional patch antenna.

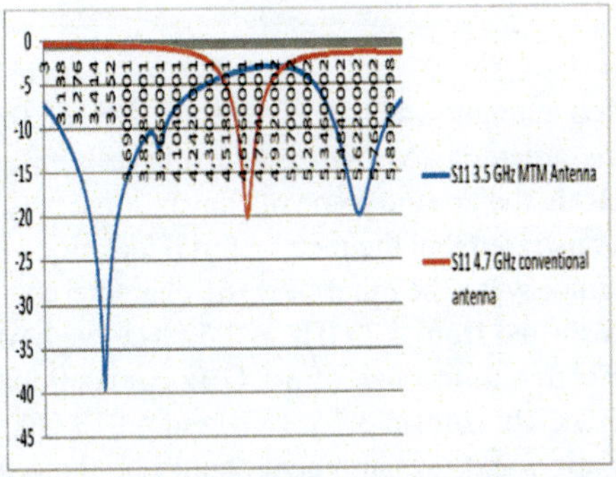

Fig. 5. Reflection coefficient (S11) of the MTM antenna and the conventional antenna.

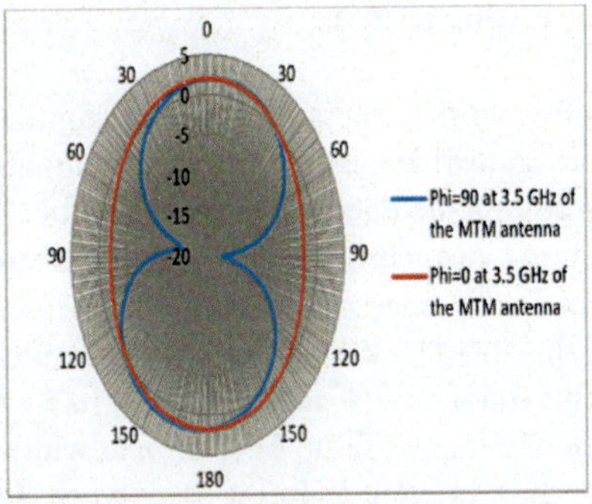

Fig. 6. Radiation pattern of the proposed metamaterial (MTM) antenna at 3.5 GHz

By following the same steps to calculate and simulate the conventional patch 4.7 GHz, we calculate the dimensions of the 3.5 GHz conventional patch, the Table 1 sums up the dimensions of both antennas; from this table we observe that we have reduced the size of the 3.5 GHz conventional patch antenna by 44, 67%.

Table 1. The dimensions of the 3.5 GHz conventional patch antenna and MTM antenna.

Antenna	Volume in mm^3
Conventional patch at 3.5 Ghz	51.54x39.48x1.6
The metamaterial antenna at 3.5 Ghz	38.58x29.18x1.6

Table 2 compares the proposed MTM antenna performance with other antennas operating at 3.5 GHz for 5G applications recently reported in literature. It shows that our conceived antenna has the best bandwidth and reduced size but it has low gain, even though it is acceptable and make with the other features of this antenna an excellent candidate for the 5G.

Table 2. Results comparison of our work with other recently published works.

Ref/Year	Antenna size (mm^3)	BW (MHz)	Gain (dB)	ε_r	Technique
9/2019	$35 \times 40 \times 1.57$	100	7	2.2	MTM
10/2015	$40 \times 45 \times 1.5$	132	4.6	2.65	MTM
11/2021	$50 \times 50 \times 3.04$	73.5	4.15	3&3	MTM
This work	$38.58 \times 29.18 \times 1.6$	872.9	2.1	4.4	MTM

4 Conclusion

In this work, a metamaterial rectangular patch antenna was conceived for 3.5 GHz, the core band of 5G. By etching two symmetric metamaterial (MTM) circular unit cells based on complementary split ring resonator (CSRR) in the ground plane of conventional patch antenna resonating at 4.7 GHz, we have demonstrated that the metamaterial technique is an effective method for miniaturization and improving the other intrinsic characteristics of the antenna, the resonance frequency was shifted from 4.7 GHz to 3.5 GHz, we have achieved 44, 67% size reduction in comparison with the size of 3.5 GHz conventional patch antenna (Table 1). The bandwidth at 3.5 GHz is from 3.1421 GHz to 4.015 GHz, the gain is 2.1 dB. In the next work, fabrication and measurement should be done to confirm the simulated results.

References

1. Gudarzi, A., Bazrkar, A., Mahzoon, M., Mohajeri, F.: Gain enhancement and miniaturization of microstrip antenna using MTM superstrates. In: International Conference on Computer and Communication Engineering (ICCCE 2012), 3–5 July 2012, Kuala Lumpur, Malaysia
2. Zamali, M., Osman, L., Regad, H., Latrach, M.: UHF RFID reader antenna using novel planer metameterial structure for RFID system. (IJACSA) International Journal of Advanced Computer Science and Appl. **8**(7) (2017)
3. Saghanezhad, S.A.H., Atlasbaf, Z.: Miniaturized Patch Antenna on Substrate Loaded with Ring-Shape Metamaterial Conference Paper, pp. 15911593 (2014). https://doi.org/10.1109/IranianCEE.2014.6999790
4. Lafmajani, I.A., Rezaei, P.: Miniaturized Rectangular Patch Antenna Loaded with Spiral/Wires Metamaterial Article **65**(11), 121130 (2011)
5. Ramzan, M., Topalli, K.: A miniaturized patch antenna by using a CSRR loading plane. Department of Electrical and Electronics Engineering, Bilkent University, 06800 Ankara, Turkey
6. Balanis, C.A.: Antenna Theory: Analysis and Design, 3rd edn. John Wiley, Hoboken, New Jersey (2005)
7. Barrou, O., El Amri, A., Reha, A. : Comparison of feeding modes for a rectangular microstrip patch antenna for 2.45 2.45 GHz applications. RITM Laboratory, CED Engineering Sciences, ESTC, Hassan II University of Casablanca, Morocco
8. Huang, Y., Boyle, K.: Antennas: From Theory to Practice. Wiley, Chichester, UK (2008)
9. Li, R., Zhang, Q., Kuang, Y., Chen, X., Xiao, Z., Zhang, J.: Design of a miniaturized antenna based on split ring resonators for 5g wireless communications. Beijing University of Posts and Telecommunications, Haidian district, 100876 Beijing, China
10. Cai, T., Wang, G.-M., Zhang, X.-F., Wang, Y.-W., Zong, B.-F., Xu, H.-X.: Compact microstrip antenna with enhanced bandwidth by loading magneto-electro-dielectric planar waveguided metamaterials. Air Force Engineering University, China
11. Dai, G., Xu, X., Deng, X.: Size-reduced equilateral triangular metamaterial patch antenna designed for mobile communications. School of Communication and Information Engineering Shanghai University, Shanghai, 200444, China

Modeling of PV Power Source and Fuel Cell Used for Energy Autonomy of Connected Objects

Wafaa Ennabirha[1(✉)], Ahmed Moutabir[2], and Abderraoouf Aboudou[1]

[1] High School of Technology, Laboratory in Computer Networks, Telecommunications and Multimedia, Hassan 2nd University, Casablanca, Morocco
wafaa.ennabirha.doc2020@ensem.ac.ma

[2] Aïn Chok Faculty of Science (FSAC), Laboratory of Electrical and Industrial Engineering, Information Processing, IT and Logistics (GEITIIL), Hassan 2nd University, Casablanca, Morocco

Abstract. A large part of the new generations of connected objects will only be able to develop if it is possible to make them completely autonomous in terms of energy. Even if the use of batteries or cells solves part of this problem by ensuring autonomy that can be important, it introduces many problems such as maintenance incompatible with some applications and environmental pollution. The recovery of solar energy is then a very attractive solution although the amount of energy obtained is generally quite low. Our objective is to propose an autonomous sensor solution based on a solar energy recovery system that can be implemented in different applications. In this article, we focus on combining renewable sources such as solar and wind to provide a backup solution when fuel cells cannot compensate for all energy needs.

Keywords: Internet of Things · PV generator · Fuel cell · Single diode · Energy recovery

1 Introduction

Currently, there are two main ways to maximize the lifetime of a communicating object. The first way is to store energy with a systematic reduction of energy consumption by the different sensor blocks (especially the radio block). While the second way consists in obtaining energy from the environment [1, 2]. Indeed, renewable energies such as wind and solar energy as well as vibratory energy have become promising sources to provide a part of the energy balance of the sensor or to totally replace the batteries used to have an Energy Autonomous Communicating Sensor. The combination of solar energy and batteries has become a key element to extend the life of wireless sensors. This technique not only allows the power of the sensor for different classes of applications but is also free from all the worries caused by the use of batteries [1, 3]. Thus, to ensure an energy-neutral operation.

Y. Farhaoui et al. (Eds.): ICAISE 2022, LNNS 635, pp. 627–634, 2023.
https://doi.org/10.1007/978-3-031-26254-8_91

This article is organized as follows: in the next section, we give a brief introduction on ITO, in Sect. 3, we did a modelling of fuel cells that are used in ITO and then in Sect. 4, We are interested in modelling the solar panel, in Sect. 5 with simulations carried out in Matlab, we present a number of results; we conclude the article in Sect. 6.

2 Background Theory

The uses of IoT technology are different because it can be applied to almost any technology and can provide insights into the response process, operations performance, and environmental variables that need to be controlled and monitored remotely [4]. Today, several organizations in different sectors and industries are using this technology to minimize, create, automate and track several processes [5].

An open architecture that uses open protocols to support many existing network applications to provide scalability, security and semantic middleware to facilitate the convergence of Internet data globally should provide an IoT architecture. IoT should then be a five-layer system: Detection and Control Layer: the basis of design and deployment, including RFID readers, smart sensor nodes and entry gateways [6]. A group of sensor nodes captures the associated ambient information, transmits it to the nearest gateway, and then transmits the data to the network via the Internet. The energy supply system consists of three main parts: energy sources included in upstream refining processes, energy transfer processes, energy demand in buildings and transport. Energy supply chains; distribution. Processing layer: Reasoning, semantic analysis of detection data, data interrogation, data collection, analytical data, mining. Cloud computing could be a powerful tool for data processing and detection. It is an integral part of the body of knowledge. IoT software can be classified, into network monitoring (logistics, emissions regulation), form of control (smart transport, smart home), type of scanning (motor bag, highway, no parking fees)

3 Fuel Cell model

The PEM fuel cell is made up of a cathode and anode electrode that are separated by a polymer electrolyte membrane.

This is controlled electrochemical combustion of hydrogen and oxygen, with simultaneous generation of electricity, heat, and water, based on a chemical reaction.

Generally, the electrochemical operation principle of a PEM fuel cell is described by two chemical reactions whose named oxidation reaction (1) and reduction reaction (2)

$$H_2 \rightarrow 2H^+ + 2e^- \tag{1}$$

$$\frac{1}{2}O_2 + 2H^+ + 2e^- \rightarrow H_2O \tag{2}$$

The overall reaction is :

$$H_2 + \frac{1}{2}O_2 \rightarrow H_2O \tag{3}$$

An electromotive force is generated between the anode and cathode electrodes during electrical reactions and electron transfer process. In order to satisfy the demand for energy, many simple elements can be assembled into fuel cells.

The model of PEM fuel cell proposed in [7] is considered in this paper.

The output voltage of a single fuel cell is given by:

$$V_{cell} = E_{Nernst} - V_{act} - V_{ohm} - V_{conc} \tag{4}$$

where E_{Nernst} represents the open cell voltage, V_{act} represents the activation potential, V_{ohm} represents the ohmic potential and V_{conc} represents the concentration potential.

The open cell voltage is defined by:

$$E_{Nernst} = 1,229 - 0,85.10^{-3}.(T - 298,15) + 4,31.10^{-5}.T.\left[ln(P_{H_2}) + \frac{1}{2}ln(P_{O_2})\right] \tag{5}$$

where P_{H_2} and P_{O_2} are respectively, the partial pressure of hydrogen and oxygen, and T the temperature in Kelvin.

The representative potentials of tension losses are given by the following expressions.

$$V_{act} = \xi_1 + \xi_2.T + \xi_3.T.ln(C_{O_2}) + \xi_4 ln(i_{cell}) \tag{6}$$

$$V_{ohm} = i_{cell}.(R_M + R_C) \tag{7}$$

$$V_{conc} = -B.ln(1 - \frac{J}{J_{max}}) \tag{8}$$

$$J = \frac{i_{cell}}{A} \tag{9}$$

where icell is the cellular current and ξ1, ξ2, ξ3, ξ4 are the parametric coefficient for the cellular model;B is a parametric coefficient (V) which is dependent on the cell and its operating state.J is the real current density of the cell (A/cm2), and Jmax is the max value of J.

C_{O_2} (the concentration of oxygen), R_M (the equivalent membrane resistance) and ρ_M (the specific resistively (Ωm)) are given by the following expressions:

$$C_{O_2} = \frac{P_{O_2}}{5,08.10^6.e^{(-\frac{498}{T})}} \tag{10}$$

$$R_M = \frac{\rho_M.l}{A} \tag{11}$$

$$\rho_M = \frac{181,6.\left[1 + 0,03.\left(\frac{i_{cell}}{A}\right) + 0,062.(\frac{T}{303})^2(\frac{i_{cell}}{A})^{2,5}\right]}{\left[\psi - 0,634 - 3.(\frac{i_{cell}}{A})\right].e^{\left[4,18.(\frac{T-303}{T})\right]}} \tag{12}$$

where: l, A, R_C and ψ are, respectively, the thickness (cm), the activation area (cm^2), the contact resistance between the membrane and electrodes (Ω) and the water contents of the membrane.

For a stack with n cells, the voltage V_{fc} can be calculated by:

$$V_{fc} = nV_{cell} \tag{13}$$

Finally, the power is defined by:

$$P = V_{fc}i_{cell} = nV_{cell}i_{cell} \tag{14}$$

As long as the unknown parameters are determined, the output V_{fc} corresponding to a certain current i_{cell} can be predicted.

4 Modeling of the Solar Photovoltaic Panel

4.1 Equivalent Circuit of a PV Cell

A PV module consists of several solar cells connected in series and parallel to achieve the desired voltage and current levels. A solar panel cell is essentially a semi-conducting p-n connection. If exposed to light, a continuous current is produced. For simplicity, the single diode model in Fig. 1 is used in this paper. This model offers a good compromise between simplicity and precision in relation to the base structure. The equivalent circuit of the general model consists of a photocurrent (Iph), a diode, a parallel resistance (Rsh) expressing a leakage current, and a series resistance (Rs) due to contacts between semiconductors and metal parts Fig. 1

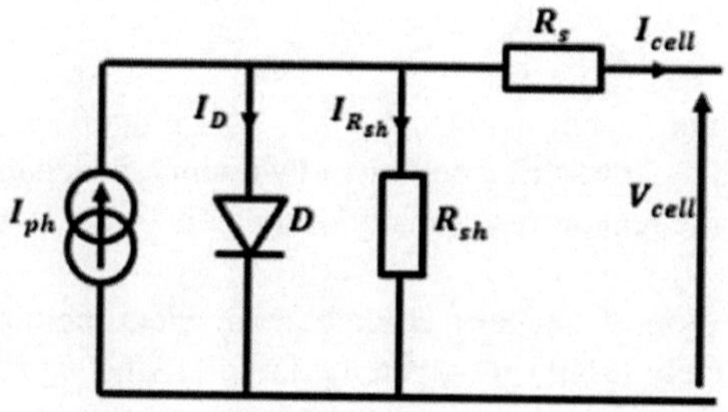

Fig. 1. Equivalent design of a PV cell.

4.2 Equations of the PV Cell

In Fig. 1, by applying Kirchhoff's law, the current will be obtained by the following equation:

$$I_{cell} = I_{ph} - I_D - I_{R_{sh}} \tag{15}$$

$$I_{R_{sh}} = \frac{V_{cell} + R_s I_{cell}}{R_{sh}} \tag{16}$$

$$I_s = I_{rr}\left[\frac{T}{298,15}\right]^3\left[exp\frac{qE_G}{\gamma k}\left(\frac{1}{298,15} - \frac{1}{T}\right)\right] \tag{17}$$

I_D: the current of the diode which is proportional to the saturation current, it is given by the following equation:

$$I_D = I_S\left(e^{\left(\frac{V_{cell}+R_S I_{cell}}{V_T}\right)} - 1\right) \tag{18}$$

with:

- I_S: is the reverse saturation current in amperes (A)
- q: Electronic load (1.6 10-19C)
- k: Boltzmann constant (1.38 10–23 J / K)
- T: is a cell temperature in Kelvin (K)
- γ: is an ideal factor
- R_s: is a series resistance (Ω)

We replace in equation (15), the characteristic voltage-current equation of a solar cell is given as follows

$$I_{cell} = I_{ph} - I_S\left(e^{\left(\frac{V_{cell}+R_S I_{cell}}{V_T}\right)} - 1\right) - \frac{V_{cell} + R_s I_{cell}}{R_{sh}} \tag{19}$$

The photo current depends mainly on the sunlight and the operating temperature of the cell, which is described by the following equation:

$$I_{ph} = \left[I_{SC} + K_I(T - T_{ref})\right]\frac{G}{G_{ref}} \tag{20}$$

- I_{SC}: is the short circuit current of the cell at 25 °C and 1000W / m2,
- K_I: the temperature coefficient of the short circuit current of the cell,
- T_{ref}: is the reference temperature of the cell, in Kelvin (K) (= 25C ° + 273),
- G: is the insolation in watts / square meter (W / m2),
- G_{ref}: is the reference insolation of the cell (= 1000W / m2)

On the other hand, the saturation current of the cell varies with the cell temperature, which is described as follows:

$$I_s = I_{rs}\left(\frac{T}{T_{ref}}\right)^3 exp\left[\frac{qE_g\left(\frac{1}{T_{ref}} - \frac{1}{T}\right)}{\gamma k}\right] \tag{21}$$

- I_{rs}: the reverse saturation current of the cell at a reference temperature and solar radiation.

- E_g: is the gap energy of the semiconductor used in the cell in electron volt (eV)

$$I_{rs} = \frac{I_{sc}}{exp\left(\frac{qV_{OC}}{N_s\gamma kT}\right) - 1} \tag{22}$$

where

- N_s: Number of cells within a series of a PV module.
- V_{OC}: is the open-circuit voltage

A typical PV cell produces less than 2W at about 0.5V; cells must be connected in series parallel on a module to produce enough power. A PV array is a group of multiple PV modules that are electrically connected in series and parallel to generate the required current and voltage.

The terminal equation for the current and voltage of a PV module becomes:

$$I_{PV} = N_p I_{cell}$$

$$V_{PV} = N_s V_{cell}$$

5 Simulation Results Obtained

5.1 Fuel Cell Simulation Results

The PEM fuel cell model has been simulated by using Matlab software with the set of parameter values depicted in the following Table 1:

Table 1. Table parameter of fuel cell.

Parameters	Values	Parameters	Values
P_{H_2}, P_{O_2}	$2, 105P_a$	ψ	23
ξ_1	$-0, 922$	T	343 K
ξ_2	$0, 00312$	A	$100\ cm^2$
ξ_3	$-9, 92.10^{-5}$	J_{max}	$2A/cm^2$
ξ_4	$7, 4.10^{-5}$	$R_M + R_C$	$0, 4.10^{-4}\ \Omega$

The fuel cell V-I characteristic is shown in Fig. 2

The fuel cell is used in the linear zone “ohmic region”, where the fuel cell deliver a stable power and a maximum efficiency.

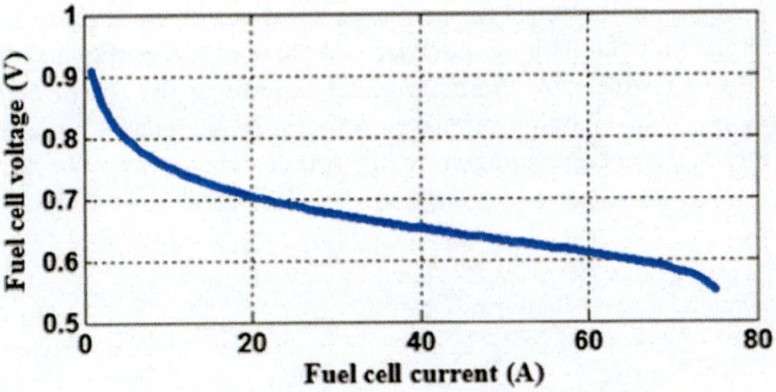

Fig. 2. The voltage fuel cell versus current fuel cell

5.2 PV Panel Simulation Results

To plot the PV and IV characteristics of the solar panel and to see the effect of the illumination on its output power, we used the electrical characteristics of a solar panel available in the laboratory. This module is composed of 8 cells in series (i.e. $N_p = 1$ and $N_S = 8$), it has respectively the electrical characteristics of P_{max}, I_{SC} and V_{OC} following: 200 W, 8,2 A and 33 V. The simulations are done under standard conditions of temperature (25°C) and illumination (1000 W/m^2).

Figure 3 shows the power-voltage characteristic (Fig. 3-a) and the current-voltage characteristic of the panel used (Fig. 3-b) .We notice that the maximum power is equal to 200 W and the voltage of open circuit is equal to 33 V) and the short-circuit current is equal to 8,2 A and the open circuit voltage is equal to 33 V.

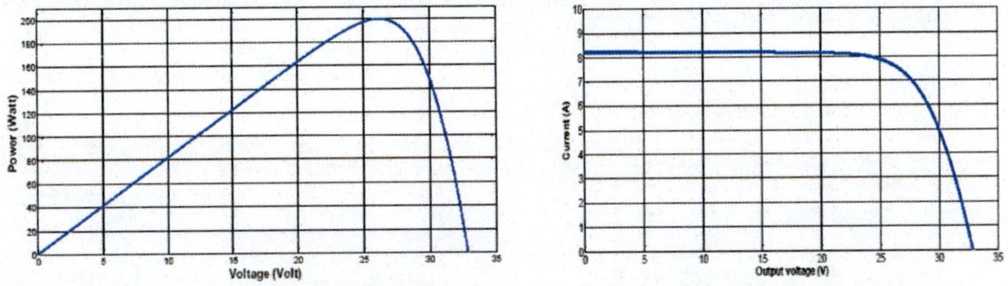

a- characteristic P-V panel

b- characteristic I-V panel

Fig. 3. Characteristics P-V and I-V panel

6 Conclusion

This article focuses on modeling of solar energy and fuel cell. In order to extend the lifetime of wireless sensors, we offer models for characterization of solar energy modeling and fuel cell. This characterization is based on mathematical equations to trace the PV power as a function of the voltage of a solar panel. Indeed, to ensure neutral operation.

This work opens the horizons to several perspectives at the level of optimization and energy storage for communicating sensors. At the energy recovery and optimization level, the aim is to characterize vibratory energy to power the sensor and free itself from the temporary unavailability of energy sources. At the energy storage level, it is proposed to model the energy storage in multi-source. Then, it is a question of model the consumption of the sensor to converge towards its neutral operating point and thus prolong its life.

References

1. Atzori, L., Iera, A., Morabito, G.: The internet of things : A survey. Computer Networks (2010)
2. Mainetti, L., Patrono, L., Vilei, A.: Evolution of wireless sensor networks towards the internet of things: a survey. In: International Conference on Software, Telecommunications and Computer Networks (SoftCOM) (2011)
3. Christin, D., Reinhardt, A., Mogre, P.S., Steinmetz, R.: Wireless sensor networks and the internet of things : selected challenges. In: Proceedings of GI/ITG KuVS Fachgespräch Drahtlose Sensornetze (2009)
4. Haji, L.M., Ahmad, O.M., Zeebaree, S.R., Dino, H.I., Zebari, R.R., Shukur, H.M.: Impact of cloud computing and internet of things on the future internet. Tech. Rep. Kansai Univ. **62**, 2179–2190 (2020)
5. Sadeeq, M.M., Abdulkareem, N.M., Zeebaree, S.R., Ahmed, D.M., Sami, A.S., Zebari, R.R.: IoT and cloud computing issues, challenges and opportunities: a review. Qubahan Academic Journal **1**, 1–7 (2021)
6. Zeebaree, S.S.R., Ameen, S., Sadeeq, M.: Social media networks security threats, risks and recommendation : a case study in the Kurdistan region. International Journal of Innovation, Creativity and Change **13**, pp. 349–365 (2020)
7. Correa, J.M., Farel, F.A., Canha, L.N., Simoes, M.G.: An electrochemical based fuel cell model suitable for electrical engineering automation approach. IEEE Transactions on Industrial Electronics **51**, pp. 1103–1112 (2011)

Monitoring Energy Consumption of Android Apps with AppsDrain

Ayyoub El Outmani[1](✉), El Miloud Jaara[1], and Mostafa Azizi[2]

[1] Faculty of Sciences (FSO), Computer Science Research Laboratory (LARI), Oujda, Morocco
ayyoub.eloutmani@gmail.com, e.jaara@ump.ac.ma
[2] MATSI Research Lab, ESTO University Mohammed First (UMP), Oujda, Morocco
azizi.mos@ump.ac.ma

Abstract. Despite the rapid and exponential evolution of innovative applications for mobile devices, their batteries still suffer from a limited capacity that cannot keep pace with new and increasingly resource-intensive apps. The gap between development rates of batteries and chips strongly requires optimizing the energy efficiency in order to meet the demands for reduced energy consumption. Therefore, we need to monitor and analyse the energy efficiency of these devices. For this purpose, we have developed an Android application, called AppsDrain. It provides a detailed analysis of battery usage, specifying the energy consumption of each component. This app could assist developers to identify instantly which apps are the sources of battery drain. In some cases, power leaks are due to poor software design, junk code running in the background, or complex code consuming a high number of CPU cycles. To validate our app, we propose to explore the energy consumption of some applications while making a comparison with their respective complexities. Among different applications that aroused our interest, we consider here most known sorting algorithms. The obtained results of our experiment confirm that more complexity implies more energy consumption.

Keywords: AppsDrain · Energy consumption · Energy efficiency · Battery drain · Energy profilers · Mobile device · Monitoring energy · Android

1 Introduction

According to statistics from 2021, there are approximately 6.25 billion smartphones in use worldwide [1]. This reflects how essential these devices have become in our daily lives. They are deployed in a variety of tasks and activities such as browsing on the web, playing games, shooting and editing videos, etc. Moreover, users prefer to continue their activities on the go. Meanwhile, these mobile devices are powered by batteries with limited capacity. Up to now, the only way to heighten the battery capacity is to increase its size, which is unfortunately conditioned by the device's dimensions. This is why researchers are constantly working on ways to improve it and avoid frequently recharging it, especially to satisfy the increasing energy requirements of modern applications [2]. Some researches focus on the hardware side as they propose a novel method for dynamic

Y. Farhaoui et al. (Eds.): ICAISE 2022, LNNS 635, pp. 635–641, 2023.
https://doi.org/10.1007/978-3-031-26254-8_92

voltage and frequency scaling (DVFS) on CPU, GPU, and RAM in a mobile equipped with multi-processor systems-on-chip (MPSoCs), with 26% more energy efficient [3]. Now-a-days, developers need to look at energy optimisation during the development process. About 18% of Android apps available on the Google Play store have energy consumption issues [4]. With millions of applications in the Play Store, energy efficiency has become as important as security. Android developers have realized that reviews left by end users on the store can have a negative effect on their Google ranking. In this sense, we design an app, named AppsDrain, to use over Android to monitor the energy consumption of each application with more details. Developers can use it to identify battery draining components in order to optimize their codes. We organize the rest of this paper as follows: we deal with some existing methodologies and related works in Sect. 2. Section 3 explains the power measurements. We describe our app in Sect. 4. Then, we present the results of a case study over some sorting algorithms in Sect. 5.

2 Related Works

There are several approaches that deal with energy measurements in mobile devices. We divide them into hardware and software approaches;

In hardware-based approaches, obtaining direct measurements either using external hardware connected to the battery interface of a device as Monsoon Power Monitor [5], or internal instrument such as NEP [6] (Nokia Energy profiler). In direct measurements, voltage and current are periodically measured and the power is calculated as the result of multiplying the two values, while total energy is calculated by integrating the power over the duration of execution. On the other hand, software-based approaches obtain the measurement indirectly with the correlates power consumption with hardware performance counters using an energy consumption model. McIntosh et al. [7] present an empirical study using a tool called GreenMiner for different machine learning algorithms that developers can use as a guide. They need to be aware of the trade-offs between these algorithms in terms of energy and accuracy. IBtryMntr is a tool developed by Gokhale et al. [8], it shows that Wi-Fi uses less power than LTE. Zou et al. [9] achieve similar results using an Arduino board, where Wi-Fi was 54% more efficient. They also find that the power consumption increases when the video quality also increases. In Android, indirect measurements are used to estimate the power consumption of various components using a utilization model, while BatteryStats tracks the power consumption values for individual hardware components [5].

3 Power Measurement

In this section, we are interested in the process of obtaining power measurements, such as the current draws with external hardware (oscilloscope). Hardware-level estimation is simple to implement and provides real-time results. However, during the execution of an application, it is not possible to know how energy is consumed inside the application and which components consume energy the most. The indirect measurement correlates the power consumption with hardware performance counters using an energy consumption model. Building an energy consumption model is a complicated task and has several levels (modelling method, generating the power models, determining the coefficients of the model, etc…). The authors in [4] and [5] categorize power models into three types:

Models based on utilization: the power consumption is correlated with the usage of hardware components. For an application that uses multiple components, the power model includes all of the active subcomponents while the application is running [5]. Android uses this method to estimate the power consumption of various applications and components based on its own energy models, while BatteryStats tracks the power consumption values for. Models based on instruction: An instruction-based model analyses the program code to be run. In this process, static analysis of the code is used to correlate energy consumption with APIs or instructions in the application. Bergel et al. [11] propose a code profiler called Pharo to measure the power consumption of any code expression. Models based on system calls: System call-based models correlate energy consumption with the number of system calls triggered by an application. However, utilization-based and instruction-based models cannot capture non-linear characteristics of energy consumption, such as tail-energy (Several components, e.g. Wi-Fi, 4G, in smartphones exhibit the tail power behaviour, where activities in one entity, e.g., a routine, can trigger a component to enter a high-power state and stay in that power state long beyond the end of the routine) [12]. Table 1 summarises the comparison between the models:

Table 1. Comparison of power models

Model type	Measures	Accuracy	Overhead
Utilization	On-device	++	++
System calls	On-device	+++	+
Instruction	Off-device	+	+++

To sum up, the instruction-based Model is rarely used because it is difficult to know what the energy consumption will be until the program code is run in a real device (On-device). For example, poor wireless connection quality can cause a file to take longer to transmit and use more energy, in this case it would be better to use a usage-based model that measures the number of bits sent over time [5]. The advantage is of not requiring external equipment to measure energy consumption or running the software on a real machine and measurements are made Off-device. The utilization-based model is great for capturing linear relationships between resources and energy consumed by the hardware being modelled. In contrast, few smartphone components have nonlinear characteristics in terms of energy consumption [13]; the case of tail-energy, for instance. The system call model can be adopted to characterize the asynchronous behaviour of the energy, using the Eprof tool, Pathak et al. presented a system-call-based approach to improve the accuracy of application energy estimation [12]. Models based on system calls provide a quantitative measure of how applications access hardware in order to overcome the limitations of utilization-based models [4], and the use of these metrics helps build energy consumption models that are more accurate [4, 5, 13]. The utilization based approaches could have an error rate as high as about 20% [14]. Finally, among these three models, the most accurate model is the one based on system calls but the latter requires an overhead.

4 AppsDrain

Sometimes, the app seems to work fine, however the battery drains quickly and the cause of this problem is unclear, this can be considered as an energy bug. Due to poor design, many energy leaks can be responsible for this drain. In this sense, we have developed an Android application called AppsDrain that allows developers to track the amount of power consumed by each component. Since 2014 (version 5.0 Lollipop), Android operating system has provided tools called BatteryStats and Battery Historian to collect information about the energy consumption of Android devices and to support the visualization and analysis of these measurements [10].This tool tracks hardware components such as the Screen, CPU, Wi-Fi, cellular data, Bluetooth, GSM radio scanning, GPS, etc. Using timestamps, BatteryStats estimates the amount of energy consumed by the various hardware [5]. It does not directly measure battery energy consumption. Rather, BatteryStats estimates the energy consumption based on the timestamps of various hardware components. In BatteryStats, statistics are collected in two ways: either different services report component status changes to it, or it periodically collects information from the process system files about the CPU and other components [5]. Many other power profilers, including GreenMiner [10], PowerBooter and AppScope [5], retrieve these statistics directly from the BatteryStats, rather than from proc system files.

Our Android app, is based on information provided by Android through BatteryStats, which provides fine-grained estimations in real time. AppsDrain extracts data periodically from the system's files: BatteryStats, ProcStats and PS (see Fig. 1). The BatteryStats class provides the statistics in mAh (milliAmpere hour) of energy consumed by the different components by UID.

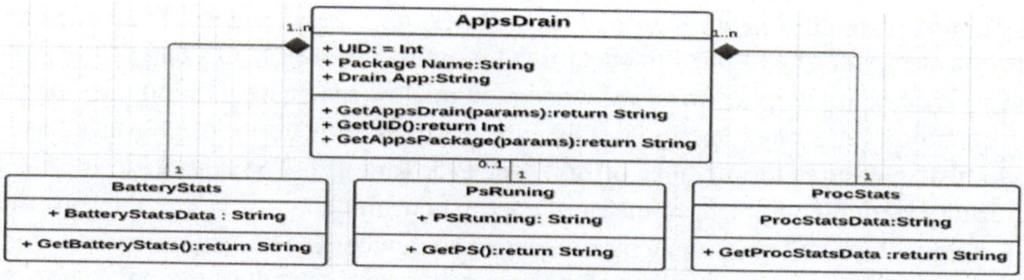

Fig. 1. The class diagram of AppsDrain.

These statistics will be used later with the ProcStats class to retrieve the names of the applications running on devices. The AppsDrain class is in composition relationship with the two classes to provide the desired result by associating to each UID its consumed energy quantity. The PS class is used to display the currently running processes.

Figure 2 shows the AppsDrain use case diagram which represents the different types of users. A detailed view of each application's energy consumption is displayed in the application, as well as the total amount of energy consumed since the battery was last recharged and its maximum capacity. Our application also shows information on the amount of energy consumed by the device's screen and standby mode.

Figure 3 shows a screenshot of our application. End-users can monitor their application to determine how much energy is being consumed and which components are consuming energy the most. For example, the Facebook application consumed more CPU power than the Wi-Fi and speaker combined.

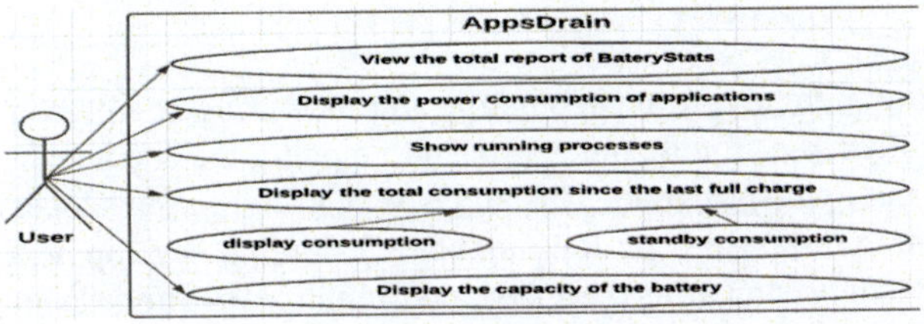

Fig. 2. The use case diagram of AppsDrain

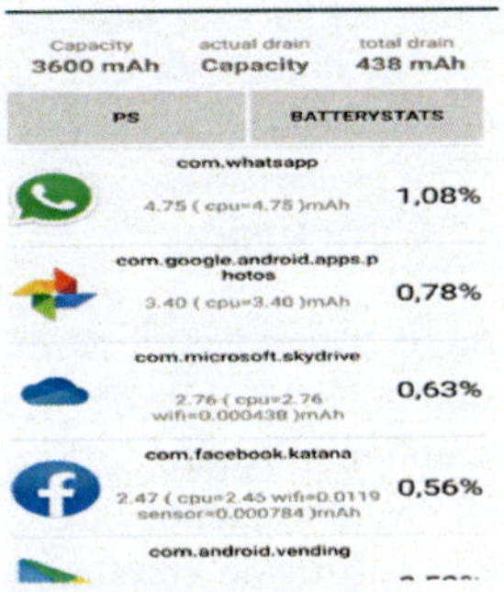

Fig. 3. Screenshot of AppsDrain

5 Case Study: Sorting Algorithms

In terms of their energy consumption, we compare different sorting algorithms. Table 2 lists all algorithms used in our study. Using AppsDrain, we measure and plot the energy consumption of different sorting algorithms (See Fig. 4). The algorithms with a complexity of (n^2) are at the top of the graph, although the graphs of nlog(n) appear identical while, so we changed the scale in Fig. 4 to make it clearer.

Table 2. The measurements of each of the sorting algorithms with several runs in mAh

Sort Algorithm	Complexity	1st Run in mAh	2nd	3rd	4th	5th	Avg
Bubble	O (n^2)	3,85	4,1	3,21	3,8	3,47	3,713
Selection	O (n^2)	1,22	1,6	1,52	1,3	1,2	1,374
Insertion	O (n^2)	1,13	1,14	1,1	1,16	1,09	1,111
Merge	O (n,log(n))	0,031	0,039	0,033	0,038	0,035	0,035
Quick	O (n,log(n))	0,031	0,033	0,035	0,033	0,037	0,033
Heap	O (n,log(n))	0,042	0,041	0,046	0,04	0,04	0,045

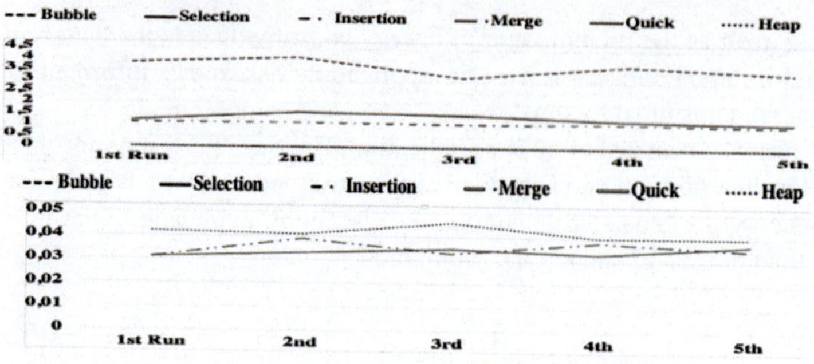

Fig. 4. The energy consumption graph.

6 Conclusion

In this paper, we present a literature review on estimating and measuring energy consumption in mobile devices; then, we propose our approach to monitor the energy consumption in Android, using our app AppsDrain. This app is very useful for developers who want to identify the source responsible of draining batteries during development. We also provide a preliminary study comparing the energy consumption of different sorting algorithms. Through the obtained results, we confirm that more complexity requires more energy consumption. In our future work, we want to investigate the energy impact of machine learning algorithms on smartphones.

References

1. (2021) Android Statistics. In: Business of Apps (2022). https://www.businessofapps.com/data/android-statistics/. Accessed 24 Aug 2022
2. Hamzaoui, K.I., Berrajaa, M., Azizi, M., Lipari, G., Boulet, P.: An experimental methodology for modeling the energy consumption of mobile devices. In: 2017 First International Conference on Embedded & Distributed Systems (EDiS). IEEE, Oran, pp. 1–6 (2017)
3. Dey, S., Isuwa, S., Saha, S., Singh, A.K., McDonald-Maier, K.: CPU-GPU-memory DVFS for power-efficient MPSoC in mobile cyber physical systems. Future Internet **14**(3), 91 (2022)
4. Zhu, C., Zhu, Z., Xie, Y., Jiang, W., Zhang, G.: Evaluation of machine learning approaches for android energy bugs detection with revision commits. IEEE Access **7**, 85241–85252 (2019). https://doi.org/10.1109/ACCESS.2019.2925350
5. Hoque, M.A., Siekkinen, M., Khan, K.N., Xiao, Y., Tarkoma, S.: Modeling, profiling, and debugging the energy consumption of mobile devices. ACM Comput. Surv. **48**, 1–40 (2016)
6. Fowdur, T.P., Hurbungs, V., Beeharry, Y.: Statistical analysis of energy consumption of mobile phones for web-based applications in Mauritius. In: 2016 International Conference on Computer Communication and Informatics (ICCCI). IEEE, Coimbatore, India, pp. 1–8 (2016)
7. McIntosh, A., Hassan, S., Hindle, A.: What can Android mobile app developers do about the energy consumption of machine learning? Empir. Softw. Eng. **24**(2), 562–601 (2018). https://doi.org/10.1007/s10664-018-9629-2
8. Gokhale, A., Wang, W,, Deuskar, G.: iBtryMntr: experimental study of power consumption characteristics for wireless multimedia communication apps in iOS devices. In: Proceedings of the 11th EAI International Conference on Mobile Multimedia Communications. EAI, Qingdao, People's Republic of China (2018)

9. Zou, L., Javed, A., Muntean, G.-M.: Smart mobile device power consumption measurement for video streaming in wireless environments: WiFi vs. LTE. In: 2017 IEEE International Symposium on Broadband Multimedia Systems and Broadcasting (BMSB). IEEE, Cagliari, Italy, pp. 1–6 (2017)
10. Pereira, R.: GreenHub: A Large-Scale Collaborative Dataset to Battery Consumption Analysis of Android Devices **26**(3), 155 (2021)
11. Bergel, A.: Power and energy code profiling in pharo. In: Proceedings of the 11th edition of the International Workshop on Smalltalk Technologies. ACM, Prague Czech Republic, pp. 1–7 (2016)
12. Pathak, A., Hu, Y.C., Zhang, M.: Where is the energy spent inside my app?: fine grained energy accounting on smartphones with Eprof. In: Proceedings of the 7th ACM european conference on Computer Systems - EuroSys '12. ACM Press, Bern, Switzerland, p. 29 (2012)
13. Pathak, A., Hu, Y.C., Zhang, M., Bahl, P., Wang, Y.-M.: Fine-grained power modeling for smartphones using system call tracing. In: Proceedings of the Sixth Conference on Computer systems - EuroSys '11. ACM Press, Salzburg, Austria, p. 153 (2011)
14. Gao, X., Liu, D., Liu, D., Wang, H., Stavrou, A.: E-Android: a new energy profiling tool for smartphones. In: 2017 IEEE 37th International Conference on Distributed Computing Systems (ICDCS). IEEE, Atlanta, GA, USA, pp. 492–502 (2017)

Multi-task Offloading to a MEC Server with Energy and Delay Constraint

Nouhaila Moussammi[1(✉)], Mohamed El Ghmary[2], and Abdellah Idrissi[1]

[1] Department of Computer Science, Faculty of Sciences, Mohammed V University in Rabat, Rabat, Morocco
{nouhaila_moussammi,a.idrissi}@um5.ac.ma

[2] Department of Computer Science, FSDM, Sidi Mohamed Ben Abdellah University, Fez, Morocco
mohamed.elghmary@usmba.ac.ma

Abstract. Nowadays, due to the large number of connected smartphones, mobile wireless communications have increased exponentially. Mobile Edge computing allows mobile devices to offload computing-intensive tasks to the Edge, which minimizes battery consumption and time delay. Edge computing is restricted by radio transmission power, bandwidth, and other limitations. The power of transmission and local CPU frequencies of mobile devices are considered constants in the majority of models. Additionally, the amount of energy used is in a positive relationship with these parameters. The values can be adjusted to accommodate multi-tasking in order to decrease the energy consumed. If several tasks have high computing density, then the methods used to offload aren't appropriate. This paper suggests an efficient single-user model that has multiple tasks and high-density computing, and presents an Offloading Decision and Energy Minimization, validated through simulation, the results achieved in terms of energy are extremely encouraging. It can improve the effectiveness of offloading and energy consumption.

Keywords: Offloading · Mobile edge computing · Energy consumption

1 Introduction

With the rapid expansion of mobile apps [1] and the IoT, there is a rising need for high-performance, secure, scalable, low latency platforms that can process all data produced at the edge. Mobile devices are limited in their resources. They could experience slow performance due to power and battery limits [10]. Mobile edge computing [2,3], which is a technology that allows computing-intensive tasks to be transferred to an edge server may assist in this case. This method allows applications to operate near their users and greatly minimizes the time needed to finish the application [9]. A lot of studies are now focusing on the single-task, multi-user type such as [11]. The power of transmission and the CPU's frequency are constant. They also considered how to allocate bandwidth

Y. Farhaoui et al. (Eds.): ICAISE 2022, LNNS 635, pp. 642–648, 2023.
https://doi.org/10.1007/978-3-031-26254-8_93

efficiently for efficient energy usage [4]. Many papers have studied the allocation of resources in an MEC infrastructure to minimize processing time [5,6]. In [7] the user has the option to delegate all or a part of any task that is dependent on the power of its transmitter and noise. The task may be divided into several time slots so that it is completed in the [8,9]. To reduce the consumption of energy, the user must alter the power of transmission and the local CPU's frequency in line with their workload [8]. The program has to be divided into smaller tasks that can be accomplished in its own manner. It is difficult to decide what tasks must be transferred and completed before the deadline to save energy. The paper uses the model of single-user multi-tasks, numerous independent tasks need to be completed before the deadline, either locally or via the server. The current efficient energy utilization model picks high-density tasks that can be offloaded. This paper addresses these issues and suggests a solution to decrease the energy consumption of mobile devices in the MEC environment. We also suggest the OFDEM algorithm that implements the minimization technique. It calculates various options for allocation and chooses the most efficient in the execution of the tasks. Simulation is utilized to test the effectiveness of OFDEM. The model includes both processes in the Mobile device and on a MEC Server. We design and modify experiments by using clearly-defined simulation software for scenarios using a MEC server and mobile device. This paper is organized as: the 1st section is the introduction and other research, in the 2nd section explains the system model and its formulation. Section 3 outlines the proposed algorithm, and a numerical simulation is presented in Sect. 4. Section 5 closes the paper.

2 System Model and Problem Formulation

This paper explores a mobile device and a multi-tasking model as illustrated in Fig. 1. The model splits code into several tasks which can be performed locally, or transferred to the Edge server.

A mobile device was supposed to contain N independent tasks, $N = N_0 + N_1$, N_0 is the number of tasks which will be executed locally and N_1 is the number of tasks which will be executed on the server, which is designated by $T = \{T_1 ... T_N\}$. A task can be described as tuple (D_i, C_i, d), which D_i is the size of the input

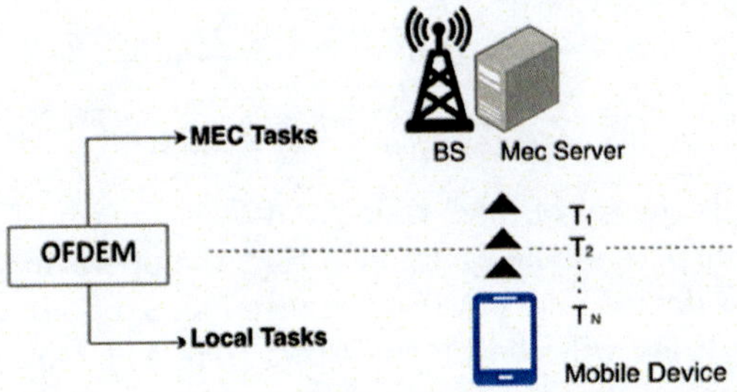

Fig. 1. Model of edge computing system

data. C_i is the amount of CPU cycles required to complete the tasks. The values can be determined by analyzing specific scenarios for task execution [10,11], the deadline for the task is known as d. It's used to determine if the task is either regular (d = 0), or non regular (d > 0). According to Shannon's equation [12] the transmission rate can be calculated as follows: $R(p) = Wlog(1 + kp)$, which W is the bandwidth of the upstream and $k = \frac{h^2}{\sigma}$ as a constant, h is the gain of the channel, p is the rate at which the user has to transmit in order to process input data onto an Edge server close to them, and σ is the background noise. When task i runs locally, the time required to process it is $t_i^{loc} = \frac{C_i}{F^{loc}}$ where F^{loc} is the CPU frequency of the device. Moreover, every CPU cycle's energy consumption can be described as $\epsilon = \omega F^{loc^2}$ (where ω is the parameter that was calculated in accordance with the architecture of the CPU chip). The formula (1) describes the total energy consumption for local task execution:

$$E_{tot}^{loc} = \sum_{i=1}^{N_0} E_i^{loc} = \sum_{i=1}^{N_0} \omega F^{loc^2} C_i \tag{1}$$

Three parts must be added to the offloading time t_i^{off} of task i which runs on the MEC Server namely $t_i^{up} = \frac{D_i}{R(p)}$ is the time of data transmission, $t_i^{exe} = \frac{C_i}{F^o}$ is the task execution time on the server (Where F^o is the server frequency), and t_i^{rep} is the time to resend the results of the calculation to a user. Generally, this time is ignored. It conforms to the considerations mentioned in [13]. Moreover, the total energy consumption can be expressed as follows:

$$E_{tot}^{up} = \sum_{i=1}^{N_1} E_i^{up} = \sum_{i=1}^{N_1} t_i^{up} * p = \sum_{i=1}^{N_1} \frac{D_i}{R(p)} * p \tag{2}$$

The following limitations are used to construct the system model in this study:

(P1) $\min_{\beta_1,\beta_2,p,F^{loc}} (\sum_{i\in\beta_1} \omega F^{loc^2} C_i + \sum_{i\in\beta_2} p * \frac{D_i}{R(p)})$

s.t

(C1) $0 < p < P^{max}$

(C2) $\sum_{i\in\beta_1} \frac{C_i}{F^{loc}} < d$

(C3) $\sum_{i\in\beta_2} \frac{D_i}{R(p)} + \frac{C_i}{F^o} < \sum_{i\in\beta_2} \frac{C_i}{F^{loc}}$

(C4) $\sum_{i\in\beta_2} \frac{D_i}{R(p)} * p + \sum_{i\in\beta_1} \omega F^{loc^2} C_i < E^{max}$

In this part, β_1 is the set of local tasks, and β_2 is the set of tasks is executed at the Edge. The purpose of the function is that the operating frequency of the chip on the mobile device and power of transmission be altered. If there is a minimal delay the Edge will choose the tasks that will require less execution time. While (C1) means that the transmit power must not exceed its maximum or fall below. The constraints (C2) and (C3) emphasize that tasks executed locally or on the Edge server must be completed before the deadline as well as when the computation offload time takes less time than local execution of a task i, and (C4) means that the energy consumption on the part of the mobile device and the Edge server is lower than a maximum energy consumption E^{max}.

3 Proposed Algorithm

In this section, we will discuss the OFDEM solution for the optimization issue. It has complexity $o(n^2)$, and allows to process of data faster. The algorithm1 shows the steps of this solution.

Algorithm 1 OFDEM : Offloading Decision and Energy Minimization

Input: $C_i \in \beta_1, D_i \in \beta_1, C_i \in \beta_2, D_i \in \beta_2$
Output: Offloading Decision
  Step 1 : Collection of data sets of mobile device and MEC Server and initialization
while true **do**
  Step 2 : Call **Algorithm 2**
  Step 3 : Tasks Status (EXECUTED or FAILED)
  if (All tasks are finished) **then**
    Break;
  end if
end while

Algorithm 2 Task Allocation

Input: $C_i \in \beta_1, D_i \in \beta_1, C_i \in \beta_2, D_i \in \beta_2, F^o, F^{loc}$
Output: $\beta_{1\_res}, \beta_{2\_res}$
for $T_i \in$ (list of non regular tasks) **do**
  for $C_i \in \beta_1$ **and** $D_i \in \beta_1$ **do**
    Calculate Mobile device execution time
  end for
  for $C_i \in \beta_2$ **and** $D_i \in \beta_2$ **do**
    Calculate the offloading time
  end for
end for
for $T_i \in$ (list of regular tasks) **do**
  for $C_i \in \beta_1$ **and** $D_i \in \beta_1$ **do**
    Calculate the Energy consumed for local execution
    Calculate the local execution time
  end for
  for $C_i \in \beta_2$ **and** $D_i \in \beta_2$ **do**
    Calculate the energy consumed by the mobile device E^{up}
    Calculate the offloading time
  end for
end for

Step 1 Collect data from mobile devices and MEC server. its calculates the operating frequency of CPUs on mobile devices and MEC server. Its complexity is y $o(n)$. Step 2 is a call to algorithm 2 that determines the task allocation. Tasks are classified first into regular and non-regular tasks. The time it takes to process tasks and the energy consumed is determined for each task. The MEC server calculates the CPUs and the energy used for data transmissions. The CPU that has the shortest total time used is allocated non-regular tasks. The non-regular tasks are first assigned because they're more urgent than regular tasks. Regular tasks shouldn't affect the efficiency of the system. it's has a complexity $o(n^2)$. Step 3: The tasks are monitored to determine if they're still in progress or completed. The cancellation of tasks is not an ideal result, however, it is an important measure to be monitored to assess the effectiveness of the algorithm proposed. And this step has a complexity $o(n)$.

4 Implementation and Evaluation

In this part, OFDEM was implemented within Eclipse IDE version 2022-03 (4.23.0). Table 1 lists three different application were defined [14].

Table 1. Simulation parameters

Parameter	Application 1	Application 2	Application 3
Task generation rate (seconds)	10	10	0,1
Computational load (millions of CPU cycles)	20	100	200
Data entry size (MB)	36,3	30,3	4
Non-regular Tasks (% from total tasks)	10	40	60
Deadline for non-regular tasks (ms)	500	400	100

The simulation considered as bandwidth $W = 20\,\text{Mhz}$, channel gain $h = 10^{-28}$, noise $\sigma = 10^{-9}$, the maximum transmitting power $P^{max} = 3.65\text{W}$, the server Edge has 20 CPUs. Each CPU has different operating frequencies. The scenarios chosen were built on 1000 tasks.

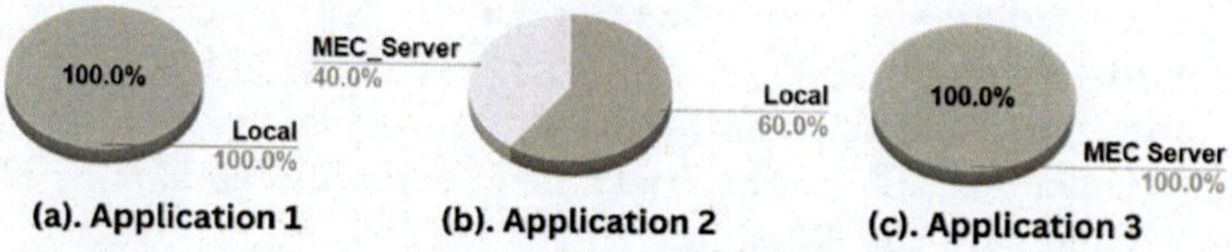

Fig. 2. Results of offloading tasks in Applications 1, 2 and 3.

Figure 2 illustrates the 3 scenarios for each app, in 2.(a) due to the small processing capacity of App1, all tasks are performed locally. In Fig. 2.(b) 60%

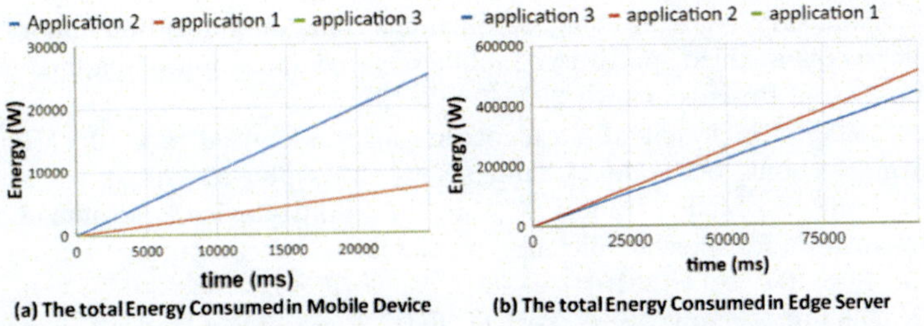

Fig. 3. Results of Energy consumed in Applications 1, 2 and 3.

of the tasks are executed locally, and 40% are transferred to the MEC Server. In Fig. 2(c) due to the volume of work the app3 is extremely high, there is no allocation made on the mobile device. The short deadlines caused the tasks to be canceled due to the fact that any setting could complete the task in an acceptable time. Figure 3.(a) shows different levels of energy consumption, it can be explained by the fact that all tasks were executed locally which can justify the high consumption concerning app2, while 60% of tasks were executed in local for app1 and none were executed in local for app3. The same reasoning can be used for Fig. 3.(b). The difference between (a) and (b) is the edge layer requires more energy than mobile devices, while the energy transfer is always equal to 0 in local.

5 Conclusion

The results allowed us to observe the behavior of the system in relation to the allocation of tasks. By adjusting the energy coefficients and the parameters utilized by the applications of the OFDEM algorithm, it was capable of reducing energy consumption. The obtained results show the effectiveness of our solution. We're planning to expand our research to include multi-user scenarios and multi-server which will reduce execution time.

References

1. Simiscuka, A.A., Bezbradica, M., Muntean, G.M.: Performance analysis of the quality of service-aware networking scheme for smart internet of things gateways. In: 13th IWCMC, pp. 1370-1374. IEEE (2017)
2. Al-Shuwaili, A., Simeone, O.: Energy-efficient resource allocation for MEC based augmented reality applications. IEEE Wirel. Commun. Lett. **6**, 398–401 (2017)
3. Chang, H., Hari, A., Mukherjee, S., Lakshman, T.V.: Bringing the cloud to the edge. In: IEEE Conference on Computer Communications Workshops, pp. 346–351 (2014)
4. Chen, M. H., Liang, B., Dong, M.: Joint offloading and resource allocation for computation and communication in mobile cloud with computing access point. In: IEEE INFOCOM Conference on Computer Communications, pp. 1–9 (2017)

5. Wu, Y., Qian, L.P., Ni, K., Zhang, C., Shen, X.: Delay-minimization nonorthogonal multiple access enabled multi-user mobile edge computation offloading. IEEE J. Sel. Top. Signal Process. **13**(3), 392–407 (2019)
6. Jošilo, S., Dán, G.: Decentralized algorithm for randomized task allocation in fog computing systems. IEEE/ACM Trans. Netw. **27**(1), 85–97 (2018)
7. Liu, K., Peng, J., Li, H., Zhang, X., Liu, W.: Multi-device task offloading with time-constraints for energy efficiency in mobile cloud computing. Future Gener. Comput. Syst. **64**, 1–14 (2016)
8. Mao, Y., Zhang, J., Song, S.H., Letaief, K.B.: Power-delay tradeoff in multi-user MEC systems. In: IEEE Global Communications Conference, pp. 1–6 (2016)
9. El Ghmary, M., Chanyour, T., Hmimz, Y., Cherkaoui Malki, M.O.: Processing time and computing resources optimization in a mobile edge computing node. In: Bhateja, V., Satapathy, S.C., Satori, H. (eds.) Embedded Systems and Artificial Intelligence. AISC, vol. 1076, pp. 99–108. Springer, Singapore (2020). https://doi.org/10.1007/978-981-15-0947-6_10
10. El Ghmary, M., Hmimz, Y., Chanyour, T., Malki, M.O.C.: Time and resource constrained offloading with multi-task in a mobile edge computing node. Int. J. Elect. Comput. Eng. (2088-8708) **10**(4), 1–10 (2020)
11. Huang, L., Feng, X., Zhang, L., Qian, L., Wu, Y.: Multi-server multi-user multi-task computation offloading for MEC networks. Sensors **19**(6), 1446 (2019)
12. Yu, H., Wang, Q., Guo, S.: Energy-efficient task offloading and resource scheduling for MEC. In: International Conference on Networking Architecture and Storage, pp. 1–4 (2018)
13. Lin, X., Zhang, H., Ji, H., Leung, V.C.: Joint computation and communication resource allocation in mobile-edge cloud computing networks. In: IEEE International Conference on Network Infrastructure and Digital Content, pp. 166–171 (2016)
14. Jansson, J.: Collision avoidance theory: with application to automotive collision mitigation. Doctoral dissertation, Linköping University Electronic Press (2005)

Multivariate Time Series Forecasting Using Recurrent Neural Network for a Complex System

Rida El Abassi[1(✉)], Jaafar Idrais[1], Yassine El Moudene[1], and Abderrahim Sabour[1,2]

[1] IMI Laboratory, Department of Mathematics, Ibn Zohr University, Agadir, Morocco
rida.elabassi@edu.uiz.ac.ma

[2] Department of Computer Science, Ibn Zohr University, Agadir, Morocco

Abstract. A complex system is a combination of interacting parameters, having the capacity to generate a new type of collective behavior through self-organization. This work focuses on the automatic processing of non-preprocessed multivariate time series obtained from a complex system, in particular the critical cases process of the SARS-Cov2 pandemic in Morocco. We have established a multivariate neural network which the target variable is the critical cases process, the system uses three input parameters, the active cases, the dead cases and the critical cases. We adopt the average error percentage (MAPE) to measure the performance of the model for which we found MAPE = 14.61% as the best performance depending on the tests performed.

Keywords: Time series · Automatic learning · Recurrent neural network

1 Introduction

A complex system is a combination of a multitude of interacting parameters, able to generate a new type of collective behavior through self-organization, including the spontaneous formation of temporal, spatial or functional structures. These interactions are not only linear, but also generate chaotic, resilient and critical behavior. The knowledge of the law of evolution allows to predict the behavior of the phenomenon at long times, indeed the law of evolution is only the rule which gives the state at the time $t + \delta t$ knowing the state at the time t.

The new SARS-COV-2 coronavirus identified in the city of Wuhan, China, in December 2019 [1], the first death reported on January 10, 2020, it has become a global pandemic [2], named COVID-19 declared by the organization World Health Organization [3]. COVID-19 has not been effectively suppressed due to the extreme irregularity of the primary stage of the epidemic and the limited understanding of the novel coronavirus by the medical community [4]. When a large-scale epidemic infectious disease emerges and a significant public health emergency is triggered, people use epidemic models to analyze and predict the trend of disease development and use the analysis results to guide disease development.

Y. Farhaoui et al. (Eds.): ICAISE 2022, LNNS 635, pp. 649–654, 2023.
https://doi.org/10.1007/978-3-031-26254-8_94

Time series analysis and forecasting is an important area where an appropriate model [5] describing the underlying relationship of the process. The model is then used to extrapolate the time series into the future. ARIMA models (traditional approach) [6] is based on Wold's representation theorem [7].

For this study we propose the neural network [8, 9, 10, 11] approach to predict saturation of intensive care beds using critical cases process. First, we have presented the data on which the study is based and then the working methodology we propose. Secondly, we will establish the results of this study and its interpretations and then finally a general conclusion which will emphasize the limits of this study and the interpretations that we can deduct from it.

2 Methodology

2.1 Data Description

The database used from September 03, 2020, until June 25, 2022, is taken daily from the official website of the Moroccan Ministry of Health, it represents respectively in Figs. 1, 2 and 3, the number of active cases, the number of deaths and the number of critical cases. These are official data that we have since the Ministry of Health has started to deposit these data on its site from September 3, 2020.

It is necessary to consider the number of tests performed daily, which has increased to reach in the period after delivery, this is one of the indicators that explains the upward trend that characterizes signals in Figs. 1, 2 and 3, reproduced by (Moroccan Ministry of Health). It should also be noted that the number of positive cases of the virus, precisely on Monday, this is due to the fact that the number of tests performed for this day is almost lower than the average performed during the week.

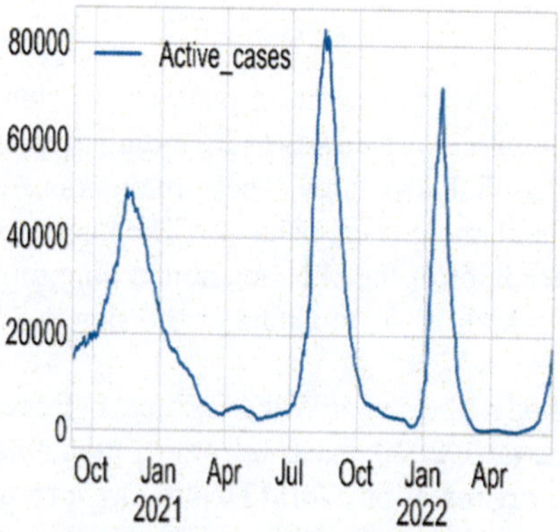

Fig. 1. Evolution of active cases

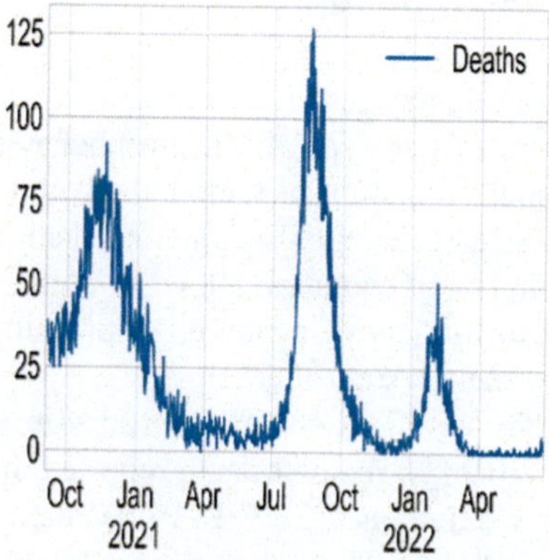

Fig. 2. Death's process

2.2 Processing Strategy

The first questions at the origin of this study were mainly related to the capacity of RNNs to model complex systems. We consider that the phenomenon of SARS-Cov2

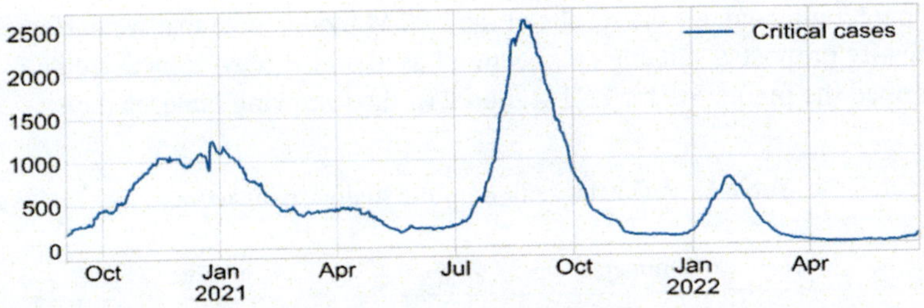

Fig. 3. Critical cases process

propagation is a complex system in which several parameters interact. The goal is to establish a model to predict the critical case process from two variables, namely, the active and dead case processes, therefore, we implement a multivariate Long short-term memory (LSTM) model where the target variable is the daily critical case process (Fig. 4).

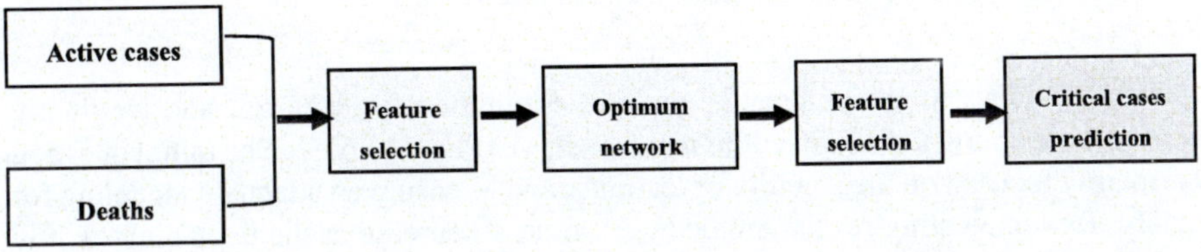

Fig. 4. An AI method for predicting critical cases combining active cases and deaths process.

It is important to note that the learning process will be carried out by a part of the signal noted (Train), which represents nearly 80% of the size of the basic signal covers the period from September 03, 2020, until February 13, 2022. However, the 20% will represent the part (Test), which will allow us to measure the performance of the learning carried out via the MAPE measure (the average error percentage) given by the following formula:

$$MAPE = \left(\frac{1}{k} \sum\nolimits_{x=1}^{k} \left| \frac{a_x - f_x}{a_x} \right| \right) \times 100 \tag{1}$$

MAPE quantifies accuracy as a rate, and can be determined as a percentage of total absolute error for each time period, as actual values minus predicted values divided by actual values.

2.3 Results and Discussion

The databases we manipulated in this study are multivariate time series. We implemented one models of recurrent neural networks in particular LSTM, using libraries such as Numpy [12], Pandas, Keras and Tensorflow [13].

Table1 shows the parameters used for both models. Note that these parameters are obtained in a heuristic way by performing a considerable number of tests.

This paragraph contains the results of the LSTM model, note that we performed six tests to finally propose a suitable architecture based on the MAPE measurement. Then we performed the prediction for LSTM model of the following hundred days.

Table 1. Parameter settings of the studied approaches.

Model	Parameter	Value
LSTM	Input	3
	Output	1
	Hidden units	2
	Neurons	64–32
	Optimizer	Adam
	Features	3
	Activation function	Relu
	Training epochs	500

Table 2 shows us the percentage of error committed for each test and we finally propose the learning done in the third test, which MAPE = 14.61%. The initial question was mainly focused on the stability of learning, and to ensure an adequate modeling for signals. Two interesting results are deduced: first, a decrease at the beginning of July after an upward trend in the number of critical cases for the next 100 days, second, Among the six tests established, we have identified three overfittings on tests 2,4 and 6.

Table 2. Parameter settings of the studied approaches.

	Test-1	Test-2	Test-3	Test-4	Test-5	Test-6
MAPE	19.55	31.47	14.61	53.86	18.11	43.27

We have established a multivariate neural network which the target variable is the critical cases process, the system uses three input parameters, the active cases, the dead cases and the critical cases. Figures 5, 6 and 7 characterize the prediction of the LSTM model, for the test part comprising nearly 20% of the basic signal. Moreover, Fig. 8 characterize the prediction of 100 future observations from June 26, 2022 until 03 October 2022. To measure the stability of the signal.

We remark a decrease at the beginning of July, and this can be explained by several factors, namely the collective immunity of the population following the vaccination. This, for a performance of 14.61%. Tests 2, 4 and 6 characterize an overfitting which system could not perform learning.

Several recent works are interested in modeling the spread of SARS-COV2 using neural network models. A. Tomar and N. Gupta established the LSTM model for modeling and predicting the number of active cases in India, the error percentage is of order

8% [14]. While N. Zheng and S. Du. Proposed a hybrid model which integrated the language processing (NLP) and the LSTM model with the improved susceptible-infected (ISI) model [15], the proposed hybrid AI model can significantly reduce the errors of the prediction results and obtain the MAPE with 0.52%, 0.38%, 0.05%, and 0.86% for the next six days in Wuhan, Beijing, Shanghai, and countrywide, respectively.

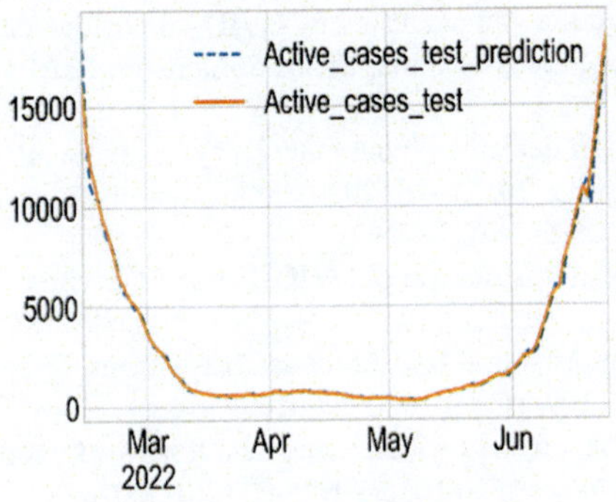

Fig. 5. Active cases prediction

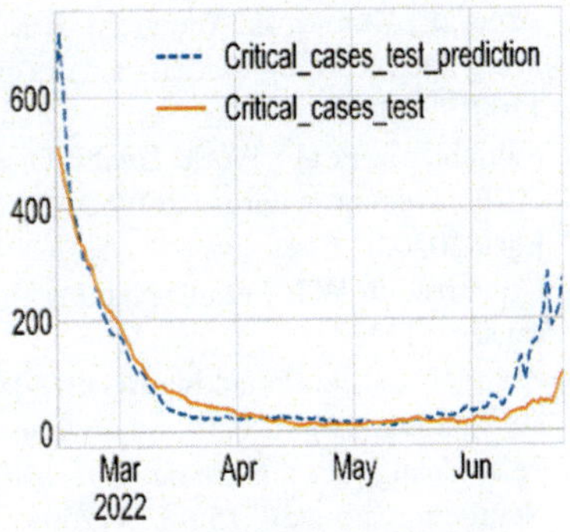

Fig. 6. Critical cases prediction

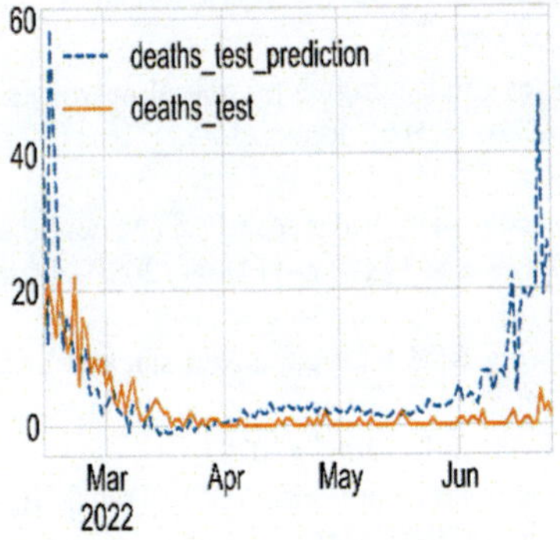

Fig. 7. Death's prediction

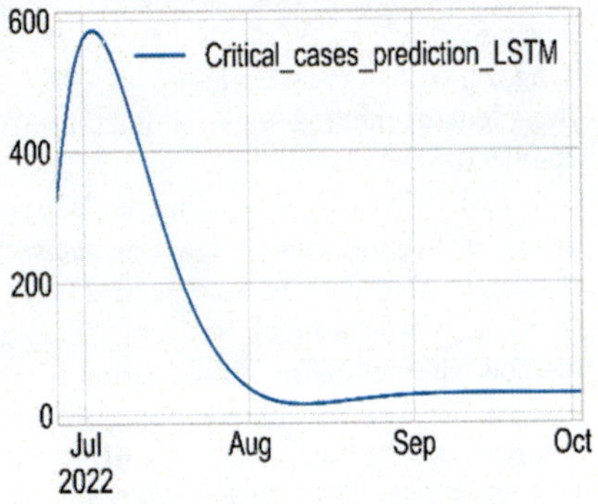

Fig. 8. 100 future observations

3 Conclusion

Analyzing and modeling time series allows the extraction of knowledge. In the present study, we have introduced the modeling using a supervised learning technique, in particular the recurrent neural networks method. For this, we have established LSTM model for multivariate time series. We conducted tests in order to test the ability of the established systems. Based on 80% of the data, the RNN extension were able to predict the rest of the series, which was then validated with the remaining 20%. The model predicts an upward trend in the number of critical cases for the next 100 days, however, note that we have not considered several determining factors regarding critical cases evolution such as, herd immunity of the population as well as the vaccination process and so on, which characterizes the complexity of the evolution dynamics of the process. Indeed, using neural network methods for signals resulting from complex processes does not allow us to establish a reliable and consistent model, and we are often faced with an overfitting as tests 2, 4 and 6. The perspective of this study is to examine the stability of the established system and comparing it with other models.

References

1. Arora, P., Kumar, H., Panigrahi, B.K.: Prediction and analysis of COVID-19 positive cases using deep learning models: a descriptive case study of India. Chaos Solitons Fractals. **139**, 110017 (2020)
2. Sohrabi, C., et al.: World Health Organization declares global emergency: a review of the 2019 novel coronavirus (COVID-19). Int. J. Surg. **76**, 71–76 (2020). https://doi.org/10.1016/j.ijsu.2020.02.034
3. Organization W.H., et al.: Naming the coronavirus disease (COVID-19) and the virus that causes it (2020a)
4. Zheng, N., et al.: Predicting COVID-19 in China using hybrid AI model. IEEE Trans. Cybern. **50**(7), 2891–2904 (2020). https://doi.org/10.1109/TCYB.2020.2990162
5. Peter Zhang, G.: Time series forecasting using a hybrid ARIMA and neural network model. Neurocomputing **50**, 159–175 (2003). https://doi.org/10.1016/S0925-2312(01)00702-0
6. Idrais, J., El Abassi, R., El Moudene, Y., Sabour, A.: Characterizing community behavior in OSNs: modeling and forecasting activity on Facebook using the SARIMA model. J. Intell. Fuzzy Syst. **43**, 1–13 (2022)
7. Tang, Z., de Almeida, C., Fishwick, P.A.: Time series forecasting using neural networks vs. Box- Jenkins methodology. Simulation **57**(5), 303–310 (2016). https://doi.org/10.1177/003754979105700508
8. Kuo, C.-E., Chen, G.-T.: Automatic sleep staging based on a hybrid stacked LSTM neural network: verification using large-scale dataset. IEEE Access **8**, 111837–111849 (2020). https://doi.org/10.1109/ACCESS.2020.3002548
9. Jahromi, A.N., Hashemi, S., Dehghantanha, A., Parizi, R.M.: An enhanced stacked LSTM method with no random initialization for malware threat hunting in safety and time-critical systems. IEEE Trans. Emerg. Top Comput. Intell. **1–11**, 2471–3285 (2020)
10. Schuster, M., Paliwal, K.K.: Bidirectional recurrent neural networks. IEEE Trans. Signal Process. **45**(11), 2673–2681 (1997). https://doi.org/10.1109/78.650093
11. Hamilton, J.D.: Stationary ARMA processes. In: Time Series Analysis 1994 by Princeton University Press. Princeton, New Jersey (1994)
12. Jiménez, F., Palma, J., Sánchez, G., David Marín, M., Francisco Palacios, M.D., López, L.: Feature selection based multivariate time series forecasting: an application to antibiotic resistance outbreaks prediction. Artif. Intell. Med. **104**, 101818 (2020). https://doi.org/10.1016/j.artmed.2020.101818
13. Oliphant, T.E.: A Guide to NumPy, vol. 1. Trelgol Publishing, USA (2006)
14. Tomar, A., Gupta, N.: Prediction for the spread of COVID-19 in India and effectiveness of preventive measures. Sci. Total Environ. **728**, 138762 (2020). https://doi.org/10.1016/j.scitotenv.2020.138762
15. Zheng, N., et al.: Predicting COVID-19 in China using hybrid AI model. IEEE Trans. Cybern. **50**(7), 2891–2904 (2020). https://doi.org/10.1109/TCYB.2020.2990162

Parallel Genetic Algorithms, Parameters and Design

Mustapha Ouiss(✉), Abdelaziz Ettaoufik, Abdelaziz Marzak, and Abderrahim Tragha

Faculty of Sciences Ben M'sik, Hassan II University of Casablanca, Casablanca, Morocco
mustapha.ouiss-etu@etu.univh2c.ma

Abstract. Genetic Algorithms are widely used in the quest of optimization of real-world complex problems. Parallel Genetic Algorithms may be considered as an evolution of the traditional GA. However, they are designed differently. Designing a Parallel Genetic Algorithm depends on many parameters other than the selection, the crossover, and the mutation parameters. In the literature, we can run into multiple and confusing Parallel Genetic Algorithms model names. These models are a sort of combination of specific parameters. Some of these parameters are the number of populations, the granularity of each population, the migration operation, the overlapping operation. This article lists these parameters used to design Parallel Genetic Algorithms so that the practitioner can design his own model, knowing the meaning and the use of each parameter.

Keywords: Parallel Genetic Algorithms · Random Heuristic Search · Parallel models · Optimization

1 Introduction

Introduced for the first time by [1], genetic algorithms (GAs) are merely a special case of Random Heuristic Search (RHS) [2], based on populations of individuals or solutions (population-based metaheuristic), used to find good solutions to complex and time consuming problems. The candidate solutions evolve according to the mechanism of selection and genetic operations as crossover and mutation. One of the strengths of GAs is that they are constructed from a number of reusable components: the chromosome encoding, the fitness function, the selection method, the recombination (crossover and mutation), and the replacement strategy. The basic form of Genetic Algorithms is often called simple genetic algorithm (SGA), and was detailed in the work of [3] and [2]. Genetic algorithms are iterated until the fitness value stabilizes or the maximum iteration number is reached.

Parallel Genetic Algorithms may be considered as an evolution of the traditional GA. However, they must be designed differently. The way in which GAs can be parallelized depends on many other parameters than the standard ones of the traditional GA. The number of populations, and the granularity of each population used by the PGA are important parameters to be considered.

Y. Farhaoui et al. (Eds.): ICAISE 2022, LNNS 635, pp. 655–660, 2023.
https://doi.org/10.1007/978-3-031-26254-8_95

The researcher may face a huge number of synonyms of parallel genetic algorithms words: subpopulations, multi populations, subpopulations with migration, overlapping subpopulations without migration, massively parallel genetic algorithm, demes, islands, and so on. PGA models can have multiple nominations, and that can make the comprehension of the PGAs quite complex for new researchers in the domain of the genetic algorithm parallelization. In order to simplify the comprehension and the design of PGAs, we listed the main parameters used to design efficient Parallel genetic algorithms.

The main goal of this article is to be a comprehensible, and concise as much as possible beginning for new researchers interested by parallel genetic algorithms. The paper presents the main parameters to design parallel genetic algorithms.

The rest of the article will be structured as follows: the first section will dress the correspondent literature in relation with taxonomies and classification of parallel genetic algorithms (PGAs). The second section will dress parallel genetic algorithms parameters.

2 Literature Review

The design of parallel genetic algorithms has many difficult and interrelated problems, despite of the benefits of using them [9]. The three main problems in the design of parallel genetic algorithms are, according to the author: the determination of the size and the number of demes (subpopulations); the efficient topology interconnecting these demes; and the choice of the best migration rate.

As many articles in the literature, the authors of [4] bring taxonomies and classification for parallel genetic algorithms.

Among parallel genetic algorithms cited in their article, we can find a massively PGA, a parallel steady-state algorithm, parallel messy algorithm. The authors proposed a version of PGA called dynamic demes which allows the combination of global parallelism, considering the algorithm as a simple master-slave distributed GA, with a coarse-grained GA. Moreover, in the cited model, the whole population is treated during evolution as a whole, hence, there is no migration operator.

Another article [10] presents some of the publications on parallel GAs. The author categorized the PGAs in four different types of parallel implementations of GAs: Master-Slave Parallelization, Multiple-Deme Parallel The First Generation Coarse-Grained Parallel GAs The Second Generation, and Fine-Grained Parallel GAs. The article gives an example of new hybrid algorithms, that tried to combine methods to parallelize GAs, called hierarchical parallel GAs. According to the author, these hybrid algorithms may add a new degree of complexity to the already complicated parallel GAs. An example of hierarchical parallel GAs cited, is the combination of a multi-deme GA and a fine-grained GA.

As we may conclude from these articles, there can be found different names for the same implementation, and no well-defined parameters to design and implement new PGA models. Furthermore, we can found a large number of names in the literature as: coarse grained GA no overlapping demes with migration, etc.

3 Designing Parallel Genetic Algorithms

To the best of our knowledge, there is no article that gives the full list of parameters and their characteristics. This article tries to the best to list the main parameters and their respective characteristics, in order to help researchers and practitioners to model their own PGA. The main parameters are: (1) number of populations; (2) the population granularity; (3) the migration operation; and (4) the overlapping operation.

3.1 Number of Populations

Parallel Genetic Algorithms with multiple population are also called multiple demes. Multiple deme PGA, are known as coarse grained genetic algorithms. In this case, the number of populations (or deme) is reduced, and the number of solutions in each deme is elevated. In the contrary, if a higher number of demes is used, and for each deme, the number of solutions is lower, the PGAs are considered as a fine-grained PGAs.

Single Population. A Single population parallel genetic algorithm is known as global parallelization, master-slave model, or distributed fitness evaluation. In the master-slave genetic algorithm, one global population is used, and the individuals' fitness evaluation is done in parallel. The genetic operations (crossover and mutation) are also done in parallel. The master stores the population, and the slaves evaluate the fitness function and execute genetic operations. The selection of parents for mating is done globally. The Master slave GA had a synchronous mode, and an asynchronous one. In the synchronous mode, the master waits until he receives the fitness from the totality of the solutions before the selection operation. The disadvantage of the synchronous mode is that the process stops until all the slaves finish their fitness evaluation operation. It may be delayed since the whole process is penalized by the slowest slave. In the asynchronous mode, the master-slave algorithm selection waits until a predefined percentage of the population has been processed. Needless to say that the percentage needs to be defined, and to be defined wisely. A high percentage will converge the asynchronous mode to the synchronous mode. In the other hand, a lower percentage may produce weaker solutions, since the choice of the parents for mating is reduced. According to [6], probably the straightforward way to parallelize GAs is to distribute the fitness evaluation among several slaves, while one master processor executes the GA operations (selection, recombination, and mutation). This observation is obvious since the Master slave GAs are easy to understand and to design. They explore the search space in exactly the same way as a serial GA.

Multiple Population. These algorithms are usually referred to as subpopulations, static subpopulations, multiple-deme GAs, coarse-grained GAs. Demes can be separated from one another, and individuals compete only within a deme if non-overlapping and non-migration are defined. The overlapping and the migration parameters are discussed in the next sections. Once dealing with multiple populations, it may be important to distinguish between a single generated population, that is divided into multiple sub-populations, and multiple generated populations. This distinction is important since the generation phase can be done randomly with certain parameters, or using a greedy heuristic.

3.2 Population Granularity

The size of each population define its granularity. Hence, there are two categories of Parallel Genetic Algorithms according to the granularity: Coarse grained GA, and fine grained GA, also called massively parallel.

Coarse Grained GA. Also known as Island GA, coarse grained algorithms are defined as a population that is divided to a small number of demes with a large number of individuals for each deme. Coarse grained parallel GAs are also known as distributed GAs since they are usually implemented on distributed memory MIMD computers, which is well suited for heterogeneous networks [4].

Fine Grained GA. Also known as cellular GA or massively parallel GA, fine grained algorithms are considered when a population is divided into a large number of demes with a few individuals in each deme. Individuals compete and mate within the neighborhoods' individuals. Due to the necessity of special hardware, fine grained GAs have not received as much attention as other types of parallel GAs [8].

Exchanging Individuals. Demes may exchange individuals in two ways: by migration or using overlapping. When neighborhoods overlap, good individuals may participate in the mating process of the overlapped demes. The migration is no needed in this case, and there is no reason to define a migration parameter. The migration should be used if demes are not overlapping.

3.3 Migration

The migration operation consists in copying (or even moving) individuals from a deme to another. There are many models of migration in the literature. The free migration between demes, whether demes are contiguous or not, are called island model migration. A stepping stone model is when individuals can only migrate to neighbor's demes. Other models exist, and the practitioner may define new one. The migration operation is controlled by a set of parameters: the migration rate, that defines the number or the percentage of individuals migrating. The migration type, that defines if the individuals are copied, or moved to the deme of destination. The interval of migration, that determines the frequency of migration (once a generation for example). Which type of individuals are migrated, and if the migrants replace individuals in the destination deme or not.

The complete isolation of the demes, that is, no migration between demes is allowed, should be avoided in practice, since the expected speedups are only marginal [9]. Its straightforward to note that migration is not necessary in the master slave model.

Each deme is connected to its neighbours. The degree of connectivity is an important property, since it determines the cost of communications, and the time of computations [7]. In the latter article, the authors deal with how to choose the size of the population, the degree of communication, that is, the number of demes that one deme communicate with, and how to define the deme size, the migration rate. The degree of the topology affects the probability that the parallel GA reaches the desired solution.

3.4 Overlapping

The overlapping operation is used when some individuals belongs to more than one deme, and compete for selection in the overlapped demes. The overlapping can be defined by a number of individuals shared between demes, or by a percentage.

3.5 PGA Design

In order to design an optimal PGA, the characteristics cited above should be defined.

An enormous number of designs can be implemented using the characteristics cited in the previous section. We can assume that a PGA implementation is equal the equation:

$$implementation = (Number\ of\ Populations\ X\ Population\ granularity) \\ +/-\ Migration\ +/-\ Overlapping$$

However, not every implementation is suitable, or applicable. For example, it is counter-intuitive to define a parallel genetic algorithm with high number of demes, and a high number of individuals for each one. Or even a low number of demes with a low number of individuals for each deme. Or even disable the migration and the overlapping operations at the same time.

While it is true that smaller demes converge faster, the quality of the solution found may be inacceptable [10]. The speedup is very modest when there is no communication between demes, and the quality of the solution degrades as smaller demes are used [5]. Adding more isolated demes to an optimal sized set of communicating demes results in a performance loss [5]. Defining the population size is one of the most important decisions needed to use a GA. If the population is too small, it will be difficult to identify good solutions. On the other hand, if the population is too large the GA will waste time processing unnecessary individuals. The parallel GAs encounter the same problems [6].

4 Conclusions and Future Work

Parallel genetic algorithms are defined considering a set of parameters as: the population number, their granularity, the use or not of migration and overlapping operations. The combination of these characteristics allows the creation of wide variety of PGAs. However, some combinations are counter intuitive, or useless. This article tries to give an overview of these parameters in order to help new PGA practitioners. In the future articles, we may develop to the best the full list of characteristics of every parameters of PGA. Also, we may compare different models with different parameters and characteristics.

References

1. Holland, J.H.: Adaptation in Natural and Artificial Systems. Univ. Michigan Press, Ann Arbour (1975)
2. Vose, M.D.: The Simple Genetic Algorithm: Foundations and Theory. Complex Adaptive System, Complex ad. (1999)

3. Goldberg, D.E.: Genetic Algorithms in Search, Optimization and Machine Learning. Addison-Wesley Publishing Company, Inc., Boston (1989)
4. Nowostawski, M., Poli, R.: Parallel genetic algorithm taxonomy. In: 1999 Proceedings of Third International Conference on Knowledge-Based Intelligent Information Engineering Systems (Cat. No.99TH8410), pp. 88–92. IEEE (1999)
5. Cantú-Paz, E., Goldberg., D.E.: predicting speedups of idealized bounding cases of parallel genetic algorithms. In: Proceedings of the Seventh International Conference on Genetic Algorithms, pp. 113–120 (1997)
6. Cantú-Paz, E., Goldberg, D.E.: Efficient parallel genetic algorithms: theory and practice. Comput. Methods Appl. Mech. Eng. **186**, 221–238 (2000).
7. Cantú-Paz, E., Goldberg, D.E.: On the scalability of parallel genetic algorithms. Evol. Comput. **7**, 429–449 (1999)
8. Wahib, M., Munawar, A., Munetomo, M., Akama, K.: Optimization of parallel genetic algorithms for nVidia GPUs. In: 2011 IEEE Congress of Evolutionary Computation (CEC), pp. 803–811. IEEE (2011)
9. Cantú-Paz, E., Goldberg, D.: Modeling idealized bounding cases of parallel genetic algorithms. In: 1997 Proceedings of Second Annual Conference on Genetic Programming, pp. 353–361 (1996)
10. Cantú-Paz, E.: A Survey of parallel genetic algorithms. Calc. Paralleles. **10**, 141–171 (1998).

Particle Swarm Optimization to Design Microwave Components Based Substrate Integrated Waveguide Technology

Souad Akkader[1](✉), Hamid Bouyghf[2], and Abdennaceur Baghdad[1]

[1] Laboratory of Electronics, Energy, Automatics, and Data Processing, Faculty of Sciences and Techniques of Mohammedia, Hassan II University of Casablanca, Casablanca, Morocco
souad.akkader-etu@etu.univh2c.ma

[2] Laboratory of Engineering Sciences and Biosciences, Faculty of Sciences and Techniques of Mohammedia, Hassan II University of Casablanca, Casablanca, Morocco

Abstract. The development of smaller and reconfigurable microwave components with lower losses is the main objective in the manufacturing of recent smart technologies, substrate integrated waveguide (SIW) structure looks like a very suitable approach. This study contributes to the design and optimization of a substrate-integrated waveguide resonator using Particle Swarm Optimization (PSO) Algorithm. The attenuation constant results are very encouraging with a 0.069 dB/m optimal value, the validation results done by Ansys HFSS also show good agreement with lower loss (Return Loss: S11 = −24 dB and Insertion Loss: S21 = −1.4 dB). The SIW technology is therefore a suitable solution to satisfy the requirements of the future IoT and smart systems.

Keywords: Substrate integrated waveguide · Particle swarm optimization · Microwave devices · Attenuation constant

1 Introduction

Among the various solutions for implementing microwave and millimeter-wave components, SIW technology looks like a very suitable approach [1]: it allows for planar integration of active and passive components, and it exhibits low loss and compact size, as a result, interference, and cross-talk are avoided. One of the major issues in the design of SIW components is related to the minimization of losses. Similar to traditional rectangular waveguides, SIW structure is also affected by conductor and dielectric losses, recently a new type of loss is studied named radiation loss [2] due to radiation leakage between gaps illustrated below in the geometry of SIW shown in (Fig. 1).

Previous works have analyzed and studied the influence of dielectric materials and the height thickness on the Quality factor of SIW resonators [3, 4]. However, is there an impact of other geometric factors like diameter of vias, spacing between two adjacent vias, or SIW width on the performance of the overall SIW resonator in the presence of a dielectric material with high performance in reducing losses? This is what this study will go through in depth.

Y. Farhaoui et al. (Eds.): ICAISE 2022, LNNS 635, pp. 661–667, 2023.
https://doi.org/10.1007/978-3-031-26254-8_96

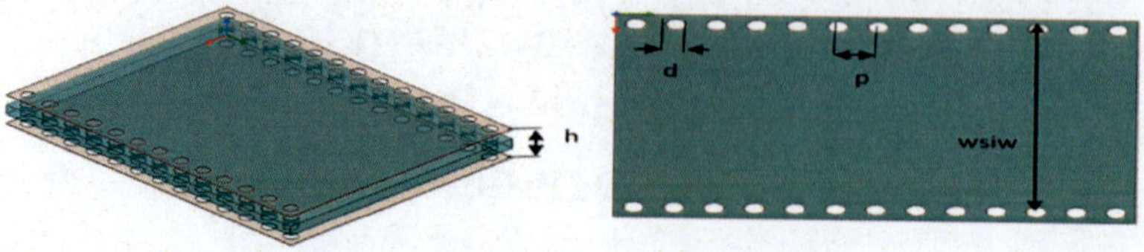

Fig. 1. Geometric parameters of SIW

In the last few decades, new optimization techniques-based meta-heuristic algorithms are intended to achieve practical designs of SIW like Ant Colony Optimization (ACO) [5], as well as in other fields like RF Integrated circuits [6]. The appeal of employing these algorithms to solve complicated problems is that they get the best and optimal solutions in a short amount of time, even for very huge issue sizes. In this paper, an optimal design of SIW resonator is done using Particle Swarm Optimization (PSO) Algorithm by minimizing the attenuation constant and taking into consideration all types of losses: conductor, dielectric, and radiation losses. Additionally, obtained results by Matlab programing language are validated using Ansys HFSS software.

2 Modeling of Losses in Substrate Integrated Waveguide

The propagation characteristics of SIW structures are similar to traditional rectangular waveguides. Only TEn0 (n ¼ 1, 2,...) modes can exist [7] and the fundamental mode is the TE10. Due to this similarity, the SIW structure exhibits too, two identical losses: conductor and dielectric losses. However, there is an additional loss in SIW called radiation losses caused by radiation leakage between two adjacent vias holes.

Conductor loss is caused by the finite conductivity of the metal planes and vias, dielectric loss is related to the loss tangent (tan δ) of the dielectric substrate, and finally, radiation loss is due to the gaps between pairs of metal vias. The attenuation constant related to each type of loss, conductor, dielectric, and radiation, is given in [8] as shown respectively in the Eqs. (1), (2), and (3).

$$\alpha_C(f) = \frac{\sqrt{\pi f \varepsilon_0 \varepsilon_r}}{h\sqrt{\sigma_c}} \frac{1 + 2\left(\frac{f_0}{f}\right)^2 \frac{h}{w_{eff}}}{\sqrt{1 - \left(\frac{f_0}{f}\right)^2}} \tag{1}$$

$$\alpha_D(f) = \frac{\pi f \sqrt{\varepsilon_r}}{c\sqrt{1 - \left(\frac{f_0}{f}\right)^2}} \tan\delta \tag{2}$$

$$\alpha_R = \frac{\frac{1}{w}\left(\frac{d}{w}\right)^{2.84} \left(\frac{s}{d} - 1\right)^{6.28}}{4.85\sqrt{\left(\frac{2w}{\lambda}\right)^2 - 1}} \tag{3}$$

where h is the substrate thickness, ε_r is the dielectric permittivity, f_0 is the cut-off frequency of the SIW, f is the frequency of operation, σ_c is the metal conductivity, and λ is the wavelength in the dielectric medium at the operation frequency.

From (1), to (2), and (3). the attenuation constant of the SIW is calculated as in [8] in Eq. (4).

$$\alpha_0 = \alpha_C + \alpha_D + \alpha_R \tag{4}$$

3 Optimization of SIW Using Particle Swarm Optimization

3.1 Overview of Particle Swarm Optimization (PSO)

PSO is an intelligent optimization algorithm and it belongs to the class of Metaheuristic optimizations. It is based on the paradigm of swarm intelligence inspired by the social behavior of animals like fish and birds, PSO is successfully applied to enormous applications like machine learning mesh processing data mining and operational research [9]. Initially, PSO is introduced by James Kennedy and Russell Eberhard in 1995 [10]. They were working to develop a model to describe the social behavior of animals like a flock of birds and a school of fish. However, they realized that their model is capable of doing optimization tests so they proposed a new optimizer, which is called particle swarm optimization. For 20 years, PSO became one of the most useful and most popular algorithms to solve various optimization problems in various fields.

3.2 Objective Function and Design Variables

We look for optimizing design variables: h, d, p, w_{siw} indicated in (Fig. 1), and minimizing the attenuation constant α_0 introduced in Eq. (4) which is the objective function. The dimensions of SIW resonator are set subject to the constraints $d/p \geq 0.5$ and $d/\lambda_0 \leq 0.1$ to substantiate minimum discharge of energy [7]. With $\varepsilon_r = 2.2$, tan $\delta = 0.0009$, $\sigma_c = 5.8e7$ S/m, the height of metal layer h1 = 0.09 mm, λ = 1.78e-3 mm, and w_{eff}= 20 mm at 5 GHz frequency. Once the cutoff frequency f_0 of the SIW is known, the equivalent width w_{eff} can be calculated as in [3] so w_{eff}= 20 mm.

PSO method is said to be simple and efficient since it can result in faster convergence. Particle location and velocity are both initially assigned random values. These settings are then changed regularly to get the best possible result. Figure 2 depicts the flowshart of the PSO method. $V_i(j)$ and $X_i(j)$ are the velocity and coordinates of the particle number i at iteration j, respectively, and then revised into their updated values at iteration j + 1 as V_i (j + 1) and $X_i(j + 1)$ by utilizing the Eqs. (5) and (6).

$$V_i(j+1) = W * V_i(j) + c1 * rand1() * [pb_i(j) - X_i(j)] + c2 * rand2() * [gb(j) - X_i(j)] \tag{5}$$

$$X_i(j+1) = X_i(j) + V_i(j+1) \tag{6}$$

$pb_i(j)$ is the self-best solution of particles, and gb(j) is the best solution of the swarm of particles at iteration j.. Both rand1() and rand2() take random values in the range of 0 to 1. c1 is a Cognitive parameter and it specifies to what extent a particle is conditioned on itself, c2 specifies to what extent it is conditioned on the swarm. To obtain the best convergence for the PSO technique, c1, and c2 is set to 2. The weight (also called inertial weight) of the process is 1 and its damping Ratio is 0.99.

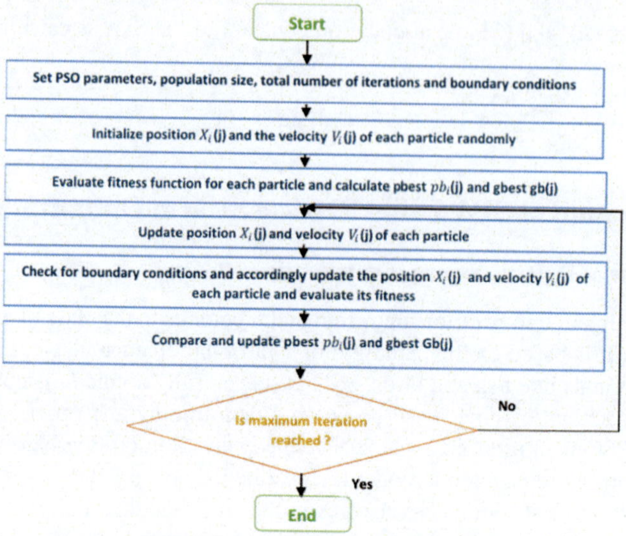

Fig. 2. Flowchart of PSO Algorithms

4 Results and Discussion

Optimized parameters obtained from the algorithm are used to achieve the required SIW design; the best solution is obtained when the algorithm converges. The attenuation constant obtained using the PSO technique on simulation was found to be 0.069 dB/m around 5 GHz frequency as represented in Fig. 3. Table 1 illustrates optimal dimensions obtained using the PSO algorithm.

In order to better appreciate the accuracy of the proposed Particle Swarm Optimization method, the variation of the attenuation constant α_0, as well as the Scattering parameters using Ansys HFSS software are shown in (Fig. 4) and (Fig. 5) respectively.

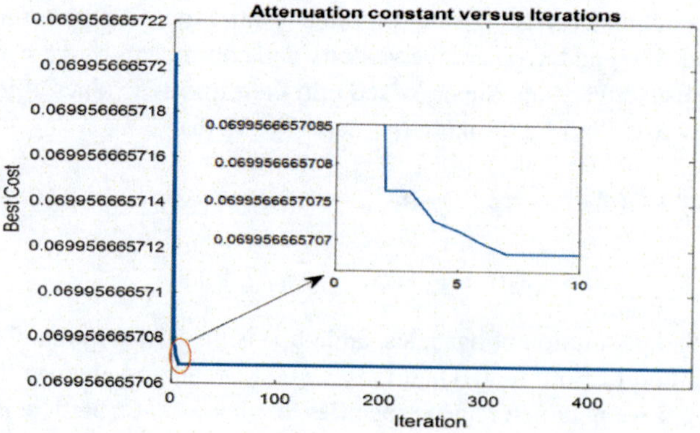

Fig. 3. Objective function (Attenuation constant α_0) versus Iterations

From the simulation results, it can be seen that $\alpha_0 = 0.8$ dB/m is approximately near the value obtained using the PSO algorithm: 0.06 dB/m. Moreover, the detail of scattering parameters show good results: Return Loss S11 = −24 dB and Insertion Loss S21 = −1.4 dB at 5.1 GHz. Obtained results indicate also that the proposed SIW design achieves better results than previous work [3, 11] with good achievements in compactness and lossless. It is reasonable to conclude that increasing the substrate thickness h reduces conductor loss and using low-loss laminates (with minimal tan δ) reduces dielectric loss. While the quantity of leakage field is controlled by the pitch between two vias holes. In the case of HFSS, a frequency sweep is necessary, and the time per frequency point was 180 s whenever the frequency response is calculated in 100 points. It should be noted that the PSO method produces a big reduction in computing time (20 s), making it a powerful tool for the efficient design of complex SIW circuits.

Table 1. Parameter ranges and their optimal values.

Variables	Minimum (mm)	Maximum (mm)	Optimal value (mm)
d	0.8	1.2	0.8
p	1.6	2.2	1.6
w_{siw}	18	20	20
h	0.9	1.2	1

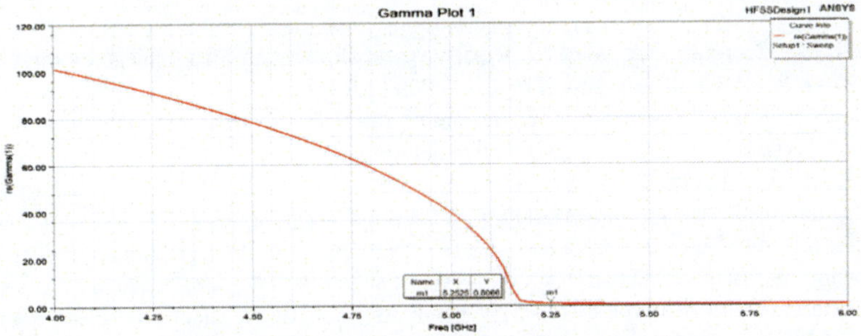

Fig. 4. Attenuation constant α_0 versus frequency

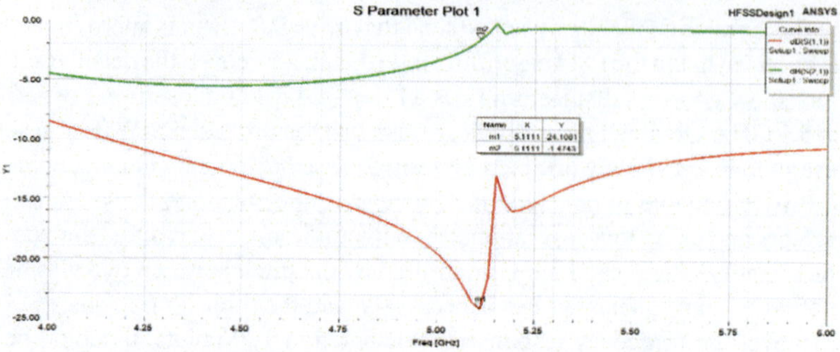

Fig. 5. Return Loss (S11) and Insertion Loss (S21)

5 Conclusion

We have presented in this paper an optimal design of substrate integrated waveguide resonator based on the expression of its attenuation constant α_0 and taking into consideration physical limitations and constraints. Simulation results obtained using particle swarm optimization in Matlab programing language and Ansys HFSS software show good agreement, a minimal attenuation constant α_0 equal to 0.069 dB/m while keeping a smaller size and lower losses. The present work is therefore an optimal method for fabricating microwave components based on SIW technology and evaluating their physical, electronic, and electromagnetic properties suitable for IoT and smart technologies.

References

1. Bozzi, M., Georgiadis, A., Wu, K.: Review of substrate-integrated waveguide circuits and antennas. IET Microw. Antennas Propag. **5**, 909 (2011)
2. Pasian, M., Bozzi, M., Perregrini, L.: Radiation losses in substrate integrated waveguides: a semi-analytical approach for a quantitative determination, 3p. (2013)
3. Ali Khan, A., Mandal, M.K., Sanyal, S.: Unloaded quality factor of a substrate integrated waveguide resonator and its variation with the substrate parameters. In: 2013 International Conference on Microwave and Photonics (ICMAP), pp. 1–4. IEEE, Dhanbad, India (2013)
4. Rhbanou, A., Bri, S., Sabbane, M.: Analysis of substrate integrated waveguide (SIW) resonator and design of miniaturized SIW bandpass filter. Int. J. Electron. Telecommun. **63**, 255–260 (2017). https://doi.org/10.1515/eletel-2017-0034
5. Lalbakhsh, P., Zaeri, B., Lalbakhsh, A., Fesharaki, M.N.: AntNet with reward-penalty reinforcement learning, 5p. (2010)
6. Bouyghf, H., Benhala, B., Raihani, A.: Artificial bee colony technique for a study of the influence of impact of metal thickness on the factor of quality-Q in integrated square spiral inductors. In: 2018 4th International Conference on Optimization and Applications (ICOA), pp. 1–7. IEEE, Mohammedia (2018)
7. Xu, F., Wu, K.: Guided-wave and leakage characteristics of substrate integrated waveguide. IEEE Trans. Microwave Theory Techn. **53**, 66–73 (2005)
8. Bozzi, M., Pasian, M., Perregrini, L.: Modeling of losses in substrate integrated waveguide components. In: 2014 International Conference on Numerical Electromagnetic Modeling and Optimization for RF, Microwave, and Terahertz Applications (NEMO), pp. 1–4. IEEE, Pavia, Italy (2014)

9. Zhang, Y., Wang, S., Ji, G.: A comprehensive survey on particle swarm optimization algorithm and its applications. Math. Probl. Eng. **2015**, 1–38 (2015)
10. Eberhart, R., Kennedy, J.: A new optimizer using particle swarm theory, 5p. (1995)
11. Khalil, M., Kamarei, M., Jomaah, J., Ayad, H.: Compact wide-band and low loss substrate integrated waveguide. In: 2015 Third International Conference on Technological Advances in Electrical, Electronics and Computer Engineering (TAEECE), pp. 130–134. IEEE, Beirut, Lebanon (2015)

PI Controller and Quadratic Feedback of Synchronous Machine

Hafid Ben Achour[1](✉), Said Ziani[2], and Youssef El Hassouani[1]

[1] Department of Physics, Faculty of Science and Technology Errachidia, Moulay Ismail University, BP-509 Boutalamine, 52 000 Errachidia, Morocco
h.benachour@edu.umi.ac.ma, y.elhassouani@umi.ac.ma

[2] Laboratory of Networks, Computer Science, Telecommunication, Multimedia (RITM), Department of Electrical Engineering, High School of Technology ESTC, Hassan II University, Casablanca, Morocco
SAID.ZIANI@univh2.ma

Abstract. The Synchronous Machine is used in many industrial sectors because it works much better than other kinds of motors. In this paper, we use the Lyapunov stability theory to show a vector control that is combined with a PI controller and dynamical feedback control of a PMSM to make it more stable. The main results of each controller are talked about and compared to each other.

Keywords: Nonlinear control · PI controller · PMSM · Backstepping technique

1 Introduction

In automatic mode, control by state feedback modifies the closed-loop behavior of a dynamic system whose state is represented by a state representation. This method implies the current state is known. When this is not the case, a state observer can be employed to rebuild the state from the data given. One goal of state feedback control might be to reduce (or maximize) a performance index (Optimal control, LQ control). It may also be possible to achieve a closed-loop system in which the poles, or eigenvalues of the state matrix, are appropriately positioned. In reality, these poles govern the system's behavior, but only if the system is univariable; in the case of multivariable systems, it is crucial to also consider the eigenvectors1. In most instances, the control must perform other key servo activities, including monitoring by producing a reference signal and rejecting disturbances. Then, the essential state increases must be made initially. A synchronous machine SM is an electrical machine with a rotating field and a rotor that both move simultaneously. For example, the rotor of a synchronous machine powered by an electrical network rotates precisely in time with the rotating field, which is set by the frequency of the network. A So, the synchronous machine is different from the asynchronous machine, whose rotor is behind the rotating field when the machine is working as a motor and in front of the field when it is working as a generator. Another difference is that, unlike the asynchronous machine, the synchronous machine needs an extra excitation field to work. [1, 2]. It has a high nonlinear dynamic. The implementation of vector-controlled

Y. Farhaoui et al. (Eds.): ICAISE 2022, LNNS 635, pp. 668–673, 2023.
https://doi.org/10.1007/978-3-031-26254-8_97

speed requires the design ofa PI controller of d-q axis current, and speed control. Also, the PI controller reqires a technique that assured the coupling of input-output [3]. Due to its complexity and dynamic nonlinear, several studies elaborated on many nonlinear technologies such as sliding mode control [4], backstepping control, predictive control, artificial intelligence control, adaptive control, and input-output linearization control. In this paper, we present a PI controller and quadratic feedback of Synchronous Machine.

2 Mathematical Model of the PMSM

The Synchronous Machine model is characterized by the following [10].

$$\begin{cases} \frac{di_d}{dt} = -\frac{R_s}{L_d} i_d + p\frac{L_q}{L_d}\Omega i_q + \frac{v_d}{L_d} \\ \frac{di_q}{dt} = -\frac{R_s}{L_q} i_q - p\frac{L_d}{L_q}\Omega i_d - p\frac{\Omega\varphi_f}{L_q} + \frac{v_q}{L_q} \\ \frac{d\Omega}{dt} = \frac{p}{J}(L_d - L_q) i_d i_q - \frac{f}{J}\Omega + \frac{P}{J}\varphi_f i_q - \frac{C_r}{J} \end{cases} \quad (1)$$

where: R_s: Stator resistance (Ω); L_d, L_q: : d, q axis self-inductance (H); φ_f: Mutual flux due to permanent magnetic (Wb); i_d, i_q: : d, q axis currents (A); Ω: Angle speed (rad/s); θ: Rotor position; J: Moment of inertia (kg.m^2); f: damping constant (N/rad/s); p: Number of pole pairs; C_r: : load torque (N.m).

3 Field-Oriented Control of Synchronous Machine SM

Usually, the stator axis component acts as the excitation and adjusts the value of the flux in the machine. The q-axis component acts as the armature current and controls the torque. The application of vector control requires the axis of the component I_q to be in quadrature relative to the rotor flux.

For a surface mounted SM, $L_d = L_q$ is fulfilled. When i_d is set to 0 in field-oriented control, also when the system works at optimal linear torque, the direct current must be $i_d = 0$, so:

$$i_d = 0; \quad i_q = i_s; \quad \varphi_d = \varphi_f \quad (2)$$

The electromagnetic torque formula, since the flux is constant:

$$C_e = \frac{3}{2}\varphi_f i_q = K_f i_q \quad (3)$$

4 Application of PI Controller

The control law of PI is:

$$u(t) = K_p e(t) + K_i \int e(t) \quad (4)$$

Therefore:

$$k_{id} = \frac{R_s^2}{L_d};\ k_{pd} = k_{id} \cdot T_{ds} \tag{5}$$

The regulator parameters for id current are:

$$k_{iq} = \frac{R_s^2}{L_q};\ k_{pq} = k_{iq} \cdot T_{qs} \tag{6}$$

Those of the Speed Are:

$$K_i = j\omega_0^2;\ K_p = 2j\omega_0 - f \tag{7}$$

5 Backstepping Design

5.1 Backstepping Designee

- **i_d current control**

The id error regulator defined by:

$$e_d = i_{d,ref} - i_d \text{ And } \dot{e} = \dot{i}_{d,ref} - \dot{i}_d \tag{8}$$

where:

$$\dot{e_{id}} = \dot{i}_{d,ref} + [\frac{R_s}{L_d} i_d - p\frac{L_q}{L_d}\Omega i_q - \frac{v_d}{L_d}] \tag{9}$$

Lyapunov energy function:

$$V_d = \frac{1}{2} e_d^2 \tag{10}$$

Its derivate is:

$$\dot{V_d} = e_d \dot{e_d} \tag{11}$$

According to (15) and (17):

$$\dot{e_d} = -k_d e_d = \dot{i}_{d,ref} - \frac{R_s}{L_d} i_d - p\frac{L_q}{L_d}\Omega i_q - \frac{v_{d,ref}}{L_d} \tag{12}$$

k_d : Is a positive scalar

The convergence condition of the error is:

$$\dot{V_d} = -k_d e_d^2 \leq 0 \tag{13}$$

the backstepping control law is:

$$v_{d,ref} = L_d [k_d e_d + i_{d,ref} + \frac{R_s}{L_d} i_d - p\frac{L_q}{L_d}\Omega i_q] \tag{14}$$

- **Speed control:**

$$i_{q,ref} = \frac{J}{p\varphi_f}[\dot{\Omega}^* + \frac{C_r}{J} + \frac{f}{J}\Omega + K_\Omega e_\Omega] \tag{15}$$

- **Control of i_q current:**

$$v_{q,ref} = L_q\left[K_q e_q + \dot{i}_{q,ref} + \frac{R_s}{L_q} i_q + \frac{p\Omega}{L_q}\left(L_d i_d + \frac{\varphi_f}{L_q}\right)\right] \tag{16}$$

6 Simulations Results

The following figures show the simulations of PI Controller of the PMSM (Fig. 1), and those of backstepping design controller (Fig. 2).

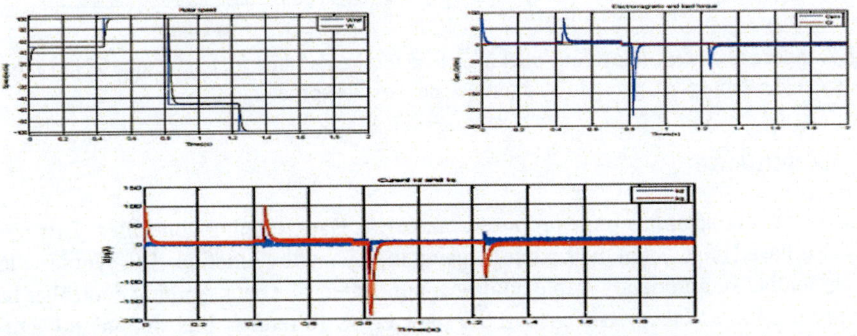

Fig. 1. Speed, (i_d, i_q) current, and torque with inversion of direction of rotation at 0.7 s and with load at 1 s

7 Discussion

The results (Fig. 1) show that the PI controller contains an overshoot in speed tracking. The application of the load leads to the speed decrease. The i_q currents take a high value when we apply the load. The electrical torque compensates the load with a remarkable exceed in the moment of load application. When the speed is inverted the electrotechnical torque and i_q current face a high drop. The variation of machine's setting affects the i_d and electro-mechanical torque.

For the backstepping controller (Fig. 2) the results show that during no-load torque we have a fast response time, the speed is tracking the reference speed without overshoot. we noticed that there are good tracking dynamics and rejection of the disturbances. The electromagnetic torque oscillates during power-up reaching a maximum value and disappears once the steady state is reached. When the load is applied, the electromagnetic torque increases so as to instantly compensate the load torque with some additional ripples in the electromagnetic torque. The characteristics of the stator currents i_d and i_q at start-up the machine draws a large current afterwards we notice a decrease as the machine has the normal operating regime. The decoupling introduced by backstepping control (id = 0) applied of PMSM illustrated by stator current i_d and i_q.

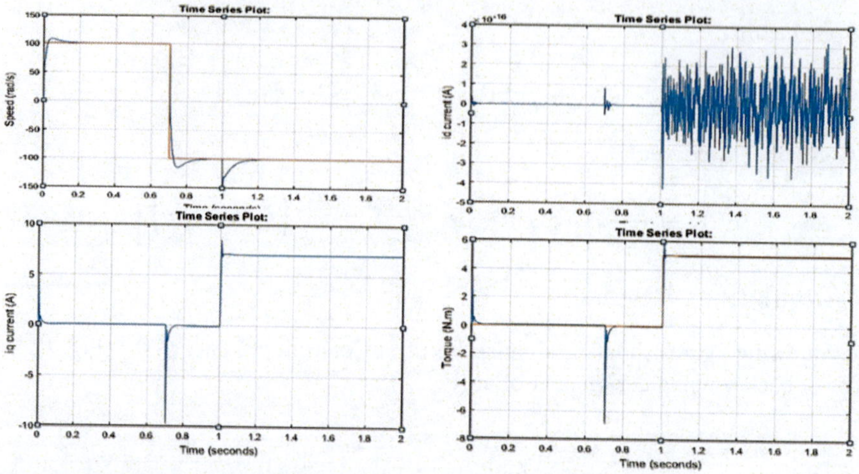

Fig. 2. Speed id current, and torque tracking a load torque variation for load torque 5 N.m at time t = 1 s at constant speed, variable load torque and variable speed

8 Conclusion

A non-linear control has been proposed, namely a Backstepping controller. This controller is based on a recent methodology using the Lyapunov function. The synthesis led to a globally asymptotically stable nonlinear controller. The backstepping controller has the disadvantage of the persistence of the static error. To remedy this, an integral action has been associated to eliminate this error. Finally, a comparison between the methods used allows us to underline the advantage of the Backstepping control because it does not lead to the cancellation of useful non-linearities and allows us to pursue stabilization or tracking objectives, rather than linearization objectives. This optimization method can also be used to avoid the heating of LEDs in optics [5, 6] and this study's findings also emphasize the need to use wavelets [7–10], integrated with data analysis techniques often employed in biomedical signal processing, such as ICA-NMF-SVD-PCA [11–13], to further improve the aforementioned techniques' efficacy.

References

1. Wang, Z.: Challenges faced by electric vehicle motors and their solutions. IEEE Access **9**, 5 (2021)
2. Achour, H.B., Ziani, S., Chaou, Y., ElHassouani, Y., Daoudia, A.: Permanent magnet synchronous motor PMSM Control by combining vector and PI controller. WSEAS Trans. Syst. Control **17**, 244–249 (2022)
3. Bimal, K.B.: Modern Power Electronics and AC Drives. Machine Press, Beijing (2006)
4. Chaou, Y., Ziani, S., Achour, H.B., Daoudia, A.: Nonlinear control of the permanent magnet synchronous motor PMSM using backstepping method. WSEAS Trans. Syst. Control **17**, 56–61 (2022). https://doi.org/10.37394/23203.2022.17.7
5. Ouhadou, M., El Amrani, A., Messaoudi, C., Ziani, S.: Experimental investigation on thermal performances of SMD LEDs light bar: Junction-to-case thermal resistance and junction temperature estimation. Optik. **182**, 580–586 (2019)

6. Ouhadou, M., El Amrani, A., Ziani, S., Messaoudi, C.: Experimental modeling of the thermal resistance of the heat sink dedicated to SMD LEDs passive cooling. In: Proceedings of the 3rd International Conference on Smart City Applications (2018)
7. Ziani, S., El Hassouani, Y.: A new approach for extracting and characterizing fetal electrocardiogram. Traitement du Signal. **37**(3), 379–386 (2020). https://doi.org/10.18280/ts.370304G
8. Ziani, S., Hassouani, Y., Farhaoui, Y.: An NMF based method for detecting RR interval. In: Farhaoui, Y., Moussaid, L. (eds.) Big Data and Smart Digital Environment. SBD, vol. 53, pp. 342–346. Springer, Cham (2018). https://doi.org/10.1007/978-3-030-12048-1_35
9. Ziani, S., El Hassouani, Y.: Fetal-maternal electrocardiograms mixtures characterization based on time analysis. In: 2019 5th International Conference on Optimization and Applications (ICOA), pp. 1–5 (2019). https://doi.org/10.1109/ICOA.2019.8727619
10. Ziani, S., Jbari, A., Bellarbi, L.: QRS complex characterization based on non-negative matrix factorization NMF. In: 2018 4th International Conference on Optimization and Applications (ICOA), pp. 1–5 (2018). https://doi.org/10.1109/ICOA.2018.8370548
11. Ziani, S., Jbari, A., Bellarbi, L., Farhaoui, Y.: Blind Maternal-Fetal ECG separation based on the time-scale image TSI and SVD – ICA Methods. Proc. Comput. Sci. **134**, 322–327 (2018). https://doi.org/10.1016/j.procs.2018.07.179
12. Ziani, S., El Hassouani, Y.: Fetal electrocardiogram analysis based on LMS adaptive filtering and complex continuous wavelet 1-D. In: Farhaoui, Y. (ed.) Big Data and Networks Technologies, vol. 81, pp. 360–366. Springer, Cham (2020). https://doi.org/10.1007/978-3-030-23672-4_26
13. Ziani, S., Jbari, A., Belarbi, L.: Fetal electrocardiogram characterization by using only the continuous wavelet transform CWT. Int. Conf. Elect. Inf. Technol. (ICEIT) **2017**, 1–6 (2017). https://doi.org/10.1109/EITech.2017.8255310

Practical Implementation of Pseudo-random Control in Step-Down Choppers and Its Efficiency in Mitigating Conducted Electromagnetic Emissions

Zakaria M'barki(✉), Kaoutar Senhaji Rhazi, and Youssef Mejdoub

Laboratory of Networks, Computer Science, Telecommunication, Multimedia (RITM), Higher School of Technology ESTC, Hassan II University, Casablanca, Morocco
mbarki.ensem@gmail.com

Abstract. Recently, pseudo-random modulation has become popular for DC-DC converter control. When compared to typical PWM technology, this approach is relatively simple to implement and has a high EMI suppression efficiency. This is because the harmonic power is spread in such a manner that the peak harmonics at discrete frequencies are substantially smaller. Many topologies of the pseudo-random method will be described in this study, including random pulse width modulation (RPWM), random carrier frequency modulation (RCFM), and random pulse position modulation (RPPM). Following that, we will provide an innovative hardware implementation of the various spread spectrum PWM approaches utilizing a low-cost solution based on the Atmega328p microcontroller. Finally, we will evaluate the impact of this random control in its many aspects on the electromagnetic emission reduction process performed in a Buck converter considered as a harmful source for its surrounding environment.

Keywords: Buck converter · Random PWM · Conducted EMI · Microcontroller ATMEGA328P

1 Introduction

In recent years, semiconductor-based power converters have become commonplace in various applications, particularly in power electronics, power engineering, and drives. However, these converters are hazardous to their surroundings and constitute a significant EMC risk. Indeed, the higher frequency and fast switching speed of semiconductor materials encourage the creation of large di/dt current and dv/dt voltage differentials, which cause significant electromagnetic disturbances [1] in the conducted and radiated modes in the frequency range [150 kHz, 1 GHz].

To address this shortcoming, a traditional approach provided in EMC filters [2, 3] demonstrates the efficient suppression of electromagnetic disturbances. However, where size is a critical consideration, such as in aerospace and automotive applications, another solution for minimizing conducted EMI noise should be available [4, 5].

Y. Farhaoui et al. (Eds.): ICAISE 2022, LNNS 635, pp. 674–682, 2023.
https://doi.org/10.1007/978-3-031-26254-8_98

This study introduces a critical spectral spreading approach whose primary goal is to reduce or even eliminate EMI noise levels by dispersing harmonic energy over a broad frequency range [6, 7]. Because pseudo-random modulation [8] is based on the randomization of three parameters (duty cycle, delay time, and period), we refer to it in three distinct ways [9]: random pulse width modulation (RPWM), random pulse position modulation (RPPM), and random carrier frequency modulation (RCFM).

In this regard, a low-cost solution based on the Atmega328p microcontroller has been presented for implementing different pseudo-random PWMs. The development of such a digital implementation strategy is prompted by the fact that analog implementations have a number of limitations, including low reliability, environmental sensitivity, and aging.

This paper will overview the three most used spread spectrum PWM techniques. The three randomized algorithms will then be implemented using the Atmega328p microcontroller [10]. Finally, given the three methodologies already offered, he will concentrate on the spectrum analysis of the conducted electromagnetic disturbances created by a Buck converter to execute a worldwide optimization to comply with the mandated EMC requirements [11].

2 The Fundamental Framework for Pseudo-random Modulation

The theory of pseudo-random modulation is based on developing a random bit sequence of 0 and 1 using linear feedback shift registers (PRBS). This sequence (P) is used to operate a multiplexer 2/1, which will also address two forms of triangle input signals 180° out of phase (c and $\overline{c}$) to the carrier output (R). The procedure of comparing the reference signal to this random carrier to create the control signal for the switch follows (Fig. 1).

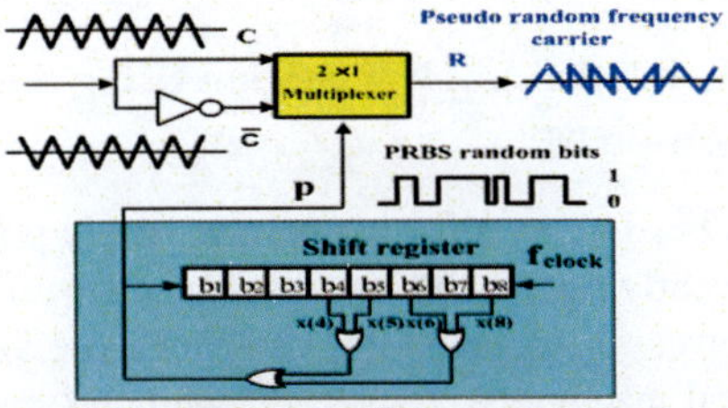

Fig. 1. The fundamental structure of pseudo-random modulation.

A typical PWM control signal for forced-switching converters such as choppers and voltage inverters is shown in Fig. 2. Three parameters entirely define it: the modulation period (T), the duty cycle (d) and the delay time (δ).

The approach of pseudo-random modulation is based primarily on the randomization of three parameters: the pulse position $\boldsymbol{\varepsilon_k}$, the switching time $\boldsymbol{T_k}$, and the duty cycle $\boldsymbol{d_k}$. . In reality, the randomization of these factors permits the output harmonic spectra to be spread and disseminated continually. Random Pulse Width Modulation (RPWM) with variable duty cycle, Random Carrier Frequency Modulation (RCFM) with variable switching period, and Random Pulse Position Modulation (RPPM) with variable pulse position are the resulting kinds. Table 1. outlines all that has come before.

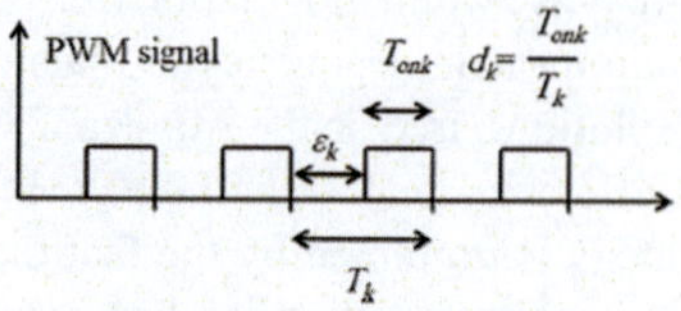

Fig. 2. Pulsed control signal.

Table 1. Different pseudo-random modulation patterns.

Configuration	T_k	d_k	δ_k
DPWM	Constant	Constant	Constant
RPWM	Constant	Random	Constant
RPPM	Constant	Constant	Random
RCFM	Random	Constant	Constant

3 Method Proposed

3.1 Pseudo-random Modulation Implementation Using the Arduino Uno Board in Real-World Applications

Random modulation has become an effective mechanism in power electronics system control. Furthermore, the early uses of this novel approach in power electronics focused on noise reduction in power semiconductors. Later, this use was expanded to reduce acoustic noise in electrical machinery. Thus, this technique ensures that the more the spectrum is uniformly dispersed, the less its amplitude is significant and the better the random PWM is in terms of EMC and noise mitigation.

Many research have been conducted on the digital application of the pseudo-random modulation approach. Indeed, FPGAs (Field-Programmable Gate Array),which are programmable circuits that typically incorporate a pseudo-random flow generator and a PWM modulator, are valuable in this regard. However, they have a substantial downside, which is the high cost of deployment. A solution based on the Atmega328p microcontroller is offered to address this. This new approach is reliable, easy to implement and economic. The designed implementations are built around an interrupt function that uses the microcontroller's Timer1.

To create the pseudo-random PWM signal in this work, the Arduino board with the Atmega328p microcontroller is employed. Timer1 has two operational modes: "Fast PWM Mode" and "Phase Correct PWM Mode". As the name implies, the Fast PWM Mode permits a PWM signal with a greater frequency than the Phase Correct PWM Mode but with a lower resolution.

Indeed, the 14 Fast mode selected in the setup of our TCNT1 timer is based on a single-slope operation, which means that the timer is only increased from BOTTOM = 0 to TOP = ICR1, then restarts from BOTTOM = 0. (overflow). It eventually achieves a comparative value of OCR1x ("x" meaning A or B). When there is equality, the comparison unit (Waveform Generation block) outputs a PWM signal on pin OC1x with the frequency:

$$F_{out} = \frac{F_{clk}}{\mathbf{N}*(1+\mathbf{ICR}1)}$$

where F_{clk} is the 16 MHz clock frequency for the Arduino Uno, and N is the Prescaler value (see Fig. 3).

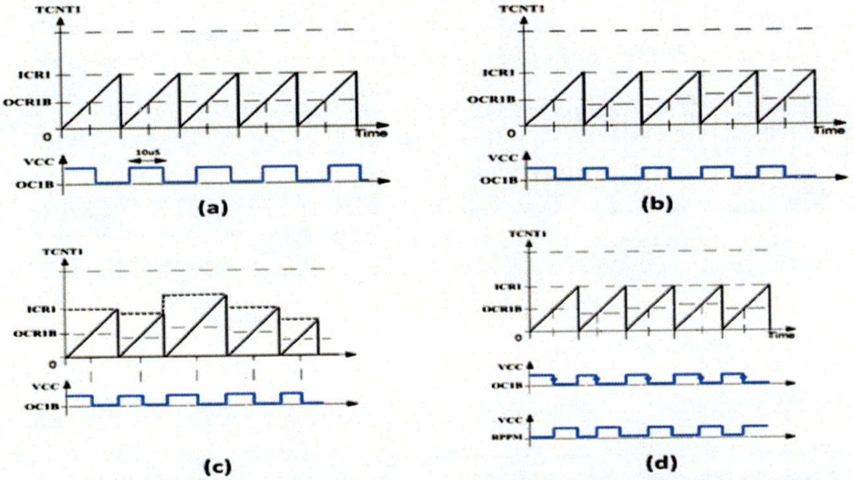

Fig. 3. Time diagram for (a) Deterministic Pulse Width Modulation (DPWM), (b) Random Pulse Width Modulation (RPWM), (c) Random Carrier Frequency Modulation (RCFM), (d) Random Position Pulse Modulation (RPPM).

The pseudo-random modulation approach devised for our application seeks to provide an RPWM control signal for the serial chopper with a duty cycle of 45 to 55 percent and a fixed switching frequency of 100 kHz. (Fig. 3.b). Then, we'll look at generating an RCFM control signal with a switching frequency between 80 and 120 kHz, with an average of 100 kHz and a constant duty cycle of 50%. (Fig. 3.c). Finally, using the NE555 circuit and the RPWM control signal, we will produce an RPPM control signal with a pulse position ranging from 0 us to 5 us. (Fig. 3.d). The "random" function is used to produce a variable z for the three previously predicted pseudo-random methods.

This function creates a pseudo-random integer number between a lower and higher band. These restrictions can be changed to get the required range of fluctuation.

The proposed deterministic pulse width modulation (DPWM) approach entails creating a control signal with a defined frequency of 100 kHz and a constant duty cycle of 0.5. (Fig. 4).

As previously stated, to produce an RPWM signal, we must randomly modify the duty-cycle parameter ($\boldsymbol{d_k}$) by operating on the value of register OCR1B using the "random" function. To maintain the required average of d = 0.5, we use an interval $\boldsymbol{d_k} \in [0.45, 0.55]$. After running the program in the Arduino IDE, timer1 generates the RPWM signal on the Uno module's output pin 10 (OC1B) (Fig. 5).

Similarly, the production of an RCFM signal is based on the random fluctuation of the period parameter ($\boldsymbol{T_k}$) around an average by operating on the value of register ICR1 with the "random" function while maintaining a fixed duty cycle $\boldsymbol{d_k} = 0.5$. In our scenario, we select an interval where $\boldsymbol{T_k}$ = [8.34 us, 12.5 us] (Fig. 6).

The RPPM signal is derived from the RPWM approach, in which the randomized falling edge controls a monostable based on the NE555 circuit to create an output pulse with a duration of 5 us (Fig. 7).

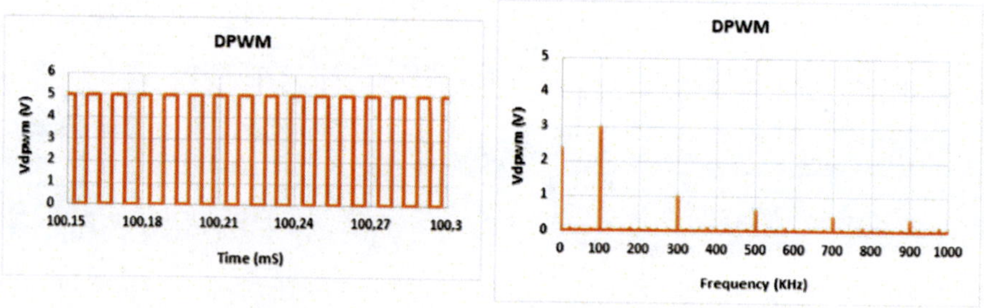

Fig. 4. DPWM command signal and its spectra.

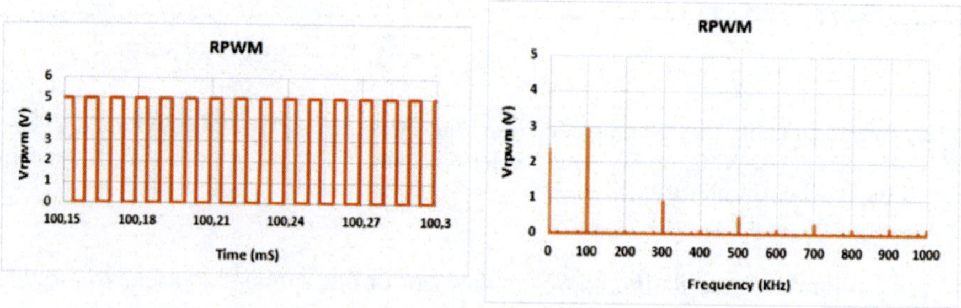

Fig. 5. RPWM command signal and its spectra.

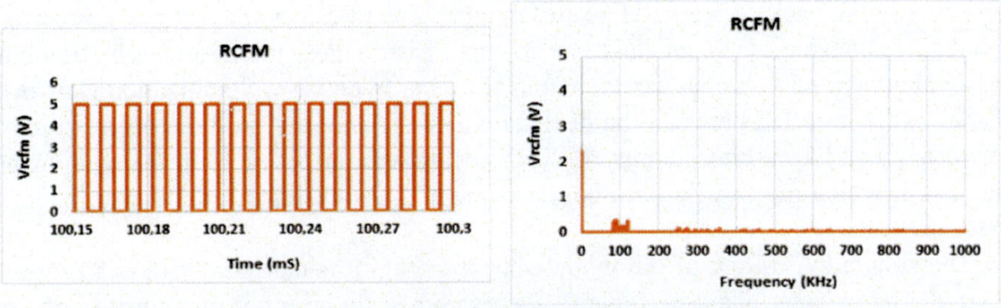

Fig. 6. RCFM command signal and its spectra.

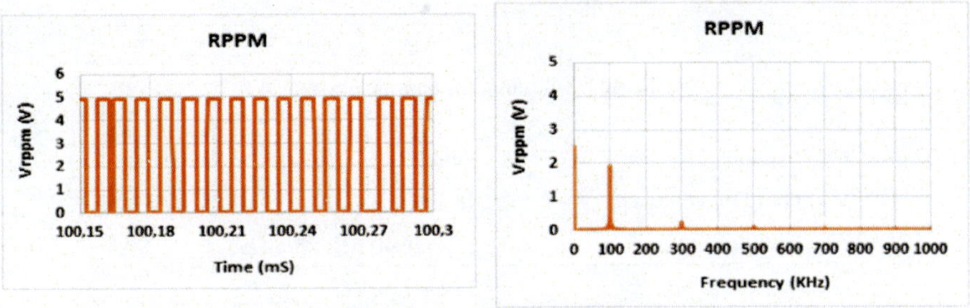

Fig. 7. RPPM command signal and its spectra.

4 Results and Discussion

4.1 Conducted Electromagnetic Interference Measurement in a Buck Converter

The static converter under consideration is a voltage step-down chopper placed between the load R_{load} and the solar generator V_{dc}, which is represented by a DC voltage source V_{dc} (Fig. 8). We used a chopper model comprised of an 15ETH06 freewheeling diode and an IRFP460 MOSFET transistor for this application (fast: used at very high frequencies and operated with a Power around 1KW). A PWM logic signal drives the MOSFET transistor.

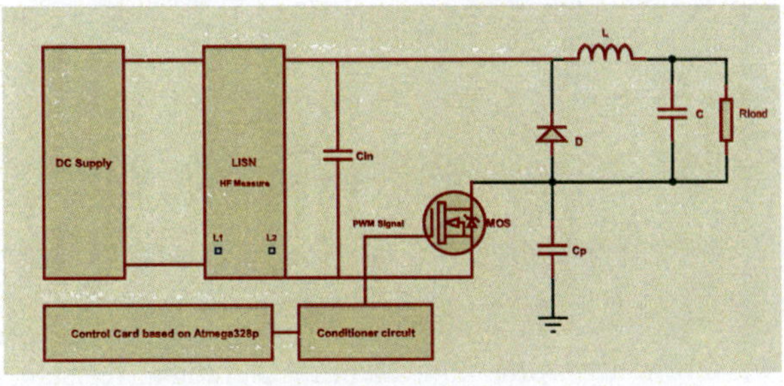

Fig. 8. The general framework of the conducted emissions testing for the Buck converter.

Because the buck converter is a substantial source of electromagnetic disturbances in conducted mode, it is beneficial to be able to measure these interferences by through the employment of a mechanism known as the Line Impedance Stabilization Network "LISN". It is installed between the device under test "chopper" and the power supply network, much like a filter.Its purpose is to separate the network from the equipment under test, which may cause both common-mode and differential-mode disturbances (buck converter).

The simulation design of the model shown in Fig. 8 is carried out in Isis Proteus using the parameters listed in Table 2. We examined the spectral composition of the voltage V_{lisn} image of the perturbations in differential and common modes. The primary goal is to keep the frequency content of this voltage (Power Spectral Density) as low as possible in the [10 kHz, 30 MHz] range.

Table 2. Simulation parameters

Parameter	Value
V_{in} input DC voltage	48 V
C_{in} capacitor	2.2 mF, 40 nH, 25 mΩ
L inductance	330 uH, 0.5 Ω
C capacitor	560 nF
R_{load}	12 Ω
C_p parasitic capacitor	130 pF, 520 nH, 0.3 mΩ

The interest in these projected techniques (DPWM, RPWM, RCFM, and RPPM) and their performances in terms of decreasing conducted electromagnetic disturbances created by the static converter will now be discussed. The simulation findings are as follows:

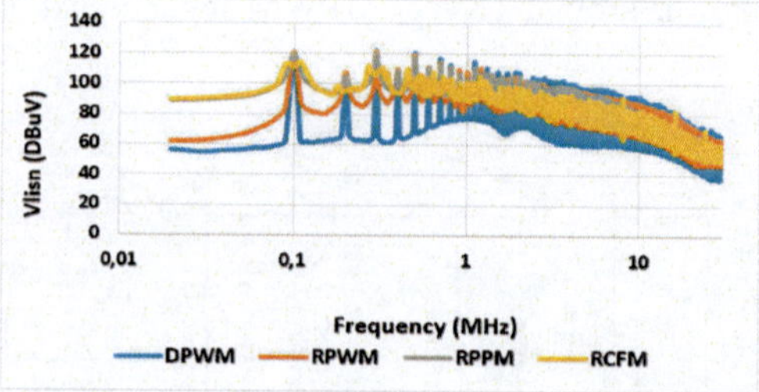

Fig. 9. Spectral content of the voltage V_{lisn} for the different methods already predicted.

The voltage waveform across the equivalent measurement resistance V_{lisn} swings rapidly in amplitude, as seen in Fig. 9. As a result, the spectral variations analysis of conducted disturbances, which is composed chiefly of discrete power harmonics, displays a certain richness in frequencies and exceeds the EMC normative limitations for DPWM approach. As a result, we conclude that this converter type must be supplied with devices to mitigate these disturbances; otherwise, it will endanger systems running in the same environment. Indeed, using pseudo-random modulation is proving to be a practical approach. After carefully examining the other techniques' results, it is evident that the RCFM approach provides an equal power spectrum distribution across a more comprehensive frequency range than the others. This is why its vital feature is the efficient elimination of harmonic peaks, which leads to a gain (approximately 20 dBuV compared to the DPWM approach) in terms of electromagnetic compatibility.

5 Conclusion

In conclusion, static DC-DC converters are especially harmful to their surroundings because they offer significant electromagnetic compatibility limitations. As a result, a pseudo-random technique has been proposed and digitally implemented using the well-known Atmega328p microcontroller, resulting in a very dependable and cost-effective solution. In this regard, we produced the various pseudo-random control topologies by adding a random component to the features of the typical PWM control signal (duty cycle, period, and delay time). Following that, we could notice the major interest of this pseudo-random approach by evaluating the spectrum content of the conducted electromagnetic emissions generated by the studied Buck converter. Indeed, the RCFM method has demonstrated an enormous efficiency in spreading the power spectrum of disturbances, which contributes to obtain a significant EMC gain.

References

1. Muttaqi, K.M., Haque, M.E.: Electromagnetic interference generated from fast switching power electronic devices. Int. J. Innov. Energy Syst. Power. 3(1) (2008)
2. Kotny, J.L., Duquesne, T., Idir, N.: Design of EMI filters for DC-DC converter. In: 2010 IEEE Vehicle Power and Propulsion Conference, pp. 1–6, 1–3 September 2010
3. Chen, W., Zhang, W., Yang, X., Sheng, Z., Wang, Z.: An Experimental study of common- and differential-mode active EMI filter compensation characteristics. IEEE Trans. Electromagn. Compat. **51**(3), 683–691 (2009)
4. M'barki, Z., Rhazi, K., Mejdoub, Y.: A proposal of structure and control overcoming conducted electromagnetic interference in a buck converter. Int. J. Power Electron. Drive Syst. **13**, 380–389 (2022). https://doi.org/10.11591/ijpeds.v13.i1.pp380-389
5. M'barki, Z., Rhazi, K., Mejdoub, Y.: A novel fuzzy logic control for a zero current switching-based buck converter to mitigate conducted electromagnetic interference. Int. J. Electr. Comput. Eng. (IJECE) **13**(2), 1423–1436 (2023)
6. M'barki, Z., Rhazi, K.: Optimization of electromagnetic interference conducted in a devolver chopper. In: Hajji, B., Mellit, A., Marco Tina, G., Rabhi, A., Launay, J., Naimi, S. (eds.) Proceedings of the 2nd International Conference on Electronic Engineering and Renewable Energy Systems, ICEERE 2020. LNEE, vol. 681, pp. 523–529. Springer, Singapore (2021). https://doi.org/10.1007/978-981-15-6259-4_55
7. Marsala, G., Ragusa, A.: Spread spectrum in random PWM DC-DC converters by PSO&GA optimized randomness level, pp. 1–6 (2017)

8. Lim, Y., Wi, S., Kim, J., Jung, Y.: A pseudorandom carrier modulation scheme. IEEE Trans. Power Electron. **25**(4), 797–805 (2010)
9. Boudouda, A., Boudjerda, N., Bouzida, A., Aibeche, A.: Dual randomized pulse width modulation technique for buck converter fed by photovoltaic source. Revue Roumaine des Sci. Techn. - Serie Électrotechnique et Énergétique. **63**, 289–294 (2018)
10. Chariag, D., Sbita, L.: Implementation of different randomized PWM schemes using Atmega328p microcontroller for EMI reduction in boost converter. Electr. Eng. **102**(2), 1063–1071 (2020). https://doi.org/10.1007/s00202-020-00934-7
11. Fang Lin, L., Hong, Y.: Investigation of EMI, EMS and EMC in power DC/DC converters. In: The Fifth International Conference on Power Electronics and Drive Systems, vol. 1, pp. 572–577 (2003)

Prediction of Sudden Death Due to COVID-19 Using Machine Learning Models

Ibtissam Chouja[1](✉), Sahar Saoud[2], and Mohamed Sadik[1]

[1] Engineering Science Laboratory, Ibn Zohr University, Agadir, Morocco
ibtissam.chouja@edu.uiz.ac.ma

[2] Higher School of Technology, Ibn Zohr University, Dakhla, Morocco

Abstract. The early classification of COVID-19 patients severity can help save lives by giving to doctors valuable instructions and guidelines for the cases that may need more attention to survive. This paper aims to classify cases depending on their severity into three classes: "survivor", "sudden death" and "death" using electronic health records (HER). The first class represents positive cases discharged from the hospital after being treated for COVID-19. While the second and the third classes are describing the level of cases severity based on the interval of death. We called the highest severity class "sudden death" to identify critical cases with a high risk of death in the first two days of admission, while the "death" class includes severe cases with an interval of death beyond two days. The sudden death class represents the biggest challenge for this classification as the number of samples representing this case is very small. This paper presents a triage system for COVID-19 cases using four machine learning algorithms (KNN, Logistic Regression, SVM, and Decision tree). The best classification results were obtained using Logistic Regression and SVM models.

Keywords: Electronic health records · EHR · Machine learning · COVID-19 · Imbalanced data

1 Introduction

The COVID-19 pandemic has caused multiple damages in different aspects and challenged medical services on several occasions especially when the toll of cases and deaths has reached the peak in some countries like France, Italy, and the USA. The health system has been overwhelmed by the huge number of infected people and with the absence of predictive systems, doctors had to decide by themselves which case should be saved. Millions of dollars have been spent inefficiently on controlling and combating the outbreak of coronavirus [1], hence, the WHO estimated 14.9 million excess deaths in 2020 and 2021, calculated as the difference between the deaths that have occurred and expected deaths, this difference could be associated directly or indirectly to the pandemic [2]. Several actions have been taken to ensure the continuity of health system services. Therefore, Countries must invest in stronger health information systems to generate data for situation analysis and better decisions to avoid future crises. Sudat et al.[3] highlighted

Y. Farhaoui et al. (Eds.): ICAISE 2022, LNNS 635, pp. 683–689, 2023.
https://doi.org/10.1007/978-3-031-26254-8_99

the importance of the electronic health record (HER) in supporting pandemic response and proposed working in collaborative teams including informatics and data analytics to support medical staff.

One use of the EHR is to build a triage system for cases to light the burden on the health system in a crisis and to reduce mortality [4]. Many studies have been done for this reason, but most of them classify cases only into two classes: survivor (non-severe cases) or death (severe cases). For example, to predict the mortality risk, Mahdavi et al. have used a model based on SVM [5]. While Yan et al. have developed an XGBoost machine learning-based model for the same prediction [6]. On other hand, Amini et al. proposed a comparative study between multiple classifiers to predict mortality risk where Random Forest has shown the best results [7].

However, the excess of death could be caused by sudden death [8] that may happen in COVID-19 [9] or severe cases that needed an emergent intervention in the few hours or first days of admission. What gives importance to the early prediction of high severe cases (sudden death).

Sudden death has been described in different manners. Yang et al. expect sudden death as death within 24 h after admission [10]. While Coleman et al. and Giudicessi [8, 11] Suggested a possible causal link between the rise in the incidence of out-of-hospital sudden death and COVID-19.

In this paper, we propose a system of triage able to classify patients depending on their emergency by taking into consideration sudden death risk, in order to help guide doctors to cases with a higher emergency.

2 Material and Methods

2.1 Data Collection

In this study, we used electronic health records published by Mahdavi et al. [12], collected from 492 patients who were admitted between February 20th, 2020, and May 4th, 2020 to Masih Daneshvari Hospital the largest respiratory and pulmonary care center in Tehran, Iran. The data contain demographic information, vital signs, and laboratory results (Coagulation, Kidney, Liver, Complete blood count, and blood gas). The laboratory results were collected on the first day of admission. We enrolled this study on 436 patients after deleting missing values.

To classify the data, we divided the outcome into three classes: Survivor ($n = 272$) to describe people who survived after being treated for COVID-19, sudden death ($n =$ 39 critical cases) to describe deceased patients at the first two days of admission, and death ($n = 125$) to describe the late death above the second day of admission.

2.2 Proposed Model

Figure 1 illustrates the proposed model architecture of this work. The data used in this study is imbalanced data with a skewed distribution of classes which is a common problem when dealing with real-world data [13]. To avoid biased results toward the majority class we decided to apply a data-level approach to uniform the classes distribution by oversampling the minority classes using the synthetic minority over-sampling technique (SMOTE), which is a data augmentation technique for tabular data, it consists of duplicating samples from the minority class without providing additional information. But before that, we must select the best features and parameters to build a useful model for classification.

We chose the P-value and principal component analysis (PCA) to analyze and select features according to their importance. While the Gridsearchcv was employed to guarantee the best parameters for our model.

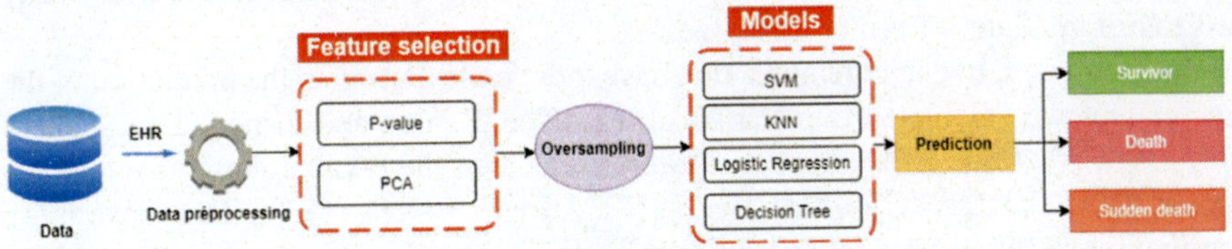

Fig. 1. Illustration of the model framework

Both the data oversampling and the feature selection steps have been applied only to the training set to avoid over-fitting.

In this study, we used four models to classify the data into three classes "Survivor, death, and sudden death". The first model is Support Vector Machine for classification (SVC), it's one of the most popular algorithms used to classify data by finding the best hyper-plan that separates classes by maximizing the margin between support vectors and thc hyper-plan. The second model is K-nearest neighbors (KNN), which assumes that similar things are close to each other. The similitude of points is calculated by the distance between samples. The third model is Logistic Regression, which is used to predict the probability of the dependent variable. Finally, Decision tree that consist of dividing the dataset into smaller subsets and building a tree-structured classifier.

2.3 Validation Metrics

To measure the efficiency of a classification model we generally calculate the accuracy, but in the case of imbalanced data, the accuracy wouldn't be a good metric of validation and could be misleading [13]. Instead, we decided to use the ROC AUC score and the F1 score which balances the precision and recall.

- Precision: ratio of the total correctly classified positive observations (TP) to the total positive predictions (TP and FP).
- Recall: ratio of total correctly predicted positive observations to the total of all observations in the actual class.
- F1 score is the harmonic mean of precision and recall.
- AUC-ROC score: (Area Under the Receiver Operating Characteristics) evaluate the global performance of a model. The near to the 1 means the model has a good measure of separability.

3 Results and Discussion

As we see before, this is imbalanced data where most of the cases are survivors (62.39%) or death (28.67%) only a few cases are representing sudden death (8.94%). A special pre-treatment was necessary because the use of this data in its native form for the predictive model might cause a lot of errors as it will classify most of the cases as non-sudden death. What we want from the model is to be certain when a non-sudden death occurs. The challenge of the proposed model is to detect patterns that give signs of sudden death to ensure a good classification and to avoid the overfitting caused by the imbalanced data. So, the first task of the proposed model was to learn the characteristics of the minority class to differentiate it from other classes.

On other hand, to detect features that have more contribution to the prediction of the three classes we calculated two statistical tests The p-value and Principal component analysis (PCA). Table 1 Describes the features selected. The P-value shows that features like 'SPo2', 'CVD', 'MCV', 'MCHC', 'PTT', 'PH', 'BUN', 'Cr', 'LDH' have more contribution on this classification with p-value less than 0.05. Whilst to explain 95% of variance we need 19 principal components. We computed the correlation between each variable and the generated components, and those are the most representative variables: HTN, SPO2, BP min, PTT, Ph, Abs Lymph, PR, T, Sex, DM, CVD, Abs Neutr, ESR, WBC, Age.

Table 1. Description of selected features

	Feature	Description	P-value
Demographic information	Age	Age of patient	0.057
	Sex	Sex of patient	0.642
	HTN	Presence or absence of hypertension	0.688
	DM	Presence of Diabetes mellitus	0.847
	CVD	Presence of cardiovascular disease	0.000
Vital signs	BP min	Blood pressure min	0.138
	PR	Pulse rate	0.456
	T	Temperature	0.158
	SPO2	Blood oxygen saturation level	0.000
CBC Complete blood count	WBC	White blood cell count	0.062
	Abs Neutr	Absolute Neutrophil	0.123
	Abs Lymph	Absolute Lymphocyte count	0.051
	MCV	Mean corpuscular volume	0.016

(*continued*)

Table 1. *(continued)*

	Feature	Description	P-value
	MCHC	Mean corpuscular hemglobin concentration	0.014
Coagulation	ESR	Erythrocyte sedimentation rate	0.121
	PTT	Partial thromboplastin time	0.046
Kidney	BUN	Blood urea nitrogen	0.000
	Cr	Creatinine	0.000
General	LDH	Lactate dehydrogenase	0.005
Blood gas	Ph	potential hydrogen	0.007

The Impact of Comorbidities on COVID-19 Severity: To better understand how comorbidities are affecting the case's severity we calculated the contribution of each comorbidity for each class. The presence of comorbidities can decrease the odds of survival by increasing the risk of death, actually, 43.59% of sudden death had diabetes Mellitus, while 46.15% of them had cardiovascular disease and 46.15% had hypertension problems. On another hand, the combination of comorbidities can increase in advantage the risk of mortality (death and sudden death), for example, 67.01% of deceased patients had two comorbidities, but the couple of CVD and HTN is more likely to lead to mortality due to COVID-19 (72.72%).

Table 2. Best models results

Model	Roc		F1		Precision		Recall	
	Train	Test	Train	Test	Train	Test	Train	Test
Logistic regression	78,34%	80%	73%	71%	70%	72%	70%	72%
KNN	73%	67%	63%	66%	65%	61%	65%	61%
SVC	79%	79%	68%	70%	70%	70%	70%	70%
Decision tree	64%	62%	51%	61%	63%	44%	63%	44%

By ranking the four algorithms based on the selected performance metrics, we can see in Table 2 and Fig. 2 that Logistic Regression and SVC are the best classifiers for these three classes.

We conclude that there is a difference between normal death and sudden death in COVID-19 that can be predicted by machine learning models, although, the sudden death needs to be presented by more samples to confirm and improve the results. By comparing this work to previous studies, it's obvious that our model represents a new classification approach by considering sudden death, while most present models classify the cases only into two classes "Survivor" and "death".

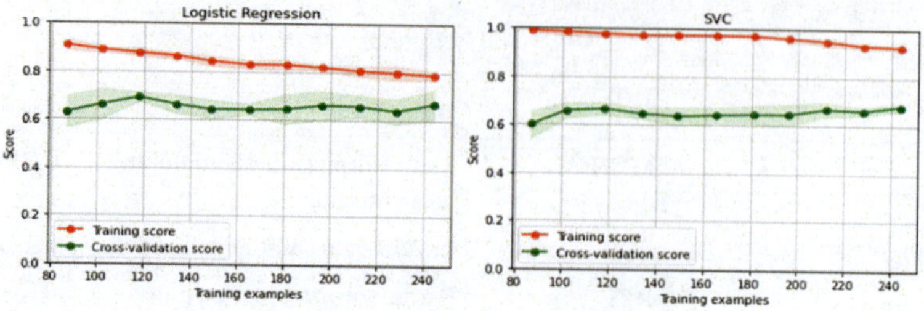

Fig. 2. ROC-AUC score for the best models

4 Conclusion

In this paper, we have proposed a system of triage for COVID-19 patients capable to predict the severity of cases with four machine learning algorithms that we have applied to electronic health records of COVID-19 patients from a respiratory hospital in Iran. The main problem of this work was to predict critical cases that may cause sudden death because of the rarity of data representing sudden death although we employed some strategies to overcome this issue by oversampling the data. The model has shown reliability but still needs to be confirmed with extra data.

References

1. Cutler, D.M., Summers, L.H.: The COVID-19 pandemic and the $16 trillion virus. JAMA **324**(15), 1495 (2020). https://doi.org/10.1001/jama.2020.19759
2. 14.9 million excess deaths associated with the COVID-19 pandemic in 2020 and 2021. https://www.who.int/news/item/05-05-2022-14.9-million-excess-deaths-were-associated-with-the-covid-19-pandemic-in-2020-and-2021. Accessed 10 Sep 2022
3. Sudat, S.E.K., Robinson, S.C., Mudiganti, S., Mani, A., Pressman, A.R.: Mind the clinical-analytic gap: electronic health records and COVID-19 pandemic response. J. Biomed. Inform. **116**, 103715 (2021). https://doi.org/10.1016/j.jbi.2021.103715
4. Estiri, H., Strasser, Z.H., Klann, J.G., Naseri, P., Wagholikar, K.B., Murphy, S.N.: Predicting COVID-19 mortality with electronic medical records. NPJ Digit Med. **4**(1), 15 (2021). https://doi.org/10.1038/s41746-021-00383-x
5. Mahdavi, M., et al.: A machine learning based exploration of COVID-19 mortality risk. PLoS ONE **16**(7), e0252384 (2021). https://doi.org/10.1371/journal.pone.0252384
6. Yan, L., et al.: An interpretable mortality prediction model for COVID-19 patients. Nat. Mach. Intell. **2**(5), 283–288 (2020). https://doi.org/10.1038/s42256-020-0180-7
7. Amini, N., Mahdavi, M., Choubdar, H., Abedini, A., Shalbaf, A., Lashgari, R.: Automated prediction of COVID-19 mortality outcome using clinical and laboratory data based on hierarchical feature selection and random forest classifier. Comput. Methods Biomech. Biomed. Eng. **26**(2), 160–173 (2022). https://doi.org/10.1080/10255842.2022.2050906
8. Giudicessi, J.R.: Excess out-of-hospital sudden deaths during the COVID-19 pandemic: a direct or indirect effect of SARS-CoV-2 infections? Heart Rhythm **18**(2), 219–220 (2021). https://doi.org/10.1016/j.hrthm.2020.12.010
9. Sudden cardiac death in COVID-19 patients, a report of three cases. https://www.futuremedicine.com/doi/epub/10.2217/fca-2020-0082. Accessed 01 Dec 2021

10. Yang, N., et al.: Sudden death of COVID-19 patients in Wuhan, China: a retrospective cohort study. J. Glob. Health **11**, 05006 (2021). https://doi.org/10.7189/jogh.11.05006
11. Coleman, K.M., Saleh, M., Mountantonakis, S.E.: Association between regional distributions of SARS-CoV-2 seroconversion and out-of-hospital sudden death during the first epidemic outbreak in New York. Heart Rhythm **18**(2), 215–218 (2021). https://doi.org/10.1016/j.hrthm.2020.11.022
12. Mahdavi, M., Chubdar, H., Zabeh, E., Lashgari, R., Kamrani, E.: Processed_Model_Data.zip. figshare, p. 123874 Bytes (2021). https://doi.org/10.6084/M9.FIGSHARE.14723883.V2
13. Tanha, J., Abdi, Y., Samadi, N., Razzaghi, N., Asadpour, M.: Boosting methods for multi-class imbalanced data classification: an experimental review. J. Big Data **7**(1), 1–47 (2020). https://doi.org/10.1186/s40537-020-00349-y

Profiles Behavior Analysis in Blockchain Social Network

Fatima Anter(✉), Fatna Elmendili, Mohammed Fattah, and Nabil Mrani

Image Laboratory, EST, Moulay Ismail University, Meknes, Morocco
f.anter@edu.umi.ac.ma

Abstract. Currently, a significant portion of the population connects to social networks. Websites such as Facebook, Twitter, LinkedIn, and Instagram, among others, enable users to exchange ideas, information, and messages as well as build friendships. These online social networking sites are becoming a likely avenue for malevolent users who create fake accounts, hijack normal accounts, or propagate fake news. Conversely, social network companies exploit the value of user-generated content for for-profit purposes. Social networks have advantages and disadvantages, but privacy and security are among the most significant weaknesses. An alternate solution is proposed to address the existing problems with social networks. In this regard, Blockchain is one of the most promising technologies and can potentially improve social networking in various ways. In this paper, we present a study of existing research on analyzing profiles' behavior in Blockchain-based social networks, focusing on Steemit, and explain the significance of Blockchain technology in social networks.

Keywords: Blockchain technology · Social network · Security · Decentralized social networks · Profiles behavior · Malicious user

1 Introduction

Social networks play a very important role in our lives today. They share ideas, information, and messages and create personal links between them, the most popular such as Facebook, Twitter, and Instagram. According to a recent study by Hootsuite and We Are Social[1] on the use of the Web and social networks, in January 2022, there are 4.62 billion users, 62% of the world's population use the Internet, and 58.4% are active on social networks. Compared to 2021, the number of Internet users has evolved by 10%, representing a significant increase over one year, or 424 million additional individuals.

Users make most of the content on social networks, making them a target for malicious attacks [1]. The growth in network users has resulted in increased threats and the spread of attacks. Social networks that are centralized always are easier to attack. Everyone who uses centralized social networking have a serious concern about user privacy. Its policies let them to access, retrieve, and store data from individual users in a central server, which increases the risk and effects of a data breach. One such case occurred in July 2020 when the Twitter[2] accounts of influential people were hacked in

Y. Farhaoui et al. (Eds.): ICAISE 2022, LNNS 635, pp. 690–695, 2023.
https://doi.org/10.1007/978-3-031-26254-8_100

a cryptocurrency-related scam. Also, the same thing happened when Facebook had the biggest crisis in its history in March 2018, Cambridge Analytica[3] collected the personal information of millions of Facebook users without their consent.

An alternative solution has been suggested to resolve the current problems in social networks. In this regard, Blockchain is among the most promising technologies.

The remainder of the paper is structured as follows: Sect. 2 contains a list of related works. Section 3 highlights the significance of Blockchain technology in social networks. Section 4 focuses on detecting fraudulent profiles on Steemit. Section 5 examines the most recent research on spam detection algorithms in Blockchain Social Network. Finally, the conclusion is presented in Sect. 6.

2 Background and Related Works

Several techniques have been used to analyze the profiles' behavior. Some of the techniques that we studied: Liu, Jiang, and Zhang [2] propose an iForest method for detecting anomalous activity in Blockchain-based social networks. Carnegie Mellon University data [3] is used. It includes login information for around 500 users, including login data, device, and location. However, only the user's login is used in this experiment. The isolation forest algorithm aims to isolate anomalous behaviors as an outlier. The smart contract receives the user data for classifying user behaviors into two categories normal or abnormal. In Blockchain, abnormal behavior is identified and recorded in the abnormal chain. Aji et al. [4] present a system for classifying spam messages utilizing Blockchain technology and machine learning with a 98.2% accuracy rate. Once 1143 data have been gathered, pre-processed, and cleaned using Tokenization, Stop Words, Bag of Words (BoW), and TF-IDF, they are ready for analysis. After the Logistic Regression algorithm classifies messages, the spammer's number will be recorded on the Blockchain. Proof of Credibility (PoC) is a novel Blockchain consensus developed in this research [5] to identify and prevent the propagation of fake news on social networks. Using the Twitter R library tool, the functionality of the PoC protocol was simulated on two datasets of noteworthy tweets (1003 tweets) obtained from various news sources on Twitter. The results of this experiment have clarified the effectiveness of PoC with an accuracy of about 89%. Another work [6] provides a solution aligned with a decentralized rumor detection system, and voting rules are used to judge the kind of message released on social media networks. The significant nodes, known as miners, can validate information; they determine whether a message is a rumor or not. Once a message is identified as a rumor, it will be marked with a red label, and rebroadcast will be forbidden. The message and its detection result are written to the block. Guidi and Michienzi [7] present an analysis to comprehend the behaviors of ordinary users and bots using Steemit as a case study to identify characteristics that may benefit bots detection techniques. They studied a dataset of around 29 million blocks comprising the activities of ordinary users as well as an initial group of bots. Overall, the study revealed that bots are much more active than regular users, especially regarding comments. The number of comments is a characteristic feature that can aid bot detection. They also follow a few accounts. However, they are popular and have a consistent number of accounts. Almost bots receive more SBD than STEEM. In addition, they utilize terms linked with the Steemit platform. In addition, presence network extensions and file extensions.

3 Blockchain and Social Network

Blockchain has emerged as a viable solution for addressing a variety of issues, including social networks. In this way, several Decentralized social networks have been proposed [8, 9]. Their primary aim is to protect the privacy of their users. Their data and their content are always under their control. There are helpful surveys of decentralized social networks [10, 11], which highlight the following characteristics:

- Decentralized social networks give users more privacy because no centralized entity can access their consumption or interactions. In traditional social networks, host businesses track users through cookies or analytics to serve personalized ads.
- The Blockchain allows the user to maintain anonymity. Instead of using real identities, Blockchain accounts are identified by public keys and addresses. This benefits those who are afraid to censor authorities, and publishing opinions or facts could put a journalist or author at risk.
- Thanks to the decentralization of data, a decentralized social network stores its data via a distributed ledger system, it means that in this system, no data are held by a central authority. Each node participant in the peer-to-peer has a copy of a ledger. That makes it more difficult to hack and tamp.
- One of the objectives of Blockchain-based social networks is to reward content creators for valuated content to limit the spread of fake news. Unlike traditional social networks, companies earn enormous profits from user-generated content and advertising.
- Another purported advantage of decentralized social networks is the absence of a central censorship authority, which enables free speech. Decentralized consensus processes guarantee user privacy and free speech. Current traditional social networking sites eliminate content that conflicts with business interests or serves another purpose.

4 Profiles Detection in Bosms: the Steemit Case Study

Steemit is the most popular decentralized social network with one million users, and it serves as an important case study compared to other BOSMs. The official launch of Steemit as a social platform occurred on July 4, 2016; it operates a blogging and social networking website on the Steem Blockchain database. It relies on a consensus mechanism referred to as delegated proof of stake (DPOS). Its primary goal is to incentivize users to be socially active, particularly in the production of information, by rewarding Those who are positively evaluated by the community.

As a Blockchain-based network, Steemit is incredibly secure. This is due to the remarkable characteristics of Blockchains. However, nothing is perfect. Some sophisticated, irreparable secrecy attacks [13] are emerging on a Blockchain. Unfortunately, no etude addresses malicious behavior on Steemit. However, as a result of the platform's incentive system for content sharing, users are able to get revenue immediately. This attracts some users to get a fast profit by posting unrelated comments to the topic of discussion and requesting money, views, upvotes, followers, or resteems. Or offer to buy/sell votes, resteems, or follows, and schemes that facilitate this. These offerings could be frauds or scams. A second type of dangerously malicious user sends malicious links or advertisements via message, post, or comment, directing victims to malware or phishing sites. Another malicious behavior on Steemit, some users post inappropriate content, such as when a user claims that she was reading her feed when she noticed a porn video posted by another user. In order to reduce spam, Stemeet limited the time between comments to 20 s and maintained posts and comments active for 7 days; after that, no one can modify or delete a post/comment. Moreover, there is no detection system for spam content. Steemit relies more heavily on community members to downvote spam content or reports it by email. Steemit allows users to post anything, but users nowadays are willing to share a significant amount of personal information, making them easy prey for reconnaissance attacks. Malicious users can gather and analyze users' profiles, relationships, and behaviors and then use this information to obtain passwords or impersonate users. Some users create multiple accounts for various reasons, including increasing their upvotes or number of followers, impersonation, spreading false information and advertisements, or utilizing these multiple accounts to launch an attack on Sybil [14] which could affect elected witnesses. Steemit allows users to create one free account to counteract this malicious behavior but must pay to create additional accounts.

5 Comparative Analysis

This section analyzed the current research on spam detection algorithms released between 2020 and 2021. Through the use of Table 1, an attempt has been made to offer researchers with a comparative analysis of various spam review detection methods and their estimated accuracy. MNB [15] is a novel method for classification algorithm for detection of fake news using Blockchain with heigh accuracy 95.20%, 91.92% as Precision, Recall F-measure and 4.87% as error rate. This system provides greater accuracy than traditional machine learning classification algorithms. Such as Naïve Bayes SVM, CNN, ANN, KNN, and Naïve Bayes. Paper [16] presented a method for detecting fake news and predicting fake user accounts and posts using Natural Language Processing, Reinforcement Learning (RL) and blockchain. The proposed method is more efficient than other models such as XGBoost, Random Forest, RNN, and LSTM.

Paper [2] introduced a new blockchain-based framework for the detection of user behavior anomalies. It could successfully determine which user exhibited anomalous behavior at what time. Paper [5] proposed a novel protocol called Proof of Credibility (PoC) that can validate shared information and detect fraudulent information through social networks. The results of this experiment have clarified the effectiveness of PoC with an accuracy of about 89%. Fall-Out, and False Discovery Rate about 10% and 9% respectively.

Table 1. Comparative analysis of various techniques for detection profile behavior.

Paper	Advantages	Disadvantage
[7]	Highlight some features that might be helpful in bot identification methods	Number of bots are small in the dataset
[2]	Detection anomaly behavior on Blockchain social media	A large number of dimensions remain unused. The user's login data only used for detecting abnormal user
[5]	Verification of shared information and detection of fraudulent information in social networks with a novel Blockchain consensus	The lack of a benchmark and standard datasets for examining the efficiency of PoCs
[4]	Classification of Spam messages using Blockchain technology and machine learning with an accuracy of 98.2%	
[12]	Detection of rumor using Blockchain technology	The proposed approach doesn't allow news resources to use Blockchain because writing and reading should be fast
[15]	mNB novel method for classification algorithm for detection of fake news using Blockchain with heigh accuracy 95.20%	
[16]	Using Blockchain and machine-learning techniques, detect fraudulent news and predict fake user accounts and postings	

6 Conclusions and Future Work

The future of social networks is probably in blockchain-based social networking platforms and application. Its advantages are centered on the protection of data and personal information. All interactions, storage, device security, and transaction verification are encrypted from end-to-end. They enable users to have more control over their data.

This paper's objective is to examine current research of profiles' behavior in Blockchain-based social networks, focusing on Steemit, it's a secure platform based on Blockchain. This does not mean that all its users are safe; malicious users will exploit the naivety of certain users using social engineering techniques, resulting in severe damage like data loss, account comprises, ransomware, and malware infection into the personal systems. Unfortunately, there is no detection system for spam content on Steemit. It relies more heavily on community members to downvote spam content or reports it by email. This presents an opportunity for us to plan, as future works, to examine further in depth the malicious actives on Steemit and to provide a method for detecting malicious users.

References

1. Xin, Y., Zhao, C., Zhu, H., Gao, M.: A survey of malicious accounts detection in large-scale online social networks. In: 2018 IEEE 4th International Conference on Big Data Security on Cloud (bigdatasecurity), IEEE International Conference on high Performance and Smart Computing, (HPSC) and IEEE International Conference on Intelligent Data and Security (IDS), pp. 155-58. IEEE, Omaha, NE (2018)
2. Liu, X., Jiang, F., Zhang, R.: A new social user anomaly behavior detection system based on blockchain and smart contract. In: 2020 IEEE International Conference on Networking, Sensing and Control (ICNSC), pp. 1–5. IEEE, Nanjing, China (2020)
3. Zyskind, G., Nathan, O., et al.: Decentralizing privacy: using blockchain to protect personal data. In: Security and Privacy Workshops (SPW), pp. 180–184. IEEE (2015)
4. Aji, G.S., Farizi, T.S., Farhan, M., Sari, R.F.: Machine learning based spam message classification system using blockchain technology. In: 2021 17th International Conference on Quality in Research (QIR): International Symposium on Electrical and Computer Engineering, pp. 129-34. IEEE, Depok, Indonesia (2021)
5. Torky, M., Nabil, E., Said, W.: Proof of credibility: a blockchain approach for detecting and blocking fake news in social networks. Int. J. Adv. Comput. Sci. Appl. (IJACSA), **10**(12) (2019)
6. Liu, X., Jiang, F., Zhang, R.: A new social user anomaly behavior detection system based on blockchain and smart contract. In: 2020 IEEE International Conference on Networking, Sensing and Control (ICNSC), pp. 1–5 (2020)
7. Guidi, M.: Users and bots behaviour analysis in blockchain social media. In: 2020 Seventh International Conference on Social Networks Analysis, Management and Security (SNAMS), pp. 1–89 (2020)
8. Le, J., Zhang, X.: BCOSN: a blockchain-based decentralized online social network . IEEE Trans. Comput. Soc. Syst. **6**(6), 1454–1466 (2019)
9. Li, C., Palanisamy, B. : Incentivized blockchain-based social media platforms: a case study of steemit . In: Proceedings of the 10th ACM Conference on Web Science, pp. 145-54. ACM, Boston Massachusetts USA (2019)
10. Guidi, B.: An overview of Blockchain online social media from the technical point of view. Appl. Sci. **11**(21), 9880 (2021)
11. Guidi, B.: When blockchain meets online social networks. Pervasive Mob. Comput. **62**, 101131 (2020)
12. Shahbazi, Z., Byun, Y.C.: Fake media detection based on natural language processing and blockchain approaches . IEEE Access **9**, 128442–128453 (2021)
13. Aggarwal, S., Kumar, N.: Attacks on blockchain. In: Advances in Computers, vol. 121, pp. 399–410. Elsevier (2021)
14. Swathi, P, Modi, C., Patel, D.: Preventing Sybil attack in blockchain using distributed behavior monitoring of miners. In: 2019 10th International Conference on Computing, Communication and Networking Technologies (ICCCNT), pp. 1–6, IEEE, Kanpur, India (2019)
15. Waghmare, A.D., Patnaik, G.K.: Social media fake news detection using MNB in blockchain. In: 2022 International Conference on Sustainable Computing and Data Communication Systems (ICSCDS), pp. 1198–1204. IEEE, Erode, India (2022)
16. Gao, H., Gao, T.: Prevention of rumor spreading based on blockchain. In: 2020 IEEE 20th International Conference on Communication Technology (ICCT), pp. 1174–1178 (2020)

PSO Optimization Algorithm for QoS Enhancement in IoT-Enabled WSNs

Abdelkader Benelhouri[1](✉), Hafida Idrissi-Saba[1], and Jilali Antari[2]

[1] Laboratory of Metrology and Information Processing, Faculty of Science, Ibn Zohr University, B.P. 8106, Agadir, Morocco
a.benelhouri@gmail.com

[2] Laboratory of Computer Systems Engineering, Mathematics and Applications, Ibn Zohr University, Polydisciplinary FacultyB.P. 271, 83000 Taroudant, Morocco
j.antari@uiz.ac.ma

Abstract. Through the recent progress in the field of microelectronics and the emergence of wireless communication technologies, wireless sensor networks (WSNs) have seen the light of day. However, one of the major problems of this type of networks is the energy efficiency. The overage is also a key problem as it determines the ability of a set of sensors to monitor a geographic area efficiently. In order to improve network deployment and throughput, this paper proposes a PSO-based approach to deploy nodes and select cluster heads efficiently. The suggested work has undergone significant QoS experimentation in a variety of scenarios. The results of this algorithm are investigated by implementing the Mod-LEACH algorithm in the data routing process. It is concluded that the PSO based clustering algorithm enhances the network throughput by improving nodes emplacement.

Keywords: WSNs · PSO · Coverage optimization · QoS enhancement

1 Introduction

The paradigm transition from static or wired networks to wireless networks has occurred through the development in micro electromechanical systems (MEMS) combined with the emergence of new information and communication technologies. A wireless system is a particular kind of system that transmits data or information between network devices through wireless communication [1, 2]. The fast evolution of wireless networks allowed the integration of the functionalities of data collection, processing and transmission in a single tiny device which is the wireless sensor [3, 4]. These tiny devices are connected as an Internet of Things (IoT) network to satisfy the crucial need for universal access and real-time monitoring. The IoT-enabled WSNs, which fosters good relationships between connected individuals and objects, improves the next generation of internet services. A base station that serves as a wireless hub and is known as an access point is what makes up an infrastructure-based wireless network. All wireless nodes can communicate through this access point, which is controlled by the network operators. The access point connects to other devices or nodes via a fixed wireless link, a fiber network connection, or cable

Y. Farhaoui et al. (Eds.): ICAISE 2022, LNNS 635, pp. 696–700, 2023.
https://doi.org/10.1007/978-3-031-26254-8_101

connections [1, 5]. The wireless sensor networks now offer a host of advantages over traditional networks, particularly in terms of simplicity and cost of deployment. This has led to the development of a wide range of applications for wireless sensor networks in the fields of health, the environment, industry, infrastructure, space activities, the military and many other fields [2, 3]. The energy efficiency of wireless networks is increased by using bio-inspired algorithms and evolutionary computing. The concepts of biological evolution form a basis for evolutionary algorithms [6, 7]. The process of flocking or swarming in vertebrates serves as the native inspiration for swarm intelligence (SI) as a new distributed intelligent technique for tackling optimization challenges. Particle swarm optimization (PSO) takes into account the phenomenon of swarming, which is seen in animal groups like fish schools, bee swarms, and flocks of birds, among others [8]. The quick convergence of PSO is one of its main benefits. In this work, wireless sensor networks' coverager optimization is carried out using Particle Swarm Optimization (PSO) algorithm.

This paper presents a PSO-based routing protocol to improve coverage and throughput in a wireless sensor network. To reduce the number of holes produced by random deployment, the proposed work employs the PSO algorithm for nodes distribution. To perform network data routing, a developed fitness function is involved in the PSO algorithm to select suitable nodes as cluster heads to aggregate data from other nodes. The rest of this paper is divided into the following sections. The problem formulation is detailed in Sect. 2. Proposed work is provided in Sect. 3. The simulation results is given in Sect. 4. Section 5 presents the conclusion and future work.

2 Problem Formulation

This section describes the problem formulation of coverage issue in WSNs.

2.1 Coverage Area of a Sensor

The coverage is a critical issue in wireless sensor networks as it affects the outcome of the perception spot exerted by a network. Random deployment, where the position of the sensors is not known in advance, is the most widely used mode for installing a wireless sensor network in difficult or dangerous environments such as battlefields or natural disasters zones. In this deployment mode many areas in the network remain uncovered which presents holes areas for the distributed network which affect its performance. The total area covered by the network is the union of the areas monitored by all nodes, it is noted:

$$\bigcup_{i=1}^{\mathrm{n}} S(i) \tag{1}$$

where $S(i)$ is the surface observable by a sensor i .

To overcome the coverage issue and enhance the network QoS, it is preferable to place sensor nodes in locations with the largest possible area coverage, for this purpose we will implement the Particle Swarm Optimization (PSO) algorithm to maximize the coverage of a distributed WSN.

2.2 Particle Swarm Optimization

The PSO technique uses a set of achievable solutions called "particles" that are deployed in the search space with random initial locations. The objective function values corresponding to the particle locations are evaluated. Then, the particles are moved in the search space by obeying rules inspired by the flight behavior of birds. Each particle is moved to the randomly weighted average of the best position the particle has encountered so far (P_{best}) and the best position encountered by the entire population of particles (g_{best}). Let $X_i = (x_{i1}, x_{i2}, \ldots, x_{iM})$ be the N-dimensional vector representing the position of the i_{th} particle in the swarm, $g_{best}i = \left[g_1, g_2, \ldots, g_n\right]$ is the position vector of the best particle in the swarm (i.e., the particle with the smallest objective function value). $P_{best}i = [p_{i1}, p_{i2}, \ldots, p_{iN}]$ is the position vector of the i_{th} particle and the velocity of the particle $V_i = [v_{i1}.v_{i2}, \ldots, v_{iN}]$ is the velocity of the ith particle. The particles evolve according to the equations:

$$V_{i;j}(k+1) = w \times V_{i;j}(k) + c_1 \times r_1 \times \left(P_{best_{i,j}} - X_{i,j}(k)\right) + c_1 \times r_1 \times \left(g_{best_j} - X_{i,j}(k)\right) \quad (2)$$

$$X_{i,j}(K+1) = X_{i,j}(k) + V_{i,j}(K+1) \quad (3)$$

3 Proposed Work

Our purpose is to find a set of coordinates (x_i, y_i) $(i = 1, 2\ldots, n)$ which gives the widest covered surface S_t by the network. In another saying, we will try to minimize the holes h_i of the network, so our fitness function will be defined as:

$$f(x, y) = \frac{1}{n_h}\sum_{i}^{n_h} h_i \quad (4)$$

where n_h denotes the total number of coverage holes.

For efficient data routing, we implement the PSO algorithm in the cluster head selection process, the objective function of the optimization algorithm will be designed to reduce the total energy intended for data communication, the average distance between the cluster node and its cluster members, and the transmission distance between the selected cluster head and the base station, it is defined as follows:

$$f(s) = \alpha * AvgDist_{CH-CM} + \beta * AvgDist_{CH-BS} + \gamma * E_{tot} \quad (5)$$

$\alpha, \beta,$ *and* γ are predefined wights assigned to have a direct effect on the fitness score.

4 Simulation and Results

With the aim of evaluating the proposed work, we will apply nodes emplacement produced by the PSO algorithm and compare the network performance with the case where sensor nodes are distributed randomly. For nodes clus- tering and data communication we will count on the clustering hierarchy with incorporating the PSO-based CH selection algorithm, and compare the throughput of PSO-network with the random-Network.

4.1 Coverage Optimization

Figure 1 presents the initial emplacement of nodes with random distribution mode for both scenarios, the green circles illustrate the number of holes that are present in the network. Figure 2 shows the estimated network emplacement by the PSO algorithm. It is evident that new emplacement obtains a great coverage than the random distribution, the PSO was successful to reduce the number of holes by moving the nodes into new better locations. The new estimated.

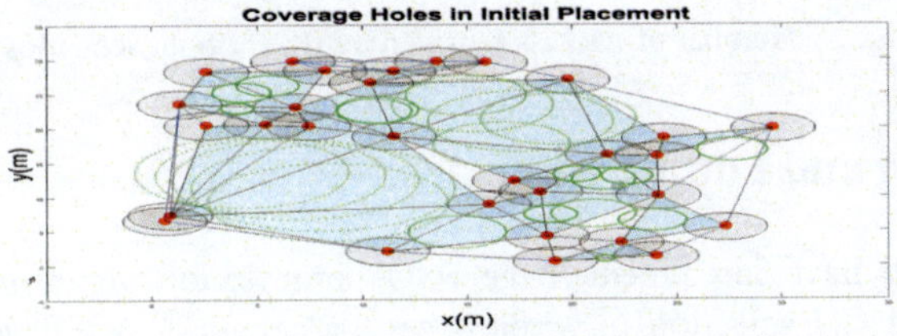

Fig. 1. Coverage Holes for 30 nodes distributed in a 30 × 30 m^2 Area

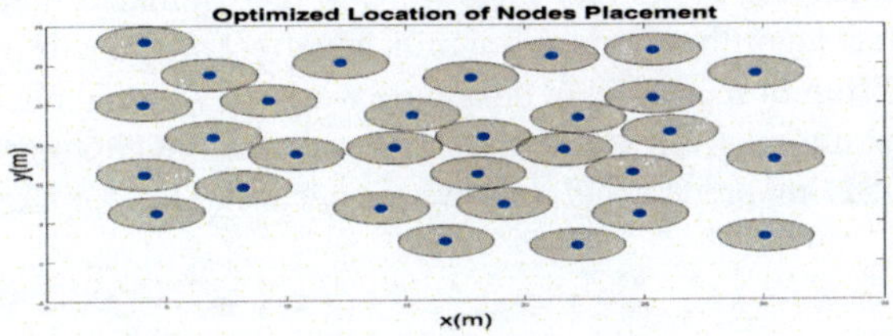

Fig. 2. Final nodes emplacement for 30 nodes distributed in a 30 × 30 m^2 Area

emplacement of network nodes will have a significant effect on the number of packets received by the base station.

4.2 Throughput

Figure 3 demonstrates the number of packets received by the base station for the Random-Network and the PSO-Network in both cases (20 nodes and 30 nodes). It is obvious that the throughput produced by the PSO-Networks is improved as compared to the Random-Networks. In the PSO- networks, nodes location are enhanced, each sensor node is surrounded by at least one node, therefore, the collected data can achieve the base station by different scenarios, clustering or direct transmission. This is what explain the differences in the throughput of both networks. Besides the improvement of node location, the efficient cluster.

head selection supported the network to avoid electing nodes with low energy and high distance to the base station as cluster heads, the PSO-based CH election process ensured the selection of nodes with high residual energy and low distance to the base station, which saves the energy of the nodes and thus the energy of the network, thus, the network lifetime is improved, as well as the throughput.

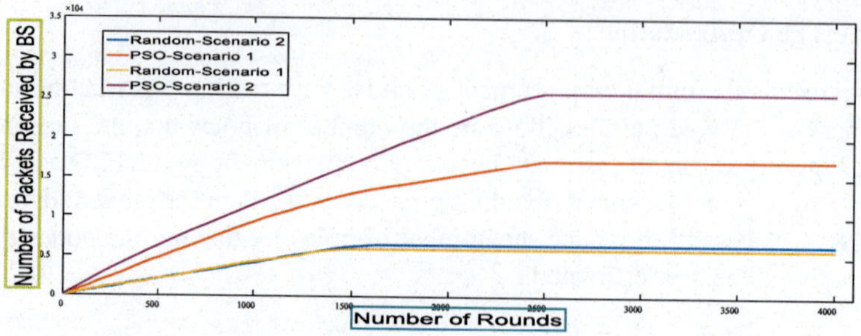

Fig. 3. Number of packets received by BS for both scenarios

5 Conclusion and Future Work

In this paper, we have implemented the PSO optimization algorithm for coverage enhancement and CH selection in wireless sensor networks. We have improved the network deployment by reducing the number of holes produced by the random deployment, in data routing phase we have incorporated a devel- oped PSO-based CH selection algorithm. The simulation results have shown that the estimated nodes distributions provided by the PSO algorithm leads to a significant effect on the wireless network performance in two different scenarios. In our future work we will to reduce the number of nodes needed to monitor a given area by implementing the PSO optimization algorithm to enhance the clustering process and improve the network QoS for large scale WSNs.

References

1. Malik, A., Bhushan, B., Kumar, A., Chaganti, R.: Opportunistic internet of things (OIoT): elucidating the active opportunities of opportunistic networks on the way to IoT. In: Sharma, R., Sharma, D. (eds) New Trends and Applications in Internet of Things (IoT) and Big Data Analytics. Intelligent Systems Reference Library, vol. 221, pp. 209–224. Springer, Cham (2022). https://doi.org/10.1007/978-3-030-99329-0_14
2. Ali, A., Ming, Y., Chakraborty, S., Iram, S.: A comprehensive survey on real-time applications of wsn. Future Internet **9**(4), 77 (2017)
3. Benelhouri, A., Idrissi-Saba, H., Antari, J.: An improved gateway-based energy-aware multi-hop routing protocol for enhancing lifetime and throughput in heterogeneous wsns. Simul. Model. Pract. Theory **116**, 102471 (2022)
4. Shafiq, M., Ashraf, H., Ullah, A., Tahira, S.: Systematic literature review on energy efficient routing schemes in wsn–a survey. Mob. Netw. Appl. **25**(3), 882–895 (2020)
5. Srivastava, A., Mishra, P.K.: A survey on wsn issues with its heuristics and meta-heuristics solutions. Wireless Pers. Commun. **121**(1), 745–814 (2021)
6. Chaganti, R., Gupta, D., Vemprala, N.: Intelligent network layer for cyber-physical systems security. Int. J. Smart Secur. Technol. (IJSST) **8**(2), 42–58 (2021)
7. Abdel-Basset, M., Abdel-Fatah, L., Sangaiah, A.K.: Metaheuristic algorithms: a comprehensive review. Comput. Intell. Multimed. Big Data Cloud Eng. Appl., 185–231 (2018)
8. Kong, H., Yu, B.: An improved method of wsn coverage based on enhanced pso algorithm. In: 2019 IEEE 8th Joint International Information Technology and Artificial Intelligence Conference (ITAIC), pp. 1294–1297. IEEE (2019)

Road Accident Forecast Using Machine Learning

Jihad Satri[1(✉)], Chakib El Mokhi[2], and Hanaa Hachimi[1]

[1] Systems Engineering Laboratory, Sultan Moulay Slimane University, Beni Mellal, Morocco
{jihad.satri,hanaa.hachimi}@usms.ac.ma

[2] Systems Engineering Laboratory, Cadi Ayyad University, Marrakesh, Morocco
c.elmokhi@uca.ma

Abstract. Accidents and mishaps keep happening every day and every now and then. Some mishaps are unnoticeable while others get serious and unavoidable. Statistics needs good data to foster change. Scattered and unorganized information can have a significant impact on decision makers, resulting in inefficient processes, a waste of time and money, and delay in making right decision. Reacting quickly to an accident scene can save many lives. The solution proposed compiles together different datasets, anatomize it thoroughly for a better and deep understanding of the different accident rates in Béni Mellal city and their visual representation for an easy yet sophisticated manner, so that the insights can be easily understood by everyone. For that a multiple dataset was gathered and integrated data related to different types of accidents. Next we will process this data and conduct analysis to get better insights, and visualizing the analysis for a clear understanding that helps the decision makers. Finally, a prediction of accidents in the upcoming years will be thrown.

Keywords: Decision making · Artificial intelligence · Machine learning · Time series · Road accident

1 Introduction

In Morocco, according to the Ministry of Equipment, Transport, Logistics and Water, traffic accidents cause an average of 3500 deaths and 12000 serious injuries per year, or an average of 10 deaths and 33 serious injuries per day. Despite efforts to reduce the number of road accidents, the situation remains alarming. In September 2021, Morocco counted 10028 road traffic accidents with injuries at the national level, an increase of + 18.6% compared to September 2020, and + 13.8% compared to September 2019, according to figures from the National Road Safety Agency (NARSA). Faced with this worrying balance sheet, initiatives are emerging and multiplying to defeat this road war, including the National Road Safety Strategy (SNSR) (2017–2026), which defines a more demanding vision and focused on the long term to develop responsible behavior and safer roads in Morocco. Serious injuries and causalities reported every year are highly caused by road accidents around the world. Law enforcement authorities are facing a challenge in determining how to address this issue.

Y. Farhaoui et al. (Eds.): ICAISE 2022, LNNS 635, pp. 701–708, 2023.
https://doi.org/10.1007/978-3-031-26254-8_102

Traffic accident has emerged as a topic of discussion and analyzing traffic accident data, becoming a major concern for researchers in search of coherent methods for road accidents forecasting [1]. Machine learning (ML) is now being applied in many different fields. It is very encouraging to know that this field finds application in improving road safety. Taking advantage of road accident data to improve decision making is an idea that has found its way to public decision makers. Open Data allows us today to revolutionize accident risk prevention. It is much cheaper and smarter to rely on real-life data collected by law enforcement. ML's predictive models and statistics models allow us to evaluate the risk according to the circumstances of the accident and the profile of the driver(s). Indeed, our contribution aims at predicting and analyzing the different accidents and helps the officials to take appropriate decisions.

The rest of this paper is organized as follows: in the Sect. 2 are some related work presented. Section 3 describes the dataset collected. Section 4, a data analysis is provided. Section 5, the development of our prediction model. Finally, a general conclusion with future work possibilities is given in the Sect. 6.

2 Related Work

Several studies have proposed ML methods to extract safety-critical information from historical data and predict the outcomes of accidental events [2]. For example KRJAJoH [3] presents some models to predict the seriousness of the injury that happened during traffic accidents utilizing ML paradigms, which is an extraordinary way to deal with taking a precise choice with the experience to deal with the current circumstance and the discoveries of the investigation part can be recommended to traffic experts for reducing the number of accidents. Vipin and Rahul [4] used a time series regression analysis methodology to forecasts time zone's-specific road traffic accident mortality reported during 2005–2018 in Kerala State, India. Sangare et al. [5] proposed a novel framework that combines the descriptive strength of the Gaussian Mixture Model with the high-performance classification capabilities of the Support Vector Classifier. Li et al. [6] incorporate the visualization method into machine learning based on Light Gradient Boosting Machine-Tree-structured Parzen Estimator (Light GBM-TPE) to analyze the traffic accidents data of the UK in 2017, this research tries to explore an innovative way to understand and evaluate feature importance of road traffic accidents, which can help suggest effective solutions to improve traffic safety. Avuglah et al. [7] used the ARIMA models to forecast road accident in Ghana. Of course, our solution uses the same prediction model with different data related to Beni Mellal, moreover our work is not limited just in the application of the ARIMA model, but we plan to exploit several models for our analysis such as ML and deep learning (DL) models to make a comparative study and choose the best model. Concerning the DL models which tends to work well with a large amount of data while the more classical ML models stop improving after a saturation point, in this respect we are working on obtaining all the road accident data at the national level and applying the DL models which remains little used. Several solutions are based on this model in other different fields, for example Chaganti et al. [8] have proposed an effective, robust, and platform independent DL model Bi-GRU-CNN based approach to perform the IoT malware classification. As well Ravi et al. [9] worked on an end-to-end model for cyber–physical systems network intrusion detection.

3 Data Collection

For the data related to road accidents, there are several sources from which we can retrieve them. In Morocco exist different administrative departments that deal with this subject such as NARSA, department of Road Transport and Road Safety, local authorities, the courts of the kingdom. Initially the data collection is done through the reports of the police department. At present, these data are not accessible to the general public at their level, so we based ourselves on the data produced by Beni Mellal court in Morocco since it has an information system SAJ (a legal information system) which manages all the data of road accidents produced by the police department. These data are recovered from an SQL SERVER database, then transformed into a CSV file. The dataset includes road accidents from 2010 to 2021, and contains numerous attributes such as the date of the accident, its object, the persons involved with their postal addresses, the judgment and so on.

4 Data Analysis

In this section, we try to go through the steps to understand and prepare the data for time series modeling. In fact, a road accident dataset is used, which describes the number of accidents over 12 years (2010–2021) in Beni Mellal city. First, a data preprocessing is performed. This step includes the verification of missing values, the aggregation of data by date, the setting of the index column on the date column, and so on. Then a visualization of the total number of accidents per year and per month, as well as the monthly relation (see Fig. 1). It can be seen that there is an increase in the number of accidents from the year 2010 to the year 2021 with a few hollow drops between these two years. But generally, since the year 2016 an increase without significant decrease is observed. The *seaborn.barplot()* method is used to draw a bar plot as shown in Fig. 1, that represents the estimated annual trend for the variable "number of accidents" with the height of each rectangle. In addition to this, a box plot shows the distribution of quantitative data in a way that facilitates comparisons between variables or across levels of a categorical variable (see Fig. 2). The box shows the quartiles of our dataset while the whiskers extend to show the rest of the distribution.

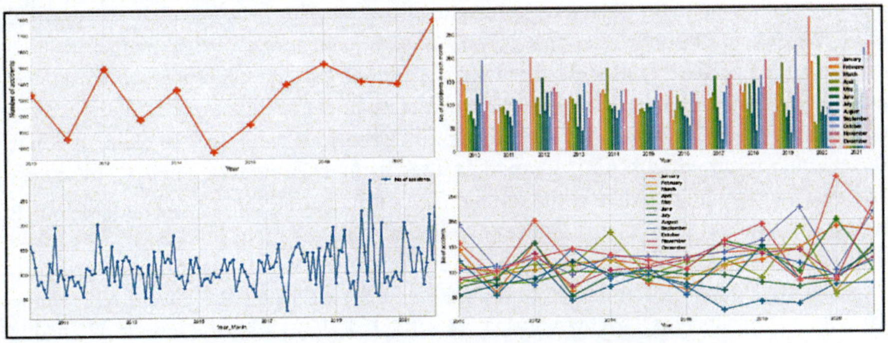

Fig. 1. Total road accidents in Beni Mellal (2010 to 2021).

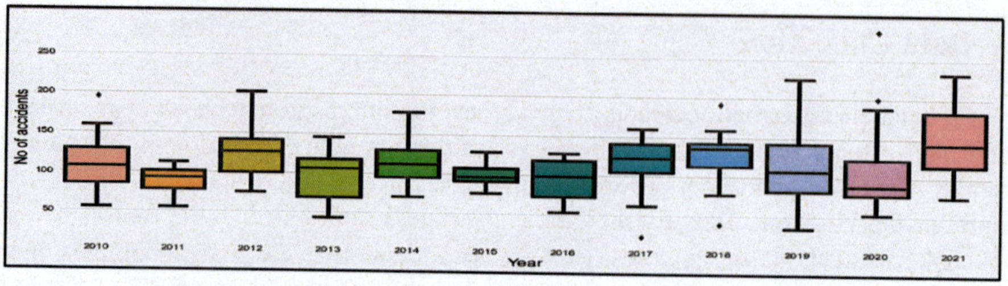

Fig. 2. Box plot of road accidents in Beni Mellal (2010 to 2021).

5 Prediction

The main objective of the analysis of a time series is the prediction of its future realizations. In order to achieve this objective, a first step of modeling the series is necessary. This step consists in selecting, among a family of models corresponding to approximations of reality, the one that best describes the series in question. A time series T_t is commonly decomposed into a trend T_t, a seasonality S_t, and an error ε_t. Sometimes, we add another component, the cycle C_t which corresponds to a regular repetitive phenomenon (predictable) of unknown or changing period.

The time-series decomposition allows us to decompose our time series into three distinct components: trend, seasonality, and noise. The plot in Fig. 3 clearly shows that the number of accidents records are unstable and it has an overall increasing trend, along with their obvious seasonality.

5.1 Arima Time Series Model

In the field of ML, there is a specific set of methods and techniques that are particularly well suited to predicting the value of a dependent variable as a function of time. For this purpose, the ARIMA model gives the possibility to make predictions based on historical observations, which creates a competitive advantage. ARIMA is considered one of the most commonly used methods for time-series modeling and forecasting. ARIMA stands for Autoregressive Integrated Moving Average. The ARIMA models are denoted with the notation $ARIMA(p, d, d)$. The three parameters take into consideration for seasonality, trend, and noise in data. The next step is the parameter selection for our ARIMA Time Series Model. The main objective here is to use a "*grid search*" to find the optimal set of parameters that gives the best performance for our model. When evaluating and comparing the statistical models fitted with different parameters, each model can be ranked against another based on its fit to the data or its ability to accurately predict future data points. The AIC (Akaike Information Criterion) value will be used, which is conveniently returned with ARIMA models fitted using *statsmodels*. The AIC measures how well a model fits the data while taking into consideration the overall complexity of the model. A model that fits the data very well while using a lot of features will be given a higher AIC score than a model that uses fewer features to achieve the same good fit. Consequently, we are interested in finding the model that yields the lowest AIC value. The output suggests that SARIMAX(0, 1, 1) × (0, 1, 1, 12) yields the lowest AIC value of 1204.89. Hence, we should consider this option as optimal.

When fitting seasonal ARIMA, it is important to run the model diagnostics to ensure that none of the assumptions made by the model have been violated (see Fig. 4). The *plot_diagnostics* object allows us to quickly generate model diagnostics and look for any unusual behavior.

Our main concern is to ensure that the residuals in our model are uncorrelated and normally distributed with a zero mean. If the seasonal ARIMA model does not satisfy these properties, it is a good indication that it can be improved. In this case, the diagnosis of our model proposes that the model residuals are normally distributed. This can be validated by the autocorrelation plot on the bottom right, which shows that the residuals of the time series have a low correlation with lagged versions of itself. Hence, the model produces a satisfactory fit that could help us to better understand our time series data and forecast future values. Although we have a satisfactory fit, some of the parameters of our seasonal ARIMA model may be modified to enhance the fit of our model. As an example, our grid search only considered a small set of parameter combinations, so we could find better models if we expanded the grid search. Indeed, it's not really perfect, nevertheless, our model diagnostics suggest that the model residuals are near normally distributed.

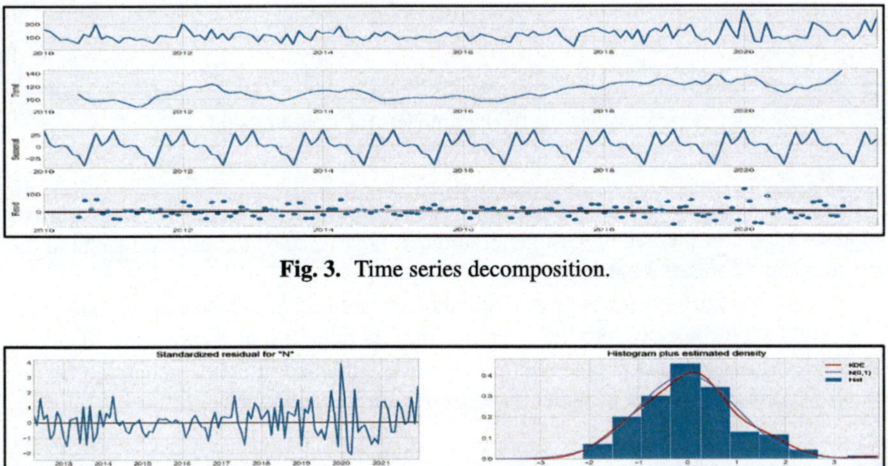

Fig. 3. Time series decomposition.

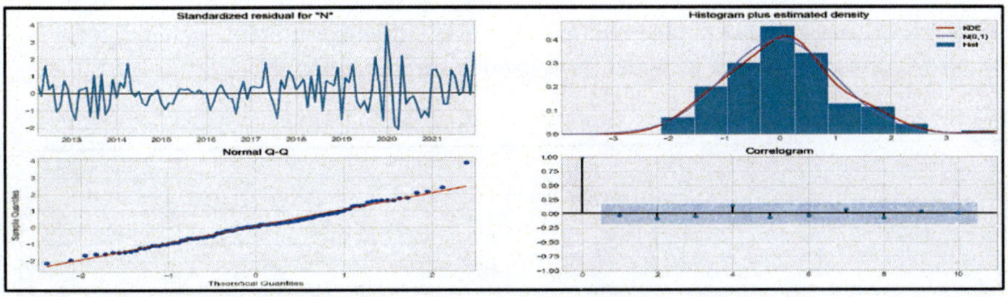

Fig. 4. Model diagnostics.

5.2 Forecast

With the purpose to understand the accuracy of our predictions (see Fig. 5), we compare the number of predicted accidents to the actual number of accidents in the time series, and set the forecasts to start at 2018–01–01 to the end of the data. It is always useful and important to quantify the accuracy of our forecasts. Therefore, the MSE will be used (Mean Squared Error), which summarizes the average error of our forecasts. For each predicted value, we calculate its distance from the actual value and square the result. The results must be squared so that the positive/negative differences do not cancel out when we calculate the overall mean. The MSE of our one-step ahead forecasts yield a value of 3321.48, which should be closer to 0. An MSE of 0 would mean that the estimator is predicting observations of the parameter with perfect accuracy, which would be an ideal scenario but is typically not possible. Concerning the value of the RMSE (Root Mean Squared Error), it is equal to 57.63, which indicates that our model was able to forecast the number of accidents within 57.63 of the real number.

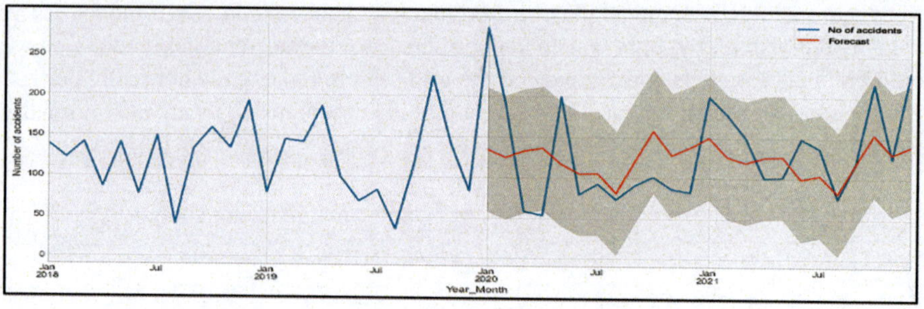

Fig. 5. Validating forecast models for 12 last months.

Finally, we can produce and visualize forecast future values (see Fig. 6). The forecasts and associated confidence interval generated can now be used to better understand the time series and foresee what to expect.

Our forecast shows that the time series should continue to increase at a steady pace. As we forecast further out into the longer term, it is natural that we lose confidence in our values. However, this is not reflected by the confidence intervals generated by our model, which are expected to widen as we move further into the future.

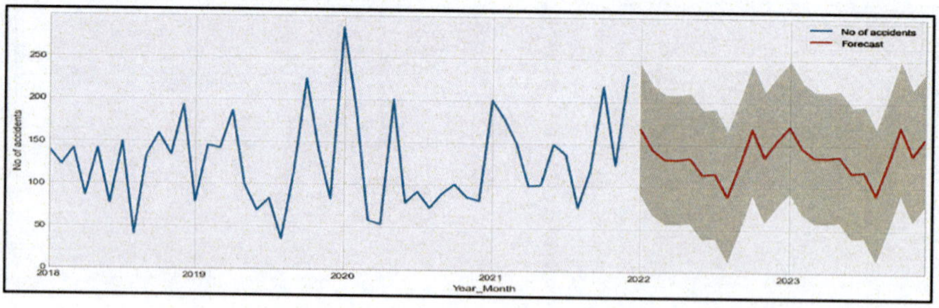

Fig. 6. Forecast of accidents up to 2023.

6 Conclusion and Future Work

Getting accurate insights from the raw road safety datasets not only require excessive human efforts and time but also is a cumbersome task if feasible at all. The challenge becomes even more complex with the arrival of Big data, where the major bottlenecks lie in data cleaning, processing, fusion and subsequently building AI (artificial intelligence) models to get the predictions and valuable insights on scale. In fact, time series forecasting has a wide application in data analysis. It can be argued that time series ML principles basically outperform the classical time series forecasting approach. Thus, traditional methods are limited to processing only previously collected, readily available data history. In turn, ML automatically defines points of interest in the unlimited data stream and then aligns them with available customer data and performs *what-if* analysis. In our case, this results in particularly effective catches to help the authorities make the right decision with regard to road safety policy. The objective in our paper is to create a prediction model for the number of accidents in the future in the city of Beni Mellal with the available data. This model will need to be improved to get the best results. Currently the police department has begun the second phase of generalization of an integrated information system, related to the management of traffic accident files at the national level. This system includes a set of functionalities that enables the completion of all procedures related to the inspection of traffic accidents. Therefore, it will allow us to have more data which will be used in our research. Indeed, after improving this current model, other models will be developed with ML and DL algorithms to predict the number of accidents, deaths and injuries from road accidents, and predict the occurrences of road accidents at a given time, at a given place. These models will provide law enforcement an assistive AI prediction system so that they take the necessary safety measures in order to control and arrange traffic with a serious framework to diminish the number of road accidents.

References

1. Chand, A., Jayesh, S., Bhasi, A.B.: Road traffic accidents: an overview of data sources, analysis techniques and contributing factors. Mater. Today: Proc. **47**, 5135–5141 (2021)
2. Tamascelli, N., Solini, R., Paltrinieri, N., Cozzani, V.: Learning from major accidents: a machine learning approach. Comput. Chem. Eng. **162**, 107786 (2022)
3. Ballamudi, K.R.: Road accident analysis and prediction using machine learning algorithmic approaches. Asian J. Hum. Art Literat. **6**(2), 185–192 (2019)
4. Vipin, N., Rahul, T.: Road traffic accident mortality analysis based on time of occurrence: evidence from Kerala, India. Clin. Epidemiol. Glob. Health **11**, 100745 (2021)
5. Sangare, M., Gupta, S., Bouzefrane, S., Banerjee, S., Muhlethaler, P.: Exploring the forecasting approach for road accidents: analytical measures with hybrid machine learning. Expert Syst. Appl. **167**, 113855 (2021)
6. Li, K., Xu, H., Liu, X.: Analysis and visualization of accidents severity based on LightGBM-TPE. Chaos Solitons Fractals **157**, 111987 (2022)
7. Avuglah, R.K., Adu-Poku, K.A., Harris, E.: Application of ARIMA models to road traffic accident cases in Ghana. Int. J. Stat. Appl. **4**(5), 233–239 (2014)

8. Chaganti, R., Ravi, V., Pham, T.D.: Deep learning based cross architecture internet of things malware detection and classification. Comput. Secur., 102779 (2022)
9. Ravi, V., Chaganti, R., Alazab, M.: Recurrent deep learning-based feature fusion ensemble meta-classifier approach for intelligent network intrusion detection system. Comput. Electr. Eng. **102**, 108156 (2022)

Security of Hadoop Framework in Big Data

Youness Filaly[1(✉)], Nisrine Berros[1], Hamza Badri[1], Fatna El mendil[2], and Younes El Bouzekri EL Idrissi[1]

[1] Engineering Sciences Laboratory, Ibn Tofail University, National School of Applied Sciences, Kenitra, Morocco
youness.filaly@uit.ac.ma

[2] Image Laboratory, School of Technology, Moulay Ismail University of Meknes, Meknes, Morocco

Abstract. A diversity of hardware and software solutions with distinct infrastructures make up big data. Huge volumes of data may now be stored and handled using the Hadoop architecture. It delivers rapid and economical Big Data solutions and is utilized across a range of industries, including social media, healthcare, and insurance. Built on a distributed computing architecture, Hadoop is an open source framework for processing and storing data on a cluster of shared equipment. Due to the framework's adaptability, vulnerabilities develop. These weaknesses put the data at danger and attract attackers. Numerous vulnerability categories are studied in this research, along with various mitigation approaches. Then, we create essential security modules and a Hadoop security management system.

Keywords: Big data · Hadoop · Security threats · Data security · Security architecture

1 Introduction

The worldwide amount of data has expanded owing to the fast growth of the Internet and cloud computing. Big Data and associated applications are increasing quickly. People working in the subject of Big Data often concentrate on data analysis and mining, machine learning, high speed computing, etc. The value of megadata has been gradually expanding with the fast maturity of organizations and linked technologies, and its security issues are becoming more and more critical [1]. Various models, frameworks, and new technologies have been generated as a consequence of the many Big Data operations that have been undertaken throughout the globe to offer higher storage capacity, parallel processing, and real-time analysis of many heterogeneous sources [2]. In addition, new solutions for data privacy and security have been devised. The cornerstone of megadata security protection is developing a safe and trustworthy megadata platform. However, builders of megadata systems, notably Hadoop platforms, have not addressed security challenges from the outset. Designers think that massive data clusters work in a trusted environment, which indicates that anybody may do anything in the cluster and that all users and servers are lawful. There is no security model in this mechanism at the

Y. Farhaoui et al. (Eds.): ICAISE 2022, LNNS 635, pp. 709–715, 2023.
https://doi.org/10.1007/978-3-031-26254-8_103

beginning of the Big Data platform. Due to a lack of security settings, the Big Data global platform server resulted in a large data breach in early 2017 [3].

Here is how the paper is organized: Sect. 2 describes the Big Data Hadoop components. Section 3 examines the security challenges of the Hadoop framework, and Sect. 4 is a presentation of remedies to the difficulties discussed in the previous section, and we develop a Big Data security architecture by merging current Big Data security key technologies. Section 5 brings us to a conclusion.

2 Hadoop Framework

2.1 Definition

The Apache Software Foundation created Hadoop, an open source batch processing system. This depends significantly on distributed computing and parallel processing approaches. HDFS for storage, MapReduce for data processing and YARN for cluster resource management make up Hadoop. Hadoop was designed in 2005 and originally published in 2011 to allow Yahoo's Nutch search engine project [5]. This version was developed with the most basic security features and has been carefully tested in a secure environment.

2.2 Distributed File System of Hadoop (HDFS)

This distributed file system was built to handle massive files and operate on commodity hardware clusters. It is made up of a master node known as NameNode and one or more slave nodes known as DataNode [5]. The name node handles the namespace, which is a collection of files and directories, and it also offers access to clients. Namenode executes file activities such as open, rename, and close. Only this node binds blocks to data nodes [5].

2.3 The Hadoop Map Reduce

This distributed file system was intended to manage huge files and operate on basic hardware clusters. It comprises of a master node called a NameNode and one or more slave nodes called DataNodes. The NameNode administers the namespace, which is a collection of files and folders, and also offers access to clients. Namenode performs file functions such as opening, renaming and shutting. Only this node binds blocks to the data nodes [5].

2.4 The Hadoop yet Another Resource Negotiator (YARN)

The scheduler and the application manager are the two most crucial resource management compo- nents. The scheduler allocates resources to the different running programs and executes scheduling depending on the applications' resource needs [5]. Task submissions are authorized by the resource manager and each task is allocated to an application manager [5]. The application manager allo- cates a container to execute the program, and each application has an application-specific master container responsible for resuming the application master container in case of failure [5].

3 Hadoop Security Issues

Hadoop security has gone a long way in only a few years, and most of the common vulnerabilities can now be handled with minimal time and effort on the side of IT and security teams. For data center administrators and security specialists, Hadoop provides significant security issues. The security threats are mentioned here [6–9]:

- Data Access and Ownership: most RDBMS and data warehouse security solutions rely on role-based access, and Hadoop is no different. Roles, groups, schemas, label security, and other aspects of relational and quasi-relational systems restrict user access to subsets of accessible data. However, authentication and permission necessitate collaboration between the application designer and the IT team in charge of the cluster.
- Protection of the Data: encryption is the industry standard for safeguarding data at rest, as it safeguards against efforts to access data outside of defined application interfaces. We are worried about someone stealing archives or reading files directly from disk in Hadoop systems, and file-level or HDFS layer encryption assures customers data secured against direct access because only file services are supplied with encryption keys.
- Communication of Hadoop nodes: by default, Hadoop and the vast majority of distributions (Cassandra, MongoDB, Couchbase, etc.) do not interact securely: they employ unencrypted RPC over TCP/IP. TLS and SSL capabilities are given in Big Data deployments, although they are seldom utilized for inter-node communication and are rarely used between client applications and the cluster resource controller. This exposes data in transit, as well as application requests, to scrutiny and alteration.
- Authentication and authorization In hadoop system: identification and authentication are crucial aspects of any security activity, since they enable us to determine who should have access to data. Fortunately, the most important advancements in Hadoop security are around identity and access control. This is primarily owing to the Hadoop distribution suppliers, who have done much of the integration and setup work. From default installs without authentication choices to fully integrated solutions based on LDAP, Active Directory, Kerberos, and X. 509, we've gone a long way.
- Audit and logging: Can you identify or trace the source of a suspected breach in your cluster? A record of activity, often from an event log, is necessary. You can use the cluster to store its own logs, but many security experts are concerned that an attacker could muddy the waters by deleting or modifying log entries. As a result, most organizations depend on additional specialized systems, such as SIEM or Splunk, to quickly route logs to another source.
- Monitor, filter, and block of Hadoop activity: no monitoring tools are built in to identify or halt harmful requests. We are just now seeing workable versions of Hadoop activity monitoring tools. As the usage of SQL queries has risen in recent years, we can now deploy database activity monitoring technology to detect and even prevent abuse. These tools are still in their infancy, but the principle has proven useful on relational systems, thus NoSQL implementations should improve with time.

4 Hadoop Security Solution

Members of the open-source community and makers of commercial Hadoop distributions discuss security as a set of fundamental characteristics. The generally stated"pillars" upon which cluster security is built are authentication, authorization, encryption, key management, and logging. These are the fundamental components of an effective Hadoop security model. And you'll very certainly implement all of these functions to some extent. However, integrating these technologies into a unified security strategy necessitates further forethought. It is not enough to just have these tools installed; how and where they are deployed is critical. The simplest method to discuss security plans is to show which security technologies are deployed. It assists businesses in piecing together a comprehensive plan, often by using what they already have and are comfortable with initially, then adding the necessary pieces afterward.

4.1 Threat-Response Models

The following diagram in Fig. 1 displays particular choices available to you to assist you in selecting'preventative' security measures:

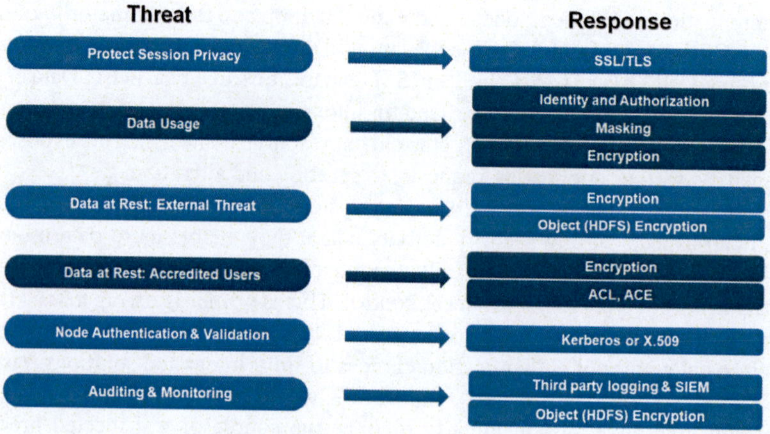

Fig. 1. Hadoop security solution.

4.2 Security Architecture

The security architectures depicted in Fig. 2 are quite useful for envisioning how you wish to approach cluster security. And they're great for determining resource allocation: which technique is your IT staff most comfortable with, and which tools do you have the cash to acquire? However, the fact is that enterprises no longer adhere only to one model; instead, the majority adopts a combination of two. On the back end, some businesses utilize application gateways to validate requests and IAM and transparent encryption to offer an administrative division of tasks. In another scenario, because the cluster was extremely multi-tenant, they relied significantly on TLS security for session privacy and built dynamic controls (masking and tokenization) for fine-grained data control.

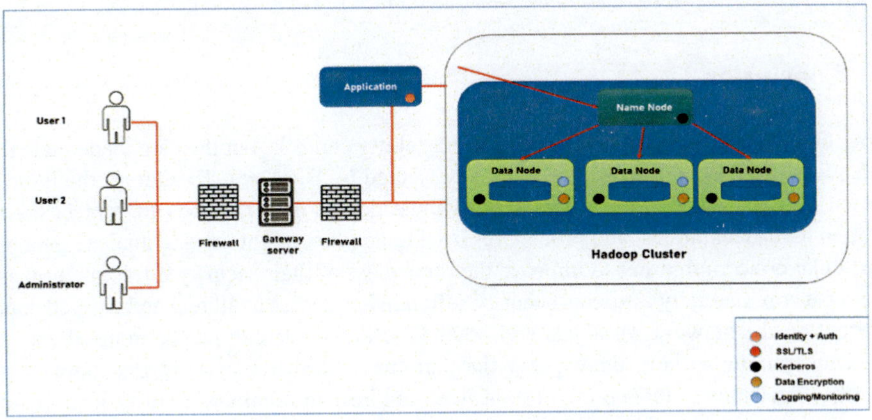

Fig. 2. Security architecture of hadoop.

Unlike conventional databases, which serve as black boxes, Hadoop exposes its innards to the network. Inter-node communication, replication, and other cluster processes take occur across numerous nodes, utilizing different kinds of services. For best protection, it is necessary to embed security into cluster operations. Our approach security strategy is akin to the "moat" model of mainframe security: install the whole cluster on its own network and tightly limit logical access through firewalls or API gateways, utilizing access restrictions for user or application authentication. In truth, this paradigm provides essentially no security within the Hadoop cluster. The security of the data and infrastructure rely on the outside "protective shell" of the network and the applications surrounding it. The benefit is simplicity: any organization may implement this paradigm with existing tools and capabilities, without affecting the performance or functionality of the Hadoop cluster. On the other side, security is fragile: if the firewall or application fails, the system is exposed. For this reason, our technique contains a second side to safeguard our application that includes the Hadoop cluster. This side of our plan employs built-in security measures or third-party products coupled with the Hadoop cluster. This security is systemic and built to be part of the underlying cluster architecture. Tools may include SSL/TLS for secure login, Kerberos for node authentication, transparent encryption for data security at rest, and identity and permission management (groups/roles), to mention a few. This technique is difficult since there are far more moving components and locations where some knowledge is required. Implementing multiple security functions targeted at unique dangers takes time. Additionally, third-party security systems might be pricey. However, they can effectively defend clusters against attackers, malevolent administrators, and foolish application programmers. This is the most effective and complete solution for Hadoop security.

This method is harder to implement since there are many more moving components and locations where some skill is necessary. It takes time to set up numerous security functions that are aimed at distinct hazards. Third-party security technologies might also be costly. However, they can successfully protect clusters from attackers, unscrupulous administrators, and inept application programmers, even if the firewall or application fails. It is the most thorough and successful solution to Hadoop security.

5 Conclusion

Traditional business security solutions offer security controls, but they are inadequate to effectively address the security concerns presented by Big Data. To address the issues raised by data aggregation, organizations must devise new methods to protect their critical tools, techniques, and procedures used to acquire, maintain, and analyze data of particular concern in order to improve the robustness of their security infrastructure, as the most commercially available security software has access to all real and derived data present on the network. Analysts and other IT users are searching for more effective security training to help them grasp the particular hazards they confront, how their security decisions influence the bottom line, and how to combine security goals with mission objectives. In this research, we examined the security issues and needs of Big Data platforms and then identified significant technologies for Big Data security. As a consequence, we built a complete security management solution for Hadoop systems coupled with big data security modules. The security management solution safeguards Hadoop systems and enables unified control of security services on a Big Data platform. According to traditional security applications and literature, the data traffic should be examined and encrypted in an endurable way. Therefore, the field of cryptography will be the topic of our next paper in order to establish an approach to developing data encryption within Hadoop because no one can claim today that there are guaranteed encryption security mechanisms for information technology. First, because the vulnerabilities are unknown, and secondly, because systems and technologies are rapidly evolving.

References

1. Vatsalan, D., Sehili, Z., Christen, P., Rahm, E.: Privacy-preserving record linkage for big data: current approaches and research challenges. In: Zomaya, A.Y., Sakr, S. (eds.) Handbook of Big Data Technologies, pp. 851–895. Springer, Cham (2017). https://doi.org/10.1007/978-3-319-49340-4_25
2. Braun, T.D., et al.: A comparison of eleven static heuristics for mapping a class of independent tasks onto heterogeneous distributed computing systems. J. Parallel Distrib. Comput. **61**(6), 810–837 (2001)
3. Shang, T., Zhuang, H., Liu, J.: Comprehensive security management system for Hadoop platforms. In: 2018 IEEE International Conference on Intelligence and Safety for Robotics (ISR), pp. 172–177. IEEE (2018)
4. Suman, A.K., Gyanchandani, M., Jain, P.: A survey on miscellaneous attacks in Hadoop framework. In: 2018 2nd International Conference on Inventive Systems and Control (ICISC), pp. 1043–1048. IEEE (2018)
5. Bhathal, G.S., Singh, A.: Big data: hadoop framework vulnerabilities, security issues and attacks. Array **1**, 100002 (2019)

6. Vormetric. Securing Big Data: Security Recommendations for Hadoop and NoSQL Environments (2012)
7. Zettaset. The Big Data Security Gap: Protecting the Hadoop Cluster (2014)
8. Vinay, S.S.: Hadoop Security: Today and Tomorrow (2014)
9. Narayana, S.: Securing Hadoop-Implement Robust End-to-End Security for Your Hadoop Ecosystem. Packt Publishing, Birmingham (2013)

Simple Inter-stage Impedance Matching Technique for 5G mm-Wave Systems

Abdelhafid Es-Saqy[1(✉)], Maryam Abata[1], Mohammed Fattah[2], Said Mazer[1], Mahmoud Mehdi[3], Moulhime El Bekkali[1], and Catherine Algani[4]

[1] Laboratory, Sidi Mohamed Ben Abdellah University, Fez, Morocco
abdelhafid.essaqy@usmba.ac.ma
[2] EST, Moulay Ismail University, Meknes, Morocco
[3] Microwaves Laboratory, Lebanese University, Beirut, Lebanon
[4] ESYCOM Lab, University Gustave Eiffel, CNRS, Le Cnam, Paris, France

Abstract. In this paper, a simple inter-stage impedance matching technique for the communication system is presented. This technique allows to choose the most suitable matching circuit between stages. It is applied to match the output impedance of an LO, with an oscillation frequency of 27.27 GHz and an optimum output power of 5 dBm, to the impedance of the LO input of an up-conversion mixer, with an intermediate frequency IF of 2 GHz and a conversion gain of 10 dB for an IF power of −20 dBm. The circuits are designed in MMIC technology based on the 0.15 µm GaAs pHEMT transistor from the UMS foundry.

The integration of these two RF blocks, which are essential in any wireless communication system, allows the optimization of the chip size. The resulting Self-Oscillating Mixer (SOM) circuit has a conversion gain of 6.4 dB for an IF power of −15.4 dB.

Keywords: Impedance matching · Millimeter waves · Mixer · SOM

1 Introduction

In order to reduce the power consumption, one of the most important challenges of the next generation of mobile communication -5G-[1–6], communication systems must be as compact as possible. This leads to reduce the time and costs of MMIC manufacturing [7–10]. On the other hand, the integration of several elements of a communication system in a single circuit, and on a single substrate, presents several difficulties. First of all, it requires the choice of a suitable semiconductor technology, which presents the right compromise between the performances of the different blocks of the system [11]. Then it is necessary to apply an adequate matching technique. Which allows to optimize the power transmission between the stages [12].

Y. Farhaoui et al. (Eds.): ICAISE 2022, LNNS 635, pp. 716–722, 2023.
https://doi.org/10.1007/978-3-031-26254-8_104

In this paper we propose a simple impedance matching technique. It allows a good matching of the output impedance of the local oscillator (LO) to the impedance of the LO input of the up-conversion mixer.

Thus, the rest of this paper is organized as follows: In the second part we explain the principle of the proposed impedance matching technique. In the third part this technique is applied to match the output impedance of the local oscillator (LO) to the impedance of the LO input of the up-conversion mixer. In the fourth part we present the performance of the resulting circuit. In the fifth part we present the conclusion and perspectives of the work presented in this paper.

2 Impedance Matching Technique

The impedance matching technique adopted in this paper is composed of three steps:

- The first step is to determine the load resistance R_{0OPT} for which the output power of the LO is of optimal value. For this reason, the LO is loaded by a variable resistance R_0 (Fig. 1).

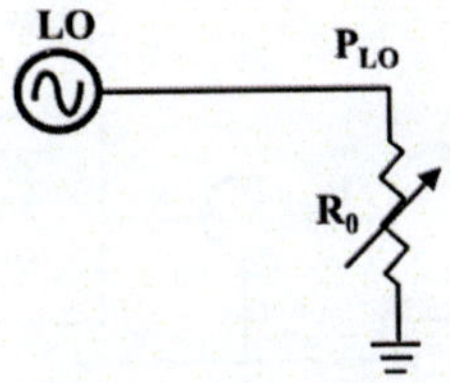

Fig. 1. LO loaded by a variable resistor R_0.

- After determining R_{0OPT} the optimum value of R_0, the second step is to match the impedance of the LO input of the up-conversion mixer to this optimum resistance R_{0OPT}. The impedance of the two other inputs of the up-conversion mixer (IF and RF) are matched to an impedance Z_0 of 50 Ω (Fig. 2).

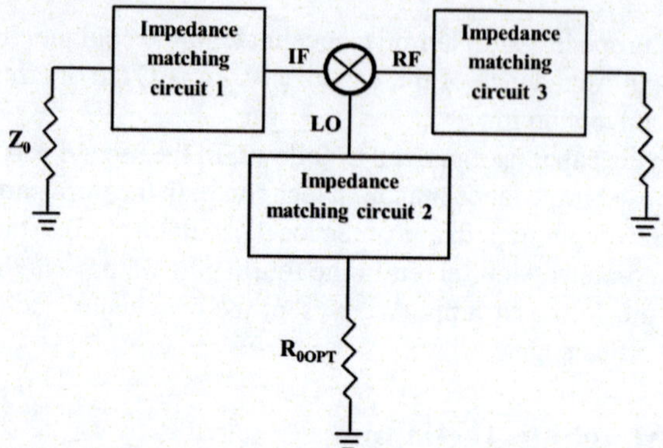

Fig. 2. High conversion mixer with matched impedances of the three ports.

- Then we combine the LO circuit and the mixer circuit (Fig. 3). The resulting circuit is a Self-Oscillating Mixer (SOM) allowing the frequency conversion of the IF signal, to the RFs frequency band, using a self-generated LO signal.

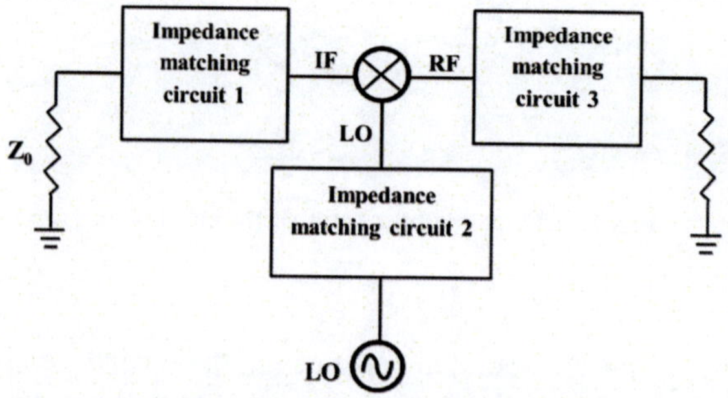

Fig. 3. Self-oscillating mixer.

3 Impedance Matching Circuit

In order to validate the impedance matching technique presented before, we present in this section the design of a SOM in pHEMT technology. This SOM circuit consists of a voltage-controlled oscillator (VCO), studied in detail in the paper [13–15], and a double balanced mixer studied in detail in the paper [16]. Both circuits are designed in Monolithic Microwave Integrated Circuit MMIC technology based on 0.15 μm GaAs pHEMT from the PH15 process of the UMS foundry.

The VCO is characterized by a center oscillation frequency of 27.7 GHz and a tuning range of 2.44 GHz. The up-conversion mixer is characterized by a conversion gain of

10 dB for an IF power of −20 dBm. The 1 dB compression point is obtained for an IF power of −16.5 dBm. The variation of the LO output power P_{LO} as a function of the load resistance value is shown in Fig. 4. The curve in Fig. 4 shows that the LO output power is maximum for an optimum load resistance value R_{0OPT} of 96 Ω. To verify if the impedance of the LO input of the up-conversion mixer is well matched to R_{0OPT}, the reflection coefficient at the LO input of the mixer S_{22} is plotted (Fig. 5).

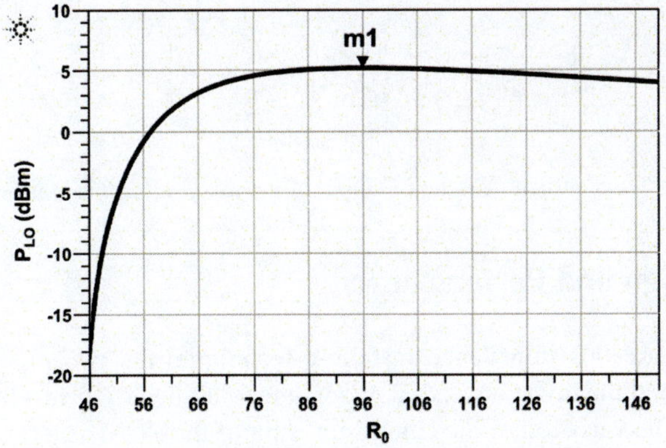

Fig. 4. Variation of LO output power as a function of load resistance R_0.

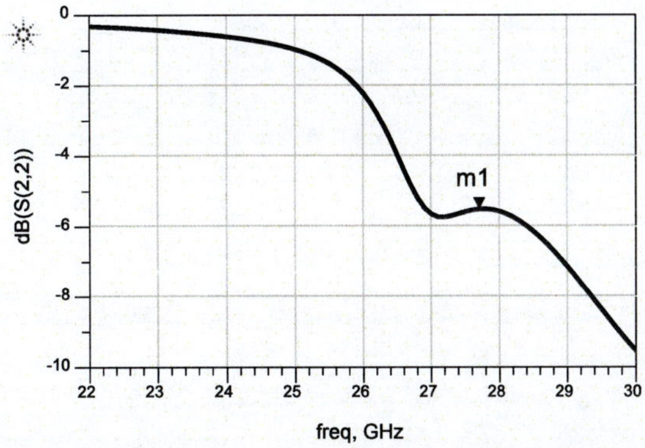

Fig. 5. Reflection coefficient at the LO input of the mixer before impedance matching.

From this figure we can deduce that the LO input impedance of the mixer is initially not well adapted. In order to improve the power transmission, an impedance matching circuit is inserted at the LO input of the mixer. This impedance matching circuit is designed with the using the "Smith Chart Utility" tool of the ADS software. The reflection coefficient at the LO input of the mixer after the impedance matching is shown in Fig. 6. From this figure we can deduce that at the frequency of 27.7 GHz the impedance of the LO input of the mixer is well matched to 96 Ω.

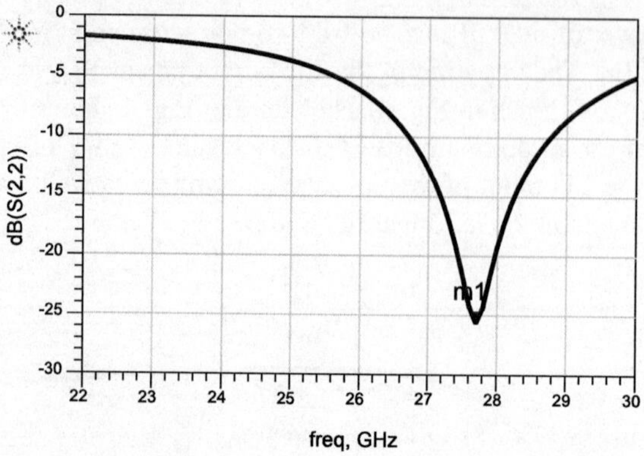

Fig. 6. Reflection coefficient at the LO input of the mixer after impedance matching.

4 SOM Design and Performance

In order to validate this impedance matching technique, the two previously studied circuits, the LO and the high conversion mixer, are combined to form a Self-Oscillating Mixer (Fig. 7). The variation of the conversion gain of the SOM as a function of the IF

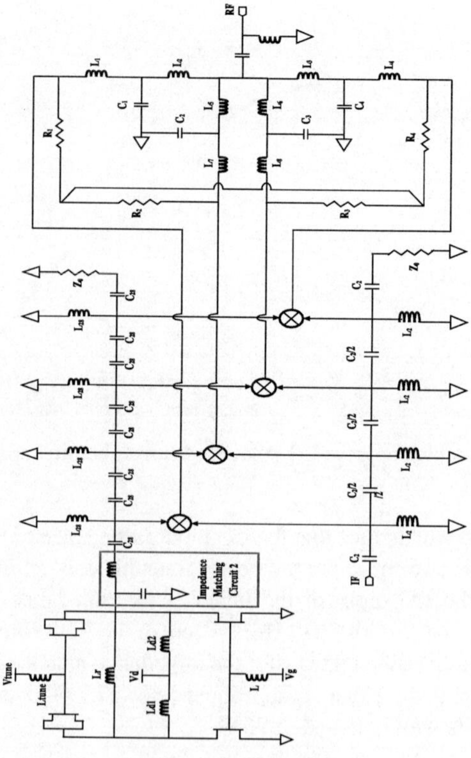

Fig. 7. SOM circuit.

input power is shown in Fig. 8. It has a conversion gain of 6.4 dB for an IF power of −15.4 dBm.

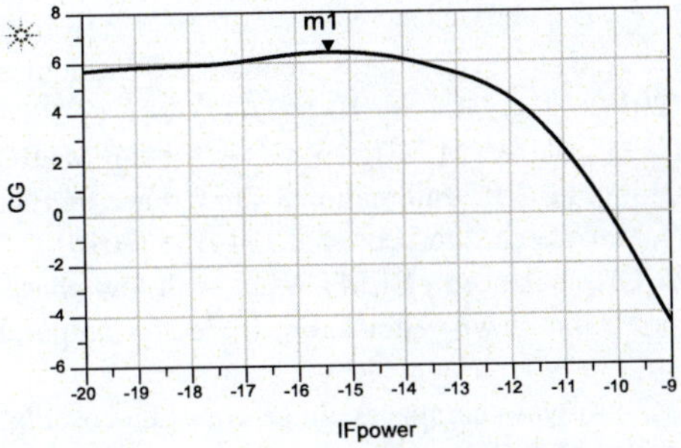

Fig. 8. Variation of the conversion gain versus IF power.

5 Conclusion

In this paper we have presented an inter-stage impedance matching technique. This technique allows to match the output impedance of a local oscillator to the impedance of the LO input of an up-conversion mixer. This technique allowed us to integrate the two circuits, the LO and the up-conversion mixer, to obtain a SOM. This circuit has a conversion gain of 6.4 dB for an IF power of −15.4 dBm.

References

1. Mamane, A., Fattah, M., El Ghazi, M., El Bekkali, M.: 5G enhanced mobile broadband multi-criteria scheduler for dense urban scenario. Telecommun. Syst. **80**(1), 33–43 (2022). https://doi.org/10.1007/s11235-022-00885-3
2. Chafi, S.-E., Balboul, Y., Mazer, S., Fattah, M., El Bekkali, M.: Resource placement strategy optimization for smart grid application using 5G wireless networks. IJECE **12**(4), 3932 (2022)
3. Balboul, Y., Fattah, M., Mazer, S., El Bekkali, M.: 5G uplink interference simulations, analysis and solutions: the case of pico cells dense deployment. IJECE **11**(3), 2245 (2021)
4. Daghouj, D., Fattah, M., Mazer, S., Balboul, Y., El Bekkali, M.: UWB waveform for automotive short range radar. IREA **8**(4), 158 (2020)
5. Moutaib, M., Fattah, M., Farhaoui, Y.: Internet of things: energy consumption and data storage. Procedia Comput. Sci. **175**, 609–614 (2020)
6. Mamane, A., Fattah, M., El Ghazi, M., Balboul, Y.E., Bekkali, M., Mazer, S.: Proportional fair buffer scheduling algorithm for 5G enhanced mobile broadband. IJECE **11**(5), 4165 (2021)
7. Chen, Y., Chen, C.-N., Chiong, C.-C., Wang, H.: A compact 40-GHz doherty power amplifier with 21% PAE at 6-dB power back off in 0.1-μ m GaAs pHEMT process, IEEE Microw. Wireless Compon. Lett. **29**(8), 545–547 (2019)
8. Wang, H., et al.: Towards energy-efficient 5G mm-wave links: exploiting broadband mm-wave doherty power amplifier and multi-feed antenna with direct on-antenna power combining. In: 2017 IEEE Bipolar/BiCMOS Circuits and Technology Meeting (BCTM), Miami, FL, pp. 30–37 (2017)

9. Es-saqy, A. et al.: High Conversion Gain Self-Oscillating Mixer for 5G mm-wave Applications, vol. 17, p. 8 (2022)
10. Didi, S. et al.: Study and design of printed rectangular microstrip antenna arrays at an operating frequency of 27.5 GHz for 5G applications. J. Nano- Electr. Phys. **13**(6), 06035–1–06035–5 (2021)
11. Bera, S.C.: Microwave Active Devices and Circuits for Communication. LNEE, vol. 533. Springer, Singapore (2019). https://doi.org/10.1007/978-981-13-3004-9
12. Abata, M., El Bekkali, M., Mazer, M., Algani, C.: A V-band MMIC frequency quadrupler with active input matching. In: 2015 International Conference on Electrical and Information Technologies (ICEIT), Marrakech, Morocco, pp. 134–138 (2015)
13. Es-Saqy, A., et al.: 28 GHz balanced pHEMT VCO with low phase noise and high output power performance for 5G mm-wave systems. Int. J. Electr. Comput. Eng. **10**(5), 4623–4630 (2020)
14. Es-saqy, A., et al.: A 5G mm-wave compact voltage-controlled oscillator in 0.25 μm pHEMT technology. IJECE **11**(2), 1036 (2021)
15. Es-Saqy, A. et al.: Very low phase noise voltage controlled oscillator for 5G mm-wave communication systems. In: 2020 1st International Conference on Innovative Research in Applied Science, Engineering and Technology (IRASET), pp. 1–4 (2020)
16. Es-saqy, A. Abata, M., Fattah, M., Mehdi, M., El Bekkali, M., Algani, C.: A pHEMT double-balanced up-conversion mixer for 5G MM-wave communication systems. IJMOT **17**(4), 403–411 (2022)

Smart Learning Using Autonomous Chatbot Based on NLP Techniques

Khadija El azhari[1(✉)], Imane Hilal[2], Najima Daoudi[2], and Rachida Ajhoun[1]

[1] Smart Systems Laboratory, ENSIAS, Mohammed V University, Rabat, Morocco
khadja_elazhari@um5.ac.ma
[2] LyRica Labs, School of Information Sciences, Rabat, Morocco
{ihilal,ndaoudi}@esi.ac.ma

Abstract. Artificial Intelligence (AI) technologies have shown exponential growth in different areas. Many companies and organizations adopt AI in their workflow to perform tasks efficacy and efficiency. Schools and universities were not an exception. They integrate several AI solutions to help students in their learning processes. AI chatbots are among the AI technologies widely used in the education domain. They understand human language and interact correctly with it through two main components: (1) Natural Language Processing techniques to understand users' requests and generate the appropriate answers. (2) Knowledge Base (KB) to store and centralize all chatbot knowledge. Chatbot's KB is the brain behind the chatbot. It is the central component behind any AI Chatbot. However, several researchers still construct the chatbot's KB manually which consumes time, energy, and cost. In our paper, we propose an autonomous solution to collect a chatbot KB, preprocess it, and store it through the Python programming language. Specifically, to demonstrate the feasibility of our solution, we focus on creating a solid KB for an educational chatbot that answers students' questions to offload teachers and automate repetitive tasks. In our paper we make four main contributions: (1) Collect the chatbot's KB through our autonomous solution. (2) Preprocess the chatbot's KB and makes it clean to use. (3) Understand students' requests by applying four main techniques: Latent Dirichlet Allocation (LDA), Term Frequency-Inverse Document Frequency (TF-IDF), BERTOPIC and KEYBERT. (4) Develop an AI chatbot to explore the KB and satisfy students' needs.

Keywords: AI · Chatbots · E-learning · Knowledge base · Machine learning · Natural language processing

1 Introduction

In the last couple of years, schools, universities, and learning platforms widely adopt educational chatbots to answer repetitive and massive questions asked by students. Artificial Intelligence (AI) chatbots address many issues that the education domain is facing: (1) Students cannot filter the information they need in the available information sources [1]. (2) Teachers are not able to answer all repetitive and massive students' questions [2].

Y. Farhaoui et al. (Eds.): ICAISE 2022, LNNS 635, pp. 723–728, 2023.
https://doi.org/10.1007/978-3-031-26254-8_105

(3) Students and teachers suffer from insufficient interaction in distance learning and online courses [3]. (4) Students hesitate to ask their questions in public in front of their peers because they fear being judged [4]. To fill this gap, we propose an autonomous solution based on Natural Language Processing (NLP) techniques to build an intelligent educational chatbot that answers massive and repetitive questions about the Machine Learning (ML) domain. Hence, satisfying students' needs related to mastering the ML skills and helping them find the information they need rapidly without wasting time. First, we started by automatically scraping the chatbot's Knowledge Base (KB), then preprocessing it and extracting keywords, and finally implementing the educational chatbot to answer the repetitive and massive students' questions.

The remainder of this paper is as follows: The next section presents a literature review related to educational chatbots, the third section proposes our autonomous solution and its implementation process, the fourth section presents the conclusion and future works, and the last section provides references.

2 Literature Review

Researchers propose various ways to use AI chatbots in the educational field. Several efforts have been carried out to collect and extract the chatbot's KB. However, there are some limitations: (1) Many papers proposed the manual collection of the chatbot's KB which consumes time and energy [1, 2, 5–10]. (2) Few papers [11–13] tried to automatically construct the chatbot's local KB, however, they limit their approaches to collecting pairs Q&As without focusing on automatically detecting the specific need of each question (Intention), thus, generating appropriate responses rapidly. Authors in [11, 12] propose Dialog flow as an AI platform to perform this task, they integrate their Q&As in Dialog flow, then, it automatically detects intent for each question. On one hand, AI platforms help create AI chatbots rapidly, however, there are some drawbacks, namely: (1) The few inquiries and responses that are present in the training sentences are the only ones the chatbot can handle. (2) AI platforms are expensive, and the number of requests available in their free editions is constrained. (3) AI platforms restrict the channels a chatbot can be integrated into. Additionally, Authors in [13] propose an automatic approach to collect Q&As, but they use a generative chatbot which can lead to bad quality responses because of applying prediction models, rather than using a local KB. To address these issues, we propose our autonomous solution to automatically construct the local KB of an educational chatbot that answers repetitive and massive questions asked by students. Thus, saving energy, and time, and reducing costs. We aimed to construct an AI chatbot able to detect the user intent automatically without the need for AI platforms or prediction models, thus, we ensure the quality of responses given to users.

3 Our Autonomous Solution and Its Implementation

In this section, we will present the methodology adopted to automatically create and construct the chatbot's KB (See Fig. 1.).

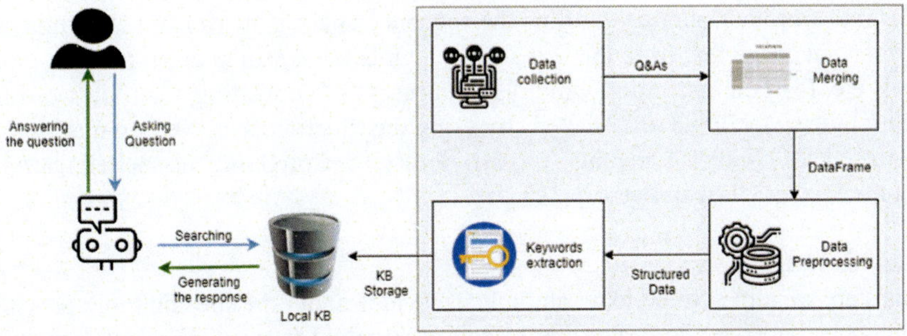

Fig. 1. The methodology adopted to create the chatbot's KB

We started by (1) Identifying relevant and reliable sources that propose pairs of Q&As about the ML domain. (2) Creating a Python Script to successfully retrieve the Q&As. And (3) Merging the data into a Pandas DataFrame [14] to facilitate the Data Analysis. (4) Preprocessing and structuring the pairs of Q&As. (5) Extracting Keywords from Q&As by applying various techniques such as Latent Dirichlet Allocation (LDA), Term Frequency-Inverse Document Frequency (TF- IDF), BerTopic, and Keybert. (6) Integrating the KB on our educational chatbot.

Data Collection

In this step, we made a comprehensive search of well-known educational sources in the ML domain. We focused our work on relevant and reliable sources which are mainly in English to ensure the high quality of Q&As. We integrated the following sources: Reddit, Stack Overflow, Json files, Edureka, JavaTpoint, Simplilearn, Analytics Vidha, FreeCodeCamp. Then, we created a Python script to scrape and extract Q&As available on each source. Especially, we used the python library BeautifulSoup [15] and Request [16] to extract the content of sources. We succussed in retrieving 1476 Q&As automatically and rapidly. Thus, reducing time, energy, and cost.

Data Merging

After retrieving 1476 Q&As from several sources, we decided to merge all Q&As to facilitate data structuring and data analysis. Thus, we merged them into a Pandas DataFrame which is a matrix containing the Q&As in rows and the variables/attributes in columns. For our case, we have two attribute questions and their answers. Thanks to the DataFrame, we can successfully perform many filtering operations, pre-processing, and data analysis.

Data Preprocessing

After scraping Q&As from different sources, we stored them in a Python DataFrame Then, we developed a Python script to clean it by applying the following NLP techniques: (1) Converting the Q&As to lowercase. It is an important task in NLP to allow the machine to process two different words like "Machine Learning" and "MACHINE LEARNING" in the same way. We used the built-in lower () [17] function provided by python to convert the text to lowercase. (2) Tokenization is the process of dividing the text of

Q&As into words or phrases to allow the sequence analysis of words. (3) Eliminating the irrelevant characters from the text such as numbers, punctuation, stopwords, empty spaces, etc. (4) Lemmatization which is the process of converting a certain word into its form base to facilitate text analysis by using the function WordNetLemmatizer [18] from the NLTK (Natural Language Toolkit) library. (5) Correcting misspelled Q&As by using the Speller Class in Python [19].

Keywords Extraction

In this step, we aimed to build a comprehensive idea about the specific topic presented in each question (Intention). For that reason, we applied four techniques that meet our needs by extracting keywords from retrieved questions: LDA, TF-IDF, BerTopic, and Keybert. After extracting keywords from Q&As, we randomly selected 100 Q&As and three members of the team evaluated the performance of each method to choose the method that generates the most representative keywords (See Fig. 2.). For the evaluation step, the members' team evaluates the generated keywords as follows: (1) If 70% of the generated keywords represent the question's subject, then, the classification is considered good. (2) If just 50% of the generated keywords represent the question's subject, then, the classification is considered partly good. (3) If 70% of the generated keywords don't represent the question's subject, then, the classification is considered bad.

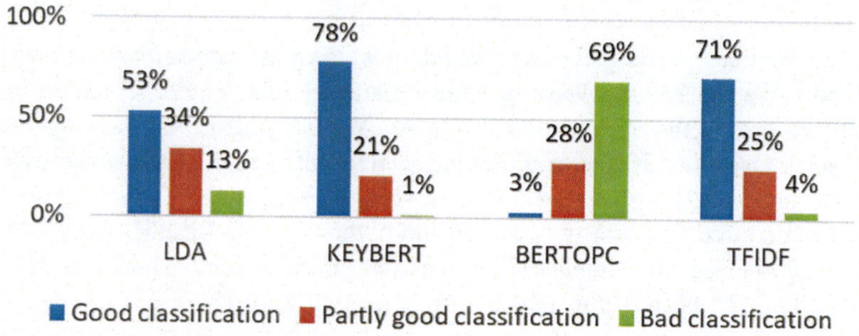

Fig. 2. Comparison between the performance of keywords extraction methods

As shown in Fig. 2. More than 50% of the given Q&As were well classified through LDA, TF-IDF, and KEYBERT. However, 69% of questions were badly classified by the BERTOPIC technique. Based on Fig. 2. We considered KEYBERRT as the most appropriate method for our approach for two main reasons: (1) 78% of questions were well-classified by the KEYBERT technique. (2) Just 1% of questions were badly classified, which demonstrates that there is a low chance to make a bad classification for new questions. Unlike LDA and TF-IDF methods that present respectively 13%, 4% of bad classified questions, and 34%,25% partly good classification. Thus, they increase the error rate. After completing the process of keyword extraction, we implemented our educational chatbot to test the local KB and automate responding to students' questions. Figure 3. Shows an example of a conversation with our chatbot.

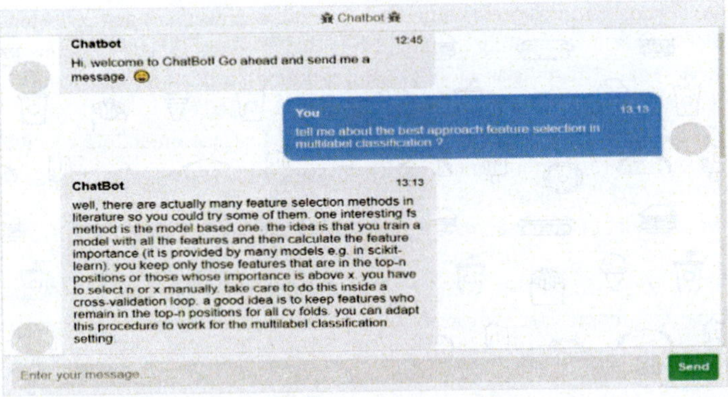

Fig. 3. Example of a conversation with our chatbot.

4 Conclusion and Future Works

Various works are carried out to create educational chatbots to automate many tasks. They offer an increasing number of services to satisfy several educational needs efficiently and efficacy. Previous papers focused their works on creating chatbots to answer repetitive and massive questions by manually collecting the local KB which is time and energy-consuming. In this paper, we propose an autonomous solution to automatically create an educational chatbot that answers repetitive and massive students' questions related to the ML domain. We succeeded in (1) Capitalizing on the knowledge available in eight well-known educational sources. (2) Extracting keywords from the retrieved Q&As. (3) Developing an educational chatbot able to interact with users and help them find the information they need rapidly. Our proposed approach can guide researchers and teachers that are interested in constructing the chatbot's KB automatically without the need to manually collect and structure pairs of Q&As. In future works, we plan to (1) add more advanced AI technologies to personalize the chatbots' answers and motivate students in their learning processes. (2) integrate external sources to extend the chatbot's KB and make it richer. (3) update the local KB and makes it evolve over time.

References

1. Kasthuri, E., Balaji, S.: Natural language processing and deep learning chatbot using long short term memory algorithm. Mater. Today: Proc., 1–4 (2021)
2. Hien, H., Cuong, P., Nam, L.,Nhung,H.,Thang,L.: Intelligent assistants in higher-educa- tion environments: the FIT-EBot, a chatbot for administrative and learning support. In: The Ninth International Symposium, pp.69–76. (2018)
3. EL azhari, K.,Hilal, I., Daoudi, N., Ajhoun, R.: Chatbots in E-learning : advantages and limitations. In: Colloque sur les Objets et systèmes Connectés - COC'2021, HAL, MARSEILLE, France (2021)
4. El Azhari, K., Hilal, I., Daoudi, N., Ajhoun, R.: "AskBot": The AI chatbot that enhances the learning process. In: Motahhir, S., Bossoufi, B. (eds) Digital Technologies and Applications. ICDTA 2022. Lecture Notes in Networks and Systems, vol 454, pp 388–396. Springer, Morocco (2022). https://doi.org/10.1007/978-3-031-01942-5_39

5. Chang, C., Hwang, G.: Promoting students' learning achievement and self-efficacy: a mobile chatbot approach for nursing training. Br. J. Edu. Technol. **53**(1), 171–188 (2021)
6. Neumann, A., et al.: Chatbots as a tool to scale mentoring processes: individually supporting self-study in higher education. Front. Artif. Intell. **4,** 1–7(2021)
7. Souali, K., Rahmaoui, O., Ouzzif, M., El Haddioui, I.: Recommending moodle resources using chatbots. In: 15th International Conference on Signal-Image Technology & Internet-Based Systems (SITIS), pp.677–680 (2019)
8. Dahiya, M.: A Tool of Conversation: Chatbot. Int. J. Comput. Sci. Eng. (IJCSE) **5**(5), 158–161 (2017)
9. Ranoliya, R., Raghuwansh, B., Singh, S.: Chatbot for university related FAQs. In: 2017 International Conference on Advances in Computing, Communications, and Informatics (ICACCI), pp. 1525–1530. IEEE (2017)
10. Santoso, H., et al.: Dinus intelligent assistance (DINA) chatbot for univer- sity admission services. In: 2018 International Seminar on Application for Technology of Information and Communication, pp. 417–423 (2018)
11. Wang, L., Wang, W.: Research and construction of junior high school subject Q&A sys- tem model based on deep learning. In: 2018 International Conference on Information Systems and Computer Aided Education (ICISCAE), pp. 504–508 (2018)
12. Carlander-Reuterfelt, D., Carrera, A., Iglesias, C., Araque, O., Sanchez Rada, J., Munoz, S.: JAICOB: a data science chatbot. IEEE Access **8**, 180672–180680 (2020)
13. Göschlberger, B., Brandstetter, C.: Conversational AI for corporate e-learning. In: Proceedings of the 21st International Conference on Information Integration and Web-Based Applications & Services, pp. 674–678 (2019)
14. pandas documentation — pandas 1.5.0 documentation. https://pandas.pydata.org/docs/index.html. Accessed 17 June 2022
15. beautifulsoup4. https://pypi.org/project/beautifulsoup4/. Accessed 20 June 2022
16. Requests: HTTP for Humans™ — Requests 2.28.1 documentation. https://requests.readthedocs.io/en/latest/. Accessed 20 June 2022
17. Built-in Functions — Python 3.10.7 documentation. https://docs.python.org/3/library/functions.html. Accessed 19 June 2022
18. NLTK:: nltk.stem.api module. https://www.nltk.org/api/nltk.stem.api.html. Accessed 18 June 2022
19. Python Speller Examples, spelling. Speller Python Examples – HotExamples. https://python.hotexamples.com/examples/spelling/Speller/-/python-speller-class-examples.html. Accessed 15 June 2022

Speech Emotion Recognition Using yet Another Mobile Network Tool

Mayowa Oyedepo Oyediran[1(✉)], Olufemi Samuel Ojo[1], Salil Bharany[2], Abidemi Emmanuel Adeniyi[3], Agbotiname Lucky Imoize[4], Yousef Farhaoui[5], and Joseph Bamidele Awotunde[6]

[1] Department of Computer Science, Ajayi Crowther University, Oyo, Oyo, Nigeria
{mo.oyediran,os.ojo}@acu.edu.ng
[2] CET Department, GNDU, Amritsar, India
[3] Department of Computer Sciences, Precious Cornerstone University, Ibadan, Nigeria
adeniyi.emmanuel@lmu.edu.ng
[4] Department of Electrical and Electronics Engineering, Faculty of Engineering, University of Lagos, Akoka, Lagos 100213, Nigeria
aimoize@unilag.edu.ng
[5] STI Laboratory, T-IDMS Faculty of Sciences and Techniques, Moulay Ismail University of Meknès, Errachidia, Morocco
y.farhaoui@fste.umi.ac.ma
[6] Department of Computer Science, Faculty of Information and Communication Sciences, University of Ilorin, Ilorin 240003, Kwara State, Nigeria
awotunde.jb@unilorin.edu.ng

Abstract. Recognizing and reacting to the emotions of others is a key necessity for successful human engagement since emotions are such an important and vital component of being human. Speaking is the most intuitive method by which individuals ex-press themselves. It follows that this communication medium should be extended to computer programs. Affective computing focuses on the automated perception, understanding, and synthesizing of emotional responses, consequently, speech emotion recognition seeks to recognize a speaker's core emotional state derived from the voice signal. The paralinguistic information conveyed by speech expressions has been identified to be beneficial in speech processing in a variety of ways, most notably as a significant component of computer emotional intelligence and contributing to Human – Computer Interaction. In this regard, this study uses a spoken emotion recognition technology to deliver "Yet Another Mobile Net-work Tool" where 1295 primary dataset acquired from Bells University of Technology. The dataset consists of five emotions (anger, fear, neutral, happy, and sad) that were used for the experiment. The experimental results show that the study obtained a recognition accuracy of 85.5% and a rejection rate of 10.5%.

Keywords: Emotion · Speech · Human-computer interaction · Recognition

1 Introduction

The word "emotion" is derived from the Latin phrase "Emovere," which means "to stir up" or "to excite," and a person's feelings literally shake them about. Because emotion

Y. Farhaoui et al. (Eds.): ICAISE 2022, LNNS 635, pp. 729–739, 2023.
https://doi.org/10.1007/978-3-031-26254-8_106

and sensation are so connected, the quantity of emotions grows with the rise in sensation, and the body is provoked or pushed. This heightened state is referred to as 'emotion' [1]. The authors of [2] define emotion as a complicated psychological condition with three unique elements (sensory perception, physiological reaction, and cognitive or communicative reaction). Emotions are very essential in our everyday communications, and there has been a lot of research effort done in recent years to build effective emotional identification systems based on many types of information sources such as audio and video [3, 4].

Emotion recognition is a technique for detecting and recognizing human feelings that incorporates technical skills including image recognition, audio and voice identification, bio-sensing, machine learning, and pattern matching [5]. Emotion detection refers to applications and services that can recognize fundamental emotions (rage, disgust, fright, happiness, sorrow, and astonishment). However, this current work will put into consideration the Speech Emotion Recognition.

Speech emotion recognition is simply the identification of emotion from speech signals. Speech Emotion Recognition is an already existing topic in the realm of Human-Computer Interaction Cognition and is utilized in a variety of applications including learning and gaming software, as well as stress management for call center employees. In the realm of e-learning, recognizing students' emotions early and providing appropriate therapy can improve the quality of instruction [6].

The capacity to perceive emotional states can aid in the identification of some agreements, such as "marry that woman, be on your boss's good side, and enjoy a life of leisure.". In psychology, emotion is a tough term to grasp and define. In the scientific literature, there are numerous diverse definitions of emotions. Mood is described as any fairly brief conciousness characterized by considerable intellectual states and a high level of satisfaction or distress or dissatisfaction in common discourse [7].

Speech emotion identification not only improves human-human contact but also bridges the gap between human-computer communication (HCI). The primary goal of HCI is to enhance the relationships between humans and computer systems, resulting in a more natural communication relationship. This may be accomplished by making technology even more sensitive to the requirements and desires of end users [6].

Numerous speech emotion recognition systems have been proposed and implemented and some scholars adopted artificial neural networks (ANN). ANNs are easier to develop and retrain than other neural network-based approaches, however a single-layer ANN cannot address particularly complicated and nonlinear issues, which is where ANN ends [8]. As a result, while they are speedy, their powers are restricted [9]. Deep Artificial Neural Networks outperform their single-layer equivalents in modeling challenges. Nonetheless, its simulation remains limited to nonlinear polynomial equations, at the expense of an exponential increase in the number of parameters to train, necessitating the use of more processes and memory. To overcome these snags, this study presents speech emotion recognition using "Yet Another Mobile Network Tool".

This study consists of five sections. The next section describes the related works. The materials and methods used as described in Sect. 3. Section 4 presents the result and discussion. Section 5 concludes the study.

2 Related Works

Speech-based emotion recognition has several applications, including interpersonal language, online gaming, and care delivery assistance. While deep learning-based techniques attain high accuracy, they frequently come at a substantial computational and time cost. True, pattern training algorithms must sift through a vast amount of voice data to find essential features. In [10], authors suggest speech sections with similar format qualities are grouped together and labeled as the same morpheme during the pre-processing step to decrease time and computational expenses. The phoneme recurrence frequencies in expressive speech are subsequently used as input attributes by classifiers. The degree of accuracy is equivalent to that of contemporary cutting-edge techniques, and the needed training period was reduced from hours to minutes by utilizing six databases (EmoDB, RAVDESS, IEMOCAP, ShEMO, DEMoS, and MSP-Improv).

The authors in [11] introduced a dynamic multi-level features extraction model based on an encryption framework for improving voice emotion detection classification performance. Four public speech emotion databases (RAVDESS Speech, Emo-DB (Berlin), SAVEE, and EMOVO) are used to validate the model's performance. Using RAVDESS, Emo-DB (Berlin), SAVEE, and EMOVO corpora, the created twine-shuf-pat, and INCA-based technique yielded 87.43%, 90.09%, 84.79%, and 79.08% classification accuracies, respectively, with a 10-fold cross-validation strategy. Four public speech emotion databases were combined to generate a mixed database with a classification accuracy of 80.05%.

Similarly, authors in [12] presented an end-to-end authentic SER based on a single-dimensional compressed convolutional neural network model (DCNN). A multi-learning strategy was employed by the program to obtain spatially salient emotional cues components from voice signals while also learning long-term contextual relationships. Using the benchmark IEMOCAP and EMO-DB datasets, the model achieved excellent recognition accuracy of 73% and 90%, respectively. The experimental findings validate the significance and effectiveness of our proposed paradigm, which has shown to be extremely useful in the establishment of a meaningful SER mechanism. As a result, the model can handle original voice data for emotion recognition using a minimal inflated Cnn model using the multi-learning method (MLT) method.

The implications of limited voice frequency and the -low companding approach employed in transmission networks on speech emotion detection performance are investigated by authors in [13], who provide a detailed explanation of a real-time voice emotion detection approach utilizing AlexNet, a pre-trained image analysis system. The benchmark approach produced an accuracy level of 82% when trained on Berlin Emotional Speech (EMO-DB) data encompassing seven classified moods. The SER accuracy was reduced by roughly 3.3% when the average sampling rate of 16–8 kHz was lowered (i.e., bandwidth was reduced from 8 to 4 kHz, respectively). When compared to the baseline data, the companding process reduced The cumulative impact of combining knowledge and banding increased accuracy level by 3.8% shortening decreased efficiency by about 7%. With emotional labeling, the SER was conducted in real-time being created every 1.033–1.026 s.

In [14], the authors introduced a unique multimodal EEG mood identification technique based on a dynamical graph convolutional neural network (DGCNN). The experimental findings demonstrate that the proposed model outperformed state-of-the-art methodologies, with a mean total accuracy of 90.4% for particular matter research on the SEED directory and 79.95% for pertaining cross-validation research, and 86.23, 84.54, and 85.02% for valence, vigilance, and dominance categorisation on the SEED directory DRE, respectively.

The dataset used in each of the literature is secondary dataset that was obtained from various online repositories. Also, some of the experimental results give a low accuracy, also some of the literature was prone to low training time of the dataset. Another limitation of the study is that the nature of the dataset used in most of the literature was not indicated. For more accurate and efficient results, there is a need to use real and actual data for the experiment with accurate recognition of emotion. Therefore, this study generated an audio file primary dataset from the Bells University of Technology for the experiment.

3 Methodology

3.1 Dataset Collection

The model was experimented with 1295 primary dataset gotten from Bells University of Technology; the dataset consists of Rage, panic, neutrality, happiness, and sadness all emotions. The dataset was divided into two categories which aree datasets for training and validation, with 70% of the dataset used for training and 30% for validation. Table 1 describe the description of dataset used in this study while Fig. 1 shows the flow diagram of the emotion recognition system.

Table 1. Dataset description

Classes	Number of audio files	Number of training	Number of testing
Anger	259	181	77
Fear	259	181	77
Neutral	259	181	77
Happy	259	181	77
Sad	259	181	77

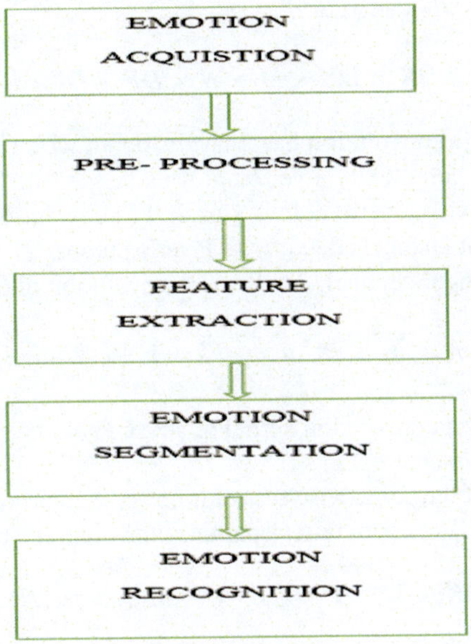

Fig. 1. The flow diagram of the emotion recognition model

3.2 The Model: Yet Another Mobile Network

"Yet Another Mobile Network" (YAMNets) is an audio classification model that includes a classification network (using Mobilnet) and feature extraction. A lightweight kind of factorized convolution is complexity detachable convolution, which converts a regular convolution into a depth-wise convolution and an 11 convolution (point-wise convolution, weight factor). Note that because the sizes of the parameters are different, depth-wise separable convolution is an incompatible fork of traditional convolution; additionally, this factorization can speed up a large amount of computation with a small loss in accuracy, which will be discussed further in the following sections. The traditional one, it takes a $DF \times DF \times M f$eature map F as input and outputs a $DG \times DG \times N$ feature map G, where DF represents the spatial width and height of a square input feature map, M represents the number of input channels, DG represents the spatial width and height of a square output feature map, and N represents the number of output channels, the convolution kernel K of size $DK \times DK \times M \times N$, where DK is the square dimension of the kernel, is used to parameterize the conventional convolution layer. This method's output feature map (with stride 1 and padding) is as follows:

$$Gk, l, n = \sum i, j, mKi, j, m, n \cdot Fk + i - 1, l + j - 1, m \tag{1}$$

With cost of computation as,

$$DK \times DK \times M \times N \times DF \times DF \tag{2}$$

Depth wise separable convolution has 2 layers: depth wise convolutions and point-wise convolutions. The complexity layer employs a single filtering for each input signal, whereas the point-wise layer employs a basic 11 convolution to construct a linear combination of the input channels depth wise layer's output.

With one filter per input channel, depth wise convolution may be stated as follows:

$$G^k, l, m = \sum i, jK^i, j, m \cdot Fk + i - 1, l + j - 1, m \tag{3}$$

where $\hat{K}$ is the depth wise convolution kernel of size $DK \times DK \times M$ and the mth filter of $\hat{K}$ *is applied to the mth channel in F.*

It's worth noting that the point-wise convolution is the same as a regular convolution with a kernel size of 11. The depth wise separable convolution's total computing cost is

$$(DK \times DK \times M \times DF \times DF) + (M \times N \times DF \times DF) \tag{4}$$

And the number of parameters (without bias vector) is $.DK \times DK \times M + M \times N$

3.3 Experimental Setup

The development tool used is MATLAB version R2020a on Windows 10 pro-64bit operating system, Intel core ™ i5-73000 HQ @2.50 GHz Central Processing Unit, 8 GB Random Access Memory and 500 GB hard disk drive. The decision taken to recognize or classify the emotions as true positive and false positive according to the categories of the five emotions. Results generated based on True Positive (TP) and False Negative (FN), that is, the recognition accuracy was determined by the overall recognised True Positive of all five categorised emotions. It was discovered that, the higher True positive value generated high recognition accuracy and lower True Positive value generated low recognition accuracy. Figure 2 display the flowchart of the developed system from loading of audio files to the recognition of emotion from the audio file analyser.

4 Experimental Result

The implementation of the study shows various emotions from the inputted dataset. Figure 3 recognized anger emotion. Figure 4 recognized sad emotion. Figure 5 recognized fearful emotion. Figure 6 display the neural emotional data analysed while Fig. 7 display the data analysis of happy emotion detection.

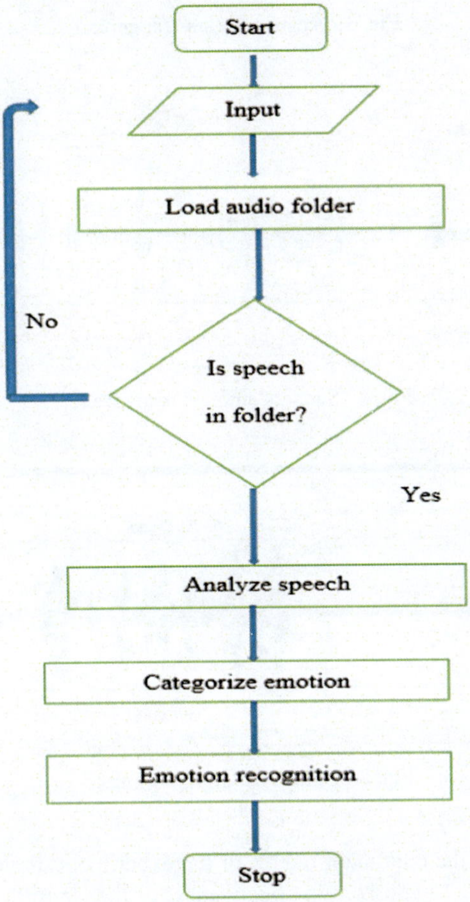

Fig. 2. Flowchart of the developed system.

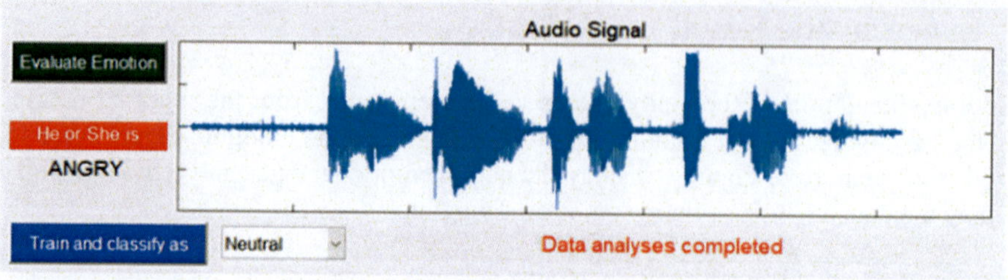

Fig. 3. Anger emotion recognized

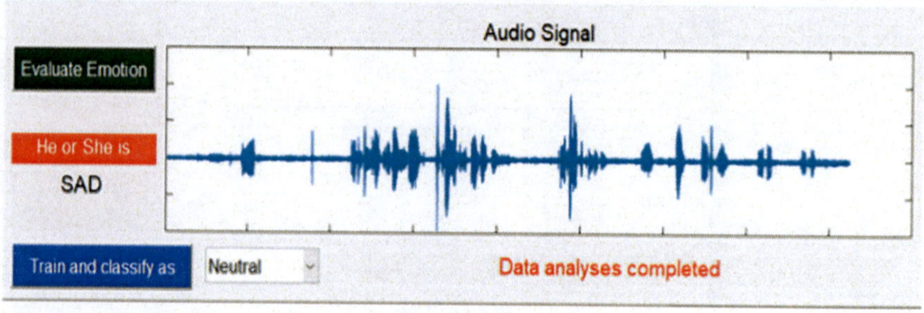

Fig. 4. Sadness emotion recognized

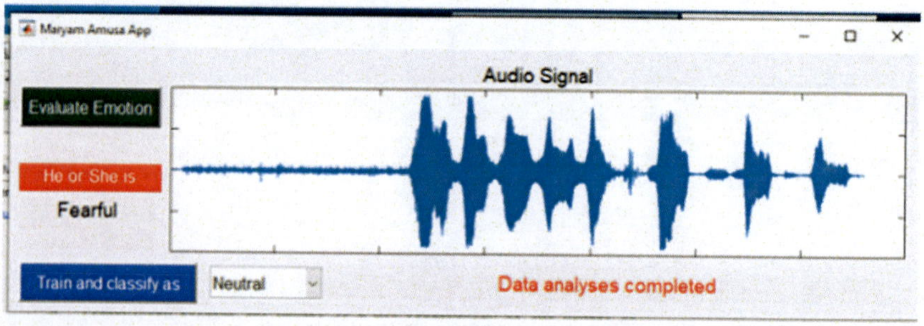

Fig. 5. Fearful emotion recognized

Table 2 presented the confusion matrix of the experiment; the confusion matrix is employed to measure a categorization effectiveness of the algorithm on a collection of data for which the real values are available. The confusion matrix presented in table two indicates a relatively high True Positive Rate per class (Anger = 234 (90.7%), Fear = 219 (85%), Neutral = 226 (87.2%), Happy = 241 (93%), and Sad = 239 (92.3%)), hence, the overall recognition accuracy is 89.5%. Thus, making the result significantly reliable as none of the classes is below the True Positive Rate of 85%, this implies that the developed model can correctly recognize these specified emotions in the ratio of 90:10.

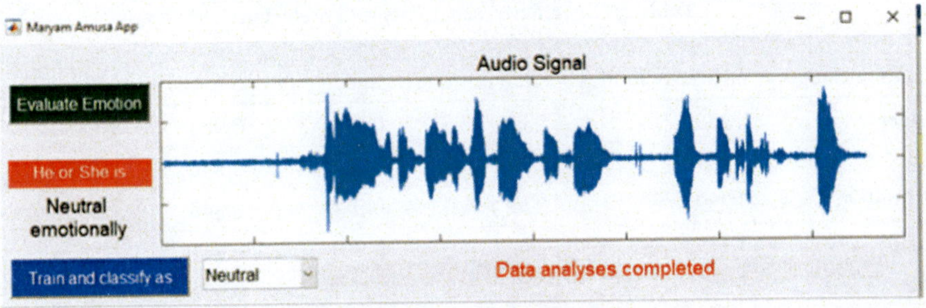

Fig. 6. Neutral emotion recognized

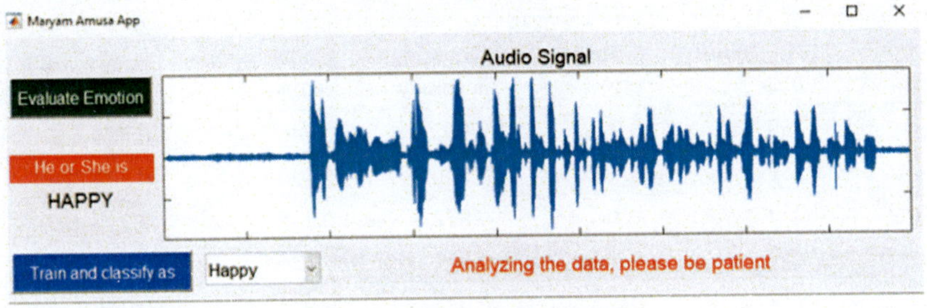

Fig. 7. Happy emotion recognized

Table 2. Confusion matrix

N=125	Anger	Fear	Neutral	Happy	Sad	Total
Anger	234	11	6	5	3	259
Fear	13	219	10	16	1	259
Neutral	9	7	226	6	11	259
Happy	13	4	1	241	0	259
Sad	5	8	0	7	239	259
Total	274	249	243	275	254	**Σ1295**

Further analysis as specify in Table 3 indicate the evaluation of the developed model, where three important metrics (specificity, sensitivity and precision) are considered. The results thus further establish the solidity and accuracy of the model.

Omission error is regarded as an error that originates from a partial or total absence of an operation from the log, whereas commission happens as a result of inaccurate transaction recording to the log. Both omission and commission error are put into consideration in this study and the result is presented in Table 4.

Table 3. Specificity, sensitivity and precision

Emotion category	No. Audio files	True positive	Specificity	Sensitivity	Precision
Anger	259	234	0.952	0.903	0.850
Fear	259	219	0.964	0.846	0.879
Neutral	259	226	0.984	0.848	0.930
Happy	259	241	0.961	0.931	0.876
Sad	259	239	0.982	0.922	0.941

Table 4. Omission and commission

Emotion categories	No. Audio files	Omission error	Omission accuracy	Commission error	Commission accuracy
Anger	259	0.096	0.903	0.146	0.854
Fear	259	0.154	0.846	0.120	0.879
Neutral	259	0.127	0.872	0.075	0.930
Happy	259	0.069	0.931	0.124	0.876
Sad	259	0.0772	0.923	0.057	0.941

5 Conclusion

Advances in neural nets, as well as the on-demand necessity precise and close to real Speech Emotion Recognition in human-computer interactions, necessitate a comparison of possible techniques and databases in Speech Emotion Recognition in order to accomplish suitable solution and a stronger comprehension of this open-ended challenge. In that regard, this study adopted YAMNets Tool to build a model to recognize emotions through speech. The developed model achieved a recognition accuracy of 89.5% with rejection rate of 10.5% and further analysis reveals a True Positive Rate of 85%. This study contributed to the body of knowledge by using a primary audio file dataset to recognize individual emotion status. Also the study used a new model "Yet Another Mobile Network Tool" to carry out the experiment compare to the start-of-the-art study which focuses only on machine learning techniques. This study only consider five emotions (anger, fear, neutral, happy, and sad), thus, it cannot be use to make a generalization in a more complex situation. It is therefore recommended that future work should look into recognition of more classes of emotions outside the five focused in this research. Also, this study uses a supervised learning mechanism, hence future work can consider unsupervised or semi supervised learning.

References

1. Akshaykumari, J.: Emotions (Psychology) (2015). https://www.researchgate.net/publication/320621079_Emotions_Psychology/citations

2. Hockenbury, D.H., Hockenbury, S.E.: Discovering Psychology. Worth Publishers, New York (2007)
3. Uddin, M.Z., Nilsson, E.G.: Emotion recognition using speech and neural structured learning to facilitate edge intelligence. Eng. Appl. Artif. Intell. **94**, 103775 (2020). https://doi.org/10.1016/j.engappai.2020.103775
4. Awotunde, J.B., Ogundokun, R.O., Ayo, F.E., Matiluko, O.E.: Speech segregation in background noise based on deep learning. IEEE Access **8**, 169568–169575 (2020)
5. Mordor Intelligence: Emotion Detection and Recognition (EDR) Market - Growth, Trends, COVID-19 Impact, and Forecasts (2021 - 2026) (2021). https://www.mordorintelligence.com/industry-reports/emotion-detection-and-recognitionedr-market
6. Sharma, A., Kumar, R., Mansotra, V.: Proposed stemming algorithm for hindi information retrieval. Int. J. Innov. Res. Comput. Commun. Eng. (An ISO Certified Organization). **3297**(6), 11449–11455 (2016). https://doi.org/10.15680/IJIRCCE.2016
7. Goyal, M., Reeves, N.D., Rajbhandari, S., Yap, M.H.: Robust methods for real-time diabetic foot ulcer detection and localization on mobile devices. IEEE J. Biomed. Health Inform. **23**(4), 1730–1741 (2019). https://doi.org/10.1109/JBHI.2018.2868656
8. Oladipo, I.D., AbdulRaheem, M., Awotunde, J.B., Bhoi, A.K., Adeniyi, E.A., Abiodun, M.K.: Machine learning and deep learning algorithms for smart cities: a start-of-the-art review. In: Sur, S.N., Balas, V.E., Bhoi, A.K., Nayyar, A. (eds.) IoT and IoE Driven Smart Cities. EICC, pp. 143–162. Springer, Cham (2022). https://doi.org/10.1007/978-3-030-82715-1_7
9. Abbaschian, B.J., Sierra-Sosa, D., Elmaghraby, A.: Deep learning techniques for speech emotion recognition, from databases to models. Sensors. **21**(4), 1249 (2021). https://doi.org/10.3390/s21041249
10. Liu, Z.T., Rehman, A., Wu, M., Cao, W.H., Hao, M.: Speech emotion recognition based on formant characteristics feature extraction and phoneme type convergence. Inf. Sci. **563**, 309–325 (2021)
11. Tuncer, T., Dogan, S., Acharya, U.R.: Automated accurate speech emotion recognition system using twine shuffle pattern and iterative neighborhood component analysis techniques. Knowl. Based Syst. **211**, 106547 (2021). https://doi.org/10.1016/j.knosys.2020.106547
12. Mustaqeem, S.K.: MLT-DNet: speech emotion recognition using 1D dilated CNN based on multi-learning trick approach. Expert Syst. Appl. **167**, 114177 (2021). https://doi.org/10.1016/j.eswa.2020.114177
13. Lech, M., Stolar, M., Best, C., Bolia, R.: Real-time speech emotion recognition using apretrained image classification network: effects of bandwidth reduction and companding. Front. Comput. Sci. **2**, 14 (2020). https://doi.org/10.3389/fcomp.2020.00014
14. Song, T., Zheng, W., Song, P., Cui, Z.: EEG emotion recognition using dynamical graph convolutional neural networks. IEEE Trans. Affect. Comput. **11**(3), 532–541 (2018)

Study and Design of a 28/38 GHz Bi-band MIMO Antenna Array Element for 5G

Salah-Eddine Didi[1(✉)], Imane Halkhams[2], Mohammed Fattah[3], Younes Balboul[1], Said Mazer[1], and Moulhime El Bekkali[1]

[1] IASSE Laboratory, Sidi Mohamed Ben Abdellah University, Fez, Morocco
salaheddine.didi@usmba.ac.ma
[2] SIEDD Laboratory, UPF, Fez, Morocco
[3] IMAGE Laboratory, Moulay Ismail University, Meknes, Morocco

Abstract. The technological progress in the telecommunications domain is undoubtedly due to the increasing needs of users in terms of throughput and bandwidth. With such a high demand, new systems should develop with the aim of offering solutions to the new requirements in this field. It is within this framework that the 5G standard has been established and based on the use of massive MIMO offering new frequency bands (millimeter waves ranging from 24 GHz to 60 GHz). MIMO systems, characterized by multiple antenna elements, have demonstrated their potential for improving communication performance in multipath environments. However, the two major obstacles to the design of MIMO antennas are their size and their mutual coupling, and there is a need to find an intermediate solution between their size and their mutual coupling. In this work, we propose the design to create a compact bi-band (28/38 GHz) MIMO antenna element for 5G applications. This proposal consists of an inverted L-shaped patch antenna element printed on an Isola Gigaver 210 type substrate and reduced in size ($7.5 \times 6 \times 0.3$ mm^3) with a relative permittivity of 3.75 and a loss tangent of 0.003. The results show that this MIMO antenna element has bandwidths of 1.91 GHz and 1.07 GHz, gains of 3.3 dB and 3.4 dB, directivities of 3.6 dB and 3.8 dB, S11 reflection coefficients of −28.99 dB and −26.08 dB, and efficiencies of 96% and 96.2% for the 28 GHz and 38 GHz frequencies respectively.

Keywords: MIMO · 5G · Dual-band · Millimeter-wave · Microstrip antenna · Slot

1 Introduction

Today, the number of wireless devices has evolved significantly, as well as the unsatisfactory bandwidth and low-speed transmission capabilities, which strongly support the development of new technologies in the domain on telecommunications. This development has also contributed to the creation of new next-generation communication systems (5G) at the millimeter-wave spectrum, which offers much better coverage and higher throughput [1, 2]. 5G technology offers improved performance, high data rates, and energy savings to meet the needs of the significant growth in connected devices. In

Y. Farhaoui et al. (Eds.): ICAISE 2022, LNNS 635, pp. 740–746, 2023.
https://doi.org/10.1007/978-3-031-26254-8_107

addition, it will also provide better business opportunities for existing technologies such as smart cities and virtual reality [3, 4]. However, the major problems posed by limitations in the millimeter-wave spectrum, including noise fading, atmospheric effects, and reductions related to path losses, which are increasingly important when using a single antenna, must be addressed [5].

The "massive MIMO" (Multiple Input Multiple Output) smart antennas deployed as part of the arrival of 5G are composed of multiple transmitters, integrated together in a single piece of equipment, and which, controlled by software, allow the radio wave beam to be focused on a given area. In other words, they will direct the mobile network signal to only those people who need it at a given time. The radio signal is therefore not directed everywhere and all the time continuously, but as the antennas of previous generations do, focus only on the equipment that requires it at the moment they need to access it. This technique, known as beamforming, increases the capacity of the mobile network through a more efficient allocation of the waves to the communications of each user connected to the cell [6]. Nevertheless, multi-element antenna arrays also offer a similar level of capability as the individual antenna, because the antenna arrays have a common power source. On the other hand, MIMO antennas offer the ability to propagate data across multiple paths and provide higher data rates, higher capacity, and better link quality, which is the key feature of 5G technology. Several types of MIMO antenna designs for 5G VHF applications have been published in the literature recently [7].

In paper [8], a PIFA antenna operating in two frequency bands for 5G applications was designed. This antenna offers two bandwidths: a bandwidth of 3.34 GHz and 1.395 GHz for frequencies corresponding to the 28 GHz and 38 GHz bands, as well as gains corresponding to 3.75 dB and 5.06 dB for these same bands. Millimeter waves (mm-wave) have received increasing universal interest across various methodologies due to the wide bandwidth available. In the paper [9], he created a SIW antenna, this antenna offers bandwidths of 0.45 GHz and 2.2 GHz for respective operational frequencies that are 28 GHz and 38 GHz, and its gains are 5.2 dBi and 5.9 dBi for these same frequencies. In the work [10], we find the design of an AiP-shaped antenna operating at two different frequency bands, 28 and 38 GHz; it has a gain of 4 dBi and 6 dBi as well as two bandwidths, 1.02 GHz and 3.49 GHz, respectively. In addition, the work in [11] proposes a small antenna, fed by the coplanar waveguide (CPW). This antenna printed on a Roger RT5880 substrate with dimensions of 5×5 mm^2 and a height of 0.254 mm, presents a maximum gain of 6.6 dB at 28 GHz. The paper published by the author [12], proposes to design a three-band MPA antenna with gains of 7.02 dBi and 5.05 dBi for 28 and 38 GHz respectively, but a very narrow bandwidth (0.9 GHz and 0.48 GHz).

In this paper, we will perform a design and simulation of an inverted L-shaped patch antenna MIMO array element, we add a stub and a hole to it these play an important role to improve the performance of this antenna-like bandwidth and size reduction in the millimeter band of 5G which operates in the bands (28 GHz and 38 GHz) The designs and simulations are performed using HFSS software. This work is structured as follows: Sect. 2 details the design process of this antenna. Then, in Sect. 3, we simulate this antenna using the HFSS (High-Frequency Structure Simulator) software, so we analyze and compare the various results obtained for this new antenna with those which currently exist. Finally, we conclude in Sect. 4.

2 Modeling Method and Antenna Design

2.1 The Choice of the 28/38 GHz Band

The 28 GHz and 38 GHz frequency bands described in our work are contained in the Ka-band (27 GHz to 40 GHz) established by the ITU and have long been used for satellite communications. The Ka-band offers indeed, many prospects in terms of bandwidth: twice the Ku band (12 GHz to 18 GHz) and four times the C band (4 GHZ to 8 GHz). The frequency bands chosen by different countries for 5G are in the Ka-band and most often around the frequencies 28 GHz and 38 GHz. All these advantages make us think that the frequencies chosen in our work will be the most used in the future. The new 5G standard, one of the main objectives of which is to increase the throughput of 4G by a factor of 100, provides for the use of millimeter waves in its second frequency domain FR2 (27 GHz–60 GHz). The antennas used in this frequency range are very small, which is a great advantage for mobile equipment manufacturers. In addition, studies have shown that Ka-band beams are more directional than C- and Ku-band beams, which results in a considerable reduction in power consumption in the network [13, 14].

2.2 Structure of This Proposed Antenna

The proposed structure is a MIMO array antenna element, as shown in Fig. 1. This figure has three layers: the ground plane in red, the substrate in gray, and the radiating element in blue. This element is an inverted L-shaped radiating patch consisting of a conductive bar of thickness $Cu = 0.035$ mm that operates at 38 GHz. Another band is added to its upper end to also have two bands operating at 28 GHz. A slot is added to the patch to improve the performance of this antenna, especially the bandwidth. In this work, the Isola Gigaver 210 (tm) type substrate is chosen to have a relative permittivity of $\varepsilon r =$ 3.75 and a loss tangent of 0.003 with dimensions of $7.5 \times 6 \times 0.3$ mm^3. The physical parameters of this proposed antenna are shown in the table below. They are calculated by the formulas [15, 16] (Table 1).

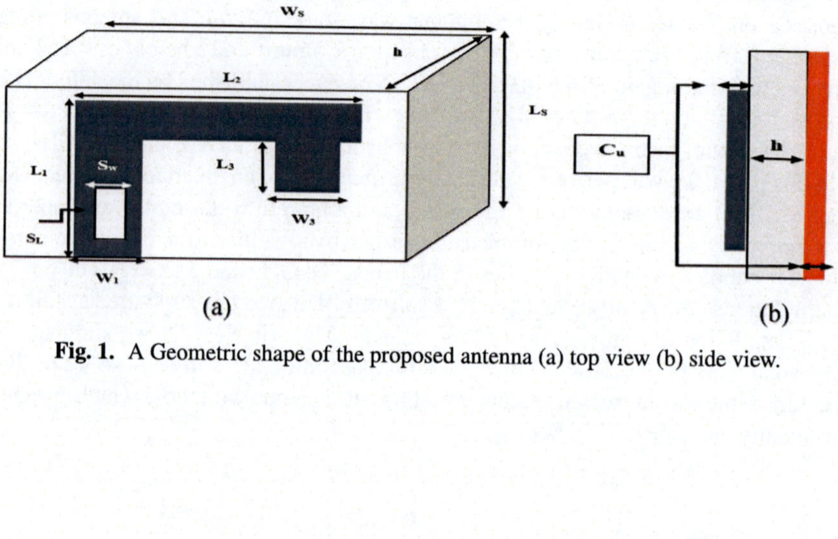

Fig. 1. A Geometric shape of the proposed antenna (a) top view (b) side view.

Table 1. Table captions should be placed above the tables.

Parameters	Values (mm)	Parameters	Values (mm)
L_1	5.5	h	0.3
W_1	1.04	S_L	0.6
L_2	4.25	S_W	1.8
L_3	1.5	W_S	7.5
W_3	0.946	L_S	6
S_W	1.8		

3 Results and Discussion

3.1 Results of the Simulations Carried Out During This Work

i. Return loss S11 and VSWR

The plane between the transmission source and the antenna has a discontinuity characterized by the reflection coefficient S11. By definition, this coefficient (usually expressed in dB) is presented by the ratio between the reflected wave and the incident wave at the antenna. The Standing Wave Ratio (SWR) expresses the quality of the antenna's match to a transmission line, coaxial or bifilar. Impedance matching is a technique used in electricity to optimize the transfer of electrical energy between a transmitter (source) and an electrical receiver (load) and to optimize the transmission of telecommunications signals. The results obtained by this proposed antenna are excellent in terms of its S11 reflection coefficient (−28.99 dB and −26.08 dB), VSWR (1.074 and 1.105), and bandwidths (1.91 GHz and 1.07 GHz) for the frequencies of 28 GHz and 38 GHz, respectively, as shown in Figs. 2(a) and (b).

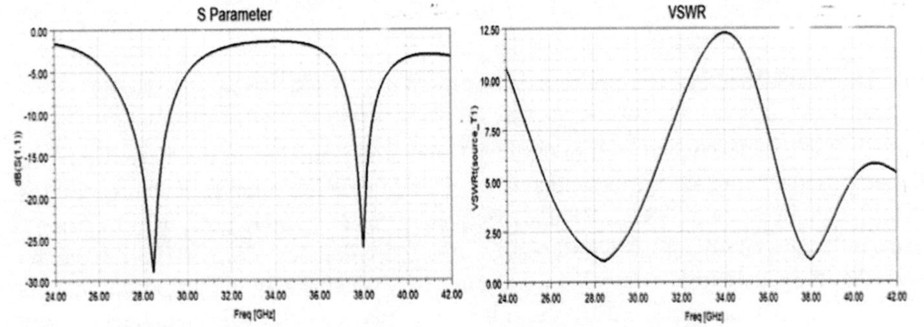

(a) (b)

Fig. 2. **(a)** S11 curve and **(b)** VSWR curve

ii. **Gain and Directivity**

The directivity of an antenna is the ratio of the power density S (radiated power per unit area) of the real antenna in its main direction to that of a hypothetical, but in reality nonexistent, an isotropic transmitter that would radiate uniformly in all directions. The power density of the isotropic transmitter is therefore the radiated power uniformly distributed over a spherical surface. Each real antenna has a more or less pronounced directivity. The directivity is defined by the quotient of the power density of the real antenna in its main direction and the power density of the isotropic transmitter. Directivity is an essential part of antenna gain. For a real antenna gain, gains and losses must be compensated. The power radiated by an antenna is, for each antenna, proportional to the injected transmit power, which can be measured quite easily on the antenna feed line. Some of this transmit power is however lost inside the antenna, at its ohmic resistors. These losses are defined as the efficiency of an antenna. If an ideal antenna existed without these losses, the gain of the antenna would be equal to its directivity. The simulation results performed in this work regarding the gain and directivity of the proposed antenna are remarkable (G = 3.3 dB and 3.4 dB) and (D = 3.6 dB and 3.8 dB) respectively for the frequencies of 28 GHz and 38 GHz, as shown in Fig. 3.

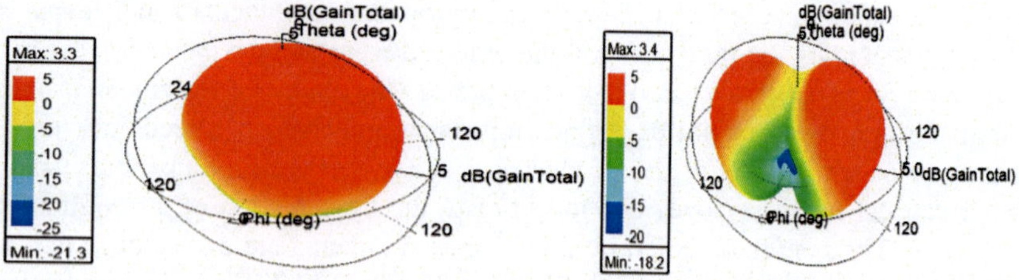

(a) (b)

Fig. 3. (a) and (b) are the graphical representations of the 3D gain, respectively for the frequencies of 28 GHz and 38 GHz.

iii. **Comparisons between the results obtained and similar research works**

Finally, Table 2 clearly and legibly outlines simulation results from work conducted on the proposed antenna design as well as those found in the literature for antenna designs that operate in the two 28 GHz/38 GHz frequency bands, namely antenna dimensions, gain, reflection coefficient, VSWR, bandwidth, and efficiency. From Table 2 it can be seen that the bandwidth, return loss, and efficiency obtained by the proposed antenna are better than those obtained by the other antennas are. The gain generated by the antenna [18] is larger than that obtained with the other antennas [17, 19, 20] as well as the proposed antenna. In addition, the antenna [19] has a smaller size than the other antennas, as shown in Table 2.

Table 2. Comparison between the simulation results and previous research works

Ref	Size (mm^2)	Freq (GHz)	S11(dB)	BW(GHz)	Gain (dB)	Eff (%)
[17]	13 × 11.25	28	−23.6	1.49	5.41	90.3%
		38	−27.1	1.01	4.89	84.3%
[18]	3.8 × 5.5	28	–	0.7	5	–
		38	–	0.3	5.5	–
[19]	3 × 7	28	–	3.34	3.75	–
		38	–	1.395	5.06	–
[20]	55 x 110	28	−13,79	1	5,126	91,02
		38	−30,36	1.3	4,607	92,81
This work	7.5 × 6	28	−28.99	1.91	3.3	96
		38	−26.08	1.07	3.4	96.2

4 Conclusion

The authors of this paper proposed the design of a 5G millimeter band MIMO antenna element. This antenna is resonant at two different frequency bands: 28 GHz and 38 GHz. The operational frequency ranges are between 25.5 and 29.6 GHz, and their bandwidths are 1.91 GHz and 1.07 GHz for 28 GHz and 38 GHz respectively. The other performances are also: gains of 3.3 dB and 3.4 dB, directivities of 3.6 dB and 3.8 dB, and S11 reflection coefficients of −28.99 dB and −26.08 dB for 28 GHz and 38 GHz respectively. Indeed, the better radiation performance attributed to the envisaged antennas demonstrates that they are compatible with the 5G telecommunications system. My next goal will be to develop this antenna proposal to design a 5G millimeter band MIMO antenna array.

References s

1. Halkhams, I., Mazer, S., Mehdi, M., El Bekkali, M., El Hamdani, W.: Design of a dual-band bandpass filter using active inductor principle. Int. J. Commun. Ant. Propag. (IRECAP) **7**(4), 342–347 (2017)
2. Didi, S.-E., et al.: Study and design of the microstrip patch antenna operating at 120 GHz. In: El Ghzaoui, M., Das, S., Lenka, T.R., Biswas, A. (eds.) Terahertz Wireless Communication Components and System Technologies, pp. 175–190. Springer, Singapore (2022). https://doi.org/10.1007/978-981-16-9182-9_12
3. Didi, S.-E., Halkhams, I., Fattah, M., Balboul, Y., Mazer, S., Bekkali, M.E.: Design of a microstrip antenna two-slot for fifth generation applications operating at 27.5 GHz. In: Motahir, S., Bossoufi, B. (eds.) ICDTA 2021. LNNS, vol. 211, pp. 1081–1089. Springer, Cham (2021). https://doi.org/10.1007/978-3-030-73882-2_99
4. Abdelhafid, E., et al.: A 5G mm-wave compact voltage-controlled oscillator in 0.25 μm pHEMT technology. Int. J. Electr. Comput. Eng. (IJECE) **11**(2), 1036–1042 (2021)
5. Salah-Eddine, D., Imane, H., Mohammed, F., Younes, B., Said, M., Moulhime, E.L.: A review of the design of a single antenna and antenna arrays at 60 GHz for 5G applications system. In: international Conference on Big Data, Modelling and Machine Learning (BML 2021), Kenitra, Morocco, pp. 103–107 (2021)

6. Mohammad, L.H., Mohammed, J.U., Hoque, M.D.J.: 28/38 GHz dual-band microstrip patch antenna with DGS and stub-slot configurations and its 2 × 2 MIMO antenna design for 5G wireless communication. In: 2020 IEEE Region 10 Symposium (TENSYMP), Dhaka, Bangladesh, pp. 56–59 (2020)
7. Mohammed, M., Mohammed, F., Yousef, F.: Energy consumption and data storage. Procedia Comput. Sci. **175**, 609–614 (2020)
8. Zhang, Y., Deng, J., Li, M., Sun, D., Guo, L.: A MIMO dielectric resonator antenna with improved isolation for 5G mm-wave applications. IEEE Ant. Wirel. Propag. Lett. **18**(4), 747–751 (2019)
9. Ahmad, W., Khan, W.T.: Small form factor dual-band (28/38 GHz) PIFA antenna for 5G applications. In: 2017 IEEE MTT-S International Conference on Microwaves for Intelligent Mobility (ICMIM), pp. 21–24. IEEE (2017)
10. Ashraf, N., Haraz, O., Ashraf, M.A., Alshebeili, S.: 28/38-GHz dual-band millimeter-wave SIW array antenna with EBG structures for 5G applications. In: 2015 International Conference on Information and Communication Technology Research (ICTRC) , Abu Dhabi, pp. 5–8 (2015)
11. Chu, C., Zhu, J., Liao, S., Zhu, A., Xue, Q.: 28/38 GHz Dual-band dual-polarized highly isolated antenna for 5G phased array applications. In: 2019 IEEE MTT-S International Wireless Symposium (IWS), Guangzhou, China, pp. 1–3 (2019)
12. Ali, M.M.M., Sebak, A.R.: Dual-band (28/38 GHz) CPW slot directive antenna for future 5G cellular applications. In: 2016 IEEE International Symposium on Antennas and Propagation (APSURSI), pp. 399–400. IEEE (2016)
13. Abdelhafid, E., et al.: Very low phase noise voltage controlled oscillator for 5G mm-wave communication systems. In: 1st International Conference on Innovative Research in Applied Science, Engineering and Technology, Meknes, Morocco, pp. 1–4. IEEE (2020)
14. Kamal, M., Islam, M., Uddin, M., Imran, A.: Design of a tri-band microstrip patch antenna for 5G application. In: 2018 International Conference on Computer, Communication, Chemical, Material and Electronic Engineering (IC4ME2), Rajshahi, Bangladesh, pp. 1–3 (2018)
15. Salah-Eddine, D., Imane, H., Mohammed, F., Younes, B., Said, M., Moulhime, E.L.: Design of a microstrip antenna patch with a rectangular slot for 5G applications operating at 28 GHz. J. TELKOMNIKA Telecommun. Comput. Electron. Control **20**(3), 527–536 (2022)
16. Salah-Eddine, D., Imane, H., Mohammed, F., Younes, B., Said, M., Moulhime, E.L.: Study and design of printed rectangular microstrip antenna arrays at an operating frequency of 27.5 GHz for 5G applications. J. Nano- Electron. Phys. **13**(6), 06035 (2021)
17. Muhammad, A., Muhammad, I.K., Syed, M.O., Abbas, A.K., Asif, S.: 2020 Design and analysis of millimeter wave dielectric resonator antenna for 5G wireless communication systems. Prog. Electromagn. Res. C **98**, 239–255 (2020)
18. Hashem, Y.A., Haraz, O.M., El-Sayed, E.D.M.: 6- element 28/38 GHz dual-band MIMO PIFA for future 5G cellular systems. In: 2016 IEEE International Symposium on Antennas and Propagation (APSURSI), Fajardo, PR, USA, pp. 393–394 (2016)
19. Ahmad, W., Khan, W.T.: Small form factor dual band (28/38 GHz) PIFA antenna for 5G applications. In: 2017 IEEE MTT-S International Conference on Microwaves for Intelligent Mobility (ICMIM), Nagoya, Japan, pp. 1–5 (2017)
20. Marzouk, H.M., Ahmed, M. I., Shaalan, A.A.: A novel dual-band 28/38 GHz AFSL MIMO antenna for 5G smartphone applications. In: Journal of Physics: Conference Series, Fourth International Conference on Advanced Technology and Applied Sciences (ICaTAS 2019), Cairo, Egypt, 10–12 September 2019, vol. 1447 (2019)

Temporal Blockchains for Intelligent Transportation and Autonomous Vehicles Management

Zouhaier Brahmia[1](✉), Fabio Grandi[2], and Rafik Bouaziz[1]

[1] University of Sfax, Sfax, Tunisia
zouhaier.brahmia@fsegs.rnu.tn, rafik.bouaziz@usf.tn
[2] University of Bologna, Bologna, Italy
fabio.grandi@unibo.it

Abstract. Nowadays, blockchain technology is being used in several domains, and in particular for managing intelligent transportation systems and autonomous vehicles. However, since block timestamps, put by miners, are inaccurate (with about one hour of delay), they could not be useful for many real-time applications that require to deal with exact timing of events, like traffic laws enforcement and insurance accident investigation. Hence we propose to add new timestamps, put by data originators, with the meaning of valid time as defined in temporal databases and which can represent the accurate time of data recording within a transaction. Consequently, the new enhanced blockchain data model, named BiTchain, can be considered, from a temporal database perspective, as a bitemporal database. In order to let users and applications enjoy the potentialities of BiTchain, we also propose an expressive temporal query language, named BiTEQL and defined as a TSQL2-like temporal extension of the EQL blockchain query language.

Keywords: Data model · Query language · Blockchain · Block timestamp · Traffic management · Autonomous vehicle · Temporal database · Transaction time · Valid time · Bitemporal data model · Ethereum query language · TSQL2 · Temporal query language

1 Introduction

Blockchain [1, 2] is a public distributed ledger that stores committed transactions in an ordered list, or a chain, of blocks. The ledger is replicated and maintained by multiple distributed nodes (computers), which are linked in a peer-to-peer network (i.e., without a server node that has full control and central authority) and possibly do not trust each other, through a consensus mechanism (i.e., an agreement among these nodes on the truth of data stored in the blockchain) and cryptography (essentially hash algorithms and digital signatures). A transaction is a sequence of operations applied on some state respecting the ACID properties as in classical database systems. It cannot be modified once it is recorded in a block of the blockchain. A block is made up of a block header and a block body: the former contains, among others, the parent block hash that points to

Y. Farhaoui et al. (Eds.): ICAISE 2022, LNNS 635, pp. 747–756, 2023.
https://doi.org/10.1007/978-3-031-26254-8_108

the previous block and a timestamp; the latter essentially contains transactions. From a database point of view [3], a blockchain can be considered as a distributed and replicated *temporal database* [4] that stores a large list of blocks.

Recently, the deployment of blockchains has been proposed for smart traffic management and autonomous vehicles support [5–9]. An autonomous vehicle [10] is a vehicle that can, without interaction with a human driver: (i) sense its environment (detecting objects like other vehicles, analyzing their behaviour and predicting their evolution) and (ii) efficiently drive itself along several types of roads having different characteristics and possibly presenting complex and unexpected situations. Exchange of information about the environment is also carried out via communication with other vehicles and roadside devices (e.g., smart grids, hotspots). The autonomous vehicle technology aims at having a safe and comfortable driving, avoiding road accidents or reducing their damage, improving mobility for disabled and elderly persons, reducing fuel consumption, resolving traffic congestion, and efficiently exploiting available infrastructures. Although the autonomous vehicle technology is still in its youth, there are some interesting commercial solutions like the Waymo One [11] of Waymo LLC and the Volvo XC90 SUV of Uber [12], and semi-autonomous driving can be rather considered the state-of-the-art (e.g., Tesla cars equipped with the Autopilot feature [13]).

Autonomous vehicles take their driving decisions in real time, based on data acquired by their sensor network or transmitted by surrounding vehicles and roadside devices. In order to *ex post* assess their decisions (e.g., to allege traffic infringements or evaluate liability in case of an accident), a trustworthy system is needed to record such data. As already proposed in [5–9], blockchains are an elective way to implement such a trustworthy recording. Moreover, an accurate timestamping of data is a strong requirement to reconstruct the premises of a real-time decision and is also needed for keeping track of real time actions performed for traffic control and safety enforcement by road control authorities and intelligent transportation support systems.

To this end, the basic blockchain timestamp support is not sufficient. Due to the mining effort, block timestamps are not accurate [14–16] and at least a 10 ÷ 15-min discrepancy with the events originating the data is usually considered. For example, in the Bitcoin blockchain [17], the timestamp of a block is required to be accurate only within two hours of the network-adjusted time and some solutions were proposed to reduce such a discrepancy, like the Median Past Time Rule and Future Block Time Rule which are used to prevent timestamp abuse [18].

Moreover, the temporal dimension of the timestamp put in the block header by the miner, and representing when the block was added to the blockchain, can be considered *transaction time* from a temporal database point of view [19]. In order to satisfy the application requirements recalled above, we think that also a timestamp representing *valid time* of data is required to be inserted by the data originator (corresponding to the validity of facts and occurrence of events in the real world, at least according to the local/network clock of the device that recorded the data), giving rise to a *bitemporal* database. The resulting new data model that we will propose in this work is named BiTchain (Bitemporal Blockchain).

Existing query languages proposed for blockchains, like EtherQL [20] and EQL [21], obviously do not provide a built-in support for querying and modifying bitemporal data and, thus, in this work, we will complete our proposal with a new temporal query language, named BiTEQL (BiTemporal EQL). BiTEQL is an expressive TSQL2-like temporal extension of EQL for querying BiTchain blockchains. TSQL2 [22] is a consensus temporal query language specified as a temporal extension of the SQL92 standard designed by a committee of temporal database experts chaired by R. T. Snodgrass in 1995. Several concepts and constructs of TSQL2 are supported in the SQL:2011 standard [23].

The rest of this paper is structured as follows. Section 2 proposes BiTchain, a data model for representing and managing bitemporal blockchain. Section 3 introduces BiTEQL, a temporal query language for querying BiTchain databases, and illustrates its functioning with some examples. Section 4 concludes the paper and provides some remarks about our future work.

2 BiTchain: A New Data Model for Bitemporal Blockchain Management

Thanks to its advantages (including distribution, absence of central parties, neutral support of security and reliability of transactions), the blockchain technology is being adopted for autonomous vehicles support. However, some problems have not been resolved yet, in particular those involving safety issues and those related to vehicle-to-vehicle communications and between vehicles and transport infrastructure components for refueling, charging, parking or repair services. Approaches to the solution of these problems could benefit from the exploitation of the temporal nature of information which can be made available to the involved agents. To this purpose, owing to a decades-long experience in temporal database research [4], we propose to extend a blockchain with valid time in order to enable the exploitation of a new bitemporal blockchain data model.

In this section, we introduce BiTchain, our bitemporal blockchain data model that extends conventional blockchains to the valid time dimension support.

2.1 Temporal Extension of the Conventional Data Model

Both in the conventional data model and in our extension, a blockchain is a sequence of blocks, where each block is composed of a header and a body and is linked to its parent block (i.e., its predecessor in the sequence) via a cryptographic hash pointer located in the block header. The first block of a blockchain is called "genesis block" and does not have a parent block.

The block header contains six components [2], including the *Timestamp*, whose value is set by the miner to represent when the block has been validated and which gives to the blockchain the semantics of a transaction-time database. The block body is composed of a transaction counter and a set of transactions (TRi, $i \in \{1, 2, \ldots, n\}$). As in conventional database management systems, a transaction in a blockchain is the result of the execution of a sequence of (low-level) operations applied on some (consistent) database state respecting the ACID properties. The number of transactions that can be packaged by a block depends on two parameters: the size of the block and the size of

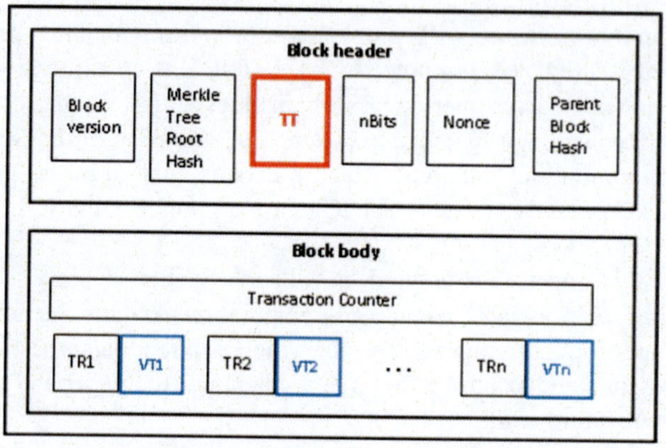

Fig. 1. The structure of a block in the BiTchain data model.

each transaction. An asymmetric cryptography mechanism is used by the blockchain to validate the transaction authentication.

Our proposed structure for a block is shown in Fig. 1. The extension consists of the addition of valid timestamps (VT1, VT2, ..., VTn in Fig. 1) to the transactions in the block body. The purpose of our proposed blockchain extension with valid time consists of allowing the data originators to specify, for each transaction, the time the data have been actually produced and representing their validity, according to the local/network clock of the system that recorded the data (or to an internet or satellite-based time server). Such time only depends on the "will" of the data originator through the execution of the recording transaction and is not affected by the blockchain mining effort.

Therefore, we obtain a bitemporal blockchain data model that maintains the history of data along both transaction time (block timestamp, provided by the miner) and valid time (data validity timestamp, supplied by the data originator). All the sets of data produced by the transactions contained in each block have the same transaction timestamp, since there is only one block timestamp (TT in Fig. 1), whereas each set of data produced by the same transaction have the same valid timestamp, since there is a (different) recording timestamp per transaction (and, thus, several valid timestamps per block). Thus, with respect to the conventional blockchain data model, the proposed valid timestamp will be associated as a new property named ValidTime to each transaction (TR1, TR2, ..., TRn) packaged in the block body.

Hence, the first part of our contribution is the definition of the BiTchain data model, which follows. In the BiTchain data model, the blockchain BC can be regarded (and queried) as a bitemporal table in the TSQL2 data model: BC.transactions(A1, A2, ..., Am | TT, VT), whose tuples correspond to the blockchain Transactions. Explicit attributes A1, A2,..., Am are the transaction data components of the blockchain (which we assume for simplicity to be flat values in order to have a 1NF temporal relation) and implicit attributes TT and VT are the transaction timestamp extracted from the blockchain Block containing the Transaction, and the valid timestamp extracted from the blockchain Transaction property ValidTime, respectively. In particular, in the TSQL2 terminology, BC is always an event table with respect to transaction time, since TT timestamps are timepoints, but can be an event or state table with respect to valid time, if valid timestamps are added to the blockchain Transactions are timepoints or periods, respectively. In practice, the table BC plays the role of a relational view, and can be implemented as a SQL foreign data wrapper, encapsulating the underlying blockchain data as follows:

BC.Blocks.BlockBody.Transactions.Aj → BC.transactions.Aj (j ∈ {1, 2, …, m}).
BC.Blocks.BlockBody.Transactions.ValidTime → VT(BC.transactions).
BC.Blocks.BlockHeader.Timestamp → TT(BC.transactions).

We also assume the BiTchain model to be a consistent extension of the EQL data model, that is the values of all the other components of the blockchain structure can be accessed from BC via explicit path expressions as defined in the EQL data model [21] (e.g., the properties Nonce, TransactionCounter, ParentBlockHash of the blockchain BC can be referenced in queries via EQL path expressions BC.blocks.header.nonce, BC.blocks.body.amountOfTransactions, BC.blocks.header.parentHash, respectively). They can be dealt with as additional explicit attributes as they are uniquely defined for each tuple of the BC.transactions bitemporal table.

3 BiTEQL: A Temporal Query Language for Querying BiTchain Blockchains

Since a blockchain is storing massive amounts of data, there is an increasing demand for effectively and efficiently querying blockchains in order to make their information contents available to applications by means of a powerful and user-friendly query language. To this purpose, EQL [21] is a SQL-like language which has been proposed for the Ethereum blockchain. Its "Select" statement is similar to that of SQL and includes five clauses: *select*, and *from*, which are mandatory, *where*, *order by* and *limit*, which are optional. The EQL Select statement allows blockchain users to specify structural and semantic filters, to reformat and transform returned data, to order query results and to limit the number of returned blocks.

However, there is no consensual proposal or standardization effort concerning temporal query languages for blockchains, although they can be considered temporal (and more precisely, transaction-time) databases. Therefore, taking into account these remarks and in order to allow blockchain users and applications to efficiently exploit our BiTchain model, we introduce a suitable temporal extension of EQL named BiTEQL (Bitemporal EQL), inspired by the well-known temporal query language TSQL2 [22]. In practice, the syntax and semantics of BiTEQL are exactly the same as the syntax and semantics of TSQL2, extended with the specific EQL constructs to deal with blockchains (in the same way that EQL extends SQL).

In the following, first we describe time representation and manipulation facilities in BiTEQL, then we describe temporal selection and temporal projection support in BiTEQL, and finally we illustrate the BiTEQL expressive power through some sample queries.

3.1 Time Representation and Manipulation in BiTEQL

As in the TSQL2 model, time is considered discrete, with a minimal system-dependent representation unit called chronon [19]. A mono-temporal chronon corresponds to an elementary interval on the time axis, whereas a bitemporal chronon corresponds to the Cartesian product of a transaction-time and a valid-time chronon. Three base temporal types have been defined for TSQL2 at the conceptual level: date-time, period and interval. The first one corresponds to an instantaneous event, without duration, which can be conventionally represented via a single chronon. The second one corresponds to a set of consecutive chronons along the time axis and is characterized by two datetime constants which represent its boundaries. The third corresponds to a pure duration, non anchored on the time axis, and can be represented as a multiple of the chronon. Union of disjoint time periods are also called temporal elements.

Functions and operators defined to manipulate time and duration datatypes are assumed to be available. We further assume that, as it happens for TSQL2, casting from another temporal datatype to a duration can be used to calculate the overall duration of a time period or element by means of, for instance, type constructor functions.

The TSQL2 language, which is based on a bitemporal data model, provides functions to access the valid and transaction time timestamps, which are implicit attributes of tuples. We assume similar functions to be available also for BiTEQL: if T is a bitemporal tuple, the expressions VALID(T) and TRANSACTION(T) can be used to extract the valid and transaction timestamps of T, respectively.

3.2 Temporal Selection in BiTEQL

Temporal selection is the most qualifying feature of a temporal query language, as it allows selecting data on the basis of their temporal properties. In order to add temporal selection capabilities to the EQL language, we extend the syntax of the WHERE clause of the SELECT statement.

As each transaction in a block is correspondingly augmented with the new valid timestamp in the data model, the BiTEQL WHERE clause is extended with variables and functions in order to specify temporal selection conditions based on data validity. In particular, in the BiTEQL WHERE clause, TSQL2 temporal (binary infix) predicates can be used, with the same semantics, to specify constraints over timestamp variables.

The available comparison operators are presented below with their semantics:

- A PRECEDES B: END(A) is earlier than BEGIN(B).
- A = B: A and B are identical (i.e., they contain the same chronons).
- A OVERLAPS B: the intersection of A and B is not empty.
- A MEETS B: END(A) immediately precedes BEGIN(B).
- A CONTAINS B: each chronon in B is also contained in A.

They can be used to compare (monodimensional) temporal elements, periods and time points. Since all temporal types can be reduced to sets of chronons, such operators can also be used to compare operands with different temporal types [22]. For instance, if A is an element and B is a period, then the expression "A PRECEDES B" is true if the last chronon belonging to A precedes the left boundary of B. All the comparison operators can be implemented on the basis of a primitive operator "Before()" which defines the relation order on the time axis. It can also be easily checked that such operators guarantee the temporal completeness of the resulting language, as they allow users to check the occurrence of all the possible relationships between two periods or events. Such operator set has been chosen for TSQL2 also considering the user-friendliness of the language among the design principle. This leads to a non minimal set of comparison operators which are closer to their meaning in natural language than the artificial definition of operators which equip other temporal languages (e.g., based on Allen's algebra [24]).

The OVERLAPS and CONTAINS operators can also be defined to work on multi-dimensional timestamps in a straightforward way.

3.3 Temporal Projection in BiTEQL

Temporal projection is the operation which specifies the value of the timestamps to be assigned to the retrieved data. TSQL2 supports an optional VALID clause to specify valid-time projection, as the transaction time assigned to query results is always the current time and cannot be changed by the user. In case a VALID clause is not specified, the data timestamps are used by default (if more than one temporal relation R1, R2, ..., Rn appear in the FROM clause, the valid time assigned to the result is computed by default as the intersection of the timestamps of the data in R1, R2, ..., Rn in order to implement an implicit temporal join). BiTEQL also supports such a clause.

3.4 Examples of BiTEQL Queries

In this section, we illustrate the use of BiTEQL by providing some examples of queries written in this language.

First of all, the basic EQL syntax of BiTEQL can be used to express standard queries over blockchains. For instance, Fig. 2 below shows a BiTEQL query over blocks of the blockchain BC similar to that presented in Sect. 4.2 of [21].

```
SELECT block.parent.number, block.hash, block.timestamp,
       block.number, block.amountOfTransactions
FROM BC.blocks AS block
WHERE TRANSACTION(block) = '2021-04-12'
  AND block.transactions.size > 10
ORDER BY block.transactions.size
LIMIT 100;
```

Fig. 2. BiTEQL query over blocks.

The only difference is in the temporal selection: whereas in EQL the transaction timestamp of a block is an explicit property that can be referenced via the path block.timestamp, in BiTEQL it is an implicit attribute that can be accessed through the TRANSACTION() function as for TSQL2.

Furthermore, in order to unleash the BiTEQL full expressive power, the TSQL2-like extensions can be used to query the transactions of a blockchain as if they were relations in a temporal database. For example, we assume to deal with a blockchain named `traffic`, whose transactions are used to track the position and speed of moving vehicles. Its schema is: `traffic.transactions(VID,X,Y,VX,VY|TT,VT)`, where VID is the vehicle identifier, X and Y are the coordinates of its position, VX and VY are the components of its velocity (TT and VT are the implicit timestamps). We further assume that built-in functions are available to compute the scalar speed of a vehicle and the distance between two vehicles (or between a vehicle and a given point). If V1 and V2 are transactions stored in blocks of `traffic`, such functions can be defined as `speed`(V1) = (V1.VX^2 + V1.VY^2)^{½} and `distance`(V1,V2) = ((V1.X-V2.X)^2 + (V1.Y-V2.Y)^2)^{½}. Then we can see some examples of TSQL2-like temporal queries issued with BiTEQL on the `traffic` blockchain.

Figure 3 shows a query doing a temporal selection exploiting the valid time of `traffic` transactions.

```
SELECT VPOS.X, VPOS.Y, speed(VPOS)
FROM traffic.transactions AS VPOS,
WHERE VPOS.VID = 'AX932EU'
  AND VALID(VPOS) CONTAINS TIMESTAMP '2021-06-01 10:15';
```

Fig. 3. BiTEQL query example.

This query retrieves the position and the speed of vehicle AX932EU at quarter past 10 in the morning on June 1, 2021.

4 Conclusion

In this paper, we proposed a bitemporal blockchain data model, named BiTchain, extending a conventional blockchain based on transaction time with the valid-time dimension of data, and BiTEQL, an expressive query language for BiTchain-based blockchains. BiTEQL is a SQL-like query language equipped with both (i) the basic constructs for querying conventional blockchains supported by EQL language and (ii) the expressive temporal constructs which have been designed for the consensual temporal database query language TSQL2.

In our future work, we will design and implement a query engine that supports the BiTEQL language, possibly via the extension of a canonical EQL engine and the adoption of suitable index and storage structures to facilitate the execution of BiTEQL temporal queries on BiTchain-based blockchains.

References

1. Nofer, M., Gomber, P., Hinz, O., Schiereck, D.: Blockchain. Bus. Inf. Syst. Eng. **59**(3), 183–187 (2017)

2. Zheng, Z., Xie, S., Dai, H.N., Chen, X., Wang, H.: Blockchain challenges and opportunities: a survey. Int. J. Web Grid Serv. **14**(4), 352–375 (2018)
3. Mohan, C.: Blockchains and databases: a new era in distributed computing. In: Proceedings of the 34th IEEE International Conference on Data Engineering (ICDE 2018), Paris, France, 16–19 April 2018, pp. 1739–1740 (2018)
4. Grandi, F.: Temporal databases. In: Khosrow-Pour, M. (ed.) Encyclopedia of Information Science and Technology, 3rd edn., pp. 1914–1922. IGI Global, Hershey (2015)
5. Guo, H., Meamari, E., Shen, C.C.: Blockchain-inspired event recording system for autonomous vehicles. In: Proceedings of the 1st IEEE International Conference on Hot Information-Centric Networking (HotICN 2018), Shenzhen, China, 15–17 August 2018, pp. 218–222 (2018)
6. Fu, Y., Yu, F.R., Li, C., Luan, T.H., Zhang, Y.: Vehicular blockchain-based collective learning for connected and autonomous vehicles. IEEE Wirel. Commun. **27**(2), 197–203 (2020)
7. Gupta, R., Tanwar, S., Tyagi, S., Kumar, N.: Blockchain-based security attack resilience schemes for autonomous vehicles in industry 4.0: a systematic review. Comput. Electr. Eng. **86**, 106717 (2020)
8. Narbayeva, S., Bakibayev, T., Abeshev, K., Makarova, I., Shubenkova, K., Pashkevich, A.: Blockchain technology on the way of autonomous vehicles development. Transp. Res. Proc. **44**, 168–175 (2020)
9. Rehman, M., et al.: A blockchain based distributed vehicular network architecture for smart cities. In: Barolli, L., Amato, F., Moscato, F., Enokido, T., Takizawa, M. (eds.) WAINA 2020. AISC, vol. 1150, pp. 320–331. Springer, Cham (2020). https://doi.org/10.1007/978-3-030-44038-1_29
10. Anderson, J.M., Nidhi, K., Stanley, K.D., Sorensen, P., Samaras, C., Oluwatola, O.A.: Autonomous Vehicle Technology: A Guide for Policymakers. Rand Corporation, Santa Monica (2014)
11. Waymo One. https://waymo.com/waymo-one/. Accessed 10 Sept 2022
12. Hawkins, A.J.: Uber has resumed testing its self-driving cars in San Francisco, 10 March 2020. https://www.theverge.com/2020/3/10/21172213/uber-self-driving-car-resume-testing-san-francisco-crash. Accessed 10 Sept 2022
13. Tesla Support: Autopilot and Full Self-Driving Capability (2022). https://www.tesla.com/support/autopilot. Accessed 10 Sept 2022
14. Szalachowski, P.: Towards more reliable bitcoin timestamps. In: Proceedings of the 2018 Crypto Valley Conference on Blockchain Technology (CVCBT 2018), Zug, Switzerland, 20–22 June 2018, pp. 101–104 (2018)
15. Ma, G., Ge, C., Zhou, L.: Achieving reliable timestamp in the bitcoin platform. Peer-to-Peer Netw. Appl. **13**(6), 2251–2259 (2020). https://doi.org/10.1007/s12083-020-00905-6
16. Zhang, Y., Xu, C., Cheng, N., Li, H., Yang, H., Shen, X.: Chronos+: an accurate blockchain-based time-stamping scheme for cloud storage. IEEE Trans. Serv. Comput. **13**(2), 216–229 (2019)
17. Nakamoto, S.: Bitcoin: a peer-to-peer electronic cash system (2008). https://bitcoin.org/bitcoin.pdf. Accessed 10 Sept 2022
18. BitMEX Research: Bitcoin's Block Timestamp Protection Rules, 30 October 2019. https://blog.bitmex.com/bitcoins-block-timestamp-protection-rules/. Accessed 10 Sept 2022
19. Jensen, C.S., et al.: The consensus glossary of temporal database concepts – February 1998 version. In: Etzion, D., Jajodia, S., Sripada, S. (eds.) Temporal Databases – Research and Practice. LNCS, vol. 1399, pp. 367–405. Springer, Berlin (1998). https://doi.org/10.1007/BFb0053710
20. Li, Y., Zheng, K., Yan, Y., Liu, Q., Zhou, X.: EtherQL: a query layer for blockchain system. In: Candan, S., Chen, L., Pedersen, T.B., Chang, L., Hua, W. (eds.) DASFAA 2017. LNCS,

vol. 10178, pp. 556–567. Springer, Cham (2017). https://doi.org/10.1007/978-3-319-55699-4_34

21. Bragagnolo, S., Rocha, H., Denker, M., Ducasse, S.: Ethereum query language. In: Proceedings of the 1st IEEE/ACM International Workshop on Emerging Trends in Software Engineering for Blockchain (WETSEB 2018), Gothenburg, Sweden, 27 May–3 June 2018, pp. 1–8 (2018)
22. Snodgrass, R.T., et al.: The TSQL2 Temporal Query Language. Kluwer Academic Publishers, Norwell (1995)
23. Kulkarni, K., Michels, J.E.: Temporal features in SQL:2011. ACM SIGMOD Rec. **41**(3), 34–43 (2012)
24. Allen, J.F.: Maintaining knowledge about temporal intervals. Commun. ACM **26**(11), 832–843 (1983)

Terahertz Antennas: Application, Research Challenges and Future Directions

Amraoui Youssef[1](✉), Imane Halkhams[2], Rachid El Alami[1], Mohammed Ouazzani Jamil[2], and Hassan Qjidaa[1]

[1] LISAC Laboratory, Sidi Mohamed Ben Abdellah University, Fez, Morocco
youssef.amraoui2@usmba.ac.ma
[2] LSEED Laboratory, UPF, Fez, Morocco

Abstract. The increasing demand for unoccupied and unregulated bandwidth for wireless communication systems will inevitably lead to the extension of operating frequencies to the THz frequency range. Terahertz wireless communications appear to be an attractive and complementary technology compared to other less flexible and more expensive technologies. However, most THz antennas suffer from relatively high loss and low manufacturing accuracy due to their small size in the high frequency bands of THz waves. Therefore, this paper presents a detailed overview of THz antennas and the latest research on improving their performance. Research directions of THz antenna in 6G technology are presented.

Keywords: THz applications · Microstrip antenna · 6G

1 Introduction

THz electromagnetic waves are generally defined at the 0.1–10 THz [7] (1 THz = 10^{12} Hz) frequency band with a wavelength of 0.03–3 mm. Figure 1 shows this spectrum as a function of frequency. The THz wave is between microwave and infrared light. The THz wave has the following excellent characteristics [1]:

- Wide bandwidth: The THz wave can be the electromagnetic wave at the highest frequency band in electronics. If the THz wave is used as the signal carrier transmitted by the antennas, the rate of information transfer may reach a new level, even at the rate of Tbps.
- Penetration: Terahertz radiations have a strong penetrating power. They potentially allow us to see through many non-conductive materials (such as skin, clothing, paper, wood, cardboard, and plastics). They are very low energy radiations; this low energy has no ionizing effect on biological tissues [1].
- Resolution: The resolution of an image increases with decreasing wavelength, and the resolution in the THz band is better than in the microwave band.

This paper is structured as follows: The first section gives a general introduction; the second part is devoted to the different applications THz frequencies, the third section

Y. Farhaoui et al. (Eds.): ICAISE 2022, LNNS 635, pp. 757–762, 2023.
https://doi.org/10.1007/978-3-031-26254-8_109

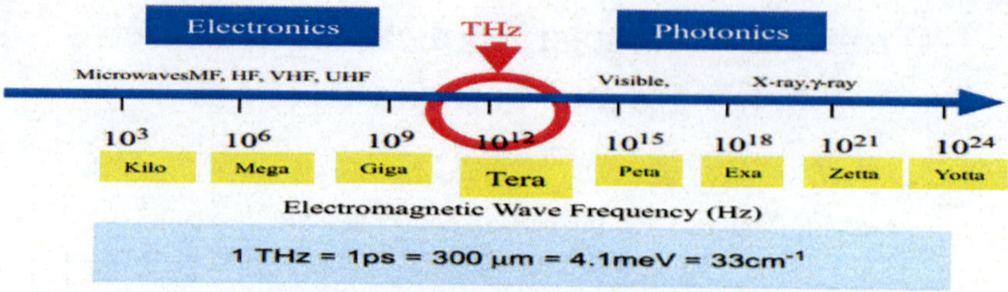

Fig. 1. The terahertz wave in the electromagnetic spectrum

presents the sources and detectors of THz radiation, the fourth part is devoted to THz antenna developments. After that, we will present the basic THz antennas, the positive impact of the use of these waves in 6G, and the future research directions of THz antennas. Finally, a brief conclusion is given in Sect. 8.

2 Applications of THz Frequencies

2.1 Spectroscopy

Generally, THz spectroscopy is among the most widely used techniques in the biomedical field [2]. Using this technique, chemical analyses can be performed to identify species and molecules present in a sample. It also allows the study of the dielectric properties of materials. Compared to those observed in the ultraviolet, visible spectrum, or infrared, the THz spectroscopy technique provides a complementary signature and fingerprint of the molecule under study [3].

2.2 Medical Application

THz radiation will be used in a variety of medical applications, including protein structure identification, diagnosis of diseases, etc. Furthermore, THz radiation, characterized by its low non-ionizing energy per photon, is safe for biological media, making it more attractive for biological and medical applications. THz imaging will allow us to visualize objects in different ways with a better spatial resolution. [4].

2.3 Telecommunications

The high frequencies of THz waves make them particularly suitable for the deployment of very high-speed and low latency telecommunications systems. The targeted applications are those related to very high speed and short range networks inside buildings, in particular those required for the deployment of 6G. Currently, several experiments have demonstrated the possibility of communication using THz waves.

3 Terahertz Radiation Sources and Detectors

3.1 Sources

There are two types of current THz sources that are extensively employed:

- Electronic sources with frequency multiplication: The principle of electronic sources with frequency multiplication, manufactured in particular by Virginia Diodes Inc. (VDI). It consists of generating continuous waves (CW) by multiplying the frequencies of the microwave domain to obtain frequencies more than 200GHz.
- Quantum cascade lasers (QCL): It is a semiconductor laser with a single type of carrier (unipolar) in a sequence of coupled quantum wells (superlattice) that has step band energy. QCL is based on optical transition of an electron between quantized levels of the conduction band [12].

3.2 Detectors THz

There are several types of THz detectors on the market and in laboratories.

- Transistors: The channel of a field effect transistor; can act as a plasma wave resonator. The resonance frequency of the plasma depends on the dimensions of the channel and for gate lengths of a few hundred nanometers. The interest in using FETs for THz detection began in the early 1990s with the work of Dyakonov [5].
- Schottky barrier diodes: They are generally used either in direct detection or as non-linear elements to implement heterodyne detection. This barrier has a very fast switching time and a low direct voltage threshold. This diode uses a metal-semiconductor junction instead of the P-N junction traditionally used. It is a non-linear element realized by epitaxy on a GaAs substrate. This detector is commonly used since it has a quick reaction time and a spectral range that ranges from 0.2 to 7 THz. The physical principle of Schottky barrier diodes is based on the transport of electrons through a potential barrier of a metal-semiconductor junction [6].
- Pyroelectric detectors: Are based on the properties of ferroelectric materials. The variation of temperature, caused by the incident electromagnetic wave generally modulated, causes the generation of an electric charge proportional to the illuminated surface, thus to the incident THz signal [7].

4 Development of Terahertz Antennas

Since the beginning of the 21st century, wireless communication technology has developed rapidly; the demand for information and the increase of communication equipment have imposed stricter requirements on the transmission rate of communication data. However, the human demand for communication capacity and speed is endless. Table 1 shows some of the previous work of the THz antenna.

In this paper [13], the proposed antenna simulates at 600 and 800 GHz for the surveillance system. Simulation results show that the proposed antenna has a radiation efficiency of 96.1%, 95.7% at 600GHz and 800GHz, respectively.

Table 1. Comparison of several THz antennas performances

Refs	Frequency (THz)	Gain	Other	Application
[13]	0.6 and 0.8	9.8dB	D = 10.7dBi	Security
[14]	4.9	4.254dB	D = 4.084dBi	Medical
[15]	0.7	6.92dB	Efficiency = 86.4%	Communication

In [14], a microstrip THz patch antenna with a rectangular shape and slotted ground is proposed. It has an impedance bandwidth of 444.5 GHz and resonates at a frequency of 4.952 THz with a minimum return loss of −55.31 dB.

5 Basic Terahertz Antennas

This section studies the different types of THz antennas. Table 2 summarizes the different THz antenna types.

Table 2. Different THz antenna types

Antenna type	Frequency	Gain	Others	Refs
Planar dipole	120GHz	25dB	10Gbps, 0.5–1 m	[8]
Patch microstrip	960GHz	3.8dB	BW = 240GHz	[9]
Horn	300GHz	18dBi	PW = 100GHz	[10]
Lens	300GHz	31dB	–	[11]

The very first 120 GHz link is established by a planar dipole and slot antenna integrated with a photodiode and Schottky barrier diode and receiver [8]. Simulation results show that the proposed antenna has a gain of 25dB at 120 GHz.

This paper [11] proposes a high gain circularly polarized (CP) lens antenna operating at 300 GHz. The measured gain bandwidth of 1 dB and axial ratio bandwidth of 3 dB of the THz CP can be as high as 13.3% and 18.8%, respectively.

6 Impact of THz on 6G Wireless Communication

The THz frequency band will be used in the future sixth generation (6G) wireless communication system to support the user demand for higher data rates and ultra-high-speed communication for many future applications. In the United States, the Federal Communications Commission (FCC) decided to open the 95 GHz–3THz band as an experimental spectrum for the validation of 6G technology. Terra Nova [19] is supported by the European Commission's Horizon 2020 framework program since July 2017; some countries contributing are Greece, Finland, and Germany. It targets for embedding broadband THz wireless links into fiber optic links for B5G. Finland launched the 6Genesis [19], the next-generation (6G) flagship project hosted by the University of Oulu, that spans 8 years starting from 2018.

6.1 Overview of the 6G Antenna

Table 3 presents some of the previous works of the 6G antenna. In [16], the paper presents a circularly polarized (CP) conical horn antenna for the 6G wireless communications. The measurement of 300 GHz CP conical horn antenna shows a measured impedance bandwidth of 60 GHz with a reflection coefficient ≤ -15 dB.

The article [18] presents a linearly polarized sub-THz FPC (Fabry–Perot cavity) antenna with high gain and low cross-polarization. The measured radiation pattern shows a highly directive pattern with a cross-polarization level below -25 dB.

Table 3. Summary of 6G antenna previous work

Refs	Frequency	Gain/ Directivity	others
[16]	300GHz	D = 18.3dBi	BW = 60GHz
[17]	300GHz	G = 16dBi	Eff = 88%
[18]	300GHz	G = 17.7dB	BW = 22GHz

7 Future Research Directions of THz Antennas

- Massive MIMO antenna arrays: The objective of MIMO is to maximize the utilization of the THz band over long distances. However, such MIMO structures require efficient antennas with low RF cross talk and mutual coupling, having the capability of choosing the carrier frequency of their choice.
- Fabrication and testing: The cost associated with the fabrication of terahertz antennas is immense, as it requires very high precision. The research community is putting its efforts to resolve this issue.
- High gain: A high antenna gain is required to overcome the path loss experienced at terahertz band. The High gain THz antennas can keep operating in a variety of harsh environments and reduce the impact of transmission loss.

8 Conclusion

The objective of this paper was to present the state of the art of terahertz antenna technology, detectors THz, typical THz antennas, and their application on 6G. Through analysis, it's found that THz antennas currently encounter high cost and other challenges. Terahertz waves may resolve the issue of lower throughput of data or latency that 5G will not be able to handle, as they are fully deployed and integrated with cellular networks. The next generation, 6G is expected to support a gigantic amount of data traffic to cope with the rise of connection wireless devices.

References

1. Guan, K., Kürner, G.L., Molisch, A.F.: On millimeter wave and THz mobile radio channel for smart rail mobility. IEEE Trans. Veh. Technol. **66**(7), 5658–5674 (2017)
2. Lampin, J.F., Mouret, G., Dhillon, S., Mangeney, J.: THz spectroscopy for fundamental science and applications. Photoniques **101**, 33–38 (2020)
3. Globus, T.R., Khromova, T., Bykhovskaia, M., Gelmont, B., Hesler, J.: THz-spectroscopy of biological molecules. Spectrosc. Biol. Mol. **29**, 89–100 (2003)
4. Mueller, E.R.:Terahertzradiation: applications and sources. Physicist., 27–29 (2003)
5. Dyakonov, M., Shur, M.: Detection, mixing, and frequency multiplication of terahertz radiation by two-dimensional electronic fluid. IEEE Trans. Electr. Devices **43**(3), 380–387 (1996)
6. Hesler, J.L., Liu, L., Xu, H., Duan, Y., Weikle, R.M.: The development of quasioptical THz detectors. In: 33rd International Conference on Infrared, Millimeter and THz Waves, pp. 1–2 (2008)
7. Hossain, A., Rashid, M.H.: Pyroelectric detectors and their applications. IEEE Trans. Ind. Appl. **27**(5), 824–829 (1991)
8. Nagatsuma, T., et al.: A 120 GHz integrated photonic transmitter. In: International Topical Meeting on Microwave Photonics MWP (Cat. No.00EX430), pp. 225–228 (2000)
9. Singh, A., Singh, S.: A trapezoidal microstrip patch antenna on photonic crystal substrate for high speed THz applications. Photonics Nanostruct. – Fundam. Appl. **14**, 52–62 (2015). https://doi.org/10.1016/j.photonics.2015.01.003
10. Tajima, T., Song, H.J., Ajito, K., Yaita, M., Kukutsu, N.: 300-GHz step-profiled corrugated horn antennas integrated in LTCC. IEEE Trans. Antennas Propag. **62**(11), 5437–5444 (2014)
11. Wu, G.B., Zeng, Y., Chan, K.F., Qu, S., Chan, C.H.: High-gain circularly polarized lens antenna for terahertz applications. IEEE Antennas Wirel. Propag. Lett. **18**, 921–925 (2019)
12. Faist, J., Capasso, F., Sivco, D.L., Sirtori, C., Hutchinson, A.L., Cho, A.Y.: Quantum cascade laser. Science **264**, 553–556 (1994)
13. Jha, K.R., Singh, G.: Dual-band rectangular microstrip patch antenna at terahertz frequency for surveillance system. J. Comput. Electron. **9**(1), 31–41 (2009)
14. Gurleen, K., Vishal, M.: Rectangular terahertzmicrostrip patch antenna design for vitamin k2 detection applications. In: 1st International Conference on Electronics, Materials Engineering and Nano-Technology (IEMENTech) (2017)
15. Saraereh, O.A., Al Tarawneh, L., Ali, A., Al Hadidi, A.M.: Design and analysis of a novel antenna for THz wireless communication. Intell. Autom. Soft Comput. **31**(1), 607–619 (2022)
16. Aqlan, B., Himdi, M., Vettikalladi, H.: Sub-THz circularly polarized horn antenna using wire electrical discharge machining for 6g wireless communications. IEEE Access **8**, 117245–117252 (2020). https://doi.org/10.1109/ACCESS.2020.3003853
17. Xu, R., et al.: A review of broadband low-cost and highgain low-terahertz antennas for wireless communications applications. IEEE Access **8**, 57615–57629 (2020)
18. Aqlan, B., Himdi, M., Vettikalladi, H.: A 300-GHz low-cost high-gain fully metallic Fabry-Perot cavity antenna for 6G THz wireless communications. Sci. Rep. **11**, 7703 (2021)
19. Salameh, A.I., ElTarhuni, M.: From 5G to 6G—challenges, technologies, and applications. Future Internet **14**(4), 117 (2022)

The Artificial Intelligence for Modeling and Predicting SMALL's Cities Attractiveness

Khalid Sohaib[1](✉), Effina Driss[1], and Mohamed Salem Chaabane[2]

[1] MASAFEQ Laboratory, Department of Economics, National Institute of Statistics and Applied Economics Rabat, Rabat, Morocco
sohaibxkhalid@gmail.com

[2] Hassania School of Public Works, Hassan II University of Casablanca, Casablanca, Morocco

Abstract. The present manuscript is intended to analyze the behavior of Moroccan small cities' residential attractiveness through several statistical modeling. For this purpose, we have built in first place a structural econometric model using binary logistic stepwise backward regression in order to extract the most relevant variables that may influence the net migration rate in small Moroccan cities. We have implemented those variables in a predictive model via artificial neural networks. Modeling results revealed many evidence such as, residential attractiveness is a rather complex phenomenon that depends on different factors such as employment supply and housing conditions. These findings might enlighten policy makers for better decision making especially since the ANN prediction algorithm's accuracy exceeded 80 %.

Keywords: Artificial neural network · Binary logistic regression · Residence attractiveness · R language

1 Introduction

Until the turn of the century, Morocco was only moderately urbanized, with only 400,000 urban residents, barely 8% of the country's total population. The cities were small, with few modestly sized urban centers. However, the last population census revealed a very marked change in the urbanization rate, which has now reached 60.3%, enabling Morocco to achieve a fairly sustained rate of growth over a relatively short period (by way of comparison, Algeria has an urbanization rate of 71.3%, Tunisia 67.05%, Mauritania 60.45%, France 79.75%, Germany 75.75%, the United Kingdom 82.84%, Brazil 85%).

The challenge of the future development of Moroccan cities is multidimensional and therefore requires the adoption of a systemic approach, which is essential for understanding the complexity of the urban phenomenon.

From this point of view, the urban framework is a system of interconnected centers linked to the territories with which they interact by a set of intense flows on which urban economies are based and which organize the systems of actors and territorial systems [1].

Y. Farhaoui et al. (Eds.): ICAISE 2022, LNNS 635, pp. 763–769, 2023.
https://doi.org/10.1007/978-3-031-26254-8_110

Cities are centers that bring together people, economic and social activities in small portions of territory. They constitute a real system whose entities are differentiated by their contents and functions,with vital interactions with their close and distant environments and which interact between them in a complex relational system. Because of these characteristics, cities are really places where social and economic interactions are maximized. In this context, and despite the fact that few studies have focused on the problems of small cities, small cities nevertheless play a crucial function in territorial development through its functional role of canalizing population flows and the diffusion of economic growth.

It is within this conceptual frame that the challenge of the residential attractiveness of small cities arises. If it is apparently a question of territorial rebalancing and consolidation of certain strata of the urban hierarchy, it actually goes beyond these dimensions to acquire a strategic scope in relation to the future of Morocco, its insertion in the inclusive territorial economy, its project of society, of territory and of its good governance.

2 Definition and Estimation

2.1 Target Population

In the same spirit as the diversity of territories and the heterogeneity of urban facts, the concept of small city does not refer to any unique and universal finish.

There are often several different approaches to defining urban formation, each of these approaches is based on several criteria such as administrative criteria (municipality, town, the community of agglomerations), functional criterion (urban areas defined by a number of jobs…), or morpho-statistical criterion (agglomerations defined by the continuity of the built-up area, by a minimum threshold of population, density,).

In this article, we use the morpho-statistical criterion to define a small city, in fact, according to the High Commission for Planning a small city is defined as any urban entity whose population is between 1000 and 50,000 inhabitants. According to this statistical definition, Morocco's urban network contains exactly **292** cities based on the most recent census of 2014.

2.2 Estimating the Net Migration Rate

Moroccan cities grow above all through their populations, either thanks to natural births or the population that immigrate. The demographic and migratory dynamics which are an engine of growth constitute an important element in the structuring of the space. The demographic past of cities is explained by the simultaneous analysis of the natural and the migratory balances. These two indices make it possible to determine the overall variation in the population over a given period, and they reflect the attraction and dynamism of the city [3].

The net migration rate is considered to be one of the most accurate indicators for measuring residential migration dynamics. However, the High Commission for Planning does not publish its own migration data. In view of this, it seems necessary to estimate the net migration rate. Net migration rate, which represents a surplus or deficit of arrivals over departures of the city's population, is calculated as the difference between overall growth and natural growth rates. The analytical estimation equation is written as follows [4].

For each city i :

$$NetMigrationRate_i = OverallGrowthRate_i - NaturalGrowthRate_i \quad (1)$$

$$NaturalGrowthRate_i = RateBirths_i - RateDeaths_i \quad (2)$$

3 Mathematical Modeling

3.1 Methodology

The aim of this work is to conduct structural and predictive modeling of the residential attractiveness of Moroccan small's cities.

In order to determine the factors that can significantly affect the residential attractiveness rate and successfully elaborate our predictive model. We will first start by building a synthetic reduced model using logistic backward regression modeling aiming to determine the most relevant factors that can positively or negatively affect the attractiveness of small cities in Morocco. Secondly, using key variables selected in the previously reduced model, we will elaborate predictions on attractiveness rate using Artificial Neural Network modeling. This method would provide us with the most precise prediction of a city's attractiveness probability.

For all that follows, the probability of a city being attractive is defined as:

$$y_i = \begin{cases} 1 \text{ if Net Migration Rate} > 0\text{(Attractive)} \\ 0 \text{ if Net Migration Rate} < 0\text{(Repulsive)} \end{cases}$$

Note: No city obtained the value zero when estimating the net migration rate.

3.2 List of Variables

Given the complexity of the concept of residential attractiveness, the modeling of this economic issue requires detailed research in literature and empirical works. This investigation aims to identify the key variables that may influence the capacity of a given territory to attract people and their income.

To carry out successfully our modeling, a set of independent variables was selected according to the literature results and the characteristics of the Moroccan urban network. In our modeling, thirteen variables were carefully selected to appreciate the territorial attractiveness of small cities along six dimensions namely: **economic and housing conditions, demographic dynamics, education, employment supply, spatial dynamics and territorial connectivity, and territorial mobility of workers**.

4 Results

4.1 Results of the Logistic Regression Using Stepwise Backward Regression

Before proceeding with the ANN modeling we should first determine key variables that influence a city's attractiveness. Therefore we'll proceed by logistic regression analysis. The final model called reduced model is considered after the estimation and the application of the backward elimination method.

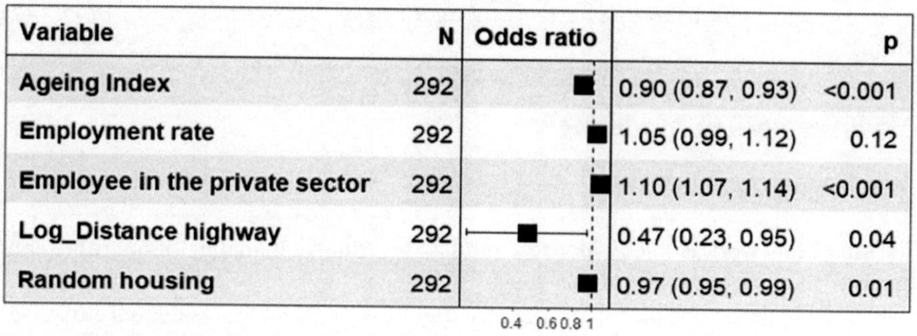

Fig. 2. Results of the reduced model using "forestmodel" library. Source: Author's manipulations using R language.

-Attractive cities are younger

The aging index negatively influences the probability of the city being attractive. In this sense, the probability of attractiveness of the city decreases when its population is more likely to be old. This fact highlights the fact that the migratory phenomenon is more related to the young population. Indeed, young people in Morocco are considered the most socio-demographic stratum affected by migration.

-Job offer, a key factor of residential attractiveness

The analysis of the results of the reduced model shows that the reduced model contains only five remaining variables, two variables among them belong both to the labor supply dimension. One variable is significant at the 5% level while the second one is significant at the 15% level. This finding is quite interesting, in the sense that it clearly shows that labor supply, especially that of the private sector, is the key driver of the residential attractiveness of small cities in Morocco.

In this context, it is worth noticing that all the variables related to labor supply have shown a positive impact on the probability of a small city being attractive to the population in both reduced and global models.

In conclusion, the labor supply, especially the private sector, is one of the key drivers of the territorial attractiveness of small cities in Morocco, in the sense that the population seeks to settle in an environment that guarantees the best chance of employment, economic and social integration.

-Better accessibility, greater gravitation

The accessibility of the city has been proven to influence its level of attractiveness, in fact, the distance between the city and the highway has shown a negative effect on the probability of the city being attractive. This finding shows that the proximity of the city to the highway network influences its level of attractiveness; in the sense that the more accessible the city is the more likely to be attractive.

-Good housing conditions means better are the immigrant flows

Good housing conditions are essential in promoting the attractiveness of any urban environment especially cities. In fact, the results of our model show that higher the level of inadequate housing in the city, the less likely it is to be attractive.

4.2 Results of the ANN

Before moving on to the performance measurement of our model, let's recall that our dataset has been divided into two parts. The first part contains 70% of the data and has been used to train the model. The second partition contains 30% of the data and was used for the predictive accuracy test.

In order to measure the accuracy of our model, we calculated the confusion matrix; this matrix is defined in the binary case as a table of four values containing all possible combinations between the actual and predicted values [5].

Table 1. Confusion matrix

	1-Attractive	**0-Repulsive**
$\hat{1}$	37	5
$\hat{0}$	6	40

Source: Author's manipulations using R language

The accuracy of the model is calculated as the fraction of the diagonal of the confusion matrix over the total of all cells:

$$\text{Accuracy} = \frac{37 + 40}{88} = 0,875 = \mathbf{87,\ 5\%}$$

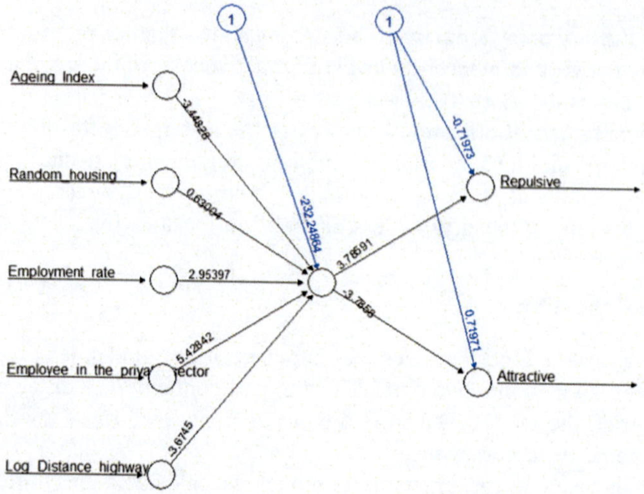

Fig. 3. Architecture of the Artificial Neural Network. Source: Author's manipulations using R language.

Note: We have calculated the accuracy in terms of prediction of both logistic regression model and ANN algorithm. The accuracy of the natural network largely exceeds that of the logistic regression model (62%).

5 Conclusion

As we have seen, residential attractiveness is a rather complex phenomenon that depends on several determinants. Each factor has its own way of influencing the capacity of the city in terms of population attractiveness.

Indeed, young people are the socioeconomic stratum most affected by the migration phenomenon. Also, housing conditions and accessibility play a crucial role in the city's ability to retain its residents. However, the supply of employment remains the most determining dimension of geographic attractiveness in the sense that the population always seeks the environment where it can obtain maximum integration into social and economic life.

This article has succeeded in producing a predictive econometric model of territorial attractiveness using advanced methods of artificial intelligence namely ANN, whose accuracy exceeds 80 %.

References

1. Bell, D., Jayne, M.: Small Cities? Towards a Research Agenda. Int. J. Urban and Regional Res. 33(3), 683–699 (2009)
2. Brabazon, T.: Unique Urbanity? SG, Springer, Singapore (2015). https://doi.org/10.1007/978-981-287-269-2

3. Sohaib, K., Effina, D., Youness, J.: an econometric analysis of the determinants of small's cities' attractiveness: evidence from moroccan case. International Conference on Optimization and Applications. The 8th Edition - ICOA2022 (2022)
4. Sohaib, K., et al.: The effect of Employment mobility and spacial proximity on the Residential attractiveness of Moroccan small cities: Evidence from Quantile regression analysis (2022)
5. Rasmy, L., Zhu, J., Li, Z., et al. : Simple recurrent neural networks is all we need for clinical events predictions using EHR data (2021). arXiv. (published online Oct 3.) (preprint). https://arxiv.org/abs/2110.00998
6. Niedomysl T.: Migration and place attractiveness. Geografiska regionstudier 68, Department of social and economic geography, Uppsala University, Uppsala (2006)
7. Rogerson, R.J.: Quality of life and city competitiveness. Urban Stud. **36**(5–6), 969–985 (1999)

The Transformation Method from Business Processes Models by BPMN to Class Diagram by UML: CIM to PIM

Mohamed Achraf Habri(✉), Redouane Esbai, and Yasser Lamlili E.L. Mazoui Nadori

Mohammed First University, Oujda, Morocco
hmohamedachraf@gmail.com

Abstract. The first step to understanding the need at the level of the information system in a company is to find tools, means of communication, or a simple modeling language that provides notations such as diagrams that are Clearly readable called BPMN (Business Process Model and Notation), the language that can satisfy the needs of internal organizations and facilitate collaboration between the components of the company. Our paper proposes a transformation based on the (model-driven architecture) MDA approach that will help us to save time, and money and to have a solid information system, up to date and consistent which ensures the passage of information between simple users and IT developers. As a start, we focus on a transformation carried out with QVT (views query transformations), between the BPMN diagram and the class diagram of UML (unified modeling language) which is a more sophisticated, specific modeling language and which will allow us to carry out other transformations to achieve other goals.

Keywords: BPMN · Model transformation · QVT · The transformation rules · Class diagram

1 Introduction

In each company there are several departments, divisions, or services which consist of human resources which have several specialties generally they are simple computer users, they can hardly express their needs at the level of the information system, The American consortium OMG (Object Management Group) has proposed a solution to this problem and has created a modeling language whose strong point is the ease of use, and which facilitates collaboration between the business side and the IT side by a description of the internal needs in the form of diagrams designed with simple, clear and easy to read symbols, it is a tool that provides graphical notations that easily describe the internal needs of the establishment and their business processes, this language based on models is BPMN (business process model and notation).

In Speaking of modeling languages, the passage between models and all that is modeling we are obliged to talk about the MDA standard created by OMG well before BPMN, this standard is based on three models which are the independent calculation

Y. Farhaoui et al. (Eds.): ICAISE 2022, LNNS 635, pp. 770–775, 2023.
https://doi.org/10.1007/978-3-031-26254-8_111

model (CIM), Platform Independent Model (PIM) and Platform Specific Model (PSM), the performance of the MDA standard manifests itself in the protection of investments, the updating of information systems in a minimized time and allow them to migrate to adaptable platforms to new technologies, which is largely beneficial for businesses [1].

This The idea is to use BPMN as a starting point to design the business process of the company as much as an independent model computation (CIM) as a business process diagram and transform it into a platform-independent model (PIM), a technically specific model that allows us to communicate with the computer side and complete the cycle of transformations later, we have chosen a model from the same OMG family which is the class diagram of the modeling language unified modeling language (UML), we opted for the first passage of the transformation of the MDA from CIM to PIM and we intend to do the other passages thereafter, that is to say from the PIM to the PSM until being able to automate the code of the desired software, this transformation will highlight the strengths of MDA, that is to say automate the trans-formations starting with the purely business side of understanding its needs and also to design its information system in order to have as an optimal result which is the suitable software, among several transformation languages we chose a language created and developed by the OMG group which is the (QVT) query view transformation) [2].

Our article contains five sections, the first is the introduction the second is the related work where we talk about the papers that inspired us to do this work, and the background is presented in the third section which offers definitions of the bases used, The fourth section is the part that reveals the added value of our work, by drawing the BPMN diagram according to the case study in the first step, then by establishing the transformation rules by respecting the standards of the QVT language and finally by having a class diagram as the desired result.

The last component of our work is a conclusion which is at the same time a future introduction for the tasks that we intend to accomplish in the future, which will enrich our work and even take it as a reference. Either.

2 Related Works

In [3] the authors propose two transformations the first is horizontal and called refinement CIM model between the business process diagram and the use case diagram UML, the second is vertical to the UML sequence diagram as a PIM model, using the specification that defines the rules and vocabulary: SBVR (semantics business vocabulary rules).

In [4] Yassine Rhazali et al. present a method of transformation based on the modernization of MDA, it starts by creating the transformable model at the CIM level and transforms it afterward using ATL atlas transformation language to obtain the PIM model as a result.

The authors propose in [5] a transformation between BPMN and the diagram of UML namely the activity diagram, practically there is a convergence between the business process diagram and the activity diagram which allows having precise and direct transformation rules.

In the work [6] the authors propose an idea of transformation that resembles the previous one, starting with the diagram of the business process BPMN and obtaining the activity diagram of UML but what characterizes this transformation is the use of an XML metadata exchange to represent the two models and to specify the document style as well as the vocabulary at the input and output level they used the XSLT transformation type.

In [7] the authors have done a work that starts by designing a BPMN business process diagram that helps to create a java ee 6 business platform, in three steps starting with a transformation between the BPMN diagram and class diagram of UML, after transforming it into a diagram UML based on java and at the end get MOF script based on java ee profiles.

We were also inspired by the transformation carried out from BPMN towards graphical and mathematical tools (COLORED PETRI NET) allowing us to model and verify the dynamic behavior of discrete event systems in [8] and from the Web Ontology Language obtained by a transformation of BPMN, OWL2 is a language with components that include an overview of the document, which serves as an introduction to OWL 2, describes the relationship between OWL 1 and OWL 2, and provides an entry point to deliverables remaining via a worksheet [9], after observing the two papers we find that BPMN remains the clearest and practical language.

3 Method

Our approach is a solution that will highlight a relationship between simple users and IT analysts, starting by translating the need, vision, and situation of a company into a BPMN business process diagram and transforming after this initial model into a more specialized model, and more technically sophisticated, namely the UML class diagram to complete our global.

In our article we are inspired by other works for example the transformation called refinement at the beginning of CIM (BPMN) towards CIM (the use case diagram of UML) after their passage to carry out another CIM to PIM transformation two diagrams of the same UML language use case diagram to UML sequence diagram, the passage between BPMN and the UML CIM to CIM activity diagram (refinement) and one can notice that there is a great resemblance between these two diagrams, in the dedescription transformation of two diagrams from UML, CIM to PIM (see Fig. 1).

We chose to start with a case study represented by a BPMN diagram and to go directly to the UML class diagram which allowed us to go more on the IT development side without going through other transformations for example case diagram usage, sequence diagram, or activity diagram of UML [10–18].

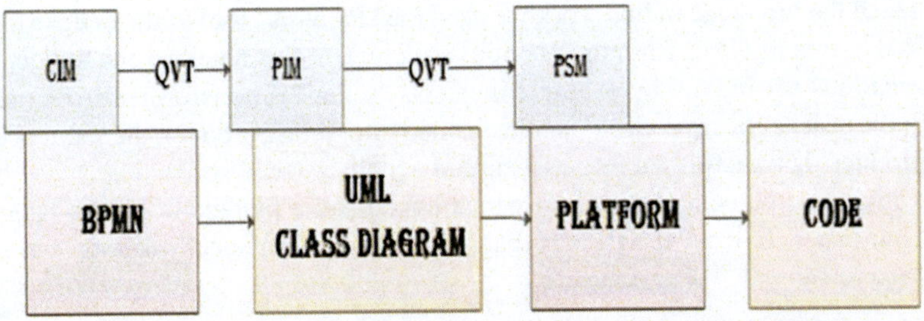

Fig. 1. Our proposed transformation compared to the MDA models.

3.1 Case Study

After reading the transformation rules between BPMN and the use case diagram, sequence diagram, or activity diagram we managed to write the rules of transformation between BPMN and class diagram, in our case, we have a simple language (BPMN) (see Fig. 2), and more efficient language (UML class diagram) so we need to use conditions to direct the transformation properly, for example, the "Task" element according to the condition can be transformed into class or method.

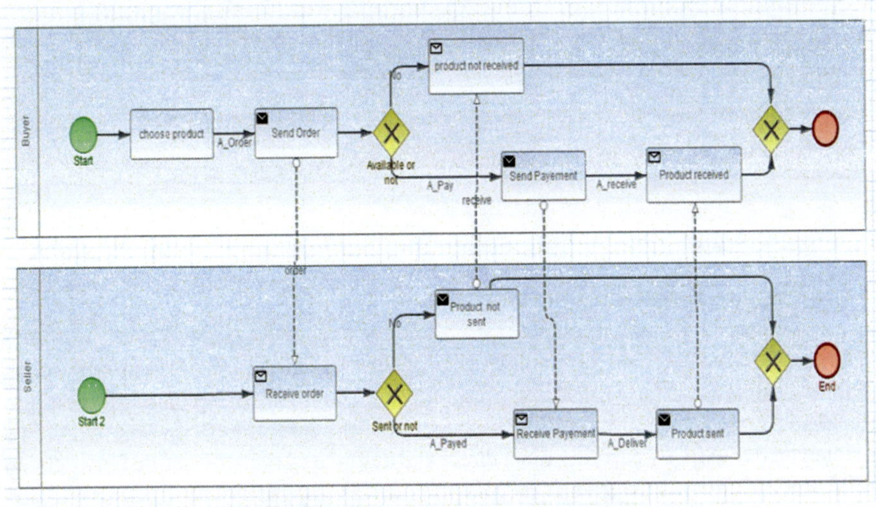

Fig. 2. The Process diagram of BPMN.

3.2 The Result

After executing the QVT code using eclipses that shows the result obtained in the form of a UML class diagram structure composed of target elements transformed from source BPMN elements.

To reach the last stage in this work we used the "Papyrus" tool of the eclipse to draw the class diagram of UML from the previous structure, at this level we can notice all the transformations carried out by example the class "seller" is transformed from the pool "seller", the task "choose product" is transformed into a class product, the sequence flow "send product" is transformed into association a_deliver (see Fig. 3).

We opted for the eclipse platform which contains the plugins of all the languages used in our paper. To draw the BPMN diagram and develop the eclipse code and obtain the desired result.

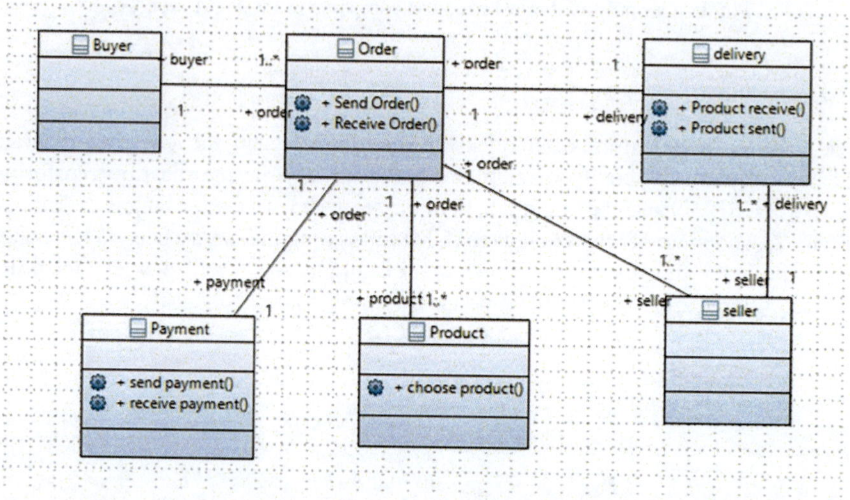

Fig. 3. Our result as a class diagram

4 Conclusion and Perspectives

The use of MDA shows us the added value of this standard for information systems and their design, the transformations of the models provided by MDA allow having a homogeneous sequence between the components of the information system, which we projected on our work by carrying out a direct and automated transformation between the "collaboration" process diagram of BPMN and the class diagram of UML assuming that we create a gateway between the simple users on the business side with the developers in the computer engineering to achieve the company's objective, which is investment protection, saving time and, above all, having a solid and up-to-date information system.

References

1. Object Management Group. https://www.omg.org/bpmn/lncs. Accessed 1 Apr 2022
2. Xavier Blanc, MDA en action Ingénierie logicielle guidée par les modèles, Eyrolles, 7 Apr 2005

3. Addamssiri, N., Kriouile, A., Balouki, Y., Taoufiq, G.: Generating the PIM Behavioral Model from the CIM using QVT. J. Comput. Sci. Inf. Technol. Vol. 2, p. 3 & 4 December (2014)
4. Rhazali, Y., Hadi, Y., Mouloudi, A.: A new methodology CIM to PIM transformation resulting from an analytical survey. In: 4th International Conference on Model-Driven Engineering and Software Development (MODELSWARD) (WCCS) (2015)
5. Bảo, N.Q.: A proposal for a method to translate BPMN model into UML activity diagram, Vietnamese-German University - BIS (2010)
6. Macek, O., Richta, K.: The BPM to UML activity diagram transformation using XSLT, International Workshop on Databases, Specifications and Objects, 15 Apr 2009
7. Debnath, N., Martinez, C.A., Zorzan, F., Riesco, D.: Transformation of business process models BPMN 2.0 into components of the java business platform. IEEE Transactions on Industrial Informatics. https://ieeexplore.ieee.org, December (2013)
8. Dechsupa, C., Vatanawood, W., Thongtak, A.: Transformation of the BPMN design model into a colored petri net using the partitioning approach, IEEE ACCESS.2853669, July 6 (2018)
9. Kchaou, M., Khlif, W., Gargouri, F., Mahfoudh, M.: Transformation of BPMN Model into an OWL2 Ontology. In: 16th International Conference on Evaluation of Novel Approaches to Software Engineering, January (2021)
10. Oualid, B., Saida, F., Amine, A., Mohamed, B.: Applying a model-driven architecture approach: transforming CIM to PIM Using UML. Int. J. Online Bio-med. Eng. (iJOE) 14(9), 170–181, 30 Sept (2018)
11. Ambara, R., Nurul Fajar, A.: business process management for system development. Int. J. Eng. Adv. Technol. (IJEAT) 9(2), 1331–1334 December (2019)
12. Braun, R., Esswein, W.: Classification of domain-specific BPMN extensions. In: 7th IFIP Working Conference on the Practice of Enterprise Modeling (PoEM), Manchester, United Kingdom, pp.42–57, Nov (2014)
13. http://www.bpmodeling.com/faq/lncs. Accessed 2 Nov 2021
14. Argañaraz, M., Funes, A.: An MDA approach to business process model transformations. SADIO Electronic J. Informat. Oper. Res. 9, 24–48 (2010)
15. Gotti, S., Mbarki, S.: IFVM Bridge: a model-driven IFML execution. Int. J. Online Biomed. Eng. (iJOE). **15**(4), 111–126 (2019)
16. Sbai, R., Arrassen, I., Meziane, A., Erramdani, M.: QVT transformation by modeling. (IJACSA) Int. J. Adv. Comput. Sci. Appl. 2(5) (2011)
17. Esbai, R., Elotmani, F., Belkadi, F.Z.: Model-Driven transformations: toward automatic generation of column-oriented NoSQL databases in big data context. Int. J. Online Biomed. Eng. (iJOE) 15(09), 4–16 (2019)
18. Habri, M.A., Esbai, R., Lamlili El Mazoui Nadori, Y.: BPMN to UML class diagram using QVT. In: Ben Ahmed, M., Teodorescu, H.-N. L., Mazri, T., Subashini, P., Boudhir, A.A. (eds.) Networking, Intelligent Systems and Security. SIST, vol. 237, pp. 593–602. Springer, Singapore (2022). https://doi.org/10.1007/978-981-16-3637-0_42

The Use of a Genetic Algorithm to Alleviate the Limited Content Issue in a Content-Based Recommendation System

Oumaima Stitini(✉), Soulaimane Kaloun, and Omar Bencharef

Computer and System Engineering Laboratory, FSTG, Cadi Ayyad University, 40000 Marrakesh, Morocco
oumaima.stitini@ced.uca.ma

Abstract. Recommendation systems today present a technology designed to make things easier for consumers. They can be used to suggest ideas, solutions, products, and even people in various fields, including medicine, education, tourism, e-commerce and industry. The most present study we identified is in the e-commerce industry, supporting people in finding what they desire. A recommender system can, however, present consumers with helpful information on topics that may be important to them. The user may become dissatisfied receiving suggestions that are identical to their preferences, resulting in the problem of over-specialization. Limited content data leads to over-specialization since content-based techniques promote products closely connected to the client profile instead of novelties. In this research, we are concerned explicitly with suggesting novel, unpredictable, and startling objects that may be loved by consumers and may help make up for the lack of content. Genetic algorithms may be used to incorporate innovations and a sense of unpredictability, which we need to accomplish in order to propose both unfamiliar and unforeseen products alongside more familiar ones. This article discusses an innovative method to overcome the limited content problem and enhance the results of a content-based recommendation system.

Keywords: Artificial intelligence · Genetic algorithm · Recommender system · Content-based filtering

1 Introduction

The World Wide Web, in particular, has seen incredible advancements in cloud computing. With this update, additional online benefits are available, including documentation, stories, and articles to study as well as videos to watch and things to buy. Currently, it is challenging for customers to make the right choice from the enormous array of products offered by these platforms due to the emergence of IoT devices and the tremendous rise of e-commerce platforms. People gain from recommender system capabilities inadvertently [1]. In this increasingly interconnected world, users have difficulty locating things they are genuinely interested in. Several of the most excellent online shopping,

Y. Farhaoui et al. (Eds.): ICAISE 2022, LNNS 635, pp. 776–782, 2023.
https://doi.org/10.1007/978-3-031-26254-8_112

such as Bestbuy.com, for example, has employed recommender systems to help buyers locate elements to purchase. Most popular websites, like Netflix, have long employed recommender systems to assist their customers in deciding which films to view based on their particular interests. These apps deliver personalized search results based on the user's choices. When users go to a website, they are typically looking for information that is relevant to them. These objects of interest might cover a wide range of themes. A recommender system can also give consumers helpful information about the goods they are passionate about [2]. Such applications benefit from the ability to respond swiftly to emerge in user tastes. A recommender system calculates the chance a specific user selects a particular item [3]. A customer model was proposed via a content-based recommendation engine by analyzing particular articles characteristics in estimating and offering recommendations. Consequently, the things presented are frequently similar to those the buyer has previously liked. A critical approach for assessing anything unique is absent from the content-based recommendation system. Compared to the profile page, the algorithm will propose goods with a perfect score. It is also called the over-specialization, which limits content-based guidance. In this respect, this study aims to apply a genetic algorithm to improve suggestions, provide diversity, and propose new items that the customer could appreciate. This study looks for proposing innovative products that the user could enjoy. The genetic algorithm (GA) is also utilized to build a better recommendation list for a specific person depending on their interests. Limited content difficulties are two main challenges of content-based filtering that the suggested RRSGA can address. The rest of this paper is categorized in the subsequent manner: We discuss our theoretical background and related works in Sect. 2. The research problem is described in Sect. 3. The suggested approach, RRSGA, is detailed in Sects. 4, 5. Section 6 concentrates on displaying and examining RRSGA results. Finally, Sect. 7 concludes the paper and suggests potential future directions for research.

2 Theoretical Background

Although addressing novelty and diversity remains a research and development priority, there has been an evolution in improvement techniques, evaluation metrics, and methodology. Novelty may be defined as the distinction between the current and a previous experience, whereas variety refers to the intrinsic distinctions between knowledge sections. The distinction between these two notions is complex, and intimate connections can be made depending on one's point of view, as we will see in the two sub-problems, we will look at. Diversity is a relatively recent concept in the recommender literature that has sparked an interest. It has emerged as a crucial consideration in developing a high-quality recommender system. Information retrieval research has impacted the notion of novelty in the recommender system literature. In the research on recommender systems, definitions of novelty often focus on two aspects: an item is unfamiliar to the user, and an item is distinct from what the user has seen before. Some studies have concentrated entirely on the latter component, proposing unique techniques for determining the distance between each user profile and a specific element. According to [4], the term "novelty" [5] refers to products that were new to the user before the proposal. Novelties are typically disliked and depart from user profiles to reduce the likelihood of being recognized.

3 Research problematic

Traditional recommender systems focus mainly on the two notions: accuracy and relevance, as objectives to satisfy user needs. In many application areas, the principles of diversity and novelty, in particular, are increasingly recognized as essential aspects of information value. The property of a system that avoids duplication and redundancy is characterized as novelty. For example, in information retrieval, the marginal value of one document over another is minimal if the system suggests two documents with similar content to consumers.

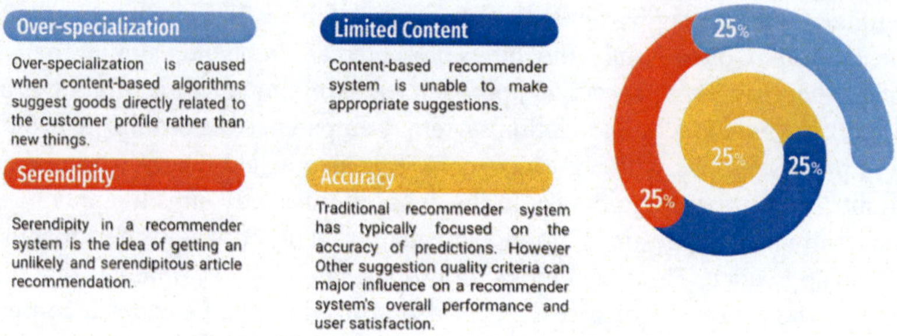

Fig. 1. Content-based filtering drawbacks.

A helpful recommendation system must be accurate and provide consumers with novelty and a variety of products. The capacity of recommendation algorithms to offer goods that users would not find on their own is suggested novelty and variety. Novelty and variety are becoming more prevalent in today's evaluation processes. They are increasingly being included in the reported efficacy measurements of new recommendation systems, and algorithmic breakthroughs are expressly targeting them. Other aspects to consider while evaluating recommender systems include novelty, user demands, and expectations [6]. Diversity and originality are critical traits for providing automated suggestions. The novelty of an item relates to how it differs from information previously viewed by specific users or groups of users [7]. Figure 1 mentions drawbacks of content-based filtering. Over-specialization, limited content, and serendipity all converge towards the same problem.

4 Research Methodology

We are interested in providing a comprehensive solution to deal with content-based filtering drawbacks mentioned in Fig. 1. The whole purpose is to provide a general scheme to be followed through 5 phases. Figure 2 describes the solution proposed for our research problematic. We offer a five-phase approach in our proposed technique to deal with this research problem. RRSGA focuses on the entire content of the recommendation list rather than selecting particular goods to construct a recommendation list. The fundamental premise is to go through the complete recommendation list and display new goods to clients who could be interested.

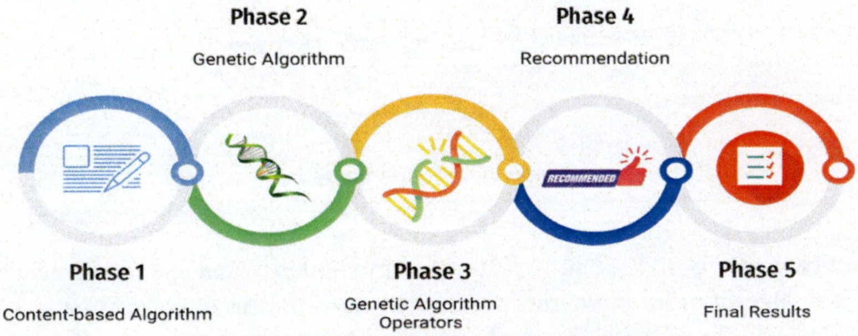

Fig. 2. Our global proposed approach.

5 Our Proposed Approach

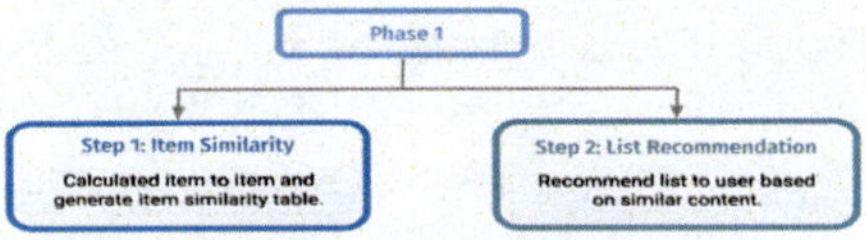

Fig. 3. Our first phase.

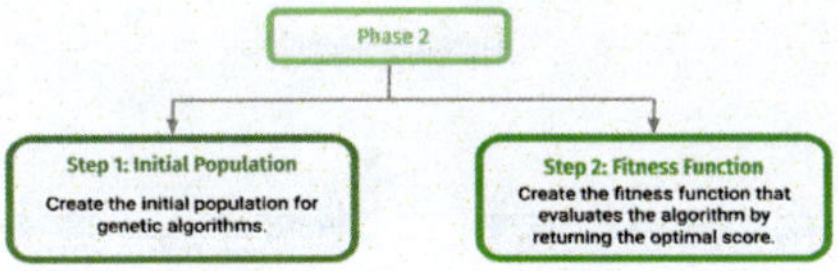

Fig. 4. Our second phase.

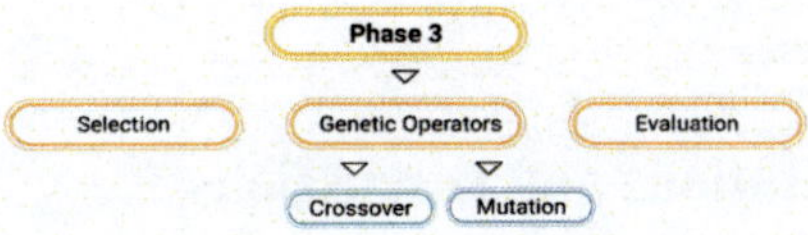

Fig. 5. Our third phase.

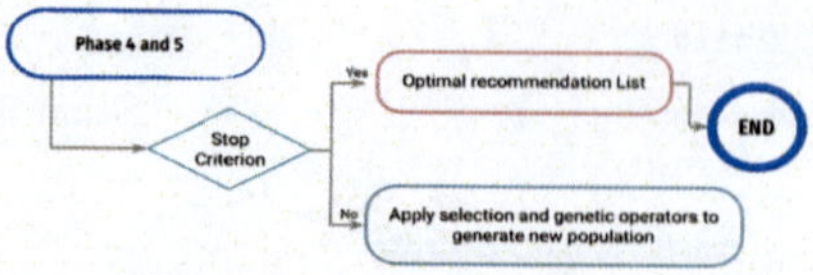

Fig. 6. Our fourth phase.

The first phase is splitted in two steps: Item similarity and list recommendation. We used an algorithm for measuring similarity between films to construct a similarity matrix, then find the highest degree of similarity by checking the matrix for all films with a value greater than 0.80 in the table (Fig. 3). The second phase is also divided into two steps: initial population and fitness function. The GA model is randomly selected among the entire population. Therefore, the algorithm starts with real data and works its way to the optimal answer. Since we will need to discover neighbors among them, our first population needs to be predicated on real existing films. The starting population is created randomly (Fig. 4). The third phase contains four steps: Selection, Crossover, Mutation, Evaluation (Fig. 5). The last phases of our proposed algorithms. This is the final phase that generates the best suggestion list; it displays the chromosome with the most excellent fitness. The user is presented with an unexpected and serendipitous suggestion in this last phase (Fig. 6). The active user has a better chance of discovering new goods that are likely to interest them (Fig. 7).

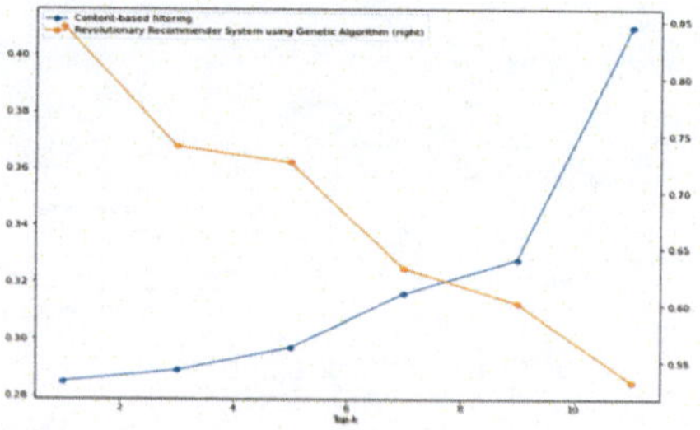

Fig. 7. Novelty results of the recommendation methods.

6 Results, Interpretations and Discussions

Table 1 shows a comparative study between the novel proposed approach and other existing ones. The limited content effect is very high in content-based filtering, by using our suggested solution to become less as shown in Table 2.

Table 1. Comparative study with our proposed approach RRSGA.

Works	Over-specialization	Limited content	Serendipity	Accuracy
[8]	✔			
[9]			✔	
[10]			✔	
[11]			✔	
[12]				✔
[13]		✔		
[6], RRSGA	✔	✔	✔	✔

Table 2. Comparison of our proposed approach with other approaches.

Properties	Content-based filtering	Clustering	Proposed approach
Result diversity	Less	Medium	High
Scalability	Low	Low	Good
Limited content	High	Medium	Less

7 Conclusion

Recommender systems have grown increasingly important as a result of information overload. Instead of studying items and then creating a suggestion list, the basic intention following the suggested approach is to assess prospective recommendation lists. As a consequence, RRSGA looks for a list of suggestions that meets three criteria starting by the proposed things that are essentially parallel to the subject that was searched for, moving to the suggested items that cover a wide range of the user's desires. Finally, customers must be offered surprising and serendipitous things. RRSGA uses innovative techniques to diversify the suggestion list to obtain a collection that fits those criteria. RRSGA takes an innovative approach to the problem of limited content by viewing each person in the population as a possible recommendation list. More characteristics should be examined and evaluated using the Boruta algorithm in future research to ensure that the recommendation quality is sufficient and that just the most significant features are kept.

References

1. Oumaima, S., Soulaimane, K., Omar, B.: Artificial intelligence in predicting the spread of coronavirus to ensure healthy living for all age groups (2021)
2. Stitini, O., Kaloun, S., Bencharef, O.: The recommendation of a practical guide for doctoral students using recommendation system algorithms in the education field (2021)

3. Stitini, O., Kaloun, S., Bencharef, O.: Latest trends in recommender systems applied in the medical domain: a systematic review. In: Proceedings of the 3rd International Conference on Networking, Information Systems & Security (2020)
4. Kotkov, D.: Serendipity in recommender systems (2018)
5. Stitini, O., Kaloun, S., Bencharef, O.: Integrating contextual information into multi-class classification to improve the context-aware recommendation. In: EUSPN/ICTH (2021)
6. Stitini, O., Kaloun, S., Bencharef, O.: An improved recommender system solution to mitigate the over-specialization problem using genetic algorithms. Electronics **11**(2), 242 (2022). https://doi.org/10.3390/electronics11020242
7. Vargas, S.: Novelty and diversity enhancement and evaluation in recommender systems and information retrieval. In: Proceedings of the 37th international ACM SIGIR conference on Research & development in information retrieval (2014)
8. Adamopoulos, P., Tuzhilin, A.: On over-specialization and concentration bias of recommendations: probabilistic neighborhood selection in collaborative filtering systems. In: RecSys '14 (2014)
9. Ziarani, R.J., Ravanmehr, R.: Serendipity in recommender systems: a systematic literature review. J. Comput. Sci. Technol. **36**, 375–396 (2021)
10. Grange, C., Benbasat, I., Burton-Jones, A.: With a little help from my friends: cultivating serendipity in online shopping environments. Inf. Manag. **56**(2), 225–235 (2019). https://doi.org/10.1016/j.im.2018.06.001
11. Kotkov, D., Veijalainen, J., Wang, S.: How does serendipity affect diversity in recommender systems? A serendipity-oriented greedy algorithm. Computing **102**, 393–411 (2018)
12. Raza, S., Ding, C.: News recommender system: a review of recent progress, challenges, and opportunities. Artif. Intell. Rev. **55**(1), 749–800 (2021). https://doi.org/10.1007/s10462-021-10043-x
13. Iaquinta, L., Degemmis, M., Lops, P., Semeraro, G., Filannino, M., Molino, P.: Introducing serendipity in a content-based recommender system. In: 2008 Eighth International Conference on Hybrid Intelligent Systems, 168–173 (2008)

The Use of the Surface Response Methodology of Water Pre-treatment for a Better Demineralization

Mariem Ennouhi(✉), Sanaa El Aggadi, Jamal Mabrouki, Amale Boutakiout, and Mohammed Alaoui El Belghiti

Laboratory of Spectroscopy, Molecular Modeling, Materials, Nanomaterial, Water and Environment, CERNE2D, Mohammed V University in Rabat, Faculty of Science, Avenue Ibn Battouta, Agdal, BP1014 Rabat, Morocco
mariem.ennouhi90@gmail.com

Abstract. In surface water reverse osmosis (RO) demineralization processes, pre-treatment is a key step in achieving high performances and avoiding frequent membrane fouling. The plant studied includes conventional pretreatment and RO process. The objective of this work is to optimize the conventional pretreatment for a better operation of the demineralization plant. The jar-test was optimized by applying the Doehlert experimental design. The results of the tests obtained showed that to obtain a better optimization of pretreatment it is necessary to reach a pH of 6.5, a dose Aluminum sulphate is about 18.5 $mg{\cdot}L^{-1}$.

Keywords: Pretreatment · Coagulation · Flocculation · Residual · Aluminium · Plan experience · Reverse osmosis

1 Introduction

Clean water is considered to be the most fundamental resource for human utilization, and at present, almost 11% of the world's total population have no access to pure potable water [1, 2]. Although water occupies more than 70% of the earth's surface, 97% of this water is too salty for human consumption [3, 4]. In addition, majority of the rest of the 3% is caught in ice caps and glaciers or underground, leaving lesser than 1% of the world's total supply of water accessible to human use [5]. Because of the growing world population and steady economic increase, we have reached a potentially critical situation in adjusting the supply and availability of freshwater assets [6]. Predictions up to 2030 demonstrate an enhancement in worldwide water uptake of almost 40% of the present available and dependable supply sources, which means a heightening of water demand [7]. With pure water resources rapidly depleting, the need for additional effective and inexpensive water treatment strategies has become extremely important to meet our constantly developing water needs. Subsequently, the requirement for technological advancement to empower innovative desalination and wastewater/water treatment technologies cannot be exaggerated [8, 9]. The transformation of brackish water, seawater, and wastewater into drinking water requires higher water treatment.

Y. Farhaoui et al. (Eds.): ICAISE 2022, LNNS 635, pp. 783–787, 2023.
https://doi.org/10.1007/978-3-031-26254-8_113

Desalination now becomes a main method to produce drinking water [10–12]. Desalination is a general term for the process of removing salt from water to produce fresh water [13, 14]. The major problem that limits this process at the plant level is the clogging of the membranes. It is a harmful phenomenon that results in less efficient systems and reduces the reliability of membrane processes [6, 15], thus increasing energy demand, cleaning frequency, chemical consumption, membrane replacement and high water treatment cost [7, 9]. In this regard, it is necessary to optimize the pretreatment process in order to avoid clogging and to achieve a successful desalination system [16–18].

The objective of this study is to optimize the conventional pretreatment, a case study the treatment plant of the city of Khenifra in Morocco, using the methodology of the response surface that was exploited to determine the optimal dose of reagents to obtain a better desalination.

2 Research Methodology

2.1 Presentation of the Study Area

The first surface water treatment plant in Morocco (Fig. 1), which combines a conventional treatment with a Reverse Osmosis process was set up in Khenifra in 2013 [8]. It will cover drinking water needs until 2030 with a production of 36.290 m3•d-1 of water. The plant was supplied from the Oum Rbiâa River [7]. This surface water is characterized by strong variations in the concentration of total dissolved solids (TDS) and high chlorides (due to the geological nature of the land crossed by these waters) [20].

The water from Oum Rbiâa River is treated at the treatment plant of the city of Khenifra by a conventional treatment followed by a demineralization by reverse osmosis.

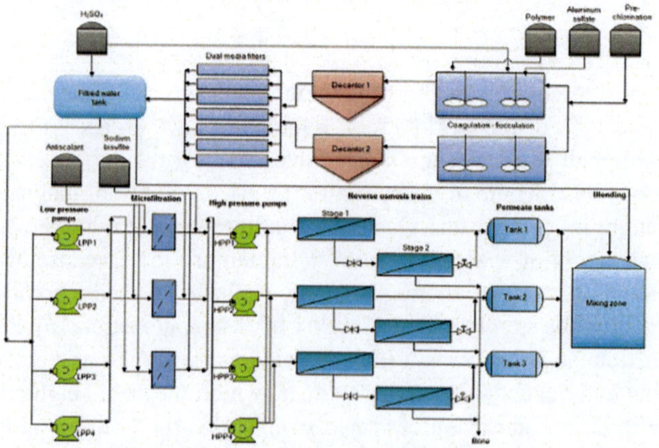

Fig. 1. General plant design of Khenifra Plant

2.2 Experimental Research Methodology

Response surface methodology is part of the method of design of experiments. This technique is intended to determine the variations of the function response to factors of significant influence in a quantitative way. Once the variables with a statistically significant influence on responses have been identified, Doehlert experimental model is used to optimize the levels of these variables. The number of required experience (N) is given by the following equation:

$$N = n^2 + n + n_0 \quad (1)$$

where n is the number of variables and n0 is the number of central points.

Answers to the central level of the variables are also performed to validate the model using experimental variance estimation. The experimental variables Xi were coded as xi according to the following equation:

$$x_i = (X_i - X_{0i})/\Delta X_i \quad (2)$$

with x_i is the coded value of the variable I^{ème}, X_i the natural value, X_{oi} the value at the center point, ΔX_i, the value of level change.

3 Results and Discussion

3.1 Determination of the pH and the Optimal Dose of Aluminum Sulphate

The experimental design method has been applied to optimize the parameters of the process. The experimental matrix proposed by Doehlert (1970) was used. The study was conducted first on two variables (pH = X1) and the dose of Aluminum sulphate (AS = X2). A model with two variables of Doehlert forms a regular hexagon where every corner and the center represent an experiment. From this study, it has been possible to express the optimization of residual Aluminum depending on the pH and the dose of Aluminum sulphate by the following equation:

$$Y = (-0.035)X1 + 0.05X2 + 0.029X3 + 0.115X12 + 0.087X22 + 0.089X32$$
$$+ 0.07X1(-0.04)X1X3 + 0.196X2X3.$$

Using this model, the iso-response curve (Fig. 2A) and the response surface curve (Fig. 2B) are plotted, showing the effect of pH and aluminum sulphate dose on residual aluminum reduction. The interaction effect between pH and aluminum sulphate dose indicates that proper adjustment of these two factors optimizes residual aluminum. The minimum value of residual aluminum was obtained near the central values of pH and aluminum sulphate dose. Optimal conditions for maximum residual aluminum reduction were observed at pH 6.5 and aluminum sulphate dose of 18.5 mg·L-1. The minimum value for residual aluminum is 0.03 mg·L-1. This value meets the requirements of the membrane suppliers (Al < 0.05 mg·L-1).

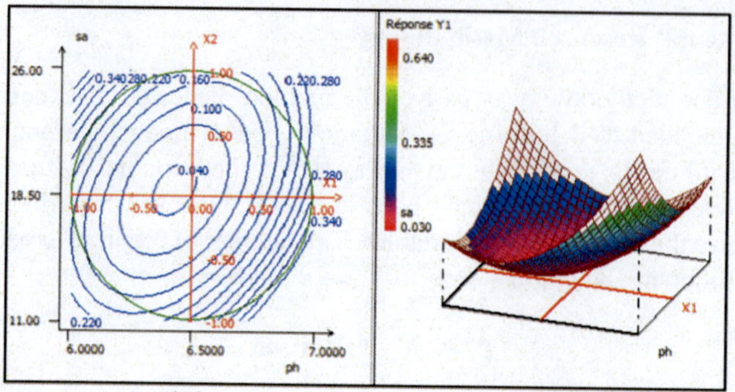

(A) (B)

Fig. 2. Isoreponse curve (A) and the response surface (B) of residual Aluminum depending on the pH and Aluminum sulphate.

4 Conclusion

This article represents the optimization of the pretreatment process for a better operation of the demineralization unit of khénifra during one year of exploitation.

Several conclusions can be drawn:

- The lowest residual Al value was obtained at pH 6.5 based on actual surface water quality.
- The recirculation of sludge minimizes the rapid clogging of sand filters and therefore reduces the number of cloggings.

References

1. Syafrudin, M., et al.: Pesticides in drinking water—a review. Int. J. Environ. Res. Public Heal. **18**, 468 (2021)
2. Abu Hasan, H., Muhammad, M.H., Ismail, N.I.: A review of biological drinking water treatment technologies for contaminants removal from polluted water resources. J. Water Process Eng. **33**, 101035 (2020)
3. Angelakis, A.N., et al.: Desalination: from ancient to present and future. Water 2021 13, 2222 (2021). https://doi.org/10.3390/W13162222
4. Issac, M.N., Kandasubramanian, B.: Effect of microplastics in water and aquatic systems. Environ. Sci. Pollut. Res. **28**(16), 19544–19562 (2021). https://doi.org/10.1007/s11356-021-13184-2
5. Al A'assam, A., Ahamd, M.W.: photonic cloning of seas and oceans. J. Phys. Conf. Ser. **2114**, 012019 (2021)
6. Saleem, H., Trabzon, L., Kilic, A., Zaidi, S.J.: Recent advances in nanofibrous membranes: production and applications in water treatment and desalination. Desalination **478**, 114178 (2020)
7. Mabrouki, J., Moufti, A., Bencheikh, I., Azoulay, K., El Hamdouni, Y., El Hajjaji, S.: Optimization of the coagulant flocculation process for treatment of leachate of the controlled

discharge of the city Mohammedia (Morocco). In: Ezziyyani, M. (ed.) AI2SD 2019. LNEE, vol. 624, pp. 200–212. Springer, Cham (2020). https://doi.org/10.1007/978-3-030-36475-5_19

8. Boulahfa, H., Belhamidi, S., Elhannouni, F., Taky, M., Hafsi, M., Elmidaoui, A.: Pretreatment process optimization and reverse osmosis performances of a brackish surface water demineralization plant. Morocco. Desalin. Water Treat. **206**, 189–201 (2020). https://doi.org/10.5004/DWT.2020.26297
9. Ikonnikova, K.A., Teslyuk, A.B., Bobkov, S.A., Zolotarev, S.I., Ilyin, V.A.: Reconstruction of 3D structure for nanoscale biological objects from experiments data on super-bright X-ray free electron lasers (XFELs): dependence of the 3D resolution on the experiment parameters. Procedia Comput. Sci. **156**, 49–58 (2019). https://doi.org/10.1016/J.PROCS.2019.08.129
10. Ennouhi, M., Khadir, A., Hasnaoui, L., Hourch, A.E., Elazzouzi, M.: Autopsy of membranes of the reverse osmosis demineralization plant in the surface waters of khénifra city morocco. Wulfenia. **27**, 1–12 (2020)
11. Mabrouki, J., Azrour, M., Hajjaji, S.E.: Use of internet of things for monitoring and evaluating water's quality: a comparative study. Int. J. Cloud Comput. **10**(5–6), 633–644 (2021)
12. Ayaz, M., Namazi, M.A., Din, M.A. ud, Ershath, M.I.M., Mansour, A., Aggoune, el H.M.: Sustainable seawater desalination: Current status, environmental implications and future expectations. Desalination. 540, 116022 (2022). https://doi.org/10.1016/J.DESAL.2022.116022
13. Darre, N.C., Toor, G.S.: Desalination of water: a review. Curr. Pollut. Rep. **4**(2), 104–111 (2018). https://doi.org/10.1007/s40726-018-0085-9
14. Abd-almohi, H.H., Alismaeel, Z.T., M-Ridha, M.J.: Broad-ranging review: configurations, membrane types, governing equations, and influencing factors on microbial desalination cell technology. J. Chem. Technol. Biotechnol. (2022)
15. Khayet, M., García-Payo, M.C., García-Fernández, L., Contreras-Martínez, J.: Dual-layered electrospun nanofibrous membranes for membrane distillation. Desalination **426**, 174–184 (2018). https://doi.org/10.1016/J.DESAL.2017.10.036
16. Boulahfa, H., Belhamidi, S., Elhannouni, F., Taky, M., El Fadil, A., Elmidaoui, A.: Impact of the Raw Water Seasonal Variations on the Reverse Osmosis Performance: Khenifra Plant. Morocco. J. Water Environ. Technol. **17**, 359–374 (2019). https://doi.org/10.2965/jwet.19-028
17. Mabrouki, J., Benbouzid, M., Dhiba, D., El Hajjaji, S.: Simulation of wastewater treatment processes with Bioreactor Membrane Reactor (MBR) treatment versus conventional the adsorbent layer-based filtration system (LAFS). International Journal of Environmental Analytical Chemistry, 1–11 (2020)
18. Lee, W.J., et al.: Fouling mitigation in forward osmosis and membrane distillation for desalination. Desalination **480**, 114338 (2020). https://doi.org/10.1016/J.DESAL.2020.114338
19. Mabrouki, J., et al.: Study, simulation and modulation of solar thermal domestic hot water production systems. Model. Earth Syst. Environ. **8**, 1–10 (2021). https://doi.org/10.1007/s40808-021-01200-w
20. Shaaban, S., Yahya, H.: Detailed analysis of reverse osmosis systems in hot climate conditions. Desalination **423**, 41–51 (2017)

Towards a Smart Photovoltaic Panel: Numerical and Experimental StudY

Youness Atifi[1](✉), Abdelhadi Raihani[1], Mohammed Kissaoui[1], Mohamed Hajjaj[2], and Abella Bouaaddi[2]

[1] Electrical Engineering and Intelligent Systems Laboratory (EEIS), ENSET Mohammedia, Hassan II University of Casablanca, Casablanca, Morocco
atifi.youness@gmail.com

[2] GEMS Laboratory Department of Physics ENSA, Ibn Zohr University, Agadir, Morocco

Abstract. In the sense of supervision and in the context of the development of photovoltaic installations, it seems very important to share the data or the physical and electrical characteristics that by turning the solar panels, these data will be exploited to locate the defects or to have a database that will be exploited for predictive studies… However, this sharing must be secure and on a large scale…

The objective of this study is to produce a prototype of a smart solar panel capable of enriching us with data and information on location and climate as well as production. In addition, the smart PV solar panel is equipped with a network of sensors.

Finally, the smart PV solar panel must share the data of the mentioned parameters with the cloud service.

An IoT based cloud monitoring system is proposed and developed using the Arduino for remote PV plant.

Keywords: PV panel · IOT · Embedded systems · Arduino · ESP 8266 · Renewable energy · Smart systems

1 Introduction

In Morocco and worldwide, the consumption of electrical energy is very accelerated [1]. It seems that there is a trend of energy consumption, which will continue to increase, driven by economic growth on the one hand and by the increase of electricity consumption per capita on the other hand, whatever the scenarios considered. There is periodic talk of various technical solutions to maintain this pace of life.

The depletion of fossil resources in the long term, the soaring prices of crude oil, and the fight against greenhouse gas emissions make it urgent to control consumption and diversify energy sources: the use and development of renewable energies [1].

One of the most widely used renewable energies, especially in Morocco, is solar photovoltaic energy [2]. However, to make photovoltaic technologies more competitive with other energy sources, the efficiency of the PV conversion system must be continuously improved. In this way, the supervision and control of the PV system can improve the overall increase of the total energy production of a PV installation [3].

Y. Farhaoui et al. (Eds.): ICAISE 2022, LNNS 635, pp. 788–793, 2023.
https://doi.org/10.1007/978-3-031-26254-8_114

This study consists in transforming a simple solar panel into a smart panel allowing the collection of data related to the place, the climate as well as the electric production.

For this purpose, the smart solar PV panel will be equipped with a smart sensor network. This network shall enable the required measurements of the panel and climate parameters, namely temperature, current and voltage of the panel as well as ambient temperature, solar radiation, wind speed and direction. Finally, the smart PV panel must share the data of the mentioned parameters with the cloud service [6].

The paper is organized as follows: related work is described in Sect. 2; the map modeling is developed in Sect. 3; The implementation and results are checked in Sect. 4.

2 Related Work

Photovoltaic energy is a clean, green, and virtually inexhaustible energy. To improve the production, transport and even consumption of this energy, a control and supervision system can be a solution to limit production losses and improve the performance of installations [6]. Thanks to an alert system, a fault analysis is possible to reduce the cost of maintenance and avoid unnecessary travel [6].

There are many systems for evaluating the performance of a photovoltaic installation. These systems, usually specific to inverters, evaluate the performance of the installation by measuring the power provided by the panels and the associated energy. Some systems allow, for example, the comparison between the actual production of an installation and the theoretical production. An alert is sent to the user warning of anomaly on the PV installation [4, 6]. This methodology allows the detection of an anomaly on the installation but has limitations, on the location of defects and is generally carried out retrospectively [3–5].

We can distinguish two types of monitoring strategies: classic strategies based on wired logic and advanced strategies including the integration of IOT systems [3, 6, 8].

Lately several research aimed at the integration of IOT systems in the field of solar energy, this immigration to IOT systems is justified by several reasons namely the high cost of supervising conventional installations, the problems of continuity of service and the lack of data sharing as well as in old installations it is difficult to continue production and locate defects [4, 6, 7, 9–12].

3 Map Modeling

3.1 Characteristics of the Panel Studied

For this study we chose the BP Solar 340 photovoltaic module. The BP Solar 340 provides cost-effective photovoltaic power for general use by direct operation with direct current loads, or with alternating current loads on inverter systems.

3.2 Constituents of Our IOT System

And as this study is part of IOT projects, it is necessary to collect all the technological systems that can intervene to build a smart and connected solar system [10]. Figure 1 shows the minimum environment to make the system studied smart and communicating.

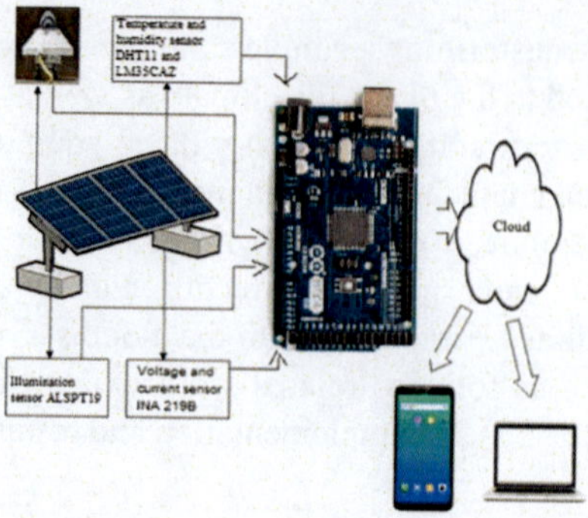

Fig. 1. Hardware implementation of proposed work.

The choice of the different elements was carefully made we take in consideration the characteristics of the solar panel and the cost the reliability as well as the speed of each component and the storage memory of the processing unit. This IOT system is organized into four parts, the first part is a map that receives the physical and electrical quantities (temperature, humidity, voltage, current and sunshine) and the second part is a data processing unit, the third is a remote data communication unit and the last part is the exploitation of data on the cloud using the ThingSpeak website [8, 9]. Figure 2 present the final electrical diagram of the map under ISIS Proteus.

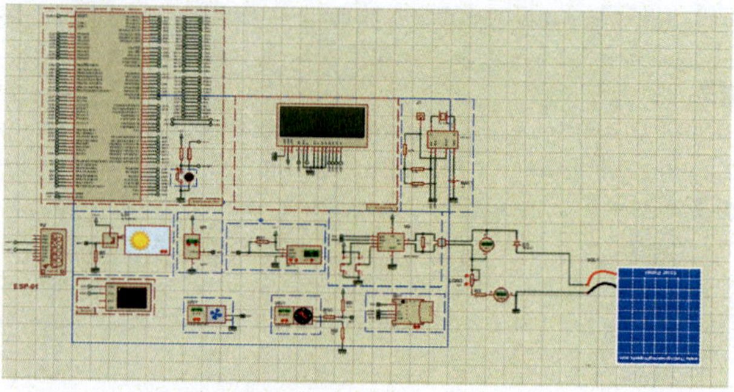

Fig. 2. Electrical diagram of the board.

4 Implementation and Results

The design and simulation of the Arduino board card and sensors firstly were done on Proteus 8 professional platform as shown in the Fig. 2 and t he final simulation was done on the real board card Arduino ATMEGA 2560.

4.1 Operating Algorithm and Simulation Results

The flowchart in Fig. 3 describes the operating algorithm of the electronic board.

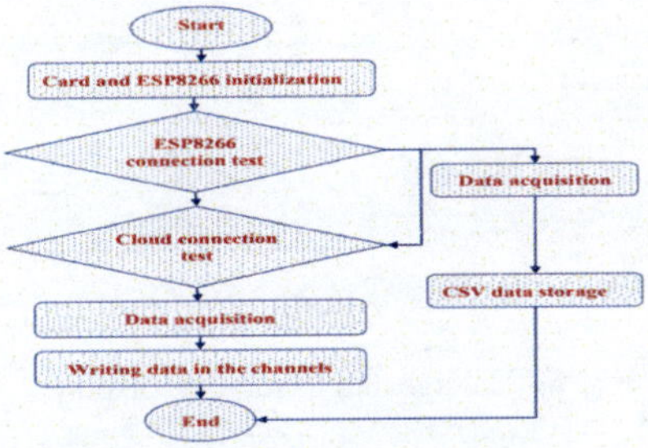

Fig. 3. Operating flowchart.

The microcontroller updates the results after every 5 min and returns the new values to the display and storage organs.

4.2 Prototype of the Smart Panel

Figure 4 shows the final view of the prototype, all the components of which are grouped in a suitable box.

Fig. 4. Prototype of the smart panel.

Once the connection of ESP and middleware are established, the owner or user of the card can access his channel in private or public mode to follow the operation of his card or to make interventions if necessary. The data is displayed in real time and in the form of graphs. Figures 5, 6, 7, 8 show the results of measured magnifier displays.

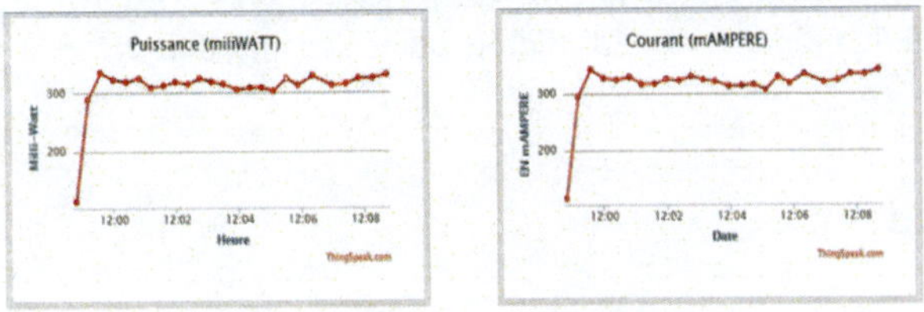

Fig. 5. Power and Current Display.

The experiment is carried out during one day of the month Mail 2021 with different time intervals. The electrical parameters of the PV module are analyzed by continuous

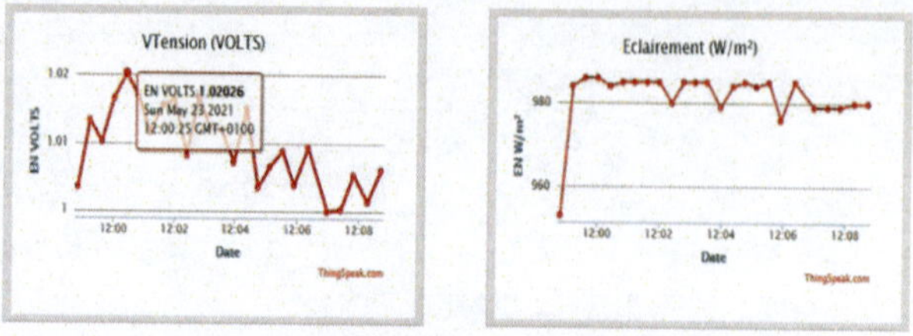

Fig. 6. Voltage and irradiation display.

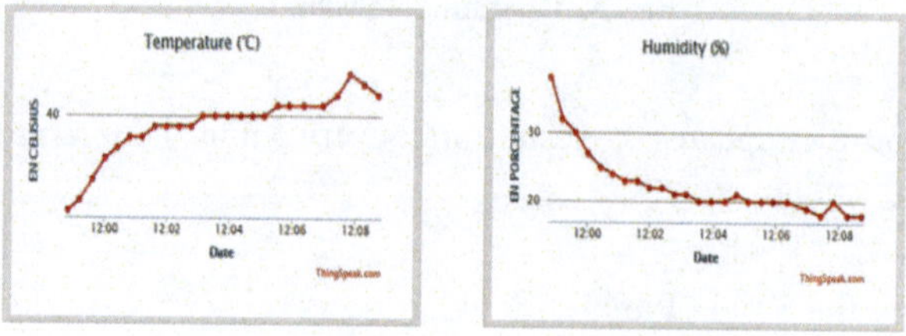

Fig. 7. Temperature and Humidity Display.

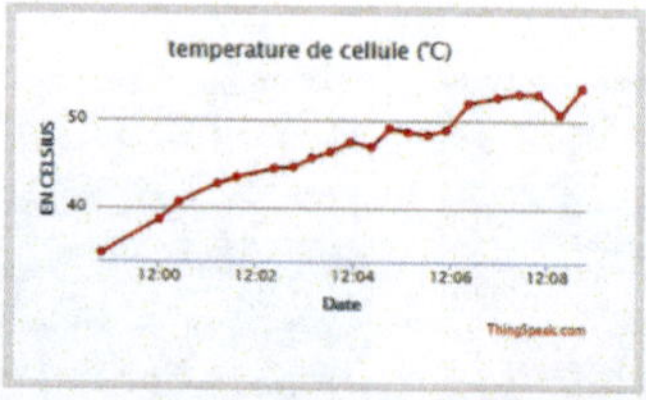

Fig. 8. Displaying the Cell Temperature.

monitoring to estimate the behavior of the solar panel. The figures show the temperature variation due to climatic conditions that influences the voltage generation, and the irradiance affects the current parameter of the PV module. The standard operating temperature of the solar panel is 25 °C. As you see the user and thanks to the integration of IOT, he can monitor and observe the state of his panel remotely and in real time, which avoids any until displacement to perform maintenance tasks or condition observation. The work presented in this study can always be improved by integrating more sophisticated middleware and adding other sensors that cover a wide range of measurement.

5 Conclusion

Thanks to the Internet of Things, today solar PV can be controlled and supervised remotely and in real time. The realization of an intelligent and communicating electronic board with middleware for the photovoltaic system will certainly and undoubtedly participate in the improvement of the production of PV energy by increasing the efficiencies of PV installations and their reliability and security as well as participating in their remote maintenance. This electronic board can be developed soon so that it can be adapted with renewable energies other than solar PV. In our next studies, we aim to create our own middleware to respond correctly to user needs.

References

1. Omri, A.:. Analysis of the transition to renewable energies in Tunisia: Risks, challenges, and strategies to adopt. Savings and finance. University of Côte d'Azur; University of Sfax (Tunisia). Faculty of Economics and Management (2016). French. NNT: 2016AZUR0030ff
2. Daoudiab, B., Laarabia, B., Barhdadia, A.: Toward a wide development of photovoltaic electricity production in the RSK region in Morocco. In: Conference: The 2nd International Materials Science and Engineering for Green Energy Conference (IMSEGEC-2018), April 25 - 27, Rabat, Morocco (2018)
3. Coulibaly S.: IoT applied in Agriculture, University Badji Mokhtar Annaba (2019)
4. Kavitha, V. Malathi, V.: A smart solar PV monitoring system using IOT, SRAIC-2019, pp. 01–33 (2019). © CS & IT-CSCP 2019
5. Xiao, W., Dunford, W.G., Palmer, P.R., Capel, A.: Regulation of Photovoltaic Voltage. IEEE Trans. Industr. Electron. 54(3), 1365–1374 June (2007)
6. Gupta, V., Soni, A., Priyadarshi, H., Banerjee, R.: Design and development of IoT-Based PV cleaning system, energy conversion and green energy storage, 1st edition (2022)
7. Gunawan, A.F., Wibawa, A.G., Linggarjati, J.: Computer engineering department, faculty of engineering, Bina Nusantara University, Jakarta, Indonesia, 11480 (2020)
8. Bressan, M.: Development of a supervision and control tool for a photovoltaic installation. University of Perpignan, Electronic (2014)
9. Nagalakshmi, R., Kishore Babu, B., Prashanth, D.: Design and development of a remote monitoring and maintenance of solar plant supervisory system. Int. J. Eng. Comput. Sci. **3**, 9382–9385 (2014)
10. Papageorgas, P., Piromalis, D., Antonakoglou, K., Vokas, G., Tseles, D., Arvanitis, K.G.: Smart Solar Panels: In-situ monitoring of photovoltaic panels based on wired and wireless sensor networks. Int. Conf. Adv. Renew. Ener. Clean Environ. **36**, 535–545 (2013)
11. Haider-e-Karar, Khuwaja, A.A., Sattar, A.: Solar power remote monitoring and controlling using Arduino, Labview and web browser, In: IEEE International Conference on Power Generation System and Renewable Energy Technologies, pp. 1–4 (2015)
12. Pereira, R.I.S., Dupont, I.M., Carvalho, P.C.M., Juca, S.C.S.: IoT embedded linux system based on raspberry Pi applied to real-time cloud monitoring of a decentralized photovoltaic plant. Int. J. Measur. Elsevier **2**, 1–18 (2017)

Towards an Effective Anomaly Detection in Solar Power Plants Using the AE-LSTM-GA Approach

Mohamed Khalifa Boutahir(✉), Yousef Farhaoui, and Mourade Azrour

Engineering Science and Technology Laboratory, IDMS Team, Faculty of Sciences and Techniques, Moulay Ismail University of Meknes, Errachidia, Morocco
moha.boutahir@edu.umi.ac.ma

Abstract. Solar energy infrastructure has been transformed into an essential part of our daily lives due to the wide spread use of electric appliances. Therefore, the performance estimation and equipment fault or anomaly detection is a challenging task requiring early knowledge to carry out early fixes. Accurate anomaly detection is essential for any system to function dependably and effectively. Machine learning approaches showed impressive quality and accuracy in identifying the various power system vulnerabilities. In this paper, we applied an AutoEncoder Long Short-Term Memory (AE-LSTM) method based on the Genetic Algorithm (GA) as a hyperparameter tuner to detect anomalies in two power plants. The AE-LSTM-GA approach demonstrated excellent performance in detecting anomalies compared to the Isolation Forest (IF) and the Local Outlier Factor (LOF) methods.

Keywords: Solar energy · Power plants · Anomaly detection · Machine learning

1 Introduction

Solar energy has become an essential solution for residential, commercial, and industrial energy production. Fossil fuels currently account for more than 80% of the planet's energy supplies [1]. Since renewable energy is cheap, plentiful, green, and ecological, its use has increased dramatically in recent years. Photovoltaic (PV) technologies revealed their potential to offer a global surplus of solar energy.

Monitoring the dynamic evolution of system parameters and informing decision-makers of anomalies is a must for PV systems. Interactive PV monitoring facilitates operation, maintenance, and economic integration into smart grids [2]. The inability to recognize photovoltaic (PV) malfunctions will lead to decrease the generated electricity. Detecting anomalies in solar panels earlier can prevent future power loss. [3].

Artificial Intelligence advancements led to new anomaly detection solutions. However, most of these methods are based on supervised models, which require significant data sources [4]. Annotating massive datasets for energy applications is complex, time-consuming, and expensive, and it requires domain expertise. In addition, the lack of labels has made anomaly detection an excellent challenge for researchers and practitioners.

Y. Farhaoui et al. (Eds.): ICAISE 2022, LNNS 635, pp. 794–799, 2023.
https://doi.org/10.1007/978-3-031-26254-8_115

Various research has examined PV systems anomalies. Mariam I. et al. [5] studied LSTM, Facebook-Prophet, and the Isolation Forest to detect anomalies in two solar power plants. The AutoEncoder Long Short-Term Memory shows promising results and effectively identifies healthy signals and detected anomalies. Maitreyee D. et al. [6] used CLARA to identify events from large datasets for two solar power stations in England. CLARA recognizes site-specific patterns, voltage dips, and spikes. This analysis included data from seven summer and seven winter days. Seven hundred twenty-five million voltage data points were analyzed for anomalies. A. Purarjo-Mandlangrudi et al. [7] presented a unique anomaly detection method using data mining and machine learning. Kurtosis and NGS were used to evaluate sensitivity and accuracy; they compared their results with the Support Vector Machine (SVM), which had a 75% accuracy rate. J. Balzategui et al. [8] develops a trustworthy solar cell inspection system using a Generative Adversarial Network-based anomaly detection model. The model locates and finds aberrations in solar cells. To detect errors, a fully convolutional network was trained to use first-stage anomalies in the following step. D. Kim et al. [9] suggested a novel technique for recognizing PV power anomalies. PV system groups were divided up using K-means. When using the normal distribution and 95% confidence interval, their suggested approach yields notable results. It should facilitate PV system diagnosis in renewable energy housing assistance schemes.

The contribution of this study is an examination of the accuracy and performance of three anomaly detection models for two solar power plants: Autoencoder LSTM with Genetic algorithm (AE-LSTM-GA), Isolation Forest, and Local Outlier Factor (LOF).

The remainder of this study describes the utilized artificial intelligence algorithms in Sect. 2. Section 3 displays the datasets used. The outputs and findings of the experiment are provided in Sect. 4. The last section concludes the paper.

2 Data Source

We evaluated the proposed models using a public dataset from Kaggle website [11]. Over 34 days, this dataset was collected from two solar power plants in India. The dataset consists of two axes, one for displaying power generation and the other for presenting sensor data. The power generation is measured using 22 inverter sensors connected at each plant's inverter and plant levels. The sensors data was collected at the plant level, where a single array of sensors is placed in the best place. The information was published, authorized, and made available according to [12].

There could be a variety of anomalies at the inverter level. To deal with that, a test was run to compare the produced Ac power and the irradiation for the various inverter sensors, as seen in Fig. 1. For inverter number 18, there was a drop in AC power on June 7 and June 9. This drop can indicate an inverter failure.

3 Results and Findings

In this section, we cover the various models and techniques for anomaly detection in the power generation for the two power plants and assess the internal and external causes of the inverter sensors data for the two power plants. In our experiments, we utilized

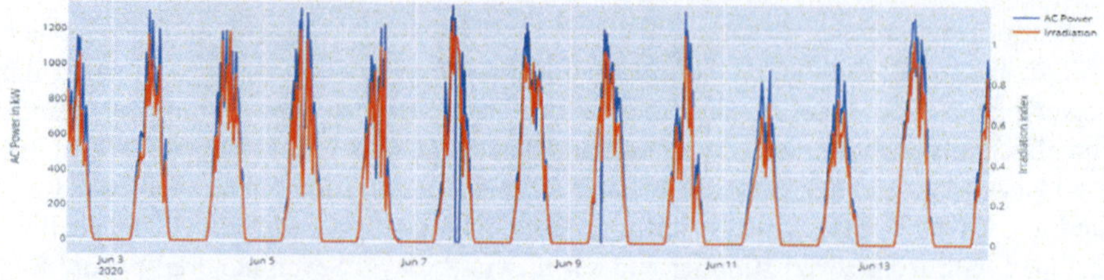

Fig. 1. AC Power and Irradiation signal comparison for inverter 18.

Python 3.7.13 in a Google Colaboratory environment [13], and also, we implemented Scikit-Learn [14] and Keras [15] packages for the proposed models.

3.1 Isolation Forest (IF)

The Isolation Forest was built to address the issue of the typical iForest algorithm's single machine experiencing a bottleneck in computation [16]. Although those algorithms have several benefits, like excellent accuracy, linear time complexity, and little memory consumption, there is no denying that they struggle when dealing with large amounts of data.

Table 1. The best parameters for the isolation forest algorithm

n_estimators	max_samples	contamination	max_features	bootstrap	n_job	random_state	verbose
1000	auto	.03	2	False	-1	42	0

The IF method categorizes the data points as normal or abnormal. Using the **Grid-Search** method, we improved the algorithm's parameters, and the optimal ones are listed in Table 1.

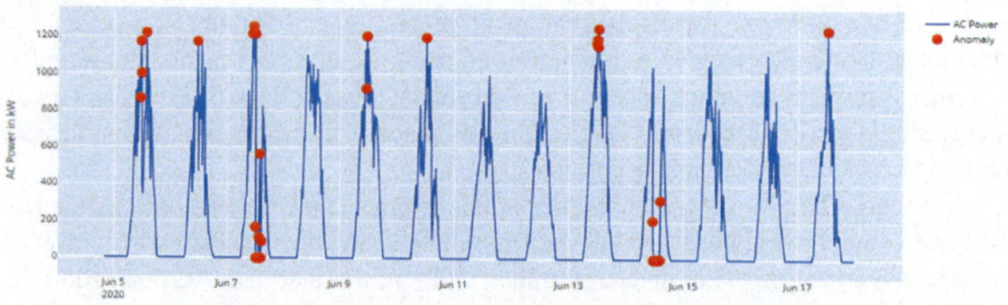

Fig. 2. The detected anomalies using Isolation Forest

Figure 2 shows the results of the IF algorithm in identifying anomalies in power generation. As seen in Fig. 1, the model determines the drop areas on June 7 and 9, but it also recognizes anomalies in the top peaks of most days. Which perhaps aren't anomalies.

3.2 Local Outlier Factor (LOF)

LOF was one of the test methods used to find abnormalities in the dataset. Using the grid-search methodology, we optimized their parameters; the optimum set of parameters is shown in the Table 2.

Table 2. The best parameters for the Local Outlier Factor algorithm

n_neighbor	leaf_size	metric	contamination	p	novelty
25	35	Minkowski	auto	2	True

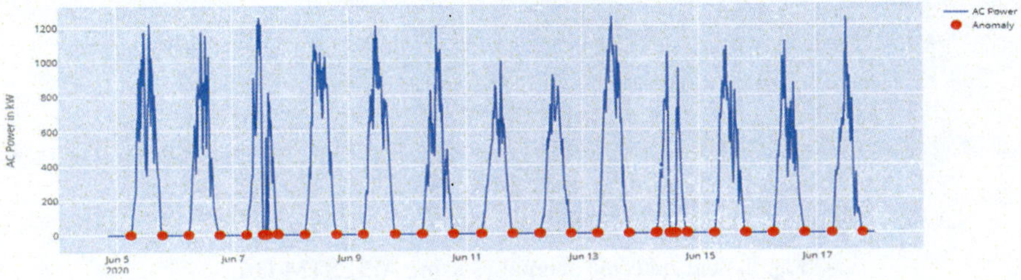

Fig. 3. The detected anomalies using the Local Outlier Factor

The outcomes of the LOF algorithm in recognizing anomalies are shown in Fig. 3. Most of the anomalies were located on the illustration's lower peaks, indicating that the algorithm didn't perform well in this case study.

3.3 AutoEncoder Long Short-Term Memory with Genetic Algorithm (AE-LSTM-GA) Approach

AE-LSTM as a form of recurrent neural network (RNN) [10] was suggested for improving the performance of detecting anomalies using the GA as a hyperparameter tuner. Therefore, the AE-LSTM is also put to the test, and the results are shown in Fig. 4. These results demonstrate a more effective performance in detecting anomalies when compared to the IF and LOF models, taking into consideration the period of June 7 and June 9, which has shown in Fig. 1.

The grid search technique, formerly employed in parameter tuning, was replaced by the genetic algorithm (GA) in recent researches. GA is a metaheuristic search technique inspired by the theory of evolution. It can be utilized to choose values within predetermined restrictions that minimize the loss of an established objective function [17]. In this approach, GA was used to optimize the AE-LSTM performance in anomalies detection.

The differences between the AC power and the irradiance are adjusted for the 12 anomalies that the AE-LSTM-GA detected compared to 8 anomalies detected using the AE-LSTM. The approach also successfully located the signal that the inverter module needed to function correctly as shown in Fig. 5.

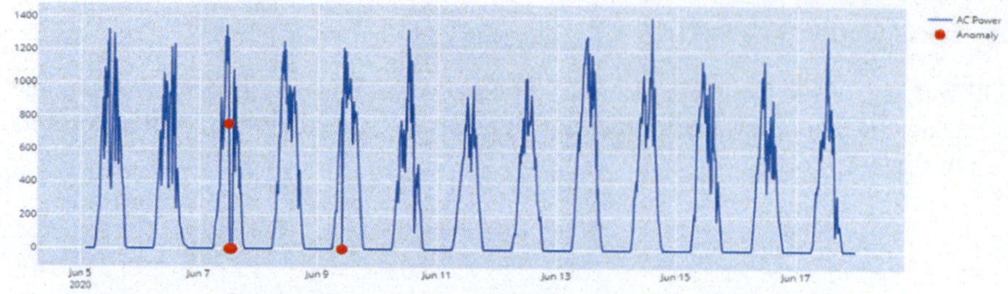

Fig. 4. The detected anomalies using AE-LSTM

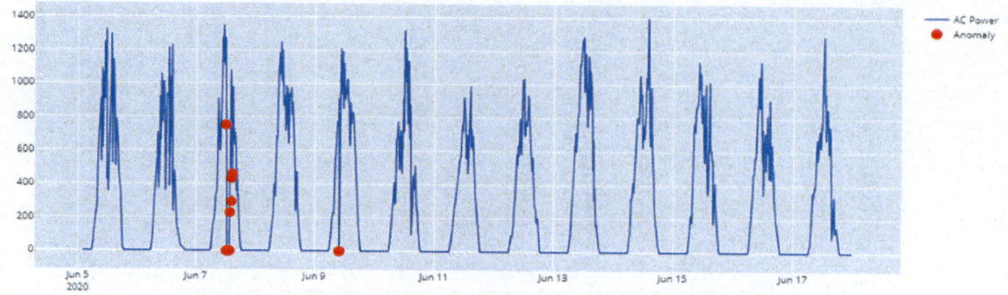

Fig. 5. The detected anomalies using AE-LSTM-GA

4 Conclusion

Solar system anomaly detection provides various advantages, including a reduction in downtime and an improvement in the equipment's efficiency. To examine some artificial intelligence algorithms' performances and choose the best model, this research introduces a new method for detecting anomalies in solar power plants. The effectiveness of machine learning models in detecting anomalies was examined using the association between the internal and external features of the power plants. The AE-LSTM-GA method outperforms the isolation forest and local outlier factor in terms of detecting anomalies in the AC-Power generation signals.

References

1. Laurentiu F., Alexandru D., Dan C., and Silvian F.: Forecasting of energy production for photovoltaic systems based on ARIMA and ANN advanced models. Int. J. Photoener. 1110–662X (2021)
2. Benninger, M., Hofmann, M., and Liebschner, M.: Online monitoring system for photovoltaic systems using anomaly detection with machine learning. In: Conference on Sustainable Energy Supply and Energy Storage Systems, Hamburg, Germany, pp. 1–6 (2019)
3. Hu, B.: Solar Panel Anomaly Detection and Classification. Master's Thesis, University of Waterloo, Waterloo, ON, Canada (2012)
4. Pereira, J., Silveira, M.: Unsupervised anomaly detection in energy time series data using variational recurrent autoencoders with attention. In: 17th IEEE International Conference on Machine Learning and Applications (ICMLA) (pp. 1275–1282). IEEE (2018)

5. Ibrahim, M., Alsheikh, A., Awaysheh, F.M., Alshehri, M.D.: Machine learning schemes for anomaly detection in solar power plants. Energies **15**(3), 1082 (2022)
6. Dey, M., Rana, S.P., Simmons, C.V., Dudley, S.: Solar farm voltage anomaly detection using high-resolution μPMU data-driven unsupervised machine learning. Appl. Ener. **303**, 117656 (2021)
7. Purarjomandlangrudi, A., Ghapanchi, A.H., Esmalifalak, M.: A data mining approach for fault diagnosis: an application of anomaly detection algorithm. Measurement **55**, 343–352 (2014)
8. Balzategui, J., Eciolaza, L., Maestro-Watson, D.: Anomaly detection and automatic labeling for solar cell quality inspection based on generative adversarial network. Sensors **21**(13), 4361 (2021)
9. Dawon, K., Sung-Min, K., Jangwon, S., Yosoon, C.: Anomaly detection of photovoltaic systems installed in renewable energy housing support project sites by analyzing power generation data (2022)
10. Boutahir, M.K., Farhaoui, Y., Azrour, M.: Machine learning and deep learning applications for solar radiation predictions review: morocco as a case of study. In: Yaseen, S.G., (eds) Digital Economy, Business Analytics, and Big Data Analytics Applications. Studies in Computational Intelligence, vol 1010. Springer, Cham (2022). https://doi.org/10.1007/978-3-031-05258-3_6
11. Kaggle website: www.kaggle.com
12. Kannal, A.: Solar power generation data. Kaggle.com. https://www.kaggle.com/anikannal/solar-power-generation-data. Accessed 09 Juin 2022
13. Google Colab website: https://colab.research.google.com
14. Cournapeau, D.: Scikit-Learn, www.scikit-learn.org (2007)
15. Chollet, F.: Keras, https://keras.io (2015)
16. Tao, X., Peng, Y., Zhao, F., Zhao, P., Wang, Y.: A parallel algorithm for network traffic anomaly detection based on isolation forest. Int. J. Distrib. Sens. Netw. **14**(11), 1550147718814471 (2018)
17. Whitley, D.: A genetic algorithm tutorial. Stat. Comput **4**, 65–85 (1994)

Towards an Integrated Rough Set and Data Modelling Framework for Data Management and Knowledge Extraction

Salem Chakhar[1,2] and Zouhaier Brahmia[3](✉)

[1] Portsmouth Business School, Faculty of Business Law, University of Portsmouth, Portsmouth, UK
salem.chakhar@port.ac.uk
[2] Centre for Operational Research and Logistics, University of Portsmouth, Portsmouth, UK
[3] MIRACL Laboratory, Faculty of Economics and Management, University of Sfax, Sfax, Tunisia
zouhaier.brahmia@fsegs.rnu.tn

Abstract. Data models and database systems are excellent tools to store and manage data. However, most of available data models and database systems lack effective techniques to extract relevant knowledge form raw data. Combining data modelling approaches and machine learning techniques represent a promising road to design and develop integrated data management and knowledge extraction systems. In this paper, first we propose a Rough Semantic data Model (RSM) based on a coupling between semantic data modelling and rough sets concepts. Then, we introduce the design of a framework that supports RSM and provides data management and knowledge extraction functionalities.

Keywords: Database · Rough set theory · Data model · Knowledge extraction

1 Introduction

In the database world, there are three main data models to design a database: the relational data model [1], the object-oriented data model [2], and the semantic data model [3, 4]. A relational model is a set of relations. Each relation has a distinct name and is a grid-like mathematical structure composed of columns, called attributes, and rows, called tuples. Each attribute (or a column of data) has a name and a data type. Each tuple (or a row) consists of a set of values such that each value per attribute. A relation is also known as a "table" in the standard SQL language and existing relational database management systems (DBMSs).

Data models and database systems are excellent tools to store and manage data. However, most of available data models and database systems lack effective techniques to extract relevant knowledge form raw data. Combining data modelling approaches and machine learning techniques represent a promising road to design and develop integrated data management and knowledge extraction systems. In fact, since the last

Y. Farhaoui et al. (Eds.): ICAISE 2022, LNNS 635, pp. 800–805, 2023.
https://doi.org/10.1007/978-3-031-26254-8_116

decade, several research papers (like [5–14]), which have been published in well-known conferences and journals, have dealt with either machine learning in DBMSs or data management in machine learning systems.

In this paper, we propose a Rough Semantic data Model (RSM) based on a coupling between semantic data modelling and rough sets [15] concepts. The rough sets theory is a mathematical tool for the analysis of a vague description of objects. It operates on a decision table composed of a set of objects described by a set of attributes. It procedures a collection of decision rules derived from rough approximations of subsets identified with decision classes. Notice that rough sets have been used in some research work that deal with database or data management problems, like [16–19], but to the best of our knowledge they have not been used in a work that combines machine learning and database management.

The paper also introduces a framework that supports the RSM model and provides data management and knowledge extraction functionalities.

The rest of the paper is organized as follows. Section 2 presents some basic concepts of the rough set theory. Section 3 proposes RSM. Section 4 provides a general overview of an integrated data management and knowledge extraction framework. Section 5 concludes the paper.

2 Principles of Rough Set Theory

Let U be a non-empty set of objects (the universe) and D be a non-empty, finite set of attributes such that $q{:}U \rightarrow Vq$, where Vq is the domain of attribute $q \in D$. With any subset $K \subseteq D$ there is an associated equivalence relation, called K-indiscernibility relation $IND(K)$ such that:

$$IND(K) = \left\{(x, y) \in U^2 \middle| q(x) = q(y), \forall q \in K\right\}$$

This relation $IND(K)$ is partitioning U into a set of equivalence classes which is denoted by $U/IND(K)$ or simply U/K. The equivalence classes induced by relation $IND(K)$ are denoted $[x]_K$. Shortly, $[x]_K \in U/K$ is the equivalence class containing x. In RST, any subset $M \subseteq U$ is defined in terms of the elementary sets (equivalence classes) of the partition U/K by lower and upper approximations as follows:

$$K_*(M) = \{x \in U | \, [x]_K \subseteq M\}$$
$$K^*(M) = \{x \in U | \, [x]_K \cap M \neq \varphi\}$$

The sets $K_*(M)$ and $K^*(M)$ (or simply M_* and M^*) are called the lower and the upper approximations of M respectively. Therefore, $M_* \subseteq M \subseteq M^*$. The difference between the upper and lower approximations is called the boundary of M and is denoted by $BN_K(M) = M^* - M_*$.

3 Rough Semantic Data Model

3.1 Basic Concepts

Let E be the universe of discourse. A rough entity e in E is a natural or artificial entity that one or several of its properties are rough. At the extensional level, a rough class K in E is a collection of rough entities having some similar properties: K = $\{(e, La(e)){:}e \in E \wedge La(e) \in T\}$, where $La{:}E \rightarrow T$ is a mapping from E to a set $T = \{Cl_h{:}h = 1, 2, \ldots p\}$ of labels.

Each RSM attribute is characterized by its name, data type and domain. A data type may be exact (e.g. integer, char) or rough. The domain of an attribute is the set of values the attribute may take. Attributes may be single-valued, i.e., the attribute cannot have more than one value at a given time, or multi-valued, i.e., the attribute can have several values at a given time. In general, the values of a multi-valued attribute may be related with different logical connectors (e.g. AND, OR, XOR) but this is will not be dealt with here.

Two crisp rough classes can be extracted from rough class K:

- K_L Lower approximation class containing objects that certainly belong to K;
- K_U Upper approximation class containing objects that may belong to K.

The boundary K_B of set rough class K is defined as the set difference between the lower and upper approximations of K, i.e., $K_B = K_U - K_L$.

3.2 Quality of Classification and Accuracy of Approximation

The quality of classification of partition T by means of condition attributes set is defined as the ratio of all correctly classified objects to all objects in E. The accuracy of approximation of decision classes is thus computed as the ratio between the number of objects in the lower approximation and the number of objects in the upper approximation.

3.3 Attribute Reducts and Core

A reduct is a subset of condition attributes that can, by itself, fully characterize the knowledge in the decision table. A reduct is minimal (with respect to inclusion) subset of condition attributes in the sense that no attribute can be removed from the reduct without deteriorating the quality of approximation. The intersection of all reducts is called core.

3.4 Data Structuring

The data used within RSM care organized into different subsets:

- Information Table: the original dataset with unlabelled objects.
- Decision Table: Information table with labelled objects.
- Learning Set: A subset from the Decision Table used for training purposes.
- Validation Set: A subset from the Decision Table used for validation purposes.
- Testing Set: A subset from the Information Table with unlabelled objects or new used testing purposes.

3.5 Knowledge Extraction

Different induction algorithms can be used to extract knowledge from the Decision Table. The most popular rule induction algorithm is LEM which generates a minimal set of rules. The extracted knowledge take the form of a set of If-then decision rules. A decision rule is a consequence relation relating a set of conditions (premise) and a conclusion (decision). Each elementary condition is built upon a single condition attribute while a conclusion is defined as assignment to decision classes.

Decision rules are evaluated through a set of quantitative measures including support, strength, accuracy, coverage and length. The obtained rules need to be validated before put into practice. Three validation techniques are commonly used: direct analysis, reclassification and cross-validation. The validated decision rules can then be used to classify unseen decision objects.

4 Towards an Integrated Data Management and Knowledge Extraction Framework

In Fig. 1, we propose the high-level architecture of a framework that supports RSM and provides data management and knowledge extraction functionalities. As shown in this figure, two categories of users are considered: DBMS oriented users and Machine Learning (ML) oriented users. This categorisation is not strict as the same user may be concerned by both data management and knowledge extraction.

The core of the proposed architecture is an extended SQL - Machined Learning (SQL-ML) language that supports the conventional SQL query language used by DBMS with additional operators and operations devoted to support knowledge extraction. The main new operations are as follows:

- Approximate <decision_table>: it generates the lower and upper approximations.
- Infer <decision_table>: it infers the decision rules.
- Reduct <decision_table|information_table>: it calculates all attribute reducts.
- Core <reducts_set|decision_table|information_table>: it calculates the core which is the intersection of all attribute reducts.
- Classify <information_table> with <classifier>: it applies the decision rules to classify the objects.
- Cross-Validate <classifier> <decision_table> <number_of_folders>: it applies the cross-validation technique.

5 Conclusion

In this, we propose (i) RSM, a Rough Semantic data Model that relies on coupling between semantic data modelling and rough sets concepts, and (ii) a framework that supports RSM and provides both data management and knowledge extraction functionalities. In our future work, we will develop a layered system that supports our framework and shows its feasibility. Such a system will help us to experimentally evaluate the usability and the performances of our framework.

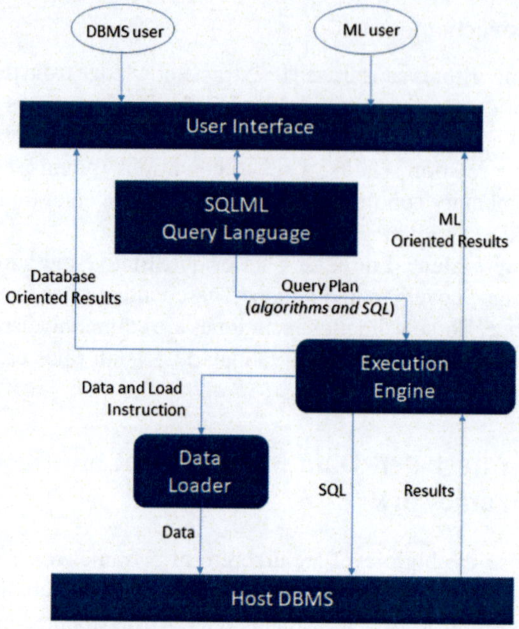

Fig. 1. The architecture of an integrated data management and knowledge extraction framework

References

1. Embley, D.W.: Relational model. In: Liu, L., Özsu, M.T. (eds.) Encyclopedia of Database Systems, 2nd edn., pp. 3149–3154. Springer, New York (2018). https://doi.org/10.1007/978-1-4614-8265-9_306
2. Urban, S.D., Dietrich, S.W.: Object data models. In: Liu, L., Özsu, M.T. (eds.) Encyclopedia of Database Systems, 2nd edn., pp. 2525–2532. Springer, New York (2018). https://doi.org/10.1007/978-1-4614-8265-9_249
3. Bouaziz, R., Chakhar, S., Mousseau, V., Ram, S., Temouldi, A.: Database de-sign and querying within the fuzzy semantic model. Inf. Sci. **177**(21), 4598–4620 (2007)
4. Embley, D.W.: Semantic data model. In: Liu, L., Özsu, M.T. (eds.) Encyclopedia of Database Systems, 2nd edn., pp. 3391–3393. Springer, New York (2018). https://doi.org/10.1007/978-1-4614-8265-9_105
5. Li, X., Cui, B., Chen, Y., Wu, W., Zhang, C.: MLog: towards declarative in-database machine learning. Proc. VLDB Endow. **10**(12), 1933–1936 (2017)
6. Domingos, P.: Machine learning for data management: problems and solutions. In: Proceedings of the 2018 International Conference on Management of Data, p. 629, May 2018
7. Fard, A., Le, A., Larionov, G., Dhillon, W., Bear, C.: Vertica-ML: distributed machine learning in Vertica database. In: Proceedings of the 2020 ACM SIGMOD International Conference on Management of Data, pp. 755–768, June 2020
8. Jankov, D., et al.: Declarative recursive computation on an RDBMS: or, why you should use a database for distributed machine learning. ACM SIGMOD Rec. **49**(1), 43–50 (2020)
9. Jasny, M., Ziegler, T., Kraska, T., Roehm, U., Binnig, C.: DB4ML-an in-memory database kernel with machine learning support. In: Proceedings of the 2020 ACM SIGMOD International Conference on Management of Data, pp. 159–173, June 2020

10. Abdennebi, A., Elakaş, A., Taşyaran, F., Öztürk, E., Kaya, K., Yıldırım, S.: Machine learning-based load distribution and balancing in heterogeneous database management systems. Concurr. Comput. Pract. Exp. **34**(4), e6641 (2022)
11. Schule, M., Lang, H., Springer, M., Kemper, A., Neumann, T., Gunnemann, S.: In-database machine learning with SQL on GPUs. In: Proceedings of the 33rd International Conference on Scientific and Statistical Database Management, pp. 25–36, July 2021
12. Png, A., Helskyaho, H.: Oracle machine learning in autonomous database. In: Extending Oracle Application Express with Oracle Cloud Features, pp. 139–191. Apress, Berkeley (2022)
13. Chai, C., Wang, J., Luo, Y., Niu, Z., Li, G.: Data management for machine learning: a survey. IEEE Trans. Knowl. Data Eng. (2022). https://doi.org/10.1109/TKDE.2022.3148237
14. Villarroya, S., Baumann, P.: A survey on machine learning in array databases. Appl. Intell., 1–24 (2022). https://doi.org/10.1007/s10489-022-03979-2
15. Pawlak, Z.: Rough Sets: Theoretical Aspects of Reasoning About Data, vol. 9. Springer, Dordrecht (1991). https://doi.org/10.1007/978-94-011-3534-4
16. Beaubouef, T., Petry, F.E., Buckles, B.P.: Extension of the relational database and its algebra with rough set techniques. Comput. Intell. **11**(2), 233–245 (1995)
17. Lin, T.Y.: Rough set theory in very large databases. In: Symposium on Modeling, Analysis and Simulation, IMACS Multi Conference (Computational Engineering in Systems Applications), Lille, France, vol. 2, pp. 936–941, July 1996
18. Suraj, Z., Grochowalski, P.: The rough set database system. In: Peters, J.F., Skowron, A. (eds.) Transactions on Rough Sets VIII. LNCS, vol. 5084, pp. 307–331. Springer, Heidelberg (2008). https://doi.org/10.1007/978-3-540-85064-9_14
19. Kovalenko, I., Shved, A., Antipova, K., Davydenko, Y.: Structuring of a transaction database using the rough set theory. In: CMIS, pp. 278–287 (2020)

Towards Cd-Free Sb_2Se_3 Based Solar Cells Using SCAPS-1D

Abdelmajid El Khalfi[1(✉)], Lhoussayne Et-taya[1], Abdellah Benami[1], Mustapha Sahal[2], and Lahoucine Elmaimouni[2]

[1] LM3ER-OTEA Department of Physics, Faculty of Sciences and Techniques, Moulay Ismail University of Meknes, BP 509, Boutalamine, 52000 Errachidia, Morocco
elkhalfi-abdelmajid@hotmail.fr, l.ettaya@edu.umi.ac.ma, a.benami@fste.umi.ac.ma

[2] ERMAM, Ouarzazate Polydisciplinary Faculty, University of Ibn Zohr, 45000 Agadir, Morocco

Abstract. Antimony selenide (Sb_2Se_3) solar cells are becoming the backbone of global photovoltaic industry due to their remarkable low cost, optoelectronic and mechanical properties. In this work, the performance of Sb_2Se_3 based solar cells with WS_2, CdS, In_2S_3 and ZnSe as buffer layers is simulated using SCAPS. The effect of the active layer thickness and acceptor doping concentration, as well as different buffer layers and the thickness of WS_2, were studied. The efficiency for WS_2 buffer layer was approximately 27% higher than that of the other buffer layers. This finding will aid researchers in the development of CdS-free high performance Sb_2Se_3 based solar cells.

Keywords: Sb_2Se_3 solar cell · SCAPS-1D · Cd-free · Buffer layer

1 Introduction

Photovoltaic solar cells are electronic devices that convert light energy into electricity [1]. Because of its excellent properties such as narrow band gap (1.1–1.3 eV) [1–3], the high absorption coefficient ($>10^5$ cm^{-1}) [4], low toxicity and comparatively abundant constituents in the earth. All of these properties make Sb_2Se_3 a highly promising long-term alternative to CIGS and CdTe for the long-term development of solar cells.

In this work, we have used the SCAPS program to simulate the performance of the device structure Sb_2Se_3/ETM with different buffer layers: WS_2, CdS, In_2S_3 and ZnSe. The effect of various parameters on device performance, such as V_{oc}, J_{sc}, FF and PCE, has been investigated, including absorber thickness, acceptor doping concentration, and WS_2 buffer layer thickness.

Y. Farhaoui et al. (Eds.): ICAISE 2022, LNNS 635, pp. 806–811, 2023.
https://doi.org/10.1007/978-3-031-26254-8_117

2 Device Structure and Simulation

SCAPS is an efficient tool for simulating the performance of solar cells, which is developed by the University of Gent that can provide good concordance between the results of both the experiment and the simulation [5]. This software uses Poisson and continuity equations for holes and electrons, as shown by the equations listed below.

$$\frac{\partial^2 \varphi}{\partial x^2} = -\frac{q}{\varepsilon}[p(x) - n(x) + N_D - N_A + \rho_P - \rho_n] \tag{1}$$

$$\frac{dJ_p}{dx} = q(G - R) \tag{2}$$

$$\frac{dJ_n}{dx} = -q(G - R) \tag{3}$$

where φ is the electric potential, q is the electron charge, ε is the dielectric constant, n is the electron concentration, p is the hole concentration, J_n is the electron current density, J_p is the hole current density, G is the carrier generation rate, and R is the carrier recombination rate.

The device structure in our simulation is glass substrate/ITO/ETM/Sb_2Se_3/Au as shown in Fig. 1. ITO is used as a front electrode for electron collection. ETM is an n-type layer that is used as an electron transport material. Sb_2Se_3 is a p-type material that is used as an active layer to generate electron-hole pairs. Back electrode holes are collected using Au. Table 1 displays the SCAPS setup parameters used as starting data for the simulations; these values were obtained from the literature [6–9].

Where A is a constant that varies depending on the layer in the simulation.

In the simulation, the working temperature is set as 300 K and the device is illuminated from the ITO side by the standard AM1.5 spectrum.

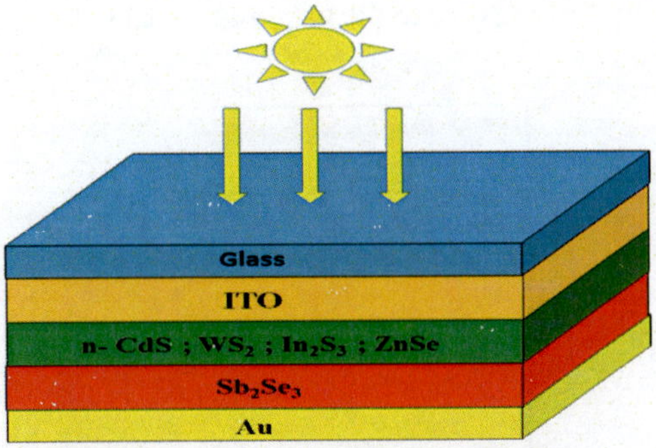

Fig. 1. Structure of Sb_2Se_3 thin film solar cell.

Table 1. Parameters used in the simulation [6–9]

Parameters	Sb_2Se_3	CdS	WS_2	In_2S_3	ZnSe	ITO
W (μm)	0.5	0.05	0.05	0.05	0.05	0.2
Eg (eV)	1.2	2.4	1.8	2.9	2.7	3.6
χ (eV)	4.04	4	3.95	4.25	4.09	4.1
ϵ	18	9	13.6	13.6	10	10
N_C (cm^{-3})	2.2×10^{18}	2.2×10^{18}	2.2×10^{18}	2.2×10^{18}	2.2×10^{18}	2×10^{19}
N_V (cm^{-3})	1.8×10^{19}	1.8×10^{19}	1.8×10^{19}	1.8×10^{19}	1.8×10^{19}	1.8×10^{19}
Vn (cm/s)	10^7	10^7	10^7	10^7	10^7	10^7
Vp (cm/s)	10^7	10^7	10^7	10^7	10^7	10^7
μ_n (cm^2/v.s)	15	100	100	100	100	50
μ_p (cm^2/v.s)	5.1	25	25	25	25	75
N_D (cm^{-3})	0	10^{17}	10^{17}	10^{17}	10^{17}	10^{20}
N_A (cm^{-3})	10^{17}	0	0	0	0	0

3 Results and Discussion

3.1 Current Density-Voltage Characteristic for Different Buffer Layers

Buffer layers are incorporated into the thin film solar cells to extract photogenerated electrons from absorber layer to front electrode. Figure 2 shows the J-V characteristic curve simulated with the data given in Table 1 for the different buffer layers. As can be seen, the device with WS_2 as ETM shows better performance as compared to others (inset of Fig. 2), which can be attributed to the fact that WS_2 has higher transmittance and higher conductivity. The Sb_2Se_3 with WS_2 as ETM shows a higher PCE of 19.55%, hence WS_2 can be a good alternative to the CdS traditional buffer layer.

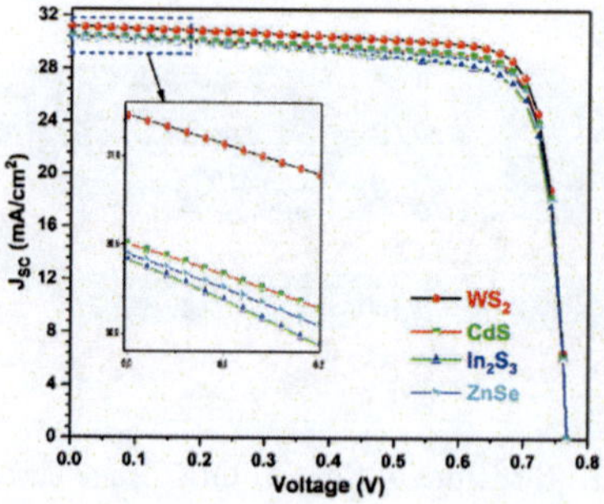

Fig. 2. J-V characteristic of Sb_2Se_3 based solar cells at deferent buffer layers.

3.2 Effect of the Sb_2Se_3 Absorber Layer's Thickness

Absorber layer has a direct effect on the performance of the thin film solar cells due to the fact that absorbed photon number depend on the absorber thickness and absorber layer thickness. We have investigated the impact of absorber layer thickness on device performance in the range of 0.5–4.1 μm. As shown in Fig. 3 (a–d), all the parameters first increased up to 2.5 μm, and then keep nearly constant when the absorber layer thickness is further increased. The obtained result is in good agreement with the Beer-Lambert's law, which relate the increase of absorber layer with increasing exciton photogenerated thus enhance thin film solar cell performance. Keeping in mind higher defect density in thicker Sb_2Se_3, a promising Voc, Jsc, FF and PCE of 0.801 V, 36 mA/cm^2, 84% and 24% were obtained at thickness of 1.5 μm, respectively.

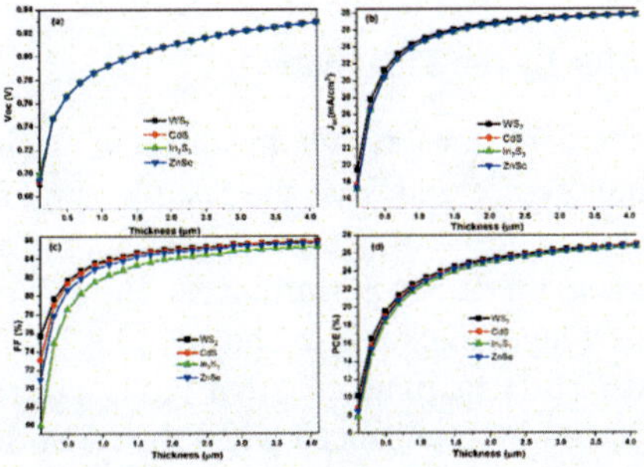

Fig. 3. Sb_2Se_3 thickness impact on cell performance: (a) Voc (b) Jsc, (c) FF and (d) PCE with different ETMs

3.3 Active Layer Acceptor Doping Concentration Optimization

The acceptor doping concentration plays a very significant role in optimizing thin film solar cell performance. Sb_2Se_3 has a p-type acceptor concentration where the impurity of the carrier depends on the imperfection in the synthesized material. Figure 4 (a–d) illustrates the impact of doping concentration on device performance with four different buffer layers. It is clear from Fig. 4 that PEC, V_{oc} and FF rise as acceptor doping concentration increases, whereas J_{sc} initially remained constant and then decreases when the concentration is greater than 10^{15} cm^{-3}. The increase of the acceptor doping concentration in the absorber p-type results in the increment of the holes density, which generate more hole trap in the absorber layer. This lead to enhance the recombination of the excitons due to the Coulomb interactions, and hence decreases of Jsc with increase of the absorber doping. The PCE increases from 22.5% to 26.9% when the doping concentration rise from 10^{13} cm^{-3} to 10^{18} cm^{-3}.

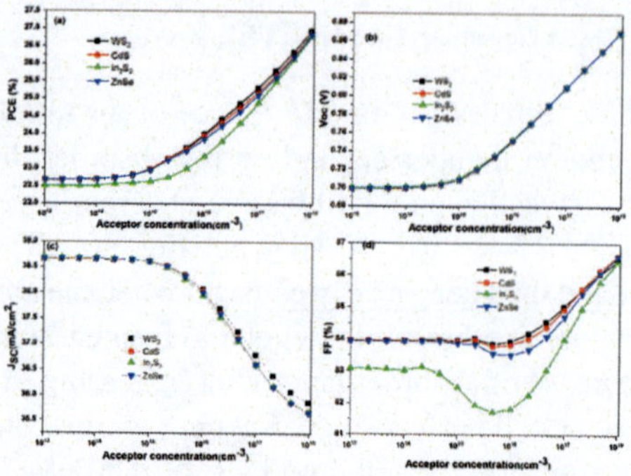

Fig. 4. Effect of acceptor doping concentration of Sb_2Se_3 absorber layer on photovoltaic parameters (a) PCE, (b) Voc, (c) Jsc and (d) FF with different ETMs.

3.4 Impact of WS2 Buffer Layer Thickness

The main function of ETM is to transport photogenerated electrons to the front metal electrode. It is of great importance and from the result comparative study of different buffer layers, the higher PCE consist of WS_2 as ETM. So, we investigated the impact of WS_2 buffer layer thickness on device performance. The key parameters of solar cell change by varying the buffer layer thickness as shown in Fig. 5. The V_{oc} was almost the same for all thickness up to 0.30 μm, indicating that V_{oc} is independent of the thickness of ETM. Thus, the FF, Jsc and PCE enhanced when the ETM thickness is raised from 0.03 μm to 0.30 μm.

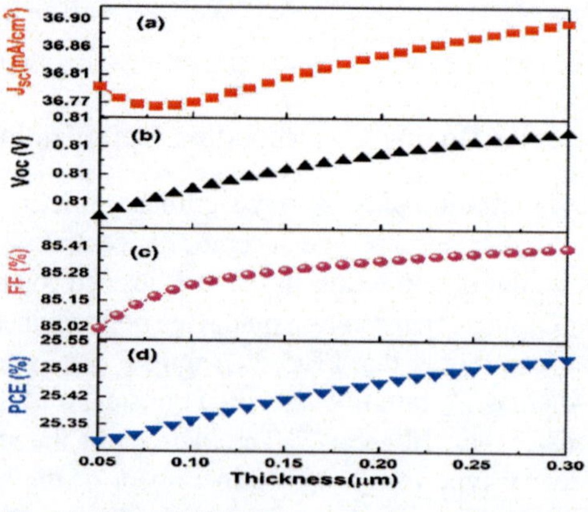

Fig. 5. (a) Jsc, (b) Voc, (c) FF and (d) PCE as a function of WS_2 buffer layer thickness.

4 Conclusion

SCAPS software was used to simulate the performance of the Sb_2Se_3 based solar cells with a structure ITO/ETM/Sb_2Se_3/Au. Four different devices with buffer layers of WS_2,

CdS, In_2S_3 and ZnSe were investigated in order fabricate highly efficient Cd-free Sb_2Se_3 based solar cells. The active layer thickness, acceptor doping concentration and buffer layer thickness affect key parameters such as the PCE, the FF, the V_{oc} and the J_{sc}. Due to its nontoxic properties and conductivity, we discovered that WS_2 is a more effective layer than other buffer layers for the fabrication of low-cost, high-efficiency Cd-free Sb_2Se_3 heterojunction solar cells. The highest PCE of around 27% was achieved for Sb_2Se_3 based solar cells with WS_2 as an ETM. Our simulation results can help to produce high performance Sb_2Se_3 based solar cells.

References

1. Ouslimane, T., Et-taya, L., Elmaimouni, L., Benami, A.: Impact of absorber layer thickness, defect density, and operating temperature on the performance of $MAPbI_3$ solar cells based on ZnO electron transporting material. Heliyon **7**, e06379 (2021)
2. Cheng, Y., et al.: Constructing lateral sulfur-gradient $Sb_2(S_xSe_{1-x})_3$ heterostructures for Sb_2Se_3 nanorod photocathodes with enhanced photoelectrochemical properties. Electrochimica Acta **403**, 139610 (2022)
3. Li, Z.-Q., Ni, M., Feng, X.-D.: Simulation of the Sb_2Se_3 solar cell with a hole transport layer. Mater. Res. Express **7**, 016416 (2020)
4. Zhao, Y., et al.: Regulating energy band alignment via alkaline metal fluoride assisted solution post-treatment enabling $Sb_2(S,Se)_3$ solar cells with 10.7% efficiency. Adv. Energy Mater. **12**, 2103015 (2022)
5. Et-taya, L., Ouslimane, T., Benami, A.: Numerical analysis of earth-abundant $Cu_2ZnSn(S_xSe_{1-x})_4$ solar cells based on Spectroscopic Ellipsometry results by using SCAPS-1D. Sol. Energy **201**, 827–835 (2020)
6. Ngoupo, A.T., Ouédraogo, S., Zougmoré, F., Ndjaka, J.M.B.: Numerical analysis of ultrathin Sb_2Se_3-based solar cells by SCAPS-1D numerical simulator device. Chin. J. Phys. **70**, 1–13 (2021)
7. Khatun, M.M., Sunny, A., Al Ahmed, S.R.: Numerical investigation on performance improvement of WS_2 thin-film solar cell with copper iodide as hole transport layer. Sol. Energy **224**, 956–965 (2021)
8. Mamta, Maurya, K.K., Singh, V.N.: Efficient Sb_2Se_3 solar cell with a higher fill factor: a theoretical approach based on thickness and temperature. Sol. Energy **230**, 803–809 (2021)
9. Et-taya, L., Benami, A., Ouslimane, T.: Study of CZTSSe-based solar cells with different ETMs by SCAPS. Sustainability **14**(3), 1916 (2022). https://doi.org/10.3390/su14031916)

Using MPI to Compare Two Protocols That Allow Paillier's Cryptosystem to Perform Homomorphic Multiplication

Hamid El Bouabidi[1](✉), Mohamed El Ghmary[2], Sara Maftah[1], Mohamed Amnai[1], and Ali Ouacha[3]

[1] Department of Computer Science, Faculty of Sciences, Ibn Tofail University, Kenitra, Morocco
{hamid.elbouabidi,sara.maftah,mohamed.amnai}@uit.ac.ma

[2] Department of Computer Science, Faculty of Sciences, Dhar El Mahraz, Sidi Mohamed Ben Abdellah University Fez, Fez, Morocco
mohamed.elghmary@usmba.ac.ma

[3] Department of Computer Science, Mohammed V University in Rabat, Rabat, Morocco
ali.ouacha@um5.ac.ma

Abstract. Cloud computing delivers hitherto unknown storage and computation outsourcing possibilities. But the potential of exposing essential data to privacy and security issues prevents many users from transferring vital data to the cloud. Several encryption techniques have been created and studied in cloud security literature. In this paper, we explore Paillier's encryption and its application to privacy-preserving computing outsourcing and securing the cloud, which may be helpful for various applications such as improving election, medicine, finance and banking, etc. We present a new implementation of Paillier's cryptosystem that uses two protocols to perform homomorphic multiplications so that Paillier's cryptosystem can be considered as a fully homomorphic encryption.

Keywords: Cloud computing · Homomorphic encryption · Cloud security · Paillier cryptosystem · Sockets

1 Introduction

The cloud is a concept that has gained popularity in recent years because it has many advantages that make computing easier and more affordable. People use the cloud to store data, process information, and make applications accessible and manageable. Traditional computers are expensive, slow, and difficult to upgrade. The cloud solves this problem by making computing more convenient. Cloud computing provides several advantages to businesses that operate in the digital age [1]. It enables them to boost efficiency and productivity and save space. Cloud services are accessible over the internet from multiple locations with high speed and reliability. This allows businesses to quickly expand their operations without worrying about server storage costs or setup times. Additionally, cloud services are affordable due to economies of scale that would enable pricing models suitable for small businesses as well as large enterprises. The biggest

Y. Farhaoui et al. (Eds.): ICAISE 2022, LNNS 635, pp. 812–817, 2023.
https://doi.org/10.1007/978-3-031-26254-8_118

worry of cloud users is security [2], according to surveys [3]. We will mention related works in the next section. Section three will be about two protocols to improve Paillier's cryptosystem. In section four we will conclude the paper.

2 Related Works

Cloud computing is gaining popularity because it allows customers to access resources at an affordable cost. However, there are issues regarding the security of user data that attracted many research works and studies. The authors of [4] proposed using Homomorphic Encryption (HE) to tackle privacy difficulties. Paillier (1978 to 2008) produced various HE techniques that allowed encrypted data to be processed concurrently with just one kind of operator [5]. In 2009, Gentry suggested the first not-yet broken Fully Homomorphic Encryption (FHE) system [6]. FHE refers to cryptosystems that perform both adds and multiplications on encrypted data. It was a breakthrough after 30 years of great work since it opened the path to many more powerful practical applications than earlier. However, because of its significant algorithmic complexity, big key size, and ciphertext expansion, modern FHE systems remain today not efficient in reality. Darko and Stjepan [7] researched the significance of homomorphic encryption in cloud computing. The study discussed IBM's involvement in the algorithm's open-source library. Authors in [8] analyze the state-of-art methods used to implement homomorphic encryption in cloud security. They emphasize homomorphic problems and constraints for utilizing HE techniques on encrypted data for the cloud. The authors of [9] provided a comprehensive analysis for HE that highlights current demand applications and prospects such as security and privacy. The authors of [10] discuss the obstacles, possibilities, and future improvements of using HE for Big data computations. In this paper, we will focus on the improvement of Paillier's cryptosystem for the first time by suggesting two protocols that allow multiplication on encrypted data and compare them using MPI.

3 Protocols that Enable Homomorphic Multiplication

Since Paillier's approach [11] does not enable homomorphic multiplication, i.e., to obtain the product of two encrypted values. We propose to enhance this algorithm by utilizing two protocols to perform multiplication. The idea is to transform the product into a sum so that the cloud server may add the encrypted data.

3.1 Paillier Encryption and the Russian Multiplication Protocol

This multiplication has the significant benefit of simply requiring the use of the table of twos [12]. To advance from one line to the next, multiply the partial products on the left by two and the partial products on the right by half. This is related to writing the multiplier in base 2 and then doing multiplications by 2 and additions. It is therefore a version of the multiplication method used in ancient Egypt. The algorithm for this approach is as follows (Fig. 1):

Algorithm 1 Russian multiplication

```
Input: m1,m2,table tab
Output: m1 × m2
while m1 > 0 do
    if (m1%2 = 1 ) then
        e2 = encrypt(m2, pubKey)
        Add e2 to tab
    end if
    m1 = m1//2
    m2 = m2 * 2
end while
return tab
```

Fig. 1. Russian multiplication algorithm

3.2 Paillier Encryption and Continuous Logarithm

In this part, we will propose a protocol that is more natural in the manner it changes a multiplication into an addition, using the well-known continuous logarithm function (Fig. 2):

$$log(a \times b) = log(a) + log(b) \tag{1}$$

Algorithm 2 Continuous logarithm multiplication

```
Input: m1,m2
Output: m1 × m2
l1 = log(m1)
l2 = log(m2)
e1 = encrypt(m1)
e2 = encrypt(m2)
e = e1 + e2
decrypt(e)
prod=exp(m)
```

Fig. 2. Continuous logarithm multiplication algorithm

3.3 Comparison Between Multiplication Protocols Using MPI

MPI (Message Passing Interface) [13] is designed to maximize scalability, maintainability, and portability of applications. It provides a foundation that research programs can develop to help them achieve their goals. MPI enables researchers in different institutions to collaborate across locations without fear of incompatibility between tools. To simulate the two protocols, we used two virtual machines, a master machine that simulates the (cloud) and a slave machine that simulates the (client).

Continuous Logarithm Protocol: Process 0, which simulates the client, will apply the logarithm on clear messages and then encrypt them by a public key before sending them to the cloud (process 1). The cloud will perform addition on the encrypted values and return the result to the client (process 0). The client will decrypt the result using his private key and apply exponential to see the plaintext as is shown in the Algorithm below (Fig. 3):

Algorithm 3 Continuous logarithm multiplication protocol

```
Input: m1,m2
Output: m1 × m2
{Client Side(processus 0)}
l1 = log(m1)
l2 = log(m2)
e1 = encrypt(l1, pubkey)
e2 = encrypt(l2, pubkey)
MPI_Send(e1,e2,destination=processus 1)
r=MPI_Recv(e,source=processus 0)
m=decrypt(r)
message=exp(m)
{Cloud Side(processus 1)}
sum=0
c1=MPI_Recv(e1,source=processus 0)
c2=MPI_Recv(e2,source=processus 0)
sum = c1 + c2
MPI_Send(e,destination=processus 0)
```

Fig. 3. Continuous logarithm multiplication protocol algorithm

Russian Multiplication Protocol: The client (Processus 0) encrypted the values of m_2 corresponding to the odd values of m_1 and put these values in a table. Send the table so that to get the value of the multiplication, the cloud (Processus 1) sums up those values as shown in the Algorithm below (Fig. 4):

Algorithm 4 Russian multiplication protocol

```
Input: m1,m2,table tab
Output: m1 × m2
{Client Side}
while m1 > 0 do
    if (m1%2 = 1 ) then
        e2 = encrypt(m2, pubKey)
        Add e2 to tab
    end if
    m1 = m1//2
    m2 = m2 * 2
end while
send tab
R = MPI_recv(sum,source=1)
{Cloud Side}
R_tab = MPI_recv(tab,source=0)
sum = 0
for x in R_tab do
    sum = sum + x
end for
MPI_send(sum,destination=0)
```

Fig. 4. Russian multiplication protocol algorithm

3.4 Implementation and Results

To evaluate our proposed solution and for better security, we created a TinyDB [14] database, and we encrypted the name of the database, tables and fields names. This technique will allow us to perform several operations without disclosing the slightest clue about what we want to do or about the content of our database. As an example, consider the following table before encrypting the database (Fig. 5):

nom	note	cne
Salah	17.5	2365
Ahmed	14	2541
Rajae	12.3	9856
Hamid	16	5874

Fig. 5. Displaying clear values from the database

After encrypting all the database by the client with the public key, it shows as follow (Fig. 6):

nom	note	cne
39f453ed708a5788225400b0d1fe2535a389eeed47e8856020fe2683e589c858ad	8432272260435046738662663667586946582013248148094801494354022471514758173241	11229679835971761764794605727448890348776791417820051894049069461875906833895
11d1069cc2bd4bf4683afeb8b0effc03052e567ac4acb4b0d464dc08ba4774d4fc	45785263804957006217764673301927578932976911199554544210035409949465213391238	29669388642339150147000897496796592694891327251657200423633936148743412468454
758735fde75e58e22c33438d75c5633ddfbe7a73449a97622d9abe0a6a7b696449	13974137691280782338656695035590789829133639173636604535795638260050954909080	34862167145836576234708229202029560247519590888246657429829468612411066788419
6c7b8f2d6ffa6b01a438e0e8d309b5e5462fe8b76fe771f9cd3393b8fa048fb7aa	32074423316829008730452875681372891650981000943867478556666952432716500264630	19867834178605928178963397126545190719451682303898462314109833966372230048209

Fig. 6. Database after being encrypted

We mention that the columns nom and note fields appear in plaintext for visibility, but in reality, in the cloud Server, they are encrypted. The figure below describes the comparison of the two protocols, using MPI, in terms of time and key size (Fig. 7):

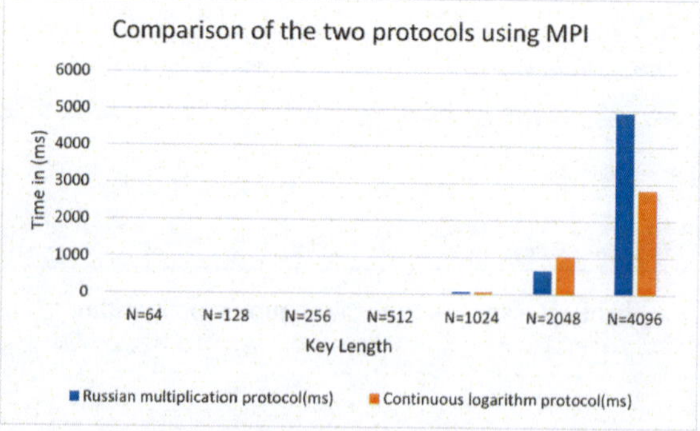

Fig. 7. Result and comparison of the two protocols using MPI

With a tiny key size, we see that both algorithms provide the same results in about the same amount of time. However, when the key size increases, the product encryption protocol employing the Russian multiplication protocol takes a long time, but it's more accurate than the continuous logarithm protocol. As a result, if the operations performed on the cloud need precision, we should use the Russian multiplication protocol. While we may count on the continuous logarithm protocol if we want quick and approximate results.

4 Conclusion

Security is among the top concerns of companies when it comes to adopting cloud computing. With sensitive data and critical business processes running on the cloud, companies cannot afford to have their information and data stolen or their systems defaced by hackers. Our proposed solution allows users to conduct sensitive mathematical operations on the encrypted data without decrypting it. This provides users with the ability to store, process, and analyze sensitive information without having to sacrifice the security of their data.

References

1. Modisane, P., Jokonya, O.: Evaluating the benefits of cloud computing in small, medium and evaluating the benefits of cloud computing in small, medium and enterprises (SMMES). Procedia Comput. Sci. **181**(2019), 784–792 (2021)
2. Gaur, A., Mallikarjuna Shastry, P.M.: Enhanced approach for secure stored data in cloud (2020)
3. Sun, P.J.: Privacy protection and data security in cloud computing: a survey, challenges, and solutions. IEEE Access **7**, 147420–147452 (2019)
4. Aguilar-Melchor, C., Fau, S., Fontaine, C., Gogniat, G., Sirdey, R.: Recent advances in homomorphic encryption: a possible future for signal processing in the encrypted domain. IEEE Signal Process. Mag. **30**(2), 108–117 (2013)
5. Fontaine, C., Galand, F.: A survey of homomorphic encryption for nonspecialists. EURASIP J. Inf. Secur. **1–10**, 2007 (2007)
6. Gentry, C.: Fully homomorphic encryption using ideal lattices. In: Proceedings of the Forty-First Annual ACM Symposium on Theory of Computing, pp. 169–178 (2009)
7. Hrestak, D., Picek, S.: Homomorphic encryption in the cloud. In: 2014 37th International Convention on Information and Communication Technology, Electronics and Microelectronics (MIPRO), pp. 1400–1404. IEEE (2014)
8. Biksham, V., Vasumathi, D.: Homomorphic encryption techniques for securing data in cloud computing: a survey. Int. J. Comput. Appl. **975**, 8887 (2017)
9. Alloghani, M., et al.: A systematic review on the status and progress of homomorphic encryption technologies. J. Inf. Secur. Appl. **48**, 102362 (2019)
10. Alaya, B., Laouamer, L., Msilini, N.: Homomorphic encryption systems statement: trends and challenges. Comput. Sci. Rev. **36**, 100235 (2020)
11. Paillier, P.: Public-key cryptosystems based on composite degree residuosity classes. In: Stern, J. (ed.) EUROCRYPT 1999. LNCS, vol. 1592, pp. 223–238. Springer, Heidelberg (1999). https://doi.org/10.1007/3-540-48910-X_16
12. Tee, G.J.: Russian peasant multiplication and Egyptian division in Zeckendorf arithmetic. Technical report, Department of Mathematics, The University of Auckland, New Zealand (2002)
13. Gropp, W., Lusk, E., Doss, N., Skjellum, A.: A high performance, portable implementation of the MPI message passing interface standard. Parallel Comput. **22**(6), 789–828 (1996)
14. Madden, S.R., Franklin, M.J., Hellerstein, J.M., Hong, W.: TinyDB: an acquisitional query processing system for sensor networks. ACM Trans. Database Syst. (TODS) **30**(1), 122–173 (2005)

Using the Ethereum Blockchain to Secure a Crowdfunding System in the Real Estate Sector

Hibatou Allah Boulsane(✉), Karim Afdel, and Salma El Hajjami

Faculty of Sciences, Ibn Zohr University, Agadir, Morocco
hibatouallah.boulsane@edu.uiz.ac.ma

Abstract. The covid-19 crisis has severely affected the dynamics of the real estate sector, which is facing various financial and structural problems. The current real estate world is complicated by the lack of transparency in transactions such as rental, purchase, and sale, and it does not reach the level of confidentiality and authenticity of operational data. In addition, real estate financing brings together several players such as banks, notaries, and others, which makes the acquisition of real estate very expensive. With the advent of blockchain technology, many fields such as finance, accounting, and real estate have received a positive impact using the benefits of this technology. This article aims to reorganize real estate into a next-generation digitized system based on blockchain technology, by proposing a crowdfunding model that aims to eliminate intermediate, costs. Moreover, this model can allow to customers, who do not have immediate financing, the possibility of acquiring real estate. We also present the implementation of this model through smart contracts and the blockchain to set up a decentralized platform that ensures the security, traceability, and transparency of transactions.

Keywords: Crowdfunding · Blockchain Ethereum · FinTech · Smart contract · Real estate · Dapp

1 Introduction

The real estate sector is an essential element of Morocco's economic system, and the acquisition of real estate is very important for Moroccan citizens and the state. There are different categories of real estate in Morocco: Residential real estate includes all built assets with the aim of housing people. The business real estate, this category includes a variety of buildings that do not have a residential use but are intended for the exercise of a professional, commercial or industrial activity. Tourist real estate includes buildings of leisure and tourist buildings.

Indeed, the real estate sector in Morocco faces several problems, such as the lack of traceability of transactions between the real estate promoter, the real estate fund, the investor, and the client. In addition, the high cost of the buying or hiring process (real estate agency fees) and the existence of an intermediary (virtual entity or human)

Y. Farhaoui et al. (Eds.): ICAISE 2022, LNNS 635, pp. 818–823, 2023.
https://doi.org/10.1007/978-3-031-26254-8_119

makes transactions longer and more expensive. The main problem is that the majority of Moroccans do not have sufficient financial funds to invest in real estate.

Fintech (Financial technology) has an important role in improving the real estate sector. [1, 2] considered Fintech as a distinctive taxonomy that mainly describes the areas of financial technology in a broad set of operations for businesses or organizations, that improve the quality of service using information technology (IT) applications.

Blockchain technology [3–6] is a series of decentralized registries and systems that provide accurate, transparency, consistent, traceable, and secure services to users at low transaction costs and without intervention. To provide infrastructure for decentralized applications, the blockchain system uses cryptographic technologies, consensus models (proof of work, proof of stake, etc.), and P2P networks. Furthermore, blockchain technology is based on the smart contract which is a self-executing program that is characterized by resistance to falsifica tion, autonomous execution, and precision.

One of the financial technologies is crowdfunding [7–10] which is the financing of projects by a large number of people, regardless of whether they are individuals or businesses. Crowdfunding is a novel and creative phenomenon in entrepreneurial finance that allows project owners to ask investors for funding.. The crowdfunding system includes three actors which are the funding portals (intermediaries), the contributors (funders/investors), and the creators of crowd- funding campaigns.

The objective of this study is to show the impact of crowdfunding using blockchain technology in real estate projects. A crowdfunding platform can reduce the cost of financing real estate projects, and facilitate the acquisition of housing for a significant number of Moroccan citizens who find difficulties in financing their housing. In this context, we present a blockchain-based real estate management model which is developed to enable a more reliable and transparent real estate sector. We then discuss the technical and operational contributions of our model in this research work

The rest of the work is presented: Section 2 comprises related work. Section 3 explains our funding approach as well as the planned application's architecture. Section 4 summarizes the outcome. The last section is the conclusion.

2 Related Work

According to [11], some countries, like the Netherlands, are developing their first blockchain applications in the real estate sector to document leasing contracts. This project aims to make it possible to digitize building data, digitize ownership status, close rental contracts, and unlock contractual information for third parties; thanks to Blockchain, all building information is gathered.

In [12] suggested a buyer and seller's platform based on the Ethereum blockchain, that allows using smart contracts which can store the data related to lands, users and transactions between customer and others participants.

In [13], the authors showed how blockchain technology can increase user trust in crowdfunding platforms by reducing asymmetric information and fraud. They use smart contracts to guarantee the transparency of transactions and to know where and how funds are spent.

In [14], the authors analyzed the functions of intermediaries's real estate in European Union and how blockchain can ensure security of transactions and reducing costs and times of these transactions. They deleted the notary to eliminate intermediary, but the problem is how to identify the official ID of the parties, further more they should add a legal clauses to control the legality.

In [15], the authors produced two conceptual models of real estate which are the foreign and South Africa real estate transaction process based on Blockchain technology. They highlighted 5 areas of application for blockchain in real estate include Smart contracts, costs, speed, immutable record keeping and transparency. On the other hand they mentioned that trust and support from stake-holders must be developed, getting correct information migrated into blockchain is a challenge that must be resolved and legal context for transactions is required to adopt in international.

All these articles are focused on the blockchain, which is considered to have great potential for the financial sector, especially for crowdfunding. However, the most important issue that is not discussed in these articles is how to add legal clauses in crowdfunding based on blockchain technology by ensuring reducing the cost of real estate for the customer and getting correct information from all the participants of the blockchain system. The general idea of this article is to integrate blockchain technology in the real estate sector to optimize the work tasks and especially minimize the costs, duration of each transaction performed and to add legal clauses to this system. The proposed approach is to design a real estate financing process based on crowdfunding and blockchain, in addition to that, having a kind of traceability and transparency of the transactions made. To deploy this solution, we have developed a web platform of Dapp type (decentralized application) accessible by customers to buy or hire a property even if they do not have sufficient financial funds.

3 Proposed Model

In our proposed system, we must choose to adopt a specific and intelligent contractual relationship between all the actors who are part of the blockchain network, by specifying their position, whether it is the investment fund, the final client, the investor, or the real estate promoter. Smart contracts will be used in each of the operations of fundraising from investors, buying from customers, and also subcontracting with real estate promoters.

1. Investors will choose the project that is suitable for them to invest their funds, and they will specify the amount of their participation. The crowd-funding platform will automatically create a contract "Modarabah [16]".
2. Once we have raised the necessary funds for the investment, the investment fund will put out a call for offers on our website to attract customers who wish to hire or buy a house there.
3. The customer chooses the desired house from the crowdfunding platform. After that he will choose the type of contract for example "Murabaha [17]", "Ijara Montahiya bitamlik [18]", or a simple ordinary sales contract.
4. Once the investment fund is collected more than half of the clients' purchase commitment, construction will be started. The real estate investment fund will issue a call

for applications, providing potential applicants with all the information and details about the project.
5. The real estate promotor look for projects that are in accordance with his budget, his estimates of bearable costs, and the ranking of his company. In order to start the house building works, the investment fund will set up an "Istisnaa [19]" with the real estate promotor as subcontracting contract that will manage the operation of the house building works.
6. Once the project has reached the final state, the promoter will hand over the project to the investment fund and the procedure of sale and transfer of ownership to the final clients will start under the agreed conditions.
7. The profits resulting from this sale transaction will be transferred to the real estate fund, which will then transfer the capital plus the profits to the investors, in accordance with the agreed return percentages at the time of sale.

4 Result and Discussion

In this section, we present a comparative study between the cost of financing through traditional financing methods and financing through blockchain technology. We present this example to have better understanding of our model: We want to have an apartment of 70 m2 with an acquisition price of 800 000 MAD.

The acquisition cost using the real estate credit

1. Property price: 800 000 MAD
2. Notary fees: 8400 MAD
3. File fees: 800 MAD
4. Interest: 414 687 MAD
5. Loan insurance: 59 200 MAD

Total cost: 1 283 087 MAD

240 monthly payments of 5233 MAD with a total amount due of 1 256 005 MAD (i.e. a profit margin of 456 005 MAD)

The acquisition cost using the client' funds

1. Property price: 16000MAD
2. Registration fee: 8400 MAD
3. Notary: 8000 MAD
4. File fees: 700 MAD
5. Agency fees: 20 000 MAD

Total cost: 844 700 MAD

The acquisition cost using Blockchain is 800 000 MAD

By comparing the total cost of acquisition using credit and the total cost using blockchain we can conclude that the percentage of gain for the customeris 38%. From the results of this comparative study, we can conclude that it is totally clear that the integration of blockchain technology in the real estate financing process is very relevant on the side of reducing the cost of financing housing by eliminating any type of intermediation and additional purchase cost.

5 Conclusion

Blockchain has proven its effectiveness in different sectors. It has proposed new way of designing financial procedures without intermediaries and raises the level of trust between the various stakeholders, based on transparency and traceability of transactions, using strong algorithms of cryptography to guarantee the security and consistency of the financial system. The general idea of this article is to try to solve the problems faced in real estate crowdfunding in Morocco, such as the lack of insurance and traceability of transactions, and adding a legal clause to improve trust of participants. The proposed solution is the implementation of the financial model of real estate acquisition based on blockchain technology and crowdfunding by guaranteeing transparency, security, decentralization, and facilitating access to information for all actors who are part of our network. Our interest is also minimizing the cost of financing real estate in Morocco whether it is to invest for people who have more money or to purchase a home for the end customer.

References

1. Gai, K., Qiu, M., Sun, X.: A survey on FinTech. J. Netw. Comput. Appl. **103**, 262–273 (2018). https://doi.org/10.1016/j.jnca.2017.10.011
2. Bollaert, H., Lopez-de-Silanes, F., Schwienbacher, A.: Fintech and access to finance. J. Corp. Finan. **68**, 101941 (2021)
3. Shrimali, B., Patel, H.B.: Blockchain state-of-the-art: architecture, use cases, consensus, challenges and opportunities. J. King Saud Univ. Comput. Inf. Sci., S131915782100207X (2021). https://doi.org/10.1016/j.jksuci.2021.08.005
4. Hewa, T., Ylianttila, M., Liyanage, M.: Survey on blockchain based smart contracts: applications, opportunities and challenges. J. Netw. Comput. Appl. **177**, 102857 (2021)
5. Cai, W., Wang, Z., Ernst, J.B., Hong, Z., Feng, C., Leung, V.C.: Decentralized applications: the blockchain-empowered software system. IEEE Access **6**, 53019–53033 (2018)
6. Mohanta, B.K., Panda, S.S., Jena, D.: An overview of smart contract and use cases in blockchain technology. In: 2018 9th International Conference on Computing, Communication and Networking Technologies (ICCCNT), pp. 1–4. IEEE, July 2018
7. Cai, W., Polzin, F., Stam, E.: Crowdfunding and social capital: a systematic review using a dynamic perspective. Technol. Forecast. Soc. Change **162**, 120412 (2021). https://doi.org/10.1016/j.techfore.2020.120412
8. Böckel, A., Hörisch, J., Tenner, I.: A systematic literature review of crowdfunding and sustainability: highlighting what really matters. Manag. Rev. Q. **71**(2), 433–453 (2020). https://doi.org/10.1007/s11301-020-00189-3
9. Alhammad, M.M., AlOthman, R., Tan, C.: Review of crowdfunding regulations across countries: a systematic review study. J. Inf. Syst. Eng. Manag. **6**(4), em0145 (2021). https://doi.org/10.21601/jisem/11395

10. Belleflamme, P., Omrani, N., Peitz, M.: The economics of crowdfunding platforms. Inf. Econ. Policy **33**, 11–28 (2015)
11. Morena, M., Truppi, T., Pavesi, A.S., Cia, G., Giannelli, J., Tavoni, M.: Blockchain and real estate: Dopo di Noi project. Prop. Manag. **38**(2), 273–295 (2020). https://doi.org/10.1108/PM-01-2019-0005
12. Kumar, P., Dhanush, G.A., Srivatsa, D., Nithin, A., Sahisnu, S.: A buyer and seller's protocol via utilization of smart contracts using blockchain technology. In: Luhach, A.K., Jat, D.S., Hawari, K.B.G., Gao, X.-Z., Lingras, P. (eds.) ICAICR 2019. CCIS, vol. 1075, pp. 464–474. Springer, Singapore (2019). https://doi.org/10.1007/978-981-15-0108-1_43
13. Nik Ahmad, N.A., Syed Abdul Rahman, S.A.H.: Applying Ethereum smart contracts to blockchain-based crowdfunding system to increase trust and information symmetry. In: 2021 7th International Conference on Computer Technology Applications, pp. 53–59, July 2021
14. Garcia-Teruel, R.M.: Legal challenges and opportunities of blockchain technology in the real estate sector. J. Prop. Plan. Environ. Law (2020)
15. Tilbury, J.L., de la Rey, E., van der Schyff, K.: Business process models of blockchain and South African real estate transactions. In: 2019 International Conference on Advances in Big Data, Computing and Data Communication Systems (icABCD), pp. 1–7. IEEE, August 2019
16. Rammal, H., Gulzar, G., Graeme, P.: Mudaraba in Islamic finance: principles and application. Doctoral dissertation, Franklin Publishing Company (2003)
17. Miah, M.D., Suzuki, Y.: Murabaha syndrome of Islamic banks: a paradox or product of the system? J. Islam. Account. Bus. Res. (2020)
18. Fares, L.I., Habachi, M.: Contrat Ijara: cadre juridique et comptable marocain et rapprochement avec le crédit-bail. Recherches et Applications en Finance Islamique (RAFI) **4**(2), 230–251 (2020)
19. Al-Bashir, M., Al-Amine, M.: Istisna and its application in Islamic Banking. Arab LQ **16**, 22 (2001)

Vigenère Implemented in Two Chaotic Feistel Laps for Medical Images Encryption Followed by Genetic Mutation

Abid Abdellah[1](✉), Mariem Jarjar[1], Abdellhamid Benazzi[1], Abdellatif Jarjar[1], Youness Qobbi[1], and Samir El Kaddouhi[2]

[1] MATSI Laboratory Mohamed First University, Oujda, Morocco
abdellahab90@gmail.com

[2] IEVIA Team, IMAGE Laboratory, Department of Sciences, Ecole Normale Supérieure, Moulay Ismail University of Meknès, Meknes, Morocco

Abstract. In this work, we will propose a new method for medical image encryption based on Vigenère injection into two greatly improved Feistel rounds, followed by a chaotic mutation applied to the integrity of the vector coming out of the first encryption stage. Chaotic maps most commonly used in cryptography were used to create this design. After vectorizing the original image, two improved Feistel circuits will be used by implementing a new Vigenère replacement function using two S-boxes generated from the chaotic maps. A genetic mutation will be applied to the resulting vector at the end of the two rounds. Simulations performed on a large number of images of different sizes and formats guarantee that our approach is not subject to known attacks.

Keywords: Image encryption · New Vigenère's replacement functions · Feistel cipher · Chaotic map · Chaotic genetic mutation · Encryption function

1 Introduction

Many standards, including Hill [1, 2], César, Vigenère [3, 4] and Feistel [5, 6], have been improved by the use of chaotic maps.

1.1 The Classical Feistel Technique

The classical Feistel [7–10] technique is described by Eq. (1).

$$\begin{cases} G_i' = D_i \\ D_i' = G_i \oplus f(D_i) \end{cases} \quad (1)$$

Y. Farhaoui et al. (Eds.): ICAISE 2022, LNNS 635, pp. 824–830, 2023.
https://doi.org/10.1007/978-3-031-26254-8_120

1.2 Vigenère Classical Technique

The classical Vigenère [8] method is based on a static matrix (V), the encryption and decryption functions are defined by Eq. (2)

$$\begin{cases} C_i = V(P_i, K_i) = (P_i + K_i) \ \ mod\ 26 \\ P_i = V(C_i, K_i) = (P_i - K_i) \ \ mod\ 26 \end{cases} \quad for\ each\ i\ in\ [1 \ldots l] \tag{2}$$

With (P) : plain text, (C) : cypher text; (K) : Encryption key, (V) Vigenere matrix and (l) : length of clear text.

1.3 Genetic Mutation

A genetic mutation [9] occurs when two DNA genes switch nucleotides from a randomly determined place to a stop codon specified by RNA. In our case, the mutation is a little change in the rank of two blocks.

1.4 Our Contribution

Our contribution is based on injecting a new Vigenère substitution function ensured by large dynamic replacement S-box in two improved Feistel laps with a new linear transformation provided by a large size matrix, which increases the number of possibilities of brutal attack. Finally, a chaotic mutation will be applied to the integrity of the image.

2 The Proposed Method

Based on chaos, our new method is based on the use of the two chaotic maps most used in cryptography, namely the tent map and the PLWCM map, which are generated by a large secret key.

2.1 Chaotic Vectors Design

Our work requires the construction of five chaotic vectors $(PL1)$, (PL2), (PL3), (PL4) and (PL5) with a coefficient of (G_{256}) using the skew Tente map and PLWCM map.

2.2 Encryption Process

Original Image Vector Transition. Prior to the application of the proposed method an operation of vectorization of the original image of size (n,m,3) into a vector (X) of size (3nm) is necessary, then a decomposition of the vector (X) into (3m) blocks of size (n) each is established. We admit that (m) is even.

Advanced Vigenère Methods. This technique requires the establishment of two substitution matrices $(SB1)$ and $(SB2)$ of size (256, 256) (algorithm 1), through the process described by the following steps

- a permutation ($Q1$) obtained by descending ordering the first $256 values$ of the sequence ($PL1$)
- a permutation ($Q2$) obtained by increasing the ordering the first $256 values$ of the sequence ($PL2$),

Algorithme 1 : substitution matrices design	

Input: Q1, Q2, PL3, PL4
Output: SB1, SB2
Begin
For i=1 **to** 256
$SB1(1, i) = Q1(i)$
$SB2(1, i) = Q2(i)$
Endfor
	For i =2 **to** 256
	For j=1 **to** 256
	$SB1(i,j) = SB1(i-1, mod(j + PL3(i),256))$
	$SB2(i,j) = SB2(i-1, mod(j + PL4(i),256))$
	Endfor
	Endfor
	End

The Eq. (3) illustrates the effective expression of the $Y(i)$ image of the pixel through the new advanced Vigenère (AV) method.

$$AV(X(i)) = Y(i) = SB1(PL1(i), SB2(PL2(i), X(i) \oplus PL3(i))) \quad (3)$$

Enhanced Feistel Methods. The Eq. (4) illustrates the effective expression of the $(B_i^{'}, B_{i+1}^{'})$ image of the block (B_i, B_{i+1}) by one round of our enhanced feistel method.

$$FT(B_i, B_{i+1}) = \left(B_i^{'}, B_{i+1}^{'}\right) = \begin{cases} B_i^{'} = B_{i+1} \\ B_{i+1}^{'} = AVB(B_i) \oplus f_i(B_{i+1}) \end{cases} \quad (4)$$

With (f_i) a function described by Eq. (5)

$$f_i(D_i(j)) = M(i,j) \oplus D_i(j) \text{ where } j \text{ in } [1 \ldots n] \quad (5)$$

(M) is chaotic matrix of size ($3m, n$) genereted by reshaping the chaotic vector ($PL5$). And (AVB) is Vigenère function for a block of size (n) defined by Eq. (6) and (7):

If we note

$$Y = AVB(X) \quad (6)$$

So

$$Y(j) = AV(X(j)) \forall j in [1..n] \quad (7)$$

The Genetic Mutation. After two rounds of feistel the blocks will be concatenated into a single vector (Y) of size (3nm). Then, it will be subdivided into (m) blocks of size (3n). Then, a genetic mutation operation is applied on the (m) blocks. This operation consists in inverting some elements controlled by the vector (VP) defined by Eq. (8)

$$VP(i) = mod\left(\sum_{j=1}^{i+5} PL2(i), E(\frac{n}{4})\right) \text{ where } i \text{ in } [1, m] \tag{8}$$

The operation of the inversion on a block B_i is described by the Eq. (9)

$$B_i^{'}(j) = B_i(j) \oplus 255 \text{ where j in [VP(i)} \ldots \text{3n]} \tag{9}$$

The Diffusion Function. At the end of each Feistel round, a CBC scheme is used to link the encrypted element with the next original pixel using the Eq. (10)

$$\begin{cases} Z(1) = IV \oplus Y(1) \oplus PL5(1) \\ Z(i) = Z(i-1) \oplus Y(i) \oplus PL5(\text{i}) \text{ } for \text{ } i \text{ } in[2 \ldots 3nm] \end{cases} \tag{10}$$

with IV an initial value calculated by applying the XOR operation on the Y vector excluding the first pixel.

2.3 Decryption of Encrypted Image

Our algorithm is a symmetric encryption system, so the same key will be used in the decryption process by applying the reverse encryption operations in the reverse order.

3 Examples and Simulations

3.1 Statistics Attack Security

Simulation and Histogram Analysis. We notice that the cipher image is visually different from the clear image and has no similarity (Fig. 1).

we notice that all the histograms of the images encrypted by our algorithm are uniformly distributed which ensures better protection against any statistical attack (Fig. 1).

Entropy Analysis. The entropy of an image of size (n, m) is given by Eq. (11)

$$H(MC) = \frac{1}{t}\sum_{i=1}^{t} - p(i)log_2(p(i)) \tag{11}$$

$p(i)$ is the probability of occurrence of level (i) in the original image.

Correlation Analysis. The correlation of an image of size (n, m) is given by Eq. (12)

$$r = \frac{cov(x, y)}{\sqrt{V(x)}\sqrt{V(y)}} \tag{12}$$

We noticed that the entropy of all images tested by our algorithm is close to 8 (Table 1).

All the correlation measures of the images tested by our system are very close to zero. This can protect our methods from statistical attacks (Table 1).

3.2 Differential Analysis

To assess the algorithm's performance against differential attacks, the Number of Pixels Change Rate (NPCR), Unified Average Changing Intensity (UACI), and avalanche effect are used.

The NPCR Constant. It is determined by Eq. (13), (14)

$$NPCR = \left(\frac{1}{nm}\sum\nolimits_{i,j=1}^{nm} D(i,j)\right) * 100 \tag{13}$$

$$WithD(i,j) = \begin{cases} 1 & if \quad C_1(i,j) \neq C_2(i,j) \\ 0 & if \quad C_1(i,j) = C_2(i,j) \end{cases} \tag{14}$$

The UACI Constant. The *UACI* mathematical analysis of an image is given by Eq. (15)

$$\text{UACI} = \left(\frac{1}{255nm}\sum\nolimits_{i,j=1}^{nm} Abs(C_1(i,j) - C_2(i,j))\right) * 100 \tag{15}$$

Avalanche Effect. The mathematical expression of this avalanche effect is given by Eq. (16)

$$\text{AE} = \left(\frac{\sum_i bit\ change}{\sum_i bit\ total}\right) * 100 \tag{16}$$

(Table 1) shows that the differential parameters results obtained in the desired ranges (NPCR close to 100%, UACI > 33% and EA > 50%). This can ensure that our method resists differential attacks.

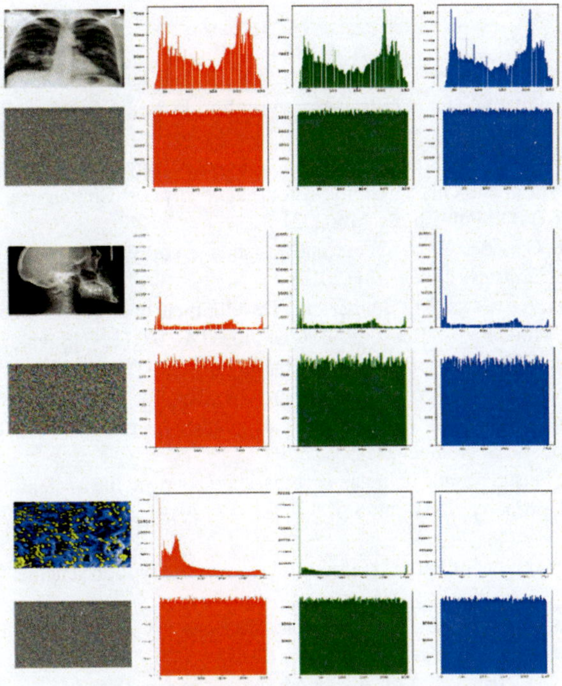

Fig. 1. Plain and cipher images histogram

Table 1. Correlation, Entropy, NPCR, UACI and EA of some tested images

Name	Vertical correlation	Horizontal correlation	Diagonal correlation	Entropy	NPCR (%)	UACI (%)	EA (%)
Img1	0.00024	0.00031	0.00033	7.99974	99.6034	33.4838	50.0138
Img2	-7.0663e-05	-0.00066	-0.00021	7.99885	99.6213	33.5232	50.0402
Img3	0.00060	-0.00019	-0.00118	7.99966	99.6133	33.4381	50.0214

4 Conclusion

Chaos theory has taken its place in cryptography due to the high sensitivity to the initial state of dynamic systems. Based on chaos, this new technology relies on two Feistel rounds implementing a new Vigenère replacement function provided by two substitution matrices generated from the most used chaotic maps in the world of image encryption. The F function of the Feistel network has been replaced by a linear transformation provided by a chaotic matrix. The mutation applied at the end of these two rounds has improved the statistical and differential parameters while increasing the complexity of the attack of our system.

References

1. Hraoui, S., Gmira, F., Abbou, M.F., Oulidi, A.J., Jarjar, A.: A new cryptosystem of color image using a dynamic-chaos hill cipher algorithm. Procedia Comput. Sci. **148**, 399–408 (2019)
2. Ibada, A.J., Ehkan, P., Ngadiran, R., Hammood, D.A., Alkhayyat, A.: RGB image encryption using Hill algorithm and chaos system. In Journal of Physics: Conference Series, vol. 1962, No. 1, p. 012061. IOP Publishing, July 2021
3. Zhu, S., Wang, G., Zhu, C.: A secure and fast image encryption scheme based on double chaotic S-boxes. Entropy **21**(8), 790 (2019)
4. Qobbi, Y., Jarjar, A., Essaid, M., Benazzi, A.: Development of large chaotic S-boxes for image encryption. In: Motahhir, S., Bossoufi, B. (eds.) ICDTA 2021. LNNS, vol. 211, pp. 847–858. Springer, Cham (2021). https://doi.org/10.1007/978-3-030-73882-2_77
5. Shams Mahmoud Abd Ali » Novel Encryption Algorithm for Securing Sensitive Information Based on Feistel Cipher"Test engeenering managenment Page Number: 10 - 16 Publication Issue:19 Volume: 80 September-October 2019
6. JarJar, A.: Two Feistel rounds in image cryptography acting at the nucleotide level exploiting DNA and RNA property. SN Appl. Sci. **1**(11), 1–17 (2019). https://doi.org/10.1007/s42452-019-1305-7
7. JarJar, A.: Improvement of Feistel method and the new encryption scheme. Optik **157**, 1319–1324 (2018)
8. Jarjar, M., Hraoui, S., Najah, S., Zenkouar, K.: New cryptosystem using two improved vigenere laps separated by a genetic operator (2022)
9. Niu, Y., Zhou, Z., Zhang, X.: An image encryption approach based on chaotic maps and genetic operations. Multimedia Tools Appl. **79**(35–36), 25613–25633 (2020). https://doi.org/10.1007/s11042-020-09237-2
10. Zhang, X., Zhou, Z., Niu, Y.: An image encryption method based on the feistel network and dynamic DNA encoding. IEEE Photonics J. **10**(4), 1–14 (2018)

Word Embedding Methods of Text Processing in Big Data: A Comparative Study

Lahcen Idouglid(✉) and Said Tkatek

Computer Sciences Research Laboratory, Faculty of Sciences, IBN Tofail University, Kenitra, Morocco
{lahcen.idouglid,said.tkatek}@uit.ac.ma

Abstract. One of the biggest challenges any NLP data scientist faces is choosing the best numeric/vector representation of a text string for running a machine learning model. This research paper provides a comprehensive study of big data and its impact on improving the performance of word embedding techniques for text processing. Therefore, we propose a method for text processing using word embedding techniques and machine learning (ML) algorithms to improve the performance of analyzing data and decision-making. For this reason, it is possible to use several word embedding methods for text processing, especially the most popular ones like CountVectoriser, TF-IDF, and HashingVectorizer, and combine them with ML algorithms like decision trees, random forest classifiers, and logistic regression for implemented text Regression and other supervised machine learning algorithms are combined with classification. Compared with recent work, our comparative study shows the impact of dataset size on the performance of text classification algorithms and gives good results.

Keywords: Word Embedding · NLP · Big Data · Machine Learning

1 Introduction

The big data term refers to vast, complex, and real-time data that necessitates sophisticated management, analytical, and processing approaches to extract management insights. There are a little of people who can think critically about big data problems and who have the skills and knowledge to tackle big data problems [14].

Machine learning (ML) is a science focused on understanding and developing learning algorithms [13]. It is considered to be a component of artificial intelligence [7]. Without being expressly taught to do so, machine learning algorithms create a model using sample data, also referred to as training data, to make predictions or judgments.

The study of theories and techniques that enable information exchange between humans and computers through natural language is known as "natural language processing" [9]. NLP combines linguistics, computer science, and mathematics [2]. NLP is an area of AI that aids computers in manipulating and interpreting human languages. Text mining is a technique used in NLP to extract relevant information from text. The goal of NLP is to glean knowledge from natural language. One of the most common

Y. Farhaoui et al. (Eds.): ICAISE 2022, LNNS 635, pp. 831–836, 2023.
https://doi.org/10.1007/978-3-031-26254-8_121

document vocabulary representations is word embedding. It can record a document's context, the semantic and syntactic similarity of its words, their relationships to one another, etc.[2].

The paper is organized as follows. Section 2 contains related work Sect. 3 discusses methodologies and the system architecture, we present the word embedding techniques and algorithms classifiers used in this work. The experimental results are discussed in Sect. 4. Finally, the conclusion is presented in Sect. 5.

2 Related Work

This section introduces related work from methods of evaluating word embeddings, and existing studies of evaluating embeddings in downstream tasks.

The performance of word embedding approaches was compared, examined, and evaluated using machine learning algorithms in a Turkish sentiment analysis study that was conducted using a dataset derived from user comments shared on multiple shopping websites in recent years. This study will serve as our starting point for the discussion of word embeddings [3]. The second study named "BnVec: Towards the Development of Word Embedding for Bangla Language Processing" approaches for Bengali word embedding. Six well-known word embedding techniques CountVectorizer, TF-IDF, Hash vectorizer, Word2vec, fastText, and GloVe are included in the first one, which highlights their well-known functionality [9]. Various qualitative and quantitative tests were conducted on a few tasks to show each technique's ability to infer word proximity and, in addition, to examine how well it performs in comparison to the other word embedding techniques [9]. The third work is about Sentiment analysis on film review in Gujarati language using machine learning this paper is a comparative study between TF-IDF vectorizer and CountVectorizer features after applying sentiment analysis [12]. Comparing the results of two different machine learning algorithms based on Accuracy, Recall, Precision, and F-score performance parameter. The last work cited is "Measuring associational thinking through word embeddings" it's an investigating various way of incorporating existing embeddings to decide the semantic or non-semantic acquainted strength between words so relationship with human decisions can be augmented [11].

3 Methodology

3.1 System Architecture

There are four steps for the text classification process and initial steps regarding collecting and preparing the datasets. Preprocessing techniques play an important role to improve the performance of the models. Three key steps of data preprocessing namely, tokenization, stop words removal and stemming. The technique of tokenizing involves separating a stream of text into recognized tokens, such as words, phrases, symbols, or other practical elements. Tokenization's objective is to analyze each word in a statement. Stemming is a technique for getting a word's numerous forms to look similar to its stems. Bag of Words is one of the most popular methods. It is a text representation that indicates the order of words in a text. After Stemming and Lemmatization, the step that becomes is the division of the Data Set into Train Set and Test Set.

3.2 Word Embedding Techniques:

TfidfVectorizer: Each word is represented as a weighted vector of the terms discovered in the super vector after the most significant terms in the super vector have been selected. Each document assigns a weight to each word. It determines the significance of a word in a corpus document. It's calculated by multiplying the term frequency (TF) by the document frequency's inverse (IDF). TF counts the number of times a phrase appears in a document, while IDF counts the phrase's significance concerning the corpus as a whole [1].

$$TF - IDF = TF(t) * IDF(t) \quad (1)$$

CountVectorizer: Count Vector is a simple yet incredibly effective method used in language processing [9]. The number of information components (N) and the number of unit components (M) present in the information components are combined to create an N * M grid [6]. The frequency of the unit elements supplied in each data component serves as a representation of that component [9]. Text is transformed into a vector by marking the presence (1) or absence (0) of a word of a given input [12].

HashingVectorizer: The hashing trick is used by the hashing vectorizer to identify the mapping from the token string name to the feature integer index. This vectorizer converts text documents into matrices by creating sparse matrices out of the collection of documents that contain the token occurrence counts [5].

3.3 Text Classifiers

In this paragraph, we present the machine learning algorithms implemented in our work.

Decision tree Classifier: A decision tree is an induction approach that has been applied to a variety of classification problems. It is based on separating features and determining their worth [4]. The splitting procedure continues until each branch can only have one classification labeled on it. It generates decision trees from random data samples, assigns expectations to each tree, and chooses the best solution.

Random Forest Classifier: The Random Forest (RF) decision tree ensemble is a well-known decision tree ensemble that is often utilized in categorization. The popularity of RF stems from its superior performance when compared to other classification methods [8].

Logistic regression Classifier: Is one of the most often used classification techniques. It's employed in a variety of fields since it's easy to understand and the results are interpretable, allowing for what-if scenarios. It's a classification method based on the Bayes Theorem and the assumption of predictor independence [10].

4 Results and Discussion

In this experiment, we evaluated the classification results, based on standard evaluation metrics of accuracy that were used to compare the state-of-the-art three machine learning algorithms, namely Decision Tree, Random Forest classifier, Logistic Regression, and their compatibility with three methods of Word Embedding, namely Tf-IDF Vectorizer, ContVectorizer, and HashingVectorizer. In the first part of this comparative study, we will see the evolution of the accuracy of the algorithms regarding the size of the dataset, for this, we will take parts of our example dataset (2000, 5000, 20000, 50000, 150000 inputs) from my dataset. And we will present all this in graphs.

The Dataset was downloaded from GitHub "Large Question Answering Datasets"; All experiments are realized, executed and tested on Google Collaboratory.

Table 1. Comparative Accuracy of ML algorithms with word embedding techniques for Text Processing

ML Algorithms	Dataset Size	Word Embedding Methods		
		Tf-idf Vectorizer	ContVectorizer	Hashing Vectorizer
Decision Tree	2000	82	82	97,5
	5000	82,6	82,6	98,1
	20000	84,42	84,42	97,88
	50000	85,47	85,48	97,77
	150000	80,26	80,26	97,47
Random Forest Classifier	2000	82	82	98,5
	5000	82,50	82,50	98,1
	20000	84,45	84,48	98,32
	50000	85,32	85,33	98,32
	150000	80,26	80,60	98,02
Logistic Regression	2000	78,75	82	97
	5000	77,70	81,70	97,20
	20000	78,85	82,82	97,75
	50000	80,12	83,86	97,7
	150000	78,76	70,59	97,49

This table has shown the results of the comparative study used the most famous machine learning algorithms with such us Decision Tree, Random Forest Classifier and Logistic Regression with Tf-idf Vectorizer, ContVectorizer and Hashing Vectorizer word embedding methods. The evaluation process is conducted using evaluation parameters such as accuracy. The Hashing Vectorizer word embedding method archived the highest scores 98,5% in term of accuracy in Random Forest Classifier ML algorithm with dataset size 2000.

The graphical representation (a) shows the evolution of the accuracy of the TfidfVectorizer method with the number of entries in the training DataSet. As the graph shows, we conclude that the accuracy increases with increasing number of entries in the

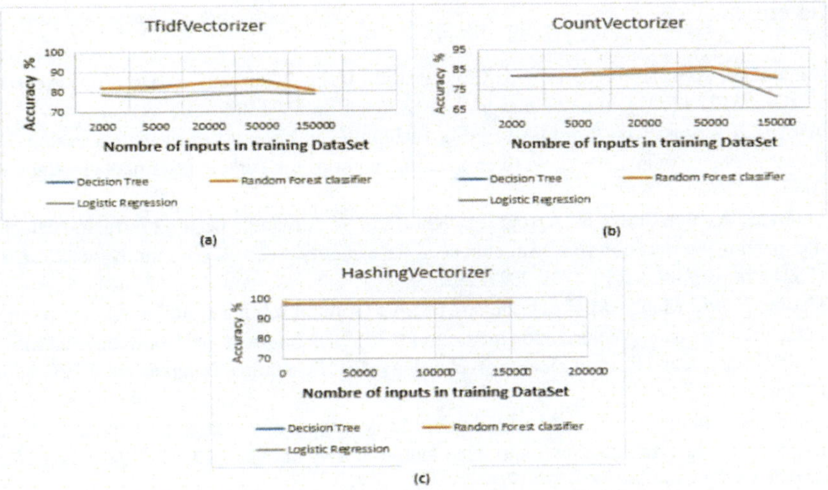

Fig. 1. Performance comparison for word embedding methods

training dataset and we arrive at the best result when the size of our training dataset is 5000 entries with the best performance obtained for this model 85%.

The second graph named (b) shows *CountVectorizer performance evolution with increasing dataset size,* we conclude that the accuracy of CountVectorizer evolves and increases with the increase in the number of inputs in the training Dataset, but there is a degradation when the Dataset size is 50000 with the best performance achieved for this model 85,47%. And the best algorithm that gives good result are DT and RF.

As shown in the graph (c) that presente *HashingVectorizer performance evolution with increasing dataset size,* we conclude that the accuracy evolves and increases with the increase in the number of inputs in the Training Dataset and we arrive at the best result when the size of our training Dataset is 15000 inputs with the best performance achieved for this model 98%. And this method HashingVectorizer work efficacelly with all algorithms tested.

5 Conclusion

This work presents the result of a comparative study of Word Embedding Methods for text processing as Tf-idf Vectorizer, ContVectorizer and Hashing Vectorizer.

The results show the impact and influence of big data and the size of training data on the performance of machine learning algorithms with the three Word Embedding Methods. When more than the dataset size and larger the performance becomes higherThe performance achieved for our model 98%, and the best method is HashingVectorizer witch work efficacelly with all algorithms tested. Int the futures work we will research the similarity semantic for text categorization using genetics algorithms in subsequent study. We'll also create a larger model for several languages.

References

1. Alammary, A.S.: Arabic Questions Classification Using Modified TF-IDF. IEEE Access **9**, 95109–95122 (2021). https://doi.org/10.1109/ACCESS.2021.3094115
2. Al-Ansari K Survey on Word Embedding Techniques in Natural Language Processing. 7
3. Aydoğan, M.: Comparison of word embedding methods for Turkish sentiment classification. 21
4. Guezzaz, A., Benkirane, S., Azrour, M., Khurram, S.: A reliable network intrusion detection approach using decision tree with enhanced data quality. Secur. Commun. Netw. **2021**, 1–8 (2021). https://doi.org/10.1155/2021/1230593
5. Haque, F., Md Manik, M.H., Hashem, M.M.A.: Opinion mining from bangla and phonetic bangla reviews using vectorization methods. In: 2019 4th International Conference on Electrical Information and Communication Technology (EICT), Khulna, Bangladesh. IEEE, pp. 1–6 (2019)
6. Haque, R., Islam, N., Islam, M., Ahsan, M.M.: A comparative analysis on suicidal ideation detection using NLP, machine, and deep learning. Technologies **10**, 57 (2022). https://doi.org/10.3390/technologies10030057
7. Hilbert, S., et al.: Machine learning for the educational sciences. Rev. Educ. **9**(3), e3310https://doi.org/10.1002/rev3.3310
8. Jain, A., Sharma, Y., Kishor, K.: Financial administration system using ML Algorithm. 10
9. Kowsher, M., et al.: BnVec: towards the development of word embedding for Bangla language processing. IJET **10**, 95 (2021). https://doi.org/10.14419/ijet.v10i2.31538
10. Mahesh, B.: Machine Learning Algorithms - A Review **9**, 7 (2018)
11. Periñán-Pascual, C.: Measuring associational thinking through word embeddings. Artif. Intell. Rev. **55**(3), 2065–2102 (2021). https://doi.org/10.1007/s10462-021-10056-6
12. Shah, P., Swaminarayan, P., Patel, M.: Sentiment analysis on film review in Gujarati language using machine learning. IJECE **12**, 1030 (2022). https://doi.org/10.11591/ijece.v12i1.pp1030-1039
13. Tkatek, S.: A hybrid genetic algorithms and sequential simulated annealing for a constrained personal reassignment problem to preferred posts. IJATCSE **9**, 454–464 (2020). https://doi.org/10.30534/ijatcse/2020/62912020
14. Tkatek, S., Belmzoukia, A., Nafai, S., Abouchabaka, J., Ibnou-ratib, Y.: Putting the world back to work: an expert system using big data and artificial intelligence in combating the spread of COVID-19 and similar contagious diseases. WOR **67**, 557–572 (2020). https://doi.org/10.3233/WOR-203309

Yolov2 Implementation and Optimization for Moroccan Traffic Sign Detection

Imane Taouqi[1] (✉), Abdessamad Klilou[1], Kebir Chaji[1], and Assia Arsalane[2]

[1] MiSET Team, Faculty of Sciences and Technologies, Sultan Moulay Slimane University, Beni Mellal, Morocco
imane.taouqi@gmail.com

[2] High School of Technologies, Sultan Moulay Slimane University, Beni Mellal, Morocco

Abstract. Image and video processing is a major contributor of object's detection, since only video sensor can analyze a lot of visual information specific of road scene. Notably road signs identification in images. Which is an important issue for Advanced Driver Assistance Systems (ADAS system), seeing that the result of detection is given to another ADAS modules i.e., planning (behavior planning and prediction), to make decision, thus the accurate and real time detection is required for vehicle safety, limitation of accidents and road management applications. The purpose of this study is to implement an algorithm for Moroccan traffic sign detection, what for solving the problems that appear in the road scene such as an occlusion, changes in luminosity and response on real time. We have proposed and optimized an implementation of the Yolov2 algorithm based on convolutional neural network (CNN) for Moroccan traffic sign. The obtained results are a precision of 100% for stop and mandatory signs, and more than 65% for the other signs, with a frame per second FPS of 10.

Keywords: Traffic sign detection · Recognition · yolov2 · CNN · Realtime · MKL-DNN

1 Introduction

Vision and image processing is the predominant method used by ADAS system to understand the on-road environment, detection of objects, and especially the detection of traffic sign. Then, the main objective of the research is to avoid false detections that may lead to a wrong alert. Recently, many algorithms of detection and recognition of traffic sign have been proposed in the literature. These algorithms must provide a good precision while detecting despite the challenges that may appear in road scene such as changes in luminosity, occlusion, presence of multiple signs in a frame and detection of inappropriate signs as road signs (e.g., commercial signs). Moreover, they must be adapted and trained on Moroccan traffic sign use case. In addition, detection must be embedded on a vehicle calculator and performed in real time, although the limited memory and processing resources. **Related work** [1], Sun et al. used Hough transform and Douglas-Peucker algorithm for detection and CNN for recognition, which

Y. Farhaoui et al. (Eds.): ICAISE 2022, LNNS 635, pp. 837–843, 2023.
https://doi.org/10.1007/978-3-031-26254-8_122

achieves 98.1% for triangular signs and 99.18% for circular signs, in [2] Islam used the masking process to detect red and blue signs, then CNN to classified them. However, both methods aren't robust enough, because the proposed detection method may generate false detection by detecting advertisement signs for example which are similar in shape and color to traffic signs. Wei et al. [3], used transfer learning based on Convolutional Neutral Network to detect stop sign, so the proposed method can only detect one class traffic sign. Shen et al. [4], used a group multi-scale attention module, that allows to focus only on the regions containing traffic sign. Tabernik et al. [5] proposed mask R-CNN with adaptation and data augmentation, both proposed methods have reached a good accuracy but there is a missing of real time notion, which is a necessary information for an embedded application. In this paper, we use optimized Yolov2 based on CNN algorithm to detect five traffic signs classes, to solve the above-mentioned problems. Then generate C/C++ code (which is a universal language, fast, optimizes memory and execution time) based on Library MKL-DNN that allows the optimization of the sign detection model for a real time execution on hardware. In summary, the major contributions in this Paper are:

- adaptation of the detection algorithm on Moroccan road scene.
- split the database into five classes based on their forms, for accurate detection if signs not included in training images. Those classes are described in the Table 1.
- optimize architecture of yolov2 to performed in real time.
- generate, optimize and test yolov2 C/C++ code using MKL-DNN and open cv.

The rest of this paper is organized as follow, in the first section, description of our algorithm, datasets, and models architecture, the second section, a global experimental results and discussions of our traffic sign detection algorithm using MATLAB and Visual Studio. In the end, a conclusion is presented to sum up this work.

2 Proposed Method

2.1 Architecture of Proposed Method

The detection algorithm proposed in this paper is Yolov2, which is both accurate and fast, you only look once (YOLO) like her name shows, it allows to do the prediction and categorization of the whole image bounding boxes simultaneously for all classes. It is useful for applications that require real-time speed like self-driving cars. The YOLO detector splits the Input image into an $S \times S$ grid. Each grid cell predicts bounding boxes based on Anchors boxes, beside its own confidence scores. Hence, all grids consist of five predictions: x, y, W, H, and confidence [6]. The coordinates (x, y) represent the center of the box, W is the width, H is the height, and the confidence prediction presents the IOU of the predicted box to ground truth box. However, each cell predicts probabilities of class C of bounding box. We proposed two yolov2 architecture, the first one is based on feature Extraction Network resnet50 (Residual Neural Network) [7], which has 48 Convolution layers. The second optimized proposed Architecture of yolov2 is shown in Fig. 1, composed of 6 convolution layers. Each architecture was trained with a database of Moroccan traffic signs.

Table 1. Classes of used traffic signs.

Classes	Exemples	Description
Mandatory		Rectangular& circular blue, white
yield		Triangular,Red, White
Prohibitory		Circular,Red, White
Warning		Triangular,Red, White
Stop		Red, White

Fig. 1. Second optimized Architecture of Yolov2.

Every convolution layer is followed up by batch normalization layer. Then activation.

Function Rectified Linear Unit (RELU) that replaces all negative values with zero to optimize and speed up the learning process. Besides, Maxpooling layer with stride equal to 2 reduce progressively the size of the input data (416x416). Finally, yolov2 layers to predict class, bounding box position and scores of detections. Before training, we estimated the anchors box from training datasets using k medoids clustering, which is more robust to outliers and noisy, and suggests the size (W, H) of the objects we'd detect for the best prediction of detected bounding boxes [8]. The training parameters are MaxEpochs = 80, MiniBatchSize = 16, Verbose Frequency = 50, learning rate = 0,001 and Optimization SGDM.

2.2 Datasets and Working Environment

In this paper, we use « The German Traffic Sign Detection Benchmark (GTSDB)» datasets [8]. Since their road signs are the same as ours, except for the yellow road sign and STOP sign. Then, we delete these unwanted images, and we add other images taken in the Moroccan road scene in different weather conditions, brightness, to adapt our database that contain 842 images. We divided the signs in 5 classes: stop, yield, mandatory, prohibitory, and warning. We use MATLAB version 2021a to label traffic sign datasets and train the Yolov2 detector, then generate the C/C++ code from yolov2 function. Finally, we use Visual studio version 2022 to test the generated C/C++ code. All this work is done in "Processor Intel(R) Celeron(R) CPU N3060 @ 1.60 GHz, 1601 MHz, 2 Core(s), 2 Logical Processor(s)".

2.3 Optimization of Generated Code C/C++

While we are generating code C/C++ from MATLAB, it is necessary to use a way to optimize the heavy computations of deep learning, in order to accelerate the detection process, especially convolutional neural networks. Among the existing optimization methods, we found the Math Kernel Library for Deep Neural Networks (MKL DNN), recently named ONE API: it's an open-source performance and optimization library from intel. That can optimize forward and backward convolution, normalization, and activation that exist in the used yolo architecture (Fig. 1) [9]. Also, it is used in implementing low accuracy numerical operations in DNN learning and inference on Intel's related hardware architectures [10].

3 Experimental Results

For each input image the trained model outputs a set of objects bounding boxes. The evaluation of our detector it's done by using Average Precision and Miss rate metrics, which are calculated for each category independently, result is exhibited in the Table 2 for both architectures. Notice that the detector is better if AP tends to 100% and Miss Rate leads to 0. The test proposed for detection methods on 29 images of traffic signs, reached a precision of 100% over four classes and execution time (25 s) using Resnet feature extractor. Then the optimized architecture gives a precision of 100% for stop and mandatory signs and more than 66% for the other signs, with a frame per second FPS of 10 (0,1s). This leads us to choose the optimized architecture for our work, that is both accurate and performed in real time. Testing optimized method on more images (74 images) gives the next results in the Table 3. That shows the precision of our algorithm despite a lot of images in road scene.

Rem: The use of yolov2 in our case is better than using YOLOv3 [11] in term of execution time, in other words calculation charge is much important with YOLOv3 than YOLOv2. Thus, embedded on a vehicle calculator can performed in real time.

Table 2. Evaluation of both architecture.

		stop	Warning	Mandatory	Yield	Prohibotory
Resnet	Accuracy	100%	100%	100%	100%	63%
	Miss Rate	0	0	0	0	0, 43
optimized	Accuracy	100%	67%	100%	88%	75%
	Miss Rate	0	0, 33	0	0, 13	0, 28

Table 3. Evaluation on more test images.

classes	Stop	Warning	Mandatory	Yield	Prohibotory
AP	90%	40%	86%	62%	82%
Miss rate	0, 1	0, 6	0, 12	0, 38	0, 17

3.1 Testing Using MATLAB

The obtained results of optimized yolov2 detector are shown in Fig. 2 for Mandatory sign class and Fig. 3 for stop sign. All those signs are contained in the training datasets. So, they are correctly localized and categorized in all tested images. While testing our detector on traffic signs (they are not included in training datasets), it tries to find their classes by looking for similitude with existing traffic signs. Because each class has its proprieties. However, our detector is capable to detect all those signs that are not contained in training datasets with accuracy of 70%. In other words, the proposed yolov2 detection has a good generalization detection ability. The evaluation considered different traffic-signs sizes and different party of the day. The results are presented in Fig. 4 and Fig. 5, when the detector localized correctly the warning and prohibitory signs respectively, with her bounding box and category traffic sign.

Fig. 2. Mandatory sign detection

Fig. 3. Stop sign detection

Fig. 4. Warning sign detection

Fig. 5. Prohibitory sign detection

3.2 Testing Generated C/C++ code Using Visual Studio

To test the generated code, it is important to note that MATLAB uses column format, while open cv uses row format. Thus, the stacking of an image's pixels start by R, G, B respectively column by column in MATLAB, and by B, G, R respectively row by row in open cv. The test in visual studio gave the same result. So, the main objective is to check out the generated code using open cv. This makes it easier and ensure prochain implementation in embedded board tasks. Therefore, the results founded are displaying in Fig. 6, 7 and 8 that's correctly localized the stop, warning, mandatory signs respectively and Fig. 9 where is the detection of both yield and mandatory signs.

Fig. 6. Stop sign detection

Fig. 7. Warning sign detection

Fig.8. Mandatory sign detection

Fig. 9. Yield and mandatory signs detection

4 Conclusion

It's clear that traffic sign detection and recognition is a needed component of ADAS system, that can limit the accidents in road scene, so the accurate results are important, especially the response in real time. As the research has been demonstrated, the work done in this paper reached a good accuracy and execution time. Therefore, we will be able to localize, categorize traffic sign in Morocco road scenes, and generate C/C++ code from MATLAB, also optimize the generated code using MKL-DNN then test it using visual studio to implement it in embedded card in future work.

References

1. Sun, Y., Ge, P., Liu, D.: Traffic sign detection and recognition based on convolutional neural network. In 2019 Chinese Automation Congress (CAC), Hangzhou, China, pp. 2851–2854 (Nov 2019). https://doi.org/10.1109/CAC48633.2019.8997240
2. Islam, M.T.: Traffic sign detection and recognition based on convolutional neural networks. In: 2019 International Conference on Advances in Computing, Communication and Control (ICAC3), Mumbai, India, pp. 1–6 (Dec 2019). https://doi.org/10.1109/ICAC347590.2019.9036784
3. Wei, L., Runge, L., Xiaolei, L.: Traffic sign detection and recognition via transfer learning. In: 2018 Chinese Control And Decision Conference (CCDC), Shenyang, pp. 5884–5887 (Jun 2018). https://doi.org/10.1109/CCDC.2018.8408160
4. Shen, L., You, L., Peng, B., Zhang, C.: Group multi-scale attention pyramid network for traffic sign detection. Neurocomputing **452**, 1–14 (2021). https://doi.org/10.1016/j.neucom.2021.04.083
5. Tabernik, D., Skocaj, D.: Deep learning for large-scale traffic-sign detection and recognition. IEEE Trans. Intell. Transport. Syst. **21**(4), 1427–1440 (2020). https://doi.org/10.1109/TITS.2019.2913588

6. Redmon, J., Divvala, S., Girshick, R., Farhadi, A.: You Only Look Once: unified, real-time object detection. In 2016 IEEE Conference on Computer Vision and Pattern Recognition (CVPR), Las Vegas, NV, USA, pp. 779–788 (Jun 2016). https://doi.org/10.1109/CVPR.2016.91
7. He, K., Zhang, X., Ren, S., Sun, J.: Deep residual learning for image recognition. In 2016 IEEE Conference on Computer Vision and Pattern Recognition (CVPR), Las Vegas, NV, USA, pp. 770–778 (Jun 2016). https://doi.org/10.1109/CVPR.2016.90
8. Valentyn, S.: Traffic Signs Detection by YOLO v3, OpenCV, Keras, In: Kaggle (2021). https://www.kaggle.com/code/valentynsichkar/traffic-signs-detection-by-yolo-v3-opencv-keras/data
9. Li, M., et al.: The deep learning compiler: a comprehensive survey. IEEE Trans. Parallel Distrib. Syst. **32**(3), 708–727 (2021). https://doi.org/10.1109/TPDS.2020.3030548
10. Azim, A.: Low precision arithmetic operations in deep neural networks: an overview, p. 7
11. Du, L., Ji, J., Pei, Z., Zheng, H., Fu, S., Kong, H., et al.: Improved detection method for traffic signs in real scenes applied in intelligent and connected vehicles. IET Intell. Transp. Syst. **14**(12), 1555–1564 (2020)

τJSONPath: A Temporal Extension of the JSONPath Language for the τJSchema Framework

Zouhaier Brahmia[1(✉)], Fabio Grandi[2], Safa Brahmia[1], and Rafik Bouaziz[1]

[1] University of Sfax, Sfax, Tunisia
zouhaier.brahmia@fsegs.rnu.tn, safa.brahmia@gmail.com, rafik.bouaziz@usf.tn
[2] University of Bologna, Bologna, Italy
fabio.grandi@unibo.it

Abstract. JSONPath is a language for locations in a standard JSON document, like the XPath language for XML documents. Its definition is based on the specification of the standard JSON format. Moreover, many JSON query languages, like JSONiq, have been inspired from JSONPath. However, JSONPath allows to navigate only in the current version of a JSON document, since it does not support temporal aspects. Besides, as several modern platforms, like IoT-based applications, cloud computing, and e-health systems, are creating, accessing, updating, and querying temporal JSON documents, it will be interesting to have a temporal query language that allows locating and retrieving temporal JSON data from such documents. Moreover, the state-of-the-art of JSON data management does not include any temporal JSON query language. Hence, to fill this gap, we propose in this paper τJSONPath (Temporal JSONPath), a temporal extension of JSONPath, to query time-varying JSON documents in our temporal JSON framework τJSchema. τJSONPath extends both the syntax and the semantics of JSONPath to support transaction time. We also provide some examples to motivate and illustrate our proposal.

Keywords: JSON · JSONPath · JSON query language · τJSchema · Temporal JSON document · Temporal database · Transaction time · Temporal query language

1 Introduction

A temporal database [1] is a database with built-in support for managing time-varying data. Two temporal dimensions have been proposed for timestamping temporal data: transaction time [2], which denotes when some datum is current in the database, and valid time [3], which denotes when some datum is valid in the modelled reality. The applications' requirements concerning temporal data management and querying are omnipresent regardless of both the application field (e.g., business, weather monitoring, health) and the underlying database model (e.g., relational, document-oriented (XML or JSON), graph-oriented, key-value, …).

Y. Farhaoui et al. (Eds.): ICAISE 2022, LNNS 635, pp. 844–853, 2023.
https://doi.org/10.1007/978-3-031-26254-8_123

Besides, since JSON [4] is a light-weight format that makes easy data storage, manipulation and exchange, JSON documents are being widely used in today's applications. Moreover, several platforms, like smart cities, Internet of things, online social networks, cloud computing, mobile applications, and e-health systems, use temporal JSON documents [5, 6] as they allow (i) to recover any JSON document version, (ii) to track document changes, (iii) to execute temporal queries [7, 8], (iv) to discover previously unknown knowledge (e.g., patterns, rules, relationships) implicitly present in time-varying data [9], and (v) to apply learning techniques on temporal data [10], in order to acquire novel useful information. Notice that a temporal JSON document is a sequence of versions of the same JSON document, each one having a distinct timestamp [5, 6]. Notice also that in our previous work [5], we have proposed a temporal JSON framework, called τJSchema (Temporal JSON Schema), to create and validate time-varying JSON documents via the use of a temporal JSON Schema. This latter is constructed from a conventional JSON Schema, i.e., a JSON Schema [11] file, and a set of temporal characteristics, such that each characteristic is associated to a component of such a conventional schema. Temporal characteristics are of two types: temporal logical characteristics, which specify which components of a JSON document can vary over transaction time and/or valid time, and temporal physical characteristics, which specify where timestamps should be placed and how the temporal aspects should be represented. The main strengths of τJSchema are: (i) ensuring logical and physical data independence, as it separates, from one hand, conventional schema, temporal logical characteristics, and temporal physical characteristics, which allows them to be changed independently, and, from the other hand, conventional JSON data from temporal JSON data and conventional JSON schema from temporal JSON schema; (ii) providing a low-impact solution, as it requires neither updates to existing JSON documents/schemas, nor extensions to JSON technologies (e.g., JSON format, JSON Schema language, available tools supporting JSON or JSON Schema).

Besides, JSONPath [12, 13] is a JSON query language inspired by the XPath [14] language of the XML World. It allows selecting and extracting values within a conventional (i.e., non-temporal) JSON document. Its definition is based on the specification of the standard JSON format. However, JSONPath allows to navigate only in the current version of a JSON document, since it does not support temporal aspects (e.g., transaction time, valid time). Moreover, the state-of-the-art of JSON data management (e.g., [15–17]) does not include any temporal JSON query language. Also, it is worth mentioning that the τJSchema framework is not yet equipped with a temporal JSON query language. Furthermore, as several modern platforms, like those mentioned above, are creating, managing, updating, and querying temporal JSON documents, it will be interesting to have a temporal query language that allows locating and retrieving temporal JSON data from such documents. So, to fill this gap, we propose in this paper τJSONPath (Temporal JSONPath), a temporal extension of JSONPath, to query time-varying JSON documents in our temporal JSON framework τJSchema. τJSONPath extends both the syntax and the semantics of JSONPath to support transaction time [2]; due to space limitations, we deal only with this temporal dimension in the present work, as we consider it sufficient for the design of a temporal extension of JSONPath. We also provide some examples to motivate and illustrate our proposal. To the best of our knowledge, we are the first to propose such a temporal extension to JSONPath.

The rest of the paper is organized as follows. Section 2 presents the temporal JSON data model on which our proposal is based. Section 3 proposes τJSONPath, the temporal JSONPath query language for the τJSchema framework, and illustrates its use via some examples of queries. Section 4 summarizes the paper and gives some remarks about our future work.

2 Temporal JSON Data Model

The temporal JSON data model on which our language is based is that of the τJSchema framework. In this latter, temporal JSON instance data are stored in JSON documents called "squashed JSON documents". In the following, we explain how a squashed JSON document is built.

Recall that in τJSchema [5], the JSON database administrator or designer (JDBA) starts by defining the conventional JSON schema (i.e., a JSON Schema file); to each conventional schema corresponds a set of conventional JSON documents which store JSON instances conforming to that JSON schema. After that, he/she specifies temporal aspects of this conventional JSON schema by annotating the components of this latter with temporal logical and physical characteristics; the whole set of temporal characteristics is stored in a standard JSON file. Finally, once the JDBA asks the system to commit his/her temporal annotation of the conventional schema, the system automatically generates the following documents (in the same order):

(i) the temporal JSON schema which associates the conventional JSON schema to its temporal characteristics document;
(ii) a temporal JSON document that links each conventional JSON document, which is conformant to the conventional JSON schema, to its temporal JSON schema and, therefore, to its set of temporal characteristics; notice that the temporal JSON document includes all the versions of the conventional JSON document with their corresponding transaction-time timestamps and indicates the temporal JSON schema associated to these JSON document versions;
(iii) the squashed JSON document corresponding to each generated temporal JSON document; it is created from this latter and it stores the temporal JSON instances that are produced from the conventional JSON instances (stored in conventional JSON documents) and the temporal JSON schema, i.e., via the application of temporal characteristics on these conventional instances.

2.1 Illustrative Example

In order to illustrate this model, let us consider the example of a squashed JSON document that stores the data of books (i.e., their ISBN, title, authors, publisher, and price), which belongs to the JSON database of an online library. Moreover, let us suppose that the JDBA, with the aim of maintaining a full history of each book, has chosen to manage the evolution over time of book data along the transaction time dimension, which means that any change to each book data is recorded. Figure 1 shows an instance of our temporal JSON data model, which stores the information of two books: the first has two versions, and the second has one version. The modification of the title, the authors, the publisher or the price of a book gives rise to a new version of this book; we suppose that the ISBN of a book could not be modified once assigned. Transaction-time timestamps are presented in blue type and changes to a book version w.r.t. the previous one are presented in red type.

Notice that in the τJSchema environment if the JDBA declares as transaction-time a conventional JSON schema component having the following form `"x":{ y }`(like the object member "book" in our example of Fig. 1), the conventional JSON instances will become, in the squashed JSON document, as follows:

```
"x": [ { y_version1, "TTbegin":temp_val1, "TTend":temp_val2},
       { y_version2, "TTbegin":temp_val3, "TTend":temp_val4}, ...]
```

The contents y is replaced by an array representing a collection of timestamped versions of y (with an "implicit" structuring of squashed JSON documents).

It is important to note that transaction-time timestamps, represented through the object members TTbegin and TTend, are automatically added by the system as an answer to the declaration of the object book as evolving over transaction time. Moreover, their values are also automatically assigned by the system, in a transparent manner to the user. It should be mentioned also that if a book version has a TTend with a value equal to "UC" ("Until Change") [2], this means that it is the current version of such a book. In the examples, we assume a granularity of a day for timestamps.

```
{ "books":[
   { "book":[ { "ISBN":"111-22-33333-44-5",
                "title":"Blockchain 2.0 Technologies",
                "authors":["Layla Ahmad", "Mario Rossi"],
                "publisher":"Zpringer",
                "price":181.99,
                "TTbegin":"2020-03-01",
                "TTend":"2022-01-26" },
              { "ISBN":"111-22-33333-44-5",
                "title":"Blockchain 2.0 Technologies",
                "authors":["Layla Ahmad", "Mario Rossi"],
                "publisher":"Zpringer",
                "price":168.99,
                "TTbegin":"2022-01-27",
                "TTend":"UC" } ] },
   { "book":[ { "ISBN":"333-44-55555-66-7",
                "title":"NoSQL Database Systems",
                "authors":["Jamila Jamal"],
                "publisher":"IZI Global",
                "price":95.99,
                "TTbegin":"2020-03-01",
                "TTend":"UC" } ] } ] }
```

Fig. 1. A squashed JSON document representing an instance of our temporal JSON document

2.2 Desiderata

The τJSONPath language is designed to satisfy mainly the four following goals:

- τJSONPath has to be fully compatible with JSONPath. Both existing and future JSONPath queries should work exactly the same in τJSONPath.
- The extensions to JSONPath must be both minimal and efficient. So, our proposal should reuse JSONPath whenever possible.
- The extensions must include all constructs that could be useful to query temporal JSON data. Such constructs should be easy to write and to memorize (by users).
- The completeness and the expressiveness of τJSONPath must be guaranteed. Due to space limitations, we will deal with these two aspects in a future work.

3 Temporal JSONPath Query Language

In this section, we propose the τJSONPath language and illustrate its use via some examples of queries.

3.1 Syntax and Semantics of tJSONPath

The temporal extension of JSONPath consists in adding, to the JSONPath's syntax, some transaction-time functions, as shown below, in order to query transaction-time JSON data. Notice that when defining these functions, we have been inspired (among others) from both the Allen's interval algebra [18], and the work of Zaniolo's team [19] on querying transaction-time relational databases.

Added Transaction-Time Functions
We have added a set of functions in order to make a separation between (i) temporal JSON queries which are supposed to be specified at a high level, via a high-level query language, and (ii) the temporal JSON data model that includes low-level information concerning the representation of time (e.g., if a transaction-time interval is closed at the end or not, the representation of "UC"). In fact, the implementation aspects should be hidden/transparent to the end user.

The following transaction-time functions have been added to JSONPath:

- tt-current-version: it returns the current version of an information;
- tt-current-between(b,e): it returns a slice of consecutive versions of an information, which were current consecutively between the temporal value b (for beginning time) and the temporal value e (for ending time);
- tt-current-asof(x): it returns the version of an information, which was current at the temporal value x (i.e., whose transaction-time contained the temporal value x);
- tt-first-version: it returns the first transaction-time version of an information;
- tt-history: it returns the full transaction-time history of an information (e.g., an employee salary, a product price) by providing each value of each version of this information, with its transaction-time timestamp;

- tt-includes(t): to return the version of an information whose transaction-time interval includes the temporal value t;
- tt-precedes(b,e): to return the version of an information whose transaction-time interval precedes the temporal interval [b, e] such that b for its beginning and e for its ending;
- tt-meets(b,e): to return the version of an information whose transaction-time interval meets the temporal interval [b, e];
- tt-overlaps(b,e): to return the version of an information whose transaction-time interval overlaps with the temporal interval [b, e];
- tt-starts(b,e): it returns the version of an information whose transaction-time interval starts the temporal interval [b, e];
- tt-during(b,e): it returns the version of an information whose transaction-time interval during the temporal interval [b, e];
- tt-contains(b,e): it returns the version of an information whose transaction-time interval contains the temporal interval [b, e];
- tt-finishes(b,e): it returns the version of an information whose transaction-time interval finishes the temporal interval [b, e];
- tt-equal-to(b,e): it returns the version of an information whose transaction-time interval is equal to the temporal interval [b, e].

Notice that:

- the tt-current-asof(x) function is used to write rollback queries (like Query 2 in Sec. 3.2), with TTend set to x in the result;
- the difference between tt-overlaps/tt-includes and tt-current-between/tt-current-asof is that the formers only select data preserving their timestamps in the results, whereas the latters select the same data but restrict their timestamps to the values of the arguments (as shown in the results of Query 2 and Query 3 below);
- some functions could be expressed via some other functions while using default arguments like the following ones:
- tt-history = tt-current-between(0,UC);
- tt-current-version = tt-includes(UC)).

3.2 Examples of τJSONPath Queries

The main types of temporal queries are [7]: history selection, rollback (with transaction time only), snapshot (with valid time only), time-point selection, temporal slicing, temporal join, temporal aggregate, and restructuring. In this subsection, we show how to use τJSONPath to express, on the squashed JSON document of Fig. 1, a query example of each one of the three following types: history selection, rollback, and temporal slicing. Notice that, for each query, we provide three things: the τJSONPath query, the implementation of this latter in JSONPath, and the result of the execution of this query. Notice also that we have tested and evaluated each JSONPath query while using an online tool that supports JSONPath, called JSONPath Online Evaluator [20]. Recall that the result of the execution of a JSONPath query is always a JSON array of values. It is worth mentioning that we have slightly changed the presentation of the query results provided below (e.g., by putting price versions between brackets, adding the corresponding object member name just before the extracted value) in order that they become more useful and easily interpretable by both application programs and humans.

History Selection Query

Query 1: Retrieve the price history of the book titled "Blockchain 2.0 Technologies".

This query could be expressed using τJSONPath as follows:

```
tt-history.$.books[*].
    book[?(@.title=="Blockchain 2.0 Technologies")].price
```

The translation of this τJSONPath query to JSONPath is as follows:

```
$.books[*].
  book[?(@.title=="Blockchain 2.0 Technologies")].price,TTbegin,TTend
```

The result of this query is as follows:

```
[  {  price: 181.99,
      TTbegin: "2020-03-01",  TTend: "2022-01-26"  },
   {  price: 168.99,
      TTbegin: "2022-01-27",  TTend: "UC"   }  ]
```

Rollback Query

Query 2: Retrieve the books edited by Zpringer and available on 2022–02-01.

This query could be expressed using τJSONPath as follows:

```
tt-current-asof("2022-02-01").$.books[*].
    book[?(@.publisher=="Zpringer")]
```

The translation of this τJSONPath query to JSONPath is as follows:

```
$.books[*].book[?((@.publisher=="Zpringer") &&
            (@.TTbegin <= "2022-02-01") && (@.TTend >= "2022-02-01"))]
```

The result of this query is as follows:

```
[ {  "ISBN": "111-22-33333-44-5",
     "title": "Blockchain 2.0 Technologies",
     "authors": ["Layla Ahmad", "Mario Rossi"],
     "publisher": "Zpringer",
     "price": 168.99,
     "TTbegin": "2022-01-27", "TTend": "2022-02-01"  }  ]
```

Temporal Slicing

Query 3: Find the price history of all books between 2020–09-01 and 2021–09-01.

This query could be expressed using τJSONPath as follows:

```
tt-current-between("2020-09-01","2021-09-01").$.books[*].
    book[*].ISBN,price
```

The translation of this τJSONPath query to JSONPath is as follows:

```
$.books[*].book[?(
(@.TTbegin <= "2021-09-01") && (@.TTend >= "2020-09-01")
)].ISBN,price,max(TTbegin,"2020-09-01"),min(TTend,"2021-09-01")
```

The result of this query is as follows:

```
[ {  "ISBN":"111-22-33333-44-5",  "price":181.99,
     "TTbegin":"2020-09-01",  "TTend":"2021-09-01"  },
  {  "ISBN":"333-44-55555-66-7",  "price":95.99,
     "TTbegin":"2020-09-01",  "TTend":"2021-09-01"  }  ]
```

4 Conclusion

In this paper, we have proposed a temporal JSON query language, named τJSONPath, as an extension of the JSONPath query language to transaction time, in our already established temporal JSON framework τJSchema. τJSONPath augments JSONPath with several transaction-time functions which allow users to query, in a user-friendly manner, transaction-time JSON data that are stored in squashed JSON documents; recall that these latters are the logical storage space of temporal JSON instances in a τJSchema environment. The proposed language fills a gap in the state of the art of temporal JSON data management and querying, and completes our τJSchema framework with an appropriate temporal query language.

In the near future, we plan to develop a tool that supports τJSONPath and shows the feasibility of our proposal, to be integrated with the τJSchema-Manager tool that has been developed to support the whole τJSchema framework. Furthermore, we intend to extend τJSONPath to also support the valid time [3], so that it will allow to query not only transaction-time JSON data but also valid-time JSON data and bitemporal ones (i.e., temporal JSON data evolving over both transaction time and valid time).

References

1. Jensen, C.S., Snodgrass, R.T.: Temporal database. In: Liu, L., Özsu, M.T. (eds.) Encyclopedia of Database Systems (2nd edition), pp. 3945–3949. Springer, New York, NY, USA (2018)
2. Jensen, C.S., Snodgrass, R.T.: Transaction time. In: Liu, L., Özsu, M.T. (eds.) Encyclopedia of Database Systems (2nd edition), pp. 4200–4201. Springer, New York, NY, USA (2018)
3. Jensen, C.S., Snodgrass, R.T.: Valid time. In: Liu, L., Özsu, M.T. (eds.) Encyclopedia of Database Systems (2nd edition), pp. 4359–4360. Springer, New York, NY, USA (2018)
4. Internet Engineering Task Force (IETF): The JavaScript Object Notation (JSON) Data Interchange Format, Internet Standards Track document, December 2017. https://tools.ietf.org/html/rfc8259. Accessed 10 Sept 2022
5. Brahmia, S., Brahmia, Z., Grandi, F., Bouaziz, R.: τJSchema: a framework for managing temporal JSON-based NoSQL databases. In: Hartmann, S., Ma, H. (eds.) DEXA 2016. LNCS, vol. 9828, pp. 167–181. Springer, Cham (2016). https://doi.org/10.1007/978-3-319-44406-2_13
6. Goyal, A., Dyreson, C.: Temporal JSON. In: Proceedings of the 5th IEEE International Conference on Collaboration and Internet Computing (CIC 2019), Los Angeles, CA, USA, 12–14 December 2019, pp. 135–144 (2019)
7. Snodgrass, R.T. (ed.), et al.: The TSQL2 Temporal Query Language. Kluwer Academic Publishers, Norwell (1995)
8. Zheng, X., Liu, H.K., Wei, L.N., Wu, X.G., Zhang, Z.: Timo: in-memory temporal query processing for big temporal data. In: Proceedings of the 7th International Conference on Advanced Cloud and Big Data (CBD 2019), Suzhou, China, 21–22 September 2019, pp. 121–126 (2019)
9. Tansel, A.U., Imberman, S.P.: Discovery of association rules in temporal databases. In: Proceedings of the 4th International Conference on Information Technology (ITNG'2007), Las Vegas, NV, USA, 02–04 April 2007, pp. 371–376 (2007)
10. Zhao, J., Papapetrou, P., Asker, L., Boström, H.: Learning from heterogeneous temporal data in electronic health records. J. Biomed. Inform. **65**, 105–119 (2017)
11. Internet Engineering Task Force (IETF): JSON Schema: A Media Type for Describing JSON Documents, Internet-Draft, 19 March 2018. https://json-schema.org/latest/json-schema-core.html. Accessed 10 Sept 2022
12. Gössner, S.: JSONPath – XPath for JSON, 21 February 2007. http://goessner.net/articles/JsonPath/. Accessed 10 Sept 2022
13. Internet Engineering Task Force (IETF): JSONPath: Query Expressions for JSON, Internet-Draft, 25 April 2022. https://datatracker.ietf.org/doc/draft-ietf-jsonpath-base/. Accessed 10 Sept 2022
14. XML Path Language (XPath) 3.0, W3C Recommendation, 8 April 2014. http://www.w3.org/TR/2014/REC-xpath-30-20140408/. Accessed 10 Sept 2022
15. Liu, Z.H., Hammerschmidt, B., McMahon, D.: JSON data management: supporting schema-less development in RDBMS. In: Proceedings of the 2014 ACM SIGMOD International Conference on Management of Data, Snowbird, UT, USA, 22–27 June 2014, pp. 1247–1258 (2014)
16. Bourhis, P., Reutter, J.L., Vrgoč, D.: JSON: data model and query languages. Inf. Syst. **89**, 101478 (2020)

17. Brahmia, Z., Brahmia, S., Grandi, F., Bouaziz, R.: JUpdate: a JSON update language. Electronics **11**(4), 508 (2022)
18. Allen, J.F.: Maintaining knowledge about temporal intervals. Commun. ACM **26**(11), 832–843 (1983)
19. Wang, F., Zaniolo, C., Zhou, X.: Archis: an xml-based approach to transaction-time temporal database systems. VLDB J. **17**(6), 1445–1463 (2008)
20. JSONPath Online Evaluator (2022). https://jsonpath.com/. Accessed 10 Sept 2022

Virtual Communities Homophily Based on Users Posts in Online Social Networks. Case Study: 198 Facebook Public Groups

Yassine El Moudene[1(✉)], Jaafar Idrais[1], Rida El Abbassi[1], and Abderrahim Sabour[2]

[1] IMI Laboratory, Faculty of Sciences of Agadir, Ibn Zohr University, Agadir, Morocco
yassine.elmoudene@gmail.com

[2] Department of Computer Sciences, High School of Technology of Agadir, Ibn Zohr University, Agadir, Morocco

Abstract. The multidimensional user profiling analysis is based on the definition of a set of attributes to describe the behavioral characteristics of users within a complex system like the online social network (OSN) Facebook. An analysis that extracts user characteristics from textual elements (text of posts and comments with dictionary analysis) and non-textual elements (time periods of posts, time periods of comments, reactions to the content). Another dimension is the measurement of the virtual communities' distance (VCs) via the analysis of publications. In this paper we analyze 2.259.249 publications made by 59.020 users in 198 several groups, the general goal is to study the kinetics of those VCs over the years to detect correlations between them and to classify them, in order to contribute to the multidimensional profiling Process.

Keywords: Online social network · Classification · Visualization · Big data · Virtual community · Facebook · Knowledge extraction

1 Introduction

People enter into relationships with people of the same characteristics as themselves. This trend is called homophily. The word "homophily" can etymologically be broken down from the Greek into -homo (similar, the same) and -phily (love). Homophily means that individuals with similar traits are more likely to form social bonds with each other, which often impacts their actions. Recent popular Online Communities (OCs) like Usenet and common discussion forums were structured by interest or according to modern hierarchies, but OSNs (as a complex system [1]) are structured as individual networks [2], with the person at the center of the community [3]. On Facebook [4], users acquire the freedom to join several VCs in order to publish, comment, or make other actions [5]. The activities made by a user in several VCs create indirectly a relationship between themes. The publication-based relationship studies the users' behavior [6] of publishing over different VCs, this relationship is visualized in order to extract the knowledge, see the evolution of groups over the years, and measure the distance between those VCs.

Y. Farhaoui et al. (Eds.): ICAISE 2022, LNNS 635, pp. 854–860, 2023.
https://doi.org/10.1007/978-3-031-26254-8_124

2 Related Works

From 2005, new complex systems [7] appeared under the name of OSNs. They are OCs whose only objective [8] is the creation and maintenance of social relationships. [9] presents What distinguishes a VC from its physical, and Shows that technological mediation filters the psychological experiences of places and people represented in virtual environments. [10] argue that OCs can be classified based on 4 dimensions: attributes, supporting software, relation to physical communities, and boundedness. However, extensive research on correlation and distance calculation between VCs has received very limited attention [11].

3 Methodology

3.1 Publishers-Based Relationship

Of course, two groups cannot interact, but how can we define the relationship between two groups? Based on the freedom to join several groups of different themes, the user has the right to publish to anyone, this action defines indirectly the relationship R(U, G) between the user U and the group G, by transitivity between those several groups R(U, G1), R(U, G2), R(U, G3), etc. The relation R(G1, G2) is illustrated in the Fig. 1.

3.2 Treatment Process

After defining the relationship between VCs, we start by identifying the active users who have published in multiple VCs, then we draw all their publications which we sort by creation time to allow their time scheduling. After we define the relationship intervals (simple and extended) to analyze the evolution of R and visualize the graph that represents all the relations.

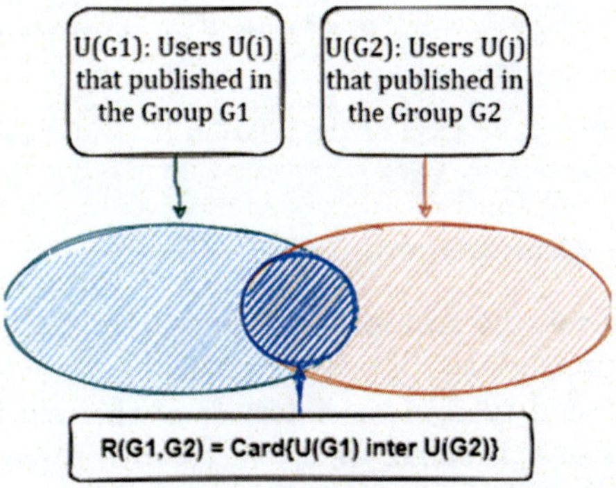

Fig. 1. Definition of relationships between groups

3.3 Interval of the Relationship Between the User and the Group

Each publication is indexed by the timestamp attribute "created_Time" (time in milliseconds) which automatically defines the creation time of the relationship R(U, G). This field let us pass from a static study to a dynamic evolutionary study of the relationships between groups. For the implementation of this dynamism, some basic questions arise: When does the relationship between a user and a group start? Over what interval does it spread? When should it disappear?

- **Definition of the simple interval of the relation R**
  Let the publication P1 be published by the user U1 at the group G1, the relation between U1 and G1 starts at the created_time of P1 (T1), and ends at the date of the last comment (T2) of P1.
- **Definition of the extended interval of the relation R**
  During the simple interval, it is possible that the user U1 launches a second publication (P2) at the level of the same group, at this moment the end time of the interval is recalculated, which will be the maximum of the times of the last comments for two publications P1 and P2. The Fig. 2 show the new interval: **[T1, max (T3, T4)]**.

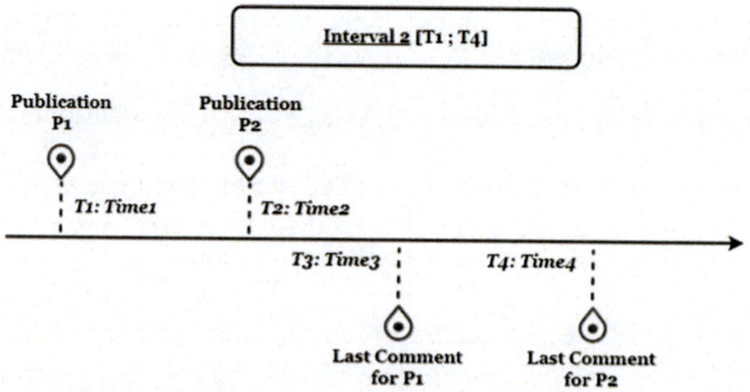

Fig. 2. Extended interval of the relation R (U, G) ($T4 \geq T3$)

3.4 Special Cases Management

- **Publication without comments:** During the definition of intervals, we were opposed to publications that didn't receive a comment, hence the problem of defining the end time of the intervals of the relationships created by these publications. At this stage, we are faced with three possibilities:
  - Eliminate these publications from the study and lose an information rate.
  - Define a cut-off time "t" to be used as the end of the interval for every relation created by these publications.

  - Define a constant "t" to add it to the end of all the intervals, and these publications will have as interval: [publication_time, publication_time+t].

 Even if the publications have not received any comments, we opt for the third choice which allows shifting the intervals with a constant "t".
- **Publication with modified "created_time"** Facebook allows for the publisher to choose the publication date that will appear to other users, this opportunity creates two possibilities:
  - **Large Interval:** Publication updated time is so old.
  - **Reduced Interval:** Publication updated time between the first and the last comment time.
  - **Wrong Interval:** Publication updated time is greater than the date of the last comment.

4 Results and Discussion

4.1 Processing Result

The treatment process has as input the 363 crawled Facebook groups [12], which includes a total of 5.251.859 publications, to have at the end of the treatment 198 interconnected groups (54,55% of the studied groups) thanks to 59.020 users (12.74% of the initial users) who have published 2.259.249 publications (42,43% of the initial publications). For the special cases, we detect 2.170.912 publications without comments, which represent (96%) of the publications at the end of the treatment process. In terms of calibration, we found only one publication with a creation time higher than that of the last comment.

4.2 Featured Users

For the different users who published in several groups, we notice that 71,63% of the users linked two VCs, followed by 16% who linked 3, and 6% for linking 4, the Table 1 displays the percentage of user's contribution in linking a given number of groups. We notice that only 2% of users have linked more than 7 groups. The Table 2 shows that 44,5% of users have published from 2 to 6 publications, a production less than 24% who have made between 7 and 16 publications, the same percentage is calculated for the third category which have made from 17 to 99 publications, while a minority of users (7,5%) have made a significant number of publications (between 100 and 27.871). The 2,32% in Table 1 and the 7,5%

Table 1. Number (K) and percentage of users who linked a Number (N) of groups

N Groups	2	3	4	5	6	≥7	Total
K Users	42.271	9.420	3.514	1.594	849	1.372	59.020
%	71,63	15,96	5,95	2,70	1,44	2,32	100

Table 2. Users who have published an interval number (Goal) of publications

Goal	02–06	07–16	17–99	100–27	Total
Users	26.297	14.155	14.113	4.455	59.020
%	44,5	24	24	7,5	100

in Table 2 have an interesting activity that leaves us wondering about the nature of the behavior of these users, and the purpose behind this huge production. The previous measurements have led to the definition of two types of users: the first type (about 4.455 users) generates a large number of relationships between these users and groups, and the second type represents users who have linked a large number of different groups (users have linked 36, 29, 27).

5 Visualisation the Relations

Using the Gephi software, we visualize the graph that represents the relation between these VCs in order to measure the degree, the modularity, and all the characteristics. We use the ForceAtlas2 [13] spatialization layout algorithm which allows having a disposition aimed by force. Nodes are regimented as charged particles, while edges attract their nodes, as springs. This force creates a movement that converges in a balanced state. The Fig. 3 shows the distribution of the 5 calculated communities, the nodes' degrees, and the balanced state.

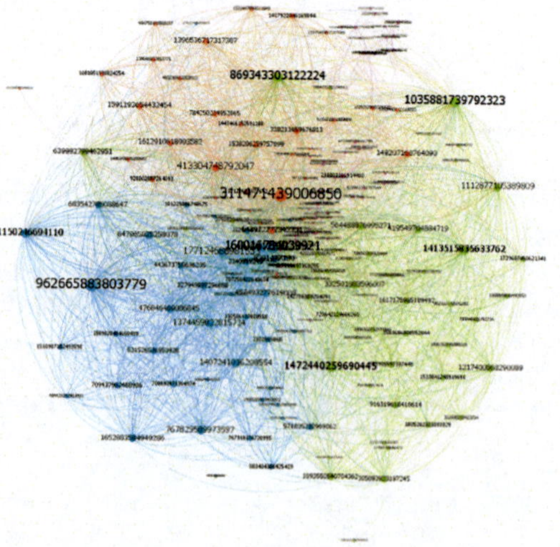

Fig. 3. Different virtual communities class visualisation

6 Conclusion and Perspective

The aim of this work is to classify the virtual communities (VCs) in order to measure the distance between them and visualize the different classes, this work is the continuation of other preliminary works on the multidimensional user profiling process: it is based on the definition of a set of attributes to describe the behavioral characteristics of users within Facebook, as a component it intends to measure VM coefficient based on the calculated distances. VCs homopholy based on comments and reactions ar perspectives of future works.

References

1. El Moudene, Y., Idrais, J., Sabour, A.: The effect of keywords used on content attraction in complex networks. In: 2020 Complex Systems, pp. 689-709 (2020)
2. Boyd, D.M., Ellison , N.B., Social network sites: definition, history, and scholarship. J. Comput.-Mediated Commun. 210–230 (2007)
3. Idrais, J., El Abassi, R., El Moudene, Y., Sabour, A.: Characterizing community behavior in OSNs: modeling and forecasting activity on Facebook using the SARIMA model. J. Intell. Fuzzy Syst. 3757–3769 (2022)
4. Idrais, J., El Moudene, Y., Sabour, A.: Characterizing user behavior in online social networks: study of seasonal changes in the Moroccan community on Facebook. In: Third International Conference on Intelligent Computing in Data Sciences (ICDS) 2019, pp. 1–5 (2019)
5. Idrais, J., El Moudene, Y., Sabour, A.: Online social networks: study and validation of the regular behaviors of a targeted community on a complex network using a time series approach. In: 2019 4th World Conference on Complex Systems, pp. 1–7 (2019)
6. Idrais, J., El Moudene, Y., Sabour, A.: Characterizing user behavior in online social networks: analysis of the regular use of Facebook. Int. J. Electr. Comput. Eng. 3329–3337 (2021)
7. El Moudene, Y., Idrais, J., Sabour, A.: Principal component analysis (PCA) use for keywords research characterization. Case study: environment and renewable energy community interest. In: 2019 4th World Conference on Complex Systems, pp. 1–7 (2019)
8. Iriberri, A., Leroy, G.: A life-cycle perspective on online community success. In: 2009 ACM Computing Surveys (CSUR), pp. 1–29 (2009)
9. Shen, K., Khalifa, M., Yu, A.: Supporting social interaction in virtual communities: role of social presence. In: AMCIS 2006 Proceedings, p. 527 (2006)
10. Lazar, J., Preece, J.: Classification schema for online communities. In: AMCIS 1998 Proceedings, p. 30 (1998)
11. Lyu, X., Wang, H., Ma, A., Wang, X., Zhao, L.: The relationship between the sense of virtual community and knowledge-sharing: the mediating role of trust. In: 2019 Human Behavior and Emerging Technologies, pp. 245–60 (2019)

12. Idrais, J., El Moudene, Y., Sabour, A.: Introduction to sociology of Moroccan online social networks: evolution analysis of the Moroccan community Activity on Facebook. In: Mizera-Pietraszko, J., Pichappan, P., Mohamed, L. (eds.) RTIS 2017. AISC, vol. 756, pp. 395–408. Springer, Cham (2019). https://doi.org/10.1007/978-3-319-91337-7_35
13. El Moudene, Y., Idrais, J., Sabour, A.: Introduction to sociology of online social networks in Morocco. Data acquisition process: results and connectivity analysis. In: Mizera-Pietraszko, J., Pichappan, P., Mohamed, L. (eds.) RTIS 2017. AISC, vol. 756, pp. 409–418. Springer, Cham (2019). https://doi.org/10.1007/978-3-319-91337-7_36

Prediction Student Performance in E-Learning Environment: Challenge and Opportunity

Hanae Aoulad Ali[1](✉), Chrayah Mohamed[2], and Bouzidi Abdelhamid[1]

[1] FS, TIMS Laboratory, Abdelmalek Essaadi University, Tetuan, Morocco
hanae.aouladali@etu.uae.ac.ma
[2] ENSATE, TIMS Laboratory, Abdelmalek Essaadi University, Tetuan, Morocco

Abstract. With advancements in e-learning technology, students may power them by interacting with the eLearning environment, such that the teacher is no longer the gatekeeper of instruction. This research attempts to examine students' prediction performance based on their interaction with educational activities in MOODLE and MOOCs; this was accomplished via the use of student log files and some extra data about the specific student. In order to discover the best approach for the student's prediction, the performance prediction was explored using Decision Tree (C4.5 algorithm), Artificial Neural Network, Support Vector Machine (SVM), and K-Nearest Neighbor (KNN) algorithm techniques. Furthermore, log file analysis shows that the rate of interaction with the e-learning context has a substantial influence on their performance, with students with the highest interaction on the MOODLE performing better than someone with low interactivity rates. According to the data, students are more interested on e-learning MOODLE than MOOCs, and as a result, they are missing out on the benefits of the available resources on MOOCs, such as viewing lecture videos and participating in quizzes, which may help them with their studies.

Keywords: Artificial intelligence · AIEd · Education · Corona virus · Online learning · MOOC · Student performance · Prediction · E-learning · Data mining · Online education · Machine learning

1 Introduction

The efficacy of web-based education (e-learning) systems has increased as information technology has advanced. The e-learning system allows students to access and participate in various learning activities such as reading, downloading and uploading papers, presentations, and assignments from anywhere and at any time. The activities are typically held on a platform known as a Learning Management System (LMS), which employs various technologies, usually on the internet, to enhance student communication between students and instructors or among students. The LMS delivers a large number of diverse data for students and instructors to use, as well as data on the students' personal data, learning methods, practices, habits, and usability preferences [1]. The LMS delivers a large number of diverse data for students and instructors to use, as well as data on the

Y. Farhaoui et al. (Eds.): ICAISE 2022, LNNS 635, pp. 861–867, 2023.
https://doi.org/10.1007/978-3-031-26254-8_125

students' personal data, learning methods, practices, habits, and usability requirements [1]. Furthermore, Massive open online courses (MOOCs) enable students all over the world to easily have access to high-quality instructional content. One of the most noticeable characteristics of MOOCs is that, owing to the accessibility of enrollment, many students may calmly enroll in a course, peruse a couple of recordings or discussion and debate gatherings, and then discontinue involvement [2–4]. Academic performance of students is an important aspect in every educational institution; therefore, strategic programs in continuous enlightenment or guiding the students should be created. Student achievement is a vital aspect in every educational institution; thus, strategic programs in continuously enlightening or directing the students for an enhanced efficiency that may lead to a brighter future should be prepared [4]. According to [5], e-learning is an important tool for assisting and encouraging the teaching and learning process. It provides tools for students to communicate with friends and teachers outside of the classroom. It also encourages students to cope with their unique learning style and the best path for each learner. Students learn in a variety of methods, including reading, observing, researching, inquiring about, cooperating, communicating, working together, studying, and sharing knowledge and experiences. Several educational data mining (EDM) methodologies have been employed. Variety of educational data mining (EDM) approaches, including as classification, regression, and density estimation, have been employed in forecasting student performance and correlating student engagement in the LMS environment. It should be noted that current studies using EDM for student performance prediction have mostly focused on situations of University of high schoolers. Students may have access to a wide range of learning tools through e-Learning, and learning can take place anywhere, at any time, because there are no longer any geologic or geographical restrictions to learning [5, 6]. ELearning has empowered the student to the point that the instructor is no longer the information gatekeeper. E-learning has increased the importance of informal learning by bringing together study, employment, correspondence, and entertainment [7]. With the evolution of e-learning, and as an e-service product, interactivity is becoming an increasingly important component of e-learning, making it feasible to investigate students' learning processes based on their interaction within the platform. [8] defines "interactivity" as "reciprocal activity between both the educated man and the e-learning system." [8] defines "interactivity" as "reciprocal activity" between the learner and the e-learning system. The learning man's activity/response is determined by the system's action and reaction. However, with the advancement of Internet and PC innovation, online instruction is becoming an increasingly important instructional environment, and known for its influence has been proclaimed by many as one of the highlights of this innovation. Intelligence, as a tool for developing the understudies' capacities and skills, is unquestionably an acceptable supplementation inside an e-learning backing of training, and it is a critical benefit of mixed media e-learning guideline. The purpose of this study is to overviews of prediction student performance in E-Learning Environment. Second section present overviews of e-learning and MOOC. Furthermore challenge and opportunity of education and intelligence artificial. Finally, we present a general conclusion.

2 E-Learning and MOOC

This section contains concepts and approaches associated with an e-learning framework by various researchers, as well as past work by various researchers on mining e-learning and MOOCs activities, interaction, and performance prediction in the virtual learning environment. This section also contains a review of similar efforts on predicting student success in eLearning and MOOCs.

2.1 E-Learning

Online learning, often known as e-learning, has grown in popularity as a result of technological advancements that allow for remote debate [11, 12]. The majority of research focuses on students' learning experiences and outcomes of successful online learning, but the assessment of the process of learning is also required. E-Learning is a training technique that employs a variety of technological improvements, primarily online or PC-based, in the learning process [16]. Learning Management Systems (LMS), also known as Course Administration Systems (CMS) or Virtual Learning Environments (VLE), provide strong support for individual courses and stimulate communication between professors and students. According to [9–13], there was a significant success in the Internet that led to spontaneous expansion in online services, e-learning, which has been described as a type of online administrations have stood out amongst the others. According to [9], there was a massive accomplishment in the Internet that brought about a spontaneous increase in digital services, e-learning, which has been depicted as a kind of online administrations has been portrayed as one of the colossal late advancements in both establishments and partnerships, the online comfort and opportunity at all times and everywhere helps give users access to date data, sought answers to research complications, and exchange the knowledge gained in the study These eLearning systems tend to increase the circulation of educational resources to students; course content functioning, exam and coursework preparedness, dialogues, and different administration of classes [14, 15]. Furthermore, these frameworks include a variety of collaborative learning capabilities, such as debates, visits, news, record storage, and so on. Unlike regular face-to-face interactions. In contrast to traditional face-to-face training methodologies, PC and e-learning environments allow for storing a large amount of available data. These frameworks hold student activity data and linkages to log documents/databases. Reading materials, assignment submission techniques, tests and quizzes, methods of carrying out various duties, interactions with other classmates in virtual classes utilizing chat, forum, and so on. Databases may also contain data such as demographic or personal information on students or tutors [1].

3 Main Trends and Challenges in the Future Development of Educational Artificial Intelligence

3.1 Development Trend

Will encourage instructors' roles to be transformed in order to adapt to the changing educational environment. The implementation of AIEd will significantly alter the job of instructors in the future. To begin with, the flattening of information access implies that instructors no longer have a competitive advantage in terms of expertise and information. Teachers become leaders of instructional activities rather than power brokers, content creators rather than disseminators, and teachers and students share the same knowledge acquisition platform [7]. The top-down education approach is no longer appropriate; second, instructors who have been relieved of simple and repetitive tasks like heavy homework revision and examination judgment must concentrate more on students' individualized teaching plan creation and comprehensive development strategy study. When a result, as education evolves with the times, education training goals and learning paradigms shift, instructors must stay up with the trend and adapt to the new teaching environment in order to perform their crucial role as an education subject. Will hasten the deployment of high-quality Al education and encourage the completion of closed-loop training. School administration is in desperate need of restructuring in light of the upcoming college admission exams. Quality education, as one of the key aspects of education reform, urgently requires a new stage of evolution of comprehensive performance assessment system to achieve full business process coverage from indicator development, evaluation methods, collection plans, integrity systems, and results announcements. Through the scientific integration and application of contemporary teaching theories and AI technology, AIEd will be able to break free from the constraints of exam-oriented education in the future and become a strong supporter of inclusive and intelligent quality education [8]. Create a data collection plan that is merged with the school's regular education and teaching activities; collect student growth data using a scientific and effective content supervision and integrity system; conduct process and summative evaluations; and introduce multiple assessments such as students, parents, schools, and third parties. The main body, and can provide intelligent trial calculations of performance evaluation based on national and regional standards, forming a closed-loop system of timely feedback and effective intervention, assisting education authorities in scientific decision-making, and achieving comprehensive training of students' exploration, analysis, and learning abilities. Conduct a scientific and fair examination of students' overall quality, and advise and support overall improvement and growth of students' overall quality. It will encourage the improvement of educational research methodologies and the advancement of educational research toward science. The arrival of the AI technological revolution may mark a "tipping point" in the transformation of educational research techniques. To overcome the current disciplinary crises and achieve "selfsalvation," educational disciplines are urged to use Al-oriented research techniques, as well as methodological reconstruction, research technology, and tool creation. The method problem is to achieve the revolutionary innovation of educational methods of research, and lastly to realize the rebuilding of educational disciplines, and to promote educational scientific research. The advantages of large data collecting and analysis in educational research are obvious. It will be feasible to develop a series of scientific research findings if it can strengthen strategic collaboration with educational institutions and technological businesses such as New Oriental and iFlytek, and promote the integration of data and technology. And immediately use big data research findings to educational activities. On this basis, the AI technique represented by machine learning has the characteristics of big data, strong computing, good at tracking and repeatability, the ability to learn independently, the ability to form a large database with self-renewal function, and the ability to develop better teaching strategies.

3.2 The Key Challenge

- ←AI technology is not yet developed enough. AI's issues and controversies have created a slew of hurdles while also opening up new potential for Aed. Technological study based on empathy, emotion, and long-term interaction is still at a relatively primitive stage due to technological maturity; in regarding product service objects, many AI technologies are only applied in the field of children's education; at the same time, AIEd is applied in many fields. The majority of them are still in the planning stages. The commercial application variety potential has yet to be realized.
- There are no legal system rules in place to address AI's possible moral and ethical difficulties [9]. At this point, AI is progressing toward strong AI that can fully reason and solve issues; yet, strong AI requires policy direction and authoritative oversight. The ethical and moral difficulties raised by supervision are concerning. The implementation of AI technology, for example, is based on a huge number of data training sets and automatic classification and screening, particularly in the field of education with different personality traits. AIEd solutions are developed using massive amounts of educational data. And because this is the fundamental data foundation, the risk of students' and instructors' social communication, teaching behavior and habits, and other privacy leaks is greatly enhanced, and data security is called into doubt. Any technical innovation Moreover, as the core data basis As a result, the danger of students' and instructors' social communication, instructional behavior and habits, as well as other privacy leaks is greatly enhanced, and data security is called into doubt. Any technical advancement should not jeopardize human privacy. As a result, data protection must be developed to ensure that now the teacher is protected. Students have complete ownership and control over personal data, and they work to increase private data management and protection in regulations and ethics to reduce the chance of data leakage and misuse.
- Overreliance on AI may result in education vassalization. Because of the rapid development of AIEd, current education has entered a comprehensive technology period, and education's reliance on technology has also increased; research and development of educational products is still biased towards advanced technologies, and the phenomenon of product homogeneity is serious, and the production of high-quality educational content cannot be considered. At the same token, over-reliance on AI and disregard for instructors' experience, knowledge, and mental labor, as well as students' learning capacity and inventive thinking, may lead to educators' loss of teaching skills [10]. No more vigilance in teaching tactics and intentions: Students lose desire and training for independent thinking, indulge in the fullness of tools, and are unable to grow a healthy and thorough personality and behavior as a result of the lack of care in teaching tactics and intentions. The goal of education is no longer to grow individuals, but rather AI. Education may become a total slave and auxiliary to AI.

4 Conclusions

AIEd, on average, presents more possibilities than difficulties. This has had a significant impact on the direction and pace of world wide education reform. It has significantly enhanced the intellectual level of educational informatization. At the same time, education must keep up with the times and revise educational aims and implications on a regular basis. More core skills must be developed in order to strengthen the reserve force for future AI technologies.

References

1. Mosavi, A., et al.: Ensemble boosting and bagging based machine learning models for groundwater potential prediction. Water Resour. Manag. **35**(1), 23–37 (2021)
2. Ali, H.A., et al.: A course recommendation system for MOOCs based on online learning. In: 2021 XI International Conference on Virtual Campus (JICV). IEEE (2021)
3. Rincy, T.N., Gupta, R.: Ensemble learning techniques and its efficiency in machine learning: a survey. In: 2nd International Conference on Data, Engineering and Applications (IDEA). IEEE (2020)
4. Polikar, R.: Ensemble learning. In: Zhang, C., Ma, Y. (eds.) Ensemble Machine Learning, pp. 1–34. Springer, Boston (2012). https://doi.org/10.1007/978-1-4419-9326-7_1
5. Dalipi, F., Imran, A.S., Kastrati, Z.: MOOC dropout prediction using machine learning techniques: review and research challenges. In: 2018 IEEE Global Engineering Education Conference (EDUCON). IEEE (2018)
6. Edalati, M., Imran, A.S., Kastrati, Z., Daudpota, S.M.: The potential of machine learning algorithms for sentiment classification of students' feedback on MOOC. In: Arai, K. (ed.) IntelliSys 2021. LNNS, vol. 296, pp. 11–22. Springer, Cham (2022). https://doi.org/10.1007/978-3-030-82199-9_2
7. Adnan, M., et al.: Predicting at-risk students at different percentages of course length for early intervention using machine learning models. IEEE Access **9**, 7519–7539 (2021). HI, USA, 27 January–1 February 2019, **33**, 517–524
8. Kuzilek, J., Hlosta, M., Zdrahal, Z.: Open university learning analytics dataset. Sci. Data **4**(1), 1–8 (2017)
9. Costa, E.B., et al.: Evaluating the effectiveness of educational data mining techniques for early prediction of students' academic failure in introductory programming courses. Comput. Hum. Behav. **73**, 247–256 (2017)
10. Hew, K.F., et al.: What predicts student satisfaction with MOOCs: a gradient boosting trees supervised machine learning and sentiment analysis approach. Comput. Educ. **145**, 103724 (2020)
11. Ali, H.A., et al.: COVID-19 and distance learning. In: 2022 XII International Conference on Virtual Campus (JICV). IEEE (2022)
12. Al-Shabandar, R., Hussain, A., Laws, A., Keight, R., Lunn, J., Radi, N.: Machine learning approaches to predict learning outcomes in massive open online courses. In: IEEE International Joint Conference on Neural Networks, Anchorage, USA, 14–19 May 2017 (2017)
13. El Habti, F.E., Chrayah, M., Bouzidi, A., Ali, H.A.: Blended learning platform model. In: 2022 XII International Conference on Virtual Campus (JICV), pp. 1–4. IEEE, September 2022

14. Ali, H.A., Mohamed, C., Abdelhamid, B.: Challenges of e-learning in COVID 19 pandemic. In: 2022 XII International Conference on Virtual Campus (JICV), pp. 1–3. IEEE, September 2022
15. Ali, H.A., et al.: A comparative evaluation use bagging and boosting ensemble classifiers. In: 2022 International Conference on Intelligent Systems and Computer Vision (ISCV). IEEE (2022)
16. Ali, H.A., et al.: A novel hybrid classification approach for predict performance student in e-learning. In: Ben Ahmed, M., Abdelhakim, B.A., Ane, B.K., Rosiyadi, D. (eds.) Emerging Trends in Intelligent Systems & Network Security, pp. 45–52. Springer, Cham (2022). https://doi.org/10.1007/978-3-031-15191-0_5

Routing Protocol Based on Artificial Neural Network for Vehicular Ad Hoc Networks

Bouchra Marzak[1](✉), Soufiane El Moumni[1], Zaki Brahmi[2], and Mohamed Talea[1]

[1] Laboratory of Information Processing, Faculty of Sciences Ben M'Sick, Hassan II University, Casablanca, Morocco
marzak8bouchra@gmail.com

[2] College of Science and Arts at Al-Ola Taibah University Medinah, Medina, Kingdom of Saudi Arabia

Abstract. In VANET networks, the clustering protocol consists of dividing the network into a set of geographically close nodes. It presents an interesting solution to optimize and simplify network services and functions. Notably, it allows the protocol to operate more efficiently by minimizing control traffic and simplifying the data orientation process. In this paper, we will propose a clustering protocol. This protocol is a generalization of a k-hop artificial neural network (ANN) algorithm. The main contribution of the proposed algorithm is to improve the stability of the cluster topology when the network topology changes frequently while keeping the data routing reliability. To improve the reliability of the network, we use the reliability rate factor introduced in our proposal.

Keywords: Cluster-head · Artificial neural network · VANET · Routing protocol · Clustering

1 Introduction

Vehicular ad-hoc network (VANET) represents a new area of research that aims to improve road safety, navigation, and congestion. This kind of networks aims also to improve driver safety while reducing the number of deaths and injuries caused by traffic, by implementing collision avoidance and warning systems. In addition, these VANETs allow the broadcast of traffic and road conditions to help with navigation and traffic congestion by providing drivers with routes that avoid hazards and bottleneck areas [1].

VANET is characterized by high node mobility making the network topology highly dynamic. This characteristic makes the lifetime of the communication links between the nodes quite short. Additionally, the density of nodes in VANET is unpredictable during peak periods, especially when the roads are congested with vehicles. We also mention that some roads have more traffic than other roads [2].

In general, clustered routing protocols are more suitable for large-scale network topology and guarantee better energy savings [3]. Clusters are conceptual structures where multiple nodes self-organize into a group around their currently selected representative called cluster-head (CH). This node acts as a coordinator for the other cluster

Y. Farhaoui et al. (Eds.): ICAISE 2022, LNNS 635, pp. 868–875, 2023.
https://doi.org/10.1007/978-3-031-26254-8_126

members (CM). The cluster head is responsible for intra- and inter-cluster management. Intra-cluster nodes interact with each other through direct links, while inter-cluster interaction is handled by the cluster heads. In cluster routing protocols, vehicles close to each other form a cluster. However, the configuration of the clusters and the choice of the cluster head are important issues. Given the high mobility of VANETs, dynamic cluster configuration becomes a major process [4].

Several clustering routing approaches have been proposed to create the most efficient and intelligent data aggregation system. Various intelligent algorithms have been used to build efficient data traffic aggregation systems, such as fuzzy logic [5, 6] which improves the efficiency of permanent cluster construction. In this paper, we use artificial neural networks ANN for node classification and their stability and efficiency to divide the network into clusters.

Our proposed approach is based on a generalization of a K-hop artificial neural network algorithm. The main objective of our contribution is to improve the stability of the cluster structure while keeping the reliability of data routing. To improve the network reliability, we employ the stabilization rate factor introduced in our proposal. Furthermore, we evaluate our achieved performances through experiments and simulations.

This paper is organized as follows: related works are presented in Sect. 2. Section 3 presents the algorithm overview. In Sect. 4, the proposed approach is explained. Achieved performance results are presented in Sect. 5. Finally, Sect. 6 concludes the paper.

2 Related Works

Different clustering protocols have been proposed for solving the routing problems in VANET. However, most of these algorithms focus on the creating clusters, none of them considers the stability rate value for creating clusters. Several examples of clustering protocols are described in the literature; we present below the most usable in VANET environment.

The k-means algorithm [7] aims to divide a data set into groups by minimizing the mean square error of the relative intra-group distances. The k-means clustering method is very simple to implement as it is very sensitive to the initial selection of CHs, the algorithm must be run several times with different initial CHs and the best result must be chosen. Therefore, this algorithm is not scalable due to the large number of repetitions required [8].

The APROVE algorithm [9] is a clustering method based on the k-means and MOBIC algorithms. Clustering by APROVE can be negatively impacted by stored memory, which can lead to delayed responses to abrupt changes in network topology. Therefore, APROVE is not able to capture changes in the network quickly enough through its calculations, as they are still partially affected by the old affinity values. It also cannot guarantee the stability of the network when confronted with various vehicle mobility conditions, where connections are of short duration.

The AMACAD algorithm [10] tried to accurately follow the network mobility pattern and extend the cluster lifetime and reduce the overall overhead. This protocol takes location, speed, and relative and final destination into consideration as a parameter for the node grouping decision.

In addition, all proposed approaches in this field were focused on the development of clustering protocols to [11–17] support efficient and reliable communication between vehicles. However, we found that several challenges have not been resolved and therefore require further research. For example, in terms of cluster-head stability, we need to focus on what can allows us to calculate a factor that determines node stability. Therefore, in this paper we seek to identify parameters that allow more stable clusters to be formed.

3 Algorithm Overview

The goal of this scientific work is to develop an algorithm capable of communicating to neighboring vehicles to the damaged vehicle an alert message to warn them of the imminent danger. This algorithm will consider several incoming parameters that will be developed later in this study.

The different objectives of our proposal are revolved around maximizing: stability of broadcaster nodes, number of clusters, number of exchanges between vehicles, latency of alerts from all vehicles, or lifetime of the cluster structure. Therefore, we have combined several metrics as: the stability rate, the speed of the knot, the geographical coordinates of the node (x, y), the difference in distance between neighboring nodes.

4 Proposed Approach

The proposed routing protocol is a clustering protocol based on vehicle-to-vehicle communication to reduce the network maintenance cost. The proposed approach should ensure optimal network operation by maximizing stability and minimizing information delivery time, as illustrated in Fig. 1.

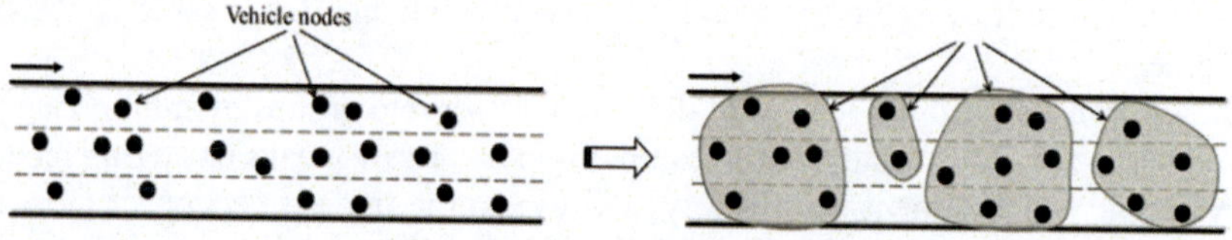

Fig. 1. Cluster formation procedure

The idea is to develop a cluster routing protocol that considers the best performance for vehicle networks. However, it is not able to classify these nodes in traditional ways because clustering protocols require large processing information. Therefore, this processing requires a long delay to make a simple decision without losing data in a mobile and dense network.

The proposed algorithm can identify a stable group structure and consider the mobile node movements, to form clustering structures resistant to vehicle mobility. The proposed algorithm uses the fundamental idea of an artificial neural network system to tackle pattern recognition and classification problems. The node collects the distance between its neighbor node, the speed, and the stabilization rate, sending it to ANN to decide on which nodes in the cluster where each node can run the algorithm.

At the first step in creating clusters, the nodes calculate their stability values, afterwards, the node collects information from neighboring nodes in a transmission zone of 800 m by exchanging a request message. The request message includes an excessive node identifier, its direction, its speed, and its stability value. The data collected by each node will make it possible to form clusters based on the ANN algorithm. As a result, the node with the highest stabilization rate will be selected as cluster-head, and their neighboring nodes in a transmission zone that does not exceed 800 m become a cluster member.

In this proposed approach, it is assumed that each vehicle is equipped with an IEEE 802.11 wireless transceiver to perform local broadcasting. In addition, each vehicle is assisted by a GPS which makes it able to obtain its own geographical coordinates at any time. We assume also that the network nodes adopt a communication that periodically samples data at regular intervals. Once the data is ready, the nodes select a cluster-head (CH). Once CH is selected, nodes belonging to the same cluster communicate their data directly to their CH.

The main objective of the proposed algorithm is to delimit the list of neighborhoods to create groups of nodes. A vehicle sends a request message to its neighbors within its transmission range to collect and store the latest data. These nodes respond with a response containing the id, speed, and current geographical coordinates. If the vehicle has not found neighboring nodes within a short time x, it will increase its transmission range until it finds its neighbors. However, in the opposite case, it will reduce its transmission range to reduce the number of jumps to avoid the loss of information in a time t. Once the message has been exchanged between the vehicles, each vehicle calculates the distance between its neighbors and the stabilization rate (2) using its list of neighbors. The formula below shows how this parameter can be calculated:

$$\|A - B\| = \left[(x_A - x_B)^2 + (y_A - y_B)^2\right]^{1/2} \tag{1}$$

The distance formula is used to find the distance between two vehicles A and B.

$$\%S = (\text{number of received message/number of answered messages}) * 100 \tag{2}$$

ANN algorithm is a particular computer modeling of complex real-world problems. It can perform parallel computations for data processing and knowledge representation. ANN is chosen to implement the structure selection of stable clusters. We use the speed, the distance difference between two vehicles, the stabilization rate as input parameters of our artificial neural network system, and the determination of the group structure of each vehicle at its output.

After the cluster selection process, CH broadcasts a message to all the members of the clusters, containing its identifier. Each new member must inform the CH of his group by sending an affiliation message. Thus, the node becomes a member of this cluster and builds its routing table.

In the next section, the proposed routing protocol will be compared to the AMACAD protocol in terms of two parameters: the average lifetime of a cluster and the average lifetime of a CH.

5 Performance Results

The performance simulation of our proposed approach is assessed using the NS2 simulator. We compare our achieved performance results to those of the AMACAD algorithm. Which use the same node mobility. The proposed approach is evaluated on different parameters such as: the average lifetime of a cluster and the average lifetime of a CH.

We use the IEEE 802.11 standard as the connection layer model. The Table 1 below summarizes the values of parameters used in our simulation. The time for each simulation was 700 s in a space of 40 m × 3000 m.

Table 1. Simulation parameters

Parameters	Scenario
Highway length	40 m × 3000 m
Transmission range	110–550 m
Speed	80 km/h–120 km/h
Package size	1000 bytes
MAC protocol	IEEE 802.11
The maximum packet size in the queue	60
Type of traffic	CBR
Message sending interval	3 s
Simulation time	700 s

The simulation parameters are a transmission range between 110 and 550 m with a data rate of 10 Mb/s, a density of 100 to 400 vehicles, and a speed between 80 and 120 km/h. CBR traffic has been used as an application layer to evaluate the performance of routing protocols. The existence time of a cluster on the route is directly related to the lifetime of its CH. Therefore, we specify the lifetime of a CH as the time interval between when a node becomes the CH node and when it relinquishes the role of CH. The lifetime of a CH makes it possible to evaluate the overall stability of the clustering algorithms. Figure 2 below illustrates the variation of the average lifetime of a CH of the two clustering algorithms for the transmission range of the nodes.

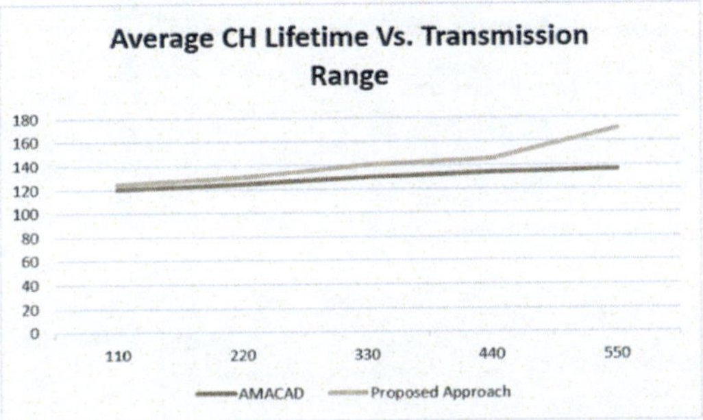

Fig. 2. Average CH lifetime (s) vs. Transmission range (m)

Interestingly, as the transmission range increases, the average CH lifetime follows the same trend as shown in Fig. 3 below. This effect is due to the greater transmission range allowing for greater diversity of nodes in terms of destinations and speed, and will improve the stability of the CH.

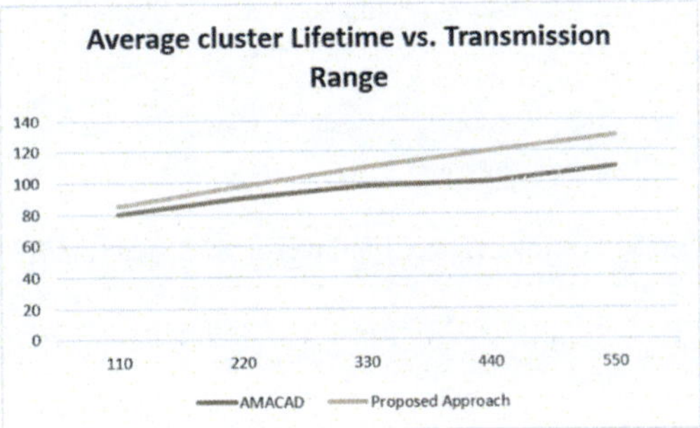

Fig. 3. Average cluster lifetime (s) vs. Transmission range (m)

We can notice that the proposed approach provides the best results in terms of cluster stability. Whereas, the AMACAD approach had a short cluster lifetime.

The average lifetimes of a CH and a cluster make it possible to assess the overall stability of the clustering algorithms. Figure 4 and Fig. 5 show the performance of the proposed approach compared to other techniques in terms of the average lifetime of a CH and a cluster respectively. Since vehicles move faster, the vehicle network topology is more dynamic, the average lifetime of a CH and the cluster decreases as vehicle speed increases.

The Fig. 4 below shows the performance of the proposed approach compared to AMACAD technique in terms of the average lifetime of a cluster.

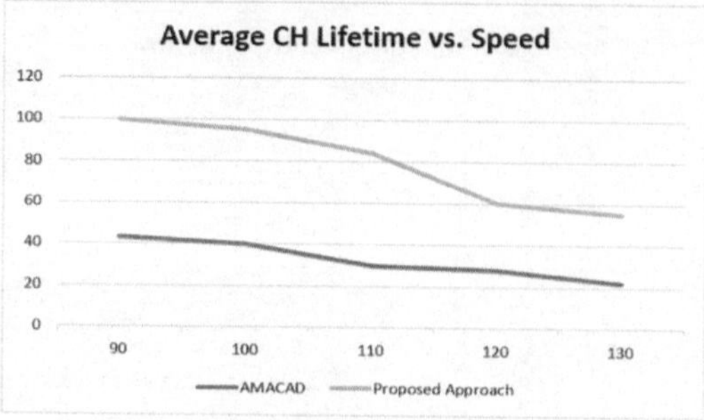

Fig. 4. Average CH lifetime (s) vs. Speed (km/h)

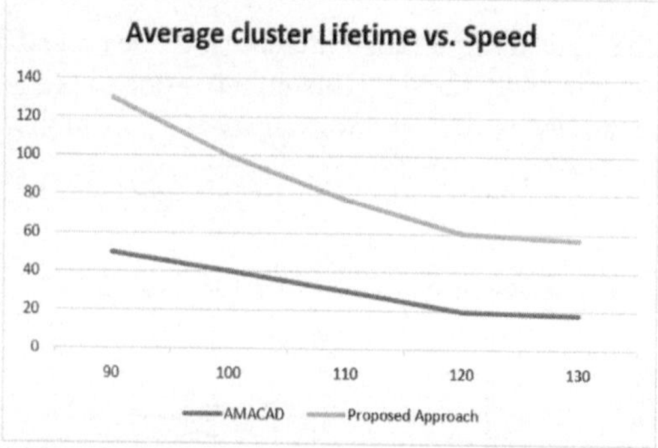

Fig. 5. Average cluster lifetime (s) vs. Speed (km/h)

6 Conclusion

Clustering protocol is based on dividing network into a set of geographically close nodes. This protocol allows a significant optimization in network services.

In this paper, we have proposed a clustering algorithm that can be integrated into a large-scale routing protocol for use in VANETs networks. The objective of our algorithm is to maintain the cluster structure as stable as possible overtime as well as routing reliability. The stability of the cluster structure makes it able to guarantee the stability of the routes thus minimizing the cost of their maintenance. The proposed routing protocol will be compared to the AMACAD protocol in terms of average lifetime of a cluster and CH (Cluster Head).

References

1. Marzak, B., Toumi, H., El Guemmat, K., Benlahmar, A., Talea, M.: A survey on routing protocols for vehicular ad-hoc networks. Indian J. Sci. Technol. **9**(S1) (2016)
2. Hamdi, M.M., Audah, L., Rashid, S.A., Mohammed, A.H., Alani, S., Mustafa, A.S.: A review of applications, characteristics, and challenges in vehicular ad hoc networks (VANETs). In: 2020 International Congress on Human-Computer Interaction, Optimization and Robotic Applications (HORA), pp. 1–7. IEEE (2020)

3. Qi, W., Landfeldt, B., Song, Q., Guo, L., Jamalipour, A.: Traffic differentiated clustering routing in DSRC and C-V2X hybrid vehicular networks. IEEE Trans. Veh. Technol. **69**(7), 7723–7734 (2020)
4. Alves Junior, J., Wille, E.C.: Routing in vehicular ad hoc networks: main characteristics and tendencies. J. Comput. Netw. Commun. **2018** (2018)
5. Marzak, B., El Moumni, S., Benlahmar, E., Ait-Mlouk, A., Talea, M.: Stable routing protocol based on fuzzy logic system in vehicular ad hoc networks. Int. J. Commun Syst **31**(10), e3587 (2018)
6. Dietzel, S., Bako, B., Schoch, E., Kargl, F.: A fuzzy logic based approach for structure-free aggregation in vehicular ad-hoc networks. In: Proceedings of the Sixth ACM International Workshop on VehiculAr InterNETworking, pp. 79–88 (2009)
7. Sinaga, K.P., Yang, M.S.: Unsupervised K-means clustering algorithm. IEEE Access **8**, 80716–80727 (2020)
8. Ahmed, M., Seraj, R., Islam, S.M.S.: The K-means algorithm: a comprehensive survey and performance evaluation. Electronics **9**(8), 1295 (2020)
9. Shea, C.: APROVE: a stable and robust VANET clustering scheme using affinity propagation. University of Toronto (2009)
10. Morales, M.M.C., Hong, C.S., Bang, Y.C.: An adaptable mobility-aware clustering algorithm in vehicular networks. In: 2011 13th Asia-Pacific Network Operations and Management Symposium, pp. 1–6. IEEE (2011)
11. Marzak, B., El Guemmat, K., Benlahmar, E.H., Talea, M.: Clustering in vehicular ad-hoc network using artificial neural network. Int. Rev. Comput. Softw. (IRECOS) **11**(6), 548–556 (2016)
12. Marzak, B., Toumi, H., Talea, M., Benlahmar, E.: Cluster head selection algorithm in vehicular Ad Hoc networks. In: 2015 International Conference on Cloud Technologies and Applications (CloudTech), pp. 1–4. IEEE (2015)
13. Cao, Y., Xu, S., Chen, X., He, Y., Jiang, S.: A forward-secure and efficient authentication protocol through lattice-based group signature in VANETs scenarios. Comput. Netw. 109149 (2022)
14. Shafi, S., Venkata Ratnam, D.: An efficient cross layer design of stability based clustering scheme using ant colony optimization in VANETs. Wirel. Pers. Commun. 1–19 (2022)
15. Azhdari, M.S., Barati, A., Barati, H.: A cluster-based routing method with authentication capability in vehicular ad hoc networks (VANETs). J. Parallel Distrib. Comput. (2022)
16. Nath, H.J., Choudhury, H.: A privacy-preserving mutual authentication scheme for group communication in VANET. Computer C (2022)
17. Kandali, K., Bennis, L., Bennis, H.: A new hybrid routing protocol using a modified K-means clustering algorithm and continuous hopfield network for VANET. IEEE Access **9**, 47169–47183 (2021). https://doi.org/10.1109/ACCESS.2021.3068074

Towards Secure SDN-Based VANETs Paradigm

Nabil Nissar[1(✉)], Najib Naja[1], and Abdellah Jamali[2]

[1] RAI2S-Team, National Institute of Posts and Telecommunications, INPT Rabat, Rabat, Morocco
nabil.nissar@gmail.com, naja@inpt.ma
[2] RI2M, Hassan I University Settat, Settat, Morocco

Abstract. In recent years, VANETs (Vehicular Ad Hoc Networks) technology has been considered as one of the most promising technologies for providing traffic management, road safety and comfort applications. Thus, security issues in VANETs are of high importance because of the sensitivity of delivered information between vehicles. Unfortunately, traditional VANETs architecture lacks a lot of flexibility and security because of the high mobility of nodes. To solve this problem, in this paper we are proposing a novel framework in order to incorporate the principles of software-defined network SDN (Soft-Defined Networking) with VANETs, which will expand the flexibility of the network and allow us to add new range of services especially in terms of security services for the introduction of VANET technology in industrial enterprises.

Keywords: Vehicular ad hoc networks · Software-defined networking · Cryptography · Digital signatures · Secure road messages · Vanet attacks

1 Introduction

In recent years, VANETs (Vehicle Ad Hoc Networks) have been witnessing growing interests with extensive research efforts among the industry and academic research community. Such interest toward this field of research appeared due to the possibility of improving road safety and the introduction of this technology in various industrial enterprises. Therefore, VANET has evolved as part of Intelligent Transportation System (ITS) whose main goal is to improve road safety and to provide different applications to the drivers and the passengers [1]. However, such technology also needs to be highly secured in order to prevent attacks which can prove to be very dangerous and can lead to loss of life and revenue. Many researchers in the literature have been working on improving security schemes in VANETs, but most of them are still unable to ensure safety because most existing security schemes couple data forwarding with control. To solve these problems, it is proposed to use a software-defined network (SDN) [2] which is a new worldwide spread networking paradigm. Using SDN technology, VANETs can benefit from many advantages as SDN provides more flexibility and programmability that makes designing and implementing security measures and applications such an easy and dynamic task. In this context, our work comes up with a novel secure framework of software-defined VANETs. Specifically, as security in VANETs is a major concern, we

Y. Farhaoui et al. (Eds.): ICAISE 2022, LNNS 635, pp. 876–883, 2023.
https://doi.org/10.1007/978-3-031-26254-8_127

investigate how these two technologies can be implemented together in order to improve VANETs security. This paper is organized as follows: Sect. 2 presents the well-known security issues in VANETs while presenting the main layer 3 attacks, and provide a big picture of the SDN architecture; Sect. 3 sum-up some of the related works in the field of securing VANETs; In Sect. 4 we introduce our approach to secure VANETs based on SDN; Sect. 5 discusses the main benefits of our presented scheme against some well-known attacks in VANETs; and finally Sect. 6 gives a general summary of the actual paper.

2 VANETs and SDN: Background Overview

2.1 VANET Architecture

Nowadays, the research of ITS has become more attractive for researchers, in which VANETs play a key role in safety, commercial, convenience, and productive applications. In the traditional VANETs, communication can be established through three ways (see Fig. 1): Vehicle to Vehicle (V2V), Vehicle-to-Infrastructure (V2I) [3], and infrastructure to infrastructure (I2I) communications. VANETs hold the same characteristics of Mobile Ad Hoc Networks [4]. However, VANETs have some distinctive features in aspect of structure, high-mobility and special traffic application.

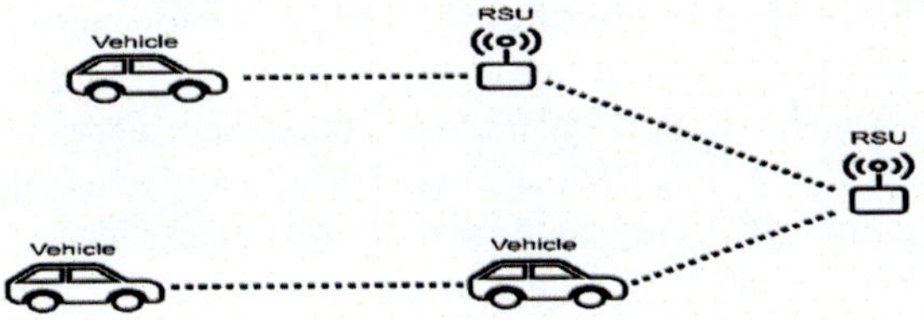

Fig. 1. Traditional VANET architecture

Security Problems in VANET. Nowadays, unlike the other networks, VANETs performance is more challenging in term of security due to many constraints such as the high mobility, network size, and the geographic relevancy. The design of VANETs architecture must take into consideration many encountered security challenges such as: optimal cryptographic algorithms, efficient security protocols, real time constraint, keys distribution, and high mobility etc.

Different Types of Attacks in VANETs. Security and privacy issues in VANETs [5] have still met a lot of challenges in order to prevent attacks which can prove to be very dangerous and can cause both loss of life and revenue. Some of the well-known attacks in VANETs are: Eavesdropping [6], Black hole attack [7], Denial of service (DoS) attacks, Impersonation attacks, Man-in-the-Middle attacks, Sybil attacks.

Overview of Software-Defined Networking SDN. SDN technology provides programmability and flexibility for network devices by decoupling control from data planes.

Moreover, SDN offers separation of control plane and data plane in the network architecture (Fig. 2). Based on this model, the network management can be done independently on the control plane without impact on the data flows.

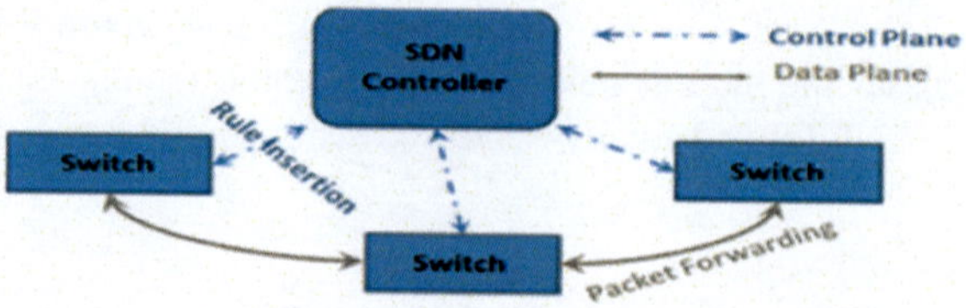

Fig. 2. SDN architecture

3 Related Work

Different approaches used in literature for securing VANETs using traditional techniques. Chen et al. [8] have proposed many techniques in order to prevent Sybil and reply attacks based on GPS time spoofing. In [9], Purohit et al. introduced a solution to secure VANETs by modifying the original AODV protocol by appending an encrypted random number to the RREQ packet of the source node. In [10], Tyagi et al. proposed a secure algorithm based on asymmetric public-key cryptography to mitigate some attacks against VANETs (Enhanced Secure AODV). In [11], Zardari et al.introduced a dual-attack detection algorithm in order to detect malicious nodes based on Intrusion Detection System (IDS). In [12], Cherkaoui et al. used a variable control chart in their proposed scheme to mitigate against some well-known attacks in.

VANETs. In [13], Hassan et al. suggested an algorithm called: Intelligent detection of Black Hole Attack in order to mitigate against Black Hole Attack. The authors in [14] introduced a secure AODV based on the RREQ and RREP control packets enhancement. Pattanayak et al. [15] suggest a mechanisms based on a fixed key infrastructure for detecting impersonation and Sybil attacks in VANETs based on different methodologies such as PKC, PVM, and RTM techniques.

The previously cited solutions are based on traditional techniques to secure and authenticate messages, but most of them does not take into consideration the design of a secure and flexible framework by preserving the security and network requirements of VANETs and simultaneously protecting the whole system from attacks, including protecting the data flow. But recently, Software defined VANETs have attracted much attention as SDN becomes one the new worldwide spread networking paradigms. Balamurugan et al. [16] discussed the benefits of SDN in VANETs in terms of Path selection, Frequency/Channel selection, Power selection management, and proposed a framework for the deployment of SDN over VANETs. However, their framework lacks flexibility and mobility management due to the absence of a central intelligence, along with the lack of security mechanisms. Software Defined Vanet Architecture Software-Defined VANET Architecture has the advantage of distributing the processing load between different entities of the network, under the supervision of a main Controller. RSUs are part of the Fog computing environment to minimize latency, real-time responses and improves user experience while meeting the frequent and variable service demands of drivers.

4 Our Proposed Software-defined VANET Model

Our suggested framework is based on the major progress in the traditional VANET architecture; therefore, in order to keep pace with the new demands and fulfill the new generation requirements, our Software-Defined VANET architecture's main components are discussed next and shown in Fig. 3: SDN Controller: Is a software defined base-station of the fourth generation 4G/LTE for mobile and control management. It's a form of certificate authority and central intelligence of the VANET network responsible for policy control, traffic management, and security enforcement. Road-Side Units (RSUs): are intelligent fixed units for direct connection with the vehicle in their working range. They provide flexibility, fast processing, scalability, and rule management. Vehicles: are intelligent wireless mobile nodes, which are considered as wireless hosts with V2V and V2I capabilities. The vehicles are controlled by the RSUs in their coverage, and they collect all necessary information and exchange safety data with the RSUs through local agents onboard.

QoS is guaranteed because the SDN controller that uses RSA security algorithm based on 3 steps: key generation (Public Key, Private Key), encryption and decryption. The public key is known to everyone and is used for encryption, and the decryption can only be done using the private key in a reasonable interval of time.

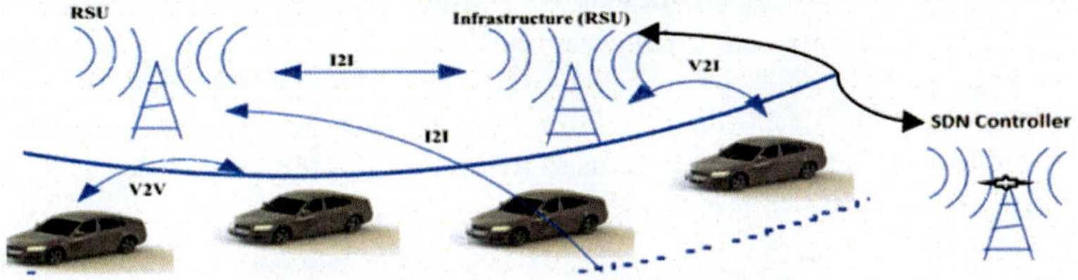

Fig. 3. SDN VANET security architecture

Join Request Phase of a New Vehicle. In our proposed scheme, we assume that the communication channel between SDN_Controller and RSUs is safe for data transmission. In addition, every vehicle(i) has to register first to the SDN_Controller in order to get its own RSA pair key (PuK_Vi, PrK_Vi) and the list of public keys for both RSUs and vehicles that have been already registered,SDN_Controller stores the drivers' personal information such as the car plate number, the contact information, the address and the assigned pair key as well during the registration phase. Similarly, every RSU in our network is pre-assigned a RSA pair key (PuK_u,PrK_u). Every vehicle(i) wants to join our software_defined VANET sends a join request message to the working range of a fixed RSU, the JOIN_Req message format is the following:

Payload	SignatureV
{JOIN_Req,Req_ID,Veh_ID}	Signature_Vi(JOIN_Req,Req_ID,Veh_ID; PrK_Vi)
The signature_Vi contains the JOIN_Req fields sent by the Vehicle(i) and is signed using its private key PrK_Vi	

The *JOIN_Req* represents the type of the packet request, the *Req_ID* represents the JOIN_Req ID, the *Veh_ID* represents the unique vehicle ID such as the car plate Number, and the *SignatureV* denotes the signature of the JOIN_Req Field using PrK_Vi based on RSA Public-Key Cryptosystem. The payload is sent without any encryption. The RSU checks first the received packet type; if it is a new JOIN_Req; then the RSU uses its own private key to generate its signature according to RSA Public-Key Cryptosystem and appends it to the received join request message before forwarding it to the SDN_Controller. Next is the new JOIN_Req message format:

Message		SignatureU
Payload	SignatureV	Signature_U(JOIN_Req,Req_ID,Veh_ID; PrK_U)
Signature_U has the fields of JOIN_Req sent by the vehicle(i) and is signed using the private key PrK_U		

Once the SDN_Controller receives the double signed JOIN_Req message, it will first check the freshness of the received JOIN_Req message, and then it will validate the RSU's signature "SignatureU" and then validates the signature of the vehicle "SignatureV". If the validation succeeds and matches with the stored registration information in its database, the integrity of the message is guaranteed and non-repudiation has been ensured (vehicle(i) actually sent the message).

Join Reply Phase. SDN_Controller replies with a JOIN_Rep packet that contains a group private key K_Gp which will be used to encrypt the transmitted Road Messages during communications, also it contains the list of valid Public Keys of the other members including RSUs and the already joined Vehicles. The group private key K_Gp is encrypted using the vehicle public key PuK_Vi, and the message is signed by the SDN_Controller using its own private key, then it is broadcasted back to all RSUs. Each RSU authenticates the signature of the SDN_Controller and then appends its signature to the message and forwards it to all vehicles in its working range. Each vehicle that receives the JOIN_Rep message authenticates the two signatures using the same approach described previously in the case of the SDN_Controller. If the vehicle is the originator of the JOIN_Req message, then it will use its private key PrK_Vi in order to decrypt the attached group private key K_Gp which will be used to encrypt any transmitted Road Messages, the other vehicles that received JOIN_Req will add the attached public key of the new vehicle(i) to their public key repository.

Road Message Dissemination. We consider the scenario in which a specific vehicle(j) with identity ID is within the working range of a RSU. We assume that the vehicle has successfully passed the authentication and acquired the group key K_Gp. For instance, when the driver passes through the road accident scene, he could consider this accident as a road message and report it to the RSU and the other vehicles, the vehicle road message is encrypted using the distributed group key K_Gp. A time stamp TS and the vehicle ID will be signed using it private key PrK_Vj and will be attached to the road message. The message will be broadcasted to vehicle(j) neighbors (RSU and vehicles), each neighbor will validate the freshness, authentication and integrity of the message then it will decrypt it using the distributed group key K_Gp.

Public Key Revocation. SDN_Controller sends periodically a HELLO message; signed using its private key; to the list of RSUs and to be forwarded to the registered vehicles in our Software Defined VANET. HELLO message contains the recently revoked public keys for some reason, for example: if a vehicle leaves the network, a vehicle is misbehaving etc. The Hello message may contain also a new symmetric Group Key K_Gp when needed, and it will be sent encrypted using the old one.

5 Security Evaluation and Analysis for Our Proposed Scheme

Securing VANETs using our enhanced Software Defined VANET approach is fully distributed, because it does not rely on a single unit in order to validate signatures. Our secure system model ensures many advantages with regard to the SDN_Controller, RSUs and vehicles such as: integrity of transmitted messages as it guarantees that messages content have not been tampered by a malicious entity, and non-repudiation of transmitted messages by ensuring that the originator unit has actually constructed and sent the received messages, because the mutual signature authentication prevent attackers from spoofing another unit's identity to send a message. Finally, our scheme guarantees the data confidentiality and transmission security by encrypting road messages and important information sent by SDN_Controller using the private Group session key. Consequently, our proposed system model can prevent many possible well-known network layer routing attacks based on the solutions summarised in the following table (see Table 1):

Table 1. Our Security Solutions against well-known routing attacks

Attack	Our proposed solution
DoS: sleep deprivation	Our scheme based on signatures verification guarantees the Non-repudiation security service. However, reactive intrusion detection and prevention systems; for instance, can be suggested as a better solution against DoS attacks in the future
Replay attack	Replay attack can be prevented using our system model because the previous messages of the past authentication sessions cannot pass the current authentication process in RSU side

(*continued*)

Table 1. (*continued*)

Attack	Our proposed solution
Black hole	In addition to the mutual authentication used by our scheme, the SDN Controller provides another solution against this attack because it controls the flow tables and prevents attacker the process of manipulating the routes in the traditional form
Eavesdropping	Confidentiality service is one of our priorities in our scheme for road messages; therefore, thanks to our adopted privacy scheme an attacker cannot insert false information while using eavesdropping attack
Sybil	SDN controller provides a centralized digital certificate distribution structure prevent the attacker from declaring himself more than a vehicle. The SDN controller also adds these digital certificates to his revoked digital certificate record and publishes this information to RSUs and vehicles
Wormhole	Thanks to SDN controller, trusted routing is done using the control plane and based on authentication, it is impossible for an attacker to manipulate routes and routing tables, which means that the system acts automatically as an IPS intrusion prevention system for this attack
Man-in-the-Middle	The existence of a digital certificate system and encryption of communication when needed between different units through public and private keys prevents such an attack

6 Conclusion and Future Work

In this paper, we presented a general and effective mechanism to implement security services (authentication and privacy) in a VANET based on the integration with SDN technology. We have tried to add security services that are supposed to be suitable to address the important security issues that were facing traditional VANETs while supporting their specific characteristics. The proposed model ensures security services during joining phase of a new vehicle to the VANET and during the dissemination of road messages. The efficiency of the suggested model and its security properties against many well-known attacks have been clearly presented. Finally, many ideas can be enhanced by our approach leading to the main direction for our future work as to integrate Intrusion Detection System in the SDN Controller to prevent DoS attacks and to simulate our approach and compare its efficiency with other existing solution.

References

1. Khan, S., Sharma, I., Aslam, M., Khan, M.Z., Khan, S.: Security challenges of location privacy in VANETs and state-of-the-art solutions: a survey. Fut. Internet **13**(4), 96–122 (2021)
2. Ghonge, M.M.: Software-defined network-based vehicular ad hoc networks: a comprehensive review. In: Software Defined Networking for Ad Hoc Networks. EAI/Springer Innovations in

Communication and Computing. Springer, Cham (2022). https://doi.org/10.1007/978-3-030-91149-2_2
3. Sanghavi, P., Sanghvi, S., Mangrulkar, R.S.: Software-defined networks a brief overview and survey of services. In: Software Defined Networking for Ad Hoc Networks. EAI/Springer Innovations in Communication and Computing. Springer, Cham (2022). https://doi.org/10.1007/978-3-030-91149-2_1
4. Yuen, C., Seah, R., Soon, S., Jia, T.: Mobile Ad Hoc Networking (2019). http://dsta.gov.sg/
5. Goumiri, S., Riahla, M.A., Hamadouche, M.: Security issues in self-organized ad-hoc networks (MANET, VANET, and FANET): A survey. In: Lejdel, B., Clementini, E., Alarabi, L. (eds.) Artificial Intelligence and Its Applications: Proceeding of the 2nd International Conference on Artificial Intelligence and Its Applications (2021), pp. 312–324. Springer International Publishing, Cham (2022). https://doi.org/10.1007/978-3-030-96311-8_29
6. Saini, J.I., St Amour, B., Jaekel, A.: Intelligent adversary placements for privacy evaluation in VANET. Information **11**, 443 (2020)
7. Kumar, A., Gupta, N.: A secure RSU based security against multiple attacks in VANET. In: 2020 3rd International Conference on Intelligent Sustainable Systems (ICISS), Thoothukudi, India, pp. 1156–1163 (2020)
8. Lyu, C., Gu, D., Zhang, X., Sun,S., Tang, Y.: Efficient, fast and scalable authentication for VANETs. In: IEEE Wireless Communications and Networking Conference (WCNC): Networks, China (2013)
9. Purohit, K.C., Dimri, S.C., Jasola, S.: Mitigation and performance analysis of routing protocols under black-hole attack in vehicular ad-hoc network (VANET). Wirel. Pers. Commun. **97**(4), 5099–5114 (2017). https://doi.org/10.1007/s11277-017-4770-6
10. Tyagi, P., Dembla, D.: Advanced secured routing algorithm of vehicular ad-hoc network. Wirel. Pers. Commun. **102**(1), 41–60 (2018). https://doi.org/10.1007/s11277-018-5824-0
11. Zardari, Z.A., et al.: A dual attack detection technique to identify black and gray hole attacks using an intrusion detection system and a connected dominating set in MANETs. Future Internet **11**, 61 (2019). https://doi.org/10.3390/fi11030061
12. Cherkaoui, B., Beni-hssane, A., Erritali, M.: Variable control chart for detecting black hole attack in vehicular ad-hoc networks. J. Ambient Intell. Humaniz. Comput. **11**(11), 5129–5138 (2020). https://doi.org/10.1007/s12652-020-01825-2
13. Hassan, Z., Mehmood, A., Maple, C., Khan, M.A., Aldegheishem, A.: Intelligent detection of black hole attacks for secure communication in autonomous and connected vehicles. IEEE Access. **8**, 199618–199628 (2020). https://doi.org/10.1109/ACCESS.2020.3034327
14. Kumar, A., et al.: Black hole attack detection in vehicular ad-hoc network using secure AODV routing algorithm. Microprocess. Microsyst. **80**, 103352 (2021). https://doi.org/10.1016/j.micpro.2020.103352
15. Pattanayak, B.K., Pattnaik, O., Pani, S.: Dealing with sybil attack in VANET. In: Mishra, D., Buyya, R., Mohapatra, P., Patnaik, S. (eds.) Intelligent and Cloud Computing. SIST, vol. 194, pp. 471–480. Springer, Singapore (2021). https://doi.org/10.1007/978-981-15-5971-6_51
16. Balamurugan, V.: An intelligent framework for vehicular ad-hoc networks using SDN architecture. IJCSN Int. J. Comput. Sci. Netw. **3**(6) (2014)

Solving a Proposed Traffic Flow Model Using Deep Learning and Physical Constraint

Mourad Haddioui[1](✉), Youssef Qaraai[1], Said Agoujil[2], and Abderrahman Bouhamidi[3]

[1] MMIS Team, MAIS Laboratory, Departement of Computer Sciences, Faculty of Sciences and Technics Errachidia, University Moulay Ismail, Meknes, Morocco
mou.haddioui@edu.umi.ac.ma, y.qaraai@fste.umi.ac.ma
[2] MMIS Team, MAIS Laboratory, ENCG El Hajeb, Meknès, University Moulay Ismail, Meknes, Morocco
s.agoujil@fste.umi.ac.ma
[3] Department of Mathematics, Laboratory LMPA, University Littoral Cote d'Opale, Dunkirk, France
bouhamidi@lmpa.univ-littoral.fr

Abstract. First order macroscopic model such as Lighthill-Whitham-Richards (LWR) has been extensively studied and applied to various homogenous traffic problems. Although numerical methods have been widely used with good performance, researchers are still searching for new methods for solving partial differential equations. In recent years, deep learning has achieved great success in many fields, such as image classification and natural language processing. Several studies have shown that deep neural networks have powerful function-fitting capabilities and have great potential in the study of partial differential equations. In this paper, we introduce an improved Physics-Informed Neural Network (PINN) for solving one dimensional proposed macroscopic model compared to B-spline collocation method. The experimental results show that PINN is effective in solving partial differential equations and deserves further research.

Keywords: PDE · Deep learning · PINNs · LWR · Trafic flow · B-spline

1 Introduction

Growth in the population and the economy raises the demand for transportation, which frequently results in higher levels of congestion, which causes delays, pollution and a decline in safety. There are many ways to reduce congestion. Encouragement to travel by means that are less taxing on the transportation system is one example. It is crucial to understand how the actual flow of traffic

Y. Farhaoui et al. (Eds.): ICAISE 2022, LNNS 635, pp. 884–889, 2023.
https://doi.org/10.1007/978-3-031-26254-8_128

will look: where and when will there be jams, what are the bottlenecks, and where is there already enough road capacity? In this paper we have proposed an exponential model that links velocity and density, which was solved using the PINNs and the B-spline methods. The paper is structured as follows: Sect. 2 introduces the proposed model used to describe the evolution of density throughout a given section of road. Section 3 presents a summary about physic informed neural networks. In Sect. 4, we give a detailed description and implementation of PINN for solving the LWR model, while Sect. 5 provides some concluding remarks and future works.

2 Proposed Model

A significant turning point in the evolution of continuum traffic flow modeling occurred in the 1950s, when Lighthill and Whitham (1955) and Richards (1956) independently proposed a straightforward continuum model to describe the characteristics of traffic flow. One of the most basic continuum traffic flow models ever described in a literature is the LWR model. The LWR describe the traffic as a compressible fluid by a first-order nonlinear hyperbolic PDE:

$$\frac{\partial \rho(x,t)}{\partial t} + \frac{\partial q(\rho(x,t))}{\partial x} = 0 \tag{1}$$

The Greenshields model is referred to as a univariate one. Greenberg [1] proposed a logarithmic formula representing the velocity-density relation whose main drawback is the inability to predict the velocity in the case of low densities, i.e. the velocity tends to infinity when the density tends to zero. To overcome the problem of the Greenberg model, Underwood [2] proposed an exponential model, in which the speed tends to zero when the density tends to infinity, which does not allow the prediction of the traffic speed when the density is high, as in the case of congested traffic. Drake [3] improved Greenberg's model by proposing a new velocity-density relation. In the case of free-flowing traffic, it was noticed that this model gives a better fit than the previous models. Yet, in the case of congested traffic, the fit is not good, as in the case of Underwood's model. In the same line of thought, Pipes [4] proposed a generalized version of the Greenshield model. By varying the parameter n, a family of models is obtained. Drew [5] slightly modified Pipes' model by replacing n with $n + 1$. However, it has been noticed that this model underestimates the speed in the case of fluid traffic, and overestimates the speed in the case of congested traffic. To avoid the problems as mentioned in the previous models, we propose an exponential model where the speed is defined according to the relationship

$$V(\rho) = \begin{cases} V_{max} \quad if \qquad \rho = 0 \\ V_{max}(1 - e^{\rho - \rho_{max}}) \qquad \rho \in]0; \rho_{max}] \end{cases} \tag{2}$$

where ρ is the traffic density, ρ_{max} the maximum density and V the traffic speed. The density maximum is calculated when the latter is in the congested state. In the next we will introduce the PINNs for solving such a model and compare it with B-spline collocation approach [8].

3 Methodologies

in this section we begin by giving an overview on physic informed neural networks that incorporate physical laws. The PINNs implementation framework are introduced.

3.1 Physics-Informed Neural Networks for Solving PDEs

Consider the following PDE parameterized by λ for the solution $u(x)$ with $x = (x_1, ..., x_d)$ defined on a domain $\Omega \subset R^d$:

$$f(x; \frac{\partial x}{\partial u_1}, ..., \frac{\partial x_d}{\partial u_d}; \frac{\partial^2 u}{\partial x_1 \partial x_1}; ... \frac{\partial^2 u}{\partial x_1 \partial x_d}; \lambda) = 0, x \in \Omega \quad (3)$$

with boundary conditions $B(u, x) = 0$ on $\partial\Omega$ where u is the unknown solution and $B(u, x)$ could be Dirichlet, Neumann, Robin, or periodic boundary conditions. In order to approximate the solution $u(x)$ of a partial differential equation, we first build a neural network which accepts the input x and produces an output vector with the same dimension as $u(x)$. The neural network u should meet two requirements: first, it should be able to reproduce the observations when given a dataset of $u(x)$ observations when x is used as input. Second, it should adhere to the physics underlying the partial differential equation. Currently, many deep learning frameworks, including Tensorflow [6] and PyTorch, have extensively incorporated automatic differentiation technics. As a result, automatic differentiation has been widely used in research on PINNs [7].

4 Implementation and Discussion Results

In this section we study the one-dimensional LWR equation using physics-informed neural networks. The neural network models are constructed for these equation based on the given initial and boundary conditions. The approximation result of the neural networks are compared with the result obtained by the B-spline collocation method. The LWR equation is given by:

$$v(\rho) = \begin{cases} \partial_t \rho(x,t) + \partial_x Q(\rho(x,t)) = 0 & \forall x \in [a,b], \forall t \in [t_0, T] \\ \rho(a,t) = f_1(t); \rho(b,t) = f_2(t) & \forall t \in [t_0, T] \\ \rho(x, t_0) = \rho_0(x) & \forall x \in [a,b]. \end{cases} \quad (4)$$

where the flow is defined as

$$Q(\rho) = V_{max}\rho(1 - exp(\rho - \rho_{max}))$$

The neural network model is trained using training data that includes the initial conditions, boundary conditions, and some randomly generated data in the space-time domain. We use the neural network model to generate multiple predictions in order to evaluate the performance of the training model, and then compare it with the true solution of the partial differential equation. A neural network is first created to approximate the solutions of partial differential equations, represented by $\hat{\rho}(x,t)$. Six hidden layers with 100 neurons each make up the neural network's architecture. A hyperbolic tangent tanh is used as the activation function. For introducing equation control information, a physics-based neural network called $f(x,t)$ is built, and TensorFlow is used in our experiments to build the neural network. A physical information neural network is defined as follows:

$$f(x,t) := \partial_t \rho(x,t) + V_{max}\partial_x \rho(x,t)(1 - e^{(\rho(x,t)-\rho_{max})} - \rho(x,t)e^{(\rho(x,t)-\rho_{max})}) - F(x,t) \quad (5)$$

where:

$$F(x,t) = \partial_t \rho_{exact}(x,t) + V_{max}\partial_x \rho_{exact}(x,t) - V_{max}\partial_x \rho_{exact}(x,t)e^{\rho_{exact}-\rho_{max}}(1+\rho_{exact})$$

Note that the expression of function F is obtained using a test solution (what we call exact solution $\rho_{exact} = e^{-tx^2}$). This choice is not the real exact solution, but it is used quite simply to be numerically guided (to have a reference for the comparison). After numerically validating our approach, the behavior of the actually approximated solution will be deduced.

The next main task is to evaluate our PDE in a certain number of "collocation points" (N_f) inside our domain (x,t). Then we iteratively minimize a loss function related to f:

$$MSE = MSE_u + MSE_f = \frac{1}{N_u}\sum_{i=1}^{N_u} \left|\hat{\rho}^i - \rho\left(x_u^i, t_u^i\right)\right|^2 + \frac{1}{N_f}\sum_{i=1}^{N_f} \left|f\left(x_f^i, t_f^i\right)\right|^2 \quad (6)$$

where MSE_u is a loss function built from observations of the initial and boundary conditions. MSE_f is a loss function that introduces physical information by using partial differential equations. We chose the data in the spatio-temporal domain to train the neural network in this study in order to fully account for the physical information contained in the equations. The spatio-temporal domain's training data are chosen at random, and there are 10000 total training data points. Additionally, the initial and boundary conditions' training data total is relatively small, and the expected outcome can be obtained when N_u is 100. We set the learning rate to 10^{-3} during the optimization process in order to balance convergence speed and global convergence. After the neural network model had been trained, we evaluated its effectiveness. The prediction made by the neural network model after training is shown in Fig. 1. In addition, the root-mean-square error was calculated (Table 1) which confirms the efficacy of this approach. The neural network model can still approximate a result from the training data that is very close to the true solution, despite the complexity of the solutions to the chosen partial differential equations. This shows that the neural network with physical information has great potential and value, and is deserving of further study. Figures 1, 2 and 3 show the results obtained for the solution of LWR equation using physics-informed neural networks compared to the result obtained from collocation method with cubic B-splines. From Fig. 4, it can be seen that the predictions of the neural network model and exact solutions are very consistent, indicating that the constructed neural network model has a good ability to solve LWR model.

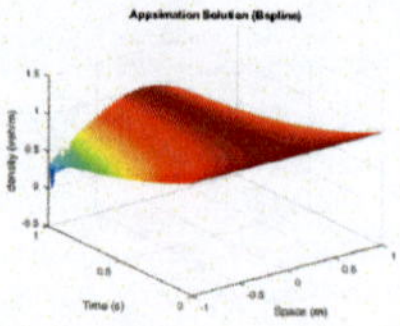

Fig. 1. Approximation solution (Bspline)

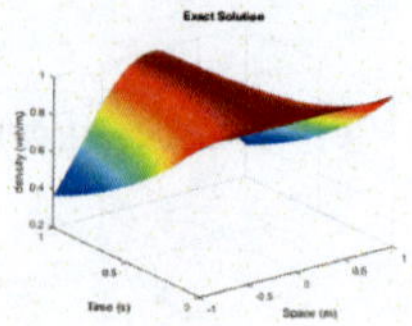

Fig. 2. Exact solution

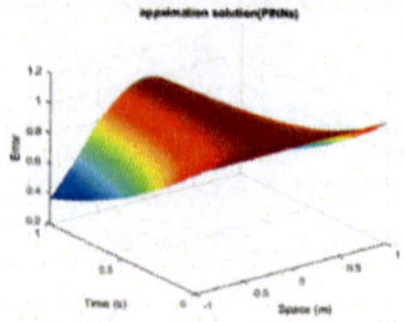

Fig. 3. Approximation solution (PINNs)

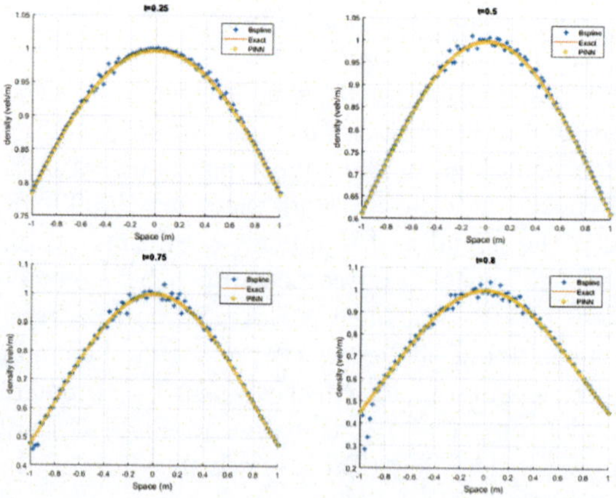

Fig. 4. Comparaison of approximate solution obtained with PINNs, Bspline and exact solution at different times $t = 0.25$, 0.5, 0.75, 0.8

Table 1. Root-mean-square error

t(s)	PINNs	B-spline method
0.25	1.1560e−05	1.3084e−05
0.5	1.1680e−06	1.1545-05
0.75	1.15e−05	9e−04
0.8	1.4969538e−06	0.001917256006619

5 Conclusion

In this paper, a macroscopic model for traffic flow is proposed and solved using physical informed neural networks, which achieves good experimental results due to the powerful function approximation ability of neural networks and the physical information contained in the partial differential equations. The produced results also seem to be more accurate than results obtained using a collocation bspline method. Among our perspectives is to extend the model to the two-dimensional domain and to use the data to make a platform for collision prevention using sensor networks and multiagent systems.

References

1. Greenberg, H.: An analysis of traffic flow. Oper. Res. **7**, 79–85 (1959)
2. Underwood, R.T.: Speed, Volume, and Density Relationships: Quality and Theory of Traffic Flow, pp. 141–188. Yale Bureau of Highway Traffic, New Haven (1961)
3. Drake, J.S., Schofer, J.L., May, A.D.: A Statistical Analysis of Speed Density Hypotheses. Highway Research Record 154, Highway Research Board, NRC, Washington D.C., USA (1967)
4. Pipes, L.: Car following models and fundamental diagram of road traffic. Transp. Res. **1**, 21–29 (1967)
5. Drew, D.: Deterministic aspects of freeway operations and control. Highway Res. Rec. **99**, 48–58 (1965)
6. Abadi, M., et al.: TensorFlow: a system for large-scale machine learning. In: Proceedings of the 12th fUSENIXg Symposium on Operating Systems Design and Implementation (fOSDIg 2016), Savannah, GA, USA, 2–4 November 2016, pp. 265–283 (2016)
7. Raissi, M., Perdikaris, P., Karniadakis, G.E.: Physics-informed neural networks: a deep learning framework for solving forward and inverse problems involving nonlinear partial differential equations. J. Comput. Phys. **378**, 686–707 (2019)
8. Agoujil, S., Bouhamidi, A., Chergui, S., Qaraai, Y.: Implementation of the vehicular occupancy-emission relation using a cubic B-splines collocation method. Discret. Continuous Dyn. Syst. Series-S **13**(3) (2020). https://doi.org/10.3934/dcdss.2020022

Check for updates

The Use of Online Social Networks: Comparing Moroccan and French Communities on Facebook

Jaafar Idrais[1](✉), Yassine El Moudene[1], Rida El Abassi[1], and Abderrahim Sabour[1,2]

[1] IMI Laboratory, Faculty of Sciences of Agadir, Ibn Zohr University, Agadir, Morocco
jaafar.idrais@edu.uiz.ac.ma
[2] Department of Computer Sciences, High School of Technology of Agadir, Ibn Zohr University, Agadir, Morocco

Abstract. The impact of online social networks (OSNs) has become a great social, political and economic impact on people's lives. This space has provoked multidisciplinary research questions. In this work, we compare the use of OSNs by Moroccan and French users through a study of public social interactions on Facebook. We use a collected database that contains more than 3 million profiles. By analyzing the average number of interactions, the number of subscribers and the average response rate. Results suggest that the usage characteristics provided an overview of Facebook usage in both communities, but were stronger indicators in a developed country than in a developing one, consistent with the focus on the average number of interactions and response rate. Future research is needed to determine the effect of uses on the lives of users who are still present on Facebook for both communities based on content analysis and motivations to study human behavior in the virtual world.

Keywords: Data mining · OSNs · User behavior · Social media · Facebook · Tracking communities · Data analysis

1 Introduction

Online social networks have become a virtual space that gives more freedom for users. Human behaviour in the internet has undergone a lot of change characterized by the emergence of new habits and led to many new activities such as advertising, e-commerce and digital marketing [1,2].

Studying online communities' behavior provides several insights into the use of social networks. In this work and given the nature of collected data, we will compare the use of Facebook by Moroccan and French communities. The relations between both populations are very strong and span several areas, historical relations dating back to the period of colonization in the early twentieth century [3], and have generated a set of strong economic and social links. France is

Y. Farhaoui et al. (Eds.): ICAISE 2022, LNNS 635, pp. 890–895, 2023.
https://doi.org/10.1007/978-3-031-26254-8_129

the preferred destination for Moroccan immigrants [4], and for economic reasons both countries use the same time zone [5]. Morocco is a French-speaking country, the French language is used in the state administrations, education, press [6]. To this end, and thanks to the existing commonalities between the two populations, this comparative study will allow to analyze interactions similarities and differences.

The rest of the paper is organized as follows: Sect. 2 presents related works, third section covers the methodology and working method. In Sect. 4, we present our results and the discussion of important points. Section 5 presents a conclusion and the perspectives of the work.

2 Related Work

Facebook as the world's leading social network has been the great beneficiary of technological revolution by allowing users to navigate through a personal and personalized profile in a rich and dynamic content system. These networks have become a very important area for the scientific community [7].

Previous works on the analysis of OSNs are an introduction to the digital sociology of OSNs in Morocco, the aim was to characterize the use of the Moroccan community and to set up a database for the benefit of researchers in the sociology of virtual social networks [8,9]. The findings found through the exploration of the collected data [10] have allowed to draw morphological characteristics related to the use of Facebook and behavioral regularities. The work on the study of users' behavior on the social network Facebook has allowed to analyze the main trends of the communities [11]. Many studies have been conducted to explore and analyze user behavior in OSNs [12]. In the area of behavioral learning of human personality on OSNs, researchers attempted to build a system capable of predicting personality from information about Facebook users [13]. Characterizing human behavior on OSNs has led to important findings on the interaction of individuals with these platforms [8]. Social interactions of Facebook are characterized by a remarkable periodicity among users, as well as some signs of dependency are reported in the studied community [14–17].

3 Measures and Methodology

3.1 Comparing Communities

In this work, the collected data is exploited and we will compare the use of OSNs in the two communities through the following measurements:

- **The number of subscribers:** we measure the number of users who follow a page on Facebook and groups subscribers.
- **The average number of monthly interactions:** this refers to the number of interactions published on Facebook per month.
- **The average response rate:** it constitutes the number of response interactions divided by the number of triggering interactions

3.2 Data Collection

The study of human interactions with OSNs constitutes one of the important research axes aiming at modeling human data and phenomena detected on virtual social platforms. The data used in this study are taken from the social network Facebook via a process of collection of online public data using a semi-supervised approach based on a Crawling agent [9,10]. The objective is to crawl the graph by targeting the most popular public pages and groups in both communities. The semi-supervised crawling process allows for parallel data collection via an improved crawling system (Fig. 1). The semi-supervised crawling system allows for the management and infiltration of agents based on the targeted community, as well as the exploitation and maintenance of connection links within the destination network. The web resources browsing algorithm allows to establish the connection with the server according to the agent and the timestamp. Once the connection is established, an access token will be generated for a secure access to the remote resources. At each iteration the existence and the lifetime of the token are checked, then the algorithm saves the collected nodes.

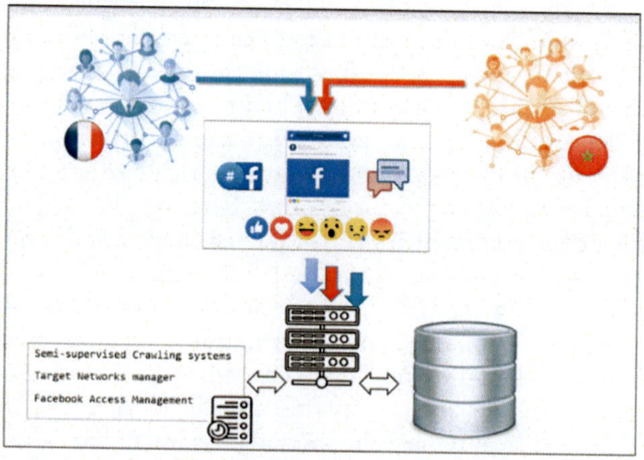

Fig. 1. Crawling process using semi-supervised agent system

4 Results and Discussions

The study of human behavior on OSNs is an area requiring more investigation and credible data and reliable processes. Comparing the use of OSNs through the exploration of some indicators of the collected database will allow the extraction of knowledge and set up a comparative analysis of both communities. In order to deduce characteristics of each population, Fig. 2(A) shows that the number of French subscribers is double compared to Moroccans, the average number of interactions remained stable during the first seven months of 2016 near 2 million subscribers, while the number of Moroccan fans underwent an increase of 0.2

million subscribers between January and July. The same behavior is noticed in the Moroccan community in the last month of the year with a variation around 0.9 million.

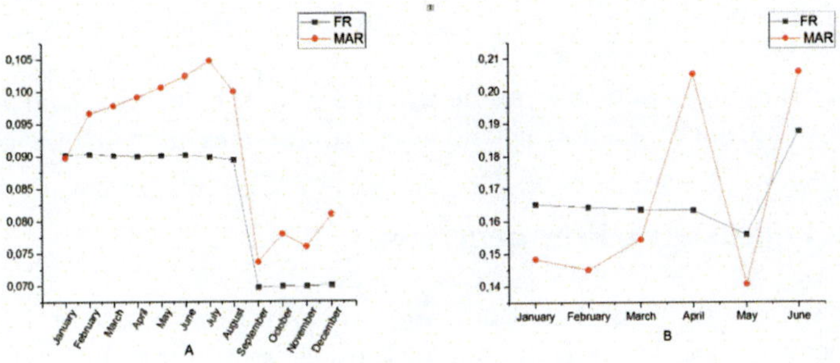

Fig. 2. The number of subscribers in 2016 (A) and in 2017 (B)

During 2017, Fig. 2(B) indicates that the first sex months are characterized by a large number of French subscribers around 1.6 million with a decrease on May (1.47 Million). Dividing the average number of subscribers by the sum of the averages throughout the year, Fig. 2(A) shows a similar behavior between communities in 2016. From January to July, the evolution of the average number of subscribers is almost stable for the French community, while a slight increase characterizes Moroccan users (1.5 in July). From September to December, a decrease appears very well for both populations, which represents a period of recovery from usual activities such as studies, work, etc.

Fig. 3. The average response rate in 2016 (A) and 2017 (B)

The response rate is a key to the quality of the information, regarding the representation's of the target population. Thus, while it is very important to check the response rate of a survey, it is not the only indicator of the quality of the results. Figure 3(A) gives a general view of the average response rate, in 2016 we see that in March and August Moroccans are very interactive with an average response rate of 67%, but in October and November a drop in value to 32% was noticed in this period of the year, for France we observe an important activity in March as the maximum value of the response rate, which means that our studied communities respond more, May and October are known by a cessation of activity. Thus, we can say that the Moroccan community responds more than the French community and this can be explained by the new role that Facebook plays in daily life such as following the news, communicating with people ... etc.

5 Conclusion and Perspectives

In general, the results suggest that in European countries, the use of the internet and specifically OSNs has become one of the needs of life, while in developing countries, people are becoming familiar with the new intruder in their lives, the importance of access and connection as well as interaction is important to ensure high usage of Facebook in Morocco. The usage characteristics provided an overview of Facebook usage in both communities, but were stronger indicators in a developed country than in a developing one, consistent with the focus on the number of average interactions and the response rate and type of interactions and reaction groups in the latter. Future research is needed to determine the effect of uses on the lives of users who are still present on Facebook for both communities based on content analysis and motivations to study behavior in the virtual world. As an outlook, we will try to study and compare our communities using content analysis to extract more similarities and differences between these populations and to see how the evolution of reactions is influenced.

References

1. Ketelaar, P.E., Janssen, L., Vergeer, M., van Reijmersdal, E.A., Crutzen, R., van't Riet, J.: The success of viral ads: social and attitudinal predictors of consumer pass-on behavior on social network sites. J. Bus. Res. **69**(7), 2603–2613 (2016)
2. Tiago, M.T.P.M.B., Veríssimo, J.M.C.: Digital marketing and social media: why bother? Bus. Horiz. **57**(6), 703–708 (2014)
3. Atouf, E.: Les Marocains en France de 1910 à 1965: l'histoire d'une immigration programmée. Ph.D. thesis, Perpignan (2002)
4. Merizak, M.: Immigration, militantisme politique et mouvement associatif des marocains en France: des origines aux évolutions. Ph.D. thesis, Paris 8 (2006)
5. Koulayom, H., Noussair, Z., et al.: Nouvelles technologies de l'information et de la communication et performances des entreprises françaises dans le cadre d'une coopération euro-méditerranéenne: Quelles pistes méthodologiques de recherche. Technical report (2008)
6. Benzakour, F.: Le français au maroc. enjeux et réalité. Le français en Afrique, Revue du Réseau des Observatoires du Français Contemporain en Afrique **25**, 33–41 (2010)

7. Kaya, T., Bicen, H.: The effects of social media on students' behaviors; Facebook as a case study. Comput. Hum. Behav. **59**, 374–379 (2016)
8. Idrais, J., El Moudene, Y., Sabour, A.: Introduction to sociology of Moroccan online social networks: evolution analysis of the Moroccan community activity on Facebook. In: Mizera-Pietraszko, J., Pichappan, P., Mohamed, L. (eds.) RTIS 2017. AISC, vol. 756, pp. 395–408. Springer, Cham (2019). https://doi.org/10.1007/978-3-319-91337-7_35
9. El Moudene, Y., Idrais, J., Sabour, A.: Introduction to sociology of online social networks in Morocco. Data acquisition process: results and connectivity analysis. In: Mizera-Pietraszko, J., Pichappan, P., Mohamed, L. (eds.) RTIS 2017. AISC, vol. 756, pp. 409–418. Springer, Cham (2019). https://doi.org/10.1007/978-3-319-91337-7_36
10. Moudene, Y.E., Idrais, J., Sabour, A.: Introduction to sociology of online social networks in morocco. data acquisition process: results and connectivity analysis. Int. J. New Comput. Archit. Appl. **1**, 56 (2017)
11. Nasim, M., Ilyas, M.U., Rextin, A., Nasim, N.: On commenting behavior of Facebook users. In: Proceedings of the 24th ACM Conference on Hypertext and Social Media, pp. 179–183 (2013)
12. Wahab, A.A., Purrohman, P.S., Ampry, E.S., et al.: Internet user behavior and social media in learning. In: 4th International Conference on Research of Educational Administration and Management (ICREAM 2020), pp. 50–55. Atlantis Press (2021)
13. Tandera, T., Suhartono, D., Wongso, R., Prasetio, Y.L., et al.: Personality prediction system from Facebook users. Procedia Comput. Sci. **116**, 604–611 (2017)
14. Idrais, J., El Moudene, Y., Sabour, A.: Online social networks: study and validation of the regular behaviors of a targeted community on a complex network using a time series approach. In: 2019 4th World Conference on Complex Systems (WCCS), pp. 1–7. IEEE (2019)
15. Idrais, J., El Moudene, Y., Sabour, A.: Characterizing user behavior in online social networks: Study of seasonal changes in the Moroccan community on Facebook. In: 2019 Third International Conference on Intelligent Computing in Data Sciences (ICDS), pp. 1–5. IEEE (2019)
16. Idrais, J., Moudene, Y.E., Sabour, A.: Characterizing user behavior in online social networks: analysis of the regular use of Facebook. Int. J. Electr. Comput. Eng. (2088-8708) **11**(4) (2021)
17. Idrais, J., El Abassi, R., El Moudene, Y., Sabour, A.: Characterizing community behavior in OSNs: modeling and forecasting activity on Facebook using the SARIMA model. J. Intell. Fuzzy Syst. **43**(3), 3757–3769 (2022)

Performance Study of Linear and Circular Arrays Based on Wideband DOA Estimation

Hassan Ougraz[1](✉), Said Safi[1], and Miloud Frikel[2]

[1] Sultan Moulay Slimane University, LIMATI Laboratory, Beni Mellal, Morocco
hassan.ougrazfpb@usms.ac.ma

[2] Caen Normandie University, LIS Laboratory, Caen, France

Abstract. Recently, direction-of-arrival (DOA) has become one of the most essential terms in the array signal processing field. In this paper, an analysis study is made for uniform linear array (ULA) and uniform circular array (UCA) geometries based on wideband DOA estimation by using the test of orthogonality of frequency subspaces (TOFS) approach. In addition, an analytical model for each geometry is produced. Finally, simulation results have demonstrated the performance of each geometry in both DOA estimation precision and root mean square error (RMSE) considering numerous signal-to-noise ratios (SNRs).

Keywords: Wideband DOA estimation · Wideband sources · Uniform linear array · Uniform circular array · TOFS method · Localization

1 Introduction

Array signal processing has been employed in a variety of wireless communication systems for many years. As a result of advancements in digital signal processing, this sector has grown significantly in recent years. Radar [1], sonar [2], microphone array systems [2], detection and ranging [3], cellular communication [4], and others are popular uses of this field. To increase overall performance and minimize computing complexity, many DOA estimation approaches have been devised. Thus, when the arriving sources are wideband, the most common technique is to divide the wideband source into narrowband components. Wideband direction-of-arrival (DOA) estimation is largely concerned with discovering a creative solution to combine several correlation matrices at different frequencies to achieve reliable DOA estimation [5,6].

Yu et al. [7] proposed the test of orthogonality of frequency subspaces (TOFS) approach, that collects data from several frequencies at the same time and necessitates no pre-processing for initial value systems [6]. Considering TOFS is an incoherent approach [7], the resolution is better when signal-to-noise ratio (SNR) is mid-to-high. However, this algorithm is unable to detect the wanted DOAs when SNR is lower [7].

The use of array geometries is important task in many radar purposes. Thus, by using the TOFS algorithm, we present a performance comparison of the uniform linear array (ULA) and uniform circular array (UCA) geometries. The ULA

Y. Farhaoui et al. (Eds.): ICAISE 2022, LNNS 635, pp. 896–902, 2023.
https://doi.org/10.1007/978-3-031-26254-8_130

geometry is the most basic and widely utilized array configuration in array signal processing and smart antenna systems [8]. In comparison to alternative geometries, the benefits of employing a ULA are its simplicity, superior directivity, as well as the formation of the sharpest main lobe in a particular direction. On the other hand, ULA cannot perform consistently well in all azimuth angles, and the DOA prediction resolution is weak in sources close to the linear array end-fire [8]. This significant disadvantage may be overcome by using various configurations, such as UCA, although this configuration may result in increased complexity of array structure and computations, as well as a bigger array aperture. As a result, it is preferable to create basic array layouts that function equally well in all azimuth orientations. The UCA has played an important role in the wireless communication field [9] due to its uniform beams spanning 360° azimuthal direction.

2 Signal Model and Array Geometries

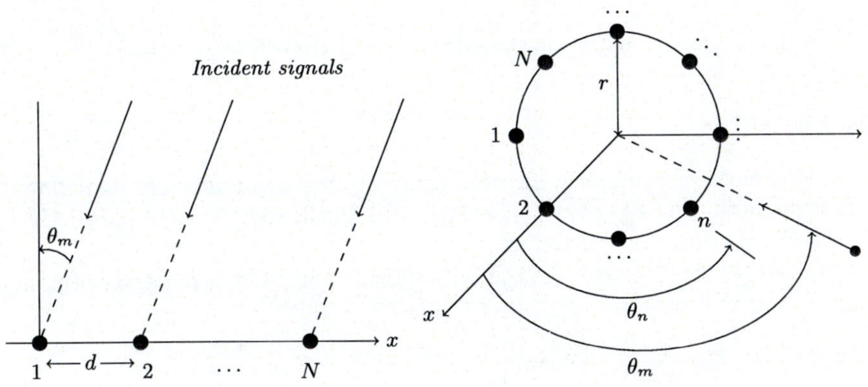

(a) ULA of an N-elements [10, 11].

(b) UCA of N-elements [11, 12].

Fig. 1. The DOA estimation problem using ULA and UCA.

2.1 Signal Model for ULA

In a uniform linear array (ULA) geometry, as illustrated in Fig. 1a, N components are positioned along a single line with a uniform inter-element distance, $d = \lambda/2$, where λ is the wavelength. These antennas are used to receive and estimate M wideband signals $s_m(\omega)$, $1 \leq m \leq M$, which radiate from different directions $\{\theta_1, \theta_2, \ldots, \theta_M\}$. We assumed that all sources are decorrelated and reside in the bandwidth between the lowest frequency (ω_L) and the largest frequency (ω_H) of the sources. The angle to be predicted is θ_m. Thus, the wideband signal is divided into K narrowband signals. The discret fourier transform (DFT) of the received data at n^{th} antenna for ULA geometry is [10]:

$$x(\omega) = \sum_{m=1}^{M} s_m(\omega) \exp\left(-j\omega v_n \sin\theta_m\right) + n(\omega) \tag{1}$$

where $s_m(\omega)$ is the mth signal source, the term $n(\omega)$ represent an additive white gaussian noise (AWGN) at the nth sensor. The delay is expressed as:

$$v_n = 2\pi \frac{(n-1)d}{c}, \tag{2}$$

where d represents the separation between nearby antennas, and c indicates the speed of light.

The DFT output signals may then be expressed in vector form as:

$$x(\omega_k) = A(\omega_k, \theta)s(\omega_k) + n(\omega_k), \quad 1 \leq k \leq K, \tag{3}$$

where $\omega_L < \omega_k < \omega_H$, and for $k = 1, 2, \ldots, K$, we have:

$$A(\omega_k, \theta) = [a(\omega_k, \theta_1),\ a(\omega_k, \theta_2),\ \ldots,\ a(\omega_k, \theta_M)], \tag{4}$$

$$a(\omega_k, \theta_m) = \Big[1, \exp\left(-j\omega_k v_2 \sin\theta_m\right), \cdots, \exp\left(-j\omega_k v_N \sin\theta_m\right)\Big]^T \tag{5}$$

2.2 Signal Model for UCA

The components of a uniform circular array (UCA) are uniformly distributed along a circle in the x-y plane. The Fig. (1b) illustrates an example of UCA shape of N components and radius r.

The DFT of the received data at n^{th} antenna for UCA geometry expressed as [12]:

$$x(\omega) = \sum_{m=1}^{M} s_m(\omega) \exp\left(-j\omega \frac{2\pi r}{\lambda} \cos(\theta_m - \theta_n)\right) + n(\omega), \tag{6}$$

where θ_n is the angular position at n^{th} component of the array and is computed as follows:

$$\theta_n = 2\pi \left(\frac{n-1}{N}\right) \tag{7}$$

In the x-y axis, the steering vector is represented as [13]:

$$a(\omega_k, \theta_m) = \left[\exp\left(-j\omega_k \frac{2\pi r}{\lambda} \cos(\theta_m - \theta_1)\right), \cdots, \exp\left(-j\omega_k \frac{2\pi r}{\lambda} \cos(\theta_m - \theta_N)\right)\right]^T \tag{8}$$

2.3 Formulation of Array Correlation Matrix

The $N \times N$ array correlation matrix is obtained by [10]:

$$\begin{aligned} R_{xx}(\omega_k) &= E[x(\omega_k)x^H(\omega_k)] \\ &= A(\omega_k, \theta)R_{ss}(\omega_k)A^H(\omega_k, \theta) + \sigma^2 I, \end{aligned} \tag{9}$$

where $E[.]$ represents the mean value operator, H is the complex conjugate transpose, $R_{ss}(\omega_k) = E[s(\omega_k)s^H(\omega_k)]$, σ^2 is the power of noise, and I is the $N \times N$ matrix of identity. The signal subspace $U_s(\omega_k)$ and the noise subspace $U_n(\omega_k)$ at each frequency ω_k are obtained by the eigen-value decomposition (EVD) of $R_{xx}(\omega_k)$ as [10]:

$$U_s(\omega_k) = [e_1(\omega_k), e_2(\omega_k), \ldots, e_M(\omega_k)], \tag{10}$$

$$U_n(\omega_k) = [e_{M+1}(\omega_k), e_{M+2}(\omega_k), \ldots, e_N(\omega_k)], \tag{11}$$

where $e_1(\omega_k), \ldots, e_N(\omega_k)$ are the orthogonal eigen-vectors of $R_{xx}(\omega_k)$.

3 TOFS Method

In this part, we will go through the various steps of the TOFS technique, an incoherent approach for estimating wideband DOA sources that takes use of the orthogonality between the steering array and the noise subspace created by EVD of the correlation matrix for every frequency. The estimator array $D(\theta)$ by TOFS can be expressed as follows [5]:

$$\begin{aligned} D(\theta) = [&a^H(\omega_1,\theta)U_n(\omega_1)U_n^H(\omega_1)a(\omega_1,\theta) \\ &a^H(\omega_2,\theta)U_n(\omega_2)U_n^H(\omega_2)a(\omega_2,\theta) \\ &\ldots a^H(\omega_K,\theta)U_n(\omega_K)U_n^H(\omega_K)a(\omega_K,\theta)] \end{aligned} \tag{12}$$

If θ is the actual angle of the source, the following equation is fulfilled for every arbitrary frequency bin ω_k [14]:

$$a^H(\omega_k,\theta)U_n(\omega_k)U_n^H(\omega_k)a(\omega_k,\theta) = 0 \tag{13}$$

We may calculate the DOA of arriving wideband sources from the spatial spectrum derived by the following formula since the rank of the matrix $D(\theta)$ reduces when θ is the one DOA of arriving wideband sources. Then, the estimated DOAs are achieved by:

$$\hat{\theta} = arg\ \underset{\theta}{max}\ \frac{1}{\sigma_{min}(\theta)}, \tag{14}$$

where $\sigma_{min}(\theta)$ denotes the shortest singular value of the estimator $D(\theta)$.

4 Simulation Results

To prove the performance of ULA and UCA geometries using the wideband TOFS approach, computer simulations are investigated in detail, which are in terms of resolution and root mean square error (RMSE). Thus, we assume that the number of sensors in both geometries is the same and equal to $N = 10$ elements. Moreover, the distance between the elements of ULA d is $\lambda/2$ and the radius of the circle r is considered to be $\lambda/(4\sin(\pi/N))$. The snapshots are set to 400.

4.1 Resolution

For the first simulation, as shown in Fig. 2, we compare ULA and UCA using TOFS algorithm for four directions that are specifically selected to evaluate the performance of each configuration. Two angles are located near the limits of a ULA $\{-85°, 85°\}$, whereas the others are chosen close to the center $\{-10°, 10°\}$. The SNR is fixed at 0 dB.

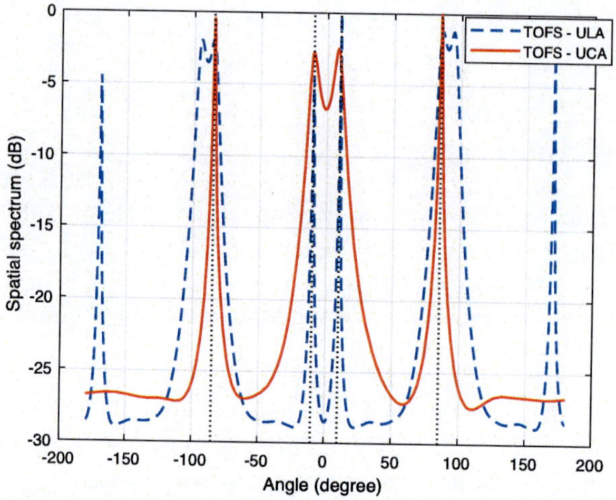

Fig. 2. DOA resolution of ULA and UCA geometries.

When the source comes from larger degrees near the linear array limits ($\theta = \pm 90°$), it is obvious that UCA offer sharper peaks better than ULA. In the other hand, the TOFS algorithm using ULA geometry works better when the sources are impinging on the center of the array. However, the ULA generates symmetric angular spectra, i.e. if it generates wanted peaks at $-10°$ and $10°$ to indicate the real DOAs, then it also provides unwanted peaks at $-170°$ and $170°$ which causes an ambiguity in the interpretation of the results. This can be handled with a UCA configuration that just demonstrates true peaks at the wanted directions.

4.2 RMSE

For the second simulation, two examples are plotted to show the impact of SNR on the performance of the linear and circular arrays for the TOFS algorithm. The definition of RMSE is given by [15]:

$$\mathrm{RMSE} = \sqrt{\frac{1}{100}\sum_{i=1}^{100}\left|\hat{\theta}(i) - \theta\right|^2}, \tag{15}$$

where $\hat{\theta}(n)$ is the predicted direction and θ is the incidence angle. In this part, 100 Monte Carlo experiments are implemented in order to calculate the RMSE in terms of SNR.

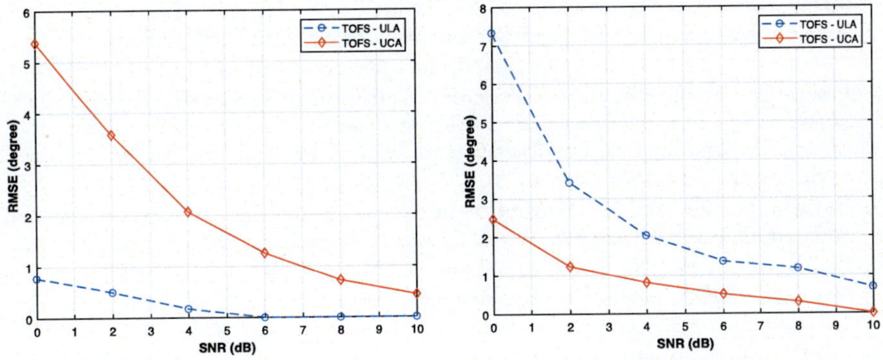

(a) $\theta = 10°$ (b) $\theta = 85°$

Fig. 3. RMSE performance against SNR.

For the first example, it is observed from Fig. 3a, where the incidence angle is $\theta = 10°$, that ULA has a lower RMSE in comparison with UCA. Therefore, when the incident signals are impinging on the center of the linear array, the TOFS algorithm has a better performance in terms of SNR, especially a lower SNR. In the second example, we consider the incidence angle is $\theta = 85°$. The Fig. 3b shows an evident benefit realized in the whole SNR range with the TOFS using UCA geometry near to zero average error in the DOA estimation of the source.

5 Conclusion

The paper introduces a comparative study of two various configurations for 1D and 2D direction finding, using TOFS approach, as a part of wideband DOA estimation are investigated in detail. We analyzed the constraints of each geometry and used several metrics such as resolution and RMSEs to compare the configurations. The TOFS using ULA is unable to identify the right DOAs at the angles that far from the center of linear array, due to that the power distribution becomes weaker at the margins of this array architecture. Finally, for the UCA, TOFS has the opposite effect; it produces stronger peaks in end-fire directions while suffering in the middle.

References

1. Jin, H.: Improved direction-of-arrival estimation and its implementation for modified symmetric sensor array. IEEE Sens. J. **21**(4), 5213–5220 (2021)
2. Shabir, K., Al Mahmud, T.H., Zheng, R., Ye, Z.: A low-complexity RARE-based 2-D DOA estimation algorithm for a mixture of circular and strictly noncircular sources. Turkish J. Electr. Eng. Comput. Sci. **26**(5), 2234–2245 (2018)
3. Adewale, A.: Embedded system based radio detection and ranging (RADAR) system using Arduino and ultra-sonic sensor. Am. J. Embedded Syst. Appl. **5**(1), 7–12 (2017)
4. Zhang, Y., Ye, Z., Liu, C.: An efficient DOA estimation method in multipath environment. Signal Process. **90**(2), 707–713 (2010)
5. Hayashi, H., Ohtsuki, T.: DOA estimation for wideband signals based on weighted Squared TOPS. EURASIP J. Wirel. Commun. Networking **2016**(1), 1–12 (2016)
6. Bai, Y., Li, J., Wu, Y., Wang, Q., Zhang, X.: Weighted incoherent signal subspace method for DOA estimation on wideband colored signals. IEEE Access. **7**, 1224–1233 (2018)
7. Yu, H., Liu, J., Huang, Z., Zhou, Y., Xu, X.: A new method for wideband DOA estimation. In: International Conference on Wireless Communications, Networking and Mobile Computing, pp. 598–601 (2007)
8. Akbari. F., Moghaddam, S.S., Vakili, V.T.: MUSIC and MVDR DOA estimation algorithms with higher resolution and accuracy. In: 5th International Symposium on Telecommunications, pp. 76–81 (2010)
9. Abdulsattar, A.M.: Evaluation of circular array antenna for mobile base station to increase reverse-link capacity. AL-mansour J. **17** (2012)
10. Ougraz, H., Safi, S., Frikel, M.: Analysis of several algorithms for DOA estimation in two different communication models by a comparative study. In: International Conference on Business Intelligence, pp. 219–230. Springer, Cham (2022). https://doi.org/10.1007/978-3-031-06458-6_18
11. Kulaib, A.R., Shubair, R.M., Al-Qutayri, M.A., Ng, J.W.: Performance evaluation of linear and circular arrays in wireless sensor network localization. In: 18th IEEE International Conference on Electronics, Circuits, and Systems, pp. 579–582 (2011)
12. Ougraz, H., Safi, S.: Minimum array elements and angular separation based on several wideband direction-of-arrival estimation approaches. Int. J. Tech. Phys. Probl. Eng. **14**(2), 228–233 (2022)
13. Ihedrane, M.A., Bri, S.: Direction of arrival using uniform circular array based on 2-D MUSIC algorithm. Indonesian J. Electr. Eng. Comput. Sci. **12**(1), 30–37 (2018)
14. Zeng, Y., Lu, G.: Efficient wideband signals direction of arrival estimation method with unknown number of signals. Int. J. Distrib. Sensor Networks **12**(11) (2016)
15. Zhang, X., Liu, X., Yu, H., Liu, C.: Improved MUSIC algorithm for DOA estimation of coherent signals via toeplitz and fourth-order-cumulants. Int. J. Control Autom. **8**(10), 261–272 (2015)

The Realization of 5G Network Booster Using Various Types of Antennas

Ouafae Elalaouy[1(✉)], Mohammed El Ghzaoui[2], and Jaouad Foshi[1]

[1] Faculty of Sciences and Techniques, Moulay Ismail University, Errachidia, Morocco
ou.elalaouy@edu.umi.ac.ma
[2] Sidi Mohamed Ben Abdellah University, Fes, Morocco

Abstract. Given the fact that the network in some regions is weak, this paper aimed at realizing a 5G network booster circuit. To achieve this objective, we suggested resolving this shortage by adopting an adequate circuit in which we implemented different types of antennas in numerous iterations to increase the network strength and reduce the response time. The core of this idea is a simple scheme that makes use of two helical antennas. The first one is excited by the mobile phone while the second is induced. Unwanted noises decrease and the signal (magnetic field) strength at the output of the system are obtained by the stability of the circuit due to the good polarization of the transistor, its functioning in a narrow band of high frequencies, and the best selection of coupling capacitors. The major finding of our study is that the use of a strong magnetic field would help improve the quality of the network. This has been done to ensure good transmission of data among devices, especially those using 5G. Lots of features, then, are enhanced, namely directivity, gain, and selectivity.

Keywords: Weak signal · Network booster · 5G · Mobile phone · Response time · Magnetic field · Antenna · Directivity · Gain

1 Introduction

The fifth generation cellular network (5G) is becoming an increasingly important r search topic interesting and its evolution is necessary to meet the exponential growth of traffic of data in all its forms [1–3]. If the 2G (GSM) networks that constituted the first system of digital communication were centered on voice and the exchange of SMS with data rates offered hardly exceeding 280 Kb/s, which remained low to meet the demand, 3G (UMTS) by its evolution (3G+ then 3G++) and with new techniques such as SIMO (Single Input Multiple Output) and MIMO (Multiple Input Multiple Output) made it possible to achieve theoretical data rates of 42 Mbit/s thus allowing the sending of data packets and the reductionthe phenomenon of fading of signals caused by multi-path environments. Then the arrival of 4G, which is based on the LTE (Long Term Evolution) standard, made it possible to offer higher bit rates (up to 150 Mbit/s under certain conditions) and many architectural improvements such as voice over IP. Nowadays, terahertz technology is also under test [4, 5]. This new technology has many advantages

Y. Farhaoui et al. (Eds.): ICAISE 2022, LNNS 635, pp. 903–910, 2023.
https://doi.org/10.1007/978-3-031-26254-8_131

but suffer also from many drawbacks [6, 7]. The idea that is beginning to be forged behind the notion of 5G is that it would not correspond to a simple increase in data rates, as has been the case for previous generations. Large mobile communication audience, downloading videos and using mobile applications represent the main the current use of radio resources in 4G networks. 5G, on the other hand, will have to meet many needs and constraints (reduced cost, better spectral efficiency, flow rates even higher, even greater number of users, lower latency, better coverage, ...). On the other hand, 5G targets very varied sectors, which would not necessarily have any other common element than this technology, but which are important pillars of our societies current: energy, health, media, industry or transport. In addition, it now seems a given that the objectives of 5G will not be able to be achieved using of a single technology but using several. Therefore, many research efforts and many proposals are beginning to emerge in order to meet these challenges. Among the devices that will be used for 5G to enhance the coverage is the booster circuit. This device helps weak cellular signal to be used by mobile phone. Especialy, this device will improve indoor signal coverage to connect one room to the entire home as well as communication between two machines.

The network booster circuit plays a significantly important role in the world of communication. It assists in rectifying the poor signal due to fading, interference, and longer range in low coverage areas, and makes it available to everyone regardless of the thickness of walls, location, etc. [8]. The main components of the network booster circuit are an input antenna a bipolar transistor used as an amplifier device, and an output antenna. The function of the system is to grab the signal via the input antenna, which is then transferred to the transistor, and finally, the output antenna without any hitches or interruption radiates the provided signal across a long distance to meet the user's requirements [9]. The proposed network booster circuit has been constructed by taking numerous iterations. The process starts by using a helical antenna in both the input and the output. Then, it ends by utilizing the patch antenna in the output to compare them based on the response time, the properties of the used antenna, and the strength of the network signal. The paper comprises three sections in addition to an introduction. The first will be devoted to presenting the study's design and functioning. The second section will be allocated to the construction and testing of the booster. In section three, the results of the study will be presented and thoroughly discussed. The paper will end up with a very brief conclusion.

2 Design and Functioning

2.1 Design of the Mobile Phone Network Booster Circuit

This section deals with the design of the mobile phone network booster. The circuit consists of a transistor and two antennas [10–12]. They are respectively the input and the output. Other materials are outlined in Table 1.

The aforementioned components were chosen based on their durability, availability, and reliability criteria. The designed circuit is depicted in Fig. 1.

Table 1. Mobile phones network booster circuit's components.

Components	Value
R1	1 KΩ
R2	470 Ω
R3	1.5 KΩ
R4	15 KΩ
C1	1 μF

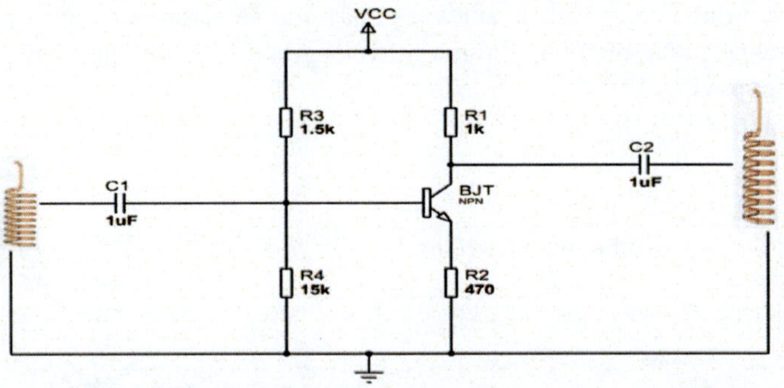

Fig. 1. Design of the mobile phone network booster circuit.

2.2 Functioning

Mobile phones enable their users to send data and make calls at large reaches. However, lack of network signal and the high cost is generally the major problems associated with them. Hence, to overcome these drawbacks, a mobile phone booster network is used. The latest is based on the following functioning. The wireless signal is trapped through the input antenna, then it enters a processing procedure to be amplified via the bipolar transistor that exhibits, in high frequencies, a unique behavior. This results in two low-frequency and two high-frequency, forming a narrow frequency fixation range suitable for telecommunications to eliminate signals that constitute undesirable noises. Additionally, the stabilization of the circuit is controlled by an ideal polarization with the aid of auxiliary components, particularly capacitors and base bridge polarization. The gain in particular becomes more powerful as a result of this. The last step consists of transmitting the result magnetic field to users via the output antenna characterized by their radiation pattern, gain, bandwidth, radiation efficiency, and operating frequency.

3 Test

This section highlights tests of the signal booster circuit using numerous types of antennas. Initially, a breadboard instead of PCB is employed because it doesn't have to be soldered its components can be either changed or removed, it's cheaper and different scenarios can be tested.

3.1 Two Helical Antenna Mobile Phone Network Booster Circuit

The core of the mobile phone network booster circuit is a simple scheme constructed by using an input helical antenna and an output helical antenna and other components as shown in the figure below. Helical antenna polarisation and radiation properties depend on the diameter, pitch, number of turns, wavelength, excitation, and spacing between the helical loops (Fig. 2).

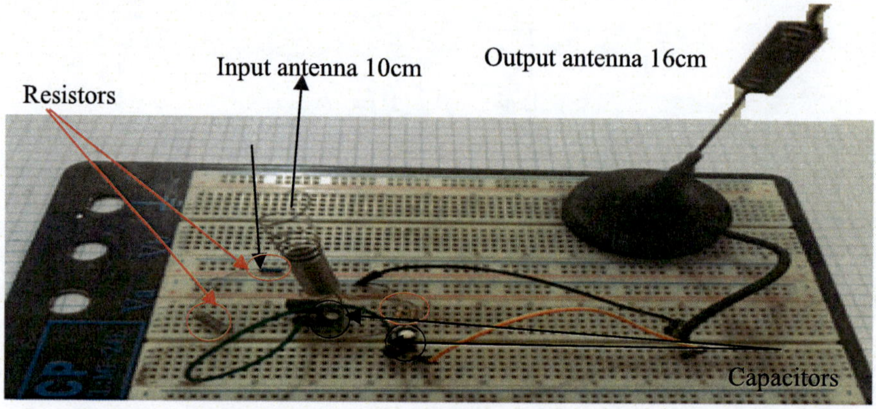

Fig. 2. Two helical antenna mobile phone network booster circuit.

3.2 Aluminium Antenna Mobile Phone Network Booster

Before installing the signal booster in the cellar, the measurement results of the signal quality are very poor, namely zero bar, as seen in Fig. 3. Hence, we cannot receive or make calls and send data, etc.

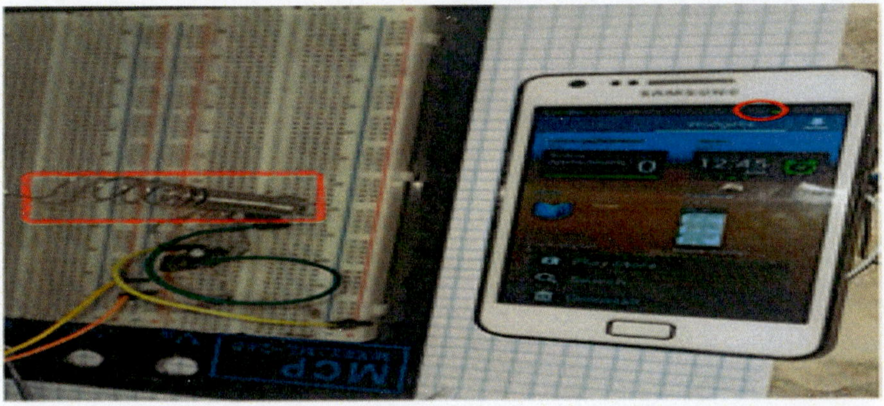

Fig. 3. Mobile phone before integrating network booster circuit.

After the signal booster circuit was installed, the measurements were carried out in the cellar. We noticed that the signal quality received on the cell phone screen shows an increase in the signal from zero to four bars, with a response time that equals 5 s. The measurement results are shown in Fig. 4.

Fig. 4. Aluminum antenna mobile phone after integrating network booster circuit.

3.3 Cover Antenna Mobile Phone Network Booster Circuit

In this sub-section rather than using the aluminum output antenna, we employed the cover material to see the changes. As we can see from the figure below, we obtained a strong signal. The handphone screen shows an increase in the signal from zero to four bar, with a response time that equals 1 s as shown in Fig. 5.

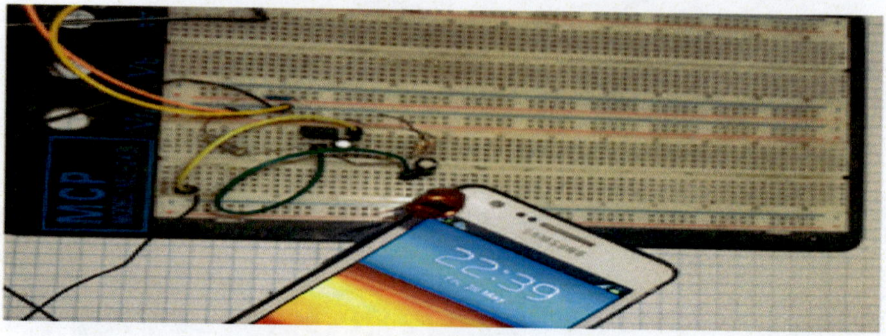

Fig. 5. Cover antenna mobile phone after integrating network booster circuit.

3.4 Patch Antenna Mobile Phone Network Booster

In the final iteration, we replaced the cover helical output antenna with a patch antenna. Hence, we noticed an increase from zero to two bars with a response time that equals 1 s as shown in Fig. 6.

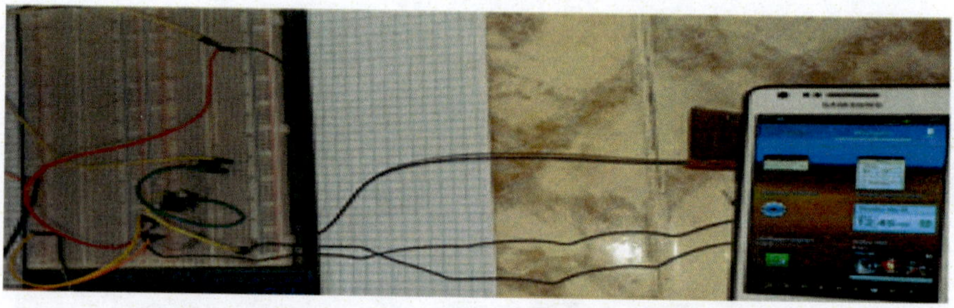

Fig. 6. Patch output antenna mobile phone network booster.

4 Results and Discussion

Based on our findings, we believe that these boosters can help improve signals in low-coverage locations to such an extent that everyone can receive signals without any disturbance. When the booster is turned on and off, we can see how the mobile signal is amplified. Further, we observed that the type of the output antenna and affects directly the signal strength and response time, for example using a cover reduced the latency of receiving data. On the other hand, the geometry of the antenna plays a vital role nowadays, for instance, patch antennas are preferable due to their light size, low cost and the possibility to be integrated into phones. As a result, the value of a repeater in signal enhancement was demonstrated effectively.

5 Construction of the Mobile Phone Booster Circuit

After validating the tests, we built the signal booster circuit as depicted in the following figure. The mobile phone network booster is made from affordable local components mentioned in Table. 1. All components are brought and soldered using the previously mentioned design shown in Fig. 1. Ample attention was taken here because any incorrect component placement could lead to the malfunctioning of the unit (Fig. 7).

Fig. 7. Mobile phone network booster circuit.

6 Conclusion

The ultimate goal of this paper has been to find ways to enhance the strength of the cell phone's signal by designing a network booster circuit and further testing its impact on phones using numerous types of antennas in either the input or output. Overall, we observed that the signal can be increased in remote areas with the help of these boosters irrespective of the sort of antenna utilized. However, we noticed slight changes, especially when using a specific antenna's material or type in terms of response type, the radiation, the number of bars, etc. Fortunately, lots of drawbacks are overcome like the expensive cost of boosters available in markets and the reduced cell phone signal across long reach. To make the booster much more efficient and effective, we will try to make use of MIMO antenna technology in our future projects.

References

1. Jadhav, M., Deshpande, V., Midhunchakkaravarthy, D., Waghole, D.: Improving 5G network performance for OFDM-IDMA system resource management optimization using bio-inspired algorithm with RSM. Comput. Commun. **193**, 23–37 (2022). https://doi.org/10.1016/j.comcom.2022.06.031
2. El Ghzaoui, M., Hmamou, A., Foshi, J., Mestoui, J.: Compensation of non-linear distortion effects in MIMO-OFDM systems using constant envelope OFDM for 5G applications. J. Circuits Syst. Comput. **29**(16), 2050257 (2020)

3. Ali, M.S., Islam, M.S., Memon, M.H., Asif, M., Lin, F.: Optimum DCT type-I based transceiver model and effective channel estimation for uplink NB-IoT system. Phys. Commun. **48**, 101431 (2021). https://doi.org/10.1016/j.phycom.2021.101431
4. Bawuah, P., Axel Zeitler, J.: Advances in terahertz time-domain spectroscopy of pharmaceutical solids: a review. TrAC Trends Anal. Chem. **139**, 116272 (2021). https://doi.org/10.1016/j.trac.2021.116272
5. Ajayan, J., et al.: A critical review of design and fabrication challenges in InP HEMTs for future terahertz frequency applications. Mater. Sci. Semiconduct. Process. **128**, 105753 (2021). https://doi.org/10.1016/j.mssp.2021.105753
6. Aghoutane, B., El Ghzaoui, M., El Faylali, H.: Spatial characterization of propagation channels for terahertz band. SN Appl. Sci. **3**(2), 1–8 (2021). https://doi.org/10.1007/s42452-021-04262-8
7. El Ghzaoui, M., Das, S.: Data transmission with terahertz communication systems. In: Biswas, A., Banerjee, A., Acharyya, A., Inokawa, H., Roy, J.N. (eds.) Emerging Trends in Terahertz Solid-State Physics and Devices, pp. 121–141. Springer, Singapore (2020). https://doi.org/10.1007/978-981-15-3235-1_9
8. Bethram Chibuzo, O., Philip Omoniyi, A.: Network and complex systems developing a signal booster for improved communication in remote areas (2019). https://doi.org/10.7176/NCS
9. Onwuka, E.N., et al.: Survey of cellular signal booster. Int. J. Inf. Eng. Electron. Bus. **10**(6), 21–31 (2018). https://doi.org/10.5815/ijieeb.2018.06.03
10. El Alami, A., Ghazaoui, Y., Das, S., Bennani, S.D., El Ghzaoui, M.: Design and simulation of RFID array antenna 2 × 1 for detection system of objects or living things in motion. Procedia Comput. Sci. **151**, 1010–1015 (2019)
11. Aghoutane, B., Das, S., El Ghzaoui, M., Madhav, B.T.P., El Faylali, H.: A novel dual band high gain 4-port millimeter wave MIMO antenna array for 28/37 GHz 5G applications. AEU Int. J. Electron. Commun. **145**, 154071 (2022)
12. Caimi, D., et al.: III-V-on-Si transistor technologies: performance boosters and integration. Solid State Electron. **185**, 108077 (2021). https://doi.org/10.1016/j.sse.2021.108077

Efficiency in the Availability of Indicators of a Datamart Based on the Kimball Methodology for a Baking Company

Samuel Roncal Vidal, Roger Huanca Díaz, and Omar L. Loaiza(✉)

Escuela Profesional de Ingeniería de Sistemas, Universidad Peruana Unión, Carretera Central Km. 19.5, Lurigancho-Chosica, Lima, Peru
{samuel.roncal,rogerhuanca,omarlj}@upeu.edu.pe

Abstract. The aim of this research was to streamline the availability of strategic indicators of the financial perspective of a bakery company in Peru, through a Datamart based on Kimball's methodology. This study is relevant because it can serve as a reference for organizations in the process of growing their operations and market coverage to monitor their performance through indicators. Methodologically, this research consists of 5 phases: i) planning, ii) e-laboration of requirements, iii) analysis and design, iv) construction and deployment and v) evaluation of availability efficiency. The conclusions are: availability is more efficient because obtaining indicators goes from 3.49 min to 16 s, which represents an improvement of 93.1%, given that there is less sending of bytes from the client from 1,176 to 362 (efficiency of 69.21%) when server; and of a lower generation of bytes in this from 780,627 bytes to 186,968 bytes (efficiency of 417.52%); that is, with fewer bytes the recoverability is better.

Keywords: Kimball methodology · Datamart · Business intelligence · Financial indicators

1 Introduction

Today, technology has become an indisputable and inescapable ally of organizations [1], ranging from the most operational tasks to the complex tasks that executives require to make decisions [2]. Years ago, organizations collected volumes of information from their processes, but computing power and software were expensive and limited; it relied more on human expertise, added to the natural possibilities of error, while it was costly and scarce [2]. Currently, there is computing power, there is software for analyzing large volumes of data and experts now focus on contextualizing and interpreting said information with a greater strategic sense [3].

The data analysis and exploitation tools for the construction of Datamart over time have been complementing strategic management tools such as the Balance Scorecard (BSC) [4] allowing assertive and timely decisions to be made [5] and based on the measurement of indicators, whether simple or complex, to continuously improve an organization [4, 6] It can be said that the BSC helps them to define and clarify their

Y. Farhaoui et al. (Eds.): ICAISE 2022, LNNS 635, pp. 911–919, 2023.
https://doi.org/10.1007/978-3-031-26254-8_132

strategy [6, 7], in the context of business goals and objectives, so a Datamart would allow feeding desired indicators for measurement [8]. On the other hand, the BSC represents a performance evaluation system [8, 9] that ensures a unitary representation of the management vision in relation to performance control and assumes the relational approach in four dimensions [4]: financial perspective, customer perspective, internal business processes, and organizational learning and growth perspective.

An integrated system with shared metadata, common security, and a common technology platform ensures that a wide range of business intelligence capabilities is easy to develop, deploy, while reducing cost, complexity, and risk [10]. In addition, Business Intelligence (BI) applications represent an information architecture [8, 10, 11], in a context of a concert of technologies under the umbrella of a business architecture [12]; which is based on software applications, integrated multidimensional database suites and decision support systems that provide rapid access to strategic information. It is necessary to indicate that the achievement of "Competitive Intelligence" [9, 13, 14] based on information represents a powerful weapon that provides advantages in trade in general.

On the other hand, BI allows users to better analyze and understand their organization plans and results [9]. This gives an idea of what is working correctly, while the identification of areas with potential problems in a certain term, allows corrective actions to be taken [13]. In addition, a BI solution allows you to discover how to keep a medium-sized company ahead of the competition [13].

As the conditions for acquiring Business Intelligence technologies are now more accessible [10], small and medium-sized companies that previously used spreadsheets as a primary BI tool have realized that it is a temporary and limited solution, and progressively they are betting on more advanced BI solutions. On the other hand, a Datamart is part of a Business Intelligence solution [9], and it is not necessary to implement the entire solution in this case since it will be a repository with processed information that can simplify access to historical points of view, current and predictive operations necessary to make better decisions [10].

The investigation is contextualized to a bakery company, the subject of this article, it manufactures products of great nutritional value, its vision is to ensure that its products are massified and preferred in the market. In this sense, the measurement of indicators in the financial perspective of its strategic management is essential. Its most critical indicators are: sales volume and profit. The sources of information come from transactions made from its information systems for its distribution channels such as: warehouses, supermarkets, internal distribution and others. This perspective becomes one of the most affected due to the inconveniences generated by the handling of a large volume of structured data, generating slowness and delays in obtaining the already existing information, which by not being prepared for the analysis and exploitation of data makes it difficult to measure indicators.

The aim of this research was to make the availability of prioritized strategic indicators of the financial perspective more efficient in the distribution channels of a bakery company in Peru, for which a Datamart based on the Ralph Kimball methodology was implemented. to process the necessary information, rather than traditional, structured queries. Table 1 shows the initial situation in terms of the speed of obtaining data for a pilot query of 2,968 rows with 4 crosses of operational tables with their respective filter and grouping with a waiting time of 03:49 min, Understanding the experimental nature of the query, but with access to the total data as a million records or more, a waiting time would be generated that needs to be improved.

Table 1. Example of basic requirements.

Type	Description	Attempt 1
Query statistics	Number of sentences SELECT clause	2
Query statistics	Rows returned by SELECT clause	2,968
Query statistics	Number of transactions	0
Network statistics	Bytes sent from client	1,176
Network statistics	Bytes received from the server	780,627
Time statistics	Hour executed	5:19:56
Time statistics	Customer processing time (seconds)	80
Time statistics	Total execution time (milliseconds)	227,040
Time statistics	Waiting time per server request (milliseconds)	226,960
Time statistics	Time in minutes	3.49

2 Methodology

A Datamart is a copy of the transactional data structured in a special way for analysis, according to the dimensional model, which includes the analysis dimensions and their attributes, their hierarchical organization, as well as the different facts of business to be analyzed [11]. On the one hand, there are tables to represent the dimensions and on the other hand tables for the facts. A dimension can be the Time that will consist of each transactional date of an established event.

For projects where you want to ensure the usability of users that allows rapid and incremental development of the solution where the global picture is not clear, Kimball's approach is appropriate [11]. The 5 phases of this methodology are grouped into 4 and a fifth is added to show the achievement of the objective of the study; Therefore, the phases of this research are: i) planning, ii) requirements development, iii) analysis and design, iv) construction and deployment; and, v) evaluation of the efficiency in the availability.

Regarding stage 1 of planning, Business Intelligence (BI) projects provide support for decisions [5, 13], in addition to providing the entire organization with the capacity for analysis. This phase is where the objectives of the project, the scope of the solution, the factors of success and risks are defined, who will make up the work team, the estimation of times and the resources to be used are determined.

For stage 2, requirements development, end users and their requirements always have an impact on the implementations of a Datamart. The requirements of the area determine the scope of the BI project where it is determined what data it should contain, how the information should be organized, how the information should be updated and who will have access. Kimball [11] provides techniques to effectively discover business requirements through information gathering through interviews, among others.

While stage 3 refers to the analysis and design of the solution, the technical architecture is determined, when the requirements are already correctly established, the architect determines which is the appropriate tool, the analyst is in charge of elaborating the dimensional model where the dimensions and fact tables to be used with their respective crosses, the objective of this phase is to have the model, the source metadata and the necessary tool to carry out the ETL process.

For stage 4 related to the construction of the Datamart, the construction of a Data staging that serves as an intermediary for data cleaning is proposed, after finishing with the data cleaning, the ETL process is carried out to fill the dimensions and Fact tables that must have already been created in the physical model. This process requires a lot of care and can be very extensive depending on the amount of information.

Regarding the last phase, the construction of the BI project solution passes to the real scope of the company, executing according to the user's request (frequency).

3 Results

3.1 Requirements

This section corresponds to phase 1 and 2 of the Kimball methodology. Thus, the work breakdown structure prepared for the project was estimated and executed in 20 days; the identification of requirements 7 days; analysis and design of the solution 32 days, construction 13 days and deployment 3 days.

The datamart to be built, so that the availability of indicators with current information systems and databases is efficient, requires taking into consideration the functional requirements that are: i) provide an intuitive and easy-to-use system to generate reports and analyzes with respect to time; ii) Provide a single version of the information avoiding data redundancy; iii) Allow users to access the information at any time; iv) Improve the response time to the end user by reducing the volume of information; and, v) Create an independence between the technical knowledge of the users and their ability to use these tools. The non-functional requirements: i) The solution must offer the necessary information to measure the following indicators (The set of reports must be specified in detail): Reduce the operational and administrative costs of Production, increase the volume of sales rates and reduce the % of returns; ii) The solution must support a connection to MS SQL Server; and, iii) The solution will allow the integration of different data sources.

3.2 Datamart Construction

This section responds to phases 3 and 4 of Kimball's methodology. Regarding the product of phase 3, on analysis and design of the solution, it is evidenced by the formulation of a dimensional model that must cover the necessary requirements for the elaboration of the

Datamart considering the specific indicators, the level of detail and the attributes of the operational database that allows measuring the indicators. Figure 1 shows the proposed dimensional model used.

Phase 4 of the Kimball methodology guides the construction of the Datamart. In this part of the work, tools such as Pentaho are used, which according to García P. y Gonzáles is a platform that includes all the main components required to implement process-based solutions and has been conceived from the beginning to be a process-based solution. It is an infrastructure of analysis tools and reports integrated with a business process workflow engine that will allow the execution of business rules.

Through the Pentaho suite, it is intended to integrate information visually from multiple information sources, visualize and analyze the company's information interactively, have an end-user web interface developed to allow self-service BI, reducing IT dependency.

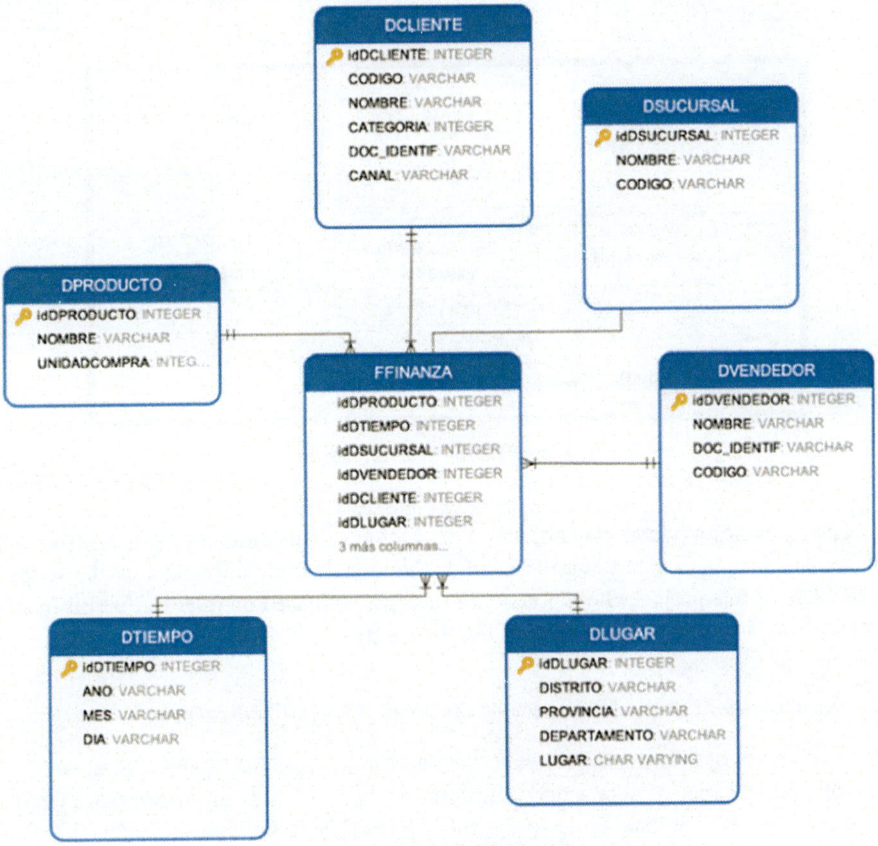

Fig. 1. Dimensional model

Pentaho is made up of a set of tools, one for each stage of your BI project, these are: Data Integration, Pentaho Reporting, Pentaho Analysis, Pentaho Dashboard and

Pentaho Datamining. For the project it will not be necessary to reach the level of offering a dashboard, the solution is oriented towards offering a source of structured information but that can be expanded and use the modules that offer decision support.

The module that performs the ETL is known as PDI or Pentaho Data Integration, it allows data cleaning (Data cleaning), Datamart loading, data integration, migration and data cleaning, data export. Figure 2 shows the ETL Process for filling in the time dimension, in which the column that refers to the sale is used, we break it down according to the requirement and level of detail desired to divide it into year, month and day; In the same way, the filling of the other dimensions specified in the model is carried out to finally fill the Datamart.

A fundamental characteristic of the Pentaho PDI module is the possibility of working based on a process, if the organization has its processes aligned, it can represent the filling as it is the process in the organization, the Jobs are a group of transformations that can be programmable Depending on the necessary periodicity, it is possible to create a Job for each dimension so that the design can be explained to end users.

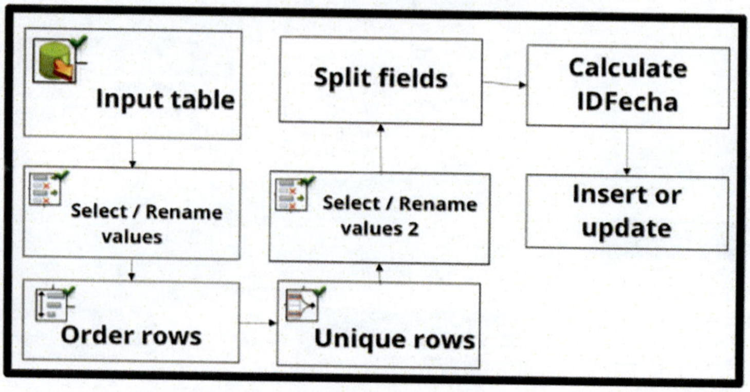

Fig. 2. ETL for time dimension

The elaboration of reports consists of exploiting the information obtained from all the crossings, always satisfying the need of the organization and based on the scope. In the case of this project, the reports are prepared, but based on queries that fill in the necessary data for the measurement of the indicators.

3.3 Improvement in the Efficiency of the Availability of Indicators

The availability of information refers to the degree of accessibility, speed of obtaining, ease of retrieval, ease of access and use. In the section on obtaining speed, As a result, to obtain the data corresponding to the Fiber Bread Sales Volume indicator, the data corresponding to the month of June 2011 will be used. Table 2 shows the statistics of two Requests based on operational data, making the necessary crosses that contain the necessary information for the measurement of the indicator, which in this case are the total amount, the product code, the place of sale, branch, client and date. The difference in waiting time is equivalent to 7% of the execution time in minutes, that is, there is a saving of the time resource corresponding to the difference.

Table 2. Obtaining time for measurement for sales volume indicator

Type	Description	Attempt 1	Attempt 2
Query statistics	Number of sentences SELECT clause	2	2
Query statistics	Rows returned by SELECT clause	2,968	2,968
Query statistics	Number of transactions	0	0
Network statistics	Bytes sent from client	1,176	362
Network statistics	Bytes received from the server	780,627	186,968
Time statistics	Hour executed	5:19:56	5:35:19
Time statistics	Customer processing time (seconds)	80	32
Time statistics	Total execution time (milliseconds)	227,040	2,783
Time statistics	Waiting time per server request (milliseconds)	226,960	2,751
Time statistics	Time in minutes	3.49	0:16

The tendency is to decrease the time of obtaining both in the Datamart as if it were done from the operational Database, but the difference in the total execution time is still noticeable, with the Datamart being the one that shows the fastest. Shows how the operational Database takes longer to perform the query.

The execution plan shows the cost of each action carried out in a SQL query, when performing the table crossings through their respective joins, it leaves the execution plan as shown in Fig. 3, the greater the number of steps, the greater the possibility of an increase in cost in actions such as search table, index scan, etc. [15], and that even the Map Reduce method is not as efficient. It should be considered that the operational database has physical indexes (non-clustered) in most of its tables, so the query becomes efficient.

The execution plan in the Datamart (Fig. 4) includes fewer steps therefore lower cost, likewise the creation of an index would facilitate having a pre-ordered table and would avoid scanning the entire table, scanning only the index in - included in the respective table.

The results in speed of obtaining will affect the measurement of the indicators, granting an improvement in terms of the availability of information, at the end of the development of the technological solution it will be possible to contrast the influence of the Datamart in the measurement not only in speed of obtaining but also in other factors such as recoverability, ease of use and access.

The filtering of information necessary to measure the indicator can be programmed according to the periodicity required in the organization, the query will be executed and depending on the frequency in which the ETL process is carried out, the information will be updated and structured.

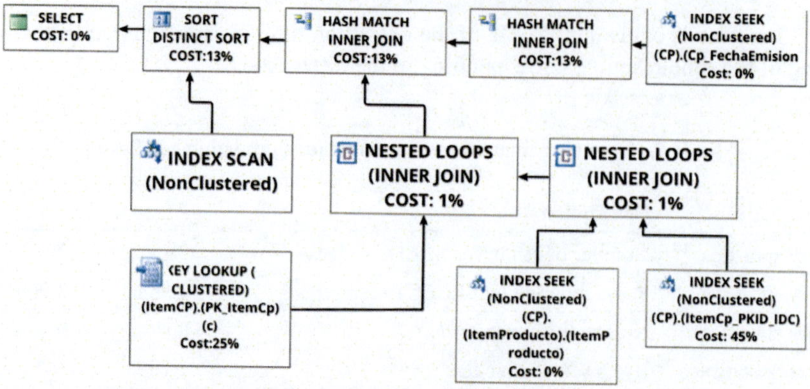

Fig. 3. Execution plan for the fiber bread sales volume indicator

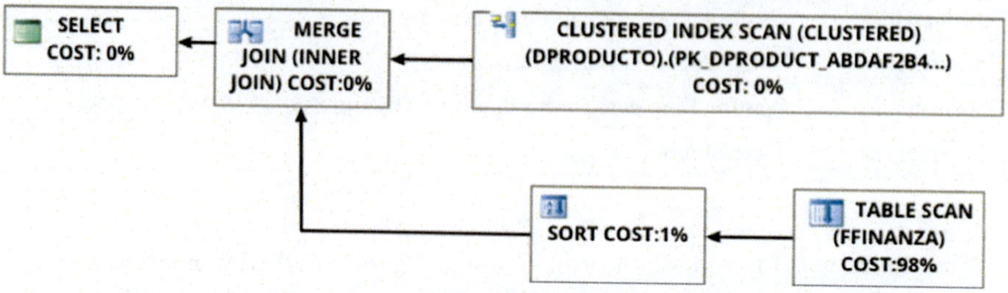

Fig. 4. Execution plan for the fiber bread sales volume indicator - Datamart

4 Conclusions

The implementation of an efficient Datamart the availability of information in obtaining the strategic indicators of the financial perspective in the distribution channels of a bakery company in Metropolitan Lima in Peru as a function of time, due to the fact that the waiting time of obtaining for the measurement of the fiber bread sale indicator went from 3.49 min to 16 s, an efficiency of 93.1%. Project planning manages to cover the scope of the project according to the reality of the environment with the estimated resources.

Well-defined requirements ensure an adequate structuring of the Datamart, while a correct analysis and design ensures the concordance of the requirements, the analysis, and the existing information; being denoted in a better structuring and performance of the queries that allowed a lower sending of bytes from the client from 1,176 to 362 (efficiency of 69.21%) to the server; and of a lower generation of bytes in this from 780,627 bytes to 186,968 bytes (efficiency of 417.52%); that is, with fewer bytes the recoverability is the same.

A Datamart offers greater efficiency in the use of stored information to be analyzed by having a structure properly prepared to serve the strategic level through a Balanced Scorecard. The technological solution, when completed in its entirety, will allow obtaining the necessary information for a larger cloud of indicators with all the data of the organization studied in the availability of information.

References

1. Damanpour, F., Sanchez, F., Chiu, H.: Internal and external sources and the adoption of innovations in organizations. Br. J. Manag. **29**(4), 712–730 (2018)
2. Dankova, P.: Main aspects of enterprise architecture concept. Econ. Altern. **1**, 102–114 (2009)
3. Brous, P., Janssen, M., Herder, P.: The dual effects of the Internet of Things (IoT): a systematic review of the benefits and risks of IoT adoption by organizations. Int. J. Inf. Manag. **51**, 101915 (2020)
4. Aryani, Y., Setiawan, D.: Balanced scorecard: is it beneficial enough? A literature review. Asian J. Account. Perspect **13**, 65–84 (2020)
5. Vanegas, E., Guerra, L.: Sistema de inteligencia de negocios para el apoyo al proceso de toma de decisiones. Rev. Ing. UC **20**(3), 25–34 (2013)
6. Lueg, R.: Strategy maps: the essential link between the balanced scorecard and action. J. Bus. Strat. **36**, 34–40 (2015)
7. Rosado, A., Rico, D.: Inteligencia de negocios Estado del arte **1**, 1–7 (2010)
8. Aspin, A. KPIs and Scorecards BT – Business Intelligence with SQL Server Reporting Services, pp. 19–66. Apress, Berkeley (2015)
9. Rozalia, N., Capatina, A., Costal, N., Cristea, D.: Performance assessment of a business intelligence system acquisition using balanced scorecard software. Recent Adv. Bus. Adm. **1**, 16–20 (2011)
10. Jaleel, R., Abbas, T.: Design and implementation of efficient decision support system using data mart architecture. In: 2020 International Conference on Electrical, Communication, and Computer Engineering (ICECCE), Istanbul, pp. 1–6. IEEE Xplore (2020)
11. Raphl, K., Margy, R.: The Data Warehouse Toolkit (2), pp. 1–27 (2002)
12. Saboya, N., Loaiza, O., Lévano, D.: Diseño de un modelo de arquitectura empresarial para publicaciones científicas basado en ADM - TOGAF 9.0. Apuntes universitarios **8**, 1–16 (2018)
13. Caseiro, N., Coelho, A.: Business intelligence and competitiveness: the mediating role of entrepreneurial orientation. Compet. Rev. **28**, 213–226 (2018)
14. Porter, E.: Competitive Advantage of Nations: Creating and Sustaining Superior Performance. Free Press, New York (2011)
15. Afrati, N., Ullman, J.: Optimizing multiway joins in a map-reduce environment. IEEE Trans. Knowl. Data Eng. **23**, 1282–1298 (2011)

Author Index

Y. Farhaoui et al. (Eds.): ICAISE 2022, LNNS 635, pp. 921–925, 2023.
https://doi.org/10.1007/978-3-031-26254-8

Lecture Notes in Networks and Systems 635

The series "Lecture Notes in Networks and Systems" publishes the latest developments in Networks and Systems—quickly, informally and with high quality. Original research reported in proceedings and post-proceedings represents the core of LNNS.

Volumes published in LNNS embrace all aspects and subfields of, as well as new challenges in, Networks and Systems.

The series contains proceedings and edited volumes in systems and networks, spanning the areas of Cyber-Physical Systems, Autonomous Systems, Sensor Networks, Control Systems, Energy Systems, Automotive Systems, Biological Systems, Vehicular Networking and Connected Vehicles, Aerospace Systems, Automation, Manufacturing, Smart Grids, Nonlinear Systems, Power Systems, Robotics, Social Systems, Economic Systems and other. Of particular value to both the contributors and the readership are the short publication timeframe and the world-wide distribution and exposure which enable both a wide and rapid dissemination of research output.

The series covers the theory, applications, and perspectives on the state of the art and future developments relevant to systems and networks, decision making, control, complex processes and related areas, as embedded in the fields of interdisciplinary and applied sciences, engineering, computer science, physics, economics, social, and life sciences, as well as the paradigms and methodologies behind them.

Indexed by SCOPUS, INSPEC, WTI Frankfurt eG, zbMATH, SCImago.

All books published in the series are submitted for consideration in Web of Science.

For proposals from Asia please contact Aninda Bose (aninda.bose@springer.com).